CAMBRIDGE LIBRARY COLLECTION

Books of enduring scholarly value

Botany and Horticulture

Until the nineteenth century, the investigation of natural phenomena, plants and animals was considered either the preserve of elite scholars or a pastime for the leisured upper classes. As increasing academic rigour and systematisation was brought to the study of 'natural history', its subdisciplines were adopted into university curricula, and learned societies (such as the Royal Horticultural Society, founded in 1804) were established to support research in these areas. A related development was strong enthusiasm for exotic garden plants, which resulted in plant collecting expeditions to every corner of the globe, some-times with tragic consequences. This series includes accounts of some of those expeditions, detailed reference works on the flora of different regions, and practical advice for amateur and professional gardeners.

Flora Capensis

This seminal publication began life as a collaborative effort between the Irish botanist William Henry Harvey (1811–66) and his German counterpart Otto Wilhelm Sonder (1812–81). Relying on many contributors of specimens and descriptions from colonial South Africa – and building on the foundations laid by Carl Peter Thunberg, whose *Flora Capensis* (1823) is also reissued in this series – they published the first three volumes between 1860 and 1865. These were reprinted unchanged in 1894, and from 1896 the project was supervised by William Thiselton-Dyer (1843–1928), director of the Royal Botanic Gardens at Kew. A final supplement appeared in 1933. Reissued now in ten parts, this significant reference work catalogues more than 11,500 species of plant found in South Africa. Containing the remaining polypetalous orders of the Calyciflorae, Volume 2 covers Leguminosae to Loranthaceae.

Cambridge University Press has long been a pioneer in the reissuing of out-of-print titles from its own backlist, producing digital reprints of books that are still sought after by scholars and students but could not be reprinted economically using traditional technology. The Cambridge Library Collection extends this activity to a wider range of books which are still of importance to researchers and professionals, either for the source material they contain, or as landmarks in the history of their academic discipline.

Drawing from the world-renowned collections in the Cambridge University Library and other partner libraries, and guided by the advice of experts in each subject area, Cambridge University Press is using state-of-the-art scanning machines in its own Printing House to capture the content of each book selected for inclusion. The files are processed to give a consistently clear, crisp image, and the books finished to the high quality standard for which the Press is recognised around the world. The latest print-on-demand technology ensures that the books will remain available indefinitely, and that orders for single or multiple copies can quickly be supplied.

The Cambridge Library Collection brings back to life books of enduring scholarly value (including out-of-copyright works originally issued by other publishers) across a wide range of disciplines in the humanities and social sciences and in science and technology.

Flora Capensis

Being a Systematic Description of the Plants of the Cape Colony, Caffraria & Port Natal

VOLUME 2:
LEGUMINOSAE TO LORANTHACEAE

WILLIAM H. HARVEY *ET AL.*

CAMBRIDGE
UNIVERSITY PRESS

University Printing House, Cambridge, CB2 8BS, United Kingdom

Cambridge University Press is part of the University of Cambridge.
It furthers the University's mission by disseminating knowledge in the pursuit of education, learning and research at the highest international levels of excellence.

www.cambridge.org
Information on this title: www.cambridge.org/9781108068079

This edition first published 1894
This digitally printed version 2014

ISBN 978-1-108-06807-9 Paperback

FLORA CAPENSIS.

FLORA CAPENSIS:

BEING A

Systematic Description of the Plants

OF THE

CAPE COLONY, CAFFRARIA, & PORT NATAL.

BY

WILLIAM H. HARVEY, M.D., F.R.S.

PROFESSOR OF BOTANY IN THE UNIVERSITY OF DUBLIN, ETC., ETC., ETC.

AND

OTTO WILHELM SONDER, PH. D.

OF HAMBURGH.

MEMBER OF THE IMPERIAL LEOP.-CAROLINE ACADEMY NATURÆ CURIOSOLUM, ETC., ETC., ETC.

VOLUME II.

LEGUMINOSÆ TO LORANTHACEÆ.

L. REEVE & CO., LTD.

THE OAST HOUSE, BROOK, NR. ASHFORD, KENT

1894.

PREFACE.

THIS second volume contains the remaining polypetalous Orders of South African CALYCIFLORÆ.

Since the completion of the first volume, our correspondents in South Africa have continued to send us large and valuable collections of dried plants, including specimens of a considerable number of new species, and of some new genera. Such of the novelties as fall under any of the Orders contained in Vol. I., as well as those belonging to, and received during the printing of, the present volume, will be found described in the ADDENDA, &c., at p. 583.

These additions to the Flora Capensis, though considerable, comprise only a portion of the new species which we have recently received, and which will find their proper place, as our work proceeds. But they sufficiently prove, not only the botanical richness of the country, but also the continued and increasing interest felt in our work by colonial botanists and collectors of plants.

We have already, in the preface to Vol. I., recorded our obligations to a numerous list of colonial contributors of specimens, to most of whom our thanks are again due for parcels received within the last two years. We have now to express similar indebtedness to the following new correspondents.

To W. T. GERRARD, Esq., of Natal, and to M. J. MCKEN, Esq., Curator of the Botanical Garden at D'Urban, jointly and severally, for large collections (already numbering nearly nine hundred species) made in the Natal Colony and in Zululand. Several of Messrs. Gerrard and McKen's discoveries will be found in the Addenda to this volume, and others will appear in the second volume of "THESAURUS CAPENSIS," now in preparation for press.

To W. S. M. D'Urban, Esq., for a collection of plants from British Caffraria, containing several rare species.

To Miss Elliott, daughter of the late Rev. W. Elliott, of the Paarl, an early collector of Cape plants, for a most interesting parcel of the plants of Damaraland, a region as yet but very imperfectly explored, not ordinarily accessible to botanists, but known to possess a highly curious Flora. From this region has recently been sent to Kew, a singular plant, the "*Tumboa*" of the natives, (*Welwitschia mirabilis*, Hook. f., —a stemless conifer!*) of which we are most anxious to receive specimens of various ages.

To the Rev. Henry Whitehead, for a collection of plants from Namaqualand, carefully dried and well selected, containing some new and many rare species, in better condition than any which we have previously received from that arid district.

To our friends Dr. Atherstone and Henry Hutton, Esq., we are more especially indebted for their voluntary and zealous exertions in making known throughout the frontier districts, both the "Flora" and the "Thesaurus," and kindly acting as our agents in distributing both works. Nor can we close this imperfect expression of thanks to our many kind friends, without again recording the obligations we are under to the Hon. Rawson W. Rawson, Secretary to Government, for his continued care in transmitting parcels of plants sent to his office by our various correspondents. Without this aid, and the privilege of post-transit for small parcels afforded to us by the Colonial Government, we should suffer under many delays and expenses which are now saved to us.

W. H. H.

Trinity College, Dublin, 1st October, 1862.

* Nat. Ord. *Gnetaceæ*.

SEQUENCE OF ORDERS CONTAINED IN VOLUME II. WITH BRIEF CHARACTERS.

Continuation of Sub-Class II. CALYCIFLORÆ. Orders XLVIII.-LXXII.

XLVIII. LEGUMINOSÆ (page 1). *Calyx* free, 5-cleft, unequal. *Petals* inserted in the base of the calyx. *Stamens* (except in gen. 80–82) 10, variously combined. *Ovary* of 1 carpel, with a terminal style. *Fruit* a legume; seeds 1 or several. *(A vast Order, including all leguminous plants.)*

XLIX. ROSACEÆ (page 285). *Calyx* free or nearly so, equal. *Petals* inserted on the margin of the calyx-tube, or none. *Stamens* mostly indefinite. *Ovary* (except in *Grielum*) apocarpous, of several (or rarely only 1) carpels; styles mostly lateral. *Fruit* of dry achenia; rarely succulent. *Albumen* none. *(Leaves alternate, compound or cut, stipulate.)*

L. SAXIFRAGACEÆ (page 305). *Calyx* 4–5-cleft or parted, adnate or free, *Petals* 4–5, regular, marginal. *Stamens* as many or twice as many as the petals. *Ovary* of 2–5 connate carpels; styles 2–5, terminal. *Fruit* a many-seeded capsule. splitting through the middle. *Seeds* albuminous. *(Herbs, shrubs, or trees. Leaves simple or compound, alternate or opposite, exstipulate or stipulate.)*

LI. BRUNIACEÆ (page 309). *Flowers* small, perfect, regular. *Calyx* 4–5-cleft, adnate. *Petals* 4–5, epigynous. *Stamens* 4–5, alternating with the petals. *Ovary* inferior or nearly so, 1–3-celled, with solitary, pendulous ovules. *Fruit* dry, 1–2-coccous. *Seeds* albuminous. *(Heath-like shrubs. Leaves small, mostly linear, entire, crowded, subsessile, exstipulate. Flowers mostly white, sessile, spiked or capitate; very rarely red.)*

LII. HAMAMELIDEÆ (page 324). *Flowers* small, in heads or spikes. *Calyx* 3–5-cleft, or subentire, adnate. *Petals* 4–5, epigynous, valvate. *Stamens* 5–10; anthers opening by a valve! *Ovary* 2-celled, with solitary, pendulous ovules. *Fruit* dry, capsular or nut-like. *(Shrubs with petiolate, entire, simple leaves. Stipules minute, deciduous.)*

LIII. CRASSULACEÆ (page 327). *Calyx* free, 4–5–7-cleft or parted, imbricate. *Petals* inserted in the bottom of the calyx, as many as its lobes, regular, free or connate in a monopetalous corolla, imbricate. *Stamens* inserted with the petals, as many or twice as many. *Ovary* of 4–5–7 carpels, nearly apocarpous; styles separate, subulate, terminal. *Fruit* of 4–7 follicles. *Seeds* albuminous. *(Succulent plants, with fleshy, opposite or alternate, mostly entire, exstipulate leaves. Flowers in cymes, regular.)*

LIV. PORTULACEÆ (page 381). *Calyx* free, 2-leaved. *Petals* 4–6, inserted in the base of the calyx, dissolving. *Stamens* numerous. *Ovary* free, 1-celled, with several (or 1) long-stalked ovules, rising from a free, central placenta; stigmata several. *Fruit* capsular or nut-like. *Embryo* curved round a central albumen. *(Succulent plants, with fles hy, entire leaves. Flowers in cymes. Readily known by the 2-lobed calyx.)*

LV. MESEMBRYACEÆ (page 386). *Calyx* 4–5-cleft, adnate or free. *Petals* very numerous, or none. *Stamens* definite or indefinite, perigynous. *Ovary* 2–5–20-celled, inferior or superior, with long-stalked ovules attached to the base or inner angle of the cell. *Stigmata* as many as the carpels. *Fruit* capsular or nut-like. *Embryo* curved round a central albumen. *(Succulent plants, with fleshy, entire leaves. Flowers showy or minute.)*

LVI. CACTEÆ (page 479). *Flowers* bisexual. *Perianth* many-leaved, imbricating

Calyx-tube adnate. *Stamens* indefinite. *Ovary* inferior, 1-celled, with several parietal placentæ. *Style* terminal, filiform; stigmata as many as the placentæ. *Fruit* succulent. *(Succulent, mostly leafless plants. Only one S. African species.)*

LVII. BEGONIACEÆ (page 480). *Flowers* monœcious. *Male: Perianth* 4-leaved. *Stamens* indefinite; anthers adnate. *Female: Perianth* 4–9-parted, with a 3-winged tube. *Ovary* inferior, 3-celled, with many axile ovules. *Stigmata* 3, subsessile, fleshy. *Capsule* 3-winged, membranous, opening by slits. *(Herbaceous plants, with juicy stems and foliage. Nodes swollen. Leaves unequal-sided, with membranous stipules.)*

LVIII. CUCURBITACEÆ (page 482). *Flowers* unisexual. *Calyx* 5-lobed, adnate. *Corolla* mostly monopetalous, 5-lobed or parted, continuous with the summit of the calyx-tube. *Stamens* inserted in the bottom of the calyx, 5–3–2, free or monadelphous; anthers extrorse, linear. *Ovary* inferior, unilocular, with 6–10 prominent placentæ, many-ovuled. *Stigma* 3–5-lobed or parted. *Fruit* succulent; seeds lying in pulp. *(Herbaceous, rarely shrubby, with prostrate or climbing stems. Leaves alternate, petioled, palmate-nerved. Tendrils.)*

LIX. PASSIFLOREÆ (page 498). *Flowers* perfect or unisexual. *Perianth* tubular, free, 3–5- (or 8–10-) lobed or parted. *Stamens* as many or twice as many as the lobes of the perianth, monadelphous or free; anthers introrse. *Ovary* superior, stipitate or subsessile, unilocular, with 3–5-parietal placentæ. *Fruit* succulent or capsular. *Seed*-coats furrowed and ridged. *(Herbaceous or shrubby, mostly climbing by tendrils. Leaves alternate, stipulate.)*

LIX* TURNERACEÆ (page 599). *Flowers* perfect. *Calyx* tubular, 5-fid. *Petals* 5, twisted in æstivation, deciduous. *Stamens* 5, alternate with the petals. *Ovary* free, unilocular, with 3-parietal placentæ; styles 3, distinct; stigmata fimbriate. *Flowers* capsular. *(Herbs, with alternate, exstipulate leaves. Only one S. African species.)*

LX. LOASACEÆ (page 502). *Flowers* perfect, regular. *Calyx* 5-lobed, adnate. *Petals* epigynous, 10, in two rows, those of the outer row concave. *Stamens* indefinite, polyadelphous. *Ovary* 3-celled. *(But one S. African species.)*

LXI. ONAGRARIEÆ (page 503). *Flowers* perfect. *Calyx* adnate; its limb 4–5-lobed, valvate. *Petals* epigynous, 4–5 or none. *Stamens* 4–8 or 5–10. *Ovary* inferior, 4–5-celled. *Style* filiform; stigma 4–5-lobed. *Fruit* a capsule or berry. *(Herbaceous (or shrubby), with exstipulate, simple leaves and axillary flowers.)*

LXII. COMBRETACEÆ (page 507). *Flowers* perfect or unisexual. *Calyx* adnate; its limb 4–5-lobed. *Petals* inserted on the summit of the calyx-tube. *Stamens* as many or twice as many as the petals. *Ovary* inferior, unilocular; ovules few, pendulous. *Style* filiform. *Fruit* a winged or ribbed drupe. *(Trees or shrubs. Leaves simple, entire, exstipulate. Flowers in spikes, racemes, or heads.)*

LXIII. RHIZOPHOREÆ (page 513). *Flowers* perfect. *Calyx* adnate; its limb 4–12-parted, valvate. *Petals* 4–12, epigynous. *Stamens* twice as many as the petals, opposing them in pairs. *Ovary* inferior, 2–4-celled; ovules in pairs. *Style* 1. *Fruit* leathery, one-seeded. *(Seaside shrubs and trees, with opposite branches and leaves. "Mangroves.")*

LXIV. LYTHRARIEÆ (page 514). *Flowers* perfect. *Calyx* free, tubular, 4–12-toothed, ribbed. *Petals* on the summit of the calyx-tube, deciduous. *Stamens* inserted at or below the middle of the calyx-tube, as many or twice as many as the petals. *Ovary* free, 2–6-celled, with many ovules. *Style* single. *Capsule* inclosed in the base of the calyx-tube. *(Herbs or shrubs, with simple, entire, exstipulate leaves.)*

LXV. MELASTOMACEÆ (page 517). *Flowers* perfect. *Calyx*-tube enclosing the ovary and partly adnate to it; limb 4–6-parted, or subentire. *Petals* 4–6, on the summit of the calyx-tube, with twisted æstivation. *Stamens* inserted with the petals, twice their number; anthers opening by terminal pores! *Ovary* plurilocular, many-ovuled. *Style* simple. *Fruit* capsular or fleshy. *(Shrubs or herbs, with opposite, 3–5-ribbed, exstipulate leaves.)*

OLINIEÆ (page 519). *Flowers* perfect. *Calyx* tubular, its base adnate with the ovary; limb 5-toothed. *Petals* 5, on the summit of the calyx-tube, with 5 alternating scales. *Stamens* 5, opposite the scales; anthers splitting. *Ovary* inferior, 4-5-celled; cells with 3 pendulous ovules. *Style* simple. *Drupe* 3-4-celled. *(A tree, with opposite, petioled, penninerved, entire, shining leaves, and densely cymose, panicled, white flowers.)*

LXVI. MYRTACEÆ (page 520). *Flowers* perfect. *Calyx* adnate; its limb 4-5-parted. *Petals* 4-5, epigynous. *Stamens* indefinite, epigynous. *Ovary* inferior, 2 or many-celled, with several axile ovules. *Style* filiform. *Fruit* a capsule or berry. *(Trees or shrubs, with mostly opposite, entire, penninerved, pellucid-dotted leaves.)*

LXVII. UMBELLIFERÆ (page 524). *Flowers* perfect, small, in umbels. *Calyx* adnate. *Petals* 5, epigynous. *Stamens* 5, alternate with the petals. *Ovary* inferior, 2-celled, with solitary, pendulous ovules. *Styles* 2, divergent. *Fruit* dry, separating into 2 one-seeded pieces. *Albumen* copious. *(Mostly herbaceous. Leaves alternate, with sheathing petioles, mostly cut or lobed.)*

LXVIII. ARALIACEÆ (page 568). *Flowers* nearly as in the last Order; but fruit fleshy, with a crustaceous or bony endocarp. *(Mostly shrubs or trees. Only one S. African genus.)*

LXIX. CORNEÆ (page 570). *Flowers* perfect, small, (panicled). *Calyx* adnate; its limb 4-toothed. *Petals* 4, epigynous, valvate. *Stamens* 4, alternate with the petals. *Ovary* inferior, 2-4-celled; ovules solitary, pendulous. *Style* single. *Fruit* fleshy, 2-4-celled. *(Trees and shrubs. Only one S. African genus.)*

LXX. HALORAGEÆ (page 571). *Flowers* minute, often unisexual. *Calyx* adnate; its limb 2-4-toothed. *Petals* 2-4, epigynous, valvate. *Stamens* as many or twice as many as the petals. *Ovary* inferior, 1-4-celled; ovules pendulous, few. *Styles* as many as the carpels, short; stigmata long, feathery. *Fruit* nut-like. *(Herbaceous, often marsh or water plants, with minute flowers.)*

LXXI. BALANOPHOREÆ (page 572). *Flowers* unisexual, in dense spikes or panicled-heads. *Perianth* 3-parted, valvate. *Stamens* opposite the segments of the perianth. *Ovary* inferior, 1-celled, 1-ovuled. *Fruit* dry or succulent. *(Fleshy, leafless or scaly, coloured, root-parasites.)*

LXXII. LORANTHACEÆ (page 574). *Calyx* adnate; its limb obsolete. *Petals* 4-8, separate or cohering, epigynous, with valvate æstivation. *Stamens* opposite the petals. *Ovary* inferior, 1-celled, with 1-3 pendulous ovules. *Fruit* succulent. *(Parasitical shrubs. Leaves entire, coriaceous, exstipulate, or none. Inflorescence various.)*

FLORA CAPENSIS.

ORDER XLVIII. LEGUMINOSÆ, Juss.

(By W. H. HARVEY).

Calyx free, 5-toothed, cleft or parted, equal or unequal, the odd segment in front. *Petals* 5 (some or all occasionally wanting), usually unequal. *Stamens* definite or indefinite, variously combined. *Ovary* simple (formed of one carpel), with one or many ovules, attached by cords to the ventral suture; style proceeding from the upper margin, i.e. continuous with the ventral suture; stigma simple. *Fruit* a legume, rarely drupaceous. Seeds one or many, without albumen; embryo either straight, or with the embryo bent back on the cotyledons.

A vast Order, found in all parts of the globe, and including huge trees, shrubs, and small or minute herbaceous plants of extremely different aspect. Leaves mostly alternate, and generally compound; sometimes with pellucid dots; sometimes resin-dotted. Stipules 2 or 1 at the base of the petiole; and often 2 *stipellæ* at the base of each leaflet. Pedicels commonly jointed and bibracteate below the flower. The three Sub-Orders, as characterised below, are readily distinguished from each other by the *æstivation* of the corolla, or the mode in which its petals are folded together in the bud. In the 1st Sub-Order, which comprises by far the largest number of the S. African genera, the *corolla* is "*papilionaceous*," that is, shaped like the blossom of the common garden pea. In such a corolla the uppermost or back petal, which is usually turned or rolled backwards, equal sided, and broader than the others, is called the *vexillum* or standard; the two lateral petals are the *alæ* or wings; and the two front petals, whose laminæ are often partly or completely connate into a boat-shaped piece, together make up the *carina* or keel.

The distribution into Tribes, here adopted, is that proposed by Mr. Bentham, who has studied this most natural Order more successfully than any living botanist, and whose admirable papers on the South African Leguminosæ, published in Hooker's London Journal of Botany, I have taken as the groundwork of my descriptions; verifying every description, however, when possible, with authentically named specimens. An examination of Thunburg's *Leguminosæ*, preserved at Upsal and Stockholm, has enabled me to correct several names, and clear up some doubtful species of early writers; and even to add some species, collected but not described by Thunberg, and which have escaped the notice of subsequent travellers. The student, in using the following Table of Genera, will pay particular attention in examining *Papilionaceæ*, to the combination of the stamens (diadelphous, monadelphous, or free); to the nature of the foliage; and the condition, whether *twining* or not, of the stem. Thus, if the stamens be *free*, the plant must belong either to *Podalyrieæ* or to *Sophoreæ*. If the stamens be united, and the leaves *simple, without stipules*, it will fall either under *Liparieæ* or the first three genera of *Genisteæ*: if the leaves be simple, and *stipulate*, other characters must be looked to. *Psoraleæ* (with *solitary* ovules), and *Indigofera* (with apiculate anthers) include every variety of foliage. Palmately 3-5-foliolate, stipulate leaves chiefly characterize *Genisteæ* (with *monadelphous* stamens) and *Trifolieæ* (with *diadelphous* stamens). Pinnato-trifoliolate leaves and *twining* stems are characteristic of *Phaseoleæ*. Pinnate-plurijugate leaves characterise *Galegeæ*, *Astragaleæ*, and *Dalbergieæ*, which all differ in their legumes; and occur also in *Abrus*, among *Phaseoleæ*, and in some *Hedysareæ*. Pinnate-*tendril tipped* leaves are peculiar to *Vicieæ*. *Hedysareæ* are only to be known by their more or less perfectly jointed or *lomentaceous* pods; but several of the genera have *one-seeded* (and of course inarticulate) pods. On the whole, in the absence of *absolute* distinctive characters to the Tribes, the student will do well, if in doubt, to try under more than one, if unsuccessful in his first guess. The Tribes and genera are all natural groups, and not difficult to learn, when the eye has become accustomed to their "*habit*," or external characters.

TABLE OF THE SOUTH AFRICAN GENERA.

Sub-Order 1. PAPILIONACEÆ. (Gen. 1–68).

Corolla papilionaceous; *petals* imbricated in æstivation; the upper-petal (*vexillum*) exterior, folding over the lateral-petals (*alæ*) and the anterior (*carina*). *Stamens* 10; either diadelphous, monadelphous, or free.

Tribe 1. PODALYRIEÆ. *Filaments* free. *Legume* continuous. Shrubs. *Leaves* either simple or palmately compound. (Gen. 1–2).

I. **Cyclopia.**—*Legume* strongly compressed. *Leaves* sessile, 3-foliolate. *Fl.* yellow.
II. **Podalyria.**—*Legume* turgid, woolly. *Leaves* simple, petiolate. *Fl.* purple.

Tribe 2. LIPARIEÆ. *Stamens* diadelphous (except in *Coelidium* and *Walpersia*). *Legume* continuous, bivalve. *Ovary* 1 or pluri-ovulate.—Shrubs or suffrutices. Leaves simple, without stipules. (Gen. 3–8).

* *Stamens diadelphous.*

III. **Liparia.**—*Flowers* yellow. Lowest calyx-segment very large, petaloid, coloured.
IV. **Priestleya.**—*Fl.* yellow. Lowest calyx-segment equal to the rest, or scarcely longer. *Corolla* conspicuous. *Ovary* several ovuled.
V. **Amphithalea.**—*Flowers* purple or purplish-white, the carina dark coloured.
VI. **Lathriogyne.**—*Fl.* yellow, inconspicuous, the corolla scarcely exceeding the calyx. *Legume* 1 seeded, hidden in the calyx.

** *Stamens united at base into a short tube, or nearly free.*

VII. **Coelidium.**—*Flowers* purplish; the *petals* not adhering to the staminal tube. *Leaves* sessile, with involute margins.
VIII. **Walpersia.**—*Flowers* yellow; the *petals* adnate at the base to the staminal tube. *Leaves* petiolate, with reflexed margins.

Tribe 3. GENISTEÆ. *Stamens* completely monadelphous. *Ovary* 2 or several ovuled.—Shrubs, suffrutices, or herbs. Leaves either simple, or palmately 3- or pluri-foliolate. (Gen. 9–24).

* *Leaves uniformly simple, without stipules.*

IX. **Borbonia.**—*Calyx* equally 5–fid, the segments pungent. *Vexillum* villous. *Legume* linear, compressed.
X. **Rafnia.**—*Calyx* 5–fid, the lowest segment much narrower than the others. *Corolla* (and the whole plant) glabrous. *Legume* lanceolate or linear.
XI. **Euchlora.**—*Calyx* and *corolla* of *Rafnia*. *Legume* ovate, turgid. Plant densely hairy.

** *Leaves stipulate, either simple or palmately compound.*

XII. **Crotalaria.**—*Carina* sharply rostrate. *Legume* very turgid. Flowers in racemes, or scattered (not umbellate).

*** *Leaves constantly palmately compound (sometimes in some* Lebeckiæ *and in* Lotononis monophylla, *reduced to a single leaflet, quasi simple).*

† Lateral and upper calyx-segments *connate in pairs;* the front segment separate, and narrower.

XIII. **Pleiospora.**—*Vexillum* erect, straight, *concave;* carina *straight*. *Style* short, straight.
XIV. **Lotononis.**—*Vexillum* reflexed; carina *inflexed,* obtuse or rostrate. *Style* geniculate. *Legume* subcompressed or subturgid, straight or falcate.
XV. **Listia.**—*Corolla* of *Lotononis*. *Legume* linear, repeatedly folded and twisted from side to side.

†† Calyx distinctly *bilabiate,* the upper lip bifid, the lower tridentate, trifid or tripartite. *Legume* linear, compressed.

XVI. **Argyrolobium.**—*Calyx* deeply divided. *Vexillum* ample, longer than the carina. *Legume* silky, not glandular.
XVII. **Dichilus.**—*Calyx* of *Argyrolobium*. *Carina* obtuse, rather longer than the vexillum. *Legume* subtorulose, not glandular.
XVIII. **Melolobium.**—*Calyx* tubular, shortly bilabiate. *Corolla* small. *Carina* obtuse. *Legume* subtorulose, generally glandular or viscoso-pubescent.

††† Calyx campanulate, *hollow or intruse at base,* shortly 5–fid, sub-bilabiate. Glabrous shrubs or suffrutices.

XIX. **Hypocalyptus.**—*Vexillum* much longer than the carina. *Legume* linear.
XX. **Loddigesia.**—*Vexillum* much shorter than carina and alæ. *Legume* ovato-lanceolate.

†††† Calyx *oblique*, 5-toothed or 5-fid.

XXI. **Lebeckia.**—*Carina* longer than the alæ, mostly than the vexillum. *Legumen* linear, flat, terete, or turgid, several seeded.
XXII. **Viborgia.**—*Petals* with long claws. *Legume* stipitate, ovate or oblong, *indehiscent*, one seeded, *winged* on the upper side.

††††† Calyx *subequally* 5-toothed or 5-fid.

XXIII. **Buchenroedera.**—Leaves *petiolate*, 3-foliolate, mostly *with stipules*.
XXIV. **Aspalathus.**—Leaves *sessile;* leaflets as if fascicled leaves, *without stipules*.

Tribe 4. PSORALIEÆ. *Stamens* diadelphous (or the vexillary filament cohæring in the middle). *Ovary* 1-ovuled. Leaves never *stipellate*, variously compound, *very generally* sprinkled with resinous, glandular dots. Infl. axillary or terminal.

XXV. **Psoralea.**

Tribe 5. TRIFOLIEÆ. *Stamens* diadelphous. *Ovary* 2 or several ovuled. *Stem* erect or trailing, but *not twining or climbing*. Herbaceous, or suffruticose. *Leaves* 3-foliolate, very rarely 5-foliolate, never stipellate. (Gen. 26–30).

XXVI. **Lotus.**—*Carina* very acute. *Legume* cylindrical, many seeded. *Flowers* umbellate.
XXVII. **Trifolium.**—*Carina* obtuse, adnate to the alæ, persistent. *Legume* small, *concealed in the calyx*. *Flowers* in dense spikes or heads.
XXVIII. **Melilotus.**—*Carina* obtuse, free, deciduous. *Legume* small, but longer than the calyx. *Flowers* in lax racemes.
XXIX. **Trigonella.**—*Carina* obtuse, free, very short. *Legume* linear, slightly curved, many seeded, much longer than the calyx. *Flowers* in very short (sub-umbellate) racemes.
XXX. **Medicago.**—*Carina* obtuse, free. *Legume* much incurved or spirally twisted, often bordered with spinous teeth. *Flowers* racemose or sub-solitary.

Tribe 6. INDIGOFEREÆ. *Stamens* diadelphous ; *the connective of the anthers apiculate*. *Ovary* 2 or several ovuled. *Stem* never twining or climbing. *Leaves* variously compound, rarely simple. *Pubescence* very frequently strigose.

XXXI. **Indigofera.**—*Vexillum* roundish, reflexed. *Carina* with a *spur* or prominence at each side, near the base. Flowers red or purple, rarely white.

Tribe 7. GALEGEÆ. *Stamens* monadelphous or diadelphous. *Ovary* 2 or several ovuled. *Legume* bivalve, one celled. *Stem* never twining. *Leaves* pinnate, rarely unifoliolate, *sometimes* stipellate. *Flowers* racemose. (Gen. 32–37).

* *Legume coriaceous or rigid, not membranous.*

XXXII. **Tephrosia.**—*Legume* linear, compressed, coriaceous. Small shrubs, suffrutices, or herbs. *Flowers* purple, pink or white.
XXXIII. **Milletia.**—*Legume* elliptic or lanceolate, few seeded, hard and woody, with thickened margins. *Trees* or large shrubs. *Flowers* purple, or purplish.
XXXIV. **Sesbania.**—*Legume* very long and slender, contracted between the seeds, but not jointed, many seeded. *Flowers* yellow.

** *Legume membranous, semi-translucent, either compressed or bladdery.*

XXXV. **Sutherlandia.**—*Leaves* pinnate. *Vexillum* shorter than the *acute* carina. *Style* bearded at back. *Flowers* scarlet, racemose.
XXXVI. **Lessertia.**—*Leaves* pinnate. *Vexillum* expanded, longer than the *obtuse* carina. *Style* bearded in front. Flowers purple, pink, or rosy white, racemose.
XXXVII. **Sylitra.**—*Leaves* simple. *Flowers* axillary, minute.

Tribe 8. ASTRAGALEÆ. *Stamens* diadelphous. *Legume* completely or incompletely *longitudinally 2-celled*, by the introflexion of one of the sutures. *Stem* never twining. *Leaves* pinnate.

XXXVIII. **Astragalus.**—*Legume* with its lower (*carinal*) suture introflexed.

Tribe 9. HEDYSAREÆ. *Legume* more or less completely *jointed*, usually separating at maturity into *indehiscent, one seeded* articulations : sometimes reduced to a *single articulus*. *Leaves* variously compound, or simple. (Gen. 39–49).

* *Leaflets 2–4, from the apex of a common petiole (pellucid dotted).*

XXXIX. **Zornia.**

** *Leaves pinnate; leaflets in two or many pairs.*

XL. **Æschynomene.**—*Leaves* multi-jugate. *Stamens* 10, connate in two 5–androus parcels. *Legume* many jointed.

XLI. **Arachis.**—*Leaves* bijugate. *Stamens* monadelphous. *Pod* subterraneous, indehiscent.

*** *Leaves pinnately trifoliolate.*

XLII. **Stylosanthes.**—*Calyx* tube very long; the *corolla* inserted in its throat. *St.* monadelphous. *Legume* short, of one or two joints.

XLIII. **Desmodium.**—*Calyx* 2-lipped, the corolla inserted in its base. *St.* diadelphous. *Legume* many jointed.

XLIV. **Anarthrosyne.**—*Flower* of *Desmodium.* *Legume* compressed, linear-subfalcate, *imperfectly inarticulate,* not spontaneously separating. *Fl.* panicled.

**** *Leaves simple or unifoliolate.*

XLV. **Alysicarpus.**—Suffruticose or herbaceous. *Legume* terete, separating into numerous joints.

XLVI. **Requienia.**—A suffrutex with *obcordate* leaves, and minute, axillary flowers. *Legume* oval, compressed, one-seeded, hook-pointed.

XLVII. **Hallia.**—Suffrutices, with cordate or lanceolate leaves, and adnate stipules. *Flowers* axillary, violet-coloured, small. *Legume* compressed, one-seeded.

XLVIII. **Alhagi.**—*Spinous* shrubs. *Flowers* red, racemose. *Legume* stipitate, terete, irregularly constricted, not perfectly jointed, indehiscent.

Tribe 10. Vicieæ. *Stamens* either diadelphous or monadelphous in the middle. Herbaceous plants, with *abruptly pinnate* leaves, *the common petiole produced into a tendril* or excurrent point. Flowers axillary.

XLIX. **Vicia.**

Tribe 11. Phaseoleæ. *Stamens* either diadelphous, or monadelphous in the middle. *Legume* bivalve. *Stem* frequently twining, prostrate or diffuse. *Leaves* usually *pinnately-trifoliolate,* sometimes unifoliate (in *Abrus* pinnate-multi-jugate), stipellate. (Gen. 50–60).

* *Leaves trifoliolate. Ovary with more than two ovules.*

L. **Dumasia.**—*Calyx* tubular, *obliquely truncate, entire.*

LI. **Teramnus.**—*Calyx* 4–5 fid. *Vexillum* with a long claw, not callous at base. *Stamens* monadelphous. *Ovary* sessile. *Stigma* capitate. *(Flowers minute).*

LII. **Galactia.**—*Calyx* sharply 4–fid, *Vexillum* suborbicular, not callous at base. *Stamens* diadelphous. *Ovary* subsessile. *Stigma* minute. *(Flowers small).*

LIII. **Erythrina.**—*Calyx* various. *Vexillum* oblong, incumbent, not callous at base, *very much longer than the alæ and carina.* *Stamens* exserted. *(Shrubs or trees, with large, showy flowers).*

LIV. **Canavalia.**—*Calyx* bilabiate; upper lip very large, with 2 rounded lobes, lower small. *Vexillum* bicallous within. *Stigma* terminal. *(Twiners).*

LV. **Vigna.**—*Calyx* subequally 4–5 fid. *Vexillum* bi-callous within. *Style* compressed, channelled on one side, with a hooked point and an oblique stigma. *(Erect or twining).*

LVI. **Dolichos.**—*Calyx* bilabiate; upper lip bifid or subentire, lower trifid. *Vexillum* bi-callous within. *Style* channelled or terete, with a straight point and terminal, truncate or capitate stigma. (*Twining or prostrate. Fl. purple or green.*)

LVII. **Fagelia.**—*Calyx* sharply 5–cleft beyond the middle. *Carina* longer than the alæ, very obtuse. *Style* straight-pointed. *Legume* turgid, constricted between the seeds. (*A viscidly-hairy twiner, with yellow flowers*).

** *Leaves trifoliolate, or unifoliolate (or pinnate). Ovary 2 ovuled.*

LVIII. **Rhynchosia.**—*Ovary* glabrous or pubescent. *Seeds* globose-reniform, with a short hilum, and subcentral seed cord.

LIX. **Eriosema.**—*Ovary* very hairy. *Seeds* oblong, obliquely transverse; the seed-cord attached at one end of a linear hilum.

*** *Leaves pinnate, multi-jugate. Ovary many-ovuled.*

LX. **Abrus.**—*Leaves* abruptly pinnate. *Seeds* globose, scarlet and black.

Tribe 12. DALBERGIEÆ. *Stamens* monadelphous or variously connate, *Legume* either perfectly indehiscent, or rarely splitting eventually into rigid-ligneous valves. *Stem* woody, either arborescent, shrubby, or twining. *Leaves* pinnate, 5 or many foliolate, or unifoliolate. (Gen. 61–63).

LXI. **Lonchocarpus.**—*Leaflets opposite.*

LXII. **Pterocarpus.**—*Leaflets* alternate. *Anthers* versatile. *Legume* orbicular.

LXIII. **Dalbergia.**—*Leaflets* alternate. *Anthers* small, terminal; loculi erect or divergent. *Legume* oblong or linear.

Tribe 13. SOPHOREÆ. *Stamens* free. *Stem* woody, arborescent, shrubby or suffruticose. *Leaves* pinnately multifoliolate, rarely (in *Bracteolaria*) unifoliolate. (Gen. 64–67).

LXIV. **Sophora.**—*Leaves* pinnate. *Carina* obtuse, straightish. *Legume* moniliform.

LXV. **Virgilia.**—*Leaves* pinnate. *Carina* incurved, acutely rostrate. *Legume* oblong, compressed, coriaceous, with very blunt edges. *(Flowers purple).*

LXVI. **Calpurnia.**—*Leaves* pinnate. *Carina* incurved, obtuse. *Legume* oblong-linear, very flat, sharp-edged. *(Flowers yellow).*

LXVII. **Bracteolaria.**—*Leaves* unifoliolate.

Sub-Order 2. CÆSALPINIEÆ. (Gen. 68–75).

Corolla irregular or subregular, not papilionaceous; *petals* imbricated in æstivation, the upper petal interior. *Stamens* 10 or fewer, free or monadelphous.

* *Leaves simply pinnate.*

LXVIII. **Parkinsonia.**—*Stamens* 10; anthers versatile, splitting. *Ovary* sessile. *Legume* linear, very long, many-seeded.

LXXIII. **Cassia.**—*Stamens* (fewer than ten perfect) *opening by terminal pores.*

LXXIV. **Schotia.**—*Stamens* 10; anthers versatile, splitting. *Ovary* stipitate. *Legume* short, broadly-oblong, coriaceous, few-seeded.

** *Leaves bi-pinnate.*

LXIX. **Guilandina.**—Arborescent. *Ovary* stipitate. *Legume* ovate, covered with sharp prickles.

LXX. **Melanosticta.**—Half-herbaceous: all parts sprinkled with convex, *black, resinous dots.*

LXXI. **Peltophorum.**—Arborescent. *Flowers* pedicellate. *Filaments* as long as the petals, hairy at base. *Style* equalling the petals. *Legume* unarmed.

LXXII. **Burkea.**—Arborescent. *Flowers* sessile. *Filaments* and *style* very short. *Legume* unarmed.

*** *Leaves simple, 2-lobed (formed of 2 terminal, confluent leaflets).*

LXXV. **Bauhinia.**

Sub-Order 3. MIMOSEÆ. (Gen. 76–82).

Flowers minute, in dense heads or spikes. *Corolla* irregular, its petals free or united in a tube, *valvate* in æstivation. *Stamens* definite or indefinite.

Tribe 1. EU-MIMOSEÆ. *Stamens* definite; *pollen* powdery.

LXXVI. **Entada.**—*Flowers* uniform, spicate, sessile. *Legume* breaking up into 1-seeded joints.

LXXVII. **Elephantorhiza.**—*Fl.* uniform, racemose, pedicellate. *Legume* continuous.

LXXVIII. **Dichrostachys.**—*Flowers* of two kinds in the spike; the upper perfect; the lower neuter, with long slender filaments without anthers.

LXXIX. **Xerocladia.**—*Flowers* uniform, capitate, sessile. *Legume* one-seeded, semiorbicular.

Tribe 2. ACACIEÆ. *Stamens* indefinite; *pollen* collected into 4–6 masses in each anther cell.

LXXX. **Acacia.**—*Corolla* small, tubular. *Stamens* free.

LXXXI. **Albizzia.**—*Corolla* funnel-shaped. *Stamens* shortly tubular at base.

LXXXII. **Zygia.**—*Corolla* narrow funnel-shaped. *Stamens* connate into a long, much exserted tube.

I. CYCLOPIA, Vent.

Calyx subequally 5-cleft, with the base indented. *Petals* subequal; *vexillum* roundish, plaited at base, with a short recurved claw; alæ oblong, with a transverse fold; *carina* incurved, obtusely rostrate. *Stamens* separate, or slightly connate at base; filaments dilated. *Ovary* glabrous, several-ovuled. *Legume* oblong, compressed, uni-locular, bivalve, coriaceous; seeds strophiolate. *Benth. in Hook. Lond. Journ.* 2. *p.* 432. *DC. Prod.* 2. *p.* 101.

Erect, dark-coloured or rusty-olivacescent South African shrubs. Leaves sessile, palmately trifoliolate; leaflets narrow, linear, lanceolate or rarely ovate, glabrous or pubescent, frequently with revolute margins. Stipules none. Peduncles axillary, 1-flowered, 2-bracted at base. Flowers bright yellow. Name from κυκλος a *circle*, and πους, a *foot*, because there is a circular depression of the base of the calyx, round the pedicel.

ANALYSIS OF THE SPECIES.

Glabrous or nearly so:
  Leaflets *flat*, with slightly *recurved* or plane margins:
    Bracts *smooth*, with *straight* points:
      Leaflets cordate-ovate, acute or acuminate, flat ... ... (1) **latifolia.**
      Leaflets elliptic-oblong or sub-lanceolate-linear, obtuse (2) **Vogelii.**
    Bracts *striate*, with *recurved* points ... ... (3) **longifolia.**
  Leaflets linear or filiform, with *strongly revolute* margins:
    Bracts very broad, *obtuse;* leaflets *linear-filiform* ... ... (4) **tenuifolia.**
    Bracts boatshaped, *acute;* leaflets linear ... ... (5) **genistoides.**
Hairy or pubescent, *at least* on the twigs and young parts:
    Bracts *ribbed*, with *recurved* points ... ... (7) **pubescens.**
    Bracts *smooth*, with *straight* points:
      Flowers pedicellate; calyx *glabrous;* its lobes *falcate* (6) **galioides.**
      Fl. sessile; calyx softly *hairy;* its lobes *falcate* ... ... (8) **sessiliflora.**
      Fl. subsessile; calyx-lobes *broadly-oblong, ciliate* ... ... (9) **Bowieana.**

1. C. latifolia (DC. Prod. 2. p. 101.); glabrous; leaflets *cordate-ovate or ovate-acuminate, truncate or cordate at base, acute;* outer bract as long as the pedicel; calyx-lobes deltoid, acute. *Benth.! in Lond. Journ.* 2. *p.* 432. *B. cordifolia, Benth! Ann. Mus. Vind.* 2. *p.* 67.

HAB. South Africa, *Scholl!* in Herb. Mus. Vind. (Herb. Benth.!)

Densely branched, robust; the branches and ramuli somewhat winged along the angles. Leaflets ½-inch long or more, the middle one longest, more than ¼-inch wide at base, the upper ones narrower: none ever *tapering* at base. Peduncles very short. A rare species, not found by Ecklon or Drege.

2. C. Vogelii (Harv.); glabrous; leaflets *flat, with reflexed or sub-revolute margins,* elliptic-oblong, oblong-linear, lanceolate, or spathulate, either acute or obtuse at one or both ends; bracts navicular, broadly ovate, acute or acuminate, *erect, smooth,* longer or shorter than the pedicel; calyx-lobes deltoid or broadly falcate, acute or obtuse, shorter, or rather longer than the tube.

VAR. α. **subternata**; pedicel as long as, or exceeding the bract; calyx lobes deltoid, acute; leaflets elliptic-oblong or lanceolate. *C. subternata, Vog. Linn.* 10. *p.* 595. *Benth. in Lond. Journ.* 2. *p.* 432. *C. latifolia, E. Mey.! Com. p.* 3. *C. grandifolia, A. DC. Burch. Cat. No.* 5519.

VAR. β. **brachypoda**; pedicel much shorter than the bracts or calyx; calyx lobes deltoid, acute; leaflets narrow, tapering at base. *C. brachypoda, Benth. Lond. Journ. l. c, C. sessiliflora, E. & Z. No.* 1147. *non E. M. Rafnia retroflexa β. Thunb.* Herb. (excl. var. *a*).

VAR. γ. **intermedia**; pedicel and the obtuse bracts of equal length; calyx lobes deltoid, obtuse; leaflets oblong-linear or lanceolate. *C. intermedia, E. Mey. Com. p.* 3. *excl. lit. c. Benth. l. c. Burch. Cat.* 4929.

VAR. δ. **laxiflora**; pedicel much longer than the *small*, acute or obtuse bracts; calyx lobes deltoid, obtuse or subacute; leaflets tapering much at base. *C. laxiflora, Benth! l. c. C. latifolia, E. & Z.! En. No.* 1149. (*non DC.*)

VAR. ε. **falcata**; pedicels scarcely equalling the ovato-lanceolate, *acuminate*, very acute bracts; calyx-lobes *longer than the tube, falcate, acute;* leaflets varying from broadly elliptic-oblong with straight edges, to lanceolate and narrow-linear with strongly reflexed margins.

HAB. Mountains in the districts of Stellenbosch, Swellendam, and George. Var. α. Drakenstein & Bosjesveld, *Drege, Mundt. and Maire, Wallich.* β. Puspas valley, Swellendam, *E. & Z.* Grootvadersbosch, *Zeyher.* γ. Near Swellendam and on the Kaureboom River, George, *Drege.* δ. The Knysna and Plettenberg Bay, *E. & Z.! Pappe* (36), *Mundt & Maire,* Georgetown, *Dr. Prior.* ε. Witsenberg, *Zeyher!* 354. *Pappe* (39). French Hoek, *W.H.H.* Caledon, *Dr. Prior.* (Herb. Th. Bth. Hk. Sd. D.)

Very variable in the breadth, flatness and form of the leaves; the length of the pedicel, of the bracts, and in the form of the calyx lobes. If the four first varieties be held separate, so must the fifth, which differs from all the rest in its very large and much acuminate bracts and strongly falcate calyx lobes. In other respects it is so like var. α. that I find it named "subternata" in Herb. Benth. Stems angular, 3-4 feet high, robust, except in var. β., which is a weak, straggling plant. Leaflets ¾-1½ inch long, 1-3 lines broad, *rarely* quite flat, usually with the margins recurved, but never covering the lower surface. Bracts in α. 1-1½ lines, in β. ¾-1 line, in γ. 2-3 lines, in δ. 1 line, in ε. 3-5 lines long, always smooth and somewhat shining, with straight points: acute, except in var. γ.

3. C. longifolia (Vogel, Lin. 10. p. 595.); glabrous; leaflets flat, with reflexed margins, elongate, linear-spathulate, tapering at base, obtuse; bracts sharply keeled, *striate, with strongly recurved points,* much shorter than the pedicel; calyx lobes broadly falcate, acute, about as long as the tube. *Benth. l. c. p.* 433.

HAB. South Africa, *Mundt & Maire.* (Herb. Hook.)

A laxly-branched, straggling shrub, with angular branches. Leaflets 1-1½ inches long, 1-1½ lines wide, with obvious petiolules. Pedicels 3-4 times as long as the strongly hooked bracts.—This comes near *C. Vogelii, var.* δ., but differs from all states of that species, in its bracts.

4. C. tenuifolia (Lehm. Linn. 5. p. 373.); glabrous; leaflets *linear-filiform,* with revolute edges, elongate; bracts *very broad, obtuse, strongly keeled,* smooth, much shorter than the pedicels; calyx lobes deltoid, obtuse. *Benth. l. c. p.* 433. *E. & Z. No.* 1150. *C. laricina, E. Mey.! Com. p.* 153. *Zeyher* 2257. *Burch. Cat.* 7522.

HAB. Mountains of Swellendam and George, *E. & Z., Mundt, Drege,* Stellenbosch, *W.H.H.* River Zonderende, *Zeyher,* 2257. (Herb. D. Hk. Sd.)

Shrub, 3-6 feet high, erect, much branched: readily known by its very narrow, *pine-like* leaflets, which are 1½-2 inches long, and ¼ line in diameter. Flowers in dense subterminal clusters, with remarkably large, very obtuse or truncate, boat-shaped bracts. Pedicels ½-¾ inch long.

5. C. genistoides (Br. Hort. Kew, Ed. 2. vol. 3. p. 5); *glabrous or nearly so;* leaflets narrow-linear, *with strongly revolute margins,* acute or mucronate; bracts boatshaped, acute, smooth, erect, shorter than the pedicel; calyx lobes broadly *falcato-subulate,* acute or mucronate. *Benth. l. c. p.* 434. *DC. Prod.* 2. *p.* 101. *Andr. Bot. Rep. t.* 427. *C. genistoides and C. teretifolia, E. & Z. No.* 1143, 1145! *C. genistoides and C.*

galioides, E. Mey. in Herb. Drege. C. No. 353, 2256, *Zeyher. Ibbetsonia genistoides, Bot. mag. t.* 1259. *Galega genistoides, Thunb. Cap. p.* 600. *excl. var. β.*

VAR. *β.* **heterophylla**; leaflets, especially of the lower leaves, broader, imperfectly revolute. *C. heterophylla, E. & Z.* 1148.

HAB. Common on hills and mountain-sides, throughout the S. W. districts. (Herb. Th., Sd., D., Hk., Bth.)

A virgate, much-branched shrub, 3–5 feet high, with angular branches; the young twigs sometimes minutely downy. Leaves closely set, patent; leaflets ½–¾ or 1 inch long, ½ line in breadth, the under surface generally concealed, except the midrib, by the revolution of the margins. In var. *β.* the leaflets are twice as broad, and imperfectly revolute.

6. C. galioides (DC. Prod. 2. p. 101); young branches and *twigs densely hirsute;* leaflets (except the young ones) *glabrous,* linear-terete, strongly revolute, acute or mucronate; bracts ovato-lanceolate, cuspidate-acuminate, smooth, erect, the outer one longer than the pedicels; calyx *glabrous,* its lobes broadly falcato-subulate, cuspidate. *Benth. l. c. p.* 434. *Galega genistoides β. Thunb.*

HAB. Summit of Table Mountain; Muysenberg and on the Cape Town ranges, *Burke, W.H.H., Wright* (554, 572), &c. (Herb. Th., Bth., Hk., D.)

A stouter but shorter growing bush than *C. genistoides,* with dark-coloured blackish-brown densely hairy branches, more erect leaves, and thicker and more revolute leaflets. The young parts are covered with long, villous, deciduous hairs. Leaflets ½–¾ inch long, nearly 1 line in breadth, the under surface with a slender medial furrow; no midrib visible. The specimens of E. & Z. under this name in Herb. Sond. belong to *C. genistoides.*

7. C. pubescens (E. & Z.! Enum. No. 1146); branches and ramuli, pedicels, and calyx, tomentose or pubescent; leaflets glabrous, narrow-linear, with revolute margins; bracts *acuminate, hooked-backwards, ribbed and furrowed,* much shorter than the villous pedicel; calyx furrowed, pubescent or tomentose, its lobes from a broad base falcato-subulate, very acute. *Benth. in Lond. Journ.* 2. *p.* 433.

HAB. Among shrubs on the Krakakamma plains, and declivities of V. Staadensberg Mountain, Uitenhage, *E. & Z.!* (Herb. Bth., Sd.)

A robust, densely branched and probably tall growing species; readily known by its strongly ribbed, hooked bracts. The pubescence varies much in different specimens; in some the hairs are very short, thinly scattered; in others dense, somewhat tomentose and even canescent. The ribs on the calyx tube are equally variable: those on the bracts are more constant. The leaflets are ¾–1 inch long, not a line wide. The pedicels ½–¾ inch. Vexillum acuminate.

8. C. sessiliflora (E. Mey. Com. p. 4. non E. & Z); young branches *densely villous,* older glabrous; leaflets linear-terete, strongly revolute, softly hairy or glabrescent, sub-acute; bracts very broadly ovate, cuspidate, smooth and shining, villous at the edges; calyx sessile, *softly hairy, its lobes falcato-subulate,* acute; vexillum apiculate. *Benth. l. c. p.* 434. *Burch. Cat.* 7770. *(fide Benth.)*

HAB. Moist rocks, Dutoits Kloof Mountains. *Drege.* (Herb. Bth., Hk.)

A rigid, stoutgrowing, much-branched shrub, more hairy than *C. galioides,* but with softer, and more silky pubescence; the young foliage is densely clothed, the older becoming nearly glabrous. The leaflets ¾ inch long, ¼ line in diameter, the margins completely revolute.

9. C. Bowieana (Harv.); young twigs thinly villous, older glabrous; leaflets linear-terete, strongly revolute, glabrous, acute; bracts broadly cymbiform, obtuse or subacute, smooth; calyx subsessile, *glabrous, its lobes broadly-oblong, obtuse or subacute, ciliolate.*

HAB. S. Africa, *Bowie.* (Herb. Hook.)

This remarkable looking *form,* or species, resembles *C. galioides* in aspect, but differs in the much shorter pedicels; the large very broad and obtuse, or scarcely acute bracts, and the calyx lobes. I am unwilling to pass it over, although nothing more is known of it than the three specimens preserved in Herb. Hooker.

II. PODALYRIA, Lam.

Calyx widely campanulate, subequally 5-cleft, with the base indented. *Vexillum* ample, rounded-emarginate, with a short recurved claw; *alæ* obovate, oblique, rather shorter than the vexillum, longer than the broad, obtuse *carina. Stamens* separate, or slightly connate at base. *Ovary* sessile, pubescent, many-ovuled. *Legume* turgid, leathery, villous. *Benth! in Hook. Lond. Journ. vol.* 2, *p.* 434. *DC. Prod.* 2, *p.* 101.

Silky or silvery-pubescent South African shrubs, with simple, alternate, expanded leaves. Stipules subulate, deciduous. Peduncles axillary, 1-2, rarely 3-4 flowered. Bracts solitary, falling off before the opening of the flowers. Flowers purple, rosy, or blush-white. Named from *Podalyrius,* a son of Æsculapius.

ANALYSIS OF THE SPECIES.

1. Nitidæ. *Adult-leaves* quite glabrous and shining on the upper surface, silky or villous on the lower. (Sp. 1-6.)

Leaves veinless: peduncles *shorter* than the leaves ... ... (1) **speciosa.**
Leaves veinless: pedunc. equalling or exceeding the leaves:
  Lateral calyx segments as long as the carina ... ... (2) **glauca.**
  Lateral calyx-segments much shorter than the carina:
    Leaves *orbicular,* with revolute margins ... ... (4) **orbicularis.**
    Leaves ovate or obovate:
      Leaves ½ inch long, 2 lines wide ... ... (6) **microphylla.**
      Leaves about uncial, densely silky-lanose beneath (3) **buxifolia.**
Leaves reticulately veined on the lower surface ... ... (5) **reticulata.**

2. Villosæ. *Leaves* tomentose on both surfaces; the toment on the *upper* surface mostly silky, that on the lower less abundant; the primary veins and often the netted veinlets conspicuous and prominent. Bracts broad, but not cap-shaped. (Sp. 7-11.)

Leaves orbicular or very broadly ovate, obtuse:
  Lvs. densely woolly beneath, not obviously nerved ... (7) **cordata.**
  Lvs. shortly tomentose and *reticulately nerved* beneath ... (8) **canescens.**
Leaves oblong or ovate-elliptic, or lanceolate, acute or subacute:
  Under surface densely hairy; lateral veins inconspicuous:
    Lvs. narrow-oblong, 1-1¾ inch long, 3-4 lines wide (9) **velutina.**
    Lvs. ovate or elliptical, ¾-1 inch long, 8-9 lines wide (10) **Burchellii.**
  Under surface thinly pubescent, conspicuously netted-veined ... ... (11) **lanceolata.**

3. Calyptratæ. *Leaves* pubescent on both sides, netted with veins and veinlets on the lower. *Bracts* very broad, *connate into a cap,* which separates at base and falls off from the opening flower. (Sp. 12.)

Leaves obovate-elliptical, obtuse, 1-2 inches long ... ... (12) **calyptrata.**

4. Sericeæ. *Leaves* silky or silky-villous on both sides; the veins concealed under the *shining* hairs, or rarely, in old leaves somewhat prominent. Bracts broad or narrow, but not cap-shaped. (Sp. 13-17.)

Peduncles *much longer* than the leaves, 2-1 flowered:
  Calyx shaggy with curled, rusty, coarse hairs ... ... (14) **argentea.**
  Calyx silky, with appressed and glossy hairs ... ... (15) **biflora.**

Peduncles *shorter* or *scarcely longer* than the leaves:
Calyx shaggy with curled, rusty hairs ... ... (13) **myrtillifolia.**
Calyx silky; its lobes *deltoid-acute*, much shorter than the carina ... ... (16) **cuneifolia.**
Calyx silky; its lobes *subulate*, as long as the carina ... (17) **sericea.**

I. **NITIDÆ** (Sp. 1–6.)

1. P. speciosa (Eck. & Zey.! No. 1164); leaves glabrous and shining above, densely silky beneath, *veinless*, the lower ones elliptic-oblong or sub-lanceolate, flattish, upper linear-elongate, with strongly revolute margins, retuse; *peduncles much shorter than the leaves;* calyces and pods densely rufo-villous. *Benth. Lond. Journ.* 2, *p.* 435.

HAB. Summits of Hott. Holl. Mts.; near Palmiet River and Klynriversberg, and near "Hemel and Aarde." *E. & Z.! Pappe* 41. (Herb. Bth., Sd., D.)

Stem 1½–2 feet high, erect, with virgate branches, appressedly pubescent. Leaves 2–3 inches long, the lower ones ½–1 inch broad, with slightly rolled margins, the upper 1–2 lines wide, perfectly linear, the revolute margins nearly concealing the whole under surface. Pubescence of the under surface generally ferruginous, coarsely silky, close-pressed. Flowers in pseudo-racemes towards the ends of the branches. Pods very hirsute.

2. P. glauca (DC. Prod. 2. p. 102); leaves glabrous and shining above, silky beneath, *veinless*, either elliptical, obovate, oblong, or lanceolate, obtuse, mucronulate, with slightly reflexed margins; peduncles generally *much longer* than the leaves; calyces rufo-sericeous, *the lateral segments about as long as the carina;* pods softly villous. *Benth. l. c. p.* 495. *P. buxifolia, Lam. Dic.* 2. *t.* 327. *f.* 4. *E. Mey. Com. p.* 7. *P. Mundiana, E. & Z.!* 1162. *P. sparsiflora, E. & Z.!* 1166. *P. racemulosa, E. & Z.!* 1165. *Zeyher, No.* 2268. *Burch. Cat.* 5118.

HAB. Mountains of Swellendam and George, *Drege.! Mundt.! Bowie, Burchell.* Zwarteberg at Klynriver; and in the Langekloof, *E. & Z.!* (Herb. Bth., Sd., Hk., D.)

Sub-erect or decumbent, 1–2 feet high, not much branched. Leaves ½–¾ inch, seldom exceeding an inch long, ¼–½ inch wide, the silky hairs of the lower surface either rufous or pale. Peduncles 2–6 times as long as the leaves, 1–2 flowered.

3. P. buxifolia (Willd.? Sp. 2. p. 505, ex parte, DC. Prod. 2, p. 102); "leaves *broadly ovate or obovate*, flattish, shining above, densely sericeolanose beneath, veinless; peduncles as long as the leaves or longer, villous, 1–2 flowered; calyces rufo-villous, the lateral segments cultrate, *much shorter than the carina.*" *Benth. l. c. p.* 436. *P. glauca, β. biflora, E. & Z.!* 1167. *in Herb. Benth.!*

HAB. Between Swellendam and Kochman's Kloof, *E. & Z.!* (Herb. Benth.)

Of this I have only seen a single specimen. The upper surfaces of the young leaves are thinly sprinkled with hairs; the older glabrous and glossy. The species requires further examination, with better specimens.

4. P. orbicularis (E. Mey.! Com. p. 8, *non E. & Z.*); leaves (becoming) glabrous and shining above, densely ferrugineo-villous beneath, *veinless, orbicular, with revolute margins*, very obtuse or mucronulate; peduncles 1–2 fl., as long as the leaves or longer; calyces rufo-hirsute, the lateral segments much *shorter* than the keel. *Benth. l. c. p.* 436. *E. & Z. No.* 1159!

HAB. Stony mountain sides near Caledon, *E. & Z.!* Gnadendahl, *Drege!* Bavians' Kloof, *Krauss.* (Herb. Bth. Sd.)

The subrotund leaves are ¾–1 inch long, and nearly as wide; the young ones thinly sprinkled with hairs, the older glabrous and shining, the under surface densely shaggy with dark, reddish-brown hairs. When dry the upper surfaces of the leaves are frequently reticulated, from a shrinking of the parenchyma.

5. P. reticulata (Harv.); leaves glabrous and shining above, appressedly pubescent, with *prominent midrib and netted veins* beneath, orbicular or obovate, very obtuse; peduncles 2-flowered, as long as the leaves or shorter; calyces rufo-sericeous, the lateral segments cultrate, acute, as long as the carina. *Zey. No.* 2269.

HAB. On the Zwarteberg, Caledon, *Zeyher*, *Pappe* 40. (Herb. D., Hk., Sd.)

Branched from below, the branches virgate, 1–2 feet long, thinly pubescent. Leaves about ¾ inch long, and nearly as broad, the young ones sprinkled with a few deciduous, short hairs, the older quite glabrous and shining above, and obviously reticulated below, green or fulvous. Calyces appressedly pubescent, rusty-brown. Peduncles mostly shorter than the leaves, 1–2 flowered. Known from others of this section by the netted venation of the leaves.

6. P. microphylla (E. Mey. in Linn. vol. 7. p. 147); leaves (very small) obovate, mucronulate, glabrous above, appressedly silky beneath, veinless; peduncles much longer than the leaves, minutely-silky, one-flowered; calyx rufo-sericeous, its segments deltoid, acute, shorter than the tube and much shorter than the carina. *E. & Z.! En. No.* 1174.

HAB. Among shrubs on the Tigerberge and Paardeberge, Stellenbosch. *Eck. & Zey.! Pappe* 48. (Herb. Sd., D.)

A woody, divaricately, much-branched shrub, the branches and twigs minutely silky, with close-pressed, very short hairs. Leaves about ¼ inch long, 2 lines broad, with a recurved point, thickish, midribbed below, but veinless. Peduncles 1–1½ inches long, with close-pressed minute pubescence. Pubescence of the calyx very short and close-pressed.

2. **VILLOSÆ** (Sp. 7–11.)

7. P. cordata (R. Br. Hort. Kew. 2. vol. 3. p. 8); thickly *villoso-tomentose in all parts;* leaves orbicular or broadly ovate, rounded or cordate at base, *densely woolly on both surfaces;* peduncles shorter than the leaves; calyces very shaggy, with reddish-brown hairs, their lobes broadly cultrate, nearly as long as the keel. *Benth. in Lond. Journ.* 2. *p.* 437. *DC. Prod.* 2. *p.* 102. *E. & Z. En. No.* 1151, *pro pte. Zey.* 2264. *P. hirsuta, Willd.? Sp.* 2. *p.* 505.

HAB. Moist mountain situations in the western districts. Hott. Holl. Mts. *E. & Z.! Pappe.!* Dutoitskloof, *Drege!* Klynhauhoek, *Zeyher!* (Herb. Hk., Bth., Sd., D.)

A stout bush, 2–3 feet high or more, densely shaggy with reddish brown or fulvous shining pubescence; feeling like coarse cloth to the touch. Leaves seldom more than 1–1½ inch long, often shorter, varying from circular, through elliptical, to oblong and ovate; very obtuse or subacute; the veins generally concealed beneath the thick pile of hairs, in old leaves somewhat apparent. Peduncles ¼–½ inch long, shaggy. Flowers purple.

8. P. canescens (E. Mey.! Comm. p. 9, not of E. & Z. 1177); branches hairy or tomentose; leaves orbicular, broadly ovate or sub-ovate, appressedly pubescent above, *reticulately veined* and shortly tomentose beneath; peduncles 1–3 flowered, shorter or scarcely longer than the leaves; calyces rufo-velvetty, their lobes broadly cultrate, scarcely as long as the keel. *Benth. l. c. p.* 437. *P. Thunbergiana, P. amoena, P. intermedia, and P. Meyeriana, E. & Z.* 1152–55! *Zey.* 2266!

HAB. Mountain sides in the districts of Stellenbosch and Swellendam, frequent. *E. & Z.! Drege! Burchell, W.H.H., &c.* Hott. Holl. Mts., French Hoek, and Paarl. (Herb. D. Hk., Bth., Sd.)

Very near *P. cordata*, but less shaggy; generally with more ovate leaves, and the under surface always obviously and mostly *strongly* reticulated with veins. Leaves 1–1½ inches long; in young vigorous shoots 1½-2½ inches. Pubescence of the stem, branches, undersides of leaves, and calyx rusty-brown, shining.

9. P. velutina (Burch. Cat. Geogr. No. 3565 & 6984); branches tomentose; *leaves narrow-oblong, acute or mucronate*, thickish, thinly pubescent, with appressed hairs above, *densely hairy*, with a prominent midrib beneath; peduncles one-flowered, shorter than the leaves; calyx shaggy with rusty hairs, its lobes cultrate-acuminate, shorter than the keel. *Benth. l. c. p.* 437. *Zeyher*, 2262!

HAB. Among shrubs by banks of rivulets, Western Mts., Grahamstown, *Zeyher*. (207 in Herb. T.C.D.); also at Howison's Poort; and near Sidbury, *Zey.* Albany, Burchell, *T. Williamson! H. Hutton!* (Herb. D. Bth., Hk., Sd.)

An erect, virgate shrub, 2–3 feet high. Leaves 1–1¾ inch long, 3–4 lines wide, somewhat acute at base, and thus sub-lanceolate-oblong, distinctly mucronate when young. Occasionally the nervation is partly obvious on the lower surface. Pods very shaggy with long, rufous hairs. Very near *P. Burchellii*, but with longer and proportionably narrower, never ovate leaves. I retain it, with considerable doubt: the form of *P. Burchelli* called "*lancifolia*" by *E. & Z.*, being almost exactly intermediate.

10. P. Burchellii (DC. Prod. 2. p. 101); branches tomentose; leaves thickish, *ovate or elliptical*, acute or mucronate, thinly pubescent with appressed hairs above, *densely sericeo-villous* beneath, with a prominent midrib; peduncles 1–2 flowered, much shorter than the leaves; calyces shaggy with rusty hairs, the lobes broadly cultrate, shorter than the keel. *Benth. l. c. p.* 431. *P. Burchellii and P. lancifolia, E. & Z.* 1157, 1158.

HAB. Mountain situations in Uitenhage and Albany, *Burchell, E. & Z.! &c.* Common. (Herb. D. Bth., Hk., Sd.)

2–3 feet high, with virgate branches. Leaves ½–¾–1 inch long, ⅓–¾ inch broad, seldom exceeding twice their breadth in length, the margin often slightly revolute; the veins sometimes apparent, but usually, except the midrib, concealed under the rusty, glossy pubescence. The peduncles are generally 1-flowered, but not always.

11. P. lanceolata (Benth. An. Mus. Vind. 2. p. 68); twigs tomentose or pubescent; leaves *oblongo-lanceolate*, or *elliptic-oblong*, sub-acute or obtuse at each end, mucronulate, *thinly pubescent on both surfaces, prominently netted-veined* on the under side; peduncles 1-flowered, shorter or longer than the subtending leaf; calyx rufo-villous, its lobes acute, equalling the keel. *Benth. Lond. Journ.* 2. *p.* 438. *P. calyptrata, β. lanceolata, E. Mey. Comm. p.* 10.

HAB. District of Swellendam, *Mundt.* Riv. Zonderende, *Zeyher*, 2272! *Pappe* (47). River side near Spaorbosch, *Drege!* (Herb. Bth., Sd. D.)

A tall shrub, 2–4 feet high, erect and virgate. Leaves 1½–2 inches long, ½–¾ inch wide; generally obtusely lanceolate, but varying on the same bush to broadly-elliptical, scarcely thrice as long as broad. The veins on the under surface are always well marked and netted. Bracts (according to Bentham) distinct. More like *P. calyptrata* than any species of this section.

3. **CALYPTRATÆ.** (Sp. 12.)

12. P. calyptrata (Willd. Sp. 2. p. 504); branches thinly pubescent; leaves obovate-elliptical, obtuse, mucronulate, thinly pubescent on both surfaces, prominently netted-veined on the underside; calyces velvetty with rusty pubescence, the lobes cultrate, nearly equalling the keel; *bracts connate into a hood. Benth. Lond. Journ.* 2. *p.* 438. *P. sty-*

racifolia, Bot. Mag. t. 1580. *DC. Prod.* 2. *p.* 102. *E. & Z.* 1160; *also P. myrtillifolia, E. & Z.* 1161.

HAB. Mountains of the western districts, common: Table Mt., Muysenberg; Howhoeksberg and Zwarteberg, &c. *E. & Z.* and others. (Herb. D., Hk., Bth. Sd.)

A tall, strong growing, much-branched, erect shrub, 3-6 feet high. Leaves 1-1¾ inch long, ¾-1 inch broad, green on both surfaces, the pubescence very short and thinly set: in no part rufous, except on the peduncle and calyx. Peduncle 1-1½ inches long, 1- rarely 2-flowered. Pods thinly villous with whitish hairs. Known from all others by its calyptrform bracts.

4. **SERICEÆ** (Sp. 13-17.)

13. P. myrtillifolia (Willd. Sp. 2. p. 505); branches pubescent; leaves elliptical or obovate, rarely sub orbicular, thickish, thinly sericeous above, more densely sericeo-pubescent beneath, the younger leaves veinless, the older netted-veined beneath; peduncles 1-2 flowered, shorter, or scarcely longer than the leaf; calyx rufo-villous, its lobes cultrate, acute, longer than the tube, but shorter than the carina. *Benth. Lond. Journ.* 2. *p.* 439. *P. buxifolia, E. & Z.! No.* 1163. non Willd.

Var. β. **parvifolia**; leaves small (4 lines long), ovate or elliptic-ovate, acute, silky on both sides, rufo-sericeous on the lower; peduncles 1 fl., equalling leaves. *E. Mey. com. Drege. p.* 8. *(not P. parvifolia, E. Mey. in Linn.)*

HAB. Tigerberg and Klein Drakensteinberg, *Drege.* Near Caledon Baths, *E. & Z.! Dr. Prior!* River Zonderende, *Zeyher,* 2273, 2277. (Herb. Bth., Sd. Hk.)

A variable species, less silky than others of this section, and approaching *P. Burchelli* in many respects. The pubescence is very variable; on some of Eck. & Zey. specimens the under surfaces of the leaves are nearly as bare as in *P. calyptrata.*

14. P. argentea (Salisb. Par. Lond. t. 7. non E. & Z.); branches pubescent; leaves ovate, elliptical, obovate, oblong or lanceolate, obtuse or acute, silky on both surfaces, especially the under side; peduncles *much longer* than the leaves, 1-2 flowered; calyces *shaggy with curled, red-brown, coarse hairs,* the segments *deltoid,* shorter than the tube or about equal to it *DC. Prod.* 2. *p.* 102. *P. biflora Sims. Bot. Mag. t.* 753. *E. M. Com. p.* 6. *P. subbiflora, Benth. Lond. Journ.* 2. *p,* 439. *P. liparioides, P. angustifolia, and P. cuneifolia, (pro parte) E. & Z. !* 1170, 1171, 1156.

HAB. Hills round Capetown and Simon's Bay, common. Hott. Holland, *E. & Z.!* &c. (Herb. Hk., Bth., Sd., D.)

Suberect or decumbent, 1-2 feet high, subsimple or much branched. Pubescence satiny, often fulvous or rust coloured. The long peduncles, and very *roughly hairy* calyces, with short and broad lobes sufficiently mark this species. The calyx lobes vary in length, and, though generally shorter than the tube, sometimes a little exceed it; their shape and pubescence are more constant.

15. P. biflora (Lam. Illustr. t. 327. f. 3); branches pubescent; leaves ovate, obovate, oblong, or suborbicular, obtuse, silky on both surfaces; *peduncles much longer than the leaves,* two flowered; calyx *silky,* its segments cultrate, acute, longer than the tube, but shorter than the carina. *Benth. l. c. p.* 439. *P. liparioides, DC. Prod.* 2. *p.* 102. *P. myrtillifolia β. liparioides, E. M. Com. p.* 8. *P. argentea and P. pedunculata. E. & Z. ! No.* 1168, 1169. *P. subbiflora ? and P. racemulosa ? DC. Prod. l. c.*

HAB. Cape district, *Sieber, Burchell, E. & Z. !* Kl. Drakenstein and Paarl, *Drege* Near the waterfall, Tulbagh, *E. & Z. !* (Herb. D., Hk., Bth., Sd.)

Suberect or decumbent, 1–3 feet high; silky in all parts. Leaves $\frac{3}{4}$–1 inch—rarely 1$\frac{1}{2}$ inches long, $\frac{1}{2}$–$\frac{3}{4}$ inch wide, very variable in form, often rusty or fulvous. Pubescence of the calyx close-pressed, very unlike that of *P. argentea.*

16. P. cuneifolia (Vent. Hort. Cels. t. 99. non E. & Z.); branches silky; leaves obcordate, obovate or cuneate-oblong, acute at base, retuse or recurvo-mucronulate, silky on both surfaces; peduncles mostly one-flowered and *shorter than the leaf,* (rarely 2 fl. somewhat longer); calyces *appressedly pubescent,* its lobes *deltoid-acuminate, half as long as* the carina; pod villous (not shaggy). *Benth. l. c. p.* 440. *DC. Prod.* 2. *p.* 101. *P. pallens, albens, hamata, splendens, and patens, E. & Z.!* 1172, 1173, 1175, 1178, 1179.

Hab. A common species from Capetown through Swellendam and George to Uitenhage; also in Worcester and Stellenbosch, &c. (Herb. Bth., Hk., D., Sd.)

Erect or procumbent, varying much in habit, in the form of leaves, hairiness and colour of flower. It is more silky and shining than the preceding species, but much less so than the following, for which it is often mistaken.

17. P. sericea (R. Br. in Hort. Kew, Ed. 2. vol. 3, p. 6.); branches silky; leaves obovate or cuneate-oblong, acute at base, recurvo-mucronulate, *silky and shining* on both surfaces; peduncles 1 flowered, shorter than the leaves; calyces silky, their lobes *subulate,* acute, *as long as* the carina; pod silky. *Benth. l. c. p.* 440. *DC. Prod.* 2. *p.* 101. *E. & Z. No.* 1176, *and P. canescens, E. & Z.* 1177. *P. anomala,* Lehm.

Hab. Cape flats and Table Mountain, *E. & Z.! Drege, &c.* Saldanha Bay, *E. & Z.!* (Herb. D., Hk., Sd., Bth.)

Erect or procumbent; the whole plant with a satiny or silvery lustre, sometimes fulvous. It much resembles the preceding species, from which it is known by its narrow and elongate calyx lobes, and (when they can be seen) by its narrower bracts.

III. LIPARIA, Linn.

Calyx indented at base, with a short tube, 5-lobed; the four upper lobes lanceolate, acute, the lowest very large and broad, petaloid. *Corolla* glabrous; *vexillum* oval-oblong; the *alæ* oblong, one infolding the other in æstivation; carina straight, acute, narrow. *Stamens* 9 & 1. *Ovary* sessile, few-ovuled. *Legume* ovate, few-seeded. *DC. Prod.* 2. *p.* 121. *Benth.! Lond. Journ.* 2. *p.* 413.

South African shrubs, with alternate, lanceolate, rigid, pungent, exstipulate leaves. Flowers bright yellow, in terminal heads. Name from λιπαρος, *brilliant.*

ANALYSIS OF THE SPECIES.

Branches glabrous:
- Calyx glabrous, its upper segments alone lanato-ciliate ... (1) **sphærica.**
- Calyx everywhere hairy, all the segments hairy within ... ... (3) **Burchellii.**

Branches thinly or densely villous: calyx hairy:
- Branches thinly villous, soon glabrous; leaves imbricated ... (2) **comantha.**
- Branches densely villous; leaves spreading or reflexed ... (4) **parva.**

1. L. sphærica (Linn. Mant. 268); branches *glabrous;* leaves erect, lanceolate-oblong, cuspidate, pungent, 3–5–7 nerved; bracts glabrous, not ciliate; calyx *glabrous,* its upper segments alone fringed with woolly hairs. *DC. Prod.* 2. *p.* 121. *Benth. Lond. Journ.* 2. *p.* 443. *E. & Z. En. No.* 1215. *Thunb. Fl. Cap. p.* 121. *Andr. Bot. Rep. t.* 568. *Bot. Mag. t.* 1241. *Lod. Cab. t.* 642.

HAB. Mountains round Capetown and Simonsbay; Muysenberg, &c. *Thunberg, E. & Z., Drege, Wright* (550), *Pappe, &c. &c.* (Herb. Th., Bth., Hk., D., Sd.)

A rigid shrub 2–6 feet high or more, perfectly glabrous in all parts except on the pedicels and along the edges of the smaller calyx-lobes. Leaves 1½–1¾ inch long, ⅓–¾ inch wide, closely imbricating, sessile, varying from lanceolate to oblong, tapering into a hard sharp point. Flowers in dense, nodding-heads, 3–4 inches in diameter; each flower on a short villous pedicel, in the axil of a large leafy bract, thinner, softer, and paler than the leaves, but of similar shape; pedicels about a ¼ inch long, softly hairy. Lower segment of the calyx obovate-oblong, acute, quite glabrous, twice as long as the narrow-subulate, woolly-edged upper segments. Flowers bright yellow or orange.

2. L. comantha (Eck. & Zey. En. No. 1216); branches *thinly villous,* soon becoming glabrous; leaves imbricated, narrow-oblong, acute, pungent, 3–5–7 nerved; bracts fringed with hairs; calyx hairy externally, all the segments fringed, the lowest, which is nearly as long as the vexillum, densely hairy on the inside also. *Benth. l. c. p.* 443.

HAB. Rocks above Hott. Holland Kloof, *E. & Z.!* (Herb. Bth., Sd., D.)

A smaller plant than *L. sphærica,* which it much resembles, but is well marked by its hairy calyx and ciliated bracts. Leaves 1¼–1½ inch long, ¼–½ inch wide, densely imbricated. Heads of flowers nodding, scarcely 2 inches in diameter. Lowest calyx-segment lanceolate, softly villous on both surfaces. Fl. bright yellow.

3. L. Burchellii (Benth. Lond. Journ. 2. p. 443); "branches *glabrous;* leaves lanceolate-oblong, mucronate-acute, thick, obscurely about 5 nerved; bracts ciliate; *calyx everywhere hairy,* all the segments ciliate, hairy within, the lowest half as long as the vexillum." *Bth.*

HAB. Cape colony, *Burchell, No.* 6881.

Said to resemble the preceding species in aspect, but having longer and narrower leaves, and a more hairy and differently proportioned calyx.

4. L parva (Vogel, ex Walp. Linn. 13. p. 468); stem slender, *densely villous;* leaves *spreading or reflexed,* elliptical or oblong, acute and pungent, 3–5 nerved, sparsely villous; bracts broadly elliptical, acuminate, fringed with silky hairs; calyx densely hairy externally, all the segments well-fringed, the lowest which is rather shorter than the vexillum, hairy on the inside also. *Benth. l. c. p.* 443. *Bot. Mag. t.* 4034. *L. crassinervia, Meisn. Lond. Journ. 2. p. 63 ?*

HAB. S. Africa, *Bowie, Forbes.* Near Simon's Bay, *Mr. C. Wright,* 549. (Herb. Bth., D., Hk.)

Stem 1–2 feet high, suberect or trailing, irregularly branched, all the younger twigs patently hairy. Leaves scattered, more distant than in the preceding species, ½–¾ inch long, about ¼ inch wide, the young ones thinly villous. Heads of flowers erect, 1–1½ inch in diameter; the bracts ½ inch wide or more, suborbicular or elliptical; their fringe rusty-brown. Calyx ½–¾ inch long, its smaller segments broader and blunter than in others. Fl. yellow. This looks like a *Priestleya.*

IV. PRIESTLEYA, DC.

Calyx subequally 5–cleft, the lowest lobe equal to the rest or scarcely longer. *Corolla* glabrous; *vexillum* subrotund, shortly stipitate; *alæ* obtuse, subfalcate; *carina* incurved, without lateral appendages. *Stamens* 9 & 1. *Ovary* sessile, several-ovuled. *Legume* plano-compressed, 4–6 seeded. *Benth! in Hook. Lond. Journ.* 2 *p.* 444. *DC. Prod.* 2, *p.* 121. *Priestleya and Xiphotheca, E. & Z.*

South African shrubs, with alternate, simple, exstipulate leaves. Flowers yellow, in terminal heads or racemes, or axillary. Name in honour of ***M. Priestley***, a physiological botanist.

ANALYSIS OF THE SPECIES.

1. **Isothea.**—*Calyx* at length indented at base. *Carina* rostrate. *Glabrous or thinly villous, rarely hirsute, shrubs, mostly turning black in drying.* (Sp. 1–11.)

Flowers *subcapitate* or very densely *capitato-racemose:*
  Branches glabrous :
    Leaves linear-lanceolate, 1-nerved ; flowers pedicellate, racemose; bracts *lanceolate* .. .. (1) **graminifolia.**
    Leaves linear or oblong-linear ; fl. subsessile, capitate ; bracts broadly ovate ... ... (5) **capitata.**
  Branches villous or hirsute :
    Leaves narrow-lanceolate ; bracts lanceolate ... (2) **angustifolia.**
    Lvs. narrow-oblong, 1-nerved ; bracts broadly ovate (4) **umbellifera.**
    Lvs. obovate-oblong, 3–5 nerved ; bracts ovate ... (6) **hirsuta.**
    Lvs. ovate-orbicular, many nerved, shaggy ... ... (11) **vestita.**
Flowers conspicuously pedicellate, *sub-umbellate :*
  Bracts and calyces glabrous :
    Pedicels downy ; pods *densely villous* ... ... (8) **myrtifolia.**
    Pedicels glabrous ; pods *glabrous and glossy* ... ... (9) **leiocarpa.**
  Bracts or calyces pubescent or hairy :
    Branches *hirsute;* leaves broadly ovate or obovate (10) **latifolia.**
    Branches glabrous, or soon becoming glabrous :
      Leaves *all narrow,* lanceolate-linear ... ... (3) **teres.**
      Leaves *broad,* the lower oblong, upper lanceolate (7) **Thunbergii.**

2. **Anisothea.**—*Calyx* not indented at base. *Carina* shortly curved, not rostrate, rounded above. *Silky or appressedly villous shrubs, not blackening.* (Sp. 12–15).

Flowers in terminal heads :
  Branches subcorymbose ; pubescence villous ; leaves elliptic-ovate ... ... (13) **villosa.**
  Branches flexuous, divaricate ; pub. silky ; leaves squarrose, ovato-lanceolate ... ... (14) **sericea.**
Flowers axillary, scattered :
  Diffuse ; pedicels shorter than the broadly ovate, *concave* leaves ... ... (15) **tecta.**
  Suberect ; pedicels longer than the bracts ; leaves ovate or elliptic, *flat* ... ... (12) **elliptica.**

I. **ISOTHEA** (Sp. 1–11.)

1. P. graminifolia (DC. Prod. 2. p. 122); branches glabrous ; leaves linear-lanceolate, rigid, glabrous, pungent-mucronate, one nerved ; flowers on hairy pedicels, in an oblong, terminal, headlike raceme ; bracts lanceolate, hairy ; calyx hairy, the segments thrice as long as the tube, fringed, the upper ones lanceolate, the lowest oblong, rather longer than the rest, membranous. *Benth. Lond. Journ.* 2. *p.* 444. *Liparia graminifolia, Linn. Mant. p.* 268. *Thunb! Fl. Cap. p.* 566.

HAB. South Africa, *Thunberg! Forbes!* (Herb. Thunb., Benth.)

A small shrub, 12–18 inches high, chiefly branched from the base, glabrous except on the inflorescence. Leaves imbricating, 1–1½ inch long, 1–2 lines wide, slightly concave, of a thickish substance, midribbed, but veinless. Heads of flowers 1½ inch long, dense, with copious rusty pubescence. Upper calyx segments very acute, broadly subulate, or semi-lanceolate. Apparently a rare species.

2. P. angustifolia (Eck. & Zey! En. No. 1222); branches villous; leaves narrow-lanceolate, rigid, the young ones sparsely villous, the older glabrous, pungent-mucronate, rather concave, one-nerved ; flowers on hairy pedicels, in a short, terminal, headlike raceme ; bracts lanceolate, hairy; calyx hairy, the upper segments lanceolate or deltoid-acuminate, the lowest cuspidate, 1½ as long as the rest. *Benth. Lond. Journ.* 2. *p.* 444. *P. umbellifera, E. M. Comm. p.* 17. *P. lævigata, E. & Z. No.* 1221. *(non Benth.) Borbonia villosa, Thunb! Fl. Cap. p.* 560.

HAB. Hott. Holland and Klynriverberg, *E. & Z.!* Capetown range, *W.H.H.* (Hb. Thb. Bth., Hk., Sd., D.)

A small shrub, 10–12 inches high, erect or ascending, chiefly branched from the base. Leaves 1–1¼ inch long, 1–2 lines wide, sometimes with obscure lateral nerves. Flowers 6–10, in a globose cluster.—Resembles *P. graminifolia* in aspect, but is more hairy, with a different calyx. Pods softly villous.

3. **P. teres** (DC. Prod. 2. p. 122); much branched, branches glabrous; leaves narrow lanceolate-linear, narrowed at base, rigid, acute, obscurely one nerved, glabrous; flowers on downy pedicels, somewhat exceeding the downy ovato-lanceolate, deciduous bracts; calyx downy, its segments ovate or ovate-oblong, acute or mucronulate. *P. lævigata, DC. Mem. t.* 30. *DC. Prod.* 2. *p.* 121 *(excl. syn. Thunb.) Benth.* l. c. *p.* 445. *Liparia teres, Thunb.! Cap. p.* 566.

HAB. S. Africa, *Thunberg!* Swellendam, *Dr. Thom!* (Hb. Thb., Hk., Bth.)

An erect, much branched shrub, 1½–2 feet high, glabrous except on the inflorescence, which is downy, with very short, soft, patent hairs. Leaves 1–1¼ inch long, 1–1½ lines wide, scattered, more dense toward the ends of the erect, virgate branches. Bracts falling away soon after the flowers open: the lowest flower only from the axil of a persistent leaf, or leaf-like glabrous "bract," longer than the pedicel. I have compared Thunberg's original specimen with that of Dr. Thom.

4. **P. umbellifera** (DC. Prod. 2. p. 122); much branched, branches softly hairy; leaves narrow-oblong or lanceolate-linear, rigid, acute, concave, obscurely one-nerved, appressedly silky on both surfaces, the older ones becoming sub-glabrous; pedicels densely silky, shorter than the broadly ovate, concave, mucronate, hairy bracts; calyx densely hairy, its lobes obliquely ovate, obtuse, of nearly equal length. *P. cephalotes, E. Mey! Comm. p.* 18. *excl. syn. Benth. l. c. p.* 445. *Liparia umbellifera, Thunb! Fl. Cap. p.* 568.

HAB. S. Africa, *Thunberg!* Mountains of Stellenbosch; also Piquetberg, and Cedarberg, Clanwilliam, *Drege!* (Hb. Th., Bth., Hk., D.)

12–18 inches high, erect, corymbosely-branched; all the younger parts softly silky, with long, appressed hairs. Leaves about ¾ inch long, 1–2 lines wide, between oblong and lanceolate, acute but scarcely pungent. Flowers 4–6, on very short, hairy pedicels, subcapitate. Calyx segments remarkably broad and blunt. Thunberg's original specimen in Herb. Upsal. quite agrees with those from Drege.

5. **P. capitata** (DC. Prod. 2. p. 121); much branched, glabrous (or thinly villous); leaves linear or oblongo-linear, rigid, acute, thick, channelled, nerveless, glabrous; bracts broadly ovate, concave, obtuse or mucronulate; flowers capitate, subsessile; calyx rigidly hirsute, the four upper segments oblong, obtuse, the lowest acute, longer than the others, glabrous and shining near the point. *Benth! Lond. Journ.* 2. *p.* 445. *P. lævigata, E. Mey. (non DC.) Liparia capitata, Thunb! Fl. Cap. p.* 566.

VAR. β. **pilosa** (E. Mey.); branches and leaves more or less clothed with soft, deciduous hairs.

HAB. Summit of Table Mountain, *Thunberg!* Dutoit's-kloof and Gnadendahl, *Drege! Burchell* (591). (Herb. Th., Bth.)

A shrub, 2–3 feet high or more, somewhat umbellately branched, usually glabrous in all parts, except on the inflorescence. Leaves crowded, erect or appressed, ½–¾ inch long, a line broad, shining, very concave or convolute. Flowers 5–6 in terminal heads. Calyces densely clothed with long, rigid, foxy bristles. Pods few seeded, short, very hairy.

6. **P. hirsuta** (DC. Prod. 2. p. 121); branches virgate, villous; leaves obovate-oblong, mucronate-acute, somewhat rigid, 3–5 nerved, glabrous, or the younger ones villous; bracts ovate, rigidly cuspidate, hirsute; calyx hairy, its segments with glabrous, rigid points, ovato-lanceolate, acute, the lowest subulato-acuminate; legume shaggy. *Benth. l. c. p.* 446. *E. & Z. En. No.* 1220. *Liparia hirsuta, Thunb! Fl. Cap. p.* 567.

HAB. In moist places, on hill sides, George and Uitenhage, *Thunberg! Drege! E & Z.! Pappe* (80), &c. (Herb. Th., Bth., Sd., Hk., D.)

A stout, branching shrub, 2–3 feet high, erect; the older branches becoming glabrous. Leaves 1–1½ inch long, ¼–½ inch broad, more or less acute or acuminate, narrowed toward the base, erect, somewhat concave, the nerves scarcely visible, except in the dried plant. Head-like racemes 6–8 or more flowered; the pubescence rust coloured. Pods erect, few seeded, very hairy.

7. **P. Thunbergii** (Benth. Lond. Journ. 2. p. 446); much branched, branches thinly villous, becoming glabrous; lower leaves broadly oblong, upper lanceolate, rigid, acute or acuminate, flat, one-nerved, glabrous, the younger ones thinly villous; flowers about four in an imperfect umbel, on silky pedicels longer than the ovate, coloured, concave, pubescent and ciliate bracts; calyx silky, its segments broadly ovate, mucronate, the lowest acuminate. *Priestleya umbellifera, E. & Z. No.* 1219. *Walp. Lin.* 13. *p.* 469. *(non DC.) Borbonia lævigata, Lin. Mant. p.* 100. *Liparia lævigata, Thunb! Fl. Cap. p.* 566. *L. villosa, Sieb! No.* 162.

VAR. β. **villosa**; branches hairy; leaves thinly villous on both surfaces. *Liparia villosa, Thunb! in Herb.* Upsal.

HAB. Moist places, East side of Table Mt., *E. & Z.!* Near Capetown, *Sieber*, &c. (Herb. Th., Bth., Sd., Hk.)

A stout shrub, 2–3 feet high, corymbose or umbellately branched, the old branches naked, the young densely leafy. Leaves uncial, ¼ inch wide, mostly lanceolate. Pedicels an inch long or less, twice as long as the bracts, or barely longer. Flowers large.

8. **P. myrtifolia** (DC. Leg. Mem. p. 194. t. 29); stems glabrous, or downy near the summit; leaves broadly ovate, obovate or ovato-lanceolate, acute, margined, glabrous, one-nerved and sub-penniveined; racemes sub-umbellate, few flowered, pedicels minutely downy, longer than the glabrous, convolute, acute bracts; calyx glabrous, with deltoid, sub-acute segments; pod densely villous. *DC. Prod.* 2. *p.* 121. *Benth. l. c. p.* 446. *E. & Z. En. No.* 1217. *Liparia myrtifolia, Thunb! Fl. Cap. p.* 565.

HAB. Hott. Holl. Stellenbosch Mts., *E. & Z.!* &c. (Herb. Th., Bth., Hk., Sd., D.)

Stem 1½–3 feet high, sparingly branched, branches virgate; the whole plant except the inflorescence, and occasionally the young ends of branches, glabrous. Leaves crowded, variable in shape, breadth and obviousness of veins, 1–1¼ inches long, ¼ to ⅝ inch wide, erect, flat; the lateral veins very erect. Pedicels 1–1½ inches long, 1–2 flowered, twice or thrice as long as the glumaceous, narrow-oblong bract. Pods uncial, ½ inch wide, suberect. The bracts and calyx are sometimes fringed with very soft, white hairs, and the young calyx sparingly villous.

9. P. leiocarpa (Eck. & Zey. En. No. 1218); glabrous; leaves lanceolate, acute, margined, one-nerved; racemes subumbellate, few flowered; pedicels longer than the bracts, glabrous as well as the calyx and pod; calyx-lobes deltoid, subacute. *Benth. l. c. p.* 447.

HAB. Grootvadersbosch, Swellendam, *Mundt* in Herb. Ecklon. (Herb. Sd.)

This precisely resembles *P. myrtifolia*, except that the pedicels and pods are quite glabrous; the latter shining, 1½ inch long.

10. P. latifolia (Benth. Lond. Journ. 2. p. 447); branches hirsute; leaves broadly ovate, obovate or elliptical, sharply mucronate, one-nerved and penniveined, softly villous and ciliate, becoming sub-glabrous; racemes short, densely umbellate; pedicels as long as the ovate, densely hairy bracts; calyx densely and softly villous, its lobes tapering, subacute.

HAB. S. Africa, *Scholl! Burchell* (8025); *Gueinzius!* (Herb. Bth., Sd.)

A robust shrub, 2–3 feet high, umbellately branched; all the younger branches densely though softly hairy. Leaves imbricating, ½–¾ inch long, ⅜–⅝ broad, flat, obviously veined, only the old ones smooth. Whole inflorescence very densely villous; racemes 6–8 flowered; the hairy pedicels ½–¾ inch long.

11. P. vestita (DC. Prod. 2. p. 122); branches densely hirsute; leaves ovate-orbicular, very concave, obtuse, imbricating, many-nerved, sparsely hispid on the upper, very densely hirsute and shaggy on the under (outer) surface; flowers sub-capitate; bracts broadly ovate, longer than the pedicel, densely hirsute as well as the acutely lobed calyx. *Benth. l. c. p.* 447. *E. & Z. En. No.* 1223. *Liparia vestita, Thunb.! Fl. Cap. p.* 568. *Bot. Mag. t.* 2223. *L. villosa, Andr. Bot. Rep. t.* 382. *(non Thunb!)*

HAB. Hott. Holl. Mountains, *Thunberg!* &c. (Herb. Th., Bth., Hk., Sd., D.)

A tall, stout shrub, 2–4 feet high, with long, erect, virgate branches closely imbricated with very concave, almost cymbiform leaves, whose outer surfaces are thickly covered with long, white, straight, rigid, coarse hairs. Leaves ½ inch long and broad, the veins visible on the upper or inner surface, which is green, thinly sprinkled with a few hairs. Hairs of the inflorescence often rust coloured. Flowers 3–4 or more, subsessile at the ends of the branches.—Quite unlike any other species.

2. **ANISOTHEA** (Sp. 12–15).

12. P. elliptica (DC. Leg. Mem. t. 33); densely much branched, twigs angular, appressedly pubescent; leaves ovate or elliptical, minutely petiolate, calloso-mucronulate, flat, thickish, one-nerved, thinly appresso-sericeous on both sides; peduncles axillary, 1 flowered or umbellately 2–3 flowered, the pedicels longer than the bracts; calyx appressedly puberulent, its lobes triangular, much shorter than the tube; legume linear-oblong, thinly pubescent. *Benth! Lond. Journ.* 2. *p.* 447. *Ingenhoussia? verticillata, E. Mey.! Herb. Drege. Com. p.* 21.

HAB. Dutoit's-kloof Mts. *Drege!* Gnadendahl, *Dr. Alexander Prior!* C. B. S. *Gueinzius!* Zwarteberg, *Zeyher!* (Herb. Bth., Sd, D.)

Erect or ascending, 1½–2 feet high or more, much branched, the whole plant thinly clothed with very short, closely appressed, rather shining, whitish hairs. Leaves either scattered, or often opposite, especially on the upper branchlets, ⅜–½ inch long, 3–4 lines wide; the petiole ½–1 line long, mucro not always obvious. Flowers from the axils of the upper leaves, on short or longish peduncles. Calyx tube conical-campanulate, somewhat ribbed at base; its teeth very short and broad.

13. P. villosa (DC. Prod. 2. p. 122); much branched, twigs villoso-hirsute; leaves ovate or elliptical, or oblong, acute at each end, sessile, flat, one-nerved, densely and softly villous on both surfaces; flowers in terminal heads, subsessile; outer bracts lanceolate, shaggy, inner setaceous; pedicels and calyx densely viiloso-hirsute, the calyx lobes acuminate, about equalling the tube. *Benth. l. c. p.* 448. *Xiphotheca villosa, and X. tomentosa, E. & Z.! No.* 1226, 1227. *Borbonia tomentosa and Liparia villosa, Linn. fide Benth. (not L. villosa, Thunb.) Crotalaria lanata, Thunb.!* Cap. p. 571.

HAB. Common on the Capetown mountains. (Herb. Th., Bth., Hk., Sd., D.)

A densely branched, somewhat corymbose bush, 2–3 feet high and wide, thickly clothed with soft, long, whitish, appressed hairs. Leaves imbricating, ½, ¾, or 1 inch long, ¼–½ inch wide, erect or spreading. Heads 6–8, several flowered; their pubescence fulvous. In Thunberg's Herbarium (Upsal), this is marked "*Crotalaria lanata*;" the specimen marked "*Liparia villosa*" in the same collection is without flowers, but appears to be a villoso-pubescent variety of *P. Thunbergii.* If not so, it is a new species of the *Isothea* section.

14. P. sericea (E. Mey.! Comm. p. 19); divaricately branched, branches villoso-hirsute; leaves spreading or reflexed, ovate, oblong, or lanceolate, flat, sessile, acute, sericeo-villous on both sides, one-nerved; flowers spicato-racemose or sub-capitate, from the axils of the upper leaves; inner bracts setaceous; calyx softly villous, the segments very narrow, acuminate, equalling the tube or longer; pods patently hairy. *Benth.! l. c. p.* 448. *Xiphotheca sericea, X. axillaris and X. lanceolata, E. & Z.! En. p.* 167. *Crotalaria reflexa, Thunb.! Fl. Cap.* 571.

HAB. Mts. round Capetown and Simonsbay, common. (Herb. Th., Bth., Sd., Hk., D.)

A prostrate or diffuse, straggling shrub, with very patent, flexuous branches, clothed with long, white, spreading hairs. Leaves ¾–1 inch long, 2–4–5 lines wide, squarrose. Flowers often in pairs, the pedicels subtended by a small bract: sometimes confined to the tips of the branches, and then capitate; sometimes distributed in longish pseudo-racemes. Thunberg's specimen of "*Crotalaria reflexa,*" belongs to this, although in Fl. Cap. he says the flowers are "*purple.*"

15. P. tecta (DC. Prod. 2. p. 122); branches divaricate, hairy; leaves broadly ovate or subrotund, acute or obtuse, concave, patently villous on both surfaces, especially the under, obscurely one-nerved; flowers axillary, on short pedicels; calyx tomentose, its segments cultrate, about as long as the tube. *Benth. l. c. p.* 448. *Liparia tecta, Thunb.! Fl. Cap. p.* 568. *Xiph. polycarpa, E. & Z.! No.* 1225.

VAR. β. **rotundifolia**; leaves suborbicular, obtuse or mucronate, flattish, somewhat penninerved. *Xiph. rotundifolia, E. & Z.! No.* 1224. *P. rotundifolia, Walp.*

HAB. Hott. Holland; Stellenbosch; the Paarlberg; Piquet Bay, &c. *Drege!* Tulbagh and Klapmuts, *E. & Z.!* (Herb. Th., Bth., Sd., Hk.)

A procumbent or diffuse shrub, with flexuous, widely spreading branches. Leaves imbricated, patent or reflexed, scarcely half an inch long, 3–5 lines broad; the nerve sometimes scarcely visible, sometimes well marked. Pods hirsute and velvetty, fulvous, 1½ inches long, ⅜ wide.

V. AMPHITHALEA, E. & Z.

Calyx subequally 5-cleft. *Vexillum* roundish, shortly stipitate, reflexed; *alæ* oblong; *carina* straightish, obtuse, spurred on each side. *Stamens* diadelphous (9 & 1). *Ovary* 1-4 ovuled. *Legume* ovate, 1-2 seeded, rarely oblong, 3-4 seeded. *Benth.! in Hook. Lond. Journ.* 2. *p.* 449. *Endl.* 6465. *Ingenhoussiæ, sp. E. Mey. Cryphiantha, E. & Z. Epistemium, Walp.*

Small, heathlike South African shrubs, with alternate, simple, entire, sessile, exstipulate leaves, frequently with revolute margins. Flowers purple or rosy, with the carina darkly tinted, axillary and subsessile, or crowded in a leafy spike. Name from αμφιθαλης, *flowering round the branch.*

ANALYSIS OF THE SPECIES.

1. *Ovary* with 2-4 ovules. *Leaves* ovate or obovate, silky and silvery, flat. (Sp. 1-2).

Leaves mucronate or obtuse; pod 5-6 lines long, *falcate* (1) **cuneifolia.**
Leaves acute; pod *ovate,* acuminate, 3-4 lines long ... (2) **densa.**

2. *Ovary* with a solitary ovule. (Sp. 3-9).

Leaves flat or nearly so:
Pubescence of the branches, leaves, and calyx *silky:*
Leaves silky on *both* surfaces; the margin slightly recurved:
Divaricately branched; legume tomentose (3) **violacea.**
Virgate; legume silky ... ... (4) **intermedia.**
Leaves glabrous above; the margin slightly involute ... ... (8) **Williamsoni.**
Pubescence of the branches *viscid; calyx glabrous* (9) **micrantha.**
Leaves with strongly revolute margins:
Leaves linear-oblong or lanceolate; pod turgid ... (5) **ericæfolia.**
Leaves very narrow linear-terete; branches virgate; pod not turgid ... ... (6) **virgata.**
Leaves *short,* spreading, ovato-lanceolate, tomentose (7) **phylicoides.**

* *Ovary 2-4 ovuled.* (Sp. 1-2).

1. A. cuneifolia (Eck. & Zey. En. No. 1231); leaves broadly obovate, *calloso-mucronulate* or *obtuse,* flat, penninerved, densely clothed on both surfaces with silky, closepressed, silvery or fulvous hairs; flowers axillary, crowded, sessile; pod *twice as long as the calyx, oblong-falcate, subacute, compressed,* with foxy pubescence, 2-4 seeded. *Benth. Lond. Journ.* 2. *p.* 450. *Priestleya axillaris, E. Mey. p.* 20. *Epistemum ferrugineum, Walp. Linn.* 13. *p.* 473.

HAB. Hottentots' Holland, *Mundt.! Bowie, E. & Z.!* Burchell (8162). Babylonisch Toornberg, *Zeyher!* 2285. (Herb. Hk., Sd.)

A robust shrub, 2-3 feet high, with virgate, tomentose branches. Leaves 5-7 lines long, 4-5 lines broad, imbricated, thickish, satiny to the touch. Flowers 5 lines long, bright purple; the calyx softly villous; a small, subulate bract. Pods 5-6 lines long, curved. Flowers twice as large as in the following; leaves suddenly mucronulate, not gradually acute.

2. A. densa (Eck. & Zey.! No. 1232); leaves ovate, elliptical or obovate, *acute,* flat, one-nerved, or obscurely penninerved, densely clothed on both surfaces with silky, close-pressed, silvery or fulvous hairs; flowers subsessile, axillary, solitary or crowded; calyx silky, its teeth subulate, rather longer than the tube; ovary 2-ovuled; *pod ovate, acuminate. Benth. l. c. p.* 450. *Indigofera sericea? L. Borbonia tomentosa, L. Crotalaria imbricata, L. Thunb.! Fl. Cap. p.* 571. *Priestleya sericea* *and P. axillaris, DC. Prod.* 2. *p.* 122. *P. elliptica, E. Mey.* (*non DC.*) *P. Meyeri, Meisn. Ld. Jrn.* 2. *p.* 65. *Lathriogyne candicans, E. & Z. !* 1245.

HAB. Mountains round Capetown; Devil's and Table Mountain summit. *E. & Z.! W. H. H., &c.* Hottentots' Holland. (Herb. Th., Bth., Hk., Sd., D.)

A stout shrub, much branched, with virgate or ramulose branches, decumbent or spreading, 2–3 feet long. Leaves densely imbricated, 3–7 lines long, 2–5 broad, very soft and satiny. Flowers nestling among the leaves, light purple, either crowded and somewhat capitate at the ends of short branches, or distributed in leafy pseudo-spikes 6–8 inches long. Pods 3–4 lines long, densely villous, very acute. The leaves vary in size and shape, and in the faintness of the ribs and veins.

** *Ovary one-ovuled* (Sp. 3–9).

3. A. violacea (Benth. Lond. Journ. 2. p. 451); *branchlets alternate, divergent, rigid,* at length bare of leaves, silky; leaves spreading, ovato-lanceolate, acute, flat, with slightly reflexed margins, one-nerved, appressedly silky and pale on both surfaces; spikes terminal, leafy, flowers sub-sessile, solitary or clustered; calyx silky, its lobes triangular, acute, half as long as the tube, legume ovate, *tomentose. Ingenhoussia violacea, E. Mey.! Com. p.* 21. *Burch. Cat. No.* 7436.

HAB. Outeniqua Mts., George, *Drege! Bowie! Burchell.* (Herb. Bth., Hk., Holm.)

2–3 feet high, woody, divaricately much branched; the larger branches closely set with erecto-patent branchlets 2–3 inches long. Leaves not very densely set, 3–4 lines long, 1–2 wide; the margin very slightly recurved. Flowers 2½ lines long, the keel dark-purple, other petals lilac.

4. A. intermedia (Eck. & Zey. No. 1234); branches slender, *virgate, silky;* leaves oblong or oblongo-lanceolate, acute or mucronate, *nearly flat,* obscurely one-nerved, silky on both surfaces; flowers axillary, solitary or crowded in terminal spikes; calyx lobes shorter than the tube; legume ovate, acute, *silky. Benth.! l. c. p.* 451, *also A. humilis, E. & Z.!* 1233, *and Zey.* 2286, *b. &c. Burch.* 5971 *and* 6667.

HAB. Stony hill-sides, in Caledon and Swellendam, *E. & Z.! Bowie, Burchell.* Klynriviersberg, *Zeyher!* (Herb. Bth., Hk., Sd., D.)

A small, sub-erect or ascending many-stemmed shrub, 8–16 inches high; branches simple or sparingly divided, virgate. Leaves scattered or crowded, erect, 3–5 lines long, 1–3 lines broad, shining and silvery. Flowers 2 lines long; the keel dark purple, other petals lilac.

5. A. ericæfolia (E. & Z. 1239); much branched, branches virgate or ramuliferous, silky; leaves erect, incurved or spreading, *linear-oblong or lanceolate,* acute, *with strongly revolute margins,* silky on one or both surfaces; calyx lobes shorter than the tube; legume ovate, *turgid,* acuminate. *Ingenhousia ericæfolia, E. Mey.! p.* 21. *Priestleya ericæfolia, DC. Prod.* 2. *p.* 122.

VAR. α. **glabrata**; adult leaves *glabrous and shining* on the upper surface. *A. ericæfolia, E. & Z.!* 1239, *and A. hilaris, E. & Z.!* 1238. *Indigofera sericea, Thunb.!*

VAR. β. **multiflora**; adult leaves *silky on both sides. A. multiflora, E. & Z.!* 1236. *A. densiflora, E. & Z.* 1237, *and A. incurvifolia, E. & Z.!* 1235. *A Vogelii ? Walp. Linn.* 13. *p.* 472. *Indigofera sericea, Thunb.! in Herb. Holm. Zey.!* 367.

HAB. Common on hills, &c. Cape and Stellenbosch. (Hb. Th., Bth., Hk., Sd., D.)

Erect or sub-erect, many-stemmed, 1–2 feet high; branches either quite simple or having lateral, erect, short twigs. Leaves 3–4 lines long, 1 line wide, more or less silky. Flowers in dense, terminal pseudo-spikes, similar to those of the preceding species. Intermediate states connect the two varieties, which I cannot always distinguish. *A. Vogelii* seems to be merely a starved state of var. β.

6. A. virgata (Eck. & Zey. No. 1240); branches slender, virgate or ramuliferous, the young twigs thinly silky; leaves incurvo-patent or erect, *very narrow, linear-lanceolate or terete,* acute, with strongly revolute margins; the adult glabrous and shining on the upper, silky on the under surface; calyx-lobes shorter than the tube; pod ovate, with a long point, *scarcely turgid, silky. Benth. l. c. p.* 452. *Amp. Kraussiana, Meisn. Lond. Journ.* 2. *p.* 65.

HAB. Rocky and sandy places near the mouth of Klynrivier, Caledon, E. & Z.! (Herb. Bth., Sd., D.)

A much more slender plant than *A. ericæfolia,* with much narrower, and rather longer leaves, and a different pod. Many-stemmed, 6–12 inches high, chiefly branched from the base; the branches curved, mostly clothed with short ramuli. Leaves 3–5–6 lines long, not ½ line wide, the lower surface generally quite concealed by the rolling back of the margins: the young leaves thinly silky. Flowers smaller than in *A ericæfolia.*

7. A. phylicoides (E. & Z.! No. 1243); *divaricately* much branched, robust; branches *tomentose;* leaves *short, spreading,* ovato-lanceolate, or sub-linear, callous-pointed or obtuse, or acute, with strongly revolute margins, tomentose at first, afterwards glabrescent above, densely villous beneath; calyx-teeth very unequal, rather shorter than the tube. *Benth. l. c. p.* 452. *Zey. No.* 2287.

HAB. Vanstaadensberg, Uitenhage, *E. & Z.!* (Herb. Bth., Hk., Sd., D.)

A coarse bush, 2–3 feet or more high, with widely-spreading, much divided branches; the older ones bare and rough with cicatrices, the younger canescent and villoso-tomentose. Leaves 2–3 lines long, generally horizontally patent; the old ones only glabrescent. Flowers solitary in the upper axils, two lines long.

8. A. Williamsoni (Harv.); branches virgate, pubescent; leaves ovate or ovato-lanceolate, callous-pointed, *nearly flat, with slightly involute* margins, the adult glabrous above, thinly appresso-pubescent beneath, somewhat 3-nerved; calyx silky, its teeth ovate, acute, shorter than the tube.

HAB. Albany, *T. Williamson!* (Herb. T.C.D.)

Of this apparently very distinct species, I have seen but a few fragments, and know not to what sized bush they may belong. The margins of the leaves are *inflexed,* not *reflexed,* as in most others of the genus. This led me at first to refer it to *Cœlidium;* but the upper stamen is quite free. Leaves 5 lines long, 2–3 lines broad, at length glabrescent on *both* surfaces; the nerves then plainly visible, and even reticulate veins obscurely so. Flowers axillary, scattered, or two together, 2½ lines long. Calyx-teeth very short. Found by *Thomas Williamson,* a soldier in the 72nd Regiment, formerly employed by me to collect plants in Albany and at Port Natal, and whose intelligence and diligence deserve honourable commemoration.

9. A. micrantha (Walp. Linn. 13. p. 471); densely much branched, *branchlets viscoso-pubescent;* leaves broadly ovate or cordato-ovate, acute, flat, slightly concave, glabrous and shining above, villous or glabrous beneath, one-nerved; calyx *glabrous,* its teeth short, *very obtuse. Benth. l. c. p.* 452. *Ingenhoussia micrantha, E. Mey.! com. p.* 21. *Cryphiantha, imbricata, E. & Z.! No.* 1247.

HAB. Vanstaadensberg, Uitenhage, *E. & Z.!* Zwarteberg, *Drege!* Also gathered by Bowie. (Herb. Bth., Hk., Sd., D.)

A much branched small shrub, 1–2 feet high; the lesser branches with a thick short coat of viscid hair, mixed with long soft white hairs. Leaves 4 lines long, 3½ lines wide, erecto-patent, imbricating, minutely dotted, the under side sometimes densely villous, more commonly quite smooth; margins when young, fringed with long soft hairs. Flowers few and small, hidden among the upper leaves; the calyx teeth round topped. Easily known by its pubescence and calyx; in its foliage it comes near *A. Williamsoni*, but the leaves are shorter and broader.

VII. COELIDIUM, Vogel.

Calyx nearly equally 5–fid. *Vexillum* obovate, shortly stipitate, reflexed, *alæ* oblong; *carina* oblong, straight, obtuse, bluntly spurred at each side. *Stamens* monadelphous, the tube often very short. *Ovary* uniovulate. *Legume* ovate, one-seeded. *Benth. in Ld. Jrn. Bot.* 2. *p.* 453.

Small, much-branched S. African shrubs or suffrutices, with simple, entire, sessile exstipulate leaves, with the margin more or less involute, appressedly pubescent on the upper side, either glabrous or silky on the lower, often twisted or transversely rugose. Flowers geminate, in the axils of the upper leaves sub-sessile, except in *C. spinosum*. Name from κοιλος, *hollow;* alluding to the frequently *concave* leaves.

TABLE OF THE SPECIES.

Leaves broadly ovate or cordate ovate:
  Leaves *very concave*, with strongly inflexed margins; flowers sub-sessile ... ... (1) **bullatum.**
  Leaves *flattish*, silky on both sides; fl. sub-sessile ... (2) **Thunbergii.**
  Leaves flattish, silky; peduncles elongate, 2-flowered (3) **spinosum.**
Leaves lanceolate, ovato-lanceolate or subulate, straight or twisted:
  Floral-leaves not broader than the ordinary leaves ... (7) **Vogelii.**
  Floral-leaves *broad*, ovate or ovato-lanceolate:
    Lobes of the calyx *acute or acuminate:*
      Leaves *straight* and erect, glabrous externally (4) **ciliare.**
      Leaves *twisted*, densely and softly villous on both sides ... ... (5) **roseum.**
      Leaves twisted, glabrous externally ... ... (6) **Bowiei.**
    Lobes of the calyx very short and obtuse ... ... (8) **muraltioides.**

1. **C. bullatum** (Benth. Lond. Journ. 2. p. 453); "leaves *broadly ovate, bullato-concave, nearly closed, with the margin strongly involute*, appressedly silky on the inside, bearded at the apex, thinly hairy on the outside; calyx silky-pilose, its segments longer than the tube; stamens shortly monadelphous." *Benth. l. c.*

HAB. S. Africa, *Burchell*, Cat. No. 7115. (Unknown to me.)

"Branchlets softly hairy. Leaves 3–4 lines long. Bracts subulate. Calyx 3 lines long, with narrow-lanceolate segments. Corolla not seen. Filaments persistent after flowering, connate to a fourth of their length. Legumes obliquely ovate, shortly acuminate, compressed, very villous, scarcely longer than the calyx." *Benth.*

2. **C. Thunbergii** (Harv.); leaves (small) *ovate or cordato-ovate*, acute, *flattish*, with the margin slightly inflexed, one-nerved, appressedly silky on both surfaces; *flowers sub-sessile;* calyx silky, its segments shorter than the tube, deltoid; stamens shortly monadelphous. *Crotalaria parvifolia, Thunb.! Fl. Cap. p.* 571.

HAB. S. Africa, *Thunberg!* (Herb. Thunb.)

An erect branching shrub, 1–2 feet high; branches virgate, ribbed and furrowed, thinly and appressedly silky. Leaves scattered, 2½–3 lines long, 1–2 lines broad, varying from broad to narrow, but always on an ovate type. Flowers on pedicels shorter than the calyx, one or two together in the axils of the leaves, scattered along the branches. Calyx 1–1½ lines long, thinly silky. Corolla not seen. Filaments persistent after flowering, connate at base for ¼ their length, the stamen-tube adnate to the calyx. Legume obliquely oblong, acute, twice as long as the calyx, thinly silky. This remarkable plant does not seem to have been found by any one save Thunberg, who has omitted to record the locality.

3. **C. spinosum** (Benth. Lond. Journ. 2. p. 455); branches divaricate, often spine-pointed; leaves *ovate-elliptical*, acute, with sub-incurved margins, silky on both sides, at length glabrescent; *peduncles axillary, capillary, two-flowered, much longer than the leaves; calyx teeth very short. Ingenhousia spinosa, E. Mey. Com. p.* 22.

HAB. Mountains between Hex River and Draai, 2800 f., *Drege!* (Herb. Sond.)

In its branching and foliage, this resembles *C. Thunbergii:* but differs from all the species, by its long, 2-flowered peduncles, bibracteate under the flowers.

4. **C. ciliare** (Vog. ex Walp. Lin. 13. p. 472); leaves lanceolato-subulate, acute, *straight and erect*, involute, hairy within and somewhat bearded at the apex, *glabrous and shining on the outside;* floral leaves broader, ovate or ovato-lanceolate; calyx glabrescent, its lobes setaceo-acuminate, somewhat bearded, much shorter than the keel; staminal tube elongated. *Benth. l. c. p.* 454. *Amphithalea ciliaris, E. & Z.! No.* 1241. *Ingenhoussia rugosa, E. Mey.! Comm. p.* 22.

HAB. Stony hills, Klynriver and Zonderende, *E. & Z.!* Klyn Drakenstein, *Drege!* Caledon, *Dr. Prior!* (Herb. Bth., Hk., Sd., D.)

Root thick and woody, throwing up many slender, erect or ascending stems, sparingly branched. Branches virgate, angular, minutely downy. Leaves generally close pressed, 4–5 lines long, not a line wide. Flowers axillary, sub-sessile; calyx not half as long as the corolla. Filaments united for nearly half their length, the staminal tube and petals perigynous.

5. **C. roseum** (Benth. Lond. Journ. vol. 2. p. 454); branches densely villous; leaves ovato-lanceolate or lanceolate, acute, *twisted and involute, densely and softly villous on both sides;* floral leaves shorter and broader, more ovate; calyx silky-villous, the segments acute; staminal tube elongate. *Ingenhoussia rosea, E. Mey! Comm. p.* 153. *Amphithalea perplexa, E. & Z.! En. No.* 1242.

HAB. Dutoitskloof and Winterhoeksberg, *Drege! E. & Z.!* (Herb. Bth., Hk., Sd., D.)

A much branched, densely and softly hairy bush, 1–2 feet high or more; the branches well covered with leaves. Leaves spreading, more or less twisted, 2–3 lines long, 1–1½ broad. Flowers nearly sessile, axillary. Calyces fulvous, with long silky hairs. Legume obliquely oblong, acute, villous.

6 **C. Bowiei** (Benth. l. c. p. 454); leaves lanceolate, involute, acute, *twisted*, tomentose within, *glabrous and shining on the outside;* floral leaves *broader;* calyx glabrous, or with downy margin, *its lobes very acute, as long as the carina, or longer;* stamens *very shortly* monadelphous. *Benth.*

HAB. S. Africa, *Bowie!* (Herb. Hook.)

Densely ramulose; the branches downy, soon glabrous, furrowed. Leaves 3–5 lines long, spreading, sub-pungent, the floral ovate-acuminate, longer than the calyx. Flowers sessile, rather smaller than in *C. Vogelii.* Stamens very nearly free to the point where they are adnate to the calyx. The foliage resembles that of *C. ciliare,* but the floral characters are very different.

7. **C. Vogelii** (Walp. Linn. 13. p. 472); dwarf, divaricately branched; leaves lanceolate, with incurved margins, twisted, tomentose on the upper surface, thinly pubescent, afterwards glabrous and shining on the outer side; *floral leaves similar;* calyx thinly silky and canescent, the teeth subulate; stamens very shortly monadelphous, and petals perigynous. *Benth. l. c. p.* 455. *Ingenhoussia tortilis, E. Mey. Comm. p.* 22.

HAB. Dutoitskloof Mountains, *Drege!* Also (fide *Benth.*) a narrow-leaved variety in Herb. Burchell (No. 6687). (Herb. Bth., Hk., D.)

A small, scrubby plant, 6–8 inches high, much branched; the twigs flexuous, furrowed and thinly silky. Leaves spreading or squarrose, 3–4 lines long, ½–1 line broad, the younger ones silky. Flowers small.

8. C. muraltioides (Benth. l. c.); branches rigid, more or less tomentose; leaves linear-sub-lanceolate, or the lower ones short, obtuse, strongly involute, spreading or squarrose, straight or somewhat twisted, tomentoso-villous on the upper side, cano-villous, becoming sub-glabrous on the outer; floral leaves *broadly ovate*, acute, one-nerved; calyx *pubescent, its teeth very short and obtuse;* stamens very shortly monadelphous; petals perigynous. *Benth.*

HAB. Pinaar's Kloof, *Burke and Zeyher!* Witsenbergsvlakte, *Dr. Pappe!* (Herb Hk., Bth. D.)

Much branched, 1–2 feet high, canescent and tomentose; the branches virgate, furrowed. Lowest leaves, 2–3, middle and upper ones 4–5 lines long, ¼ line wide, so strongly involute as to be nearly cylindrical. Flowers in capitate, five-flowered spikes, near the ends of the branches, about 2 lines long, purple. Young plants are very villous, with silvery-white hairs.

VIII. WALPERSIA, Harv.

Calyx campanulate, 5–cleft; two upper segments broader than the three lower. *Petals* sub-æqui-long, all adnate at base to the staminal tube; *vexillum* ovate, with a small callosity at the summit of the claw; *alæ* oblong, eared at base; *carina* sub-incurved, bluntly spurred at each side. *Stamens* shortly monadelphous, 5 longer. *Ovary* bi-ovulate. Style subulate.

A small shrub, closely allied to *Cœlidium*, but differing in foliage and in the soldering of the petals to the short staminal tube. Leaves *petiolate*, linear, with reflexed margins and a prominent midrib beneath. Flowers axillary, yellow. Calyx bibracteate at base. This genus is inscribed to the memory of *W. G. Walpers*, author of the useful "*Repertorium Botanices Systematicæ*," &c., &c., who commented learnedly on S. African *Leguminosæ* in the "Linnæa," vol. 13. p. 453, et seq. The genus *Walpersia, Reiss.* is the same as *Phylica, L.*

W. burtonioides (Harv.)

HAB. Glassenbosch, *Zeyher*. Feb.–Apr. (Herb. Sond.)

A small shrub, 6–12 inches high, erect, much branched; branches erect, angular, villous. Leaves spirally inserted, imperfectly whorled, 4–5 lines long, ½ line wide, exactly linear, acute, sub-mucronulate, thickish, plano-convex and muricated on the upper surface, the margins revolute and the broad midrib very prominent beneath. Petioles slender, 1 line long. Flowers axillary, towards the end of the branches, on hairy pedicels, 1–2 lines long. Bracts leaf-like at the base of the calyx, 4–5 lines long. Calyx silky-villous, all its segments subulate-attenuate, sub-aristate, the two upper ovate at base, the lower lanceolate. Corolla 4–5 lines long, the claws of the petals attached to the short staminal tube. Vexillum, with a slender, channelled claw, callous-tubercled at the summit, suddenly passing into the ovate limb. Alæ eared at base and corrugate at the sides. Carina obovate, incurved, sub-acute, scarcely rostrate, eared at base, and furnished with a small, blunt, pouch-like spur. Staminal tube scarcely exceeding the claws of the petals; filaments filiform, elongate. Ovary sessile, silky, with a long style. Legume unknown. This has more the look of an Australian *Burtonia* than of any S. African species known to me.

IX. BORBONIA, Linn.

Calyx acute at base, equally 5-cleft, the segments pungent. *Vexillum* hairy, emarginate, *carina* obtuse. *Stamens* 10, monadelphous, with a split tube. *Ovary* 2 or several ovuled; *style* filiform; *stigma* capitate. *Legume* linear, compressed, longer than the calyx, several seeded, (rarely 1–2 seeded. *Endl. Gen.* 6461. *DC. Prod.* 2. *p.* 120.

Shrubs or suffrutices, with alternate, simple, very rigid, many-nerved sessile or amplexicaul, exstipulate leaves; flowers yellow, axillary, or ending the branches, scattered or shortly racemose. Name in memory of Gaston de Bourbon, Duke of Orleans, son of Henry IV. of France, a great patron of botany. Some authors say it means "Farmers' (Boers') beans."

ANALYSIS OF THE SPECIES.

* All the petals hairy:
- Robust, densely branched, with densely imbricate leaves:
  - Leaves cordate-ovate; calyx-lobes glabrous ... .. (1) **cordata.**
  - Leaves broadly subulate; cal.-lobes villous ... ... (2) **barbata.**
- Slender, virgate, with *scattered*, linear-lanceolate leaves (3) **lanceolata.**

** The alæ and carina glabrous:
- Leaves lanceolate, 3–5 nerved:
  - Lvs. villous, ½-uncial; pedicels shorter than the leaf (5) **villosa.**
  - Lvs. glabrous, 1½-uncial; pedicels shorter than the leaf; legume 1-seeded ... ... (4) **monosperma.**
  - Lvs. glabrous uncial; pedicels equalling the leaf; legume many-seeded ... ... (6) **trinervia.**
- Leaves ovato-cordate, cordate-amplexicaul, or perfoliate:
  - Branches sharply angular:
    - Leaves cordate-ovate, 7–11 nerved, ciliato-papillate (8) **parviflora.**
    - Leaves orbicular, 11–15 nerved, entire or rough edged ... ... (9) **latifolia.**
  - Branches roundish or scarcely angular, leaves very entire:
    - Lvs. pungent-mucronate, peduncles 2–3 flowered (7) **alpestris.**
    - Lvs. taper-pointed, netted; pedunc. many-flowered (10) **complicata.**
  - Branches roundish; leaves more or less calloso-ciliate:
    - Robust, bushy; lvs. amplexicaul; calyx-teeth short ... ... (11) **crenata.**
    - Diffuse; lvs. amplexicaul; calyx teeth subulate, long ... ... (12) **undulata.**
    - Diffuse; lvs. perfoliate; cal. teeth short ... ... (13) **perforata.**

* *All the petals hairy.* (Sp. 1–3.)

1. B. cordata (Lin. Sp. p. 994); robust, densely much branched; the branches very villous; leaves densely imbricated, *cordato-ovate*, acuminate and pungent, many-nerved, glabrous, with very entire rib-like margins; flowers sub-capitate; calyx-tube densely barbate, segments glabrous. *DC. Prod.* 2. *p.* 120. *Benth. l. c. p.* 461. *Jacq. Schoenb. t.* 218. E. & Z.! No. 1210.

HAB. Mountains round Capetown, common. (Herb. Thb., Bth., Hk., Sd., D.)

A robust shrub, 1–2 feet high, with very rigid and sharply pungent, perfectly smooth leaves; ¾–1 inch long, ½ inch wide, spreading to all sides. The hairiness is confined to the branches and calyces.

2. B. barbata (Lam. Dict. 2. p. 436. Ill. t. 610. f. 2.); robust, with corymbose branches; leaves very densely imbricated, *broadly subulate*, broader at base and sub-cordate, villoso-ciliate, many-nerved; flowers sub-capitate, on very short pedicels; calyces densely villoso-barbate. *DC. Prod.* 2. *p.* 120. *Benth. in Lond. Journ.* 2. *p.* 460. *E. & Z.!* 1211.

HAB. Common on Table mountain, &c. (Herb., Hk., Bth., Sd., D.)

A coarse, furze-like bush, 1–2 feet high, with bare, woody stems, branched chiefly at the summit. Branches crowded, fastigiate, much divided, the young twigs hairy. Leaves ¾ inch long, 1–2 lines wide at base, whence they taper to a very slender point, densely crowded, their bases 1 line apart; upper leaves broadest, the young ones villous, older glabrescent, except on the edges, especially towards their base. Flowers 3–4, sub-sessile at the ends of the branches, both calyx and corolla densely hairy. Lower leaves 5–7, upper 12–15 nerved.

3. **B. lanceolata** (Linn. Sp. p. 994); glabrous, or nearly so, slender, with virgate branches; leaves scattered, lanceolate or linear-lanceolate, 5–7 (rarely 3–9) nerved; flowers solitary, in pairs, or sub-corymbose, on pedicels shorter than the leaves; calyces glabrous. *DC. Prod.* 2. *p.* 120. *Benth. l. c. p.* 460. *E. & Z.! No.* 1212. *Jacq. Schœnb. t.* 217. *Lodd. Cab. t.* 81. *B. angustifolia, Lam. Dict. E. & Z.! No.* 1213. *B. trinervia, DC.* (*non Thunb.*) *B. decipiens, E. M.! Comm. p.* 15.

Var. β. **gracilis**; leaves 3-nerved; flowers *small*, on uncial pedicels.

HAB. Common from Capetown to Uitenhage, and to the Camiesberg: in all collections. Var. β. Tulbagh Waterfall, *Dr. Pappe!* (Herb. Thunb. T.C.D., &c.)

A slender, slightly-branched shrub, 1–3 feet high; branches curved, glabrous or occasionally villous, with long soft white hairs. Leaves 2–4 lines apart, ¾–1½ inch long, ½–2½ lines broad, broadest in the middle. Flowers 5–6 lines long, sub-terminal, either on a short, 2-flowered peduncle, or in a corymbose, 4–6 flowered raceme. Calyx-lobes very variable in length, strongly 3-nerved. Corolla silky-villous. The leaves vary much in comparative breadth; and the inflorescence as above stated; but the species is easily limited by the characters assigned. β. is a very slender variety, with small (3 lines long) pale flowers, and 3-nerved leaves.

** *The vexillum hairy; the alæ and carina glabrous.* (Sp. 4–13)

4. **B. monosperma** (DC. Prod. 2. p. 120); leaves oblongo-lanceolate, three-nerved, glabrous, as well as the stem, or slightly villous; peduncles one-flowered, longer than the small flower, hairy, as is also the calyx; ovary 2–ovuled; legumes ovate-oblong, acute, 1–2 seeded. *Benth. l. c. p.* 461.

HAB. Cape Colony, *Bowie!* (Herb. Hook.)

Leaves 1½ inch long, two lines broad, very acute, nerveless on the upper, with 3 prominent distant nerves on the under side. Flowers I have not seen.

5. **B. villosa** (Harv.); slender, diffuse, villoso-pubescent; leaves scattered, patent or deflexed, lanceolate-oblong, acute, three-nerved, *villous on both surfaces;* peduncles one-flowered, shorter than the leaf; calyx-teeth subulate, longer than the tube; ovary several-ovuled, legume linear-oblong, acute, 1–2 seeded, *very villous.*

HAB. South Africa, *Wallich!* Between Witsenberg and Schurfdeberg, *Zey.* 437! *Pappe!* 128. (Herb. Hk., Bth., Sd., T.C.D.)

Stems 2 feet long or more, decumbent, not much branched; the branchlets angular, thinly covered with soft, spreading hairs. Leaves ½ inch long, 2½–3 lines broad, the lower ones more oblong, the upper more lanceolate; all acute, ¼–½ inch apart, patent. Vexillum pubescent; as long as the glabrous carina. Ovary and pod densely covered with long, soft hairs, by which character it differs from all other species of *Borbonia.* It is in other respects allied to *B. monosperma.*

6. **B. trinervia** (Thunb. Prod. p. 122); glabrous; branches slender; leaves scattered, linear-lanceolate, pungent, 3 (rarely 5) nerved; peduncles filiform, equalling the leaves, 2–3-bracteate below the flower;

legume oblong-linear, acute, many-seeded. *E. & Z.! No.* 1214. *B. pungens, Mundt.! Benth. l. c. p.* 461. *B. monosperma, E. Mey. Comm. p.* 15.

HAB. Cape of Good Hope, *Thunberg!* Mountains of southern region, *Drege!* Subalpine places near Gauritz Hoogte, *Mundt!* Between Riversdale and Gauritz River, *Pappe!* (Herb. Thb., Bth., Hk., Sd., T.C.D.)

A slender, diffuse shrub, 1–2 feet high, much branched above; the flowering branches twice as thick as hog's bristle. Leaves scarcely an inch long, 1–1½ lines wide, patent or squarrose, almost always 3-nerved; occasionally the marginal rib is removed inward, and becomes a nerve. Flowers 3 lines long, glabrous; the calyx-teeth deltoid-acuminate, one-nerved. The *B. trinervia* of Linnæus is said to have been founded on an imperfect specimen of *Cliffortia ruscifolia.* However this may be, an excellent specimen in Thunberg's Herbarium, marked "*Borbonia trinervia,*" by Thunberg himself, belongs to the plant now described. I think it right, therefore, to restore the early and appropriate trivial name.

7 **B. alpestris** (Benth. Lond. Journ. 2. p. 461); "glabrous; the branches scarcely angular; leaves *ovato-cordate, pungent-mucronate, very entire,* many-nerved; peduncles 2–3 flowered, longer than the calyx." *Benth. l. c.*

HAB. Subalpine bushy places, near Kochman's Kloof, *Mundt!* (Herb. Hook.)

"A small, divaricately-branched shrub, with short and slender branches. Leaves 4–6 lines long, 2 lines wide, 7–9 nerved on the lower side, veinless between the nerves. Flowers not seen. Fruit-stalk 3–4 lines long, slender, divided near the apex into 2–3 short pedicels. Bracts under the calyx subulate, striate. Calyx-tube 1½ line long, with narrow, setaceous, pungent segments as long as the tube. Pod nearly an inch long, 2 lines wide, acute, glabrous." *Benth. l. c.*. Of this I have only seen the single, imperfect specimen described by Bentham, and preserved in Herb. Hook.

8. **B. parviflora** (Lam. Dict. 1. p. 437); glabrous; branchlets sharply angular; leaves broadly *cordate-ovate,* acuminate, pungent, minutely ciliato-papillate, 7–11 nerved on both surfaces, faintly netted-veined between the nerves; flowers sub-capitate, on short pedicels; calyx-tube shorter than the narrow subulate segments. *DC. Prod.* 2. *p.* 120. *Benth. l. c. p.* 462. *E. & Z.! No.* 1209. *B. ruscifolia, Bot. Mag. t.* 2128. *DC. l. c. E. & Z.! No.* 1208. *B. alata, Willd. B. serrulata, Thunb.! Herb.*

HAB. Mountains of the Cape and Stellenbosch districts, common, *Thunberg! E. & Z.! Drege! Pappe! &c.* (Herb. Th., D., Bth., &c.)

A robust, much branched, very rigid shrub, 2–3 feet high. Leaves ¾–1 inch long, ½–¾ inch wide, spreading, very rigid and sharp, flat, with cartilaginous, minutely denticulated or tuberculated margins. The intermediate veins are more obvious in dried specimens. Flowers in dense, terminal, capitate racemes; the bracts setaceous, longer than the pedicels. Vexillum hairy. Pods an inch long.

9. **B. latifolia** (Benth. Lond. Journ. 2. p. 462); "branchlets sharply angular; leaves *orbicular,* mucronulate, cordate at base, very entire, or rough-edged, 11–15 nerved, obsoletely veined between the nerves; peduncles very short, many-flowered." *Benth. l. c.*

HAB. Cape Colony, Burchell, No. 8087.

"A fruiting specimen. Leaves an inch long and wide, concave, nerved on both sides. Peduncles very short, 8–12 flowered. Pedicels rigid, 2–3 lines long. Flowers not seen. Legumes 8–9 lines long, 3 lines wide, glabrous, coriaceous, reticulated." *Benth.*

10. **B. complicata** (Benth.! Lond. Journ. 2. p. 462); "branchlets round, glabrous; leaves amplexicaul, broadly cordate-ovate, *taper-*

pointed, pungent, *very entire*, many-nerved, closely and *delicately-netted between the nerves;* racemes many-flowered, shorter than the leaf; calyx teeth rather shorter than the tube." *Benth. l. c. B. parviflora, E. Mey.! Comm. p.* 16. (*non Lam.*)

HAB. Along rivulets on the Piquetberg, *Drege!* (Herb. Hk., Bth.)

A rather slender, much-branched bush, 1–1½ foot high; the branchlets short, crowded, spreading, obtusely angular or terete. Leaves horizontally patent, 6–7 lines long, 5 lines wide, with a long, tapering point. Calyx-teeth falcato-subulate. Flowers small.

11. B. crenata (Linn. Sp. 994); erect, robust; twigs roundish, glabrous or sparsely pilose; leaves amplexicaul, orbicular-cordate, mucronulate, calloso-ciliate, many-nerved, strongly netted between the nerves; racemes densely many-flowered; calyx sparsely villous, its teeth shorter than the tube. *Benth. l. c. p.* 462. *Thunb.! Prod. cap. p.* 122. *Bot. Mag. t.* 274. *DC. Prod.* 2. *p.* 120. *B. undulata, E. & Z.!* 1204 (*non Thunb.*) *Zey.!* 362.

HAB. Winterhoek and Dutoitskloof, *Drege! Pappe!* Tulbagh, *E. & Z.!* Erste R., Stell. *W.H.H.* (Herb. Th., Bth., Hk., Sd., D.)

A strong-growing, much-branched, and densely leafy, nearly glabrous and very rigid bush, 3–4 feet high. Leaves 1 inch broad and long, pale green, very cordate at base, the margin wavy, and set with rigid cilia. Racemes terminal, 12–15 flowered; fl. 4 lines long.

12. B. undulata (Thunb.! Prod. p. 122); diffuse or trailing; branchlets terete, thinly pilose, or glabrescent; leaves amplexicaul, deeply cordate at base, orbicular-ovate, undulate, with a reflexed sharp point, many-nerved, netted-veined, the margin ciliato-papillate, or villoso-ciliate, and when young pilose; peduncles 1–3 or many flowered; *calyx pilose, its segments subulate, longer than the tube. Benth. l. c. Borbonia perforata, E. Mey.! Comm. p.* 16. (*non Thunb.*) *B. ciliata, Willd. DC. Prod.* 2. *p.* 120, *ex parte. B. Candolleana, E. & Z.!* 1207. *B. commutata, Vog. Linn.* 10, *p.* 596.

Var. β. **multiflora**; peduncles 7–8 flowered; calyces very hairy. *B. crenata, E. & Z.! No.* 1203, *non Thunb.*

HAB. Among shrubs and tall grass, Dutoit's Kloof, *Drege!* Tulbagh Waterfall, *E. & Z.!* Witsenberg, *Zey.!* 363 (more pilose than usual). (Herb. Thb Bth. Hk. Sd. D.)

A slender, spreading, weak and often decumbent shrub; either nearly glabrous, or more commonly sprinkled with long, soft, horizontally patent hairs. Leaves closely clasping the stem, but their margins always free, ½–¾ inch long, ½ inch wide, laxly inserted, flat or folded together, the margins not always wavy.

13. B. perforata (Thunb. Prod. p. 122); diffuse or trailing; branchlets sub-terete, glabrous or pilose; leaves *perfoliate*, undulate, orbicular or elliptical, obtuse or recurvo-mucronulate, many-nerved, netted-veined, the margins pilose-ciliate, sub-papillate or entire; pedunc. 2–6 flowered, shorter than the leaf; calyx-teeth shorter than the tube. *Benth. l. c. B. ciliata. Willd. DC. Prod.* 2. *p.* 120, *ex parte. E. & Z.* 1206.

Var. α. **pluriflora**; robust, ascending or sub-erect, much-branched; leaves broader than long; peduncles 5–6 flowered. *E. & Z.!* 1205 & 1206.

Var. β. **pauciflora**; slender, trailing on the ground, the branches long and simple, leaves longer than broad; peduncles 1–2 flowered.

HAB. Var. α. Grootvadersbosch, Swell., *Mundt.!* Tulbagh, *E. & Z.!* β. Camps Bay, *W.H.H.* (Herb. Th., Bth., Hk., Sd., D.)

Very similar to *B. undulata;* but here the edges of the leaf-lobes are more or less united or connate round the stem: the flowers are smaller, and the calyx-lobes shorter.

X. RAFNIA.

Calyx unequally 5–fid, the lowest segment narrowest. *Corolla* glabrous; *vexillum* roundish; *carina* incurved, either rostrate or obliquely truncate. *Stamens* 10, monadelphous. *Ovary* sessile or stipitate, many ovuled; stigma capitate. *Legume* lanceolate or linear, the upper suture sharp or somewhat winged. *Endl. Gen.* 6459, *and Pelecynthis,* 6460. *Benth.! in Lond. Journ.* 2. *p.* 463. *Vascoa, DC. Œdmannia, Thunb.*

Glabrous, and frequently glaucous shrubs or suffrutices, with simple, very entire, alternate, exstipulate leaves and yellow flowers. Named in memory of C. G. Rafn, a Danish botanist.

ANALYSIS OF THE SPECIES.

1. **Vascoa.**—*Carina* rostrate. *Legume* many-seeded. Leaves *broadly amplexicaul, strongly netted* with veins. (Sp. 1–3.)

Leaves very obtuse, membranaceous :
 Calyx-teeth *longer* than the tube ... ... (1) **virens.**
 Calyx-teeth *shorter* than the tube ... ... (2) **amplexicaulis.**
Leaves acute or mucronate, rigid ... ... (3) **perfoliata.**

2. **Eu-Rafnia.**—*Carina* rostrate. *Legume* many-seeded. Leaves *never amplexicaul,* veinless, or with obsolete veins. (Sp. 4–15.)

* Latifoliæ : Leaves *broad;* either ovate, obovate, elliptical, or ovato-lanceolate.
 † Peduncles axillary, *leafless:*
  Leaves very broad, *cuspidate-acuminate* ... ... (4) **ovata.**
  Leaves acute or mucronate:
   Upper calyx-teeth *broader* than the lateral... (5) **triflora.**
   Upper calyx-teeth similar to the lateral ... (6) **fastigiata.**
 †† Peduncles axillary; with a pair of leafy bracts under the flower ... ... (7) **elliptica.**
 ††† Peduncles in a *terminal,* leafless raceme ... ... (8) **racemosa.**
** Angustifoliæ : Leaves *narrow;* either linear-oblong, lanceolate or linear.
 † Peduncles axillary, one-flowered, simple :
  Bracts minute, setaceous, close to the flower ... (9) **lancea.**
  Bracts leafy, remote from the flower ... ... (10) **crassifolia.**
 †† Peduncles forked, bearing flowers in the fork, and at the end of each arm:
  Branches roundish. *Legume* sessile, broad at base (11) **axillaris.**
  Branches angular. *Legume* narrowed at base and stipitate:
   Upper and lateral calyx-lobes lanceolate-falcate, acute, as long as the tube ... ... (12) **angulata.**
   Upper calyx-lobes *broader* than the lateral; both acuminate and twice as long as the tube (13) **humilis.**
   Upper and lateral calyx-lobes *dilated, obtuse* (14) **Ecklonis.**
   Calyx-teeth not half as long as the tube and separated by wide interspaces ... ... (15) **Thunbergii.**

3. **Pelecynthis.**—*Carina* somewhat fornicate, *broadly and obliquely truncate* or emarginate. Legume many-seeded. *Leaves* of sec. 2. (Sp. 16–18.)

† Flowering branchlets forked, a flower in the forks between a pair of opposite leaves :
 Leaves *narrow-oblong or lanceolate,* scarcely veiny (16) **opposita.**
 Leaves *ovate-elliptical or broadly-oblong,* veiny ... (17) **affinis.**
†† Flowers sub-corymbose, on terminal, leafless peduncles.
 Leaves *broadly* obovate, rhomboid or oblong ... . . (18) **cuneifolia.**

4. **Caminotropis.**—*Carina* completely fornicate, its petals united to the extreme point, obtuse or truncate. *Legume* on a long stipe, *one or two seeded.* (Sp. 19–22.)

† Leaves, at least the lower ones, broadly ovate or obovate:
 Erect, robust: leaves broad and *rounded at base* ... (19) **dichotoma.**

Erect, slender: leaves obovate, *narrowed at base* ... (20) **retroflexa.**
Procumbent, slender; lower leaves obovate, upper small, lanceolate ... ... (21) **diffusa.**
†† Leaves linear-lanceolate, very acute ... ... (22) **spicata.**

1. **VASCOA** (Sp. 1–3).

1. R. *(Vascoa)* **virens** (E. Mey.! Comm. p. 11); leaves orbicular or reniform, cordato-amplexicaul, *very obtuse;* four upper calyx-teeth *deltoideo-cuspidate,* as long as or *longer than* the tube, not very dissimilar. *Benth. Lond. Journ.* 2. *p.* 464.

HAB. Rocky places, Dutoit's-kloof, *Drege!* (Herb. Benth.)

A tall shrub. Leaves 1½ inch long, 2 inches wide, densely crowded, membranaceous, reticulately veined. Flowers terminal, concealed between two oblate bracts; pedicels 2–3 lines long. Calyx 2½ lines long, with a wide sinus between the upper and lateral segments. Very similar to the more luxuriant forms of *R. amplexicaulis,* from which this is merely distinguished by the different proportions of the calyx-teeth: I fear a variable character.

2. R. *(Vascoa)* **amplexicaulis** (Thunb.! Fl. Cap. p. 563); leaves orbicular or reniform, cordato-amplexicaul, *very obtuse;* four upper calyx-teeth *deltoid-acute,* shorter than the tube, not very dissimilar. *Vascoa amplexicaulis, DC.* 2. 119. *E. & Z.* 1200.

HAB. Mountains of Cape and Stellenbosch districts; Paarlberg, Cederberg and Giftberg, *Drege!* Tulbagh, *E. & Z.!* Witsenberg, *Zey.* 36. *Pappe* (55). Caledon, *Dr. Prior!* (Herb. Th., Bth., Sd., D., Hk.)

A much branched, densely leafy shrub, 2–4 feet high. Leaves ¾–1½ inch long, broader than their length, glaucous, thinly membranous, reticulated: rarely mucronate or sub-apiculate. Flowers concealed between two leafy bracts. Calyx 2 lines long, with a wide, *rounded* sinus between the upper and lateral teeth: by which character it is known from all varieties of the following.

3. R. (*Vascoa*) **perfoliata** (E. Mey. Comm. p. 12); leaves *rigid,* ovate or orbicular, cordato-amplexicaul, *acute* or scarcely obtuse; two upper calyx-teeth broadly falcato-cultrate, *broader* than the *triangular-acute* lateral ones; all nearly equalling the tube in length. *Benth. l. c. p.* 464. *Vascoa perfoliata, DC. Prod.* 2. *p.* 119. *Borbonia perfoliata, Thunb.! E. & Z.* 1202; *also V. acuminata, E. Mey.! E. & Z.* 1201. *Zeyher,* 2283!

HAB. Rocky hills in the Western Districts. Kochmanskloof, *Mundt.* Tulbagh and Hott. Holl. *E. & Z.!* Dutoit's-kloof, &c. *Drege!* (Herb. Bth., Hk., Sd., D.)

Smaller, more slender and diffuse than *R. amplexicaulis,* with much thicker, more rigid, and more evidently netted leaves. The leaves vary greatly in size and shape, sometimes they are nearly as orbicular and obtuse as in *R. amplexicaulis;* but more commonly they are longer than their breadth, decidedly cordate-ovate or cordate-oblong and acute or acuminate, ½–¾, rarely 1 inch long. Flowers smaller than in the preceding, similarly placed on short pedicels, and hidden between a pair of leafy bracts. The sinus separating the upper from the lateral calyx-teeth is *sharp* and narrow; and the lateral teeth are narrow (acute-angled) triangular, not *deltoid*

2. **EU-RAFNIA** (Sp. 4–15).

4. R. ovata (E. Mey.! Comm. p. 12); robust, branches roundish; leaves broadly *elliptic-ovate, sharply acuminate,* sub-petiolate; upper leaves more lanceolate, peduncles axillary, leafless; four upper calyx-teeth *triangular-acuminate,* as long as the tube, lowest narrow-subulate, of equal length to the rest; pod stipitate, with a broad wing. *Benth.! l. c. p.* 465. *R. cordata, E. & Z.* 1180.

HAB. Alpine, rocky places of the Western Districts. Cederbergen, *Drege!* Klapsmuts, *E. & Z.!* Hott. Holland, *Pappe!* (59). Witsenberg, *Zeyher!* 356. (Herb. Bth., Hk., Sd., D.)

The largest and most luxuriant of the genus, 3-4 feet high, very densely leafy. Leaves 2–3 inches long, 1½–2 inches broad, the lower ones much the broadest, sharply acuminate or cuspidate, penninerved, but not remarkably veiny. Flower 7-8 lines long, on axillary pedicels ¾–1 inch long.

5. R. triflora (Thunb. ! Fl. Cap. p. 563); robust, branches angular or two edged; leaves roundish-obovate, elliptical or ovato-lanceolate, acute or mucronulate, obtuse at base; peduncles axillary, 1–3 together, leafless (or branched and leaf bearing); upper calyx-teeth broadly *falcato-cultrate*, lateral acutely triangular, lowest narrow-subulate, as long as the rest; keel about twice as long as the calyx-tube; pod stipitate, with a narrow wing. *Benth. l. c. p.* 465, *DC. Prod.* 2. *p.* 118. *E. & Z. No.* 1181; *also R. diffusa, E. & Z.* 1183, *and R. alpina, E. & Z.* 1184. *Sieb. No.* 51.

HAB. Common in sub-alpine places, near Capetown and throughout the western districts. (Herb. Th., Bth., Hk., Sd., D.)

2–4 feet high, densely leafy. Leaves 1½–3 inches long, ½–2½ broad, very variable in shape, the upper ones narrowest and most acute, the lower often obtuse. Peduncles nearly an inch long, from the axils of the upper leaves. Flowers 5–6 lines long.

6. R. fastigiata (E. & Z. ! En. No. 1182); branchlets angular; leaves ovate, ovato-lanceolate or oblong, cuneate or rounded at base, acute; peduncles 1–3, axillary, leafless (or branched and leaf bearing); *upper* and lateral calyx-teeth *triangular-acuminate*, lowest setaceo-subulate, rather shorter than the rest; keel thrice as long as the calyx tube; pod scarcely winged. *Benth. l. c. p.* 466.

HAB. High mountains near Puspas Valley, Swellendam, *E. & Z.! Burchell*, 7177. (Herb. Sd., D.)

Very similar to *R. triflora*, but with more acute, more lance-shaped leaves; more isosceles-triangled *upper* calyx-teeth, and a longer and more rostrate vexillum. I retain the species with much doubt. Dr. Pappe (No. 58), unites it to *R. triflora.*

7. R. elliptica (Thunb. ! Prod. p. 123); branches angular; leaves broadly obovate, or elliptical, oblong, or ovato-lanceolate, acute or obtuse and mucronate; the upper ones narrow and more lanceolate, all narrowed at base; peduncles axillary, one-flowered, with a pair of leafy bracts under the flower; calyx-segments as long as or longer than the tube, the two uppermost cultrate or oblong-acuminate, much broader than the lateral which are broadly subulate; the lowest narrow-subulate, longer than the rest; pod sessile, linear-oblong, broader at base. *R. elliptica and R. intermedia, Benth. l. c.*

VAR. *α.* **erecta**; lower leaves obovate, mucronate; upper oblongo-lanceolate, acute; calyx tube dorsally umbonate, as long as the segments; the upper segments broadly cultrate. *R. elliptica, Thunb.! Herb. R. erecta, E. & Z.! No.* 1168. *R. intermedia, Walp., Benth. l. c.*

VAR. *β.* **intermedia**; foliage as in var *α*; calyx as in *γ*. *R. retroflexa, E. & Z.! En. No.* 1187. *Zey.* 359. *R. cuneifolia, litt. b. E. Mey.! Com. p.* 12, non Thunb. *R. intermedia, Walp. (partim).*

VAR. *γ.* **acuminata**; leaves oblongo-lanceolate, acuminate; calyx tube shorter than the segments; the upper segments oblong-acuminate. *R. elliptica, E. & Z.!* 1185. *Benth. l. c.*

HAB. Mountain sides and grassy slopes, Northern and Eastern Districts. Langekloof, George; Vanstaadensberg and Adow, Uitenhage, *E. & Z.!* Howison's Poort, *Mr. Hutton.* (Herb. Th., Bth., Sd., D.)

A stout, leafy, erect or spreading, slightly branched shrub, 1–3 feet high; branches virgate, more or less angular, sometimes obtusely so. Leaves 2–3 inches long, ½–1½ broad, varying much in shape on different parts of the plant. On young root shoots they are generally broadly obovate, obtuse or mucronulate: on older parts more or less lanceolate, and often much acuminate and very narrow. The calyx-teeth vary in length, as compared with the tube, rather than in shape. The original specimen in Herb. Thunb. has exactly the calyx of var. *a*, but the narrow and sharp foliage of var. *γ*: these forms are brought together through var. *β*, which is the commonest in Uitenhage and Albany.

8. R. racemosa (Eck. & Zey. ! No. 1188); leaves elliptical or oblong, mucronulate, somewhat cuneate at base, thick, midribbed, *veinless; flowers* 3–4 *in a short, terminal raceme;* bracts small, subulate; calyx lobes equalling the tube, acute, the upper broadly cultrate, lateral lanceolate-subulate; lowest setaceo-subulate, equalling the rest; carina shortly rostrate.

HAB. Assegaiskloof and Breederiver, Swellendam, *E. & Z.!* (Herb. Sd.)

This has the calyx of *R. elliptica;* but a different inflorescence, a much shorter, though rostrate, carina, and leaves of denser substance, leathery and not obviously veined. It might be conceived to be a cross between *R. elliptica, a*, and *R. cuneifolia, γ*. Branches roundish, or somewhat angled. Leaves 1–1½ inch long, ½–¾ inch wide, not much crowded, alternate.

9. R. lancea (DC. Prod. 2. p. 119); stem angular; leaves linear-oblong or oblongo-lanceolate, acute or mucronulate, somewhat veiny; peduncles axillary, one flowered, *setaceo-bracteate below the flower;* upper and lateral calyx-lobes *connate in two opposite pairs, connivent,* triangular-acuminate, lowest setaceous, all much shorter than the carina; legume cultrate, tapering at base. *Oedmannia lancea, Thunb.! Fl. Cap. p.* 561. *E. & Z.! En. No.* 1194. *Harv. Thes. t.* 72. *Benth.! l. c (pro parte).*

HAB. C. B. S. *Thunberg! Dr. Thom!* In sandy places, on the flats near Tigerberg, Cape district; also at Klynriver, Caledon, *E. & Z.!* Stellenbosch and the Paarl, *Drege!* Tulbaghskloof, *Dr. Pappe!* (Herb. Th., Bth., Hk., D., Sd.)

A small, ascending or spreading suffrutex, 6–12 inches high; simple or branched from the base. Leaves 1½–2 inches long, ¼–⅓ inch broad, with obvious midrib, more or less feather-veined. Peduncles an inch long, shorter than the subtending leaf, articulate near the summit, and there furnished with a pair of setaceous or subulate bracts. Calyx different from that of any other species, more deeply cloven between the two upper segments than between each of the upper and its lateral; so that the calyx may be said to be 3-lobed, the two larger lobes sharply bifid, with connivent teeth, the smaller lobe setaceous. The vexillum is strongly revolute, and it and the sharply rostrate keel are nearly twice as long as the calyx.

10. R. crassifolia (Harv. Thes. t. 71); branches sharply angular, or slightly winged; leaves narrow-oblong, or cuneate-oblong, obtuse, mucronate, thick and veinless; peduncles axillary, one flowered, bibracteate at base, deflexed after flowering; calyx lobes separate, twice or thrice as long as the tube, and as long as the carina, the four upper ones lanceolate-acuminate; legume oblong-cultrate, cuneate at base. *R. axillaris, E. & Z.!* 1192. *(non Thunb.) R. lancea (pro parte). Benth.! l. c. p.* 467. *R. angulata, litt. f., Thunb.! in Herb.*

HAB. Hottentot's Holland, near Palmietriver and Klynriversberg, *E. & Z.!* Klein-How-Hoek, *Zeyher!* 2281. *Pappe!* 60. Simon's Bay, *C. Wright!* 564. Capetown Hills, *Dr. Hooker! Bowie! Dr. Alexander Prior!* (Hb. Th., Bth., Hk., Sd. D.)

1½–2 feet high, many stemmed, simple, or branched from the base; branches virgate, curved, sub-trigonous or compressed. Leaves scattered, 1–2 inches long, nearly ½ inch broad, tapering more or less to the base, of a thickish substance, one nerved, but without obvious veins, more uniform in shape than in most species. Peduncles an inch long, shorter than the subtending leaf, articulate about a line from their base, and there furnished with a pair of linear, leafy, long or short bracts. Corolla scarcely protruding beyond the attenuated calyx lobes; keel rostrate. Legume pendulous. This seems to me to be a well marked species.

11. R. axillaris (Thunb.! Prod. p. 123. non Benth.); branches subterete; leaves narrow-oblong or lanceolate, acute, somewhat veiny; flowering branches forked, with opposite leaves, a flower in the fork and on each arm; peduncles shorter than the flower; upper and lateral calyx lobes separate, falcate, acuminate, rather longer than the tube, lowest setaceo-subulate; legume pendulous, *sessile, broad at base;* carina rostrate. *DC. Prod.* 2. *p.* 119.

HAB. Cape, *Thunberg! Bowie!* Sidbury, near Grahamstown, *Burke & Zeyher!* Swellendam, *Mundt.* (Herb. Th., Hk., Bth.)

A small, slightly branched suffrutex, 1–2 feet high, diffuse or ascending, with the habit and foliage of *R. lancea,* for which it is frequently taken; but with a very different calyx, more nearly resembling that of *R. elliptica.* The inflorescence is a three flowered cyme; the pedicels rarely more than ⅛ inch long. The calyx lobes are sometimes as long as the carina, sometimes shorter.

12. R. angulata (Thunb. Fl. Cap. p. 564); densely much branched, branchlets angular; leaves oblongo-cuneate, lanceolate, linear-lanceolate or lineari-filiform, sub-obtuse or acute; flowering branchlets forked, leafy; upper and lateral calyx lobes *lanceolate or falcate,* acute, subequal, nearly as long as the calyx tube or somewhat longer, the lowest setaceo-subulate, slightly shorter than the rest; legume much *narrowed at base into an evident stipe. Benth. l. c. p.* 467. *R. angulata, angustifolia and filifolia, Thunb.! Prod.* 123. *DC. Prod.* 2. 119.

VAR. *α.* **latifolia**; leaves cuneate-oblong, or lanceolate; calyx lobes longer than the tube.

VAR. *β.* **angustifolia**; leaves linear-lanceolate or filiform; calyx lobes shorter than in var. *α. E. & Z. No.* 1196, 1197, 1199.

HAB. Very abundant on the Cape Flats, and Capetown Hills; also Hott. Holland and Drakenstein Mts. &c. (Herb. Th., Hk., Bth., Sd., D.)

12–18 inches high, suffruticose, erect or sub-erect, generally much branched; the branches crowded, virgate or somewhat corymbose. Leaves ¾–1½ inch long, varying from half a line to nearly half an inch in breadth, of thickish substance and veinless. Flowers with a very long, sharply rostrate keel, twice as long as the calyx, by which character, as well as that of the pod, the broad-leaved forms differ from *R. crassifolia.*

13. R. humilis (Eck. & Zey.! No. 1198); slender, branching; branchlets angular; leaves linear-lanceolate, attenuate at base, acute; flowering branchlets leafy, one-flowered; calyx lobes *twice as long as the tube, falcate-acuminate, the upper pair broadest and somewhat cultrate,* lowest setaceo-subulate; legume *narrowed at base* into a long stipe.

HAB. Eastern side of Table Mt. near Constantia, *E. & Z.!* Near "Paradise," *W. H. H.* (Herb. Sd., D.)

A smaller, weaker, and less branched plant than *R. angulata,* with a differently proportioned calyx; in other respects similar and perhaps a mere variety. My specimen precisely agrees with that of Ecklon, in Herb. Sond. There are very few flowers, and they are of smaller size, with a much shorter corolla (in proportion to the calyx) than in *R. angulata.*

14. R. Ecklonis (E. Mey. Com. p. 13); "leaves linear, four upper calyx lobes *dilated, obtuse.*"

HAB. Cape Flats, *Ecklon.* (Unknown to us).

This can hardly be intended for *R. humilis;* but nothing like it occurs in Ecklon's (now Sonder's) private herbarium.

15. R. Thunbergii (Harv.); branches virgate, angular; leaves linear-lanceolate, attenuate; flowering branchlets (in a long pseudo-thyrsus) forked, leafy, 1–3 flowered; upper and lateral calyx teeth deltoid-acuminate, not half as long as the tube, *with rounded interspaces,* the lowest setaceous, shorter than the rest; carina falcato-rostrate, 3–4 times as long as the calyx; pod tapering at base into a stipe. *Crotalaria virgata, Herb. Holm.!*

HAB. South Africa, *Thunberg!* (Herb. Upsal, Holm.)

Apparently a tall shrub; branches 1–2? feet long, densely leafy below; the upper half, for the space of ten or twelve inches, converted into a dense thyrsoid inflorescence. Individual flowering branchlets axillary, 1½–2 inches long, bearing 1–3 flowers and a pair or two of leaves. Leaves ½–1 line broad, 1–1½ inch long, tapering to each end. Carina sharply bent upwards, much longer than the vexillum.—This species is founded on one of the sheets marked "*Rafnia filifolia*" in Thunberg's Herbarium. The sheet holds three specimens, each of the full length of the paper (14 inches), and seemingly but broken tops of much longer branches. The thyrsus reminds one of *Lebeckia Simsiana.* The calyx is so unlike that of any other of this section, and the thyrsoid habit so peculiar, that I have no hesitation in proposing the species, which I recommend to the notice of our South African friends. Possibly a native of Groenekloof? Another specimen also from Thunberg, exists in Herb. Holm. marked *Crotalaria virgata.*

3. **PELECYNTHIS** (Sp. 16–18.)

16. R. *(Pelecynthis)* **opposita** (Thunb. Fl. Cap. p. 564); branches somewhat angular; cauline leaves scattered, oblong or oblongo-lanceolate, narrowed at base, acute, mucronulate; flowering branches once or twice forked, with opposite leaves; pedicels shorter than the leafy bracts; upper and lateral calyx lobes triangular, acute, of equal size, with a wide interspace; the lowest subulate, slightly longer; carina very broad, truncate and emarginate; legume stipitate. *Benth. l. c. p.* 468. *DC. Prod.* 2. *p.* 119. *E. & Z. En. No* 1191, *also R. spicata,* 1193, *and R. pauciflora,* 1195.

HAB. Cape Flats and Muysenberg, Falsebay; also in Stellenbosch and Swellendam districts, *E. & Z.! Pappe* (61), &c. (Herb. Th., Bth., Hk., Sd., D.)

1–1½ feet high, many stemmed, ascending or sub-erect, branches not much divided. Leaves 1½ inch long, ¼–½ inch broad, thick and nearly veinless. Flowering branchlets 2–4 inches long, slender, compressed. Carina 2–3 times as long as the calyx, the truncate extremity nearly two lines broad.

17. R. *(Pelecynthis)* **affinis** (Harv.); branches sub-terete, the flowering branchlets once or twice forked, compressed; leaves ovate-elliptical or broadly oblong, acute at each end, mucronulate, somewhat veiny; peduncles short, one flowered, in the forks of the branchlets; upper and lateral calyx-lobes triangular-acuminate, distant, lowest subulate, somewhat longer; legume shortly pedicellate, narrowed at base, lanceolate. *R. axillaris, Benth.! Lond. Journ. non Thunb.*

HAB. Cape Flats and hill sides, *Bowie! W. H. H., Gueinzius,* &c. (Herb. Bth., Hk., Sd., D.)

I retain this species with much doubt, as distinct from *R. opposita,* to which it is very nearly allied, but has much broader, thinner, and more veiny leaves; more taper-pointed calyx lobes and somewhat larger flowers. Leaves 1–1½ inch long, ¾–⅞ inch broad.

18. R. *(Pelecynthis)* **cuneifolia** (Thunb. ! Fl. Cap. p. 563); branchlets more or less angular or compressed; leaves either broadly-obovate, ovato-rhomboid, elliptical, oblong, or lanceolate, shining, one nerved, somewhat veiny or veinless; peduncles several at the ends of the branches, sub-corymbose, leafless; upper and lateral calyx lobes broadly triangular, acute or acuminate, shorter than the subulate lowest lobe. *DC. Prod.* 2. *p.* 118. *E & Z. No.* 1189.

VAR. *α.* **rhomboidea**; leaves smaller, more rhomboid, thicker and less veiny than in the following. *R. rhomboidea, Walp. Linn.* 13. *p.* 464. *Benth. ! l. c. p.* 499.

VAR. *β.* **obovata**; leaves larger, more obovate or ovate-elliptical (the upper ones rhomboid), thinner and more veiny than in *α. R. cuneifolia, Benth. ! l. c. p.* 468.

VAR. *γ.* **lanceolata**; leaves oblong-elliptical or lanceolate, somewhat veiny. *R. corymbosa, Walp. Linn.* 13. *p.* 484. *Benth. ! l. c.*

HAB. Near Tulbagh, *E. & Z. !* Piquetberg, *Drege, β.* & *γ.* About the Paarlberg and Dutoit's-kloof. (Herb. Th., Bth., Sd., Hk., D.)

An erect or ascending, slightly branched, rather robust undershrub, 2–3 feet high. Leaves 1–2 inches long, of thickish substance, ½, ¾, or nearly 1 inch broad, subacute or mucronulate. This is readily known from the preceding species by its inflorescence. In Thunberg's Herb. are two specimens glued on one sheet; one of them referable to *R. rhomboidea,* Walp., the other to *R. cuneifolia,* E. Mey. In Ecklon's private Herb. the same intermixture occurs. Our three varieties come from the same district, and appear to me to differ merely in the foliage, which is notoriously variable throughout the genus.

4. **CAMINOTROPIS** (Sp. 19–22).

19. R. *(Caminotropis)* **dichotoma** (E. & Z. No. 1190); robust, the flowering branches repeatedly forked, angular or compressed, with opposite leaves; leaves broadly ovate or elliptic-oblong, acute or mucronate, *rounded at base and quite sessile,* thick and veinless; flowers solitary in the forks of the branches, on short pedicels; calyx oblique, upper and lateral segments triangular, acuminate, distant, lowest rather broadly subulate, scarcely shorter than the tube; carina fornicate, truncate, rectangular above; legume ovate-oblong, obtuse, on a long stipes. *Benth. ! l. c. p.* 469. *Pelecynthis gibba, E. Mey. Comm. p.* 14.

HAB. South Africa, *Burchell, No.* 7742, *Thom !.* Mountains near Gnadendahl, *E. & Z. !* Cederberg, *Drege.* Appelskraal, Riv. Zonderende, *Zeyher !* 2280. (Hb. Bth., Hk., Sd.)

A stout, strong growing, sub-erect undershrub, 1½–2 feet high, with sub-simple, terete stems, bearing towards the summit a long pseudo-panicle of leafy flowering branches. These latter are 4–6 inches long, 3–4 times patently dichotomous, with a pair of leaves at each fork. Leaves ½–¾ inch long, about ½ inch wide, somewhat fleshy, broad at base. Flowers 3–4 lines long, the two petals of the carina united to the very point; vexillum with involute edges, strongly bent.

20. R. *(Caminotropis)* **retroflexa** (Thunb. ! Prod. p. 123); erect, divaricately much branched, the branches nodoso-articulate, dichotomous; leaves mostly opposite, narrow-obovate, cuneato-attenuate at base, very obtuse, thick and fleshy, veinless; flowers terminal, on short pedicels, small; upper and lateral calyx segments triangular, acute, lowest subulate, about equalling the tube; carina fornicate, broadly truncate. *Thunb. ! Fl. Cap. p.* 564. *DC. Prod.* 2. *p.* 119.

HAB. South Africa, *Thunberg !* (Herb. Thunb.)

A distinctly woody, though slender shrub, 1–1½ feet high, with the aspect of a *Zygophyllum.* Leaves ½ inch long, of uniform size and shape, 2–3 lines wide; the lower ones (wanting in Thunberg's specimen) appear to have been alternate. The upper branches are very distinctly articulate, bent and flexuous, and irregularly forked. The plant turns black in drying. Not found since Thunberg's time. There are good specimens of it in the Upsal Herbarium, marked *R. retrofracta.*

21. R. *(Caminotropis)* **diffusa** (Thunb. ! Prod. p. 123); diffuse or procumbent, with slender, filiform branches; leaves scattered, the lowest obovate, cuneate at base, sub-obtuse or mucronate, upper (much smaller and sometimes opposite) ovato-lanceolate or oblong, acute; flowers at the ends of the branchlets, on short pedicels, small; calyx oblique, the upper and lateral segments triangular, acuminate, distant, lowest subulate, as long as the tube; carina fornicate; legume broadly oblong, on a long stipes. *R. retroflexa and R. diffusa, Benth. ! Lond. Journ.* 2. *p.* 469, 470. *E. Mey. ! Comm. p.* 15.

HAB. S. Africa, *Thunberg ! Zeyher !* (357). Zwartland, at Malmesbury, *Pappe !* 64. Under Bokkeveld and Cederbergen, *Drege !* (Herb. Th., Sd., Bth., Hk., D.)

Root thick and woody, deeply descending, emitting from the crown many trailing or diffusely sub-ascending, slender, terete, patently much branched stems. Lower leaves broadly cuneato-obovate, 1–1½ inch long, ½–¾ inch wide, thin, midribbed and somewhat veiny; upper and especially the uppermost mostly lanceolate, ¼–½ inch long, and 1–3 lines wide. Flowers sub-terminal, 2–3 lines long. Our specimens from Zwartland precisely agree with those in Herb. Thunberg, and also with "*R. retroflexa,*" Drege ! non Thunb.

22. R. *(Caminotropis)* **spicata** (Thunb. ! Fl. Cap. p. 564); slender; stems angular, sub-simple, incurvo-erect; leaves scattered, linear-lanceolate, acute or acuminate, veinless; flowers small, in a dense, leafy, pseudo-raceme; peduncles axillary, one flowered, nearly as long as the leaf, with a pair of leafy bracts near the summit; upper and lateral calyx-teeth triangular, acute, lowest subulate, shorter than the tube; carina fornicate, obtuse; legume stipitate, 2 seeded. *DC. Prod.* 2. *p.* 119.

HAB. South Africa, *Thunberg !* (Herb. Thunb.)

Many stemmed, 12–14 inches high, simple or branched from the base only. Peduncles (simple, one flowered, two-leaved flowering *branches*) about an inch long, crowded toward the end of the stem, for the space of 2–3 inches, each from the axil of a leaf. Leaves 1–1¼ inch long, 1–2 lines broad. The foliage is nearly that of *R. angulata;* the inflorescence that of *R. Thunbergii;* but the calyx and corolla those of a *Caminotropis.* A broad leaved variety, mentioned by Thunb. in *Fl. Cap.* does not now exist in his Herbarium.

(Doubtful Species).

R. *(Caminotropis ?)* **erecta** (Thunb. Fl. Cap. p. 565); "leaves oblong; flowers lateral; stem erect." *Thunb. l. c. DC. l. c.*

HAB. S. Africa, *Thunberg.* (Herb. Thunb.; *a battered fragment only !*)

"Stem shrubby, terete, branching, a foot or more high; branches alternate. Leaves sessile, ovate, entire, an inch long. Flowers axillary, pedunculate." *Thunb. l. c.* It is impossible to say what this may be. The specimen in Herb. Thunb. has but half a dozen leaves, and a broken flower remaining.

XI. EUCHLORA, E. & Z.

Calyx deeply 5-cleft; the lowest segment much narrower than the rest. *Corolla* glabrous; vexillum long-clawed, roundish, reflexed; alæ obtuse, longer than the sub-truncate carina. *Stamens* monadelphous, the tube slit above. *Ovary* few-ovuled, hairy; style glabrous. *Legume* swollen, ovate, few-seeded. *Endl. Gen. No.* 6484.

Only one species known, viz.,

E. serpens (*E. & Z.!* En. No. 1246). *Benth. Lond. Journ.* 2. *p.* 470. *Crotalaria serpens, E. Mey.! Lin.* 7. *p.* 153. *Ononis hirsuta, Thunb.! Fl. Cap. p.* 584. *Microtropis hirsuta, E. Mey.! Comm. p.* 65.

HAB. Sandy plains, &c., Cape Flats, near Salt River; also in Zwartland and at Saldanha Bay, *E. & Z.!* Drege, Thunberg! (Herb. Th., Bth., Hk., Sd.)

A small prostrate suffruticose plant, with filiform underground stems, at intervals throwing up leafy branches; the places below the soil glabrous, those above densely clothed with long, sub-ferruginous hairs. Overground branches 1–2–3 inches long, ascending, imbricated with leaves. Leaves lanceolate, sessile, about ½ inch long, and 1–1½ lines wide, hispid on both sides, but especially on the under: the hairs appressed. Peduncles terminal, patently hairy, supporting an oblong, dense, sub-capitate spike of small, purplish flowers. Calyx very hairy; its 4 upper segments semi-lanceolate, lowest narrow subulate. Vexillum broader than long, emarginate, about as long as its claw; alæ and carina subtruncate. Pod short and hairy.

XII. CROTALARIA, Linn.

Calyx sub-bilabiate, the upper lip bifid, the lower trifid. *Vexillum* large, cordate; *carina* falcate-acuminate. *Stamens* monadelphous. *Ovary* 2 or many ovuled; style elongate, knee-bent, often laterally pubescent. *Legume* turgid, with very convex valves, sessile or stipitate, few or many seeded. *Endl. Gen.* 6472. *Benth.! in Lond. Journ.* 2. *p.* 472.

Herbs or shrubs, common throughout the tropics and sub-tropics of both hemispheres. Leaves either simple, or palmately 3–5–7 foliolate, commonly stipulate; bracts and stipules sometimes wanting. Flowers either racemose or sub-solitary, but not umbellate. The sharply rostrate carina (in *C. purpurea* short and rather blunt) and the inflated pod mark this genus. Some species of *Lotononis*, especially in the section "*Oxydium*," approach it in the form of corolla; but differ by their umbellate inflorescence and unswollen pod. In other *Lotononides*, when the pod is more turgid, the carina is not sharp. The name is derived from κροταλον, a *castanet;* because the seeds rattle in the inflated pods, when shaken.

ANALYSIS OF THE SOUTH AFRICAN SPECIES.

1. Simplicifoliæ. *Leaves* simple, sessile. (Sp. 1–2.)

Leaves narrow-oblong or sub-lanceolate; pedicels shorter than the calyx ... ... (1) **virgultalis.**

Leaves linear-subulate or setaceous; pedicels *longer* than the calyx ... ... (2) **spartioides.**

2. Oliganthæ. *Leaves* digitately 3-foliolate. *Peduncles* opposite the leaves (or rarely terminal), 1–2 flowered, or distantly 2–6–8 flowered. (Sp. 3–12.)

Diffuse or prostrate herbs or suffrutices:

Leaflets narrow-lanceolate or subulate:

Stipules minute; leaflets linear-subulate, downy on the lower surface ... ... (9) **angustissima.**

Stipules linear-lanceolate; leaflets lanceolate, silky (10) **Ecklonis.**

Leaflets obovate or ovato-lanceolate:

Dwarf; thinly silky or glabrescent:

Leaflets silky-canescent on *both* surfaces; stem canescent ... ... (3) **sparsiflora.**

Leaflets appressedly pubescent on the *under* side; stem glabrescent

Peduncles 2–3 inches long, 1 rarely 2-flowered ... ... (4) **humilis.**

Peduncles 3–5 inches long, 3–6-flowered (5) **effusa.**

Dwarf; densely and softly silky-villous; pedunc. 3–8 flowered ... ... (6) **mollis.**

Larger; hispid with coarser, rusty hairs:
Suffruticose, slender, very hispid; peduncles 4–6 inches long, 2–5 flowered … … … (8) **obscura.**
Shrubby, much branched; pedunc. not much exceeding the leaves, 1–3 flowered … … (7) **lotoides.**
Erect or sub-erect herbs, scarcely suffruticose:
Leafl. oblong-linear or lanceolate; racemes 2–6 flowered (12) **distans.**
Leafl. cuneate, obtuse; pedunc. 1 rarely 2-flowered … (11) **Grantiana.**

3. Racemosæ. *Leaves* digitately 3–5 foliolate. *Racemes* mostly terminal, densely or laxly many or several flowered. (Sp. 13–24.)

Stipules small, obsolete or none:
Leaflets narrow-lanceolate, acute:
Branches puberulent; stipules obsolete; calyx-teeth *short* … … … … … … … … … (13) **lanceolata.**
Branches and petioles *densely hispid;* stipules subulate; calyx-lobes long … … … … … (14) **Burkeana.**
Leaflets obovate, oblong or linear-oblong;
Calyx-lobes as long as the tube or longer, lanceolate:
Leafl. elliptic-oblong; legume subsessile, long, many-seeded … … … … … … … (15) **striata.**
Leafl. obovate; legume large, oblong, on a long stipe … … … … … … … (20) **macrocarpa.**
Leafl. cuneate-oblong; legume stipitate, small sub-globose or ovoid … … … … … (16) **globifera.**
Leafl. oblong-linear; *fl. minute;* legume sessile small, obovoid … … … … … … (17) **Nubica.**
Calyx-lobes longer than the tube, *oblong, truncate, and cuspidate* … … … … … … … (18) **platysepala.**
Calyx shortly 5-toothed:
Whole plant densely velvetty-tomentose; racemes very long … … … … … … … (19) **elongata.**
Sub-glabrous or thinly silky:
Leaflets ¾–1 inch long; flowers *purple* … (21) **purpurea.**
Leaflets *minute,* 2 lines long, 1 line wide; racemes lax … … … … … … … (22) **aspalathoides.**
Stipules leaf-like, petiolulate (occasionally wanting to some leaves)
Branches terete; *stipules* and leaflets obovate; carina glabrous … … … … … … … … … (23) **Capensis.**
Branches angular; stipules *lanceolate;* carina woolly on upper edge … … … … … … … … … (24) **Natalitia.**

I. **SIMPLICIFOLIÆ** (Sp. 1–2).

1. C. virgultalis (Burch. in DC. Prod. 2. p. 128); nearly glabrous, or minutely silky; branches virgate, rush-like, striate; stipules none; leaves *narrow-oblong, or lanceolate-linear*, thickish; racemes terminal, 8–10 flowered, lax, bracts and bracteoles minute, subulate; calyx appressedly silky, *longer* than the pedicel; legume sessile, elliptic-oblong, minutely pubescent. *Benth. Ld. Journ. 2. p. 561. C. spartioides, E. Mey.! Comm. non DC.*

HAB. S. Africa, *Burchell* (No. 1752). On the Gariep, near Verleptram, *Drege! Pappe* (68)! *A. Wyley, Esq.* (Herb. T.C.D.)

1½–2 feet high, slender, the younger parts very minutely and appressedly downy; older glabrescent. Leaves few and distant, obtuse or acute, 1 inch long, 1 line broad, twice or thrice as long as their petiole. Flowers yellow, ½ inch long; the vexillum silky; carina very acute and slender, nearly twice as long as the alæ. Pods ¾ inch long, 3 lines in diameter.

2. C. spartioides (DC. Prod. 2. p. 128); nearly glabrous, or thinly and minutely silky; branches virgate, rush-like, striate; stipules none;

leaves *linear-subulate or setaceous;* racemes elongate, distantly pluri-flowered, bracts and bracteoles very minute, setaceous; calyces *shorter* than the pedicels; legume shortly oblong, sessile, glabrous or minutely puberulous. *Benth.! l. c. p.* 561.

HAB. S. Africa, Burchell, No. 2336. Rhinoster River, *Burke! and Zeyher! Pappe*, No. 69. Zoolu country, *Miss Owen!* (Herb. Hk., Sd., D.)

2–3 feet high, broom-like, much branched; the branchlets slender and wiry. Leaves few and distant, scarcely thicker than hog's bristle, 1–1½ inch long. Flowers larger than in *C. virgultalis*, on longer stalks; the calyx-lobes more taper-pointed, and both calyx and vexillum less hairy. Pods smaller; in my specimens quite glabrous.

2. FOLIOLATÆ (Sp. 3–24).

3. C. sparsiflora (E. Mey.! Comm. p. 26); "dwarf, diffuse; stipules small or obsolete; leaflets obovate, silky-canescent on both sides, as is also the stem; pedicels opposite the leaves, one-flowered; carina with a straight beak; ovary many ovuled; legume sessile, oblong, silky, not longer than the calyx." *Benth. l. c. p.* 573.

HAB. On the Gariep, near Verleptram, *Drege!* (Unknown to me.)

"Herbaceous, dichotomously much branched, many-flowered. Flowers small. Legume 2–3 lines long." *Benth.*

4. C. humilis (E. & Z.! En. No. 1263); dwarf, diffuse, or trailing, much branched from the base, branchlets filiform, minutely pubescent or glabrous; stipules minute, subulate; leaflets scarcely as long as the petiole, obovate or oblong, or the uppermost ones linear, appressedly pubescent beneath; peduncles opposite the leaves, 1–2 flowered; carina with a straightish beak; ovules numerous; legume sub-sessile, oblong-cylindrical, minutely and appressedly pubescent. *Benth.! l. c. p.* 574. *C. diffusa, E. Mey. Linn.* 7. *p.* 151, *non Link. & C. effusa, E. Mey, (ex parte). Lotononis diffusa, E. & Z.! No.* 1274 *and L. perplexa, E. & Z.!* 1275 (*non Benth.*) *Ononis excisa, Thunb.! Fl. cap. p.* 586.

HAB. In clayey soil. Common near Capetown, *E. & Z.! &c.* Paarl and Groenekloof *Drege!* Sandy places on Oliphant's River, Clanwilliam, *E. & Z.!* (Herb. Hk., Sd., D.)

A small spreading plant, the size of *Lotus corniculatus.* Root deeply descending, perennial; many stemmed from the crown. Petioles ¾–1 inch long, leaflets ½–¾ inch: sometimes the petioles are proportionably shorter. Pod ¾ inch long. Peduncles 2–3 inches, more commonly one than two flowered. The solitary specimen marked "*Lot. perplexa*" in Herb. Ecklon, belongs to this plant, and not to *L. perplexa* of *E. Mey.!* and *Benth.*, which Ecklon probably confounded with it.

5. C. effusa (E. Mey.! Comm. p. 25, ex parte); rather dwarf, diffuse, much branched from the base, branchlets terete, glabrescent; stipules minute; leaflets shorter than the petiole, obovate or oblong, minutely appressedly-pubescent beneath; peduncles terminal or (at length) opposite the leaves, elongate, distantly 3–6 flowered; bracts *oblong, blunt;* carina with a straightish beak; ovules numerous, legume sub-sessile, oblong, much inflated, appressedly pubescent. *Benth.! l. c. p.* 574. *Ononis racemosa, Thunb.! Fl. cap. p.* 587.

HAB. Sands near Krakkeelskraal, *Drege!* (Herb. Th., Hk., D.)

Herbaceous, more robust and less branching than *C. humilis*, with a different inflorescence; and differing from *C. mollis* by its pubescence. Peduncles 3–6 inches long. Leaflets broadly obovate or narrow oblong, the uppermost narrowest. *Thunberg's* specimen has rather larger flowers and narrower leaves than *Drege's*

6. C. mollis (E. Mey.! Comm. p. 23); diffuse, much branched from the base, *densely and softly hairy;* stipules subulate, small; leaflets obovate, glabrescent *above;* peduncles terminal, (at length) opposite the leaves, elongate, distantly 3–8 flowered; bracts *subulate, acute;* carina with a straightish beak; ovules numerous; legume sub-sessile, oblong, much inflated, softly hairy. *Benth. l. c.* 575.

HAB. Dry hills and islands at the mouth of the Gariep, *Drege!* (Herb. Hk., Sd., D.)

Resembles *C. effusa*, but easily known as well by its thick coat of long, silky hairs, which cover all parts except the upper surfaces of the leaflets, as by the bracts, &c. Petioles ¾, leaflets ½ inch long. Peduncles 4–5 inches.

7. C. lotoides (Benth.! Lond. Journ. 2. p. 575); shrubby, sub-dichotomously much branched; branches with dense rusty pubescence and spreading hairs; stipules linear-lanceolate; leaflets broadly obovate, hairy beneath, or on both surfaces; peduncles opposite the leaves and not much exceeding them, 1–3 flowered; calyx pubescent; carina with a straightish beak; legume sub-sessile, cylindrical, hairy.

HAB. Magalisberg and Aapges River, *Burke and Zeyher!* Near Grahamstown, *Mr. Ward*, in Herb. Hook. (Herb. Hook., Sond.)

A rigid, somewhat woody, much-branched undershrub, either prostrate or spreading widely over the ground; branches 1–1½ feet long, their pubescence variable, dense or rather thin, but always *rough* and patent. Leaflets about equalling the petioles, ½ inch long, and nearly as broad. Legume 1–1¼ inch long. Flowers rather small, 3–4 lines long.

8. C. obscura (DC. Prod. 2. p. 134); diffuse, branched from the base, hispid with long, very patent, rusty-coloured hairs; stipules linear-lanceolate; leaflets broadly obovate or ovato-lanceolate, hairy beneath, or on both surfaces; peduncles terminal or opposite the leaves, elongated, 2–5 flowered; calyx hispid and ciliate; carina with a falcate beak; legume sessile, oblong, hispid. *Benth. l. c. p.* 575. *Cr. pilosa, Thunb.! Fl. Cap. p.* 572 (*non Mill.*) *E. & Z.! En. No.* 1260.

HAB. Eastern districts and Caffraria. Grassy pastures of Uitenhage, "Adow," and Krakakamma, *E. & Z.!* Albany, *Mrs. Barber!* Omsamwubo, *Drege!* (Herb. Th., Hk., Sd., D.)

Slender, suffruticose, 1–2 feet high, decumbent or spreading; less branching, less woody and more slender than *C. lotoides;* with larger leaves on longer petioles, much longer peduncles, and longer, stiffer, and more spreading hairs. Petiole 1–1½ inch long, rather longer than the lamina. Peduncles 4–6 inches long. Legumes 1–1½ inch long, nearly ½ inch wide.

9. C. angustissima (E. Mey.! Comm. p. 26); very slender, diffuse, much branched from the base; stipules minute; leaflets on very long petioles, *linear-subulate*, or those of the lower leaves linear-cuneiform, appressedly and thinly downy underneath, as is the stem; peduncles elongate, 1–2 flowered at the summit; legume short-stalked, oblong, downy. *Benth. l. c. p.* 576.

HAB. Sandy hills near Ebenezer, Stellenbosch, *Drege!* (Hb. Sond.)

"Flowers rather large, flesh-coloured. Legume 9–10 lines long," *E. Mey.!* Leaflets an inch long, not ½ line wide, folded together; petioles 2 inches long. A very imperfect fragment only seen.

10. C. Ecklonis (Harv.); diffuse, slender, pubescent; stipules linear-lanceolate; leaflets on long petioles, *lanceolate*, those of the upper leaves linear-lanceolate, appressedly-pubescent on both surfaces; peduncles opposite the leaves, 2–flowered; carina with a falcate beak. *C. stenophylla, E. & Z.!* 1261 *(non Vog.)*

HAB. In rocky places, Oliphant's River, Clanwilliam, *E. & Z.!* (Herb. Sond.)

This may possibly be a broad-leaved state of the preceding. The petioles are 1½ inch long, frequently deflexed; the leaflets of the upper leaves 1¼–1½ inch long, 1 line wide, acute at each end, with reflexed margins; those of the lower leaves shorter and broader, but lanceolate (not cuneate). Whole plant silky, with somewhat the habit of an *Argyrolobium*. I have only seen one specimen, with imperfectly preserved yellow flowers.

11. **C. Grantiana** (Harv.); herbaceous, erect; branches appressedly-pubescent; stipules minute, setaceo-subulate; leaflets cuneate, obtuse or emarginate, glabrous above, appressedly-pubescent beneath; peduncles filiform, opposite the leaves or terminal, bracteolate beyond the middle, not much longer than the petiole, 1, rarely 2–flowered; calyx segments lanceolate, longer than the tube; carina with a long straight beak; vexillum downy; legume sessile, broadly-oblong, pubescent, many-seeded.

HAB. Port Natal, *Dr. W. B. Grant.* (Herb. Hook.)

A small, slender, probably annual species, a foot in length, branching, leafy, many flowered. Leaves trifoliolate, on shortish petioles; leaflets ½–¾ inch long, not 2 lines wide. *Peduncles* threadlike, about an inch long, articulate and minutely bracteolate beyond the middle; flowers inclined, small, yellow, striate. Legume 5–6 lines long, much inflated, 2–3 lines wide. Except in inflorescence, this much resembles *C. Senegalensis*, and possibly it may be only a dwarfed form of that species.

12. **C. distans** (Benth. Lond. Journ. vol. 2. p. 582); herbaceous, minutely puberulous, slender, erect or ascending, with sub-terete, virgate branches; stipules bristle-shaped; leaflets oblong-linear or lanceolate, obtuse or mucronate, glabrous above, very minutely and appressedly pubescent beneath; racemes opposite the leaves, distantly 2–6 flowered; ovules numerous; legume sessile, oblong-oval, minutely downy.

HAB. Lay Spruit and Tal River, N. East, *Burke and Zey.!* (Herb. Hk., Sd., D.)

A slender, much branched, slightly suffruticose or herbaceous species, thinly covered with extremely minute, close-pressed hairs. Petioles shorter than the leaflets, ⅓–½ inch long; leaflets 1–1½ inch long, 1–2 lines wide, the middle one longer than the others. Peduncles 3–5 inches long, the flowers an inch apart, yellow. Legumes turgid, ½–¾ inch long.

13. **C. lanceolata** (E. Mey.! Comm. p. 24); erect, herbaceous; branches virgate, angular and striate, appressedly downy; stipules obsolete; leaflets narrow-lanceolate, elongate, appressedly puberulent beneath; racemes terminal, elongating, many flowered; calyx-teeth triangular, not half as long as the tube; legume sessile, elongate, appressedly downy. *Benth. l. c. p.* 585.

HAB. Between the Omsamculo and Omcomas, Caffraria, *Drege!* Port Natal, *T. Williamson! Krauss* (469)! *Gueinzius!* (Herb. Hk., Sd., D.)

Two or three feet, with the habit of a *Lupin*. The pubescence *extremely minute.* Leaflets 1½–2½ inches long, 2–3 lines wide, acute at each end, the upper surface, except along the midrib, quite glabrous: petioles shorter than the leaflets. Racemes 6–8 inches long, 20–30 flowered. Flowers yellow, with dark purple veins, 3–4 lines long. Pods more than an inch long.

14. **C. Burkeana** (Benth. Lond. Journ. 2. p. 593); herbaceous or suf-

fruticose, erect; branches, petioles, and racemes densely hispid, with long, patent, rusty hairs; stipules linear-subulate; leaflets 3–5, linear-lanceolate, acute, glabrous above, pilose beneath; racemes terminal, lax, several flowered; bracteoles lanceolate; calyx deeply cut, its segments lanceolate, nearly as long as corolla; legume sub-sessile, oblong, very hairy.

Var. β. **sparsipila**; much less hairy, with longer petioles, leaflets and racemes.

HAB. Magaliesberg and Aapjes River, *Burke and Zeyher!* β. in the Zooloo country, *Miss Owen!* (Herb. Hk., Sd., D.)

One to two feet high, suffruticose at base, with many herbaceous, slightly branched stems. Pubescence copious, rusty brown, and harsh. Petioles 1–1½ inches long, leaflets as long, often 5 together, 1–2 lines wide, acute at each end. Racemes pedunculate, terminal, 10–15 flowered. Flowers yellow, ¾ inch long. Legumes 1½ inches long. Var. β. is a more luxuriant and less hairy form, and probably grew in richer, alluvial soil.

15. C. striata (DC. Prod. 2. p. 131); herbaceous or suffruticose, erect, divaricately branched; branches angular and striate, thinly canescent; stipules none; leaflets on long petioles, elliptic oblong, or obovate, obtuse, mucronulate, glabrous or very minutely strigoso-puberulent bebeneath; racemes terminal, densely many-flowered, elongate; calyx appressedly pubescent, its teeth acuminate, about equalling the tube; carina falcate, twice as long as the alæ; legume sub-sessile, elongate, hook-pointed, many-seeded, minutely downy. *Benth. l. c. p.* 586. *Bot. Mag. t.* 3200.

HAB. Port Natal, *Mr. Hewitson! Mr. Sanderson!* (Herb. Hk., D.)

A large species, but with moderate flowers. Petioles 2–3 inches long, leaflets 2–2½ inches long, 1–1½ inch broad, petiolulate, somewhat acute at base, tipped with a minute bristle-like mucro, mostly obtuse, rarely ovato-lanceolate and sub-acute. The very young buds and branchlets are somewhat woolly; the mature very thinly pubescent and whitish. Racemes 6–8 inches long, 40–50 flowered; flowers yellow, streaked with purple, pendulous: the vexillum and alæ much shorter than the carina. Legume 1–1½ inch long, 2–2½ lines in diameter, the upper suture depressed. This species occurs throughout the tropics of both hemispheres.

16. C. globifera (E. Mey. Comm. p. 24); suffrutescent, many-stemmed, more or less puberulent or canescent, with angular, virgate branches; stipules bristle-shaped or wanting; leaflets cuneate-oblong, glabrous above, appressedly pubescent beneath; racemes terminal, densely many flowered; ovules four; legume *stipitate, obliquely obovoid-subglobose,* appressedly pubescent. *Benth. l. c. p.* 581. *C. macrostachya, Sond.! Linn.* 23. *p.* 26.

Var. β. **brachycarpa**; legume very oblique, depressed-globose. *Benth.! l. c.*

Var. γ. **glabra**; petioles much shorter than the lamina; leaflets nearly glabrous below; flowers smaller.

HAB. Between the Omsamcaba and Omsamwubo, *Drege.* Port Natal, *Krauss!* 341 and 440, *Gueinzius!* β. Magaliesberg, *Burke!* γ. Port Natal, *Gueinzius!* (Herb. Hk., Sd., D.)

Root with a thick crown throwing up several sub-erect or ascending, curved stems, a foot or eighteen inches in height; with several lateral virgate branches. Pubescence close-pressed, thin and minute. Petioles ½–¾ inch long, scarcely equalling the lamina, in var. γ. much shorter; leaflet cuneate at base, obtuse or mucronate, the midrib very prominent on the lower surface. Raceme short, with 12–20 yellow flowers: carina taper-pointed; vexillum thinly silky. Pods scarcely as large as peas; in var. β. smaller. I cannot distinguish Sonder's *C. macrostachya* from *Krauss's*

No. 440, which differs from his 341 merely in what may be referred to luxuriant growth; namely, a longer raceme, somewhat longer petioles (but this varies), and more evident stipules. In Gueinzius' specimens, referred by Sonder to *C. globifera*, the flowers are smaller and the foliage more glabrous; these answer well to E. Meyer's var. *a, glabra.* I have not seen any Dregean specimens.

17. C. Nubica (Benth.! l. c. p. 581); herbaceous, spreading, much branched, piloso-pubescent; stipules minute, lanceolate; leaflets *oblong-linear*, obtuse, glabrous above, or nearly so; pubescent beneath; racemes opposite the leaves, elongate, slender, laxly many flowered; *flowers minute;* calyx-lobes subulate, longer than the tube; ovules 4–6, legume *sessile, small, obovoid,* pubescent. *C. sphærocarpa, var. angustifolia, Hochst.! Hb. Un. It. Pl. Arab.* 282.

HAB. Delagoa Bay, *Forbes!* (Herb D.)

Annual? 1–2 feet high; branches divaricating or angularly bent, or sub-dichotomous. Leaves sub-distant. Petioles uncial; leaflets 3, about as long, 2–3 lines wide, tapering at base, either quite glabrous above, or minutely pubescent. Racemes 5–6 inches long or more, the flowers 2–4 lines apart, on slender pedicels longer than the bracts. Flowers 2 lines long, with a very sharp, slender carina. Pod 2–3 lines long. The flowers are much smaller than in any other South African species.

18. C. platysepala (Harv. Thes. t. 29); shrubby; branches thinly downy; stipules setaceous; leaflets about equalling the petiole, elliptic-oblong, mucronulate, glabrous above, appressedly pubescent beneath; racemes terminal, elongate, many flowered; calyx pubescent, deeply lobed, *its segments obovate-oblong, cuspidate-mucronate, the lowest narrower than the rest;* vexillum pubescent, alæ very broad, nearly as long as the much acuminate carina; ovary pubescent, *stipitate,* 6–8 ovuled; legume?

HAB. Sandy soil between the Rivers Tamulahan and Zougha, beyond the Lake Ngami, *Jos. M'Cabe!* (Herb. Hook.)

"A shrub." The flowering ends of branches only seen by me. Leaves 3-foliolate, petiole and leaflets ¾–1 inch long, the middle leaflet longest, 2–3 lines broad. Racemes 5–6 inches long, lengthening, 20–30 flowered; flowers bright yellow, ⅓ inch long. *Calyx* very different from that of any other S. African species.

19. C. elongata (Thunb. Fl. Cap. 571); suffruticose, erect, *densely velvetty-tomentose* in all parts; branches terete, virgate; stipules wanting; leaflets ovate or cuneate-oblong, *densely lanato-sericeous on both sides,* thickish, obtuse; racemes terminal, very long, densely many-flowered; calyx villous, its lobes deltoid, short; ovules three; legume sub-sessile, oblong, acute, densely woolly, one-seeded.

HAB. Karroo, near Bockeveld, *Thunberg!* (Herb. Thunb.)

Apparently two feet high, erect, robust, every part clothed with soft, silky and silvery dense pubescence. Branches long and simple. Petioles ¾–1 inch long. Leaflets inch long, 5 lines wide at top, cuneate at base. Racemes 8–12 inches long, the flowers patent, imbricating, small, pale yellow. Petals glabrous. Carina rather blunt. Legumes 3 lines long, not very turgid.

20. C. macrocarpa (E. Mey. Comm. p. 24); "Shrubby; stipules minute; leaflets obovate, glabrous above, minutely downy beneath, as are also the branchlets; racemes laxly several-flowered; calyx-segments lanceolate, thrice as long as the tube; legume (large) oblong, faintly veined transversely, minutely downy, *on a stipes somewhat longer than the calyx.*" *Benth. l. c. p.* 592.

HAB. Marshy places between the Omcomas and the Omblas, Caffraria, *Drege.* (Unknown to me.)

E. Meyer compares this species with *C. purpurea*, stating that it differs in the calyx and legumes. Bentham places it in a different sub-section.

21. C ? purpurea (Vent. Malm. t. 66); shrubby; the twigs appressedly pubescent; stipules subulate or obsolete; leaflets longer than the petioles, elliptical or obovate, glabrous, or minutely canescent beneath; racemes terminal, many-flowered, elongating; calyx silky, its teeth shorter than the tube, *the two uppermost truncate;* carina with a short, obtuse beak; legume glabrous, stipitate, oblong-inflated, transversely veined. *Benth. l. c. p.* 590. *DC. Prod.* 2. *p.* 133. *Eck. & Zey.! En. No.* 1257. *Bot. Reg. t.* 128. *Bot. Mag.* 1913. *C. coluteoides, Lam. C. elegans, Hort.*

HAB. In moist, sub-alpine places, from Caledon to Uitenhage, *E. & Z.!* George, *Drege.* Gnadendahl, *Dr. Prior!* Zwarteberg, *Pappe!* &c. (Herb. Hk., Sd., D.)

A shrub, 3–4 feet high, erect and much branched. Leaflets ¾–1 inch long, ⅓–½ inch wide, slightly glaucous, often drying blueish or livid, as if they contained *indigo.* Flowers bright purple, half an inch long. Calyx sub-truncate or intruse at base, ¼ as long as corolla. Flower-buds abruptly ovoid. A highly ornamental plant, cultivated in English green-houses.

22. C. aspalathoides (Lam. Dict. 2. p. 202); "shrubby, rigidly much branched; stipules obsolete; leaflets small, cuneate-oblong, glabrous on both sides, or silky canescent, like the twigs, on the lower surface; racemes lax, several-flowered; calyx 5–toothed; carina . . . . ; legume shortly stipitate, oblong-inflated, glabrous, veinless." *Benth. l. c. p.* 591. *DC. Prod.* 2. *p.* 134.

HAB. Onder Bokkeveld, on the Grasberg River, *Drege.* (Unknown to me.)

A small, but woody species, with quite a distinct habit. Leaflets on very short petioles, 2 lines long, 1 line wide. Racemes few-flowered. Flowers (not seen). Calyx of the fruit downy, with a ten-nerved tube and acuminate teeth. Legume half an inch long.

23. C. Capensis (Jacq. Hort. Vind. t. 64); shrubby, with *terete*, appressedly silky branches; stipules when present *petiolulate, obovate and leaf-like,* obsolete or wanting on many petioles; leaflets broadly obovate, obtuse or mucronulate, glabrous or very minutely pubescent on one or both surfaces; racemes terminal or opposite the leaves, lax, many-flowered; flowers (large) in longish pedicels; calyx pubescent, its segments longer than the tube; legume stipitate, appressedly pubescent. *Benth. l. c. p.* 590. *Thunb.! Fl. cap. p.* 572. *C. arborescens, Lam. DC. Pd.* 2. *p.* 130. *Cr. incanescens, Linn. f.*

HAB. Eastern districts, extending to Port Natal; common. (Herb. Th., Sd., D.)

A tall, much branched, stout bush, thinly canescent or sub-glabrous, 4–5 feet high. Petioles very variable in length, ½–1½ inch long, some of them furnished with a pair of leafy stipules, others on the same branch exstipulate; leaflets ¾–1 inch long, obtuse or emarginate or acute or mucronate. Flowers bright yellow, an inch long, the alæ transversely rugulose and pitted. Flower-buds lanceolate. Carina glabrous.

24. C. Natalitia (Meisn. Lond. Journ. 2. p. 67); shrubby, with *angular*, appressedly pubescent branches; stipules *narrow-lanceolate, petiolulate, leafy;* leaflets cuneate-oblong, obtuse or mucronulate, minutely pubescent beneath; racemes terminal, few-flowered; calyx *nearly glabrous,* its teeth scarcely as long as the wide tube; carina *woolly* on the upper edge; legume shortly stipitate, *glabrous. Benth. l. c. p.* 590.

HAB. Port Natal, *Krauss!* (339) *Gueinzius.* (Herb. Hk., Sd., D.).

Shrubby, much branched, glabrescent except the younger branches and peduncles. Petioles variable, ½–1 inch long, almost all furnished with a pair of leafy stipules of about half their length; leaflets ¾–1 inch long, 2–3 lines wide, cuneate at base. Racemes 5–6 flowered, the bracts and bracteoles caducous; pedicels longer than the calyx, glabrous above the bracteoles. Calyx cup-shaped, with wide rounded spaces between the teeth. Carina covered with white, woolly hairs on its upper half; the other petals glabrous. Flowers yellow, ½ inch long.

XIII. PLEIOSPORA, Harv.

Calyx ovoid, 5–fid; 4 upper segments approaching in lateral pairs, the lowest narrowest. *Vexillum* straight, vaulted; *alæ* patent; *carina* straight. *Stamens* monadelphous, with a split tube. *Ovary* sessile, tapering into a subulate, straight style; *ovules* numerous; *stigma* simple. *Legume?*

A shrub, with the habit of a *Psoralea*, the calyx and stamens of a *Lotononis*, and a corolla very different from either. All the petals are uniform in colour. Name from πλειον, *more*, and σπορα, *a seed;* the numerous ovules distinguishing it from every *Psoralea.*

P. cajanifolia (Harv. Thes. t. 81); *Psoralea cajanifolia, Benth.! in Herb. Hook.*

HAB. Magaliesberg and near the Crocodile River, *Burke & Zeyher! Zey. No.* 448. *Pappe!* 161 (Herb. Hk., Bth., Sd., D.)

A tall shrub, branching and densely leafy. Branches and twigs silky with fulvous, shining hairs. Petiole ¾ inch long. Leaves trifoliolate; leaflets 2–2½ inches long, ¾–1 inch wide, broadly lanceolate, or lanceolate-obovate, acute at base, mucronate, opaque, prominently ribbed and closely penninerved beneath, fulvo-sericeous. Stipules setaceo-subulate, equalling the petiole, free. Peduncles terminal and axillary, sub-corymbose towards the end of the branches, 1–2 inches long, bearing globose or oblong, spicate heads of flowers: each flower sub-tended by a subulate bract, and having a pair of setaceous bracteoles at the base of the calyx. Calyx ovoid, sub-inflated, with a narrow mouth, silky; its segments of nearly equal length, the four upper connate in pairs, the lowest subulate, narrower. Vexillum spathulate, narrow, silky on the outside, and on the claw within, concave, not bent backwards. Alæ and carina of nearly equal length, the alæ widely spreading; the carina not curved upwards at point, its petals nearly free, and of the same colour as the rest. Colour of the flower uncertain. Ovules 6–8.

XIV. LOTONONIS, DC. Benth.

Calyx sub-unequally 5–fid, the lowest segment narrower than the rest, and unconnected with them; the four upper approaching in pairs, and more or less connate into two bifid lobes, rarely quite separate, and then all the segments sub-equal. *Vexillum* obcordate or oblong, commonly pubescent; carina obtuse or acute. *Stamens* monadelphous. *Ovary* many ovuled. *Legume* oblong or linear, more or less compressed, many seeded. *Benth.! in Lond. Journ.* 2. *p.* 594. *Leobordea,* Del. *Leptis, Lotononis, Krebsia and Polylobium,* E. & Z. *Acanthobotryæ sp.,* E. & Z. *Crotalariæ sp., Aulacinthus, Telina, Lipozygis and Capnitis,* E. Mey.

A large genus, chiefly South African (a few inhabit Southern Europe and Central Asia), very various in habit, shrubby, suffruticose or herbaceous. Leaves very generally 3-foliolate, rarely 5-foliolate, and in one case uni-foliolate. Stipules frequently solitary, more rarely in pairs, or none, always free. Inflorescence various, racemose, umbellate, capitate, or with solitary flowers. Fl. commonly yellow. The species are here arranged after Bentham, under eight sections, readily distinguishable by the characters given in the following

ANALYSIS OF THE SPECIES.

1. Aulacinthus.—Racemes terminal. Flowers small. Legume short, turgid, with the carinal suture inflexed. *Small rigid shrubs.* (Sp. 1–3).

Leaves on long petioles; calyx thinly silky ... ... (1) **gracilis.**
Lvs. on short petioles:
Stipules none or very minute; calyx patently hairy ... (2) **rigida.**
Stipules solitary, lanceolate; calyx minutely puberulous (3) **viborgioides.**

2. Krebsia.—Flowers solitary, on short peduncles, or few, in a terminal raceme. Vexillum ample. Carina obtuse. Legume compressed, or scarcely turgid. *Shrubs or suffrutices.* (Sp. 4–11).

Erect, with rod-like simple or sub-simple branches:
Silky; leafl. cuneate-obovate; calyx-lobes acute ... ... (4) **cytisoides.**
Minutely puberulent; leafl. narrow-linear, glabrescent; calyx teeth short ... ... (5) **carnosa.**
Thinly silky-*canescent;* leafl. linear-cuneate; calyx lobes *much acuminate* ... ... (6) **genuflexa.**
Villoso-pubescent; leaflets pellucid-dotted; racemes 4–6 flowered, sub-paniculate ... ... (11) **dichiloides.**
Diffusely or divaricately much branched:
Sub-spinescent; leafl. cuneate-oblong; legume pubescent (7) **divaricata.**
Unarmed, albo-sericeous; fl. subsolitary; legume silky (8) **sericophylla.**
Unarmed, thinly canous; leafl. 3, fl. racemose; vex. glabrous (9) **densa.**
Unarmed, glabrous; leafl. 5, on long petioles; leg. *falcate* (10) **digitata.**

3. Telina.—Peduncles elongate, one-flowered, or rarely sub-umbellately 2–4 flowered. Vexillum ample. Carina obtuse. *Slender, diffuse or decumbent suffrutices or herbs.* (Sp. 12–19).

Peduncles one flowered (in *L. prostrata* sometimes 2–3 flowered):
Stipules in pairs; peduncles one flowered:
Peduncles short; bracts long, *obovate,* deciduous ... (12) **bracteata.**
Peduncles long; bracts minute, *setaceous* ... ... (13) **azurea.**
Stipules solitary: leafl. obovate, or lanceolate:
Leafl. *glabrous above* ... ... (14) **prostrata.**
Leafl. densely *hirsute* on both sides ... ... (15) **villosa.**
Leafl. *cuspidate,* rigid, with *netted veins,* thinly pilose (16) **acuminata.**
Leafl. acuminate, *veinless,* densely silky on both sides (17) **argentea.**
Peduncles sub-umbellately 2–4 flowered, or interruptedly racemose, pluri-flowered:
Calyx ebracteolate, the *lanceolate* segments long ... ... (18) **varia.**
Calyx bibracteolate, the *deltoid* teeth shorter than the tube (19) **macrocarpa.**

4. Polylobium.—Peduncles elongate, umbellately many flowered, or imperfectly racemose, several flowered. Vexillum of moderate size. Carina *obtuse. Diffuse or rarely sub-erect suffruticose-herbaceous plants.* (Sp. 20–25).

Stems diffuse or procumbent; leaflets elliptical or obovate:
Bracts minute: (stipule mostly solitary)
Appressed-pubescent, or sub-glabrous ... ... (20) **umbellata.**
Patently pubescent; leafl. pilose beneath and ciliate (21) **debilis.**
Bracts obovate or orbicular, longer than the pedicel ... (22) **pallens.**
Stems diffuse or procumbent; leafl. narrow, linear or lanceolate:
Loosely hairy; leafl. hairy; peduncles shortish ... ... (23) **involucrata.**
Pubescent; leafl. glabrescent; pedunc. long ... ... (24) **peduncularis.**
Stems ascending or erect; leafl. linear lanceolate; peduncles elongate ... ... (25) **angustifolia.**

5. Oxydium.—Peduncles umbellately many flowered, rarely 1–2 flowered. Vexillum moderate. Carina *acute. Procumbent or ascending, slender, sub-herbaceous plants.* (Sp. 26–31).

Lvs. trifoliolate: peduncles slender, 1–2 flowered ... ... (27) **perplexa.**
Lvs. 3-foliolate; peduncles umbellately many flowered:
Leafl. broadly-obovate or obcordate, glaucous; cal. lobes subulate... ... ... (26) **trichopoda.**
Leafl. cuneate-oblong or linear; cal. teeth triangular acuminate ... ... (28) **micrantha.**

Leafl. obovate or oblong, silky ; fl. capitate ; cal. tomentose, its lobes deltoid acuminate ... ... (30) **oxyptera.**
Lvs. (usually) 5-foliolate ; peduncles shorter than the leaf (29) **acutiflora.**
Lvs. *unifoliolate ;* peduncles umbellately several flowered ... (31) **monophylla.**

6. Lipozygis.—Heads of flowers sub-sessile, terminal. Vexillum usually oblong. Carina obtuse, or rarely sub-acute. *Ascending, erect or prostrate, villous or silky suffruticose plants.* (Sp. 32–37).

Procumbent, much branched :
Leafl. oftener 5 than 3 ; bracts very narrow ... ... (32) **pentaphylla.**
Leafl. 3, softly hairy on both sides ; bracts broad ... (33) **polycephala.**
Leafl. 3, glabrous and green above ; bracts broad ... (34) **anthylloides.**
Erect or sub-erect, sub-simple, many stemmed :
Leafl. oblong-elliptical, or sub lanceolate, *very acute ;* heads pluri-flowered ... ... (35) **eriantha.**
Leafl. ellipt. oblong or obovate, *sub-obtuse ;* heads *densely* many flowered ... ... (36) **corymbosa.**
Leafl. *linear-lanceolate,* acute ; heads laxly flowered ; vexillum glabrous ... ... (37) **lanceolata.**

7. Leobordea.—Flowers small, sub-sessile, opposite the leaves or few together in the forks of the stem. Lowest segment of the calyx very slender, minute. Carina obtuse. *Prostrate herbs or suffrutices, some annual.* (Sp. 38–41).

Much branched, silky ; stipule falcate ; leaflets glabrous above ; petals longer than the calyx ... ... (38) **porrecta.**
Dichotomous ; calyx teeth short and broad ; carina *straight, twice as long as the vexillum,* much longer than the calyx (39) **carinalis.**
Dichotomous, tomentoso-canescent ; stip. minute ; petals shorter than the calyx ... ... (41) **clandestina.**
Dichotomous, silky-canescent ; stip. oblong ; cal. lobes lanceolate ; petals longer than the calyx ... ... (40) **Leobordea.**

8. Leptis.—Flowers small, sub-sessile, opposite the leaves, solitary or few together. Carina elongate, obtuse, or rarely acute. Small suffrutices, rarely shrubs, erect or more commonly diffuse or prostrate. (Sp. 42–58).

Leaves 5-foliolate ... ... (48) **quinata.**
Lvs. 3-foliolate ; carina acute :
Stipules cordate-ovate or orbicular ; bracts very broad, sub-reniform ... ... (42) **Burchellii.**
Stipules lanceolate or oblong :
Leaflets obovate or cuneate-oblong :
Flowers solitary ; cal. shorter than the corolla (44) **lenticula.**
Fl. 3–4 together ; calyx lobes lanceolate, longer than corolla ... ... (43) **crumanina.**
Leaflets linear or linear lanceolate ; flowers solitary :
Villous ; calyx deeply divided, its subulate lobes shorter than corolla ... ... (45) **pungens.**
Silvery ; calyx lobes shorter than corolla ; legume *piloso-villous* ... ... (46) **depressa.**
Silvery ; cal. lobes sub-equalling corolla ; legume *canescent* ... ... (47) **laxa.**
Leaves 3-foliolate ; carina obtuse :
Suffruticose or herbaceous, dwarf :
Leaflets broadly obovate or obcordate :
Flowers 2–4 together, terminal or opposite the leaves :
Stems very dwarf, tufted ; petioles uncial ; petals villous ... ... (53) **mollis.**
Prostrate, much branched ; petioles ½ uncial ; petals nearly glabrous ... ... (54) **pumila.**
Flowers solitary.
Flowers *sub-sessile ;* leafl. *minute* (1 line long) (55) **microphylla.**
Flowers *pedunculate ;* leaflets obovate or obcordate ... ... (52) **humifusa.**

Leaflets cuneate-oblong or linear, or *somewhat obovate* :
Thinly appressedly silky or sub-glabrous :
Stipules in pairs ; petioles very short ;
petals silky ... ... (51) **carinata.**
Stipule solitary ; petioles rather long :
Procumbent ; fl. 2-5 together ; legume thrice as long as calyx ... (49) **brachyloba.**
Diffuse or sub-erect ; fl. 1-2 ; legume many times as long as the calyx (50) **falcata.**
Patently hairy or densely villous :
Calyx-lobes nearly as long as corolla ... (56) **tenella.**
Calyx-lobes much shorter than corolla ... (57) **versicolor.**
Shrubby ; stems erect, virgate, sub-simple ; lvs. sessile (58) **sessilifolia.**

1. **AULACINTHUS** (Sp. 1-3).

1. **L. gracilis** (Benth. ! Lond. Journ. 2. p. 597); shrubby, erect, much branched, silky and canescent ; branches slender ; leaflets on long *petioles*, linear or oblong, obtuse, acute at base, appressedly silky on one or both sides; racemes terminal, laxly 6–8 flowered; bracts subulate, deciduous ; *calyx* thinly silky ; legume appressedly downy, about twice as long as the calyx. *Aulacinthus gracilis, E. Mey. ! Comm. p.* 156.

VAR. β. **anomala** ; legumes (abortive ?) an inch long, linear-attenuate.

HAB. Rocky places, on the Roodeberg and Ezelkop Mountain, Kamiesberg, *Drege!* Var. β. Kamiesberg, *Dr. Pappe!* (100). (Herb. Hk., Bth., D.)

A woody bush, a foot or more in height, densely much branched, silvery. Petioles 1-2 inches long, channelled ; leaflets ½-¾ inch long, ½-1 line wide ; sometimes the lower ones are but 2 lines long, and nearly 1 line wide. Racemes 3-4 inches long, the flowers half an inch apart. The pods are ordinarily 3-4 lines long, 1½ lines broad ; but in var. β, which in all other respects is identical with Drege's specimens, the pods are over an inch long, not a line wide, and nearly glabrous.

2. **L. rigida** (Benth. Lond. Journ. 2. p. 597); shrubby, spinescent, densely much branched, rigid, silky-canescent ; stipules *minute or none ; leaflets on short petioles*, linear, obtuse, acute at base, appressedly silky ; racemes short, few-flowered; calyx *patently hairy ;* legume roughly pubescent, more than twice as long as the calyx. *Aulacinthus rigidus, E. Mey. ! Comm. p.* 157.

HAB. Zwartland and at Mortkuil, *Drege !* (Herb. Benth.)

Very like *L. gracilis*, but smaller and more stunted, with short petioles, and a roughly hairy calyx and legume. I have only seen very imperfect specimens.

3. **L. viborgioides** (Benth. l. c.) ; shrubby, slender, diffuse or decumbent, much branched, very thinly and appressedly downy or glabrescent ; branches slender, here and there hardened into spines ; *stipules solitary, lanceolate ;* leaflets oblongo-cuneate, longer than the petiole ; racemes terminal, laxly 4–6 or 8 flowered ; calyx *minutely puberulous ;* ovary glabrous. *Zey.* 2319.

HAB. Cape Colony, *Bowie! Thom!* Hassagaiskloof, *Zeyher!* (Herb. Hk., Bth., Sd.)

A low growing, somewhat trailing, slender shrublet, woody at base ; the pubescence scant and sometimes excessively minute. Leaves, including the petiole, about ½ inch long. Stipule as long as the petiole, or much shorter. Flowers 2 lines long. Legume not known.

2. **KREBSIA** (Sp. 4-11).

4. **L. cytisoides** (Benth. Lond. Journ. 2. p. 598) ; shrubby, branches

virgate, *softly hairy or silky ;* leaflets *cuneate-oblong or obovate,* acute or acuminate or obtuse, silky and villous on both sides ; stipules oblong, leafy ; peduncles axillary or terminal, short, one flowered ; calyx-teeth acute, equalling the tube ; vexillum ample, silky or glabrescent ; legume pubescent or hairy. *Telina cytisoides & T. eriocarpa, E. Mey ! Krebsia stricta, E. Z.* 1284.

HAB. Mountains of Uitenhage, *E. & Z. ! Drege,* and extending through Cafferland to near Natal. (Herb. D., Hk., Bth., Sd.)

A stout shrub, 2–3 feet high, with rod-like branches 12–16 inches long. Leaflets very variable in size, and sometimes fascicled, 1–2 to 6–8 lines long ; the petiole as variable. Stipules in pairs, resembling the leaflets and varying like them. Flowers mostly axillary ; the peduncle scarcely longer than the calyx. Pods compressed, ¾ inch long, acute. The habit is nearly that of an *Aspalathus.*

5. **L. carnosa** (Benth. l. c. p. 598) ; minutely and *appressedly puberulous ;* branches slender, virgate ; leaflets and leafy stipules *narrow-linear, fleshy,* glabrescent ; raceme terminal, laxly several flowered, and short 1–2 flowered peduncles opposite the leaves ; calyx teeth *shorter than the tube ;* legume nearly glabrous. *Krebsia carnosa, E. & Z. !* 1287. *Telina striata, E. M. Comm. p.* 68.

HAB. Caffirland, near Silo, *E. & Z. !* Between Omtata and Omsamwubo, *Drege!* (Herb. Bth., D., Hk., Sd.)

Slender, erect, not much branched, 12–18 inches high ; the pubescence very scanty. Leaves, including petiole, about ½ inch long ; leaflets commonly hooked at the point, of thick substance and veinless. Calyx teeth subulate, widely separated. Legume ¾ inch long.

6. **L. genuflexa** (Benth. l. c.) ; thinly *canescent,* with short, silky, close pressed hairs ; branches slender, filiform ; leaflets *linear-cuneate,* or narrow-oblong ; stipules *solitary,* narrow-oblong ; peduncles one flowered, jointed and bent beyond the middle, opposite the leaves and longer than them, or two or three in a terminal raceme ; calyx lobes much *acuminate ;* legume minutely and appressedly silky. *Telina genuflexa, E. Mey. Comm. p.* 69.

HAB. Grassy spots, between the Kliplatt and Key Rivers, Caffraria, and dry hills near Gaatje, *Drege !* (Herb. Bth., D., Hk.)

A slender, upright suffrutex, 1½–2 feet high ; branches erect, scattered, 6–8 inches long. Leaves, including petiole, about ½ inch long. Peduncles nearly an inch long, angularly bent ; calyx lobes subulate from a broad base. Pods an inch long.

7. **L. divaricata** (Benth. l. c. p. 599) ; shrubby, *divaricately much branched, somewhat spiny,* thinly canescent with short closely pressed hairs ; stipule *solitary,* shorter than the petiole ; leaflets cuneate-oblong, appressedly puberulous ; peduncles one flowered, jointed and bent beyond the middle, opposite the leaves, or terminal ; calyx lobes much acuminate ; legume appressedly pubescent or nearly smooth. *Krebsia divaricata, E. & Z.* 1285.

HAB. Caffirland, near Silo, on the Kliplaat River, *E. & Z.* Zuureberge, *Burke.* Albany, *Mrs. F. W. Barber.* (Herb. Hk., Sd., Bth., D.)

A scraggy, woody, densely branched dwarf bush, different in *aspect* from *L. genuflexa,* but so nearly allied in essential characters, that I suspect it to be a mere variety ; and the more so as they come from the same part of the country. Burke's specimens are much more woody than Ecklon's, and may be different.

8. **L. sericophylla** (Benth. ! l. c. p. 599) ; dwarf, diffusely much

branched, *everywhere thinly silky*, with short, white, closepressed hairs; leaflets small, linear or cuneate; stipule solitary, lanceolate-linear; peduncles one-flowered, *short; vexillum and legume densely silky;* calyx teeth acuminate, as long as the tube. *Zeyher!* 399.

HAB. Wolf-kop, near Caledon, *Burke and Zeyher*. (Herb. Bth., Hk., Sd.)

A small, much branched suffrutex, 8–12 inches high; stem woody at base, breaking up into many, slender, flexuous, vaguely divided branches. Leaves, including petiole, not half an inch long, the leaflets mostly cuneate, ½ line in breadth; stipules generally as long as the petiole, sometimes wanting. Flowers sub-solitary toward the ends of the branches. Pods ¾ inch long, 1½ lines wide.

9. **L. densa** (Harv.); shrubby, diffuse or decumbent, much branched, unarmed, *thinly canescent*, stipule . . . ; leaflets 3, on short petioles, narrow cuneate-oblong, folded, *glabrous above, thinly silky beneath; racemes terminal, laxly few-flowered;* calyx thinly canescent, campanulate, its teeth triangular, about equalling the tube; vexillum *glabrous, acute;* legume . . . ? *Lebeckia densa, Thunb.! Fl. Cap. p.* 562. *Acanthobotrya decumbens, E. & Z.* 1345.

HAB. South Africa, *Thunberg!* Kochmanskloof, Swell., *E. & Z.* (Herb. Th., Sd.)

A rigid, woody, thick stemmed and densely much branched, small bush, 12–18 inches high, very thinly covered with short, closepressed, whitish hairs. Petioles 3–4 lines long; leaflets about equalling them, very narrow and folded together. Flowers yellow; 4 or 5 in a raceme, nodding, 3 lines long.

10. **L. digitata** (Harv.); slender, ascending, much branched, *nearly or quite glabrous;* stipule solitary, falcate; leaves on very long petioles, 5-foliolate, leaflets linear or lin. lanceolate, acute, very narrow, complicate; racemes opposite the leaves and terminal, laxly 3–4 flowered; calyx narrow, acute at base, the upper segments lanceolate, sub-connate, the lower subulate; legume *falcate*, glabrous.

HAB. South Africa, *Capt. Carmichael!* (Herb. T.C.D.)

In foliage and flower this is puzzlingly like *L. quinata*, but the inflorescence and legume are quite different. Stems filiform, 6 inches high, flexuous. Petioles 1–1½ inches long; leaflets ½ line wide, ½ inch long, pale green. Legume 1½ as long as the calyx, not a line wide, strongly curved backwards. The only specimens yet seen are in a collection made by the late Captain Carmichael, in some part of the eastern provinces, about the year 1814.

11. **L. dichiloides** (Sond. in Linn. 23. p. 28); suffruticose, *villoso-pubescent;* stem (or main branches?) *very long, straight and rod-like*, densely beset with short, filiform branchlets; leaves subtending the ramuli fascicled; leaflets narrow-spathulate, mucronate, *pellucid-dotted;* stipules leaf-like, linear-lanceolate, as long as the petiole; *racemes laxly 4–6 flowered*, terminating the branchlets; calyx teeth shortly subulate, the four upper approaching in pairs, the lowest longest; petals glabrous, the vexillum stipitate; legume?

HAB. Port Natal, *Gueinzius!* (Herb. Sond.)

Two or more feet high, erect, simple? densely clothed with sub-fascicled leaves and short axillary branchlets. Lower branchlets 2–3 inches long, upper gradually shorter, almost all ending in racemes of yellow flowers. Leaves about 3 together; petiole ½ inch, leaflets ½–¾ inch long, the middle one longest, all tapering at base, scarcely 1 line broad. Stipules nearly as long as the petioles or shorter. Racemes erect; the peduncle an inch long, pedicels rather shorter than the calyx. Lower calyx tooth widely separated from the rest: all the teeth shorter than the tube. This has the habit of *Dichilus lebeckioides*, but the calyx of a *Lotononis*.

3. **TELINA** (Sp. 12–19).

12. L. bracteata (Benth. ! Lond. Journ. 2. p. 600); decumbent, suffruticose, silky-pubescent; stipules in pairs, *small, linear, acute;* leaflets linear-oblong or lanceolate, acute, longer than the petiole; peduncles *scarcely as long as the leaf,* one flowered; bracts and bracteoles *obovate, truncate,* erect, deciduous; calyx *sub-inflated;* legume *linear, sub-compressed, silky. Zeyher,* 385.

HAB. Mooije River, *Burke and Zeyher !* (Herb. Bth., Hk., Sd., D.)

The whole plant pallid, thinly clothed with short, closepressed hairs. Stems trailing, 6–12 inches long, the points ascending. Petioles $\frac{1}{4}$–$\frac{1}{2}$ inch long; leaflets $\frac{3}{4}$–1 inch long, 1–2 lines wide, thickish, often incurved, midribbed, sometimes mucronate. Peduncles $\frac{1}{2}$–$\frac{3}{4}$ inch long, jointed and bracteate near the summit, with two bracteoles also under the calyx; bracts and bracteoles 1–1$\frac{1}{2}$ line long, membranous, the two upper segments curved, connate with the lateral lanceolate ones, the lowest deeply divided, subulate. Corolla pubescent, not much exceeding the calyx. Legume 1 inch long, 1 line wide.

13. L. azurea (Benth. ! l. c. p. 600); decumbent, suffruticose, glabrescent or villoso-pilose; stipules in pairs, *obliquely ovate or lanceolate;* leaflets cuneate-oblong, sub-glabrous or silky, thickish, longer than the petiole; peduncles *much longer than the leaves* and opposite them, one flowered; bracts *minute, setaceous;* vexillum densely pubescent; legume *turgid, patently hairy. Crotalaria azurea, E. & Z. !* 1262. *Zey. !* 2297. *Telina heterophylla, E. Mey.! Comm. p.* 69. *excl. syn.*

VAR. β. **lanceolata**; densely hairy; leaflets and stipules linear-lanceolate, acute; calyx more deeply divided, with narrower, lanceolate segments. *Ononis villosa, Thunb. ! Herb.; Fl. Cap. p.* 585.

HAB. Sandy hills. Krakakamma and Port Elizabeth, *E. & Z. !* Between Eschenbosch and the Gamtoos River, *Drege !* (Herb. Hk., Bth., Sd., D.)

Nearly glabrous, or thinly or thickly clothed with long soft hairs. Stems 6–12 inches long, spreading on the ground. Petiole $\frac{1}{4}$–$\frac{1}{3}$ inch long, channelled; leaflets $\frac{1}{2}$–$\frac{3}{4}$ inch long, tapering at base, broader upwards, blunt or mucronate, midribbed. Stipules nearly as long as the petioles, sometimes but one, lanceolate. Peduncles 2–4 inches long, articulate and bracteate below the flower. Calyx 3 lines long, softly villous, deeply cut, the lowest segment subulate, the rest lanceolate, connate beyond the middle. Vexillum densely hairy; the keel shorter than the wings. Legume 1 inch long, 2$\frac{1}{2}$ lines wide, the ventral suture elevated and often tuberculate. *Drege's* specimens are much less hairy than *E. & Z.'s.* Var. β. chiefly differs in the narrower and more acute leaflets and longer calyx lobes.

14. L. prostrata (Benth. ! l. c. p. 600); diffuse or prostrate, silky-pubescent; stipule mostly *solitary,* ovate or lanceolate, much shorter than the petiole; leaflets obovate or obcordate, *glabrous above,* silky beneath; peduncles elongate, opposite the leaf, 1 (rarely 2–3) flowered; bracts minute, setaceous; vexillum very ample, *silky along the midrib, otherwise glabrous;* legume turgid, thinly and minutely pubescent. *Ononis prostrata, Linn. Thunb.! Cap. p.* 586. *Ononis heterophylla, Thunb.! and O. elongata, Thunb. Telina prostrata, E. Mey. Comm. p.* 69. *Crot. vexillata, E. Mey.! Lin.* 7. *p.* 153. *Lot. vexillata, E. & Z.!* 1270. *Zey.* 2312.

VAR. β. **glabrior**; peduncles sometimes 2–3 flowered; plant glabrescent; legume smaller. *Tel. excisa, E. M. !*

VAR. γ. **heterophylla**; leaflets of the lower leaves obovate; of the upper lanceolate, inch long, and thinly silky on both sides. *L. heterophylla, E. & Z. !* 1273, *non Thunb.*

HAB. Mountains round Capetown, *Thunberg ! E. & Z. ! Pappe, Drege, W.H.H.,*

&c. Between Howhoek and Potrivier, *Zeyher!* Var. γ. near the Waterfall, Tulbagh, *E. & Z.!* (Herb. Th., Hk., Bth., D., Sd.)

Root woody, deeply descending; stems many, filiform, trailing. Petioles varying much in length, ¼–1 inch long; leaflets ⅓–½ inch long, 2–3 lines wide, mucronulate or emarginate, thickish, midribbed. Stipules variable in length and breadth, sometimes linear-subulate. Peduncles 2–4 inches long, jointed below the calyx. Calyx deeply parted, thinly silky, with lanceolate lobes. Legume ¾–⅞ inch long, 2 lines in diameter. Flowers yellow. Of var. γ. I have only seen a single small specimen, from which I cannot determine whether it be more than a mere variety.

15. L. villosa (Steud.); slender, diffuse, *densely hirsute with fulvous, patent hairs;* stipule mostly solitary, lanceolate, acute; leaflets *obovate,* mucronulate, shorter than the petiole, *densely hairy on both sides;* peduncles very long, opposite the leaves, mostly one flowered; bracts minute, setaceous; calyx rufous, deeply cut, nearly equalling the uniformly silky vexillum; legume hairy. *Benth. l. c. p.* 601. *excl. syn. Thb. Telina villosa, E. Mey.! Comm. p.* 70.

HAB. Mountain thickets. Riebeck's Kasteel, Stellenb. *Drege!* (Hb. Bth., Hk., D.)

Root perennial. Stems numerous, weak and trailing. *Petioles* inch long or more; leaflets ½ inch long, 2½ lines wide. Stipules ¼–½ inch long. Peduncles 3–4 inches long. Calyx 4 lines long, densely hirsute, the segments lanceolate, longer than the tube. I have not seen the legume. This is by much the most hirsute of the section; the hairs long, foxy or golden.

16. L. acuminata (E. & Z. No. 1269); diffuse, the branches pubescent; stipule mostly solitary, *lanceolate, acute, midribbed;* leaflets obovate, acute or cuspidate, *rigid, with netted veins,* thinly pilose, the young ones silky; peduncles elongate, terminal or opposite the leaves, 1 (rarely 2) flowered; bracts setaceous; calyx pubescent, deeply cut, much shorter than the pubescent vexillum; legume turgid, downy. *Benth.! l. c. p.* 601.

HAB. Fields near the Zwartkop River, Uit., *E. & Z.! Pappe!* (Hb. D., Sd., Bth.)

Root thick and woody. Stems 6–8 inches long, thickly clothed with very short hairs. Petioles ¼–½ inch long; leaves ⅓–½ inch long, 2–3 lines wide, when dry distinctly netted, especially on the under side. Peduncles 1½–3 inches long. Calyx lobes lanceolate, 3 nerved, longer than the tube. Legume nearly 1 inch long, 1½ lines wide, sub-terete.

17. L. argentea (E. & Z. No. 1272); diffuse, the whole plant silky, with closely appressed hairs; stipule solitary, lanceolate; leaflets *obovate, acute or acuminate, midribbed, but veinless, densely silky on both sides;* peduncles elongate, opposite the leaves, one flowered; bracts minute, subulate; calyx deeply cut, shorter than the silky vexillum; legume (young) sericeous. *Benth. l. c. p.* 601.

HAB. Barren hills between Kochmanskloof and Gauritz River, Swell., *E. & Z.!* (Herb. Sond.)

Root thick. Stems slender, 6 inches long, decumbent or prostrate. Petioles and leaflets each ½ inch long. Leaflets gradually acuminate. Peduncles 3 inches long. Calyx lobes lanceolate, longer than the tube, not obviously ribbed. Vexillum hairy along the midrib and lateral nerves, glabrous between the nervures. The pubescence is close and glossy, but scarcely silvery, rather somewhat fulvous.

18. L. varia (Steud.); diffuse or sub-erect, glabrescent or thinly pubescent, with appressed hairs; stipules in pairs, leaf-like, obliquely ovate or oblong, acute; leaflets obovate or obcordate or cuneate, mucronulate, becoming glabrous; peduncles elongate, opposite the leaves, *subumbellate or interruptedly racemose, several flowered;* bracts linear-spathulate, nearly as long as the pedicel; calyx ebracteolate, the *lanceolate segments longer than the tube;* legume . . . ? *Telina varia, E. Mey. Comm. p.* 70.

HAB. S. Africa, *Thom.* Drakenstein Hills, *Drege!* Near Capetown, *Dr. Alexander Prior!* (Herb. Bth., Hk., D.)

Many stemmed, ascending, more erect than most others of this section. Stem 6–8 inches long, pubescent when young, flexuous. Petioles ¼ inch long; leaflets as long or somewhat longer, 2–3 lines wide, frequently emarginate, with a minute mucro. Calyx deeply cut, the lowest segment subulate, the four upper broader, connate below, lanceolate above. The legume may afford further characters to distinguish it from *L. macrocarpa,* from which at present it is most easily known by the calyx.

19. L. macrocarpa (E. & Z.! 1271); diffuse, thinly silky or glabrescent; stipules in pairs, lanceolate or oblong or obovate, acute; leaflets broadly obovate, mucronulate, *nearly glabrous,* the younger puberulous beneath; peduncles elongate, opposite the leaves, *sub-umbellately* 2–4 *flowered;* bracts like the stipules; calyx bibracteolate, the deltoid teeth shorter than the tube; corolla glabrous, twice as long as the calyx; legume much compressed, glabrous and veiny, widening upwards, acute. *Benth. l. c. p.* 601. *Zey.* 403.

HAB. Near Brackfontein, Clanwilliam, *E. & Z.!* Kalebasskraal, *Zeyher! Pappe!* (Herb. Sd., D., Hk.)

Stems 6–12 inches long, decumbent or prostrate, the young ones clothed with short, appressed hairs, as are also the young leaves. Afterwards the hairs frequently disappear. Petioles ⅓ inch long; leaflets ½ inch long; 3–3½ lines broad, with a very small mucro. Calyx campanulate, with very short, triangular teeth, minutely pubescent. Legume an inch long, 2½ lines wide near the point, narrowing to 1½ near the base. In drying it becomes dark.

4. **POLYLOBIUM** (Sp. 20–25).

20. L. umbellata (Benth.! l. c. p. 602); diffuse or decumbent, *appressedly pubescent or sub-glabrous;* stipule mostly solitary, oblong or lanceolate, shorter than the petiole; *leaflets obovate;* umbels (or umbellate-racemes) on long peduncles, many-flowered; bracts minute; flowers cernuous; legume sub-falcate, compressed, nearly glabrous. *Ononis umbellata, Linn.? (non Thunb.) O. strigosa, Thunb.! Cap.* 588. *O. anthylloides, DC. Prod.* 2. *p.* 168. *Lipozygis umbellata, E. Mey.! Comm. p.* 76. *Polylobium truncatum, E. & Z.* 1292, *and P. filiforme, E. & Z.!* 1291.

HAB. Moist places on the Capetown and Stellenbosch hills, and to River Zonderende, *E. & Z.! Drege, Pappe, W.H.H., &c.* (Herb. Th., Bth., Sd., Hk., D.)

Root thick and woody. Stems many from one crown, spreading 1–2 feet in all directions. Leaves an inch apart, patent; petioles ¼–½ inch long; leaflets rather longer, 2–2½ lines wide, very obtuse, puberulous on the underside. Peduncles terminal, becoming lateral and opposite the leaf, 8–12 flowered; umbels sometimes dislocated, and passing into short, dense racemes. Calyx thinly silky, deeply cut, the segments broadly lanceolate. Legumes an inch long, slightly curved, 1½ line wide.

21. L. debilis (Benth.! l. c. p. 604); procumbent and slender, *patently pubescent,* stipule oblongo-lanceolate, sub-falcate; leaflets elliptical or obovate, glabrescent above, *pilose beneath, ciliate;* peduncles longer than the leaf, umbellately 4–6 flowered; bracts minute; calyx silky; carina short and very blunt; ovary linear, pubescent. *Polylobium debile, E. & Z.!* 1290.

HAB. Barren hills, near Assagaiskloof, Swell., *E. & Z.!* near River Zonderende, *Zeyher*, 2316. (Herb. Sond.)

2–4 inches high, many-stemmed, slightly-branched, with spreading pubescence. Leaves densely set; petiole ⅓–½ inch long; leaflets shorter, 1–2 lines wide, the upper ones often acute. Peduncles 1½ inch long; pedicels very short. The legume is probably, judging by the shape of the ovary, elongate.

22. L. pallens (Benth. l. c. p. 605); procumbent, thinly pilose; stipules solitary or in pairs, obliquely ovate or oblong, small; leaflets obovate or obcordate, glabrous above, ciliolate and sparsely pilose beneath; peduncles much longer than the leaves, sub-capitately several-flowered; bracts *obovate or orbicular*, longer than the short pedicel; calyx 10-ribbed, piloso-ciliate on the ribs and margin of the falcate segments; carina shorter than the vexillum; legume ?; ovary glabrous. *Polylobium pallens, E. & Z.!* 1294.

HAB. Mountain-sides, near Brackfontein, Clanwilliam, *E. & Z.!* (Herb. Sond.)

Stems decumbent, pale and weak, 1–2 feet long, thinly sprinkled with long, horizontally patent hairs. Peduncles 3–4 inches long, similarly hairy. Leaves sub-distant, the petiole shorter than the laminæ; leaflets 3, not quite ½-inch long, 3 lines wide, usually emarginate. The calyx has ciliate ribs, but the inter-spaces are naked. Bracts very broad. The carina is short, and not at all acuminate.

23. L. involucrata (Benth.! l. c. p. 602); diffuse or decumbent, *loosely hairy in all parts;* stipules in pairs, linear or lanceolate, mostly longer than the petiole; leaflets *linear or lanceolate, the lower ones narrow-cuneate;* peduncles mostly terminal, rather short, umbellate or sub-racemose, many-flowered; bracts leaflike, lanceolate, longer than the pedicel, or sometimes equalling the flowers; calyx-lobes *subulate, longer than the tube;* legume turgid, not twice as long as the calyx. *Ononis involucrata, Lin. f. Thunb.! Cap. p.* 587. *Polylobium involucratum, E. & Z.!* 1296; *P. tenuifolium, E. & Z.!* 1295, *and P. angustifolium, E. & Z.!* 1297. *Lipozygis involucrata, E. Mey.! Comm. p.* 80. *Ononis aspalathoides, DC. Zeyher*, 2388, 2389.

HAB. Common on dry hill-sides in Cape and Stellenbosch districts. Witsenberg, *Zeyher!* (Herb. D., Bth., Hk., Sd., Th.).

Variable in the amount of pubescence; sometimes rather thinly, sometimes very densely hairy or shaggy, with pale or foxy hairs. Leaves thickly set; petiole ¼–⅓ inch long; leaflets ½–¾ inch, sometimes very narrow, not ½-line wide, sometimes 1 line. Peduncles ½–1 inch, rarely 2 inches long, mostly terminating short, ascending, leafy branches. Bracts, like the leaves, broad or narrow, ⅓–½ inch long. Flowers yellow, generally in umbels, but varying to racemes, on the same root. Pod hairy or glabrescent.

24. L. peduncularis (Benth.! l. c. p. 602); diffuse, more or less hairy, the flowering branches ascending; stipules in pairs, linear or lanceolate; leaflets linear or linear-cuneate, glabrescent; *peduncles terminal, elongate,* umbellate or sub-racemose, several flowered; bracts lanceolate, linear, or ovate-oblong, about equalling the pedicel; calyx silky, *its lobes lanceolate, shortly acuminate, equalling the tube;* carina very obtuse; legume sub-compressed, *glabrescent, more than twice as long as the calyx.*

VAR. *α.* **Meyeri**; less branching and more hairy, with narrower leaflets and linear or lanceolate bracts. *Lipozygis peduncularis, E. Mey. Comm. p.* 79. *Ononis umbellata, Thunb.! Cap. p.* 587.

VAR. *β.* **secunda**; flowering branches numerous, short, secund, sub-erect, with few leaves; leaflets more glabrous, cuneate; bracts ovate or oblong; flowers 4–8, umbellate. *Ononis secunda, Thunb.! Cap.* 588, *(but not Lotononis secunda, Benth.)*

HAB. Sandy ground near the Paarl and at Groenekloof, *Drege!* round Capetown, *Thunberg.* β. at Gnadendahl, *Dr. Alexander Prior!* (Herb. Th., Bth., Hk.)

Allied to the more glabrous forms of *L. involucrata*, but with longer flower-stalks, different calyx and legume, and rather smaller flowers. *Thunberg's* specimen of our var. β, exactly agrees with *Dr. Alexander Prior's.*

25. L. angustifolia (Steud.); stems *ascending or sub-erect, elongate, sub-simple, laxly hairy;* stipules linear-lanceolate, longer than the petiole; leaflets *lanceolate-linear,* acute, sparsely pilose; peduncles terminal, elongate, umbellate or sub-racemose, several-flowered; bracts lanceolate equalling or exceeding the pedicel; calyx silky, its lobes lanceolate; carina arched, obtuse, vexillum pilose on the dorsal ridge; legume (young) *very hairy. Telina angustifolia, E.Mey.? Herb. Drege. Polylobium fastigiatum, and P. Mundianum, E. & Z.!* 1298, 1299. *Lotononis secunda, Benth. l. c. p.* 603, *(excl. syn. Thunb.)*

HAB. Cape flats, *E. & Z.! W. H. H., &c.* Koeberg, *Dr. Pappe,* 105. Swellendam, *Mundt.!* (Herb. Bth., Hk., Sd., D.)

Many-stemmed. The stems rigid, curved, or sub-decumbent at base, then ascending or erect, 12–14 inches long, densely clothed with narrow, erect leaves. Petioles ¼–½ inch long; leaflets ½–¾ inch, ½-line wide, acute at each end. Pubescence more or less copious, the hairs long and white. Peduncles 3–4 inches long, 6–12 flowered, the umbel sometimes breaking into a short raceme. Flowers larger than in any, except *L. involucrata.* The habit is quite unlike that of the rest. I cannot find it in Thunberg's Herbarium. His *O. secunda* is a branchy form of *L. peduncularis.* The specific name here adopted, though not the oldest, is perhaps the most appropriate.

5. **OXYDIUM,** (Sp. 26–31.)

26. L. trichopoda (Benth. l. c. p. 603); procumbent, effuse, glaucous and glabrescent or minutely canescent; branches filiform; stipule solitary, small, ovate, or sub-rotund; leaflets broadly elliptical, obovate or obcordate; peduncles slender, elongate, *umbellate, many-flowered;* bracts minute; calyx thinly silky, semi-5-fid, the lobes subulate; petals glabrous, *on long claws;* the carina acute; legume *linear, compressed, minutely puberulent, with close-pressed hairs. Crotal. trichopoda, E. Mey.! Comm. p.* 154. *Polylob. typicum and P. intermedium, E. & Z.!* 1288, 1289. *Ononis glabra, Thunb.! Cap. p.* 588.

HAB. Uitenhage districts, from Van Staadensberg to Sondag River, *Drege, E. & Z. &c.* Algoa Bay, *Dr. Alexander Prior!* (Herb. Th., D., Bth., Hk., Sd.)

Root woody; stems spreading on the ground in a circle of 2–3 feet diameter. The whole plant looks glabrous and pale, but under a lens is seen to be thinly clothed with very minute, white, close-pressed hairs. Leaves scattered; leaflets ½–¾ inch long, 2–3 lines wide. Peduncles 3–6 inches long; umbel 10–20 flowered. Flowers bright yellow. Legume ¾-inch long, 1–1¼ line wide, 3–4 times as long as the calyx.

27. L. perplexa (E. & Z. (ex parte); Benth.! l. c. p. 605); procumbent, glabrous or sprinkled with very minute appressed hairs; branches filiform; stipule solitary, small, obliquely ovato-lanceolate; leaflets cuneate-oblong or obovate, those of the upper leaves narrower; peduncles *slender, elongate,* 1–2-*flowered;* bracts minute; calyx thinly silky, with broadly subulate teeth; petals glabrous, on long claws, the carina rostrate; *legume oblong, turgid, about twice as long as the calyx. Crotalaria perplexa, E. Mey. Linn.* 7. *p.* 151. *Lotononis strigosa, Pappe!* 96.

HAB. Mountain-sides round Capetown, common. (Herb. D., Hk., Bth., Sd.)

Root woody, sinking deeply. Stems trailing, 2 feet long or more, slender. Pubescence very minute, scanty, and close-pressed, the hairs stiff. Leaves scattered, leaflets variable in breadth. Peduncles 2–3 inches long. Flowers small. Legume 2–2½ lines long, very turgid. E. & Z. confounded this plant with *Crotalaria humilis.*

28. L. micrantha (Thunb. & Harv.; non E. & Z.); procumbent, thinly sprinkled with very minute, appressed hairs; branches filiform; stipule solitary, small, oblong or lanceolate; leaflets cuneate-oblong or linear, or those of the lower leaves obovate; peduncles slender, elongate, *umbellate, many-flowered;* bracts minute; calyx thinly silky, with triangular-acuminate teeth; petals glabrous, on long claws, the carina rostrate; *legume ovoid, turgid, scarcely longer than the calyx. Ononis micrantha, Thunb.! Cap. p.* 587. *Crotalaria micrantha, E. Mey. Comm. p.* 27. *C. tenuiflora, Steud. Loton. rostrata, Benth. l. c. p.* 604.

HAB. Cape, *Thunberg!* On shrubby hills, Roodesand, *Drege!* (Herb. Th. D. Bth. Hk.)

Stems extensively trailing. Leaflets 3, narrow; those of the upper leaves especially. Peduncles 1–1½ inch long; *flowers very small.* Peduncles not quite twice as long as the leaves; the umbel somewhat like that of *Trifolium repens.* Stipule sometimes falcate. Legume very short. This is the original "*Ononis micrantha*" of Thunberg, a very appropriate name and worthy of being preserved.

29. L. acutiflora (Benth. l. c. p. 604); procumbent, thinly canescent or glabrescent; ramuli filiform; stipule solitary, small, "orbicular" or oblong-lanceolate; *leaflets usually five,* narrow-cuneate or sub-linear; peduncles *shorter than the leaf,* umbellately several-flowered; bracts shorter than the longish pedicel; calyx appressedly and thinly silky, its segments lanceolate; legume silky, oblong, falcate, at length turgid, scarcely longer than the calyx. *Crotalaria quinata, E. Mey. Com. p.* 27*!*

HAB. Near Krakkeelskraal, Clanw., *Drege!* (Herb. Benth. Hook.)

A small, half-herbaceous, slender species, 2–4 inches high; known from its neighbours by the usually quinate or *digitate* leaves. Petioles 1 inch, lamina ½–¾ inch long. The pubescence is very scanty, the hairs minute and closely appressed.

30. L. oxyptera (Benth.! l. c. p. 605); procumbent, thinly silky or villoso-pubescent; stipule solitary, small, obliquely ovate or lanceolate; leaflets 3, *ovate or oblong;* peduncles *long or short,* villous, capitately several-flowered; bracts longer than the very short pedicel; calyx tomentose, with deltoid-acuminate, sub-falcate teeth; legume ("turgid, not much exceeding the calyx" ?)

VAR. *α.* **longipes**; peduncles longer than the leaf. *Pol. sparsiflorum, E. & Z.!* 1293*!*

VAR. *β.* **brevipes**; peduncles shorter than the leaf. *Crotalaria oxyptera, E. Mey.! Comm. p.* 28.

HAB. *α.* Tulbagh on moist hill sides, *E. & Z.!* *β.* Drakensteen hills, *Drege!* (Herb. Bth., Hk., Sd.)

Stems trailing, not much branched, thinly or thickly clothed with short, curly, soft hairs. Petioles ½-inch long; leaflets as long, 2–3 lines wide, very blunt or emarginate. Peduncles ¾-inch long, softly hairy. Flowers 5–6, sub-sessile. Legume unknown to me. Bracts linear or oblong, small. The two varieties scarcely differ, except in the length of the peduncle; var. *β.* is rather the most hairy.

31. L. monophylla (Harv. Thes. t. 63); suffruticose, slender, ascending, appressedly puberulous and silvery; stipules none; leaves *unifoliolate,* leaflet ovate or oblong, mucronulate, glabrous above, silvery beneath; peduncles elongate, umbellate, 4–5-flowered, bracts minute; calyx appressedly and minutely silvery-puberulous, upper segments triangular-acute, lowest subulate; petals on short claws, the vexillum sub-rotund, silky, the carina glabrous, rostrate; legume? *Zey. No.* 2313!

HAB. Stony places, on the Vanstaadensberg, Uit., *C. Zeyher!* (Herb. D., Sd.)

Root woody. Stems 2–4 inches long, scarcely branched, decumbent, then erect. Leaves an inch apart; petiole ½–1 inch long; leaflet ¾-inch long, 4–5 lines wide. Peduncles terminal or opposite the leaf, 3–4 lines long. Flowers like those of *Lotus corniculatus*. The pubescence is very minute, and closely appressed. Legumes not known.

6. **LIPOZYGIS.** (Sp. 32–37.)

32. L. pentaphylla (Benth. l. c. p. 605); procumbent, much-branched, softly and densely silky, and fulvous; stipule solitary, small, lanceolate; leaflets *often five, obovate;* heads sessile, densely many-flowered; *bracts very narrow;* calyx densely and softly hairy; petals hairy, the vexillum oblong, obtuse, carina oblong, incurved. *Lipozygis pentaphylla, E. Mey. Comm. p.* 79.

HAB. Karakuis, *Drege!* (Herb. D., Bth., Hk.)

Stems perhaps prostrate, 6–12 inches long or more, pale, softly silky: the pubescence of the whole plant fulvous or foxy. Petioles ¼–½ inch long; the leaflets not longer, broadly obovate, 2–2½ lines wide. Heads of flowers densely hairy, globose, terminal and lateral.

33. L. polycephala (Benth.! l. c. p. 605); decumbent, branching, *densely and very softly silky-villous;* stipule solitary, oblongo-lanceolate; leaflets three, obovate, softly and densely villous on both sides; heads sessile, densely many-flowered; *bracts broadly ovate;* calyx shaggy, somewhat inflated; vexillum *broadly obovate,* it and the incurved helmet-shaped carina silky. *Lipozygis polycephala, E. Mey. Comm. p.* 79.

HAB. Kamiesberg, *Drege!* (Herb. D., Hk., Bth.)

The whole plant densely clothed with long and soft, pale or fulvous hairs. Petioles ½–1 inch long; leaflets rather shorter. Heads of flowers at the ends of short axillary ramuli; the bracts very broad, acute or acuminate; by which character it differs from all of the present section, except the following.

34. L. anthylloides (Harv.); decumbent, flexuous, branching; stems *thinly pilose;* stipule solitary, broadly oblong or oval; leaflets 3, obovate or obcordate, *glabrous above, silky-pilose beneath;* heads sessile, densely many-flowered; bracts *broadly oblong or ovate;* calyx shaggy, its segments long and subulate, nearly equal; vexillum *narrow*-oblong, it and the blunt carina silky-pilose; *alæ and carinal petals eared at base.*

HAB. Namaqualand, *A. Wyley, Esq.* (Herb. T.C.D.)

Allied to *L. polycephalum,* of which it has the inflorescence and calyx, but from which it remarkably differs in pubescence, in the shape of the vexillum, the long earlike appendages of the lower petals, and in minor characters. Stems 6–8 inches long, curved, purplish. Leaflets 4–5 lines long, 3 lines wide at the very obtuse or emarginate summit, as long as the petiole or shorter. Stipules and bracts very broad. Heads very hairy, the calyx tube somewhat swollen, scarcely equalling the laciniæ. Legume not seen. The habit is that of an *Anthyllis.*

35. L. eriantha (Benth. l. c. p. 605); ascending or sub-erect, slightly branched, thinly and softly villous; *stipule leaflike, longer than the short petiole;* leaflets *oblong-elliptical or broadly lanceolate, very acute,* pilose on both sides; heads sessile, laxly several-flowered; bracts small, setaceous; the calyx, the oblong-acuminate vexillum, and the obtuse, arched carina, all silky; legume compressed, at length sub-turgid, silky, not quite twice as long as the calyx.

HAB. Magaliesberg, *Burke and Zeyher!* (Herb. D., Hk., Bth.)

Many-stemmed, from a woody crown; the stems 4–6 inches high, with a few erect branches. Leaves pale, trifoliolate, on short petioles; the leaflets often with a much acuminate point. Stipules as large as the leaflets and of similar form. Nearly allied to *L. corymbosa*, but the heads have much fewer and larger flowers; the leaflets are differently shaped and very acute, and the stems more branching.

36. L. corymbosa (Benth.! l. c. p. 606); ascending or erect, sub-simple, pilose; stipule leaflike, longer than the petiole; *leaflets elliptic-oblong or obovate, obtuse or sub-acute,* pilose on both sides; heads sessile, very-many-flowered; bracts setaceous; calyx hirsute; the oblong-acuminate vexillum, and the obtuse, arched carina pubescent. *Lipozygis corymbosa, E. Mey.! Comm. p.* 79.

HAB. Grassy hills, near the Umsata, *Drege!* Top of Table Mt., Natal, *Krauss! Gueinzius!* (Herb. D., Hk., Bth., Sd.)

Many-stemmed from a thick, woody crown. Stems 4–5 inches high, in our specimens quite simple, terminated with a somewhat corymbose-capitate cluster of shortly pedicellate flowers. Petiole ¼–⅓ inch long; the leaflets longer, 2–3 lines wide, and generally blunt: by which character and the smaller and more numerous flowers this species is easily known from *L. eriantha.*

37. L. lanceolata (Benth! l. c. p. 606); thinly villous, stems ascending, simple or slightly branched; the leaflike stipule and leaflets *linear-lanceolate, acute;* heads laxly corymbiform, terminal, sub-sessile, many-flowered; bracts setaceous; calyx thinly silky; the oblong-acuminate vexillum and the incurved, obtuse carina, *glabrous* or minutely pilose. *Aspalathus lanceolatus, E. Mey.! Comm. p.* 37.

HAB. Witbergen, on grassy hills, Leewenspruit, *Drege!* (Herb. D., Bth., Hk., Sd.)

A foot or more in height, turning dark in drying, thinly sprinkled with long, patent, very slender and soft hairs. Petioles 1–2 lines long; leaflets ¾–1 inch long, 1–1½ lines wide, tapering to each end. Flowers clustered rather than capitate, not large.

7. LEOBORDEA. (Sp. 38–41).

38. L. porrecta (Benth. l. c. p. 606); procumbent or prostrate, much branched, silky-pubescent; stipule falcate, small; petioles short, leaflets (small) cuneate-oblong or obovate, *glabrous above, appressedly pubescent beneath;* flowers in pairs or solitary, sub-sessile; calyx thinly silky, oblong, cleft to the middle, the lowest lobe subulate, very narrow; *petals exserted;* legume pubescent, scarcely longer than the calyx. *Capnitis porrecta, E. Mey.! Comm. p.* 81. *Leptis prolifera and L. debilis, E. & Z.!* 1264, 1265! *Pappe,* 98.

HAB. Zwartbulletje, on stony hills, and at Gamke River, *Drege.* Between Grahamstown and Bothasberg; Zwartkops River, *E. & Z.! Pappe!* (Herb. D., Bth., Sd.)

A very dwarf suffruticose plant, the stems 6–8 inches long, spreading flat over the ground in all directions and densely leafy, with patent, prostrate branches. Leaves 3-foliolate, including the petiole not ½-inch long; the leaflets 2–2½ lines long, 1 line broad, very blunt. Flowers small, yellow. The upper calyx lobes are connate in pairs for ½–¾ of their length, the lowest is separated by a much deeper sinus, and is very narrow, but nearly as long as the rest. As well as I can make out from a very bad specimen of Drege's plant in Herb. Bentham, it is the same as that of E. & Z.! from which I chiefly describe.

39. L. carinalis (Harv.); procumbent, *dichotomous, silky-canescent;* stipule solitary, falcate, shorter than the petiole; leaflets cuneate-oblong, silky canescent on both sides, complicate; flowers 1–3, sub-sessile; calyx thinly silky, tubular, *the segments much shorter than the tube*, the lowest very small, subulate; petals nearly twice as long as the calyx, silky, *the carina straight, very long.*

HAB. Namaqualand, *A. Wyley, Esq.* (Herb. T.C.D.)

A slender (perhaps annual) species, with sub-distantly forked stems. Petioles 3–4 lines long; leaflets rather shorter, hoary on both sides. Flowers 7 lines long; the carina nearly twice as long as the vexillum. Upper calyx segments connate in pairs for half their length, much broader than the lowest segment, and nearly twice as long. Ovary many-ovuled. This has a different habit and calyx, and a much longer carina than *L. porrecta;* and differs equally (except in ramification) from *L. Leobordea.* It has larger flowers than either.

40. L. Leobordea (Benth.! Lond. Journ. 2. p. 607); procumbent, sub-dichotomous, silky-canescent; stipules *oblong* or lanceolate, small; leaflets obovate-oblong, silky-villous; flowers 2–5 together, sub-sessile; *calyx deeply and sharply cut, its upper segments lanceolate, acuminate, the lowest setaceous;* petals exserted, villous, the carina arched; legume oblong or oval, as long or twice as long as the calyx, turgid, sparsely puberulous. *Leobordea lotoides, Del. &c., fide Benth. l. c. Leptis sp., Zeyher*, 409!

HAB. Springbokkeel and Bitterfontein. Feb.–May, *Zey.!* (Herb. Sd., Hk., Bth.)

Root annual? stems prostrate, 2–4 inches long, the whole plant clothed with long and soft whitish hairs. Petioles ¼–½ inch long; leaflets about the same length. Flowers 2–3 lines long, the carina most prominent. In the S. African specimens the legumes are about as long as the calyx, but sometimes ⅓ longer; in those from Arabia and N. Africa, which are in other respects very similar, the legume is sometimes short as in the Cape individuals, and sometimes fully twice as long. The habit is very like that of *L. clandestina*, but the calyx segments are much longer and more tapering; the *lowest* is always very slender, but variable, being either as long as the rest, or very much shorter.

41. L. clandestina (Benth.! l. c. p. 607); procumbent, dichotomous, tomentoso-canescent; stipules very minute; petioles short, leaflets (small) obovate-oblong; flowers in pairs or solitary, sub-sessile; calyx tubular, tomentose, *with short teeth*, the lowest subulate, very narrow; *vexillum and alæ shorter than the calyx*, the carina exserted, arched, tomentose; legume tomentose, scarcely longer than the somewhat enlarged calyx. *Benth.—Capnitis clandestina, E. Mey.! Comm. p.* 81.

HAB. Plains of the Gariep, between Verleptram and Natvoet, *Drege.* (Herb. Bth. Sd.)

Slender, repeatedly forked, with leaves and flowers at the forks; the internodes longer than the leaf. Petiole 1–2 lines long; leaflets 2–2½ lines. Flowers 3 lines long, the calyx ½-line wide, in fruit 1½ lines. I have only seen the single specimen in Herb. Benth.

8. **LEPTIS.** (Sp. 42–58.)

42. L. Burchellii (Benth. Lond. Journ. 2. p. 612); "dwarf, decumbent? much branched, densely silky-villous, greyish or silvery; *stipules cordate-ovate or orbicular;* leaflets obovate; heads of flowers dense, sessile, few-flowered; *bracts very broad, sub-cordate-reniform;* corolla shorter than the calyx; the oblong, acuminate vexillum and the arched carina nearly glabrous; legume silky, at length turgid, as long as the calyx.

HAB. S. Africa, *Burchell, No.* 2539.

This may be easily known by its bracts and stipules.

43. L. crumanina (Burch. Cat. 2445); procumbent, slender, silky and somewhat silvery; stipules solitary, lanceolate, much shorter than the petioles; leaflets cuneate-oblong; *flowers 3–4–together in small, subsessile heads* opposite the leaves; calyx very deeply cleft, densely silky, its narrow-lanceolate sub-equal segments longer than the corolla; the oblong vexillum and the acute carina glabrous, legume obliquely ovate, silky, at length somewhat turgid, *scarcely as long as the calyx. Benth. l. c. p.* 612.

HAB. South Africa, *Burchell.* Caledon River, branch of the Gariep, *Burke and Zeyher!* (Herb. Bth., Hk.)

Root woody; stems several, 6–8 inches long, slender, branched near the base, trailing. Petioles an inch or more in length; leaflets about ½-inch long, 1–1½ lines wide. Heads of flowers mostly sessile, occasionally on a very short peduncle.

44. L. lenticula (Benth. l. c. p. 611); prostrate, dwarf, silky; stipules solitary, small; leaflets small, obovate or cuneate-oblong; flowers solitary, opposite the leaves, subsessile; calyx silky, the lobes acute, the lowest smaller; vexillum oblong, acuminate, pubescent on the dorsal ridge, rather shorter than *the narrow, rostrate carina;* legume turgid, slightly curved, silky, 1½ to twice as long as the calyx. *Crotalaria lenticula, E. Mey.! Herb. Drege. Zey.* 410.

VAR. *β.* **brachycarpa**; legume oblong, densely silky, not much longer than the calyx. *Zey.* 411.

HAB. S. Africa, *Burchell,* 1455. In the Sneeuwbergen, on the flat between Rivertje and Kieuwkerkshoogte, *Drege!* Bitterfontein, *Zeyher, β.* Springbokkeel, *Zeyher.* (Herb. Hook., Bth. D.)

Very small; the stems 1–3 inches long. Petioles ¼–¾ inch long; leaflets 2–3 lines. Flowers small; the taper-pointed carina very conspicuous.

45. L. pungens (E. & Z.! 1277); slender, suberect, or diffuse, silky and *villous;* stipule solitary, *linear-lanceolate;* leaflets ternate, linear-lanceolate, acute, subpungent; flowers solitary, pedicellate, opposite the leaves; calyx deeply divided, silky, its lobes subequal, *subulate, shorter than the corolla;* vexillum obovate, acuminate, rather shorter than the broad, falcate, glabrous carina; legume *villoso-pilose,* twice as long as the calyx, at length turgid. *Also L. affinis, E. & Z.!* 1279, *and L. decidua, E. & Z.!* 1281.

HAB. On barren hills, at Koonabshoogde, Ceded Territory; also between Bosjesman's and Karrega Rivers, Albany, and near Gauritz R., Swell., *E. & Z.* (Hb. Sd.)

Very similar to *L. laxa,* but the leaflets are narrower, more lanceolate, and sharper; the pubescence is not silvery but villous, or inclining to pilose, the calyx is more deeply divided and the lobes much narrower; and the corolla is larger and more exserted. The three Ecklonian species here united, differ chiefly in the more or less abundant villosity; the character of the pubescence is the same in all. Stems 6–12 inches long. Leaflets ½ inch long, not a line wide.

46. L. depressa (E. & Z.! 1278); suffruticose, small, diffuse, silky and silvery; stipule solitary, *linear or cuneate,* leaflike; leaflets ternate, linear, or cuneate, or sublanceolate, silvery; flowers solitary, opposite the leaves, on very short pedicels; calyx deeply cut, silvery, its lobes subequal, lanceolate, shorter than the corolla; vexillum subciliate, equalling the broadly falcate glabrous carina; legume *densely pilloso-villous,* twice as long as the calyx, subturgid.

HAB. Between the Gauritz River and the Langekloof, *E. & Z.!* (Herb. Sd. D.)

Densely cæspitose, the stems 3–4 inches long, branched from the base. Pubescence of stem and leaves copious, silvery and close pressed; that of the legume loosely but softly hairy. Leaflets variable in shape. It is nearly intermediate between *L. laxa* and *L. pungens*, having the pubescence of the former and the legumes of the latter.

47. L. laxa (E. & Z.! 1276); suffruticose, small, diffuse or decumbent, *silky and silvery;* stipule solitary, *lanceolate;* leaflets ternate, rarely solitary, *oblongo-lanceolate or sublinear, acute;* flowers solitary, opposite the leaves, on short pedicels; calyx thinly silky, its lobes subequal, nearly as long as the corolla; vexillum obovate, acuminate, silky on the dorsal ridge, as long as the shortly rostrate carina; *legume* canescent, 2–3 times as long as the calyx, at length sub-turgid. *L. diversifolia, Benth.! Lond. Journ.* 2. 611. *Crot. diversifolia, E. Mey.! Comm. p.* 77.

HAB. Near Silo, on the Klipplaat River, Caffraria, *E. & Z.! Drege.* Thaba Uncka and Caledon River, *Burke & Zey.* Graff Reynet, *Mrs Barber.* (Herb. D., Hk., Sd., Bth.)

Stems very many from the crown, trailing, 6–12 inches long or more; the whole plant silvery-white and shining. Leaves varying in size and shape; stipules as long as the petiole or longer. Legume 4 lines long, 1 line wide.

48. L. quinata (Benth.! l. c. p. 608); suffruticose, prostrate, either glabrous or thinly silky and canescent; stipule solitary, small; *leaves quinate,* leaflets minute, cuneate-oblong, obtuse or sub-acute; flowers solitary or in pairs, opposite the leaves, sub-sessile; upper calyx-lobes approaching in pairs, lowest subulate, rather shorter; vexillum obovate-oblong, silky on the dorsal ridge, equalling the oblong, *obtuse* carina; legume compressed, appressedly downy, twice or thrice as long as the calyx. *Benth. Ononis quinata, Thunb.! Fl. Cap. p.* 586.

VAR. β. **minor**; very dwarf; the foliage minute, thinly silky; leaflets 1½–2 lines long. Lipozygis quinata, *E. Mey. Comm. p.* 77

HAB. South Africa, *Thunberg,* var. β, Kamiesberg, on rocks near Modderfontein, *Drege!* (Herb. Th., Bth.)

Small but robust and woody, much branched, appressed to the soil. Stems 3–12 inches long. Petioles 2–6 lines long; leaflets digitate, 2–5 lines long and very narrow. Flowers in α, 4–5 lines long, rufescent; in β, 2–2½ lines long, pale yellow. Thunberg's specimen is twice as large in all parts as Drege's, and nearly glabrous; otherwise the two agree.

49. L. brachyloba (Benth.! l. c. p. 608); *procumbent,* thinly silky and canescent, the branches filiform; stipules solitary, linear, small; leaflets on longish petioles, cuneate-oblong or linear, the lower ones broader; flowers 2–5, subsessile, opposite the leaves; bracts minute; calyx-lobes of nearly equal length, half as long as the petals; vexillum obovate-oblong, acuminate, scarcely silky at back, carina sub-fornicate, obtuse, glabrous; legume *thrice as long as the calyx, appressedly hairy, at length somewhat tumid. Benth. Lipozygis brachyloba, E. Mey.! Comm. p.* 78. *Burch. Cat.* 1273.

HAB. S. Africa, *Burchell.* Karakuis, *Drege!* (Herb. Bth.)

Very slender, trailing, the stems 6–8 inches long, spreading from a centre. Petioles of the lowest leaves ¾–1 inch long, of the upper ¼–⅓ inch; leaflets 3–4 lines long, 1–1½ broad. The hairs of the pubescence are very short, closely appressed and silvery. Flowers 2–3 lines long.

50. L. falcata (Benth. ! l. c. p. 608); *diffuse*, thinly sprinkled with minute, close-pressed, white hairs; stipule solitary, small; leaflets cuneate-oblong or linear, obtuse; flowers *solitary or in pairs*, subsessile, opposite the leaves; calyx teeth acutely triangular, subequal; petals glabrous, 1½ as long as the calyx; vexillum oblong, acute, scarcely as long as the oblong, subfalcate, obtuse carina; *legume several times as long as the calyx, falcate*, appressedly canescent. *Lipozygis falcata, E. Mey.! Comm. p.* 78. *Zeyher*, 390.

HAB. On the Gariep, near Verleptpram, and hills near Ebenezer, *Drege*. Bitterfontein, Bosjesland, *Zeyher !* (Herb. D., Bth., Sd., Hk.)

Drege's original specimens are prostrate, branched from the base, the stems 4–5 inches long; *Zeyher's* are erect, but diffusely much branched, stiff and wiry, somewhat dichotomous, with distant nodes, and are 10–12 inches high. In the characters of leaf, flower and fruit, the two forms nearly agree. The legumes are ¾–1 inch long, 1 line wide. Flowers 2–2½ lines long.

51. L. carinata (Benth. ! l. c. p. 609); procumbent, glabrous, or microscopically puberulous; branches filiform, ascending, elongate; stipules *in pairs, linear, unequal, rarely solitary;* leaflets on very short petioles, cuneate-oblong or linear; flowers subsessile, 1–3 together opposite the leaves; calyx-lobes from a broad base, subulate; *the obovate-oblong vexillum, and the obtuse carina silky;* legumes . . ? *Lipozygis carinata, E. Mey.! Comm. p.* 80.

HAB. Caffraria, between Omsamculo and Omcomas, *Drege!* (Herb Bth.)

A very slender and nearly glabrous plant; the pubescence, where it exists, is extremely minute, thin, and closely appressed. Flowers 2–3 lines long. Leaflets 2–3 times as long as their petiole, acute or mucronulate, 5–6 lines long.

52. L. humifusa (Burch. Cat. 3927); branches slender, prostrate, *pubescent;* stipules solitary, small, oblong or subulate; leaflets *obovate or obcordate*, longer or shorter than the petiole, glabrous above, thinly puberulous beneath; *flowers shortly pedunculate*, opposite the leaves; calyx tubular, thinly and appressedly pubescent, its segments shorter than the tube; the obovate, acute vexillum and the obtuse carina minutely puberulous (or sub-glabrous). *Benth. l. c. p.* 609. *Lipozygis humifusa, E. Mey.! Comm. p.* 77.

VAR. β. **Radula**; stems more rigid; petioles longer; flowers nearly twice as large. *Lipozygis Radula, E. Mey ! Comm. p.* 77.

HAB. S. Africa, *Burchell*. Foot of the Witberg, near Shiloh, *Drege!* β., Nieuwe Hantam, *Drege!* (Herb. Sond.)

A very small plant. Stems 2–3 inches long, depressed, branching. Leaflets 3–4 lines long, 1–2 lines wide, nearly glabrous. Peduncles equalling the petioles or a little longer. β. is larger in all its parts, with somewhat longer petioles and peduncles, but is otherwise so similar, that I hesitate to keep it apart.

53. L. mollis (Benth.! l. c. p. 609); suffruticulose, *very dwarf, softly silky and canescent;* branchlets short, ascending; stipules solitary, small; *petioles long;* leaflets cuneate-obovate, blunt; flowers 1–4, terminal or opposite the leaves, pedicellate; lowest segment of the puberulent calyx setaceous, shorter than the rest; the orbicular vexillum, and the obtuse, equally long carina softly villous. *Lipozygis villosa, E. Mey. Comm. p.* 79. *Leptis mollis, Steud.*

HAB. Liliefontein, Kamiesberg, *Drege*. (Herb. Bth.)

Root thick and woody; stems 1–2 inches high, densely cæspitose. Leaves crowded, the petioles nearly uncial, the leaflets ¼ inch long. Flowers 2–3 lines. The pubescence is short, but copious, very soft, and greyish white.

54. L. pumila (E. & Z. 1283); suffruticose, *prostrate*, much branched, thinly silky or silvery; stipules small, oblong, solitary; leaflets *very short, obovate-orbicular or cuneate;* flowers *sub-umbellate,* 2–4, *on a short peduncle* opposite the leaf; calyx-lobes sub-equal; vexillum sub-ciliate, nearly as long as the arched, obtuse, glabrous carina; *legume linear, sub-compressed,* 2–3 *times as long as the calyx, silky. Benth. l. c. p.* 609. *Lipozygis erubescens, E. Mey. Comm. p.* 76.

VAR. β. **micrantha**; Stems longer and more slender; leaflets cuneate or cuneate-oblong; flowers 1–3, umbellate, on a short peduncle. *L. micrantha, E. & Z.* 1280 *(non Thunb.)*

HAB. Stony places on the little Fish River, *Drege!* Somerset, *Mrs. Barber.* Near the Gauritz River, *E. & Z.!* Var. β. near the Sondag River, *E. & Z.!* (Hb. Bth., Sd., D., Hk.)

Root woody and thick, many stemmed from the crown; the stems rigid, and spreading over the soil. Petioles ¼–½ inch long; leaflets 2–3 lines long, and 1½–2 lines broad, appressedly silky on both sides. Peduncles ¼–⅓ inch long. Calyx-lobes narrow-triangular, acute. Fl. yellow. β. is more slender, with narrower leaflets and somewhat smaller flowers.

55. L. microphylla (Hv.); suffruticose, prostrate, *very much branched;* stipules minute; leaves *very small, on short petioles;* leaflets ternate, obovate, glabrous above, thinly silky below; flowers solitary, opposite the leaves, sub-sessile; calyx appressedly silky, semi-quinquefid, the segments lanceolate, lowest narrower; vexillum and the blunt carina thinly silky; legume . . ?

HAB. S. Africa, *Zeyher!* (Herb. Sond.)

Mixed with *L. tenella* in Zeyher's collection. Stems thick and woody, but dwarf; 3–4 inches long, the branchlets closely crowded, ¼–½ inch long, densely covered with minute leaves. Petioles 1 line long; leaflets not longer, concave, thinly canescent on the lower surface. Flowers yellow, 2 lines long. Apparently a distinct species.

56. L. tenella (Eck. & Zey. 1282); suffruticose, dwarf, either prostrate, decumbent or sub-erect, *patently hairy or densely hirsute;* stipules solitary, linear; leaflets oblong, obovate, or linear or linear-lanceolate, acute or sub-obtuse; flowers 1–3, sub-sessile; *calyx pilose, deeply cut, its lobes lanceolate, nearly as long as the corolla, the lowest subulate;* vexillum obovate, about equalling the carina, but more or less silky or villous; legume turgid, as long as the calyx or scarcely longer, piloso-hispid.

VAR. α. **angustifolia**; leaflets narrower, *oblong-linear or sub-lanceolate;* pods rather longer than the slightly increased calyx. *L. tenella, E. & Z. Benth. l. c. p.* 610. *Lipozygis tenella,* γ *piloso-villosa, E. M. p.* 78. *Zey.* 408.

VAR. β. **hirsutissima**; very hairy; sub-erect; leaflets lanceolate; pods somewhat longer than the calyx. *L. calycina, var. Herb. Benth.!*

VAR. γ. **calycina**; leaflets *obovate or oblong,* broader and shorter than in α. & β.; pod not quite as long as the calyx. *L. calycina, Benth. l. c. L. divaricata, E. & Z.* 1266. *Lipozygis calycina, E. Mey. p.* 78.

HAB. Uitenhage and Albany, *Drege! E. & Z., &c.* β. Magaliesberg, *Burke & Zey.* γ. Katberg and Kliplaat River, *D.!* Thaba Uncka and Vaal River, *Burke & Zey.* (Herb. D., Hk., Bth., Sd.)

Certainly a very variable species. Vars. α. and γ. are held for species by *Meyer* and *Bentham;* but to me they appear to run together by insensible gradations. Stems 2–6–8 inches long, the shorter sub-erect, the longer trailing. Some of *Drege's* specimens may belong to different species. *Zeyher's No.* 2311 is more robust than usual, but scarcely different.

57. L. versicolor (Benth. l. c. p. 610); suffruticose, minute, diffuse, much-branched, villoso-pubescent; stipules oblong, solitary; leaflets small, on short petioles, obovate or cuneate-oblong, thinly silky; flowers solitary, on very short pedicels, opposite the leaf; *calyx silky, deeply cut, its lobes acuminate, much shorter than the petals,* nearly of equal length, the lowest subulate; vexillum obovate-acuminate, sub-ciliate or villous, about equalling the arched, obtuse, glabrous or villous carina; pod thinly hairy, scarcely longer than the calyx, somewhat turgid. *Crot. versicolor, E. Mey. Linn. p.* 152. *Leptis versicolor, E. & Z.!* 1267. *Lept. filicaulis, E. & Z.!* 1268. *Lipozygis Kraussiana, Meisn.?—fide Benth. l. c.*

HAB. Uitenhage and Albany, *E. & Z.!* (Herb. D., Sd., Bth., Hk.)

Root woody; stems 3–6 inches long, spreading on the ground. Pubescence very variable. Leaflets mucronulate. Nearly allied to *L. tenella,* but with a shorter calyx, pedicels, and more glabrous corolla.

58. L. sessilifolia (Harv.); shrubby, *erect, densely rufo-sericeous; branches virgate,* leaves crowded, *sessile or nearly so,* trifoliolate, the leaflets lanceolate, acute or subpungent, complicate; *stipules none;* flowers 3 or 4 together, subsessile towards the ends of the branches; calyx silky, its 4 upper teeth very short, triangular, the lowest longer, subulate; petals glabrous, the vexillum narrow-obovate, the rest with long claws; legume rufo-sericeous.

HAB. Magaliesberg, *Burke.* (Herb. Hk., Bth., Sd.)

Erect, woody, branched from the base, the whole plant densely clothed with soft, silky, close-pressed, reddish hairs. Leaves closely set, imbricate; leaflets ½–1 inch long, 1–1½ line wide, tapering to each end, when dry the leaves of the laminæ folded together. *Flowers* small, often crowded near the ends of the branches. Legume ¾–1 inch long, 1½ lines wide, linear, acute, many seeded. This has the foliage of one of the flat-leaved triphyllous *Aspalathi;* but the pod of a *Lotononis.*

XV. LISTIA, E. Mey.

Calyx trifid, the anterior segment subulate, the lateral broader, bidentate. *Carina* obtuse, longer than the vexillum and alæ. *Stamens* monadelphous. *Legume* linear, compressed, many-seeded, repeatedly folded and twisted from side to side. *E. Mey. Comm. p.* 81. *Endl. Gen.* 6491. *Benth. in Hook. Lond. Journ.* 3. *p.* 338.

The only known species has the habit of a *Lotononis* of the section *Polylobium;* but differs from all the other Genisteæ in the remarkable twisting and folding of its legume. The trifid calyx is formed of 5 sepals, the four uppermost of which are broadest and connate in pairs; the anterior is much narrower, and stands apart. The name is in memory of *Fr. L. List,* a German botanist.

1. L. heterophylla (E. Mey. l. c.). *Benth. Lond. Journ.* 3. *p.* 338. *Zeyher,* 413!

HAB. Dry stony hills near Gaatje, 4600 ft., and on table lands near Wildschutshoek, 4000 ft. *Drege!* Stormberg's spruit, *Burke and Zeyher! Burchell,* 2297. Queenstown District, *Mrs. Barber!* (Herb. D., Hk., Sd., Bth.)

Root perennial. Stems slender, procumbent, not much branched, 6–12 inches long; the whole plant nearly glabrous. Stipules in pairs, unequal, small. Leaves trifoliolate, scattered, leaflets narrow-cuneate or lanceolate-oblong, 5–7 lines long, 1 line wide, on petioles ⅛–⅜ inch long. Peduncles 1½–2 inches long, bearing a short sub-umbellate 6–8 flowered raceme. Calyx-teeth short. Vexillum oblate; carina very obtuse. Flowers yellow.

XVI. ARGYROLOBIUM, E. & Z.

Calyx campanulate, deeply cleft, bilabiate, the upper lip bifid or bipartite, the lower trifid or tridentate. *Vexillum* ample, longer than the carina. *Stamens* monadelphous. *Ovary* many-ovuled. *Legume* linear, compressed, silky, not glandular, many-seeded. *Benth. in Lond. Journ.!* 3. *p.* 339. *Endl. Gen.* 6504. *Chasmone, E. Mey. Trichasma, Gamochilum and Argyrolobium, Walp.*

Small shrubs or more commonly suffrutices, natives of the Mediterranean region as well as of the Cape, generally with silky or silvery, copious pubescence. Leaves trifoliolate, with stipules. Flowers yellow, pedunculate or pedicellate, solitary or racemose, or subumbellate, bracteate. Name from *αργυρος, silver,* and *λοβιον, a legume.*

ANALYSIS OF THE SOUTH AFRICAN SPECIES.

Section 1. **Chasmone.** *Legume* without divisions between the seeds, the valves convex, not constricted. (Sp. 1–29.)

§ 1. SPICATA. *Stems* virgate, simple, from a perennial root.
Flowers in a terminal spike ... ... (1) **crinitum.**

§ 2. RACEMOSA. *Stems* herbaceous, erect, from a perennial root. *Flowers* in terminal, pedunculated racemes.

Leaflets obovate or cuneate-oblong;
Glaucous, sub-glabrous; stipules ovato-lanceolate; lower calyx-lip 3-cleft ... ... (2) **speciosum.**
Softly hairy; stipules linear or subulate; lower cal.-lip 3-toothed ... ... (3) **baptisioides.**
Leaflets linear-lanceolate:
Silky-villous; stipules equalling the petiole; legume thickly hairy ... ... (4) **longifolium.**
Glabrescent; stip. much shorter than the petiole; legume glabrescent ... ... (5) **tuberosum.**

3. FRUTICOSA. Much-branched *shrubs;* flowers in short, terminal racemes.

Thinly or minutely silky, with short hairs:
Stipules lanceolate-setaceous; vexillum silky ... (6) **polyphyllum.**
Stipules very minute; vexillum glabrous ... ... (7) **crassifolium.**
Densely tomentose or silky-villous: vexillum glabrous.
Leafl. silky on *both* sides; racemes 3–4 flowered (8) **incanum.**
Leafl. glabrous *above;* racemes 10–15 flowered (9) **sericeum.**
Silky-villous; vex. silky; cal. bluntly bilabiate (10) **obsoletum.**

§ 4. BREVIPEDES. Suffrutices. Flowers solitary or in pairs, subsessile; rarely subumbellate, on very short peduncles, opposite the leaves.

Stipules large and leaflike, equalling or nearly equalling the leaflets:
Subsimple, *albo-pilose;* stipules and leafl. uncial; flowers *several;* calyx green, pilose ... ... (13) **pilosum.**
Sub-simple, *silky;* stipules and leafl. *uncial;* flowers few; calyx densely silky ... ... (12) **stipulaceum.**
Branching; stipules and leafl. *semi-uncial* or less, thick, *nerveless,* woolly ... ... (11) **candicans.**
Stipules *ovate* or ovato-lanceolate, *shorter than the petiole, free* ... ... (14) **velutinum.**
Stipules deltoid, *connate-perfoliate;* petiole very short (15) **connatum.**
Stipules lanceol.-subulate, small; leafl. obovate or lanceolate, hairy; petiole very short ... ... (16) **collinum.**

Stipules very minute; petiole very long ... ... (17) **petiolare.**
Stipules minute; leaflets *linear-acute,* glabrescent, petiole *short* ... ... (18) **uniflorum.**

§ 5. PEDUNCULARES. Suffrutices. Peduncles elongate, or sometimes short, one or several flowered.

Leaflets linear, very narrow: *peduncles* shorter than the leaf.
Slender, sub-simple; vexillum and legumes *densely silky* ... ... (19) **filiforme.**
Suffruticose, much-branched; vex. and leg. glabrescent ... ... (20) **tenue.**
Leaflets obovate, oblong, or lanceolate: peduncles 1–2–3 flowered, shorter or not much longer than the leaf ... ...
Pubescence scanty, appressed; stipules small; leafl. cuneate-oblong or lanceolate, rigid, *veiny* (26) **patens.**
Pubescence copious, silky-villous; stipules long, subulate; leafl. oblong or lanceolate ... ... (23) **pauciflorum.**
Pubescence copious, silky-villous; stipules small; leafl. obovate or broadly oblong ... . (25) **pumilum.**
Pubescence *pilose;* leafl. *piloso-ciliate;* stipules connate-amplexicaul, ovato-lanceolate ... ... (24) **barbatum.**
Leaflets obovate or oblong; peduncles *elongate,* umbellately 2–3–5 flowered:
Diffuse, branching; stipules small, subulate; petiole uncial ... ... (21) **rupestre.**
Erect, *dwarf,* sub-simple; stipules linear-lanceolate, exceeding the *short* petiole ... ... (22) **molle.**
Ascending; branches *long;* stipules *ovate,* exceeding the very short petiole; leafl. *puberulous* (27) **ascendens.**
Leaflets ovate or elliptical; petioles long; *racemes* on long petioles, few or many flowered ... ... (28) **Andrewsianum.**

§ 6. INVOLUCRATA. Suffruticose. Peduncles elongate, umbellately several flowered; umbels subtended by a leaf-like bract. Stipules adnate to the petiole, connate-perfoliate. ... ... (29) **involucratum.**

Section 2. **Eremolobium.** *Legume* strongly compressed, *sub-torulose,* with transverse divisions between the seeds.

Suffruticose, decumbent, silky-canescent; stipules falcate; leafl. lanceolate; peduncles elongate, 1–3 flowered ... (30) **lanceolatum.**

Section 1. **CHASMONE.** (Sp. 1–29.)

1. A. crinitum (Walp. Linn. 13. p. 506); "very villous; leaflets and stipules nearly similar, oblong; spike terminal, elongate." *E. Mey. Chasmone crinita, E. Mey. Comm. p.* 71.

HAB. Trado, in the Zwarteberge range, *Drege.* (Unknown to me.)

Root perennial, many stemmed. Stems, as the whole plant, densely villous, virgate, subsimple, nearly a foot high to the spike. Petiole ½ inch long. Leaflets ovate-oblong, twice as long as the petiole, the lateral ones a little oblique. Stipules very large. Spike 6 inches long, the bracts oblong, scarcely shorter than the calyx. Petals yellow. Legume unknown. Habit peculiar; nearly that of a *Thermopsis.* Calyx a little more closed than in the rest. Possibly the type of a new genus." *E.M.*

2. A. speciosum (Eck. & Zey.! 1320); *glaucous, nearly glabrous* or thinly silky-villous; stem erect, triangular, flexuous, rigid; stipules *ovato-lanceolate,* acute, *the upper ones longer than the petioles;* leaflets obovate or oblong, or the upper ones lanceolate, setaceo-mucronulate; raceme elongate, terminal; the lower lip of the calyx *deeply trifid;* legume silky. *Benth. in Lond. Journ.* 3. *p.* 341. *Chasmone heterophylla, E. Mey. Comm. p.* 71. *Zeyher,* 360.

HAB. North-east of Colony, and Caffraria. Katberg and between Zandplaat and Coega, *Drege!* Winterberg, *E. & Z.!* Magaliesberg, *Burke and Zeyher!* Natal, *Sanderson!* (Herb. D., Bth., Hk., Sd.)

Stem 1–2 feet high, sharply angular, not much branched. Leaves very variable in shape and size, turning dark in drying, rigid, 2 inches long, from ½ to 1½ inch in breadth, the uppermost always narrowest. Stipules ¾–1 inch long. Racemes 4–6 inches long, many-flowered; the rachis, pedicels, leaves and calyces appressedly silky. Flowers greenish yellow, with purple striæ. Pods 2½–3 inches long, narrow-linear.

3. A. baptisioides (Walp. Linn. 13. p. 306); thinly and *softly hairy;* stem ascending, branched, angular; stipules *linear or subulate, the upper ones shorter than the petiole;* leaflets cuneate-oblong, narrow; raceme lax, terminal; lower lip of the calyx *shortly 3-toothed;* legume ? *Benth.! l. c. p.* 341. *Chasmone baptisioides, E. Mey. p.* 71.

HAB. Katberg, *Drege!* Winterberg, *E. & Z.!* (Herb. Hk., Bth., Sd.)

Similar in habit to *A. speciosum,* but much more hairy, with narrower leaves and smaller stipules, &c. Stems 1–2 feet high, sub-erect, patently hairy. Leaflets 1–1½ inch long, seldom half inch wide, mucronulate. Calyces and petals covered with fulvous, shining, appressed hairs.

4. A. longifolium (Walp. Repert. 2. p. 844); silky-villous; stem erect; branches slender, round, and striate; stipules *setaceous, about equalling the short petiole;* leaflets very long, *linear-lanceolate,* racemes pedunculate, remotely few flowered; pedicels very short; lower lip of the calyx *longer than the upper, incurved,* 3-toothed; *legume thickly hairy. Benth. l. c. p.* 341. *Chasmone longifolia, Meisn! Lond. Journ.* 2. *p.* 74.

HAB. Summit of the Tafelberg, Port Natal, *Krauss!* 214, *Gueinzius.* (Herb. D., Hk., Bth., Sd.)

Stem 1–2 feet high, branching. Petioles not ½ inch long; leaflets 2–2½ inches long, 1–3 lines wide, acute at each end, infolding. Peduncles 3–4 inches long, bearing 4–5 sub-distant flowers.

5. A. tuberosum (E. & Z.! 1322); glabrescent or *minutely silky;* stem erect, slender, slightly branched, 3-cornered near the top; stipules lanceolate-setaceous, *much shorter than the petiole* (or rarely longer); leaflets *linear-lanceolate,* the lowest ones sometimes obovate-cuneate; racemes pedunculate, laxly few-flowered; the lips of the silky calyx *about equal,* the lower 3-toothed, the teeth acuminate; legume *minutely appressedly puberulous,* at length glabrescent. *Benth. l. c. p.* 341. *Chasmone tuberosa, Meisn! Zeyher, No.* 384. *Argyr. angustifolium, E. & Z.!* 1321.

HAB. Krakakamma and Winterberg, *E. & Z.!* Howison's Poort, *Mr. Hutton!* Magaliesberg, *Burke and Zeyher!* (Herb. D., Bth., Hk., Sd.)

Root tuberous. Stems 2–3 feet high, very slender, straggling, often simple. Petioles 1–2, or sometimes 3 inches long; those of the upper leaves much shorter, leaflets 1–2 inches long, 1–2 lines wide. The pubescence is always scanty and very closely appressed, the hairs scarcely visible but with a pocket lens. Flowers pubescent, dark yellow, turning brown in drying. Pods 2 inches long, straight or slightly curved, 1 line wide. Allied to *A. filiforme, No.* 19.

6. A. polyphyllum (Eck. & Zey.! 1302); thinly silky, with short hairs, leafy, much branched; stipules *lanceolate-setaceous;* leaflets *cuneate-oblong* or obovate, mucronulate; lower lip of the silky calyx 3-toothed; *vexillum silky;* legume densely silky. *Benth. l. c. p.* 342. *Chasmone cuneifolia, E. Mey.! Comm. p.* 71.

HAB. Chumie and Winterberge, *E. & Z.!* Katberg, and between Keiskamma and Buffalo river, *Drege!* (Herb. Hk., Bth., D.)

A densely branched, robust, erect shrub, 1–3 feet high, well covered with leaves. Pubescence appressed, short, fulvous. Petioles ¼–½ inch long, leaflets 1–1½ inch long, 2–4 lines broad. Racemes terminal, few-flowered, rather dense. Flowers yellow. Occasionally more glabrous.

7. A. crassifolium (E. & Z ! 1305); thinly and minutely silky, much branched; stipules *very minute*, setaceo-subulate; leaflets *short*, broadly *obovate or obcordate; glabrous on the upper*, thinly silky on the lower side; lower lip of the silky calyx sub-entire or minutely 3-toothed; *vexillum glabrous;* legume silky. *Benth! l. c. p.* 342. *Chasmone crassifolia, E. Mey. Comm. p.* 72. *Ch. Goodioides, Meisn ? fid. Benth. Zey. No.* 2303.

HAB. Among bushes, &c. Eland's River, Uit., *E. & Z.! Drege.* (Herb. D., Hd. Sd.)

A densely branched, closely and shortly pubescent shrub, 1–1½ foot high, erect. Stipules, except on the young shoots, where they are sometimes ¼ inch long and lanceolate ! very short. Leaflets 3–4 lines long, 2–3 lines wide, mucronulate. Peduncles short, terminal, 3–5 flowered.

8. A. incanum (E. & Z.! 1306); densely *tomentose*, much branched; stipules setaceous, small; leaflets short, broadly obovate or obcordate, *densely silky-villous on both sides;* racemes sub-terminal (becoming lateral), pedunculate, subumbellately 3–4 flowered; lower lip of the tomentose calyx sub-entire; carina rostrate, *vexillum glabrous;* legumes silky. *Chasmone obcordata, E. Mey? Comm. p.* 72.

HAB. Mountain sides. Bothasberg, near Grahamstown, *E. & Z.!* Assigaisbosch, *Zey.!* 2302, *ex parte.* (Herb. Sond.)

In ramification this resembles *A. crassifolium*, to which it is united by Walpers; but the pubescence is different and much more copious.

9. A. sericeum (E & Z.! 1304); *densely silky-villous*, branching, *virgate;* stipules setaceous, *equalling the petiole or shorter;* leaflets broadly obovate or obcordate, *glabrous above*, densely silky and fulvous beneath; racemes dense, short, several flowered; the lower lip of the densely silky calyx *sub-entire; vexillum glabrous;* legume silky-villous. *Benth.! l. c. p.* 342. *Dichilus sericeus, E. Mey.! Linn.* 7. *p.* 154. *Chasmone holosericea, E. Mey.! Comm. p.* 72. *Gamochilum sericeum, Walp. Galega trifoliata, Thunb.! Cap. p.* 600. *Zey.* 2300.

HAB. Grassy hills, Vanstaadensberg, Uit., *E. & Z.! Drege!* (Herb. Th., D., Hk., Sond., Benth.)

A stout shrub, 1–3 feet high, more or less branched; the branches virgate 8–16 inches long. Pubescence copious, fulvous and rather glossy; the hairs sometimes short, sometimes long and loosely spreading. Petioles very short; leaflets rarely ½ inch long, 2–4 lines wide. Racemes 2 inches long, 10–16 flowered. Legumes fulvous and densely silky, 1½ inch long.

10. A ? obsoletum (Harv.); shrubby, much branched and ramulous, silky-villous; stipules small, *deltoid-amplexicaul;* leaves sub-sessile, leaflets lanceolate-oblong, short, with strongly revolute margins, densely silky on both sides; flowers 2–3 subterminal, on short pedicels; calyx silky, *campanulate, very short, obsoletely 2-lipped*, the short upper lip minutely bidentate, the lower tridentate; petals silky, more than twice as long as the calyx; ovary linear, many ovuled, densely silky-villous. *Aspalathus sericea, Thunb! Fl. Cap. p.* 574 (*excl. syn.*)

HAB. South Africa. (Herb. Thunb.!)

A woody shrub, densely branched and leafy, 1–2 feet high; twigs flexuous, silky. Stipules scarcely a line long, clasping the stem and somewhat perfoliate, triangular. Petiole equalling the stipules; leaflets 3–4 lines long, 1–1½ wide, bluntish, the rolled back margins nearly closing over the under surface. Flowers subumbellate at the ends of the branches. Calyx 2 lines long, widely campanulate, two-lipped; upper and lower lips both very short, with a wide, rounded sinus between, bluntly and minutely toothed. Vexillum and carina silky. The calyx, though formed on a two-lipped type, is very different from that of any other species.

11. A. candicans (E. & Z.! 1312); suffruticose; stems short, erect, *branching*, densely silky-villous; stipules ovate or ovato-lanceolate, leaf-like, equalling the leaflets or smaller; petiole short; leaflets obovate-cuneate, *densely silky-villous on both sides, veinless;* flowers *solitary, sub-sessile;* vexillum and legume densely silky-villous *Benth. l. c. p.* 343. *Chasmone sessiliflora, E. Mey.! Comm. p.* 72.

HAB. Winterberg, *E. & Z.!* Windvogel-berg, Wittberg and at Moozplatz, *Drege!* (Herb. Bth., Sd.)

Taller, more woody, and stiffer and more branched than *A. stipulaceum*, with much smaller, thicker, and more densely hairy leaves, &c. Pubescence fulvous or whitish. Leaflets ½ inch long, shorter or somewhat longer than the leaf-like stipules. Flowers yellow.

12. A. stipulaceum (Eck. & Zey.! 1318); suffruticose; stem short, erect, *sub-simple*, densely silky-villous; stipules *leaf-like, broadly ovate or ovato-lanceolate*, equalling the leaflets and much longer than the very short petiole; leaflets elliptic-oblong or cuneate, *thinly or densely silky-villous*, indistinctly veiny; peduncles *short, subumbellately* 1–5 *flowered;* lower lip of the *silky* calyx 3-toothed; vexillum densely silky. *Benth.! l. c. p.* 343. *Chasmone verticillata, E. Mey! Comm. p.* 72.

HAB. Winterberg, near Philipstown, *E. & Z.! Mrs. F. W. Barber.* Katberg and between the Kei and Basch, *Drege!* (Herb. Bth., Hk., D., Sd.)

Stems 3–6 inches high. Stipules uncial, 4–5 lines wide, broader than the leaflets. The pubescence is sometimes very copious and silky, sometimes rather scant on the upper surface of the leaves; in the former case the venation is in a great measure concealed. Flowers dull yellow.

13. A. pilosum (Harv.); suffruticose; stem short, erect, sub-simple, patently *albo-pilose;* stipules leaf-like, broadly ovate, equalling or exceeding the elliptic-oblong, *albo-pilose, membranaceous, green*, veiny leaflets; peduncles short, shortly sub-racemose, 7–9 flowered; lower lip of the *green* albo-pilose calyx 3-toothed; vexillum silky.

HAB. Trans Kei Country, *H. Bowker, Esq.* (106). (Herb. D.)

Very like *A. stipulaceum*, except in pubescence. In this the surface is *glabrous*, except for *scattered, long, white, patent hairs;* in *A. stipulaceum* there is a more or less dense undercoat of appressed, short, silky hairs. The flowers are more numerous and smaller.

14. A. velutinum (E. & Z.! 1313); erect, branched, everywhere densely silky-tomentose; stipules *ovate or ovato-lanceolate, much shorter than the petiole;* leaflets obovate, oblong, or oblongo-lanceolate, obtuse or acute, thick; flowers in pairs or solitary, sub-sessile; lower lip of the calyx tridentate; vexillum villous; legume densely silky-villous. *Benth. l. c. p.* 344. *Zeyher*, 386, 387.

HAB. Sides of Table Mountain, Capetown, *Mundt.* Near Simonstown, *E. & Z.!* Paalen, Brandenberg, and Doornhoogde. *Zeyher!* (Herb. Hk., Bth., Sd.)

Stems 1–1½ feet high, branched above, flexuous. Leaves scattered; petioles ¾–1 inch long, patent; leaflets as long, 2–3 lines wide, thick and veinless. Pubescence pallid, the hairs long and close. Stipules 2 lines long. Pods 1–1½ inches long, densely, but softly hairy.

15. A. connatum (Harv.); shrubby, branched, densely silky-canescent; stipules *deltoid, connato-perfoliate;* petiole very short; leaflets short, narrow-lanceolate, concave, silky on both sides; flowers in pairs, sub-sessile, terminal; lower lip of the calyx three toothed; vexillum and carina villous.

HAB. S. Africa, *Dr. Thom!* (Herb. Hooker.)

I venture to found a species on a single specimen preserved in Herb. Hooker, relying on the very remarkable character of the stipulation. The stipules are broad based, deltoid, clasping the stem, and connate at the side opposite the insertion of the leaf, so as to be completely "perfoliate:" the compound stipule is sharply bifid. In other respects this plant resembles narrow leaved forms of *A. collinum,* to which it was doubtfully referred by Bentham.

16. A. collinum (E. & Z. 1311); suffruticose; stems short, sub-simple or branched, densely silky or silky-villous; stipules *narrow-lanceolate or subulate, small;* petiole *very short;* leaflets obovate-apiculate or lanceolate, silky on one or both sides; flowers solitary or in pairs, sub-sessile, the lower ones sometimes pedunculate; lower lip of the silky calyx sharply trifid; vexillum densely silky-villous; legume appressedly silky, erect. *Benth.! l. c. p.* 343. *Chasmone apiculata, E. Mey.! Comm. p.* 73.

VAR. *α.* **vestitum**; leaflets obovate, densely silky on both sides.

VAR. *β.* **seminudum**; leaflets obovate, glabrous above. *Zey.!* 2301, 2302 *pte.,* 2306. *Argyr. obovatum, E. & Z.!* 1307. *Dichilus obovatus, E. Mey.! Linn.* 7. *p.* 154.

VAR. *γ.* **angustatum**; leaflets narrow lanceolate, silky on one or both sides. *A. argentum, E. & Z.* 1303.

HAB. Uitenhage district in many places, *E. & Z.! Drege! &c. γ.* near Gauritz R., Swell.. *E. & Z.!* (Herb. Hk., Bth., D., Sd.)

A variable plant, from 2–12–14 inches high, sometimes copiously branched. Leaflets ½–¾ inch long, 2–4 lines wide, mostly much longer than the petiole. Pods erect, 1½ inch long, densely silky.

17. A. petiolare (Walp. Rep. 1. p. 632); rigidly branched, appressedly silky; stipules *very minute;* leaflets obovate-oblong, *much shorter than the rigid petiole;* peduncles one flowered, shorter than the petiole; the lower lip of the silky, bipartite calyx three-toothed; vexillum pilose. *Benth.! l. c. p.* 344. *Chasmone petiolaris, E. Mey.! Comm. p.* 75. *Crotalaria argentea, Jacq. Hort. Schoenb. t.* 220.

HAB. Hills near Uitkomst, Clanw., *Drege!* (Herb. Bth., Hk.)

Erect, 1–1½ foot high, dichotomously much branched, bushy, remarkably rigid. Petioles 1–1½ inch long, spreading; leaflets about ⅓ inch long. Pubescence close pressed, shining and fulvous. Stipules obsolete, tooth-like. Calyx parted nearly to the base.

18. A. uniflorum (Harv.); slender, erect, sub-simple, *glabrescent* or thinly appresso-puberulent; stipules small, subulate; petioles shorter than the linear, acute, infolded leaflets; flowers solitary (rarely in pairs) opposite the leaf, sub-sessile; bracts setaceous; lower lip of the thinly pubescent calyx trifid; vexillum silky; legume sparsely appresso-pubescent. *Zey.!* 387. *A filiforme β. uniflorum, Harv. in Hb. Hook.*

HAB. Berg River and on the Vanstaadensberg Mts., *Zeyher!* Albany, *T. Williamson! Mrs. F. W. Barber!* (Herb. T.C.D., Sd., Hk.)

Closely allied on the one hand to *A. tuberosum*, and on the other to *A. filiforme*, from both which it differs in inflorescence, and is more glabrous than either. It appears to be not uncommon in Albany, and constant to the differences noted. Stems 10–18 inches long. Leaves few and distant. Flowers smaller and much less hairy than in *A. filiforme.*

19. A. filiforme (E. & Z. ! 1315); *slender*, erect, *sub-simple*, thinly silky-canescent; stipules small, subulate; petioles shorter than the linear, acute, infolded leaflets; *peduncles shorter than the leaf*, umbellately 2–4 flowered; bracts setaceous; lower lip of the silky-villous calyx trifid; vexillum and *legumes densely silky. Benth.! l. c. p.* 345. *Galega filiformis, Thunb.! Cap. p.* 600. *Chasmone angustissima, E. Mey.! Comm. p.* 75.

HAB. Hills round Capetown, *Pappe! W.H.H., &c.* Doornhoogde, *E. & Z.!* Hott. Holl. Mounts., *Thunberg!* Paarl, *Drege!* (Herb. Th., D., Hk., Bth.)

Root thickened, fusiform, 2–3 inches long, 2–3 lines in diameter. Stems 6–8 inches high, filiform. Petioles ¼–½ inch long; leaflets 1½ inch long, and 2 lines wide. Peduncles from ¼ inch to 1 inch long; flowers very hairy. Pubescence pallid.

20. A. tenue (Walp. Rep. 1. p. 632); *suffruticose, much branched,* thinly silky; stipules minute or obsolete; petioles very short, leaflets linear, apiculate, convolute, the lower ones cuneate-oblong; peduncles *shorter than the leaf,* 1–2 flowered; calyx thinly appressedly-silky, the lower lip shortly 3-toothed; vexillum and legume *glabrescent or puberulous. Benth.! l. c. p.* 345. *Chasmone tenuis, E. Mey.? Comm. p.* 75.

HAB. Cape Colony, *Bowie!* (Herb. Hooker.)

Rather ligneous, somewhat corymbosely branched; branches erect, slender. Petioles 1–2 lines long; leaflets 6–7 lines long, with hooked points, pale green, puberulous. Young legume silky; the ripe one minutely pubescent.

21. A. rupestre (Walp. Linn. 13. p. 508); diffuse, slender, branching, silky-villous; stipules *small*, deltoid-subulate; leaves on longish petioles; leaflets obovate or oblong, apiculate, densely silky on one or both sides; peduncles *much longer than the leaf*, umbellately 2–3 flowered, or one flowered; the lower lip of the silky calyx trifid; vexillum and legume silky. *Benth.! l. c. p.* 345. *E. Mey.! Comm. p.* 74.

HAB. Rocky places. Stormberg, *Drege!* Graaf Reynet, *Mrs. F. W. Barber* (Herb. Bth., D., Hk.)

Stems spreading, 12–14 inches long, alternately divided. Pubescence copious on the ramuli, under surfaces of leaves and calyx. Leaflets ¼–½ inch long, or somewhat longer, 2–3 lines wide, commonly oblong or elliptical. Peduncle 1½–3 inches long, rarely abbreviated.

22 A. molle (E. & Z. ! 1319); dwarf, slender, erect, *sub-simple, thinly covered with long, patent, silky hairs;* stipules linear or subulate, *longer than the short petiole;* leaflets broadly obovate-oblong, *netted with veins;* peduncles much longer than the leaves, umbellately 3–5 flowered; lower lip of the calyx deeply trifid; vexillum silky. *Benth.! l. c. p.* 347. *Chasmone venosa, E. Mey.! Comm. p.* 73 (excl. var. β.)

HAB. Albany, *Drege!* On the Winterberg, *Eck. & Zey.!* Port Natal, *Gueinzius.* (Herb. Bth., D., Sd.)

Stems 4–5 inches high. Petioles 1–2 lines long; leaflets ½–1 inch long, 5–7 lines wide, strongly veined. The pubescence on *E. & Z.'s* specimens is fulvous, and much more patent and less silky than on *Drege's*. The peduncles are almost shaggy, 3–4 inches long.

23. A. pauciflorum (E. & Z.! 1314); small, slender, sub-erect or diffuse, sub-simple, *densely silky and villous, with long, soft hairs;* stipules linear-lanceolate or subulate, sub-falcate; leaflets either obovate, oblong or lanceolate, silky and pilose; peduncles 1–2 flowered, longer than the leaf, the lower sometimes short; upper lip of the calyx bipartite, lower trifid; vexillum and legume silky. *Argyr. strictum, Steud. Benth.! l. c. p.* 346. *Chasmone stricta, E. Mey.! Comm. p.* 75.

VAR. β. **semiglabrum**; leaves glabrous on the upper surface. *A. biflorum, E. & Z. No.* 1317.

HAB. Among Acacias, on the Stormberge Mts., near the source of the Key River, *E. & Z.!* Near Grahamstown and on the Caledon and Mooje R., *Burke and Zey.!* Between Klipplaat and Black Key River, *Drege, Burchell,* 5035. Queenstown district, common, *Mrs. F. W. Barber.!* Var. β., Winterberg, *E. & Z.!* (Herb. Hk., Sd., D.)

3–6 inches high, branched from the base. Pubescence copious, long, somewhat fulvous and shining. Leaflets very variable in length and breadth, ½–1½ inch long, 2–4 lines wide. Peduncles long or short, rarely 2 flowered. Calyx nearly equalling the petioles; the upper lobes lanceolate.

24. A. barbatum (Walp. Rep. 2. p. 845); dwarf, densely leafy, the *branches, stipules, petioles, leaflets, calyces and legumes bearded and ciliated with long white, patent hairs;* stipules ovato-lanceolate, *connato-amplexicaul, persistent;* leaflets obovate-oblong, acute or obtuse, longer than the petiole; peduncles sub-terminal, shorter than the leaf, or a little longer, umbellately 2–3 flowered; bracts lanceolate; upper lip of the calyx bipartite, lower trifid; vexillum silky. *Benth. l. c. p.* 345. *Chasmone barbata, Meisn. Lond. Journ.* 2. *p.* 77. *Zeyher!* 2305.

HAB. Near the Koega River, Uit., *Krauss.* Near Salem, and at Zoutpanshoogte, *Zeyher!* (Herb. Sd.)

Very small, but thick and woody. Stems 3–5 inches long, diffuse or prostrate, not much branched, the branches short and close together, densely covered with leaves. Except for the long white scattered hairs, the surface is glabrous. Leaflets ½ inch long, 2–2½ lines wide. Stipules clasping the stem and connate at the side opposite the leaf. Calyx nearly as long as the petals. I have not seen *Meisner's* plant, but *Zeyher's* specimens, here described, agree well with his diagnosis. It is a very well marked species.

25. A. pumilum (E. & Z.! No. 1308); *dwarf, densely much branched,* silky-villous; stipules small, from a broadish base *subulate;* petioles shorter than the *obovate or obovate-oblong leaflets;* peduncles shorter or somewhat longer than the leaf, 1–2 flowered (the lower flowers sometimes sub-sessile); lower lip of the silky calyx 3 toothed; vexillum and legumes silky-villous. *Benth.! l. c. p.* 346. *Arg. venustum, E. & Z!* 1310. *Chasmone argentea, β. pilosa, E. Mey. Comm. p.* 75. *and C. pusilla,* 600. *Galega sericea, Thunb! Fl. Cap.* 601.

VAR. α. **verum**; leaflets glabrous on the upper, silky-villous on the lower side. *Zey.* 2302, 2306.

VAR. β. **pilosum**; leaflets ovato-lanceolate, densely hairy on both sides. *Chasmone argentea, β. pilosa, E. Mey. Comm. p.* 75.

HAB. Bockeveld, *Thunberg!* District of Uitenhage in several places. Nieuweveld and Camdebo, *E. & Z.! Drege! Burchell*, 3491. (Herb. Th., Bth., Sd., Hk, D.)

Root and stems woody. Branches diffuse, 3–4 inches long. Pubescence variable in amount, sometimes rather scanty. Petioles very short; leaflets ¼–½ inch long, 2–3 lines wide, veinless, mostly blunt. Peduncles ½–1½ inch long: occasionally some or all the flowers are sub-sessile.

26. A. patens (E. & Z.! No. 1309); dwarf, slender, much branched, *thinly and appressedly silky;* stipules from a broadish base subulate; petioles much shorter than the *narrow, cuneate-oblong or lanceolate, acute, rigid, veiny, thinly-silky leaflets;* peduncles somewhat longer than the leaf, one flowered; lower lip of the silky calyx deeply 3-fid; vexillum and legume puberulent. *Benth. l. c. p.* 347. *Zey.* 2304.

HAB. Cape Colony, *Bowie!* Grassy pastures near the Zwartkops River, *E. & Z.* (Herb. Bth., D., Hk., Sd.)

Nearly allied to *A. pumilum*, but with narrower, more acute and rigid and veiny leaves, and a much more scanty pubescence. Leaflets 5–7 lines long, 1–1½ broad, glabrous on the upper side. Peduncles 1 inch long.

27. A. adscendens (Walp. Linn. 13. p. 507); slender, appressedly silky; *branches elongate, ascending;* stipules small, *ovate, acute, longer than the very short petiole;* leaflets ovate or elliptical, puberulous on both sides, silky-margined; peduncles much longer than the leaves, umbellately 1–3 flowered; lower lip of the calyx deeply trifid. *Benth. l. c. p.* 347. *Chasmone ascendens, E. Mey. Comm. p.* 73.

HAB. Caffraria, between the Omcomas and Omblas Rivers, *Drege!* (Herb. Bth.)

Branches 1–2 feet long, curved. Leaves 2–3 inches apart; petioles 1 line long; leaflets ½–¾ inch long, 3–4 lines wide, with a border of fulvous, silky hairs. Peduncles 2–3 inches long.

28. A. Andrewsiana (Steud.); *tall, flexuous, laxly-branched,* sparsely silky or glabrescent; stipules small, subulate; leaflets *ovate or elliptical, on long petioles,* silky-ciliate, and sparsely silky; racemes on long pedicels, few or many flowered; lower lip of the calyx trifid; vexillum silky; legumes glabrescent. *Benth. l. c. p.* 349. *Cytisus tomentosus, Andr. Rep. t.* 237. *Goodia? polysperma, DC. Prod.* 2. *p.* 118. *Chasmone Andrewsiana, E. Mey.! Comm. p.* 74. *Dichilus ciliatus, E. & Z.* 1301.

VAR. α. **racemosum**; raceme fasciculately many flowered.

VAR. β. **pauciflorum**; peduncles abbreviate, sub-umbellately 2–3 flowered.

VAR. γ. **helvolum**; stems, petioles and peduncles patently hispid, with yellow tawny hairs; peduncles few flowered.

HAB. Eastern Districts; extending from Uitenhage to Port Natal, *Drege! E. & Z.! Hutton, &c.* Var. γ. Natal, *Gueinzius!* (Herb. D., Bth., Hk., &c.)

Stems 2–3 feet high, angularly bent, straggling. Pubescence variable, sometimes very scanty. Petioles 1–1½ inch long; leaflets as long, or shorter, ¾–1 inch wide, thin, pale green. Not much like any of the other species.

29. A.? involucratum (Harv.); procumbent, much branched; branches, peduncles and pedicels woolly; stipules broadly ovate, acute, leafy, *connate-perfoliolate and adnate to the very short petiole,* glabrous; leaflets (small) ovato-oblong, acute, complicate, glabrous; peduncles terminal, longer than the leaves, umbellately 4–5 flowered; umbels subtended by a leaflike (*connato-stipulate and trifoliolate*) *bract;* pedicels equalling the calyx; calyx villous, tubular, its upper lip bipartite, lower deeply trifid; vexillum and carina silky. *Psoralea involucrata, Thunb.! Fl. Cap. p.* 607.

HAB. South Africa, *Thunberg!* (Herb. Upsal).

Diffuse or procumbent; stems woody at base, flexuous, much branched, robust, terete, densely fulvo-tomentose. Leaves close together; petioles 1–3 lines long, altogether adnate to the broad, leafy stipules which unite round the stem into a perfoliolate stipule. Leaflets 3–4 lines long, 2 lines wide, smaller than the stipules. Peduncles 1–2 inches long, ending the branches, crowned with a floral leaf or bract exactly similar to the ordinary leaves, and similarly stipulate. This floral leaf forms an involucre to the umbellate flowers. Pedicels ½–¾ inch long. Calyx 4 lines long, the tube nearly twice as long as the limb. Carina nearly equalling the vexillum Ovary, silky, linear, 6–ovuled. The calyx tube is longer than in other species; somewhat similar stipules are found in *A. barbatum* and *A. connatum.*

SECT. 2. **Eremolobium.** *Legume* strongly compressed, subtorulose within, between the seeds, transversely septate. *(The species of this section are chiefly Asiatic or European).*

30. A. lanceolatum (Eck. & Zey.! 1316); suffruticose, decumbent, silky and silvery; stipules oblique, falcato-lanceolate or ovate, much shorter than the petioles; leaflets of the lower leaves obovate, elliptical or oblong, of the upper lanceolate, nearly or quite glabrous on the upper, silky and silvery on the lower side and at the margin; peduncles elongate, 1–5 flowered; calyx and vexillum densely silky; legume *furrowed between the seeds. Benth.! l. c. p.* 349. *Ononis sericea, Thunb.! Crotalaria lunaris, Linn. Dichilus lanceolatus, E. Mey.! Linn.* 7. *p.* 154. *Chasmone lanceolata, E. Mey.! Com. p.* 75. *Zey.* 388, 389.

HAB. Mountains of the Cape District, &c., common. (Herb. D., Hk., Th., &c.)

Root woody; stems many from the crown, decumbent, 2 feet long or more, slender, not much branched; the branches ascending. Leaflets of the lowest leaves short, blunt and broad; of the upper narrow, and acute at each end. Peduncles 4-6 inches long. Legume 2–2½ inches long, 1½ line wide, somewhat falcate.

(Doubtful Species.)

A. umbellatum (Vogel, MSS.); leaves trifoliolate, petiolate; leaflets oval-subrotund, acuminate, silky tomentose on both sides, about twice as long as the petiole; flowers umbellate, terminal. *Walp. in Linnæa,* 13. *p.* 509.

A. splendens (Walp.); suffruticose, densely much branched, ascending, branches silky villous; leaves densely set, on short petioles; leaflets coriaceous, oblong, acute, with revolute margins, midribbed beneath, the young densely silky on both sides, the adult becoming *glabrate* and shining above; stipules obliquely ovate, acute, about equalling the petioles; peduncles terminal, elongate, umbellately 2–4 flowered; calyx silky-villous, little shorter than the corolla, the upper lip bipartite, the lower longer, trifid; legumes lanceolate, silky. *Meisn.—Meisn. in Hook. Lond. Journ.* 2. *p.* 78.

HAB. Mountain sides near "Hemel and Aarde," Swell., *Krauss!* 927. (Hb. D.)

Of this plant I have seen neither perfect flowers nor legumes, and am uncertain where to place it. *Meisner* regards it as nearest to *A. lanceolatum,* of which it has the stipules and in some respects the foliage; but it is more robust, more densely branched, with much more frequent leaves, shorter petioles, and revolute margins, &c. Still *A. lanceolatum* varies considerably, and after all, this may be one of its extreme forms.

XVII. **DICHILUS**, DC.

Calyx as in *Argyrolobium*. *Vexillum* oblong, shorter than the obtuse carina. *Stamens* monadelphous. *Ovary* many-ovuled; stigma minute. *Legume* linear, compressed, sub-torulose. *Endl. Gen.* 6480. *Benth.! in Hook. Lond. Journ. vol.* 3. *p.* 353. *Calycotome, E. Mey. Melinospermum, Walp.*

A few S. African suffrutices, having the habit of some *Lebeckiæ*, and the calyx, but not the corolla of *Argyrolobuim*, constitute this genus. The name is compounded of δις, *twice*, and χειλος, *a lip*; alluding to the two-lipped calyx.

TABLE OF THE SPECIES.

Erect, virgate: calyx *much shorter* than the alæ, with *short, ovate teeth*	(1) **strictus.**
Erect, virgate: calyx nearly equalling the alæ; with lanceolate, acuminate teeth	(2) **lebeckioides.**
Diffuse or trailing, slender	(3) **gracilis.**

1. D. strictus (E. Mey.! Comm. p. 36); erect, virgate, all the leaves alternate; leaflets linear-oblong; racemules 2–3 flowered, terminating short ramuli, sub-paniculate; calyx *scarcely cleft to the middle, half as long as the alæ, the teeth short, ovate. Benth.! l. c. p.* 353.

HAB. On the Witbergen, *Drege!* (Herb. Bth., D.)

A slender, fruticose, many-stemmed plant; stems 12–15 inches high, simple, ramulose toward the top, appressedly pubescent. Leaflets minutely and appressedly puberulent, 4–5 lines long, 1–2 lines wide, sub-acute, tapering at base. Stipules none. Racemules on sub-uncial ramuli, crowded round the ends of the branches. Flowers yellow. Carina prominent and very blunt.

2. D. lebeckioides (DC. Leg. Mem. p. 202. t. 35); erect, virgate, all the leaves alternate; leaflets linear-oblong or linear; racemules 2–3 flowered, teminating short ramuli, sub-paniculate; calyx *cleft beyond the middle, not much shorter than the alæ, the teeth lanceolate, acuminate. DC. Prod.* 2. *p.* 136. *Benth.! l. c.*

HAB. S. Africa, *Burchell,* 2614. Macallisberg, *Burke!* (Herb. Bth., D.)

Only to be known from *D. strictus*, by its deeply parted calyx, with longer and more acuminate lobes. It is questionable whether this character be constant.

3. D. gracilis (E. & Z.! 1300); *slender, diffuse, or procumbent,* branches spreading; floral leaves mostly opposite; leaflets obovate; peduncles short, filiform, 1–2 flowered, terminal or spuriously lateral; calyx deeply bilabiate, the teeth lanceolate, acuminate. *Benth.! l. c. p.* 354. *D. patens, E. Mey.! Comm. p.* 36. *Indigofera sarmentosa, Herb. Holm.!*

VAR. β. **pusillus**; leaflets narrower; flowers rather smaller. *Calycotome pusilla, E. Mey.! Comm. p.* 113. *excl. syn. Thunb. Dichilus pusillus, Benth.! l. c.*

HAB. Chumieberg and near Fort Beaufort, Kat River settlement, *E. & Z.!* Fish River, *Drege!* Albany, *Mrs. F. W. Barber!* (Herb. Holm., Bth., Hk., D., Sd.)

Stems 1–2 feet long, ascending, diffusely branched. Leaves sub-distant, spreading; leaflets 2–3 lines long, 1–2 lines wide, palish green, appressedly and minutely pubescent. Peduncles very slender, jointed and bracteolate above the middle, terminal between a pair of leaves, or in the forkings of the branches. Flowers pale yellow, nodding. Our var. β., retained as a species by Bentham, appears to me to be a very trifling variety.

XVIII. **MELOLOBIUM**, E. & Z.

Calyx tubular, shortly or deeply bilabiate, the upper lip bipartite, the lower trifid or tridentate. *Corolla* not much longer than the calyx;

vexillum oblong; carina blunt. *Stamens* monadelphous. *Ovary* several ovuled. *Legume* linear, compressed, torulose, glandular or hairy. *Benth.! in Lond. Journ. Bot.* 3. *p.* 350.

South African shrubs or suffruticose plants, more or less viscid and glandular, variously pubescent. Leaves petiolate and stipulate, 3-foliolate. Flowers yellow, in terminal spikes or spicate racemes, with 3 bracts under each flower. Name from μελος, a *joint*, and λοβιον, a *legume;* the legumes are constricted, as if jointed, between the seeds.

TABLE OF THE SPECIES.

* Armed with spines, divaricately much-branched. Spicate-racemes few-flowered.
  Bracts longer than the cal.-tube. Calyx equalling the corolla (1) **calycinum.**
  Bracts short. Calyx shorter than the corolla.
    Viscidulous, but scarcely *glandular*; twigs canescent.
      Legume straight ... ... (2) **candicans.**
      Legume falcate-curved ... ... (3) **canescens.**
    Glandular, scabrous, and pubescent; not canescent (4) **microphyllum.**
* Unarmed, suffruticose. Spicate-racemes or spikes many-flowered:
  Spicate-racemes *lax,* flowers not imbricating:
    Leaflets linear-cuneate, or obovate-oblong, much narrower than long:
      Rough in all parts with stalked *glands:*
        Branches and twigs nearly without hairiness (5) **adenodes.**
        Branches and twigs silky, *subcanescent* ... (6) **humile.**
      Without *prominent* glands, but more or less viscid:
        Nearly glabrous; calyx deeply parted, lower lip trifid ... ... (7) **exudans.**
        Thinly pubescent; calyx shortly bilabiate, lower lip subentire ... ... (8) **alpinum.**
        Densely and softly pubescent; calyx deeply parted, lower lip trifid ... ... (9) **cernuum.**
    Leaflets oblong-obcordate, *sharply emarginate,* glandular and pubescent ... ... (10) **obcordatum.**
  Spikes very dense, cylindrical, the flowers imbricating; stipules leafy ... ... (11) **stipulatum.**

1. **M. calycinum** (Benth.! in Hook. Lond. Journ. 3. p. 350); divaricately much-branched, scarcely glandular, spiny; twigs *hairy-pubescent;* leaflets glabrous or nearly so; stipules semi-cordate, equalling the petiole; bracts ovate-acuminate, *longer than the calyx-tube;* calyx villous, *equalling the corolla,* its segments ovato-lanceolate, acute; legume villous. *Zey.!* 394.

HAB. Sand River, Betchuanaland, *Burke & Zeyher!* (Herb. Hk., Bth., Sd., D.)

A depressed, excessively and intricately branched bush, every where bristling with sharp, spinous ramuli, $\frac{1}{2}$–$\frac{3}{4}$ inch long, leafy at base. Leaflets linear-cuneate, folded together, green, 2–3 lines long, a line wide, minutely impresso-punctate. Flowers 1–3, sub-sessile, at the base of rigid ramuli, which are prolonged into naked spines; bracts and calyces hairy—one or two of the upper bracts barren.

2. **M. candicans** (E. & Z.! 1323); divaricately much-branched, scarcely glandular, spiny; twigs *velvetty-canescent;* leaflets glabrescent; stipules semi-cordate-ovate; bracts shorter than the calyx-tube; calyx villous or glabrescent, *shorter than the corolla,* its segments ovate; legume *straight,* 4–6 seeded, villous-hirsute. *Benth.! l. c. p.* 351. *Dichilus candicans, E. Mey.! Linn.* 7. *p.* 154. *Sphingium velutinum, E. Mey.! Comm. p.* 67. *M. squarrosum, E. & Z.!* 1325. *Zey.!* 2307, 391, 395.

HAB. Cape, *Bowie.!* Fields by the Zwartkops River, *E. & Z.!* Adow, *Zeyher.* Sneeuberg and Uitvlugt, *Drege!* Bitterfontyn, Bechuana Land, *Zeyher!* (Herb. Bth., Hk., D., Sd.)

A rigid, spreading or depressed, spiny bush, laxly or densely branched: all the younger branches canescent. Leaflets narrow-obovate, obtuse, folded together when dry, 2–3 lines long. Flowers 2–3 sub-sessile, near the base of rigid, spine-tipped ramuli. The calyces are sometimes viscidulous. Legumes ¾-inch long, sub-moniliform, pointed, nearly straight.

3. M. canescens (Benth.! l. c. p. 351); divaricately much-branched, scarcely glandular, spiny; twigs velvetty-canescent; leaflets glabrescent; stipules and bracts semi-cordate, small; legume *falcate-curved*, 4–6 seeded, silky-canescent, as well as the calyx. *Sphingium canescens, E. Mey.! Comm. p.* 67.

HAB. Rhinoster Kop, *Drege!* Gamke River, *Burke.* (Herb. Hk., Bth., Sd.)

So like *M. candicans*, that, except by its sickle-shaped pods, it can scarcely be distinguished.

4. M. microphyllum (E. & Z.! 1324); divaricately much-branched or decumbent, spiny, *glandular-sub-viscid, scabrous or pubescent-hairy;* stipules and bracts cordate-ovate, small; legume *incurved or arched,* 2–4 seeded, *glandularly scabrid,* either hispid or glabrescent. *Benth.! l. c. p.* 351. *Ononis microphylla, Thunb.! Fl. Cap. p.* 585. *Dichilus microphyllus, E. Mey.! Linn.* 7. *p.* 155. *Sphingium microphyllum, E. Mey.! Comm. p.* 67.

VAR. α. **Thunbergii**; dwarf, very spiny, glabrescent; legume slightly curved, 2–4 seeded. *M. microphyllum, Benth. l. c.*

VAR. β. **collinum**; lax, less spiny, glabrescent; legume arched, 4–6 seeded, hispid *M. collinum, E. & Z.* 1326. *Benth. l. c. Trigonella armata, Thunb.! and T. villosa, subinermis, tenuior.*

VAR. γ. **lampolobum**; legume arched, 4–6 seeded, shining. *Sphingium lampolobum, E. Mey.! Comm. p.* 66.

VAR. δ. **decumbens**; densely *pubescent-hairy* in all parts; legume sub-curved or falcate, 2–3 seeded, glandular-hispid or scabrous. *M. decumbens, Benth.! l. c. Sphingium decumbens, E. Mey. Comm. p.* 67.

HAB. Eastern and north-eastern districts, and Caffraria. *Thunb., E & Z., Drege.* β. Gauritz R., *E. & Z.* γ. Kendo, *Drege.* δ. Sneeuberg, and near Graaf Reynet, *Drege.* Somerset, *Dr. Atherstone.* Near Orange and Caledon River, *Burke.* Zooloo Country, *Miss Owen.* Namaqualand, *Wyley.* (Herb. Th., Bth., Hk., Sd., D.)

A dwarf, thorny, depressed or spreading *glandular* and viscidulous bush, varying in pubescence and number of spines, but the varieties indicated seem to run into one another, and all are from the same districts. Leaflets 1–2 lines long, cuneate or obovate. Flowers small, cernuous, pedicels 1–2 lines long.

5. M. adenodes (E. & Z.! 1327); suffruticose, *scarcely* spinous, *rough in all parts with stipitate glands, glabrescent;* leaflets linear-oblong; stipules semi-cordate-acuminate; flowers in a lax pseudo-spike, the rachis *rigid* and at length spinescent; bracts lanceolate, nearly equalling the calyx; legume *arched,* 4–5 seeded, *glandular and hispidulous. Benth. l. c. p.* 352.

HAB. Cape District, at Berg River and Zwartland, *E. & Z.* (Herb. Sd., D.)

A rigid, slightly-branched, erect, scabrous suffrutex, 6–8 inches high; the slightly divided branches and the naked, barren tips of the flowering ramuli, hardening and sub-pungent. Leaflets 3–4 lines long, 1-line wide, obtuse. Legumes falcate, 8–9 lines long. Known from *M. cernuum*, by the stalked glands that roughen stem, branches, leaflets, and legumes.

6. M. humile (E. & Z.! 1330); suffruticose, *scarcely* spinous, much-branched; branches and twigs *appressedly silky, sub-canescent;* petioles, leaflets, bracts, and calyx *rough with stipitate glands;* leaflets obovate-oblong, sub-pilose; stipules semi-cordate, acute; fl. in a lax, pseudo-spike, the rachis flexuous; bracts ovato-lanceolate, equalling the calyx tube; legume ?

HAB. Sandy places near Brackfontein, Clanw., *E. & Z.* (Herb. Sd.)

Larger, more robust and branching than *M. adenodes*, with broader leaflets and *silky-canescent* twigs. Petioles as long as the leaflets or longer. Leaflets 4–5 lines long, 2–3 wide.

7. M. exudans (Harv.); unarmed, shrubby, much branched, *viscid*, but scarcely glandular, nearly glabrous; leaflets linear-cuneate, obtuse or retuse, glabrous; stipules semi-sagittate, much shorter than the petioles; flowers in a lax pseudo-spike, the rachis rigid (and at length subspinescent ?); bracts lanceolate, longer than the calyx-tube; calyx deeply parted, the upper lobes ovato-lanceolate, the lower lip sharply trifid.

HAB. Cape, *Dr. Thom!* (Herb. Hook.)

A dwarf shrub, distinctly woody at base, and much branched, the branches and twigs suberect. All parts exude a gummy-resinous matter, but the glands are not prominent. Petioles ⅓ inch long; leaflets as long, 1 line broad, quite glabrous. Flowers suberect, nearly sessile. Legume unknown. The young twigs have a few short hairs.

8. M. alpinum (E. & Z.! 1331); suffruticose, unarmed, ascending, thinly pubescent, *viscidulous*, but not glandular; leaflets linear-oblong, mucronulate, *glabrescent* or sparsely pubescent; stipules semi-cordate, acuminate, much shorter than the petiole; flowers in a dense pseudo-spike, rachis flexuous; calyx shortly cleft, its upper lobes bluntly ovate, *lower lip sub-entire, tridenticulate;* legume ?

HAB. On the Winterberg, *E. & Z.!* Sunny spots half way up the Mt., Autumn. *Mrs. F. W. Barber.* (Herb. Sd., D.)

Much-branched from near the base, the branches 8–12 inches long, curved, simple, ending in the inflorescence. Flowers bright yellow. Much less hairy than *M. cernuum*, with a different calyx.

9. M. cernuum (E. & Z.! 1328); suffruticose, unarmed, diffuse or ascending, *loosely and softly hairy* and viscidulous; leaflets narrow-cuneate, obtuse, equalling or exceeding the petioles; stipules semi-cordate-lanceolate, flowers in a longish pseudo-spike with flexible rachis; bracts ovato-lanceolate, equalling the calyx-tube; lower lip of the calyx sharply trifid; legume straightish, curved or falcate, 5–6 seeded, *softly-hairy. Benth.! l. c p.* 352; *also M. spicatum, E. & Z.!* 1329. *Benth.! l. c. Ononis cernua, L. Cytisus Æthiopicus, L. Dichilus spicatus, E. Mey.! Linn.* 7. *p.* 154. *Sphingium spicatum, E. Mey.! Comm. p.* 66. *Trigonella hirsuta, Thunb.! Cap. p.* 611. *Zey.!* 396.

HAB. Cape and neighbouring districts, common, *E. & Z.! W.H.H., Pappe, &c.* (Herb. Th., D., Hk., Sd., Bth.)

Root deeply descending. Branches many from the crown, sub-erect or spreading, sub-simple or ramulous, densely or sparsely, but always *softly and loosely hairy*. Leaflets 2–4 lines long, 1 line wide. Spikes 2–4 inches long, curved. Flowers small, yellow. Legumes deflexed, at first nearly straight, becoming more curved as they ripen.

10. M. obcordatum (Harv.); suffruticose, unarmed, *sprinkled with stalked glands, and pubescent;* leaflets *oblong-obcordate, sharply emargin-*

ate; stipules semi-cordate-ovate; flowers numerous in a longish-spiked raceme; bracts obliquely ovate; legume 2–3 seeded, sub-incurved, villous.

HAB. On plains in Graaf Reynet, *Mrs. F. W. Barber.* (Herb. Hk.)

Spreading, 1–1½ foot high; branches ascending, alternately divided. The whole plant is covered with stalked glands, mixed with soft hairs. Leaflets 3–4 lines long, 2 lines wide, flat, viscid. Flowers small, yellow. Spikes 3 inches long. Legumes deflexed, 4 lines long. A more branching and woody plant than *M. cernuum,* but glandular, with shorter and broader pods and distinctly obcordate leaflets.

11. **M. stipulatum** (Harv.); robust, suffruticose, unarmed, densely covered with *sub-sessile glands, and hirsute;* stems hirto-tomentose; leaflets linear, cuneate at base, and hairy; stipules foliaceous, semi-hastate, nearly equalling the petiole; flowers *imbricated, in a short, dense, pseudo-spike;* bracts lanceolate, rather shorter than the deeply-parted, densely-glandular calyx; legumes lanceolate, erect, 2–3 seeded, glandular and pubescent. *Ononis stipulata, Thunb.! Cap p.* 585.

HAB. Hex River, *Thunberg!* (Herb. Th., Holm.)

Stems strong and woody at base, erect or curved, 12–14 inches high, densely leafy, branched chiefly near the root. Branches with short, but patent and loose pubescence. Stipules nearly half inch long, the uppermost largest, 2 lines wide. Leaflets 7–8 lines long, 1½ line wide, often infolded. Colour pale green. Spikes cylindrical, very dense. A well-marked plant, not found since Thunberg's time.

(Doubtful Species.)

M. parviflorum (Benth.!l. c. p. 351); "divaricately much-branched, spiny, scarcely glandular, sub-glabrous" (*twigs velvetty-canescent*); "stipules and bracts semi-cordate, minute; calyx glabrescent-viscidulous;" legume unknown.

HAB. Dwaka R., *Burke & Zeyher,* 392. (Herb. Bth., Hk., Sd., D.)

I cannot distinguish this satisfactorily from *M. candicans.* The half ripe legumes are straight and villous, and the twigs constantly hoary, except where the indument has been casually abraded.

M. viscidulum (Steud.); "unarmed, glandularly-pubescent, green; leaflets oblong-cuneate, spikes few-flowered; bracts equalling the calyx-tube; legumes curved, glandular-scabrid." *E. Mey. Comm. p.* 66.

HAB. Sandy hills near Ebenezer, *Drege.* (Unknown to me.)

M. canaliculatum (Benth.); "unarmed, viscoso-scabrid; leaflets linear, channelled; spikes short; bracts lanceolate, equalling the calyx-tube." *E. Mey. Comm. p.* 66.

HAB. Karroo, 2000–2500 f., *Drege.* (Unknown to me.)

XIX. HYPOCALYPTUS, Thunb.

Calyx widely campanulate, shortly 5-toothed, hollowed at base. *Vexillum* roundish, reflexed, longer than the alæ and carina. *Stamens* 10, monadelphous. *Ovary* lanceolate, many ovuled. *Legume* linear, flat, the upper suture thickened, many-seeded. *Benth.! in Hook. Ld. Journ.* 3. *p.* 354. *Endl. Gen. No.* 6477.

A glabrous, densely leafy shrub, with palmately trifoliolate, stipulate leaves, and purple flowers. Name from *υπο, under,* and *καλυπτω, to veil;* not applicable to the genus as now limited. Thunberg included under this name the species of *Podalyria,* which have large bracts under which the young flowers are hidden or veiled.

1. H. obcordatus (Thunb. Prod. 124); *Thunb. Fl. Cap. p.* 570. *DC. Prod.* 2. *p.* 135. *Deless. Ic.* 3. *t.* 63. *Bot. Mag. t.* 3894. *Maund. Bot.* 4. *t.* 198. *Eck. & Z.!* 1255. *Benth. Lond. Journ.* 3. *p.* 354. *Crotalaria cordifolia, Linn. Mant. p.* 266. *Spartium sophoroides, Berg. Cap.* 198.

HAB. Cape, *Thunberg!* Kochman's Kloof, *Mundt!* Puspasvalley and Kochman's Kloof, Swell., *E. & Z.!* Cederberg and Dutoit's Kloof, *Drege!* Schurfdeberg, *Pappe!* (Herb. Th., Bth., Hk., Sd., D.)

A large shrub or small tree, densely branched, ramulous and closely covered with leaves, glabrous, except the growing parts. Twigs angular. Leaves trifoliolate; leaflets 8–10 lines long, 6–8 lines wide, obcordate or obovate, deeply emarginate or truncate, mucronate, or sub-cuspidate, folding together when dry, reticulate with prominent ribs and veins beneath. Stipules subulate, deciduous. Racemes terminal, ovoid or oblong, densely many-flowered. Flowers reddish purple or lilac. Calyx rufescent. Legumes 1½ inch long, 1½ lines wide, linear-lanceolate, tapering at base, with a short stipe. A very handsome shrub.

XX. LODDIGESIA, Sims.

Calyx of *Hypocalyptus*. *Vexillum* much shorter than the alæ and carina. *Legume* ovato-lanceolate, acute at each end, flat, the upper suture thickened, few-seeded. *Benth. in Hook. Lond. Journ.* 3. *p.* 355. *Endl. Gen. No.* 6476.

A small glabrous suffrutex, with palmately trifoliolate leaves, and small, purple and white flowers in short, terminal racemes. The name is in honour of *Conrad Loddiges*, the celebrated nurseryman at Hackney, who died in 1820; and who, with his sons, greatly contributed to the progress of horticulture in England, during the last and present century.

L. oxalidifolia (Sims, Bot. Mag. t. 965); *DC. Prod.* 2. *p.* 136. *E. & Z.!* 1256. *Benth. in Lond. Journ.* 3. *p.* 355.

HAB. Among shrubs, on mountain sides. Zwarteberg and Klynriviersberge, *E. & Z.! Pappe!* Near George and Gnadendahl, *Drege! Alexander Prior, &c.* (Herb. Bth., Hk., Sd., D.)

A slender, erect or diffuse, laxly branched, small, glabrous shrub. Twigs terete, rufescent, flexuous, patent. Leaves trifoliolate, on very slender, setaceous petioles, equalling the laminæ or longer. Leaflets obcordate or roundish-obovate, deflexed, mucronulate, netted-veined. Racemes terminal, few or several flowered, lax, lengthening. Bracts minute. Pedicels nodding, bibracteolate above the middle. Flowers 3–5 lines long, the vexillum and alæ white; the keel dark purple at the point. The vexillum varies much in size in different specimens, sometimes scarcely exceeding the calyx, sometimes 3–4 times as long, but it is always shorter than the other petals. Legume stipitate, ovato-lanceolate, compressed, pointed at each end.

XXI. LEBECKIA, Thunb.

Calyx obliquely-campanulate, shortly 5-toothed with rounded interspaces, rarely 5-cleft. *Carina* obtuse or sub-rostrate, longer than the alæ, and usually than the vexillum. *Stamens* monadelphous. *Ovary* linear, sessile or stipitate, many ovuled. *Legume* linear, either flat, subcompressed, terete or turgid. *Benth. in Hook. Lond. Journ.* 3. *p.* 355. *Endl. Gen. No.* 6478. *Stiza, Lebeckia, and Sarcophyllum, E. Mey. Acanthobotrya, Calobota and Lebeckia, E. & Z.*

Small shrubs (often spiny) or suffrutices, very diverse in habit. Leaves either unifoliolate or palmately 3-foliolate, glabrous or silky, without stipules. Flowers racemose, yellow. Name in memory of *Lebeck*, an obscure botanist.

ANALYSIS OF THE SPECIES.

1. Stiza. *Carina* longer than the vexillum. *Legume* flat. Rigid, much branched, *spiny shrubs*. *Leaves* few, flat, unifoliolate. *Flowering branches* (peduncles) ending in a rigid, spiny, naked point. (Sp. 1–3.)

Branches and twigs thinly tomentose:
Legume quite glabrous; calyx-teeth *cuspidate* ... ... (1) **macrantha.**
Legume albo-tomentose; calyx-teeth *blunt* ... ... (2) **pungens.**
Branches and twigs glabrous; legume glabrous .. ... (3) **psiloloba.**

2. Phyllodiastrum. *Carina* acute or sub-rostrate, shorter than the vexillum. *Legume* flat. Glabrous and sub-glaucous, unarmed suffrutices or herbs, with *filiform* leaves. Racemes terminal, many-flowered. (Sp. 4–7.)

Calyx-teeth deltoid, shorter than the tube:
Legume sub-sessile, *linear-falcate* ... ... (4) **Plunkenetiana.**
Legume shortly stipitate, *broadly oblong*, straight ... (5) **Meyeriana.**
Legume *on a long stipe*, *linear*, straight or curved ... (6) **Candolleana.**
Calyx-teeth lanceolate-acuminate, longer than the tube ... (7) **grandiflora.**

3. Eu-Lebeckia. *Carina* acute or sub-rostrate, equalling or exceeding the vexillum and alæ. *Legume* narrow-linear, terete or sub-terete. Glabrous and glaucous suffrutices, with *filiform* leaves. Racemes terminal. (Sp. 8–11.)

Calyx-teeth subulate, longer than the tube ... ... (8) **pauciflora.**
Calyx-teeth deltoid, shorter than the tube:
Legumes deflexed; *rachis* furrowed:
Flowers 5–6 lines long; stems tall, sub-erect ... (9) **Simsiana.**
Flowers 2–2½ lines long; stems diffuse ... ... (10) **sepiaria.**
Legumes ascending-erect; *rachis* smooth ... ... (11) **ambigua.**

4. Calobota. *Carina* obtuse, longer than the alæ, equalling or exceeding the vexillum. *Legume* terete or turgid. *Suffrutices* or *shrubs*, variably *pubescent*, or canescent. *Leaves* trifoliolate, rarely simple, never filiform. *Flowers* racemose. (Sp. 12–22.)

Leaves simple (unifoliolate):
Canescent; racemes laxly few-flowered ... ... (12) **linearifolia.**
Appressedly pubescent; racemes elongate ... ... (13) **subnuda.**
Leaves trifoliolate:
Twigs not spiny; *petals* glabrous:
Leaflets oblong or sub-linear; young parts pubescent ... ... (14) **mucronata.**
Leaflets very narrow-linear; whole plant glabrous (15) **leptophylla.**
Twigs not spiny; vexillum and keel more or less silky
Calyx and legume glabrous ... ... (17) **cytisoides.**
Calyx *silky* (or glabrescent); legume silky or hairy:
Petioles short; racemes few flowered ... ... (16) **cinerea.**
Petioles long; racemes many-flowered:
Flowers 4–5 lines long; calyx-teeth shortly subulate ... ... (18) **sericea.**
Flowers 3 lines long; leaflets *narrow*; cal.-teeth acuminate ... ... (19) **multiflora.**
Twigs *spinous-pointed*:
Petiole short; calyx-teeth blunt; legume hairy ... (20) **microphylla.**
Petiole longish; calyx-teeth acute; legume glabrescent ... ... (21) **spinescens.**

5. Vibergioides. *Carina* sub-rostrate, scarcely longer than the alæ. Legume terete or turgid. Rigid, unarmed shrubs. Leaves *sub-sessile* or shortly petiolate, trifoliolate. (Sp. 23–25.)

Whole plant nearly glabrous; calyx-teeth short ... ... (22) **humilis.**
Twigs silky pubescent; leaflets sub-glabrous ... ... (23) **sessilifolia.**
Twigs silky pubescent; leaflets densely cano-pubescent ... (24) **Bowieana.**

I. **STIZA,** (Sp. 1–3.)

1. L. macrantha (Harv.); divaricately much branched, spiny; twigs thinly *tomentose-canescent;* leaves few and distant, unifoliolate, obovate or oblong; calyx ample, puberulous, *the tube from a turbinate base, campanulate, the teeth deltoid-cuspidate;* legume *quite glabrous.*

HAB. Zooloo Country, *Miss Owen!* (Herb. T.C.D.)

A rigid, nearly naked, spiny shrub with the habit of *L. pungens* and *L. psiloloba*, but with much larger flowers than either, and a very different calyx. Leaves subsessile, 4–5 lines long, 2 lines wide, puberulous, thick, sub-coriaceous. Flowers racemose, on rigidly pungent, robust rachides; pedicels bibracteolate near the base, shorter than the calyx. Calyx sub-contracted and rugulose at base, then widened; its teeth sub-equal, deltoid, suddenly tapering into a subulate acumination. Corolla 7–8 lines long; the keel bluntly rostrate, longer than the erect vexillum. Legume (half grown) shortly stipitate, flat, several seeded, perfectly glabrous. The flowers are nearly as large as in *L. cytisoides.*

2. L. pungens (Thunb. Cap. p. 561); divaricately much branched, spiny; *twigs thinly tomentoso-canescent;* leaves few and distant, unifoliolate, obovate or oblong; calyx-teeth *very short* and broad; *legume albo-tomentose. Benth.! l. c. p. 356. Stiza erioloba, E. Mey.! Comm. p. 31.*

HAB. Near Olifant's River and in Cannaland, *Thunberg!* Near Klaarstroom, at the foot of the Great Zwarteberge, *Drege!* (Herb. Thb. Bth.)

A very rigid, nearly leafless, much branched, densely ramulose bush; the ramuli 1–2 inches long, patent, very pungent, all the younger parts thinly tomentulose. Known from the following, which is much commoner, by its pubescence and tomentose legumes.

3. L. psiloloba (Walp.! Linn. 13. p. 478); divaricately much branched, spiny; *glabrous;* leaves few and distant, mostly unifoliolate, obovate or oblong; calyx-teeth *very short* and broad; legume membranaceous, *quite glabrous. Benth. l. c. p. 356. Stiza psiloloba, E. Mey! Comm. p. 32. Acanthobotrya pungens, E. & Z.* 1340. *Spartium cuspidosum, Burch. vol.* 1. *p.* 348. *Genista cuspidosa, DC. Prod.* 2. *p.* 147.

HAB. Near Uitenhage, *E. & Z.! Alexander-Prior! &c.* (Herb. Bth., D., Sd., Hk.)

A rigid, spiny, nearly leafless bush, closely resembling the preceding, but with nearly glabrous twigs, and perfectly glabrous legumes. Flowers yellow, nodding, laxly racemulose, on spine-tipped branchlets. Legumes oblong, flat, 8–14 lines long, 2 lines wide.

2. PHYLLODIASTRUM. (Sp. 4–7.)

4. L. Plukenetiana (E. Mey.! Comm. p. 33, excl. syn. Lam. & Willd.); glabrous, decumbent or ascending, suffruticose; leaves filiform, secund, patent, the older deflexed, acute, continuous or obscurely articulate; racemes laxly many-flowered; legume *sub-sessile, linear, falcate,* flat, the ventral suture margined, valves membranous. *Benth. l. c. p. 356.*

Var. *β.* **brachycarpa**; leaves ¾ inch long; racemes few-flowered; legume shortly linear, sessile, 6–7 lines long, nearly straight. *Zey. No.* 378.

HAB. Hex River, *Drege!* Camps Bay, *Ecklon.* Sandy places at the foot of Table Mountain, *Dr. Pappe* (115), *β.* at Klipfontein, *Zey.!* (Herb. Bth., D., Sd., Hk.)

Root deeply descending. Stems several from the crown, spreading, 1–2 feet long, sub-simple or branched; branches curved, virgate. Leaves 1–1½ inch long, turned to one side, pale green. Racemes 2–4 inches, 10–20 flowered; bracts minute, pedicels shorter than the calyx. Flowers yellow, often secund. Legumes 1–1½ inch long, linear, curved, acute at each end, sessile or *minutely* stipitate. *β.* is more slender in all parts, with smaller flowers, shorter leaves, and much shorter legumes.

5. L. Meyeriana (E. & Z.! 1339); glabrous, decumbent or ascending, suffruticose; leaves filiform, secund, patent or deflexed, acute, articulate or continuous; racemes laxly several-flowered; legumes *shortly stipitate, broadly oblong, straight,* the ventral suture margined, valves membranous.

HAB. Sandy places near Constantia, and at the Waterfall, Tulbagh, *E. & Z.! Gueinzius!* Near Simon's Bay, *M'Gillivray!* (Herb. Sd., D., Hk.)

Very like *L. Plukenetiana,* but with much broader. straighter and distinctly, though shortly stipitate legumes. *E. & Z.'s "L. contaminata"* partly belongs to this species. Bracts deciduous, equalling the pedicel. Flowers 3–4 lines long. Calyx-teeth triangular, shorter than the tube. Legume an inch long, 4 lines wide, sub-obtuse at each end; the stipes 1–2 lines long.

6. L. Candolleana (Walp. Repert. 1, p. 607); glabrous, ascending, suffruticose; leaves filiform, secund, sub-erect, acute, distinctly jointed; racemes pedunculate, laxly several-flowered; calyx teeth *triangular, shorter than the tube;* legumes *on long stipites, linear,* straight or sub-falcate, flat, the ventral suture with a narrow margin, valves membranous. *L. contaminata, Benth.! non Thunb. E. & Z.* 1335, *ex parte. Sarcophyllum carnosum, E. Mey.! Comm. p.* 32. *L. pauciflora, Benth.? non E. & Z.*

HAB. Cape flats and neighbouring districts, common. (Herb. Bth., Hk., D., Sd.)
A more slender and less densely leafy plant than *L. Plukenetiana,* with smaller flowers, and readily known when in fruit, by the long stalk to the slender legume. Racemes few or several flowered, lengthening. Flowers 2–2½ lines long, pale yellow. Pods 1–1½ inch long, 1 line wide; the stipes ½ inch long or more. This appears to be *L. contaminata, DC. Prod.* 2. *p.* 136, and is certainly the plant of Bentham and partly of E. & Z.! and others; but it is not Thunberg's plant, which, according to his Herbarium, is *Aspalathus corymbosa, E. Mey.! and A. tenuifolia, DC!* The name therefore had better be suppressed altogether. It is impossible to say whether Bentham's "*L. pauciflora*" be a good species or not until the fruit be known.

7. L. grandiflora (Benth.! l. c. p. 357); glabrous, sub-erect, suffruticose, sub-simple; leaves filiform, crowded, erect, acute, distinctly jointed; racemes pedunculate, many-flowered; calyx-lobes *lanceolate, acuminate, sub-falcate, longer than the tube;* legume stipitate, linear, flat, straight or slightly curved, the ventral suture scarcely margined. *Sarcophyllum grandiflorum, E. Mey? (fide Benth. l. c.)*

HAB. Cape Colony, *Bowie!* Witsenberg, *Zeyher!* Clanwilliam, *Drege.* (Herb. Hk., Sd.)
Robust, many-stemmed; stems 6–14 inches high, simple, densely leafy below, produced above into a naked peduncle, ending in a many-flowered raceme. Leaves 2–2½ inches long. Racemes 4–8 inches long; the bracts subulate, longer than the pedicel; bracteoles equalling the calyx-tube. Flowers 5 lines long. Unripe legumes only seen.

3. EU-LEBECKIA. (Sp. 8–11.)

8. L. pauciflora (E. & Z.! 1337); glabrous, sub-erect, suffruticose, sub-simple; leaves filiform, sub-secund, erect, acute, distinctly jointed; racemes pedunculate, laxly few-flowered; calyx-teeth *subulate, longer than the tube;* legume (unripe) stipitate, very slender (terete ?).

Hab. Sandy places near the Howhoek, Caledon, and in the Langekloof, George, *Ecklon and Zeyher!* Zwarteberge, Witsenberge and Scurfdeberge, *Zeyher, No.* 377. (Herb. Sd., Bth., Hk.)

1½–2 feet high, many stemmed, unbranched. Leaves sub-distant, 3–4 inches long, jointed above the middle, rarely shorter and then without joint. Flowers 4–5 lines long, erect; the bracts and bracteoles much shorter than the pedicel. Carina equalling or exceeding the vexillum, longer than the alæ. Legume, so far as can be determined, like that of *L. Simsiana*, from which this is readily known by its long and acuminate calyx-teeth. This may be *Sarcophyllum grandiflorum*, E. Mey., which is unknown to me; but it differs from Bentham's *L. grandiflora*.

9. L. Simsiana (E. & Z.! 1338); glaucous, *sub-erect*, suffruticose, sub-simple; leaves filiform, crowded, erect, jointed; raceme elongate, many-flowered; rachis angular or furrowed; *flowers large; calyx-teeth shorter than the tube;* legumes deflexed. *L. sepiaria, Benth. l. c. p.* 357, *non Thunb. L. Sarcophylloides, E. Mey.! in Linn.* 7, *p.* 155. *L. contaminata, E. Mey.! Comm. p. p.* 34, *non Thunb. Sarcophyllum carnosum, Sims, Bot. Mag. t.* 2502, *non. Thunb. Zey.* 2294.

HAB. Cape and Stellenbosch districts, *E. & Z.! Drege! W.H.H., &c.* Winterhoek, Tulbagh; and Grootvadersbosch, Swell., *Pappe!* (Herb. Bth., Hk, D., Sd.)

The largest and handsomest of this section. Stems 12–18 inches high, robust, simple, densely leafy at base, racemose above. Raceme 6–12 inches long, dense. Flowers 5–6 lines long, nodding, bright yellow, conspicuous. Bracts and bracteoles minute, setaceous. Calyx-teeth from a broad base, acuminate. Legumes shortly stipitate, 1½ inch long, scarcely 1 line in diameter, mostly pendulous.

10. L. sepiaria (Thunb. Fl. Cap. 561); glaucous, *ascending*, suffruticose, sub-simple; leaves filiform, crowded, sub-erect, jointed; raceme elongate, densely many-flowered; rachis furrowed; *flowers small;* calyx teeth short, deltoid; legumes deflexed. *E. & Z.!* 1334, *also L. gracilis, E. & Z.!* 1336. *Benth. l. c. p.* 358. *L. ambigua, litt. c. E. Mey., fide Benth.*

HAB. West side of Lion's mountain and elsewhere, *Thunberg!* Near the Berg River, Cape District, *E. & Z.!* also in Uitenhage, *E. & Z.! Pappe!* Knysna, *Pappe!* (Herb. Th., D., Sd., Hk.)

Many stemmed, diffuse or decumbent, the ends of the branches ascending. Leaves densely set, 2–2½ inches long, jointed at about 1 inch from the base, mucronulate. Racemes very densely flowered, cylindrical, 3–4 inches long; the flowers 2–2½ lines long, bright yellow. Legumes 1–1½ inch long, not a line wide, sub-compressed, acute at each end, margined on the ventral suture. This is Thunberg's "*L. sepiaria*," according to the original specimens in Herb. Upsal.

11. L. ambigua (E. Mey.! Comm. p. 34, excl. litt. c.); glaucous, ascending, suffruticose, sub-simple; leaves filiform, *scattered*, erect, jointed; racemes elongate, *laxly several flowered;* rachis *smooth;* flowers *mediocre;* calyx-teeth triangular, acute; legumes slender, *ascending-erect*, sub-torulose. *Benth. l. c. p.* 358.

HAB. Saldanha Bay and near Uienvalley, Clanwilliam, *Drege!* (Herb. Bth., D., Hk.)

12–18 inches high, many stemmed, sub-erect. Stems curved, ending in a raceme of half their length. Leaves 3 inches long, not very numerous, jointed in the middle. Racemes 8–10 inches long, the flowers ½ inch apart, 3–3½ lines long. Legumes 1–1½ inch long, ½ line diameter, turned upwards.

4. **CALOBOTA.** (Sp. 12–21.)

12. L. linearifolia (E. Mey.! Comm, p. 33); suffruticose, *canescent*, minutely puberulous; leaves narrow-linear, acute at each end, channelled, sub-erect; racemes laxly few-flowered; calyx-teeth triangular, short; petals silky; legume sub-terete, thinly silky. *Benth. l. c. p.* 358.

HAB. On the Gariep, near Verleptpraam, *Drege! Mr. A. Wyley!* (Herb. Bth., Sd., D.)

A slender, slightly branched suffrutex, whitish in all parts, with very minute, sub-silky pubescence. Branches straight, erect, virgate, striate. Leaves few, scattered, 1–1½ inch long, ½ line wide, on minute petioles. Racemes 4–6 flowered; the flowers sub-sessile, or on pedicels shorter than the calyx, 4 lines long.

13. ? L. subnuda (DC. Prod. 2, p. 136); "leaves very few toward the end of the branches, linear, deciduous, appressedly pubescent, as well as the ramuli; flowers in a long raceme!"

Unknown to us. Referred to this place by Bentham. By Ecklon and Zeyher considered a synonym of *L. pungens*.

14. L. mucronata (Benth.! l. c. p. 359); unarmed, suffruticose, ramulous, *at first thinly adpresso-pubescent*, afterwards glabrate; twigs furrowed; leaves petiolate, trifoliate; leaflets *cuneate-oblong or sub-linear*, mucronate, glabrous or minutely puberulous underneath; racemes *short*, several flowered; calyx campanulate, puberulous, its teeth broadly triangular, acute; petals glabrous; legume sub-compressed, acute, quite glabrous. *Zey.!* 2318 (*and* 344 *in Hb. T.C.D.*)

HAB. Stony places, Vanstaadensberg, Uit., *Zeyher!* Albany, *T. Williamson*. (Herb. Hk., D., Sd.)

An erect, densely branching and ramuliferous suffrutex, woody below, 6–12 inches high or more. Branches and twigs furrowed, rufescent. Petioles ½ inch long; leaflets 4–6 lines long, 1–1½ line wide, the lateral smaller. Flowers 3 lines long, nodding, turning dark when dry. The ripe legume unknown.

15. L. leptophylla (Benth.! l. c. p. 359); unarmed, suffruticose, *glabrous*; twigs furrowed; leaves petiolate, trifoliolate; leaflets *very narrow-linear*, mucronate; racemes *elongate*, many-flowered; calyx campanulate, its teeth short, acute; legume compressed, at length sub-turgid, quite glabrous..

HAB. Subalpine places near Swellendam, *Mundt.!* 87. (Herb. Hk., Sd.)

Very nearly allied to *L. mucronata*, but more glabrous, with much narrower leaflets, longer racemes and broader and more compressed legumes. Leaflets 6–8 lines long, not ½ line wide, slightly channelled above, prominently ribbed beneath. Flowers scarcely 3 lines long.

16. L. cinerea (E. Mey.! Comm. p. 35); thinly tomentulose or subsilky-canescent, shrubby; branches *rigid*; leaves *few*, petiolate, trifoliolate, leaflets oblong-cuneate; racemes few-flowered; calyx campanulate, *silky*, its teeth broadly triangular, acute; legume linear-terete, *softly hirsute*, deflexed.

HAB. Hills near Noagas, and near Aris, on the Gariep, *Drege!* (Herb. Bth. Hk. D.)

A rigid, but not spiny bush, greyish white in all parts, with very minute pubescence. Twigs sulcato-striate. Leaflets ½ inch long or rather more, 1–2 lines wide. Flowers nodding, shortly pedicellate; the petals downy.

17. L. cytisoides (Thunb.! Prod. p. 122); silky-canescent, shrubby; twigs virgate, terete; leaves on long petioles, trifoliolate, leaflets linear-oblong, acute at base, mucronulate; racemes elongate, many-flowered; flowers *large*; calyx widely campanulate, *glabrous*, somewhat coloured, its teeth short, *triangular*, sub-acute, distant, with rounded interspaces, puberulous; vexillum and keel silky; legume terete, *glabrous*, septate within. *Benth.! l. c. p.* 360. *Thunb.! Fl. Cap. p.* 562. *Crotalaria pulchella, Andr. Bot. Rep. t.* 417. *Bot. Mag. t.* 1699. *Calobota pulchella, and C. cytisoides, E. & Z.!* 1332, 1333. *Zey.!* 2320.

HAB. Cape, Stellenbosch and Swellendam Districts, on hills, *Drege., E. & Z.! &c.* Bergriver, *Pappe* 112. (Herb. Th., Bth., Hk., D., Sd.)

A much-branched shrub, 2–4 feet high, with whitish bark, and pale grey foliage. Petioles 9–10 lines long; leaflets 10–1? lines long or more, acute at base and sometimes attenuate, obtuse or acute at point, sometimes narrow-cuneate. Racemes several inches long; flowers 8–10 lines long, bright yellow, on spreading, longish pedicels. Legumes deflexed, 1½ inch long, mucronulate, terete, 1–1¼ lines in diameter.

18. L. sericea (Thunb.! Prod. p. 122); silky-canescent, shrubby, ramulose; leaves on long petioles, trifoliolate; leaflets linear-oblong or sub-cuneate, acute at base; racemes secund, many-flowered; calyx widely campanulate, silky, its teeth *shortly subulate*, with rounded interspaces; vexillum and keel silky near the point; legume sub-terete, silky-canescent, when old glabrate. *Bth.! l. c. p.* 360. *Thunb.! Fl. Cap. p.* 562. *L. decipiens, and L. flexuosa, E. Mey.! Crotalaria angustifolia, Jacq. Hort. Schoenb. t.* 219.

HAB. Karroo, near Mieren's Kasteel; rocks on the Kwek River; Zilverfontein and Platberg, *Drege!* (Herb. Th., Hk., Bth., Sd., D.)

A smaller, more densely branched, more silky or silvery shrub than *L. cytisoides*, with shorter racemes and smaller flowers. Leaflets 1–1½ inch long, 1–2 lines wide, obtuse or acute, generally much narrowed to the base. Petioles 1½–3 inches long, generally exceeding the leaflets. Flowers 4–5 lines long, deep yellow.

19. L. multiflora (E. Mey.! Comm. p. 34); silky-canescent, shrubby, ramulous; leaves on long petioles, trifoliolate, leaflets *narrow-linear;* raceme secund, many-flowered; calyx campanulate, silky, its teeth *triangular-acuminate*, short; vexillum and keel thinly silky near the point, or glabrous; legume thinly silky or quite glabrate. *Benth.! l. c. p.* 360. *L. decutiens, L. canescens, E. Mey.! Comm. p.* 34. *L. angustifolia, E. Mey. Acanthobotrya angustifolia, E. & Z.!* 1346.

VAR. β. **glabrata**; thinly silky-canescent or *glabrate;* raceme *elongate, lax;* calyx glabrescent; legume *quite glabrous. L. decutiens, β. glabrata, E. Mey. l. c.*

HAB. Mouths of the Gariep; Olifant's River, and between Holrivier and Mieren Kasteel, *Drege.! Mr. Wyley.!* β. Karroo, between Goedman's Kraal and Kaus, *Drege!* Sands between Berg R. and Zwartland, *E. & Z.* (Herb. Bth., Hk., Sd., D.)

Very near *L. sericea*, but smaller in all parts, with much smaller flowers, and narrower leaves. Leaflets 1 inch long, ¼–½ line wide, in α. canescent, in β. nearly glabrate. Flowers about 3 lines long. Pods uncial.

20. L. microphylla (E. Mey.! Comm. p. 155); silky-canescent, shrubby; twigs rigid, at length spine-tipped; leaves on *short* petioles, trifoliolate, leaflets small, linear-obovate or oblong, folded; calyx villoso-pubescent, its teeth very short and blunt; vexillum thinly pubescent near the point; legume *hairy. Benth.! l. c. p.* 361. *Aspalathus cærulescens, E. Mey.! Comm. p.* 54. *Krebsia argentea, E. & Z.!* 1286.

HAB. Rocky situations, Camdebosberg and Klein Bruintjeshoogte, *Drege!* near the Klipplaat R., *Drege! E. & Z.!* (Herb. Bth., Hk., D., Sd.)

A small, woody, rigid, and densely branched bush, with thinly silky twigs and silvery leaves. Leaflets ¼–⅓ inch long, on petioles half that length, ¼–½ line wide. The habit is that of a *Buchenroedera*. The calyx teeth are remarkably short and sometimes nearly obsolete; the calycine pubescence is rufescent.

21. L. spinescens (Harv.); silky-canescent, at length sub-glabrescent, divaricately branched, rigid, spiny; leaves on *longish* petioles, trifoliolate, leaflets small, oblong-cuneate or linear, folded, canescent-puberulous; calyx thinly silky or glabrescent, its teeth *triangular-acute;* vexillum silky above; legume sub-terete, glabrescent. *L. armata, E. Mey.! Comm. p.* 35. *Benth.! l. c. p.* 361, *non Thunb., nec. E. & Z.*

HAB. Rocks of Driekoppe, *Drege!* Dwaka R., *Burke.* Great Fish R., and Zout Rivier, Beaufort, *Zeyher!* 397. (Herb. Bth., Hk., Sd.)

Very rigid, robust, and woody, about a foot high, with patent branches and twigs. Leaflets generally shorter than the petiole, very narrow, 2–4 lines long, thinly silvery.

5. VIBORGIOIDES. (Sp. 22–24.)

22. L. humilis (Thunb.! Prod. p. 122.); sub-glabrous, shrubby, rigidly much-branched; leaves sub-sessile, sub-fasciculate, trifoliolate, leaflets small, *cuneate-oblong, acute or emarginate;* racemes short, *laxly* few-flowered; calyx *shortly, but sharply 5 toothed,* with rounded interspaces; legume sessile, linear, turgid, quite glabrous. *Thunb. Fl. Cap. p.* 562. *L. fasciculata, Benth. Lond.! Journ.* 3. *p.* 361.

HAB. Bockland, *Thunberg!* Cape Colony, *Bowie!* (Herb. Th., Hk.)

A small, densely-branched, woody bush, 6–12 inches high; the younger parts minutely appresso-puberulent, the older sub-glabrous. Leaves on very short petioles, the leaflets 1–3 lines long, quite or nearly glabrous. Racemes 1–3 inches long, laxly several flowered. Calyx 1 line long, very oblique. Petals on long claws. Legume 6–8 lines long, many-seeded. Turns black in drying.

23. L. sessilifolia (Benth.! l. c. p. 362); robust, shrubby, rigidly much-branched; twigs silky-pubescent, leaves on very short petioles, leaflets *obovate or cuneate-oblong, thickish,* glabrous or nearly so; racemes short, dense, secund; calyx campanulate, puberulous, *its teeth lanceolate, acute, nearly equalling the tube;* legume sessile, oblong, acute, turgid, glabrescent. *Acanthobotrya sessilifolia, E. & Z.! No.* 1344. *Viborgia grandiflora, E. Mey. Comm. p.* 31.

HAB. Calcareous hills between Breede and Duivenhoeksrivier, *E. & Z.!* Caledon, *Mundt.!* (Herb. Sd., Hk.)

A thick-stemmed, woody, small bush, turning blackish in drying, the younger parts minutely and appressedly puberulous, the older mostly glabrous. Twigs becoming naked and rigid, but scarcely spiny. Leaflets 2–3 lines long, or a little longer, of thickish, fleshy texture. Racemes 1–1½ inch long, densely several-flowered. Calyx 4 lines long. Petals glabrescent, their claws shorter than the calyx-teeth. It turns blackish in drying.

24. L. Bowieana (Benth.! l. c. p. 362); robust, shrubby, rigidly much-branched; twigs silky-pubescent; leaves sub-sessile, sub-fasciculate; leaflets small, cuneate-oblong or obovate, *densely cano-pubescent on each side;* racemes short, few-flowered; calyx widely tubular, *quite glabrous, shortly,* but sharply 5-toothed, with rounded interspaces; legume sessile, oblong, acute, turgid, quite glabrous.

HAB. Cape Colony, *Bowie! Zeyher,* No. 2345. (Herb. Hk., Sd., Bh.)

A stout, small bush, with curved, sub-virgate branches, beset with short ramuli, and somewhat crowded or tufted leaves. Petioles very short. Leaflets 2 lines long. Flowers 1–3 at the ends of the branches, nodding. Calyx turning black in drying, 2–2½ lines long. Petals glabrous, with longish claws. Keel shortly rostrate, obtuse. Young legume turgid.

(Doubtful Species.)

L. marginata (E. Mey. Comm. p. 35); "silky-canescent; leaves trifoliolate, leaflets spathulate, margined; racemes short, lax; calyx 5-fid; vexillum pilose externally." *E. Mey.*

HAB. With *L. cinerea, Drege.* (Unknown to me.)

The plant distributed under this name by Ecklon (94.10) is merely *C. cytisoides.*

XXII. VIBORGIA, Thunb.

Calyx oblique, shortly 5-toothed. All the petals with long, slender claws; vexillum ovate; carina incurved or rostrate, longer. *Stamens* monadelphous. *Ovary* stipitate, few-ovuled. *Legume* stipitate, ovate or rarely oblong, indehiscent, winged on the upper suture, sharp and thin along the lower. *Benth.! in Hook. Lond. Journ.* 3. *p.* 363. *Endl. Gen.* 6479.

Rigid, slender, sometimes spiny shrubs, with palmately trifoliolate leaves and yellow flowers: all natives of the Cape. Name in honour of Eric Viborg, an acute Danish botanist.

TABLE OF THE SPECIES.

Branches, twigs, and leaves glabrous:
Branches and twigs *virgate*, scarcely spiny:
Leaflets linear-cuneate, mucronate ... ... (1) **flexuosa.**
Leaflets obovate; calyx-teeth blunt ... ... (3) **fusca.**
Branches and twigs *divaricating*, more or less spiny:
Racemes few-flowered; legume *narrow-oblong*, subturgid ... ... (4) **cuspidata.**
Racemes many-flowered; twigs *spinous*; legume compressed, winged above ... ... (6) **armata.**
Racemes many-flowered; legume broadly winged all round and crested on the sides ... ... (7) **tetraptera.**
Branches and twigs more or less silky or pubescent:
Young leaflets *thinly silky*, older glabrate; legume *narrow*-winged ... ... (2) **obcordata.**
Leaflets (and all parts) *densely silky-canescent*; legume *broad*-winged ... . (5) **sericea.**

1. V. flexuosa (E. Mey.! Comm. p. 28); *quite glabrous;* branches and racemes erect, elongate, lax; leaflets *linear-cuneate* or linear-oblong, subglaucous, mucronate, on longish petioles; legume? *Benth.! l. c. p.* 363.

HAB. Rocks on the Kamiesberg, between Pedroskloof and Leliefontein, 3000–4000f., *Drege!* (Herb. Bth., D., Hk.)

A tall, laxly-branched, slender shrub, glabrous in all parts. Twigs virgate. Leaves scattered; the petiole 4–5 lines long; leaflets as long, 1–1½ line wide, turning blackish when dry. Racemes 5–6 inches long, many-flowered. Calyx-teeth acuminate. Young legume on a long stipe, winged on the upper margin.

2. V. obcordata (Thunb.! Fl. Cap. p. 560); branches virgate, striate, *silky;* leaflets cuneate-oblong, obtuse or emarginate, the young ones thinly silky, the older glabrate; legume stipitate, turgid, transversely wrinkled, with a very narrow wing. *DC. Prod.* 2. *p.* 136. *E.Z.* 1347; *also W. fusca, E. & Z.!* 1348, *and W. sericea, E. & Z.!* 1349. *W. sericea, Benth.! l. c. p.* 363, *non Thunb. W. floribunda, E. Mey.! Comm. p.* 28. *Crotalaria floribunda, Lodd. Bot. Cab. t.* 509. *Zey.!* 380, 381.

HAB. Common in sandy soil, on the Cape flats and throughout the western districts. (Herb. Th., Bth., Hk., D., Sd., &c.)

A slender, slightly-branched shrub, 3–6 feet high, with long, virgate, graceful branches; all the younger parts thinly silky, the older glabrescent. Leaves sub-distant, turning black in drying. Racemes terminal or spuriously lateral, 3–8 inches long, densely many-flowered. Calyx-teeth very short. Legume oval, turgid, wrinkled, 4 lines long, 2½ wide.

3. V. fusca (Thunb.! Fl. Cap. p. 560); *glabrous;* branches rigid, twigs subvirgate; leaflets *obovate* or cuneate-oblong, mucronulate or rarely obtuse; racemes *many flowered;* calyx teeth short and blunt; carina subobtuse; legume stipitate, *compressed, with a broad wing along* *the ventral margin,* the sides delicately netted-veined. *Benth.! l. c. p.* 364. *V. oblongata, E. Mey.! Comm. p.* 29. *excl. var. β.* *V. incurvata, E. Mey. Comm. p.* 30. *Pterocarpus peltaria, DC. Leg. Mem. p.* 394. *t.* 57. *f.* 2.

VAR β. **microphylla**; petioles very short; leaflets much smaller. *V. parvifolia, E. Mey. p.* 31.

HAB. Sands of Zwartland, *Thunberg!* Between Groenekloof and Saldanha Bay, and between Zilverfontein and Kaus, Namaqualand, *Drege!* β. Near the Breede River, Swell., *Mundt.!* (Herb. Th., Bth., Hk.)

More densely leafy, and much more glabrous than *W. obcordata,* with very different legumes. Racemes 3–4 inches long, dense, the rachis rather rigid, but scarcely spinous. Leaves turn dark or black in drying. Legume ½ inch long, and nearly as broad, very thin and flat, the wing 1½–2 lines wide.

4. **V. cuspidata** (Benth.! l. c. p. 364); glabrous; branches rigid, *twigs short, flexuous or divaricate,* rarely spinescent; leaflets small, obovate or cuneate-oblong, mucronulate or obtuse; racemes lax, *few flowered;* carina subrostrate; legume stipitate, *narrow-oblong, subturgid,* winged on one margin, *the sides quite smooth.* *V. oblongata, β. cuspidata, E. Mey.! Comm. p.* 29.

HAB. Between Uitkomst and Geelbekskraal, 2000–3000f, *Drege!* (Herb. Bth. Hk.)

A rigid, stunted or depressed, robust, densely ramulose bush; the twigs intricate and much divided. Leaves on short petioles; leaflets 2–3 lines long. I have only seen immature legumes.

5. **V. sericea** (Thunb.! Fl. Cap. p. 560); *densely silky-canescent* in all parts; branches and twigs spreading, rigid, at length spinescent; leaflets cuneate-oblong, obtuse or subacute, *densely silky;* racemes short, many flowered; calyx widely campanulate, pubescent; carina subobtuse; legume stipitate, broadly winged at one side, the wing truncate below the style, the sides delicately netted-veined. *E. Mey. Comm. litt. a & b?* *V. lanceolata, E. Mey! Comm. p.* 29. *Benth. l. c. p.* 365. *Acanthobotrya cinerea, E. & Z.!* 1343.

HAB. Cape, *Thunberg!* Dry rocky mountains of Mierenkasteel and Kasparskloof, near Uitkomst and Wupperthal, *Drege!* Brackfontein, Clanw., *E. & Z.!* (Herb. Th., Bth., Hk., D., Sd.)

Robust, densely ramulous, and everywhere silky or silvery, with copious appressed pubescence. Leaves closely set, on longish petioles; leaflets 4–5 lines long, usually blunt, sometimes emarginate, drying pale. Flowers 3 lines long, yellow.

6. **V. armata** (Harv.); glabrous, or nearly so; branches and twigs rigid, *divaricate, spinous-pointed;* leaflets cuneate-oblong or obovate, subobtuse, drying pale; racemes many flowered; calyx teeth acuminate; carina rostrate; legume *compressed,* winged on one edge, the sides netted-veined, the lower suture narrow. *V. spinescens, E. & Z.! No.* 1351. *Benth.! l. c. p.* 365. *V. heteroclada, E. Mey.! Comm. p.* 30. *Lebeckia armata, Thunb.! Fl. Cap. p.* 562. *Acanthobotrya armata, E. & Z.! No.* 1342. *Ac. disticha, E. & Z.!* 1341. *Aspalathus mucronata, Thunb.! Fl. Cap. p.* 573.

VAR. β. **puberula**; leaflets and calyces minutely and appressedly pubescent. *V. monoptera, E. Mey.! l. c. p.* 30.

HAB. Kochman's Kloof, *Mundt.!* 24 Rivers, *Drege.!* Near Groenekloof, *E. & Z.!* *Pappe!* Klipfontein, *Zey.* 382. Namaqualand, *V. Schlicht!* Tulbaghskloof, *Zey.* 379. (Herb. Th., Bth., Hk., D., Sd.)

A very diffuse, laxly-branched, spiny bush; the twigs horizontally patent or reflexed, rigid, and mostly tipped with spines. Foliage always pale, quite glabrous, except in var. β. Flowers small, pale yellow. Calyx-teeth remarkably tapering. Wing of the legume variable, sometimes wide, sometimes comparatively narrow.

7. V. tetraptera (E. Mey.! Comm. p. 29); glabrous; branches and twigs patent, rigid, flexuous or divaricate, subspinescent; leaves obovate-oblong or cuneate, obtuse or mucronate; calyx-teeth acute; carina subrostrate; *legume broady winged on the upper and under edges, and wing-crested on the sides. Benth.! l.c. p. 365. Zey.! No.* 2299.

VAR β. **angustifolia**; leaflets mostly linear. *V. tenuifolia, E. Mey.! Linn.* 7. *p.* 153. *E. & Z.!* 1350, *ex pte.*

HAB. Berg River, at the Paarl, Riebeckskasteel, and between Langevalei and Olifant's River, &c. *Drege!* Cape District, *Bowie.* Breede River and R. Zonderende, *Zeyher!* β. Mouth of the Breede River, *Mundt.!* (Herb. Bth., Hk., D., Sd.)

Rigidly much-branched, flexuous; readily known, when in fruit, by its 4-winged legumes, the face of the valves rising in the middle into a sharp, wing-like crest; and both margins, ventral and carinal, being broadly winged. The leaves turn dark in drying.

XXIII. BUCHENROEDERA, E. & Z.

Calyx campanulate, subequally 5-fid. *Petals* villous, on long claws; the *carina* roundish, short and blunt. *Stamens* monadelphous. *Ovary* 8–10 ovuled. *Legume* obliquely ovate, somewhat turgid, 1–3 seeded. *Benth! Lond. Journ.* 7. *p.* 580.

Densely silky or silvery shrubs or suffrutices, with petiolate and stipulate, 3-foliolate leaves, and white or purple flowers. The floral characters are nearly those of *Aspalathus*, but the presence of stipules, and the distinctly petiolate leaves indicate a group naturally separable. The legume is very short, scarcely 1½ times as long as the calyx. The generic name is in honour of *W. L. V. Buchenroeder*, a South African botanist, and friend of Ecklon and Zeyher.

TABLE OF THE SPECIES.

Stipules leaflike, equalling or exceeding the petiole; flowers white or yellowish:
  Leaflets cuneate or obovate:
    closely silky; bracts ovate, longer than the calyx ... (1) **holosericea.**
    loosely silky; bracts short; leaflets minutely apiculate (2) **Meyeri.**
    closely silky; bracts short; leafl. *recurvo-mucronate* ... (3) **multiflora.**
  Leaflets linear-lanceolate or linear-acuminate:
    loosely silky; *spikes* cylindrical, dense ... . (4) **spicata.**
    closely silky; *umbel* many flowered; fl. pedicellate ... (5) **umbellata.**
Stipules leaflike, equalling or exceeding the petiole; flowers purple.
  closely silky; much branched; bracts shorter than the calyx (6) **tenuifolia.**
  loosely silky, depressed or decumbent; bracts equalling the cal. (7) **trichodes.**
Stipules none or inconspicuous; stems simple, virgate ... ... (8) **viminea.**

1. B. holosericea (Benth.! in Lond. Journ. 7. p. 581); *closely silky;* leaflets cuneate, apiculate; heads of flowers dense, *the ovate or oblong* bracts longer than the very villous calyx; all the petals villous, of nearly equal length; legume densely villous. *Benth.!—Aspalathus holosericea, E. Mey.! Comm. p.* 37.

HAB. Caffraria, near Omgaziana and Morley, *Drege!* (Herb. Bth.)

A slender, virgate shrub, the branches ramuliferous near the ends; every part fulvous with closely-appressed, short, silky hairs. Petiole 1–2 lines; leaflets 3–4 lines long. Stipules leaflike, 1½ as long as the petiole. Bracts 2 lines long, often equally broad.

2. B. Meyeri (Presl. Bot. Bem. p. 47); *loosely silky*; leaflets cuneate, minutely acuminate; heads dense, the narrow oblong bracts *shorter* than the very villous calyx; all the petals villous; the vexillum 1½ as long as the alæ and carina; legume very villous. *Benth. l. c. p.* 581. *Asp. cuneata, α, retusa, E. Mey.! Comm. p.* 37.

HAB. Caffirland, between the Buffel and Kei Rivers, *Drege!* (Hb. Bth., D., Hk.)

1–2 feet high, virgate, afterwards ramulous, straight, canescent, with copious, long and softly silky whitish or fulvous hairs. Leaflets like those of *B. holosericea*, or smaller. Bracts much shorter than the calyx, or rarely taper-pointed and nearly as long. Calyx 2–2½ lines long. Vexillum twice as long.

3. B. multiflora (E. & Z. ! 1354); closely silky; leaflets cuneate, *recurvo-mucronulate; spikes oblong, loose or rarely sub-capitate;* bracts cuneate or oblong, shorter than the calyx; petals villous, the vexillum longer than the others; legume villous. *Benth.! l. c. p.* 582. *B. gracilis, E. & Z.!* 1353. *Asp. cuneata, β. hamulosa, E. Mey.! Comm. p.* 37. *A. polyantha, Walp. Burch.* 3864. *Buch. alpina, E. & Z.!* 1352.

HAB. Eastern Province, Uitenhage and Albany. Zuureberg, Gekau and Assagaisbosch, *Drege!* Near Grahamstown and on the Fish River, *E. & Z.!* Vanstaadensberg, *Zeyher*, 2335. (Herb. Bth., Hk., Sd., D.)

Smaller, more slender, and less virgate than *B. Meyeri*, with shorter and more closely pressed pubescence, more lax inflorescence, and hook-pointed leaflets. In E. & Z.'s *B. alpina*, from the Winterberg, the leaflets are rather broader than usual, with shorter and less recurved points.

4. B. spicata (Harv.); *loosely* silky and silvery, virgate; leaflets *linear-lanceolate, acute; spikes cylindrical,* densely many flowered, the lanceolate bracts equalling the calyx; all the petals villous; calyx teeth acuminate; legume densely villous.

HAB. Sides of the Winterberg, among rocks and long grass, *Mrs. F. W. Barber!* 43. (Herb. T.C.D.)

Chiefly branched near the base, 1½ feet high, the branches long, rodlike, ending in a dense spike of white flowers, slightly tinged with greenish-yellow. Whole plant silvery, with long, soft and loose hairs. Leaflets 3–4 lines long, 1 line broad, acute at each end. Stipules similar to the leaflets and nearly as long, longer than the petiole, erect. Bracts broader than the leaflets, but similar in shape. Spikes 2–3 inches long. Calyx teeth lanceolate.

5. B. umbellata (Harv.); appressedly silky and silvery, subsimple; leaflets linear-acuminate, acute at base, nearly twice as long as the petiole; stipules and bracts linear-lanceolate; umbels subsessile, many flowered; pedicels of the flowers at least as long as the calyx, shorter than the bracts; petals villous.

HAB. Transkei Country, on the Plains, *Mrs. F. W. Barber!* 35. (Herb. D.)

Many stemmed, tufted, erect; stems simple, 6–8 inches high, densely leafy, ending in a shortly pedunculate or subsessile umbel of cream coloured flowers. Flowers 12–15 in each umbel, on pedicels 2–3 lines long. Leaflets 7–9 lines long, 1 line wide, tapering at the base, and almost cuspidate at the apex.

6. B. tenuifolia (E. & Z. ! 1355); closely *silky;* leaves narrow cuneate or linear; bracts lanceolate, shorter than the calyx; flowers either interruptedly racemose or umbellate-subcapitate; petals villous, the vexillum somewhat longer than the others. *Benth. l. c. p.* 582.

VAR. β. **pulchella**; taller and more robust, with shorter, broader, more cuneate leaves. *Asp. pulchella, E. Mey.! Comm. p.* 38.

HAB. On the sunny slopes of mountains near Silo, on the Klipplaat River, Tambukiland, *E. & Z.! Mrs. F. W. Barber!* Katberg and Stormberg, *Drege!* (Herb. Hk., Bth., Sd., D.)

A much branched, silky and fulvescent shrub, 1 foot high, with blueish-purple flowers, in pedunculate umbel-like 4–6 flowered heads, or in interrupted racemes. Lower stipules small. Peduncle 1–1½ inch long. Stems 10–15 inches high.

7. B. trichodes (Presl. Bot. Bem. p. 47); *softly hairy with long hairs*; leaflets linear-cuneate, acuminate; bracts lanceolate, equalling the calyx; flowers umbellato-capitate; petals villous, the vexillum not much longer than the rest. *Benth.! l. c. p.* 582. *Aspalathus trichodes, E. Mey.! Comm. p.* 38.

HAB. Summit of the Katberg, *Drege!* (Herb. Bth., Hk., D.)

A small, depressed shrublet, 6–12 inches long, with short, ascending branches, densely clothed with long, pale hairs. Umbels shortly pedunculate, 4–8 flowered. Calyx 3 lines long; pedicel nearly 2 lines. Flowers blueish-purple. The smallest and most hairy of the genus; resembling a *Lotononis.*

8. B. viminea (Presl. Bot. Bem. p. 47); softly hairy, *stems simple, virgate;* leaflets oblong-cuneate, mucronate, the upper somewhat lanceolate; bracts nearly similar; flowers terminal, spicate or lateral, and clustered. *Benth. l. c. p.* 583. *Asp. viminea, E. Mey.! Comm. p.* 38.

HAB. Caffraria, between the Omsamcaba and Omsamwubo, *Drege!* (Herb. Bth., Hk., D.)

Stems 1–2 feet high, simple and rodlike, leafy throughout, from a perennial root. Petiole very short, or scarcely any; leaflets 4–6 lines long, varying from cuneate to lanceolate, always acute. Calyx-lobes short, triangular. Very similar in habit to *Lotononis sessilifolia.*

XXIV. ASPALATHUS, Linn.

Calyx subequally 5–toothed or 5–cleft, or the two upper lobes shorter and broader. *Vexillum* short-clawed, erect, keeled at back, spreading; *carina* incurved or rarely straight. *Stamens* monadelphous, with a split tube. *Ovary* 2–4–8, rarely many ovuled; style glabrous, incurved. *Legume* obliquely ovate or sublanceolate, subcompressed, acute, one or few seeded. *Benth.! Lond. Journ* 7. *p.* 583.

A large and very natural South African genus of shrubs or undershrubs, with heathlike rarely flat and broadish, sessile, entire leaves, without stipules. The leaves are rarely solitary, commonly three together (*ternate*), on a more or less prominent, sometimes spine-pointed or spurred callosity of the stem (the *leaf-tubercle*); and in very many cases numerous additional leaves spring from the axils of the exterior three, and then the leaves are said to be *tufted* or fascicled. The inflorescence is properly *terminal* and racemose or spiked; but when the flowers are solitary from within a tubercle, as they seem to be axillary, they are here called *lateral.* The corolla is yellow, or rarely blueish-purple, red or white. I have adopted the general arrangement of Mr. Bentham, who groups the species under twelve tolerably natural, but not very absolutely limited sections, as set forth in the following key:

KEY TO THE SECTIONS OF *Aspalathus.*

Flowers either *sessile or short-pedicelled,* solitary or spiked, capitate or racemose:
  Leaves *flat,* broad or narrow:
    Lvs. glabrous or roughly villous, not *silky* … … (1) **Cephalanthæ.**
    Lvs. *silky* or *very softly and closely* villous … … (2) **Sericeæ.**
  Lvs. terete or trigonous (linear or subulate):
    Claws of *carina and alæ* adnate to the staminal tube (3) **Synpetalæ.**
    Claws of carina and alæ *quite free* from staminal tube:

Legume obliquely *ovate*, short, (flowers small) (4) **Leptanthæ.**
Legume *villous*, turgid, ovate or lanceolate, *reflexed*. Flowers *sessile*, *lateral*, mediocre ... (5) **Laterales.**
Legume *linear-lanceolate*, many seeded ... ... (6) **Macrocarpæ.**
Legume *thick*, *villous*, *obliquely lanceolate*, several seeded:
Flowers *lateral*, or subterminal, 1–2 together (7) **Grandifloræ.**
Fl. *terminal*, subcapitate ... ... (8) **Pachycarpæ.**
Legume *glabrous or silky*, obliquely lanceolate:
Lvs. fleshy. Flowers mediocre or large, subsessile, capitate, spiked or solitary and lateral, *mostly glabrous* ... ... (9) **Carnosæ.**
Lvs. fleshy. Fl. small, lateral, sessile; pet. mostly glabrous ... ... (10) **Pingues.**
Lvs. not fleshy. Flowers at the ends of short branches, solitary, in pairs or racemose ... ... (11) **Terminales.**
Flowers one or few on the end of a long, filiform peduncle (12) **Pedunculares.**

ANALYSIS OF THE SPECIES.

1. **Cephalanthæ.**—Leaves ternate, or scarcely tufted, *flat*, coriaceous, glabrous, or villous, *but not silky*. Flowers terminal, sessile or on short pedicels. Legume commonly obliquely ovate, shorter than the calyx, rarely lanceolate, exserted, erect. (Sp. 1–19.)

* *Lowest calyx-lobes longer and broader than the rest, concave, boat-shaped.*
† *Carina* glabrous: *floral leaves* large:
Lvs. oblong, undulate, squarrose. Ovary glabrous (1) **undulata.**
Lvs. linear-lanceolate, flat. Ovary villous ... (2) **suaveolens.**
†† *Carina* villous; *floral leaves small:*
Lvs. glabrescent. *Calyx*-lobes as long as tube (3) **venosa.**
Lvs. villoso-hirsute. *Cal.*-lobes very long ... (4) **polycephala.**
** *Calyx nearly equally 5-lobed. Flowers capitate, or 2–4 terminal.*
† *Leaves* orbicular, obovate, broadly spathulate, or cuneate-oblong.
Bracts roundish-obovate, concave:
Calyx-lobes triangular-acuminate, glabrous (8) **truncata.**
Calyx-lobes round-ovate, obtuse, pubescent (15) **psoraleoides.**
Bracts small, linear or setaceous:
Lvs. thickened at the margin ... ... (10) **marginata.**
Lvs. not thickened at margin; calyx glabrous ... ... (9) **myrtillæfolia.**
Lvs. not margined; calyx pubescent or villous:
Lvs. orbicular; ovary 8-ovuled ... (5) **orbiculata.**
L. obovate or cuneate-oblong; ovary 4–5 ovuled ... ... (6) **securifolia.**
L. cuneate-spathulate, puberulous; ovary 4-ovuled ... ... (7) **exigua.**
†† *Lvs.* narrow-spathulate, linear, lanceolate or oblongo-lanceolate.
Flowers 2–6 together, in *small* heads or subsolitary.
Lvs. narrow-spathulate, 2 lines long; calyx hairy ... ... (11) **inops.**
Lvs. linear, 4–6 lines long; *cal.* hirsute (12) **stenophylla.**
Lvs. very narrow-linear; *cal.* glabrescent (13) **angustissima.**
Lvs. oblongo-lanceolate; cal. very hairy (14) **stellaris.**
Flowers densely capitate; heads many flowered.
Lvs. oblongo-lanceolate, rather broad; bracts subulate ... ... (16) **anthylloides.**
Lvs. linear-lanceolate, narrow; bracts ovate acute ... ... (17) **linearifolia.**
*** *Calyx nearly equally 5-lobed. Flowers somewhat racemose.*
Lvs. linear-acute; cal. downy, its lobes triangular (18) **rugosa.**
Lvs. oblongo-lanceolate, or spathulate; cal. villous, its lobes lanceolate ... ... (19) **cytisoides.**

2. Sericeæ.—Lower or all the leaves tufted or rarely ternate, *flat, silky* or *softly and densely villous.* Flowers sessile or shortly pedicellate. Legume obliquely ovate, shorter than the calyx, or taper-pointed and a little longer. (Sp. 20–33).

* *Leaf-tubercles neither prominent nor aculeate. Flowers in terminal heads or spikes, or 2–4 in terminal tufts.*

† *Ovary* 2-ovuled.

Diffuse or procumbent:
Heads 2–5 flowered ... ... (20) **villosa.**
Heads densely many flowered ... ... (24) **jacobæa.**
Erect and virgate:
Pubescence silvery; bracts minute ... ... (21) **sericea.**
Pubescence villous; bracts leaflike:
Calyx teeth broad and shorter than the tube (25) **elongata.**
Calyx teeth subulate, equalling the tube:
Pubescence scanty; carina short ... (22) **Meyeri.**
Pub. copious; carina nearly eq. to vex. (23) **virgata.**

† *Ovary* 4–8 ovuled.
Procumbent; bracteoles lanceolate ... ... (26) **lotoides.**
Erect, virgate; bracteoles filiform ... ... (27) **heterophylla.**

** *Leaf-tubercles conspicuous, and mostly aculeate. Flowers sessile, solitary or few together in small heads.*

† Ovary 4–8 ovuled; tubercles rarely pointless.
Flowers in terminal few flowered heads:
Involucral bracts lanceolate, acute ... ... (28) **tridentata.**
Invol. bracts obovate, much shorter than calyx (29) **argentea.**
Flowers lateral; petals *purple* within:
Calyx-lobes short, triangular ... ... (30) **ferruginea.**
Calyx-lobes taper-pointed, elongate ... ... (31) **purpurea.**
Flowers lateral; petals pale or yellow within ... ... (32) **dasyantha.**

†† Ovary 2–ovuled; tubercle blunt; leaves silvery ... (33) **æmula.**

3. Synpetalæ.—Leaves tufted, terete or trigonous. Flowers subsessile, lateral or terminal. *Claws of the carina and alæ adnate to the staminal tube.* Legume obliquely ovate, shorter or scarcely longer than the calyx. (Sp. 34–44).

1. *Flowers capitate.* (*A. ciliaris* has occasionally *solitary, terminal fl.*)
Leaf-tufts subtended by a sharp spine ... ... (34) **aculeata.**
Leaf-tufts unarmed; *cal.-lobes* subulate.
Lvs. rigid, straight, pungent; ovary glabrous ... (35) **chenopoda.**
Lvs. soft, curved, mucronate; ovary villous ... (36) **araneosa.**
Leaf-tufts unarmed; *calyx-lobes* lanceolate, 3–nerved (37) **ciliaris.**

2. *Flowers either lateral, or 1–3 at ends of short branchlets.*
Calyx-lobes subulate, or lanceolate, acute:
Flowers 1–3 at the end of short branchlets ... (38) **Benthami.**
Flowers lateral; *calyx-lobes* lanceolate, 3 nerved.
Calyx-lobes twice as long as tube ... ... (39) **incurva.**
Calyx-lobes shorter than tube ... ... (41) **nervosa.**
Flowers lateral; *calyx-lobes* subulate.
Albo-plumose; *cal.-lobes* thrice as long as tube (40) **leucophæa.**
Pubescent; *cal.-lobes* recurved, equalling tube (44) **comosa.**
Calyx-lobes short, broad and obtuse:
Bracts linear; *cal.-lobes* concave, boat-shaped (42) **uniflora.**
Bracts ovate; *cal.-lobes* flat, ovate ... ... (43) **prostrata.**

4. Leptanthæ. Leaves tufted, terete or trigonous. Flowers subsessile (small). Claws of the petals free from the staminal tube. *Legume obliquely ovate,* shorter or *not much longer* than the calyx. (Sp. 45–57.)

Flowers spicate or capitate:
Calyx-teeth ovate or oblong-lanceolate, *shorter than the tube:*
Cal. lobes blunt; fl. many, hairy; ovary 4–ovuled (45) **nigra.**
Cal. lobes acute; fl. few, pubesc.; ovary 2–ovuled (46) **parviflora.**
Calyx-teeth lanceol. or subulate, *longer* than the tube:
Robust, rigid; lvs. short, blunt, fleshy, 2 lines long (47) **Forbesii.**
Slender, virgate; lvs. subulate, slender, 3–4 lines (48) **spicata.**

Flowers lateral, or *interruptedly* subspicate:
Diffuse or prostrate, much branched:
Leaf-tubercles neither prominent nor spiniferous:
Leaves densely imbricated, straight:
Pubescent; fl. lvs. not much longer than calyx-tube; cal. lobes twice as long as the tube ... ... (49) **ericifolia.**
Softly pilose; fl. leaves much longer than calyx-tube; cal.-lobes twice as long as the tube ... . (50) **mollis.**
Pubescent or subglabrous; fl. lvs. very short; calyx lobes not longer than the tube ... ... (51) **thymifolia.**
Villoso-canescent; fl. lvs. short; cal.-lobes equalling the villous tube ... (52) **diffusa.**
Leaves very slender, patent, incurved, 5–6 lines long ... ... (53) **asparagoides.**
Leaf-tubercles prominent and armed with short spurs:
Cal.-lobes subulate; vexillum ovate, acute, thinly silky ... ... (54) **calcarata.**
Cal.-lobes broadly lanceolate; vex. obcordate, silky-villous ... ... (55) **Pappeana.**
Erect shrubs, virgate or ramulous:
Densely covered with *silky-white hairs;* calyx-teeth short ... ... (56) **albens.**
Roughly hairy or pilose, with dark hairs; cal.-lobes long ... ... (57) **rubrofusca.**

5. Laterales. Leaves tufted, terete or trigonous. Flowers subsessile, lateral. Ovary 2–4 ovuled. Legume *villous, exserted,* obliquely ovate or lanceolate, commonly *turgid, at length horizontally patent or deflexed.* (Sp. 58–74.)

Leaf-tufts not spine-bearing:
Leaves *straight and rigid,* pungent-mucronate (rarely pointless), scarcely more than half inch long:
Carina pubescent:
Adult leaves glabrous; cal--lobes subulate; ovary 4–ovuled ... ... (58) **teres.**
Cinereo-pubescent; calyx-lobes lanceolate; ovary 2–ovuled... ... ... (61) **rigescens.**
Carina glabrous:
Calyx-lobes setaceo-pungent or needleshaped:
Adult leaves glabrous; flowers 6–9 lines long ... . (59) **hirta.**
Leaves *albo-tomentose;* flowers 5 lines long ... ... (60) **hystrix.**
Cal.-lobes broadly lanceolate, ribbed, longer than the tube ... ... (62) **opaca.**
Leaves *scarcely pungent,* more than ½ inch long; ovary 2–ovuled:
Leaves glabrous (the young pubescent).
Calyx-teeth lanceolate-falcate, mucronate, longer than the tube ... ... (63) **acanthes.**
Calyx-teeth much shorter than the tube ... (64) **Burchelliana.**
Leaves silky *tomentose;* calyx truncate, with short, setaceous teeth ... ... (65) **glomerata.**
Leaves silky and *silvery;* cal.-teeth very short; vexillum with a long claw ... ... (66) **longifolia.**
Leaves silky and silvery; calyx-teeth subulate, equalling the tube ... ... (67) **eriophylla.**
Leaves not pungent, rarely more than 4 lines long:
Leaves glabrous or *very thinly* puberulous, 3–4 lines long ... ... (68) **laricifolia.**

Leaves *canescent or appressedly silky*, acute. 3–4 lines long ... ... (69) **canescens.**
Leaves canescent and *tomentose*, obtuse, 1–2 lines long ... ... (70) **frankenioides.**
Leaves dense, *slender, commonly setaceous*, mucronate, incurved, 4–6 lines long ; ovary 2–ovuled.
Leaves villoso-canescent ; calyx-lobes lanceolate-subulate, needlepointed, longer than the tube, (71) **setacea.**
Leaves hairy ; calyx-lobes setaceous, 1½–2ce as long as the tube ... ... (72) **Alopecurus.**
Leaves glabrous or nearly so ; calyx-lobes short, deltoid ... ... (73) **incurvifolia.**
Leaf-tufts armed with a sharp, central spine ... ... (74) **Chamissonis.**

6. Macrocarpæ. Leaves tufted, terete or flattish. Ovary *many-ovuled.* Legume *linear-lanceolate, elongate.* (Sp. 75–78.)

Suffruticose, glabrescent ; branches subsimple ... ... (75) **filicaulis.**
Virgate shrubs ; leaves villous or glabrous :
Leaf-tubercle calcarate ; calyx *downy*, its teeth short (76) **macrocarpa.**
Leaf-tubercle pointless ; calyx *villous*, its teeth long (77) **pinea.**
Shrubby, *divaricate and spiny ;* leaves *silky and silvery* ... (78) **Garipensis.**

7. Grandifloræ. Leaves tufted, terete, or trigonous. Flowers lateral or subterminal, solitary or in pairs, large. Ovary several-ovuled. Legume *broadly-lanceolate, thick*, subturgid. (Sp. 79–84.)

Lvs. setaceo-*filiform*, ½–¾ uncial ; calyx-teeth long, subulate-acuminate :
Leaf-tubercle *unarmed*, or minutely mucronulate ... (79) **Willdenowiana.**
Leaf-tubercle *sharply calcarate ;* stigma decurrent ... (80) **leptophylla.**
Lvs. linear-terete, bluntish, *canescent*, short ... ... (81) **rostrata.**
Lvs. setaceo-*subulate*, glabrescent, bracteoles simple ... (82) **macrantha.**
Lvs. keeled, pilose or glabrous ; bracteoles trifoliolate :
Calyx-lobes *broadly-lanceolate ;* branches hairy ... (83) **grandiflora.**
Calyx-lobes *subulate-acuminate*, pungent ... ... (84) **galeata.**

8. Pachycarpæ. Leaves tufted, terete, or trigonous. Flowers terminal, subcapitate, *large.* Ovary several-ovuled. Legume thick, obliquely broadly-lanceolate or ovate, rhomboid, villous. (Sp. 85–87.)

Leaves acute or mucronate, glabrous or pilose :
Lvs. 4 lines long ; cal.-lobes short ; ovary 6–ovuled (85) **densifolia.**
Lvs. 2–2½ lines long ; cal.-lobes equalling tube ; ovary 4–ovuled ... ... (86) **triquetra.**
Leaves obtuse, *cano-tomentose*, 1 line long ... ... (87) **propinqua.**

9. Carnosæ. Leaves tufted (rarely ternate), terete or trigonous, generally *fleshy.* Flowers (moderate or large) sessile or short-pedicelled. Calyx subcarnose. Petals mostly glabrous. Legume *glabrous*, obliquely lanceolate or acute, usually exserted. (Sp. 88–101.)

Flowers spicate or capitate ; leaves obtuse or acute, *but not pungent.*
Glabrous or *very minutely* puberulous :
Lvs. ternate, flat or concave above, acute ; fl. spiked ... ... (88) **callosa.**
Lvs. short, fleshy, *blunt* or mucronulate ; cal.-teeth acuminate ... ... (90) **variegata.**
Lvs. acutely *mucronate ;* fl. 1–2 terminal ; cal.-teeth broadly ovate, *obtuse* ... ... (93) **sarcodes.**
Twigs pubescent or hairy :
Lvs. glab. { Lvs. 4–6 lines long, incurved, ternate, *acute ;* cal. lobes lanceolate, longer than the tube ... ... (89) **erythrodes.**
Lvs. 2–3 lines long, blunt ; cal.-lobes lanceolate, acute, longer than the tube ... ... (92) **Priori.**
Lvs. 2–3 lines long, blunt ; cal.-lobes oval, obtuse, shorter than tube ... ... (91) **carnosa.**
Lvs. 3–5 lines long, blunt, tufted ; cal.-lobes ovate-acuminate, equalling the tube ... ... (94) **sarcantha.** }

Lvs. 4–5 lines long, villoso-ciliate; fl. densely capitate; calyx villous ... ... (95) **capitata.**

Flowers capitate or solitary: leaves *pungent-mucronate;* glabrous.

Fl. subcapitate; vexillum glabrous; lvs. mucronate ... (96) **subulata.**

Fl. solitary, terminal; vexillum silky ... ... (97) **collina.**

Fl. 2–4 together; vexillum glabrous; lvs. subulate (98) **aciphylla.**

Flowers solitary and mostly lateral; leaves *pointless:* ovary 2–ovuled:

Glabrous; (flowers very variable in size, *glabrous*) ... (99) **arida.**

Branches densely *tomentose;* leaves glabrous, legume glabrous ... ... (100) **pachyloba.**

Branches tomentulose; leaves cano-puberulous; legume pubescent .. .. (101) **pallescens.**

10. Pingues. Leaves tufted, terete or trigonous. Flowers (small) lateral, solitary, sessile or short-pedicelled. Petals glabrous or rarely silky. Legume glabrous or silky, scarcely turgid, exserted, obliquely lanceolate. (Sp. 102–117.)

Unarmed; ovary with 4–6 ovules:

Leaves half inch long or more, much longer than the flowers ... ... (102) **verrucosa.**

Lvs. *linear,* 1½–3 lines long:

Flowers yellow; vexillum with a basal callus; ovules 6 ... ... (103) **affinis.**

Fl. reddish; vexil. naked at base; ovules 4; calyx 15–striate ... ... (105) **costulata.**

Lvs. *ovoid* or subrotund, ½–1 line long:

Flowers yellow; ovules 6 ... ... (104) **pinguis.**

Flowers red; ovules 4–6 ... ... (106) **sanguinea.**

Unarmed; ovary with only two ovules:

Calyx-teeth very short, obtuse, or obsolete:

Lvs. 2–3 lines long; calyx glabrous ... ... (107) **adelphea.**

Leaves ½–1½ lines long; calyx downy ... ... (108) **microdon.**

Cal.-teeth *triangular,* with strongly recurved margins (109) **recurva.**

Cal.-teeth *linear* or *subulate,* as long as the tube or longer ... ...

Leaves and petals glabrous ... ... (110) **Wurmbeana.**

Leaves appressedly puberulous; vexillum thinly silky ... ... (112) **incomta.**

Calyx-teeth much shorter than the tube:

Pubescence scanty; leaves glabrous or subcanescent:

Calyx-teeth subulate ... ... (111) **lactea.**

Calyx-teeth triangular ... ... (113) **lepida.**

Pubescence copious; whole plant white, with silky hairs ... ... (114) **argyrea.**

Armed with rigid spines:

Leaves very short (not 1 line long), fleshy, obtuse ... (115) **spinescens.**

Leaves 3–8 lines long, linear, terete or *compressed,* glabrous or subcanescent ... ... (116) **spinosa.**

Leaves *linear-spathulate,* quite *flat,* canescent ... . . (117) **obtusata.**

11. Terminales. Leaves tufted, terete or trigonous (rarely solitary or ternate), not carnose. Flowers (small or mediocre) *at the ends of the twigs,* either solitary, in pairs, or racemulose or spicate. Petals silky or glabrous. Calyx turbinate or widely campanulate. Legume obliquely lanceolate, glabrous or silky, scarcely turgid. (Sp. 118–137.)

Leaves ternate or tufted, the adult glabrous; fl. 1–3, pedicellate at the ends of short twigs:

Leaves subulate; calyx-lobes pungent; carina rostrate

Cal.-lobes twice as long as the tube; carina arched and rostrate ... ... (118) **abietina.**

Cal-lobes as long as tube; carina fornicate, strongly inflexed ... ... (119) **fornicata.**

Leaves linear, fleshy, blunt or mucronulate; carina straight and blunt ... ... (120) **pedicellata.**

Leaves tufted, glabrous or silky; fl. 1–2, sessile among the uppermost leaves:

Nearly glabrous, or minutely and thinly canescent:

Calyx-lobes longer than the tube or equalling it, subulate; leaves acute:

Lvs. puberulous, the floral shorter than the calyx, ovary silky ... ... (121) **retroflexa.**

Lvs. glabrous, the floral equalling the calyx; ovary glabrous ... ... (122) **galioides.**

Cal.-lobes longer than the tube, lanceolate, with reflexed margins ... ... (123) **marginalis.**

Cal.-lobes shorter than the tube; leaves acute, subpungent; petals pubescent ... ... (124) **exilis.**

Canescent; leaves white and silky ... ... (125) **rubens.**

Leaves tufted, glabrous or downy; fl. pedicellate, 2–3 or more in a short raceme; branches often spiny:

Calyx-lobes short, triangular, not pungent:

Unarmed; lvs. minute, obtuse ... ... (133) **vermiculata.**

Spinous; lvs. slender, filiform ... ... (132) **ferox.**

Calyx-lobes *pungent* or needle-pointed:

Lvs. subulate-acicular, pungent, very patent, ½–¾ inch long or more:

Erect, robust; vexillum pilose on the ridge, other petals glabrous ... ... (126) **astroites.**

Spreading, more slender; vexillum and carina silky; (fl. small, pale yellow) ... ... (127) **vulnerans.**

Lvs. terete, *pungent*, short or very short, 1–3 lines long:

Spinescent; divaricately much branched:

Leaves slender, 2–3–4 lines long ... ... (128) **pungens.**

Leaves short and fleshy, 1–1½ lines long ... (130) **acuminata.**

Unarmed or scarcely spinescent, divaricate; lvs. short ... ... (131) **divaricata.**

Lvs. linear-terete, obtuse or mucronulate, not pungent fl. bright yellow ... ... (129) **genistoides.**

Leaves tufted or ternate, *silky and canescent;* flowers subsessile, in a terminal spike:

Lvs. ternate, lanceolate, acute, *flat;* cal.-teeth ovate, acute ... ... (134) **Agardhiana.**

Lvs. tufted. subulate, *pungent;* cal.-teeth needle-pointed ... ... (135) **armata.**

Lvs. *solitary* or subsolitary, glabrescent, uncial or longer; racemules few-flowered:

Vexillum pubescent, as long as the *pubescent* carina (136) **corymbosa.**

Vexillum puberulous, longer than the much arched *glabrous* carina ... ... (137) **tenuifolia.**

12. Pedunculares. Leaves ternate or fascicled (rarely solitary), linear-subulate or flat. Flowers at the end of *a long, capillary peduncle* (or leafless ramulus), solitary or 2–3 together. (Sp. 138–148.)

Leaves slender, terete or semiterete, nearly glabrous, not pungent:

Ovary sessile, 6-ovuled:

Procumbent; lvs. *compressed,* acute at each end (138) **capillaris.**

Ascending; lvs. linear-filiform, *subterete;* pedunc. 2–3 flowered ... ... (139) **pedunculata.**

Ovary stipitate, 2–4 ovuled:

Lvs. solitary; bracts *minute,* subremote toothlike (140) **nudiflora.**

Lvs. tufted; bracts *leaflike,* equalling cal. tube (141) **bracteata.**

Leaves slender, terete (and the whole shrub) *densely silky-canescent:*

Lvs. acute; fl. subtended by slender bracts; legume ovate ... ... (145) **longipes.**

Lvs. blunt; fl. naked; legume lanceolate ... ... (146) **nivea.**

Leaves slender, terete or subulate, *pungent-mucronate, glabrescent:*
Lvs. short (2–3 lines), *linear*, mucronate ... ... (147) **suffruticosa.**
Lvs. longish (6–8 lines) *subulate*, aristato-pungent ... (148) **ulicina.**
Leaves *flat*, linear or linear-lanceolate :
Lvs. clothed with long, white hairs ... ... (142) **lanata.**
Lvs. ternate or tufted, thinly villous, soon glabrous and glossy ... ... (143) **falcata.**
Lvs. *solitary*, glabrous ; stems angular and furrowed (144) **alternifolia.**

I. CEPHALANTHÆ. (Sp. 1–19.)

1. A. undulata (E. & Z. No. 1368); branches hairy-villous; leaves *oblongo-lanceolate*, acute at each end, *undulate*, *squarrose*, glabrous or ciliate, midribbed and veiny below; the floral leaves orbicular-acute, villoso-ciliate, veiny, longer than the flowers; the four upper segments of the hairy calyx linear-lanceolate, the lowest longest, cymbæform, 2–3 times as long as the tube; vexillum silky, longer than the glabrous keel; ovary 4–ovuled, *glabrous;* legume obliquely lanceolate, obtuse. *Benth.! in Hook. Lond. Journ. vol. 7. p. 594. A. involucrata, E. Mey.! Comm. p. 38. Ononis fasciculata, Thunb.! Fl. Cap. p. 589.*

HAB. On the Winterhoeksberg, 2–3000 f. *E. & Z.! Drege.!* (Hb. Th., Bth., Hk.)

A rigid, robust, densely-branched, erect bush, 1–2 feet high; the branches densely tomentose and hairy. Leaves closely set, ternate, rigid, pale green, 4–5 lines long, 1–2 lines wide, often recurved, with minutely calloso-denticulate edges, the younger ciliate, older glabrous. Floral leaves solitary, many-nerved, ½ inch long, 4 lines wide, yellowish, fringed with long, woolly hairs. Calyx 4 lines long, densely hairy. Bracteoles linear, 4–5 lines long. Flowers pale yellow.

2. A. suaveolens (E. & Z. No. 1369); pilose, with long hairs; leaves *linear-lanceolate,* subfalcate, very acute, midribbed and veiny below; the the floral leaves obovate, acuminate, hairy, equalling the flowers; the four upper segments of the hairy calyx lanceolate, acute, thrice as long as the tube, the lowest longer, cymbæform; vexillum silky, longer than the glabrous keel; *ovary villous,* 5–6 ovuled. *Benth. l. c. p. 594.*

HAB. Upper regions of the Winterhoeksberg, Worcester, *E. & Z.!* (Herb. Bth., D., Sd., Hk.)

A small, slender, suberect or ascending, shrubby plant, thinly covered with long, softish hairs. Leaves ternate, subdistant, 5–6 lines long, ½–¾ line wide, flat, erecto-patent, more or less falcate. Floral leaves purplish or green, veiny, densely hirsute and fringed with woolly hairs. A smaller, more slender, less erect, and less densely branched, more hairy bush than *A. undulata,* with narrower leaves, and smaller heads and flowers.

3. A. venosa (E. Mey.! Comm. p. 39); branchlets villoso-tomentose; leaves oblongo-lanceolate, pungent-mucronate, rigid, *thinly villous, afterwards glabrous, midribbed and veiny below;* floral leaves *small, much shorter than the calyx,* obovate, villous; the four upper lobes of the densely hirsute calyx lanceolate, *equalling the tube,* the lowest cymbæform, longer; vexillum silky-villous, longer than the calyx and the *silky-villous carina;* alæ little shorter than the carina; ovary 2–ovuled. *Benth.! l. c. p. 595.*

HAB. Among rocks on the Giftberg, Cederberg, 1500–2500 f. *Drege.!* Nov. (Herb. Benth.)

A robust, rigid, divaricately branched, subdichotomous bush, 1–2 feet high; the old branches with rough, ashen bark, the twigs fulvous and densely hairy. Leaves subdistant, very rigid, 4–5 lines long, 1–2 lines wide, subrecurved: the floral leaves much shorter, smaller and of thinner substance than the cauline. Calyx 3½–4 lines long, as well as the orange flowers, thickly villous.

4. A. polycephala (E. Mey.! Comm. p. 39); *densely villoso-hirsute;* leaves obovate or oblong, recurvo-mucronate, pungent, squarrose, subcanescent; floral rather shorter than the calyx; four upper calyx-lobes lanceolate, acuminate, *thrice as long as the tube,* the lowest broadly-obovate-cymbæform; vexillum villous, as long as the calyx and longer than the villous carina; *alæ very small;* ovary hairy, 2–ovuled. *Benth.! l. c. p.* 595.

HAB. On the Blaauberg and Kaudeberg, 3–4000 f., *Drege.* Nov.-Dec. (Herb. Bth., D., Hk.)

A rigid, spreading, divaricately branched, very hairy bush. Leaves 2–3 lines long, 1–1½ line broad, ternate, patent or reflexed, pale green. Heads of flowers very numerous, terminating short branchlets. Floral leaves yellowish, larger than the cauline, but shorter than the flowers, very hairy. Calyx 5 lines long, the lowest segment much broader and longer than the rest. Flowers yellow.

5. A. orbiculata (Benth. Lond. Journ. 7. p. 595); twigs tomentose; leaves *orbicular,* cuneate at base, thick, concave, glabrous; flowers capitate; calyx thinly *villous,* its lobes subequal, lanceolate, acute, as long as the tube; petals silky, the vexillum orbicular, scarcely longer than the carina; ovary pubescent, 8–*ovuled. Benth.! l. c.*

HAB. South Africa, *Scholl!* (Herb., Vind., Benth.)

A robust, rigid, erect shrub, 2–3 feet high, with suberect branches. Leaves 3–4 lines long, equally wide or wider, mucronulate, pale green, shining, obscurely ribbed and veiny below, tapering at base into an imperfect petiole. Floral leaves small, spathulate. Calyx 2½ lines long. Vexillum twice as long as the calyx, on a longish claw, densely villous; the other petals thinly silky. Flowers yellow.

6. A. securifolia (E. & Z.! No. 1363); twigs tomentose; leaves obovate, orbicular, or suboblong, cuneate at base, thick, concave, *the lateral oblique,* glabrous or minutely puberulous, acute or mucronulate; flowers capitate, the floral leaves *very small, linear;* calyx *puberulous,* its lobes *lanceolate-subulate, scarcely equalling the tube;* petals pubescent, vexillum broadly ovate, slightly longer than the carina; ovary villous, 4–5 *ovuled;* legume obliquely ovate. *Benth. l. c.*

VAR. β. **spathulata**; leaves broader, less acute; calyx larger and more hairy. *A. spathulata, E. & Z.!* 1364.

HAB. Rocky hills on the Zonderende R., Swell., *E. & Z.!* Var. β. on Babylon's Toorensberg, Caledon, *E. & Z.! Mundt. and Maire!* (Herb. Bth., D., Hk., Sd.)

A robust, rigid, diffusely branched shrub, 1–2 feet high. Leaves 3–4 lines long, 1–2½–3 wide, shining, pale green, crowded, the young ones microscopically downy, the older punctate. Heads of flowers very short, the floral leaves very small; the bracts setaceous. Calyx 2 lines long. Petals twice as long, pale yellow. Near *A. orbiculata,* but the flowers are smaller, the petals more minutely downy, the calyx lobes narrower, and the ovules fewer.

7. A. exigua (E. & Z.! No. 1365); branches rufo-tomentose; leaves *cuneato-spathulate-oblong,* obtuse or mucronulate, *puberulous,* at length glabrous, shining; flowers capitate, *the fl. leaves linear and bracteoles setaceous;* calyx pubescent, its lobes subequal, lanceolate-subulate, acute, *longer than the tube;* petals silky, the vexillum broadly ovate, longer than the carina; ovary 4–ovuled, villous; legume shortly ovate, acuminate. *A. conferta, Benth. l. c. p.* 596.

HAB. Mountain sides near River Zonderende, Swell., *E. & Z.!* Zwarteberg, *Mundt!* Also collected by *Bowie and Burchell, Cat. No.* 6956. (Herb. Hk., Sd.)

An erect, robust, much branched shrub, 1–1½ feet high. Leaves 4–5 lines long, 1–1½ broad, the lateral oblique; veins inconspicuous or faint. Calyx-lobes nearly as long as the orange corolla. This resembles some of the narrow-leaved forms of *A. securifolia*, but the leaves are still narrower and longer, and less coriaceous; the calyx-lobes longer and flowers deeper coloured.

8. **A. truncata** (E. & Z.! No. 1359); branches tomentose; leaves *cuneate-oblong*, acute and mucronulate, glabrous, shining, plurinerved and veiny below; flowers capitate, *the floral leaves and bracteoles broadly obovate, mucronate, petiolate, concave; calyx-lobes quite glabrous, triangular-acuminate*, scarcely longer than the pubescent, obconic tube; petals silky, the vexillum longer than the carina; ovary 4–ovuled; legume cultrate, acute, strigillose. *Penœa, No.* 1220, *Drege! Zey. No.* 425.

HAB. Witsenberg, *Zeyher!* Mountain sides near the Waterfall, Tulbagh, *E. & Z.!* S. Africa, *Drege!* (Herb. Bth., Sd., Hk.)

A very robust and rigid dwarf shrub, with pale green or yellowish foliage. Leaves 5–7 lines long, 1–2 lines wide, densely crowded, slightly concave. Heads of flowers ending the branches, 1–1½ inch in diameter, pale yellow. The specimens are more or less altered by insect punctures; but one from Zeyher in Herb. Sond. has some seemingly normal flowers; and these have the bracts and calyx-lobes which mark the species. In the diseased specimens the bracts are more expanded, but the calyx-lobes not materially altered. By Bentham all are referred to *A. exigua*.

9. **A. myrtillæfolia** (Benth. l. c. p. 597); branchlets rufo-pubescent; leaves small, obovato-spathulate, mucronate, immarginate, glabrous; flowers 3–4, subcapitate; *bracteoles very minute, toothlike;* calyx *glabrous*, its lobes falcato-lanceolate, very acute, longer than the wide tube; petals minutely downy; ovary glabrous, 2–ovuled.

HAB. Cape Colony, *Bowie!* (Herb. Hook., Benth.)

A minute (probably alpine), rigid and woody, depressed shrub, 6–8 inches broad, glabrous, except on the branches and petals. Leaves 2–3 lines long, 1–1½ wide, spreading, glossy, pale green, midribbed, dotted. Flowers 2–3 lines long, pale yellow. Lowest calyx-lobe narrower and longer than the rest.

10. **A. marginata** (Harv.); branchlets canescent; leaves obovate or ovate, *spathulato-petiolate*, acute or mucronulate, coriaceous, *margined and ribbed*, puberulous, becoming glabrous, somewhat concave, the lateral oblique; flowers few, capitate, subsessile; bracteoles very minute, setaceous; calyx appressedly downy, its lobes subulate, as long as the tube; petals silky, twice as long as the calyx; *ovary villous, 2–ovuled. A. exigua, Benth.! l. c p.* 597, *non E. & Z.*

HAB. On the Hott. Holl. berge; Babylonische Tooren and Zwarteberge, *Zeyher!* 2346. (Herb. Hook., Sd.)

A depressed, subtrichotomously branched, small woody shrub. Leaves crowded, ternate, 2–2½ lines long, 1½–2 lines wide, tapering at base into a more or less distinct petiole and generally mucronate; the margin thickened, and the midrib and often two lateral ribs prominent below. Flowers small, hidden among the leaves at the ends of the branches, 3–4 together.

11. **A. inops** (Eck. and Zey.! 1362); slender; branches pubescent; leaves *spathulate-oblong*, pubescent or glabrous, acute, rigid, one-nerved

beneath; flowers 2–4 together, terminal, subsessile, bracts very minute; *calyx hairy, its lobes subulate,* as long as the tube; petals silky, not much longer than the calyx; ovary *downy,* 2–ovuled; legume obliquely ovate. *A. capitella, Burch. Cat.* 7148, *Benth. l. c. p.* 597.

HAB. Mountains near Swellendam, *Mundt!* in Herb. *E. & Z.!* Cape Colony, *Burchell.* (Herb. Sond., Bth.)

A very slender, depressed fructiculus, 4–6 inches high. Leaves ternate or fascicled, 2 lines long, ½ line wide, becoming glabrous. Flowers 2 lines long, yellow, usually 2 together, but in one of *E. & Z.'s* specimens, 4. Calyx covered with long soft hairs.

12. A. stenophylla (E. & Z.! 1361); twigs puberulous; *leaves linear,* mucronulate, acute at each end, *incurved, channelled,* strongly ribbed and keeled beneath, rigid, becoming glabrous; flowers 3–4 together, capitate, bracts small, subulate; calyx hirsute, the lobes subulate, distant, scarcely longer than the wide tube; petals densely silky; ovary *villous,* 2-ovuled. *Benth. l. c. p.* 598. *A. canaliculata, E. Mey.! Comm. p.* 44.

HAB. Mountain sides near Riv. Zonderende, *E. & Z.!* Gnadenthal, *Drege!* (Herb. Benth. Sond.)

A small, slender, diffuse or procumbent fruticulus. Leaves 4–6 lines long, ½ line wide, sometimes slightly spathulate, full green. Flowers few together, at the ends of short branchlets, 3–4 lines long, yellow. This is more robust than *A. inops,* with longer and narrower leaves, and larger flowers; and smaller and less rigid than *A. stellaris.* All are closely related to each other.

13. A. angustissima (E. Mey. Comm. p. 44); "leaves fascicled, very narrow-linear, rather obtuse, glabrous; flowers terminal, solitary or in pairs, shortly pedicellate; segments of the glabrescent calyx acuminate subulate; vexillum pubescent." *E. M. l. c.*

HAB. Drakensteensberg, *Drege.*.

Said to resemble *A. stenophylla* in habit. Leaves 2 lines long, flat.

14. A. stellaris (E. & Z.! 1360); branchlets pubescent or villous; leaves *oblongo-lanceolate,* the lateral incurved, mucronate, scarcely pungent, rigid, glabrous or thinly villous, flat, midribbed; flowers capitate; bracts subulate; calyx *very hairy,* its lobes subulate, as long as the tube; petals *densely silky;* ovary villous, 2-ovuled. *Benth. l. c. p.* 598.

HAB. Hills of Kannaland, near Gauritz R., Swell., *E. & Z.!* Also in *Bowie's* Coll. (Herb. Hk., Sd., D., Bth.)

A small but robust shrub, 6–8 inches high, branched near the base Leaves 4–5 lines long, 1 line wide, the young ones villous. Heads 6–8 flowered; flowers 3–4 lines long, fulvous, with purplish tints. Calyx-lobes broadly subulate, the tube shaggy.

15. A. psoraleoides (Benth.! l. c. p. 598); branchlets pubescent; leaves spathulato-lanceolate, mucronate, puberulous, convex below and midribbed, the lateral ones incurved; flowers capitate, *bracts roundish, obovate, 3–toothed;* calyx *pubescent, its lobes ovato-subrotund, obtuse,* half as long as the tube; corolla silky; ovary 2–ovuled, villous. *Paraspalathus psoraleoides, Presl. Bot. Bem. p.* 134.

HAB. Cape Colony, *Sieber!* (Herb. Hook.)

A small shrub resembling *A. stellaris* or *A. exigua,* from both which it differs in the bracts and calyx; and (as suggested by Bentham) may, as well as *A. truncata,* be an abnormal diseased state of either species.

16. A. anthylloides (Linn. Sp. p. 1002); branches villous or tomentose, erect, virgate; leaves crowded, *oblongo-lanceolate*, acute, the lateral oblique, softly hairy, becoming glabrescent, flat, veiny or veinless; *flowers densely capitate, with lanceolate-subulate bracts;* calyx villose or hirsute, its lobes lanceolate-subulate, longer than the tube; petals silky villous; ovary villous, 2-ovuled; legume obliquely ovate, shorter than the calyx. *Benth. l. c. p.* 599. *Thunb.! Fl. Cap. p.* 574. *E. & Z.! No.* 1358.

VAR. β. **Kraussiana;** branches tomentose (not villous); leaves thicker, less hairy, and more distinctly nerved; the upper ones 3-nerved. *A. Kraussiana, Meisn.! in Hook. Lond. Journ.* 2. *p.* 69. *Benth.! l. c.*

HAB. Common in the mountains of Cape, Stellenbosch, and Swellendam Districts. *Thunberg! Eck. & Zey.! W.H.H., &c.* β. Klein River, Swellendam, *Krauss, Bowie, Thom, &c.* (Herb. Th., Bth., Hk., D., Sd.)

1-2 feet high, chiefly branched near the base; the branches erect and somewhat virgate. Leaves 6-8 lines long, 1½-2 lines wide, variable in pubescence and in the degree of prominence of the nerves and veins, when dry. Flowers bright yellow, in large dense heads, surrounded by somewhat longer leaves. Thunberg's original specimens in Hb. Upsal, as well as *E. & Z.'s* in Hb. Sond. belong to the form called *Kraussiana,* which is scarcely distinct from the ordinary varieties found on the Capetown range.

17. A. linearifolia (DC. Prod. 2. p. 142); branches softly villous; leaves ternate, *linear-lanceolate, acute at each end,* midribbed, shortly villous, at length glabrous; flowers densely capitate, *bracts ovato-lanceolate,* silky-villous; calyx softly villous, its lobes lanceolate, more than twice as long as the tube; vexillum pubescent, scarcely longer than the calyx or the pubescent carina; ovary villous, 4-ovuled. *Benth.! l. c. p.* 599. *A. linifolia, E. Mey.! Linn.* 7 *p.* 162. *E. & Z.! No.* 1370.

HAB. Tulbaghskloof, Worcester, *E. & Z.!* Berg Rivier, Nieuwekloof and Dutoitskloof, *Drege!* &c. Oct. (Herb. Bth., Sd., D. Hk.)

A tall, much branched bush, 2-3 feet high, with flavescent branches. Leaves 1-1¼ inch long, scarcely a line wide, pale green, flat and patent. Heads many-flowered, fulvous, subtended by fulvous and silky bracts, half as long as the calyces. Flowers 3-4 lines long, tawny yellow. Legume scarcely longer than the calyx.

18. A. rugosa (Thunb.! Fl. Cap. p. 574); branchlets canescent, puberulous; leaves linear-sublanceolate, the lateral incurved, depressedly downy, then glabrous, acute, midribbed; flowers shortly racemose, *with small linear bracts;* calyx appressedly pubescent, its lobes *triangular-acute, about equalling the tube;* petals silky; ovary pubescent, 4-5 ovuled; legume silky, obliquely ovato-lanceolate, turgid, nearly thrice as long as the calyx. *A. Plukenetiana, E. & Z.! En. No.* 1371. *Benth. l. c. p.* 600. *Zey. No.* 430.

HAB. Near the Hex River, *Thunberg!* Tulbagh Valley, Worcester, *E. & Z.!* (Herb. Th., Bth., Hk., D., Sd.)

A much branched, erect shrub, 3-4 feet high, with pale foliage. Leaves ternate, ½ inch long, 1-1½ lines wide, densely set. Racemes short, 6-8 flowered. Flowers pale yellow. There are excellent specimens of this plant in Herb. Thunb., agreeing in all respects with those collected by recent travellers.

19. A. cystisoides (Lam. Dict. 1. p. 392); branches villous; leaves *oblongo-lanceolate or subspathulate,* the lateral recurvo-falcate, pungent-mucronate, rigid, appressedly canescenti-villous, becoming glabrous; flowers *racemulose, with setaceous bracts;* calyx villous, *its lobes lanceolate,* acute, rather longer than the tube; petals villous; ovary villous, 4-ovuled; legume obliquely oblong, longer than the calyx. *Benth.! l. c. p.* 600. *E. Mey.! Comm. p.* 39. *A. cinerea, Thunb.! Fl. Cap. p.* 575., *E. & Z.! No.* 1366.

HAB. Tulbagh Valley, *E. & Z.!* Dutoitskloof, *Drege!* Hott. Holl., *Pappe!* French Hoek, *Dr. Alexander Prior!* (Herb. Hk., Bth., Thb., Sd., D.)

A low but woody, much branched, somewhat corymbose bush, a foot or so in height. Leaves 2–4 lines long, 1 line broad, densely set, canescent when young, afterwards yellowish. Racemes subcapitate, few-flowered. Flowers yellow, densely silky. A more slender variety sometimes occurs, less densely branched, with somewhat longer leaves, and more laxly set flowers.

2. **SERICEÆ** (Sp. 20–33).

20. A. villosa (Thunb.! Fl. Cap. p. 574); *slender, procumbent;* leaves subfasciculate, small, narrow-oblong, obtuse or subacute, silky and canescent; heads few flowered, with linear-oblong bracts; calyx densely silky-villous, *its lobes oblongo-lanceolate, acute,* scarcely shorter than the tube; legume oblique at base, ovate, long-beaked. *Benth.! l. c. p.* 601.

HAB. Cederberg, near Honigvalley, *Drege!* Bockland, *Thunberg!* Simonsbay Hills, *C. Wright!* (Herb. Th., Bth., Hk., D.)

Root woody. Stems several, procumbent or ascending or trailing, 3–8 inches long, filiform, appressedly silky. Leaves 2 lines long, ½ line wide, ternate or fascicled, carnose, the older ones glabrescent. Heads 2–5 flowered. Flowers 3 lines long, the vexillum and keel densely silky. Ovary very hairy, 2 ovuled.

21. A. sericea (Berg. Pl. Cap. p. 212); *erect, robust,* virgate; leaves densely tufted, short, linear-oblong or subcuneate, obtuse, *silky and silvery;* heads densely many flowered, ovoid; bracts minute; calyx campanulate, densely villous, *with very short, deltoid teeth;* petals densely villous, subequal; legume ovate, hairy, with a long beak. *Benth. l. c. p.* 601. *A. argentea, Thunb.! Fl. Cap. p.* 576.

HAB. Cederberg, in various collections. Near Ezelsbank and Giftberg, *Drege! Wallich!* (Herb. Th., D., Bth., Hk.)

A strong-growing, somewhat divaricately branched shrub, 3–4 feet high, with long, subsimple, densely leafy branches. Leaves very glossy, of thickish substance, clothed with appressed, white or fulvous hairs. Heads of flowers 1–1½ inch long, 1 inch diameter, fulvous; the flowers yellow, 5 lines long. Calyx and petals very hairy.

22. A. Meyeri (Harv.); erect, virgate; branches *tomentose;* leaves ternate or tufted, small, oblong or lanceolate, acute, thinly silky-puberulous or glabrescent; heads densely many-flowered, ovato-globose; lower bracts stipitate, ovato-lanceolate, upper subulate; calyx densely villous, *its teeth lanceolate-subulate,* equalling the tube; *vexillum and alæ half as long again as the carina. A. lotoides, E. Mey.! Comm. non Thb. Benth.! l. c. p.* 602.

HAB. Cederberg, near Ezelsbank, *Drege!* At the 24 Rivers, *Zeyher!* Dec. (Hb. Bth., Hk., Sd.)

1–1½ foot high, with the habit of *A. virgata,* to which it is nearly allied; but the pubescence is more scanty, the heads larger and more ovoid, and the flowers conspicuously larger, with a shorter carina Heads 1½ inch long, 1 inch in diameter, fulvous.

23. A. virgata (Thunb.! Fl. Cap. p. 576); erect, robust, shrubby, virgate; leaves ternate or tufted, small, oblong or linear, obtuse or acute, *silky;* heads oblong or globose, densely many-flowered; bracts *orbicular, ovate or obovate, acute;* calyx densely villous, its lobes plumose, subulate, equalling the tube; corolla not much exceeding the calyx, and the *carina not much shorter than the vexillum;* legume ovate, turgid, acuminate. *Benth.! l. c. p.* 603. *E. & Z.! No.* 1388. *A. quinquefolia, Thunb.! Cap. p.* 575.

VAR. β. **globosa**; heads globose, or depressed. *Benth. l. c.*

VAR. γ. **leucocephala**; heads globose; calyx-lobes longer and more plumose; corolla somewhat shorter. *A. leucocephala, E. Mey.! Comm. p.* 41. *Benth.! l. c. p.* 602. *Zey.! No.* 424.

HAB. Picketberg, Stellenb., *E. & Z.!* Jackall's River and Piquetberg, *Drege!* Longvalley and Bergvalley, *Zeyher!* var. γ. on the Giftberg, *Drege!* (Herb. Th., Bth., Sd., Hk., D.)

A strong-growing, or rarely slender, erect, virgate bush, 1–2 feet high, chiefly branched near the base. Leaves 1–2 or 3–4 lines long, ¼–½ line broad, sometimes almost tomentose, sometimes silky. Heads ½ inch to 1½–2 inches long, dense, and very hairy, pale or fulvous; flowers yellow. Var. γ. has longer and more plumose calyx-lobes than usual, but as Bentham remarks, is connected through var. β. with the common form, from which it does not appear to me to be specifically distinct.

24. A. jacobæa (E. Mey.! Comm. p. 41); slender, *diffuse or ascending;* leaves ternate or fascicled, narrow-oblong or linear-lanceolate, acute, silky-villous; spikes oblong, dense; *bracts* lanceolate, substipitate; *bracteoles setaceous;* calyx very hairy, its lobes lanceolate-subulate, plumose, shorter than the corolla; vexillum and alæ ⅓ longer than the carina. *Benth.! l. c. p.* 603. *A. sericea, E. & Z. fide Walp.*

HAB. Cape Flats and Hills round Capetown. Paarlberg and Dutoitskloof, *Drege!* (Herb. Bth., Hk.)

Root simple; stems numerous, rigid, ascending or curved, branched chiefly from near the base. Leaves 1–2 or 4–5 lines long, green, with lax, but long pubescence, variable in shape. Spikes 1–2 inches long, less dense than in *A. virgata,* with narrower, longer and more lanceolate bracts; the bracteoles setaceous.

25. A. elongata (E. & Z.! No. 1387); erect, virgate, the branches tomentose or silky; leaves fascicled or ternate, linear or oblong, obtuse, silky pubescent; *spikes loose, elongating, several flowered;* bracts orbicular or ovate, acute, bracteoles lanceolate or ovato-lanceolate; calyx densely villous, its teeth lanceolate, *shorter than the tube;* vexillum a third longer than the carina. *A. ascendens, E. Mey.! Comm. p.* 41. *Benth. l. c. p.* 604. *excl. syn. E. & Z.! A. stricta, Steud.*

HAB. In sandy and stony places, on mountain sides near Tulbagh, *E. & Z.!* Near the Paarl, *Drege!* Also *Zeyher,* 434 (424 in Hb. Sond.) from "Predikstael." (Herb. Bth., Sd., Hk., D.)

1–2 feet high, slender, erect or ascending. Leaves variable in pubescence and size, 1–2, or 3 lines long, ½ line wide, green or silvery, shorter or longer than the internodes. Spikes at first 1–2 inches long, with somewhat imbricating flowers, afterwards 3–5 inches long, the flowers ¼ inch apart, alternate on a filiform rachis. Bracts shorter than the calyx; the lateral bracteoles leafy, variable in breadth; by which characters and the much shorter and broader calyx teeth this species is easily known from *A. jacobæa.* From some states of *A. virgata* it is best known by its short calyx-lobes.

26. A. lotoides (Thunb.! Fl. Cap. p. 575); procumbent or ascending, rather slender; leaves ternate or fasciculate, linear or lanceolate, acute, hairy or silky; spikes oblong, *lax, or elongating and sparsely flowered; bracts and bracteoles lanceolate,* leaflike; calyx hairy, its teeth lanceolato-subulate, equalling or exceeding the tube; vexillum a third longer than the keel; ovary substipitate, 6–8 ovuled. *A. quinquefolia, Linn. sp. p.* 1002. *non Thunb. A. anthylloides, Berg. Fl. Cap. p.* 211. *A. procumbens, E. Mey.! in Linn.* 7. *p.* 162. *Benth.! l. c. p.* 604. *E. & Z. No.* 1384. *E. Mey.! Comm. p.* 40.

VAR. β. **sparsiflora**; suberect, woody and much branched; spikes elongating, laxly many flowered; calyx segments rather exceeding the tube, often recurved. *Asp. heterophylla, Thunb.! in Herb. Upsal, non E. Mey. nec Benth. A. procumbens, β. squarrosa, E. Mey.*

VAR. γ **stachyera**; spikes very lax; leaves linear-lanceolate, 6–7 lines long; calyx segments twice as long as the tube, equalling or exceeding the corolla. *A. stachyera, E. & Z.! No.* 1386. *Ononis Lagopus, Thunb.!*

HAB. Hills round Capetown, common. Various collectors. Var. β. Piquetberg, *E. & Z.!* Oliphant's R., *Zeyher!* (Herb. Th., Bth., Sd., Hk., D.)

Stems numerous, 6–12 inches long, spreading or procumbent. Pubescence copious and long, rather hairy than silky. Inflorescence rather variable; the spikes sometimes short, with imbricating flowers, sometimes 3–6 inches long, the flowers ¼ inch apart. Fl. yellow, 3 lines. Bracts varying from ovate to lanceolate; bracteoles more constantly lanceolate. This resembles *A. jacobœa* in habit, but has laxer inflorescence, and differs essentially in the calyx teeth and bracteoles and numerous ovules. I have examined Thunberg's original specimen in Hb. Upsal, and find it completely to agree with the *A. procumbens*, E. Mey. The flower examined had 6 ovules. Var. γ. (from Oliphant's R.) has the aspect of a plant *drawn* in tall grass; it is slender, pale green, with much longer leaves and calyx-lobes; but is connected through the Piquetberg specimens, with var. β.

27. A. heterophylla (E. Mey! non Thunb.); erect, virgate; leaves *scattered or subternate or tufted, linear-lanceolate,* oblong or linear, acute, silky or silvery; *spikes lax,* oblong or elongate; bracts *linear-lanceolate*; bracteoles setaceous; calyx densely and softly hairy, its segments falcato-lanceolate, about equalling the tube; carina nearly equalling the vexillum; ovary densely hairy, 4–5 ovuled. *Benth.! l. c. p.* 605. *E. Mey.! Comm. p.* 40. *excl. syn. Thunb. Ononis spicata, Thunb.! Cap. p.* 584. *A. linifolia, Steud.?*

HAB. Cape Flats, *Drege! Wallich!* Klipfontein, *Zeyher! No.* 435. Simonsbay, *Dr. Alexander Prior.* (Herb. Th., Sd., Hk., Bth., D.)

Slender, 1½–2 feet high, branched from the base; branches simple, rodlike. Lower leaves short, crowded, often densely fascicled, 2–3 lines long and ½ line wide; upper often ternate, 4–5 lines long, 1 line wide; uppermost generally scattered, solitary, 4–7 lines long, ½–1½ line wide. The pubescence is nearly as variable. The habit is most like *A. virgata*, from which this is readily known by its different bracts and numerous ovules. I retain the name '*heterophylla*,' as it is generally adopted by botanists and sufficiently characteristic, though this is not the plant so called by Thunberg.

28. A. tridentata (Linn. sp. p. 1002. non E. Mey.); *diffusely much branched,* the twigs slender and virgate; leaves ternate or fascicled, short, lanceolate, oblong, or sublinear, acute, silky or glabrescent, the leaf-tubercle mostly aculeate; heads small, shortly globose or subumbellate, 5–8 flowered, the involucral bracts *broadly lanceolate, acute,* nearly equalling the calyx, *bracteoles filiform;* calyx densely shaggy, its segments lanceolate, acute, equalling the tube; keel much shorter than the alæ; ovary 6-ovuled. *Thunb.! Fl. Cap. p.* 575. *Benth.! l. c. p.* 606.

A. argentea, var. glabriuscula, E. Mey.! Comm. p. 43. *A. pilosa, E. & Z.! No.* 1381. *fide Drege. Zeyher, No.* 2338.

HAB. Cederberg, on the Giftberg, *Drege!* Tulbaghskloof and Vogelvalley, *Zey. No.* 423. Cape Flats, *W.H.H., &c.* Zwartland, *E. & Z.!* Under Table Mount., *Dr. Pappe!* (Herb. Th., Bth., Sd., Hk., D.)

A much branched, spreading bush, 3–4 feet in diameter, with twiggy, rodlike branches directed to all sides. Spine under the leaf-tuft ¼–½ line long, rarely wanting. Pubescence scanty on the leaves; very dense, long and fulvous on the bracts, calyces and petals. Flowers 3–4 lines long. *Zeyher's* 2338, in Herb. Sond. has blunt leaf-tubercles, and very small, tufted leaves on its lower twigs; but in other characters is like some of *Drege's* specimens.

29. A. argentea (Linn. sp. p. 1002? non Thunb.); *divaricately much branched*, the twigs *recurved;* leaves ternate or tufted, small, narrow-obovate or oblong, silky or glabrescent, the leaf-tubercle sharply aculeate; heads small, shortly globose or subumbellate, few flowered; *invol. bracts obovate-orbicular, or subovate, much shorter than the calyx;* calyx densely shaggy, its segments lanceolate, acute, shorter than the tube; keel not much shorter than the alæ; ovary 5–8 ovuled. *Benth! l. c. p.* 607. *A. argentea, α, E. Mey. Comm. p.* 43. *A. staurantha, E. & Z.! No.* 1383. *Burchell, Cat.* 7455.

HAB. Sandy hills in the Cape district, *Drege!* Zwartland, *E. & Z.!* Klipfontein, *Zeyher*, 422. Gnadendal, *Dr. Roser!* (Herb. Bth., Hk., Sd., D.)

A more densely branched, divaricate and less virgate bush than *A. tridentata*, which it much resembles and from which it chiefly differs in the shorter and broader bracts. The pubescence is too variable to afford a character. Both occur in the same district, and I fear are scarcely sufficiently distinct.

30. A. ferruginea (Herb. Banks.); robust, rigid, divaricately much branched; branchlets often spine-tipped; leaves tufted or ternate, oblongo-linear or sub-cuneate, subacute, silky-pubescent; the leaf tubercle aculeate; flowers 1–3 together; calyx silky-villous, widely campanulate, *its teeth small, triangular, much shorter than the tube;* ovary 4–5 ovuled; legume obliquely ovato-lanceolate. *Benth.! l. c. p.* 607. *A. tridentata, E. & Z.! En. No.* 1378. *E. Mey. Comm. p.* 43, *et auct. pl. Galega ternata, Thunb.! Fl. Cap. p.* 601.

HAB. Sands near the sea side, Cape District. Rietvalley and Cape Flats, *E. & Z.! Zeyher, No.* 421. (Herb. Thb., Bth., Sd., Hk., D.)

A strong growing, woody and spiny, spreading bush, 2–3 feet high, with pale or yellowish bark, and canescent or fulvous foliage. Leaves 3–5 lines long, ½ line wide, sometimes longer and broader. Flowers either solitary or 2–3 together, sessile in the axils of the upper leaves, or at the ends of short branchlets, 5–6 lines long. Calyx 4 lines long, 2½–3 wide, fulvous. Petals purple within, densely silky and fulvous externally.

31. A. purpurea (E. & Z.! No. 1379); robust, rigid, divaricately branched; leaves fascicled or subternate, cuneate-linear, subobtuse, densely silky, the leaf-tubercle aculeate or pointless; flowers lateral, solitary; calyx villous, widely campanulate, *its teeth from a triangular base taper-pointed, scarcely shorter than the tube;* ovary 4–5 ovuled, the legume obliquely ovato-lanceolate. *Benth.! l. c. p.* 608. *A. purpurascens, E. Mey. Comm. p.* 44.

HAB. Sandhills near the sea. Clanwilliam, near Haartebeetskrall, *Drege!* Near Berg-vallei, *E. & Z.!* (Herb. Bth., Hk., Sd.)

Very nearly allied to *A. ferruginea,* but with longer, more densely silky leaves, and much longer, narrower, and taper-pointed calyx-teeth. Flowers purple within; densely fulvous without.

32. A. dasyantha (E. & Z.! No. 1377); robust, rigid, with patent, virgate branches; leaves ternate or fasciculate, short, *obovate or cuneate-oblong, pilose, afterwards glabrescent;* the leaf-tubercle sharply aculeate; flowers lateral, subsolitary; calyx *densely shaggy,* widely campanulate, *its segments lanceolate,* shorter than the tube; ovary 4–5 ovuled; legume obliquely ovate, acute. *Benth.! l. c. p.* 608.

HAB. Maritime Sands, Plettenberg Bay, *Mundt.! Bowie! Thom!* (Herb. Hk., Sd.)

A rigid, much branched, woody bush, resembling *A. ferruginea,* but with longer and straighter branchlets, and broader, more roughly hairy and afterwards glabrescent leaves. Leaves 2–3 lines long, 1 line broad. Flowers externally densely clothed with fulvous hairs; internally yellowish (Eck.), or perhaps a pale rosy? (judging from dried specimens.)

33. A. æmula (E. Mey. Comm. p. 42); robust, virgate; leaves fascicled or subternate, small, oblong or obovate, pointless, *silky and silvery, the leaf-tubercle blunt;* flowers sessile, lateral, 1–3 together; calyx densely hirsute, its segments falcato-lanceolate, scarcely shorter than the tube; *ovary 2–ovuled;* legume lanceolate, acuminate, densely villous. *Benth. l. c. p.* 608. *A. argentea, E. & Z.! No.* 1390, *non Thunb., nec. Linn. Zey. No.* 2342.

VAR. β. **ramulosa**; branches densely ramuliferous; leaves obovate; calyx-lobes rather shorter. *A. ramulosa, E. Mey.! Linn.* 7. *p.* 163. *E. & Z.! No.* 1380.

HAB. Caledon and Swellendam Districts, *E. & Z.!* Klynriviersberg, *Zeyher!* Also in *Bowie's and Mundt's* collections. Hexriver, *Drege!* β. on the Zwarteberg, *E. & Z.! Zeyher! No.* 2343. (Herb. Hk., Bth., Sd.)

A large, strong growing, robust bush. Branches 1–2 feet long, simple, densely leafy; in β. emitting multitudes of ramuli, ½–1 inch long. Leaves 1½–2 lines long, ⅓–¾ lines wide, with shining, white, appressed hairs. Flowers yellow, often extending a foot or more along the branches. A very distinct and easily known species.

3. **SYNPETALÆ.** (Sp. 34–44.)

34. A. aculeata (Thunb.! Fl. Cap. p. 584); leaves in dense tufts, *subtended by a long spine,* linear, keeled, pilose, *the upper ones and floral hooked and mucronate;* flowers capitate; calyx villous, its lobes hooked, as long as the tube; vexillum villous; carina glabrous. *Benth.! l. c. p.* 609. *E. & Z. No.* 1497. *E. Mey. Comm. p.* 43. *Zey.* 417. *DC. Prod.* 2. *p.* 138.

HAB. Zwartland and Paardeberg, *Thunberg! E. & Z.!* Paarlberg and Daal Josaphat, *Drege, &c.* (Herb. Thb., D., Bth., Hk., Sd.)

A robust, very densely branched, thorny and hairy bush, with pale yellow bark. Branchlets villous. Spines 3–4 lines long, subhorizontally spreading, yellow, sharp. Leaves 3–5 lines long, ⅓ line wide, thickish, the lower ones quite blunt, the upper strongly hooked, the older glabrescent. Heads 3–5 flowered; fl. yellow, 5–6 lines long. Vexillum and keel nearly equal; alæ shorter.

35. A. chenopoda (Linn. sp. p. 1000); leaves tufted or subternate, *subulate-triangular, rigid, pungent-mucronate, straight,* sparsely pilose, the floral densely plumoso-pilose; flowers capitate; calyx hirsute, its lobes subulate, rigid, pungent, much longer than the tube; vexillum villous, carina glabrous; *ovary glabrous;* legume obliquely ovato-falcate, acuminate. *Benth.! l. c. p.* 610. *Thunb. Fl. Cap. p.* 577. *Bot. Mag. t.* 2225. *Lodd. Cap. t.* 316. *DC. Prod.* 2. *p.* 138. *E. & Z.! No.* 1454.

HAB. Sides of Table Mountain and the neighbouring hills, common. (Herb. Thunb., Bth., Hk., Sd., D.)

A rigid, rough, strong-growing, furze-like bush, 3–4 feet high and wide, with very pungent leaves, but no prickles. Branches roughly hairy, with brown hairs. Leaves 6–7 lines long, or 3–4 lines, all generally straight and spreading. Heads several flowered; the petals yellow, the vexillum fulvous externally. Calyx and its segments thickly covered with long, foxy, rigid, straight hairs.

36. **A. araneosa** (Linn. Sp. p. 1001); *pilose, with long pale hairs;* leaves tufted, *filiform, curved or squarrose,* mucronate, the floral densely plumose; flowers capitate; calyx hirsute, its segments from a broader base, subulate-attenuate, plumose, much longer than the tube; vexillum villous, carina glabrous; *ovary villous;* legume obliquely ovato-rhomboid, falcate-acuminate. *Benth.! l. c. p.* 610. *Thunb. Fl. Cap. p.* 577. *DC. Prod.* 2. *p.* 141. *Bot. Mag. t.* 829 *(bad). Sieb. No.* 48. *Zey.* 426. *E. & Z.!* 1463, *and A. Simsiana,* 1464.

HAB. Cape, Stellenbosch, and Worcester districts, on the hill sides, common. (Herb. Th., Bth., Hk., D., Sd.)

A tall, but not densely branched shrub, 3–5 feet high, more or less densely covered with long, stiffish, spreading hairs; the upper and floral leaves especially. Leaves 6–8 lines long or more, very slender, terete, spreading and flexuous. Heads dense, globose, several flowered; fl. 4–6 lines long, yellow, sometimes pale or whitish, or rufescent. Zeyher's 426 is a much depauperated form.

37. **A. ciliaris** (Linn. Mant. p. 262); pilose; leaves fascicled, subulate, linear-trigonous, or filiform, short or long; flowers terminal, capitate or 2–3 together, or sometimes solitary; calyx hirsute, *its segments narrow-linear-lanceolate, ribbed, sub-trinerved,* acute, much longer than the tube; vexillum villous, carina glabrous or villous; ovary very hairy. *Benth.! l. c. p.* 611. *E. & Z.* 1469. *A. appendiculata, E. Mey.! E. & Z.!* 1470. *A. dubia, E. Mey.! E. & Z.!* 1475. *A. Meyeriana, E. & Z. No.* 1473. *A. papillosa, E. & Z.!* 1472. *A. oresigena, E. & Z.* 1471. *A. aulonogena, E & Z.!* 1465.

HAB. Mountains and hills of Cape, Stellenbosch, Worcester, and Caledon, common. Krum River, Uitenhage, *E. & Z.! No.* 1471. Klynhowhoek, *Zeyher,* 2341. (Herb. Th., Bth., Hk., D., Sd.)

A very variable plant, sometimes robust, tall and woody, with densely tufted, short, straight or curved leaves; sometimes slender, decumbent or diffuse, with scarcely tufted or ternate, very long, filiform, curved leaves, as in *A. araneosa,* from which, in all cases, the present may be known by its *broader, ribbed and nerved* calyx-lobes. The young leaves are pilose, the older glabrescent or glabrous; they vary extremely in length, often on the same bush. In *A. dubia,* E. M., all but the floral are 2–3 lines long; commonly they are 4–5 lines; and in slender specimens 6–7 lines. Flowers deep yellow or fulvescent, 5–6–8 lines long. The carina varies much in pubescence, and is often quite glabrous.

38. **A. Benthami** (Harv.); branches hairy; leaves tufted, linear-carinate or subulate, mucronate, hispid; flowers *sessile on the tips of very short, leaf-crowned branchlets,* 1–3 together; calyx-tube turbinate, hirsute, its segments lanceolate, three-nerved, twice as long as the tube, but shorter than the hirsute, suborbicular vexillum; carina silky or glabrous; legume obliquely ovate, acuminate. *A. comosa, E. Mey.! Comm. p.* 63. *Benth.! l. c. p.* 612, *non Thunb. nec. E. & Z.*

HAB. Cape and Stellenbosch districts. Paarlberg and Dutoit's Kloof, *Drege!* *Zeyher*, 427, *ex parte.* (Herb. Hk., Bth., D., Sd.)

A stout, much branched, spreading shrub, 2–4 feet high; the twigs rusty-red and hairy, 6–8 inches long, patent. Leaves 4–6 lines long, close or subdi tant, spreading, curved or squarrose, rufescent or brownish, the old becoming subglabrous. Flower bearing branchlets $\frac{1}{8}$–$\frac{1}{4}$, rarely half inch long, crowned with a tuft of leaves. surrounding the flowers; sometimes the branchlet is abortive and the fl. lateral. Flowers yellow, $\frac{1}{4}$ inch long. Cal.-lobes 3 lines long. Vexillum very broad. Bentham attributes to this species *solitary* flowers, but I find them to vary from one to 2–3 on specimens not otherwise differing; and these last come near some forms of *A. ciliaris.* Quite unlike Thunberg's "*comosa.*"

39. A. incurva (Thunb.! Fl. Cap. 578); branches villoso-tomentose; leaves tufted, linear-triangular, mucronulate, villoso-pubescent or subglabrous; flowers *sessile in the lateral leaf-tufts,* solitary; calyx-lobes lanceolate, acute, three-nerved, *twice as long as the tube,* and nearly equalling the suborbicular, villous vexillum; carina glabrous; ovary villous. *A. elongata, E. Mey.! Comm. p.* 63, *non E. & Z. Benth.! l. c. p.* 612. *Zey. No.* 433. *A. Dregeana, Walp. Linn.* 13, *p.* 586.

HAB Cape and Stellenbosch District, *Drege, Caley, Pappe, &c.* Between Paalen and the Blauwberg, *Zeyher!* (Herb. Th., Bth., Hk., Sd., D.)

A rigid bush, but smaller and less branching than *A. Benthami,* from which it is easily known by its different inflorescence, shorter petals, smaller and paler flowers, and less copious and softer pubescence. Leaves pale green or fuscous, 3–4 lines long; the young rough with tubercle-based hairs, or minutely and appressedly pubescent: the older becoming naked. Flowers pale yellow.

40. A. leucophæa (Harv.); *covered with long, white, spreading hairs;* leaves tufted, *triangular-filiform, slender, flexuous,* mucronate, *plumoso-pilose;* flowers sessile in the lateral leaf-tufts, solitary; *calyx-lobes subulate, 3-nerved,* mucronate, 2–3 times as long as the hirsute tube and nearly equalling the villous vexillum; carina thinly silky, longer than the alæ; ovary densely hirsute.

HAB. Mountains near 24 Rivers, *Zeyher.* (Herb. Sond.)

An erect shrub, with the foliage of *A. araneosa,* but still more copiously hairy, with longer and whiter hairs; and with a different inflorescence and calyx. I have only seen a single specimen. The leaves are 6–8 lines long, as thick as hog's bristle; their hairs 1–1$\frac{1}{2}$ lines long. The flowers are smaller than in *A. araneosa,* the corolla scarcely, if at all, exceeding the calyx, and are scattered along the branches. Occasionally there is an imperfect, 2–3 flowered capitulum. If it be a "divarication" of *A. araneosa* or of *A. ciliaris,* it is a singular one.

41. A. nervosa (E. Mey.! Comm. p. 62); branches *thinly tomentose;* leaves tufted, linear-carinate, mucronate, *appressedly-pubescent, subcanescent;* flowers lateral and terminal, solitary or in pairs; calyx turbinate, tomentose, furrowed, its lobes lanceolate, subfalcate, three-ribbed, *nearly equalling the tube, much shorter than the villous vexillum;* carina glabrous; ovary villous. *Benth.! l. c. p.* 613.

HAB. Grassy hills near Swellendam, *Mundt. Thom!* Rocky places in the Nieuwekloof, *Drege!* (Herb. Hk., Bth.)

A much branched, densely ramuliferous robust bush, with rusty or foxy, short, close pubescence. Leaves 2–3 lines long, spreading or squarrose. Flowers either solitary on the ends of very short, leaf-crowned ramuli, or in sessile lateral leaf-tufts, or sometimes 2 together at the ends of the branches, very abundant, pale yellow, with rusty tomentum outside. Alæ very narrow, much shorter than the keel, which nearly equals the vexillum.

42. A. uniflora (Linn. Sp. p. 1001); branches villoso-tomentose; leaves tufted, short, linear-terete, blunt, glabrous or hairy; flowers subsessile, solitary or in pairs; *calyx-lobes leafy, 3-nerved, oblongo-cymbæform, concave, very obtuse,* as long as the pubescent tube; vexillum villous; carina glabrous; legume pubescent, obliquely rhomboid, scarcely longer than the calyx. *Benth. l. c. p.* 613. *A. cymbæformis, DC. Prod.* 2. *p.* 140. *E. & Z.! No.* 1408. *A. scaphoides, E. & Z.!* 1409. *Sieb. Fl. Cap.* 160. *Zeyher,* 2685.

HAB. Hills round Capetown and in the Cape District, common. (Herb. Bth., Hk. Sd., D.)

A much branched, divaricate, prostrate or spreading, ramuliferous shrub, with pale green foliage, variable in its pubescence, but easily known by the boat-shaped calyx-lobes. According to Bentham, this is the original *A. uniflora* of Linnæus (but not of Thunberg).

43. A. prostrata (E. & Z.! 1410); *prostrate, slender,* ramulous, pubescent; leaves tufted, very short, linear-terete, blunt, pubescent; flowers solitary, sessile at the ends of very short, leaf-crowned ramuli; *bracts and bracteoles ovate;* calyx campanulate, shortly hairy, *its lobes ovate, obtuse, flat,* shorter than the tube; vexillum pubescent, longer than the glabrous carina and alæ; ovary pedicellate, silky, 2-ovuled.

HAB. Mountains near Swellendam, *Mundt!* (Herb. Sond.)

A small, slender suffrutex, quite prostrate, with spreading, filiform branches, subdivided, and thickly set with erect ramuli, ¼–½ inch long; each ramulus ending in a tuft of leaves and usually tipped with a flower. Leaves 1–1½ line long. Flowers yellow, 4 lines long; the broad and short, flat bracteoles appressed to the calyx. Calyx-tube 15 striate, the striæ darker. All parts of the plant shortly pubescent. Quite distinct from *A. uniflora.* The claws of the lower petals are adnate with the staminal tube; the vexillum hairy at summit of claw, inside, and silky externally.

44. A. comosa (Thunb. Fl. Cap. p. 577, non Benth.); branches villous; leaves tufted, *setaceo-filiform, curved,* hairy or subglabrous, acute; flowers lateral, solitary, sessile in the leaf-tufts; calyx-tube campanulate, pubescent, 13–15 nerved, *its segments setaceo-subulate, mucronate, recurved, about equalling the tube;* vexillum silky, narrow-obovate, carina *much shorter than the alæ,* glabrous; ovary 2-ovuled, silky. *A. incurva, E. & Z.! No.* 1419. *E. Mey.! Comm. p.* 55. *Benth. l. c. p.* 619, *non Thunb.! A. thymifolia, Thunb.! non Linn.*

VAR. β. **Namaquana**; villoso-canescent, with shorter leaves; carina silky; calyx-lobes twice as long as the tube.

HAB. S. Africa, *Thunberg!* Paarl and Hott. Holl., *E. & Z.! Pappe!* 207. *Alexander Prior!* Tulbagh, *Drege!* β. Namaqualand, *Wyley.* (Herb. Th., Bth., D., Hk., Sd.)

A rigid, suberect or spreading, small shrub, 6–12 inches high; the branches curved, with grey bark; the older ones naked and tubercled, the younger hairy. Leaftufts densely crowded or imbricate. Leaves 3–4 lines long, very slender, subterete, the younger softly pilose. Flowers small, slender, 3–4 lines long, pale. Cal. 2 lines long, the base of the teeth 3-ribbed, the ribs continued, and partly confluent on the tube. Vexillum twice as long as the calyx, tapering to a cuneate base. Claws of the lower petals adnate with the staminal tube for at least ⅓ of their length! Alæ nearly equalling the vexillum; carina not half so long. This has quite the aspect of one of the *Leptanthæ* section, among which it is placed by Bentham, but its petals are distinctly adnate to the stamens, and the 15-ribbed calyx accords with the 3-nerved sepals of the *Synpetalæ.* It is the *A. comosa* of Thunberg's Herbarium, and also his *A. thymifolia,* on the same authority.

4. **LEPTANTHÆ** (Sp. 45–57.)

45 A. nigra (Linn. Mant. p. 262); leaves tufted, short, *linear-terete*, blunt, glabrous or thinly pubescent; *flowers spiked or capitate;* calyx villous, its teeth *broadly ovate or sublanceolate, shorter than the tube, obtuse or acute;* petals hirsute, the ovate vexillum longest; *ovary 4–ovuled;* legume obliquely rhomboid, scarcely longer than the tube. *Benth.! l. c. p. 614. Thunb.! Fl. Cap. p. 581. E. & Z.! No. 1430. A. nigrescens, E. Mey.! E. & Z.! No. 1431. A. melanoides, E. & Z.! 1432. A. pallens, E. & Z.! 1433. A. deciduifolia, E. & Z.! 1434. A. globulosa, E. Mey.! E. & Z.! 1424. Zey.! 2333.*

VAR. β. **involucrata** (Pappe); flowers capitate, *heads involucrated with numerous broadly ovate or suborbicular, striate bracts. A. involucrata, Pappe, MSS.*

HAB. Hills &c. in the Cape and Stellenbosch Districts, common. Extending to Swellendam & George, *E. & Z.!* River Zonderende, *Zeyher.* β. Kuilsriver, Cape, *Dr. Pappe!* (Herb. Th., Bth., Hk., D., Sd.)

A small, ramuliferous bush, 1–2 feet high, erect or spreading, variable in several characters, but readily known from others of this section by the broad, short, calyx lobes, and 4–ovuled ovary. Leaves 2–3 lines long, turning dark or black in drying. Flowers either in globular heads, or in oblong spikes, or scattered thinly toward the ends of the branches. Bracts sometimes (especially in var. β.) orbicular or ovate, and ribbed, entire or trifid; sometimes linear or oblong. Calyx either glabrescent or densely hirsute; its segments either very obtuse and short, or longish and acute. Flowers blue or pale. β. is a remarkable form with highly developed bracts, but connected by many intermediate links with the ord' ary state of the species.

46. A. parviflora (Berg. Pl. Cap. p. 208); small and depressed, divaricate; leaves tufted, filiform-subulate, short, blunt, pubescent or subglabrous; *flowers 3–4 in a short spike;* calyx pubescent, *its teeth ovate or sublanceolate, acute,* shorter than the tube; petals pubescent, carina shorter than the alæ; *ovary 2–ovuled;* legume obliquely ovate, acute, rather longer than the calyx. *Benth.! l. c. p. 615. E. & Z.! No. 1429. Thunb.! Fl. Cap. p. 579. Zey.! 2332. Burch. 6359.*

HAB. Mountains of Swellendam. Gnadendahl, *Drege! Alexander Prior!* Puspasvalley and Kochmanskloof, *E. & Z.!* Grootevadersbosch, *Zeyher!* (Herb. Th., Hk., Bth., Sd., D.)

A small, depressed or prostrate, much-branched, ramuliferous shrub, 6–8 inches long. Leaves 2 lines long, drying green. Flowers 2–4 toward the ends of the branches in a spurious spike, or subcapitate, blue or pale. Like some of the smaller forms of *A. nigra*, but the branches are more slender, the flowers smaller and much less hairy, the calyx lobes longer and more acute, and the ovary constantly 2–ovuled.

47. A. Forbesii (Harv.); robust, rigid, ramuliferous; leaves tufted, *short, blunt, linear, fleshy,* pubescent or glabrous; flowers capitate; calyx hairy, its lobes lanceolate, rather longer than the tube; vexillum oval-oblong, villous, longer than the glabrous or pubescent carina; ovary 2–ovuled. *A. cephalotes, Benth.! l. c. p. 615. non Thunb. nec. E. & Z.*

HAB. S. Africa, *Forbes!* (Herb. Hk., Bth.)

A much-branched, woody bush (2–3 feet high?). Branches tomentose. Leaves 2 lines long, very obtuse, subtrigonous or terete, shrinking when dry. Heads globose, terminating short branchlets 1–2, rarely three inches long, the calyces much longer than the involucrating leaves. Calyx teeth 2 lines long, its tube rather roughly but shortly pubescent. Petals externally silky villous.

48. A. spicata (Thunb. Fl. Cap. p. 578); virgate, slender; leaves tufted, *linear-subulate, long or short,* submucronate, pilose or glabrous;

flowers spiked or capitate; calyx *very hairy, its segments lanceolato-subulate,* longer than the tube; petals villoso-pubescent, the ovate vexillum longer than the carina; ovary 2–ovuled; legume oblique, ovate, acute, shorter than the calyx. *Benth. l. c. p.* 615. *E. & Z.! No.* 1421.

VAR. β. **cephalotes**; heads globular or depressed, more hairy than usual. *A. cephalotes, Thunb.! Fl. Cap. p.* 578. *E. & Z.! No.* 1423. *Zey.!* 2336. *A. cerrantha, E. & Z.!* 1422, *and A. globosa, E. & Z.!* 1425, *vix Andr. Rep. t.* 510?

HAB. Common on the hills of Cape and Stellenbosch Districts. β. French Hoek, *Thunberg!* Zwarteberg, *E. & Z.!* Klynriviersberg, *Zeyher.!* (Herb. Th., Bth., Hk., D., Sd.)

An erect or ascending, small shrub, with long, curved, simple, pale-barked, branches, occasionally ramuliferous. Leaves 2–4 lines long, slender, becoming darker in drying. Flowers blue, in terminal, ovoid, oblong or cylindrical, dense spikes; in β. in short heads. The pubescence of the calyx and bracts is sometimes very copious, sometimes scanty. Thunberg's *A. cephalotes,* judging by his specimen, is merely a stunted form of the present species with short heads and more hirsute flowers. It is quite different from *A. cephalotes,* Benth. (our *A. Forbesii*).

49. A. ericifolia (Linn. Sp. p. 1000); diffusely much-branched; leaves tufted, short, linear, blunt, villous or glabrescent, *the floral ones scarcely longer than the calyx tube;* flowers lateral, in interrupted spikes toward the ends of the branches; calyx villous, its segments linear-subulate, bluntish, *twice as long as the tube;* the ovate-oblong vexillum and the keel villous; ovary 2–ovuled; legume obliquely ovate, acute, villous, equalling the calyx. *Benth.! l. c. p.* 616. *A. ericoides, E. Mey. Linn.* 7. *p.* 160. *Thunb.! Fl. Cap. p.* 579. *A. varians, E. & Z.!* 1428. *A. kannaensis, E. & Z.! No.* 1412.

HAB. Abundant on dry hills in the Cape and Stellenbosch Districts, &c. (Herb. Th., Bth. Hk., Sd., D.).

Diffuse or prostrate, robust, 3–4 feet long, much and divaricately branched. Pubescence generally copious and soft, rusty or fulvous, sometimes scanty. Leaves 1–1½ line long, sometimes drying black. Flowers extending from 1–2–3 inches along the branches, small and pale, 3 lines long. This is the plant of Thunberg's Herbarium; by Linnæus *A. mollis and thymifolia,* were confounded, fide *Benth.! l. c.*

50. A. mollis (Lam. Dict. 1. p. 290); every where *covered with long, soft hairs;* leaves tufted, setaceo-filiform, mucronulate, pilose, *the floral much longer than the calyx tube;* flowers lateral, subsessile, solitary; calyx pilose, its segments subulate, *twice or thrice as long as the tube;* the ovate-oblong vexillum and often the carina villous; ovary 2–ovuled, villous. *Benth.! l. c. p.* 617. *A. flexuosa, Thunb.! Fl. Cap. p.* 579. *E. & Z.! No.* 1466. *Zeyher, No.* 2339.! *A. muraltioides, E. & Z.! No.* 1427?

HAB. Paarlberg, *Drege!* Near Tulbagh Waterfall, *E. & Z.!* Voormansbosch, *Zey.!* (Herb. Th., Bth., Hk., D., Sd.)

A spreading or procumbent, much or little branched, softly pilose shrub, varying in habit, length of leaves, and amount of pubescence. Leaves very slender, 2–4, or sometimes 6 lines long, imbricated. Flowers 3 lines long, nestling among the leaves like those of a *Muraltia,* pale, with a fulvescent, either glabrous or pubescent carina, nearly equalling the alæ. Thunberg's specimens have shorter leaves than usual, but are otherwise identical. The "*A. hispida, α.*" of his Herb. is *A. mollis,* but his "*A. hispida β.*" is *A. thymifolia.*

51. A. thymifolia (Linn. Sp. p. 1000); branches thinly tomentose; leaves tufted, short, linear-filiform, blunt, glabrous or pubescent, the floral shorter than or nearly equalling the calyx tube; flowers solitary, sessile, lateral; calyx pubescent, its segments subulate, *as long as the tube or shorter;* vexillum pubescent; *alæ much shorter than the glabrous or pubescent carina;* ovary 2–ovuled; legume obliquely ovate, acute, downy, longer than the calyx. *Benth.! l. c. p.* 617. *E. Mey. Comm. p.* 57. *A. hispida, Thunb.! Fl. Cap. p.* 579, *ex parte. E. & Z.! No.* 1413. *A. frankenioides, E. & Z.!* 1416, *non DC. A. multiflora, Sieb. No.* 49, 163. *Zey.!* 2330, 431.

VAR. *α.* **tenuifolia**; leaves longer and more slender. *A. mollis, β. flexuosa, E. Mey. Comm. excl. syn. Thunb.*

VAR. *β.* **micrantha**; flowers smaller; calyx teeth shorter. *A. micrantha, E. Mey.! Linn.* 7. *p.* 161. *E. & Z. No.* 1418. *Pappe,* 242.

VAR. *γ.* **albiflora**; more glabrous, with shorter leaves, short calyx teeth and white flowers. *A. multiflora, E. Mey., and Thunb.! Herb. lit. β. non α. A. albiflora, E. & Z.!* 1417. *Zey.!* 2331.

HAB. Very abundant on dry hills and by roadsides throughout the western districts. Vars. *β.* and *γ.* in Uitenhage and Albany. (Herb. Th., Bth., Hk., D., Sd.)

A spreading or prostrate, very much-branched, and ramulose rigid shrub; stems 3–4 feet long; branches spreading or divaricate. Leaves 1–2 lines long or shorter. Calyx teeth sometimes very short, rarely equalling the tube. Flowers among the smallest in the genus, 2 lines long, slender, with a reflexed vexillum and projecting carina. Legume 1½ line long, 1 line wide, compressed.

52. A. diffusa (E. & Z.! No. 1420); *small, slender, villoso-canescent;* leaves ternate or subfasciculate, short, *linear-terete, blunt, villous;* flowers (minute) lateral, subsessile, scattered; calyx albo-villous, its lobes lanceolate, equalling the tube; petals . . . . ?; legume ovate, acute, rather longer than the calyx, silky-canescent.

HAB. Mountain sides near Plettenbergsbay, *Mundt.!* (Herb. Sd.)

A small, depressed, much-branched shrub, 6–8 inches long, with filiform branches, and softly pubescent, with whitish hairs. Leaves 1½ lines long, slightly fleshy, thicker than in *A. thymifolia,* to which this imperfectly known species seems to be allied.

53. A. asparagoides (Linn. f. suppl. 321); much-branched, ramulose; branches hirsuto-tomentose; leaves tufted, *setaceo-filiform, incurved, mucronate,* ciliato-pilose, at length subglabrous; flowers lateral, solitary, sessile; calyx-lobes subulate, acutely mucronate, ciliate, more than twice as long as the hairy tube, and nearly equalling the corolla; vexillum mucronate, pubescent, longer than the glabrous carina; alæ shorter; ovary 2–ovuled; legume obliquely-ovate, acute, pubescent, shorter than the calyx. *Benth.! l. c. p.* 619. *Thunb.! Fl. Cap. p.* 579. *Burch.* 5203 *and* 6131.

HAB. Collected by *Thunberg! Nelson, Bowie! and Burchell.* Knysna, *Dr. Pappe!* (Herb. Th., Bth., Hk., D.)

A much-branched and ramulous, rather slender bush, 1–2 feet high, with reddish bark, the old branches glabrate. Leaves in subdistant fascicles, 4–6 lines long, very slender, remarkably curved, arcuate, red-brown when dry. Vexillum with a long point, rarely muticous. Flowers rufescent.

54. A. calcarata (Harv.); diffuse, *the leaf-tubercle armed with a sharp, rigid spur;* leaves tufted, linear-terete, obtuse or mucronulate, minutely appressedly puberulous or glabrescent, the floral exceeding the calyx-tube; flowers lateral, subsessile, solitary; calyx turbinate, puberulous, its segments *subulate,* equalling the tube; the vexillum *broadly-ovate, acute, thinly silky;* carina *rostrate, glabrous;* ovary 2–ovuled, silky; legume obliquely-ovate, acuminate, thinly silky.

HAB. In the Karroo District of Swellendam, *Dr. Pappe*, 244. (Herb. D.)

A small, but woody, spreading, many-stemmed, and ramuliferous bush, 12–18 inches high. Ultimate twigs crowded, ascending, curved. Spur of the leaf-scars ½–1 line long. Pubescence of twigs, leaves, calyx, vexillum, and legume all close-pressed, very thin, of short hairs. Leaves 3–4 lines long, the younger mucronulate. Flowers 2–3 lines long, yellow, the vexillum fulvous. Legume 3–3½ lines long, 2½ broad at base, three times as long as the calyx.

55. A. Pappeana (Harv.); diffuse, much-branched, twigs tomentulose; *the leaf-tubercle armed with a short, sharp, rigid spur;* leaves tufted, linear-terete, obtuse or mucronulate, nearly glabrous, the floral equalling the calyx tube; flowers lateral, subsessile, solitary; calyx silky, its segments *broadly lanceolate,* equalling the tube; vexillum *obcordate, emarginate, densely silky-villous;* carina *oval, obtuse, straight,* pubescent; ovary 2–ovuled; legume obliquely ovate, acute, thinly silky.

HAB. Swellendam, *Dr. Pappe*, 239. (Herb. D.)

A much-branched, ramulose and rigid, depressed bush, 1 foot high, with foliage somewhat like that of *A. laricifolia.* Spur of the leaf-scars not so long as in *A. calcarata,* from which species this is readily known by its calyx and corolla. The cal.-lobes are green, with a subdefined midrib; the tube pale. Flowers yellow, 3–4 lines long; the vexillum densely silky. Leaves slender, 3 lines long, nearly or quite glabrous. Legume not much longer than the calyx. Differs from *A. opaca* in the armed leaf-tubercle, smaller flowers and calyx, more obtuse leaves, &c.

56. A. albens (Linn. Mant. p. 260, non Benth.); *canescent,* much-branched; leaves tufted, tereti-filiform, blunt, *white-silky;* flowers lateral, 1–3 together, pedicellate, crowded toward the ends of the branches; *calyx silky, with very short, acute teeth;* petals *nearly glabrous,* the alæ shortest; legume ovate, acute, cano-tomentose, longer than the calyx. *Thunb.! Fl. Cap. p.* 576. *A. candicans, Ait.! Hort. Kew Ed.* 2. *vol.* 4. *p.* 264. *Benth.! l. c. p.* 618.

HAB. In the Great Karroo, *Thunberg!* also from *Nelson* and *Oldenberg* in Herb. Banks, fide *Bentham, l. c.* Cape, *Verreaux!* (Herb. Th., D.)

A much-branched, erect shrub, covered in all parts with soft, silky, white pubescence; the flowering branches slender, flexuous, 6 inches long. Flowers lateral, but crowded in a subterminal, racemose series, small, white; the petals glabrous. Calyx 1½ lines long. Leaves 3–4 lines long, sometimes mucronulate. Legumes 2½ lines long. Of this plant there are two excellent specimens, one in flower, one in fruit, in Herb., Thunb., and I adopt it therefore as the '*albens*' of Linnæus, although in Herb. Linn. (fide Benth.!) it is confounded with *A. armata,* which in some points it resembles. It is quite unlike the plant called '*albens*' by E. Meyer; our *A. exilis.*

57. A. rubro-fusca (E. & Z.! 1467); robust, virgate, with hirsute branches; leaves tufted, imbricated, triangular-subulate, erect or incurved, rigid, mucronate, pilose, becoming glabrescent; flowers lateral, sessile, solitary; calyx lobes triangular-subulate, keeled, acute, pilose, twice as long as the hairy tube and equalling the corolla; vexillum and carina villous, subequal, longer than the glabrous alæ; ovary stipitate, 2–ovuled; legume very oblique, ovate-acute, villous, shorter than the calyx. *Benth.! l. c. p.* 619. *A. alpina, E. & Z.! No.* 1474. *Zey.!* 2340.

HAB. Hills near the Zwartkopsriver, Uitenhage, and in the Langekloof, George, *E. & Z.!* Bethelsdorp, *Zeyher.* (Herb. Bth., Hk., D., Sd.)

An erect, strong-growing shrub, 2–3 feet high; with erect, virgate branches, closely imbricated with leaves. Leaves drying a reddish-brown, 3–5 lines long, thicker and more trigonous than in most of this section. Calyx 4 and sometimes 5 lines long. Vexillum obovate, with a sharp acumination. *A. alpina, E. & Z.!* has longer leaves, rather more densely crowded, but does not otherwise differ.

5. **LATERALES** (Sp. 58–74).

58. A. teres (E. & Z.! No. 1460); branches tomentose; leaves tufted, tereti-subulate, rigid, pungent-mucronate, the adult *glabrous;* flowers solitary; calyx widely campanulate, tomentose, truncate, its lobes distant, *linear-subulate, pungent,* longer than the tube; *vexillum villous, scarcely equalling the pubescent carina;* ovary 4–ovuled; legume obliquely ovato-lanceolate, falcate-acuminate, turgid, silky. *Benth.! l. c. p.* 620. *Zey.!* 2368. *Burch.* 4640.

HAB. Vanstaadensberg Mts., Uit., *E. & Z.!* (Herb. Bth., Hk., D., Sd.)

A large, strong shrub, 8–15 feet high. Branches long, ramuliferous, pubescent. Leaves densely tufted, straight, spreading, ½ inch long, green. Flowers few, scattered, 7–8 lines long, on pedicels 1–1½ lines long. Vexillum subfalcate. Calyx teeth rigid, like the leaves, 3 lines long. Legume 8–9 lines long, 3–4 lines wide near the turgid base, compressed toward the narrowed point.

59. A. hirta (E. Mey.! Linn. 7. p. 156); branches villoso-tomentose; leaves tufted, narrow-subulate, rigid, pungent-mucronate, pubescent, afterwards glabrous; flowers lateral, solitary; calyx widely campanulate, densely villous, its lobes distant, *setaceo-pungent,* rather shorter than the tube; *vexillum silky villous, longer than the glabrous carina;* ovary 2–3 ovuled; legume obliquely ovato-lanceolate, turgid, very villous. *A. vulnerans. Benth.! l. c. p.* 621. *non Thunb. E. & Z.! No.* 1458. *A. hirta, E. Mey. Linn.* 7. *p.* 156. *E. & Z.* 1457.

HAB. Swellendam Mts., at Puspasvalley and Kochmanskloof, *Mundt., E. & Z.!* Voormansbosch, *Zeyher,* 2369! Gnadendahl, *Alexander Prior, Pappe!* Giftberg, *Drege.* (Herb. Bth., Hk., Sd., D.)

A large, strong, furzelike bush, with an abundance of softly hairy, yellow flowers and pale green foliage. Leaves ½ inch long, more slender than in *A. teres,* at first covered with whitish, silky hairs, afterwards glabrous and shining. Calyx teeth acicular. Flowers 6–9 lines long. Legume ½ inch long, covered with long white hairs. This plant does not exist in Herb. Thunberg: his *A. vulnerans* is a very different plant.

60. A. hystrix (Thunb.! Fl. Cap. p. 377); *densely tomentose and canescent;* leaves tufted, subulate, terete, rigid, pungent-mucronate, albo-tomentose; flowers solitary, *equalling the leaves;* calyx widely campanulate, softly villoso-tomentose, its teeth from broad deltoid bases *shortly acicular,* shorter than the tube; vexillum densely tomentose, longer than the glabrous carina; ovary 2–ovuled; legume obliquely lanceolate, densely woolly. *Benth.! l. c. p.* 621. *E. Mey.! Comm. p.* 51. *non E. & Z.*

HAB. Dry hills in Onderbokkeveld and Kendo, *Drege!* Ataquaskloof, *Gill!* also *Thunberg! Masson.* (Herb. Th., Hk., Bth., D.)

A large bush with white or silvery, copious pubescence. Leaves about ½ inch long, spreading, sometimes curved, needle-pointed, rarely bluntish. Flowers 5 lines long, yellow. Legume 6–8 lines long, woolly. Tubercles under the leaves bearded.

61. A. rigescens (E. Mey! Comm. p. 52); *cinereo-pubescent;* leaves tufted, terete, subulate, rigid, incurved, pungent-mucronate, pubescent; flowers solitary, mostly shorter than the leaves; calyx widely campanulate, pubescent, *its lobes linear-lanceolate, mucronate, rather longer than the tube* and nearly equalling the corolla; vexillum villous, longer than the pubescent carina; ovary 2–ovuled; legume obliquely lanceolate, appressedly villous. *Benth.! l. c. p.* 622.

VAR. β. **echinata**; branches divaricate; leaves 4 lines long; calyx lobes lanceolate-acuminate, pungent-mucronate, *about as long as the tube*. *A. echinata, E. Mey. Comm. p.* 51. *Benth.! l. c. p.* 622. *A. corrudæfolia, DC. Prod.* 2. *p.* 139. *E. & Z.! En.* 1461. *Zey.!* 2367.

HAB. Near Port Elizabeth, Uit., and Gnadendahl, George, *Drege!* Winterhoeksberg and Koega River, *Zeyher!* 2322. (Herb. Bth., Hk., D., Sd.)

A much and densely branched, leafy bush, with appressed, short, greyish, copious pubescence. Leaves 6–8 lines long, needle-pointed. Flowers yellow. This resembles *A. hystrix*, but the leaves are generally longer, the flowers smaller, the pubescence less tomentose, and the calyx lobes very different. Our var. β. is retained as a species by Bentham, and if we confine attention to the original specimens of *Drege*, it looks distinct; but on comparing several specimens from different collectors and herbaria, I do not find the differences sufficiently constant. Both plants come from the same districts, and β. looks like a plant from a drier situation and probably more stony soil.

62. ? A. opaca (E. & Z.! 1468); branches pubescent; leaf-tubercle unarmed; leaves tufted, short, trigono-subulate, rigid, acute or sub-pungent-mucronate, *the adult glabrous;* flowers solitary; calyx oblique, *its lobes broadly lanceolate, acute, midribbed, longer than the pubescent tube;* vexillum obovate-mucronulate, villous, equalling the *glabrous* carina; ovary 2–ovuled, villous; legume ?

HAB. On hills in Adow, Uitenhage, *E. & Z.!* (Herb. Sd.)

The only specimen seen is about 8 inches high, with virgate, rufescent, thinly tomentulose branches. Leaves 2½–3 lines long, scarcely equalling the internodes, patent, acute but scarcely pungent, brownish when dry. Flowers sessile, in the centres of all the upper leaf-tufts, 4–5 lines long, the calyx longer than the leaves. The calyx lobes have a prominent, keel-like rib, and two fainter, marginal nerves. Possibly this should be placed among the LEPTANTHÆ, next *A. rubrofusca*, with which, and also with *A. Pappeana*, it has several points in common.

63. A. acanthes (E. & Z.! 1459); branches virgate, tomentose; leaves densely tufted, imbricating, long, linear-terete, mucronate, rigid, glabrous, *the younger pubescent;* flowers solitary; calyx widely campanulate, villous, *its teeth lanceolate-subulate, subfalcate, rigid, mucronate, longer than tube;* vexillum subsessile, orbicular, villous, much longer than the *glabrous* carina; legume thick and very woolly. *Benth.! l. c. p.* 623.

HAB. In Kannaland, near Gauritz R., Swell., *E. & Z.! Scholl.!* (Hb. Bth., Sd.)

A robust, erect, slightly branched shrub, with long, simple branches, densely imbricated with leaves, the branch (leaves included) 1½ inches diameter. Leaves incurved 9–10 lines or an inch long, green, the young ones silky. Calyx lobes thick and rigid, 4–5 lines long. Legume turgid, acuminate, 6–8 lines long. Flowers yellow, 6–7 lines long, nestling among the leaves.

64. A. Burchelliana (Benth.! l. c. p. 623); "leaves tufted, long, linear-terete, scarcely mucronate, somewhat fleshy, *glabrous;* flowers subsolitary, shorter than the leaves; calyx widely campanulate, pubescent, *its teeth much shorter than the tube;* vexillum villous, rather longer than the *tomentose* carina; legume thick, turgid, villous." *Benth.! l. c.*

HAB. From *Burchell's* collection, No. 7456.

"At first sight, very similar to *A. verrucosa*, but easily known by its villous flowers and legume. Branchlets thick, tomentose, tuberculate. Leaves 8–10 lines long, numerous, incurved. Calyx 1½ lines long. Corolla twice as long. Carina arched, longer than the alæ."—*Benth.*

65. A. glomerata (Benth.! l. c.) "leaves tufted, long, subulate, pungent-mucronate, rigid, *silky-tomentose;* flowers lateral, glomerato-racemose, much shorter than the leaves; calyx widely campanulate, tomentose, *truncate*, with *setaceous teeth much shorter than the tube;* vexillum broadly ovate, villous, scarcely longer than the *glabrous* carina; legume obliquely ovato-lanceolate, woolly."—*Benth. l. c.*

HAB. From *Burchell's* coll. No. 5786.

"Branches thickish, softly pubescent. Leaf-tufts distant, or crowded at the ends of the branchlets. Leaves very numerous in each cluster, the larger an inch long. Flowering branch often lengthening a little from the fasciculus, bearing 4–12 flowers, separately or 2–3 together, subtended by 1–3 leaves; pedicels often a line long. Calyx 1 line long. Corolla 4 lines. Legume less deflexed than in neighbouring species, but not mature in the specimen examined." *Benth. l. c.*

66. A. longifolia (Benth.! l. c. p. 624); leaves densely tufted, elongate, subulate, subincurved, mucronulate, *silky and silvery;* flowers solitary or glomerate, much shorter than the leaves; calyx widely campanulate, softly villous, with *very short, acute teeth;* vexillum *broadly orbicular, on a long claw*, villous, longer than the glabrous carina; legume obliquely ovato-lanceolate, very villous. *Benth.! l. c.*

HAB. From *Scholl's* collection. (Herb. Bth., Vind.)

Only to be known from the following, which it greatly resembles, by its very short calyx lobes, stipitate vexillum and longer (1–1½ inch) leaves.

67. A. eriophylla (Walp.! Linn. 13. p. 499); leaves densely tufted, elongate, subulate, incurved, mucronate, *silky and silvery;* fl. solitary or glomerate, as long as the leaves; calyx villoso-sericeous, *with subulate-acuminate teeth, nearly as long as the tube; vexillum oboval, villous, narrowed into a short claw*, longer than the glabrous carina; legume obliquely ovato-lanceolate, turgid, silky and very villous. *Benth.! l. c. p.* 624. *Zey.!* 2321.

HAB. Near Grahamstown, *Zeyher! Krebs*, 892. (Herb. D., Hk., Sd.)

A large, strong growing shrub, not much branched, with long, subsimple, densely leafy branches; every part densely clothed with soft, glossy, white or silvery pubescence. Leaves 8–10 lines or an inch long. Flowers nestling among the leaves, yellow. Vexillum 6 lines long, 3 lines wide. Legume 5–6 lines long.

68. A. laricifolia (Berg. Pl. Cap. p. 204. non Lam.); leaves densely tufted, short, terete-subulate, mucronate, *glabrescent;* flowers solitary, lateral; calyx widely campanulate, villous, with subulate-acuminate teeth, shorter than the tube; vexillum pubescent, longer than the glabrous or villous carina; legume obliquely lanceolate, subacute, turgid, very villous. *Benth.! l. c. p.* 625. *E. & Z.! No.* 1452. *A. laricina, DC. Prod.* 2. *p.* 141. *A. genistoides, E. & Z.! No.* 1453. *Zey.!* 2358.

VAR. β. **sericantha**; calyx teeth nearly equalling the tube; *carina villous. A. sericantha, E. Mey.! Comm. p.* 49. *Benth.! l. c. p.* 625. *Burch.* 3485. *A. hilaris, E. & Z.!* 1451.

HAB. Hills near Capetown, and throughout the western districts, common. Var. β. in the eastern districts, Caffraria and Port Natal. (Herb. Bth., Th., H., D., Sd.)

A much branched, ramuliferous, spreading bush; branchlets 3–12 inches long, shortly tomentose. Leaves 3–4 lines long, glabrous or minutely downy, especially the younger. Flowers yellow, very numerous, toward the ends of the branches. This plant is only known from *A. canescens* by its green, not canescent leaves. I venture to reduce Meyer's *A. sericantha*, which differs by a very inconstant character: glabrous and pubescent carinæ occur indifferently in many species.

69. A. canescens (Linn. Mant. p. 262); leaves tufted, subulate, acute, *canescent or silky;* flowers solitary, lateral; calyx widely campanulate, villous, with subulate-acuminate teeth mostly shorter than the tube; vexillum pubescent, longer than the glabrous or pubescent carina; legume obliquely lanceolate, turgid, very villous. *Benth.! l. c. p.* 625. *Thunb.! Cap. p.* 577. *E. & Z.! No.* 1448.

VAR. β. **sericophylla**; leaves white and silky. *A. neanthes, E. & Z.!* 1449. *Zey.!* 2328.

VAR. γ. **Bowieana** (Benth.); larger, with longer (4–5 line) leaves; shorter calyx teeth; larger flowers, and a more acute legume, 7 lines long. *Benth. l. c.*

VAR. δ. **Joubertiana**; leaves very densely imbricated; cal. lobes as long as the tube; vexillum glabrescent; carina rostrate, acute. *A. Joubertiana, E. & Z.* 1450.

HAB. Western Districts from Capetown to Swellendam, common; the varieties chiefly in Swellendam. (Herb. Th., D., Bth., Hk., Sd.)

Erect or spreading, robust, much branched; branches virgate or ramulous. Leaves always pubescent, but sometimes thinly so; mostly canescent, and in β. almost silvery. Except in its pubescence it does not materially differ from the equally common *A. laricifolia.*

70. A. frankenioides (DC. Prod. 2. p. 139); ramulose, divaricate, with *tomentose branchlets;* leaves tufted, short, obtuse or subacute, pointless, linear-terete, canescent or tomentose, the old glabrescent; flowers solitary, sessile; calyx campanulate, pubescent or tomentose, with subulate teeth shorter than the tube; vexillum pubescent or tomentose, rather longer than the glabrous or pubescent carina; legume ovato-lanceolate, villoso-tomentose, scarcely twice as long as the calyx.

VAR. α. **chortophila**; leaves thinly pubescent, becoming subglabrous; carina villous. *A. chortophila, E. & Z.!* 1436. *Benth.! l. c. p.* 626. *A. frankenioides, E.M.*

VAR. β. **intermedia**; leaves very short; carina pubescent. *A. intermedia, E. & Z.* 1437. *Benth.! l. c. p.* 626.

VAR. γ. **albanensis**; leaves longer and more hairy; carina either minutely puberulous or quite glabrous. *A. frankenioides, Benth.! l. c. p.* 627. *A. albanensis, E. & Z.!* 1435.

VAR. δ. **poliotes**; leaves longer and more tomentose; carina glabrous or pubescent. *A. poliotes, E. & Z.!* 1446. *Benth. l. c. p.* 627. *A. leptothria, E. & Z.!* 1447. *A. tomentosa, E. Mey.! Comm. p.* 55. *Zey.!* 2329.

VAR. ε. **alpina**; leaves as in α.; vexillum and carina thinly pubescent or glabrescent; calyx-lobes as long as the tube. *A. hiatuum, E. & Z.!* 1438.

HAB. Common, in one or other of its forms, in Uitenhage and Albany, *E. & Z.! Drege! &c.* (Herb. Bth., Hk., D., Sd.)

A much branched, spreading or depressed, small bush, with canescent branches and foliage, and yellow flowers. Leaves 1–2 lines long, in δ. 2–3 lines, variable in pubescence, sometimes glabrous, but even then whitish or grey. Flowers 3–3½ lines long, the corolla thrice as long as the calyx. Vexillum broadly orbicular or subcordate; the carina blunt, hairy or smooth. Legume 2–4 lines long, more or less taper-pointed. Calyx teeth very variable in length, sometimes obsolete! Our 5 varieties are upheld by Bentham as species; otherwise I should scarcely have indicated them all. This species varies like *A. thymifolia,* of which it has the habit, but from which, and others of the *Leptanthæ* section, it differs in the broadly cordate vexillum and in the calyx.

71. A. setacea (E. & Z.! 1462); branches villoso-tomentose; leaves tufted, setaceo-subulate, pungent-mucronate, *villoso-canescent;* flowers solitary; calyx villous, campanulate, *its lobes lanceolate-subulate, needle-pointed, longer than the tube;* vexillum villous, longer than the villous carina; legume ovato-lanceolate, appressedly villous, longer than the calyx.

VAR. α. Eckloni; calyx teeth 1½ as long as tube; vexillum setaceo-mucronate. *A. setacea, Benth.! l. c. p.* 618. *A. alopecuroides, E. Mey.! Comm. p.* 52.

VAR. β. Gillii; calyx teeth 1¼ as long as tube; vexillum mucronulate. *A. Gillii, Benth.! l. c. p.* 617.

HAB. Vanstaadens Hills, Uitenhage, *E. & Z.! Drege!* and at the mouth of the Omsamcaba, *Drege!* Caffirland, *Dr. Gill.* (Herb. Bth., Hk., D.)

A densely much branched, villous shrub, 2–3 feet high; the branches erect, ramulous, densely imbricated with leaves. Leaves 3–6 lines long, the younger ones all needle-pointed. Flowers sessile, yellow, nestling among the leaves. The flowers in *A. Gillii* are a very little larger than in Ecklon's plant, with somewhat shorter, but similarly shaped and pointed calyx-lobes, and less mucronate vexillum; but in foliage and general habit both are very similar.

72. A. Alopecurus (Burch. Cat. 5561); branches woolly and very villous, densely leafy; leaves tufted, setaceo-subulate, pungent-mucronate, *hairy; calyx softly villous, its lobes setaceous,* 1½ *to twice as long as the tube;* vexillum and carina villous; legume ovato-lanceolate, woolly, much longer than the calyx. *Benth.! l. c. p.* 618. *A. comosa, E. & Z.!* 1426, *non Thunb.*

HAB. S. Africa, *Burchell, Thom!* Mountain sides, near the Howhoek Pass, Caledon, *E. & Z.!* (Herb. Hk., Sd.)

A robust, diffusely branched and very softly hairy bush, 1–2 feet high; the branches and leaf-tubercles shaggy. Leaves closely imbricated, 4–6 lines long, very slender, fulvous or dark in drying. Flowers small, pale, nestling among the leaves; vexillum ovate, stipitate, pointless. Legume 3–4 lines long.

73. A. incurvifolia (Walp. Linn. 13. p. 497); branches tomentose, densely leafy; leaves tufted, incurved, setaceo-subulate, mucronate, *glabrous* or nearly so; flowers solitary; calyx thinly pubescent, campanulate, *with short, deltoid teeth;* vexillum *glabrous or subpubescent,* scarcely longer than the glabrous carina; legume obliquely ovato-lanceolate, turgid, densely villous, much longer than the calyx. *Benth.! l. c. p.* 618.

HAB. S. Africa, *Burchell, Bowie!* (Herb. Hk., Bth.)

A large, robust, spreading bush, with long, virgate, leafy branchlets. Leaves spreading or squarrose, 3–4 lines long, all but the young ones glabrous. Flowers abundant, distributed along the branches, small and nearly glabrous.

74. A. Chamissonis (Vog. Linn. 10. p. 597); *branchlets and gemmæ spiniferous;* leaves tufted, short, linear-terete, blunt, silky-canescent; flowers solitary; calyx softly villous, widely campanulate, with subulate teeth scarcely equalling the tube; petals glabrous; ovary 2–ovuled; legume obliquely ovato-lanceolate, thick, softly pubescent. *Benth.! l. c. p.* 619. *A. acanthophylla, E. & Z.!* 1496.

HAB. Tigerberg, *Chamisso.* Swellendam, *Pappe!* 223. Sides of Table Mountain, 1–2000f. *Mundt.!* Oct. (Herb. Hk., Sd., D.)

A robust, straggling, thorny bush; the spines either ending the branches, or issuing from the middle of the leaf-tufts. Leaves 2–3 lines long, canescent, shorter than the horizontally patent spines. Flowers small and yellow. E. & Z. give Swellendam as the district where Mundt gathered this plant; but the original specimen in Hb. Hook., and of which Ecklon's seems to be a branch, is marked as above in Mundt's hand-writing.

6. **MACROCARPÆ.** (Sp. 75–78.)

75. A. filicaulis (E. & Z.! 1395); *suffruticose*, ascending, thinly pilose, *then glabrescent;* branches subsimple, incurved, angular, with prominent leaf-tubercles; leaves ternate or fascicled, unequal, linear, acute or mucronate, compressed, incurved, pilose or glabrous; *flowers solitary and lateral, or* 4–5 *in a terminal, imperfect umbel*, pedicellate, with setaceous bracteoles; calyx campanulate, striate, thinly villous, its segments subulate, distant, longer than the tube; vexillum silky, obovate, equalling the cuneate, glabrous carina; ovary linear, 16–20 ovuled; legume linear, many-seeded, 3–4 times as long as the calyx. *Benth.! l. c. p.* 619.

VAR. *β.* **subumbellata;** flowers 4–5 in an imperfect umbel.

HAB. Near the Tulbagh Waterfall, *E. & Z.! β.* Scurfdeberg, *Zeyher!* (Herb. Hk., Bth., Sd.)

A slender suffrutex, branched chiefly from the base; branches 6–12 inches long, reddish brown. Leaves 3–6 in each tuft, slender, 4–6 lines long, pale green. Pedicels 1–2 lines long. Flowers yellow, 3–4 lines long, the carina semicircular, strongly arched in front, subacute. Legume 10–11 lines long, 1–1½ lines wide, acute, subturgid or compressed.

76. A. macrocarpa (E. & Z.! 1392); *robust, woody*, glabrescent; branches rigid, virgate, the leaf-tubercle prominent, tomentose, *mucronate* (or unarmed); leaves densely tufted, linear-subulate, glabrescent; flowers solitary, pedicellate, cernuous; calyx *downy*, ribbed, *its teeth scarcely so long as the tube;* legume long and lanceolate, turgid, villoso-pubescent. *Benth. l. c. p.* 630.

HAB. Mountains near Wagenmakersbosch, *Swell.! Mundt!* (Herb. Hk., Sd.)

A robust but not densely branched shrub, resembling *A. Wildenowiana*, but with a very different pod. The leaf-tubercles on the younger twigs are armed with a minute spine, on the older unarmed, but densely albo-tomentose and pulvinate. Leaves 4–5 lines long, pale green, many in each tuft. Flowers unknown. The ripe pods are an inch long, nearly 2 lines wide, acuminate.

77. A. pinea (Thunb.! Cap. p. 582); branches rigid, virgate, the young parts villous, older glabrescent; leaf-tubercle small, *pointless;* leaves tufted, linear-subulate, acute, the young villous, the adult glabrous; flowers lateral, pedicellate, cernuous; calyx *villous*, widely and obliquely campanulate, ribbed; *its teeth from a broad base subulate, longer than the tube;* vexillum subrotund, densely villous, the carina lunate and rostrate; legume linear-lanceolate, villous, many seeded.

HAB. South Africa, *Thunberg!* Winterhoeksberg, *Pappe!* (Herb. Th., D.)

An erect, virgate, slightly branched shrub, at length nearly glabrous, when young villous; old branches glabrous, pale, with decurrent raised striæ. Leaves ½–¾ inch long, very slender, not very densely tufted, but closely set and imbricating each other; the leaf-tubercles small, dark brown, quite blunt or minutely mucronulate. Pedicels lateral, 1–2 lines long, hairy, the branch extending beyond the uppermost flower. Calyx thinly clothed with long, soft hairs. Vexillum very broad and hairy; the carina as long, very much arched, semicircular. Ovary at least 16–ovuled. Legume 8–12 lines long, 2 lines broad, linear, acute, villous. The *A. pinea*, Benth. is our *A. macrantha.*

78. A. Garipensis (E. Mey. Comm. p. 44); shrubby, divaricate, with spinous branches; leaves tufted, linear-spathulate, subinvolute, channelled, obtuse, silky-canescent; "flowers solitary, shortly pedicellate, lateral, near the ends of spine-tipped ramuli; calyx silky, its lobes lanceolate, pointless, shorter than the tube; vexillum and carina silky towards the points; legume straight, linear, silky." *E. Mey.! l. c.*

HAB. Banks of the Gariep, *Drege!*

Of this I have only seen a fragment without flower or fruit, in Hb. Sond.

7. GRANDIFLORÆ. (Sp. 79–84.)

79. A. Willdenowiana (Benth. l. c. p. 630); branches rigid, the leaf tubercle unarmed *or spinoso-mucronulate;* leaves setaceo-filiform, subglabrous, acute; flowers solitary or few, on very short pedicels; calyx villous, widely and obliquely campanulate, with subulate, acuminate teeth as long as the tube; vexillum villous; carina glabrous, strongly arched and beaked; *stigma scarcely oblique;* legume obliquely lanceolate, very villous. *Benth. A. verrucosa, Willd., non L. E. Mey.! Comm. p.* 50. *A. hystrix, E. & Z.!* 1481, *non. Linn. f. A. uniflora, Thunb. non L.*

HAB. Hill sides, Puspas Valley, Swell. *E. & Z.!* Cape Town Hills, *Drege!* Hout Bay, *Alexander Prior!* (Herb. Bth., Sd., Hk.. Th.)

A tall shrub, erect and loosely branched; branches virgate, tapering off to a fine point, the younger twigs pubescent, older nearly glabrous. Leaves densely tufted, slender, $\frac{1}{3}$–$\frac{2}{3}$ inch long. Flowers $\frac{1}{2}$ inch long, the vexillum subsessile and very wide. Ovules 6. Stigma much less oblique than in the following. Legume 6 lines long, $2\frac{1}{2}$ wide.

80. A. leptophylla (E. & Z.! 1482); branches rigid, spine-tipped, the leaf-tubercle *sharply spiniferous;* leaves setaceo-filiform, mucronulate, subglabrous; flowers solitary or in pairs, shortly pedicellate; calyx pubescent, widely campanulate, with lanceolato-subulate teeth as long as the tube; vexillum villous, carina glabrous, strongly arched and beaked; *stigma obliquely decurrent;* legumes obliquely lanceolate, villous. *Benth.! l. c. p.* 31. *A. laricifolia, Lam. non Berg. A. verrucosa, litt. b. & c. E. Mey.!*

HAB. Hills and mountain sides, Tulbagh, *E. & Z.!* The Paarl and Dutoitskloof, *Drege! Alexander Prior!* (Herb. Bth., Hk., Sd., D.)

Very like *A. Willdenowiana*, from which it is known by its sharply-aculeate leaf-tubercles and decurrent stigma. Branches reddish yellow, shining; leaves pale, $\frac{3}{4}$ inch long.

81. A. rostrata (Benth.! l. c. p. 361); divaricate, with rigid, spine-tipped branchlets; the leaf-tubercle spinoso-mucronulate; leaves linear-terete, bluntish, *cano-puberulent;* flowers solitary, shortly pedicellate; calyx pubescent, with subulate teeth *shorter than the tube;* vexillum villous, *shorter than the arched, long beaked, glabrous carina;* stigma oblique.

HAB. From *Scholl's* collection. (Herb. Benth.)

Allied to *A. leptophylla*, but with much shorter, minutely canescent, blunter and thicker leaves, shorter calyx-lobes, and a longer and more tapering carina.

82. A. macrantha (Harv.); branches virgate, villous; the leaf-tubercle mucronulate; leaves densely tufted, setaceo-subulate, mucronulate, at length glabrous; flowers towards the ends of the branches lateral or terminal, few together, shortly pedicellate; *bracteoles simple;* calyx widely campanulate, *its teeth from a broad, ovate base shortly subulate-acuminate;* vexillum villous, oblong-ovate, acute, equalling the glabrous arched and beaked carina. *A. pinea, Benth.! l. c. p.* 631, *non Thunb. A uniflora, E. & Z.!* 1483, *non L. nec. Thunb. Sieb. Fl. Cap. p.* 161.

HAB. Hott. Holl., *E. & Z.! Sieber! Reeves! Forbes!* (Herb. Bth., Hk., Sd.)

A tall, densely branched and densely leafy shrub, with erect, rod-like branches, crowned with 2–4 large, fulvous-yellow flowers. The younger leaf-tubercles are mucronate, but not spinous. Leaves very slender, ½–¾ inch long, the young ones softly pilose, older glabrous and pale green. Vexillum tapering to an acute point. Ovules about 7. Legume thick, an inch long, 3–4 lines broad at base, somewhat falcate.

83. A. grandiflora (Benth.! l.c. p. 32); branches ramuliferous, hairy; leaves tufted, linear-trigonous, keeled, acute, pilose or glabrous, shining; flowers solitary or few, subsessile, at the end of short branchlets; *bracteoles trifoliolate;* calyx villoso-hirsute, *its lobes broadly lanceolate,* rather longer than the tube;* vexillum villous, obovate, obtuse; carina glabrous, arched and rostrate; *stigma very oblique.*

HAB. From *Thom's* collection. (Herb. Hook.)

Probably a large shrub. Branches thick and woody; the younger roughly hairy; ramuli closely set, 2–3 inches long, ending in 1-2-3, uncial, fulvous, yellow flowers. Leaves on the branches 2–3 lines, on the ramuli ½ inch long, reddish brown, glossy. Calyx 6 lines long. Nearly allied to *A. galeata*, but a coarser, more roughly hairy bush, with longer leaves, larger flowers, and an oblique stigma, &c.

84. A. galeata (E. Mey.! Comm. p. 49); branches ramuliferous, pilose, soon glabrous; leaves tufted, short, linear-keeled, acute, soon glabrous, shining; flowers few, ending the branches, shortly pedicellate; *bracteoles trifoliolate;* calyx thinly pilose, *its lobes subulate, acuminate, pungent, longer than the tube;* vexillum villous, obovate, obtuse; carina glabrous, arched and rostrate; *stigma straight;* legume obliquely lanceolate, falcate, villous. *Benth. l. c. p.* 632.

HAB. Clanwilliam, between Pikenierskloof and Olifant's River, *Drege!* (Herb. Bth., Hk., D.)

A much branched, leafy bush, with large, 8–9 lines long, fulvescent flowers. Leaves 2–3 lines long, rufescent. The leaves are shorter than in any other of this section.

8. PACHYCARPÆ. (Sp. 85–87.)

85. A. densifolia (Benth. l. c. p. 632); branches virgate, hairy; leaves very densely tufted, subulate, mucronate, glabrescent, those under the 2–4 flowered heads plumoso-ciliate; calyx piloso-villous, obliquely campanulate, the two upper teeth broadly ovate-acuminate, the 3 lower lanceolate, *all shorter than the tube;* vexillum very villous, longer than the incurved, obtuse, glabrous carina; *ovary 6-ovuled;* legume obliquely rhombic-ovate, acuminate, thick, villous. *Benth.!*

HAB. Witsenberg, *Zeyher*, 428! Dec. (Herb. Hk., Bth., Sd.)

A robust, erect, densely leafy, not much branched bush; the branches very erect, 8–14 inches long, ending in small, subglobose heads of flowers, encircled with white, hairy, floral leaves. Flowers deep yellow or orange, half inch long. Calyx inflated, unequal; the lowest lobe narrow. Leaves 4 lines long, the upper ones ½ inch, glabrous and glossy, brownish. Very near the following, but with longer leaves and larger flowers.

86. A. triquetra (Thunb.! Fl. Cap. p. 578); branches ramuliferous, pilose; leaves densely tufted, *short*, incurved, acute, linear, strongly keeled, glabrous or pilose; flowers 2–4 together, capitate; calyx villoso-pilose, obliquely campanulate, the four upper teeth triangular, acute, the lowest lanceolate, *all nearly equalling the tube;* vexillum pubescent, equalling the arched and beaked, glabrous carina; *ovary 4-ovuled;* legume obliquely rhomboid, thick, hairy. *Benth. l. c p.* 633. *E. & Z.* 1480.

HAB. Cederberg and Dutoitskloof, *Drege.!* Tulbagh, *E. & Z.!* (Herb. Th., Bth., Hk., D., Sd.)

Robust, tall, densely branched, and leafy. Leaves scarcely equalling the internodes, 2–2½ lines long, incurved, the younger piloso-ciliate, the old glabrous and glossy, all flat on the upper side and keeled below; triangular. Flowers 5 lines long. Legume woody, 4 lines long and wide. A handsome shrub, with foliage resembling that of *Erica cerinthoides.*

87. A. propinqua (E. Mey.! Comm. p. 53); branches tomentose; leaves tufted, *small, trigonous, obtuse, cano-tomentose;* flowers terminal, shortly pedicellate, 2–3 together; calyx softly villous, campanulate, the four upper teeth triangular, lowest lanceolate, all rather shorter than the tube; vexillum pubescent, as long as the glabrous, arched and beaked carina; *ovary 6-ovuled;* legume obliquely ovate, acute, silky-villous. *Benth.! l. c. p.* 633.

HAB. Cederberg, *Drege!* (Herb. Benth.)

A small ramulous shrub, thinly tomentose and canescent in all parts. Leaves about a line long, few in each fascicle. Flowers 5 lines long, in pairs (or threes) at the ends of the branchlets, on pedicels 1 line long. Bracteoles trifoliolate. Very much smaller than either of the preceding.

9. **CARNOSÆ.** (Sp. 88–101.)

88. A. callosa (Linn. Sp. 1002); quite *glabrous,* or with the twigs tomentulose; leaves *ternate, linear, mucronate-acute, or obtuse, flat above, round-backed;* flowers spiked or capitate; calyx-lobes *ovate or ovato-lanceolate,* acute or acuminate, rather shorter than the tube or equalling it; ovary 4–5 ovuled, glabrous or subcanescent; legume obliquely lanceolate, twice as long as the calyx. *Benth.! l. c. p.* 634. *Bot. Mag. t.* 2329. *E. & Z.!* 1272, 1273. *Thunb.! Cap.* 573.

VAR. β. **brevifolia**; twigs tomentulose; leaves 1–2 lines long; flowers smaller and fewer. *A. tylodes, E. & Z.!* 1374.

VAR. γ. **fusca**; dwarf, robust, with tomentulose twigs; leaves 4–5 lines long; flowers 2–3 together, terminal; bracts subulate; calyx-lobes lanceolate, fully as long as the tube. *A fusca, Thunb.! Cap. p.* 574. *Benth. l. c. p.* 599.

HAB. Common on Table Mountain and on the Cape Flats, &c. Hott. Holl. and the Western Districts. (Herb. D., Bth., Hk., Sd.)

Erect, somewhat virgate, 1–2 feet high; branches erect, angular. Leaves pale green, erect, 4–6 lines long, ½ line wide. Flowers either laxly spiked or densely subcapitate, minutely pedicellate, the bracts ovato-lanceolate, equalling the calyx-tube; bracteoles lanceolate, acute at each end. Flowers yellow, quite glabrous. Var. γ. (in Hb. Upsal) is much stunted, with few flowers, and rather longer and more tapering calyx-lobes; it may be distinct.

89. A. erythrodes (E. & Z.! 1375); branches *rufo-tomentose;* leaves *ternate,* linear, mucronate, semiterete, carnose, incurved; flowers spicate; calyx-lobes *lanceolate, rather longer than the tube;* ovary 7–8 ovuled, glabrous.

HAB. Hill sides near Tulbagh, *E. & Z.!* (Herb. Sond.)

A small shrub, with spreading, ramulose, slender branches, clothed with short, foxy pubescence. Leaves 4–6 lines long, sharply mucronate, shrivelling when dry, patent, with the points incurved. Bracts green, with purple tips, narrow lanceolate. Spikes few flowered. Calyx purpurascent, its lobes longer and narrower than in *A. callosa*, to which this is nearly allied.

90. A. variegata (E. & Z.! 1376); leaves ternate or tufted, short, slender, fleshy, blunt or mucronulate, glabrous; flowers *in loose racemules or subumbellate;* calyx campanulate, downy, somewhat fleshy, *its teeth ovate-acuminate, shorter than the tube;* vexillum nearly glabrous, shorter than the arched carina; ovary 4-ovuled; legume obliquely lanceolate, glabrous, twice as long as the calyx. *Benth. l. c. p.* 634.

HAB. Cape Flats, *E. & Z.! Wallich, W.H.H., &c.* Varschevalley, *Zeyher, No.* 429. (Herb. D., Bth., Sd.)

A much branched, slender, subcorymbose, ramulous bush, 1–1½ foot high; branchlets thinly and minutely puberulent, with laxly set, pale foliage. Leaves 2–3 lines long. Racemules 3–6 flowered, the flowers yellow. Legume 5 lines long, 2 lines wide. Bracts and bracteoles linear-lanceolate.

91. A. carnosa (Berg. Pl. Cap. p. 206; non E. & Z.); branchlets rufo-tomentulose; leaves tufted and ternate, short, linear-terete, fleshy, blunt or mucronulate, glabrous; flowers pedicellate, capitato-racemulose, few together; calyx campanulate, fleshy, pubescent, *its teeth broadly oval, obtuse, glabrous, half as long as the tube;* vexillum downy on the ridge, equalling the glabrous carina; legume obliquely lanceolate, about twice as long as the calyx. *Benth.! l. c. p.* 635. *Thunb.! Fl. Cap. p.* 580. *Bot. Mag. t.* 1289.

HAB. Capetown and Simon's Bay Hills, *Thunberg! Scholl! Bowie! M'Gillivray! W.H.H., &c.* (Herb. Thb., Bth., D., Hk.)

A very densely branched shrub, 2–4 feet high and wide; glabrous except on the young twigs and the flowers. Leaves pale, 2–3 lines long, thick and blunt, patent. Flowers pale yellow. Legume 4–5 lines long. Bracts oblongo-cuneate, or linear-oblong; bracteoles lanceolate or linear.

92. A. Priori (Harv.); branchlets pubescent; leaves tufted, short, linear-terete, fleshy, blunt, glabrous, *the younger ones and the floral setoso-ciliolate;* flowers sessile, capitate, few together; *calyx very oblique, setoso-pubescent, the two upper lobes deeply parted, ovato-lanceolate, the three lower lanceolate, acute, longer than the tube;* vexillum silky, longer than the very blunt carina; ovary villoso-sericeous, 2-ovuled, stipitate.

HAB. Table Mountain, Capetown, *Dr. Alexander Prior!* (Herb. Benth.)

A much branched shrub, in aspect very like *A. carnosa;* but the young leaves are setulose; the floral longer, lanceolate, pubescent and rigidly ciliate; the flowers very much smaller, with differently shaped petals, and the calyx very dissimilar.

93. A. sarcodes (Vog., ex Walp. Linn. 13. p. 480); *glabrous;* leaves tufted or ternate, linear, *mucronate*, fleshy, subterete; flowers pedicellate, one or two, terminal; bracts and bracteoles *broadly ovate;* calyx widely campanulate, glabrous, fleshy, *with broadly ovate, obtuse lobes, dilated at the sinuses and nearly equalling the tube;* ovary with several ovules; legume obliquely lanceolate. *Benth.! l. c. p.* 635. *Sarcophyllum carnosum, Thunb.! Fl. Cap. p.* 573. *Sarcocalyx capensis, Walp. l. c.*

HAB. Steenberg, at FalseBay, *Thunberg! Pappe! Dr. Hooker! Dr. Alexander Prior!* (Herb. Thb., Bth., D., Hk., Sd.)

A loosely branched ramulous shrub, 3–4 feet high and broad, glabrous in every part except in the axils of the leaves and round the leaf-scars. Branches pale, yellow green, somewhat succulent. Leaves incurved, patent, 5–6 lines long, ½ line in diameter. Bracts and bracteoles thick, keeled, 2½ lines long, 1½ wide. Calyx nearly 6 lines long, its lobes dilated at the base, and imbricating or recurved. Corolla 9–10 lines long, the vexillum reflexed, the alæ and carina straight, oblong, obtuse. Legume 9–10 lines or an inch long.

94. A. sarcantha (Vog., ex Walp. Linn. 13. p. 689); branches tomentulose; leaves tufted, terete, linear, obtuse or mucronulate, fleshy, glabrous, or the upper downy; flowers pedicellate, 4–5 terminal, capitato-umbellate; bracts ovate or oblong, bracteoles lanceolate, acute, equalling the calyx-tube; calyx *puberulent*, campanulate, fleshy, *its lobes ovate-acuminate, often falcate, equalling the tube;* vexillum pubescent on the ridge, equalling the glabrous carina. *Benth. l. c. p.* 636. *A. carnosa, Linn. Mant. non Berg.*

HAB. Cape District, *Mundt. Wallich! &c.* Simon's Bay, *Dr. Hooker! Dr. Alexander Prior.* (Herb. Bth., D., Hk.)

Almost intermediate in character and aspect between *A. carnosa* and *A. sarcodes;* more nearly related to the former, from the more luxuriant states of which, except by the longer and acute calyx-lobes, it is sometimes with difficulty distinguished. Foliage pale. Flowers yellow.

95. A. capitata (Linn. Amoen. Acad. 6. p. 92); branches *rufo-villous;* leaves densely tufted, linear, 3–cornered, somewhat fleshy, mucronate-acute, incurved, *villoso-ciliate,* or glabrescent; flowers *densely capitate;* calyx widely campanulate, *thinly villous, the lobes broadly ovate, acute, somewhat longer than the tube;* bracts and bracteoles ovato-lanceolate, villous; vexillum glabrous, equalling the rostrate carina, which is villous on its upper edge; ovary 2–ovuled. *Benth. l. c. p.* 636. *Lam. Ill. t.* 620. *f.* 2. *Thunb.! Cap. p.* 578. *E. & Z.!* 1479. *A. glomerata, Linn. f. Sup.* 321.

HAB. Table Mountain, &c., *Thunberg* and most recent collectors. (Herb. Th., Bth., Hk., D.)

Robust 2–3 feet high, not very densely branched; the branches a foot or more in length, erect, simple or ramulose, clothed, as well as the margins of the leaves, with long soft hairs, under which is a reddish tomentum. Leaves pale, 4–5 lines long, more slender than in the preceding. Flowers ½ inch long, in many-flowered heads, yellow or buff. Carina strongly arched and taper-pointed, woolly at the upper edge near the base. Ovary ciliate on the ventral suture, otherwise glabrous.

96. A. subulata (Thunb.! Fl. Cap. p. 583); branches tomentulose; leaves subfasciculate, short, linear-trigonous, *pungent-mucronate or mucronulate-obtuse,* glabrous; flowers *subcapitate,* few together; calyx puberulous, *its lobes lanceolate, carinate, very acute, nearly equalling the tube;* petals glabrous, of equal length, the carina arched, bluntly subrostrate; ovary 4–ovuled. *A. floribunda, Benth.! l. c. p.* 636. *Zey.* 2347.

HAB. Zwarteberg, near Caledon, *Mundt. Bowie. Pappe!* 163. (Herb. Thb., Hk., Bth., D., Sd.)

A much branched, ramulous, erect shrub, with the habit of *A. carnosa,* from which it is sufficiently distinguished by the pungent leaves and the calyx. Leaves two lines long, spreading, variably mucronate, and sometimes almost pointless.

97. A. collina (E. & Z.! 1488); divaricate, the twigs pubescent; leaves tufted, short, trigonous, rigid, pungent-mucronate, glabrous; flowers terminal, solitary, subsessile; calyx glabrous, with deltoideo-subulate, acuminate, pungent segments, rather longer than the tube;

vexillum *thinly silky*, scarcely longer than the arched-subrostrate, glabrous carina; ovary 4–ovuled, legume obliquely ovate, turgid. *Benth.! l. c. p.* 657. *A. versicolor, E. Mey.! Comm. p.* 48.

Hab. Hills near Port Elizabeth, *E. & Z.* Groot Zwarteberg? *Drege.* (Herb. Bth., Sd., D., Hk.)

A much branched, divaricately flexuous and twisted, scraggy bush, with many short ramuli, each bearing a solitary, terminal, pale yellow flower. Leaves subdistant, about 2 lines long, patent and pungent. Flowers on very short pedicels. Calyx-teeth attenuated from a deltoid base.

98. A. aciphylla (Harv.); branches tomentulose; leaves tufted, subulate, *aristato pungent*, glabrous; flowers 2–4 together, or solitary, sessile; calyx fleshy, glabrous, its lobes ovato-subulate, pungent-acuminate, equalling or exceeding the wide tube; petals glabrous, subequal, the carina arched, obtuse; ovary 6–ovuled; bracts ovate, keeled, pungent. *A. abietina, E. Mey. Comm. p.* 48. *Benth. l. c. p.* 637.

Var. β. nana; very dwarf; leaves *densely crowded, subulate, taper-pointed, and pungent-mucronate*, squarrose; flowers 1–2 terminal, depauperated. *A. batodes, E. & Z.* 1455.

Hab. Grootzwarteberg and the Krom Riv., *Drege.* Zwarteberg (but probably from higher, drier, and more rocky ground) *Ecklon and Zeyher.* (Herb. Bth., Sd.)

A rigid, much branched, robust bush, well covered with twigs and leaves. Leaves very patent, unequal in the same tuft, 3–4–5 lines long. Flowers either solitary in lateral tufts, or more frequently subcapitate at the ends of the branches.

99. A. arida (E. Mey.! Linn. 7. p. 156); quite glabrous or thinly and minutely downy; leaves tufted, linear-terete, blunt or mucronulate; flowers solitary, shortly pedicellate (*very variable in size*); calyx widely campanulate, glabrous, carnose, the lobes from a very broad, ovate base, acuminate, as long or twice as long as the tube; petals glabrous, broad, nearly equally long; ovary 2–ovuled; legume obliquely rhombic-ovate, enclosed in the marcescent and enlarged corolla. *Benth. l. c. p.* 638.

Var. α erecta; unarmed; stem and branches erect, slender, elongate; leaves shorter and flowers smaller, 2–4 lines long; calyx-lobes *more subulate* and often shorter than the tube. *A. pinguis, E. & Z.!* 1487. *Zey.!* 420.

Var. β. procumbens; rigid, procumbent, much branched, *the branches spine-tipped*; leaves longer, flowers larger, 4–5 lines long, with lanceolate bracts; calyx-lobes more *ovato-lanceolate*, overlapping at edges. *A. arida, E. & Z.* 1485. *A. spinescens, DC.?*

Var. γ. grandiflora; erect, unarmed; leaves thicker, carnose; flowers very large, 5–8 lines long; calyx-lobes *ovate-acuminate*, much longer than the tube, their edges strongly imbricating; bracts lanceolate. *Zeyher*, 2355.

Hab. Common on the Capetown hills and by road sides in dry places; also throughout the Western Districts. Var. γ. Howhoek Pass, *Zeyher; also in Forsyth's* and *Bowie's* Coll. (Herb. Bth., Hb., D., Sd.)

An extremely variable plant, especially in the size of the flowers, which in var. *grandiflora* are sometimes 8 lines long, while in the smallest varieties they are scarcely 2 lines. The calyx varies with the size of the flowers, its lobes passing from *subulate* to *broadly ovate.* But between these extreme forms a perfect gradation may be established, and though our three varieties are very distinct in their typical condition, the limits are indefinable.

100. A. pachyloba (Benth. l. c. p. 638); branches thick, densely tomentose; leaves tufted, short, fleshy, linear-terete, very obtuse or mucronulate, rarely subacute, glabrous; flowers solitary, lateral, sessile; calyx carnose, glabrous or pubescent, its segments lanceolate or deltoid, acute, nearly as long as the tube or shorter; *vexillum hairy at base behind*, scarcely longer than the glabrous, obtuse carina; ovary 2–ovuled, ciliated at back; legume obliquely rhomboid, very thick, glabrous. *Benth.! A. carnosa, E. & Z.!* 1492, *non Thunb.*

HAB. Mountains of Swellendam and near Plettenberg's Bay, *Mundt.* Voormansbosch, Swell., *Zeyher*, 2354. Gnadenthal, *Alexander Prior.* (Herb. Th., Bth., Hk., Sd.)

A strong bush, 2–4 feet high, erect, with thick, erect and densely tomentose branches. Leaves crowded, 1–2 lines long, passing from *very obtuse* to *acute* or mucronate. The calyx also varies in the length and breadth of its lobes, and its pubescence; the lobes are sometimes strongly keeled. In Thunberg's Herbarium are three specimens marked *A. affinis;* one of them belongs to the present species; one to *A. recurva*, Benth., and the other to '*A. pinguis*,' E. Mey.! non Thunb. *E. & Z.*'s '*A. carnosa*' is a variety with mucronulate leaves.

101. A. pallescens (E. & Z.! 1476); branches tomentulose; leaves tufted, carnose, linear-terete, obtuse, incurved, minutely cano-puberulous; flowers solitary, lateral, sessile; calyx carnose, thinly puberulous, ribbed, its segments oblong, subacute, midribbed, equalling the tube; petals . . . ?; ovary puberulent, 2–ovuled; legume ovate, acute, not twice as long as the calyx, pubescent. *Benth. l. c. p.* 639.

HAB. Mountain sides near Plettenberg's Bay, *Mundt.* (Herb. Sond.)

A bush intermediate in aspect and character between *A. pachyloba* and *A. costulata*, but with longer leaves and broader calyx-lobes than either. I have only seen an imperfect specimen without petals. The foliage is microscopically canescent; the leaves 3 lines long, closely set, imbricating, patent, but curving inwards.

10. PINGUES. (Sp. 102–117.)

102. A. verrucosa (Linn.! Sp. p. 1001); branches tomentulose; leaves tufted, *elongate*, linear-terete, acute or mucronate, fleshy, glabrous; flowers lateral, shortly pedicellate, solitary, *shorter than the leaves;* calyx nearly glabrous, its teeth acute, much shorter than the tube; petals glabrous; ovary 4–ovuled; legume obliquely ovato-lanceolate, sparsely puberulous. *Benth.! l. c. p.* 638. *A. succulenta, E. Mey.! Linn.* 7. *p.* 159. *E. & Z.!* 1491. *A. Mundtiana, E. & Z.!* 1490. *Zey.!* 2360 ?

HAB. Mountains near Caledon and Hott. Holl., *E. & Z.! Drege! Wallich! Masson.* Near Swellendam (with sharper or mucronate leaves) *Mundt.!* (Herb. Th. Hk. D. Sd.)

A stout bush, 2–3 feet high or more, much-branched, the older branches bare of leaves and warted with hemispherical leaf-tubercles. Leaves ½ inch or more in length, spreading and closely approximated; their point varying from subobtuse to sharply mucronate. Flowers small, 3 lines long, hidden among the leaves, yellow; pedicels 2 lines long. Legume recurved, ½ inch long.

103. A. affinis (Thunb.! Fl. Cap. p. 580, non Benth.); branches tomentulose; leaves tufted, *short*, subacute or mucronulate, linear-terete, fleshy, glabrous; flowers lateral, subsessile, longer than the leaves; calyx glabrous, obliquely campanulate, its teeth acute, much shorter than the tube; petals glabrous, *vexillum with a basal callus; ovary ciliate behind, 6–ovuled;* legume obliquely and broadly lanceolate. *A. pinguis, E. Mey.! Comm. p.* 60. *Benth.! l. c. p.* 640, *non Thunb. Zeyher*, 439.

HAB. Piquetberg and Gnadendahl, *Drege!* Oliphant's R., *Zeyher!* Kamanassie Hills, *Alexander Prior!* (Herb. Th., Bth., Hk., Sd.)

A spreading bush; its older and naked branches warted, the younger closely leafy. leaves 1–2, rarely 3 lines long, spreading. Flowers 3–3½ lines long, yellow, cernuous. legume 5–6 lines long. In Thunberg's Herbarium are three plants marked '*affinis;*'

one is the present species; another, on the same sheet, is *A. recurva*, and the third, marked *β.* is *A. pachyloba.*

104. A. pinguis (Thunb.! Fl. Cap. p. 580, non Benth.); branches tomentulose; leaves tufted, *minute*, fleshy, *ovoid or oblong*, obtuse, glabrate; flowers lateral, subsessile, solitary; calyx obliquely campanulate, glabrous, its teeth deltoid, much shorter than the tube; petals glabrous, vexillum with a callus at base; ovary glabrous, several-ovuled; legume obliquely lanceolate. *A. affinis, E. Mey.! Comm. p.* 60. *Benth. l. c. p.* 640, *non Thunb. A. minutifolia, Vog.*

HAB. On dry hills. Aasvogelsberg and Kendo, *Drege!* Also found by *Thunberg! Mundt., &c.* (Herb. Th., Hk., Bth., D.)

A large, stout, much-branched, ramulose shrub, nearly related to *A. affinis*, from which it chiefly differs in the very short, ovoid or subrotund leaves, which are ½–1 line long. Flowers yellow, chiefly toward the ends of the branches, 2½ lines long. This is the true '*A. pinguis*' of Herb. Thunb.!

105. A. costulata (Benth.! l. c. p. 641); branches tomentulose; leaves tufted, short, fleshy, terete, obtuse, at length glabrous; flowers lateral, solitary, sessile; calyx campanulate, glabrescent, *its teeth triangular-acute, shorter than the* 15*–ribbed tube*; petals glabrous; *vexillum naked at base;* ovary glabrescent, 4–ovuled; legume obliquely rhomboid-ovate, downy. *Benth.! l. c. p.* 641.

HAB. Cape Colony, *Scholl!* (Herb. Benth.)

A much-branched, ramuliferous bush, resembling *A. affinis*, and also *A. pachyloba* and *A. pallescens;* from each of which however it differs in characters of detail. I have only seen the specimen in Herb. Benth., above quoted.

106. A. sanguinea (Thunb.! Fl. Cap. p. 580); branches tomentulose; leaves tufted, very short, fleshy, obtuse, glabrous; flowers lateral, subsolitary, pedicellate; calyx turbinate, glabrous, its lobes linear or subtriangular, pointless, shorter than the tube; petals glabrous, *alæ much shorter than the incurved carina;* ovary hairy at base, 4–6 ovuled. *Benth.! l. c. p.* 641.

HAB. South Africa, *Thunberg! Bowie!* (Herb. Th., Hk., Bth.)

A densely much-branched bush, with minute, fleshy-terete leaves and purple-red small flowers. Pedicels 1–1½ lines long: flowers of same length. Leaves 1 line long, the floral ones shorter than the pedicel.

107. A. adelphea (E. & Z.! 1442); branches tomentulose; leaves tufted, linear-terete, fleshy, obtuse, glabrous; flowers solitary or in pairs, shortly pedicellate; *calyx glabrous, with very short, obtuse teeth;* petals glabrous, the alæ shorter than the arched carina; ovary 2–ovuled; legume obliquely lanceolate, glabrous, several times as long as the calyx. *Benth.! l. c. p.* 641. *A. iniqua, E. & Z.!* 1443. *A. subtingens, E. & Z.!* 1441, *and A. rubescens, E. & Z.!* 1444. *Zey.!* 755, 2352, 2351, 2353.

HAB. Hills and dry ground. Uitenhage District, in several places, *E. & Z.! Bowie, Burchell, &c.* (Herb. D., Hk., Bth., Sd.)

A densely-branched, ramuliferous, leafy bush; the upper branches studded with small, reddish-yellow or rufescent flowers, of about the same size as in *A. sanguinea*, but on shorter pedicels. Leaves 1½–3 lines long, closely set, spreading. Legume 4 lines long.

108. A. microdon (Benth.! l. c. p. 642); branches puberulent; leaves tufted, *short*, terete, obtuse, fleshy, glabrous; flowers subsolitary, minutely pedicellate, lateral; calyx *downy* or glabrescent, with very short or obsolete, distant teeth; petals glabrous; alæ not much shorter than the arched and somewhat beaked carina; ovary downy, 2–ovuled; legume obliquely lanceolate, subfalcate, thinly pilose. *Benth.! l. c. p.* 642. *A. affinis, E. & Z.!* 1440, *non Thunb. A. pinguis, litt. c. E. Mey.! Comm. p.* 60. *Zeyher*, 2350.

HAB. Swellendam hills, on Riv. Zonderende, Kars, &c. *E. & Z.! Pappe!* Klyn Fish R., *Drege!* (Herb. Bth., Hk., D., Sd.)

A small, ramulous bush, resembling the preceding, but with shorter and smaller leaves, shorter or obsolete pedicels, generally pubescent calyces and rather smaller flowers. Also like *A. affinis*, but smaller in all parts, with a 2–ovuled ovary. Leaves ½–1½ lines long. The young plant is thinly puberulous.

109. A. recurva (Benth.! l. c. p. 642); divaricate, with canescent branchlets; leaves tufted, short, terete, obtuse, glabrescent; flowers solitary, lateral, shortly pedicellate; *calyx puberulent, subcanescent, its lobes triangular, with thickened, recurved margins, nearly as long as the tube;* carina glabrous, longer than the pubescent or glabrous vexillum and alæ; ovary 2–ovuled; legume glabrous, obliquely ovate-falcate-acuminate, 2–3 times longer than the calyx. *Zey.! No.* 419.

HAB. Near the 24 Rivers, *Zeyher!* also in *Paterson's* collection. (Herb. Hk. Bth. D. Sd.)

A spreading or depressed, excessively-branched, rigid shrub, with divaricate or recurved branches, and pale and scanty foliage. Leaves 1–2 lines long. It differs from neighbouring species by its calyx, &c. Flowers pale yellow, 3–3½ lines long, their pedicels as long as the leaves or rather longer.

110. A. Wurmbeana (E. Mey.! Comm. p. 58); branchlets canescent; leaves subfasciculate, slender, filiform, curved, obtuse, *glabrous*; flowers shortly pedicellate, solitary or in pairs, lateral; *calyx puberulent-canescent, its lobes linear, longer than the turbinate tube;* corolla *glabrous*, the vexillum and carina longer than the alæ; ovary glabrous, 2–ovuled. *Benth.! l. c. p.* 643.

HAB. Wupperthal, *Drege!* (Herb. Benth.)

A small shrub with the habit of *A. incomta*, from which it differs in its glabrous leaves and petals, and turbinate calyx-tube; from *A. lactea* it further differs in the longer calyx-lobes. Leaves 3–4 lines long, few in each tuft. Flowers scarcely two lines long.

111. A. lactea (Thunb.! Fl. Cap. p. 580, ex parte); branches *puberulent;* leaves tufted, slender, linear-terete, obtuse or mucronulate, *glabrous or incano-puberulous*, the floral longer than the calyx; flowers shortly pedicellate; calyx campanulate, oblique, glabrous or puberulent, *with subulate teeth much shorter than the tube;* the glabrous or thinly silky vexillum and the glabrous carina longer than the alæ; ovary glabrous or silky; legume obliquely ovato-lanceolate, twice as long as the calyx.

VAR. α. **Thunbergii**; leaves minutely cano-puberulous; flowers subsessile; calyx puberulent; vexillum, ovary, and legumes thinly silky. *A. lactea, litt. a. Herb. Thunb.! A. incomta, E. Mey.! Benth. l. c. p.* 643.

VAR. β. **Meyeri**; leaves, calyx, vexillum, ovary, and legumes glabrous or nearly so; flowers shortly pedicellate. *A. lactea, E. Mey.! Benth.! l. c. p.* 643.

VAR. γ. **Zeyheri**; glabrous; leaves much denser, longer, and acute or mucronate. *Zey. No.* 2348.

HAB. S. Africa, *Thunberg!* Between Straat and Hex Rivier, Stell., and in the Onderbokkeveld, *Drege!* Var. γ. Swellendam, *Zeyher!* (Herb. Th. Bth. Hk. D. Sd.)

An erect, branching, and ramulous bush, 2–4 feet high, variable in habit and in pubescence. Leaves in α. and β. 2–3 lines long; in γ. 4–5 lines, and much more closely placed. Flowers numerous, small, cream-coloured; the pedicels in α. ¼ line, in β. and γ. ½–1 line long. Thunberg's original specimen quite agrees with Meyer's 'incomta,' so far as I can judge from the very imperfect scrap I have seen. Our var. β. chiefly differs from it in indument. Var. γ. looks, at first sight, much more distinct, having almost the aspect of *A. verrucosa.*

112. A. incomta (Thunb. ! Fl. Cap. p. 579, non Benth.); branches tortuous, silky; leaves laxly tufted, slender, filiform, curved, obtuse, *thinly appresso-puberulent;* flowers scarcely pedicellate, solitary, lateral; *calyx puberulent-canescent, its lobes distant, subulate, as long as the campanulate tube;* vexillum *thinly silky,* alæ much shorter than the glabrous carina; ovary 2–ovuled, silky.

HAB. S. Africa, *Thunberg.* (Herb. Thunb.)

A small, depressed, or prostrate shrublet, 6–8 inches long, densely much-branched; the branches and ramuli short, much twisted, leafy. Leaves setaceous, 3–4 lines long, few in each tuft, squarrose, thinly silky, and subcanescent. Flowers yellow, 2 lines long, the pedicel ½-line long. Described from the original specimen in Herb. Upsal. It has not been found by recent collectors.

113. A. lepida (E. Mey. ! Comm. p. 58); youngest twigs minutely canescent; leaves linear-terete, obtuse, becoming glabrous, pale; flowers lateral, solitary or in pairs, shortly pedicellate; calyx minutely cano-puberulent, *its teeth triangular-acute,* much shorter than the tube; *petals silky,* the roundish vexillum longer than the carina; ovary 2–ovuled, silky. *Benth.! l. c. p.* 643. *A. lactea, litt. B. Thunb. Herb.*

HAB. S. Africa, *Thunberg.* Sandy hills, Piquetberg, *Drege!* (Herb. Th. D. Bth.)

A spready, perhaps procumbent, loosely-branched bush, with small, subdistant leaves, and slender, rodlike branches. Leaves 1–2 lines long, microscopically puberulent when young, becoming glabrous. Flowers 3 lines long, pale buff, on pedicels 1 line long. This was included by *Thunberg* under his 'lactea.'

114. A. argyrea (DC. Prod. 2. p. 139); *whole plant covered with short, white, silky tomentum;* leaves tufted, linear-terete, blunt, elongate; flowers subsessile, solitary, lateral; calyx turbinate, the teeth shorter than tube; petals pubescent; ovary 2–ovuled; legume obliquely ovato-lanceolate, acute, silky. *Benth.! l. c. p.* 644. *E. & Z.!* 1393.

HAB. Uitenhage District, *E. & Z.!* Albany, *Dr. Atherstone.* Langekloof, *Drege,* &c. (Herb. Bth., Hk., Sd.)

Readily known from all allied species by its copious, *very white,* but short, downy pubescence. Stem erect, much-branched, and ramulous. Leaves 3–4 to 6–8 lines long, spreading. Flowers 2 lines long, primrose colour. Calyx-teeth either linear or somewhat deltoid, variable in length, but never long. Legumes 6 lines long, 2 lines wide.

115. A. spinescens (Thunb. ! Fl. Cap. p. 584); branches rigid, divaricating, spine-tipped; *leaves very short, fleshy, obtuse,* glabrescent; flowers solitary, lateral, shortly pedicellate; calyx-teeth very short, obtuse, the tube silky; vexillum and carina puberulent, longer than the wings; ovary glabrous, 2–ovuled. *Benth. l. c. p.* 644.

HAB. Cape, *Thunberg.* Near Groenekloof, *Drege.* (Herb. Th., Bth., Hk , D.)

A coarse, straggling shrub, with widely spreading branches and ramuli, whose ends harden into strong spines. Leaves scarcely 1 line long, very thick and fleshy. Flowers scarcely 3 lines long, on pedicels 1 line long, pale.

116. A. spinosa (Linn. Sp. 1000); glabrous or thinly canescent; the branchlets spine-tipped, and generally the leaf-tufts spiniferous; leaves tufted, linear-terete or plano-compressed, pointless; flowers lateral, shortly pedicellate; calyx-teeth very short, sometimes obsolete; vexillum oval, scarcely pubescent near the point, equalling the glabrous alæ and carina; ovary 2-ovuled; legume silky-pubescent, acuminate, 2–3 times as long as the calyx. *Benth. l. c. p.* 644. *E. & Z. No.* 1498. *Thunb. Fl. Cap. p.* 584. *Zey.* 2363.

VAR. β. **flavispina**; more glabrous; leaves more slender; calyx-teeth, scarcely any; a shorter vexillum and a little longer and more glabrous legume. *Benth. A. flavispina, Presl. Bot. Bem. p.* 126.

VAR. γ. **inermis**; leaf-tufts destitute of spines. *E. Mey.*

VAR. δ **horrida**; leaves thick, plano-compressed, narrower to the base, pubescent; legume shorter and more woolly. *A. horrida, E. & Z.* 1499.

HAB. Common in dry ground throughout the colony, and extending to Port Natal. (Herb. Th., D., &c.)

A very rigid, divaricate, much-branched bush, bristling at all points with sharp spines; those of the leaf-tufts ¼–½ inch long, subhorizontal, sometimes absent. Flowers sulphur yellow, 3–3½ lines long. Var. δ. is almost intermediate with *A. obtusata.*

117. A. obtusata (Thunb.! Fl. Cap. p. 574); cano-pubescent, the branchlets and often the leaf-tufts spiniferous; leaves subfasciculate, *flat, broadly-linear or obovato-linear, very obtuse, narrowed to the base,* canescent; flowers lateral, shortly pedicellate; calyx oblique, with very short teeth; vexillum silky, equalling the nearly glabrous, blunt carina; legume silky-canescent, acute, 2–3 times as long as the calyx. *A. glauca, E. & Z.* 1500.

HAB. Near Oliphant's River, *Thunberg.* Gauritz River, *E. & Z.* (Herb. Th., Sd.)

This has the habit of *A. spinosa*, but the leaves are quite flat, 1–1½ lines broad, and sometimes narrow-obovate and retuse. In Thunberg's original specimen, in Herb. Upsal, the ramuli are alone spine-tipped; in E. & Z.'s (Herb. Sond.) almost every leaf-tuft is also spiniferous. The flowers are like those of *A. spinosa*, but the vexillum is more hairy.

11. **TERMINALES** (Sp. 118–137).

118. A. abietina (Thunb.! Fl. Cap. p. 583, non Benth.); glabrous; leaves ternate or subfasciculate, *linear-subulate,* mucronate, flattish; calyx-lobes subulate, setaceo-acuminate, *pungent, more than twice as long as the tube;* the vexillum and the arched and rostrate carina *nearly twice as long as the alæ;* ovary 2-ovuled; legume obliquely lanceolate, glabrous, longer than the calyx. *A. filifolia, E. Mey.! Linn.* 7. *p.* 158. *Benth.! l. c. p.* 646. *A. retroflexa, E. & Z.!* 1397, *non Linn. Zey.!* 418.

HAB. Summit of Table Mountain, Cape, *E. & Z.! W.H.H., &c.* (Herb. Th. Bth., Hk., Sd., D.)

A small, erect or spreading shrublet, 12–14 inches high, subcorymbosely branched; the branches rubescent, striate and shining. Leaves erect, 1–6 together, the very young appressedly puberulous, unequal, the longest half inch or more long, all sharply mucronate. Flowers bright yellow, 1–2 together, at the ends of the erect branches. Legume 5 lines long. Carina straightish or somewhat bent.

119. A. fornicata (Benth.! l. c. p. 646); very minutely strigillose;

leaves subfasciculate, *subulate*, carinate or with recurved margins, mucronate; calyx-lobes subulate, setaceo-acuminate, pungent, *rather longer than the tube; the vexillum and the fornicate much incurved, carina longer than the alæ;* ovary 2-ovuled; legume obliquely lanceolate, glabrous, longer than the calyx. *Benth.!*

HAB. On Table Mountain, *Mundt! Thom! Gueinzius!* (Herb. Bth., Hk., Sd.)

Very similar to *A. abietina*, but with rather shorter, blunter and more convex leaves; shorter calyx teeth, and more exactly distinguished by the strongly inflexed, fornicate carina. Branches more rigid, a foot long, erect.

120. A. pedicellata (Harv.); shrubby, diffuse, ramulous, *glabrous;* leaves ternate or tufted, *linear, fleshy, blunt or mucronulate, glabrous;* flowers 1–3 at the ends of the branches and twigs, *pedicellate;* calyx campanulate, glabrous, its segments *lanceolate-acuminate, thickened at margin, not pungent*, equalling the tube; vexillum subrotund, thinly silky, mucronulate, equalling the short, *straight and blunt* carina; ovary 2-ovuled, legume ?

HAB. Tulbagh, *Dr. Pappe!* 245. (Herb. T.C.D.)

A strongly woody, procumbent or depressed, densely leafy and ramulous small bush; twigs quite glabrous, pale, roundish. Leaves tufted on the branches, ternate on the twigs, 3–4 lines long, quite glabrous, the older obtuse. Flowers pale yellow, in imperfect umbels; pedicels 2–3 lines long. This has the aspect of *A. galioides* and *A. marginalis*, but differs in inflorescence, petals, glabrous twigs, &c. The ovary has a line of hairs on the ventral suture, but is otherwise glabrous.

121. A. retroflexa (Linn. sp. 1001); diffuse or procumbent, slender, *minutely downy;* leaves tufted, subulate, acute, microscopically puberulous or glabrescent, *the floral shorter than the calyx;* flowers solitary or in pairs, sessile; calyx downy, its lobes green, subulate, acute, mostly longer than the tube, shorter than the corolla; petals glabrous; ovary villous, 2-ovuled; legume lanceolate, silky, twice as long as the calyx. *Benth.! l. c. p.* 646. *A. galioides, Berg.; E. Mey.! ex parte.*

VAR. β. **bicolor**; calyx-lobes not longer than the tube, strongly reflexed; vexillum and carina purple on the outside. *A. bicolor, E. & Z.!* 1404.

VAR. γ. **parviflora**; leaves not 2 lines long; flowers 2–2½ lines long, the calyx-lobes equalling the tube, very patent. *A. leptocoma, Pappe! non E. & Z.*

HAB. On the Cape Flats, &c. common. Var. β. near Doornhoogde, *E. & Z.!* Var. γ. between the Breede and Duivenhoek's River, *Pappe!* (Herb. Th., Bth., Hk., Sd., D.)

Stems numerous, 2–4 feet long, spreading over the ground, much branched; the flowering ramuli slender and filiform, 1–2 inches long, generally bare of leaves, except a single tuft just below the terminal flowers. Tufts of leaves laxly set, especially on the smaller branches. Leaves 3 lines long. Flowers 4 lines long, orange yellow; in β. purple on the outside. Legume densely clothed with long white, silky hairs. Calyx-lobes leaflike, longer or shorter. Var. γ. differs chiefly by its small flowers.

122. A. galioides (Linn. Mant. p. 260, non Thunb.); diffuse or procumbent, the branches puberulent; leaves densely tufted, subulate, acute, *glabrous, the floral about equalling the calyx;* flowers in pairs, sessile; *calyx nearly glabrous*, its lobes leaflike, keeled, acute, twice as long as the ribbed tube, and not much shorter than the corolla; petals glabrous, the incurved, obtusely rostrate carina longer than the alæ; ovary 2-ovuled, *glabrous. Benth.! l. c. p.* 647. *A. galioides, var. foliosa, E. Mey.! No.* 2357. *A. juniperina, Thunb.! Fl. Cap. p.* 583.

HAB. Subalpine situations in the Western Districts, common. (Herb. Th., Bth. Hk., D., Sd.)

A procumbent, shrubby plant, with stronger stems, much more closely set, thicker and more glabrous leaves, and smaller flowers than *A. retroflexa;* the flowering branchlets are densely leafy throughout, the flowers nestling among the terminal leaves; the calyx tube is ribbed and the ovary quite glabrous. Leaves 2–3 lines long, many in a cluster, sometimes glaucous.

123. A. marginalis (E. & Z. ! 1445); much branched, with pubescent branches; leaves tufted, linear-terete, fleshy, *obtuse or submucronulate*, glabrous; flowers subsessile, 1–2 together; calyx cano-pubescent, the lobes green, *lanceolate, with reflexed margins*, acute, longer than the tube, one half shorter than the flower; ovary villous, 2–ovuled; legume obliquely lanceolate, silky, twice as long as the calyx. *Benth.! l. c. p.* 647. *Zey.!* 2349.

HAB. Grassy fields, near the Zwartkops River, *E. & Z.* (Herb. Bth., Hk., D.)

A densely branched, cæspitose, small shrub, 8–12 inches high; the branches subcanescent, densely leafy. Leaves 2–3 lines long, mostly very blunt and slightly tapering at base. It is readily known from allied species by its calyx.

124. A. exilis (Harv.); rigid, depressed, ramulous, nearly glabrous or minutely and thinly canescent; leaves tufted, *tereti-subulate, mucronate, subpungent;* flowers solitary, or 3–4, subsessile; calyx campanulate, pubescent, its teeth tapering and *mucronate*, shorter than the tube; *petals* pubescent; ovary 2–ovuled, glabrous. *A. albens, E. Mey.! (non Linn.) Benth.! l. c. p.* 648.

HAB. Sandy Hills, in the Cape District, *Drege!* Kuilsriver, *Pappe!* 164. (Hb. Bth., D.)

A scraggy, low-growing, stunted, much branched, robust and rigid shrub. Leaves either green or thinly canescent, squarrose, unequal, 2–3 lines long. Flowers minute, 1–1½ line long, in terminal 2–4 flowered racemules. Petals changing colour, at first pale yellow, afterwards purplish or livid-red.

125. A. rubens (Thunb.! Fl. Cap. p. 576); *canescent;* leaves tufted, short, terete-subulate, incurved, subacute, *white and silky;* flowers subsessile, 1–2 together; calyx turbinate, tomentose, with acute teeth much shorter than the tube; petals silky; carina obtuse; ovary 2–ovuled; legume obliquely lanceolate, silky-villous. *Benth.! l. c. p.* 648. *Zey.!* 2325. *Burch.* 4642.

HAB. Vanstaadensberg, Uit., *Drege, Zeyher.* Sidbury, *Burke.* (Herb. Th., Bth., Hk., D., Sd.)

A slender, suberect or diffuse, much branched bush, 10–12 inches high, all parts albo-canescent. Leaves 1½–2 lines long. Flowers fulvous or rufescent, 3 lines long.

126. A. astroites (Linn. sp. p. 1000); branches villous; leaves tufted, tereti-subulate, subtrigonous, pungent-mucronate, rigid, spreading, at length glabrate; flowers shortly racemulose, subcapitate; calyx widely campanulate, sparsely pilose, its lobes subulate-pungent, needle-pointed, rather longer than the tube; vexillum pilose on the ridge, as long as the glabrous, arched and rostrate carina; ovary 2–ovuled; legume obliquely and broadly lanceolate, glabrous or scarcely hairy. *Benth.! l. c. p.* 648. *Thunb.! Fl. Cap. p.* 582.

HAB. Hill sides and mountains, Cape and Stellenbosch Districts, frequent. (Hb. Th., Bth., D., Hk.)

A large, furzelike bush, with very pungent, rigid, and widely spreading leaves, unequal in the tuft, the longest ¼ inch long. Flowers orange yellow, 3–4 lines long, several together in short, imperfect, terminal racemes; rachis and pedicels white-hairy. Legumes ½ inch long, mostly glabrous, thickened on the ventral suture.

127. A. vulnerans (Thunb.! Fl. Cap. p. 582, non Benth.); branches divaricate, pubescent, spine-tipped; leaves tufted, very patent, slender, trigono-subulate, pungent-mucronate, glabrous; flowers racemulose, 3–6 together; calyx glabrescent, its lobes setaceo-subulate, pungent, scarcely longer than the tube; petals pubescent; legume obliquely lanceolate, glabrous. *A. acicularis, E. Mey.! Comm. p.* 46. *Benth.! l. c. p.* 649.

HAB. Cape District. Hills at Ebenezer, and thence to Kamiesberg, *Drege, W.H.H.* (Herb. Th., Bth.)

Much more slender than *A. astroites*, with spreading or divaricate branches, fewer and more slender leaves, and smaller and paler flowers. Leaves acicular, ⅓–⅔ inch long. Flowers 2½–3 lines long, in a lax, terminal, subsecund raceme. Except in its *inflorescence* this very closely resembles *A. ulicina*, E. & Z.

128. A. pungens (Thunb.! Fl. Cap. p. 584); divaricately much branched, spinescent; leaves tufted, linear-terete, mucronate, glabrous; flowers racemulose; calyx pubescent, the teeth triangular-acuminate, needle-pointed, shorter than the turbinate tube; the vexillum and the arched and rostrate carina silky, longer than the alæ; ovary 2–ovuled; legume obliquely lanceolate, silky-villous. *E. & Z.! No.* 1494. *A. secunda, E. Mey.! Comm. p.* 47. *Benth. l. c. p.* 649.

HAB. Near Brackfontein, Clanwilliam, *E. & Z.!* Riebeckskasteel, *Drege, Bowie, &c.* (Herb. Th., Bth., D., Hk.)

Densely branched and ramuliferous; every branchlet ending in a needle-pointed reddish or yellow spine. Leaves 2–3, rarely 4 lines long, slender, but not so sharp as in *A. vulnerans*. Flowers on a spinescent rachis, 4–5 lines long, yellow. Legumes clothed with white hairs, 5 lines long. This is the true '*pungens*' of Thunb.! in Herb. Upsal.

129. A. genistoides (Linn. Mant. p. 261); unarmed, with virgate branches and ramuli; twigs downy; leaves tufted, linear-terete, blunt or mucronulate, glabrous or nearly so; flowers pedicellate, racemulose; the teeth of the thinly silky or glabrescent calyx deltoideo-subulate, needle-pointed, distant, shorter than the turbinate tube; petals thrice as long as the calyx, puberulent, the long and straightly rostrate carina somewhat longer than the alæ; legume obliquely ovato-lanceolate, thinly silky. *Benth.! l. c. p.* 650. *Thunb. Cap. p.* 581.

HAB. S. Africa, *Thunberg!* Waterfall, Tulbagh, *Dr. Pappe!* 240. (Hb. Th., D.)

An erect or spreading bush, with rodlike branches, and rodlike, very erect ramuli. Leaf-tufts densely set, many leaved; leaves pale green, squarrose, 3–4 lines long. Racemules 4–5 flowered, ending the lateral branches. Flowers bright yellow, like those of a *Genista*, 5 lines long; the vexillum ample; the carina bent at a right angle, with a long, straight point.

130. A. acuminata (Lam. Dict. 1. p. 287); divaricately much branched spinescent; leaves subfasciculate, *very short, trigonous*, fleshy, subobtuse or mucronate and pungent, nearly glabrous; flowers 1–3, racemulose, pedicellate; calyx turbinate, puberulous, its teeth triangular-acuminate, needle-pointed, shorter than the tube; the ovate vexillum and the arched and beaked carina silky, longer than the alæ; ovary 2–ovuled; legume obliquely lanceolate, downy-canescent. *Benth. l. c. p.* 650.

HAB. Dry ground and by way sides. About Capetown, and throughout the Cape District, common. (Herb. Bth., D., Hk., Sd.)

A rigid bush, with the aspect of *A. pungens*, from which it is at once distinguished by its short, fleshy leaves. Leaves scarcely a line long. Flowers 3 lines, yellow.

131. A. divaricata (Thunb. ! Cap. p. 582); slender, diffuse, divaricate, unarmed (or rarely subspinescent); twigs puberulous; leaves short, tufted, terete or trigonous, pungent-mucronate, erect or squarrose, glabrous or nearly so; flowers racemulose or subsolitary, shortly pedunculate; calyx turbinate, puberulous, its teeth deltoid, or deltoideo-subulate, pungent, *variable in length;* the orbicular, thinly silky vexillum and the arched and rostrate, glabrous or silky carina longer than the alæ; ovary 2–ovuled; legume obliquely ovato-lanceolate, puberulous or glabrescent, 2–4 times as long as the calyx.

VAR. α. **Thunbergii**; leaves squarrose; calyx-lobes commonly shorter than the tube; legume 3–4 times longer than the calyx. *A. divaricata, Benth.! l. c. p.* 651. *A. galioides, Sieb.!* 159. *Zey.!* 2359. *A. leptocoma, E. & Z.!* 1402. *A. divergens, E. Mey.! Comm. p.* 45.

VAR. β. **microphylla**; leaves suberect, unequal, the outer shorter and swollen at base; calyx-lobes commonly as long as the tube; legume twice as long as the calyx. *A. microphylla, DC. Prod.* 2. *p.* 143. *Benth.! l. c. p.* 650. *E. & Z.* 1401. *Zey.!* 438.

VAR. γ. **subinermis**; leaves rather slender, 1–2 lines long; branches rarely spinescent; calyx teeth commonly much shorter than the tube. *A. subinermis, Benth. l. c. p.* 650. *A. spinescens, E. & Z.* 1495, *non Thunb.*

HAB. Dry ground and by way sides, throughout the Colony. β. and γ. in the Western Districts. (Herb. Th., Bth., Sd., Hk., D.)

A straggling or procumbent, rigid, but slender, much branched bush, with very spreading branches, rarely spiniferous. Leaves variable in length and direction, 1–2, rarely 3 lines long. Calyx teeth sometimes very short, sometimes equalling the tube, but variable on the same bush. I cannot, after comparing numerous specimens of the three varieties given above, consider them as more than varieties.

132. A. ferox (Harv.); divaricately much branched, *spinous*, very rigid; leaves tufted, *linear-terete, blunt or mucronulate*, glabrous, slender; flowers 4–5, pedicellate, fascicled or racemulose; calyx turbinate, puberulous, its teeth short, *triangular, acute*, but not pungent; the obovate, silky vexillum and the bluntly rostrate, glabrous carina, longer than the alæ; ovary 2–ovuled, silky; legume ?

HAB. In barren Karroo-ground. Bosjesveld, Swell., *Pappe!* 246. (Herb. D.)

A very stout, rigid bush, with pale grey or white bark; branches spreading widely, the ramuli at nearly right angles with the branch, and straight, rigid, 1–1½ inch long, spine-pointed. Leaves 3–5 lines long, many in the tuft, slender. Umbellate-tufts of flowers from the centre of the leaf-tuft, or terminating the spinous ramuli; in the latter case racemulose. Flowers 2–2½ lines long, cream-white, the tip of the carina purple. Easily known by its purple-tipped white flowers and the non-spinous calyx teeth.

133. A. vermiculata (Lam. Dict. 1. p. 288); ramulous, the branches thinly tomentulose; *leaves tufted, minute, fleshy, obtuse*, glabrous or pubescent; flowers in pairs, pedicellate; calyx pubescent, *with short, triangular, subacute teeth; corolla silky*, the alæ longer than the keel, not much shorter than the vexillum; ovary thinly villous, 2–ovuled; legume obliquely-lanceolate. *Benth. l. c. p.* 651. *A. sanguinea, E. & Z.* 1439, *non Thunb. A. microphylla, Steud. A. multiflora, litt. α. Herb. Thunb.*

HAB. Alpine valleys in the Langekloof, George, *E. & Z.* (Herb. Bth., Hk., D.)

A densely, much-branched, erect shrub, 1–2 feet high, with very minute leaves and a profusion of reddish, silky flowers. Leaves ¼–½ line, rarely 1 line long. Flowers 3–4 lines, the corolla more than twice as long as the calyx. The calyx-teeth vary in length, sometimes they are nearly half as long as the tube, usually much less.

134. A. Agardhiana (DC. Prod. 2. p. 143); silky-canescent, with very minute white hairs; leaves *ternate, narrow-lanceolate, acute, flat,* mid-ribbed; racemules teminal, few-flowered; calyx silky-canescent, *the teeth ovate, acute,* shorter than the tube; petals villous; ovary villous, 4–ovuled. *Benth. l. c. p.* 653.

HAB. From *Forsyth's* Herbarium. (Herb. Benth.)

Leaves ½-inch long, or rather longer, usually 3, springing from a woolly leaf-tubercle, as in other *Aspalathi.*

135. A. armata (Thunb. Fl. Cap. p. 577); *densely canescent;* leaves tufted, subulate, *pungent-mucronate,* or mucronulate, *albo-sericeous,* the old ones at length glabrate; flowers in a terminal pluriflowered raceme, longer than the leaves, subsessile; calyx villous, with needle-pointed teeth scarcely shorter than the tube; petals villous, *scarcely longer than the calyx;* ovary 2–ovuled; legume shortly and obliquely sublanceolate, turgid, tomentose. *Benth. l. c. p.* 652. *E. Mey. Comm. p.* 51. *Buchenroedera teretifolia, E. & Z.* 1356.

HAB. Bergvalley, Clanwilliam, *E. & Z., Drege.* (Herb. Th., Bth., Hk., D.)

A much-branched, erect bush, 12–15 inches high, all the younger parts clothed with soft, white, slender hairs. Flowers 4–6 or 8 in a terminal, inch-long raceme, small and white. Petals less exserted than in any other species. Calyx 1½ line long. Legume more turgid than in others of this section.

136. A. corymbosa (E. Mey. Linn. 7. p. 159); leaves solitary or subfasciculate, very long, linear-terete, rigid, glabrous or thinly puberulous, acute; *flowers on short pedicels, few, in pairs or imperfect corymbose-racemules;* calyx turbinate, downy, its teeth lanceolate, shorter than the tube; vexillum pubescent, as long as the pubescent carina; ovary 2–ovuled; legume *long,* lanceolate, scarcely puberulent. *Benth. l. c. p.* 652. *E. & Z.* 1396. *A. cognata, Presl. Bot. Bem. p.* 126. *Lebeckia contaminata, Thunb.! Cap.* 561.

HAB. Alpine and subalpine situations of the western districts. Table Mountain, common, also on the Cederberg, *E. & Z., Drege, &c.* (Herb. Th., Bth., Hk., D., Sd.)

Diffuse or procumbent, not much-branched; branches ascending, curved, 1–2 feet long, with distant nodes. Leaves 1–1½ inch long, thicker than in the following species. Racemules very short, 2–4 flowered. Calyx 1 line long. Corolla 3 lines, yellow. Legume 7–8 lines long, 1½ broad.

137. A. tenuifolia (DC. Prod. 2. p. 143); leaves solitary, or subfasciculate, very long, tereti-subulate, rigid, glabrous, subacute; *flowers few, interruptedly racemose;* calyx glabrescent, obliquely-turbinate, the teeth *much shorter* than the tube; vexillum puberulous, longer than *the much arched, glabrous* carina. *Benth. l. c. p.* 653.

HAB. Piquetberg, Cape District, *Drege.!* (Herb. Bth., Hk., D.)

Very like the preceding, but more slender, with a laxer pubescence, shorter calyx teeth, and a different corolla. Leaves needleshaped, but not very sharp, 1–1½ inch long. Racemes lax, elongating, with widely distant flowers; pedicels 2–3 lines long.

12. PEDUNCULARES. (Sp. 138–148.)

138. A. capillaris (Benth.! l. c. p. 653); diffuse or procumbent, slender, nearly glabrous; leaves subternate or tufted, *narrow-linear, very acute, tapering at base, flattish; peduncles capillary, one-flowered;* calyx-lobes setaceo-subulate, distant, rather exceeding the turbinate tube; ovary sessile, 6–ovuled; legume lanceolate. *Benth. l. c. p.* 658. *Ononis capillaris, Thunb. Fl. Cap. p.* 585. *A. pedunculata, litt. b. E. Mey. in Hb. Drege.*

HAB. Summit of Table Mountain, common, *Thunberg, W.H.H., &c.* (Herb. Th., D., Bth., Hk.)

A very slender, trailing suffrutex, with many capillary branches, 2–3 feet long; the internodes from ½ inch to one or more inches apart. Leaves sometimes solitary, often ternate or quinate, ⅓–¾ inch long, obviously narrowed toward the base. Peduncles 1–2–3 inches long, bracteate just below the flower; bracts setaceous. Flowers 4 lines long, yellow: the vexillum very broad, and much longer than the keel, minutely silky.

139. A. pedunculata (l'Her. Sert. Angl. t. 26); diffuse or ascending, slender, glabrous or nearly so; leaves tufted, rarely ternate, *linear-filiform, subterete,* acute; *peduncles* 1–3 *flowered, with setaceous bracts;* calyx-teeth subulate, about as long as the turbinate tube; ovary sessile, 6–ovuled; legume lanceolate. *Benth. l. c. p.* 654. *Bot. Mag. t.* 344. *A. squarrosa, Thunb. in Hb. Upsal* (*A. squamosa, Fl. Cap. p.* 581). *A. biflora, E. Mey.? Comm. p.* 64. *Acropodium suffruticosum, Desv.*

HAB. Groenekloof, *Dr. Pappe.* (Herb. Th., Hk., D., Sd.)

More erect and stronger than *A. capillaris,* with narrower and less acute leaves, and frequently 2–3 flowered peduncles. It is known from the two following species by its sessile ovary and 6–ovules. Leaves ⅓–¾ inch long. Peduncles 1–1½ inch long. Flowers 4–5 lines long, yellow; the broad vexillum thinly silky.

140. A. nudiflora (Harv.); diffuse, slender, nearly glabrous; *leaves mostly solitary,* linear-filiform, flattish, acute or subobtuse; peduncles 1–flowered, *with minute, subremote, toothlike bracts;* calyx-teeth subulate, scarcely as long as the turbinate tube; ovary shortly pedicellate, 2–ovuled *Zey.* 2362.

HAB. Between Knoflockskraal and Kleinhowhoek, *Zeyher.* (Herb. Hk., Sd.)

A diffusely flexuous, perhaps suberect suffrutex, with long, curved branches, and distantly scattered, almost always solitary, nearly uncial leaves. Peduncles opposite the leaf, and scarcely as long; both bract and bracteoles very minute, ¼–line long. Flowers 3 lines long, yellow, purplish externally. This is included by Bentham under *A. pedunculata,* but is readily known by the solitary leaves, minute, almost obsolete bracts, and the ovary. In the ovary examined I found but 2–ovules.

141. A. bracteata (Thunb.! Fl. Cap. p. 581); erect or subdiffuse, much-branched, nearly glabrous; leaves *tufted,* rarely subternate, linear-filiform, subterete, subacute; peduncles 1–flowered, *with leaflike bracts equalling the calyx-tube;* calyx-teeth setaceous, equalling the turbinate tube; ovary stipitate, 3–4 ovuled; *legume subfalcate, acuminate, tapering at base into a long stipe. Benth. l. c. p.* 654. *A. pedunculata, litt. b. E Mey. in Herb. Drege. Sieb. Fl. Cap. No.* 46.

HAB. Paarl and Drakenstein Hills, *Drege.* Table Mountain, *W.H.H., &c.* (Herb. Th., Bth., Hk., D., Sd.)

Much-branched, either dichotomous or flexuous, and alternately decompound. Leaves more closely placed, and in denser fascicles than in any of the three preceding, 4–5 lines or ½-inch long, slender, often incurved. Peduncles as long or twice as long as the leaves: the flowers and bracts as in *A. capillaris.*

142. A. lanata (E. Mey.! Comm. p. 64); slender, suffruticose, *softly hairy; leaves tufted or ternate, linear, very acute, flat, clothed with long, white spreading hairs;* 'peduncles one-flowered; calyx-lobes twice as long as the tube, subulate, pointless; vexillum hairy.' *Benth. l. c. p.* 655.

HAB. Piquetberg, Clanwilliam, *Drege.* (Herb. Bth., Hk., Sd.)

The specimens here quoted have no flowers. The foliage resembles that of *A. capillaris*, but is pilose with copious and long hairs. Leaves ½ inch long, several in a tuft. Peduncles 1½ inch, with leafy bracts at the summit.

143. A. falcata (Benth.! l. c. p. 655); diffuse, much-branched, *thinly pilose, becoming subglabrous;* twigs terete; leaves ternate or somewhat tufted, *linear-lanceolate, acute at each end, flat, the lateral falcate;* peduncles 1–3 flowered; calyx-teeth subulate, shorter than the turbinate tube; ovary shortly stipitate, about 6 ovuled; legume long and lanceolate, puberulent, subturgid. *Benth.*

HAB. At the 24-rivers, *Zeyher*, 436. (Herb. Bth., Hk., D., Sd.)

Stems 2–3 feet long, widely spreading, ascending; branches rufescent, the young ones and the young leaves with more or less abundant, long, white hairs. Leaves 6–9 lines long, 1–1½ lines wide, yellowish-green; nearly like those of *Cliffortia strobilifera.* Legume 10–12 lines long, 2 lines wide, scarcely oblique, tapering to a sharp point.

144. A. alternifolia (Harv.); diffuse or procumbent, nearly glabrous; twigs angular, furrowed; leaves *solitary, distantly scattered,* linear-lanceolate, flat, acute, mucronate, sessile, with thickened margins, glabrous; peduncles opposite the upper leaves or terminal, one-flowered, with two minute, subdistant bracts; calyx-teeth deltoideo-subulate, about equalling the turbinate tube; ovary stipitate, 2 ovuled; vexillum broadly obcordate, puberulous, equalling the glabrous carina: legume ?

HAB. Waterfall, Tulbagh, *Dr. Pappe.* (Herb. T.C.D.)

Root woody, deep. Stems many from the crown, 1–2 feet long, decumbent, pale red-brown. Leaves ½–1 inch asunder, ¾–1 inch long, 1 line wide, erecto-patent, pale green. Flowers at or near the ends of the branches, 3½ lines long, pale-yellow; the vexillum brown-backed. Peduncle 3–5 lines long. Legume not seen. The leaves are not unlike those of *A. falcata* in shape, but longer and narrower, and always solitary.

145. A. longipes (Harv.); erect, much-branched and ramulous, *densely canescent-tomentose;* leaves tufted, tereti-subulate, acute or mucronate; peduncles capillary, elongate, 1–2–flowered, *with a tuft of bracts close to the flowers;* corolla not seen; calyx-lobes subulate, distant, twice as long as the tube; legume *obliquely ovate, silky, scarcely longer than the calyx. Zey.* 2324.

HAB. Stony places on the hillsides. Riv. Zonder Einde, *Zeyher.* (Herb. Sd.)

A tall, virgate, and ramulous bush, everywhere covered with soft, white hairs. Leaves in dense tufts, 4–5 lines long. Peduncles from the leaf-cluster, 1–1½ inch long, bearing 2 terminal, sessile flowers, surrounded by a tuft of floral leaves longer than the calyx. Legume like that of one of the *Leptanthæ.* Except in inflorescence and legume this looks like *A. rubens*, but has much longer leaves.

146. A. nivea (Thunb.! Fl. Cap. p. 576); erect, *the whole plant white and silky;* leaves tufted, filiform, blunt; peduncles capillary, 1–3 flowered; calyx-teeth linear, distant, blunt, scarcely equalling the tube; ovary 4–ovuled; legume sessile, *obliquely lanceolate, silky-canescent. Benth.! l. c. p.* 655. *E. & Z.!* 1391. *DC. Prod.* 2. *p.* 144. *Zey.!* 2323.

HAB. Near the Zwartkops River, Uit., *E. & Z.*, *Drege*, *&c.* (Herb. Th., Bth., Hk., D., Sd.)

An erect, virgate shrub. Leaves in dense fascicles and closely placed, slender, subterete, ½-inch long or more. Peduncles 8 lines to 1 inch long, spreading; bracts 3, filiform, below the flower. Legume 8 lines long, nearly 3 lines wide at base.

147. A. suffruticosa (DC. Prod. 2. p. 144); diffuse, puberulous; leaves tufted, *short, linear-terete, green, pungent-mucronulate*, glabrous or puberulous; peduncles 1–3 flowered; calyx-teeth triangular, acute, much shorter than the tube; vexillum pubescent, longer than the carina; ovary 2–ovuled; legume obliquely lanceolate, minutely and appressedly puberulous. *Benth.! l. c. p.* 655. *E. & Z.* 1405. *Zey.!* 215, 2361.

HAB. Uitenhage District, *E. & Z.*, *&c.* (Herb. Bth., Hk., D., Sd.)

Branches numerous, 2 feet long or more, spreading or trailing, with erect branchlets; the young ones strigoso-pubescent. Leaves 2–3 lines long, patent or squarrose, slightly fleshy. Flowers 3 lines long, with a very blunt carina. Legume 7–8 lines long, 1½ wide at the turgid base. This resembles *A. divaricata*, except in inflorescence.

148. A. ulicina (E. & Z.! 1407); shrubby, divaricate, with hairy twigs; leaves tufted, stellately patent, subulate, rigid, pungent-mucronate, pilose, becoming glabrous; peduncles from the centre of the leaf-tuft, racemulose, 2–4 flowered; calyx puberulous, the teeth deltoid, needle-pointed, about equalling the tube; petals silky, vexillum longer than the alæ, shorter than the carina; ovary 2–ovuled; legume appressedly downy, obliquely ovato-lanceolate, acute. *Benth. l. c. p.* 656.

HAB. Mountains near Tulbagh, and in Pikenierskloof, Clanw., *E. & Z.* Winterhoek and Kardouw, *Dr. Pappe.!* (Herb. Bth., Hk., Sd., D.)

A slender, but rigid, laxly-branched, spreading or decumbent bush, with pale foliage, and pale yellow flowers. Leaves 6–8 lines long, very slender, spreading every way, unequal in the tuft. Penduncles 1–2 inches long, leafless, or with one or two distant leaves, and sometimes with a fascicle. Flowers 3–4 lines long. Except in its inflorescence this closely resembles *A. vulnerans.*

(Doubtful Species.)

A. cinerascens (E. Mey. Comm. p. 54); "leaves tufted, terete, blunt, silky-canescent; flowers solitary, subspicate; teeth of the pubescent calyx triangular; vexillum pubescent."

HAB. Drackensteen, *Drege.*

Allied to *A. canescens*, but with the aspect of *Lebeckia microphylla*, according to E. Meyer.

A. arachnoidea (Hort. Berol.); "shrubby, erect; branches arched, subfastigiate; leaves tufted, subulate, mucronate, sprinkled with cobwebby hairs; flowers axillary, solitary; teeth of the silky calyx distant, subulate, longer than the tube, equalling the silky corolla." *Walp. in Linn.* 13. *p.* 497. Allied to *A. setacea*, according to Bentham.

A. crassifolia (Andr. Bot. Rep. t. 353); like *A. subulata*, Th., but with larger flowers and very short, blunt calyx-lobes.

A. nodosa (Vog. Linn. 13. p. 496); like *A. sanguinea*, but with leaves 1½ line long and silky petals.

XXV. PSORALEA, L.

Calyx campanulate, unequally 5-lobed, the lowest lobe longer, and commonly broader than the rest. *Vexillum* broad, with reflexed sides. *Carina* shorter, dark-coloured. *Stamens* diadelphous. *Ovary* sessile, 1-ovuled; style slender. *Legume* concealed in the calyx, one-seeded, indehiscent. *Endl. Gen. No.* 6526.

A large genus of shrubs or herbs, common in the tropics and subtropics of both hemispheres, almost always copiously sprinkled with resinous, black or pellucid dots, and strongly scented. Leaves pinnate, or trifoliolate, rarely unifoliolate. Stipules free or adnate with the petiole. Flowers blue, purple or white, variously disposed.

ANALYSIS OF THE SOUTH AFRICAN SPECIES.

1. Sparsifloræ. Flowers axillary, pedicellate or subsessile, solitary or clustered (not spiked or racemose). (Sp. 1–19).

Leaves pinnate: leaflets in 3–7 pairs.
Leafl. linear-lanceolate, in 7 pairs ... ... (1) **odoratissima.**
Leafl. linear or lanceolate-linear, in 3–5 pairs ... (2) **pinnata.**
Leaves trifoliolate, or unifoliolate, or abortive:
Leafl. mostly solitary or abortive (reduced to scales);
Stipules free from the petiole:
Leafl. *lanceolate*, complicate ... ... (3) **Gueinzii.**
Leafl. narrow-linear or none:
Variably pubescent; pedicels shorter than the cal. (4) **aphylla.**
Glabrous; pedicels much longer than the calyx (5) **oligophylla.**
Stipules adnate with the petiole, stem-clasping ... (6) **restioides.**
Leafl. uniformly three:
Virgate or filiform, slightly branched shrubs or suffrutices:
Stipules adnate with the petiole, subulate-acuminate ... ... (7) **fascicularis.**
Stipules minute, subulate, free;
Slender, straggling suffrutices;
Petioles 2–3 lines long; leafl. narrow-linear ... ... (8) **tenuissima.**
Petioles 6–12 lines long; leafl. linear-lanceolate ... ... (9) **glaucina.**
Virgate shrubs:
Leafl. lanceolate, acute at each end (10) **verrucosa.**
Leafl. linear-oblong, obtuse, mucronate (11) **axillaris.**
Much branched, ramulous and densely leafy shrubs:
Lateral leaflets very *unequal sided* at base:
Leafl. elliptic-oblong, aristate, prominently dotted ... ... (12) **obliqua.**
Leafl. linear or lin.-oblong, with immersed dots ... ... (13) **polyphylla.**
Lateral leaflets nearly equal sided at base:
Leafl. lin.-oblong, *flat*, with convex glands (14) **Mundtiana.**
Leafl. *narrow, complicate*, minutely dotted, glabrous ... ... (17) **carnea.**
Leafl. *obcordate*, the young ones silvery (15) **macradenia.**
Leafl. *cuneate*, sharply recurvo-mucronate, *glabrous* ... ... (16) **aculeata.**
Trailing or prostrate suffrutices or herbs:
Petioles 1–2 inches long; leafl. glabrous (18) **repens.**
Petioles 2–3 lines long; leafl. pubescent (19) **decumbens.**

2. Fasciculatæ. Flowers in *dense*, terminal or axillary *fascicles*, but not distinctly capitate or spiked. (Sp. 20–24).

Petioles very short; leafl. broadly, cuneate-oblong, *blunt* (20) **polysticta.**
Petioles very short; leafl. *narrow*, linear-spathulate, *acute*, hook-pointed:
Calyx-teeth short, triangular ... ... (21) **triantha.**

Calyx-lobes *tapering*, much longer than the tube :
Slender ; fascicles loose, axillary ; calyx silky ... (22) **candicans.**
Robust ; fascicles dense, terminal ; calyx hairy ... (23) **uncinata.**
Petioles longish ; leaves distant ; leafl. *linear-lanceolate* ... (24) **capitata.**

3. Racemosæ. Flowers 2-3 together, imperfectly racemose or spicate on the naked or nearly naked ends of lateral or leading branches ; the pseudo-spikes lax and often interrupted. (Sp. 25-30).

Pseudo-racemes *sessile* (the lower tufts of flowers axillary) :
Leafl. cuneate-oblong, *flat*, mucronulate ... ... (25) **racemosa.**
Leafl. obovate, *complicate*, arched backwards ... ... (26) **hamata.**
Pseudo-racemes *pedunculate* (no axillary flowers).
Robust, *divaricately* much branched, rigid shrubs :
Twigs roughly hirsute ; leafl. elliptic-obovate, *hirsute* (27) **stachydis.**
Twigs silky-canescent ; leafl. cuneate-obcordate ... (28) **hirta.**
Slender or suffruticose, erect or diffuse :
Silky-canescent ; leafl. *pellucid*-dotted, *not* nigro-punctate ... ... (29) **argentea.**
Thinly pubescent ; leafl. *nigro*-punctate ... .. (30) **striata.**

4. Spicatæ. Flowers in *dense*, terminal, *sessile* spikes. (Sp. 31-32).

Spikes of flowers ovoid, or broadly oblong ... ... (31) **bracteata.**
Spikes of flowers cylindrical, elongating ... ... (32) **spicata.**

5. Pedunculares. Flowers in *distinctly peduncled*, axillary or terminal racemes spikes or heads. (Sp. 33-41).

Inflorescence laxly racemose or subspicate.
Leaves subsessile, *digitately* trifoliolate :
Racemose ; calyx pale, subglabrous, *with purple striæ* ... ... (33) **Bowieana.**
Spicate ; calyx *silky*, *impunctate* ... ... (34) **venusta.**
Leaves petiolate, *pinnately*-trifoliolate ... ... (35) **caffra.**
Inflorescence densely spiked or capitate :
Leaves unifoliolate ; *glabrous* ... ... (36) **rotundifolia.**
Leaves unifoliolate ; *hairy* or *hispid* ... ... (37) **Thomii.**
Leaves trifoliolate :
Peduncles longer than the leaves :
Heads oblong ; leafl. narrow, glabrescent (38) **Zeyheri.**
Heads globose ; leafl. silky-silvery beneath (39) **tomentosa.**
Peduncles very short ; leaflets *toothed* ... ... (40) **obtusifolia.**
Peduncles short, *two-flowered ;* leafl. entire ... (41) **biflora.**

I. SPARSIFLORÆ (Sp. 1-19).

1. P. odoratissima (Jacq. Schoenb. t. 229) ; "leaves impari-pinnate, *leaflets* about 7 pairs, *linear-lanceolate ;* pedicels axillary, shorter than the leaf." *DC. Prod.* 2. *p.* 216.

HAB. Raised from Cape seeds, *Jacquin.*

Except in the more numerous pairs of lanceolate leaflets and short pedicels this does not materially differ from *P. pinnata, var. ε.* Having been figured and described by Jacquin from cultivated specimens, it may be only a garden variety.

2. P. pinnata (Linn. sp. 1074) ; tall, erect, densely much branched, pubescent or glabrous ; branches and twigs angular ; leaves impari-pinnate, in 3-5 pair, *leaflets linear or lanceolate-linear*, acute, very narrow ; stipules small, free, lanceolate or ovato-lanceolate ; pedicels axillary, long or short, bracteolate beyond the middle, the bracteoles connate ; calyx very *variable* in incision and pubescence (see vars.) *DC. Prod.* 2. *p.* 216.

VAR. *α.* **vulgaris** ; twigs and leaves villoso-pubescent ; pedicels shorter than the leaves calyx glabrescent externally, the *ovate-acute* lobes ciliate and more or less hairy within. *P. pinnata, E. & Z.* 1515, *excl. var. β. E. Mey.! Comm. Drege. p.* 82. *P. lævigata, E. & Z.* 1518. *P. tenuifolia, E. & Z.!* 1519.

VAR. *β.* **speciosa**; twigs and leaves villoso-pubescent; pedicels shorter than the leaves; calyx glabrescent or sparsely villous externally, *the lanceolate-acuminate*, ciliate lobes longer than the tube, the *lowest longest and incurved. P. speciosa, E. & Z* 1517.

VAR. *γ.* **subglabra**; twigs and leaves pubescent or glabrous; pedicels nearly *as long as* the leaves; calyx *nigro-hirsute* externally, its lobes short, broad and blunt. *P. pinnata, β. subglabra, E. & Z.; also P. affinis, E. & Z.* 1516.

VAR. *δ.* **glabra**; nearly or quite glabrous; leaflets linear or lanceolate-linear; pedicels either short or long; calyx *glabrous or sparsely hispid and ciliate*, its lobes either broad and blunt, or subacute or acute! *P. glabra, E. Mey.! Comm. p.* 83. *Zey.* 2380, *and* 2381 *(calyx-lobes acute!)*

VAR. *ε.* **latifolia**; subglabrous; leaflets linear-lanceolate, flat, obviously midribbed; pedicels long, calyx subglabrous, or nigro-hirsute, its lobes acute. *P. arborea, Sims. Bot. Mag. t.* 2090. *E. & Z.* 1514. *E. Mey. Comm. p.* 82. *DC. Prod. l. c.*

HAB. Common throughout the Colony, extending to Caffraria and Port Natal. (Herb. Th., D., &c., &c.)

Arborescent or shrubby, 6–12 feet high, densely branched and leafy; very variable in pubescence, sometimes densely hairy, sometimes quite glabrous, different specimens showing various intermediate states. Leaflets commonly 10–15 lines long, ½ line wide; but in var. *ε.* 1–2 lines wide, and evidently lanceolate. A garden specimen, in Herb. Hooker, shows very instructively the little value, in this species, of the form of leaflets; part of the same branch producing filiform, almost *setaceous* leaflets, and part linear-lanceolate leaflets, flat, midribbed and 1 line wide; thus altogether uniting var. *α.* and *ε.* The length of pedicels is also most variable; sometimes the flowers are subsessile, sometimes on stalks 1–2 inches long. Nor are the calycine characters more constant, as seen in the above mentioned varieties. Other minor varieties might be noticed; thus, different specimens of var. *δ. glabra*, have obtuse, subacute and very acute calyx-lobes, passing from broadly-ovate to almost lanceolate! The most singular calyx occurs in var. *β.*, but by no other character does it differ from *α.*, and in general aspect the two are identical. After a careful examination of many specimens of all the above named varieties, I cannot but regard them as local forms of *one* variable species.

3. **P. Gueinzii** (Harv.); suffruticose, diffuse, much branched, leafy, variably pubescent, the young parts villous; leaves subsessile, unifoliolate, leaflets *lanceolate, acuminate, complicate;* stipules ovato-cuspidate, striate; pedicels axillary, very short, bibracteolate; calyx-lobes lanceolate, acuminate. *Hallia lanceolata, Thunb.! Herb. Ups.*

HAB. Cape District? *Thunberg! Dr. Gueinzius! No.* 58. (Herb. Sd.)

1–2 feet high, the lower parts woody, the upper branches half herbaceous. Branches numerous and close-set, flexuous or arching, closely set with leaves. Leaflets ¾–1 inch long, 2 lines wide, the sides folded together and closely applied when dry, the point very acute, almost pungent. Calyces 2½ lines long, rather longer than the pedicel. This is near *P. aphylla*, but differs in the broadly lanceolate-acuminate, folded leaflets, and in habit. It is also less woody.

4. **P. aphylla** (Linn. Mant. 450); erect, or drooping, twiggy, *variably pubescent;* twigs virgate, very erect, either leafless (having a subulate scale instead of a leaf) or sparsely leafy, the leaves *unifoliolate* or rarely 3–foliolate; leaflets narrow-linear, semi-terete, acute; stipules small (or none) deltoid, acute; pedicels axillary, *not longer than the calyx*, bibracteolate below *the middle;* calyx-lobes acute, ovate or lanceolate. *DC. Prod.* 2. *p.* 217. *Jacq. Schoenb. t.* 223. *Bot. Mag. t.* 1727. *Lod. Cab. t.* 221. *Thunb. Fl. Cap. p.* 605. *E. & Z.* 1530. *E. Mey. Comm. p.* 84. *P. Jacquiniana, E. & Z.* 1531. *P. decidua, Berg. p.* 220, *non Thunb.*

VAR. α.; calyx-lobes ovate or ovato-lanceolate, equalling the tube.

VAR. β.; calyx lobes linear-lanceolate, subfalcate, longer than the tube. *P. filifolia, E. & Z.* 1532.

HAB. By streamlets in mountain kloofs and on hill sides. Common throughout the Western Districts. (Herb. Th., D., Sd., Hk., &c.)

Generally a tall, slender, graceful shrub, with rodlike, erect branches, bending over with the weight of the pale blue flowers. In alpine situations it grows dwarf, with very densely crowded, closely scaly, subcorymbose branches. The pubescence is scanty, and sometimes nearly deficient. The young plant usually bears leaves, regularly stipulate; the full grown rarely anything but *leaf-scales* or *phyllodia*, without stipules. These are either closely or distantly set, and vary from ovate to subulate. The pedicels are rarely as long as the calyx and mostly much shorter. The calyx is sometimes nearly glabrous, sometimes pubescent, and again nigro-hirsute; its lobes vary much in length and breadth.

5. **P. oligophylla** (E. & Z.! 1533); erect or drooping, twiggy, glabrous; twigs virgate, very erect, *laxly leafy;* lower leaves trifoliolate, upper mostly unifoliolate; leaflets narrow-linear, acute; stipules small, subulate; pedicels axillary, 1–3 together, much *longer than the glabrous calyx,* bibracteolate *near the summit;* calyx-lobes ovate or ovato-lanceolate. *P. decidua, ex pte. Thunb. Herb. non Berg.*

VAR. α. **vera**; calyx-lobes ovate, subobtuse, short. *E. & Z.!* 1533.

VAR. β. **glaucescens**; calyx-lobes ovato-lanceolate, acute. *P. glaucescens, E. & Z.* 1534. *P. fascicularis, E. M.! Comm. p.* 83.

HAB. Moist places in Uitenhage and Albany, *E. & Z.!* Near Grahamstown, *Pappe!* Howison's Poort, *Mr. H. Hutton, &c.* β. Onderbokkeveld, *Drege! E. Z.!*

A graceful shrub, very like *P. aphylla,* but quite glabrous, never *leafless,* though often with few and distant leaves; and constantly known by its long pedicels. The leaflets are sometimes flattish and narrow-*lanceolate*-linear.

6. **P. restioides** (E. & Z.! 1529); suffruticose, erect or ascending, many-stemmed, glabrous; branches incurved, filiform; leaves trifoliolate or oftener unifoliolate; leaflets linear-subulate, acute; stipules *adnate to the base of the petiole,* stem-clasping, *their points short, tooth-like;* pedicels axillary, solitary, not longer than the calyx; calyx-lobes lanceolate, rigidly ciliate, the lowest subulate.

HAB. Mountain sides, Klynriviersberg, Caledon, *E. & Z.* (Herb. Sd., D.)

12–18 inches high, branched chiefly from near the base; the branches slender, simple, rushlike, rather bare of leaves. Leaflets often wanting, or only one, ½–¾ inch long, ⅓ line wide, acute but not taper-pointed. Near *P. fascicularis,* but with much shorter stipules, shorter pedicels, fewer leaflets, &c. The free points of the stipules are only a line long.

7. **P. fascicularis** (DC. Prod. 2 p. 217); suffruticose, diffuse or ascending, glabrous; branches incurved; leaves frequent, trifoliolate; leaflets very narrow-lanceolate, *tapering to a very acute point;* stipules *adnate to the base of the petiole,* stem-clasping, their *points subulate-acuminate;* pedicels axillary, several together, longish; calyx-lobes lanceolate-acuminate, the lowest subulate, glabrous. *P. Thunbergiana, E. & Z.* 1523. *P. tenuifolia, Thunb. Cap. p.* 606.

HAB. Hott. Holland, *E. & Z.! Pappe!* About the Table Mt., *W.H.H.* (Hb. Th., Hk., D., Sd.)

A straggling suffrutex, with long, subsimple, densely leafy branches. Leaflets 1–1½ inch long, ½–1 line wide, remarkably taper-pointed. The stipules are adnate for half their length to the petiole, their free-points 3–6 lines long and very slender.

8. P. tenuissima (E. Mey.! Comm. p. 84); suffruticose, diffuse or procumbent, glabrous; branches *filiform*, subsimple; leaves distant, *on short petioles*, trifoliolate; leaflets *very narrow-linear*, acute; stipules minute, subulate; pedicels axillary, solitary, about equalling the leaves, bibracteolate near the summit; calyx-lobes narrow, oblong-lanceolate, acute.

HAB. Dutoit's Kloof, *Drege*. (Herb. Bth., D., Hk.)

A very weak, slender, almost trailing or subascending suffrutex, 1–2 feet long, with the aspect of *Hallia angustifolia*. Petioles 2–3 lines long; leaflets 1–1¼ inch long, ¼ line wide, semiterete. Pedicels 1 inch long, bibracteolate near the summit.

9. P. glaucina (Harv.); suffruticose, diffuse or suberect, glabrous, and somewhat glaucous; branches virgate, compressed-trigonous, subsimple; leaves distant, *on longish petioles*, trifoliolate; leaflets *linear-lanceolate*, flat, midribbed beneath, acute at each end; stipules minute, subulate; pedicels axillary, 1-3 together, about equalling the leaves, bibracteolate above the middle; calyx-lobes unequal, lanceolate. *P. axillaris, E. & Z.! 1524, non Linn. Zey. 2387.*

HAB. Sand hills at Doornhoogdte, Cape District, *E. & Z*, (Herb. Sond.)

More robust than *P. tenuissima*, with much broader leaflets and longer petioles; allied also to *P. verrucosa*, but not woody, and much smaller and weaker and more straggling in growth. Leaves 2–4 inches apart; petiole 6–12 lines long; leaflets ¾–1½ inch long, 1 line wide. Flowers spreading or deflexed.

10. P. verrucosa (Willd. Sp. 3. p. 1343); erect, virgate, glabrous, *leafy;* branches dotted; leaves pinnately trifoliolate or bijugate; leaflets *lanceolate* or linear-lanceolate, midribbed, *acute at each end*, pale underneath; stipules small, subulate, recurved; pedicels axillary, 1–3 together, much longer than the calyx, bibracteolate near the summit; calyx-lobes obtuse or acute. *DC. Prod. 2. p. 216. E. Mey. Comm. p. 83. E. & Z. 1520. P. triflora, Thunb. Fl. Cap. p. 606. E. & Z. 1521. P. angustifolia, Jacq. Schoenb. t. 226. Burch. Cat. 4218.*

HAB. Moist places, Langekloof, *Drege*. Near the Zwartkops River, *E. & Z. Pappe, &c.* (Herb. Th., D., Hk., Bth., Sd.)

An erect, leafy shrub, 4–6 feet high; the smaller branches rough with raised points, striate. Petioles nearly ½ inch long; leaflets petiolulate, 1–1½ inch long, 1–3 lines wide, flat, with slightly reflexed margins, prominently midribbed and faintly penninerved beneath. Flowers pale blue, scattered; pedicels ¾–1 inch long. Calyx ribbed and nigropunctate; its lobes broadly ovate and short, or acute and subfalcate.

11. P. axillaris (Linn., f., fide E. Mey.! Comm. p. 83); erect, virgate, glabrous, leafy; branches dotted; leaves pinnately trifoliolate; leaflets *linear-oblong*, acute at base, *obtuse and mucronate* at the apex, the medial longest; stipules minute, subulate, spreading; pedicels axillary, subsolitary, nearly equalling the leaves, bibracteolate near the summit; calyx-lobes ovate, obtuse. *DC. Prod. 2. p. 217? P. linearis, E. & Z.! 1522. Thunb. Fl. Cap. p. 606? P. triflora, DC. 2. p. 217?*

HAB. Banks of Vanstaadens River, *E. & Z. Drege*. (Herb. Benth., D.)

Very nearly related to *P. verrucosa* and perhaps a mere variety. It chiefly differs in the *shape* and size of the leaflets. Petioles 2–3 lines long; medial leaflet ¾–1 inch long, lateral ½ inch long, all 1½–2 lines wide, round-tipped, mostly mucronulate, sometimes quite blunt. Flowers not so numerous as in *P. verrucosa*, and rather smaller. The "*P. linearis*," Thunb is not in Herb. Upsal.

12. P. obliqua (E. Mey.! Linn. 7. p. 164); shrubby, erect, much branched and leafy, *at first villous*, afterwards glabrescent; twigs pubescent; leaves trifoliolate; leaflets *elliptic-oblong*, *aristate-mucronate*, densely gland-dotted, the lateral ones remarkably *unequal-sided;* stipules membranous, ovate, withering; pedicels axillary, in threes (or one *trifid*), much shorter than the leaves; calyx somewhat villous, its carinal segment longer than the rest, *ovate-acuminate*. *E. & Z.* 1535. *E. Mey. Comm. p.* 84.

HAB. Common near Stellenbosch and the Paarl, Dutoitskloof, &c. *E. & Z. Drege! W.H.H., &c.* (Herb. Bth., Hk., Sd., D.)

A strong growing shrub, 2–4 feet high; the numerous erecto-patent, villoso-pubescent branches densely covered with broad leaves. Petioles 2-3 lines long; leaflets about ¾–1 inch long, 3–4 lines wide, bright green, very blunt at each end, the medial one equal sided, the lateral rather shorter and very unequal sided, rounded on the outer side at base: all tipped with a slender awn or excurrent midrib. The pedicels are either in threes, or there is a short *peduncle*, bearing three pedicels. The four superior calyx-lobes are lanceolate, ⅓ shorter than the lowest or carinal one, which is broad, with a long, tapering, acute point.

13. P. polyphylla (Eck. & Zey.! 1536); shrubby, erect, much branched and leafy, pubescent or afterwards subglabrous; twigs pubescent; leaves trifoliolate, *subsessile*, leaflets *linear or linear-oblong*, mucronate, immersedly-dotted, the lateral ones remarkably unequal-sided; stipules *small*, withering, ovate-acuminate; pedicels axillary, 2–3 together, very short; calyx nigro-villous and veiny, the four upper segments subulate, the carinal *ovato-lanceolate*, longer than the rest.

HAB. Woods at Krakakamma, Uit., *E. & Z.* Cape, *Bowie*. (Herb. Sd., Bth., D., Hook?)

Very near *P. obliqua*, but with narrower leaves of much darker green colour and denser substance, with less conspicuous glands; smaller stipules and narrower calyx lobes. The branches are virgate, sparingly ramulous towards the ends. Leaflets 6–7 lines long, 1-2 lines wide, either exactly linear, or somewhat cuneate at base.

14. P. Mundtiana (Eck. & Zey.! 1537); shrubby, erect, densely leafy, the twigs hairy; petioles and leaves at first pubescent, then glabrescent; leaves trifoliolate, leafl. *broadly linear*, or *linear-spathulate* or *linear-oblong* recurvo-mucronate, acute at base, prominent, pellucid-glanded; stipules rigidly scarious, striate, ovate, pubescent; pedicels axillary, very short, pubescent; calyx villoso-hirsute, its segments lanceolate, much longer than the tube, the carinal ovato-lanceolate. *P. spathulata, E. Mey. Comm. p.* 85. *P. decumbens, Willd. Herb.* 14137, *fide Walp.* (*not Ait.*)

HAB. Groenhoogte, Swell., *Mundt.* Dutoitskloof, *Drege.* (Herb. Hk., Sd., Bth.)

A stout, rather shrubby bush, 1-2 feet high, rigid, with very short, densely leafy, persistently pubescent twigs. Petioles 2 lines long; leaflets ¾–1 inch long, 1½–3 lines wide, with large, wartlike, pellucid glands, the young ones hairy and ciliate, the older glabrescent. Stipules conspicuously scarious, earlike, broad and striate. Pedicels scarcely longer than the petiole, 2–3 together.

15. P. macradenia (Harv.); shrubby, erect, densely much branched, leafy; twigs thinly canescent; leaves minutely petiolate, trifoliolate; leaflets *cuneate-obcordate*, the younger *silvery-canescent*, the older glabrate, all with *many prominent, wart-like glands* on the lower surface; stipules minute, subulate; flowers axillary, 2-3 together, on very short pedicels (or 2–3 on a minute common peduncle); calyx silky-canescent, its teeth triangular-acute.

HAB. Zwarteberg Mountains near Wanhop, George, *Dr. Alexander Prior.* Mountains of Graaf Reinet, Jan. *Mrs. F. W. Barber.* (Herb. Bth., D.)

A densely branched shrub, 1–3 feet high, with the habit between that of *P. aculeata* and *P. bracteata;* remarkable for the great size and prominence of its glands, and the silvery pubescence of all the *young* parts. The leaves dry pale; the leaflets are 4–5 lines long, 2 lines wide at top, slightly notched at the extremity and *scarcely* or not at all mucronulate. Flowers purple, crowded in the axils of the upper leaves, occasionally subpedunculate. The foliage is very strongly scented.

16. P. aculeata (Linn. Sp. 1074); shrubby, erect, much branched and densely leafy, glabrous; leaves trifoliolate, leaflets small, *cuneate, sharply recurvo-mucronate;* stipules subulate, *persistent, rigid;* pedicels from the axils of the upper leaves, 1-2 together, with two ovate, connate bracts near the summit; calyx enlarged after flowering, its lobes ovate-acuminate. *DC. Prod.* 2. *p.* 217. *Andr. Bot. Rep. t.* 146. *Bot. Mag. t.* 2158. *E. & Z.!* 1545. *P. mucronata, Thunb. Cap. p.* 606, *fide Herb. Ups.*

HAB. On the summit and eastern slopes of Table Mountain, common, *E. & Z. W.H.H., &c.* (Herb. Th., Bth., D., Hk., Sd.)

A strong growing densely branched, woody bush, 2–4 feet high, closely covered with small leaves. Petioles 1–2 lines long; leaflets 3 lines long, 1 line wide at top, rather concave and somewhat keeled, exactly cuneate. The *stipules* remain after the leaves fall, and are sharp, but scarcely spiny. The large blue flowers are sometimes crowded together towards the ends of the branches, but do not form a true raceme

17. P. carnea (E. Mey.! Comm. p. 85); fruticose; twigs rusty, pubescent; leaves trifoliolate, subsessile, leaflets narrow-oblong or cuneate-oblong, recurvo-mucronate, sub-complicate, thickish, glabrous, the younger ciliate on margin and keel; stipules deltoid, membranous; flowers axillary, subternate, on short pedicels; calyx *pale, netted with dark veins,* the segments ciliate, acuminate-mucronate, the lowest longest, broadly ovate, the rest lanceolate.

HAB. Western Districts? no station assigned, *Drege!* (Herb. Benth.)

Shrubby, slender, ramulous, the twigs reddish brown, the youngest hairy, the oldest pubescent. Leaves ½–1 inch apart; petioles not 1 line long; leaflets ½ inch long, 1–2 lines wide, coriaceous, minutely pellucid-dotted, but not nigro-punctate. Peduncles very short, from the axils of the upper leaves, 1–flowered, 2–3 together, "subracemose," (*E. Mey.*) Calyx remarkably veiny, its lowest lobe longer and broader than the rest. Of this I have only seen a very imperfect specimen in Hb. Benth.

18. P. repens (Linn. Mant. 263); suffruticose, procumbent or prostrate, sparsely leafy; leaves *on long petioles,* pinnato-trifoliolate, leaflets cuneate-oblong, subelliptical or obovate, thickish, obtuse or retuse, *glabrous;* stipules minute, ovate, acute; pedicels axillary, 2-3 together, about equalling the petiole, bibracteolate above the middle; calyx campanulate, glabrous or villoso-ciliate, shortly 5-toothed. *DC. Prod.* 2. *p.* 217. *Thunb. Cap. p.* 607. *E. & Z.!* 1525, *also P. diffusa, E. & Z.* 1526.

HAB. Sandy flats in moist places throughout the colony. (Herb. Th., Bth., Hk., Sd., D., &c.)

Stems half herbaceous, spreading widely over or under the sandy soil, often rooting at the nodes, the whole plant nigro-punctate, either glabrous or the younger parts and inflorescence more or less pubescent. Petioles 1–2 inches long; leaflets 5–7 lines long, variable in breadth and shape. Pedicels 1–1½ inches long: the bracteoles minute, toothlike. Flowers blueish purple, the tip of the carina very dark.

19. P. decumbens (Ait. Kew. 3. p. 80); suffruticose, trailing, sparsely or densely leafy, pubescent or hairy; leaves on short petioles, pinnato-trifoliolate; leaflets *broadly cuneate or obcordate*, recurvo-mucronulate, *pubescent; stipules adnate with the petiole*, ovato-cuspidate, ciliate; pedicels very short, axillary, 2-3 together (or crowded towards the ends of short ramuli in an imperfect capitulum); calyx villous, semi-5-fid, the lobes lanceolate. *DC. Prod.* 2. *p.* 217. *E. Mey. Comm. p.* 86. *E. & Z.* 1538. *Lodd. Cab. t.* 282. *P. hirta, Th. ex pte.*

VAR. *β*. **subspicata**; flowers crowded in the axils of the upper leaves, in an imperfect leafy spike.

HAB. Dry hilly or mountain places throughout the colony. (Herb., Th., Bth., Hk., Sd., D., &c.)

Stems slender, spreading widely over the soil, frequently throwing up short, erect ramuli; the branches long and filiform, procumbent. Petioles 2–3 lines long. Leaflets 4–5 lines long, 2–3 wide at the truncate or deeply emarginate extremity, minutely black-dotted. Pubescence copious or scanty, never quite absent. The flowers either issue in small tufts from the axils of all the leaves, or only from the upper ones; pedicels ½–1 line long. DC. erroneously refers Thunberg's "*P. mucronata*" to this species, instead of to *P. aculeata.*

2. FASCICULATÆ. (Sp. 20–24.)

20. P. polysticta (Benth.!); erect, virgate, densely leafy, the twigs hairy; leaves trifoliolate, *subsessile; leaflets cuneate-oblong*, very blunt, *recurvo-mucronulate*, nigro-punctate and pellucid-dotted, glabrescent; stipules *minute*, triangular; tufts of flowers subsessile, axillary, 4-5 flowered; calyx tomentulose, campanulate, shortly 5-toothed; the teeth triangular; legume tomentose. *Benth.! in Herb. Hook. Zey.!* 449.

HAB. Doorn Kop, *Burke and Zeyher!* (Herb. Hook., Sond.)

A virgate shrub, with a good deal of the aspect of *P. obliqua*, but having different inflorescence, &c. Petioles not half a line long, with minute, fleshy, pubescent stipules. Leaflets an inch long, the medial rather the longest broadly oblong, with cuneate base, conspicuously black-dotted, slightly undulate at the margin. Tufts of flowers hidden among the leaves, on peduncles 1–2 lines long, pedicels equally short. Calyx 1½–2 lines long. Legume longer than the calyx.

21. P. triantha (E. Mey. Comm. p. 88); shrubby, slender; branches virgate, ramuliferous, tomentulose; leaves shortly petiolate, trifoliolate, *leaflets linear-spathulate*, thickish, recurvo-mucronulate, glabrous, nigro-punctate; racemules subumbellate, 3–5 flowered, on very short axillary peduncles, which are crowded in an interrupted spurious-raceme; calyx appressedly puberulent, nigro-punctate, *its teeth short, triangular*, the lowest longest and broadest.

HAB. Hills near the Berg River; near Lauwskloof; near Breedrivier, and among shrubs near the Zwartkops River, *Drege!* Paarl, *Rev. W. Elliott!* Caledon, *Dr. Alexander Prior!* Willsenberg, *Zeyher!* 446. (Herb. Hk., Bth., Sd., D.)

An erect and rigid, though slender shrub, 2–4 feet high, the branches long and straight, set with alternate, erecto-patent, straight, leafy ramuli. Petioles 2–3 lines long; leaflets ¾–1 inch long, 1–1½ line wide, with largish dots, the veins hidden. Stipules small, acuminate. Peduncles scarcely as long as the petioles, crowded into a cylindrical or interrupted pseudo-raceme. Calyx 2 lines long, its teeth very short. This is referred by Meyer to *P. hirta*, Thunb., but according to Herb. Upsal and Holm., that synonyme belongs to *P. candicans, E. & Z.*

22. P. candicans (E. & Z.! 1540); shrubby, slender; branches virgate and ramulous, *canescent;* leaves shortly petiolate, trifoliolate; leaflets *linear-oblong or cuneate-obovate*, recurvo-mucronate, minutely puberulous or glabrescent, nigro-punctate; stipules minute, deltoid-acuminate; racemules subumbellate, 3-flowered, on very short axillary peduncles, which are crowded in an interrupted, spurious raceme; calyx silky, mostly impunctate, its *segments linear-lanceolate, taper-pointed*, much longer than the tube. *P. hirta, Thunb. Fl. Cap. p.* 609, *non Linn., nec. Jacq., E. & Z.* 1539. *Zey.* 2373.

HAB. Caledon, *Mundt. Bowie.* Grootvadersbosch and Gauritz River, Swell.; and near Tulbagh, *E. & Z.!* Hassagaiskloof and Zwartkopsriver, *Zeyher.!* (Herb. Th. Sd. Hk., Bth., D.)

Nearly related to *P. triantha*, but not so straight or virgate, more canescent and with a very different calyx. Three out of four specimens of *P. hirta* in Hb. Upsal belong to this species. E. & Z.'s No. 1540 is rather whiter than 1539, but not otherwise different. Leaflets ½–¾ inch long, 1–2 lines wide, mostly acute at each end. Flowers small, pedicels equalling the calyx, and generally longer than the peduncles, which are sometimes obsolete. Calyx 2 lines long, the tube not ½ a line, rarely nigro-punctate.

23. P. uncinata (Eck. & Zey.! 1554); suffruticose, erect, rigid; branches virgate, appresso-pubescent; leaves shortly petiolate, trifoliolate, leaflets *linear* or *linear-oblong* (some of the lower ones cuneate-oblong), uncinato-acuminate, glabrescent, complicate; stipules subulate, longer than the petiole; spikes *capitate, subglobose*, densely many-flowered, on *very short*, terminal and axillary peduncles; bracts ovate, bracteoles filiform; calyx hirsute, its segments much longer than the tube, the 4 upper narrow-lanceolate, the lowest twice as broad. *P. cephalotes, E. Mey. Comm. p.* 87. *non. E. & Z.*

HAB. Mountain sides near Tulbagh, *E. & Z.!* Klein Drakensteensberg, *Drege!* Stellenbosch, *W.H.H.* 24 Rivers, *Zeyher*, No. 444! (Herb. Hk., Bth., Sd. D.)

Stems woody at base, 1–3 feet high, not much branched; branches erect, rigid, 10–12 inches long, laxly leafy. Leaflets of the lower leaves more or less cuneate-oblong or narrow-obovate, about ¾ inch long; of the upper 1–1½ inch long, 1–1½ lines wide, almost exactly linear, all of them strongly hooked at point. Heads crowded round the ends of the branches, one terminal, the rest in the axils of the upper leaves, on peduncles 2–5 lines long. Zeyher's specimens are stronger, with somewhat larger flowers.

24. P. capitata (Linn. f. Suppl. p. 339); tall, sparingly branched, laxly leafy or leafless; leaves distant, trifoliolate or unifoliolate, petiolate; leaflets *linear-lanceolate* or linear, acute at each end, sparsely pilose or glabrous, punctate; stipules subulate-acuminate; peduncles very short, many together, densely fascicled in the upper axils, and aggregated in a terminal head, or interruptedly spicato-capitate; calyx variably pubescent, its segments lanceolate-acuminate. *DC. Prod.* 2. *p.* 218. *E. & Z.!* 1528. *E. Mey.! Comm. p.* 88. *P. multicaulis, Jacq. Schoenb. t.* 230. *DC. l. c. p.* 217. *E. & Z.!* 1527.

HAB. Common in moist, sandy places throughout the colony. (Herb. Th., Hk., Bth., Sd., D.)

Stems numerous, 3–4 feet high, rushlike, striate, sometimes quite simple or branched near the base, the branches simple and curved, variably pubescent. Leaves 2–6 inches apart, the petiole variable in length; lower leaves mostly trifoliolate, the upper often unifoliolate, or (as in *P. aphylla*) reduced to a scale. Flowers very densely aggregated at the ends of the branches; the inflorescence sometimes interrupted, several fascicles closely following each other in an oblong pseudo-spike. Flowers purplish blue.

25. P. racemosa (Thunb. Cap. p. 607); suffruticose, flexuous; twigs pubescent; leaves minutely petiolate, trifoliolate, leaflets *cuneate-oblong, flat,* obtuse, mucronulate, coriaceous, nigro-punctate, glabrous; stipules minute, subulate; spikes terminal, sessile, *interrupted,* the rachis flexuous, laxly many flowered; bracts small; calyx hispidulous, nigro-punctate, its lobes lanceolate, subequal. *P. hilaris, E. & Z.!* 1552.

HAB. Mountain sides in grassy places. Langekloof, George, *E. & Z.!* (Herb. Th., Sd., D.)

A slender, slightly branched, ascending suffrutex; the branches incurved and more or less zig-zag, the principal stems and lateral branches ending in sparsely flowered, flexuous spikes. Petioles 1–2 lines long; leaflets 1 inch long, 2–3 lines wide, cuneate at base, the medial rather longer. Bracts equalling the pedicels. Flowers in threes, 2–3 lines long, the corolla nearly twice as long as the calyx, (white?) carina deeply coloured.

26. P. hamata (Harv.); fruticulose, erect, twigs appressedly subcanescent; leaves minutely petiolate, trifoliolate; leaflets *obovate, complicate,* recurvo-arcuate, mucronate, glabrous, nigro-punctate; stipules subulate; spikes terminal, sessile, *interrupted,* laxly many flowered; flowers subternate, rachis straight; bracts small; calyx appressedly puberulous, punctate, its lobes lanceolate, subequal. *P. racemosa, E. Mey. Comm. p.* 87, *non Thunb.*

HAB. Roodeberg, 3800 f., *Drege!* (Herb., Hk., Bth.)

A small shrub, with slender twigs, thinly clothed with very minute whitish hairs, rather densely leafy. Petioles 1 line long; leaflets 4–5 lines long, 3 lines wide, folded together, the midrib arched backwards and hispidulous; a few minute hairs also on the lower surface, otherwise glabrous. Dots conspicuous. Racemes 2–4 inches long, the fascicles of flowers ⅓ of an inch apart. Much more ligneous than *P. racemosa,* with closer, shorter, and differently shaped leaflets, and virgate not flexuous branches. The pubescence also differs.

27. P. stachydis (Linn. f. Sup. 335); shrubby, divaricately branched, rigid, *roughly hirsute with patent, rusty hairs;* leaves shortly petiolate, pinnato-trifoliolate, leaflets *elliptico-obovate,* obtuse or acute, recurvo-mucronulate, *hirsute;* stipules adnate, subulate; flowers 3–5 together in an interrupted, flexuous, terminal, pedunculate spike; bracts minute; rachis and calyx fulvo-hirsute, shaggy, the calyx-lobes sublanceolate, acute, the lowest broadest. *DC. Prod.* 2. *p.* 218. *P. stachyos, Thunb.! Cap. p.* 608.

HAB. Near Piquetberg, *Thunberg!* (Herb. Upsal, Holm.)

Near *P. hirta,* but with larger and more elliptical, not at all cuneate leaflets and a much coarser and more patent, dark coloured pubescence. Leaflets ¾–1 inch long, ½ inch wide, minutely pellucid-dotted, often strongly recurved. Peduncles 2 inches long to the first cluster of flowers. Calyx 2 lines long. This seems to be a rare species. I have only seen the specimens above referred to.

28. P. hirta (Linn. Amœn. 6. Afr. 35); shrubby, divaricately much branched, rigid, *cano-pubescent;* leaves shortly petiolate, trifoliolate, leaflets *broadly cuneate-obovate or obcordate,* recurvo-mucronulate, minutely appresso-pubescent, nigro-punctate; stipules small, subulate; flowers ternate, sessile, either axillary or in an interrupted, leafless, flexuous, terminal, pedunculate spike; bracts minute; calyx cano-hirsute, its lobes ovate-acuminate, the lowest broadest. *DC. Prod.* 2. *p.* 217. *Jacq. Schoenb. t.* 228. *E. Mey.! Comm. p.* 87 (*non Thunb. nec E. & Z.*) *P. Stachyos, E. & Z.!* 1555.

HAB. Dry hills in the Western Districts. Zwartland, *E. & Z.!* Malmsbury, *Pappe!* Paarlberg, *Drege!* Lion's Mountain, Capetown, *W.H.H.* (Herb. Hk., Bth., Sd. D.)

A very rigid, sparsely leafy bush, 2–3 feet high and wide, the branches and twigs issuing to all sides at right angles, all parts thinly canescent with appressed or sub-patent, short hairs. Petioles 2–3 lines long; leaflets 3–5 lines long, 2–3 lines wide, blunt or mucronulate, sometimes emarginate and often folding together when dry. Tufts of flowers rarely in the axils of the uppermost leaves, usually ½–1 inch apart on a flexuous, naked prolongation of the branch. This comes very near *P. Stachydis*, but has a very different pubescence, and shape of leaflets, &c.

29. P. argentea (Thunb. ! Fl. Cap. p. 608); shrubby, slender, diffuse or suberect; twigs minutely silky; leaves shortly petiolate, trifoliolate, leaflets obovate, recurvo-mucronulate, thinly *silky-silvery* on both sides; stipules minute, subulate; flowers ternate, shortly pedicellate, either axillary or in an interrupted, leafless, terminal pedunculate raceme; bracts minute; calyx *silky-canescent*, its lobes lanceolate, the two uppermost semi-connate. *DC. Prod.* 2. *p.* 218. *E. Mey.! Comm. p.* 87.

HAB. S. Africa, *Thunberg!* Krakkeelkraal, *Drege!* (Herb. Th., Hk., Bth., D.)

A slender, diffusely much branched, leafy shrub, with many filiform twigs, all parts clothed with very short, appressed, whitish, glossy pubescence. *Thunberg* says it is "erect." Petioles 1–2 lines long. Leaflets 3–4 lines long, 2–2½ wide at top, when dry folded together. Inflorescence nearly as in *P. hirta*, but the flowers are evidently pedicellate, and the rachis not flexuous. The leaflets are minutely pellucid dotted, but not nigro-punctate. Flowers 2–2½ lines long.

30. P. striata (Thunb. ! Fl. Cap. p. 608); suffruticose or shrubby, *thinly canescent;* twigs striate, pubescent, subvirgate, laxly leafy; leaves shortly petiolate, trifoliolate; leaflets obovate-oblong, recurvo-mucronate, appressedly pubescent, nigro-punctate; stipules subulate; pseudo-spikes terminal, pedunculate, somewhat interrupted, laxly or rather closely flowered; bracts minute; calyx cano-pubescent, punctate, its lobes ovate or ovato-lanceolate, acute, the 2 upper subconnate. *DC. Prod.* 2. *p.* 218. *E. Mey. Comm. p.* 87. *P. rupicola, E. & Z.!* 1551. *P. Eckloniana, Otto! Zey. No.* 441. (421, Hb. Sond.)

VAR. β. **gracilis**; more slender, with longer and laxer spikes and smaller flowers. *P. albicans, E. & Z.* 1556.

HAB. Doorn River, Bokkeveld, *Thunberg!* Grasbergsrivier, *Drege!* Clanwilliam, *Zeyher!* Gauritz R., Swell., *E. & Z.!* Swellendam, *Dr. Thom!* (Herb. Th., Bth., Hk., Sd., D.)

A strong-growing or slender suffrutex, 1–2 feet high or more, becoming woody in age, the branches, leaves, and inflorescence clothed with very short, whitish hairs, which give a pallid aspect to the whole plant. Petiole 2–3 lines long, sometimes produced a little beyond the lateral pair of leaflets. Leaflets 1–1½ inch long, ½ inch wide, frequently complicate, the slender veins obvious on the lower surface. Dots minute, impressed. Calyces somewhat enlarged in fruit. Var. β. from Olifant's R., Clanwilliam, is smaller and more slender; but I think not specifically different.

31. P. bracteata (Linn. Mant. 264); shrubby, erect or procumbent, densely leafy; twigs pubescent or subcanescent; leaves trifoliolate, subsessile; leaflets broadly *obovate or obcordate*, mucronulate, pellucid-dotted, glabrous or the younger subvillous; stipules obliquely ovato-lanceolate, acuminate, striate, membranous; spikes terminal, sessile, oval or oblong, *densely many flowered*, bracteate; bracts membranous, variable in size and shape; calyx villous or glabrescent, its segments much acuminate, very acute, the lowest longest and broadest. *DC. Prod.* 2. *p.* 218. *Jacq. Schoenb. t.* 224. *Bot. Mag. t.* 446. *Lodd. Cab. t.* 1559. *P. aculeata, Thunb. in Herb. Ups.*

VAR. α. **vera**; bracts obovate, cuspidate, as long as the calyx, whose lowest segment is broadly ovato-cuspidate. *P. bracteata, α., Curtisiana, E. Mey. Comm. p.* 86.

VAR. β. **bracteolata**; bracts ovato-lanceolate, much shorter than the calyx, whose lowest segment is lanceolate. *P. bracteolata, E. & Z.* 1543. *P. stachyera, E. & Z.* 1549. *P. cephalotes, E. & Z.* 1548. *P. parviflora, E. Mey. Comm. p.* 86.

VAR. γ. **Algoensis**; depressed or procumbent; leaflets shorter and broader; spikes subglobose or oblong, often few flowered; bracts small; calyx much less villous, with rather shorter segments. *P. Algoensis, E. & Z.* 1544. *P. acuminata, E. & Z.* 1546. *P. densa, E. Mey. Comm. Drege, p.* 86. *Zey.* 2376.

HAB. Moist places, from Capetown to Albany and Caffirland, both on the plains and mountains. Var. γ. near the sea coast, in George, Uitenhage and Albany. (Herb. Th., Bth., Sd., Hk., D.)

A densely branched, leafy bush, 1-4 feet high, erect when growing in sheltered places and well watered soil; depressed or procumbent in exposed or dry places. Pubescence variable, but never absent. Petioles 1 line long, the stipules obliquely adnate to their base. Leaflets ½–¾ inch long, 3–5 lines wide at the end, cuneate at base, often nigro-punctate as well as pellucid-dotted, the older almost always glabrous; mucro short, recurved. Bracts very variable in size. None of the above varieties if traced through many specimens will be found true to its characters, and more might be enumerated showing intermediate forms. After a comparison of the accumulated materials of several herbaria, I am forced to regard as synonymous the species above named. The flowers vary from purple to blue and white.

32. P. spicata (Linn. Mant. 264); shrubby, erect, *virgate*, densely leafy; twigs cano-pubescent; leaves trifoliolate, subsessile, leaflets obovate-oblong or oblong, reflexo-mucronulate, nigro-punctate, glabrous, or the younger pubescent; stipules adnate, obliquely ovato-lanceolate, striate, membranous; spikes terminal, sessile, *elongating, cylindrical*, laxly many flowered, bracteate; bracts small, membranous; calyx villous, its segments lanceolate, acute, subequal. *DC. Prod.* 2. *p.* 218. *Andr. Rep. t.* 411 *(bad). E. & Z.* 1547.

HAB. S. Africa, *Thunberg! Bowie!* Moist places, Puspasvalley, Swell., *E. & Z.!* (Herb. Th., Hk., Bth., D., Sd.)

Nearly allied to *P. bracteata*, but more virgate, with narrower, longer and less decidedly obovate leaflets, much longer and more cylindrical spikes and smaller flowers. The dots on the leaves are more opaque than in *P. bracteata*, but are translucent. Leaflets ¾–1 inch long, 3 lines wide, cuneate at base, with a very small, reflexed mucro. Calyces 2 lines long, their segments less acuminate and more equal among themselves than in *P. bracteata*. Flowers blue or purplish.

33. P. Bowieana (Harv.); suffruticose, slender, diffuse, branching; twigs terete, puberulous; leaves subdistant, trifoliolate, minutely petiolate; leaflets cuneate-obovate, recurvo-mucronate, glabrescent, pellucid dotted; stipules subulate; peduncles axillary, longer than the leaves, fasciculato-racemose; pedicels 2–3 together, equalling the calyx; bracts lanceolate; calyx subglabrous, pale, with purple striæ, its lobes narrow-lanceolate, the lowest longest.

HAB. Cape, *Bowie!* (Herb. Hk.)

Seemingly a spreading plant, woody at base, the stems probably 2 feet or more long; the branches 6–12 inches long, filiform, flexuous. Petioles 1 line long; leaflets ¾–1 inch long, 3–3½ lines wide, the younger ciliate and sparsely hispid, the older becoming glabrous. The glands are minute, pellucid, and not prominent. Racemes short or elongating, interrupted, several flowered. Calyx segments each elegantly marked with 3 purple striæ.

34. P. venusta (E. & Z.! 1553); suffruticose, flexuous; twigs appressedly pubescent; leaves subsessile, trifoliolate; leaflets cuneate-oblong, obtuse-mucronulate, pellucid-dotted, appressedly puberulent; stipules subulate; spikes terminal and axillary, cylindrical, pedunculate; bracts half as long as the calyx; calyx *silky*, its lobes lanceolate, acuminate, *impunctate*, midribbed, the lowest longest.

HAB. Sand dunes, near Saldanha Bay, *E. & Z.!* (Herb. Sd.)

A single branch in Herb. Sond. is about a foot long, curved, simple, with a few small lateral ramuli, ending in a cylindrical spike 1½ inch long, and having much shorter, pedunculate spikes in the axils of the uppermost leaves. Petioles scarcely any; leaflets 1 inch long, 3–3½ lines wide, cuneate at base. Calyx 3 lines long, its lowest segment nearly equalling the petals; two uppermost shorter than the two lateral.

35. P. caffra (E. & Z.! 1550); shrubby, coarsely glandular; twigs tomentulose; leaves petiolate, *pinnately*-trifoliolate, lateral leaflets *broadly elliptical*, terminal *much larger*, *elliptical or obovate*, all *round-topped*, but mucronulate, nearly glabrous; stipules small, deltoid-acuminate; spikes axillary, pedunculate, laxly many flowered, elongating; bracts small; calyx tomentose, nigro-punctate, its lobes lanceolate, the lowest acuminate; legume pubescent.

HAB. Grassy hills between the Kat and Key Rivers, Caffraria, *E. & Z!* (Herb. Sd., Bth.)

Not much branched, diffuse, 12–18 inches high; branches curved, densely leafy, roughly warted with prominent glands, as are also the leaves and calyces. Tomentum wearing off in age. Common petiole half inch long or more, the pair of leaflets about its middle. Leaflets 1–1½ inch long, 6–10 lines wide, very obtuse, but the younger ones with a minute mucro, penni-nerved. Glands very copious and large. *Flowers* in twos and threes on the rachis, the common bract scarcely 1 line long, about equalling the pedicels. Calyces 2 lines long. Legumes turgid, 2½ lines long.

36. P. rotundifolia (Linn. f. suppl. p. 338); suffruticose, subsimple; stems ascending, glabrous or hispidulous; leaves shortly petiolate, unifoliolate, the leaflet *ovate, elliptic-oblong or elliptico-lanceolate*, obtuse or acute, mucronulate, reticulately veined, glabrous, *nigro-punctate* and pellucid-dotted; peduncles terminal (and axillary?) longer than the leaves, hirsute; spikes oblong, many flowered, bracteate; calyx *plumoso-hirsute* with long white hairs, its segments oblong or oblongo-lanceolate, longer than the tube. *Thunb. Cap. p.* 605. *DC. Prod.* 2. *p.* 218.

HAB. S. Africa, *Thunberg! Scholl!* Drakenstein, Stellenbosch (not in fl.) *W.H.H.* (Herb. Th., Bth., D., sine fl.)

A small subsimple suffrutex, a span high, densely leafy. Leaves on very short petioles; the leaflet 1½–2 inches long, ¾–1½ inch broad, the youngest ciliolate, otherwise glabrous, the ribs and veins conspicuous on both sides. Stipules lanceolato-subulate, longer than the petiole. Spikes 1½ inch long; the calyces shaggy with long white patent straight hairs.

37. P. Thomii (Harv.); suffruticose, subsimple; stems ascending, angularly winged, hispidulous; leaves shortly petiolate, unifoliolate; the leaflet ovate, elliptic-oblong, or elliptico-lanceolate, mucronulate, reticulately veined, the younger *hairy*, the older hispid or glabrescent, *impunctate*, scaberulous at the margin; peduncles terminal and axillary, longer than the leaves, hirsute; spikes oblong, many flowered, bracteate; calyx plumoso-hirsute with long hairs, its segments narrow-lanceolate, veiny, longer than the tube.

HAB. S. Africa, *Dr. Thom! Bowie!* (Herb. Hk.)

Very near *P. rotundifolia*; but the leaves when young are densely hairy, and never quite glabrous, destitute of glands and quite opaque (at least when dry), and having a minutely roughened, cartilaginous edge. In size and shape the leaflets agree with those of *P. rotundifolia.* The stipules are broader and the calyx-segments narrower and more taper pointed than in that species. The *alæ* are much longer than the carina.

38. P. Zeyheri (Harv. Thes. t. 80); suffruticose, subsimple; stems ascending, subterete, hispidulous; leaves shortly petiolate, trifoliolate; the leaflets of the lower leaves shortly and broadly obovate, of the upper linear-oblong or linear, all mucronulate, impunctate, scaberulous at margin, the younger villous, the older glabrescent; stipules lanceolato-subulate; peduncles terminal and axillary, much longer than the leaves, hirsute; spikes oblong, many flowered, bracteate; calyx densely albo-hirsute, pale, veined with purple, its 4 upper segments linear-lanceolate, the lowest oblongo-lanceolate, much longer, reticulately veined.

HAB. Mountains over the 24 Rivers, *Zeyher, No.* 2375. (Herb. Hk., Sd.)

Stems 6–8 inches high, densely leafy. Leaflets of two shapes; those of the lower leaves 4–5 lines long, 3–4 lines wide; of the upper 1–1½ inch long, 1–2 lines wide. Peduncles 4–6 inches long. Calyces very hairy, their lowest segment conspicuous in the spike, much longer than the bracts and nearly equalling the corolla. Obviously allied to *P. rotundifolia* and *P. Thomii*, but with different foliage and calyx.

39. P. tomentosa (Thunb.! Fl. Cap. p. 606); shrubby, *silky and silvery*, densely leafy; twigs striate; leaves petiolate, pinnato-trifoliolate, leaflets elliptico-lanceolate or elliptical, glabrous on the upper, *silky and silvery* on the under surface, mucronate, penni-nerved; stipules lanceolate, equalling the petiole; peduncles terminal and axillary, much longer than the leaves; *heads globose, bracteate*, dense; bracts ovato-lanceolate, very villous bracteoles narrow-lanceolate; *calyces villoso-hirsute*, the segments lanceolate, lowest much longer than the rest; ovary glabrous. *DC. Prod.* 2. *p.* 218. *and P. sericea, Poir. Dict.* 5. *p.* 687. *DC. l. c. p.* 219. *E. & Z.!* 1541. *P. pedunculata, Bot. Reg. t.* 223. *Eriosema capitatum, E. Mey.! Comm. p.* 130.

HAB. Shrubby places. Districts of George, Uitenhage and Albany, *E. & Z.! Drege! Pappe! Mrs. Barber, &c.* (Herb. Th., Bth., Hk., Sd., D.)

A small, erect or ascending, slightly branched, half woody shrub, 1–2 feet high, the twigs, under sides of leaflets, inflorescence, and young leaves silvery and silky; the older parts glabrescent. Petioles ¼–1 inch long. Leaflets 1–2 inches long, ½–¾ inch broad; the margin slightly recurved. Peduncles 4–5 inches long, erect or spreading. Heads very villous, with white or dark hairs and the bracts silvery. Carina striate with dark purple. This has much the habit of an *Eriosema*, but is a true *Psoralea.* The gland-dots are very minute; often scarcely obvious.

40. P. obtusifolia (DC. Prod. 2. p. 221); diffuse or procumbent; branches striate, villoso-canescent; leaves on longish petioles, pinnately trifoliolate, leaflets *obovate or elliptical*, obtuse, plaited, *repando-dentate*, thinly silky on each side, minutely nigro-punctulate; stipules ovate, acute, membranous; spikes axillary, on peduncles shorter than the leaf, 4–8 flowered; calyces densely villous with white hairs; legume villous. *E. Mey.! Comm. p.* 88. *P. plicæfolia, E. & Z.!* 1557, *and P. exigua, E. & Z.!* 1558. *Trigonella tomentosa, Thunb. Fl. Cap. p.* 611.

HAB. Cape, *Burchell*, No. 1214. Near the Gariep at Verleptpram, and on dry plains near Rhinosterkop, *Drege!* Gauritz River, George; Nieuweveld, Beaufort; and Olifant's R., Clanw., *E. & Z.* Fish River, *Burke & Zeyher!* (Z. 450). Grahamstown, *Dr. Atherstone!* (Herb. Th., D., Hk., Bth., Sd.)

Stems 2–3 feet long, spreading over the soil; branches filiform, alternate, distichous, very pale, clothed with soft hairs. Petioles 1–2 inches long, prolonged 1–2 lines beyond the insertion of the pair of leaflets. Leaflets 4–5 lines long, 3–4 lines wide, cuneate at base, frequently elliptical. Spikes either capitate or interrupted, in two heads, the densely hirsute calyces enlarging after flowering and subinflated. *E. & Z.'s P. exigua* is merely a starved state of this species.

41. P. biflora (Harv.); dwarf, prostrate, much branched, glabrescent; leaves on longish petioles, pinnately trifoliolate; leaflets (small) obovate or obcordate, veinless, glabrous, nigro-punctate; stipules small, ovate, blunt, withering; peduncles axillary, shorter than the leaves, 2 (or 1) flowered and bracteate at the summit; bracts very short, truncate; flowers sessile; calyx hispid, 4 upper lobes oblong, rather blunt, lowest twice as broad, concave, obtuse.

HAB. South Africa, *Burchell*, No. 1720.

A small, depressed, distichously much branched and ramulous suffrutex. Petioles not ½ inch long; leaflets 2–3 lines long, 1–1½ wide, blunt or emarginate. Peduncles 3–5 lines long, generally bearing 2 sessile flowers at the bracteate extremity. Flowers 2 lines long. Carina adnate to the alæ. The inflorescence is peculiar; nor does this little plant seem nearly allied to any S. African species.

(Imperfectly known Species.)

P. velutina (E. Mey. Comm. p. 89); "leaves 3–foliolate, short-stalked, canescent and velvetty; leaflets obovate, retuse, pointless." *E. Mey.*

HAB. Under the Zwarteberg, in moist hollows near Klaarstroom, *Drege.*

"Flowers and fruit unknown. An erect, much branched, rigid shrub. Leaves close set; leaflets 3–4 lines long, thickish, with many yellow-brown glands. Stipules minute." Can this be *P. macradenia?*

XXVI. LOTUS, L.

Calyx campanulate, 5–cleft or 5–toothed. *Vexillum* roundish, spreading, recurved, equalling the porrect, connivent *alæ; carina* ascending, narrow, rostrate. *Stamens* diadelphous. *Style* ascending, subulate. *Legume* linear, terete or subcompressed, many seeded, one celled or having septa between the seeds; when ripe splitting into two valves. *Endl.* 6514. *DC. Prod.* 2. *p.* 209.

Herbs or suffrutices, erect or diffuse, common in Europe, and temperate Asia, with outlying species in Australia, S. Africa and S. America. Leaves trifoliolate. Stipules in pairs or connate, free, resembling the leaflets. Peduncles axillary and terminal, umbellately 1–2 or many flowered, with leafy bracts under the flowers, which are *yellow*, or rarely white, red or very dark brown purple. Name from the Greek λωτος. The English name is *Bird's-foot Trefoil.*

1. L. discolor (E. Mey.! Comm. p. 92); suffruticose, *pubescent;* branches striate; *leaflets and stipules similar*, oblong-cuneate, submucronulate, glaucous above, paler beneath; *peduncles* axillary, *elongate;* umbels 4–8 flowered; bracts leaflike, unequal, the longest equalling the flowers; legumes straight, cylindrical, glabrous, locellate within; seeds ellipsoid-subglobose, smooth, dark brown. *Zey.! No.* 453.

HAB. Grassy hills, mouth of the Omsamcaba, *Drege!* Magalisberg, *Burke and Zeyher!* Natal, *Mr. Sanderson, Krauss*, 290. *Gueinzius!* (Herb. Hk., Sd., D.)

1–2 feet high, slender, alternately branched, straggling, variably pubescent; the stems and branches pale, the foliage glaucous. Leaves an inch apart, with a pair of leaflike stipules longer than the petiole; leaflets ½–¾ inch long, 2 lines wide near the top, tapering to the base, obtuse or subacute, scarcely mucronulate, appressedly hispid on the lower surface. Peduncles 1½–4 inches long, spreading. Flowers orange-yellow, the rostrate carina darker. Legumes 1½ inch long, many seeded, with spongy septa between the seeds.

(Imperfectly known Species.)

L.? amplexicaulis (E. Mey. Comm. p. 92); "erect, suffruticose, *very hairy;* leaflets oblong-obovate; stipules *solitary*, opposite the leaves, *amplexicaul*, *cordate*, *large;* heads terminal and in the forks of the branches *sessile;* legumes straight, compresso-turgid, very hairy; seeds subglobose, with a tubercle at the micropile, smooth and shining." *E. M. l. c.*

HAB. Between Zandplaat and Komga, in grassy places, *Drege.*

Flowers unknown, and the genus so far doubtful. Stipules nerved at base, with two thick nerves, which pass upwards into a veiny network. Unknown to us.

L. anthylloides (Vent. Malm. t. 92); "shrubby; leaflets and stipules spathulate, pilose; stipules shorter than the calyx; peduncles elongate; flowers corymbose, minutely pedicellate; legumes . . ?" *DC. Prod.* 2. *p.* 210.

HAB. Formerly cultivated in France.

L. linearis (Walp. Linn. 13. p. 518); "procumbent; stem and branches filiform; leaves trifoliolate; leaflets lanceolate or linear, glabrous; stipules half as long as the leaflets, leaflike; flowers brown, subumbellate; umbels pedunculate, axillary, 2–4 flowered; bracts linear, leafy, equalling the calyx; legume unknown." *Walp. l. c.*

HAB. Cape of Good Hope, *Lalande* in Hb. Kunth, fide *Walpers.*

XXVII. TRIFOLIUM, L.

Calyx campanulate or tubular, unequally 5–cleft or toothed (sometimes inflated after flowering.) *Corolla* persistent; the *vexillum* longer than the *alæ* and the obtuse *carina*. *Stamens* diadelphous. *Legume* minute, 1–4 seeded, enclosed in the calyx, indehiscent. *Endl.* 6511. *DC. Prod.* 2. *p.* 189.

Herbaceous plants, annual or perennial, erect or procumbent, prostrate or creeping, common throughout the temperate zones, especially of the northern Hemisphere. Leaves trifoliolate, rarely 5-foliolate, the common petiole sometimes extending beyond the pair of leaflets. Stipules adnate with the petiole. Flowers small, red, white or yellow, in dense heads or spikes. Name from *tres*, three, and *folium*, a leaf; literally "*Trefoil.*" The various kinds of "*clover*" are well known examples of this genus. Besides those here described, E. Meyer notices, as having been collected by Drege, *T. pratense*, L. *(purple clover)*, and *T. repens*, L. *(white clover)*. Of these I have seen no S. African specimens, and as they are undoubted *escapes* from culture, I refrain from introducing them to the Flora.

ANALYSIS OF THE SOUTH AFRICAN SPECIES.

Prostrate or procumbent:
- Calyces after flowering bladdery, much inflated, woolly … (3) **tomentosum.**
- Calyces not inflated; their segments setaceo-subulate:
  - Whole plant nearly glabrous; leaflets obcordate … (1) **Burchellianum.**
  - More or less villoso-pubescent; leaflets cuneate-oblong … … … … … … … … … … (2) **africanum.**

Erect or diffuse, not prostrate : *root* annual.
- Flowers red or white :
  - Leaflets very long, linear-lanceolate, acute .. ... (4) **angustifolium.**
  - Leaflets obovate-oblong, retuse or emarginate ... (5) **stipulaceum.**
- Flowers yellow :
  - Heads 30–40 flowered ; vex. spreading, strongly furrowed ... ... (6) **agrarium.**
  - Heads 10–15 flowered ; vex. connivent, nearly smooth ... ... (7) **procumbens.**

1. **T. Burchellianum** (Ser. in DC. Prod. 2. p. 200) ; nearly glabrous ; stems procumbent or prostrate, rooting ; leaves on long petioles ; leaflets obcordate, denticulate, nerve-striate, *glabrous ;* stipules oblongo-lanceolate, leafy ; peduncles axillary, elongate ; heads many flowered, flowers subsessile ; segments of the subglabrous calyx setaceo-subulate, nerved, longer than the tube ; legume obovate, mucronate, 2–seeded ; seeds ovate, dark brown. *T. repens, Thunb.! in Herb. Ups.. non Linn. E. & Z.!* 1507.

HAB. S. Africa, *Burchell.* Moist places at the foot of Devil's Mt., Capetown ; Puspas Valley, Swell. ; Adow and Port Elizabeth, Uit., *E. & Z.!* Near Swellendam, *Dr. Pappe!* Howison's Poort, *Mr. Hutton!* Orange River ; near Colesberg ; near Sedbury, *Burke & Zeyher!* (Herb. Th., Hk., D.)

Root perennial. Stems numerous, proceeding horizontally on or under the soil, rooting at the nodes, 1–2 feet long. Petioles 1–3 inches long. Leaflets ¼–½ inch long, broadly obcordate, cuneate at base, the midrib and closely placed lateral nerves prominent ; the latter prolonged upwards into marginal teeth. Heads resembling those of *"white clover" (T. repens).* Calyx teeth very slender and long, bristle-pointed. There are sometimes a few villous hairs on the peduncle and calyx : such specimens almost unite with *T. africanum, β.*

2. **T. africanum** (Ser. in DC. Prod. 2. p. 200); *villoso-pubescent* or *hirsute ;* stems procumbent or prostrate ; leaflets cuneate-oblong, obtuse or emarginate, denticulate, nerve-striate, *villoso-pubescent ;* stipules oblongo-lanceolate, leafy ; peduncles axillary, elongate, hirsute ; heads many flowered ; segments of the *villous* calyx setaceo-subulate, nerved, longer than the tube ; legume ? *T. hirsutum, E. Mey.! Comm. p.* 91, *non Thunb.*

VAR. β. **glabellum** ; stem and leaves either glabrous or sparingly hirsute ; petioles, peduncles and calyces villous. *E. Mey. l. c.* T. africanum, *E. & Z.* 1508 *in Hb. D.*

HAB. Moist places near Mooyplaats and Camdeboosberg, 4–4500f., *Drege.* Near the Orange River, *Burke & Zeyher!* β. Tambukeland, *E. & Z.* Port Natal, *Drege, Sanderson, &c.* (Herb. Hk., D.)

Closely allied to *T. Burchellianum,* from which it chiefly differs by its narrower leaflets, rarely subobcordate, mostly rounded at end, and usually copious pubescence. Var. β. is, however, an intermediate state, sometimes nearly glabrous, and then only to be known by its narrow leaflets. I have seen no authentic specimen of Burchell's plant, and describe from a sp. in Hb. Hooker, marked by Bentham. *T. hirsutum,* Thunb. ! fide Hb. Upsal, is some *Lotononis* not ascertained.

3. **T. tomentosum** (Linn. sp. 1086); stems procumbent or prostrate ; leaflets broadly obovate, glabrous, sharply serrulate, striate ; stipules ovato-lanceolate, much acuminate ; peduncles *very short,* axillary ; heads globose, flowers sessile ; calyces after flowering *much inflated, membranous, netted with veins and tomentose,* the segments shorter than the minute corolla ; legumes one seeded. *DC. Prod.* 2. *p.* 203. *E. & Z.!* 1513.

HAB Sandy places near the Amsterdam Battery, Table Bay, *E. & Z.! Pappe! W.H.H.* Paarl, *Rev. W. Elliott.* Simon's Bay, *C. Wright!* (Herb. D., &c.)

Roots very fibrous, perennial or annual. Stems numerous from the same crown, spreading in a circle, 3–6 inches long, flexuous, glabrous. Leaves on long petioles, the leaflets short and broad, glabrous. Flowers minute, not conspicuous till the corolla withers, after which time the calyx enlarges greatly, becomes bladdery and veiny, and then the heads resemble small cotton balls. This plant is a native of the South of Europe; probably merely naturalized, but completely so in S. Africa.

4. T. angustifolium (Linn. sp. 1083); annual, erect, subsimple, virgate; leaflets *linear-lanceolate, acute*, ciliate; stipules setaceo-subulate; spikes solitary, terminal, oblong, becoming cylindrical; segments of the very hairy, glandular, ribbed calyx setaceo-subulate, *pungent*, about equalling the corolla, the lowest longest. *DC. Prod.* 2. *p.* 189. *E. Mey. Comm. Drege. p.* 90. *E. & Z. No.* 1511.

HAB. Naturalized (from the S. of Europe) in various places near Capetown, Simonsbay, the Paarl, &c. *E. & Z.! Pappe! &c.* (Herb. D., &c)

Stems 6 inches to 2 feet high, only luxuriant ones branching. Leaves widely separate: leaflets 2–3 inches long, 1–2 lines wide, variably silky. Stipules very long and slender. Spikes at first conical-oblong, afterwards elongating, 2–3 inches long, very hairy, fulvous; the flowers small and red.

5. T. stipulaceum (Thunb.! Fl. Cap. p. 609); annual or biennial, suberect, villous; leaflets obovate-oblong, retuse or emarginate, denticulate, striate; stipules membranaceous, striate, subulate-acuminate; heads *oblong, subsessile*, or shortly pedunculate; segments of the densely villous, 10–ribbed, campanulate calyx setaceo-subulate, erect, longer than the tube and the corolla; legume membranous, one seeded; seed reddish-brown. *E. & Z.! No.* 1512. *T. micropetalum, E. Mey.! Comm. p.* 90.

HAB. Grassy fields, Puspasvalley, and near the Zwartkops River, *E. & Z.* Near Saldanha Bay; also at Algoa Bay, in sandy places near the sea, *Drege!* (Herb. Th., Hk., D.)

A small, apparently annual species, 3–6 inches high, simple or branched from the base, clothed with long, soft, fulvous hairs. Petioles long or short; stipules remarkably taper pointed, 4–6 lines long. Leaflets 4–8 lines long, 1½–3 lines wide, closely parallel veined. Heads very hairy, the calyx slightly enlarged after flowering, and its segments becoming rigid and awnlike.

6. T. agrarium (Linn. sp. 1087); annual, nearly glabrous, ascending, rigid; leaves shortly petiolate; leaflets obovate or obcordate, denticulate, striate; stipules ovato-lanceolate, acute; peduncles axillary, long or short; heads densely many flowered, ovoid, yellow; flowers deflexed, the vexillum very broad, spreading, obcordate, striate; calyx minute, its limb very unequal, the upper segments shorter than the lower, glabrous; legume one seeded. *DC. Prod.* 2. *p.* 205. *Benth. Handb. p.* 169. *T. procumbens, Thunb. Fl. Cap. p.* 610. *E. Bot. t.* 945. *E. & Z.! No.* 1509.

HAB. Naturalized (from Europe) throughout the Colony. (Herb. D., &c.)

A branching annual, 3–12 inches high, bearing many heads of bright yellow, small flowers. Each head contains 30–40 flowers, which as they wither become somewhat enlarged and strongly deflexed, and then turn pale brown, the corolla remaining. The vexillum is very broad and distinctly marked with lines. This is the "*Hop-clover*" of England.

7. T. procumbens (Linn. sp. 1088); annual, nearly glabrous, diffuse, slender; leaves shortly petiolate; leaflets obovate or obcordate, denticulate, striate; stipules broadly ovate, subacute; peduncles axillary, longer than the leaves; heads small, 10–15 flowered, subglobose, yellow; flowers deflexed, subsessile, the vexillum folded over the alæ and keel, scarcely furrowed; calyx minute, its limb very unequal, the upper segments much shorter than the lower, glabrous; legume one seeded. *DC. Prod. l. c. Benth. Handb. p.* 170. *T. minus, E. Bot. t.* 1256. *T. filiforme, E. & Z.* 1510. *Pappe!* 134.

HAB. Naturalized (from Europe). About Table Mountain; Hott. Holland; and near Tulbagh, *E. & Z.* Cape Flats, *Dr. Pappe.* (Herb. D., &c.)

Near *T. agrarium*, but much more slender, with smaller flowers and fewer of them in the heads. *T. filiforme*, L. is a still smaller plant, with 2–3, rarely 5 flowers in the head, on pedicels as long as the calyx.

XXVIII. MELILOTUS, Tournef.

Calyx campanulate, subequally 5-toothed. *Corolla* deciduous. *Vexillum* and *alæ* longer than the obtuse carina. *Stamens* diadelphous. *Legume* oval or oblong, 1–4 seeded, longer than the unaltered calyx, indehiscent. *Endl.* 6510. *DC. Prod.* 2, *p.* 186.

Annual or biennial, rarely perennial, strongly scented herbs, chiefly found wild in the Mediterranean region, whence some of the commoner have been dispersed throughout the temperate zones. Leaves pinnately trifoliolate, the leaflets sharply and coarsely toothed. Stipules adnate with the base of the petiole. Flowers small, yellow, cream-coloured or white, in long axillary or terminal racemes. Name from *mel*, honey, and *lotus*; the flowers are frequented by bees. English name, "*melilot.*"

1. **M. parviflora** (Desf. Fl. Atl. 2, p. 192); diffuse or ascending; leaflets of the lower leaves broadly obovate, of the upper cuneate-oblong, all toothed; stipules lanceolate or subulate; flowers densely racemose, *very minute*, pale yellow; calyx teeth subequal, ovate; legumes oval, obtuse, irregularly wrinkled, one seeded. *DC. Prod.* 2, *p.* 187. *E. & Z. No.* 1506. *E. Mey. Comm. Drege. p.* 91. *M. indica, Thb. Fl. Cap. p.* 609.

HAB. Sandy places near the coast. About Table, Simon's, and Algoa Bays, *E. & Z.!* Greenpoint, *Dr. Pappe!* Boschekloof and Klipberg, &c., *Drege.* (Hb. D., &c.)

A common weed in sandy ground, introduced from Europe, and now dispersed over most parts of the globe. Stems 6–12 inches high, the branches spreading at base, and often procumbent, their ends curved upwards. Leaflets variable in shape, and in their serratures; in our specimens sharply and coarsely serrated. Racemes 1–2 inches long, on peduncles 1–1½ inch, the flowers ½ line to 1 line long, on minute, nodding pedicels. Calyx glaucous green. Whole plant very strongly scented.

XXIX. TRIGONELLA, L.

Calyx campanulate, 5-toothed. *Corolla* deciduous; *vexillum* and *alæ* spreading; *carina* obtuse. *Stamens* diadelphous. *Legume* linear or oblong-linear, compressed or terete, acuminate, many seeded. *Endl. Gen.* 6508. *DC. Prod.* 2, *p.* 181.

Strongly scented herbs, chiefly natives of the Mediterranean and Caucasian regions. Leaves pinnately-trifoliolate, the common petiole extending beyond the pair of leaflets. Stipules adnate with the base of the petiole, often toothed or lacerate. Flowers either racemose, umbellato-capitate, or subsolitary. Name from τρεις, *three*, and γωνια, an *angle*; in many species (not in *T. hamosa*), the carina is very small, and the 3 other petals conspicuous, spreading three ways.

1. **T. hamosa** (Linn. sp. 1094); procumbent; leaflets cuneate-obovate

or obcordate, coarsely toothed, nerve-striate; stipules laciniate; flowers racemose, the common peduncle longer than the leaves, ending in a spine; legumes declined, terete, falcate or hooked. *DC. Prod.* 2, *p.* 183. *Fl. Græc. t.* 764. *E. Mey. Comm. Drege. p.* 91. *T. glabra, Thunb.! Fl. Cap. p.* 610.

HAB. S. Africa, *Thunberg!* Sandy places at Zwartland, Cape; and in the Nieuweveld, Beaufort, *E. & Z.* Hills near Ebenezer; and at Verleptpram, on the Gariep, *Drege! Wyley!* Albany, *T. Williamson!* (Herb. Th., D.)

A small, glabrous or sparingly pubescent annual, resembling a melilot. Leaves scattered; the common petiole extending 1–2 lines beyond the first pair of leaflets. Stipules sometimes digitate, varying much in size and number of lobes. Flowers small, pale yellow, cernuous, on pedicels longer than the calyx. Legumes ½ inch long, arched upwards. A native also of Nubia and Egypt.

XXX. MEDICAGO, L.

Calyx campanulate, subequally 5–toothed. *Vexillum* longer than the *alæ* and the obtuse *carina*. *Stamens* diadelphous. *Stigma* capitate. *Legume* one or many seeded, spirally twisted or falcate. *Endl. Gen.* 6507. *DC. Prod.* 2. *p.* 171.

Herbaceous or frutescent plants, abundant in Central and Southern Europe and Middle Asia, from which some are widely scattered throughout the temperate zones, either as weeds or cultivated for cattle food. Leaves pinnately trifoliolate, very rarely impari-pinnate; stipules adnate to the petiole and generally deeply cut; peduncles axillary, few or many flowered; flowers minute, yellow or purple. Legumes very various in form and sculpture, very often bordered with rigid prickles, and spirally rolled together like snail-shells. Name, the μηδικη of the Greeks, so called because introduced by the Medes. English name, *Medick.*

ANALYSIS OF THE SOUTH AFRICAN SPECIES.

Legume unarmed; flowers purple ... ... (1) **sativa.**
Legume bordered with a double row of prickles; flowers yellow:
Leafl. obcordate, toothed; legumes obliquely netted-veined; prickles *curved* ... ... (2) **denticulata.**
Leafl. obcordate, toothed; legume not prominently veined; prickles long, *hooked* ... ... (3) **nigra.**
Leafl. cuneate or linear, coarsely incised or jagged ... ... (4) **laciniata.**

1. M. sativa (Linn. Sp. 1096); stem erect, glabrous; leaflets cuneate-obovate, truncate and mucronate, toothed; stipules lanceolate; peduncles many flowered, racemose; calyx-teeth subulate; legumes *unarmed*, compressed, rugose, twisted once or twice in a loose spire. *DC. Prod.* 2, *p.* 173. *Eng. Bot. t.* 1749. *Lam. Encycl. t.* 612. *Fl. Dan. t.* 2244. *E. & Z.* 1501. *E. Mey. Comm. Drege. p.* 91.

HAB. Introduced in culture: now naturalized near Capetown and Simonsbay, *E. & Z.!*

1–2 feet high, branching. Leaves 2–3 inches apart: leaflets ½–1 inch long, 2–4 lines wide. Stipules sometimes toothed. Flowers purple, 3–4 lines long. Cultivated for cattle-food, under the name of "*Lucerne.*"

2. M. denticulata (Willd. Sp. 3. p. 1414); prostrate; leaflets obovate, or obcordate, toothed; stipules laciniate; peduncles 3–5 flowered; calyx-teeth triangular; legumes bordered with a double row of curved prickles, flat and *obliquely netted-veined at the sides*, twice or thrice spirally twisted. *DC. Prod.* 2, *p.* 176. *E. Bot. t.* 2634. *E. & Z.! No.* 1503. *E. Mey. Comm. Drege. p.* 92. *T. ciliaris, β. Thunb.! Cap. p.* 612.

HAB. Sandy places near Capetown and in Hott. Holl. *E. & Z.!* Paarl, *Drege!* (Herb., D., Th.)

A small annual. Flowers minute, yellow. Known from the following by the more prominent reticulations on the sides of the legume, and the shorter, curved but not hook-pointed prickles.

3. M. nigra (Willd. Sp. 3. p. 1418); diffuse or erect; leaflets obovate or obcordate, toothed; stipules laciniate; peduncles 2–3 flowered; calyx-teeth subulate; legumes bordered with a double row of long, *setaceous hook-pointed prickles*, flat at the sides and somewhat netted-veined, several times spirally and closely twisted. *DC. Prod.* 2, *p.* 178. *E. & Z. No.* 1502. *E. Mey. ! Comm. p.* 92.

HAB. About Table Mountain, *E. & Z.* Duckervallei and Ebenezer, *Drege.* Albany, *T. Williamson!* (Herb. D.)

Very like the last, but with longer and narrower calyx-segments, and much longer and more bristle-shaped, *hook-pointed* prickles on the legume. The Cape plant agrees with European specimens in Herb. T.C.D. I describe from those collected by Williamson.

1. M. laciniata (All. Fl. Ped. n. 1159); erect or diffuse; leaflets cuneate or linear, *coarsely inciso-dentate*, truncate, mucronulate; stipules incisodentate; peduncles 1–2 flowered, short; calyx-teeth short, acute, narrow; legumes bordered with a double row of subulate, hook-pointed prickles, thickened at the margin, with flat sides, and two or more times spirally twisted. *DC. Prod.* 2, *p.* 180. *E. & Z.!* 1504. *E. Mey. in Comm. Drege. p.* 92.

HAB. Sandy ground, various parts of the colony, *E. & Z. Drege!* Greenpoint, *Dr. Pappe!* (Herb. D.)

Known by its deeply cut or laciniated leaflets, which are smaller and more rigid and narrower than in either of the preceding. The prickles on the legume are much shorter in Cape specimens than on our European examples in Herb. T.C.D.

XXXI. INDIGOFERA, L.

Calyx small, campanulate, 5-fid or 5-toothed. *Vexillum* subrotund, reflexed; *carina* with a spur or prominence at each side, near the base. *Stamens* diadelphous, the connective of the anthers apiculate. *Ovary* 2 or several ovuled. *Legume* linear, terete, compressed, or flattened, 1 or several seeded, mostly with dissepiments between the seeds. *Endl. Gen.* 6530.

Shrubs, undershrubs or annuals, very abundant in tropical and subtropical climates, Leaves imparipinnate or digitate, 3 or many foliolate, rarely unifoliolate or abortive. Hairs commonly *strigose*, i.e., rigid, fixed by a medial point, and tapering to each end, set in subparallel lines. Flowers purple, or rosy, or white. *Indigo* is obtained by fermentation from the foliage of several species of this large genus, whence the generic name.

ANALYSIS OF THE SOUTH AFRICAN SPECIES.

Subgenus I. **EU-INDIGOFERA.** Legume subterete or terete, straight or subfalcate, the *convex* valves coriaceous, veinless. Seeds separated by transverse septa. (Sp. 1–107.)

1. Juncifoliæ. Petioles very long, *filiform*, acute, mostly leafless; bearing in the young plant only small, terminal, and lateral leaflets, in impari-pinnate order. (Sp. 1–2.)

Racemes on *long* peduncles, laxly many flowered ... (1) **filifolia.**
Racemes *shorter* than the petioles, few flowered ... ... (2) **podophylla.**

2. Simplicifoliæ. *Leaves* simple, subsessile. (Sp. 3–5.)

Shrubby, divaricate, *subglabrous; lvs.* oblong or obovate (9) **denudata,** vars. δ. ε
Shrubby, virgate, canescent; *lvs.* obcordate, flat ... (3) **obcordata.**
Shrubby, flexuous, canescent; *lvs.* linear-cuneate, convolute ... ... (4) **nudicaulis.**
Suffruticose, diffuse, subglabrous; *lvs.* ovate-oblong, flat (5) **ovata.**

3. Trifoliolatæ. *Leaves* digitately *trifoliolate;* i.e., *three* leaflets springing from the apex of a common petiole. (Sp. 6–32.)

(1.) Stipules subulate or setaceous, or obsolete:
Shrubs, divaricately much branched; lvs. scattered or tufted, not imbricate:
Spine-tipped; racemes short; leafl. 1–2 lines long, strigose beneath ... ... (6) **spinescens.**
Spine-tipped; racemes long, rigid; leafl. strigoso-canescent ... ... (7) **pungens.**
Unarmed, piloso-canescent; flowers solitary (8) **dealbata.**
Unarmed; leaves sub-glabrous; racemes few flowered
Leafl. obovate or oblong-cuneate, 4–5 lines long ... ... (9) **denudata.**
Leafl. lanceolate-linear, acute, 5–10 lines long ... ... (10) **stenophylla.**
Shrubby, villoso-canescent; lvs. *closely imbricate* ... (32) **glomerata.**
Suffruticose or *herbaceous,* not woody:
Leaflets *narrow* (lanceolate or linear)
Annual; very slender; racemes capillary (11) **tenuissima.**
Perennial; *canescent* or *silvery.*
Leafl. *linear, complicate,* shorter than the thick petiole ... ... (13) **complicata.**
Leafl. *lanceolate,* mostly open:
Strigose; *stip.* short; *cal.* lobes attenuate ... ... (14) **heterophylla.**
Silky; *stip.* short; *cal.* lobes sub-lanceolate ... ... (15) **candicans.**
Strigose; *stip.* elongate; *cal.* lobes subulate ... ... (16) **psoraleoides.**
Perennial; thinly strigillose (*not canescent*).
Petioles elongate; *bracts very broad, ovate* ... ... (12) **venusta.**
Petioles elongate; bracts minute, *subulate:*
Branches and petioles sharply *triquetrous,* lvs. glabrous ... (17) **triquetra.**
Branches mult-angular, thinly *strigillose* ... ... (18) **adscendens.**
Petioles 1–2 lines long; *lfts.* linear; racemes *short* ... ... (19) **leptocarpa.**
Leaflets *broad* (elliptical or obovate).
Thinly strigillose or subglabrous:
Peduncl. setaceous; racemes few flowered *subcapitate* ... ... (20) **gracilis.**
Peduncl. robust; racemes *long,* many flowered:
Herbaceous; *leaflets* roundish-obovate ... ... (21) **procumbens.**
Suffrutescent; *lfts.* cuneate-obovate ... ... (22) **porrecta.**
Strigoso-canescent; lfts. obcordate ... (23) **cardiophylla.**
Densely hairy, with loose, whitish hairs:
Racemes dense, on *long,* hairy peduncles ... ... (24) **tomentosa.**
Racemes laxly few flowered, on *short* peduncles ... ... (25) **depressa.**

(2.) Stipules *broad;* either ovate, lanceolate, or semisagittate:
Procumbent, herbaceous or half herbaceous:
White-hairy; stipules semisagittate, acute; legumes hispid ... ... (26) **incana.**
Softly-hairy; *lfts.* and stipules roundish, obtuse; legumes glabrous ... ... (27) **mollis.**
Hairy; *lfts.* obovate; *stipules* cordate-ovate; *leg.* glabrous ... ... (28) **stipularis.**
Subglabrous; *leafl.* ovato-lanceolate; *stip.* semisagittate ... ... (29) **dimidiata.**
Erect; suffruticose or shrubby:
Suffruticose; *lfts.* narrow, linear-lanceolate (30) **monostachya.**
Shrubby; *lfts.* obovate-cuneate, or lanc. oblong (31) **cuneifolia.**

4. Digitatæ. *Leaves* digitately 5–9 foliolate (rarely but 3-foliolate); the leaflets all springing from the apex of the common petiole. (Sp. 33–41.)

(1.) Suffruticose or herbaceous, slender:
Thinly strigillose or subglabrous:
Lvs. *sessile;* pedunc. *capillary,* 3–4 flowered (33) **filicaulis.**
Lvs. *sessile;* pedunc. *setaceous,* 6–8 flowered; *vex.* silky ... ... (34) **dillwynioides.**
Lvs. *petiolate;* racemes long, laxly many flowered ... ... (36) **digitata.**
Lvs. *petiolate;* pedunc. setaceous, 1–3 flowered (35) **pentaphylla.**
Canescent or silvery white:
Thinly canescent, prostrate ... ... (37) **Burchellii.**
Very white, densely silky-strigose, diffuse ... (38) **bifrons.**
(2.) *Shrubs,* with woody stems and twigs:
Twigs stigillose; leafl. *convolute-subterete,* round-backed, hairy ... ... (39) **quinquefolia.**
Twigs strigillose; leafl. linear, with *revolute margins* ... ... (41) **sulcata.**
Twigs *hirsute;* leafl. linear-cuneate, *flat* ... ... (40) **flabellata.**

5. Pinnatæ. *Leaves* impari-pinnate, bi-multi-jugate, the leaflets *opposite;* the terminal *sessile* or *subsessile* (the common petiole *ending* at the uppermost pair of leaflets.) (Sp. 42–71.)

(1.) Stem *erect or suberect,* rigid, ligneous or lignescent.
Fruticosæ: much branched *shrubs,* with woody stems and *twigs:*
Twigs tomentose and *rigidly bristly;* cal. lobes *setaceo-subulate* ... ... (42) **hispida.**
Twigs tomentose, not bristly; cal.-lobes *broadly* subulate ... ... (43) **brachystachya.**
Twigs glabrescent; leaves *patent;* leaf-pairs close; pod glabrescent ... ... (44) **stricta.**
Twigs strigillose; lvs. *erect;* leaf-pairs subremote; pod short, strigose... ... ... (45) **foliosa.**
Vestitæ: stem tall, *quite simple,* densely imbricated with leaves.
Minutely strigose; leafl. *glabrescent;* racemes *very short* ... ... (46) **vestita.**
Strigoso-*hirsute;* racemes *shorter* than the lvs. (47) **hirta.**
Villoso-canescent; peduncles *very long,* many flowered ... ... (48) **Dregeana.**
Virgatæ: slender *suffrutices,* with *virgate* stems and branches, and scattered leaves.
Calyx 5-toothed; petals *dark brown and silky;* lfts. cuneate-oblong ... ... (49) **tristis.**
Calyx 5-toothed; petals *fulvous* and silky; lfts. lin.-lanceolate ... ... (53) **viminea.**
Cal.-lobes subulate; peduncles *very long;* petals glabrous ... ... (51) **fastigiata.**
Cal.-lobes subulate; pedunc. *short;* petals glabrous ... ... (52) **affinis.**
Cal.-lobes subulate; petals silky; pod short (50) **arrecta.**

Ramosæ; slender suffrutices, irregularly *much branched*, and twiggy.
Leafl. *elliptical*, glabrescent; racemes *few-flowered*, twice as long as leaf ... ... (56) **elliptica.**
Leafl. *cuneate-oblong*, strigillose; peduncles long, many flowered ... ... (54) **corniculata.**
Leafl. linear, canescent beneath; pedunc. *long*, many flowered ... ... (58) **Zeyheri.**
Leafl. sublanceolate, acute, *canescent;* pedunc. *short*, 3–4 flowered ... ... (57) **poliotes.**
Leafl. 2–1 jugate, linear-oblong, with *inrolled* margins ... ... (59) **concava.**
Multicaules: Stems *numerous*, suffruticose, subsimple or branched from the base, branches erect.
Pubescence scanty, *strigillose*, appressed:
Racemes on *long* slender peduncles; calyx 5-toothed ... ... (66) **hedyantha.**
Racemes subsessile, or scarcely longer than the leaves ... ... (67) **hilaris.**
Pubescence *glandular*, copious; the glands stipitate:
Lfts. 2–3 lines long; gland-bristles *long* and *unequal* ... ... (70) **heterotricha.**
Lfts. 4–5 lines long; gland-bristles *short*, *equal* ... ... (71) **sordida.**
(2.) Stem *diffuse or procumbent*, shrubby, or half-herbaceous.
Fruticosæ: Stem distinctly woody:
Villoso-tomentose; leaves 4–5 jugate, gland-stipelled ... ... (55) **malacostachys.**
Tomentose and hirsute; lvs. 2–3 jugate; leafl. obovate ... ... (63) **coriacea.**
Strigose; leaves *subdigitate*, 2 jugate; leaflets linear-oblong ... ... (62) **filiformis.**
Thinly strigose; lvs. 3–4 jugate; lfts. linear, with revolute margins ... ... (60) **angustifolia.**
Suffruticosæ: Stems either scarcely ligneous, or very slender and filiform.
Stem and branches with dark-brown bark, glabrescent:
Lvs. 3–4 jugate; leafl. linear, with *revolute* margins ... ... (60) **angustifolia,** *β.*
Lvs. 4 jugate; leafl. elliptic-oblong, *white* underneath ... ... (61) **Mundtiana.**
Lvs. 4–6 jugate; leafl. linear-lanceolate, with involute-margins ... ... (65) **capillaris.**
Lvs. 2-jugate (or *trefoil*); leafl. elliptical-obovate ... ... (64) **sarmentosa.**
Stem and branches with *pale green* bark, strigoso-canescent:
Racemes on *short* peduncles, laxly few-flowered ... ... (68) **ovina.**
Racemes on *long* peduncles, many-flowered (69) **glaucescens.**

6. Unijugæ. *Leaves* impari-pinnate, *unijugate* (or, leaflets 3, 2 of them opposite and lateral on a common petiole, which is prolonged beyond their insertion and bears the third ("*terminal*" leaflet) at its summit.) (Sp. 72–76.)

Racemes on *long* peduncles; *bracts small*, subulate:
Glabrescent; leafl. obovate-oblong; petals pubescent ... ... (72) **tetragonoloba.**
Strigoso-canescent; leafl. narrow; petals glabrous (73) **intermedia.**
Racemes on *long* peduncles; *bracts broadly-ovate*, acuminate ... ... (74) **amoena.**
Racemes *subsessile*, shorter than the leaf; annual, silky (75) **argyroides.**
Racemes subsessile, longer than the leaf; shrubby, glabrescent ... ... (9) **denudata,** var. *γ.*
Flowers *axillary*, in pairs, on short pedicels ... ... (76) **polycarpa.**

7. Productæ. *Leaves* impari-pinnate, bi-multi-jugate, the leaflets *opposite*, the terminal leaflet sensibly *petiolate* (the common petiole prolonged beyond the uppermost pair of leaflets). (Sp. 77–100).

* *Stipules* and *bracts* broad and membranous; the bracts enwrapping the unopened bud, deciduous.

Stem suffruticose; leaves 2-jugate ... ... (77) **fulcrata.**

** *Stipules* and *bracts* narrow, mostly *subulate and small:*

Fruticosæ: Stem *erect*, shrubby, distinctly ligneous:

Racemes longer than the leaf, *many flowered;* fl. 3–5 lines long

Leaves 2-jugate; leaflets silky; legumes erect, canescent ... ... (78) **cytisoides.**

Lvs. 2–4-jugate; leafl. glabrous; legumes pendulous, glabrous ... ... (79) **frutescens.**

Lvs. 5–6-jugate, villoso-pubescent; leg. turgid, very hairy ... ... (84) **eriocarpa.**

Racemes scarcely longer than the leaf, *few*-flowered; fl. minute ... ... (90) **micrantha.**

Racemes *shorter* than the leaves:

Leaflets strigoso-canescent; calyx-lobes setaceo-subulate ... ... (86) **grata.**

Leaflets nearly glabrous; calyx teeth *short;* legumes moniliform ... .. (89) **cryptantha.**

Suffruticosæ: Stem erect, *half woody*, with reddish-brown bark;

Leaves 4–5-jugate; leafl. obovate, minutely strigillose beneath ... ... (80) **cylindrica.**

Lvs. 8-jugate; leafl. oblong, attenuate at base, *glabrous* ... ... (81) **rufescens.**

Canescentes. Stem erect, suffruticose, *pale or canescent*, strigillose:

Pubescence appressed, altogether *strigillose:*

Hoary; leafl. obovate, 3–4 lines long ... (93) **adenocarpa.**

Lutescent; twigs angular; leafl. linear-oblong, 6–9 lines long ... ... (94) **Burkeana.**

Pale; twigs terete; leafl. linear-sublanceolate, 5–7 lines long ... ... (95) **macra.**

Pubescence patent, of *rigid, gland-tipped bristles* (92) **seticulosa.**

Herbaceæ. Stem *herbaceous*, erect, annual or biennial:

Peduncles much longer than the leaf:

Softly hairy; leafl. elliptic-oblong; leg. hirsute ... ... (83) **hirsuta.**

Thinly strigillose; leafl. linear-lanceolate (98) **filipes.**

Racemes *subsessile*, dense, much shorter than the leaf ... ... (97) **parviflora.**

Diffusæ. Stem *diffuse or procumbent*, suffruticose or half herbaceous:

Pubescence copious, silky or villous: (petiole mostly *gland-stipelled*).

Leaves 5–8-jugate; leafl. *fulvo-sericeous* beneath; branches *hirsute* ... . (82) **oxytropis.**

Lvs. 6–8-jugate; leafl. oval, *villous* beneath; twigs pubescent ... ... (85) **velutina.**

Lvs. 3–4-jugate; leafl. cano-sericeous beneath; twigs tomentose ... ... (87) **melanadenia.**

Lvs. 2–4-jugate; leaflets oblong, fl. few, distant, minute ... ... (88) **varia.**

Pubescence scanty, thinly strigose or strigillose, or subpilose:

Pedunc. short, filiform, laxly few-flowered (91) **pauciflora.**

Peduncles long, many-flowered:

Leaves 2–3-jugate; leafl. elliptic-oblong, strigose ... ... (96) **disticha.**

Lvs. 5–7-jugate; leafl. obovate-oblong, strigillose beneath ... ... (99) **declinata.**

Lvs. 5–6-jugate; leafl. narrow-obovate, glabrous or subpilose beneath (100) **humifusa.**

8. **Alternifoliæ.** Leaves alternately pinnate; leaflets 3 or several, *scattered* along a common petiole, one *terminal.* (Sp. 101–107).

Stem ascending; leafl. 9–16, *sublinear*, thinly strigose (101) **exigua.**
Procumbent or prostrate; leaflets *obovate* or *oblong*:
Upper surface of leaflets glabrous, or thinly strigose:
Subglabrous and glaucous; stipules *membranous*, broad ... ... (102) **endecaphylla.**
Albo-strigose; stipules subulate ... ... (103) **alternans.**
Both surfaces of leaves *albo-strigose* or *hirsute*:
Pubescence wholly strigose, *appressed*, white:
Racemes pedunculate; leafl. obovate, 4–5 lines long ... ... (104) **hololeuca.**
Racemes pedunculate; leafl. oval-oblong, 6–12 lines long ... ... (106) **auricoma.**
Racemes *sessile*; leaflets three or five ... (107) **argyræa.**
Pubescence *roughly hairy*; lvs. albo-hirsute (105) **daleoides.**

Subgenus II. **AMECARPUS.** *Legume* plano-compressed, falcate or circinnate, the flat valves netted with veins, submembranous. (Sp. 108–114.)

Rigid shrubs, without axillary spines; twigs sometimes spinescent.
Racemes much longer than the leaves:
Leaves sessile; twigs strigoso-canescent ... (108) **patens.**
Lvs. petiolate; plant glabrous or pubescent (109) **falcata.**
Racemes *short*, scarcely longer than the leaves:
Leafl. obovate, glabrous; legumes glabrous ... (110) **hedranophylla.**
Leafl. cuneate, silky; legumes pubescent ... (111) **sessilifolia.**
Rigidly shrubby, with *axillary spines*; rachis spinous (112) **circinnata.**
Suffruticose, effuse; leaves pinnate, 2-jugate ... ... (113) **torulosa.**
Annual, diffuse; leaves pinnato-trifoliolate ... ... (114) **rhytidocarpa.**

Subgenus **I. Eu-Indigofera.** *Legume* subterete or terete, straight or subfalcate, the *convex* valves coriaceous, veinless. *Seeds* separated by transverse partitions. (Sp. 1–107).

1. JUNCIFOLIÆ. (Sp. 1–2.)

1. I. filifolia (Thunb.! Fl. Cap. p. 595); suffruticose, tall, erect, glabrous; branches terete, virgate; petioles filiform, very long, acute, naked, or, *in the young plant only*, bearing 3–4 pair of obovate-oblong leaflets, the terminal petioled; racemes *on long peduncles*, erect, laxly several-flowered; flowers pedicellate, bracts subulate, deciduous; petals glabrous, or minutely puberulous; calyx teeth subulate; legumes terete, pedicellate, acute, glabrous, many-seeded. *I. juncea, DC. Prod.* 2, *p.* 225. *E. & Z.!* 1624. *E. Mey.! Comm. Drege, p.* 103. *Zey.!* 2444. (the young, *leaf-bearing* plant.)

HAB. Common in moist situations in the Cape and other western districts. (Herb. Th., Hk., D., Sd.)

Stems 3–6 feet high, somewhat woody below, half herbaceous above. Branches numerous, very erect, round, laxly set with petioles or *phyllodia*. These petioles are 3–5 inches long, subulate, somewhat pungent, glabrous, always *leafless* except in the young plant. Racemes lengthening as the flowers advance, exceeding the petioles. Flowers purple, nearly glabrous. Calyx-teeth variable in length. Legumes 1½–2 inches long, very convex. I have compared with *Thunberg's* specimens in Herb. Upsal, and can detect no difference between them and the "*J. juncea*" of *Ecklon's* and *Drege's* collections and of the Herbaria above quoted. Whether *De Candolle's* original plant be different or not, I cannot say.

2. I. podophylla (Benth.! in Herb. Hook.); suffruticose, glabrescent; branches virgate; petioles filiform, very long, naked, or supporting a terminal, obovate or oblong, flat, mucronulate, minutely stipellate leaflet; racemes *shorter than the petioles*, few-flowered; flowers . . . . . ?; legumes terete, straight, thinly hispidulous, many-seeded.

HAB. De la Goa Bay, *Forbes.!* (Herb. Hk., D.)

A slender, laxly branched, erect suffrutex, 2–3 feet high; branches long and simple, incurved. Petioles 2–3 inches long, twice as thick as hog's bristles, incurved or recurved, bi-stipellate at the summit, frequently leafless. Leaflet glaucescent, ¼–¾ inch long, the smaller obovate, the larger oblong, all very thinly and appressedly puberulous. Racemes not an inch long, 3–8 flowered. Legumes 10 lines long, ½-line in diameter, brown.

2. SIMPLICIFOLLIÆ. (Sp. 3–5.)

3. I. obcordata (E. & Z. ! 1561); shrubby; branches straight, rigid, terete, canescent; leaves scattered, simple, subsessile, *obcordate*, expanded, strigoso-canescent on both sides; stipules obsolete; "spikes" (fide *E. & Z.*) "pedunculate, few-flowered, scarcely longer than the leaves;" legumes unknown.

HAB. Karroo Hills at the Gauritz River, Swell., *E. & Z.!* (Herb. Sond.)

A single specimen, without flower or fruit, exists in Hb. Ecklon, now Dr. Sonder's. It seems to be a rigid, strong growing, divaricate, and sparsely leafy shrub, thinly canescent in all parts, and ashen-grey. The branches 8–12 inches long, are very rigid and straight. Leaves ½ inch apart, 4–5 lines long, 3 lines wide, retuse or deeply emarginate, many of them exactly heartshaped, the midrib obvious on the upper, and prominent on the under surface. No appearance of inflorescence, save very young flower-buds.

4. I. nudicaulis (E. Mey.! Comm. p. 92); shrubby, *canescent;* branches flexuous, striate; leaves few, distant, simple, subsessile, linear-cuneate, convolute, recurved at point, deciduous, *appressedly canescent;* racemes shortly pedunculate, axillary, laxly many-flowered, elongating; flowers subsessile; vexillum silky; calyx oblique, with short, blunt lobes; legumes subterete, straight, thinly canescent.

HAB. Near Verleptpram, at the Gariep, on stony hills, 500f., *Drege.!* (Herb. Hk., Bth., D.)

Whole plant pale yellowish or cream-coloured, microscopically puberulous. Stems much-branched, 1–1½ foot high, the branches angularly bent. Leaves an inch apart, ½ inch long, 1–1½ line wide, the sides infolded; apex blunt. Racemes 1½–2 inches long, subspicate. Calyx 1 line long, canescent. Vexillum 2–3 lines long. Ovary densely canescent; legume nearly uncial, spreading.

5. I. ovata (Thunb.! Fl. Cap. p. 596); suffruticose, diffuse; branches filiform, angular, nearly glabrous; leaves scattered, simple, subsessile, ovate-oblong or elliptical, *flat*, coriaceous, veinless, mucronate; stipules subulate; racemes on long, glabrous peduncles, pluriflowered; flowers pedicellate, bracts deciduous; petals thinly silky; calyx pilose, its segments subulate, acuminate; legume terete, straight, glabrous, many-seeded. *DC. Prod.* 2, *p.* 222. *Burch. Cat.* 5928, 7876.

HAB. S. Africa, *Thunberg!* Klyn Howhoek, *Zeyher*, 2414. Near Georgetown, *Dr. Alexander Prior!* (Herb. Th., Hk., Sd., Bth.)

A slender, diffuse, or somewhat trailing, glabrescent undershrub, 1–2 feet long, laxly branched. Leaves half inch apart, on petioles scarcely 1 line long, 5–8 lines long, 3–5 lines wide, thickish, minutely and appressedly puberulous on one or both sides. Peduncles 2–3 inches long, supporting a short raceme. Legumes uncial, acute at each end, the seminal suture thickened. *Thunberg's* specimen, in Hb. Upsal, has rather larger leaves than *Zeyher's*, but in other respects it agrees. It is certainly not "villous," as described.

3. TRIFOLIOLATÆ. (Sp. 6–32.)

6. I. spinescens (E. Mey.! Comm. p. 93); shrubby, rigid, divaricately branched; *the branches and twigs terete, spinescent*; leaves subfasciculate, trifoliolate, on very short petioles; leaflets (minute) obovate, concave, thick, *glabrous above*, minutely strigillose beneath; stipules obsolete; racemes subsessile, laxly few-flowered, scarcely exceeding the leaf; calyx pubescent, bluntly 5-fid; petals pubescent; legumes terete, straight.

HAB. Dry rocky, mountain ground in Namaqualand. Leliefontein, 4000f., and Witpoort, and Zwaanepoelspoort, 3600f., *Drege!* (Herb. Bth., Hk., D.)

A very scraggy, small bush. Leaflets about 1-1½ lines long, ½ line wide, pale green. Flowers very small. I have not seen legumes. The specimens above quoted are all imperfect.

7. I. pungens (E. Mey.! Comm. p. 93); shrubby, rigid, divaricately branched, *strigoso-canescent*, the terete branches and twigs spinescent; leaves solitary, trifoliolate; leaflets obovate-oblong, strigose on both surfaces; stipules obsolete; racemes *subsessile, much longer than the leaf, on rigid rachides which become spinous*; calyx canescent, shortly 5-toothed; petals pubescent; legumes straight, compressed, with thickened sutures, narrowed at base, curve-pointed, the valves somewhat keeled in the centre, strigose.

HAB. Among rocks near Verleptpram, Gariep, *Drege.!* (Herb. Sond.)

A rigid shrub, nearly allied to *I. spinescens.*

8. I. dealbata (Harv.); shrubby, rigid, divaricately much-branched, unarmed (or subspinescent?), *ashen-grey*; twigs angular; leaves short-petioled, *tufted*, trifoliolate; leaflets elliptic-oblong, subacute, complicate or keeled, *strigoso-canescent on both sides*; stipules obsolete; *flowers axillary, solitary*, on pedicels, shorter than the leaf; calyx canous, bluntly 5-toothed; petals pubescent; legumes terete, straight, few-seeded, short, canescent.

HAB. Wolvekop, *Zeyher!* (Herb. Hk., Sd., D.)

A scraggy, densely-branched bush, resembling *I. spinescens*, but not spinous (on our specimens), much more pubescent and cinereous, with denser foliage and differently shaped leaflets, &c. The twigs are angular and striate, and set at short distances with prominent tubercles or abortive ramuli, which bear several tufted leaves. The few flowers seen (in Hb. Hook.) are *solitary*, on pedicels 2 lines long. Leaflets 2-2½ lines long, 1 line wide. Legume (in Hb. Sond.) 2-2½ lines long, acute.

9. I. denudata (Thunb.! Fl. Cap. p. 597, non Jacq.); shrubby, *rigid, divaricately much-branched*, unarmed or spinous, *subglabrous*; leaves short-petioled, trifoliolate; leaflets oblong-cuneate, obovate or linear-oblong, obtuse, mucronulate, coriaceous, expanded, glabrous or puberulous beneath, (*the terminal sometimes petioled*); stipules obsolete or minute, toothlike; racemes canescent, subsessile, equalling or somewhat exceeding the leaf, laxly few-flowered; calyx canous, bluntly 5-toothed; petals pubescent; legumes terete, straight, mucronate, minutely strigillose. *I. rigescens, E. Mey.! Comm. p.* 94. *I. denudata, E & Z.!* 1567. *I. centrota, E. & Z.* 1566. *I. rechodes, E. & Z.!* 1565. *I. nigromontana, E. & Z.!* 1564. *I. athrophylla, E. & Z.!* 1563. *Zey.* 2417 (*fol. angustissimis*), 2415, 2419.

VAR. *α.* **spinosa** (E. Mey.); branches divaricate, spine-tipped.

VAR. *β.* **inermis** (E. Mey.); branches straighter and more upright, unarmed.

VAR. *γ.* **luxurians**; terminal leaflet, especially on the young shoots, *petiolulate*;

racemes longer and many-flowered. *I. centrota, E. & Z.! ex pte. Zey.!* 2416. *Burch.!* 6195, 4718.

VAR. δ. ? **simplicifolia**; *leaves simple*, shortly petioled, elliptical or obovate, *expanded. I. flexuosa, E. & Z.!* 1562. *Zey.!* 2420.

VAR. ε. ? **dumosa**; *leaves simple*, subsessile, linear-oblong, *convolute*, glabrescent. *I. dumosa, E. Mey. Comm. p.* 93. *Zey.! No.* 489.

HAB. Mountain and hill-sides, and by river banks among shrubs in Uitenhage, George, and Albany in many places, *Thunberg! E. & Z.! Drege! &c.* γ. Vanstaadenberg. δ. Winterhoeksberg and Eland R., *Zeyher!* ε. Near Grahamstown and on the Zuureberg, *Zeyher!* (Herb. Th., Hk., Bth., D., Sd.)

A very rigid, much and intricately branched bush, laxly covered with coriaceous, veinless leaves: varying slightly in pubescence, and considerably in the proportions of the leaflets, which are commonly 4–5 lines long, and 1–2½ wide. Vars. δ. and ε. seem to have constantly *simple* or unifoliolate leaves, but do not otherwise differ from the broader and narrower leaved forms. Var. γ. seems to owe its peculiarities either to richer soil, or to the effect of surface burning, producing a more luxuriant state of the plant.

10. I. stenophylla (E. & Z.! 1568); shrubby, rigid, *erect*, much-branched, subglabrous; branches *virgate*, angle-ribbed; leaves on very short petioles, trifoliolate; leaflets *lanceolate-linear or subspathulate*, acute, channelled, coriaceous, glabrous; stipules obsolete or tooth-like; racemes subsessile, scarcely longer than the leaf, few-flowered; calyx canescent, 5-toothed; petals pubescent; legumes terete, straight, glabrous. *I. angustata, E. Mey.! Comm. p.* 94. *I. Zeyheri, var. β. trifoliata, E. & Z.! Burch.* 3614, 3531.

HAB. Hills about Grahamstown, and at the Fish River, *E. & Z.! Drege! Mr. Bunbury! Zey.!* 2422, *&c.* (Herb. Sd., Bth., Hk., D.)

A densely branched, leafy shrub, with very pale bark and foliage; the young branches long and rodlike, very erect. Petioles sometimes scarcely any, sometimes 2–3 lines long. Leaflets 5–10 lines long, ½–1 line wide, mostly acute or subacute. Legumes 1½ inch long, cylindrical or slightly nodose. Much more leafy than *I. denudata*, with longer and narrower leaflets, &c.

11. I. tenuissima (E. Mey. Comm. p. 94); "annual?; stem and branches virgate, *very slender*; leaves trifoliolate, shortly petiolate, strigilloso-pubescent; *leaflets very narrow-linear, acute*; racemes axillary, *capillary*, on long peduncles; legumes subcompressed, linear, straight." *E. Mey. l. c.*

HAB. Port Natal, on grassy hills, *Drege!*

Of this I have only seen a fragment in Hb. Sond. The capillary peduncle is 4–5 inches long, having flowers an inch below the apex. Cal.-teeth very short or obsolete. Flowers minute, the petals fulvo-sericeous. Leaflets 1–1½ inch long, not a line wide.

12. I. venusta (E. & Z.! 1576); herbaceous, slender, minutely strigillose; branches *terete*; leaves on long petioles, trifoliolate; leaflets of the lower leaves short and obovate, of the medial and upper linear-lanceolate or linear, acute; stipules subulate-attenuate, erect; racemes subspicate, on long peduncles; *bracts ovato-cuspidate, enwrapping the flower-buds*, deciduous; calyx-segments setaceo-subulate; petals glabrescent; legumes (unripe) strigoso-canescent, pendulous.

HAB. Muddy soil, on hills about Brackfontein, Clanw., *E. & Z.!* (Herb. Sd. Bth. D.)

Stems 12–18 inches high, ascending, filiform. Petioles 1½–2 inches long. Leaflets shorter or longer, the lower broader and blunter, the upper very narrow, acute at each end. Peduncles 4–6 inches long, the upper half floriferous; the raceme lengthening as the flowers expand. Bracts ovate-oblong, with a long, narrow point. Flowers small.

13. I. complicata (E. & Z. ! 1577); suffruticose, *strigilloso-canescent;* branches angular; leaves on long, *thickened* petioles, trifoliolate; leaflets linear, complicate, mucronulate, shorter than the petiole; stipules falcato-subulate, recurved; racemes subspicate, on long peduncles; bracts shorter than the calyx; calyx canescent, its segments subulate-acuminate; petals puberulous; legumes linear, terete, *erect,* canescent. *I. platypoda, E. Mey. Comm. p.* 95.

HAB. Gauritz River, Swell., *E. & Z.!* South Africa, *Drege!* (Herb. Bth., Sd.)

Whole plant whitish, covered with very minute, appressed strigillæ. Petioles 1–1¼ inch long, ½–¾ line in diameter. Leaflets 5–8 lines long, scarcely 1 line wide. Peduncle 3 inches long; racemes lax; flowers small, subsessile, erect.

14. I. heterophylla (Thunb.! Fl. Cap. p. 597); suffruticose, *suberect* or *ascending*, strigoso-canescent, densely leafy; branches ribbed and furrowed; leaves trifoliolate, leaflets of the lower leaves *cuneate-obovate,* of the upper *sublanceolate,* all mucronate, expanded, *strigoso-canescent;* stipules *small,* subulate, patent; racemes subspicate, on peduncles longer than the leaves, densely many-flowered; bracts shorter than the calyx; calyx canous, whitish, its segments *subulate-acuminate;* petals pubescent; legumes terete, pendulous, thinly strigillose. *I. candicans, E. Mey.! Comm. p.* 95, *non Ait. Burch.* 3613, 5515, 5684.

HAB. Nieuweveldsberg, Kamiesberg, Kasparskloof, and on hills near Vischbay, *Drege!* Appelskraal, near the Zonder End, *Zey.!* 2439. Uitenhage, *Dr. Alexander Prior.* Albany, *T. Williamson.* Somerset, *Mrs. F. W. Barber!* (Herb. Th., Hk., Bth., D., Sd.)

Stem 6–12 inches high, erect or more or less diffuse, scarcely procumbent, thinly canescent, with short, sparse, white hairs. Leaves crowded; the petiole 3–7 lines long. Leaflets 4–7 lines long, those of the lower leaves shorter and broader. Racemes either very dense or, in weakly grown specimens, laxly flowered. The common petiole sometimes extends a short way beyond the insertion of the lateral leaflets, and then this species comes very close to *I. intermedia,* which has a more evidently petioled, terminal leaflet.

15. I. candicans (Ait. Kew. 3. p. 67); suffruticose, *procumbent,* thinly *cano-sericeous;* branches compressed-angular; leaves petiolate, trifoliolate; leaflets lanceolate, acute, prominently ribbed beneath, *silky-subsilvery on each side;* stipules *small,* subulate, recurved; racemes subspicate, on long peduncles; bracts shorter than the calyx; calyx semi-fid, *the segments lanceolate;* petals puberulous; legumes pendulous, compresso-terete, silky-canescent. *DC. Prod.* 2, *p.* 232. *E. & Z.!* 1575, *ex pte. non E. Mey. Curt. Bot. Mag. t.* 198.

HAB. Dry hills round Capetown and Camps Bay, *E. & Z.! Pappe! W.H.H., Dr. Alexander Prior!* (Herb. Th., Hk., Bth., D., Sd.)

Stems decumbent, the ends ascending, 1–2 feet long. Petioles ½–1 inch long. Leaflets ¾–1 inch long. Peduncles 4–5 inches long. Stipules much shorter than the petiole, and mostly hooked backwards. Flowers pinky-purple. Legumes 1–1½ inch long. This is scarcely more than a dwarf, and therefore procumbent, variety of *I. psoraleoides,* from drier ground. The pubescence is more minute and *glossy,* the leaves and stipules smaller, and the calyx-segments rather broader, &c.

16. I. psoraleoides (Linn. Syst. 469); suffruticose, *suberect,* thinly *strigoso-pubescent;* branches angular and furrowed; leaves on long petioles, trifoliolate; leaflets lanceolate, acute, ribbed beneath, *strigoso-pubescent* on each side; stipules *elongate*, subulate, erect; racemes subspicate, on long peduncles; bracts longer than the calyx; calyx-segments subulate, longer than the tube; petals pubescent; legumes terete, subtorulose, pendulous, strigoso-pubescent. *DC. Prod.* 2, *p.* 232. *E. & Z.* 1574. *E. Mey. Comm. p.* 95. *Lam. Ill. t.* 626. *f.* 4.

HAB. Dry ground round Capetown, on the hills and by roadsides, &c. common. (Herb. Th., D., &c.)

Stems 2–3 feet high, sparingly branched, suberect. Leaves subdistant. Petioles 1½–2 inches long; stipules 6–8 lines long. Leaflets 1½–2 inches long, 3–5 lines wide, open or complicate. Pubescence not copious, but rigid, of close-pressed, medi-fixed bristles. Peduncles 6–12 inches long or more. Flowers small, dense, purple. Legumes 1¼ inch long.

17. I. triquetra (E. Mey. Comm. p. 95); herbaceous, subsimple, *nearly glabrous;* stem *sharply triquetrous*, somewhat fistular, glabrous; leaves distant, on long, triangular petioles, trifoliolate; leaflets narrow-linear, acute, glabrous above, minutely strigillose beneath; stipules subulate, erect; racemes spicate on very long, angular peduncles, many-flowered; bracts subulate, longer than the calyx, deciduous; calyx silky, its lobes *lanceolate*, longer than the tube; petals puberulous; legumes (young) pendulous, strigilloso-pubescent.

HAB. Piquetberg, 1500–2000f., *Drege*. Steendaal, Tulbagh, *Pappe!* (Herb. D.)

Two feet or more high, slightly branched, weak and scarcely at all lignescent. The general aspect is that of *I. psoraleoides*, from which this is at once known by its very sharply 3-angled and subinflated stem and petioles, its nearly glabrous surface and different calyx.

18. I. adscendens (E. & Z.! 1578); subherbaceous, ascending, minutely strigillose; branches angular, curved; leaves on long petioles, trifoliolate; leaflets oblong-sublanceolate, much shorter than the petiole, thinly strigillose; stipules very minute, setaceous; racemes subspicate, on peduncles longer than the leaves; bracts minute; calyx semi-5–fid, the segments shortly subulate; petals pubescent; legumes?

HAB. Karroid hills between Hassaquaskloof and Breederiver, Swell., *E. & Z.!* (Herb. Sond.)

Very imperfect specimens, almost denuded of leaves, only seen. Branches 14–18 inches long, incurved. Petioles 2–2½ inches long, persistent. Leaflets deciduous, 6–8 lines long, about 2 lines wide. Peduncles 2–3 inches long, laxly spicate. Apparently allied to *I. psoraleoides*, but with minute stipules and bracts, and shorter leaflets, &c.

19. I. leptocarpa (E. & Z.! 1579); suffruticose, slender, thinly strigillose; branches flexuous, subangular; leaves on *short* petioles, trifoliolate; leaflets linear, recurved, pointed, complicate, longer than the petiole; stipules minute, toothlike; racemes laxly few-flowered, on *short* peduncles; calyx-segments subulate; legumes tereti-compressed, about 4–seeded, thinly strigillose.

HAB. Mountain sides near Eland's River, Uit., *E. & Z.!* (Herb. Sond.)

A small, angularly-branched plant, 6–8 inches long. Petioles 1½–2 lines long. Leaflets 3–5 lines long, rigid, subglaucous. Peduncles (in fruit only seen) 1–1½ inch long. Legume brown, 6–7 lines long, not a line wide. Differs from *I. complicata*, in the short petioles, stipules, racemes, legumes, &c. The foliage is not dissimilar; the pubescence much more scanty.

20. I. gracilis (Spreng. Cur. Bot.); suffruticose, decumbent, slender;

branches filiform; leaves sparse, petiolate, trifoliolate; leaflets elliptic-oblong or obovate, thin, expanded, strigose on the under surface; stipules small, subulate; racemes *subcapitate, few-flowered,* on long, setaceous peduncles; calyx-segments subulate; petals glabrous; legumes turgid, 2–5 seeded, *hispido-canescent.* *I. setacea, E. Mey! Comm. p.* 95. *I. erecta, E. & Z.!* 1588. *Lotus microphyllus, Hook. Bot. Mag. t.* 2808.

HAB. Hott. Holl. *E. & Z.!* Paarlberg, *Drege!* About Table Mountain, Lion's Mount, and on the Cape Flats, *Dr. Pappe! W.H.H.!* (Herb. D., Sd., Hk., Bth.)

Root fibrous. Stems many from the crown, 1–2 feet long or more, spreading over the soil, alternately branched, strigose or glabrescent. Leaves an inch apart, on filiform petioles, 2–3 lines long. Leaflets 2–5 lines long, varying from obovate to elliptical, sometimes glabrous on the upper side, sometimes strigillose, always rigidly strigose beneath, upper surface pale. Flowers small, purple. Legumes sometimes 2 lines, sometimes 6–8 lines long. Whether this be "*I. erecta*" of Thunberg or not, that very inappropriate name is undeserving of being retained.

21. I. procumbens (Linn. Mant. 271); *herbaceous,* procumbent, prostrate or running under the soil, sparsely strigillose; branches angular, compressed; leaves on longish petioles, trifoliolate; leaflets obovate or rhomboid, obtuse or mucronulate, glabrescent or sparsely strigillose; stipules subulate; racemes on peduncles much longer than the leaves, erect, elongating, many flowered; bracts minute; calyx-segments shortly subulate; petals glabrous or downy; legumes . . . . ? *Thunb. Fl. Cap. p.* 597. *DC. Prod.* 2, *p.* 232. *E. & Z.* 1571. *Burch.* 5687.

VAR. *α.* **concolor**; glabrescent; leaflets broadly obovate, *green* on both sides, mostly glabrous above, sparsely strigillose beneath. *I. procumbens, E. Mey. Comm. p.* 97.

VAR. *β.* **discolor**; thinly strigillose; petioles shorter; leaflets smaller, thinly strigillose on the upper surface, *glaucous* and more densely strigillose beneath. *I. discolor, E. Mey! l. c.*

HAB. Round Capetown, and in moist places on the Cape Flats, in several localities, *Thunberg! E. & Z.! W.H.H., &c.* (Herb. Th., Bth., Hk., D., Sd.)

Stems several inches long, lying on the ground or creeping under ground, throwing up leaves and peduncles above the soil. Pubescence variable, always scanty, sometimes almost absent. Petioles ½ inch to 2 inches long, erect. Leaflets in *α.* ½ inch long and broad; in *β.* 3–5 lines long, 2–3 lines wide. Flowers purple, 3–4 lines long, in a raceme 2–3 inches long. The two varieties above indicated appear to run into each other; and other varieties might be indicated equally deserving of separation. The *peduncle,* for instance, is sometimes very thick and succulent, sometimes slender, and the calyx-segments are longer or shorter, &c.

21. I. porrecta (E. & Z.! 1572); *suffruticose,* procumbent, much branched, thinly strigillose; branches *ribbed and furrowed;* leaves on longish petioles, trifoliolate; leaflets cuneate-obovate, or obovate-oblong, acute or mucronate, thinly strigillose; stipules setaceo-subulate; racemes on long, ribbed peduncles, elongate, many flowered; bracts minute; calyx cano-strigillose, its lobes setaceo-subulate; petals pubescent; legumes terete, pendulous, several-seeded, minutely strigillose. *Zey.* 2437, 2436.

VAR. *β.* **bicolor**; leaflets elliptical or oblong, subglabrous above, *very pale, glaucous* and strigillose beneath; calyx-segments shorter and less acuminate.

HAB. Fields by the Zwartkops and Koega River, in Adow, and in Albany and Kaffirland, *E. & Z.!* Cape, *Bowie!* Var. *β.* Algoa Bay, *Forbes! Koelbing!* (Herb. Sd., Hk., Bth., D.)

Stem 1–2 feet long, much branched in the upper part; the branches flexuous and often aggregated, mostly strongly ribbed. Pubescence variable, never copious, often scanty. Petioles ½–¾ inch long. Leaflets 4–5 lines long, 2–3 lines wide, sometimes very obtuse, sometimes acute, variable in shape. Flowers purple, 2–3 lines long. Peduncles 4–5 inches long, erect. Legumes 1–1½ inch long, sometimes subtorulose. In many respects allied to *I. procumbens*, but a much more rigid, less herbaceous. and more branching plant, with narrower leaflets and smaller flowers, &c. Var. *β*, has longer and more elliptical leaflets, very pale underneath, and is altogether more luxuriant. Leaflets 5–7 lines long, 3–5 wide.

23. I. cardiophylla (Harv.); suffruticose, procumbent, branching, *canescent;* branches ribbed and furrowed; leaves on longish channelled petioles, trifoliolate; leaflets *obcordate, strigoso-canescent;* stipules minute, subulate, patent; racemes on long, angular peduncles, many flowered; bracts minute; calyx canescent, *shortly* 5-toothed; petals puberulent; legumes . . ? *Burch. Cat.* 1245.

HAB. S. Africa, *Burchell!* (Herb. Burch.)

Near *I. porrecta*, but with decidedly cordate leaflets and a different calyx, and the white, close pressed and copious pubescence of *I. candicans.* Petioles uncial; leaflets 3–4 lines long, 2 lines wide at top, deeply emarginate. Stipules ¼ line long. Peduncles 3–5 inches long. Calyx-lobes deltoid, acute.

24. I. tomentosa (E. & Z.! 1585); suffruticose, procumbent, *densely albo-hirsute;* branches subangular; leaves on short petioles, trifoliolate; leaflets cuneate-obovate, mucronulate, *densely albo-hirsute* on both sides; stipules *small, subulate;* racemes subspicate, on long, hairy peduncles; bracts minute; calyx hirsute, its segments setaceo-subulate; petals puberulous; legumes terete, deflexed, *hirsute. Burch.!* 6311.

HAB. Sand hills by the seaside. Cape Recief and the mouth of Zwartkops River, Uit., *E. & Z.!* Jan.-Feb. (Herb. Sond.)

Stems 12–18 inches long, prostrate, alternately and subdistichously branched; the whole plant white with coarse, patent hairs. Petioles ¼–½ inch long, patent. Leaflets 4–5 lines long, 3 lines wide, very obtuse. Flowers small. Legumes 1–1¼ inch long, straight, white-hairy. In many respects like *I. incana*, but readily known by its small, narrow stipules.

25. I. depressa (Harv.); suffruticose, prostrate, slender, cano-hispid, branches filiform; leaves on short petioles, trifoliolate; leaflets *cuneate-obovate*, mucronate, expanded, hispid on both sides; stipules *subulate*, equalling the petiole; racemes laxly few-flowered, *on short peduncles;* bracts minute; calyx-segments subulate; petals pubescent; legumes very short, 2–3 seeded, *thinly tomentose.*

HAB. South Africa, *Mundt and Maire!* Near Georgetown, *Dr. Alexander Prior!* (Herb. Benth.)

A slender, prostrate plant, with stems 12–18 inches long, alternately much branched. Petioles 1–2 lines long. Leaflets 3–4 lines long, 1–1½ line wide, pale green, rough with short, patent hairs. Peduncles in flower scarcely uncial, in fruit 1½ inch long. Legumes 3–4 lines long. This has the look of a small *trefoil.* It is near *I. incana*, but differs in its much smaller and narrower stipules, shorter, few-flowered racemes, hairy petals, and very small, few-seeded legumes.

26. I. incana (Thunb. Fl. Cap. 596); suffruticose, procumbent, branched, *cano-hirsute;* branches flexuous; leaves on *short* petioles, trifoliolate; leaflets obovate, or obovate-oblong, mucronate, expanded, *hirsute on both sides; stipules broad, semi-sagittate, acuminate, mostly longer than the petiole;* racemes subspicate, on long peduncles; bracts minute; calyx-segments subulate-acuminate; petals *glabrous;* legume terete, pendulous, *hispid*, several seeded. *DC. Prod.* 2, *p.* 232. *E. & Z.* 1584. *E. Mey. Comm. p.* 96. *Burch. Cat.* 5108.

HAB. Common on the hills round Capetown, at Kamps Bay and Rondebosch, &c. *Dr. Pappe, W.H.H., E. & Z., &c.* Klipplatt River and Zwartkey, *Drege!* (Herb. Th., Hk., D., Sd.)

Tap-root subsimple, slightly fibrous. Stems many from the crown, 1–2 feet long, spreading over the soil in all directions, much or little branched, roughly hairy. Petioles 2–3 lines long. Leaflets 3–7 lines long, 2–4 lines wide, palish green, obtuse or subacute. Peduncles 3–6–8 inches long, ending in a short raceme of several purple, dark-tipped, nearly glabrous flowers. Corolla 2½ lines long. Legume 1–1½ inch long, densely but shortly hispid. Varying in the size of leaf, length of peduncle, &c., but generally known by its pubescence and stipules. *Burchell's* 5108 has narrower stipules than usual.

27. I. mollis (E. & Z.! 1586); herbaceous, procumbent, *densely and softly hairy;* stems filiform; leaves on longish petioles, trifoliolate; leaflets roundish-obovate, very obtuse, sub-glabrous on the upper, softly villous and paler on the under side, thin, expanded; *stipules amplexicaul. roundish-ovate, obtuse;* racemes on long peduncles, many-flowered; bracts minute; calyx-lobes lanceolate; petals subglabrous; legumes pendulous, terete, *glabrous*, several seeded. *I. mollis, E. Mey.! Comm. p.* 96.

HAB. Grassy places on the Kat River Berg, above the woods, *E. & Z.!* S. Africa, *Drege!* (Herb. Sd., Bth.)

Stems trailing, subsimple, several inches long. The whole plant (save the upper surfaces of the leaflets, the petals and the legumes) densely clothed with long, soft, patent, very slender, white hairs. Petioles ½–¾ inch long. Leaflets 4–5 lines long, 4 lines wide, pale green. Stipules 1½ line long and broad. Flowers 1½ line long, pink. Legumes 1¼ inch long, quite glabrous.

28. I. stipularis (Linn.; fide E. Mey.! Comm. p. 96); subherbaceous, decumbent, hairy, subcanescent; leaves petiolate, trifoliolate; leaflets *obovate-cuneate*, mucronate, expanded, hispid on both surfaces; stipules broadly *cordate-ovate*, acuminate, equalling or exceeding the petiole; racemes on long peduncles; bracts minute; calyx segments subulate; petals glabrous; legumes terete, pendulous, *glabrous*, several seeded. *DC. Prod.* 2. *p.* 232, *non E. & Z. Maund. Bot.* 4, *t.* 191. *I. alpina, E. & Z.!* 1581. *also Zey. Legum.* 111. 11, and 112. 11.

HAB. Among rocks at river banks between Klipplaat Riv. and Zwartekey, and on the Katberg, 4500–5000 f. *Drege! E. & Z.!* Frontier and Kaffirland in various places, *Mrs. F. W. Barber, No.* 51. (Herb. Sd, Bth., D.)

Stems several from the crown, subsimple, roughly hispid, subterete. Petioles ½–¾ inch long. Leaflets ½–1 inch long, 4–6 lines wide, more or less tapering to a cuneate base, blunt or subacute, with a small mucro, pale green. Stipules leaflike, very broad, acute or acuminate. Peduncles 6–10 inches long, ½ occupied by a slender, lengthening raceme. Flowers pink or crimson, "petals producing an indigo blue" (*Mrs. Barber*). Easily known from *I. incana* by its broader and more cordate stipules and *glabrous* legumes.

29. I. dimidiata (Vogel.); herbaceous, decumbent, *subglabrous* (very sparsely strigillose); leaves on *long* petioles, trifoliolate; leaflets *ovato-lanceolate*, or lanceolate, mucronulate, expanded, pale, glabrous above, thinly strigillose beneath; stipules broad, semi-sagittate, acuminate, shorter than the petiole; racemes on long peduncles, elongating, many flowered; bracts and calyx-segments subulate; legumes terete, pendulous, *glabrous*, many seeded. *I. stipularis, E. & Z.!* 1852, *ex pte. non E. Mey.*

HAB. Open, grassy places on the Winterberg, *E. & Z.!* (Herb. Sond.)

A slender, quite herbaceous species, drying very pale; nearly glabrous, with the exception of a very few small, appressed bristles on the stem and the undersides of the leaflets. Stems 1–2 feet long, subsimple. Leaves 2 inches apart on petioles 1–1½ inch long. Leaflets 1–1½ inch long, 4–5 lines wide, those of the upper leaves lanceolate, of the lower obovate or oblong-cuneate. Peduncles 10–12 inches long. Flowers small, dense, pink. Legumes 1–1¼ inch long. One of Ecklon's specimens in Hb. Sonder belongs to *I. cuneifolia*, β.

30. I. monostachya (E. & Z.! 1583); suffruticose, suberect, subglabrous; leaves on short petioles, trifoliolate; leaflets *linear-lanceolate* or lanceolate, very acute, *with inrolled margins*, green and sparsely strigose above, dark coloured, glabrous and veiny beneath; stipules *lanceolate-acuminate*, striate, much longer than the petioles; racemes on long peduncles, elongating, many-flowered; bracts ovato-lanceolate, longer than the flowers, deciduous; calyx semi 5-fid, the lobes sublanceolate; petals glabrous; legumes terete, peduncles glabrous. *I. oroboides, E.. Mey.! Comm. p.* 94.

HAB. Grassy hills near Philipstown, *E. & Z.!* Katberg, *Drege!* (Herb. Bth., Sd., D.)

Densely tufted 6–8 inches high, slightly woody at base. Leaves closely set, much longer than the internodes. Petioles 4–5 lines long. Leaflets 1½–2 inches long, 1–3 lines wide, tapering much to each end, a dull, rusty brown externally, with prominent veins. Stipules ¾–1 inch long, brown. Peduncles 5–8 inches long, the upper half floriferous. Flowers pinky-crimson, 3 lines long. Allied to *I. cuneifolia*, but less woody, with much narrower and longer leaflets and stipules, &c.

31. I. cuneifolia (E. & Z.! 1570); shrubby, erect, thinly strigillose or subhispid; leaves on short petioles, trifoliolate; leaflets *obovate-cuneate*, or *lanceolate-oblong*, obtuse or acute, mucronulate, strigillose on both sides; stipules obliquely ovato-lanceolate, acuminate, striate, longer than the petiole; racemes densely many flowered on subterminal or axillary short peduncles; bracts *broadly ovate*, acuminate, *enwrapping the buds*, deciduous; calyx glabrescent, its segments lanceolate; petals glabrous; legumes . . . . . ? *I. florida, E. Mey.! Comm. p.* 97.

VAR. β. **angustifolia**; leaflets smaller, narrower, less obovate and more lanceolate-oblong. *Zey.! No.* 3418. *Burch. Cat.* 3544.

HAB. Winterberg near Philipstown, *E. & Z.!* Katberg, *Drege!* Somerset, *Mrs. F. W. Barber.* β. Between Boschesman's River and Karrego, *Zeyher!* Near Grahamstown, *T. Williamson! Dr. Atherstone!* (Herb. Sd., Bth., Hk., D.)

A strong, woody, densely branched bush, a foot or more in height: variably pubescent, sometimes hispid, sometimes but sparsely strigose. Petioles 3–5 lines long. Leaflets 7–14 lines long, 3–5 lines wide, varying from obovate to narrow-oblong or sublanceolate. Racemes sometimes springing near the end of short ramuli, from the axils of depauperated leaves, and thus seemingly *terminal*, at other times axillary from the ordinary rameal leaves. Bracts broad, quite enclosing the young flower-buds, glaucous or livid-purplish. Flowers handsome, pink or crimson, 3–4 lines long. Legumes not seen.

32. I. glomerata (E. Mey.! in Linn. 7. p. 166); *shrubby, depressed or prostrate*, much branched, *villoso-canescent;* leaves *densely imbricated*, short petioled, trifoliolate; leaflets (small) obovate-oblong, mucronate, white-hairy, the older denuded; stipules *setaceo-subulate*, longer than the petiole; racemes *sessile*, few-flowered, equalling the leaves; calyx pilose, its segments setaceo-subulate; petals hairy; legumes 2–4 seeded, turgid, *tomentose. E. & Z.!* 1589. *Hb. Un. it. No.* 428. *I. nivea, E. Mey. Comm. p.* 96, *an Willd.*

HAB. Tops of Hott. Holl., in exposed places; also Potberg and Klynriviersberg, *E. & Z.! Bowie!* Howhoek, *Dr. Pappe!* Table Mountain Summit, *W.H.H.* About Simonsbay, *C. Wright!* 561, 565. Grietgesgat, *Zeyher!* (Herb. Sd., D., Hk., Bth.)

A small, woody, dwarf, spreading shrub, 6–12 inches long, hoary with long hairs, which fall off from the older leaves. Petioles 1–2 lines long. Leaflets 3–4 lines long, 2 lines wide, with a sharp, brown mucro. Stipules much acuminate, twice as long as the petiole or longer. Racemes 3–5 flowered, woolly. Flowers small. Legumes 3–4 lines long, very woolly. Sometimes confounded with *I. coriacea, var. hirta*, and sometimes with *I. sarmentosa*, but very distinct from either.

4. DIGITATÆ. (Sp. 33–41.)

33. I. filicaulis (E. & Z.! 1594); suffruticose, *very slender*, diffuse or procumbent, *subglabrous;* branches filiform; leaves *subsessile*, digitately 3–7–9 foliolate; leaflets *linear-lanceolate*, acute, with involute margins, sparsely strigose; stipules obsolete; peduncles *capillary*, much longer than the leaves, 3–4 flowered; flowers pedicellate, bracts minute, persistent; calyx-segments *setaceo-subulate;* petals *glabrous*, legume stipitate, compressed, with prominent sutures, glabrous, oblong, 2–3 seeded. *I. subtilis, E. Mey. Comm. p.* 98.

HAB. Moist, grassy places near Tulbagh Waterfall, *E.. & Z.!* Draakensteenberg, *Drege!* Table Mountain Summit, *W.H.H.!* Near Simonstown, *C. Wright!* 548. (Herb. Sd., D.)

A very slender, wiry plant, drying dark. Leaflets rising from a tubercular petiole, petiolulate, 4–6 lines long, ½ line wide, tapering to each end. Peduncles hairlike, 1½ inch long, bearing a minute raceme of 2–4 small flowers. Legumes 2–3 lines long, on a longish stipe.

34. I. dillwynioides (Benth.! in Herb.); suffruticose, slender, ascending-suberect, subglabrous; branches angular; leaves subsessile, digitately 5–7 foliolate; leaflets *oblongo-lanceolate*, acute, with involute margins, sparsely strigose, glaucous; stipules obsolete; peduncles *setaceous*, much longer than the leaves, 6–8 flowered; flowers pedicellate, bracts persistent; calyx-segments *lanceolate;* petals *silky;* legume . . ?

HAB. Klipfontein, *Zeyher! No.* 494. (Herb. Hk., Bth., Sd.)

Very like *I. filicaulis*, but more robust, more erect in growth, more densely leafy, with larger flowers and more of them in the raceme; and (judging by the ovary) probably with a different legume. Root fibrous. Stems ligneous, much branched from the base; the branches curved, simple, suberect, imbricated with leaves throughout. Leaflets 6–8 lines long, 1–2 lines wide, more or less involute. Peduncles as thick as hog's bristle, 2 inches long, bearing a short raceme; the flowers often in pairs on slender pedicels. Petals nearly 3 lines long. Ovary sessile, linear, glabrous, with 6–8 ovules.

35. I. pentaphylla (Burch.! Cat. 7366); suffruticose, diffuse, very slender, subglabrous; stems filiform; leaves on short, setaceous petioles, digitately 5–7 foliolate; leaflets lanceolate, acute, with involute margins, concolorous, glabrate; stipules minute; peduncles setaceous, much longer than the leaves, 1 (–3 ?) flowered; calyx canescent, its segments ovate-acute; petals silky.

HAB. S. Africa, *Burchell!* (Herb. Burch.)

Very slender, with something the habit of *I. filicaulis*, but with petioled leaves and much larger flowers. I have only seen a small specimen with a single flower. Petioles 2–2½ lines long; leaflets 4–5 lines long, not ½ line wide.

36. I. digitata (Thunb.! Fl. Cap. 598); suffruticose, procumbent or ascending, thinly strigillose; leaves on short, channelled petioles, digitately 7–9 foliolate; leaflets cuneate sublanceolate, narrow, acute, mostly complicate, concolorous, strigoso-hispid on both surfaces; stipules broadly subulate, equalling or exceeding the petiole; racemes on long peduncles, laxly many flowered; calyx segments subulate; vexillum sparsely hispidulous; legumes terete, subcompressed, 6–8 seeded, strigoso-puberulent. *DC. Prod.* 2. *p.* 231. *E. & Z.!* 1593. *E. Mey.! Comm. p.* 98. *Zey.* 2440.

HAB. Mountain sides, Hott. Holl., Zwarteberg and Winterhoeksberg, *E & Z.!* Paarlberg, *Drege.!* Swellendam, *Dr. Thom!* (Herb. Th., Hk., Bth., Sd.)

Stems ligneous but slender, 1–2 feet long, with ascending, subvirgate, 4-angled branches. Petioles 2–3 lines long. Leaflets 4–5 lines long, sometimes very narrow, sometimes 2 lines wide beyond the middle, always tapering at base, subobtuse or acute; the pubescence scanty on the upper, copious on the lower surface. Peduncles 4–6 inches long; racemes dense, elongating; pedicels 1 line long. Legumes ½–¾ inch long.

37. I. Burchellii (DC. Prod. 2. p. 231, non E. Mey.); suffruticose, slender, prostrate, *thinly canescent;* branches angular; leaves on channelled petioles, digitately 3–5 foliolate; leaflets obovate or obcordate, mucronate, open, short, *strigillose above, thinly canescent* beneath; flowers and fruit unknown. *Burch. Cat.* 2918.

HAB. Interior of S. Africa, *Burchell!* (Herb. Benth.!)

This may perhaps be merely a more glabrous state of *I. bifrons*, but till the inflorescence shall be found it is impossible to say. The stems are more slender than in *I. bifrons*, 6–12 inches long, alternately branched, and lying prone along the soil.

38. I. bifrons (E. Mey.! Comm. p. 97); suffruticose, diffuse, *white, with dense, minute, appressed pubescence;* branches angular; leaves on thick, angular petioles, digitately 3–7 foliolate; leaflets obovate, complicate, short, *silvery;* stipules minute, toothlike; racemes on angular peduncles longer than the leaves, several flowered; calyx shortly 5-toothed; petals puberulous; legumes short, turgid, oblong, few-seeded, hook-mucronate, canescent.

VAR. α. **trifoliata**; leaves mostly 3 foliolate. *I. Meyeriana, E. & Z.!* 1573.

VAR. β. **digitata**; leaves mostly 5–7 foliolate. *E. Mey. l. c. Zey.! Legum.* 15. *Z. n. N. n. E.* 113. 10.

HAB. Nieuweveldsbergen, 3500–4000 f. *Drege! E. & Z.!* β. Among stones on the summits of the Witbergen, 7500 f. and Los Tafelberg, 6000 f.; Camdebosberg, 4500 f., and near Graaf Reynet, 3000 f.. *Drege! Zeyher!* (Herb. Sd., Bth., D.)

A very small, depressed or prostrate, much branched plant, quite white in all parts, with minute, soft, close pubescence. Branches, petioles, and peduncles sharply angular. Petioles 2–6 lines long. Leaflets 2–4 lines long, ½–1½ broad, thick, veinless, very white. Peduncles about twice as long as the leaves; pedicels very short; flowers 2 lines long. Legumes ½–¾ inch long, 2–4 seeded.

39. I. quinquefolia (E. Mey. Comm. p. 98); *dwarf*, shrubby, ascending, much branched; leaves minutely petioled, digitately 5-foliolate; leaflets cuneate-oblong, *convolute, subterete, channelled above, round-*

backed, hairy, mucronulate; stipules minute, subulate; racemes on short peduncles, scarcely twice as long as the leaves, few-flowered; rachis capillary, flexuous; calyx-segments subulate; petals hairy; legumes . . . ?

HAB. Dry, mountain situations round Genadendal, 3000 f., *Drege!* (Herb. Hk , Bth.)

A dwarf, but woody little shrub, 2–4 inches high, densely ramulous and well covered with leaves. Petioles 1 line long. Leaflets 3–4 lines long, the margins remarkably inflexed or inrolled, and the backs rounded, shortly hispid. Flowers small, just projecting beyond the leaves. Legumes not seen.

40. I. flabellata (Harv); shrubby, erect, densely much branched; branches and twigs terete, *roughly tomentose or hirsute;* leaves on short petioles, digitately 5–foliolate; leaflets *linear-cuneate,* obtuse, mucronulate, expanded, strongly midribbed, minutely strigillose on both sides; stipules minute, toothlike; racemes subsessile, shorter than the leaves, loosely few-flowered; bracts deciduous; calyx canescent, *shortly 5-toothed;* petals cano-puberulent; legumes terete, short, few seeded, *canescent. I. hispida, Herb. Berol., non E. & Z.! Burch.!* 5174, 6915.

HAB. S. Africa, *Niven! Mundt and Maire! Miller and Thom!* Near George, *Dr. Alexander Prior.* (Herb. Sd., Hk., Bth.)

A strong, erect, much branched and ramulous, densely leafy bush, with the habit of *I. sulcata,* to which it is nearly allied, but from which it differs in the rameal pubescence, shape of leaflets, calyx and legume. Petioles 1–2 lines long. Leaflets cuneate at base, gradually widening upwards, *flat* or nearly so, 3–5 lines long, 1–1½ wide near the blunt extremity. Racemes 4–5 flowered; pedicels 2 lines long. Calyx-teeth very short, bluntish. Flowers rosy purple, 2 lines long. Legumes 4–5 lines long, densely clothed with microscopic, white pubescence.

41. I. sulcata (DC. prod. 2. p. 231); shrubby, erect, densely much branched; branches augularly furrowed, thinly strigillose; leaves subsessile, digitately 5-foliolate; leaflets *linear, acute,* furrowed above, with *revolute margins* and strong midrib, minutely strigillose, subcanescent beneath; stipules minute, toothlike; racemes subsessile, shorter than the leaves, loosely pluriflowered; bracts minute, persistent; calyx canescent, its lobes broadly subulate; petals subglabrous or puberulent; legumes terete, several-seeded, *glabrous. E. & Z.!* 1601. *E. Mey.! Comm. p.* 98. *Burch. Cat.* 4706, 5126, 5562.

HAB. Mountain sides near Kromrivier and Vanstaadensberg, *E. & Z.! Drege!* Albany, *Dr. Atherstone.* (Herb. Bth., Hk., D., Sd.)

A densely branched, strong bush, 1–2 feet high; branches and twigs erect, densely leafy. Leaves either exactly digitate on an obsolete petiole, or obliquely and imperfectly digitate on a little longer petiole. Leaflets 5–7 lines long, scarcely a line wide, with thickened and recurved margins. Flowers pinky-purple, very minutely downy, 2–2½ lines long. Legumes uncial, when ripe quite glabrous. This has much the aspect of *I. brachystachya,* but the leaves are digitate, not pinnate; and the legumes glabrous not tomentose, &c.

5. PINNATÆ. (Sp. 42–71.)

42. I. hispida (E. & Z.! 1600); shrubby, erect, much branched; branches and twigs *tomentose, and set with rigid, gland-tipped bristles;* leaves subsessile, bijugate; leaflets close, linear-subcuneate, obtuse, mucronate, acute at base, with prominent midrib, *the margins minutely recurved,* both surfaces thinly cano-sericeous; stipules minute, subulate;

racemes subsessile, equalling the leaves, canescent, pluriflowered; bracts minute; calyx canescent, *its lobes setaceo-subulate;* petals albo-sericeous; legumes . . . ? *I. lotoides, E. Mey.! Linn.* 7. *p.* 168, *non Lam.*

HAB. Heathy ground on the mountains above Uitenhage, *E. & Z.* (Herb. Sd.)

Very like *I. brachystachya*, from which (unless the legumes afford further characters) it chiefly differs in the *gland-tipped, rigid setæ*, mixed with the tomentum of the branches, the flatter and thinner leaflets, and more slender calyx-segments. Leaflets 6–8 lines long, 1½ line wide, the pairs nearly 1 line apart; the common petiole about quarter inch long.

43. I. brachystachya (E. Mey.! Linn. 7. p. 168); shrubby, erect, densely much branched; branches tomentose; leaves sessile, bi-trijugate; leaflets close, linear-subcuneate, with prominent midrib and strongly revolute margins, recurvo-mucronulate, albo-puberulous above, tomentose and canescent beneath, the terminal sessile; stipules minute, subulate; racemes *on very short peduncles*, as long or twice as long as the leaf, canescent, pluriflowered; bracts minute; calyx canescent, its lobes *broadly subulate;* petals albo-sericeous; legumes short, few seeded, turgid, *tomentose. E. Mey.! Comm. Drege, p.* 98. *E. & Z. No.* 1599. *I. angustifolia, litt. b., Herb. Thunb.! Fl. Cap. p.* 599. *I. angust. β. brachystachya, DC. Prod.* 2. *p.* 231.

HAB. Cape Flats, also about Muysenberg and Simonsbay, *E. & Z.! C. Wright*, 585. Near Capetown, *Dr. Alexander Prior!* Stellenbosch and Attaquaskloof and Kromriver, *Drege.* Mouth of Bot River and Onrust R., Caledon, *Dr. Pappe! Zey.* 2424. (Herb. Th., Bth., Hk., Sd., D.)

A strong, rigid, densely branched bush, 1–2 feet high, more or less tomentose. Common petiole 3–4 lines long, the leaflets scarcely a line apart. Leaflets 4–6 lines long, 1–2 lines wide, rarely flat, usually with strongly revolute margins. Racemes rarely longer than the leaves, few or several flowered. Flowers purple, 3 lines long. Legumes 6–7 lines long, terete, 1½ line in diameter.

44. I. stricta (L. fil.); shrubby, erect, much branched; branches straight, rigid, angular-furrowed, minutely strigillose; leaves frequent, very patent, 3–4 jugate; the short common-petiole naked or gland stipelled; leaflets *close*, cuneate-oblong, narrow, obtuse or subacute, mucronulate, concave, glabrous above, thinly strigillose beneath; stipules small, subulate; racemes *subsessile, equalling the leaves, laxly 3-4 flowered;* calyx pubescent, its lobes shortly subulate; petals fulvo-sericeous; legumes terete, spreading, acute, thinly strigillose or glabrescent. *Thunb.! Fl. Cap. p.* 599. *I. pauciflora, E. Mey! Comm. p.* 99. *Burch. Cat. p.* 5479.

VAR. β. **acuta**; petiole gland-stipelled; leaflets often subacute, somewhat lanceolate. *Burch. Cat.* 3706. *I. stricta, E. & Z.!* 1602, *excl. var. β.*

HAB. Margins of woods near George, *Drege! Dr. Alexander Prior!* β. Open places near Olifant's Hoek, Uit., *E. & Z.!* Block House Hill, Grahamstown, *Dr. Atherstone! T. Williamson!* (Herb. Th., Hk., Bth., D., Sd.)

A slender but rigid, straight-branched shrub, 1–2 feet high, with green (not hoary) foliage and thinly strigillose. Common petiole not half inch long; leaflets 3–5 lines long, in α. *rounded* at point, in β. more or less acute. Peduncles slender, very short. Flowers like those of *I. Zeyheri.* Legumes 1–1½ inch long, many-seeded, at first strigillose, then glabrescent. *Drege's* specimens entirely agree with the authentic sp. in Herb. Upsal! E. & Z.'s var. β. "*pedunculata*" seems to me to belong *I. Zeyheri.*

45. I. foliosa (E. Mey.! Comm. p. 102); shrubby, erect; dark coloured, furrowed, strigillose, *densely leafy;* leaves subsessile, *laxly* 3–4-jugate, *erect;* leaflets linear-cuneate, mucronate, glabrous above, strigillose beneath, with incurved margins, the terminal subsessile; stipules obsolete; racemes *shorter than the leaf, 3-6 flowered;* calyx-segments subulate; petals fulvo-sericeous; legumes few-seeded, subterete, acute, deflexed, thinly strigillose.

HAB. Between Omsamwubo and Omsamcaba, *Drege!* (Herb. Benth.)

Small twigs only seen by me. Common petiole uncial, the pairs of leaflets 2 lines apart. Leaflets 4–5 lines long, more or less concave. Racemes ¾ inch long; pedicels filiform, at length 3–4 lines long. Flowers 2 lines long, tawny externally. Legumes ½ inch long. The very erect leaves, much longer petiole, and more distant pairs of leaflets, the terminal sometimes petioled, and the short legumes distinguish this from *I. stricta.*

46. I. vestita (Harv.); fruticose, erect, minutely strigillose; stem subsimple, virgate, very straight, *densely leafy;* leaves imbricate, 2–3 jugate; leaflets *lanceolate-linear, with strongly revolute margins,* mucronate, *glabrescent;* the terminal sub-petioled; racemes *few-flowered, shorter than the leaves;* calyx 5-toothed; petals fulvo-sericeous; legumes terete, straight, deflexed, strigilloso-puberulent.

HAB. Coast land of Natal to 1000 f. *Dr. Sutherland!* (Herb. Hook.)

Stem 2–3 feet high, robust, quite simple, closely imbricated with leaves, among which the small inflorescences are hid; the whole plant when in flower converted into a leafy thyrsus. Leaflets very narrow, and by revolution of margins almost filiform, 5–7 lines long, acute at each end and mucronate, nearly glabrous, with a very few small strigæ. Flowers small, brownish externally. Legumes an inch long, black, with very minute, appressed pubescence. This may be Meyer's *I. hirta,* and is certainly closely allied to it; but seems to differ in the much narrower and more glabrous leaflets, and different pubescence; that of our plant could scarcely be called "*strigilloso-hirta, subcanescens,*" nor the legumes "*hirtis.*"

47. I. hirta (E. Mey. Comm. p. 101); "fruticose, *strigilloso-hirsute, subcanescent;* stem erect, subsimple; leaves 2–3 jugate, short petioled, leaflets lanceolate, subacuminate-mucronate, the terminal subsessile; racemes shorter than the leaves; petals strigillose and hairy; legumes subterete, straightish, *hairy,* spreading." *E. Mey. l. c.*

HAB. Grassy places between Omtata and Omsamwubo, *Drege.* (Unknown to me.)

48. I. Dregeana (E. Mey.! Comm. p. 100); fruticose, erect, *villoso-canescent; stem subsimple, virgate, very straight;* leaves *closely set,* short-petioled, 5–7 jugate, suberect; leaflets linear-oblong, mucronate, flat, tomentose on both sides, the terminal subsessile; racemes on long, terete peduncles, densely many flowered, elongate, *patent or pendulous;* calyx shortly 5-fid; petals fulvo-sericeous; legumes terete, straight, sharp pointed, tomentulose.

HAB. Between Omcomas and Port Natal, 300 f. *Drege!* Coast land of Natal to 1000 f. *Dr. Sutherland!* (Herb. Bth., Hk.)

Stem 2–3 feet high, erect, quite simple, rigid and rodlike, closely imbricated with leaves. Pubescence copious, whitish, soft, short, patent. Leaves 1–1¼ inch long, erect. Leaflets 4–5 lines long, 1–2 lines wide, obtuse, somewhat cuneate at base. Peduncles very patent, 2–3 inches to the base of the dense raceme, which is 4–5 inches long; pedicels much longer than the calyx. Flowers rather small (withered only seen). Legumes 1½–1¾ inch long, quite straight. A very remarkable species, quite unlike any other S. African.

49. I. tristis E. Mey.! Comm. p. 101); shrubby, *erect, slender;* stem *virgate, ramuliferous above,* thinly strigillose, subangular; leaves subremote, patent, short petioled, 3–5 jugate; leaflets cuneate-oblong, retuse, mucronulate, minutely strigilloso-puberulous on one or both sides, the the terminal subpetioled; racemes on peduncles longer than the leaves, elongate, many flowered; calyx canescent, *bluntly 5-toothed; petals densely silky with dark brown hairs;* legumes . . ?

HAB. Grassy places between Gekau and Basche, and between the Omsamcabo and Omsamwubo, *Drege!* Port Natal, *Gueinzius and Dr. Sutherland!* (Herb. Bth. Hk., D.)

Stem 2–3 feet high, straight and rodlike, pale brown, minutely puberulous, somewhat panicled above; twigs slender, erecto-patent, subsimple, 5–8 inches long. Leaves nearly an inch apart, recurvo-patent; the common petiole ½–¾ inch long. Leaflets 4–5 lines long, 1–1½ line wide, pale. Racemes loosely many flowered, 2–3 inches long; pedicels longish. Calyx very small, white-hairy. Petals 4–5 lines long, dark brown externally. Legumes not seen. I have not seen *Drege's* specimens.

50. I. arrecta (Benth.! in Herb.); suffruticose, erect, *virgate,* laxly branched; branches angular, thinly strigillose; leaves scattered, petiolate, 3–jugate; leaflets *distant,* sublanceolate-linear, narrow, mucronate, complicate, thinly strigillose; stipules minute, toothlike; racemes on very long peduncles, elongating, many flowered; calyx strigose, its segments subulate; petals albo-sericeous; legumes *very short,* oblong, tumid, mucronulate, strigose, 2–3 seeded. *Zey.* 417.

HAB. Mooje River, *Burke & Zeyher, Pappe!* (Herb. Hk., Bth., Sd., D.)

1½–2 feet high, slender, glabrescent; the whole plant of a livid, blue-green, indicating the presence of *indigo.* Leaves an inch or more apart, the common petiole often gland-stipelled; the pairs 3 lines apart. Leaflets 6–8 lines long, 1–2 lines wide. Peduncles 3–4, afterwards 6–8 inches long or more, ⅔ occupied by flowers. Flowers small, 1½–2 lines long, on short pedicels. Legumes 3 lines long, very turgid, 1–1¼ line in diameter.

51. I. fastigiata (E. Mey.! Comm. p. 102); suffruticose, erect, thinly strigillose; branches virgate, angular-furrowed; leaves *subsessile, 2–3 jugate;* leaflets *linear-oblong* or sublanceolate, mucronulate, strigillose beneath, the terminal sessile or somewhat petioled; stipules subulate; racemes on *very long,* slender, angular peduncles, many flowered; bracts minute; calyx-lobes shortly subulate, subcanescent; *petals glabrous;* legumes linear, terete, straight, pendulous, thinly strigillose.

VAR. β. **angustata**; leaflets very narrow-linear, acute, complicate; legumes straight, linear, sericeo-canescent, *Zey.!* 2442.

HAB. Between the Omsamwubo and Omsamcaba, *Drege!* Coastland, Natal, 1–1000 f., *Dr. Sutherland.* Var. β. Howison's Poort, *H. Hutton!* Zwartkops R., *Zey.!* 2442. (Herb. Hk., Sd., D.)

18 inches to 2 feet high, slender, with a few long, simple, laxly leafy branches. Leaves 2–3 inches apart, the common petiole ½–¾ inch long, bearing the lowest pair of leaflets almost at its base. Leaflets ¾–1 inch long, ½ line wide, glabrous above, paler and thinly strigillose beneath. Peduncles 6–8 inches long, erect, floriferous beyond the middle. Flowers pale, pinkish-purple; the flower-buds very acute. Ovary canescent, pluri-ovulate. Legume 1¼–1½ inch long. In *Drege's* specimens the leaflets are complicate, subacute at each end and somewhat lanceolate; in *Dr. Sutherland's* they are expanded, obtuse but mucronulate, and more oblong, the terminal leaflet not quite sessile: in other respects the plants are identical. Var. β. has very narrow leaflets, and is altogether depauperated, resembling some of the weaker forms of *I. glaucescens;* but it has the glabrous petals and long peduncles of the present species.

52. I. affinis (Harv.); suffruticose, erect, thinly strigillose; branches virgate, angular, furrowed; leaves *petiolate*, distantly 3–4-jugate; leaflets linear-spathulate, obtuse or recurvo-mucronulate, complicate, strigillose beneath; stipules obsolete; racemes on *short* peduncles, densely many flowered; bracts minute; calyx canescent, its lobes shortly subulate; petals *glabrous*.

HAB. Near Lake Ngami, *J. M'Cabe!* (Herb. Hook.)

Very like *I. fastigiata*, E. Mey., but with distinctly petioled leaves (the lowest pair of leaflets ½–¾ inch from base of common petiole) and much shorter peduncles. The naked portion of the peduncle is shorter than the leaf, the floriferous not twice as long. The foliage is a very pale green; the leaflets 6–8 lines long, 1–1½ line wide, cuneate at base.

53. I. viminea (E. Mey.! Comm. p. 102); suffruticose, erect, strigilloso-canescent; stem erecto-patent, branches virgate, angular-furrowed; leaves shortly petioled, 4–5-jugate, patent; leaflets *remote*, lanceolate-linear, with involute margins, mucronate, thinly strigilloso-puberulent, the terminal sessile; racemes on long, filiform peduncles, laxly many flowered; pedicels long; calyx shortly 5-toothed; petals fulvo-sericeous; legumes subcompressed, straight, with thick sutures, strigillose.

HAB. In grass fields between Gekau and Basche, 1500–2000 f. and at Klein Bruintjeshoogte, *Drege!* (Herb. Benth., Sond.)

Two or more feet high, slender. Leaves subdistant. Common petiole uncial, the pairs of leaflets 3 lines apart. Leaflets 9–10 lines long, 1 line wide, pale green, scarcely canescent. Peduncles setaceous, 2 inches long; the raceme 1–2 inches long. Pedicels 2–3 lines long. Both the calyx and petals are pubescent.

54. I. corniculata (E. Mey.! Comm. p 101); suffruticose; "stem erect, much branched;" leaves short petioled, 3–4-jugate; leaflets *cuneate-oblong, obtuse*, mucronulate, minutely strigillose on both sides, *the terminal shortly petioled;* racemes on long peduncles, loosely several flowered, suberect; pedicels longer than the calyx; calyx canescent; *bluntly 5-toothed;* petals fulvo-sericeous; "legumes terete, straight, cuspidate." *E. Mey. l. c.*

HAB. Between the Omtata and Omsamwubo, *Drege!* (Herb. Benth., Sond.)

I have only seen one specimen. This is curved, somewhat angular, pale, thinly strigillose. Leaves nearly an inch apart. Common petiole ¾ inch long, slightly prolonged beyond the last pair of leaflets. Leaflets 4–5 lines long, 1½ wide near the apex, cuneate and acute at base, pale green, flat or nearly so, midribbed beneath. Peduncles 2–3 inches long, the upper half bearing flowers; pedicels 2 lines long. Calyx not a line long, white. Petals 4–5 lines long, covered with tawny hairs. Legumes not seen.

55. I. malacostachys (Benth.! in Herb.); shrubby; stems procumbent, much branched, *villoso-tomentose;* branches flexuous, canescent, terete; leaves subsessile, 4–5-jugate, the common petiole *gland-stipelled;* leaflets obovate-oblong, obtuse, mucronulate, pilose above, villoso-canescent beneath, the terminal sessile or minutely petioled; stipules small, subulate; racemes *spicate, cylindrical, dense;* on peduncles equalling or exceeding the leaves; calyces and bracts *densely albo-villose and tomentose;* cal. segments setaceo-subulate; petals tomentose; legumes . . . ? *Zey.*! 478.

HAB. Magalisberg, *Burke and Zeyher!* (Herb. Hk., Bth., Sd.)

Seemingly a prostrate or procumbent, much-branched shrub, the stem and larger branches ligneous, the twigs half herbaceous; the whole plant covered with soft, white hairs, which are particularly copious on the inflorescence. Leaves 1–1½ inch long. Leaflets 3–4 lines long, 2–2½ lines wide, open or folded. Peduncle 1–1½ inch to the base of the spike, which is 1–2 inches long. Flowers small and hoary. Legumes unknown. This species is more naturally allied to *I. eriocarpa, oxytropis, &c. in Sec. 7.*

56. I. elliptica (E. Mey.! Comm. p. 99); "somewhat shrubby, strigilloso-canescent; leaves 3-jugate, recurved, on short petioles; leaflets *elliptical*, glabrescent above, sparingly strigillose and pale beneath, the terminal minutely petioled; racemes twice as long as the leaf, flexuous, few-flowered; petals nearly glabrous." *E. Mey.*

HAB. S. Africa, *Drege.* (Unknown to me.)

57. I. poliotes (E. & Z.! 1609); suffruticose, *strigoso-canescent;* branches curved, filiform, subsimple; leaves frequently reflexed, subsessile, 3–4–jugate; leaflets sublanceolate, acute, furrowed above, prominently midribbed beneath, strigoso-canescent on both surfaces; stipules minute, subulate; racemes on *filiform, angularly flexuous* peduncles *scarcely longer* than the leaves, laxly 3–4 flowered; calyx-segments setaceo-subulate; petals fulvo- or cano-sericeous; legumes short, terete, canescent. *I. adoensis, E. Mey. Comm. p.* 99 *(ex diagnosi). I. rupestris, E. & Z.!* 1603, *and I. punctata, E. & Z.!* 1604. *Burch.!* 3492, 5113.

HAB. Among shrubs on hills at Adow, Uit., and on the Winterberg, *E. & Z.! Drege.* (Herb. Sd., D.)

A slender, erect or suberect, hoary suffrutex, with long, simple, incurved whitish branches, rather densely leafy. Common petiole recurved, ½–¾ inch long; leaflets 3–4 lines long, acute at each end, somewhat keeled, and with a narrow medial furrow above. Peduncles setaceous, angularly zig-zag from flower to flower. Flowers small. Legume (imperfectly known to me) canescent, almost tomentulose. Allied to *Zeyheri*, but differing in inflorescence, &c. In Hb. Ecklon a specimen of *I. Zeyheri* is preserved, along with one of *I. poliotes,* under No. 1609. As far as very imperfect specimens enable me to judge, *E. & Z.'s "I. rupestris"* is a form of this species.

58. I. Zeyheri (Spreng); suffruticose, slender, erect, thinly canescent; branches patent, angular; leaves subremote, 3–5-jugate, the petiole naked or gland-stipelled; leaflets linear-subcuneate, complicate, mucronulate, thinly appresso-canescent beneath; stipules small, subulate; racemes on long, slender peduncles, laxly several flowered; bracts minute, pedicels longer than the calyx; calyx canescent, its lobes shortly subulate; petals fulvo-sericeous; legumes terete, spreading, acute, several seeded, thinly canescent. *E. & Z.!* 1606 *(excl. var. β.) I. cinerascens, E. & Z.!* 1607. *I. nana, E. & Z.* 1611. *I. punctata, Thb.!*

VAR. **β. leptophylla**; petiole generally *gland-stipelled;* leaflets narrow, complicate; racemes few-flowered; legumes glabrescent. *Zey.* 2431, 2433. *I. leptophylla, E. Mey.! Comm. p.* 99. *I. verrucosa, E. & Z.* 1608.

HAB. About Uitenhage, by the Zwartkops and Adow, *E. & Z.! &c.* Rhinosterkop, *Burke & Zeyher!* Port Natal, *Gueinzius!* (Herb. Th., Sd., Bth., Hk., D.)

A slender, laxly branched, more or less albescent suffrutex, varying in the breadth of its leaflets and the length and fertility of its racemes; but the varieties are scarcely definite, and all grow together. The var. β. with its pulvinate tufts of brown glands between the leaflets looks distinct; but some of Drege's specimens have these glands, and others want them. *Zeyher's* 2429, from the Zwartkops, is a more glabrous form, with laxer and larger foliage, and remarkably long peduncles; it probably grew in a very wet, perhaps shady situation.

59. I. concava (Harv.); suffruticose, ramulous; twigs glabrescent; leaves subsessile, 2–1-jugate; leaflets close, *linear-oblong, blunt, with strongly involute margins*, minutely strigillose or glabrescent; stipules minute, subulate; racemes several flowered, on peduncles rather longer than the leaves; bracts small, ovate-acuminate; calyx canescent, its lobes acute, the lowest ovato-lanceolate, the upper ovate; petals silky-villous, canescent; legumes ?

HAB. Near George, *Dr. Alexander Prior! Burchell*, 1593. (Hb. Bth., D., Bch.)

Erect or suberect, slender but lignescent; the branches and twigs brown, and the leaves drying dark. Common petiole 2–3 lines long, generally with two, sometimes with but one pair of leaflets, and an odd one. Leaflets remarkably inrolled, the upper surface nearly hidden, 3–4 lines long. Racemes 5–12 flowered, hoary. Peduncles equalling or exceeding the leaves.

60. I. angustifolia (Linn. Mant. 272); suffruticose, diffuse or ascending; branches curved, angular, glabrescent or thinly strigillose; leaves subsessile, 3–4-jugate; leaflets close, *linear*, with prominent midrib and reflexed margins, recurvo-mucronulate, puberulent above, canescent beneath, the terminal sessile; stipules small, subulate; racemes *on long, filiform peduncles*, laxly many flowered; bracts minute; calyx thinly canescent, its segments broadly subulate; petals albo-sericeous; legumes terete, *glabrous*, several seeded. *DC. Prod.* 2, *p.* 230. *E. & Z. No.* 1612. *E. Mey. Comm. p.* 99. *Zey.! No.* 2425. *I. angustifolia, litt. a., Herb. Thunb.!*

VAR. β. **tenuifolia**; more slender in all parts, with shorter and narrower leaflets, and setaceous peduncles. *I. tenuifolia, E. & Z.!* 1613, *and I. leptocaulis, E. & Z.!* 1616. *Zey.!* 493. *I. strigosa, Spr. Neu. Ent.* 3. *p.* 54.

HAB. About Table Mountain and on the Cape Flats; also in Worcester, Caledon and Swellendam, *E. & Z. &c.* (Herb. Th., Bth., Hk., D., Sd.)

A small, slender, rather *woody* suffrutex, diffuse, procumbent, or ascending, much branched, 1–2 feet long, many stemmed. Common petiole ½–¾ inch long; the pairs 1–2 lines apart. Leaflets 5–7 lines long, 1 line wide, sometimes nearly flat, mostly with strongly reflexed, thickened margins, variably pubescent. Peduncles filiform, 3–5 inches long. Flowers purple, 2½ lines long. Legumes pendulous, scarcely an inch long, quite glabrous, acute. A much more slender, less woody plant than *I. brachystachya*, with very dissimilar inflorescence and legumes. β. is often very slender in all parts, but connected by insensible gradations with *a.*

61. I. Mundtiana (E. & Z. 1617); suffruticose, slender, procumbent; branches filiform, nearly glabrous; leaves scattered, shortly petioled, 4-jugate; leaflets *elliptic-oblong*, obtuse, mucronulate, flat, rugulose and minutely puberulous above, *silvery and silky* beneath, the terminal sessile; stipules small, subulate; racemes on filiform peduncles much longer than the leaves, several flowered; calyx canescent, its segments subulate; petals silky; legumes ?

HAB. Mountains near Swellendam, *Mundt!* (Herb. Sd.)

A slender trailer. Leaves an inch apart, the common petiole an inch long. Leaflets 3–4 lines long, 2 lines wide, dark-coloured above, quite white beneath. Peduncles 3–4 inches long, very slender, glabrous. Unopened flowers only seen. Allied to *I. angustifolia*, but with much broader and flatter leaflets, &c.

62. I. filiformis (Thunb. Fl. Cap. p. 598); shrubby, diffusely much branched, the branches and slender twigs *strigoso-tomentose*, becoming glabrate; leaves on very short petioles, *imperfectly digitate* or bijugate, with an odd one; leaflets *oblong* or *linear-oblong, subcuneate,* obtuse, mucronulate, flat, with subrecurved margins, strigose on both sides; racemes on long slender peduncles, *loosely many flowered;* pedicels long; calyx albo-villous, its segments setaceo-subulate; petals canescent; legumes long, slender, terete, pendulous, glabrous. *DC. Prod.* 2, *p.* 231. *E. & Z.!* 1592. *E. Mey. Comm. p.* 98. *I. candicans, Sieb.! No.* 55. *C. Wright,* 545, 526.

HAB. Moist places among shrubs, on hill sides round Capetown, common. (Hb. Th, Hk., Sd., D., &c.)

Stem strong and robust below, the diffuse, long branches slender but ligneous. Pubescence variable, sometimes scanty, sometimes copious and long, always white. The common petiole is either very short, with the five leaflets springing obliquely, in a pedate manner from a minute distance below the summit; or rather longer, with the two pairs sensibly apart. Leaflets 4–8 lines long, 1–3 lines wide, varying from linear to elliptic-oblong, often grey-hoary, drying dark. Peduncles 3–4 inches long; pedicels 2 lines long. Legumes an inch long or rather more, half line in diameter, dark brown.

63. I. coriacea (Ait. Kew. 3. p. 68); shrubby, much branched, diffuse or erect, the branches and twigs *tomentose,* becoming nude; leaves on very short common petioles, patent, 5–7-foliolate, *imperfectly subdigitate* or closely 2–3-jugate; leaflets *broadly obovate or obcordate or elliptic-oblong,* recurvo-mucronate, flat, coriaceous, strigose above, *villoso-canescent or hirsute beneath;* racemes short, oblong, *densely subcapitate,* on long or shortish filiform peduncles; pedicels short or long; calyx albo-villous and tomentose, its segments setaceo-subulate; petals cano-tomentose; legumes long, terete, pendulous, glabrous.

VAR. *α.* **cana**; diffuse or procumbent; leaflets 5, broadly obovate or obcordate, green above, *hoary* underneath; pedicels short. *I. coriacea, DC. Prod.* 2, *p.* 231. *I. mauritanica, E. & Z.* 1590, *var. α. Sieb. Fl. No.* 56. *E. Mey. Comm. p.* 100. *Lotus mauritanicus, Linn. L. fruticosus, Berg. Cap. p.* 226.

VAR. *β.* **hirta**; erect or subdiffuse; leaflets 5, broadly obovate, recurved-pointed, concolorous, densely hairy at both sides; pedicels elongate. *I. mauritanica, β. erecta, E. & Z. Sieb. Fl. mixta,* 18 *and* 215. *I. alopecuroides, E. Mey.! Comm. p.* 100, *non DC.*

VAR. *γ.* **alopecuroides**; leaflets 5–7, elliptic-oblong or elliptical-obovate, acute, concolourous, densely hairy on both sides; pedicels short. *I. alopecuroides, DC. Prod.* 2, *p.* 231. *E. & Z.* 1591. *Zey.* 2426.

VAR. *δ.* **minor** (E. Mey.); slender, procumbent; leaflets 5–7, *minute,* elliptical or obovate, concolourous, hairy on both sides; pedicels short. *E. Mey. Comm. p.* 100. *I. alopec. β. minor, E. & Z.! Zey.* 2434.

HAB. Vars. *α.* and *β.* very common round Table Mountain and the Simonsbay range, &c. *γ.* Babylon's Toorensberg and Klynrivierberg, Caledon, *E. & Z.* *δ.* same locality and the Zwarteberg, *E. & Z.* Gnadendal and Driefontein, *Drege!* (Herb. Th., D., Hk., Bth., Sd., &c.)

A very variable, woody, small shrub; erect, procumbent or prostrate; robust or slender; hirsute or hoary. The above varieties glide insensibly into each other. Leaflets 2–4 lines long, 1½–3 lines wide; in *γ.* scarcely 1 line long and ½–¾ wide. Peduncles sometimes not much longer than the leaves; usually 2–3 times as long, or longer, tomentose. Flowers sometimes subcapitate; in *β.* long-pedicelled. Inflorescence always very hoary. Legumes 1–1¼ inch long, shining brown.

64. I. sarmentosa (Linn. f. Suppl. 334); suffruticose, *prostrate,* with erect branchlets; branches filiform, minutely strigillose; leaves subsessile, *bijugate or trifoliolate;* leaflets elliptical-obovate or oblong, obtuse, mucronulate, midribbed, paler beneath, thinly *strigilloso-puberu-*

lent on both sides; stipules minute, toothlike; peduncles capillary, much longer than the leaves, laxly 3–4 flowered; bracts minute; calyx segments setaceo-subulate; legumes terete, acute, *glabrous*, patent, 7–8 seeded. *DC. Prod.* 2, *p.* 231. *E. & Z.!* 1587, *excl. var.* γ. *E. Mey. Comm. p.* 99. *I. filiformis, var. Thunb. in Herb. Upsal.*

HAB. Table Mountain summit, *Ecklon, W.H.H., Dr. Pappe.* Riv. Zonderende, *Zey.!* 2435. Near the Omsamculo, *Drege!* (Herb. Th., Hk., Bth., D., Sd.)

Root woody. Stems numerous from the crown, very slender, but ligneous, 1–2 feet long, spreading to all sides, trailing, and throwing up short, erect branchlets and peduncles. Leaves on petioles scarcely 1 line long; leaflets 5 or 3 on the same branch, 2–3 lines long, 1½–2 lines wide, very blunt, with a minute mucro and rather prominent midrib. Peduncles 2–4 inches long, setaceous, the flowers racemose, 2–3 lines apart. Legumes ¾ inch long, dark brown, quite glabrous. *E. & Z.'s* var. γ. "*latifolia,*" is *I. ovata, Thunb.*

65. I. capillaris (Thunb. Fl. Cap. p. 599); suffruticose, diffuse or ascending, much branched, nearly *glabrous;* branches slender, angular; leaves on longish, channelled petioles, 4–6-jugate; leaflets *linear-lanceolate, with involute margins,* acute, *glabrous* or *sparsely* strigillose; the terminal sessile; stipules and bracts subulate; racemes on very long peduncles, laxly several flowered; calyx subglabrous, its segments subulate; petals puberulent; legumes terete-subcompressed, linear, deflexed, *glabrous*, many seeded. *DC. Prod.* 2, *p.* 230. *E. & Z.!* 1615. *E. Mey.! Comm. Drege, p.* 102.

HAB. Mountain sides, Zwarteberg, Caledon, *E. & Z.!* Dutoitskloof and Groenekloof, *Drege.* Cape Flats, and near Stellenbosch, *W.H.H.* (Hb. Th., Bth., Hk., D., Sd.)

Root thick and woody. Stems many from the crown, ascending or procumbent, 6–12 inches long or more, either quite glabrous or sprinkled with a few appressed setæ, brown. Leaves rather distant; the common petiole 1½–2 inches long, the first pair of leaflets generally ¾–1 inch from the base. Leaflets 4–6 lines long, ½–1 line wide, mostly involute, sometimes flattish, 2–3 lines apart. Peduncles 6–8 inches long, erect, slender. Flowers 2–3 lines long, subglabrous or puberulent. Legumes 1¼ inch long, ⅓ line in diameter, pendulous, dark brown.

66. I. hedyantha (E. & Z.! 1614); suffruticose, many stemmed, suberect, thinly strigillose; branches ligneous, angular-furrowed; leaves 3–4-jugate, patent; leaflets *linear-oblong* or *sublanceolate*, complicate, mucronulate, thinly strigillose or glabrescent; stipules small, subulate; racemes on long, slender peduncles, loosely many flowered; calyx glabrescent, shortly and sharply 5 toothed; petals externally fulvo-sericeous; legumes . . . ? *I. secunda, E. Mey. Comm. p.* 102.

HAB. Mountains on the Eastern Frontier. Near Philipstown, *E. & Z.* Between Klipplaat R. and Zwartkey, 3800 f.; Katberg, 4–500 f.; and between the Gekau and Basche, 2000 f., *Drege!* Near Grahamstown, *T. Williamson!* Winterberg, *Mrs. F. W. Barber, No.* 50. Spring and autumn. (Herb. Sd., Bth., D., Hk.)

About a foot high, tufted and densely branched from the base; the bark brown; the foliage full green. Leaves close set, ½–¾ inch long. Leaflets 4–6 lines long, ½–1 line wide. Peduncles 4–5 inches long. Flowers 5–6 lines long, "the large vexillum yellow-brown outside, splendid deep crimson within; alæ rich crimson; carina a crimson, yellow-brown near the tip. Petals very soon falling."—*Mrs. Barber.* A very beautiful species, well worth introducing to English gardens.

67. I. hilaris (E. & Z.! 1605); suffruticose, thinly strigillose; stems *short, subsimple, tufted,* suberect, compressed and angular; leaves close,

1–2–4 jugate, short petioled; leaflets lanceolate-oblong or obovate, rigid, acute or mucronate, midribbed beneath, strigillose on the under or on both surfaces, the terminal subsessile; stipules setaceo-subulate; racemes on short peduncles, densely few or several flowered, *scarcely longer than the leaves;* calyx segments setaceo-subulate; petals cano-puberulous; legumes ? *Krauss,* 439. *Zey.* 2438.

HAB. Grassy sides of the Winterberg, at Philipstown, *E. & Z.!* Mts. near Grahamstown, and Aapjies R., *Zeyher!* Port Natal, *Gueinzius! Krauss! Dr. Sutherland!* (Herb. Sd., Hk., Bth., D.)

Root thick and woody. Stems many from the crown, 6–8 inches long, simple or with one or two branches, rigidly strigillose and sometimes hispid, with more spreading bristles. Leaves crowded; the common petiole 6–7 lines long; the pairs of leaflets 1½ line apart. Leaflets 4–5 lines long, varying much in breadth, open or complicate, dull green, rather rigid. Peduncles 1–1½ inch long, somewhat hoary, as are also the flower buds. Stipules and calyx segments remarkably attenuate. Petals 3 lines long, purple-crimson. Legumes unknown.

68. I. ovina (Harv.); suffruticose, procumbent or prostrate, thinly canescent; branches 4 angled, flexuous; leaves 4–5 jugate; leaflets linear-oblong, mucronulate, glabrescent above, paler and thinly strigillose beneath, the terminal sessile; stipules minute, subulate; racemes on *short* peduncles, laxly few-flowered; bracts minute; calyx segments subulate; petals fulvo-sericeous; legumes subterete, 6–8 seeded, straight, deflexed, thinly strigose.

HAB. Summits of rocky hills in Queenstown and Cradock Districts, *Mrs. F. W. Barber, No.* 61. (Herb. D.)

A stunted plant, with thick, much branched, depressed woody stems and ascending slender branches. Leaves laxly set, 1–1½ inch long. Leaflets pale green, 5–6 lines long, ½–1 line wide, expanded. Peduncles rarely as long as the leaves; flowers rosy, on longish pedicels, the vexillum brownish behind. Bracts subpersistent, ¼ line long. Flowers 3 lines long. Legumes 1–1¼ inch long, not a line wide. "Greedily eaten by sheep and goats"—*M.E.B.*

69. I. glaucescens (E. & Z. 1610); suffruticose, strigoso-canescent, pale, ascending, erect or diffuse; branches flexuous, angular and ribbed; leaves subdistant, shortly petiolate, reflexed, 3–4 jugate; leaflets linear-oblong or narrow-obovate-cuneate, recurvo-mucronate, strigose on both sides, the terminal sessile; stipules minute, toothlike; racemes on angular peduncles, much longer than the leaves, elongating, many flowered; pedicels short; calyx canescent, its segments subulate; petals albo-sericeous; legumes subterete, straight, acute or hook-pointed, thinly canescent. *I. reflexa, E. Mey. Comm. Drege, p.* 100. *Zey.* 2441, 2443.

HAB. Hills by the Zwartkops River and on the Vanstaadensberg, *E. & Z.! Pappe.* Koega's Kopje, Algoa Bay, *Zeyher!* Between Zondag and Koega Rivers, *Drege!* (Herb. Sd., Bth., D.)

Rather variable in aspect; 12–20 inches long, more or less ligneous, and more or less erect or diffuse: but recognisable by its pale green foliage, the leaves very patent and distant, rigid pubescence, and long, subspicate racemes of small, canescent flowers. Leaves 1–2 inches apart: the common petiole robust, ½–¾ inch long. Leaflets varying from 3 to 7 lines long, and from narrow-obovate to lanceolate-linear. Raceme occupying ¾ of the peduncle, which is eventually 6–7 inches long. Legumes 1–1½ inch long.

70. I. heterotricha (DC. Prod. 2. p. 227); suffruticose, erect; *stem,* *branches and peduncles densely set with horizontal, rigid, gland-tipped bristles;* leaves scattered, 4–6 jugate; leaflets (small) cuneate-obovate, obtuse, thick, pale, often complicate, rigidly strigillose and glandular, the terminal sessile; stipules subulate; racemes subspicate, on long petioles, much longer than the leaves, laxly many flowered; calyx segments subulate; petals puberulous; legumes terete, acute, pendulous, strigillose and glandularly setose, several seeded. *Burch. Cat. No.* 2635. *Zey.!* 482.

VAR. ? β. **Eckloni**; pubescence and bristles more scanty; leaflets broader and more open, margined with stipitate glands; peduncles shorter. *I. heterotricha, E. & Z.* 1619. (*Specimens in very bad condition*).

HAB. Interior of S. Africa, *Burchell.* Magalisberg, *Burke & Zeyher! Pappe*, 18. Zooloo Country, *Miss Owen!* Var. β. Konab and Kat River, *E. & Z.* (Hb. Bth., Hk., D., Sd.)

Root thick and woody. Stems numerous, erect, 6–12 inches high, subsimple or branched from below, rigid, densely gland-bristled, pale green. Leaves subdistant, the common petiole 1–1¼ inch long. Leaflets 2–3 lines long, 1–1½ wide, usually folded together. Peduncles 5–6 inches long, the raceme occupying 2–3 inches from the extremity; pedicels scarcely 1 line long. Legumes ¾ inch long, about 8 seeded, spreading or pendulous, pale brown. Ecklon's specimens are in very bad order, without flowers, and much broken; they may possibly belong to a distinct but allied species. *Burchell's* 2157, 2432, 2526, 2538, *and* 2637, seem to be states of this species.

71. I. sordida (Benth. in Herb.); suffruticose, erect; *stem, branches, peduncles, petioles, leaflets and calyx densely covered with shortly-stalked glands*, and also strigillose; leaves scattered, petioled, 4–8 jugate, erect; leaflets obovate, obtuse, complicate, strigillose and glandular; stipules subulate, recurved; racemes elongate, on long peduncles, several times longer than the leaves; calyx segments lanceolate; petals glandular; legumes . . . ? *Zey.!* 480, 483.

HAB. Aapjies River, *Burke & Zeyher!* (Herb. Bth., Hk., Sd.)

Root thick and woody. Stems numerous, 12–18 inches high, half herbaceous, terete, subsimple or branched from below, distantly leafy. Every part of the plant covered with gland tipped *short* bristles. Common petiole 2–3 inches long, the pairs of leaflets 3–4 lines apart. Leaflets dull dark green, 4–5 lines long, 2–3 lines wide. Peduncles and racemes 6–8–10 inches long. Flowers small and dull. Legumes unknown. A taller and more robust plant than *I. heterotricha*, much more thickly and equally glandular, with shorter and more uniform glands.

6. UNIJUGÆ (Sp. 72–76).

72. I. tetragonoloba (E. Mey. Comm. p. 106); "*suffruticose, glabrescent;* leaves pinnato-trifoliolate, the petiole equalling the lateral leaflets; leaflets obovate-oblong, strigillose beneath, the terminal larger, on *a long petiole;* racemes subspicate, on long peduncles, at length very long, incurved, filiform; flowers *minute*, vexillum strigillose; legumes torulose, somewhat foursided, curved, thinly strigillose, pendulous, 8–12 seeded."—*E. Mey. l. c.*

HAB. Among shrubs at the mouth of the Omsamculo, and between Omcomas and Port Natal, *Drege.*

Only known to me by a very imperfect fragment in Herb. Sond.

73. I. intermedia (Harv.); *suffruticose, ascending*, thinly strigoso-canescent; branches angular; leaves on longish petioles, pinnately-trifoliolate; leaflets narrow-oblong or sublanceolate, setaceo-mucronate, canescent, thinly strigillose, the terminal petioled; stipules setaceo-subulate; racemes subspicate, on long peduncles, many flowerod; flowers *pedicellate; bracts* lanceolate, deciduous; calyx pilose, its segments subulate; petals glabrous; legumes terete, acute, deflexed, strigillose, many seeded. *I. amoena, E. Mey. Comm. p.* 106.

HAB. Ebenezer, *Drege!* (Herb. Benth.)

A rather slender, spreading suffrutex, with weak, straggling, long, little divided branches. Leaves an inch or two apart. Petioles uncial, the terminal leaflet 1–2 lines removed from the lower pair. Leaflets ½–¾ inch long, 2–3 lines broad, acute or obtuse at each end, thinly canescent. Peduncles 4–5 inches long; pedicels 1 line long. Flowers purple. Legumes 1–1¼ inch long. Very like *I. heterophylla*, but the common petiole is more sensibly prolonged beyond the pair of leaflets. It is much less woody than *I. amoena*, with narrower stipules and bracts, smaller flowers and longer pedicels.

74. I. amoena (Ait. Hort. Kew. 3. p. 68); *shrubby, erect;* branches angular, subcanescent; leaves on longish petioles, pinnately trifoliolate, leaflets elliptic-oblong or obovate, mucronulate, pale underneath, thinly strigillose on both sides; stipules lanceolate, adnate; spikes on long peduncles, densely many flowered, elongating; flowers *subsessile;* bracts *broadly ovate, acuminate,* deciduous; calyx pilose, its lobes broadly subulate; petals glabrous; legumes terete, acute, pendulous. *DC. Prod.* 2. *p.* 224 (*excl. syn. Thunb.*) *Jacq. Schoenb. t.* 234. *E. & Z.!* 1569.

HAB. Brackfontein, Clanw., *E. & Z.!* Cape, *Dr. Thom.* (Herb. Sd., Hk., Bth.)

Robust, woody, erect, 2–3 feet high, thinly clothed with very minute, appressed, pale rigid hairs. Petioles nearly uncial, the terminal leaflet a line or two distant from the lower pair. Leaflets ¾ inch long, 4–5 lines broad, obtuse, shortly mucronate. Peduncles 5–6 inches long, gradually lengthening as the flowers open; immature inflorescence cylindrical. Bracts broad, hispid, suddenly cuspidate-acuminate. Flowers purple. I have not seen legumes.

75. I. argyroides (E. Mey.! Comm. p. 106); *annual,* strigoso-canescent; stem *diffuse* or suberect, angular; leaves on longish petioles, pinnately bi-trifoliolate; leaflets rhomboid-obovate or ovate, mucronulate, the terminal petiolate, much larger than the lateral; racemes spicate, subsessile, *shorter than the leaf,* densely flowered; calyx white hairy, its segments acuminate; petals glabrous; legumes crowded, deflexed, terete-subcompressed, slightly curved, albo-pilose, several seeded.

HAB. Muddy banks of the Gariep, 200 f., *Drege.* Namaqualand, *A. Wyley, Esq.* (Herb. Bth., D.)

Branches flexuous, angularly compressed. Petiole uncial, prolonged 3–4 lines beyond the first pair of leaflets. Lateral leaflets 4–5 lines long, 3–4 lines wide; terminal 6–10 lines long, 5–8 lines wide, flat, pale green above, white beneath, thinly covered on both sides with appressed, white hairs. Peduncles ½–1 uncial, 10–15 flowered. Flowers small. Legumes scarcely uncial. In *Mr. Wyley's* specimens one of the lateral leaflets is frequently absent.

76. I. polycarpa (Benth.! in Herb. Hook.); shrubby, *densely and softly villoso-canescent;* leaves pinnato-trifoliolate; leaflets lanceolate-oblong, tomentose and villous, the terminal petioled; stipules *lanceolate,* longer than the petiole; *flowers axillary, in pairs, pedicellate;* calyx segments setaceo-subulate; legumes short, 2–3 seeded, tomentose.

HAB. Delagoa Bay, *Forbes!* (Herb. Hk., D.)

A much branched, shrubby, but slender bush, densely clothed in all parts with pale tomentum, mixed with longer, soft hairs. Leaves closely set. Leaflets 4–5 lines long. Pedicels 1–2 lines long. Legumes 3 lines long. Differs in *inflorescence* from all the other Cape species.

7. **PRODUCTÆ** (Sp. 77–100).

77. I. fulcrata (Harv.); suffruticose, ascending, nearly glabrous; branches terete; leaves on long petioles, bijugate; leaflets elliptical or obovate-oblong, mucronulate, one nerved, glabrous, glaucous beneath, the terminal long-petioled; stipules *large, membranous, lanceolate;* racemes on long peduncles, laxly many flowered; bracts *large, membranous, enclosing the buds,* deciduous; calyx open, shortly 5 toothed; petals puberulous; legumes straight, subterete with prominent sutures, narrow, acute at each end, glabrous, many seeded. *Zey.* 496.

HAB. At the 24 Rivers, Gelustwaard, *Zeyher.* Dec. (Herb. Sd., Hk., Bth.)

1–2 feet high or more, woody at base, herbaceous upwards, ascending-suberect; branches curved, finely striate, not angular. Common petiole 2–2½ inches long, the first pair of leaflets an inch or more from the base, the others ¾–1 inch apart. Leaflets nearly inch long and about half inch wide, dark when dry, the under surface pale blueish. Stipules ⅓ inch long, 2 lines wide, withering. Peduncles 6–8 inches long; pedicels 2–2½ lines long, erect, 1–2 lines apart. Bracts broadly ovate or subrotund-acuminate, falling off on the opening of the buds. Calyx ½ line long. Petals purple, 5–6 lines long, thinly downy. Legumes at least 2 inches long, 1 line wide, callous pointed, dark brown. A very distinct species, with *stipules* unlike any other of this section. It seems most allied to *I. amoena,* but the leaves are constantly *bijugate,* and there are other differences.

78. I. cytisoides (Thunb. Prod. 133); *shrubby, robust, erect;* branches angular, subcanescent; leaves subsessile, *bijugate;* leaflets stipellate, obovate-oblong, obtuse, mucronate, midribbed, pale beneath, thinly strigilloso-canescent on both sides, the terminal petiolate, rather larger; stipules broadly subulate, stipellæ setaceo-subulate; racemes erect, longer than the leaf, cylindrical, densely flowered, elongating; bracts broadly ovate, deciduous; calyces silky, the lobes subulate, acuminate; petals minutely silky; legumes *erect,* tereti-quadrangular, subtorulose, thinly canescent. *Thunb.! Fl. Cap. p.* 598. *DC. Prod.* 2, *p.* 230. *Jacq. Schoenb. t.* 235. *Bot. Mag. t.* 742. *E. & Z.! No.* 1598.

HAB. Sides of Watercourses, &c. Common round Table Mountain and Hott. Hollandsberg. (Herb. Th., D., &c.)

A tall, strong-growing, woody, much branched and densely leafy shrub; branches very erect and straight, with a brown bark, thinly covered with whitish hairs. Common petiole 1–1½ inch long, the lowest pair of leaflets 1–3 lines from the base. Leaflets generally 5, sometimes but 3, 1–2 inches long, 3–10 lines wide, acute at base, flat, or with slightly recurved margins. Racemes 3–6 inches long, on short peduncles, dense. Flowers purple or pink, 4–5 lines long, erect. Legumes 1½ inch long, subcontracted between the seeds, pale greyish.

79. I. frutescens (Linn. f. Suppl. 334); shrubby, robust, erect; branches terete, the young twigs silky; leaves laxly 2–4–jugate, the common petiole channelled, with a rough, dark gland between each pair of leaflets; leaflets obovate, mucronulate, flat, glaucous, minutely puberulent, becoming glabrous; stipules minute, deciduous; racemes erect, scarcely longer than the leaf, *laxly* several flowered; bracts minute, subulate, deciduous; calyx canescent, *very open, obsoletely and bluntly 5–toothed;* petals silky; legumes spreading or pendulous, terete, corrugated, glabrous, many-seeded. *Thunb.! Fl. Cap. p.* 598. *DC. Prod.* 2, *p.* 226. *E. & Z.!* 1597. *E. Mey. Comm. p.* 103.

HAB. Among shrubs on mountain sides. Near Tulbagh, Heerelogement, and Olifant's River, Worcest., and Clanw., *E. & Z.!* Barnskloof, *H. Hutton!* Drakensteenberg, Dutoits' Kloof, Cederberg and the Giftberg, *Drege!* (Herb. Th., Bth., Hk., D., Sd.)

A tall, strong, leafy shrub; the younger parts microscopically silky, the older becoming glabrous, and pale, blueish-green. Common petiole 2-4 inches long, the pairs of leaflets $\frac{1}{2}$–$\frac{3}{4}$ inch apart, the terminal at least $\frac{1}{4}$ inch beyond the last pair. Leaflets $\frac{1}{2}$–$\frac{3}{4}$ inch long, 4–6 lines wide, thickish, flat, with wholly immersed or obsolete veins. Racemes elongating, the flowers 2–3 lines apart on pedicels 2–3 lines long. Calyx remarkably open and subtruncate, repand-toothed. Legumes 1$\frac{1}{2}$–1$\frac{3}{4}$ inch long, somewhat 4-angled, with very convex valves, spongy and chambered within, wrinkled.

80. I. cylindrica (DC. Prod 2, p. 225); suffruticose, erect; branches angular, hispidulous; leaves laxly 4–5-jugate, the common petiole slightly channelled, stipellate; leaflets obovate, emarginate, glabrescent above, paler and minutely strigillose beneath, the terminal petioled; stipules small, subulate; racemes on long peduncles, *densely many-flowered,* elongating; bracts minute, subulate; calyces puberulous, very open, *shortly and bluntly 5-toothed;* petals puberulous; legumes spreading, straight, cylindrical, 8–10 seeded, glabrous, the sutures not prominent. *E. Mey.! Comm. Drege, p.* 103. *Burch.!* 6954.

HAB. Banks of the Basche River, *Drege!* Port Natal, *Gueinzius!* 289. (Herb. Bth., Hk., D., Sd.)

A slender, half-woody, virgate suffrutex, rather laxly leafy. Common petiole 3–4 inches long, patent, the pairs of leaflets $\frac{1}{2}$ inch apart. Stipellæ generally solitary at the base of each leaflet, subulate. Leaflets about $\frac{1}{2}$ inch long, 4 lines wide, either quite glabrous above or with a very few minute appressed hairs, mostly emarginate, with a very minute mucro. Racemes dense: pedicels 2 lines long. Flowers pink or purple, 2–3 lines long. A much more slender plant than *I. frutescens,* with different inflorescence, smaller flowers, and different stipules and stipellæ.

81. I. rufescens (E. Mey. Comm. p. 103); "suffruticose; stem and branches erect, glabrescent, rufescent; leaves 8-jugate, short-petioled; leaflets oblong, subobtuse, mucronate, attenuate at base, *the adult quite glabrous,* the terminal petioled; racemes subpaniculate, pedunculate, thrice as long as the leaves, filiform; vexillum externally hairy." *E. Mey. l. c.*

HAB. Natal country, *Drege!* (Unknown to me.)

82. I. oxytropis (Benth.! in Herb.); suffruticose, diffuse, *densely and softly hairy;* branches flexuous, terete; leaves subsessile, 5–8-jugate, the common petiole gland-stipelled; leaflets elliptic-oblong, mucronate, often folded, villous beneath, the terminal petioled; stipules subulate-attenuate; racemes on hirsute, divergent peduncles, longer than the leaf, densely many-flowered; calyx hirsute, its segments setaceo-subulate; petals pubescent; legumes short, terete, densely hirsute, sharp-pointed. *Zey.! No.* 477.

HAB. Rocky places near Aapjes River and Magalisberg, *Burke and Zeyher, Pappe,* 26. (Herb. Bth., Hk., D., Sd.)

Procumbent or depressed, much-branched, with copious, soft, pale pubescence. Leaves 1$\frac{1}{2}$ inch long, of several pair, the lowest pair near the base of the common petiole. Leaflets 4–5 lines long, 2–3 lines wide, green and nearly glabrous above, fulvous and softly hairy beneath. Peduncles 2–3 inches long, the upper half floriferous. Flowers purple, 3 lines long. Legumes about 8 seeded, $\frac{1}{2}$ inch long, or rather more, with a hard, sharp, black point.

83. I. hirsuta (Linn. Sp. 1062); herbaceous, erect, densely and softly hairy; branches angular; leaves short-petioled, 4–6-jugate, the common petiole gland-stipelled; leaflets elliptic-oblong, subobtuse, mucronulate, villous on both sides, the terminal petioled; stipules setaceo-subulate, long; racemes subspicate, on long, hirsute peduncles, densely many-flowered; bracts minute; calyces hirsute, the segments subulate-acuminate; legumes short, turgid, mucronate, deflexed, densely hirsute, few-seeded. *DC. Prod.* 2, *p.* 228. *Burm. Zeyl. t.* 14. *Lam. Ill. t.* 626. *I. astragalina, DC. l. c.*

HAB. Mohlamba Range, Natal, 5–6000f., *Dr. Sutherland.* (Herb. Hook.)

Stem one or two feet high; the whole plant densely clothed with long, soft, pale hairs. Leaves 1½–2 inches long, closely set. Leaflets 4–6 lines long, 2–3 lines wide. Flowers small, closely set, subsessile, on peduncles 4–6 inches long. Legumes strongly deflexed, imbricating, very hairy with dark or blackish hairs. A common plant in tropical Asia and Africa.

84. I. eriocarpa (E. Mey.! Comm. p. 103); shrubby, erect, hoary, and hirsute; leaves 5–6-jugate, the common petiole filiform, gland-stipelled; leaflets petiolulate, elliptic-oblong or sublanceolate, mucronate, thinly canescent, the terminal petiolate; stipules subulate-acuminate, long; spikes on long, rigid peduncles, densely many-flowered; bracts minute, setaceous; calyces hirsute, the segments long, setaceo-subulate and ciliate; petals silky; legumes crowded, horizontal, *short, turgid, and densely cinereo-hirsute.*

VAR. β. **Williamsoni**; more glabrous; branches villoso-pubescent; leaflets obovate-oblong, glabrous above, thinly silky-subcanescent beneath; calyx pubescent, the segments subulate-acuminate.

HAB. Hills round Port Natal, *Drege, Gueinzius.* Coast land near Natal, 1000f., *Dr. Sutherland*, (Herb. Bth., Hk., D., Sd.)

1–3 feet high, suffruticose or quite shrubby, robust and much-branched: the whole plant clothed with spreading, whitish, rather rough, short hairs. Leaves densely crowded on the more shrubby specimens, patent; the common petiole 2½–3 inches long. Leaflets horizontal, variable in width and in the bluntness and acuteness of the ends, about ½ inch long. Spikes nearly twice as long as the leaves, cylindrical, hairy. Flowers purple, 3–4 lines long. Legumes not ½ inch long, 3–4 seeded. Var. β. may possibly be a species, but no perfect specimens of it have yet been seen; it grows about Port Natal, and is in Herb. T.C.D. and Benth.

85. I. velutina (E. Mey.! Comm. p. 104); suffruticose, *decumbent, softly pubescent;* leaves 6–8-jugate, the common petiole filiform, gland-stipelled; leaflets elliptical, mucronulate, thinly pilose or glabrous above, pubescent or villous beneath, the terminal petioled; stipules long, subulate-acuminate; racemes subspicate, elongating, on long peduncles; calyx hirsute, its lobes acuminate; petals downy; ovary white-hairy; legumes? *Krauss!* 373.

HAB. At the mouth of a small stream not far from the Omsamcaba, *Drege.* Coast land near Natal. *Dr. Krauss, Gueinzius.* (Herb. Bth., Hk., D., Sd.)

Very like *J. eriocarpa*, but the stems are procumbent and scarcely suffruticose, 6–12 inches long, many rising from a woody root. Petioles 2–3 inches long, the pairs of leaflets 2–4 lines apart. Leaflets exactly oval, 3–5 lines long, 2½–3 lines wide. Peduncle 2–5 inches long.

86. I. grata (E. Mey.! Comm. p. 103); shrubby, much-branched, "flexuoso-erect," branches curved, thinly canescent; leaves 5-jugate, the common petiole filiform, minutely gland-stipelled; leaflets linear-oblong, mucronulate, strigilloso-canescent on each side, the terminal petiolate; stipules subulate, patent; racemes *shorter than the leaves, laxly few-flowered*; bracts small, lanceolate; calyces canescent, the segments setaceo-subulate; petals puberulous; legumes . . . . ?

HAB. Mouth of the Omsamculo, *Drege.* (Herb. Hk., Bth.)

A woody and twiggy, pale grayish small shrub. Leaves densely set. Common petiole 1–1½ inch long, the pairs of leaflets 2 lines apart. Stipules 1–2 lines long, remarkably patent. Leaflets 3–4 lines long, 1½ line wide, greyish. Racemes subsessile, 8–10 flowered; scarcely an inch long.

87. I. melanadenia (Benth.! in Herb.); suffruticose, diffuse, much-branched; branches flexuous, patent, *tomentoso-canescent;* leaves *horizontal,* subsessile, 3–4–jugate, the common petiole gland-stipelled; leaflets elliptic-oblong, subcomplicate, silky-canescent beneath, the terminal petioled; stipules subulate, deciduous, *the axil multi-glandular;* racemes shortly pedunculate, deflexed, little longer than the leaf, pluri-flowered; calyx hirsute, its segments subulate; petals . . ?; legumes very short, turgid, villoso-hirsute, 2–3 seeded. *Zey.! No.* 481.

HAB. Doornkop, Betchuanaland, *Burke & Zeyher*, *Pappe*, 27. (Herb. Hk., Bth., D., Sd.)

Apparently prostrate, the branches spreading to all sides and angularly bent; the whole plant more or less canescent with soft hairs. Leaves remarkably patent, uncial, with a pulvinate cluster of dark-brown glands in place of stipules, and smaller tufts of similar glands as stipellæ between the leaflets. Leaflets 2–3 lines long, 1½ wide, generally folded together. Flowers not seen. The fruit peduncles are as long or twice as long as the leaves. Legumes 3–4 lines long, very turgid and hairy; like those of *I. eriocarpa.*

88. I. varia (E. Mey.! Comm. p. 104); suffruticulose, effuse, very slender, piloso-strigose; leaves 2–4 jugate, short-petioled; leaflets oblong, mucronate, thinly pilose, the terminal petioled, the lower smaller; peduncles capillary, longer than the leaf, *distantly few-flowered;* flowers minute; calyx-lobes subulate; petals pubescent; legume subcompressed, straight, 5–6 seeded, pilose.

HAB. Basche River, *Drege.* (Herb. Sond.)

Only known to me by a fragment in Herb. Sonder.

89. I. cryptantha (Benth.! in Herb.); shrubby, erect, much-branched, thinly strigillose; branches angular and furrowed; leaves petiolate, 5–8 jugate; leaflets linear-sublanceolate (rarely subalternate), or narrow-oblong obovate, acute or subobtuse, often folded, microscopically strigillose beneath, the terminal petioled; stipules minute or obsolete; racemes subsessile, shorter than the leaf, laxly several-flowered; bracts subulate, flowers small; calyx teeth *short;* petals downy; legumes *moniliform-torulose,* much constricted between the seeds, 4–5 seeded.

HAB. Crocodile River, *Burke & Zeyher!* 473. Coast land, Natal, *Dr. Sutherland.* (Herb. Hk., Bth., Sd., D.)

A much-branched, leafy, erect shrub, turning dark in drying. Leaves 1½–2 inches long; the pairs of leaflets 2 lines apart. Leaflets ½–¾ inch long, 1–2 lines wide. Flowers small, hidden among the dense leaves. Legumes almost jointed, glabrescent. Allied to *I. tinctoria.*

90. I. micrantha (E. Mey.! Comm. p. 104); shrubby, erect, much-branched, subglabrous; twigs terete, pale; leaves 4-5 jugate, short-petioled, leaflets *broadly obovate, obtuse, thin, pale, and microscopically strigillose beneath,* the terminal petioled and largest; stipules *obsolete;* racemes very slender, laxly few-flowered, scarcely longer than the leaves; *bracts and flowers very minute;* calyx glabrescent, shortly 5-toothed; petals puberulous; legumes subcompressed, straight, glabrous, many-seeded.

HAB. Between Omtendo and Omsamculo, at the edges of the wood, *Drege.* Port Natal, *Dr. Sutherland, Gueinzius.* (Herb. Bth., Hk., D.)

A slender, but woody shrub, with many branches, and filiform, curved twigs, with a pale bark, nearly glabrous in all parts. Common petiole 1–1½ inch long, minutely gland-stipelled; the pairs of leaflets 2–3 lines apart. Leaflets 3–5 lines long, 2–4 lines wide, the terminal largest, the rest successively smaller, the lowest not 2 lines long, dark-green above, glaucous-grey beneath. Racemes 1–2 inches long, very slender. Bracts ¼ line long, subulate, subpersistent. Flowers 1 line long. Legumes nearly inch long, 1 line wide, black, linear.

91. I. pauciflora (E. & Z.! 1618); suffruticose, slender, (diffuse?) thinly strigillose; branches terete; leaves subsessile, 3–4-jugate; leaflets (small) elliptic-oblong, mucronulate, thinly substrigillose above, rigidly strigose beneath, the terminal petioled; stipules setaceo-subulate, long; racemes on filiform peduncles scarcely longer than the leaves, laxly few-flowered; bracts minute; calyx segments setaceo-subulate; petals puberulous; legumes short, *curved,* subcompressed, acute, strigilloso-canescent.

HAB. Among shrubs on the mounts. near Eland's River, Uit., *E. & Z.!* (Herb. Sd.)

I have only seen two small branches, in bad preservation. The whole plant is thinly, but rigidly, *strigose,* with appressed, white, middle-fixed hairs. Common petiole scarcely inch long; the pair of leaflets 2 lines apart. Leaflets 2½–3 lines long, 1–1½ line wide, flat. Raceme 5–8 flowered, 1–1½ inch long: peduncle setaceous. Flowers 2 lines long. Unripe legume 4 lines long, ½ line wide.

92. I. seticulosa (Harv.); suffruticose, slender, erect, *whitish,* the filiform branches and twigs, peduncles, and petioles *covered with short, gland-tipped, rigid, horizontally-patent bristles;* leaves petiolate, 3–4-jugate; leaflets (*minute*) oblong, pubescent above, rigidly strigose beneath, the terminal petioled; stipules setaceo-subulate; racemes on filiform, patent peduncles, equalling the leaves, laxly few-flowered; bracts minute; calyx strigose, its segments shortly subulate; petals puberulous; legumes terete, subtorulose, straight, acute, 5–8 seeded, thinly strigillose *and sprinkled with glandular bristles.*

HAB. Uncertain, *Armstrong.* (Herb. Hooker.)

Apparently an erect, very slender, much-branched plant, densely sprinkled in most parts with rigid, glandular setæ, ¼ line in length: the bark pale or whitish. Common petiole ¾ inch long. Leaflets 1–2 lines long, ½ line wide. Racemes scarcely inch long. Flowers 2 lines long. Legumes ½–¾ inch long, brown, with white appressed hairs and erect bristles. The habitat of this is quite uncertain, and possibly it may not be S. African. In its remarkable pubescence it is allied to *I. heterotricha* and *I. sordida,* but is very different in other respects.

93. I. adenocarpa (E. Mey.! Comm. p. 105); suffruticose, erect, *white with appressed, rigid hairs,* much-branched, the twigs spreading, rigid; leaves long-petioled, 3–4-jugate; leaflets *obovate,* submucronulate, often complicate, subdistant, the terminal petioled; stipules subulate; racemes shortly peduncled, longer than the leaves, patent, laxly many-flowered; the peduncle becoming rigid after flowering; bracts subulate, flowers subsessile; calyx-lobes subulate; petals hairy; legumes terete, straight, strigilloso-canescent, gland-dotted, spreading or deflexed.

HAB. Hills by the Gariep, 300–800f., *Drege.* (Herb. Hk., Bth., Sd.)

The whole plant is very white with rigid strigillose pubescence. Branches spreading widely. Common petiole 1½ inch long, with two reflexed, broadly-subulate stipules at base. Leaflets 3–4 lines long, about 2 lines wide. Racemes in flower 3, in fruit 5 inches long. Flowers 2 lines long. Legumes 1–1¼ inch long, scarcely 1 line wide, slightly constricted between the seeds.

94. I. Burkeana (Benth.! in Herb.); suffruticose, erect, rigid, *strigoso-canescent;* branches ribbed and furrowed; leaves subsessile, 3–4 jugate; leaflets linear-oblong or subcuneate, subacute, wavy, silky-canescent on both sides, the terminal petioled; stipules small, subulate; racemes on long, rigid, furrowed peduncles, subspicate, densely many-flowered; calyx-segments subulate; petals glabrous; legumes terete, with prominent sutures, straight, deflexed, thinly canescent, many-seeded. *Zey.!* 476.

HAB. Magalisberg, *Burke* & *Zeyher!* (Herb. Hk., Bth., Sd.)

Many-stemmed, 12–14 inches high, chiefly branched from the base; branches erect and subsimple. Leaves about an inch apart; leaflets ½–¾ inch long, 1–2 lines wide, the terminal shortly petioled. Peduncles 5–6 inches long, after flowering becoming very rigid. Flowers closely crowded, pinky-purple, 3 lines long. Legumes 1–1¼ inch long, slightly rugulose, 12–14 seeded. The whole plant is very pale, whitish-green, harsh to the touch and rigid.

95. I. macra (E. Mey.! Comm. p. 105); suffruticose, erect, *much-branched,* thinly strigillose; twigs terete; leaves 3–4-jugate, leaflets linear-sublanceolate, subobtuse, strigillose beneath, the terminal short-petioled; stipules minute, recurved, subulate; racemes *on long peduncles,* laxly subspicate; bracts minute; calyx shortly 5-toothed; petals glabrous; legumes terete, with prominent sutures, straightish, minutely strigillose, 16–20 seeded.

HAB. Banks of the River Basche, *Drege!* (Herb. Bth., Sd,)

Woody below, slender, laxly-branched, with subdistant leaves. Common petiole 1–1½ inch long, the pairs of leaflets about 3 lines apart. Leaflets ½ inch long, 1 line wide, very pale green. Racemes 3–4 inches long. Flowers small and pale. Legumes 1½–1¾ inch long, pale ochraceous. This has much of the general aspect of *I. parviflora,* but differs in inflorescence, calyx, &c.

96. I. disticha (E. & Z.! 1623); suffruticose, diffuse, thinly strigillose, subcanescent; branches flexuous, distichous, angular; leaves short-petioled, 2–3-jugate, leaflets elliptic-oblong or linear-oblong, obtuse, mucronulate, flat, one-nerved, thinly strigilloso-canescent on both sides, the terminal petiolate; stipules minute, subulate; racemes spicate on long peduncles, gradually elongating, and laxly many-flowered; flowers subsessile, bracts minute; calyx canescent, its segments *short,* acute; petals externally pubescent; legumes subterete, bluntly-angular, straight, very acute, thinly strigilloso-canescent. *I. Enonensis, E. M. Comm. p.* 105.

HAB. Among shrubs on mountain sides. Bothasberg, near Hermanskraal and at the Fish River, *E. & Z.!* Albany, *T. Williamson! Dr. Alexander Prior! Mrs. F. W. Barber!* Near Enon, *Drege.* (Herb. Sd., Bth., Hk., D.)

Scarcely erect, ascending or trailing, very laxly branched, slender; the branches long and subsimple, *distichous.* Pubescence very short, thinly and equally spread.

Leaves mostly bijugate, sometimes trijugate, the common petiole scarcely uncial; the pairs 2–3 lines apart, the terminal leaflet 1½ line distant. Leaflets 7–10 lines long, 2–4 lines wide, elliptical or oblong. Peduncles at first not twice as long as the leaf, afterwards much longer. Unopened buds silvery. Legumes 1–1½ inch long, with a sharp, hard point.

97. I. parviflora (Heyne, Wall. Cat. 5457); *annual*, erect, much-branched, thinly strigillose; branches angular and furrowed; leaves 2–4-jugate, leaflets *linear-lanceolate*, mucronulate, strigillose beneath, the terminal petioled; stipules small, subulate; racemes *sessile*, spicate, shorter than the leaf; bracts minute; calyx-segments subulate; petals glabrous; legumes subterete, with prominent sutures, falcate at the point, minutely strigillose, 16–20 seeded. *W. & A. Prod. Ind. Or.* 1. *p*' 201. *I. deflexa, Hochst. Pl. Kotschy, Nub. No.* 14. *Schimp. Abyss.* 1467

Hab. Vetrivier, *Burke & Zeyher!* 475. (Herb. Hk., Bth., Sd.)

A tall annual, 2 feet or more high, woody at base, many-stemmed, branching, pale yellow-green, glabrescent or very thinly covered with rigid, appressed strigæ. Leaves an inch or more apart; petiole 1–2 inches long. Leaflets 4–5 lines apart, ½–¾ inch long, 1–2 lines wide, very pale. Racemes sometimes very short and few-flowered; sometimes longer, always sessile. Flowers small. Legumes pale-ochraceous, thin in substance, with very delicate partitions between the seeds, the extremity usually curved upwards, acute. A native of India and North Africa.

98. I. filipes (Benth. in Herb.); annual, slender, erect, branching, thinly strigillose; branches angular and furrowed; leaves subsessile, 3–4 jugate; leaflets linear-lanceolate, acute, strigillose beneath, the terminal petioled; stipules small, subulate; racemes on *long, filiform peduncles*, laxly few flowered; bracts subulate, minute; calyx-segments subulate; petals thinly downy; legumes ?

Hab. Caledon River, *Burke, Zeyher!* 472. Zooloo Country, *Miss Owen!* (Herb. Hk., Bth., Sd., D.)

Very slender, 12–18 inches high, with many erect, laxly leafy branches. Leaves an inch apart; the common petiole about an inch long, bearing the first pair of leaflets near its base, the other pairs 3–4 lines apart. Leaflets ¾ inch long, not a line wide, acute at each end. Peduncles setaceous, 2–3 inches long, with a very few small flowers near the apex; pedicels 1–2 lines long. This has the habit of *I. macra* and *I. parviflora*.

99. I. declinata (E. Mey. Comm. p. 104); suffruticose, decumbent, sparsely strigillose; branches angular; leaves 5–7 jugate, the common petiole channelled, exstipellate; leaflets obovate-oblong, mucronulate, glabrous above, thinly strigillose beneath, the terminal *shortly* petioled; stipules subulate; racemes on long peduncles, elongating, rather lax; bracts very minute; calyx-segments subulate; petals pubescent; legumes *compressed, subfalcate, glabrous*, pendulous.

Hab. Rocky mountain places between Keurebooms River and the Langekloof, *Drege!* (Herb. Bth.)

I have only seen a small branch of *Drege's* plant. Petioles 1½ inch long, the pairs of leaflets 1½ lines apart. Leaflets 3–4 lines long, 1½ line wide. Peduncles 3–4 times as long as the leaf; pedicels scarcely 1 line long, nodding. Petals not seen. Calyces white hairy. Unripe legumes nearly an inch long, ½ line wide, curved upwards at the point. *Burchell's* No. 7907, without flowers, seems to belong to this species.

100. I. humifusa (E. & Z.! 1622); suffruticose, diffuse or procumbent, nearly *glabrous*, pale; branches terete; leaves long-petioled, 5–6 jugate; leaflets petiolulate, narrow-obovate, obtuse or acute, glabrous or thinly pilose beneath, the terminal petioled; stipules small, ovate or falcato-lanceolate, recurved; racemes on very long peduncles, laxly many flowered; bracts lanceolate; calyces glabrescent, the teeth subulate; petals minutely downy; legumes terete, straight, pendulous, glabrous. *I. calva, E. Mey. Comm. p.* 104.

HAB. Winterhoeksberg, near Tulbagh, *E. & Z.! Pappe!* (Skurfdeberg, *Zeyher!* 492. (Herb. D., Bth., Hk., Sd.)

A slender, glabrous and very pale, yellowish-green, almost herbaceous species; "*erect*" according to E. Meyer, but the specimen from *Drege*, in Hb. Benth. has the same trailing habit of those of Ecklon and Zeyher, from which our description is more especially drawn. Common petiole 1½–3 inches long, the first pair of leaflets at least ¾ inch from the base, the rest 2–3 lines apart. Leaflets 3–4 lines long, obtuse or acute, sometimes almost lanceolate. Peduncles 6–8 inches long, in fruit erect; pedicels 1–2 lines long. Legumes 1¼ inch long, subtorulose, dark brown.

8. **ALTERNIFOLIÆ** (Sp. 101–107).

101. I. exigua (E. Mey. Comm. p. 108, non E. & Z.); "fruticulose, ascending; leaves pinnate; leaflets 9–16, alternate or irregularly subopposite, *sublinear*, obtuse, sparingly strigillose; fruiting racemes much longer than the leaves, on long, incurved peduncles; legumes subterete, straight, substrigillose, pendulous."—*E. Mey. l. c.*

HAB. Under Bokkeveld, between Waterfall and Grasbergrivier, *Drege.*

102. I. endecaphylla (Jacq. Ic. Rar. t. 570); herbaceous, procumbent or prostrate, thinly *strigillose or subglabrous;* leaves pinnate; leaflets 7–11, alternate, obovate-oblong or linear-oblong, expanded, glaucescent, thinly strigillose beneath (sometimes on both sides); stipules *membranous, dimidiate, oblong acuminate;* racemes spicate, on peduncles shorter than the leaves, densely many flowered; calyx setaceo-subulate; petals glabrous; legumes subquadrangular-terete, slightly constricted between the seeds, 8–10 seeded, pendulous, thinly strigillose. *DC. Prod.* 2. *p.* 228. *I. anceps, Vahl. DC. l. c. Zey.!* 488.

VAR. *β.* **angustata**; leaflets linear-oblong.

HAB. Crocodile River, *Burke & Zeyher!* *β.* Port Natal, *Gueinzius!* 195, 283. (Herb. Hk., Bth., Sd.)

A native also of tropical Africa. Root annual. Stems 6 inches to 2 feet long, glabrous or strigillose. Pubescence sometimes scanty. Leaflets 5–10 lines long, varying much in breadth, 2–4 lines wide; sometimes drying pale, sometimes a dark glaucous green. Flowers small. Legumes 1–1¼ inch long.

103. I. alternans (DC. Prod. 2. p. 229); herbaceous, slender, prostrate, albo-strigose; stems filiform; leaves pinnate; leaflets 7–11, alternate, obovate or oblong, mucronulate, *green and thinly strigillose or glabrous above,* albo-strigose beneath; stipules subulate; racemes spicate, densely many flowered, on *longish or short* peduncles; bracts minute; calyx white, its segments setaceo-subulate; petals hoary; legumes 6–8 seeded, terete, slightly curved or straightish, pendulous, strigose. *Burch. Cat.* 2079, 1961, 1963. *I. effusa and I. arenaria, E. Mey.! Comm. p.* 107. *I. exigua, E. & Z.!* 1620. *I. enneaphylla, E. & Z.!* 1621. *Zey.! No.* 491, 486, 484. *Burch.* 1994, 2079.

HAB. S. Africa, *Burchell.* Los Tafelberg, Colesberg, Veltevrede and Ebenezer, *Drege!* Olifant's River, Clanw., and between Graaf Reynet and Uitenhage, *E. & Z.*

Gamkeriver, Cradock, and Kamos, *Zeyher & Burke!* Queenstown district, and other parts of the frontier, *Mrs. F. W. Barber*, 60. Zooloo Country, *Miss Owen!* (Herb. Sd., Bth., Hk., D.)

Root annual? Stems many from the crown, prostrate, flexuous, slightly branched, 6–10 inches long, rather rigid. Pubescence variable, more or less canescent. Upper surface of the leaflets sometimes quite glabrous, sometimes sparsely strigose, and sometimes *thinly* but equably strigose; never altogether hoary. Peduncles shorter than the leaf, or twice as long. Flowers "pink, shaded with deep crimson, handsome," *E. B.* Legumes 5–7 lines long. The characters attributed to *I. effusa* and *I. arenaria*, E.M. are very variable; even in Drege's original specimens.

104. I. hololeuca (Benth.! in Herb.); herbaceous, slender, procumbent, *wholly albo-strigose;* stems filiform; leaves pinnate; leaflets 7–9, alternate, cuneate-obovate, mucronulate, *densely cano-strigose on both sides;* stipules subulate; racemes spicate, on long peduncles, many flowered, elongating; bracts minute; calyx white, its segments setaceo-subulate; petals hoary; legumes short, 4–5 seeded, subterete, straight, pendulous, canescent. *I. alternans, E. Mey. Comm. p.* 107, *non DC.*

HAB. Gariep, near Verleptpram, 300 f., *Drege!* (Herb. Bth., Hk., D.)

Stems 1–2 feet long, terete, slender, subsimple, or branched near the base, probably prostrate, the habit resembling that of a *Tribulus.* Leaves 1–1½ inch long; leaflets 4–5 lines long, 2–3 lines wide. Flowering peduncles 3, fruiting 5 inches long; flowers small, subsessile. Legumes 4–5 lines long. Every part of the plant is clothed with *dense*, white, appressed, short bristles, by which character and the more diffuse habit it is known from *I. alternans.*

105. I. daleoides (Benth. in Herb.); suffruticose, procumbent, *albo-hirsute;* branches flexuous, *roughly hairy;* leaves pinnate; leaflets 9–15, elliptic-oblong, open or concave, albo-hirsute on both sides; stipules lanceolate; racemes spicate, cylindrical, dense, hoary, on long peduncles; bracts minute; calyx hirsute, its lobes setaceo-subulate; petals *glabrous*, small; legumes (immature) terete, slightly curved, deflexed, several seeded, *albo-hirsute. Burch.! Cat.* 2540. *Zey.!* 479.

HAB. Magalisberg, *Burke and Zeyher! Pappe*, 31. Zooloo Country, *Miss Owen!* (Herb. Hk., Bth., D., Sd.)

Stems robust, 2–3 feet long, spreading widely. Whole plant densely clothed with loosely spreading, longer or shorter, white hairs; the stems, peduncles and petioles roughly villoso-hirsute. Leaves 1–1½ inch long, patent. Leaflets 4–5 lines long, 2–3 wide. Peduncles 2–3 times as long as the leaves; the spike 1–1½ inch long. Calyces very hoary. Unripe legumes an inch long, ¾ line in breadth.

106. I. auricoma (E. Mey. Comm. p. 107); herbaceous, (annual?), procumbent, wholly albo-strigose; stem knee-bent, angular; leaves pinnate; leaflets 4–7, alternate, oval-oblong, acute at each end, densely albo-strigose on both sides, white beneath, yellowish-green above; stipules subulate; racemes subspicate, on short peduncles; bracts minute; calyx segments setaceo-subulate; petals pubescent; legumes subterete, straight, canescent, pendulous.

HAB. Stony hills near Verleptpram, Gariep, *Drege!* (Herb. Hk., Bth., D.)

Root scarcely fibrous, apparently annual. Stems many from the crown, 6–12 inches long or more, subsimple, very flexuous. Petioles 1–1¼ inch long. Leaflets 6–12 lines long, 3–5 lines wide, the terminal largest, acute or subobtuse and mucronate. Peduncle shorter than the leaf; raceme longer, many flowered. Legumes (unripe) 8–9 lines long, ¾ line diameter. This has longer leaflets than any of the allied species: the leaflets are sometimes golden-greenish, and shiny on the upper side; sometimes white above as below.

107. I. argyræa (E. & Z. 1595); suffruticose, prostrate, strigoso-canescent; branches flexuous, ramuli compressed; leaves pinnate; leaflets 3–5, alternate, *obovate or cuneate*, obtuse or mucronulate, *albo-sericeous* on both sides, paler beneath; stipules subulate; racemes nearly *sessile, few flowered;* bracts minute; calyx canescent, its segments subulate; petals *glabrous;* legumes 4–6 seeded, terete, subincurved, deflexed, strigilloso-canescent. *I. Burchellii, E. Mey. Comm. p.* 106. *non DC. Zey.!* 485. *I. collina, E. & Z.!* 1596.

HAB. Stony places on mountain sides, near Silo; at Klipplaat River, and Zwartekey, *E. & Z.* Zwartbulletje and Great and Little Fish River, *Drege!* Near Uitenhage, *Dr. Alexander Prior!* Orange River, *Burke & Zeyher!* Albany, *Mrs. F. W. Barber*. Zooloo Country, *Miss Owen!* (Herb. Sd., Hk., Bth., D.)

Root woody. Stems many, prostrate, spreading every way, flexuous, distichously much branched. Leaves ½–¾ inch long. Leaflets 2–5 lines long, 1–3 wide, pale green above, silvery white beneath, open or folded. Racemes usually quite sessile, rarely minutely pedunculate, 6–8 flowered. Flowers pink. Legumes 3–6 lines long.

Subgenus II. **Amecarpus.** *Legume* plano-compressed, falcate or circinnate, the flat valves netted with veins, submembranous. (Sp. 108-114.)

108. I. *(Amecarpus)* **patens** (E. & Z.! 1580); shrubby, rigid, divaricately much branched, the twigs thinly strigoso-canescent, at length spinescent; leaves sessile, trifoliolate; leaflets obovate-oblong or subcuneate, obtuse or mucronate, often complicate, pale and thinly strigillose beneath; stipules subulate; racemes much *longer than the leaves*, many flowered, at length rigid and spinescent; calyx thinly strigose, its segments *shortly subulate;* petals pubescent; legumes compressed, falcate, 4–5 seeded, netted with veins, thinly pubescent, often contracted between the seeds. *I. melolobioides, Benth.*

HAB. Near Silo, Klipplaat River, *E. & Z.!* Brandkraal, Grahamstown, *Zeyher!* Sand River, *Burke & Zeyher!* 499. Dec. (Herb. Hk., Bth., D., Sd.)

This seems chiefly to differ from *I. sessilifolia*, by the longer and fuller racemes, and the more taper-pointed calyx segments. It may possibly be the same as Meyer's *I. falcata, β. pubescens.* It is more slender than *I. sessilifolia.* Leaflets 2–4 lines long, 1–2 lines wide, often folded together, pale. Raceme finally 2–2½ inches long. Flowers small. Legume 1½ inch long, 2 lines wide, strongly falcate.

109. I. *(Amecarpus ?)* **falcata** (E. Mey. Comm. p. 93); "shrubby; branches flexuous, subrecurved; leaves subfasciculate, trifoliolate, on very short petioles; leaflets obovate-cuneate, subretuse; racemes *much longer than the leaf* (and the twigs) *patent, straight, at length spinescent;* petals pubescent; legumes compressed, linear, subtorulose, falcate, shining." *E. Mey. l. c.*

VAR. *α.* **glaberrima**; "whole plant, save the petals, quite glabrous." *E. Mey.*

VAR. *β.* **pubescens**; "all parts, save the upper surface of the leaflets, strigilloso-pubescent." *E. Mey.*

HAB. Dry stony hills, round Platdrift. *β.* same place, and near Bitterwater, Rhinosterkopje and Klein Bruintjeshoogte, *Drege.*

Of this I have only seen a fragment of var. *α.* in Hb. Benth., wanting both flowers and fruit.

110. I. *(Amecarpus)* **hedranophylla** (E. & Z.! 1560); shrubby, the branches flexuous, the younger minutely pubescent, at length spiny; leaves subsessile; leaflets obovate, emarginate, coriaceous, nearly glabrous; racemes *abbreviate;* flowers? legumes compressed, falcate, 4–5 seeded, netted with veins, *glabrous,* sometimes sinuous between the seeds.

HAB. Among shrubs at Korabshoogde, and near Fort Beaufort, Brit. Kaffraria, *E. & Z.!* (Herb. Sd.)

A very imperfect, almost leafless specimen only exists in Hb. Sond. More perfect specimens are required to establish the species.

111. I. *(Amecarpus)* **sessilifolia** (DC. Prod. 2, p. 231); shrubby, rigid, divaricately much branched; *twigs canescent,* at length spinous; leaves *sessile,* trifoliolate; leaflets cuneate-obovate, emarginate, *thinly silky on one or both sides;* racemes subsessile, few flowered, as long or twice as long as the leaves; calyx 5-toothed; petals canescent; legumes compressed, falcate, 4–5 seeded, netted with veins, thinly pubescent. *E. & Z.!* 1559. *Zey.* 490.

HAB. Mountains round Uitenhage, *E. & Z.!* Gamke Riv., *Zey.!* Dwaka Riv., *Burke.* (Herb. Sd., Bth., Hk., D.)

A very scraggy, intricately branched, robust, spiny and sparsely leafy bush; the the younger parts canescent with short, silky, appressed hairs. Leaflets 2–4 lines long, the upper surface, except of the young leaflets, mostly glabrous. Racemes in fruit sometimes uncial, in flower much shorter. Flowers small. Legumes 4–8 lines long, 2 lines wide, strongly falcate.

112. I. *(Amecarpus)* **circinnata** (Benth. in Herb.); shrubby, rigid, *very spiny;* branches *villoso-canescent,* virgate; spines axillary, horizontal; leaves sessile, trifoliolate; leaflets (small) oblong-obovate, obtuse or mucronulate, strigoso-canescent; stipules setaceous; racemes very short, about two-flowered, the *rachis sharply spinous;* calyx segments short, acute; petals pubescent; legumes compressed, *circularly inflexed,* with thick sutures, netted with veins, 3–4 seeded, thinly strigose.

HAB. Magalisberg, *Burke & Zeyher!* (Herb. Hk., Bth., D.)

A small, canescent bush, bristling with patent, axillary spines—(abortive ramuli and old, denuded rachides).—Spines fulvous, glabrous, ¾–1 inch long. Leaflets 3–4 lines long, about 2 lines wide. Flowers small. Legumes curved round spirally in a circle, till the apex touches or overlaps the base, pale brown. A very distinct species.

113. I. *(Amecarpus)* **torulosa** (E. Mey. Comm. p. 105); suffruticose, effuse, thinly strigoso-canescent; leaves pinnate, 2-jugate, petiolate; leaflets oblong, mucronulate, strigoso-canescent beneath, the terminal petioled; stipules subulate; racemes subspicate, subsessile, in flower equalling, in fruit exceeding the leaf; flowers minute; calyx-lobes setaceo-subulate; petals pubescent; legumes compressed, subfalcate, canescent, 3 seeded, swollen at the seeds.

HAB. Banks of the Basche River, and between Omtata and Omsamwubo, *Drege!* (Herb. Bth., Sd.)

Stem slender, subdistichously branched, probably prostrate. Common petiole 1¼ inch long, the lowest pair of leaflets ⅛ inch from its base. Leaflets 4–6 lines long, 1½–2 lines wide, pale green and thinly strigulose above, whitish beneath. Racemes 1–2 inches long, dense. Legumes 4–5 lines long, curved upwards. Nearly allied to *I. senegalensis,* but scarcely the same.

114. I. *(Amecarpus)* **rhytidocarpa** (Benth. in Herb.); *annual,* diffuse, strigoso-setose; leaves pinnately trifoliolate, petiolate; leaflets linear-lanceolate, longer than the petiole, acute, flat, midribbed, setoso-strigose, the terminal petioled; stipules subulate; racemes spicate, subsessile, shorter than the leaf, several flowered; calyx segments setaceo-subulate, longer than the pubescent vexillum; legumes compressed, falcate, *strongly wrinkled*, 4–5 seeded, hispid.

HAB. Thaba Uncka and Vet Rivier, *Burke & Zeyher!* (Herb. Hk., Bth.)

Root subsimple, with few fibres. Stem much branched from near the base, the branches angular, subsimple, spreading, 6–12 inches long, coarsely strigose. Petioles 4–5 lines long. Leaflets 1½ inches long or more, 1–2 lines wide, strigose on both sides, pale green. Flowers small, hidden among the leaves. Legumes ¾ inch long, 1½ line wide, compressed between the seeds, curved upwards, pale.

XXXII. TEPHROSIA, Pers.

Calyx ebracteolate, campanulate, subequally 5-toothed or cleft. *Vexillum* suborbicular, large, patent, silky or villous externally; *alæ* adhering to the *carina*. *Stamens* monadelphous or diadelphous. *Ovary* multi-ovulate; *style* filiform, glabrous or bearded. *Legume* linear, compressed, coriaceous, straight or curved, sessile or stipitate, continuous or with partitions between the seeds: *seeds* compressed. *Endl. Gen.* 6539. *DC. Prod.* 2. *p.* 248. *Apodynomene, E. Mey.! Endl. No.* 6538.

Trees, shrubs, suffrutices or herbs common throughout the tropics and subtropical regions of both hemispheres. Leaves imparipinnate or digitate, rarely unifoliolate. Stipules free. Stipellæ none. Flowers racemose, on terminal, axillary, or lateral peduncles, red, purple or white. Name from τεφρος, *ashen;* because many of the species have a grey or silvery pubescence. The Cape species are conveniently grouped under two sections or *subgenera*, readily distinguished by their stipules and bracts.

ANALYSIS OF THE SOUTH AFRICAN SPECIES.

Section 1. **Eu-Tephrosia.** *Stipules* 1–3 nerved, subulate. *Bracts* subulate, small, *persistent.* (Sp. 1–16.)

Leaves digitate ... ... (1) **lupinifolia.**
Leaves pinnate: *style* glabrous.
Leaflets in several pairs.
Shrubs, or rigidly ligneous:
Lfts. oblong-lanceolate, pale silky ... ... (2) **suberosa.**
Lfts. narrow-linear, green, pale beneath ... (8) **Kraussiana.**
Suffrutices: stem *erect* or ascending:
Ped. *very short*, scarcely any, *axillary* ... (4) **stricta.**
Ped. *short, opposite* the leaves; whole plant *silky* ... ... (3) **canescens.**
Ped. *longer* than the leaves and opposite them:
Lfts. 2–4 pair, linear-lanceolate, *long-petioled* ... ... (11) **Dregeana.**
Lfts. 5–9 pair, cuneate-oblong, *short-petioled* ... ...
Racemes densely many flowered ... (5) **pallens.**
Racemes interruptedly few flowered (6) **semiglabra.**
Ped. long; terminal and axillary:
Lfts. cuneate-linear, pubescent beneath; stem *slender* ... ... (7) **amoena.**
Lfts. cuneate-oblong, pubescent, pale beneath; stem straight ... ... (9) **polystachya.**
Lfts. *linear*, green above, silvery white beneath ... ... (10) **discolor.**
Suffrutices; stem procumbent or trailing ... ... (12) **Capensis.**
Leaflets in 1–2 pairs, linear-lanceolate (2–4 inches long) ... . (13) **elongata.**

Leaves pinnate ; *style* bearded :
  Leafl. linear-lanceolate, in 2–3 pairs ... ... (14) **lurida.**
  Leafl. linear-lanceolate, in 4–8 pairs ... ... (15) **longipes.**
  Leafl. elliptic-oblong, broad, in 2–5 pairs ... ... (16) **oblongifolia.**

Section 2. **Apodynomene.** *Stipules* ovate, many nerved. *Bracts* ovate, spathaceous, many nerved, enwrapping the young flower, *deciduous.* (Sp. 17–21.)

Erect suffrutices or small shrubs :
  Calyx-lobes broad-based-subulate, taper-pointed ... (17) **grandiflora.**
  Calyx-lobes, short, ovate, subacute ... . (18) **glomeruliflora.**
Procumbent or trailing suffrutices :
  Stamens monadelphous ; style bearded :
    Robust, pubescent ; lfts. broadly elliptic-oblong (19) **macropoda.**
    Slender, glabrous ; lfts. cuneate-oblong ... ... (20) **æmula.**
  Stamens diadelphous ; style glabrous ; fl. small ... (21) **diffusa.**

1. **Eu-Tephrosia.** (Sp. 1–16.)

T. lupinifolia (DC. Prod. 2. p. 255) ; suffruticose, diffuse ; stems, petioles, young leaves and peduncles *fulvo-hirsute ;* leaves on long petioles, *palmately 5-foliolate,* cuneate oblong, obtuse, mucronulate, margined, glabrous on the upper, hairy on the under surface ; stipules short, broadly subulate, ribbed ; peduncles terminal and axillary, elongate, distantly many-flowered ; flowers small, spicato-racemose ; legumes compressed, fulvo-pubescent, about six-seeded ; style glabrous. *Galega lupinifolia, Burch. Cat. No.* 2488. *Zey.! No.* 458.

HAB. S. Africa, *Burchell.* Vaal and Mooje Rivers, *Burke and Zeyher.* (Herb. Hk., Bth., D., Sd.)

Stems 3–4 feet long, patently branched, flexuous, terete, densely clothed with short, patent, rather rigid, foxy hairs, which also invest all the young portions. Peduncles uncial, patent, or divaricate ; leaflets ½ inch long, 4 lines wide, thick and opaque, obliquely nerve-striate, often complicate. Stipules 2–3 lines long. Peduncles 6–12 inches long, racemose, the small shortly pedicellate ; flowers 1½–1 inch apart. Legumes foxy, 1¼–1½ inch long, 1½ line wide. Known from all other S. African species by its digitate leaves.

2. T. suberosa (DC. Prod. 2, p. 249) ; a *shrub,* leaves shortly petiolate, 8–10 jugate ; leaflets oblongo lanceolate or lanceolate, very pale, silky-canescent on both surfaces, subacute, midribbed, the older reticulately veiny ; stipules minute, deltoid ; peduncles terminal and axillary, racemose, shorter than the leaves ; calyx-teeth from a broad base, subulate ; style glabrous ; legumes *fulvo-velutinous,* 8–10 seeded, irregularly constricted between the seeds. *Zey !* 352. *Pappe* 35.

HAB. Magalisberg, and near Aapges R., *Burke and Zeyher !* (Herb. Hk., D., Sd.)

Seemingly a large, strong-growing shrub. Twigs softly tomentose, with thickish, corky, ribstriate bark. Leaves 4–6 inches long ; leaflets 1–1½ inch long, 4–5 lines wide, the young ones densely silky and silvery on both sides, the older becoming less silky, with more obvious venation, but never glabrous. The whole plant very pale yellowish grey. Corolla 4–5 times as long as the small calyx, whose two upper teeth are connate nearly to the tip. Vexillum with a very short, callous claw, broadly oval. Alæ shorter than the obtuse carina. Stamens monadelphous, pubescent. Legume 3 inches long, with thickened sutures, here and there constricted. Quite unlike any other S African species. It occurs also in trop. Africa, and is common in trop. Asia and the Asiatic Archipelago.

3. T. canescens (E. Mey.! Comm. Drege, p. 109) ; suffruticose, erect, the whole plant *densely silky and silvery canescent ;* leaves petiolate, 4–8 jugate; leaflets obovate-oblong, obtuse or emarginate, faintly penni-nerved; stipules and bracts *minute, toothlike;* peduncles opposite the leaves and shorter than them, densely racemose, many-flowered; legumes linear, broadish, 4-6 seeded, silky.

Hab. Sandy places near the Omsamculo, *Drege.* Delagoa Bay, *Forbes!* (Herb. Bth., Hk., D.)

Stem 2-3 feet high, terete, slender, flexuous, not much branched. Leaves distant, 3 inches long, the first pair of leaflets nearly 1 inch from the base of petiole; leaflets petiolulate, ¾ inch long, 3-4 lines wide, densely covered with shining, white hairs, the veins immersed. Racemes 2 inches long, on peduncles of equal length. Flowers purple, 3 lines long. Legumes 1½-2 inches long, 2½ lines wide, strongly compressed, spreading or deflexed. Style glabrous. Vexillum silky, the other petals puberulous.

4. T. stricta (Pers. Ench. 2, p. 329); suffruticose, erect, branching, pubescent; leaves subsessile, 4-5 jugate; leaflets elliptical or oblong, recurved-pointed, glabrous and green above, rusty-pubescent and closely nerve-striate beneath; stipules and bracts subulate; peduncles *very short,* axillary, 2-4 flowered; legumes narrow, rusty-pubescent, about 6-seeded. *DC. Prod. 2, p.* 253. *E. & Z.* 1630. *Indigofera stricta, Linn. f. Suppl. p.* 334. *Jacq. Schoenb. t.* 236.

Hab. Among shrubs near the Vanstaaden River and Olifant's Hoek, Uit., *E. & Z.!* (Herb. Bth., D.)

1-2 feet high, erect or suberect; branches erect, angular and ribbed, rufescent and thinly pubescent. Leaves uncial; leaflets ½-¾ inch long, 2-3 lines wide, blunt or acute, mucronulate, the points generally recurved. Flowers subsessile or on very short peduncles towards the ends of the branches, small, purple. Calyx-teeth subulate. Legumes 1½ inch long.

5. T. pallens (Pers. Ench. p. 329); suffruticose, erect or ascending, pubescent; leaves shortly petiolate, 5-8-9-jugate; leaflets narrow-cuneate-oblong, recurved-pointed, green, striolate and thinly pubescent above, closely penninerved, silky and paler beneath; stipules and bracts subulate; peduncles opposite the leaves and longer than them, rigid, angular, and furrowed; densely racemose; many flowered; legumes narrow, pale and velvetty-pubescent, 6-7 seeded. *DC. Prod. 2, p.* 254. *E. & Z.!* 1631. *T. angulata, E. Mey.! Comm. p.* 109.

Hab. Grassy hills near Olifant's Hoek at Bushman's River, *E. & Z.!* Glenfilling, *Drege!* Albany, *Dr. Alexander Prior!* (Herb. Bth., D., Hk.)

Stem suberect, angularly bent, ribstriate, with internodes 1½-2 inches apart. Leaves very patent or recurved, the lowest pair of leaflets within ¼ inch of base of petiole. Leaflets in several pairs, ¾ inch long, 2-3 lines wide; somewhat thickened at margin. Stipules broadly subulate, ¼ inch long or more. Peduncles strongly ribbed, curved, 4-6 inches long, the upper half bearing flowers. Bracts 2-3 lines long. Flowers 4-5 lines long; the vexillum pubescent, the other petals glabrous. I have not seen ripe legumes.

6. T. semiglabra (Sond.! in Linn. 23, p. 29); suffruticose, pubescent; stem and branches angularly-bent, ascending, *rufo-tomentose below,* subglabrate toward the end; angular and rib-striate; leaves on very short petioles, 5-7-jugate; leaflets cuneate-oblong, recurved-pointed, glabrous above, silky-pubescent and closely nerve-striate beneath; stipules and bracts subulate; peduncles opposite the leaves and much longer, rigid, angular, and furrowed, interruptedly few-flowered; legumes unknown.

Hab. Magalisberg, *Zeyher! No.* 459. (Herb. Sond.)

Stem decumbent at base, 2-3 feet long, flexuous, densely rufo-tomentose; the branches sparsely hairy or subglabrous. Petioles 3–4 lines long; leaflets 10–12 lines long, 2–3 lines wide. Flowers about an inch apart. Compared by Sonder with *T. stricta* and *T. polystachya*; but it seems to me rather intermediate between *T. pallens* and *T. Capensis*, var. β. having the habit and foliage, but not the pubescence or inflorescence, of the former; and having the inflorescence and much of the pubescence of the latter.

7. **T. amoena** (E. Mey.! Comm. p. 109); suffruticulose, slender, erect, branching; leaves shortly petiolate, close-set, patent, 7–9 jugate; leaflets cuneate-linear, obtuse, margined, glabrous above, appressedly pubescent and closely penninerved beneath; stipules subulate; peduncles axillary and terminal, elongate, laxly racemose; pedicels much longer than the subulate bracts; legumes narrow, thinly downy.

HAB. Mouth of the Omsamcaba and on grassy hills near Omtendo, *Drege!* (Herb. Bth., Hk.)

Stems 12 inches high, erect, subfastigiate, ribstriate, with internodes about 1 inch apart or less; leaves recurved. Leaflets ¾ inch long, 1 line wide, very blunt or subtruncate or recurvo-mucronulate. Racemes laxly many flowered; the flowers 3 lines long, in pairs, on pedicels ½ inch long. Allied to *T. pallens*, but much smaller, more slender, less hairy, with narrower leaflets.

8. **T. Kraussiana** (Meisn. in Hook. Lond. Journ. 2, p. 87); *shrubby*, erect, straight, *densely leafy*; leaves shortly petiolate, patent, 7–10 jugate; leaflets narrow-linear, complicate, recurvo-mucronulate, glabrous above, silky-canescent beneath; stipules setaceo-subulate, longer than the petiole; peduncles axillary and terminal, elongate, laxly racemose; pedicels much longer than the subulate bracts; legumes narrow, thinly canescent, 6–7 seeded.

HAB. Tafelberg, Port Natal, *Krauss, No.* 40. Coastland, 1000 f., *Dr. Sutherland.* (Herb. Bth., Hk., D., Sd.)

A rigid, woody, densely leafy shrub, 2-3 feet high; the branches virgate or ramulous, rib-furrowed; all parts thinly covered with short, appressed whitish pubescence. The plant turns dark in drying. Leaves recurved; leaflets 1 inch or more in length; scarcely 1 line wide, acute at each end. Racemes crowded in a spurious panicle toward the end of the branches, 4-5 inches long. Flowers 3 lines long; the vexillum silky, the other petals glabrous. Legumes 1½ inch long, not two lines wide. Allied to *T. amoena*, but much more robust and woody, with narrower and longer leaves, &c.

9. **T. polystachya** (E. Mey.! Comm. Drege, p. 109); suffruticose, erect; stem and branches straight, rib-striate and angular, pubescent or hairy; leaves shortly petiolate, suberect, 5–9 jugate; leaflets subcuneate-oblong, pubescent on one or both sides, paler beneath and closely penninerved, mucronate; stipules subulate; peduncles axillary and terminal, elongate, laxly racemose; bracts subulate; legumes narrow, linear, pubescent or hairy, about 8–seeded.

VAR. β. **latifolia**; leaflets oblong, uncial, retuse, mucronate, 4–5 lines wide.

VAR. γ. **hirta**; stems, branches, and inflorescence *roughly* rusty-pubescent; racemes shorter and more densely flowered than usual, and legumes more hairy.

HAB. Flats and grassy valleys between Gekau and Basche, near Omtata, Omsamwubo, Omsamcaba, and Port Natal, 200–2000 f., *Drege! Williamson!* β. at Port Natal, *Gueinzius, No.* 616. γ. at Port Natal, *Sutherland!* (Herb. Hk., Sd., D.)

Stems 2–3 feet high, slender, strongly furrowed, with close or spreading, yellowish or foxy pubescence; branches and leaves suberect. Leaflets ½–¾ inch long, 2–3 lines wide (in β. longer, broader, thinner, and less hairy), flat or subcomplicate, sometimes recurved at point, the older often glabrate above. Peduncles longer than the leaves and ending in a distantly many-flowered, 4-6 inch long raceme. Flowers 3 lines long, the vexillum densely hairy. Pods 1½ inch long, 1-1½ line wide, straight or slightly curved, acute. β. appears to be a form from moister and more shady situations. In many respects this agrees with *T. stricta,* but differs in inflorescence.

10. T. discolor (E. Mey.! Comm. Drege, p. 111); suffruticose, erect, virgate, canescent; branches angular and rib-striate, straight; leaves very shortly petiolate, 4–6 jugate; leaflets linear, subobtuse, margined, green above, *silky and silvery* beneath; peduncles terminal and axillary, elongate, laxly many flowered, straight; stipules and bracts shortly subulate; legumes narrow-linear, biuncial, fulvous-pubescent, many-seeded.

HAB. Grassy places near Port Natal, *Drege! Williamson! Gueinzius!* (Herb. Bth., Hk., D.)

2-3 feet high, slender, with the habit and inflorescence of *T. polystachya,* but with much narrower and more linear leaflets, white hairy on the under surface. Leaflets uncial, scarcely a line wide, obtuse or mucronulate, of thickish substance, midribbed but not obviously nerved, erect or somewhat patent. Racemes 6–8 inches long, tapering; the flowers nearly an inch apart, 4 lines long. Legume 2–2½ inches long, about a line wide, somewhat tomentose-pubescent and rusty brown.

11. T. Dregeana (E. Mey.! Linn. 7. p. 169); suffruticose, much branched, rigid, flexuoso-erect; branches 4–angled, thinly appresso-puberulous; leaves *on long, rigid petioles,* 2-4 jugate; leaflets *linear-lanceolate,* pale, flat, membranous, penninerved, thinly puberulous underneath; peduncles opposite the leaves, elongate, slender, *straight,* laxly pluri-flowered; stipules and bracts shortly subulate; legumes *pale, curved,* puberulous or glabrate, 3-4 seeded. *E. & Z.! No.* 1634. *T. brachyloba, E. Mey.! Comm. Drege, p.* 110.

HAB. Near Bitterwater, 2400 f., and between Natvoet and the Gariep, 800 f. *Drege! A. Wyley!* (Herb. Sd., Bth., Hk., D.)

Stems woody at base, remarkably rigid, and angularly bent. Foliage very pale, subglabrous. Leaflets 1½-1¾ inch long, 1-1½ line wide, flat, tapering to each end or with a blunt or emarginate apex. Racemes 6–8 inches long. Flowers small, 2 lines long. Legumes somewhat scymetar shaped, ¾-1 inch long, pale-yellowish green.

12. T. Capensis (Pers. Ench. 2, p. 330); suffruticose, *procumbent,* flexuous, much branched, slender, *variably pubescent;* branches angular; leaves on longish petioles, 3-6 jugate; leaflets *elliptical,* cuneate-oblong or *lanceolate,* obtuse or acute, glabrescent or hairy, penninerved; stipules subulate; peduncles opposite the leaves, elongate, slender; raceme *interrupted,* attenuated, distantly pluri-flowered; legumes linear, narrow, minutely puberulous or subglabrate, sometimes pubescent. *E. Mey.! Comm. Drege, p.* 110. *E. & Z.!* 1633.

Var. α. **Jacquini**; leaflets *glabrescent* or thinly puberulous, oblong, obtuse or subacute. *T. Capensis, DC. Prod.* 2. *p.* 252. *Galega Capensis, Thunb. Fl. Cap. p.* 602. *Jacq. E. Rar. t.* 574.

VAR. β. **hirsuta**; leaflets densely pubescent or hairy underneath; oblong, obtuse, or acute. *T. Capensis, β. acutifolia, E. Mey.! l. c.*

VAR. γ. **angustifolia**; leaflets *linear-lanceolate,* acute or acuminate, flat or complicate, very narrow. *E. Mey.! l. c.*

HAB. Common throughout the Colony and in Caffraria. Port Natal, *Gueinzius!* (Herb. Th., D., Sd., &c.)

Stems many from a woody crown, 2–3 feet long, procumbent or trailing. Petiole an inch long to the first pair of leaflets. Leaflets very variable in breadth, size, shape, and amount of pubescence, but usually glabrous on the upper surface. Peduncles 8–10 inches long, about half occupied by the interrupted raceme. Flowers purple, 3 lines long; the vexillum pubescent. Var. β. gathered by Drege at the Zuureberg and at Glenfilling, and by Dr. Pappe in the district of George, has its short pubescence quite patent, and has rather larger flowers than the other forms. The legume is often nearly glabrous, but is sometimes patently pubescent on specimens which have subglabrous leaflets.

13. T. elongata (E. Mey. Comm. Drege, p. 111); suffruticose, ascending, flexuous, *variably pubescent or glabrescent;* branches rib-striate; leaves on longish petioles, 2–1–jugate; leaflets linear-lanceolate, 2–4 uncial, margined, obliquely nerve-striate, the terminal longest; racemes opposite the leaves, elongate, interruptedly pluri-flowered; legumes narrow-linear, 2-2½ uncial, velvetty or subglabrous, many-seeded, somewhat turgid. *Sond. in Linn. vol.* 23, *p.* 30.

VAR. α. **pubescens**; leaves mostly 2-jugate, or the lower unijugate; leaflets, stems, and legumes pubescent. *T. ensifolia, Harv. (olim) in Herb.*

VAR. β. **glabra**; leaves mostly unijugate; leaflets and stems glabrous or minutely puberulent; legumes velvetty or glabrate. *T. coriacea, Benth. in Herb.*

HAB. Between the Omsamculo and Omcomas, *Drege!* Port Natal, *Williamson! Gueinzius! Sanderson!* (28) Vaal River and Magalisberg, *Burke and Zeyher!* (Herb. Hk., Bth., Sd., D.)

Stem 6–12 inches to 2 feet long, subsimple, the shorter ones nearly erect, the longer angularly bent and either ascending or procumbent. Leaflets of the upper leaves 3–4 inches long, and 3–4 lines wide, acute or acuminate; of the lower leaves shorter, broader, and more obtuse. Pubescence variable. Flowers in pairs, about an inch apart, on a long peduncle. Legumes 2 inches or more in length. Var. β. is not always glabrous, and in other respects is so like α, that we consider it best to unite them.

14. T. lurida (Sond. Linn. 23, p. 30); suffruticose, ascending, flexuous; branches quadrangular, thinly canescent; leaves on long petioles, 2–3–jugate; leaflets linear-lanceolate, 3–5–*uncial,* margined, glabrous above, appressedly silky and paler beneath, obliquely nerve-striate; stipules setaceo-subulate, shorter than the petiole; peduncles opposite the leaves and terminal, elongate, 2–6 flowered in an interrupted raceme; legumes narrow-linear, 2½ uncial, fulvo-tomentose; style bearded. *Zey.!* 456.

HAB. Mooi River, Magalisberg, and Crocodile River, *Burke and Zeyher!* (Herb. Sd., Bth., Hk., D.)

Stems many, 12–14 inches high, subsimple, either canescent or thinly and appressedly pubescent, pale, as is also the foliage. Leaflets 5 inches long, 1½ line wide, tapering to each end, variably pubescent. Flowers 5 lines long, in distant pairs towards the extremity of the 6–8 uncial peduncles. Legumes 2–3 inches long, 2 lines wide, compressed and densely tomentose, fulvescent. Chiefly distinguished from *T. longipes* by its longer leaflets in fewer pairs. I fear the limits between the two are rather indefinite.

15. T. longipes (Meisn. in Hook. Lond. Journ. 2. p. 87); suffruticose, erect, thinly silky-canescent or appresso-pubescent; branches quadrangular; leaves on long petioles, 4–8 (10)-*jugate;* leaflets narrow-linear, acute, 2–3–*uncial,* complicate or open, glabrous above, appressedly silky and paler beneath, obliquely nerve-striate; stipules setaceo-subulate, shorter than the petiole; peduncles terminal and opposite the leaves, interruptedly racemose near the summit; legumes narrow-linear, 2–3 uncial, fulvo-tomentose, compressed, many-seeded; style bearded.

VAR. β. **uncinata**; leaflets 2 lines wide, obtuse, recurvo-mucronate. *Zey.!* 455.

HAB. Port Natal, *Krauss!* No. 20. Aapjes River and Macallisberg, *Burke and Zeyher!* Delagoa Bay, *Forbes!* (Herb. Bth., Hk., Sd., D.)

Stems 1–2 feet high, curved at base, then erect, subsimple, pale. Leaves 4–6 inches long; leaflets in several pairs, 1½–2½ inches long, 1–2 lines wide, acute at each end, but not tapering, mucronate and sometimes hook-pointed, as in var. β., which is a stronger growing plant, with broader leaflets. Flowers in pairs, 1–2 inches apart, several on a lengthening raceme. Legumes 3 inches long, 2 lines wide.

16. T. oblongifolia (E. Mey.! Comm. p. 108); "stems procumbent, terete, pubescent, flagelliform; leaves shortly petiolate, 2–5-jugate; leaflets elliptic-oblong, strigoso-pubescent or hairy, netted-veined, prominently ribbed and veined beneath, mucronulate; stipules setaceo-subulate; peduncles axillary, longer than the leaves; racemes elongated, interrupted, the flowers in subdistant pairs, shortly pedicellate; calyx densely hirsute; flowers small and hairy; style bearded; legumes coriaceous, broadish, fulvous-hairy, 3-seeded." *E. Mey. Zey.!* 520.

HAB. Grassy hills at Omsamcaba and near Omtendo, and Port Natal, *Drege!* Macallisberg, *Burke and Zeyher!* Port Natal, *Krauss*, No. 174. (Hb., Bth., Hk.)

Stems very long, trailing and subsimple, either appressedly or patently pubescent, pale or fulvous. Leaflets 1½–2 inches long, ½ line wide, obtuse or acute, flat, with slightly revolute margins, green, appressedly pubescent above, with more copious and looser pubescence beneath, green on both sides; the young leaves densely silky. *Calyx*-lobes of equal length, the four upper ovate-acuminate, the lowest lanceolate. *Corolla* 1½ as long as the calyx; petals hairy. I have not seen legumes. Burke and Zeyher's specimens from Magallisberg are of stronger growth, with more copious pubescence and shorter peduncles than usual.

2. **APODYNOMENE.** (Sp. 17–21.)

17. T. grandiflora (Pers. Ench. 2, p. 329); shrubby, erect, variably pubescent; leaves shortly petiolate, 5-7-jugate; leaflets cuneate-oblong, or linear-oblong, obtuse or acute, retuse or mucronulate, *variably pubescent* on one or both sides; peduncles terminal and opposite the leaves, angular and canescent, *fasciculato-corymbose* at the summit; bracts broadly ovate, deciduous; *calyx-teeth from a broad base subulate;* legume broad, linear, glabrescent, plano-compressed, hispid at the sutures, many seeded; style bearded; vexillum silky. *DC. Prod.* 2. *p.* 251. *E. & Z.! No.* 1629. *Galega grandiflora, Vahl. Symb.* 2, *p.* 84. *Thunb.! Fl. Cap. p.* 602. *G. rosea, Lamk. Apodynomene grandiflora, E. Mey.! Comm. p.* 111

HAB. Frequent among shrubs in the districts of Uitenhage and Albany, and in Caffraria, *E. & Z.! Drege! Pappe, &c.* Paarl, *Rev. W. Elliott!* Coastland, lat. 30° s., Natal, *Dr. Sutherland!* (Herb. Thb., Bth., Hk., Sd., D.)

A rigid shrubby plant, 1–2 feet high, very variable in the amount of pubescence; sometimes subglabrous, sometimes with densely hairy stems and leaflets canescent beneath. Leaflets also very uncertain in length and breadth; sometimes shortly cuneate and almost obcordate, sometimes long and verging to lanceolate. Peduncles shorter or longer than the leaves. Flowers the largest in the genus, 8–10 lines long, red, fulvescent on the outside. The young flowers are enwrapped in very broad ovate bracts, which fall off on the opening of the flower.

18. T. glomeruliflora (Meisn. in Hook. Lond. Journ. 2, p. 86); suffruticose, erect; stems terete, branching, thinly canescent or glabrate; leaves on short petioles, 6-8 (10)-jugate; leaflets *on longish petiolules*, narrow-oblong, obtuse or subacute, mucronate, flat, glabrous on the upper, silky-canescent on the under surface, faintly nerve-striate; stipules lanceolate or ovato-lanceolate; peduncles terminal and opposite the leaves, angular, interruptedly fasciculato-racemose near the summit; bracts ovate, deciduous; *calyx-teeth short, broadly ovate, subacute;* legume broad, stipitate, plano-compressed, glabrous, downy at the sutures, many-seeded; style bearded; vexillum thinly silky.

HAB. Port Natal, *Krauss, Gueinzius! No.* 306. *Sanderson!* (Herb. Sd., Hk.)

1–2 feet high, branching, the stem augularly bent. Pubescence in all parts pale, whitish-grey. Leaves 2½ inches long; the leaflets uncial, 2–2½ lines wide, green above, whitish beneath. Flowers in clusters, on pedicels that lengthen as the flowers advance, and in fruit are nearly ½ inch long. Corolla ½ inch long; carina sharply rostrate. This has something the habit of *T. grandiflora*, but smaller flowers, and very different calyx.

19. T. macropoda (E. Mey.! Comm. p. 112); suffruticose, procumbent, variably pubescent; leaves on long petioles, 2-3-jugate; leaflets *broadly elliptical or oblong*, obtuse or mucronulate, glabrous above, hispid or hirsute or glabrescent beneath, nerve-striate and veiny; peduncles opposite the leaves and terminal, *very long*, racemose or fasciculato-corymbose at the summit, bracts broadly ovate, deciduous; calyx-teeth subulate, alternate; stamens monadelphous; legumes broadly linear, plano-compressed, sub-glabrous, hairy at the sutures. *Apodynomene macropoda, E. Mey. l. c.*

HAB. In grassy places between Kachu and Zandplaat and between Gekau and Basche, &c., *Drege!* Natal, *Krauss! Gueinzius! Sutherland!* &c. In Kreilis country, *H. Bowker*. (Herb. Hk., Bth., Sd., D.)

Stems several, 2–3 feet long, subsimple, lying on the ground. Leaves 1–2 inches apart, on petioles 2–4 inches long. Leaflets 1–1¾ inch long, ¾–1 inch wide, dark green, quite flat, conspicuously veiny. Pubescence copious or scanty. Stipules broadly cordate-ovate, many-ribbed, 3–4 lines long. Peduncles 1–2 feet long, sometimes with a few flowers crowded at the end; sometimes laxly or interruptedly racemose. Bracts broad, involving the young flower. Flowers purple ("yellow," fide *E. M.*), ½ inch long. Legumes biuncial, 3–4 lines wide, sessile. Native name "*Itozane.*" The roots are used by the Zooloo Caffres for stupifying or poisoning fish. The flesh of the prey so captured is eaten without injurious consequences.

20. T. æmula (E. Mey.! Comm. p. 113); subherbaceous, procumbent, *glabrous* or nearly so; stems slender, filiform; leaves on long petioles, 2-3-jugate; leaflets *cuneate oblong or elliptical*, subobtuse, mucronulate, glabrous, faintly penninerved; peduncles terminal and opposite the leaves, very long, few-flowered at the summit; bracts ovate, deciduous; stamens monadelphous; calyx-teeth subulate; legumes linear, glabrous, many-seeded. *Apodynomene, E. Mey. l. c.*

HAB. Between Zandplaat and the Komga, 2500–3000 f., *Drege.* Common in Albany, *Mrs. F. W. Barber! Genl. Bolton!* Port Natal, *Dr. W. B. Grant!* Also collected by *Zeyher!* (Herb. Sd., Hk., D.)

Resembles *T. macropoda* in miniature. The whole plant is more glabrous, the stems more slender, the leaflets and flowers much smaller, and the pods narrower. Leaflets ½–1 inch long, 2–3 lines wide. Flowers pale purple ("yellow," fide *E.M.*) 4 lines long. Legumes 1½ inch long, 2 lines wide.

21. T. diffusa (E. Mey.! Comm. p. 113); suffruticose, procumbent; stem and branches subfiliform; leaves 3-4-jugate, on long petioles; leaflets subcuneate-oblong, appressedly puberulous; stipules ovate, multistriate; racemes opposite the leaves, filiform, few-flowered; vexillary stamen free; bracts deciduous; legumes narrow, subglabrous, 6-seeded. *E. M. l. c. subApodynomene.*

HAB. Rocky places near a small river between the Omsamwubo and Omsamcaba, 1500 f., *Drege!* (Herb. Bth., Sd.)

With the aspect of *T. capensis*, from which this is readily known by its broadly ovate stipules, and ovate, deciduous bracts.

XXXIII. MILLETTIA, W. & A.

Calyx urceolate, bluntly toothed. *Vexillum* recurved, emarginate, rather longer than the *alæ* which are longer than the carina. *Stamens* imperfectly monadelphous, the vexillary stamen free at base. *Legume* elliptical or lanceolate, few seeded, hard and woody, with thickened margins, tardily dehiscent. *W. & A. Prod. p.* 263. *Endl. Gen. No.* 6715.

Trees or large shrubs, natives of the hotter parts of Asia and Africa. Leaves large, abruptly or impari-pinnate; leaflets opposite, stipellate. Racemes or panicles axillary or terminal. Flowers purple or reddish. Named in honour of Dr. Millett, of Canton, China.

ANALYSIS OF THE SOUTH AFRICAN SPECIES.

Leaves 5–6-jugate; vexillum densely silky externally ... ... (1) **Caffra.**
Leaves 2–3-jugate; vexillum glabrous ... ... (2) **Sutherlandi.**

1. **M. caffra** (Meisn. Lond. Journ. 2, p. 99); young parts pubescent; stipellæ setaceous, equalling the petiolule; leaves impari-pinnate, 5–6 *jugate;* leaflets lanceolate-oblong, acute, glabrous above, paler, penninerved and *thinly silky beneath;* panicles fulvo-sericeous, terminal, fasciculato-racemose or branching; calyx thinly silky, *deeply lobed,* two upper lobes connate, three lower elliptic-oblong, very obtuse; vexillum *silky;* legume lanceolate, obtuse, 2-seeded, densely velvetty, brown. *Virgilia grandis, E. Mey. Comm. p.* 1.

HAB. Between Omgaziana and Omsamcaba, and near Port Natal, *Drege.* Port Natal, *Krauss! Plant!* (Herb. Hk., Sd., D.)

A tree, 20–30 feet high, with very hard, close grained brown wood, dark coloured, rugulose bark, and thinly downy or glabrous twigs. Leaves on channelled common petioles 6–8 inches long, the pairs of leaflets an inch apart. Stipellæ 2–3 lines long, very slender. Leaflets on hairy 2-lineal petiolules, 2–2½ inches long, ¾ inch wide, coriaceous, obtuse at base, acute and somewhat mucronate at apex, closely penninerved beneath. Panicle 6–8 inches long, robust, rusty brown; its lateral branches short or long, several flowered. Calyx shortly campanulate, with very broad and blunt lobes. Flowers 7–9 lines long, purple. Legumes coriaceous, very velvetty. The native name, fide *E. Meyer,* is *Omzambeet* (Iron wood); the fruit is a Caffir medicine.

2. **M. Sutherlandi** (Hv.); young parts fulvo-pubescent; stipellæ very minute; leaves 2–3-*jugate;* leaflets elliptic-oblong, subacute, at first minutely puberulous, afterwards *glabrous, netted veined beneath;* panicles fulvo-sericeous, terminal, much branched; calyx thinly silky, its *teeth deltoid, much shorter than the tube; corolla glabrous;* ovary linear, silky, 3-ovuled; legume?

HAB. From the "Windsor Forest," N. of S. John's River, Natal, 1000 f., *Dr. Sutherland.* (Herb. Hk., D.)

"A magnificent tree, 70–90 feet high, 3 feet or more in diameter." Full grown leaves not yet seen. Those sent by Dr. Sutherland have a common petiole about 3 inches long, the pairs of leaflets nearly an inch apart, with very minute stipellæ. Leaflets 1½ inch long, ¾ inch wide, deep green, membranaceous. Panicles terminating the lesser twigs, 4–5 inches long, ovate in outline, densely branched; branches alternate, racemose. Peduncle, pedicels and calyx clothed with minute, glossy, appressed, deep brown hairs. Flowers purple. Vexillary stamen free at base, cohering above.

XXXIV. SESBANIA, Pers.

Calyx bibracteolate, cup-shaped, subequally 5-toothed or cleft. *Petals* subequal; *vexillum* roundish, complicate, crested on the claw or naked; *alæ* oblong; *carina* long-clawed, ascending, sharply eared or toothed at base. *Stamens* 9–1, the tube wide and eared at base. *Ovary* mult-ovulate; style curved. *Legume* very long, slender, compressed or cylindrical, with thickened sutures constricted between the seeds, and transversely multi-loculate; seeds cylindrical-oblong. *Endl. Gen.* 6551. *DC. Prod.* 2. *p.* 264.

Shrubs or herbaceous plants, common throughout the tropics of both hemispheres, with outlying species in the warmer temperate zones. Leaves abruptly pinnate, multi-jugate, the common petiole prolonged into a bristle. Stipules small, deciduous. Flowers yellow, in axillary racemes. *Sesban* is the Arabic name of the original species.

1. **S. aculeata** (Pers. Ench. 2, p. 216); herbaceous, nearly glabrous; the common petiole prickly; leaflets linear, obtuse, mucronulate, 12–30 jugate; racemes few flowered; calyx-teeth very short, triangular, distant; legumes erect, terete, acute. *DC. Prod.* 2, *p.* 265. *E. Mey. Comm. Drege, p.* 114. *Æschynomene bispinosa, Jacq. Ic. Rar. t.* 564.

HAB. Mouth of the Omsamculo and Omcomas, *Drege.* Near Port Natal, *Mr. Hewetson!* (Herb. T.C.D., &c.)

A tall growing annual, several feet high, becoming almost woody below; the young parts sometimes slightly pubescent, otherwise glabrous and glaucous. Leaflets sometimes in 40 pairs, sometimes in but 10, ½–¾ inch long, 1–1½ line wide. Racemes 4–6 flowered, the peduncle and pedicels very slender. Pods a foot long, not 2 lines wide, scarcely constricted between the seeds. A native also of the East Indies.

XXXV. SUTHERLANDIA, R. Br.

Calyx campanulate, 5-toothed. *Vexillum* oblong, shorter than the oblong, boat-shaped *carina*, its sides reflexed; *alæ* very short. *Stamens* 9 1. *Ovary* stipitate, mult-ovulate; *style* bearded along the upper side, and in front below the terminal stigma. *Legume* papery, inflated, many seeded, indehiscent; seeds reniform. *Endl. Gen.* 6566. *DC. Prod.* 2, *p.* 273.

A canescent S. African shrub. Leaves impari-pinnate, multi-jugate. Stipules minute, lanceolate-subulate, withering. Racemes axillary. Flowers handsome, scarlet or bright red. Legumes bladdery, glistening. Named in honour of *James Sutherland*, one of the earliest superintendants of the Botanic Gardens, Edinburgh. The compliment may now be worthily extended to DR. PETER SUTHERLAND, Surveyor-General of Port Natal, who is careful to use all opportunities for extending our knowledge of the botany of that most interesting district.

1. **S. frutescens** (R. Br. Hort. Kew, Ed. 2, p. 327); *DC. Prod.* 2, *p.* 273. *Colutea frutescens, Linn. sp.* 1045. *Mill. Ic. t.* 99. *Bot. Mag. t.* 181

VAR. *a.* **communis**; thinly canescent; leaflets elliptical or oblong, *glabrous above*, canescent beneath; ovaries and legumes quite *glabrous*, shining. *S. frutescens, E. Z.* 1658. *S. frutescens, vars. a. and δ. E. Mey. Comm. p.* 121.

VAR. β. **tomentosa**; thickly canescent-tomentose; leaflets short and broad, obovate or obcordate, silvery white *on both surfaces;* ovaries and legumes *hispid. S. tomentosa, E. & Z.!* 1659. *S. frutescens, β. E. Mey.! l. c.*

VAR. γ? **microphylla**; thinly pubescent; leaflets oblong-linear, glabrous above, pubescent beneath; peduncles 2–3 flowered. *DC. Prod.* 2. *p.* 273. *Deless. Ic.* 3. *t.* 71. *S. frutescens, γ. E. Mey. l. c.*

HAB. Dry hills and mountain sides throughout the colony. Var. β. Seashore near Wagenhuisgrotte and Cape L'Agulhas, *Mundt!* Simonsbay, *Mc Gillivray!* Green Point, *Dr. Pappe!* (Herb. Hk., Sd., D., &c.)

Very variable in the size of the bush, the copiousness of pubescence, and the shape of the leaves; varying also in a less degree in the proportions and colours of the petals. Var. β. in its extreme form, as collected by *Mundt*, looks to be very distinct; but intermediate states connect it with α: it seems to be merely a sea-side condition, growing probably in loose sands and exposed to the sea breezes. Var. γ., judging by Delessert's figure, is a starved, weak growing state of α.

XXXVI. LESSERTIA, DC.

Calyx campanulate, shortly and subequally 5-toothed. *Vexillum* obovate, emarginate, expanded, longer than the obtuse *carina*. *Stamens* diadelphous. *Ovary* substipitate, several ovuled; style filiform, ascending, bearded in front below the apex; *stigma* terminal. *Legume* scarious, compressed or inflated, unequal sided or linear, at length opening at the apex, several seeded. *DC. Prod.* 2, *p.* 271. *Endl. Gen.* 6563.

Suffrutices or herbs, rarely annual, natives of South Africa. Leaves impari-pinnate, multi-jugate; leaflets frequently alternate and albo-pubescent, rarely glabrous. Stipules small. Racemes axillary, on long or short peduncles. Flowers pink or crimson, rarely white. The species are very difficult to define, and probably too many have been established. The best characters are to be found in the shape and pubescence of the legume, the length of the pedicels with reference to the calyx, the general pubescence, whether strigose, patent, or silky; and the nature of the raceme, whether elongating or subcapitate. The length of the peduncle is a more variable character; the shape and size of the leaflets very uncertain. Named in honour of the late Baron Benj. Delessert, of Paris, a munificent patron of botany.

ANALYSIS OF THE SPECIES.

1. **Platylobæ.** Legume inflated or subcompressed, obliquely obovate, roundish or broadly oblong. (Sp. 1–25).

(*a.*) Rigid shrubby plants; or suberect or erect and virgate suffrutices:
  Legume glabrous; stems irregularly branched or divaricate:
    Peduncles rigid, at length spine-pointed:
      Peduncles *short;* legumes broad and short ... (2) **spinescens.**
      Peduncles long; legumes ovate-oblong, acute at base ... ... (3) **fruticosa.**
    Peduncles not spine-pointed.
      Peduncles much shorter than the leaves ... (1) **brachypus.**
      Peduncles much longer than the leaves ... (4) **flexuosa.**
  Legume glabrous; stems suberect or erect, virgate:
    Nearly glabrous; pedunc. long ... ... (5) **margaritacea.**
    Densely villous, with soft, white, loose hairs ... (7) **polystachya.**
    Thinly silky, with soft, appressed, whitish hairs (6) **perennans.**
  Legume pubescent; peduncles very long ... ... (8) **macrostachya.**
  Legume pubescent; pedunc. short; leaflets lanceolate, silky white ... ... (9) **candida.**

(*b.*) Herbaceous; or scarcely suffruticose, and then *procumbent or diffuse:*
  Legume glabrous, *compressed;* pubescence none or scanty.
    Stems and foliage *glabrous* or nearly so:
      Racemes laxly many flowered; root annual; leafl. narrow ... . (10) **linearis.**

Racemes *subumbellate*; leafl. 5-6-jugate; stipules toothlike .. .. (11) **subumbellata.**
Racemes subcorymbose; leaflets 6-14 pair; stipules ovato-lanceolate ... ... (12) **pulchra.**
Stems and foliage *thinly pubescent,* racemes long, many flowered; leaflets linear-lanceolate ... (13) **Pappeana.**
Legume glabrous, *inflated*; pubescence rigid, *appressed,* strigose:
Erect or ascending; leafl. 4-6-jugate; peduncles long, subcorymbose ... ... (14) **physodes.**
Erect or ascending; leafl. 10-12-jugate; pedunc. short, laxly racemose ... ... (15) **tenuifolia.**
Procumbent or prostrate; leafl. 3-6-jugate; peduncles short, racemose ... ... (16) **depressa.**
Legume glabrous; pubescence *copious,* of short, *spreading* hairs;
Racemes *subcapitate,* not elongating, 5-8 flowered (17) **capitata.**
Racemes elongating, few or many flowered:
Peduncles *shorter* than leaf; pod inflated ... (11) **inflata.**
Pedunc. *longer* than the leaf; pod *inflated,* rigid (19) **microcarpa.**
Pedunc. long; pod subcompressed or compressed:
Raceme dense; pod roundish-elliptical (20) **diffusa.**
Raceme lax; almost half-moon shaped (21) **excisa.**
Legume *pubescent*:
Pubescence spreading; pedunc. long, 8-12 flowered (22) **tomentosa.**
Pub. spreading; pedunc. short, 2-3 flowered ... (23) **prostrata.**
Pub. *appressed,* strigose; peduncles long, several flowered ... ... (24) **argentea.**
Pub. scanty; leaflets *semi-terete,* rigid, furrowed (25) **carnosa.**

2. Stenolobæ. Legume compressed, *linear,* straight or falcate. (Sp. 26-30).
Stem suffruticose; peduncles much shorter than the leaves:
Pedicels much longer than the calyx ... ... (28) **brachystachya.**
Pedicels shorter than the calyx; stem *dwarf*; legume mostly *straight,* hispidulous ... ... (27) **pauciflora.**
Pedicels shorter than the calyx; stem *shrubby*; legume *falcate,* mostly glabrous ... ... (26) **falciformis.**
Stem herbaceous (root perhaps annual?) peduncles long or longish:
Legumes nearly straight, 4 times as long as broad (29) **stenoloba.**
Legumes much arched or annular ... ... (30) **annularis.**

I. PLATYLOBÆ (Sp. 1-25.)

1. L. brachypus (Harv.); shrubby, the branches, petioles and peduncles minutely albo-puberulous; leaflets 4-5-jugate, elliptico-obovate, mucronate, glabrous above, sparsely puberulous beneath; stipules triangular; peduncles *much shorter than the leaf,* 6-8 flowered; pedicels longer than the puberulous calyx; legume glabrous, compressed (or subturgid?) obliquely ovate-oblong, 4-6 seeded. *L. falciformis, β. glabrata, E. Mey. Comm. p.* 119.

HAB. Namaqualand, *A. Wyley, Esq.* Koussie and Zilverfontein, *Drege.* (Herb. Sd., D.)

The petioles are 2-2½ inches long; the leaflets 4-5 lines long, 2-2½ wide, thickish. The peduncles ½-¾ inch long. Flowers 4-5 lines long. The legumes on *Mr. Wyley's* specimen, though but half ripe, are sufficiently advanced to indicate form.

2. L. spinescens (E. Mey.! Comm. p. 115); shrubby, erect, divaricately branched, rigid, thinly villoso-pubescent or glabrous; leaflets 5-8-jugate, lanceolate or linear, acute or obtuse, glabrous, or thinly villous beneath; stipules subulate; peduncles *shorter than the leaf or scarcely longer,* laxly racemose, rigid, at length spinescent; pedicels pubescent, longer than the puberulous calyx; legume glabrous, subcompressed, broadly and obliquely obovate, or suborbicular, 1–2 seeded, mucronate. *Zey.!* 2390.

HAB. Cape, *Carmichael!* Goedmanskraal and Kaus, and at Modderfonteinsberg, *Drege!* Brandenburg, *Zeyher!* Kuils River, *Pappe!* 124. (Hb. Bth., Hk., Sd., D.)

A much branched, rigid, bush, 1–2 feet high, with yellowish twigs, and pale, rather scanty foliage, either nearly glabrous or thinly villous. Petioles 1½–2 inches long; leafl. 5–6 lines long, 1½–2 lines wide. Peduncles 1–3 inches long, very patent. Legumes 12–13 lines long, 9–10 lines wide.

3. L. fruticosa (Lindl.? Bot. Reg. 970); shrubby, divaricately much-branched, rigid, the branches, petioles and peduncles cano-pubescent or glabrescent; leaflets 5–8-jugate, cuneate-oblong, convolute, obtuse or retuse, glabrescent or piloso-villous beneath; stipules triangular, rigid; peduncles longer than the leaves, rigid, at length spinescent, laxly racemose; *pedicels nigro-pubescent, longer than the nigrescent calyx;* legume glabrous, compressed, obliquely obovate-oblong, tapering at base, 4–8 seeded, obtuse, the ventral suture straightish, the dorsal convex. *E. & Z.!* 1649. *L. rigida, E. Mey.! Comm. p.* 115.

HAB. Groenekloof, *E. & Z.* Lauwskloof, Groenekloof, and Saldanha Bay, *Drege!* (Herb. Hk., Bth., Sd.)

A rigid, erect or spreading bush, 1–2 feet high, with patent, more or less canescent branches. Petioles 1½ inch long; leaflets 3–5 lines long, their edges rolled in. Peduncles 3–5 inches long, spreading, the pedicels 4–5 lines apart, at length pendulous. Legume almost pyriform in outline, rigid, pale, and opaque, 1–1¼ inch long, 5–6 lines wide Whether this be the plant figured by Lindley I cannot say. *Thunberg's 'L. rigida'* seems to me to be a glabrate form of *L. flexuosa,* with which it agrees in the pedicels, calyx, filiform peduncles, and legume.

4. L. flexuosa (E. Mey.! Comm. p. 116); *shrubby,* erect, much-branched, flexuous, the striate branches, petioles, and peduncles thinly appressed-silky; leaves short, patent, closely 6–7-jugate, leaflets obovate-oblong, obtuse, glabrous above, thinly silky beneath; racemes *on long peduncles,* distantly several-flowered; pedicels *much longer* than the puberulous calyx; ovary *glabrous,* few-ovuled; legumes glabrous, compressed, substipitate, obliquely ovate, the ventral suture curved or nearly straight, the dorsal much arched. *L. macrostachya, E. & Z.!* 1643. *Zey.!* 2401.

VAR. β. **rigida**; leaflets very thinly pubescent beneath, ovato-lanceolate, acute. *Colutea rigida, Thunb. Cap. p.* 603, *fide Hb. Ups. L. rigida, DC.*

HAB. Cape, *Bowie!* In Adow and Olifantshoek, *E. & Z.! Drege!* Bed of the Tarka River, *Mrs. F. W. Barber,* 64. (Herb. Hk., Bth., Sd., D.)

A much-branched bush, 2–3 feet high, twiggy, with short, curved, half herbaceous branches. Petioles 1½ inch long, the leaflets 1–2 lines apart, 3–4 lines long, 2 lines wide. Peduncles numerous, 5–8 times as long as the leaves, curved, bearing flowers beyond the middle. Flowers deep pink, almost crimson, 2½ lines long. Legumes 10–11 lines long, 6–7 wide, nearly semicircular. A very handsome species, quite distinct from *L. macrostachya,* but very near *L. fruticosa,* differing in the longer, not spinous peduncles, the pedicels, the pubescence of the calyx, &c. VAR. β. chiefly differs in being more glabrous, with more acute leaflets.

5. L. margaritacea (E. Mey.! Comm. p. 116); shrubby, erect, *virgate,* nearly glabrous; leaflets 3–6-jugate, oblongo-lanceolate or linear-obovate, obtuse or acute, mostly infolded, glabrous; stipules lanceolate; peduncles *much longer than the leaf,* laxly racemose, rigid, at length subspinescent; pedicels puberulous, longer than the calyx; legume glabrous, subcompressed, stipitate, (*small*), broadly obovate or suborbicular, 1–4 seeded. *Zey.!* 2391.

HAB. Near Rustbank and Ezelsfontein, 3800f., *Drege.* Buffaljagdriver and Rietkuil, *Zeyher.* (Herb. Hk., Sd.)

Near *L. spinescens*, but more erect, with long, straight, ribstriate branches; much longer peduncles and smaller legumes. Petioles 1½ inch long. Peduncles 3–4 inches, erect. Legumes 6–7 lines long, 4–5 broad. Leaflets 2–5 lines long, 1–2 broad.

6. L. perennans (DC.? l. c. 271); suffruticose, erect, virgate, the ribstriate stem, the petioles and peduncles *thinly silky;* leaflets 8–10-jugate, elliptical or elliptico-lanceolate, acute or mucronate, *thinly silky-villous* on one or both sides; stipules lanceolate-acuminate, membranous; peduncles much longer than the leaves, laxly racemose, at length rigid; pedicels much longer than the puberulous calyx; *calyx-lobes taper pointed;* legume glabrous, compressed, (*small*), obliquely obovate-oblong, stipitate, mucronate, 3–4 seeded. *E. Mey.! Comm. p.* 117.

HAB. Grassy places in Albany and Caffraria. Between Kachu and Zandplaat, and between Gekau and Baasche, *Drege.* Brooker's Hill, Natal, *Sanderson.* Grahamstown, *Genl. Bolton.* (Herb. Bth., Hk., D., Sd.)

2–3 feet high, subsimple or branched from below, the branches erect, straight or incurved; the stem fistular, pale. Foliage subcanescent; petioles 1½–2 inches long, Peduncles 3–6 inches long. Flowers 2–2½ lines long, white or pale-purple. Legumes 8–9 lines long, 5–6 lines wide.

7. L. polystachya (Harv.); erect, virgate, suffruticose, the rib-striate stem, the petioles and peduncles *villoso-canescent;* leaves short, 7–8-jugate, leaflets lanceolate-oblong, *villoso-canescent on both sides;* racemes on long peduncles, floriferous beyond the middle, many-flowered, elongating; pedicels longer than the villous calyx; ovary glabrous; legumes (small) glabrous, compressed, elliptical, acute, substipitate, 2–3 seeded. *Zey.!* 460.

HAB. Magaliesberg, *Burke & Zeyher.* (Herb. Hk., Bth., Sd.)

Seemingly 2–3 feet high, rigid, subsimple, all parts hoary with loose, soft, short, and very slender hairs. Peduncles somewhat panicled toward the end of the stem, 5–6 inches long or more, at first densely flowered; fl. 2–2½ lines long, pale. Petioles 1½ inch long, the leaflets 2–3 lines apart, 4–6 lines long, about two wide. Legumes 5 lines long, 3 lines wide, pale yellowish-horn colour.

8. L. macrostachya (DC. l. c. p. 272); suberect, virgate, suffruticose at base, the striate stem, the petioles and peduncles thinly and softly silky; leaflets distantly 8–10-jugate, elliptic-oblong, *thinly silky on both sides;* racemes *distantly many-flowered, twice or thrice as long as the leaves;* pedicels scarcely equalling the thinly silky calyx; ovary villoso-canescent, few-ovuled; legumes compressed, elliptic-oblong, very obtuse, shortly stipitate, thinly pubescent, about 3-seeded. *E. Mey.! Comm. p.* 116, (*non E. & Z.!*).

VAR. β. **atomaria**; more slender, with shorter peduncles; legumes thickly spotted with small purple dots.

HAB. Cape, *Burchell, No.* 2356. On the Gariep, *Drege.* β. Namaqualand, *A. Wyley, Esq.* (Herb. Bth., Hk., D., Sd.)

Stems perhaps 2 feet long, distantly branched, thinly canescent. Petioles 4–5 inches long, the leaf-pairs half inch apart; leaflets 5–6 lines long, 2–2½ lines wide, gradually smaller upwards, but the terminal leaflet equals the lowest. More than ¾ of the long peduncle bears flowers, which are ½ inch apart, and 2–2½ lines long. Legume 9–10 lines long, 5–6 broad, very obtuse at each end. β. is chiefly remarkable for its purple-spotted legumes. Its peduncles vary from once to twice as long as the leaf.

9. L. candida (E. Mey.! Comm. p. 116); 'shrubby, erect, rigid villoso-canescent;' leaflets distantly 6–8-jugate, lanceolate, silky and white on both sides; 'racemes shorter than the leaves; pedicels shorter than the calyx; legumes turgid, obliquely subovate, pubescent, 6-seeded. *E. Mey.*

HAB. Aris and Verleptpram, on the Gariep, *Drege.* (Herb. Sond.)

Of this I have merely seen a leaf and a legume. Leaflets 5–6 lines long, 1½ broad, acute.

10. L. linearis (DC. l. c. 272); annual (or biennial?), erect or ascending, sparsely strigillose or glabrescent; leaflets laxly 5-7-8-jugate, *narrow-linear*, obtuse or retuse, glabrous above, sparsely strigillose beneath; racemes on long peduncles, *laxly many-flowered*, elongating; pedicels much longer than the nigro-puberulous calyx; ovary 2–6 ovuled; legumes compressed, glabrous, shortly stipitate, obliquely ovate-oblong, 2–6 seeded. *E. & Z.!* 1651. *L. annua, DC. l. c.* 271. *E. Mey! Comm. p.* 117. *E. & Z.!* 1648. *L. propinqua, E. & Z.!* 1650. *Colutea linearis, Thunb.! Cap. p.* 604. *Zey.!* 468, 2394.

HAB. Picketberg and Verlooren Valley, *Thunberg!* Moist spots round Capetown and Campsbay, *Pappe!* Zwartland and Saldanha Bay; Olifants R. Clanw.; Gauritz R., Swell.; Zwartkops R., Uit., *E. & Z.!* Klipfontein and Hassagaiskloof, *Zeyher.* (Herb. Th., Bth., B., Hk., D.)

A slender, wiry annual (perhaps occasionally subperennial), 12–18 inches high. Leave few and distant. Stipules minute, triangular. Leaflets 5–8 lines long, 1 line wide, often infolded. Peduncles longer, often much longer than the leaves, becoming rigid and arched in fruit. Fl. 3 lines long, purple. Legumes often veiny, ¾ inch long, 6–7 lines wide, the ventral suture straightish, the dorsal hemispherical. The figure in *Hook. Exot. Fl. t.* 84 is doubtful.

11. L. subumbellata (Harv.); herbaceous, perennial, decumbent, nearly glabrous or sparsely strigillose; leaflets 5-6-jugate, oblong or linear-oblong, obtuse or mucronulate; stipules small, triangular; peduncles longer than the leaf, subumbellate, 4–8 flowered; pedicels equalling the puberulous calyx; legumes glabrous, compressed, obliquely oblong, many seeded. *Zey.* 2392, 2393.

HAB. Grassy places and wood sides, Voormansbosch, Swell., *Zeyher! Pappe!* 119. (Herb. Hook. Sd. D.)

Root woody. Stems numerous, 12–18 inches long, subsimple, trailing, ribbed and furrowed. Pubescence very scanty, of a few appressed, rigid bristles. Petiole 1–1½ inch long; leafl. 3–4 lines long, 1 line wide. Peduncle 2–3 inches long, erect, rigid in fruit, floriferous at the summit only. Legume 1–1½ inch long, ⅓ inch wide; ovules 16–20.

12. L. pulchra (Sims. Bot. Mag. t. 2064); suffruticose at base, *diffuse or ascending,* the *multangular and striate stem, the petioles and peduncles glabrous or nearly so;* leaflets 6-14-jugate, linear-oblong, mucronate or retuse, rigid, *glabrous* (or sparsely pubescent beneath); stipules obliquely ovato-lanceolate, acute; racemes on long peduncles, densely several flowered, subcorymbose or oblong; rachis nigro-pilose; pedicels equalling the nigro-pilose calyx or longer; ovary several ovuled, glabrous; legumes broadly-elliptical, obtuse, acute at base, nearly equal-sided, glabrous, compressed, 4–6 seeded. *DC. Prod.* 2, *p.* 272. *E. & Z.!* 1640. *E. Mey.! Comm. p.* 117. *L. astragalina, Meisn. Lond. Journ.* 2, 89. *Galega striata, Thunb.! Vicia capensis, Berg! Cap. p.* 215.

VAR. β. **luxurians**; larger and stronger in all parts; leaflets puberulous beneath; racemes longer and many flowered, the rachis and calyx very dark. *Galega striata, Thunb.! Lessertia pubescens, E. & Z.! No.* 1647.

VAR. γ. **alpina**; dwarf, densely cæspitose; leaflets obovate, cuneate or obcordate; peduncles not twice as long as the leaf, laxly racemose. *L. venusta, E. & Z.!* 1641.

HAB. Sides of Table and Devil's Mt., *E. & Z.! W.H.H.* &c. Drakensteenberg, *Drege!* β. shady places on Table Mt., *E. & Z.! W.H.H.* γ. on the Cederberg, *E. & Z.! Mundt.* (Herb. Th., Hk., Bth., D., Sd.)

Root and base of the stem woody. Stems herbaceous, spreading, branched chiefly near the base. Leaves closely multi-jugate, bright green. Racemes short or slightly elongating, the peduncle becoming rigid in fruit; flowers bright purple, 4–5 lines long. Stipules larger and more leafy than in most others. Legumes 1½ inch long, 9–10 lines wide, tapering at base, but scarcely stipitate. β. is a strong-growing form, often springing up after surface burning. γ. has all the look of an alpine, stunted plant.

13. L. Pappeana (Harv.); herbaceous, ascending, the sulcate stem and peduncles and the petioles *thinly pubescent with soft, short, spreading hairs;* leaflets 10-14-jugate, *linear-lanceolate,* obtuse, mucronate, glabrous above, thinly pubescent beneath; stipules ovato-lanceolate, acuminate; peduncles 3–4 times longer than the leaves, racemose, many flowered, elongating; pedicels nigro-pubescent, longer than the *puberulous* calyx; legume (immature) glabrous, compressed, oblong, 8–12 seeded. *L. astragalina, Pappe! non Meisn.*

HAB. Tulbagh, *Pappé!* (Herb. D.)

Stems 2 feet long, pale yellowish, strongly rib-furrowed, somewhat fistular; the peduncles similar, 12–14 inches long, bearing flowers from 5–6 inches below the summit upwards. Petioles 3–4 inches long; leaflets 9–10 lines long, 2 lines wide. Pubescence scanty and soft. Flowers 5–6 lines long, the dark purple carina somewhat rostrate. The half-ripe legumes are linear-oblong, both margins nearly straight, probably the dorsal afterwards becomes arched. Most like *L. pulchra,* var. *luxurians,* but the raceme is much longer, the flowers larger, the calyx less hairy, and the leaflets different, the ovules more numerous, and the legumes probably longer and narrower.

14. L. physodes (E. & Z.! 1644); erect or ascending, suffruticose, the angular stem, the petioles and peduncles thinly albo-strigose; leaflets laxly 4-6-jugate (often alternate), linear-oblong or sublanceolate, obtuse or acute, glabrous above, strigose beneath; stipules lanceolate; peduncles longer than the leaves, corymbo-racemose near the summit; pedicels not exceeding the albo-puberulous calyx; legumes glabrous, inflated, oval-oblong, acute at each end, not very unequal sided. *L. acuminata, E. Mey! Comm. p.* 118. *L. tumida, E. & Z.!* 1646.

HAB. Philipstown; also (*L. tumida*) at Wagenhausgrotte, Swell. and Adow, Uit. *E. & Z.!* Buffel River and between Gekau and Basche, *Drege!* (Herb. Bth., Sd.)

Stem 12–18 inches long, pale. Leaves subdistant; the petiole 1½–2 inches long, the leaflets 4–5 lines apart, 7–8 lines long, 1 line wide. Peduncles 3–4 inches long, the last inch bearing flowers. Legumes about an inch long, nearly ½ inch wide. The specimens of *L. tumida,* E. & Z. examined are imperfect; they may belong to a different, but closely allied form.

15. L. tenuifolia (E. Mey.! Comm. p. 117); half herbaceous, erect or ascending, the angular stem, the peduncles and petioles thinly albo-strigose; leaflets laxly 10-12-jugate, linear-oblong or obovate, obtuse or retuse, glabrous or thinly strigose beneath; stipules triangular; peduncles scarcely equalling the leaves, laxly racemose; pedicels shorter than the strigillose calyx; legumes glabrous, much inflated, membranous, ovoid, substipitate, many-seeded.

HAB. Mooyplaats, 4600 f. and at the Compasberg, 4800 f., *Drege!* Zooloo Country, *Miss Owen!* (Herb. Sd., D.)

Suffruticose at base, 6–12 inches high. Petioles 2½–3 inches long; leafl. 3 lines long, 1½ wide. Peduncles 2–2½ inches long; flowers 3½–4 lines, crimson. Legumes 1¼ inch long, ¾ inch wide, very bladdery.

16. L. depressa (Harv.); half-herbaceous, *procumbent or prostrate*, the stem, petioles, and peduncles thinly albo-strigose or subpilose; *leaflets 3-6-jugate, oblongo-lanceolate or oblong*, acute or obtuse, thinly strigoso-pilose beneath; stipules acuminate; peduncles *shorter* than the leaves or scarcely longer, shortly racemose, several flowered; bracts broadly ovate, obtuse; pedicels shorter than the nigro-pubescent, bluntly lobed calyx; legume glabrous, inflated, obliquely ovoid, substipitate, many-seeded. *L. prostrata, E. & Z.!* 1645, *non Thunb.*

HAB. Zwartkops River, Uit., *E. & Z.!* Albany, *Mrs. F. W. Barber!* (Herb. Sond., Hook.)

Near *L. physodes* and *L. tenuifolia*, but differing in habit. The stems in E. & Z.'s specimens are 18 inches long, branched only at the base, floriferous for half their length. Leaflets 5–6 lines long. The pubescence is generally close-pressed and strigose, but that of the stem and peduncles is sometimes looser and more copious. Legumes scarcely an inch long, ½ inch wide.

17. L. capitata (E. Mey.! Comm. p. 118); herbaceous, decumbent, the stem, petioles, and peduncles hoary with short, patent hairs; leaflets closely 7-9-jugate, oblong, obtuse or retuse, patently pubescent on one or both sides; stipules ovate-acute; peduncles much longer than the leaf, *capitato-racemulose*, 5-8 flowered; pedicels unequal, shorter than the nigro-hirsute calyx; legumes (immature) *glabrous, many seeded.*

HAB. Kasparskloof, Camisberge, 2000 f., *Drege!* (Herb. Hk., Bth., Sd.)

Very like *L. tomentosa*, but with glabrous legumes, the exact form of which is unknown.

18. L. inflata (Harv.); herbaceous, diffuse or decumbent, villoso-canescent with loose hairs; leaflets 8-10-jugate, short, obovate or obcordate, *villoso-canescent* beneath; peduncles *shorter than the leaf*, laxly racemose, few-flowered; pedicels shorter than the villous calyx; legumes inflated, glabrous, ovate-subglobose, several seeded. *L. vesicaria, E. Mey.! Comm. p.* 119, *non DC.*

HAB. Steelkloof in Uitvlught, 3500 f. *Drege!* (Herb. Bth., Sd.)

Thunberg's Colutea (Lessertia) vesicaria, in Hb. Upsal and Holm., has pubescent legumes, and seems scarcely distinct from *L. tomentosa.* Leaflets 2½–3 lines long, 2 lines wide, mostly emarginate. Legumes 7–8 lines long.

19. L. microcarpa (E. Mey.! Comm. p. 119); herbaceous, procumbent, dwarf, villoso-canescent with loose hairs; leaflets 6-8-jugate, short, obovate-oblong, or obcordate, villous beneath; peduncles *about twice as long as the leaf*, laxly racemose, few-flowered; pedicels shorter than the villous calyx; legumes (small) inflated, glabrous, obliquely obovate, rather rigid, few-seeded.

HAB. Modderfonteinsberg, 4000 f., *Drege!* (Herb., Hk., Bth., Sd.)

Very like *L. inflata*, but smaller in all parts, with longer peduncles. Leaflets 2–3 lines long, 1–2 wide, often infolded. Peduncles 2–3 inches long. Legume 5–6 lines long, 4 lines wide. Flowers white?

20. L. diffusa (R. Br. Hort. Kew.); perennial, diffuse or procumbent, the stems, petioles, and peduncles albo-pubescent with short, patent hairs; leaves closely 8-12-jugate, leaflets (often alternate) elliptic-oblong, obtuse or emarginate, albo-pilose *on one or both sides;* racemes on long peduncles, densely many-flowered, elongating; pedicels scarcely equalling the albo-pilose calyx; ovary few or several ovuled, glabrous; legumes subcompressed, glabrous, elliptic-oblong, *both margins convex* and not very unequally. *E. Mey.! Comm. p.* 118. *DC. Prod.* 2, *p.* 271. *Galega dubia, Jacq. Ic. Rar.* 576.

HAB. Kasparskloof, near Koussie, Zilverfontein and little Namaqualand; and in the Nieuweveld, *Drege!* (Herb. Hk., Bth., D., Sd.)

Whole plant hoary with short white, patent hairs. Leaflets 2–4 lines long, 1½–2 lines wide, sometimes well covered with hairs above, and sometimes nearly bare, except at the edges. Racemes 5–6 inches long, on peduncles of equal length. Flowers purple, 3 lines long. Legumes scarcely uncial; sometimes semi-uncial, nearly orbicular and 1–2 seeded, but more commonly 6–8 seeded. Near *L. excisa*, but differing in the legumes.

21. L. excisa (DC. l. c. 272); perennial, diffuse or procumbent, the stems, petioles, and peduncles albo-pubescent with short, patent hairs; leaves closely 8-12-jugate, leaflets (often alternate) cuneate-oblong, truncate or emarginate, glabrous above, thinly albo-pilose beneath; racemes on long peduncles, *laxly* several or many flowered; pedicels equalling the nigro-puberulous calyx or shorter; ovary several ovuled, glabrous; legumes compressed, glabrous, at first subfalcate-ovate (the ventral margin concave), then *obliquely semi-elliptical, the ventral margin straightish, the dorsal arcuate. E. Mey.! Comm. p.* 119. *L. perennans, E. & Z.!* 1652. *L. diffusa, E. & Z.!* 1654.

HAB. Cape, *Thunberg!* Lion's Mt., *E. & Z.! Pappe!* Camps Bay, *W.H.H.* Paarlberg, *Drege!* Near Tulbagh, *Pappe!* Brackfontein, Clanw., *E. & Z.!* Klipfontein, *Zey.!* 469. (Herb. Th., Bth., Hk., D., Sd.)

Many stemmed, trailing or partly ascending, branched from the base. Pubescence patent. Petioles 2–3 inches long, leaflets 2–6 lines long, 1–2 lines wide, sometimes obcordate, sometimes elliptical, but usually abruptly cuneate. Racemes 2–3 inches long or more, on peduncles 3–5 inches long, which become rigid in fruit. Legumes commonly ¾–1 inch long, 6–8 lines wide, when ripe nearly half-moon shaped; sometimes 1¼ inch long, 9–10 lines wide.

22. L. tomentosa (DC. l. c. p. 272); perennial, diffuse or procumbent, the stems, petioles, and peduncles albo-pubescent with short, patent hairs; leaflets closely 8-12-jugate, cuneate-oblong, obtuse or emarginate, glabrous above, *albo-pilose beneath;* racemes on long peduncles, *subcapitate or shortly several flowered;* bracts deltoid; pedicels shorter than the nigro-hirsute calyx; ovary few-ovuled, hoary; legume *inflated, ovoid, albo-hirsute,* shortly stipitate. *Colutea tomentosa,* and *C. vesicaria, Th.! Fl. Cap. p.* 604. *L. excisa, E. & Z.!* 1653.

HAB. South Africa, *Thunberg! Carmichael!* Near Capetown, *Pappe!* Near Bergrivier, *E. & Z.* (Herb. Th., D., Hk., Sd.)

Many stemmed; stems subsimple, trailing, 12–14 inches long, hoary with short, white hairs. Petioles 2–3 inches long; leaflets 4–5 lines long, 2–3 wide, open or infolded, green above, hoary beneath. Racemes subcapitate, or slightly lengthening, much shorter than the peduncles, 8-12 flowered; all the hairs, except those of the calyx white. Flowers 3 lines long, purple. Legumes 5-7 lines long, bladdery.

23. L. prostrata (DC. l. c. p. 272); herbaceous, procumbent, the stem, petioles, and peduncles sparsely hispid with short, patent hairs; leaflets 8-12-jugate, lanceolate-linear, obtuse or mucronate, glabrous above, sparsely hispid beneath; stipules ovate-acuminate; peduncles *rather shorter than the leaf, 2-3 flowered;* pedicels shorter than the nigro-pubescent calyx; legumes *turgid, pubescent,* oblong, the ventral suture concave or straightish, the dorsal arched. *E. Mey.! Comm. p.* 118. *Colutea prostrata, Thunb! Cap. p.* 603.

HAB. Verlooren Valley, *Thunberg!* Cape Flats, *Pappe!* 118. S. Africa, *Drege!* (Herb. Th., D., Sd.)

Root branching, perennial. Stems pale, 10-12 inches long, trailing. Pubescence scanty. Petioles 2-3 inches long, the first pair of leaflets ½ inch from the base; leaflets 6-7 lines long, 1-2 lines wide, green. Peduncles 1½ inch long, 2- rarely 3-flowered. Legume 8-10 lines long, 4-5 wide. Less hairy than *L. tomentosa,* with narrower leaflets, shorter peduncles, and fewer flowers.

24. L. argentea (Harv.); perennial, erect or ascending, *the stems, petioles and peduncles appressedly albo-strigose;* leaves closely 8-12-jugate, leaflets oblong or linear-oblong, obtuse or emarginate, glabrous above, thinly *albo-strigose* beneath; racemes on long peduncles, shortly several flowered; bracts small; pedicels shorter than the *albo-strigose* calyx; ovary 6-8 ovuled, *silvery;* legume broadly elliptic-oblong, obtuse at both ends, compressed, subsessile, *thinly albo-strigose,* 4-6 seeded. *Coronilla argentea, Thunb.! Lessertia villosa, E. Mey.! Linn.* 7, *p.* 169. *E. & Z.* 1655; *also L. vesicaria, E. & Z.!* 1657.

VAR. β. **angustifolia** (E. & Z.); leaflets narrower, linear, mucronate.

HAB. S. Africa, *Thunberg!* Near Greenpoint, Cape, *Pappe! W.H.H.* Simon's Bay and near Hott. Holland and Vischhoek, *E. & Z.* (Herb. Th., D., Hk., Bth., Sd.)

Stems several, 1-2 feet high, subsimple, erect or spreading. Pubescence thinly spread, close pressed, the short, rigid, white hairs fixed by a middle point. Petioles 3-5 inches long; leaflets 3-5, in β. 6-7 lines long, 1½-2½ lines wide. Stipules triangular-acuminate. Flowers 3-4 lines long. Legumes 1-1¼ inch long, 6-8 lines wide.

25. L. carnosa (E. & Z.! 1642); suffruticose, diffuse or ascending, the stem, petioles, and peduncles thinly strigose or glabrescent; leaflets closely 7-9-jugate, *linear semiterete,* rigid, carnose, *complicate, furrowed above,* glabrescent; stipules lanceolate; peduncles longer than the leaves, densely corymbo-racemose at the summit; pedicels scarcely equalling the albo-puberulous calyx; ovary canescent, many ovuled; legumes *pubescent,* broadly oblong, subinflated, the central suture straight or concave, the dorsal arcuate, several seeded. *Zey.!* 2399.

HAB. Karroo-places at the mouths of the Coega and Sondag's Rivers, *E. & Z.!* Zoutpanshoogdte, *Zeyher!* (Herb. Bth., Sd.)

Root woody. Stems several, more or less procumbent, 10-12 inches long, rigid. Leaves closely set, patent, 1-1½ inch long. Leaflets 4-6 lines long, ½ line wide, with a narrow furrow on the upper side, otherwise terete, the young ones subpilose, the older glabrous. Peduncles 2-3 inches long, the raceme scarcely uncial, 8-12 flowered. Legumes 1¼ inch long, 7-8 lines wide, broadly subfalcate-oblong, acute, obtuse at base.

2. **STENOLOBÆ.** (Sp. 26-30.)

26. L. falciformis (DC.? l. c. 272); suffruticose, erect or ascending, virgate, thinly strigoso-canescent; petioles elongate, leaflets distantly 6-9-jugate, elliptic-oblong or lanceolate-linear, or obovate, obtuse or retuse, albo-strigose on one or both sides; stipules subulate; peduncles much shorter than the leaves, shortly racemose near the summit, several flowered; pedicels shorter than the puberulous calyx; legumes glabrous or *sparsely* setulose, broadly linear-falcate, compressed, 6-8 seeded. *L. falciformis, E. Mey.! Comm. p.* 120, *excl. var. β.*

VAR. *β? Thunbergii;* legumes nearly straight, setulose. *Galega humilis, Thunb. Cap.* 601, *fide Herb. Upsal.*

HAB. Little and Great Namaqualand. Zilverfontein, Koussie, Kaus and near Verleptpram, N.W., *Drege!* Bitterfontein, *Zeyher!* 470. Gariep, *Wyley!* (Herb. Hk., Bth., Sd., D.)

A virgate suffrutex, becoming shrubby in age, more or less canescent. Petioles on the young plant 6 inches, on the older 2–3 inches long; leaflets 3–8 lines long, 2–4 wide. Peduncles about an inch long, several flowered. Legumes 1½ inch long, 3–5 lines wide, glabrous or with a very few setæ. In var. *β.* the legumes are nearly straight, and as closely setulous as in *L. brachystachya*, to which, but for the short pedicels, I should have referred Thunberg's specimens. This is the largest and most woody of all the linear-fruited *Lessertiæ.*

27. L. pauciflora (Harv.); many stemmed, herbaceous or suffruticose, suberect or prostrate, *dwarf*, variably strigoso-pubescent or canescent; leaflets distantly 8-12-jugate, oblong, obtuse or retuse, with involute margins; stipules subulate or ovato-lanceolate; peduncles much shorter than the leaves, mostly 2-flowered (rarely subumbellately 4–6 flowered); pedicels *shorter than the pilose calyx;* legumes linear, obtuse, mucronate, straight or somewhat falcate, 3-4 times as long as broad, thinly strigose, 10-12 seeded.

VAR. *α.* **erecta**; stems erect; leaflets glabrescent. *Zey.!* 461.

VAR. *β.* **prona**; stems quite prostrate; leaflets glabrescent or albo-pilose.

VAR. *γ.* **diffusa**; stems diffuse, much branched; legumes often falcate; leaflets albo-pilose.

VAR. *δ.* **canescens**; leafl. canescent on both sides; peduncles 2–6 flowered; leg. straight or falcate. *Zey.* 471.

HAB. Var. *α*, Thaba Uncka, *Burke & Zeyher!* *β.* Valleys of Queenstown; also in Cradock and on the Winterberg, blossoming throughout the summer, *Mrs. F. W. Barber!* 54. *γ.* Bassoutosland, *v. Schlicht!* *δ.* Namaqualand, *Wyley!* Bitterfontein, *Zeyher.* (Herb. Hk., Bth., Sd., D.)

A very variable plant, intermediate between *L. brachystachya* and *L. falciformis*, having usually *straight* legumes, but varying to curved or even falcate ones. Flowers dull-purple and white, with the vexillum striped. Petioles 1½–2½ inches long; leafl. 3–4 lines long. Peduncle ½–1 inch long. Var. *δ.* may perhaps be a species; yet its distinctive marks are not constant.

28. L. brachystachya (DC. l. c. 272); suffruticose, diffuse or ascending, thinly albo-strigose; leaflets 7-10-jugate, linear-oblong, glabrous above, thinly strigose beneath; stipules triangular; peduncles shorter than the leaves, densely few-flowered; *pedicels much longer than the calyx;* ovary canescent; legumes linear, obtuse or acute, mucronate, 4-5 times longer than broad, nearly straight, thinly strigose, distantly 7–8 seeded. *E. & Z.* 1635. *E. Mey.! Comm. p.* 121. *Burch. Cat.* 3453.

VAR. *β.* **acutiloba**; legumes taper-pointed; slightly falcate. *Zey!* 2400.

HAB. Cape, *Bowie! Burchell.* Zwartkops River, Uit. and in Caffraria, *E. & Z.* Mouth of Gauritz R. and near Enon, *Drege.* *β.* Zoutpanshoogdte, *Zeyher.* (Herb. Hk., Bth., Sd.)

Stems 1–2 feet long, subsimple or branched, more or less ligneous. Peduncles commonly very short, the rachis, even in fruit, not an inch long. The pedicels are slender, 5–6 lines long, by which character this is best known from *L. falciformis;* the curvature of the legume is variable. Our var. β. with the long pedicels of *L. brachystachya* has legumes approaching those of *L. falciformis.*

29. L. stenoloba (E. Mey. Comm. p. 121); herbaceous, erect or diffuse, thinly or sparsely albo-strigose; leaflets 5-10-jugate, linear-oblong or sublanceolate, variably pubescent; peduncles *longer* than the leaves, shortly racemose at the summit; pedicels shorter than the calyx; ovary canescent; legumes linear, obtuse, mucronate, nearly straight, four times as long as broad, thinly strigose, many-seeded.

VAR. **Meyeri**; leaflets subacute, 5-jugate, albo-strigose on both sides. *L. stenoloba, E. Mey.*

VAR. β. **obtusata**; leafl. obtuse, 7–10–jugate, glabrous above. *L. obtusata, E. & Z.* 1636, *non Thunb.*

HAB. α. Nieuweveld, between Waschbank and Rietpoort, *Drege.* β., Sandy hills near the Zwartkops River, Uit., *E. & Z.* Cape, *Bowie.* (Herb. Sd., Bth., Hk., D.)

More herbaceous than *L. brachystachya,* with (usually) much longer peduncles, and (constantly) much shorter pedicels. It differs from *L. annularis* merely in the straight or nearly straight legumes.

30. L. annularis (Burch. Voy. 1, p. 304); herbaceous, erect or diffuse, thinly albo-strigose; leaflets 6–10-jugate, oblong-obovate or linear-oblong, obtuse or emarginate, thinly strigose on one or both sides; peduncles *longer or shorter* than the leaf, shortly racemose near the summit; pedicels *shorter* than the calyx; legumes linear, compressed, obtuse, mucronate, *arched in a semicircle or more or less completely annular. DC. Prod.* 2. 272. *E. Mey.! Comm. p.* 120. *E. & Z !* 1638. *Also L. arcuata, E. & Z. !* 1639 *and L. falciformis, E. & Z. !* 1637.

HAB. Near Bokpoort, Nieuweveld and Matjesvalei, *Drege!* Beaufort and Albany, and in Uitenhage and Swellendam, *E. & Z.! Mrs. F. W. Barber.* (Herb. Hk., Bth., Sd., D.)

Many stemmed, slightly lignescent at base, 10–15 inches high, the young plant erect. Pubescence scanty, close-pressed. Leaflets variable in size, 2–8 lines long. Peduncles very variable, even on the same specimen. Legumes commonly semicircular, but sometimes completely annular. Scarcely two specimens exactly alike of those examined.

(Doubtful or imperfectly known species.)

L. pubescens (DC. l. c. 272); *Colutea pubescens, Thunb.! Cap. p.* 603.

L. obtusata (DC. l. c.); *Colutea obtusata, Th.! l. c.* 604.

L. mucronata; *Galega mucronata, Thunb.! p.* 601.

Thunberg's specimens of the above in Herb. Upsal are too imperfect for accurate determination.

L. vesicaria (DC); *Colutea vesicaria, Th.! p.* 604, seems to be a var. of *L. tomentosa.*

L. procumbens (DC. 273); stem suffruticose, procumbent, leafl. 12-14 pairs, ovato-linear, tomentose; peduncles very long, 3-4 flowered. *Colutea procumbens, Mill. Dict. No.* 7.

L. abbreviata (E. Mey. Comm. p. 118); herbaceous, erect, glabrescent; leaflets 8-jugate, subconvolute-linear, obtuse; racemes equalling the leaves; legumes semi-obovate, glabrous, about four-seeded.

Growing with *L. linearis* (perhaps merely a short-peduncled variety) *Drege.*

L. lanata; suffruticose? densely clothed in all parts with long, soft white hairs; leafl. 6-8-jugate, elliptic-oblong (5-6 lines long, 2-2½ wide,) obtuse; stipules lanceolato-subulate; peduncles equalling the leaves, shortly racemose at the summit; pedicels equalling the calyx; legume . . . . ? *L. tomentosa, E. & Z.!* 1656, *excl. Syn.*

HAB. Gauritz River, Swellendam. *E. & Z.* (Herb. Sond.)

Of this there exists a mere fragment in Herb. Ecklon. It is much more densely villous and woolly than any recorded species.

L. sulcata (E. Mey. Comm. p. 116); suffruticose, erect, glabrous; stem furrowed, flexuous or kneebent; leaflets multi-jugate, linear-oblong, obtuse, emarginate; raceme twice as long as the leaf; the peduncle thick, furrowed, erect; legumes compressed, ovate-oblong, very large, 1-2 seeded? *E. Mey.*

HAB. Rocky places near Leeuwenkrall, in Dutoitskloof, *Drege.*

Unknown to me.

XXXVII. SYLITRA, E. Mey.

Calyx sub-bilabiate, 5–fid. *Carina* erect, round-pointed, shorter than the subequal *vexillum* and adnate *alæ*. *Stamens* monadelphous. *Style* glabrous, the stigma capitellate. *Legume* scarious, indehiscent, compressed, much broader than the seeds. *E. Mey. Comm. Drege, p.* 114.

A virgate perennial. Leaves unifoliolate. Stipules subulate, free. Flowers axillary, mostly in pairs, subsessile, very small. The name συλιτρα, was applied by Dioscorides to some *Glycyrhiza*, and by Medicus formerly given to the genus now called *Lessertia*. It is now revived in favour of the present plant, which with the legume of a *Lessertia* has a very distinct general habit.

S. biflora (E. Mey.! Comm. Drege, p. 114); *Harv. Thes. t.* 78.

HAB. Bitterwater, near the Gamke River, 2300f., *Drege.* (Herb. Benth., Sond.)

Stems 12–16 inches high, terete, slender, thinly canescent, slightly branched; branches virgate, flexuous. Leaves nearly an inch apart; petiole 1–2 lines long, articulated at the summit, with a terminal, linear-lanceolate leaflet, 1½–2 inches long, 2 lines wide, acute at each end, glabrous above, minutely canescent beneath, obliquely striate. Stipules setaceo-subulate, equalling the petiole. Flowers in axillary pairs, on pedicels not a line long. Calyx 1½ line long, silky, with sharp teeth. Corolla not twice as long as the calyx. Legume oblong, 1½ inch long, half inch wide, strongly compressed, thickened at the sutures, pubescent, membranous, and pale. A very remarkable plant.

XXXVIII. ASTRAGALUS, L.

Calyx tubular or campanulate, 5–toothed. *Vexillum* equalling or exceeding the alæ; carina obtuse. *Stamens* diadelphous. *Ovary* many-ovuled. *Legume* (variable in form) incompletely or completely divided longitudinally into two cells, by the introflexion of the *carinal* (dorsal) suture. *DC. Prod.* 2, *p.* 281. *Endl. Gen.* 6573.

An immense genus of herbs or suffrutices, natives chiefly of the northern hemisphere, very abundant in temperate Asia; a few in North Africa. Habit extremely various But one S. African species known. The name was given by the Greeks to some leguminous plant.

1. **A. Burkeanus** (Benth.! in Herb.); erect, quite glabrous, glaucescent; stipules very large, leafy, semicordate-oblong, acute, free; leaflets

8–12-jugate, oblong, mucronate; peduncles longer than the leaves; flowers racemose, patent; calyx-lobes equalling the campanulate tube; legume elongate, compressed, glabrous, on a stipe longer than the calyx. *Harv. Thes. Cap. t.* 82.

HAB. Magaliesberg, *Burke & Zeyher*. (Herb. Hk., Bth., Sd.)

This belongs to the section "*Galegiformes*," and is near *A. graveolens, tigrensis, venosus,* and *abyssinicus*, the three last natives of North Africa. Root annual. Stems 12–18 inches high, slightly flexuous, terete, pale. Lower stipules uncial, 5–6 lines wide, upper smaller and narrower. Leaflets 6–7 lines long, 2–3 wide, varying from oblong to sublanceolate, pale green, thin. Flowers small and slender, scarcely 4 lines long. Calyx-tube a line long, the segments narrow-lanceolate. Petals subequal or the alæ rather shorter. Legume on a 2–3 line long stipe, uncial, 2½ lines wide, completely bilocular, slenderly netted-veined and thin.

XXXIX. **ZORNIA**, Gmel.

Calyx bilabiate, the upper lip obtuse, emarginate, the lower trifid. *Corolla* inserted in the base of the calyx; the *vexillum* roundish, with reflexed sides; *alæ* oblong; *carina* of lunate petals, cohering in the middle. *Stamens* monadelphous; the alternate anthers small. *Legume* sessile, compressed, 3–6-jointed; the joints roundish, often hispid. *DC. Prod.* 2, *p.* 316. *Endl. Gen.* 6599.

Tropical or subtropical herbs or suffrutices. Leaves digitate, of 2–4, pellucid-dotted leaflets. Stipules broad, rigid, peltate; the upper broader, forming bracts, which enclose the sessile flowers. Flowers minute, in terminal or lateral, lax pseudo-spikes. Named from *J. Zorn*, an apothecary of Kempten, in Bavaria, and author of *Icones Plantarum medicalium*.

1. **Z. tetraphylla** (Mich. Fl. Bor. Amer. t. 41); stems diffuse; leaves 2–4-foliolate; leaflets lanceolate or linear; bracts ovate, acute, ciliolate, 5-nerved. *DC. l. c.* 317. *Hedysarum tetraphyllum. Thunb. Cap.* 595. *Zey.!* 467.

VAR. α. **Capensis**; leaflets broadly lanceolate; legumes reticulate-scabrid. *Z. Capensis, Pers. E. & Z.!* 1660. *E. Mey.! Comm.* 122.

VAR. β. **linearis**; leaflets very narrow-linear; legumes echinulate. *Z. linearis, E. Mey! l. c.*

HAB. Eastern districts and Port Natal, common. (Herb. Hk., Sd., D.)

A slender, wiry plant, 6–12 inches high, glabrous or variably pubescent. Leaflets generally 4, but varying to 3 and 2, 7–8 lines to 1–1¼ inch long, and ½ line to 2–3 lines wide. Stipules and bracts prolonged at base below their insertion; the latter enclosing the small yellow flowers, as if between a pair of "winkers." The spines on the legumes vary much; sometimes they are mere points, sometimes 1–2 lines long.

XL. **ÆSCHYNOMENE**, L.

Calyx bibracteate at base, more or less bilabiate or bipartite, the upper lip entire or bifid, the lower either entire, trifid, or tridentate. *Vexillum* subrotund or oblong, simple at base; alæ oblong, equalling or exceeding the incurved carina. *Stamens* 10, in two equal parcels of 5 each. *Ovary* stipitate. *Legume* stipitate, compressed, exserted, transversely articulated, several-jointed. *DC. Prod.* 2, *p.* 320. *Endl. Gen.* 6605.

Tropical and subtropical herbs or shrubs. Leaves impari-pinnate (rarely pari pinnate); multi- or pluri-jugate, bistipulate. Peduncles racemose or 1-flowered, axillary, rarely terminal. Fl. yellow or reddish. Name from αισχυνομαι, *to be modest*; one of the species has sensitive leaves.

ANALYSIS OF THE SOUTH AFRICAN SPECIES.

Sect. 1. **Eu-Æschynomene.** *Calyx* bi-partite, deciduous. Stipules peltate.
Peduncles racemoso-paniculate. Stems glabrous. ... (1) **erubescens.**
Peduncles 1-flowered. Stem hispid ... ... (2) **uniflora.**
Sect. 2. **Ochopodium.** Calyx persistent, subequally 5-cleft. Stipules sessile.
Leaves 4-5-jugate, ciliate; racemes axillary, few-flowered (3) **micrantha.**

1. **EU-ÆSCHYNOMENE** (Sp. 1-2.)

1. Æ. erubescens (E. Mey. Comm. p. 123); suffruticose, erect, *glabrous;* leaflets 10-15-jugate, linear-oblong, ciliolate, *the mid-rib beneath and the common petiole sparsely set with bristles;* racemes axillary, *sub-panicled,* bristly; lips of the bipartite, setulose calyx, shortly but sharply toothed; legume .... ?

HAB. Between the Omtendo and Omsamculo, *Drege.* Natal, *Gueinzius!* (Herb. Sd.)

Stems long, simple, purplish, striate, with distant nodes. Leaves spreading 1½ inch long; leafl. 3-4 lines long, 1½ wide, paler below. Racemes rather longer than the leaf, setulose above. Stipules scarious, peltate, lacerated. Bracts ovate, acute, half as long as the calyx. Legume unknown.

2. Æ. uniflora (E. Mey. l. c.); suffruticose, erect, the stem, petioles, and peduncles rigidly hispid; leaflets 10-20-jugate, linear-oblong, ciliolate; peduncles axillary, shorter than the leaf, one-flowered, scabrid; calyx hispid, bipartite, the lips entire; legume on a long, rough stipes, 7-8-jointed, the joints oblong, margined, densely warted in the middle and sparsely hispid.

HAB. Near the mouth of the Omsamculo, among tall grass, *Drege.* Coast land of Natal, *Dr. Sutherland.* (Herb. Hk., Sd., D.)

Stem virgate, "5-6 feet high," terete, rufescent, with distant nodes. Leaves 1-2 inches long; leafl. 3-4 lines long, 1 line wide. Stipules rigid, eared at base. Peduncles in flower ½, in fruit 1 inch long. Legume 2 inches long; each joint 3 lines long, 2½ wide, brown.

2. **OCHOPODIUM.** (Sp. 3.)

3. Æ. micrantha (DC. Prod. 2, p. 321); diffuse, slender, the stem, petioles, and peduncles hispidulous; leaflets 4-5-jugate, mostly alternate, obovate-oblong, ciliate, mucronate, netted-veined and sparsely hispidulous beneath; racemes axillary, distantly few-flowered; the bracts and the subequally and bluntly 5-lobed calyx ciliate, persistent; legumes stipitate, 2-4-jointed, recurved, even, thinly silky or glabrescent. *Patagonium racemosum, E. Mey.! Comm. p.* 123.

HAB. Grassy places between the Omtendo and Omsamculo, and near Natal, *Drege! Sutherland!* (Herb. Hk., Sd. D.)

Stems 1-2 feet long, subsimple, filiform, pale, weak, and trailing. Petioles uncial; leaflets 3-4 lines long, 2 lines wide, nearly glabrous. Flowers 2-3, on slender pedicels. Bracts and bracteoles rigid, ovate, ciliato-dentate. Joints of the legume suborbicular. The stamens are 10, in 2 pentandrous parcels, exactly as in others of this genus.

XLI. ARACHIS, Linn.

Flowers polygamous. STERILE: *calyx-tube* very long and slender (resembling a flower-stalk); limb bipartite, the upper lip 4-toothed, the lower slender, entire. *Corolla* inserted in the throat of the calyx; *vexillum* roundish; *alæ* oblong, free; *carina* incurved, rostrate. *Stamens* monadelphous, inserted with the petals. *Ovary* concealed in the base of the calyx-tube, subsessile, 2–3–ovuled, abortive. FERTILE: *Calyx, corolla,* and *stamens* none. *Ovary* on a quickly elongating, rigid, reflexed (pedicel-like) torus, stipitate, unilocular, with 2–3 *anatropous* ovules. *Style* very short. *Stigma* dilated. *Legume* (buried under ground) oblong, thick, reticulated, indehiscent, subtorulose, 2–3-seeded. Embryo straight, with thick cotyledons. *Endl. Gen.* 6601. *DC. Prod.* 2, *p.* 474.

A small, herbaceous plant, said to come originally from tropical America, but now common in all the warmer parts of the world, and much cultivated by half-civilized man, for its seeds; the common *earth-nut*. The name was given by Pliny to a plant with neither stem nor leaves, but all root. It is now applied to one without proper flower-stalk, or axis of inflorescence.

1. A. hypogæa (Linn. Sp. 1040); *DC. l. c. E. & Z.!* 1696.

HAB. Coast land about Port Natal, *Dr. Sutherland!* (Herb. Hk.)

Stems herbaceous, diffuse, 1–2 feet long, pubescent. Leaves abruptly bijugate, the petiole with two adnate, subulate stipules at base. Leaflets obovate-obtuse, penni-nerved, becoming glabrous, 1–1½ inch long, ¾ inch wide. Flowers solitary, axillary; the sterile from the upper, the fertile from the lowest axils. The stipe of the ovary rapidly elongates after fertilization, and forces the young fruit under the soil, where it ripens.

XLII. STYLOSANTHES, L.

Flowers polygamous, very generally sterile. *Calyx-tube* very long and slender; limb deeply bilabiate, the upper lip 4-fid, the lower elongate, entire. *Corolla* inserted in the throat of the calyx; *vexillum* roundish; *alæ* oblong, free; *carina* incurved, rostrate, shorter than the alæ. *Stamens* monadelphous, with a split tube. *Ovary* sessile, in the base of the calyx-tube (commonly abortive); the style filiform, elongate. FERTILE: *Calyx, corolla,* and *stamens* none. *Ovary* subsessile, erect, 2–ovuled; style short, hooked. *Legume* sessile, mostly 2-jointed; joints compressed, the lower often sterile, the upper one-seeded, separating. *Endl. Gen.* 6606. *DC. Prod.* 2, *p.* 317.

Weedlike herbs or undershrubs, frequently viscid-pubescent, common throughout the warmer regions of the globe. Leaves pinnately-trifoliolate. Stipules adnate, striate. Flowers crowded in dense, terminal or axillary spikes, each in the axil of a leafy bract, pedicellate; or solitary, bi-tribracteolate, or in pairs, one perfect, the other barren. Name στυλος, a *style,* and ανθος a *flower;* a flower with a very long style.

1. S. setosa (Harv.); suffruticose, dwarf, rigidly hispid and pubescent; leaflets oblongo-lanceolate, subpungent-mucronate, pubescent, rib-striate beneath, rigid; stipules subpungent; *fl. unknown. Zey.!* 404.

HAB. Aapjes River, *Burke & Zeyher!* (Herb. Hk., Sd.)

Root woody. Stems numerous, subsimple, erect, 6–8 inches high, roughly hairy. Leaflets 6–7 lines long, 2½ wide, longer than the petiole; the terminal 1–2 lines removed. Nerves prominent beneath.

XLIII. DESMODIUM, DC.

Calyx 5-parted or deeply bilabiate, the upper lip bifid, the lower trifid. *Vexillum* roundish; *alæ* oblong, longer than the straight, obtuse carina. *Stamens* diadelphous, 9 and 1. *Ovary* sessile, many-ovuled, *Legume* several-jointed, the joints compressed, one-seeded, membranous or rigid, separating at maturity. *Seeds* compressed, reniform. *Endl. Gen.* 6615. *DC. Prod.* 2, *p.* 325, *Nicolsonia, DC. l. c. E. Mey.! Comm. p.* 123–4.

Herbs or suffrutices, common in warm countries of both hemispheres. Leaves pinnately trifoliolate or unifoliolate; the terminal leaflet bistipellate, the lateral unistipellate. Racemes terminal, slender or densely flowered. Flowers small, purple or white. Name, *δεσμος*, a *bond;* the stamens are connected.

ANALYSIS OF THE SOUTH AFRICAN SPECIES.

Lvs. trifoliolate; racemes *ovoid*, very dense, short ... (1) **Dregeanum.**
Lvs. foliolate; racemes *cylindrical, elongating;* pedicels short (2) **grande.**
Lvs. trifoliolate; racemes *very lax*, paniculate; pedicels long and filiform:
Leaflets ovate, *acuminate*, the lateral unequal-sided... (3) **strangulatum.**
Lfts. obovate, *obtuse;* legume moniliform ... .. (4) **setigerum.**
Leaves unifoliolate; leaflet *cordate;* racemes slender, long (5) **natalitium.**

1. **D. Dregeanum** (Benth.!); suffruticose, erect, thinly silky-villous, with appressed hairs; leaves on short petioles, pinnately trifoliolate; leaflets obovate-oblong, obtuse, often complicate, glabrous above, thinly pilose and netted beneath, concolourous; stipules lanceolate, acuminate, stipellæ setaceous; racemes terminal and axillary, *very dense and short;* pedicels crowded, 2–3 together, short; calyx densely piloso-barbate with yellow hairs; legumes deflexed, about 4-jointed, the joints subquadrate, pubescent, the intermediate constrictions shallow. *Krauss.!* 143. *Nicolsonia caffra, E. Mey.! l. c.*

HAB. Banks of streams between the Great Cataract and Omsamcaba, *Drege!* Natal, *Krauss! Sutherland! Sanderson, &c.* (Herb. Hk., Sd., D.)

2–3 feet high, more ligneous than other S. African species. Petioles ½–¾ inch long; leaflets 1–1¼ inch long, 5–6 lines wide, the terminal largest. Racemes ovoid, shorter than the leaf, very dense, but not capitate; the pedicels 2–3 lines long. The yellow calycine hairs conspicuous. Legumes about an inch long, but little constricted between the joints.

2. **D. grande** (E. Mey. ! l. c.); suffruticose, erect, hispido-pubescent; leaves on longish petioles, pinnately trifoliolate; leaflets *elliptic-oblong, subacute*, rigid, paler beneath, penni-nerved, glabrous above, appressedly hispid along the nerves beneath; petiolules hispid; stipules scarious, ovato-lanceolate, stipellæ subulate; *racemes terminal and axillary, cylindrical, elongating, many-flowered*, pedicels scarcely longer than the calyx, 2–3 together; calyx setose, bilabiate; legume 4–6 or more jointed, joints oblong or subquadrate, coriaceous, pubescent, the intermediate constrictions variable.

HAB. Wet places in the plains between Omblas and Port Natal, *Drege! Gueinzius!* (Herb. Hk., Sd.)

Stem 3-4 feet high? branching. Leaflets 2½–3 inches long, 1–1½ wide, with prominent nerves beneath. Racemes from the upper axils, as well as ending the branches, 3–5 inches long, closely flowered, the bracts, before flowering, imbricating; afterwards deciduous, lanceolate, acuminate. Pedicels 2 lines long. Legume sometimes scarcely constricted between the joints, sometimes deeply so.

3. **D. strangulatum** (W. & Arn. Fl. Ind. 1, p. 228); suffruticose, erect, hispido-pubescent; leaves on long petioles, pinnately trifoliolate; leaflets *broadly ovate, acuminate*, paler beneath, sparsely hispidulous on both sides, the lateral unequal-sided; petiolules roughly hispid; stipules scarious, lanceolate, acuminate, stipellæ setaceous; panicle *very lax, remotely flowered;* pedicels 2–4 together, filiform, elongate; calyx setose, subequally 5-fid; legume 2–3-jointed, joints *falcato-cultrate*, mar-

gined, pubescent, joined by narrow bands. *Wight. Ic. t.* 985. *D. Caffrum, E. & Z.! No.* 1662.

HAB. Makasani-river, Caffr., *E. & Z.!* Port Natal, *Gueinzius.* (Herb. Sond.)

"3 feet high, simple." Stem weak, angular, more or less pubescent or hispidulous. Leaflets 2–3 inches long, membranous, 1½–2 inches wide. Panicle a foot or more long, terminal, very lax. *E. & Z.'s* specimens are rather more robust and hairy than the Indian, but those from *Gueinzius* exactly agree with Ceylon individuals in Herb. T.C.D. The joints of the legume are almost half-moon shaped, but narrowed to one end.

4. D. setigerum (Benth.!); subherbaceous, diffuse, the stem patently hirsute; leaves on shortish petioles, pinnately 3-foliolate; leaflets *broadly obovate, obtuse,* appressedly pilose on both sides; stipules scarious, lanceolate, acuminate, stipellæ setaceous; racemes (paniculate) terminal and axillary, very long and lax, remotely flowered; pedicels 2–3 together, filiform, elongate; calyx setose, subequally 5-fid; legume 4–5 jointed, joints ovato-subrotund, pubescent, the intermediate constrictions deep. *Nicolsonia setigera, E. Mey.! l. c.*

HAB. Grassy places near Omsamwubo, *Drege!* Coast land of Natal, *Sutherland! Sanderson!* (Herb. Hk., Sd., D.)

Stems weak, suberect or spreading, 1–2 feet long, roughly pubescent, with foxy hairs. Leaflets 1–1½ inch long, ¾–1 inch wide, scarcely paler beneath. Racemes often 12 inches long, the pedicels an inch apart, and nearly or quite an inch long, very slender. Joints of the legume 1 line or rather more in diameter.

5. D. natalitium (Sond. in Linn. vol. 23, p. 32); suffruticose, decumbent, slender, hispido-pubescent; leaves on longish petioles, *unifoliolate;* leaflet *cordate-ovate,* subacute, scaberulous above, sparsely pilose, netted veined, and paler beneath; stipules subulate, acuminate, stipellæ setaceous; racemes terminal, long and lax; pedicels sub-binate, erect, scarcely longer than the flower; calyx pubescent, bilabiate; legume 5–8-jointed, joints ovato-subrotund, pubescent, the intermediate constrictions deep.

HAB. Port Natal, *Gueinzius!* (Herb. Sond.)

The smallest and most slender of the South African species. Stems filiform. Petioles nearly uncial; leaflet an inch long, ¾ inch wide at base. Racemes 5–6 inches long; pedicels 2–3 lines long. Legumes moniliform, the ventral suture straight, the carinal deeply crenate.

XLIV. ANARTHROSYNE, E. Mey.

Same as *Desmodium,* but: *Legume* compressed, linear-subfalcate, imperfectly articulate, not spontaneously separating into one-seeded fragments. *E. Mey. Comm. p.* 124.

Tropical and subtropical suffrutices and herbs, with the habit of *Desmodium,* from which they are distinguished by the unjointed legume, as the generic name (derived from *α, privative,* and *αρθροω, to have joints*) signifies.

1. A. robusta (E. Mey. l. c.); suffruticose, erect, densely and softly tomentose; leaves on short petioles, 3-foliolate; leaflets broadly elliptical or oblong, obtuse, pilose above, paler, tomentose and penni-nerved beneath, the lateral unequal-sided; racemes in a terminal panicle, cylindrical, elongating, closely flowered, villous; pedicels short; legumes pubescent, slightly constricted.

HAB. Among tall grass near Omgaziana, *Drege!* Natal, *Plant, Gueinzius, &c.* (Herb. Hk., Sd., D.)

Stem 3 feet high, rigid, robust, furrowed, densely tomentose, pale. Petioles uncial, the terminal leafl. remote. Leaflets 3-4 inches long, 1½-2 inches wide, thick and soft. Panicle 1 foot long, of many slender racemes, each 4-6 inches long; pedicels 2-3 lines long, erect. Flowers 2½ lines long, yellow or reddish?

XLV. ALYSICARPUS, Neck.

Calyx persistent, *glumaceous*, deeply 4-parted, the upper segment emarginate or bifid. *Corolla* papilionaceous, small, scarcely longer than the calyx. *Stamens* diadelphous, 9 and 1. *Legume* terete or subcompressed, several-jointed, the joints equal-sided, separating. *Endl. Gen.* 6626. *DC. Prod.* 2, *p.* 353.

Small, weedlike herbs or suffrutices, natives of tropical and subtropical Asia and Africa. Leaves unifoliolate, bistipulate; stipules and bracts scarious. Flowers racemose, pedicellate, in pairs, purple, inconspicuous. Name from λυσις, a separation or *solution*, and χαρπος, *fruit*; because the legumes break up.

ANALYSIS OF THE SOUTH AFRICAN SPECIES.

Joints of the glabrous legume deeply furrowed and ribbed (1) **Wallichii.**
Joints of the downy legume neither furrowed nor ribbed, even (2) **Zeyheri.**

1. A. Wallichii (W. & Arn. Prod. 1. p. 234); ascending, glabrous; leaves subcordate-oblong, the upper ones linear-oblong or ovato-lanceolate; racemes terminal, *cylindrical, imbricating;* bracts broadly ovate, shorter than the flower, deciduous; legumes scarcely longer than the calyx, 4-5-jointed, tipped with the straight base of the style, subcompressed, glabrous, *the articulations deeply furrowed and ribbed transversely*, broader than long. *A. glaber, E. Mey.! Comm. p.* 125. *Zey.!* 463. *ex pte.*

HAB. Among grasses between Omsamwubo and Omsamcaba, *Drege.* Crocodile and Aapjes River, *Burke & Zeyher!* (Herb. Hk., Sd., D.)

Pale yellowish green, rigid, 12-18 inches high. Leaves very variable in shape, sometimes 1 inch long, ¾ inch wide, and very blunt; sometimes 2 inches long, ¼ inch wide, and acute: the lower leaves are always the broadest and bluntest. Racemes 3-4 inches long, densely many-flowered.

2. A. Zeyheri (Harv.); ascending, glabrous; leaves elliptic-oblong, the upper ones linear-oblong, all strongly netted-veined; racemes terminal, *lax, interrupted;* bracts broadly ovate, nearly equalling the flower, deciduous; legumes longer than the calyx, 3-4-jointed, tipped with the straight base of the style, subcompressed, *downy, the articulations beadlike, smooth,* (not wrinkled) or faintly subreticulate. *Zey.!* 463, *ex pte.*

HAB. Aapjes River, *Burke & Zeyher!* (Herb. Hk., Sd.)

The leaves are more rigid and much more strongly veiny than in *A. Wallichii;* the racemes are longer and laxer; and the legume is very different.

XLVI. REQUIENIA, DC.

Calyx campanulate, 5-fid, the segments acute, the lowest longest. *Vexillum* obovate; *carina* obtuse, dipetalous. *Stamens* monadelphous, the tube cleft above. *Ovary* sessile, uniovulate; style short, incurved.

Legume oval, compressed, mucronate, one-seeded. *DC. Leg. Mem. p.* 224. *t.* 37, 38. *Endl. Gen.* 6471. *DC. Prod.* 2, *p.* 168.

Tomentoso-canescent suffrutices, natives of Senegal and S. Africa. Leaves alternate, unifoliolate; leaflet obcordate, closely penninerved, mucronate. Stipules free. Flowers axillary, very small, subsessile, subsolitary or clustered. Named in honour of *M. Requien*, a French botanist.

1. R. sphærosperma (DC. l. c.); stipules shorter than the calyx; legumes pubescent, tapering at base; seeds globose. *DC. Prod.* 2, *p.* 168.

HAB. S. Africa, *Burchell.* Aapjes River, *Zeyher!* 368. Eastern frontier, *H. Rutherfoord!* (Herb. Hk., Sd., D.)

Root thick and woody. Stems numerous, rigid, woody at base, suberect or flexuous and diffuse, 6–8 inches long, not much-branched, densely tomentose and canescent. Petioles 1 line long; leaflet broadly obovate or orbicular, subobcordate, with a recurved or hooked mucro, complicate, silky-canescent on both sides, prominently penninerved on the lower. Stipules subulate, 1½ line long, patent. Flowers on very short pedicels, minute, 1–2–3 together. Pods thinly pubescent, tapering much at base, oblong-subobovate, 2–3 times longer than the calyx.

XLVII. HALLIA, Thunb.

Calyx subequally 5-fid. *Vexillum* ovate; *alæ* oblong, longer than the obtuse *carina*. *Stamens* completely monadelphous. *Ovary* substipitate, uniovulate. *Legume* compressed, membranous, one-seeded. *Endl. Gen.* 6469. *DC. Prod.* 2, *p.* 122.

Small, ascending or trailing suffruticose plants, natives of the Cape. Leaves alternate, simple, very entire, often nigro-punctate, bistipulate. Flowers axillary, solitary, pedunculate or subsessile, small, purple. Named in honor of Birger Martin Hall, a favorite pupil of Linnæus.

ANALYSIS OF THE SPECIES.

Leaves acute or tapering *at base*; stipules erect.
- Stem flattened; stipules adnate; peduncles very short ... (1) **alata.**
- Stem filiform; stipules adnate, subulate; peduncles very long (2) **filiformis.**
- Stem angular; stipules nearly free, toothlike, erect; pedunc. moderate ... ... (3) **virgata.**

Leaves cordate or ovate *at base*; stipules reflexed or spreading.
- Lvs. open, ovate-oblong or ovato-lanceolate, acuminate; pedunc. setaceous ... ... (4) **cordata.**
- Lvs. open, broadly ovate or oblong, mucronate; pedunc. setaceous ... ... (5) **asarina.**
- Lvs. complicate, broadly-cordiform, acute; flowers subsessile (6) **imbricata.**

1. H. alata (Thunb.! Cap. 593); stem *flattened, two-edged;* stipules adnate with the petiole throughout and longer than it; leaves lanceolate or obovate-oblong, acute at base, folded, the younger thinly silky, the older glabrous, rigid; *peduncles very short;* calyx glabrous, silky-ciliate, enlarged and ribbed in fruit. *DC. l. c. p.* 123. *E. Mey. Comm. p.* 81. *E. & Z.!* 1251.

HAB. About Capetown, and on the Cape Flats, and common generally through the Western Districts. Knysna, *Dr. Pappe!* (Herb. Th., Hk., Sd., D.)

Stems trailing or ascending, very rigid, 1–2 feet long, branched chiefly from near the base; the branches long and simple, curved. Petioles 3–4 lines long, the linear lanceolate, adnate stipule about 1 line longer. Leaves very variable in breadth, ½–1 inch long: the upper ones often abortive. Peduncle shorter than the petiole. Calyx lobes lanceolate, the lowest longest and boatshaped. Legume terete, 1½ line long, the seed completely filling the cavity.

2. H. filiformis (Harv.); stem *filiform*, striate; stipules adnate with the longish petiole throughout *and much longer than it, their points subulate;* leaves very narrow-linear, acute, folded, tapering at base, the younger silky-ciliate, the older glabrous; peduncles *setaceous, flexuous, much longer than the leaves;* calyx-lobes silky-ciliate.

HAB. Tulbagh Waterfall, *Dr. Pappe!* (Herb. D.)

This singular plant looks like a "sport" or monstrosity, and if so it must have sprung from *H. alata*, judging by the completely adnate stipules, the folded leaflets and the tendency to silky pubescence; but in characters of stem, and in the very long stipules and peduncles, it is quite unlike that species. Free portion of the stipules 5–6 lines long. Peduncles 1½–2 inches long.

3. H. virgata (Thunb.! Cap. 593); glabrous; stem angular, flaccid; stipules adnate *at base* to the very short petiole, and longer than it, *rigid, toothlike, erect;* leaves lanceolate or linear (rarely obovato-lanceolate), acute at each end, flat, rigid; peduncles filiform, shorter than the leaf or equalling it; calyx but slightly enlarged in fruit. *DC. l. c; also H. angustifolia, DC. E. & Z.!* 1253, 1254. *H. flaccida, Sieb.!* 242, *and Fl. Mixt. No.* 27, *non Thunb.*

HAB. Moist places and among grass, common. (Herb. Th., Hk., Sd., D.)

Stems trailing, 2–3 feet long, branched near the base; the branches long, simple, curved. Leaves scattered, very variable in breadth and form, always acute at base, 1–2 inches long, 1–8 lines wide, nigro-punctate beneath. Peduncles hairlike, very variable in length. Legume ovoid, wrinkled.

4. H. cordata (Thunb.! Cap. 593); stem *triangular*, villous or glabrate; stipules free, *lanceolate or ovato-lanceolate*, reflexed, mostly longer than the short petiole; leaves cordate at base, *ovate-oblong or ovato-lanceolate*, acuminate, pilose or glabrate; peduncles setaceous, nearly equalling the leaf; calyx villous or glabrate. *DC. l. c. E. & Z.!* 1250. *H. flaccida, Thunb.! Fl. Cap.* 593. *E. & Z.!* 1252.

HAB. Moist places and among grasses, common. (Herb. Th., Hk., Sd., D.)

Stem sharply 3-angled, trailing, 1–2 feet long, branched near the base; branches simple. Leaves 1–1½ inch long, 2–6 lines wide, most commonly oblongo-lanceolate, obtusely cordate at base. Pubescence variable, sometimes nearly absent. Leaves much narrower and less cordate than in the next.

5. H. asarina (Thunb.! l. c. 594); stem angular and striate, villoso-pilose; stipules free, *ovate* or ovato-lanceolate, reflexed, equalling or exceeding the petiole; leaves *cordate at base, ovate or elliptic-oblong, mucronate*, pilose or subglabrate; peduncles setaceous, nearly equalling the leaf; calyx villous. *DC. l. c. E. & Z.!* 1249. *H. convexa, Burch.* 4046.

HAB. Moist places and among grass, common. (Herb. Th., Hk., Sd., D.)

Nearly allied to *H. cordata*, but with much broader and shorter leaves, often, but not always, deeply cordate at base, and with multangular stems. The pubescence is more copious. Petioles 1–5 lines long. Leaves 1–1¼ inch long, 5–9 lines wide, not acuminate. *Burchell's H. convexa*, has exactly elliptic-oblong leaves, scarcely cordate at base, but such occur often on specimens having also deeply cordate lower leaves.

6. H. imbricata (Thunb.! l. c. 594); stem angular and striate, villoso-pilose; stipules free, ovato-lanceolate, spreading or reflexed, acute; leaves subsessile, broadly cordate, acute, *folded, the uppermost distichously* *imbricating*, villous and piloso-ciliate; *flowers subsessile;* calyx pilose. *DC. l. c. E. & Z.!* 1248.

HAB. Moist places, among grass and shrubs, common. (Herb. Th., Hk., Sd., D.) Diffuse or procumbent; stems filiform, 1–2 feet long; the branches long and flaccid. Leaves closely set, the lower ones sometimes petioled, the upper nearly sessile, closely placed, rarely an inch long, nearly 10 lines wide at base, exactly heartshaped. Flowers deep purple.

XLVIII. ALHAGI, Tournef.

Calyx shortly 5-toothed. *Vexillum* obovate, complicate; *alæ* oblong; *carina* straight, obtuse. *Stamens* diadelphous, 9 and 1. *Ovary* several-ovuled; style filiform. *Legume* stipitate, ligneous, terete, few-seeded, irregularly constricted here and there, but not articulated, indehiscent. *Endl. Gen.* 6625. *DC. Prod.* 2, *p.* 352.

Undershrubs or herbs, natives chiefly of the deserts of N. Africa and Central Asia. Leaves simple, with minute stipules. Peduncles axillary, spinous. Flowers few, red. The name is from the Arabic *Algul.* Manna is collected from these plants about the Taurus, and in other eastern countries.

A. maurorum (Tournef.); stem shrubby; leaves obovate-oblong; calyx-teeth acute. *DC. l. c.*

HAB. Karroo, near Olifant's River and Brackfontein, *Eck. & Zey.!* (Herb. Sd.) I introduce this with much hesitation. A single, spiniferous branch, without flower or fruit, which may or may not belong to *Alhagi maurorum*, exists in Herb. Ecklon. This is the only evidence for the plant in S. Africa.

XLIX. VICIA, L.

Calyx campanulate, subequally 5-cleft or toothed. *Corolla* much exserted; vexillum expanded. *Stamens* 9–1. *Ovary* subsessile; the style bent upwards at a right angle, with a tuft of hairs under the stigma. *Legume* compressed or turgid, 2 or many-seeded; seeds subglobose, with an oval or linear scar. *Endl. Gen.* 6581. *DC. Prod.* 2, *p.* 354.

Twining and climbing herbs, annual or perennial, common throughout the temperate zones of Europe, Asia, and America; only *naturalized* in S. Africa. Leaves abruptly pinnate, in several pairs, the common petiole mostly produced into a branching or rarely simple tendril; stipules mostly semi-sagittate; peduncles axillary, short or long, 1–2, or racemosely many-flowered. Flowers blue, purple, yellow or white, or parti-coloured. Name, said to be from *vincio*, to bind together; because these plants attach or bind themselves to objects by their tendrils. English name, Vetch; French, Vesce.

ANALYSIS OF THE (NATURALIZED) SOUTH AFRICAN SPECIES.

Flowers solitary or in pairs, subsessile ... ... (1) **sativa.**
Peduncles elongate, many-flowered ... ... (2) **atropurpurea.**

1. V. sativa (Linn. Sp. 1037); leaves cirrhose; leaflets 6–12, the lower ones obovate or obcordate, the upper narrower and often linear, all mucronate, pubescent or subglabrous; stipules semisagittate, sharply toothed or laciniate; *flowers solitary or in pairs, subsessile;* calyx oblong-campanulate, its teeth subulate, equalling the tube; vexillum glabrous; legume linear, compressed, suberect, pubescent or rarely glabrous; seeds globose, smooth. *DC. Prod.* 2, *p.* 61. *E. & Z.!* 1664. *E. Mey. Comm. Drege, p.* 126. *E. Bot. tt.* 334, 2614, 2708.

HAB. A weed, in cultivated and waste ground, throughout the colony. (Herb. T.C.D., &c.)

A small, weak-growing biennial, varying in the breadth of the leaflets and in pubescence. Leaves subsessile; leaflets ½–1 inch long, 2–1 line wide. Flowers purple or blueish, or a reddish-lake, on peduncles 1 line long. Legumes 2 inches long, strongly compressed. This species is now naturalized throughout the temperate zones of both hemispheres.

2. V. atropurpurea (Desf. Fl. Atl. 2, p. 164); densely villous; stems 4-angled; leaflets 10–14, oblong or linear, mucronate, softly hairy, alternate or opposite; stipules semisagittate, often one-toothed at base; *peduncles nearly equalling the leaves, many-flowered;* flowers secund, close together; calyx-teeth bristle-shaped, pilose, longer than the tube; legumes oblong, compresso-turgid, densely hairy; seeds globose, black, the scar velvetty. *DC. Prod.* 2, *p.* 359. *Vent. Hort. Cels. t.* 84. *Bot. Reg. t.* 871. *V. albicans, Lowe!*

HAB. Near the Capetown Observatory, *Zeyher!* (Herb. Sond.)

A native of Algeria, Madera, and the Azores; probably a mere escape from the Observatory Garden. I have seen but a single Cape specimen, in Herb. Sond.; it agrees in all respects with those from the Mediterranean and Azores, in Hb. T.C.D. Stem 2–3 feet high, sharply angular and rib-striate, much branched. Leaves subsessile, 2–3 inches long, with many pairs of leaflets. Flowers dark purple, the alæ paler or white at base. Young pods very hirsute; older much less so.

L. DUMASIA, DC.

Calyx cylindrical, obliquely truncate, entire, bibracteolate at base. Claws of the petals equalling the calyx; limb of the *vexillum* cordate-oval; *carina* obtuse. *Stamens* 9–1. *Ovary* few-ovuled; style filiform at base and apex, dilated beyond the middle. *Legume* attenuated at base, 2-valved, compressed, few-seeded, contracted between the seeds. *DC. Prod.* 2, *p.* 241. *Endl. Gen.* 6631. *W. & A. Prod.* 1, *p.* 206.

Twining, herbaceous or suffruticose, slender plants, common in tropical Asia. Leaves pinnato-trifoliolate. Racemes axillary. This genus is readily known by its *truncate*, shortly tubular calyx. It is named in honour of *M. Dumas*, a French naturalist.

D. villosa (DC. Leg. Mem. p. 257, t. 44); stem and leaflets more or less pubescent or villous; legumes villous. *D. pubescens, DC. l. c. t.* 45. *Prod.* 2, *p.* 241. *D. capensis, E. & Z.!* 1625. *Burch.! Cat.* 5437.

HAB. S. Africa, *Burchell!* Shady places at the Knysna, *E. & Z.!* (Herb. Sond.)

Very variable in pubescence; the S. African specimens glabrescent. Stems slender, twining, the younger parts retrorsely puberulous. Petioles filiform. Leaflets broadly ovate, obtuse, subglabrous above, minutely appresso-puberulent beneath, thin and membranous. Racemes several-flowered, equalling or slightly exceeding the leaves. Flowers 4–5 lines long.

LI. TERAMNUS, Sw.

Calyx tubuloso-campanulate, 4–5-fid. *Vexillum* obovate, with a longish claw; *alæ* narrow-oblong, oblique; *carina* shorter, oblique, obtuse. *Stamens* monadelphous, the alternate rostrate. *Ovary* sessile, with a short, thick style and capitate stigma. *Legume* linear, many-seeded, hook-pointed, septate within. *Benth. Fl. Braz. XXIV. p.* 138. *Glycine, sp.* Auct.—*Bujacia, E. Mey. Comm. p.* 127.

Slender, twining, tropical plants. Leaves pinnately-trifoliolate; leaflets stipellate, the terminal remote. Flowers minute, on slender, axillary peduncles, in pairs or fascicled, or in interrupted racemes. Pedicels short, bibracteolate under the calyx. Name from *τεραμνος, soft;* because the pods and leaves are soft.

T. labialis (Spreng. Syst. 3. p. 235); variably pubescent with reflexed hairs; leaflets ovate or oblong, obtuse or mucronate, appresso-pubescent or silky; peduncles longer than the leaves, interruptedly many-flowered; upper lip of the calyx deeply bifid; vexillum narrowed at base; alæ unidentate, longer than the obtuse carina; legume appressedly hispid or glabrescent. *Glycine labialis. Linn. Suppl.* 325. *G. parviflora, Lam, DC.* 2, *p.* 242. *G. abyssinica, and Kennedya arabica, Hochst. Bujacia gampsonychia, E. Mey.! Comm. p.* 127.

HAB. Caffraria, by the Key R. and between Omtendo and Omsamculo, *Drege!* Natal, *Gueinzius!* (Herb. Hk., Bth., Sd.)

Stems slender; in our specimens rough with reflexed, fulvous, rather rigid hairs, sometimes glabrescent. Petioles 1–2 inches long. Leaflets 1–1½ inch long, ¾–1 inch wide, varying from sparsely hispid to densely silky, the hairs appressed. Racemes 3–4 inches long, the flowers minute, 2–3 together in tufts, 3–4 lines apart, silky pubescent or subvillous. Calyx 2 lines long, the segments subequal, the two upper broader, connate at base or nearly to the middle. Vexillum 3 lines long, obovate, with a long, tapering claw. Legume 1½–2 inches long, 1½ line wide, linear, slightly falcate, with a thick, incurved style and 10–12 transverse seeds, separated by cellular septa.

(Doubtful Species.

Bujacia anonychia (E. Mey.! Comm. l. c.) "staminal tube entire; legumes subtorulose, muticous, 5-seeded; leaflets broadly ovate, acuminate." *E. Mey.*

HAB. Among shrubs near Natal, *Drege.* (Unknown to me.)

LII. GALACTIA, P. Brown.

Calyx bibracteate at base, 4-fid, the segments acute, nearly equal. *Vexillum* ovate or suborbicular, patent or reflexed; *alæ* oblong, shorter than the subincurved carina. *Stamens* diadelphous. *Ovary* several-ovuled, subsessile. *Style* filiform, incurved, glabrous; stigma small. *Legume* linear, compressed, with cellular partitions between the seeds, several-seeded. *Endl. Gen. No.* 6653.

Voluble or prostrate herbs or suffrutices, chiefly tropical. Leaves trifoliolate; leaflets stipellate, the terminal distant. Racemes axillary, few-flowered. Flowers small. Name from *γαλα, milk.*

G. tenuiflora (Wight & Arn. Prod. 1, p. 206); voluble, variably pubescent; leaflets from oval to lanceolato-oblong, glabrous and shining above, paler and pubescent beneath; peduncles equalling or exceeding the leaves, 2–4-flowered near the summit; calyx silky (or glabrescent) with linear-falcate segments. *G. tenuiflora and G. villosa, W. & A. l. c. Copisma subsericeum, Sond.! in Linn. Vol.* 23, *p.* 34.

HAB. Port Natal, *Gueinzius!* (Herb. Sd., Hk., D.)

A common coast plant in tropical Asia and Australia. Our specimens exactly agree with the "*G. villosa,*" W. & A., a hairy form that gradually passes into the subglabrous *G. tenuiflora* of the same authors. Stem 2–3 feet high, slender. Petioles 1–2 inches long. Stipules lanceolate, 3 lines long. Leaflets 1½–2 inches long, 8–10 lines wide, more or less ovate or subcordate at base. Calyx-lobes 3 lines long, ½ line wide. Corolla yellow, twice as long. Ovary silky. Legume 2 inches long, strongly compressed, slightly curved, 7–8 seeded.

LIII. ERYTHRINA, L.

Calyx either truncate or bilabiate or cleft on one side and spathaceous. *Vexillum* ovate-oblong, without basal ears or calli, incumbent, very much longer than the alæ and the dipetalous carina. *Stamens* straight, exserted, diadelphous or imperfectly monadelphous. *Ovary* stipitate, many-ovuled. *Style* straight, glabrous, with a lateral stigma. *Legume* indehiscent, compressed between the seeds, tipped with the hardened style. *Seeds* oval, with a linear scar. *Endl. Gen. No* 6667.

Trees or shrubs, natives of warm countries generally. Stem and leaves often prickly. Leaves pinnately-trifoliolate, the terminal leaflet remote. Stipellæ glandular. Stipules free. Flowers racemose, large and handsome, red or scarlet. Seeds commonly red and black. English name, "coral-tree." The generic name is from *ερυθρος, red.*

TABLE OF THE SOUTH AFRICAN SPECIES.

1. **Eu-erythrina.**—*Calyx* truncate or bilabiate, subentire or shortly 5-toothed.
   Petioles without prickles; stem *arborescent;* legume unarmed, glabrate ... ... (1) **caffra.**
   Petioles and often the nerves of the leaflets armed with *prickles.*
   Legumes unarmed; leaflets subacute or obtuse; calyx truncate ... ... (2) **Zeyheri.**
   Legumes unarmed; leafl. acuminate; calyx shortly 5-toothed ... ... (3) **Humei.**
   Legumes *prickly;* leafl. transversely elliptic, obtuse (4) **acanthocarpa.**
2. **Chirocalyx.**—*Calyx* cleft on one side, spathaceous; its lobes filiform ... ... (5) **latissima.**

1. **EU-ERYTHRINA,** (Sp. 1–4).

1. E. caffra (Thunb.! Fl. Cap. p. 559); a tree; branches prickly, *petioles unarmed;* leaflets petiolulate, broadly ovate, obtusely acuminate, glabrous; racemes densely many-flowered; calyx tomentulose, obliquely labiate or splitting, afterwards enlarged, obsoletely denticulate; vexillum minutely velvetty; vexillary stamen adnate below the middle to the split staminal tube; legumes moniliform, glabrate, unarmed. *E. & Z.!* 1691. *E. Mey. Comm. p.* 159. *Krauss!* 286.

HAB. In woods, not far from the sea, in Uitenhage, Albany, Caffraria, and Port Natal; often cultivated in colonial gardens. (Herb. Th., Hk., Bth., Sd., D.)

A tree 30–40–60 feet high. Bark of the twigs pallid, rugose. Prickles small. Leaves clustered toward the end of the twigs, on puberulous petioles 2–4 inches long, the terminal leaflet 1–1½ inch from the lateral pair. Stipellæ gland like. Leaflets 2–2½ inches long, 1½–2 inches wide, rounded at side, suddenly tapering to a blunt point, thin and membranous. Peduncles thick as a goose-quill, 4–6 inches long, floriferous from the middle. Flowers scarlet, the vexillum 1½–2 inches long, falcate-oblong, of thick substance. Ovary and young pods densely tomentose, the old ones naked, strongly constricted between the seeds.

2. E. Zeyheri (Harv.); arborescent? ; branches, petioles, and the nerves of the leaves prickly; petioles araneo-pubescent, ribstriate; terminal leaflet *broadly ovate, subacute,* lateral ovate-oblong or ovato-lanceolate, subcuneate at base, all prominently ribbed and veined beneath; stipules oblong, obtuse, or subacute; racemes on long peduncles, densely many-flowered; calyx puberulous, tubular, *obsoletely crenate;* vexillum scarcely velvetty; vexillary stamen nearly free to the base; legumes torulose, unarmed, subglabrate. *Zey.!* 531.

HAB. Between Mooje R. and Magalisberg, *Burke & Zeyher!* (Herb. Hk., Bth., Sd., D.)

Very near *E. Humei*, but a much larger and coarser plant, and perhaps arborescent; judging from the leaves which are crowded round the end of the twigs. Stipules ¾–1 inch long, thick and leathery, much larger than those of *E. Humei.* Petioles 6–8 inches long, 2–2½ lines in diameter, ribbed and furrowed. Terminal leaflet 3–6 inches long, 2–5 inches broad, sparsely prickly on both sides, subacute, but not acuminate; the lateral 2–5 inches long, 1½–3 wide, less ovate. Peduncles 10–16 inches long; the rachis tomentulose. Calyx somewhat membranous, truncate, with very obsolete lobes. Legumes 7–8 inches long, ¾ inch wide at the seeds, narrowed between. Flowers crimson.

3. E. Humei (E. Mey.! Comm. p. 150); a shrub; branches, *petioles and (often) the nerves of the leaves prickly;* leaflets petiolulate, broadly ovate, *subobtusely acuminate,* glabrous; racemes *on long peduncles,* densely many-flowered; calyx puberulous or tomentulose, tubular, shortly 5-toothed; vexillum minutely velvetty; vexillary stamen free nearly to the base; legumes unarmed. *E. caffra, Ker. Bot. Reg. t.* 736. *A. & B. Bot. Mag. t.* 2431. *DC. Prod.* 2, *p.* 412, *excl. syn. Thunb.*

VAR. β. **Raja**; smaller; the teeth of the calyx longer and more acuminate; strongly recurved; vexillary stamen adnate above the base. *E. Raja, Meisn.! in Hook. Lond. Journ.* 2, *p.* 96.

HAB. Grassy hills in Caffraria, between Kovi and Kap R., and near the Keiskamma, *Drege!* Queenstown District, *Mrs. F. W. Barber!* β. Natal, *Krauss!* 62. (Herb. Hk., Bth., Sd., D.)

A shrub, growing always alone, in open places. Bark of the half herbaceous twigs dark-coloured, glabrous, and even. Prickles pale horn-colour, glossy, triangular. Leaves scattered, on glabrous petioles, 3–5 inches long, the terminal leaflet remote. Leaflets 2–3 inches long, 1½–2½ broad, all broad at base, coriaceous, green, reticulated, glossy. Peduncles 12–16 inches long, tapering, floriferous only beyond the middle; the rachis tomentulose. Flowers crimson-scarlet, 1–1¾ inch long. Ripe legumes not seen.

4. E. acanthocarpa (E. Mey.! Comm. p. 151); shrubby; twigs, petioles, and midrib of the leaflets prickly; petioles slender, villous; leaflets *transversely elliptical, obtuse or apiculate,* glabrous and glaucous; racemes lateral or terminal, few or many flowered, shortly pedunculate or subsessile; calyx glabrous, campanulate, subtruncate, obsoletely and bluntly lobulate; vexillum scarcely velvetty; vexillary stamen adnate to the split tube; legumes clavato-stipitate, torulose, incurved, armed with prickles. *E. Humeana, E. & Z.!* 1692, *excl. Syn. Bot. Mag. &c.*

HAB. Forming low thickets in Albany, Queenstown, and Caffraria, *Drege! E. & Z.! Mrs. F. W. Barber!* (Herb. Hk., Bth., Sd., D.)

A divaricately branched, rigid shrub, 4–6 feet high, armed with sharp, subulate, reflexed prickles. The bark of the twigs is pale and rugulose. Petioles 1–2½ inches long, slender, woolly when young. Leaflets broader than long, ¾–1 inch long, 1–1½ inch broad, pale, especially beneath. Flowers 1–1½ inch long, the vexillum scarlet, tipped with green. The root, according to Mrs. Barber, is long and succulent, and when perfectly dry is extremely light, and in that state sometimes made into light summer hats (probably like those made in India of the *Neptunia*). The colonial name is Tambookie-thorn.

2. CHIROCALYX, (Sp. 5).

5. E. latissima (E. Mey.! Comm. p. 151); arborescent, velvetty-lanuginous; leaves on long petioles, pinnately 3-foliolate; leaflets broadly ovate, obtuse, densely tomentose at each side, penninerved; petioles and peduncles lanuginous; spike ovoid, densely many-flowered; calyx lanuginous, cleft down one side, the segments filiform; corolla glabrous. *E. Sandersoni, Harv. Thes. t.* 61, 62. *Chirocalyx mollissima, Meisn.! in Hook. Lond. Journ.* 2, *p.* 98.

HAB. Between the Basche, Omtata, and Omsamwubo, 1000–2000f., *Drege.* Tafelberge, Port Natal, *Krauss!* Near Sterk Spruit, *Sanderson!* (Herb. D., Hk.)

A scrubby tree, 10–12 feet high, with greyish-green foliage. Petioles 5–8 inches long to the leaf-pair; the terminal leaflet 2–3 inches distant. Leafl. 5–8 inches broad, 4–6 inches long, subtruncate at base. Peduncles 6–8 inches long, bearing a very dense spike of dull crimson flowers. Calyx spathaceous; its slender limb-segments spreading like the fingers of a hand. Pubescence woolly, whitish, deciduous. In publishing this in the *Thesaurus* as a novelty, I overlooked two previous "*discoveries!*"

LIV. CANAVALIA,

Calyx bilabiate, the upper lip very large, truncate, emarginate or bifid, with broadly rounded lobes; lower small, subentire or trifid. *Vexillum* ample, suborbicular, ridged at back, bi-callous within, with a short claw; *alæ* oblong, eared at base; *carina* equalling the alæ or longer, shorter than the vexillum, incurved. *Disc* sheathing. *Stamens* monadelphous, or imperfectly diadelphous. *Ovary* linear, multiovulate. *Style* incurved, glabrous, with a terminal stigma. *Legume* compressed, subfalcate, with transverse partitions between the seeds. *Seeds* compressed, with a linear scar. *Endl. Gen. No.* 6663. *Benth Fl. Braz. XXIV.*

Climbing or prostrate, tropical or subtropical herbs or suffrutices. Leaves pinnately trifoliolate, the terminal leaflet subdistant. Stipules small; stipellæ minute or none. Racemes axillary, subspicate; flowers solitary or in pairs, rosy, purplish or white; bracts deciduous. Name, from *Canavali*, the Malabar name.

ANALYSIS OF THE SOUTH AFRICAN SPECIES.

Leaflets suborbicular, very obtuse; lower calyx-lip 3-lobed ... (1) **obtusifolia.**
Leaflets ovate-oblong, subacuminate; lower calyx-lip minute, entire ... ... (2) **Bonariensis.**

1. C. obtusifolia (DC. Prod. 2, 404); creeping, glabrous, or when young, silky-pubescent; leaflets *obovate or orbicular, very obtuse;* upper lip of the calyx bilobed, much shorter than the tube, *lower 3-lobed;* carina erostrate. *Benth.! l.c. p.* 178, *tab.* 48. *Dolichos emarginata, Jacq. Schoenb. t.* 221. *Can. emarginata, Don. E. Mey. Comm. p.* 148.

HAB. Caffraria, between Omtendo and Omsamculo, *Drege!* Natal, *Gueinzius!* (Herb. Hk., Sd., D.)

Common near the sea, throughout the tropics, often cultivated. Leaflets 2–4 inches long, 2–3 broad, drying pale. Peduncles a foot long, floriferous near the summit, thinly silky, the subsessile flowers springing from fleshy tubercles.

2. C. Bonariensis (Lindl. Bot. Reg. t. 1199); leaflets oval-oblong, obtusely-acuminate, coriaceous; upper lip of the calyx bifid, *lower very minute, subentire;* carina incurved, erostrate, longer than the wings. *Benth.! Fl. Braz. XXIV. p.* 178. *C. monodon, E. Mey.! Comm. p.* 149. *C. cryptodon, Meisn.! Lond. Journ.* 2, *p.* 96.

HAB. Mouth of the Omtendo, *Drege!* Natal, *Gueinzius! Krauss!* (Herb. Hk., Sd., D.)

A native also of extra-tropical S. America. Leaflets rigid, 2–3 inches long, 1–1½ wide, reticulated. Peduncles short, few-flowered; the short-stalked flowers rising from fleshy tubercles. Calyx ⅜ inch long.

LV. VIGNA, Savi.

Calyx bibracteate at base, campanulate, 4–5-fid (the upper lobes separate or connate), the lowest longest. *Vexillum* ample, patent, with an arched and vaulted claw, and two callous ridges at base within; *alæ* oblong, produced at base, or eared on the claw; *carina* not twisted, inflexed or rostrate. *Stamens* diadelphous or monadelphous, nearly to the middle. *Disc* sheathing. *Ovary* linear, several-ovuled; style compressed and channelled on one side, incurved; *stigma* hooked, oblique. *Legume* terete or compressed, subfalcate, subtorulose, with cellular partitions between the seeds. *Seeds* subreniform, with a small strophiole. *Endl. Gen.* 6675. *Otoptera, DC. Prod.* 2, *p.* 240. *Scytalis, Sphenostylis and Strophostylis, E. Mey.*

Voluble or erect suffrutices or herbs, natives of the warmer parts of both hemispheres. Leaves pinnately trifoliolate. Flowers on long peduncles, floriferous at the summit or racemose. The pods of many may be eaten as Kidney beans. *V. Catjang* (*Dolichos Catjang, Linn.*) is commonly cultivated in the colony for its pods. Name, from *Domenic Vigna*, a commentator on Theophrastus.

ANALYSIS OF THE SOUTH AFRICAN SPECIES.

Peduncles 2–4-flowered at the summit:
- Rigid, suberect, ligneous, nearly glabrous undershrubs; leaves coriaceous, lanceolate or linear, mucronate:
  - Calyx 4-fid, its lobes lanceolate-acuminate, very acute ... (1) **Burchellii.**
  - Calyx-lobes very short, broad, and blunt ... ... (2) **angustifolia.**
- Voluble or prostrate, herbaceous or scarcely ligneous plants, more or less hispidopubescent:
  - Leaflets ovate, oblong, or lanceolate or linear, *not lobed:*
    - Calyx-segments short, broad, blunt ... ... (3) **marginata.**
    - Calyx-segments *acuminate;* carina *spurred at one side* ... (4) **vexillata.**
  - Leaflets hastate, 3-lobed, broad or narrow; the lateral lobes small:
    - Stipules *peltate,* ovate-oblong; cal.-lobes falcato-lanceolate (5) **triloba.**
    - Stipules *basifixed,* lanceolate; cal.-lobes broadbased subulate (6) **decipiens.**

Peduncles racemose, several-flowered.
- Leafl. ovate or ovato-lanceolate, hispidulous ... ... (7) **luteola.**
- Leafl. obovate-subrotund, retuse, glabrate ... ... (8) **retusa.**

1. **V. Burchellii** (Harv.); suffruticose, diffuse or *suberect,* much-branched, nearly glabrous; leaflets ovato-lanceolate or lanceolate, rigid, setaceo-mucronate, glabrous; stipules sagittato-peltate, ovato-lanceolate, stipellæ subulate; peduncles longer than the leaves, 2–4-flowered at the summit; calyx campanulate, *deeply 4-fid, the lobes lanceolate, acuminate, very acute; carina falcate-acute; alæ with an ear-shaped appendage to the claw;* legume . . . . ? *Otoptera Burchellii, DC. Prod.* 2, *p.* 240. *Mem. Leg. t.* 42.

HAB. Interior of S. Africa, *Burchell.* Zooloo Country, *Miss Owen!* (Herb. D.)

Stems ligneous, diffuse; the branches rigid and suberect. Petioles ¼–½ inch long. Adult leaves 1–2½ inches long, ¼–⅓ inch wide near the base, quite glabrous, the younger puberulous, all tipped with a long, setaceous mucro. Peduncles 3–5 inches long, umbellate; pedicels 3–6 lines long. The two upper segments of the calyx are completely connate to the point, all are spreading or reflexed after the flowers open. Carina very acute, like that of a *Crotalaria.* This has quite the habit of *V. angustifolia,* but differs in the calyx and corolla, &c.

2. V. angustifolia (Benth !); suffruticose, *suberect,* much-branched, glabrescent; leaflets linear-lanceolate or oblong, rigid, mucronate, glabrous; stipules and stipellæ shortly-subulate, rigid; peduncles longer than the leaves, 2–4 flowered at the summit; calyx campanulate, *its lobes very short, broad, and blunt,* the two uppermost connate; carina falcate, round-topped; legumes linear, margined, glabrous, tipped with the long, persistent style. *Sphenostylis angustifolia, Sond.! Linn.* 23, *p.* 33.

HAB. Port Natal, *Gueinzius!* 624. Magaliesberg, *Burke & Zeyher!* (Herb. Sd., Hk., Bth.)

Erect or ascending, many-stemmed, 1–3 feet high, very rigid, when young thinly pubescent, quite glabrous when adult. Stipules 1½ line long, triangular. Petioles ½–1 inch long, erect. Leaflets varying from very narrow, linear-lanceolate, acute, to broadly oblong, obtuse, 1½–2 inches long, 3–5 lines wide, full green. Peduncles long or short, subumbellately few-flowered. Flowers purple. Calyx bibracteolate at base, 2 lines long, the lobes broader than their length. Vexillum 7 lines long and wide. Legume 2–3 inches long, 2–3 lines wide, with ridged sutures and a long, yellow persistent style.

3. V. marginata (Benth.); *procumbent or voluble,* sparsely puberulous or glabrate; leaves long-petioled, leaflets elliptic-oblong, obtuse or retuse, margined, sparsely pubescent or glabrous; stipules minute, broadly subulate; peduncles very long, shortly racemose near the summit; calyx cupshaped, its lobes *very short, broad, and rounded;* carina falcate; legume straightish, compressed, linear, coriaceous, narrowed at each end, tipped with the long, persistent style. *Sphenostylis marginata, E. Mey. Comm. p.* 148.

HAB. Port Natal, *Gueinzius!* Mouth of the Omsamculo, *Drege.* (Herb. Hk.)

Stem suffruticose at base, flagelliform. Petioles 2 inches long, the terminal leaflet ½ an inch beyond the pair. Leaflets 1½–2 inches long, ¾–1 inch wide, the young ones with a few small hairs, most persistent on the ribs and veins beneath. Stipules rigid, 1 line long. Peduncles 12 inches long, incurved; the raceme commencing about an inch from the summit. Calyx very short and wide, more crenate than lobed. Corolla purple, 7–8 lines long. I describe from a specimen from Gueinzius.

4. V. vexillata (Benth.! Fl. Braz. XXIV. p. 194, t. 50, f. 1); voluble, retrorsely-hispid, or rarely glabrescent; leaflets varying from ovate to lanceolate; peduncles 2–4-flowered at the summit; calyx tubuloso-campanulate, the 5 segments acuminate, somewhat longer than the tube; carina *obliquely circularly incurved, spurred at one side. Phaseolus vexillatus, L. Ph. capensis, Thunb. Fl. Cap.* 589. *V. hirta, Hook. Ic. Pl. t.* 637. *E. & Z.!* 1682. *Strophostyles capensis, E. Mey. Comm. p.* 147. *Zey.!* 529, 528. *Vigna scabra, Sond.! Linn.* 23, *p.* 32.

HAB. In grassy places. Frequent in the eastern districts, extending through Caffraria to Natal, *Thunberg! E. & Z.! Drege! &c.* (Herb. Hk., Bth., Sd., D.)

The herbaceous, voluble stem several feet long, the petioles and peduncles rough, with rigid, reversed, foxy hairs, rarely becoming subglabrous. Leaflets extremely variable in shape and length; sometimes ovate, 1½–2½ inches long, ¾–1¼ inch wide; oftener ovato-lanceolate, 4–5 inches long, ½ inch wide at base, tapering to an acute point; sometimes linear, 1–2 inches long, 2–3 lines wide; all green at both sides, membranous, sparsely hairy above, setose, especially along the veins beneath; reticulately veined. Peduncles 3-4-12 inches long, 2-4-flowered at the summit; pedicels minute. Vexillum broad, reflexed, 9–10 lines long, with a folded claw, the limb with inflexed ears at base. Carina broadly falcate, much curved at the point, nearly forming a complete circle, and having on one side above the middle a conical, prominent spur. Legume sessile, linear, straight, hispid, 3–4 inches long, 2 lines wide. Flowers greenish-yellow, tinged with purple. This plant is common to South Africa, Australia, S. America and tropical Asia.

5. V. triloba (Walp. Linn. XIII. 534); procumbent, herbaceous; stem retrorsely-hispid or glabrescent; leaflets *hastate-trilobed,* the lobes *obtuse,* the medial lobe much the longest; lateral leaflets unequal-sided, their upper lobe obsolete; stipules *peltate, constricted at the insertion, and prolonged downwards,* ovate; peduncles elongate, 2-flowered at the summit; calyx campanulate, the lobes falcato-lanceolate, longer than the tube; carina broadly ovato-falcate, acute; legume straight, terete, hispid. *Dolichos trilobus, Thunb. Fl. Cap. p.* 590. *E. & Z.!* 1686. *Scytalis protracta, E. Mey.! Comm. p.* 146.

VAR. β. **stenophylla**; subglabrous; leaflets narrow-linear, (2–3 inches long, 2–3 lines wide), attenuate, mucronate, *obsoletely-hastate, sublobed* at base. *Zey.! No.* 529.

HAB. Sand dunes near Algoa Bay, and at Krakakamma, *E. & Z.!* Near Galgebosch, and between the Gekau and Basche, *Drege!* Mouth of Riet R., Albany, *Dr. Atherstone!* Var. β. Schoen Stroem and Vaal River, *Burke & Zeyher!* (Herb. Hk., Sd., Bth.)

Stems 1–2 feet long or more, weak, somewhat angular. Leaflets variable in breadth, and in incision; sometimes narrow-hastate, 1–2 inches long, the medial lobe 2–6 lines wide, 10–15 lines long, the lateral widely spreading, narrow-oblong, 5–6 lines long, 2 broad; sometimes the terminal leaf only is distinctly 3-lobed, the lateral leaflets being obliquely-ovate and subacute. The *stipules* are constantly peltate in insertion, vertical, rigid, 2-lobed, the lobes directed contrary ways. Pods 3–3½ inches long, 2 lines wide. Var. β. is an extravagantly narrow leaved form, but has the remarkable stipules of the species.

6. V. decipiens (Harv.); prostrate, herbaceous: stem and petioles retrorsely hispid; leaflets hastate, 3-lobed, *the middle lobe acute,* lateral short, obtuse, lateral leaflets unequal-sided, obsoletely lobed; stipules *sessile, lanceolate,* small; peduncles equalling or exceeding the leaves, 2-flowered at the summit; calyx campanulate, its lobes from a broad base subulate, equalling the tube; carina falcate, subacute; legume straight, somewhat constricted between the seeds. *Dolichos ? decipiens, Burch.* 4117. *Zey.!* 523.

HAB. S. Africa, *Burchell!* Grassy places by the Vaal R., *Burke & Zeyher!* (Herb. Hk., Bth., Sd.)

Very like *V. triloba,* but the stipules are *basifixed,* not peltate, the flowers are smaller and the calyx different. Stems flagelliform, 2 feet long or more, striate, and rather roughly hispid. Petioles uncial. Stipellæ lanceolate, equalling the petiolules. Leaflets about 1–1½ inch long, ½–¾ inch wide, green, subglabrous, except on the ribs, veins, and margin, the middle lobe mostly acute. Flowers 6 lines long, seemingly greenish-rosy. Young pods only seen.

7. V. luteola (Benth.! Fl. Braz. XXIV. p. 194); voluble, herbaceous; stem and petioles retrorsely villoso-pubescent; leaflets *ovate or ovato-lanceolate,* hispidulous; stipules minute, eared at base, ovate; peduncles very long, racemose toward the extremity, several-flowered, the rachis spirally twisted; calyx campanulate, 4-fid, its segments ovate, acute, shorter than the tube; carina falcate, acute; legume compressed, straightish, hook-pointed, subtorulose, pubescent. *Scytalis helicopus, E. Mey.! Comm. p.* 146. *Zey.!* 2412. *Vigna helicopus, Walp.*

HAB. Between Omsamculo and Omcomas, *Drege!* Thorfield, Albany, *Dr. Atherstone!* Natal, *Krauss!* 233. Zwartkops R., Uit., *Zeyher!* (Herb. Hk., Sd., D.)

Stem long, climbing, with scanty, reversed, and very soft pubescence. Petioles 2–5 inches long. Stipules and stipellæ minute. Leaflets mostly ovato-lanceolate, but varying much in size, the larger 4–5 inches long, the smaller 1–1½ inch, all pale green. Peduncles 5–10 inches long, closely many-flowered within an inch of the summit. Fl. greenish-yellow, 6–7 lines long. Pods 2–3 inches long, 2–3 lines wide, irregularly constricted. The two uppermost calyx-lobes are completely connate into one very broad one, which is shorter than the rest. A native of S. America.

8. V. retusa (Walp. Rep. 1, p. 778); procumbent or voluble, herbaceous, glabrescent; leaflets *obovato-subrotund, retuse,* the younger thinly puberulous, the old glabrous; stipules minute, triangular-acute; racemes equalling or exceeding the leaves, many-flowered, rachis at length spirally twisted; calyx small, its segments ovate, half as long as the tube; legumes *torulose or moniliform,* straightish, pendulous. *Scytalis retusa, E. Mey.! Comm. p.* 147.

HAB. Sandy, littoral hills between Omtendo and Omcomas, *Drege!* Port Natal, *Gueinzius!* (Herb. Hk., Sd., D.)

Stem long, weak, and soft, soon becoming glabrous. Petioles 1–2 inches long. Leaflets about two inches long and broad, somewhat cuneate and 3-nerved at base, very obtuse or retuse, pale green. Peduncles 3–4 inches long, the last inch floriferous. Flowers yellow-green, 5–6 lines long. Legumes strongly constricted between the seeds, forming a succession of knobs. Near *V. anomala,* Walp.

(Species unknown to me.)

V. (*Scytalis*) **tenuis** (E. Mey.! Comm. p. 145); "glabrous; stem filiform, voluble; leaflets ovate or oblong, obtuse, the lateral ones gibbous on the outer margin, the terminal on both sides; peduncles 2-flowered; calyx-lobes from an ovate base produced; legumes terete, straight, retrorsely scabrid, subserrate at the margin."

VAR. α. **ovata**; "leaflets ovate; legumes obsoletely serrate."

VAR. β. **oblonga**; "leaflets oblong; legumes evidently serrate."

HAB. Between Omtendo, Omsamculo, and Natal, *Drege.*

V. (*Scytalis*) **hispida** (E. Mey. l. c. 146); "stem voluble, retrorsely-hispid; leaflets broadly ovate, rounded at the point, ciliate, and sparingly pilose on both sides; peduncles very long, furrowed, 2-flowered; calyx-lobes lanceolate-acuminate, equalling the tube; legumes straight, terete, hispidulous."

HAB. Grassy hills near Omtata, *Drege!*

LVI. DOLICHOS, L.

Calyx campanulate, bilabiate, the upper lip bifid or subentire, the lower trifid. *Vexillum* spreading or incumbent, equalling the carina, with 2–4-ridge-like callosities within; *alæ* oblong; *carina* falcate or incurved (or nearly straight), neither twisted nor bent to one side. *Stamens* diadelphous. *Disc* sheathing. *Ovary* substipitate, several-ovuled, style channelled or terete; stigma capitate. *Legume* compressed, straight or falcate, 2 or several-seeded, with cellular partitions between the seeds. *Endl. Gen.* 6676. *Lablab, Adans. Endl.* 6677. *Chloryllis, E. Mey. Endl.* 6664.

Voluble or prostrate herbs or suffrutices, common throughout the tropics and warmer zones of both hemispheres. Leaves pinnately-trifoliolate or rarely 5-foliolate, stipellate. Flowers racemose, subcorymbose or rarely solitary, bibracteolate, red, purple, blue, or white. The name is unexplained ; it was used by Dioscorides for some similar plant, or for some species of the allied genus *Phaseolus* (*kidney-bean*). Several have edible pods.

ANALYSIS OF THE SOUTH AFRICAN SPECIES.

1. Lablab.—Upper lobe of the calyx entire. Vexillum expanded; carina elongate, taper pointed, sharply bent upwards. Style laterally flattened, broad, pubescent all round.

(1) **Lablab.**

2. Dolichos.—Upper lip of the calyx emarginate or bidentate. Vexillum expanded; carina incurved, shortly rostrate. Style channelled or terete. (Sp. 2–10).

Style channelled, bearded along its upper edge, or glabrous :
- Voluble. Leafl. rhomboid-ovate, silky. Lowest calyx-lobe very long ... ... (2) **sericeus.**
- Voluble, glabrous. Leafl. ovate-acuminate. Cal.-lobes short, subequal ... ... (3) **gibbosus.**
- Voluble. Leafl. 3-lobed-hastate, shining; pedicels thrice as long as the calyx... ... ... (4) **smilacinus.**
- Prostrate ; Leafl. hastate-ovate or hastate-linear ; pedicels short ... ... (5) **hastæformis.**
- Voluble. Leafl. linear-lanceolate, elongate, *folded;* pedunc. short, 2-4-flowered ... ... (6) **angustifolius.**

Style slender, nearly terete, pubescent all round, or round the stigma, or glabrous.
- Voluble. Leafl. linear-lanceolate, *expanded.* Peduncle short, 2-4-flowered ... ... (7) **linearis.**
- Voluble. Leafl. elliptic-ovate, membranous. Pedunc. very short, 1-3-flowered ... ... (8) **axillaris.**
- Prostrate, scabrid. Leafl. rhomb-ovate, thickish. Pedunc. long, corymbose, many-flowered ... ... (9) **decumbens.**
- Voluble, pilose. Leafl. roundish-ovate, 3-nerved. Pedunc. long, racemose ... ... (10) **falciformis.**

3. Chloryllis.—Upper lip of the calyx entire. Vexillum incumbent, oblong; carina nearly straight, boatshaped, obtuse. Style flattish below, subterete, tapering and pubescent above.
- Procumbent, roughly hispid. Pedunc. densely racemose. Fl. yellow-green ... ... (11) **Chloryllis.**

I. LABLAB. (Sp. 1.)

1. D. Lablab (Linn.) ; voluble ; leaflets broadly ovate or rhomboid, acute, glabrous, membranous ; peduncles elongate, interruptedly racemose, many-flowered ; calyx bibracteolate, its lower lobes subacute ; carina angularly incurved, rostrate ; style laterally compressed, subspathulate, equally pubescent ; legume tapering at base, broadly scimetar-shaped or subfalcate, the thickened sutures crispulate. *Benth.! Fl. Brazil. xxiv. p.* 198, *tab.* 51, *f.* 2. *Lablab vulgaris, DC. Prod.* 2, *p.* 401. *E. Mey. Comm. p.* 140.

HAB. Port Natal, *Drege, Gueinzius, &c.* (Herb. Hk., Bth., Sd , D.)

A common plant throughout tropical Africa and Asia, from whence it has passed into South America. Plant glabrous or nearly so, extensively climbing. Petioles 2–3 inches long. Leaflets 1½–2 inches long, 2 inches broad, thin, full green, finely veiny. Stipellæ subulate. Peduncles 10–12 inches long; the rachis flexuous, and subdistantly nodulose, emitting 2 or more flowers from each node. Bracteoles suborbicular, striate, equalling the calyx-tube. Corolla purple or white, the vexillum reflexed, and the long, rostrate carina upturned like that of a *Crotalaria*. Legume curled along the margin.

2. **DOLICHOS.** (Sp. 2–10.)

2. D. sericeus (E. Mey. Comm. p. 141); "stem voluble, retrorsely *silky-hirsute;* leaflets subrhomboid-ovate, the lateral ones unequal-sided, *silky-pubescent on both surfaces;* racemes 2–4-flowered, the peduncle about equalling the leaf; *carinal segment of the calyx thrice as long as the lateral and narrower;* legumes subfalcate, glabrous, many-seeded.

HAB. Omsamwubo, *Drege.* (Unknown to me.)

3. D. gibbosus (Thunb. ! Fl. Cap. p. 590); nearly glabrous, suffruticose, voluble; leaflets *ovate-acuminate,* paler beneath, the terminal gibbous at each side, the lateral oblique, gibbous on the lower margin; peduncles longer than the leaves, shortly and densely racemose near the summit; rigid, incurved, the rachis and pedicels retro-hispid; calyx-lobes very short and broad, ciliate; legumes falcate, 4–6-seeded, seeds blackish. *DC. Prod.* 2, *p.* 400. *E. & Z.!* 1683. *E. Mey. Comm. Drege, p.* 141. *Zey.!* 2413. *D. Benthami, Meisn.* ? *Lond. Journ.* 2, *p.* 95.

VAR. *β.* **uniflorus;** peduncles single-flowered. *D. capensis, Thunb.! Fl. Cap. p.* 590.

HAB. Climbing among shrubs on hillsides, from Capetown to Cafferland, common. (Herb. Th., D., &c.)

Stem woody below, several feet long, branching, the younger portions thinly pubescent, becoming glabrous. Petioles 1–3 inches long, stipulate and stipellate. Leaflets 1½–2½ inches long, 1–1½ broad, swelling out at base, thence tapering to an acute point, thin and membranous; the petiolules retro-hispid. Peduncles 3–12 inches long. Flowers bright purple, 5–7 lines long. Style channelled, bearded along its upper margin. Legume about 2 inches long, acute at base, subacute and tipped with the persistent style. The original specimen of *D. capensis. Thunb.!* in Herb. Upsal. is evidently a depauperated state of this common plant, with the raceme reduced to a single flower.

4. D. smilacinus (E. Mey. ! Comm. p. 142); "stem filiform, voluble, glabrescent; leaflets subtrilobed-hastate, glabrous, shining, the middle lobe lanceolate, acute, the lateral lobes short and obtuse; racemes 4–6-flowered, little longer than the leaf; pedicels thrice as long as the calyx; calyx-segments minutely ciliate."

HAB. Outiniquabergen, near Roodemuur, on grassy and stony hills 1500–2000f. *Drege!*

5. D. hastæformis (E. Mey.! Comm. p. 142); stem prostrate, scabro-hispidulous; leaflets thickish, veinless, with scabrous margins, the lowermost hastate-ovate, the uppermost hastato-linear, the intermediate obsoletely lobed at base; peduncles longer than the leaves, subumbellately 3–6-flowered, pedicels about equalling the calyx; calyx-lobes short and broad, ciliolate; style flattened and channelled, bearded on its upper side; legumes straightish, glabrous, 2–3-seeded. *E. Mey. l. c. D. Capensis, E. & Z.!* 1684, *non Thunb. Zey.!* 530.

HAB. Near the Zwartkops River, Uit., *E. & Z.!* Grahamstown, *Colonel Bolton!* Kookhuis at the Fish River and Modderfontein, Brach R., *Drege!* (Herb. Hk., Sd., Bth., D.)

This resembles *D. decumbens*, but differs in foliage and especially in the flattened and channelled style. Petioles ½–1 inch long, stipulate and stipellate. Leaflets ½–1½ inch long, the lowest 3–6 lines wide, the upper ½–1–2 lines. Flowers 3–5 lines long, blueish-purple. E. & Z.'s specimens are more luxuriant than Drege's, but do not essentially differ.

6. D. angustifolius (E. & Z. ! 1687); quite glabrous, slender; stem suberect at base, filiform and twining upwards; leaflets *narrow-linear-lanceolate*, very long, *complicate*, rigid, netted-veined and margined; peduncles filiform, flexuous, scarcely exceeding the short petiole, 2–4-flowered; calyx-lobes short and blunt; legumes falcate, acute at base, 6–8-seeded, seeds brown. *D. angustissimus, E. Mey.! Comm. p.* 142.

HAB. Among shrubs by the Sunday and Zwartkops R., Uit., and Klipplaat and Key, *E. & Z.!* Same places, and Stormberg and Moojeplaats, *Drege!* Magaliesberg, *Burke and Zeyher!* (Herb. Sd., Bth., Hk., D.)

Stems filiform, not much-branched, laxly-leafy, the upper portions twining. Petioles about uncial; the terminal leaflet close to the pair. Leaflets 2–4 inches long, 1–3 lines wide, either folded or open, often hook-pointed, pale-green. Peduncles about uncial, 2–4-flowered near the extremity, the flowers in pairs, on patent pedicels, longer than the calyx. Flowers 3 lines long, purple. Style channelled, bearded on the upper margin. Legumes 10–12 lines long, 2–3 wide, subacute. Some of Drege's specimens of *D. linearis*, (c. from Klipplaat R.) belong to this species.

7. D. linearis (E. Mey. ! Comm. p. 142); quite glabrous, slender; stem suberect at base, filiform and twining upwards; leaflets *linear-lanceolate or lanceolate*, *expanded*, rigid, netted-veined, and margined; peduncles filiform, flexuous, scarcely exceeding the short petiole, 2–4-flowered; calyx-lobes deltoid, half as long as the tube; legumes falcate, acute at base, 4–6-seeded, blunt; style slender, subterete, with an encircling tuft of hairs below the extremity. *Zey.* 525, 526.

VAR. β. **pentaphyllus**; leaves 3–5-foliolate, the lateral leaflets in the latter case in pairs. *D. pentaphyllus, E. Mey. l. c.*

HAB. Zwartkops and Klipplaat Rivers, Glenfilling and Stormberg, *Drege!* Crocodile River, Aapjes R., and Thaba Unka, *Burke & Zeyher!* Queenstown District, *Mrs. F. W. Barber!* (Herb. Hk., Bth., Sd., D.)

Very like *D. angustifolius*, and only to be accurately distinguished by the style, and the broader and more obtuse legumes. The leaflets are generally, but not always, expanded, not complicate, 2–3 inches long, 2–4 lines wide; those of the lower leaves are often broadly lanceolate, 1–1½ inch long, 3–4 lines wide. Flowers as in *D. angustifolius.* In var. β. leaves of the ordinary form occur on the same stem as the 5-foliolate ones, especially on Burke and Zeyher's specimens from Orange River.

8. D. axillaris (E. Mey. Comm. p. 144); suffruticose at base, voluble, pubescent or glabrescent; leaflets membranous, *elliptic-ovate, subacute;* peduncles *shorter* than the petiole, 1–3-flowered; calyx-lobes deltoid-cuspidate; legumes straightish, broadly linear, 6–8-seeded.

VAR. α. **pubescens**; all parts densely and softly pubescent.

VAR. β. **glaber**; all parts glabrous or nearly so.

HAB. Between Omtendo and Omsamculo, and at Natal, both varieties, *Drege! Gueinzius!* (Herb. Hk., Bth., D., Sd.)

Climbing among shrubs, 2 or several feet long. Petioles 1½–2½ inches long. Leaflets 1–1½ inch long, ½–¾ inch wide, thin, pale-green, not prominently veiny. Peduncles 3–6 lines long. Flowers greenish-yellow, 6 lines long. Vexillum oblong, incumbent, scarcely equalling the boatshaped, obtuse, scarcely falcate carina. Style slender, subulate, equally pubescent. Legume 2 inches long, 3 lines wide. The glabrous variety is very like *D. biflorus*, of Schimper's Abyssinian plants.

9. **D. decumbens** (Thunb.! Fl. Cap. p. 590); stem prostrate, scabro-hispidulous; leaflets thickish, immersedly veiny, with scabrous margins, rhombic-ovate or trowel-shaped, obtuse or acute; peduncles longer than the leaves, corymbo-racemulose, several-flowered, pedicels equalling the calyx; calyx-lobes short, broad, rounded, ciliolate; style slender, subterete, equally pubescent; legume tapering much at base, straight, 2–1-seeded. *E. & Z.!* 1685. *E. Mey.! Comm. p.* 143.

HAB. Common about Capetown and Hott. Holl., *E. & Z.! Pappe! W.H.H.* Near Blaawberg and Piquetberg, *Drege!* (Herb. Th., Hk., Bth., Sd., D.)

Root very thick, woody, and deeply descending. Stems many from the crown, trailing, 1–2 feet long. Petioles uncial, with ovato-rotund stipules and linear stipellæ. Leaflets ½–¾ inch broad and long, sometimes oblate, with a cartilaginous, denticulate margin. Peduncles 1–2 inches long, shortly racemose, 4–8–10-flowered. Flowers dark blue-purple, scented like violets. Pods uncial, 4–5 lines wide. Distinguished from *D. hastæformis* by the slender style, &c.

10. **D. falciformis** (E. Mey.! Comm. p. 144); stem voluble, hispid; leaflets subrotund or rhombic-ovate, hispidulous, mucronulate, prominently 3-ribbed and veiny beneath, the lateral unequal-sided; peduncles elongate, interruptedly racemose, pluri-flowered; pedicels short; calyx-lobes triangular, acute, puberulous; legumes compressed, falcate or scymetar-shaped, glabrous, 5–6-seeded; style glabrous, terete; stigma penicillate. *Zey.! No.* 521. *Burch. Cat.* 4079.

HAB. Between Omtendo and Omsamculo, in grassy places, *Drege!* Vaal River, *Burke & Zeyher!* Albany, *H. Hutton!* (Herb. Hk., Bth., Sd., D.)

Stem herbaceous, several feet long, trailing or climbing, somewhat angular, thinly hispid. Petioles 1–1½ inch long. Leaflets about 12–14 lines long, 10–15 lines wide, rather rigid. Peduncles 6–8 inches long, flexuous. Raceme at first dense, then lengthening and sparsely flowered, the fl. often in pairs, purple-blue, 3–4 lines long. Legume 1½ inch long, acute at base, septate within; seeds dark-brown. The style is quite glabrous, except immediately under the *stigma*, where it is penicellato-barbate.

3. CHLORYLLIS. (Sp. 11.)

11. **D. Chloryllis** (Harv.); stem procumbent or voluble, hispid; leaflets hispid, ribbed and veiny beneath, rhomboid-ovate, subtrilobed, the lateral lobes very short, gibbous at the sides, all setaceo-mucronate; peduncles compressed, equalling or exceeding the leaves, tortuous, densely racemose, many-flowered; calyx-lobes bluntly ovate, equalling the tube; vexillum oblong, incumbent, scarcely equalling the boatshaped, nearly straight, blunt carina; alæ much shorter than the carina; legume broadly oblong, 3–4-seeded. *Chloryllis pratensis, E. Mey.! Comm. p.* 149.

HAB. Gekau, Caffr., *Drege!* Schoenstrom and Caledon River, *Burke & Zeyher!* (Herb. Hk., Sd.)

Rather roughly hispid with the aspect of a *Phaseolus*. Peduncles 2 inches long. Leaflets 1–1½ inch long, nearly equally wide, reticulated; rather rigid, broadly cuneate at base, hispid on both sides, the mucrons conspicuous. Stipules oblong, rigid, deflexed. Raceme 3–5 inches long, with pendulous, yellowish-green flowers. Flowers 8–9 lines long. Vexillum straight, elliptic-oblong, folded over the other petals, auriculate at base, callous within, with two shallow ridges. Alæ oblong, simple, short-clawed. Carina straightish, slightly longer than the vexillum. Vexillary stamen slightly adnate in the middle. Style compressed and flattish below, subterete and villous toward the inflexed point. Calyx strumose at base. Legume 2–4 inches long, 6–8 lines wide, rounded at base, subacute, with thickened sutures, scaberulous.

LVII. FAGELIA, Neck.

Calyx 5-cleft beyond the middle, the segments linear, acute, straight, the two uppermost somewhat connate. *Vexillum* reflexed; *carina* very obtuse, longer than the *alæ*. *Stamens* diadelphous. *Ovary* sessile, several-ovuled; *style* subulate, glabrous; stigma obtuse. *Legume* turgid, about 6-seeded, constricted between the seeds, bivalve. *Seeds* ovate, strophiolate, with a linear hilum. *Endl. Gen.* 6685. *DC. Prod.* 2, *p.* 389.

A twining, strong smelling suffrutex, clothed with viscid hairs. Leaves pinnately-trifoliolate; the terminal remote. Racemes axillary; flowers yellow, the carina dark-purple at the point. Name, in honour of some unknown botanist?

1. **F. bituminosa** (DC. l. c.); *E. & Z.!* 1679. *E. Mey.! Comm. p.* 139. *Glycine bituminosa, Linn. sp.* 1024. *Lam. Ill. t.* 609, *f.* 2. *Thunb.! Fl. Cap.* 591. *Bot. Reg. t.* 261. *Fagelia flexuosa, Meisn. in Lond. Journ.* 2, *p.* 93.

HAB. Common among shrubs, &c. in the western districts. (Herb. D. Sd. Hk. &c.)
Stems several feet long, twining, woody below, viscoso-pubescent. Leaflets rhomb-ovate, obtuse or acute, pale and gland-dotted beneath, 1–1½ inch long, ¾–1 inch wide. Stipules ovate-acute. Peduncles long, laxly many-flowered. Bracts broadly ovate, deciduous. Calyx-lobes much longer than the tube. Flowers 6–7 lines long, fulvous-yellow, the apex of the carina dark-purple. Whole plant viscidly hairy. Legumes 1–1½ inch long.

LVIII. RHYNCHOSIA, Lour.

Calyx campanulate, mostly oblique, 4–5-fid, the two upper lobes more or less united, the lowest longest. *Petals* nearly of equal length, or the alæ shorter. *Vexillum* obovate or orbicular, mostly with two minute, inflexed auricles at base, naked or bicallous within. *Alæ* narrow, eared at base; *carina* broader, incurved at the apex, obtuse or subrostrate. *Stamens* 9–1; the vexillary filament quite free, mostly geniculate; *anthers* uniform. *Ovary* subsessile, with 2 ovules; style incurved beyond the middle, quite glabrous and mostly thickened above, filiform and often hairy at base. *Legume* compressed, oblique or falcate, rarely septate within; *seeds* 2–1, compressed-globose or reniform, with a lateral short or oblong hilum and a subcentral seed-cord. *Caruncle* thick or minute. *Benth.! in Mart. Fl. Braz. xxiv. p.* 199. *Endl. Gen.* 6692. *Copisma, Orthodanum, Chrysoscias, and Hidrosia, E. Mey.! Comm. Drege. Polytropia, Presl. Sigmodostyles, Meisn.*

Voluble or prostrate, rarely erect, herbs or suffrutices mostly sprinkled with resinous dots, natives of the tropics and warmer temperate zones of both hemispheres. Stipules ovate or lanceolate. Leaves pinnately-trifoliolate, rarely unifoliolate. Peduncles axillary, racemose, rarely umbellate or single-flowered. Bracts caducous. Flowers mostly yellow, the vexillum lined with brown, rarely purple. Ovary very rarely *uniovulate*. Name from ῥυγχος, a *beak*; but the carina is scarcely rostrate.

ANALYSIS OF THE SOUTH AFRICAN SPECIES.

1. **Chrysoscias.**—Stem voluble. Leaves pinnately-trifoliolate. Stipules broad. Flowers in axillary umbels or solitary, yellow. (Sp. 1–4.)

Flowers umbellate :
Fulvous ; two upper calyx-lobes separate nearly to the base (1) **Chrysoscias.**
Canescent ; upper calyx-lobes connate to, or beyond the middle ;
Stipules broadly oblong, obtuse ; indument copious (2) **leucoscias.**
Stipules ovato-lanceolate, acute ; indument scanty (3) **microscias.**
Flowers solitary, axillary ; upper cal.-lobes connate ... (4) **uniflora.**

2. Polytropia.—Stem prostrate or trailing. Leaves pinnate or bipinnate, plurijugate. Flowers racemose. (Sp. 5–6.)

Lvs. pedately-bipinnate or supra-decompound ; leaflets lanceolate or linear ... ... (5) **ferulæfolia.**
Lvs. simply pinnate, 2–3-jugate ; lfts. ovate or rhomboid (6) **pinnata.**

3. Copisma.—Stem prostrate or voluble, (in *R. Memnonia*, suberect). Leaves pinnately-trifoliolate, rarely unifoliolate. Flowers racemose. (Sp. 7–26.)

Bracts broad, ovate or oblong, persistent :
Glabrous or downy ; petioles long ; leaves trifoliolate (7) **rotundifolia.**
Hairy ; petioles very short ; lvs. 3-foliolate ... (8) **grandifolia.**
Hairy ; leaves *unifoliolate* ... ... (9) **simplicifolia.**
Bracts small or very small, deciduous :
Leaflets obovate, 3–5-lobed, 3-nerved at base ... ... (10) **ficifolia.**
Leaflets rhomb-ovate or subrotund, 3–5-nerved at base :
Stipules *broad*, ovate or oblong, acute or obtuse :
Vexillum glabrous : leaves subsessile .. ... (11) **sigmoides.**
Vexillum glabrous ; leaves petioled :
Softly pubescent ; stipules large ; pedicels short (12) **secunda.**
Vicoso-pubescent ; stip. small ; pedi. 3–4 lines long (13) **viscidula.**
Vexillum pubescent ; leaves petioled :
Patently hairy ; leaflets hispid, rigid ... ... (14) **hirsuta.**
Densely velvetty ; lfts. *thick*, softly velvetty ... (15) **crassifolia.**
Tomentose ; lfts. velvetty above, *whitish* beneath (16) **argentea.**
Stipules *narrow*, lanceolate, subulate, or minute :
Cal.-lobes lanceolate or subulate ; lvs. pubescent or silky :
Stem, petioles, and lvs. hispido-pubescent ... (17) **nervosa.**
All parts densely and softly silky-tomentose ... (18) **Memnonia.**
Cal.-lobes lanceolate or subulate ; lvs. glabrescent :
Ovary hirsute ; peduncles long ; fl. 5 lines long (19) **adenodes.**
Ovary pilose ; pedunc. long ; fl. 2–3 lines long ... (20) **minima.**
Ovary glabrate ; pedunc. short, fl. 5–6 lines long (21) **quadrata.**
Upper cal.-lobes *short and broad*, lowest subulate (22) **gibba.**
Leaflets ovate-oblong or lanceolate, one-nerved, netted-veined .
Petioles very short ; pedunc. shorter than the leaves (23) **puberula.**
Petioles long or longish :
Pubescent ; lfts. broadish ; pod oblong, pubescent (24) **Totta.**
Pilose ; lfts. narrow ; pod falcate, pilose ... ... (25) **pilosa.**
Glabrous (or downy) ; legume *stipitate*, glabrous (26) **glandulosa.**

4. Orthodanum.—Erect, virgate shrubs or suffrutices. Leaves pinnately-trifoliolate. Flowers racemulose ; peduncles short, few-flowered. (Sp. 27–29.)

Silky-villous or glabrescent ; lfts. ellipt.-oblong or lanceolate (27) **orthodanum.**
Silvery and satiny ; lfts. broadly ovate or cordate ... (28) **nitens.**
Viscidly hairy ; lfts. small, oblong, bullate ; stipules leafy (29) **bullata.**

1. CHRYSOSCIAS. (Sp. 1–4.)

1. R. Chrysoscias (Benth. !) ; stem suffruticose at base, the branches flexuous or voluble, *fulvo-villous ;* petioles very short, leaflets oblongo-lanceolate with revolute margins, puberulent above, paler and tomentulous beneath ; stipules and the deciduous bracts broadly oblong, obtuse ; peduncles umbelliferous ; calyx-tube very short, the segments lanceolate, leafy, pubescent and ciliate, scarcely shorter than the ample vexillum, the two uppermost *slightly connate at base. Glycine erecta, Thunb.! Fl. Cap. p.* 592. *Cylista lancifolia, E. & Z.!* 1690. *Chrysoscias grandiflora, E. Mey.! Comm. p.* 139.

HAB. S. Africa, *Thunberg! Bowie!* Kaymansgat, *Drege!* Between Langekloof and Plettenberg Bay, *E. & Z.!* (Herb. Th., Bth., Hk., Sd.)

A climbing suffrutex, several feet long; the younger portions clothed with golden or tawny soft hairs and resin-dotted. Petioles ½–¾ inch long. Stipules and bracts similar, each 4–5 lines long, 2–3 wide, brown, softly fringed. Leaflets 1½ inch long, ½–¾ wide, dark coloured above, fulvescent beneath. Peduncles equalling the leaf, or shorter or longer, 3–4-flowered at the summit. Flowers golden-yellow or orange, 6–8 lines long. Legumes not much longer than the calyx, pilose.

2. R. leucoscias (Benth.!); suffruticose, voluble, *albo-villous;* petioles very short, leaflets oblongo-lanceolate (or linear) with revolute margins, albo-tomentose beneath; stipules *broadly oblong, obtuse;* peduncles umbelliferous; calyx-tube very short, the segments lanceolate, leafy, silky and silvery, scarcely shorter than the vexillum, the two uppermost *connate to or beyond the middle. Cylista argentea, E. & Z.!* 1688. *Cylista angustifolia, E. Mey.! Linn.* 7, *p.* 171. *Chrysoscias calycina, E. Mey! p.* 140.

VAR. *β.* **angustifolia**; leaflets narrow-linear, strongly revolute; flowers subsolitary. *Zey.!* 2410.

HAB. Mountain sides. Vanstaadensberg, *E. & Z.!* *β.* Near the River Zonderende, *Zeyher!* (Herb. Sd., Hk., Bth.)

Readily known from *R. chrysoscias* by its much whiter, more copious and woolly pubescence, and the connate upper calyx-lobes. Var. *β.* is less woolly, with few flowers, and very narrow, almost terete-revolute leaves. Leaflets in *a.* 2 inches long, 4 lines wide; in *β.* 1 inch long, 1–2 lines wide. Flowers yellow.

3. R. microscias (Benth.!); suffruticose, voluble, silky-canescent; petioles very short, leaflets lanceolate or linear with revolute margins, whitish and tomentulose beneath; stipules and bracts *ovato-lanceolate, acute;* umbels subsessile or pedunculate, 4–6-flowered; calyx-tube very short, the segments lanceolate, leafy, cano-pubescent, shorter than the vexillum, the two uppermost connate to the middle. *Chrysoscias parviflora, E. Mey.! Comm. p.* 139.

HAB. Cape, *Bowie!* Mountain sides near George, *Drege!* (Herb. Hk., Bth., D.)

Very near *R. leucoscias,* from which it is best known by its acute stipules and bracts and somewhat smaller flowers; the indument is shorter and less copious, particularly on the calyx. The leaves vary from narrow-linear to lanceolate.

4. R. uniflora (Harv.); voluble, silky-canescent; petioles very short, leaflets lanceolate or linear with revolute margins, whitish and tomentulose beneath; stipules ovate, subacute; flowers *solitary, axillary, on short pedicels;* calyx-tube very short, the segments lanceolate, leafy, silky-villous, not much shorter than the vexillum, the two uppermost connate to the middle. *Cylista angustifolia, E. & Z.!* 1689, *non E. Mey. Glycine angustifolia, Jacq. Schoenb.* 2, *t.* 231.

HAB. Puspasvalley, Swell., and Zwarteberg, near Caledon, *E. & Z.!* (Herb. Sd.)

Very like *R. microscias,* but the flowers are constantly solitary, on pedicels shorter than the calyx. Jacquin's figure fairly represents the specimens in Hb. Sonder.

2. **POLYTROPIA.** (Sp. 5–6.)

5. R. ferulæfolia (Benth.!); prostrate, nearly glabrous; leaves *pe-*

dately-bipinnate, paucijugate or *supra-decompound,* leaflets sessile, narrow-lanceolate or linear, acute at each end, impresso-punctate; peduncles elongate, 5–12-flowered near the summit; flowers viscoso-pubescent. *Psoralea prostrata, Linn. Galega pinnata, Thunb.! Fl. Cap. p.* 602. *Polytropia ferulæfolia, Presl. Symb. t.* 13. *E. & Z.!* 1626, *and P. umbellata, E. & Z.!* 1627.

HAB. Common on the Cape Flats and in several parts of the western districts. (Herb Th., D., Hk., Sd., &c.)

Root deeply sinking. Stems many from the crown, 2–3 feet long or more, simple or slightly branched, angularly-striate. Leaves 1–2 inches apart on long petioles, variable in composition; the simplest are ternately-bipinnate, with two larger, simple leaflets (or leafy stipellæ ?) from the apex of the common petiole; in the more compound 5-7-9-pinnate or bipinnate petiolules crown the petiole. Leaflets ½–¾ inch long, ½ line to 2 lines wide, dark-green. Peduncles 4–6 inches long; pedicels nodding. Flowers yellow, at length turning brown at the tips. Legume 6–8 lines long, 2½ lines wide, subfalcate or straightish, nearly glabrous. Seeds strophiolate.

6. R. pinnata (Harv. Thes. t. 79); prostrate, puberulous; leaves *simply pinnate,* bi-tri-jugate; leaflets minutely petiolulate, *rhomboid-ovate or elliptic-oblong,* acute, puberulous, impresso-punctate; peduncles elongate, 5–6-flowered near the summit. *Polytropia pinnata, E. & Z.!* 1628.

HAB. Sandy places near Saldanha Bay, Aug.–Sept. *E. & Z.!* (Herb. Sd.)

Very similar in general habit to *R. ferulæfolia,* but with much less compound leaves and broader and shorter leaflets; the whole plant, or at least the younger parts, minutely downy. The leaves are sometimes impari-, sometimes abruptly-pinnate. Leaflets ½ inch long, 3–5 lines wide, the petiolules ½ line long. Flowers yellow, 5 or 6 in a pedunculate raceme. Stipules ovate, striate, ribbed, deflexed. Calyces and pods as in *R. ferulæfolia.*

3. COPISMA. (Sp. 7–26.)

7. R. rotundifolia (Walp. Rep. 1, p. 787); prostrate, glabrous or downy; leaves on long petioles, leaflets orbicular or ovate, obtuse, with recurved margins, netted-veined and resin-dotted; stipules broad, ovate; peduncles elongate, densely racemose near the summit; *bracts broadly-ovate, persistent;* calyx subglabrous, its segments broad, twice as long as the tube, half as long as the glabrous corolla. *Copisma rotundifolium, E. Mey.! Comm. p.* 137.

HAB. Among grass. Omsamcaba, *Drege!* (Herb. Bth.)

Known at once from *R. adenodes* by its broad, striate stipules, and broad, persistent bracts. Petioles 1½ inch long. Leaflets 6–7 lines long, nearly as broad, close together near the apex of the petiole, reflexed. Peduncles 3–4 inches long, angular, the flowering part uncial. Pedicels 1 line long. Flowers rather large, 7–8 lines long. Calyx-tube short, the segments 2 lines long, the two uppermost connate. Vexillum broadly obovate or orbicular. Keel broadly falcate.

8. R. grandifolia (Harv.); procumbent, robust, hirsuto-pubescent; petioles very short; leaflets broadly ovato-subrotund, hispido-pubescent on both sides, prominently nerved and veined beneath; stipules *broadly cordate, leafy;* peduncles longer than the leaves, closely spicato-racemose beyond the middle, many-flowered; bracts *ovate-oblong, acute, persistent;* calyx-segments lanceolate, the uppermost semi-connate, the lowest subulate, all shorter than the glabrous vexillum; (young) legume hirsute. *Copisma grandifolium, E. Mey.! Comm. p.* 138.

HAB. Hills near Kat River, *Drege!* Creeping among grass at the foot of the Winterberg, and on hills above Waterkloof and Kaalneck, *Mrs. F. W. Barber!* (Herb. D.)

Sent by Mrs. Barber among specimens of *R. hirsuta*, from which this is distinguished by its larger size, dense inflorescence, with *erect*, short pedicelled flowers, and especially by the large stipules and persistent bracts. Stem rigid, terete. Petioles ½ inch long. Leaflets nearly 2 inches long, 1½ inch broad, rigid and roughly pubescent Stipules 6–8 lines long, 4–5 lines wide. Peduncles 5–6 inches long, hairy. Flowers 6–7 lines long; vexillum brown and striped at back, yellow within; alæ bright yellow, much shorter than the brown-tipped carina. The flowers resemble those of *Fagelia bituminosa*.

9. R. simplicifolia (E. Mey. Comm. p. 138); same as *R. grandifolia*, but "the leaves are simple; stipules cordate-ovate, acuminate; bracts ovato-lanceolate."

HAB. Omtata, on grassy hills, *Drege.*

10. R. ficifolia (Benth. !); prostrate, roughly hairy and subvillous; stem and branches compressed, angular; petioles short, leaflets (large) cuneate-obovate, 3-nerved, *mostly 3–5-lobed*, reticulate above, with prominent ribs and veins beneath; stipules lanceolate; peduncles elongate, racemose beyond the middle; calyx very villous, its segments longer than the tube, scarcely shorter than the glabrous vexillum. *Zey.!* 520. (*in Hb. Sond.* 519.)

HAB. Vaal River, Mooje R., and Magaliesberg, *Burke & Zeyher!* (Herb. Hk., Bth., Sd., D.)

Readily known from other Cape species by the lobed leaflets, resembling those of a fig. Stems 2 or more feet long, trailing, robust, compressed and sharply angled and striate. Lateral leaflets near the base of the common petiole, terminal remote, an inch apart. Leaflets 2½–4 inches long, 2–2½ broad, cuneate at base, the lateral oblique, all rather rigid, softly pubescent, and very pale-green, more or less deeply cleft, the lateral lobes short, acute, or cuspidate. Racemes 3–4 inches long, several-flowered. Ovary villous. Legume unknown.

11. R. sigmodes (Benth. !); prostrate, *softly villous;* stem 3–4-angled; leaves *subsessile;* leaflets cordate-ovate or suborbicular, rugose, thick, velvetty; stipules broadly-triangular, acute; peduncles elongate, densely racemose beyond the middle; bracts lanceolate; calyx villous, its segments longer than the tube, half as long as the glabrous vexillum; legume falcate, hairy. *Sigmodostyles villosa, Meisn.! in Hook. Lond. Journ.* 2, *p.* 93.

HAB. Sides of Bosjesmansrand Mt., near P. Maritzburg, Natal, 2500f., *Krauss!* 246. (Herb. Hk., Bth., D.)

Stems robust, 2 or more feet long, sharply angled and softly hairy. Common petiole uncial, the lateral leaflets near its base, the terminal remote. Terminal leaflet 2–3 inches long, 2–2½ broad, the lateral smaller and unequal sided, the young ones lanigerous and glossy, the older more velvetty-villous and green. Stipules 4-5 lines long, 2-3 broad at base, acute. Peduncles 6 inches long. Bracts narrow, 3–4 lines long, deciduous. Flowers deflexed subsessile, 5–6 lines long. Calyx-segments narrow, the two upper connate nearly to the point. Vexillum broad, 2-callous over the claw. Alæ narrow. Legume 10–12 lines long, 3–4 broad, sessile, obtuse, hairy.

12. R. secunda (E. & Z. ! Enum. No. 1665); prostrate, *softly pubescent or velvetty;* branches flexuous; leaves petiolate; leaflets orbicular-rhomboid, obtuse, veiny beneath; *stipules broadly ovate, tomentose;*

peduncles shorter or scarcely longer than the leaves, laxly 2–4-flowered; calyx-segments longer than the tube, shorter than the glabrous vexillum; legume sessile, falcate, tomentose. *Burch. Cat. No.* 2457.

HAB. Woods at Ado and Bushman's River, Uitenhage and Chumiberg, Caffr., *E. & Z.!* Small Deel, *Zeyher!* 514, *Burke!* (Herb. Sd., Bth., Hk.)

Spreading on the ground, 1–2 feet long, branching. Stems slender, terete. Petioles ½–1 inch long. Leaflets 6–14 lines long, as wide, the terminal somewhat cuneate at base, the lateral very oblique, all round-topped and submucronulate. Peduncles ½–1½ inch long in small leaved, 3 inches long in more vigorous specimens. This has the flowers of *R. Totta* and the leaves of *R. adenodes*, but differs from both (besides other marks) in its broad stipules.

13. R. viscidula (Steud. Nom. Bot.); stem prostrate or voluble, slender, flexuous, striate, thinly viscoso-pubescent; leaves on longish petioles; leaflets ovato-rhomboid or subrotund, obtuse, viscoso-pubescent on both sides, thickly gland-dotted beneath; stipules small, ovate or ovato-lanceolate, acute; peduncles longer than the leaves, laxly 2–3-flowered; pedicels nearly as long as the calyx, whose linear-lanceolate lobes are much longer than the tube, but shorter than the glabrous, striate vexillum; legumes sessile, oblong, viscoso-pubescent. *Fagelia pubescens and F. viscida, E. & Z.!* 1680, 1681. *Copisma viscidulum, E. Mey.! Comm. p.* 134.

HAB. Summit of Table Mt. and at the Waterfall, Tulbagh, *E. & Z.!* Kasparskloof and between Koussie and the Gariep, *Drege!* (Herb. Sd., Bth.)

Stem weak, about 2 feet long. Petioles 1–1½ inch long, the terminal leaflet distant. Leaflets ½–¾ inch in diameter, scarcely longer than broad, the lateral ones very oblique. Peduncles 3–5 inches long. Flowers 5 lines long, the vexillum and tip of carina purple-lined. Legume 8–9 lines long, 3 lines wide.

14. R. hirsuta (E. & Z.! 1676); procumbent or twining; branches and *longish* petioles patently hairy; leaflets broadly ovate-suborbicular, rigid, hispido-pubescent; stipules ovate, acute; peduncles elongate, laxly several-flowered; calyx-segments longer than the tube, shorter than the *pubescent* vexillum; legume sessile, villous. *Copisma diversifolium, E. Mey.! Comm. p.* 135. *Burch. Cat.* 4163. *Zey.!* 502.

VAR. β. **angustifolia**; upper leaflets oblong, lower ovate. *Benth.*

HAB. Among shrubs on the Winterberg, Kat R., *E. & Z.!* Buffelsriver and Camdeboosberg, *Drege!* Doorn Kop and Magaliesberg, *Burke & Zeyher!* β. Port Natal, *Gueinzius!* (Herb. Sd., Bth., Hk., D.)

Stems widely spreading, 2–3 feet long, flexuous, rigid, striate, hirsute. Pubescence less copious, and often scanty on the leaves. Stipules 2–3 lines long. Petioles 1–1½ inch long. Leaflets 1½–2 inches long, and nearly as wide, very blunt at each end, the lateral oblique, strongly nerved, scabrescent above. Peduncles 6–8 inches long, above the middle distantly-flowered. Calyx 3 lines long, the tube 1 line. Vexillum more or less densely pubescent, 5 lines long. Legume 9–12 lines long, 3–4 lines wide, falcate or straightish.

15. R. crassifolia (Benth.!); stem procumbent, branches, petioles, and leaves *densely velvety;* leaflets elliptic-ovate, obtuse, thick, softly velvetty; stipules oblong-ovate, subobtuse, silky; peduncles elongate, laxly several-flowered; calyx silky-villous, the segments longer than the tube, shorter than the *pubescent* vexillum; legume sessile, villous. *Zey.!* 505.

HAB. Magaliesberg, *Burke & Zeyher!* (Herb. Hk., Bth., Sd.)

The whole plant *velvetty* with thickly set, short, soft hairs. Stems robust, terete, much-branched. Petioles 1–1½ inch long. Leaflets 1½–2 inches long, 1–1½ inch wide, prominently nerved and veined beneath, of thick substance, soft to the touch, the lateral very oblique, all round-topped and very minutely mucronulate. Peduncles 6–10 inches long, beyond the middle laxly floriferous. Calyx 2½ lines long. Petals about 5 lines, vexillum hairy. Legume falcate, softly villous, 1 inch long, 3 lines wide. Known from *R. hirsuta* by its soft, velvetty pubescence, thicker leaflets and stipules.

16. R. argentea (Harv.); voluble, tomentose; petioles longish, *the terminal leaflet approximate;* leaflets broadly ovate, thickish, velvetty-pubescent and rugulose above, *cano-tomentose* beneath; stipules ovate, acute; peduncles longer or short, laxly few-flowered; bracts caducous; calyx pubescent, its upper segments broadly lanceolate, lowest rather longer, subulate; vexillum *pubescent. Glycine argentea, Thunb. Fl. Cap. p.* 592.

HAB. S. Africa, *Thunberg! Burchell!* 5122. (Herb. Th., Burch.)

Nearest to *R. gibba β. picta,* but with more densely tomentose and *hoary* undersides to the leaflets, shorter and broader stipules, and a pubescent vexillum. Petioles ½–¾ inch long. Leaflets 10–12 lines long, 6–8 wide, very pale or whitish beneath, the terminal about 1 line apart from the pair, which are somewhat oblique, margin reflexed. Stipules 1–1½ line long, 1 line wide, bright brown, reflexed, ovate or ovato-lanceolate. Legume unknown.

17. R. nervosa (Benth.!); prostrate or twining, branches and *shortish* petioles hispido-pubescent or villous; leaflets elliptic-oblong or sub-rhomboid, rigid, hispido-pubescent, 3-nerved at base, strongly veiny beneath; stipules small, narrow-lanceolate or subulate; peduncles long (or short), 4–6-flowered; calyx-lobes lanceolate-acuminate, longer than the tube, shorter than the *glabrous* vexillum; legume sessile, villous. *Zey!* 500. *R. hirsuta β. rhombifolia, E & Z.! Zey.* 515?

VAR. *β.* **pauciflora**; smaller in all parts; peduncles shorter than the leaf, generally 2-flowered. *Zey.!* 508, 510, 511.

HAB. Sand rivier and Langspruit, *Zeyher!* Winterberg, *E & Z.! β.* Aapjes River, Magalies River and Rhinosterkop, Vaal R., *Zeyher!* (Herb. Bth., Hk., Sd.)

This resembles *R. hirsuta,* but differs in the generally shorter petioles, dense pubescence, glabrous vexillum and narrower stipules. Var. *β.* has smaller flowers and much shorter peduncles, but seems otherwise the same. Of *Zeyher's* No. 515 I have only seen imperfect specimens.

18. R. Memnonia (DC. Prod. 2. p. 386); voluble or *suberect, in all parts tomentoso-canescent or silky;* branches angular; petioles moderate, leaflets obovate or sub-rotund, or bluntly rhomboid, velvetty; stipules lanceolate, deciduous; peduncles elongate, laxly racemose, many-flowered; calyx tomentose, its segments longer than the tube, half as long as the pubescent vexillum, the uppermost connate, legume falcate, softly tomentose or villous. *Glycine Memnonia, De Lile, Fl. Æg. p.* 100. *t.* 38, *f.* 3.

VAR. *β.* **prostrata**; prostrate; thinly tomentulose. *Zey.!* 504.—*R. minima, β. calycina, Benth.! in Herb.*

HAB. Magaliesberg, both forms, *Burke & Zeyher!* 503. (Herb. Bth., Hk., Sd., D.)

Our specimens are suberect, 1–1½ foot high, with flexuous (probably afterwards voluble) branches. The whole plant is clothed with a white, soft woolly nap. Petioles 1–1¼ inch long. Leaflets about an inch long and nearly as wide, commonly roundish-obovate, varying to ovate, elliptical or rhomboid, sometimes rugose, the nerves not prominent. Stipules brown. Peduncles 4–8 inches long or more, bearing flowers for ½ or ⅝ their length; fl. 2–4 lines apart, 4 lines long, yellow. Calyx-lobes acute. Legume 9–10 lines long, curved upwards, densely woolly and white in our specimens. A native of Tropical and North Africa, Cape de Verds, Arabia, &c.

19. R. adenodes (E. & Z.! 1670); prostrate, subglabrous or minutely puberulous; leaves on longish petioles, leaflets ovate-orbicular or subrhomboid, acute or obtuse, resinous-dotted; stipules lanceolate; peduncles longer than the leaves, shortly and closely racemose near the summit; bracts narrow, deciduous; calyx pubescent, its segments lanceolate, longer than the tube, half as long as the glabrous or puberulent vexillum; ovary hirsute. *R. amatymbica, E. & Z.!* 1671. *Copisma effusum, E. Mey.! Comm. Drege, p.* 135. *Zey.!* 516.

VAR. β? **robusta**; larger, with scabro-pubescent stems; leaflets subrotund, or oblate, 10–14 lines in diameter, less conspicuously resin-dotted. *Zey.!* 509.

HAB. Winterberg and Zwartkey River, *E. & Z.!* Orange R., *Zeyher!* Queenstown District, *Mrs. F. W. Barber!* Kat River and Buffelsriver, *Drege!* Near Natal, *Krauss,* 301! β. Thaba Uncka, *Burke & Zeyher!* (Hb. Sd., Bth., Hk., D.)

Stems very numerous, 1–2 feet long, spreading every way, angular and striate, mostly downy. Stipules small. Petioles ¾–1 inch long. Leaflets (except in β.) scarcely more than half inch across, as broad as long or broader, sometimes round-topped, more commonly acute, rigidly membranous, netted, mostly glabrous. Peduncles generally longer, sometimes 3–4 times longer than the leaves, the raceme about uncial, 6–8 flowered. Bracts caducous. Fl. 5 lines long. Vexillum glandular, often minutely downy, striped with brown, carina straw colour. Legume hairy.

20. R. minima (DC. Prod. 2, p. 385); voluble, slender, thinly downy or glabrescent; petioles longish, leaflets rhomboid, membranous; peduncles longer than the leaves, laxly many flowered; bracts minute, caducous; calyx segments longer than the tube, shorter than the small corolla, the uppermost united at the base only or below the middle; legume subfalcate, 2–4 times longer than the calyx, hispidulous or pubescent. *Dolichos minimus, Linn. Copisma tenue and C. falcatum, E. Mey.! Comm. p.* 136.

HAB. Sea shore between Omtendo and Omsamculo, *Drege!* Port Natal, *T. Williamson, Dr. Grant,* &c. (Herb. Bth. Hk. Sd. D.)

A variable species, chiefly recognizable by its small flowers. Stems much branched from the base, thinly pubescent or nearly glabrous, the branches twining, filiform. Stipules small and narrow, reflexed. Petioles 1–2 inches long, channelled. Leaves ¾–1 inch long and broad, bluntly angled, broadly cuneate at base, the lateral unequal-sided, membranous, green, gland-dotted beneath. Racemes 3–4 inches long, flowers 2–3 inches long. Calyx segments narrow, tapering. Vexillum downy or glabrous. Legume thinly pubescent or glabrescent, 6–10 lines long, 2–3 wide. A common species within the tropics of both hemispheres.

21. R. quadrata (Harv. in Herb.); voluble, thinly tomentulose; petioles longish, leaflets *broadly-rhomboid,* acute, subglabrous, conspicuously resin-dotted beneath; stipules small, lanceolate; racemes lax, shorter or scarcely longer than the leaf; bracts narrow, caducous; calyx-tube short, its segments lanceolate, broad, nigro-punctate, shorter than the glabrous vexillum, the uppermost connate below the middle; ovary puberulous; legume substipitate, glabrate, falcate, twice as long as the calyx.

HAB. Port Natal, *T. Williamson!* (Herb. D., Hk, Bth.)

The whole plant is very pale green, the stems, petioles, and inflorescence softly downy. Stems slender, striate, stipules very small and withering. Petioles 1½ inch long. Leaflets 1–1½ inch long, the terminal on a petiole 3–4 lines long, wider than long, generally acute, lateral oblique, all pale green, the younger puberulous above, the old becoming glabrous. Flowers yellow, 5–6 in a lax raceme, 6 lines long. Pedicels slender, 1–2 lines long. Calyx 4–5 lines long, very pale, densely black-dotted, the segments taper-pointed. Legume sometimes minutely downy, (the unripe) 9–10 lines long, 3 lines wide, acute, tapering at base, and almost stipitate.

22. R. gibba (E. Mey.! Linn. 7. p. 170); voluble or prostrate, thinly pubescent; petioles longish, leaflets ovato-rhomboid, acute or obtuse, membranous; stipules minute; racemes longer than the leaves, laxly several flowered; upper calyx-segments *short and broad, acute, lowest conspicuously longer, subulate;* vexillum glabrous; legume falcate, much longer than the calyx, thinly pilose (or hirsute). *E. & Z.!* 1667; *also R. acuminata, E. & Z.!* 1666. *Copisma gibbum, E. Mey. Comm. p.* 137. *Burch.* 3344. *Glycine caribœa, Jacq. Ic. Rar. t.* 146.

VAR. β. **pictum** (Benth.!); more densely pubescent or tomentulose, the terminal leaflet approximate. *Cop. pictum, E. M. l. c. p.* 135. *Zey.!* 2406.

HAB. Frequent in Uitenhage and Albany, and on to Port Natal, *E. & Z.! Drege! &c. Zey.!* 501; 2407. (Herb. Bth., Hk., D., Sd.)

Stems several, climbing or trailing, elongate, slender, mostly densely and shortly pubescent. Petioles uncial, the terminal leaflet remote. Leaflets very variable in form, sometimes almost orbicular and very obtuse, sometimes rhomboid-acuminate, the lateral very unequal-sided, ½–1½ inch long and broad, pale or darkish green. Peduncles 3–4 inches long, floriferous above the middle; flowers 6–8 lines long, the ovate vexillum streaked with brown. Ovary very hairy. Legume 12 lines long, 4 lines wide, acute at each end. The common form is well characterized by its calyx. *Glycine trilobaand G. secunda, Thunb.!* both seem to belong to this species, but the specimens in Herb. Upsal are very imperfect. β. has thicker, somewhat rugulose leaflets, pale underneath, and the upper calyx-lobes are longer and narrower. It closely approaches *R. argentea,* but differs in the vexillum.

23. R. puberula (Harv.); prostrate or subvoluble, nearly glabrous or silky-puberulous; *petioles very short;* leaflets elliptic-oblong or oblong, rigid, netted-veined; stipules lanceolate, striate, patent; peduncles 1–2 flowered, shorter than the leaves; calyx villous, its segments longer than the tube, shorter than the glabrous vexillum; legume sessile, very villous. *Copisma tottum, E. Mey.! Comm. p.* 133. *Eriosema puberulum, E. & Z.!* 1677. *Burch. Cat.* 2487–4, 2433. *Zey.!* 2408, 513, 512. *Hedysarum ciliatum, Thunb.! Fl. Cap. p.* 594.

HAB. Galgebosch, *Thunberg!* Cape, *Bowie!* Caledon River and Magaliesberg, *Burke and Zeyher.* Uitenhage, Albany, and Kaffirland, *E. & Z.! &c.* Zondags R., Koega and Klein Bruintjeshoogte, Gekau, *Drege!* (Herb. Th., Sd., Bth., Hk., D.)

Root thick and woody. Stems several, spreading every way, 1–2 feet long, chiefly branched near the base, flexuous, mostly thinly covered with long, white, slender hairs. Stipules brown, narrow, 1–1½ lines long. Petioles 1–2 lines, rarely 4–6 lines long. Leaflets 1–1¼ inch long, 4–8 lines broad, always longer than broad, but variable in shape and proportions, green on both sides and conspicuously netted. Flowers 4 lines long. Legumes 6-8 lines long, 3–4 lines wide, oblique but not falcate; seed oblique, subtransverse, nearly as in Eriosema, but the hilum is medial, not linear. Strophiole small. *Benth.*

R. Totta (DC. Prod. 2, p. 388); procumbent or subvoluble, slender, hispido-pubescent or glabrate; petiole long; leaflets ovate-oblong or lanceolate-oblong, hispidulous or glabrescent; stipules small, ovato-lanceolate, rigid; peduncles equalling the leaves or longer, distantly 2–3 flowered; calyx segments longer than the tube, half as long as the glabrous vexillum; legume oblong, sessile, densely piloso-pubescent. *E. & Z.!* 1668; *also R. humilis, E. & Z.!* 1669, *and R. rigidula, E. & Z.!* 1675. *Copisma paniculatum, E. Mey.! Comm. p.* 134. *Zey.!* 506, 507. *Glycine Totta, Thunb.! Fl. Cap. p.* 591.

HAB. Cape, *Thunberg!* Near Grahamstown, and on the Winterberg, Zwartkey and Kat R., *E. & Z.! Mrs. F. W. Barber!* Magaliesberg, *Burke & Zeyher!* Uitenhage and Albany. (Herb. Th., Sd., Hk., Bth., D.)

Something like *R. glandulosa*, but more hairy, with a different legume, and wanting the large resinous dots; differing from *R. puberula* in the longer petioles, &c. Stems either short and decumbent, or 12–20 inches long and twining. Leaflets generally longer and broader than in *R. glandulosa*, rigid, strongly netted-veined, 1½–2 inches long, 4–8 lines wide, sometimes 10 lines. Flowers few, 3–4 lines long. Legume 6–8 lines long, 2–3 lines wide, quite sessile. Pubescence variable, whence the above synonyms.

25. R. pilosa (Harv.); stem *filiform*, voluble, *patently pilose*; leaflets of the lower leaves ovate-oblong, of the upper linear, all obtuse at base and acute at apex, hispidulous and ciliate, or glabrate, rigid and veiny; stipules small, subulate; peduncles setaceous, equalling the leaves, distantly 2-flowered; calyx segments setaceo-subulate, longer than the tube; vexillum glabrous; legume *falcate*, thinly pilose. *Copisma pilosum, E. Mey.? Comm. p.* 133.

HAB. Near Port Natal, *Gueinzius!* Zuureberg and between the Omtendo and Omsamculo, *Drege.* (Herb. Sd., D.)

Stems very slender, 2–3 feet long, subsimple, when young clothed with long, soft, rufous hairs. Lower leaflets 1½ inch long, 7–8 lines wide; upper as long, but only 2–3 lines wide, green on both sides, variably pubescent, sometimes glabrate and shining above, not conspicuously resin-dotted. Stipules withering. Peduncles thread-like, 2–2½ inches long, with 2 small flowers nearly an inch apart. Calyx pilose. Legume 8–9 lines long, acute at each end. I have not seen *Drege's* specimens, on which Meyer founded the species. It comes very close to *R. Totta.*

26. R. glandulosa (DC. Prod. 2, p. 388); prostrate or voluble, *glabrous* or minutely downy, slender; branches filiform; leaves on longish petioles, leaflets ovate, oblong or oblongo-lanceolate, resinous-dotted and paler beneath; stipules ovate or lanceolate, acute, persistent; peduncles slender, longer than the leaves, 1–6-flowered at the summit; calyx-tube narrow, shorter than the lanceolate segments, which are half as long as the glabrous vexillum; legume *stipitate*, glabrous or downy. *Glycine glandulosa, Thunb.! Cap. p.* 591. *Glycine heterophylla, Thunb.! Cap.* 592. *Copisma glandulosum, E. Mey. p.* 133. *Rhyn. glandulosa, riparia, glabra, E. & Z.!* 1672, 1673, 1674. *Gl. glabra, Spreng. R. glabra, DC. p.* 387. *Copisma glabrum, E. Mey.? p.* 134. *Zey.!* 518, 2404.

HAB. Common on stony hills, &c., from Capetown to Uitenhage and Albany. (Herb. Th., D., Sd., &c.)

Stems trailing or climbing, very slender; branches flexuous or twining round plants, &c. Stipules broad, ovate or sub-lanceolate, 1–2 lines long, rigid. Petioles ½–1 inch long. Lower leaflets ovate or oblong, ½–1½ inch long, ½–¾ inch wide; upper narrower and longer, sometimes nearly linear, 1–1½ lines wide. Peduncles thread-like, 1–3 inches long, rigid. Flowers 4–5 lines long, the calyx-segments narrow, the upper scarcely connate to the middle. Alæ much shorter than the carina. Legume 8–9 lines long, 3 lines wide, tapering at base into a 1–2 line long stipe.

4. **ORTHODANUM.** (Sp. 27–29.)

27. R. Orthodanum (Benth. !); erect, suffruticose, silky-villous or glabrescent; branches virgate; leaves subsessile or very shortly petiolate; leaflets *elliptic-oblong* or lanceolate-oblong, obtuse, mucronate, netted-veined; stipules lanceolate, withering; peduncles shorter than the leaves, closely few-flowered; calyx lobes lanceolate-acuminate, much longer than the tube, shorter than the nearly glabrous vexillum; legume silky-pilose. *Orthodanum latifolium, O. sordidum, and O. argenteum, E. Mey.! Comm. Drege, p.* 131–2. *Eriosema sericeum, E. & Z.! No.* 1678. *Burch. Cat.* 4674?

VAR. β. **Muhlenbeckii**; stems dwarf; leaves and flowers smaller; calyx segments rather shorter in proportion to the tube; legume less pilose. *Orthodanum Muhlenbeckii and O. glabratum, Meisn. in Hook. Lond. Journ.* 2, *p.* 91–92.

HAB. Grassy mountain sides of the Winterberg, above Philipstown, Kat River, *E. & Z.!* Between the Basche and Omtata, and on to Port Natal, *Drege! Krauss.!* 374. β. Near Kromme R., Uitenhage and in Outenequa, *Krauss., Bowie!* (Herb. Bth., Hk., D., Sd.)

Very variable in pubescence and in the size and breadth of the leaflets. Stems numerous, 1–2 feet high, subsimple, straight or incurved. Leaflets 1–2 inches long, ½–1 inch wide, sometimes silvery, especially on young shoots; mostly softly pubescent, but green; occasionally thinly pubescent. Lower petioles sometimes ½ inch long, the upper 1–2 lines long, shorter than the spreading, narrow stipules. Lower leaves often unifoliate. Peduncles 2–3-flowered. Calyx-tube 1 line, segments 2–3 lines long. Legume 6–8 lines long, 2–2½ lines wide, silky, sometimes becoming glabrescent.

28. R. nitens (Benth. !); erect, suffruticose, *every where silky-silvery*, with minute, soft tomentum; leaves on shortish petioles, leaflets *broadly ovate or cordate-ovate*, flat, penni-nerved beneath; stipules obsolete; peduncles 2–3-flowered, shorter than the leaf; calyx-lobes shorter than the wide tube, much shorter than the pubescent vexillum; (unripe) legume densely pilose. *Zey.!* 383.

HAB. Magaliesberg, *Burke & Zeyher!* (Herb. Bth., Hk., Sd., D.)

Apparently an erect suffrutex, 2–3 or more feet high, with flexuous, somewhat voluble, terete branches, everywhere softly velvetty with shining, white tomentum. Stipules very caducous and minute. Petioles 3–6 lines long, the lower ones uncial. Lower leaflets 1 inch long, ¾ inch wide, upper smaller, often cordate, but varying to ovate or suborbicular; the lateral scarcely oblique, all subacute, scarcely mucronulate. Peduncles 6–12 lines long, mostly 3-flowered at the summit; pedicels shorter than the calyx. Calyx pubescent, subcanescent, the tube very obtuse at base, 1–1½ lines long, the segments triangular-acuminate. Vexillum 8 lines long, downy; alæ much shorter than the incurved, subrostrate carina. Style conspicuously thickened beyond the middle.

29. R. bullata (Benth. !); shrubby, erect, viscidly-hairy and resin-dotted; branches virgate; leaves on very short petioles; leaflets small, oblong, bullate, with recurved margins; stipules leaf-like, taper-pointed; peduncles scarcely longer than the leaves, 1–2-flowered; calyx-lobes scarcely longer than the tube, shorter than the glabrous vexillum; legume sparsely pilose. *Hidrosia bullata, E. Mey.! Comm. Drege, p.* 89.

HAB. Wupperthal, in stony mountain places, *Drege!* (Herb. Bth., Hk., D.)

1–2 feet high, slender, not much branched; leaflets 4–5 lines long, 1–2 lines wide, thickly resinous-dotted, pale; petioles 1–2 lines long. Stipules ovate-acuminate, longer than the petiole. Calyx tube obtuse at base, 2 lines long, patently pubescent and viscid; laciniæ acuminate, scarcely longer, the uppermost slightly connate, lowest a little longer, all rigidly ciliolate. Vexillum 6 lines long, with a longish claw. Alæ narrow, shorter. Carina equalling the vexillum, broad, incurved and blunt at the point. Legume 5–6 lines long, 2–3 broad, oblong, compressed.

Species of Copisma unknown to us.

R. trichodes (E. Mey. Comm. p. 134); stem procumbent, flexuous, very hairy, with patent, yellow hairs; leaflets elliptic-oblong, subacute, netted-veined, yellow-hairy on both sides, gland-dotted beneath, equalling the petiole; peduncles one-flowered, equalling the leaf; legumes (unripe) straightish, villous. *Cop. trichodes, E. Mey.*

Collected by *Drege;* no habitat assigned.

R. nitida (E. Mey. Comm. p. 136); stem terete, filiform, voluble, glabrescent; leaflets broadly-triangular, lateral scarcely unequal-sided, glabrous and shining at both sides, ciliate; racemes equalling the leaves, few-flowered; flowers suberect, pedicels equalling the pubescent, nearly glandless calyx, legumes falcate, glabrescent, about 3-seeded. *Copisma nitidum, E. Mey.*

HAB. Grassy hills near Omsamculo, *Drege.*

LIX. ERIOSEMA, DC.

Calyx campanulate, 5-fid, the upper segments sometimes connate. *Petals* subunequal; *vexillum* obovate or oblong, with inflexed auricles at base; *alæ* narrow, longer or shorter than the wider, incurved, obtuse *carina*. *Stamens* 9–1. *Ovary* sessile, *very hairy*, 2-ovuled; style filiform, quite glabrous above the middle, incurved and often thickened upwards; stigma small or capitate. *Legume* compressed, obliquely orbicular-rhomboid, or broadly oblong, hairy; seeds 2 or 1, compressed, oblong, obliquely transverse, the seed-cord fixed at one end of a linear hilum. *Benth.! in Mart. Fl. Bras. XXIV. p.* 206. *Endl. Gen.* 6691.

Erect or prostrate, rarely voluble, herbs or suffrutices common in the warmer regions of both hemispheres, less conspicuously resin-dotted than in *Rhynchosia.* Stipules lanceolate, free or concrete. Peduncles axillary, racemose; flowers yellow or purple. Very nearly allied to *Rhynchosia,* but slightly different in habit, and essentially characterised by the obliquely transverse seeds, with excentral seed-cords: a character easily seen in the unripe fruit. Name from εριον, *wool,* and σημα, a *standard.*

ANALYSIS OF THE SOUTH AFRICAN SPECIES.

Leaves mostly unifoliolate (rarely 3-foliolate):
- Petioles rather long; pubescence *white, soft, and silky,* copious ... ... (1) **populifolium.**
- Petioles short; pubescence *red-brown,* harsh, pilose ... (2) **cordatum.**

Leaves pinnately-trifoliolate:
- Peduncles naked at base, densely racemose beyond the middle.
  - Petioles very short:
    - Leafl. (large) broadly-ovate or elliptic-ovate ... (2) **cordatum, β.**
    - Leafl. oblong, obtuse, *rugose, rufo-tomentose* beneath (3) **oblongum.**
    - Leafl. ovate or lanceolate-oblong, *whitish* beneath (6) **squarrosum.**
    - Leafl. linear-lanceolate or oblong-lanceolate, *silky-white* beneath ... ... (7) **salignum.**
    - Leafl. lanceol.-oblong, concolorous, thinly silky beneath ... ... (8) **Kraussianum.**

Petioles rather long:
Leafl. 2–3-uncial, oblong-lanceol., acute, rugose, rufo-sericeous beneath ... ... (4) **Burkei.**
Leafl. $1\frac{1}{2}$–2-uncial, ellipt.-oblong, obtuse, thinly silky beneath ... ... (5) **parviflorum.**
Peduncles bearing flowers for $\frac{3}{4}$ their length, or more (9) **cajanoides.**

1. E. populifolium (Benth.!); dwarf, ascending, *every where densely clothed with long, soft, silky, white hairs;* leaflets solitary (large), on longish petioles, *cordate,* subacute, penninerved; stipules broadly lanceolate, equalling the petiole, striate; peduncles rather shorter than the leaf, densely many flowered near the summit; calyx pilose, its lobes about equalling the tube; vexillum pilose.

HAB. Trans Vaal, Natal, *Mr. Sanderson!* (Herb. Hk.)

Stems 6–12 ? inches high, simple ? compressed or angular. Petioles uncial. Leaves 4 inches long, 3 inches wide, exactly heart-shaped. Stipules inch long, 3–4 lines wide, silky, scarious. Peduncles 3–4 inches long, bearing flowers for an inch below the end. Pedicels 2 lines long, subtended by lanceolate, deciduous bracts. Flowers 8 lines long. Calyx laxly pilose, its two upper segments connate nearly to the bifid summit, 3 lower lanceolate. Legume unknown.

2. E. cordatum (E. Mey. Comm. Drege, p. 128); the decumbent stem, petioles, peduncles and calyx *roughly pilose with spreading, red hairs;* leaves *subsessile,* either *solitary,* roundish-cordate, obtuse, or *pinnato-trifoliolate,* the leaflets elliptic-ovate, all membranous, rufo-pilose on both sides, somewhat bullate, penninerved and veiny beneath; stipules free, broad, ovate-acuminate; peduncles shorter or longer than the leaves, densely spicato-racemose above the middle, the flowers reflexed-imbricate.

VAR. β. **Gueinzii**; less densely hairy; leaves all pinnato-trifoliolate, leaflets membranaceous, elliptic-oblong, acute; peduncles much longer than the leaves, more laxly racemose. *E. Gueinzii, Sond. Linn.* 23, *p.* 34. *Dietr. Fl. Univ. cum icone.*

HAB. In grassy places between Gekau and Bache, *Drege.* Port Natal, *Gueinzius! Dr. Sutherland! Krauss,* 475. (Herb. Hk., Bth., Sd.)

Root woody. Stems 6–12 inches long or more, subsimple. Leaves either simple or pinnately trifoliolate. The simple leaves are nearly or quite sessile, 4–3 inches long, about equally wide, often nearly orbicular; the unijugate have a common petiole $1\frac{1}{2}$ inch long, and the leaflets are smaller and more ovate and subacute. Peduncles 2–6 inches long; the raceme 1–2 inches. Calyx very hairy, the tube 1 line, segments 2 lines long, lanceolate-subulate, the uppermost free. Vexillum 4 lines long, pubescent. Legume orbicular, very oblique, compressed, densely rufo-hirsute. Remarkable for its rough clothing of *foxy,* rigid, but glossy hairs. Simple and trifoliolate leaves occur sometimes on the same stem. Var. β. (Natal *Gueinzius!* in Hb. Sond.) has quite the aspect of a plant grown in a moister and more shady locality.

3. E. oblongum (Benth.!); dwarf, erect or ascending, the stems, petioles and peduncles roughly hairy with rusty hairs; petioles very short, leaflets 3, oblong, obtuse, rugose, hispid above, rufo-tomentose and penninerved beneath; stipules oblong-acute; peduncles elongate, shortly and closely several-flowered near the summit; calyx hairy, its lobes longer than the tube, much shorter than the villous vexillum. *Zey.!* 466.

Aapjes River, *Burke & Zeyher!* (Herb. Hk., Bth., Sd.)

Stems 6–8 inches high. Common petiole 3–4 lines long. Leaflets $1\frac{1}{2}$–2 inches long, $\frac{1}{2}$–$\frac{3}{4}$ inch wide, at length subglabrous above, always densely and softly pubescent beneath. Stipules equalling the petiole. Peduncles 4–6 inches long, the 8–10 flowered raceme 1–$1\frac{1}{2}$ inch long. Bracts subulate or narrow lanceolate. Flowers subsessile, 5–6 lines long, spreading (not reflexed). Calyx-tube 1½ lines long. Vexillum pubescent, about equalling the incurved, subrostrate carina. Legume 7–8 lines long, 4–5 broad, coriaceous, with long, foxy hairs. Allied to *E. cordata*, but differs in the form and size of the leaves, the softer and shorter pubescence, longer peduncles and not reflexed flowers, &c. It is in all parts smaller.

4. **E. Burkei** (Benth.!); dwarf, suberect; the stem, petioles and peduncles rufo-sericeous with short, soft hairs; petioles longest, leaflets 3, oblongo-lanceolate, acute, rugose, hispidulous above, rufo-sericeous, prominently penninerved and netted-veined beneath; stipules free, silky, lanceolate; peduncles longer than the leaves, densely racemose towards the summit, with reflexed, imbricating flowers; calyx silky-pilose, its narrow segments thrice as long as the tube. *Zey.!* 465.

HAB. Magaliesberg, *Burke & Zeyher!* (Herb. Hk., Bth., Sd.)

Stems from a woody base, a foot or more in height, incurved, branching, with soft, woolly pubescence. Common petiole about an inch long. Leaflets 2–3 inches long, ⅓–¾ inch wide, the terminal longest, green above, rufescent beneath, especially on the prominent ribs and nerves and at the margin: substance thickish. Peduncles 4–6 inches, the raceme 2–3 inches long. Flowers 6–7 lines long. Calyx clothed with long, silky, fulvous hairs, the segments nearly equalling the piloso-pubescent vexillum. Legume densely pilose, 7–8 lines long, 4–5 wide.

5. **E. parviflorum** (E. Mey.! Comm. p. 130); suffruticose, erect or ascending, the stem, petioles and long peduncles shaggy with rusty, deflexed hairs; leaves petiolate, leaflets 3, elliptic-oblong, obovate, or ovate, pubescent above, rather paler and more silky but reticulately veined beneath; stipules lanceolate, free; peduncles elongate, shortly and densely spicato-racemose at the summit; calyx segments shorter than the tube; vexillum puberulent; legume shaggy. *E. podostachyum, Hook. f. Fl. Nig. p.* 314, *Cytisus glomeratus, Boj.! Hort. Maurit. p.* 89.

HAB. In grassy places, in moist valleys between Omsamculo and Omcomas, near Omblas, *Drege!* Natal, *Gueinzius, Sutherland.* (Herb. Bth., Hk., Sd.)

Stems 1–2 feet high, branching, woody at base, more or less hairy. Stipules membranous, brown, narrow, 2–3 lines long. Lower petioles ½ inch, upper 1 inch or more long. Leaflets 1½–2 inches long, ¾–1 inch wide, mostly elliptic, varying to obovate and oblong; in some varieties ovate. Peduncles 3–6 inches long, the raceme 1–1½ inch long. Flowers subsessile, strongly reflexed, 3–3½ lines long. Calyx scarcely 1 line long, pubescent; the lobes broad, about equal in length. Vexillum minutely downy. Legume 6 lines long, 3–4 wide, very shaggy with foxy hairs.

6. **E. squarrosum** (Walp. Linn. 13, p. 536); erect or diffuse, the stem, petioles, and peduncles *fulvo-sericeous* with short, soft, hairs; leaflets 3, *ovate, obovate, oblong or lanceolate-oblong,* obtuse or acute, green above, silky-white and penninerved beneath; stipules lanceolate, free; peduncles mostly longer than the leaves, densely racemose beyond the middle with many reflexed, imbricating flowers; calyx pilose, its segments longer than the tube. *Hedysarum squarrosum, Thunb.! Fl. Cap. p.* 595. *Desmodium squarrosum, DC. Prod.* 2. *p.* 333, *E. & Z.!* 1661. *Eriosema Zeyheri, E. Mey.! Comm. p.* 129. *E. reticulatum, β. canescens, Meisn.! in Lond. Journ.* 2, *p.* 80.

VAR. β. **acuminatum** (E. & Z.!); leaflets narrower, longer, and more acute; vexillum violet.

VAR. γ. **Dregei** (Benth.!); leaflets *silky-canescent above;* vexillum concolorous. *E. Dregei, E. Mey.! Comm. p.* 129.

VAR. δ. **latifolium** (Benth. !); stem and inflorescence with longer red hairs, leaflets much larger (2½ inches long, 1½ wide), pilose hairy above, densely cano-tomentose beneath; vexillum violet.

HAB. Grassy fields beyond Camtoos R., near Galgebosch and elsewhere, *Thunberg!* Zwartkops and Vanstaadens R., and Adow, Ait., *E. & Z.!* Zuureriver and Klipplaat R., *Drege!* Slaay Kraal, *Burke & Zeyher!* β. Winterberg, *E. & Z.! Bowie! Krebs.!* γ. Omsamcubo, *Drege!* Port Natal, *Gueinzius!* (Herb. Th., Hk., Bth., Sd. D.)

Somewhat woody at base; stems erect or ascending, branched from the root, 6–14 inches long, clothed with short, reversed, rusty-coloured or dirty-white soft hairs. Stipules membranous, striate, 4–6 lines long. Petioles rarely ½ inch long. Leaflets 1–1½ inch long, ¾–1 inch wide, in δ. much larger, the lower often obovate, the rest mostly oblong or elliptical; those near the ends of the branches narrower and longer in proportion and passing into lanceolate. Peduncles 2–6 inches long, the raceme about uncial, very dense; the flowers strongly reflexed, 4–5 lines long. Calyx cano-villose, the segments lanceolate, half as long as the pubescent vexillum. Legume 6 lines long, 4–5 lines wide, shaggy with long red hairs. I follow *Bentham* in referring γ. and δ. to this place.

7. **E. salignum** (E. Mey.! Comm. Drege, p. 129); suberect, tall, the stem, petioles, and peduncles *cano-sericeous*, with short, soft hairs; leaflets 3, oblong or linear-lanceolate, elongate, acute, green above, *silky-white* and penninerved beneath; stipules oblongo-lanceolate, free; peduncles longer or shorter than the leaves, racemose beyond the middle, with many reflexed, imbricating flowers; calyx pilose, its segments longer than the tube. *Burch.* 3877.

VAR. β. **concolor**; pubescence scanty; leaflets green on both sides, pilose above, thinly silky beneath, obtuse or subobtuse, mucronulate; flowers lax.

HAB. Magaliesberg, *Burke & Zeyher!* Natal, *Sanderson, Krauss, T. Williamson, &c.* (Herb. Bth., Hk., Sd., D.)

Stems several, 6–18 inches high, subsimple. Stipules ½–¾ inch long, striate. Petioles ½–¾ inch long. Leaflets 1½–3 inches long, ½–¾ inch wide, subglabrous above, mostly white beneath; in β. green and partly naked. Peduncles 3–6 inches long, the raceme 1–2 inches long. Flowers closely reflexed, 5 lines long. Calyx-segments a little shorter than the hairy vexillum. Very near *E. squarrosum*, but the leaves are longer and narrower, and the flowers rather larger. β. from Natal (Herb. Hk.) may belong to a different species.

8. **E. Kraussianum** (Meisn.! in Hook. Lond. Journ. 2, p. 91); *dwarf*, erect, subsilky, cano-pubescent; petiole very short, leaflets 3, *lanceolate-oblong*, tapering at base, sparsely pilose above, thinly silky beneath; stipules lanceolate, free; peduncles elongate, densely racemose above the middle, many-flowered; calyx fulvo-hirsute, its lobes shorter than the tube; vexillum hairy.

HAB. Grassy places at foot of the Tafelberg, Port Natal, *Krauss!* 474, *ex pte.* (Herb. Hk., Bth.)

Root woody. Stems several, 6–8 inches high, incurved, angular, loosely cano-pubescent. Petioles 1–2 lines long. Leaflets 1½ inch long, 3–4 lines wide, thinly silky and mottled beneath, minutely netted and obliquely nerved, the margin slightly recurved. Racemes spicate, 1½ inch long, rich brown. Flowers closely set, 3–4 lines long. Stipules longer than the petioles, scarious, brown. Much smaller than *E. cajanoides*, with denser and shorter racemes, larger stipules, and a hairy vexillum, &c.

9. **E. cajanoides** (Benth.); stem erect, tall, rib-striate, fulvo-canescent, tomentose; petiole very short; leaflets 3, oblong, obtuse, acute at base, glabrescent above, cano-sericeous beneath; stipules minute,

free; peduncles longer than the leaves, floriferous for ⅔ their length, laxly many flowered; calyx-segments nearly equalling the tube; vexillum glabrous. *Rhynchosia cajanoides, Guill & Perr! Fl. Seneg. p.* 215. *Eriosema polystachyum, E. Mey.! Comm. Drege, p.* 130. *Zey.!* 464.

HAB. Port Natal, *Krauss*, 64! Crocodile River, *Burke & Zeyher!* Coastland near Natal, *Dr. Sutherland!* Caffraria, *Drege!* (Herb. Bth., Hk., Sd.)

Stems 2–3 feet high, angular and slightly flexuous, not much branched, whitish or rufescent. Stipules lanceolate, patent. Petioles 2–6 lines long. Leaflets 2–2½ inches long, ½–¾ inch wide, green and finely netted above, discoloured and penninerved beneath, the margin slightly reflexed. Racemes 3–4 inches long, many-flowered, subdistichous; flowers 2–3 lines apart, 4–5 lines long. Calyx canescent. 2½ lines long, its lobes ovate-acute. Vexillum mostly quite glabrous. Legumes 7 lines long, 5 lines wide, coriaceous, densely clothed with long, white hairs. A native also of tropical Africa. The flowers in the S. African specimens are rather longer, and the pubescence more copious and whiter.

Species unknown to us.

E. reticulatum (E. Mey. Comm. p. 129); "stem flexuous, covered with reflexed, yellow hairs; medial leaflet obovate-oblong, lateral obliquely oblong, all acute, transversely venoso-reticulate beneath, appressedly hirsute at the veins, fuscescent; racemes shorter than the leaves, few-flowered; vexillum violet." *E. Mey. l. c.*

HAB. Caffraria, *Drege.*

E. trinerve (E Mey. Comm. p. 130); "stem flexuous, retrorsely hairy; leaflets oblong-lanceolate, subacute, 3-nerved as far as the middle, green beneath, between the brownish veins, pubescent; racemes about equalling the leaves, few-flowered." *E. Mey. l. c.*

HAB. Caffraria, *Drege.*

E. capitatum (E. Mey. Comm. p. 130); "stem erect, appressedly pubescent, canescent; leaflets lanceolate-oblong, subacute at each end, white and silky beneath; heads subglobose, involucrate, on long peduncles." *E. Mey. l. c.*

HAB. Among shrubs at Ruigtavalei, *Drege.* Petals all violet-coloured,

LX. ABRUS, L.

Calyx campanulate, shortly 4-fid or 4-toothed, the upper lobe entire or bifid. *Vexillum* ovate, about equalling the subfalcate carina. *Stamens* 9, monadelphous in a split tube; no vexillary stamen. *Ovary* several-ovuled; style short, incurved, glabrous. *Legume* oblong, compressed, 4–6-seeded, with partitions between the seeds. *Endl. Gen.* 6698. *DC. Prod.* 2, *p.* 381.

Diffuse or climbing, slender, ligneous plants, chiefly from tropical Asia; naturalized in several parts of the tropics. Leaves abruptly pinnate, multi-jugate. Flowers racemose, orange. Seeds glossy, oblong, red, with a black spot round the hilum. Roots used as a substitute for liquorice. Name from *αβρος*, elegant.

ANALYSIS OF THE SOUTH AFRICAN SPECIES.

Flowering branches bare of leaves. Legume 2–2½ times longer than broad ... ... (1) **precatorius.**
Flowering branches leafy. Legume 5 times longer than broad ... ... (2) **lævigatus.**

1. **A. precatorius** (Linn. Syst. 533); racemes densely many-flowered,

terminating short, axillary, nearly leafless branches; rachis incrassated in fruit; vexillum nearly free; alæ shorter than the carina; legume quadrate-oblong, 2–2½ *times longer than broad,* very oblique at base, about 5-seeded. *Benth.! Fl. Braz. XXIV. p.* 215. *A. squamulosus, E. Mey.! Comm. p.* 126.

HAB. Port Natal, *Drege! Krauss! T. Williamson, &c.* (Herb. Hk., D., &c.)

Common throughout tropical Asia and Africa. Stems climbing, sparingly pubescent. Leaves abruptly pinnate, 10–20-jugate; leaflets oblong, blunt, pale green, glabrous above, minutely silky beneath, 4–8 lines long, 2–4 lines wide. Flowering branches axillary, longer or shorter than the leaves, very patent or divaricate, rarely quite leafless, generally with 1–2 leaves at base, and always pluri-stipulate in the lower part. Stipules subulate. Flowers red, 4–5 lines long. Legumes 1–1½ inch long, 6–7 lines wide, in our specimens squamulose and pubescent, sometimes glabrous; seeds bright scarlet with a black spot round the short hilum. The seeds are often strung as beads by children, &c.

2. A. lævigatus (E. Mey.! Comm. p. 126); racemes several-flowered, *terminating leafy, lateral branches;* rachis scarcely thickened in fruit; vexillum adnate to the base of the staminal tube, alæ shorter than the carina; legume subfalcate, 5 *times longer than broad,* 6–8-seeded, subglabrous and even.

HAB. Caffraria and Natal, *Drege!* (Herb. Hk., D., &c.)

Very near *A. pulchellus,* Wall. Cat. 5819. It is easily known, when in fruit, from *A. precatorius* by the longer and narrower, several-seeded legume. The habit and foliage are similar, but the flowers are less numerous, and the flowering branch bears leaves nearly to the base of the raceme.

LXI. LONCHOCARPUS, H.B.K.

Calyx truncate or shortly 4–5-toothed. *Alæ* slightly cohering above the claw of the carina. Petals of the *carina* slightly cohering at back above. *Vexillary-stamen* quite free at base, above connate with the rest into a complete tube. *Anthers* versatile. *Ovary* 2 or several-ovuled. *Legume* flat, oblong or elongate, membranaceous, coriaceous or ligneous, indehiscent, with terminal style; the sutures not winged. *Seeds,* if many, distant, compressed. Radicle inflexed. *Benth.! in Journ. Linn. Soc. vol.* 4, *Suppl. p.* 85.

Trees or climbing shrubs. Leaves alternate, impari-pinnate; leaflets opposite; stipellæ few or none. Racemes simple or panicled. Flowers violet, purple or white, but not yellow; petals silky or glabrous. Name from λονχη, a *lance,* and καρπος, *fruit;* alluding to the shape of the pods.

L. Philenoptera (Benth.! l. c. p. 97); young parts velvetty-canescent; leaves on long petioles, 2–3-jugate, leaflets broadly elliptic-oblong, mucronulate, at first tomentose, afterwards glabrescent or glabrous, netted-veined beneath; panicles hoary and velvetty, terminal, much-branched; calyx velvetty, its teeth deltoid, nearly equalling the tube; corolla glabrous; ovary 4-ovuled; legumes lanceolate-oblong, tapering at base, obtuse or acute, coriaceo-membranous, pale, glabrous. *Philenoptera Schimperiana and Dalbergia Schimperiana, Hochst. Pl. abyss. No.* 897, 1778. *Tapassa violacea, Kl.*

HAB. Banks of the Tamulakau R. and between the Chobu and Mabalu Rivers, near Lake Ngami, *Jos. M'Cabe.* (Herb. Hk., D.)

A tree 20–30 feet high, 2 feet in diameter. Leaves toward the ends of the branches and twigs. Common petiole 6–12 inches long, the pair of leaflets 1–1½ inch apart. Leaflets very pale green, at first densely velvetty, canescent beneath, afterwards becoming naked, and when old quite glabrous. Panicles 6 inches long, alternately branched, the branches racemose, many-flowered. Calyx silvery, 3 lines long. Corolla purplish-pink. Flowers sweetly-scented. Ovary silky. Legumes 3–4 inches long, ¾ inch wide, strongly compressed, of a parchment-like substance. A native also of Mozambique and Abyssinia.

LXII. PTEROCARPUS, L.

Calyx turbinate-campanulate, acute at base, oftener incurved, 5-toothed, sub-bilabiate. *Petals* of the carina at back near the apex shortly connate or nearly free. *Stamens* 10, sometimes monadelphous with a split tube; sometimes equally diadelphous, and sometimes diadelphous, 9–1. *Anthers* versatile. *Ovary* 6–8-ovuled. *Legume* compressed, indehiscent, orbicular or ovate, more or less oblique or falcate, with a lateral or rarely terminal style, bearing seeds in the middle, more or less indurated or thickened, with a surrounding membranous wing or sharp ridge, sometimes almost completely attenuate-coriaceous or membranous. *Seeds* 1–3, separated by hard partitions. *Benth.! in Journ. Linn. Soc. vol.* 4, *Suppl. p.* 74.

Unarmed, tropical and subtropical trees of both hemispheres. Leaves alternate, impari-pinnate. Leaflets alternate or irregularly opposite, without stipellæ. Racemes simple or laxly panicled. Flowers yellow, rarely whitish, with violet shades, often handsome. Petals glabrous. Name from πτερον, a *wing*, and καρπος, a *fruit*. The strongly astringent gum-resin called *Kino* is a natural exudation from several species of *Pterocarpus*.

1. **P. sericeus** (Benth.! l. c. p. 75); leaflets 3–5, broadly ovate, shining silky beneath or on both sides; legume shortly stipitate, ovate-suborbicular, with a coriaceous wing. *Dalbergia rotundifolia, Sond.! in Linn.* 23, *p.* 35.

HAB. Hex and Aapjes Rivers, *Burke & Zeyher!* (Herb. Hk., Sd.)

Twigs thinly and appressedly silky. Under-surfaces of the young leaflets pale, of the older concolourous, all densely silky; upper surface puberulous, at length glabrate and netted. Petioles 2–4 inches long. Leaflets 1½–2 inches long. Flowers not seen. Legume 1½ inch long, about 12–14 lines wide, minutely stipitate-elliptical, membranaceo-coriaceous, veiny.

LXIII. DALBERGIA, Linn.

Calyx campanulate, 5-toothed, the upper teeth broader, the lowest longest. Carinal petals keeled above at the apex. *Stamens* 10, monadelphous, with a split tube; or 9, the vexillary filament wanting; or equally diadelphous. *Anthers* small, erect, didymous, shortly opening at the apex; or rarely longitudinally splitting. *Ovary* stipitate, few-ovuled. *Legume* oblong or linear, rarely falcate, flat, thin, indehiscent, either one-seeded or distantly few-seeded, slightly hardened and often netted at the seed, the margins neither thickened nor winged. *Benth.! in Journ. Linn. Soc.* 4, *Suppl. p.* 28.

Trees or climbing shrubs natives of the warmer zones of both hemispheres. Leaves alternate, impari-pinnate, the leaflets exstipellate and mostly alternate. Inflorescence dichotomously cymose or irregularly panicled. axillary or terminal. Flowers small, often numerous, purple, violet or white? Named in honour of Nicholas Dalberg, a Swedish botanist.

ANALYSIS OF THE SOUTH AFRICAN SPECIES.

Leaflets small, 6–10–16-jugate.
Thinly pubescent, becoming glabrous ; adult leaves and legumes glabrous ... ... (1) **armata.**
Fulvo-villous ; adult leaves and legumes rusty-pubescent (2) **multijuga.**
Leaflets large, 2–3-jugate ; legumes rusty-pubescent ... ... (3) **obovata.**

1. D. armata (E. Mey.! Comm. p. 152); branches spinous ; leaves 6–12-jugate, leaflets subopposite or alternate (small), oblong, obtuse at each end, paler beneath, *becoming glabrous;* young petioles and inflorescence thinly-velvetty; panicles pedunculate, corymbose, shorter than the leaves, axillary or subterminal; calyx puberulous, its two upper lobes short, obtuse, 3 lower deltoid, subacute ; legumes oblong, obtuse, tapering at base, glabrate. *D. myriantha, Meisn. ! in Lond. Jrn.* 2. *p.* 100.

HAB. Between Omtendo and Omsamculo, and near Port Natal, *Drege! Krauss!* 220, *Gueinzius! &c.* (Herb. Hk., Bth., D., Sd.)

A tree with dark-coloured bark, not always spiny. Common petiole 2–2½ inches long, at first rusty-pubescent, afterwards glabrous. Leaflets 3–4 lines long, 1–1½ lines wide, the young ones thinly silky beneath, older quite glabrous, thick in substance and glossy, all midribbed and minutely marginate. Flowers very small, in dense corymbs on a peduncle 1 inch long, either from the axils of the upper leaves, or ends of the twigs. Legumes 1–1½ inches long, 5–8 lines wide, sometimes subtruncate and very obtuse, sometimes tapering to a subacute point, always tapering at base and stipitate.

2. D. multijuga (E. Mey.! Comm. p. 153); leaves 10–16-jugate, leaflets alternate (rather small), oblong, obtuse at each end, mucronulate, paler beneath, with recurved margins, *pubescent;* twigs and petioles densely fulvo-villous; racemes short, terminal and axillary, crowded toward the end of the branches *(E. M.);* legumes broadly lanceolate, netted with veins, rusty pubescent.

HAB. Morley, at the end of the wood, 1000–1500 f., *Drege!* (Herb. Bth.)

A tree. Twigs and petioles densely and persistently *foxy*, with close, short, erect hairs. Common petiole 4–5 inches long. Leaflets 5–7 lines long, 2–2½ wide, rather roughly though thinly pubescent. Legumes 2 inches long, 8–9 lines wide in the middle, acute or subacute, fulvous.

3. D. obovata (E. Mey. ! Comm. p. 152); leaves bi-tri-jugate, leaflets alternate, oblong or obovate, coriaceous, reticulated, glabrous; (young) petioles and inflorescence velvetty; panicle dense, its branches corymbose; calyx puberulous, its two upper lobes broadly oblong, obtuse, three lower narrow, acute; legumes broadly lanceolate, netted with veins, rusty-pubescent. *Podiophyllum reticulatum, Hochst. Flora,* 1841, *p.* 658.

HAB. River Basche and near Port Natal, *Drege,* Krauss! 193. (Herb. Hk., Bth., Sd., D.)

A tree. Bark dark-coloured. Common petiole 2–3 inches long, at first densely pubescent, afterwards subglabrous, bearing 5 or 7 alternate leaflets, each about 1½ inch long, ¾ inch wide, netted on each side and glossy. Panicles terminal and axillary, much branched, the branches fasciculato-corymbose. Flowers not 2 lines long. Legumes 1½–1¾ inch long, 6–8 lines wide, acute, tapering at base into a short stipe, strongly netted in the middle. Leaflets much larger and fewer than in the other S. African species.

LXIV. SOPHORA.

Calyx widely campanulate, obliquely truncate, obsoletely or shortly

5-toothed. Petals of equal length; *vexillum* obovate or roundish, erect or spreading; alæ oblong, clawed, eared at base; *carina* obtuse, straight, its petals imbricating and connate in the middle, free above. *Stamens* 10, free, glabrous. *Ovary* subsessile, many ovuled; style slightly curved, glabrous. *Legume* moniliform, indehiscent, wingless, several-seeded. *Endl. Gen.* 6738. *DC. Prod.* 2, *p.* 95.

Trees, shrubs, or herbs chiefly from the tropics of Asia and America. Leaves impari-pinnate, plurijugate, the terminal leaflet remote from the last pair. Racemes axillary or terminal, mostly simple. Flowers white, blue, or yellow. The name is an alteration of the Arabic *Sophera.*

1. S. nitens (Benth.!); shrubby (or arborescent?), densely silky-tomentose in all parts, with close-pressed, glossy hairs; leaflets 5–7, elliptic-oblong, obtuse; racemes terminal, elongating, many-flowered; bracts subulate, nearly equalling the pedicels; petals glabrous.

HAB. Near Port Natal, *T. Williamson!* (Herb. D.)

Stems and branches unknown. Petioles 3–4 inches long; leaflets alternate, 1–1½ inches long, ½–¾ inch wide, resembling the leaves of *Podalyria sericea,* but larger and blunt. Raceme 4–5 inches long; flowers white? Occasionally there are two ovaries in a flower. Not found by *Drege* or recent explorers, but well worth looking after! It must be a very ornamental plant.

LXV. VIRGILIA, Lam.

Calyx widely campanulate, shortly 2-lipped, the upper lip bifid, the lower trifid. *Vexillum* suborbicular, strongly reflexed; *alæ* oblong; *carina* incurved, rostrate. *Stamens* 10, free. *Ovary* sessile, villous, several-ovuled; style glabrous. *Legume* coriaceous, compressed, tomentose, many-seeded, stuffed between the seeds, the sutures very obtuse. *E. Mey.! Comm. p.* 1. *Endl. Gen.* 6741. *DC. Prod.* 2. *p.* 98, *ex pte.*

Only one species known. *V. grandis, E. Mey. is Milletia Caffra, Meisn.* Name in honour of the poet *Virgil.*

1. V. Capensis (Lam. Ill. t. 326 f. 2); *DC. Prod.* 2. *p.* 98. *Bot. Mag. t.* 1590. *E. & Z.!* 1141. *E. Mey. Comm. p.* 1. *Sophora Capensis, Linn. Mant.* 67. *S. oroboides, Berg. Cap.* 142. *Hypocalyptus Capensis, Thunb.! Cap.* 570. *Podalyria Capensis, Andr. Rep. t.* 347.

HAB. River-sides, &c., throughout the colony, (Herb. Th., Hk., Sd., D.)

A tree, the *Wilde Keureboom* of the colonists. Twigs furrowed, thinly tomentose. Leaves 6–20-jugate, exstipulate; leaflets linear-oblong, mucronate, coriaceous, the young ones silky on both sides, the old glabrous and glossy above, pale and tomentose beneath, with slightly revolute margins, nearly an inch long, 2–3 lines wide. Racemes lateral, longer than the leaves, many-flowered; flowers rosy-purple, half-inch long. Calyx silky, at first obtuse, then intruse at base. Bracts broadly ovate, mucronate, deciduous.

LXVI. CALPURNIA, E. Mey.

Calyx widely campanulate, shortly 5-fid, the two upper lobes semi-connate. *Vexillum* erect; *alæ* oblong; carina incurved, obtuse, bifid. *Stamens* 10, free, or connate at base, persistent. *Ovary* stipitate, several-ovuled. *Legume* membranaceous, compressed, glabrescent, reticulate, few or several-seeded, somewhat winged along the ventral suture, the valves cohering between the seeds. *E. Mey.! Comm. p.* 2. *Endl. Gen.* 6740.

Trees or shrubs, natives of the warmer parts of Africa. Leaves imparipinnate, multijugate. Racemes axillary and terminal, the peduncles often panicled. Flowers yellow. Named in honour of *Calpurnius*, an imitator of Virgil, because these plants are nearly allied to *Virgilia*.

TABLE OF THE SOUTH AFRICAN SPECIES.

Calyx convex-conical at base. Flowers 5–6 lines long.
  Leaflets *obovate*-elliptical, retuse. Ovary quite glabrous (1) **sylvatica.**
  Leaflets elliptic, obtuse. Ovary silky and silvery ... (2) **lasiogyne.**
Calyx intruse or concave at base. Flowers 2½–3 lines long.
  Ovary silky and silvery. Leaflets subglabrous, retuse ... (3) **floribunda.**
  Ovary glabrous, ciliate on the sutures. Leafl. silky on both sides, netted-veined ... ... (4) **sericea.**
  Ovary glabrous. Twigs tomentose. Leafl. glabrous and even above, silky beneath ... ... (5) **villosa.**
  Ovary glabrous (?). Racemes very long. Leafl. nearly glabrous, mucronate ... ... (6) **intrusa.**

1. C. sylvatica (E. Mey.! Comm. p. 2); leaves 3–5–10-jugate, the petiole glabrescent; leaflets glabrescent, membranous, *obovate-elliptical*, retuse or obtuse; calyx conical at base, and ribbed, glabrescent; ovary and legume *quite glabrous. Virgilia sylvatica, DC. Prod.* 2, *p.* 98. *E. & Z.!* 1142. *Sophora sylvatica, Burch. Cat.* 3138.

HAB. Woods in Uitenhage, Albany, and Caffraria, frequent. (Herb. Hk., Sd., D.)

A shrub, 6–10 feet high; the young twigs minutely pubescent. Leaves 2–6 inches long, varying much in the number of leaflets. Leaflets acute at base, very blunt or subemarginate at the point, ½–¾ inch long, 4–5 lines wide, pale green. Racemes equalling the leaves or shorter; the rachis glabrescent; bracts minute, deciduous. Flowers 5 lines long, bright yellow. Legume 2½–3 inches long, ½–¾ wide, mucronate, veiny.

2. C. lasiogyne (E. Mey.! Comm. p. 3); leaves 4–6–10-jugate, the petiole glabrescent; leaflets *thinly pubescent beneath*, coriaceo-membranaceous, *elliptical*, obtuse or retuse; calyx convex at base, glabrescent; *ovary silky, with short white hairs;* legume sparsely pubescent. *Virgilia aurea, Lam. Ill. t.* 326, *f.* 1. *DC. Prod. l. c. Schimp.! Abyss.* 453, 200, 278, 1898. *Krauss.!* 325.

HAB. Port Natal, *Drege! Krauss! Gueinzius!* (Herb. Hk., Sd., D.)

Very like *C. sylvatica*, but a taller shrub, with rather larger leaves and flowers, more coriaceous, more pubescent, and exactly elliptical or oblong (not obovate) leaflets. The *silky* ovary at once distinguishes it. It is found, perhaps more commonly, in Abyssinia.

3. C. floribunda (Harv.); leaves 6–8-jugate, the petiole, twigs, and peduncles thinly silky-canescent; leaflets (often alternate) glabrate, the young ones minutely puberulous beneath, coriaceo-membranous, obovate-elliptical, retuse; calyx intruse at base, puberulous, the lobes deltoid; ovary *silky*, with white hairs; legume . . . . . ?

HAB. Roadside near Grahamstown, *H. Hutton, Esq.* (Herb. T.C.D.)

A shrub or small tree, the younger parts slightly hoary. Leaves 3–5 inches long, the leaflets laxly set, 5–6 lines long, 3–3½ wide. Peduncles crowded toward the ends of the branches, equalling or somewhat exceeding the leaves. Flowers 3 lines long. *Mr. Hutton*, in sending this plant, says, "I believe, *not indigenous;*" if not, as it is doubtless of African origin, it may have been brought from some part of the interior as an ornamental shrub, which it certainly is.

4. C. sericea (Harv.); leaves 5–6-jugate, the petiole, twigs, and peduncles thinly silky; leaflets *appressedly pubescent on both sides*, coriaceous, *reticulated*, oblong, retuse, mucronulate; calyx intruse at base, silky, the lobes obtuse; ovary *glabrous*, ciliate along the sutures.

HAB. Bassutos Land, *v. Schlicht!* 82. (Herb. Sd.)

Seemingly a small shrub, appressedly silky in all parts. Leaves 2 inches long; leaflets 4–5 lines long, 2½–3½ wide, thickish and rigid when dry, conspicuously netted with veins. Racemes not longer than the leaves. Flowers 2½ lines long, yellow. Legume not seen.

5. C. villosa (Harv.); leaves 5–8-jugate, the petiole, twigs, and peduncles *villoso-tomentose;* leaflets glabrous and even above, appressedly silky beneath, coriaceo-membranous, elliptical or oblong, obtuse, mucronate; calyx intruse at base, pubescent; ovary *glabrous;* legume . . . . . ? *C. intrusa, Mundt.! in Herb. Reg. Berol.*

HAB. South Africa, *Mundt. & Maire!* (Herb. Hk.)

With foliage not unlike that of *Virgilia capensis*, this has the habit of a *Tephrosia*, and is the most copiously and loosely pubescent of any of the present genus. The peduncles are scarcely longer than the leaves and few-flowered; the pedicels not longer than the calyx. Leaves 2–3 inches long; leafl. 5–7 lines long, 2½–3 wide often folded.

6. C. intrusa (E. Mey. Comm. p. 2); "leaves 10–14-jugate; leaflets glabrescent, *elliptical, mucronate;* calyx intruse at base; peduncles flexuous, much longer than the leaf."

HAB. Sides of woods between Gekau and Basche, 2000f., *Drege!*

This I have not seen. *Meyer* says it has the habit of a *Tephrosia*, but the free stamens and other characters of *Calpurnia;* by his silence respecting the ovary I presume that it is glabrous. Flowers scarcely 3 lines long.

Imperfectly known Species.

C. robinioides (E. Mey. Comm. p. 3); 'leaves 3-jugate, subglaucescent; leaflets oblong, the terminal one obovate; legumes stipitate, obsoletely winged, about one-seeded? *E. Mey. l. c. Virgilia robinioides, DC. Prod.* 2, *p.* 98. *Robinia Capensis, Burm. Fl. Cap.* 22.

HAB. Rocky hill near Kraai River, Witbergen, 4500f., *Drege.*

LXVII. BRACTEOLARIA, Hochst.

Calyx bibracteolate, deeply bilabiate, reflexed. *Corolla* expanded; *vexillum* ample, suborbicular: *alæ* spreading widely; petals of the *carina* shortly connate in the middle, spurred at base. *Stamens* 10, free, glabrous, exserted. *Ovary* sessile, villous, few-ovuled; *style* short, reflexed; *stigma* simple. *Legume* unknown.

African trees or shrubs, with unifoliolate leaves. Flowers in axillary racemes or panicled. The name is in allusion to the bracts on the calyx. The genus is allied to *Baphia*.

1. B. racemosa (Hochst.); glabrescent; leaflets ovato-lanceolate, acuminate, shining above, netted veined beneath, glabrous; racemes downy, lax; lobes of the calyx toothed; bracteoles lanceolate. *Harv. Thes. t.* 20.

HAB. Port Natal. *Krauss! Gueinzius!* (Herb. Hk. Sd., D.)

A climbing (?) shrub, with dark-coloured, rough bark. Petioles ½–¾ inch long; leaflet 2 inches long, about 1 inch broad; the young leaflets downy. Stipules caducous. Racemes shorter than the leaf, several-flowered; pedicels 1–2–3-flowered, bracteolate in the middle; flowers white (?) the vexillum very wide.

Sub Order II. CÆSALPINIEÆ.

LXVIII. PARKINSONIA, Linn.

Calyx coloured, with a short, urceolate tube and 5-parted, subequal, deciduous limb. *Petals* 5, in the throat of the calyx, ovate, flat; the upper one with a long claw. *Stamens* 10, free, declined; filaments villous at base. *Ovary* sessile, many-ovuled; *style* subulate, ascending; stigma simple. *Legume* very long, acuminate at each end, compressed between seeds, 2-valved, many-seeded. *Endl. Gen.* 6775. *DC. P.* 2, 485.

Tropical and subtropical shrubs, armed with simple or 3-forked spines. Leaves pinnate, multijugate; leaflets small. Flowers racemose, yellow. Name in honour of John Parkinson, a London apothecary and botanist of the seventeenth century.

P. africana (Sond. Linn. 23, p. 38); flexuous, divaricately branched, glabrous and glaucous, spiniferous; leaves 3 or more from the axils of horizontal spines; common petiole semiterete, wingless, channelled, taper-pointed; leaflets opposite, *very minute*, oblong, acute; racemes axillary and terminal, laxly 6–10-flowered; pedicels bracteate at base; calyx puberulous; legumes linear-lanceolate, narrowed at base and apex, finely striate, 8-seeded, seeds oblong, obtuse, shining. *Zey.!* 557.

HAB. Springbokkeel, Bosjesmansland, *B. & Z.!* Namaqualand, *Wyley!* (Herb. Hk., Sd., D.)

A tall bush, called "Wilde Limoenhout" by the colonists. Branches pale yellowish, spreading subhorizontally, flexuous, tapering, the younger minutely downy. Leaves juncoid, the petiole 4–5 inches long; leaflets in 8–10 distant pairs, ½–1 line long, very narrow. Racemes 2–3 inches long, spreading; the peduncle rigid; pedicels 4–6 lines long. Calyx coloured, with a short downy tube and oblong, deciduous limb. Filaments hairy at base. Petals yellow. Legumes 4–5 inches long, 4–5 lines wide, contracted and compressed between the seeds.

LXIX. GUILANDINA, Juss.

Calyx with a short, urceolate tube, and subequally 5-parted limb. *Petals* 5, in the throat of the calyx, sessile, nearly equal. *Stamens* 10, free; filaments villous at the base. *Ovary* stipitate, several-ovuled; *style* short; *stigma* simple. *Legume* ovate, ventricose, compressed, 2-valved, 1–2-seeded, covered with straight prickles. *Endl. Gen. No.* 6763. *DC. Prod.* 2, 480. *Lam. Ill. t.* 336.

Tropical and subtropical trees and shrubs; the stem and petioles armed with hooked prickles. Leaves abruptly bipinnated. Flowers spicato-racemose. Bracts long. Name in honour of Melchior Guilandinus (or Wieland), a celebrated traveller; died in 1589.

1. **G. Bonduc** (Ait. Hort. Kew. 3, p. 32); leaves pubescent or villous and velvetty; leaflets ovate; spines subsolitary; seeds yellowish. *DC. Prod.* 2 *p.* 480. *E. Mey. Comm. p.* 158.

HAB. Mouths of the Omsamcaba, *Drege.*

A native of India and Arabia. I have not seen South African specimens. Fls. yellow.

LXX. MELANOSTICTA, DC.

Calyx deeply 5-parted, segments deciduous, the lowest largest. *Petals* 5, in the throat of the calyx; 4 lower obovate-oblong, subequal; upper one shorter and broader, with inflexed edges. *Stamens* 10, free, ascending, equalling the petals; anthers short. *Ovary* sessile, ovate-oblong,

4-ovuled; style short, straight. *Legume* compressed oblong, setose. *Endl. Gen. No.* 6772. *DC. Prod.* 2, 485. *Leg. Mem. t.* 69.

Small half-herbacous plants; the stem, petioles, leaflets, peduncles, calyx, ovary, and legumes, all parts of the plant except the petals and stamens, thickly sprinkled with black, hemispherical, resinous dots. Leaves bipinnate, stipulate, and stipellate. Flowers racemose; racemes opposite the leaves. Name from μελας, *black*, and στικτος, a dot.

ANALYSIS OF THE SPECIES.

Sparsely setose; leaves 1–2-jugate, with a long, terminal pinna; flowers drooping ... ... (1) **Burchellii.**
Densely setose; leaves equally 4–5-jugate; lat. pinnæ 14–16 foliolate; fl. erecto-patent ... ... (2) **Sandersoni.**

1. M. Burchellii (DC. Leg. Mem. t. 69); leaves 1–2-*jugate, with a long, terminal pinna;* lateral pinnæ 6–8 foliolate, terminal 16–20 foliolate; racemes scarcely as long as the leaves, few-flowered; flowers drooping; bracts minute. *DC. Prod.* 2, *p.* 485. *Harv. Thes. t.* 2. *Burch. Cat.* 2345.

HAB. S. Africa, *Burchell.* Zooloo Country, *Miss Owen.* (Herb. D.)

Stem 3–6 inches high, herbaceous, sparsely plumoso-setose, as are also the peduncles and petioles. Leaves 4–5 inches long; the lateral pinnæ uncial, the terminal 3 inches long. Leaflets oval, thickly sprinkled with gland-dots. Racemes 5–6-flowered; pedicels 2–3 lines long, recurved. Flowers purple, small. Legume oblong, acute, densely plumoso-setose.

2. M. Sandersoni (Harv.); leaves 4–5-*jugate, the terminal pinna not longer than the rest;* lateral pinnæ 14–16 foliolate; racemes longer than the leaves, many-flowered; flowers erecto-patent; bracts lanceolate, deflexed.

HAB. Transvaal, *J. Sanderson, Esq.* (Herb. Hk.)

Taller and more robust than *M. Burchellii*, much more setose, with longer racemes and very different leaves, the black dots smaller and less abundant. Stem 8–10 inches high, herbaceous, thickly setose, as well as the peduncles and petioles. Leaflets margined with small gland-dots, very few or none on the disk. Racemes 6–8 inches long, erect, 18–20-flowered; pedicels 6 lines long, straight and spreading. Ovary glabrescent. Legume not seen.

LXXI. PELTOPHORUM, Vog.

Calyx-tube turbinate; limb 5-parted, deciduous, the segments oblong, reflexed. *Petals* 5, obovate, curled at the edge, clawed. *Stamens* 10, inserted with the petals, free; filaments equalling the petals, hairy at the base, inflexed in æstivation; anthers versatile, slitting longitudinally; *ovary* sessile, compressed, few-ovuled; style filiform; stigma peltate, depressed in the centre. *Legume* broadly-oblong, much compressed, unarmed, acute at each end. 1–2-seeded; seeds oblong, with a subterminal hilum and straight embryo. *Vogel, in Linnæa, vol.* 11, *p.* 406.

Trees or shrubs. Leaves abruptly bipinnate. Racemes axillary and terminal; flowers yellow. Separated by Vogel from *Cæsalpinia*, from which it differs by the peltate stigma. Name from πελτοφορος, *shield-bearing;* alluding to the stigma.

P. africanum (Sond. in Linn. Vol. 23, p. 35); unarmed; twigs, inflorescence, and calyx thinly rusty-pubescent; leaves abruptly bipinnate, pinnæ 7–8 pair, 14–20-jugate; pinnules linear-oblong, apiculate, appressedly pubescent on both sides, pale underneath; racemes in a terminal panicle, many-flowered; bracts linear-subulate, deciduous; pedicels at length longer than the calyx; inner sepals with scarious, denticulate margins; petals broadly obovate, longer than the calyx, the claw and the base of the filaments hirsute; style glabrous, equalling the ovary, stigma from a conical base broadly peltate; legume ovato-lanceolate, acute, multistriate, puberulous, strongly compressed, 1–2-seeded. *Sond. l. c.*

HAB. Crocodile River and forests on the North side of Magaliesberg, Transvaal, *Zeyher, No.* 554. (Herb. Hk., Sd.)

A very tall tree. Twigs furrowed, rusty. Leaflets opposite, 4 lines long, 1 line wide. Racemes 3–4 inches long. Legume 2–2½ inches long.

LXXII. BURKEA, Hook.

Calyx 5-parted, the segments equal, imbricate. *Petals* 5, subequal, patent. *Stamens* 10; filaments very short, the alternate slightly longest; anthers oblong, equal, tipped with a deciduous gland. *Ovary* subsessile, 2-ovuled; *style* very short; *stigma* obliquely peltate, concave, with a wavy margin. *Legume* plano-compressed, oblong, narrowed at base, stipitate, thinly coriaceous, indehiscent. *Seeds* ovate-orbicular. *Embryo* straight. *Benth. in Hook. Ic. Pl. t.* 593–594.

A shrub or small tree. Leaves abruptly bipinnate; pinnæ in 2 pairs, opposite; leaflets about 8 on each pinna, alternate, distant. Racemes axillary, many-flowered. Named in compliment to Mr. Joseph Burke, a collector, employed by the late Earl of Derby, and who, jointly with the late Mr. Charles Zeyher, accomplished a very extensive journey beyond the Gariep, making large collections of plants.

1. B. africana, *Hook. MSS. Ic. Pl. l. c.*

HAB. Magaliesberg, *Burke & Zeyher!* (Herb. Hk., D., Sd.)

12–15 feet high. Branches thick, short, the younger covered with reddish tomentum. Stipules minute. Petioles 3–6 inches long, with two distant pairs of opposite, alternately pinnate pinnæ; pinnules petiolulate, obliquely ovate or oblong, obtuse, the younger minutely silvery, the older coriaceous and glabrate. Racemes scarcely shorter than the leaves. Flowers small, subsessile. Sepals membrane-edged. Petals twice as long as the calyx.

LXXIII. CASSIA, L.

Calyx 5-parted nearly to the base, more or less unequal. *Petals* 5, clawed, more or less unequal. *Stamens* 10, free, the 3 upper commonly sterile (sometimes wanting); fertile anthers opening by 2 terminal pores or short clefts. *Ovary* sessile or stipitate, multiovulate; style filiform; stigma simple. *Legume* terete or compressed, linear, many-seeded. *Endl. Gen.* 6781. *DC. Prod.* 2, 489.

A vast tropical and subtropical genus of trees, shrubs, suffrutices, and herbs. Leaves alternate, simply and abruptly pinnate, often having glands on the petioles. Flowers yellow or orange. Name, the κασσια of Dioscorides.

ANALYSIS OF THE SOUTH AFRICAN SPECIES.

I. Sepals obtuse:
  Shrubs or large suffrutices: leafl. pale and pubescent beneath:
    Leafl. lanceolate, acute or acuminate; jugal glands slender (1) **Delagoensis.**
    Leafl. oval-oblong, obtuse; jugal glands tubercular ... (2) **tomentosa.**
  Suffruticose. Leafl. *glabrous*, green, oblong-lanceolate, acute (3) **occidentalis.**
  Herbaceous. Leafl. glabrous, oblong-obovate, oblique. Legume oblong-orbicular, flat ... ... (4) **arachoidea.**
II. Sepals very acute. Suffruticose. Leafl. minute, linear-falcate (5) **mimosoides.**

1. **C. Delagoensis** (Harv.); a shrub; leaflets 8–12–14-jugate, lanceolate or ovato-lanceolate, acute or acuminate, puberulous or glabrate above, thinly silky and paler beneath; petiole with a slender filiform gland between each pair; stipules broadly reniform, with one lobe cuspidate, deciduous; peduncles many-flowered, racemose, in a terminal corymbose panicle; ovary silky-canescent; legume . . . ?

HAB. Delagoa Bay, *Forbes!* Port Natal, *Rev. Mr. Hewittson!* (Herb. Hk., Bth., D.)

Allied to *C. auriculata*, but with different leaflets, &c. Young parts thinly pubescent, with pale, appressed hairs. Petioles 4–8 inches long; leaflets 1–1¼ inch long, 3–5 lines wide. Anthers birimose, glabrous; 3 short, 4 mediocre, and 3 long and incurved. Petals veiny, orange, unequal. Ovary slender, multiovulate.

2. **C. tomentosa** (Lam. Dict. 1, p. 647); a tomentose shrub; leaflets 6–8-jugate, *oval-oblong*, *obtuse*, mucronulate, puberulous or glabrate above, cano-tomentose beneath; petiole with a tubercular gland between each pair; stipules inconspicuous; peduncles short, 2–4-flowered; ovary woolly; legume linear, acute, compressed, villoso-tomentose. *DC. Prod.* 2, *p.* 496. *C. multiglandulosa, Jacq. Ic. Rar. t.* 72. *Sieb. Fl. Cap. No.* 153.

HAB. Naturalized? S. Africa, *Sieber!* Simon's Bay, *C. Wright!* Grahamstown, *General Bolton!* (Herb. Sd., D.)

A South American species, often cultivated and now naturalized in several warm countries. All the young parts velvetty-tomentose. Leaflets 1–1¼ inch long, 3–4 lines wide. Flowers bright yellow. Legume 4–5 inches long.

3. **C. occidentalis** (Linn. Sp. 539); suffruticose, *ciliate;* leaflets 4–6-jugate, *ovato- or oblongo-lanceolate, acute, glabrous; petiole with a tubercular gland above the base;* peduncles very short, 2–4-flowered, the lowest axillary, the rest in a short, terminal raceme; legumes linear, elongate, glabrescent, plano-compressed, with thick margins. *DC. Prod.* 2, *p.* 497. *Sloane, Hist. Jam. t.* 175, *f.* 3, 4. *Bot. Reg. t.* 83. *C. Natalensis, Sond.*

HAB. Near Natal (probably naturalized), *Gueinzius!* (Herb. Sd.)

A common West Indian plant, now naturalized in several parts of the tropics. Nearly glabrous. Petioles pale, 4–6 inches long, bearing leaflets for half that length; leaflets 1½–2½ inches long, the lowest smallest. Flowers little conspicuous, pale yellow. Legume 3–4 inches long, brown, with pale margins.

4. **C. arachoides** (Burch. Trav. 1, p. 341); *herbaceous, procumbent,* glabrous and *glaucous,* stems flexuous, angular; leaves 6–7-jugate, without petiolar gland; *leaflets unequal-sided, broadly oblong or obovate, rounded at both ends, penninerved* and somewhat veiny; stipules small, ovato-lanceolate, rigid, ribbed; peduncles axillary, racemose, many-flowered, longer than the leaf; bracts oblong, deciduous; sepals broadly-elliptical, blunt, concave; 7 stamens fertile, of which 2 have much longer falcate anthers, 3 sterile, depauperated; legume oblong-orbicular, flat, ridged in the middle, veiny and papery, glabrous.

HAB. S. Africa, *Burchell,* 1680. Magalisberg and Vaalriver, *Burke & Zeyher!* Zooloo Country, *Miss Owen!* (Herb. Hk., Sd., D.)

Very similar to *C. obovata* of N. Africa and Tropical Asia, but differing in the shape of the legume, and slightly in minor characters. Stems pale yellowish. Petioles 3–5 inches long, channelled. Leaflets 6–12 lines long, 3–6 lines wide, prominently veiny and thickish in substance. Peduncles 6–12 inches long, 12–20 flowered. Calyx dark olive, petals yellow. Legumes nearly 1½ inches long and broad, scarcely longer than their breadth, brown.

5. C. mimosoides (Linn. Sp. 543); suffruticose, erect or diffuse, variably pubescent or glabrescent; *leaves* 10–40-*jugate, with a larger or smaller petiolar gland below the lowest pair; leaflets* very oblique or dimidiate, *linear-falcate, mucronulate, rigid,* obliquely striate beneath; stipules from a semi-cordate base, subulate, striate; peduncles axillary, 1–3 together, 1-flowered, longer or shorter; *sepals ovate, acute or acuminate,* villous, equalling or nearly equalling the petals; legumes linear, oblique at base, plano-compressed, with thicker sutures, pubescent or glabrescent, 10–25-seeded.

VAR. α. **Capensis**; diffuse; leaflets in 10–35 pairs, with a very minute petiolar gland; peduncles solitary or in pairs, equalling or exceeding the leaf. *C. Capensis, Thunb.! Cap. p.* 388. *E. & Z.!* 1698. *E. Mey.! Comm. p.* 158, *and C. plumosa, E. Mey. l. c.*

VAR. β. **stricta**; erect, virgate, subsimple; leaflets 30–40 pairs, with a large, ellipsoid petiolar gland; peduncles 1–3, unequal, much shorter than the leaf. *C. stricta, E. Mey.! Comm. p.* 159. *C. angustissima, Lam.*, and several other species of authors.

VAR. γ.? **comosa**; stem erect, glabrous; lower leaves 8-jugate, upper 20–30-jugate, with a large, oblong, bilabiate petiolar gland; peduncles subsolitary, supra-axillary, scarcely longer than the flower. *C. comosa, E. Mey. l. c. p.* 160.

HAB. Moist Sandy places in Uitenhage, Albany, Caffraria, and about Port Natal. α. & β. common. γ. (which I have not seen) between Omsamwubo and Omsamcaba, *Drege.* (Herb. Hk., Sd., D., &c.)

A common tropical and subtropical weed, which has received at least a score of names in different countries. The pubescence, number of leaflets, size and shape of petiolar-gland, length of peduncle, and size of flower, are very variable, but I find it impossible strictly to limit the variations, and all the S. African varieties inhabit the same districts, and probably often grow intermixed.

Doubtful Species.

C. Burmanni (DC. Prod. 2, p. 502); suberect? leaflets 7–9-jugate, oblong, aristate-mucronate, glabrous, subciliate; petioles with a sessile gland above the base; pedicels axillary, tufted, bracteolated; legume glabrous, downy at the sutures. *C. flexuosa, Burm.*

HAB. Cape, *Burmann.* (Herb. Deless., fide DC.) Unknown to me.

Cassia Burmanni, E. & Z.! 1697, preserved in Herb. Sonder, is founded on garden specimens of *C. revoluta*, F. Muell., (Hook. Kew Journ. 8, p. 45) an Australian species, allied to *C. australis* and *C. Schultesii.*

LXXIV. SCHOTIA, Jacq.

Calyx tube conical, limb 4-parted; the segments oval, obtuse, strongly imbricated, deciduous. *Petals* 5, in the throat of the calyx, nearly equal. *Stamens* 10, more or less connate at base; filaments free above, the alternate shorter; anthers ovate, longitudinally slitting. *Ovary* stipitate, ovate, several ovuled; *style* filiform, elongate; *stigma* capitate or simple. *Legume* coriaceous, oblong, compressed; the upper margin or both margins winged. *Seeds* 1–6, either with the hilum naked, or having a large, fleshy, cuplike arillus. *Endl. Gen.* 6785. *DC. Prod.* 2, 507.

Small trees or shrubs, natives of South Africa. Leaves pinnate; leaflets coriaceous, entire. Flowers panicled, crimson or pink, or flesh-coloured, handsome. Name, in honour of Richard Van der Schot, a travelling companion and friend of Jacquin's

ANALYSIS OF THE SPECIES.

Flowers on longish pedicels; calyx-tube conical:
 Petals conspicuous, much longer than the calyx ... (1) **speciosa.**
 Petals minute, hidden within the (crimson) calyx ... (2) **micropetala.**
Flowers subsessile, in much-branched panicles; calyx-tube short (3) **latifolia.**

1. **S. speciosa** (Jacq. Ic. Rar. t. 75); leaves *polymorphous*, 4–5 or 6–10–12–16-jugate; leaflets linear, oblong, obovate-oblong or obovate, mucronate or obtuse, pubescent or glabrous; panicles terminal, fasciculato-corymbose, many-flowered; flowers pedicellate; calyx-tube conical; stamens shortly connate at base or nearly free; petals *much longer than the calyx. DC. Prod.* 2, *p*, 508. *Thunb. Cap.* 388.

VAR. *α.* **ovalifolia**; leaflets in few pairs, oval-oblong or obovate, obtuse or mucronate, or retuse, 6–9 lines long, 3–5 lines wide. *S. stipulata, Ait. E. & Z.!* 261.

VAR. *β.* **tamarindifolia**; leaflets in 8–10 pairs, linear-oblong, oblong or elliptical, mucronate or obtuse, unequal and subtruncate, or rounded at base, 4–5 lines long, 1½–3 lines wide. *S. tamarindifolia, Afz. E. Mey.! Comm. p.* 161. *S. speciosa, E. & Z.!* 1699. *Andr. Rep. t.* 348. *Bot. Mag. t.* 1153.

VAR. *γ.* **angustifolia**; leaflets linear, mucronate, or obtuse, unequally truncate at base, and *frequently* produced on the upper margin into a small, toothlike lobe. *S. angustifolia, E. Mey.! Comm. p.* 161.

HAB. Dry, Karroo-places in the eastern districts and Caffraria, frequent; all the varieties. (Herb. Th., Hk., Sd., D.)

A small tree or large shrub, 8–12 feet high, the "*Boerboom*," of the colonists. The leaves are surprisingly inconstant in form, scarcely two of the many specimens under examination being decently similar; the extreme forms of *γ.* and *α.* looking as if they belonged to very different species. The flowers, however, are constantly the same, and our colonial correspondents seem to recognise but one, though European botanists make four species of this shrub. The legumes, half-ripe, roasted on the coals are eaten by the natives. The powerfully astringent bark is used medicinally and for tanning.

2. **S. brachypetala** (Sond.! in Linn. vol. 23, p. 39); leaves 4–5-jugate; leaflets (large) ovate-oblong, or obovate, obtuse, netted-veined; panicles axillary and terminal, many-flowered; flowers pedicellate; calyx-tube conical; petals *very minute, linear*, hidden under the calyx-lobes; stamens monadelphous; ovary on a long stipe. *Harv. Thes. t.* 32.

HAB. Near Port Natal, rare, *Gueinzius! Sanderson!* (Herb. Hk., Sd., D.)

A large shrub or small tree, sometimes blossoming from the trunk or large branches. Though less variable than *S. speciosa*, the leaflets do vary considerably in shape and size, the smallest being ¾ inch long, ½ inch wide; the larger varying from 1½–2½ inches long, 1–1½ wide, sometimes cuneate or tapering, sometimes truncate at base, always conspicuously, though not prominently, veiny. The calyces and peduncles are rich crimson. I have not seen legumes.

3. **S. latifolia** (Jacq. Fragm. 23, t. 15, f. 4); leaves 2–4-jugate; leaflets (large) obovate oblong or obovate, obtuse, coriaceous; panicles axillary and terminal, much-branched, densely many-flowered; *flowers subsessile, calyx-tube very short;* petals longer than the calyx; stamens monadelphous; seeds with a large, fleshy arillus. *DC. Prod.* 2, *p.* 508. *E. Mey.! Comm. p.* 162. *E. & Z.!* 1701. *S. diversifolia, Walp.*

HAB. Frequent in the woods of Uitenhage and Albany, &c. (Herb. Hk., D., Sd.)

A tree, 20–30 feet high. Leaflets variable in shape, 1½–2½ inches long, ½–1 inch wide, glabrous, rigid, and thick. Panicles excessively branched; flowers rosy or flesh-coloured. Legume 1–3 seeded, 1½–4 inches long, 1½–2 inches wide, very rigid; each seed sitting in a cuplike, yellow, fleshy arillus. The roasted pods are eaten.

LXXV. BAUHINIA, Plum.

Calyx-tube cylindrical or campanulate; limb 5-parted, deciduous or persistent, its segments separate or cohering in a reflexed, strap-shaped lobe. *Petals* 5, clawed, subunequal, variously inserted. *Stamens* 10, monadelphous or free, exserted, either all fertile or several, (5–7–9) sterile; filaments filiform; anthers incumbent, slitting. *Ovary* stipitate, several or many-ovuled; style curved. *Legume* stipitate, compressed, one or several-seeded. *Endl. Gen.* 6790. *DC. Prod.* 2, *p.* 513. *Lam. Ill. t.* 329, *also Casparea, Phanera, and Schnella, Auct.*

Trees, shrubs, or twining suffrutices, natives of the tropics generally, a few straggling into the temperate zone. Leaves formed of two partially connate or nearly entirely confluent leaflets, resembling a bilobed leaf. Flowers racemose. Though differing much in floral characters, chiefly of the calyx and stamens, these plants form so truly natural an assemblage, agreeing in general habit and in their very peculiar foliage, that it seems inexpedient to break up the genus, as has been proposed. If it should be broken up, our three S. African species should be referred to three different genera. The name is in honour of the brothers *Bauhin*, famous botanists of the fifteenth century, whose relationship is fancifully commemorated in the connate leaves of these plants.

ANALYSIS OF THE SOUTH AFRICAN SPECIES.

Calyx-limb spathaceous, reflexed; stam. 10; lvs. semibilobed ... (1) **tomentosa.**
Climbing; calyx-limb 5-parted, spreading; stamens 2–3; leaves deeply bilobed ... ... (2) **Burkeana.**
A rigid shrub; calyx campanulate, 5-fid; stam. 10; leaves emarginate; pedunc. one-flowered ... ... (3) **Garipensis.**

1. B. *(Pauletia)* **tomentosa** (Linn. Spec. 536); fruticose, leaves rounded at base, their underside, the twigs, petioles, peduncles, bracts, and calyx *thinly pubescent;* leaflets oval, obtuse, 3-nerved, concrete to or beyond the middle; peduncles 1–3-flowered; calyx-limb *spathaceous, reflexed;* stamens 10, fertile, unequal. *DC. Prod.* 2, *p.* 514.

HAB. Near Port Natal, *T. Williamson!* (Herb. D.)

I venture to refer to this species a fragmentary specimen in Herb. D., although the pubescence is so very thin and minute as not to be obvious without a lens. Still, the aspect, the foliage, and flower are those of *B. tomentosa*, which species varies in pubescence.

2. B. *(Phanera)* **Burkeana** (Benth. !); suffruticose, climbing; leaves deeply reniform at base, netted-veined beneath, glabrous; leaflets obliquely-elliptical, 3-nerved, obtuse, concrete for a short distance above the base; peduncles tomentose, many-flowered, some abortive and tendril-bearing; calyx-limb spreading, its lobes separate, *lanceolate;* stamens two, fertile, exserted; staminodia 5, spathulate, three broader than the rest; ovary glabrous, stipitate. *Benth.! Pl. Jungh. p.* 62.

HAB. Mooi River, Transvaal, *Burke & Zeyher! Sanderson!* (Herb. Hk., Sd., D.)

A slender climber, several feet long, the young parts thinly villoso-pubescent. Stems angular. Leaves 3–4 inches broad; each leaflet 1½–2 inches long, 2–2½ inches broad, nerves branching. Flowers small; calyx-lobes ½ inch long; petals not much longer, or uncial, striate, the vexillum with a very prominent, channelled callus. Legume not seen.

3. B. *(Adenolobus)* **Garipensis** (E. Mey.! Comm. p. 162); erect, shrubby, divaricately-branched, unarmed; leaves (small) ovate or cordate at base, emarginate, glabrous, veinless, coriaceous, glaucescent;

leaflets semicircular, concrete nearly to the summit; peduncles short, solitary or fascicled, one-flowered; calyx *campanulate, shortly 5-toothed, persistent*, its lobes deltoid, erect; petals inserted with the stamens on a perigynous disc in the base of the calyx, obovate, purple-veined; stamens 10, all perfect, 5 shorter; ovary glandular; legume stipitate, broadly and shortly falcate, few-seeded, either quite smooth or covered with small, wartlike glands.

HAB. Between Verleptpram and Natvoet, Gariep, in vallies, under 600f., *Drege!* Namaqualand, *Wyley!* (Herb. Hk., Sd., D.)

A rigid, laxly-branched shrub, with ash-colour bark, and bare, virgate, spreading branches. Leaves scattered or tufted on very short, lateral twigs. Petioles slender, ½ inch long; leaves ½ inch long, ⅜ inch wide, pale green. Flowers generally on the lateral twigs, 2–3 together. Peduncles ½–⅝ inch long. Calyx 5 lines long; petals thrice as long, elegantly veiny. The legume varies remarkably in its surface, as above stated.

Sub-Order III. MIMOSEÆ.

LXXVI. ENTADA, Linn.

Flowers sessile or shortly pedicellate. *Calyx* campanulate, shortly 5-toothed. *Petals* 5, free or nearly so. *Stamens* 10; *anthers* gland-bearing. *Legume* linear, plano-compressed, margined with thickened, persistent sutures; the valves transversely jointed, separating into 1-seeded, indehiscent, frustules. *Benth. in Hook. Journ. Bot.* 4, *p.* 332.

Shrubs, mostly scandent, armed or unarmed. Leaves bipinnate, the terminal pair of pinnæ often changed into cirrhi. Stipules small, setaceous. *Spikes* of flowers slender, solitary or in pairs, or in a terminal panicle. *Entada* is the Malabar name of *E. scandens.*

ANALYSIS OF THE SOUTH AFRICAN SPECIES.

Leaflets 2–5-jugate, 1–2 inches long, 1 inch wide ... ... (1) **scandens.**
Leaflets 8–15-jugate, 4–6 lines long, 1–2 lines wide:
  Prickly, thinly tomentose; flowers sessile ... ... (2) **Natalensis.**
  Unarmed, quite glabrous; fl. pedicellate ... ... (3) **Wahlbergii.**

1. E. scandens (Benth.! l. c.); unarmed; leaves cirrhiferous; pinnæ 1–2-jugate; leaflets 2–5-jugate (large), ovate, elliptical, or oblong-obovate, obtuse or acuminate or emarginate, often oblique, shining above, glabrous or downy beneath; spikes elongate, solitary or in pairs. *Mimosa scandens, Linn. E. Pursætha and E. monostachya, DC. Prod.* 2, *p.* 425.

HAB. South Africa, *Wahlberg!* (Herb. Holm.)

Stem climbing to the top of lofty trees, very long, ropelike. Petiole ending in a simple or branched tendril; pinnæ few, 1–2 inches apart; leaflets 1½–2 inches long, 1 inch wide. Spikes 2–8 inches long. Legume 2–3 feet long, 3–4 inches broad. This is the common sword-bean of the East and West Indies, and tropical Pacific.

2. E.? Natalensis (Benth.! l. c.); twigs and petioles thinly tomentose, here and there armed with hooked prickles; pinnæ in 5–7 pair; leaflets 9–15 pair, obliquely-oblong, obtuse, glabrous or downy, the petiole here and there gland-bearing; spikes axillary, 2–3 together in a leafy panicle. *Benth. Mimosa spicata, E. Mey.! Comm. p.* 164.

VAR. *β.* **aculeata**; stem and petioles copiously armed with hooked prickles; leaflets broader and more glabrous.

HAB. River banks about Natal, *Drege! Krauss!* 199. *β.* a garden plant raised at Capetown from Natal seeds, *Commis. Genl. J. D. Watt!* (Herb. Hk., D.)

A slender, bramble-like bush, either climbing or forming an entangled mass some feet in diameter. Prickles few or many. Petioles 4–6 inches long, bare in the lower half; leaflets 4–5 lines long, 1½–2 lines wide, pale beneath. Spikes chiefly toward the ends of the branches, pedunculate, closely-flowered, 2–2½ inches long. Unripe legume subfalcate, glabrous, shining, 5–6 inches long, nearly an inch broad, many-seeded, jointed between the seeds.

3. E.? Wahlbergii (Harv.); unarmed, slender, scandent; twigs and petioles glabrous; pinnæ in two, distant pair; leaflets in 7–10-pair, obliquely falcate-oblong, obtuse, glabrous, paler beneath, mucronulate; spikes axillary, solitary, shortly pedunculate; *flowers pedicellate;* legumes ?

HAB. South Africa, *Wahlberg!* (Herb. Holm.)

Whole plant glabrous. Twigs striate, brownish, flexuous. Petioles 2½ inches long, ending in a short, cirrhulous point, reflexed. Pinnæ in two pairs, 1¼ inch asunder, with longish petiolules. Leaflets 4–6 lines long, 1–1½ line broad, unequal-sided, scarcely curved, the edges inclining to be revolute. Spikes racemulose, 2–2½ inches long, on peduncles ½ inch long. Pedicels of the flowers ½ line long or more. Calyx cupshaped, sharply 5-toothed. Petals reflexed. Anthers tipped with a stalked, deciduous gland. Legumes unknown. In many respects this resembles a W. African species, (*Barter*, 991 in Herb. Hooker) with broadly falcate, many-seeded, crenate pods. Possibly these plants, from opposite sides of the continent, may be identical, but it would be hardly safe to pronounce them so till the fruit of our present plant be ascertained.

LXXVII. ELEPHANTORHIZA, Benth.

Flowers shortly pedicellate. *Calyx* short, 5-toothed. *Petals* lanceolate, at length free. *Stamens* 10; anther tipped with a deciduous, stalked gland. *Legume* straight, compressed, coriaceous, the sutures remaining closed, but the long-persisting, rigid valves separating (as in a *siliqua*), without transverse septa, and not pulpy within.

Small, glabrous suffrutices, with large, thick roots; the "*Elandsbontjes*" of the colonists. Leaves bipinnate, multi-jugate. Flowers densely spicato-racemose. The name signifies "*Elephant's-root.*"

ANALYSIS OF THE SPECIES.

Leaflets very narrow-linear, sharply mucronate ... (1) **Burchellii.**
Leaflets broadly-linear, blunt ... ... (2) **Burkei.**

1. E. Burchellii (Benth.! in Hook. Journ. Bot. 4, p. 344); leaflets narrow-linear, very oblique, rigid, *subacute, sharply mucronate;* racemes dense, the pedicels shorter than the calyx; legume smooth or somewhat rugulose. *Acacia elephantina, Burch. Trav.* 2, *p.* 236. *A. elephantorhiza, DC. Prod.* 2, 457. *Prosopis elephantorhiza, Spreng. E. & Z.!* 1693. *P. elephantina, E. Mey.! Comm. p.* 165.

HAB. Very common in grassy places between the Klipplaat and Zwartkey Rivers, &c. *E. & Z.! Drege!* Caledon R., *Burke & Zeyher!* Zooloo Country, *Miss Owen!* Cradock and Queenstown Districts, *Mrs. F. W. Barber!* (Herb. Hk., D., Sd.)

Root very large and thick, creeping. Stems 1–2 feet high, suberect, quite glabrous, densely leafy. Leaves 6–12 inches long, 6–12-jugate; leaflets multi-jugate, 5 lines long, not 1 line wide, slightly narrowed at base, bright green. Racemes from the axils of the lower leaves, subsessile, 2–3 inches long. Legumes 6–7 inches long, 1½ wide. Mrs. Barber says, "all grazing animals, wild and domestic, are exceeding fond of this plant. It has long, succulent roots and an underground stem. It does not shoot until rather late in the summer, seldom before December, and its stems are killed again by the first frosts of May. The seed-pods are still green when the frost comes, and the seeds not ripe, but they are so well protected by the strong, leather-like pod, that the frost cannot hurt them, and they ripen in the pod long after the stem that bore them has been killed by the frost. The roots are used for tanning leather."

2. E. Burkei (Benth.!); leaflets broadly linear or linear-oblong, not very oblique, *obtuse*; racemes lax, the pedicels as long as the calyx; legume evidently netted with veins. *Zey.!* 560.

HAB. Magaliesberg, *Burke & Zeyher!* (Herb. Hk., D., Sd.)

Similar in habit to the preceding, but with longer, broader, blunter, and paler leaflets, and fewer flowers in the raceme. Leaflets 6–7 lines long, 2–2½ wide, round-topped. Legume 6–8 inches long, 1½ wide.

LXXVIII. DICHROSTACHYS, DC.

Flowers of two kinds in the spike; the uppermost flowers hermaphrodite and sessile, as in *Entada*; the lower neuter, with calyx and corolla as in the perfect, 10 long, slender filaments without anthers, and a rudiment of ovary. *Legume* linear, twisted, compressed, membranaceo-coriaceous or subcarnose, one-celled, without pulp, indehiscent, or the valves irregularly breaking from the sutures. *Benth. in Hook Journ. vol.* 4, *p.* 353.

African and Asiatic shrubs. The twigs occasionally abortive and converted into spines. Flowering branchlets in the axils very short, fasciculately leafy, covered with imbricating stipules. Spikes pedunculate, nodding, solitary or in pairs. Flowers sessile, the perfect ones yellow, the sterile whitish or purplish. Name, *διχροος, of two colours*, and *σταχυς, a spike*.

1. D. nutans (Benth. ! l. c.); twigs, petioles, and peduncles downy or glabrate; pinnæ subdistant, 8–12-jugate; leaflets 20–30-jugate, ciliate or glabrous; glands between the pinnæ stipitate; spikes shorter or scarcely longer than the leaves. *Caillea dichrostachys, Guill. & Perr. Fl. Seng.* 1, 240. *Desmanthus nutans, divergens, trichostachys, and leptostachys, DC. Prod.* 2, 445, 446. *Dichrostachys caffra, Meisn.! Pl. Krauss*, 148. *Zey.!* 561.

HAB. Port Natal, *Krauss! Gueinzius, &c.* Magaliesberg, *Burke & Zeyher!* (Herb. Hk., D., Sd.)

A rigid shrub, armed with strong axillary spines. Petioles 3–6 inches long, pauci- or multi-jugate, mostly pubescent. Leaflets 3–4 lines long, 1 line wide. Spikes 1½ inch long, on peduncles of variable length. Legumes 2–5 inches long, 3 lines wide.

2. D. Forbesii (Benth. l. c.); "nearly glabrous; pinnæ 3–4-jugate; leaflets 10–15-jugate; glands between the leaflets stipitate; spikes slender, scarcely longer than the leaf. *Benth.!*

HAB. Delagoa Bay, *Forbes.* (Herb. Hook.)

More glabrous than *D. nutans*, with fewer pinnæ and leaflets.

LXXIX. XEROCLADIA, Harv.

Flowers capitate, sessile. *Calyx* 5-parted to the base. *Petals* 5, free. *Stamens* 10; filaments free, the 5 alternate shorter; *anthers* with a very minute, sessile gland. *Legume* sessile, semiorbicular, plano-compressed, one-seeded, indehiscent, the carinal suture arched and wing-bordered. *Seed* flattened; embryo straight.

The only species is a small, dry, and very rigid bush, with pale bark, spinous stipules, distant, bipinnate, deciduous leaves, and subsessile heads of flowers. It seems to be allied to *Prosopis humilis* and *denudans*, but differs in the legumes. The name is compounded of *ξερος, dry*, and *κλαδος, a branch*.

X. Zeyheri (Harv. in Herb. Hook.).

HAB. Springbokkeel, *Burke & Zeyher!* 558. (Herb. Hk., Sd., Bth.)

A rigid. much-branched shrub, 1–2 feet high. Branches alternate, flexuous, terete, cano-puberulous, and substriate. Stipular spines in pairs, short, recurved, glabrous, horn-colour. Leaves bipinnate, pinnæ unijugate, on a very short, gland-tipped petiole; leaflets 6–10-jugate, obliquely linear-oblong, blunt, glabrous, or minutely downy. Flowers capitate, on short, axillary peduncles, 8–12 in the head. Sepals nearly free, oblong, blunt, woolly-edged. Petals oblong, free. Filaments glabrous, not much longer than the petals, 5 opposite the petals, shorter. Anthers with a very minute or obsolete gland. The singular legumes, which look like the carpels of a *Malpighiacea*, are 6–7 lines long, and rather more in breadth, the carinal wing, 1–1½ line wide.

LXXX. ACACIA, Willd.

Flowers frequently polygamous. *Sepals* 3–5, either connate in a campanulate calyx or free. *Petals* as many, more or less united in a monopetalous corolla, rarely at length free. *Stamens* numerous (mostly more than 50), free or connate at base, rarely (in male flowers) collected in a central column. *Legumen* various, mostly dry. *Benth. in Lond. Journ. Bot. vol.* 1, *p.* 318.

Trees or shrubs, widely distributed. Leaves (at least the primordial) bipinnate. Glands on the upper margin of the petiole; *petiolar* below the lowest pair of pinnæ; *jugal*, smaller between or a little below the upper pair of pinnæ or all the pairs, and sometimes between the upper pairs of leaflets; often absent. Stipular or axillary spines often present, and the ends of branches sometimes spiny. Prickles in many. Flowers in heads or spikes, yellow. An immense genus, very abundant in Australia, where most of the species bear *phyllodia*, or leaflike petioles, in place of true leaves. *Acacia* was the Greek name of some plant of this genus.

ANALYSIS OF THE SOUTH AFRICAN SPECIES.

1. Gummiferæ.—Stipules spinous; prickles none. (Sp. 1–10).

Flowers capitate; *bracts* close under the head of flowers.

Glabrous; legume semilunate, tumid, tomentose ... (1) **erioloba.**

Glabrous; legume oval, thick, indehiscent ... ... (2) **Giraffæ.**

Pubescent; leafl. 10–15-jugate, oblong-lin.; legume lin. (3) **heteracantha.**

Cano-tomentose; spines long; leaflets 18–24-jugate, *minute;* legume falcate ... ... (4) **hæmatoxylon.**

Flowers capitate; *bracts* remote, about the middle of the peduncle.

Twigs and petioles pubescent or tomentose:

Legume oblong, obtuse, very thick, tumid, tomentose, few-seeded ... ... (5) **hebeclada.**

Legume flat, linear-moniliform, or deeply indented (6) **arabica.**

Legume stipitate, narrow-linear, flat, falcate ... (7) **hirtella.**

Glabrous; legume long and narrow, linear-falcate ... (8) **horrida.**

Glabrous; legume lanceolate-oblong, broad.

No petiolar gland; leafl. oblong-linear ... ... (9) **robusta.**

A large, petiolar gland; leafl. narrow-linear ... (10) **Natalitia.**

2. Vulgares.—Stipules not spiny. Branches armed with sharp thorns or prickles, below the nodes or scattered. (Sp. 11–17.)

Flowers capitate. Prickles in pairs just below the nodes. *Stem arborescent.*

Glabrous; pinnæ 3-jugate; leafl. unijugate, obovate (11) **detinens.**

Twigs and petioles hispid; pinnæ 4–6-jugate; leaflets 3–6-jugate ... ... (12) **ferox.**

Flowers *spicate.* Prickles in pairs, or scattered, or none.

Prickles in pairs below the nodes, or none.

Twigs hispid; leafl. 5–8-jugate, oblong or obovate, obtuse ... ... (13) **Burkei.**

Twigs subpuberulous; leafl. 15–30-jugate, narrow-linear ... ... (14) **caffra.**

Prickles scattered; twigs and petioles tomentose; leafl. silky beneath ... ... (15) **eriadenia.**

Flowers capitate. Prickles scattered. Stem fruticose, *climbing*.

Pinnæ 3–5-jugate; leafl. 6–12-jugate, oblong ... (16) **Kraussiana.**

Pinnæ 8–20-jugate; leafl. 30–35-jugate, narrow-linear (17) **pennata.**

1. A. erioloba (E. Mey. Comm. p. 171); stipular-spines straight; leaves glabrous, pallid; pinnæ trijugate, with glands at each pair; leaflets 8–10-jugate, linear-oblong, obtuse; legume woody, indehiscent, semilunate, tumid, tapering at each end, with a whitish, rough tomentum, spongy within. *Benth. in Lond. Journ.* 1, *p.* 496.

HAB. Namaqualand, *Schmeling*. (Probably *A. Giraffæ*). "*Kameeldoorn*" of the colonists.

2. A. Giraffæ (Burch. Trav. 2, p. 240, plate 6); quite glabrous; stipular-spines straight, strong, brown; pinnæ 1–3-jugate, with shield-like glands at most pairs; leaflets 8–15-jugate, oblong, linear, obtuse, thickish; peduncles crowded on abortive ramuli; legume oval, thick, indehiscent, spongy within. *Benth. Lond. Journ.* 1, *p.* 496. *A. erioloba? E. Mey.*

HAB. Dry and sandy deserts to the north of the colony, *Burchell*.

A tree, called *Kameeldoorn*, because the cameleopard browses chiefly on it, but known to the Bichuana natives as the *Mokáala*. It is much larger than the common karroo-thorn (*A. horrida*), with a thick and spreading umbrella-shaped head, and thick, brown thorns. The wood is excessively hard and heavy, of a dark, red-brown colour, and used by the Bichuanas for spoons and knife-handles, &c. *Burchell, l. c.*

3. A. heteracantha (Burch. in DC. Prod. 2, 473); branches and petioles pubescent; stipular-spines either short and hooked back or very long and straight, all pubescent; pinnæ 5–10-jugate, with few, small, jugal glands; leaflets 10–15-jugate, oblong-linear, downy or glabrate; legume linear. *Benth. Lond. Journ.* 1, *p.* 497.

HAB. Near the Gariep, *Burchell*, No. 1710., *Trav.* 1, *p.* 389.

A tree upwards of 20 feet high, with a tall trunk of 18 inches diameter, supporting a flat, wide-spreading, umbrella-like head. *Burchell, l. c.*

4. A. hæmatoxylon (Willd. Enum. 1056); branches and leaflets thinly cano-tomentose; stipular-spines mostly long, straight, subulate; pinnæ 8–19-jugate, with few, small jugal glands; leaflets very small, 18–24-jugate, closely imbricate, canescent; legume linear, falcate, thick, tomentose, spongy within, with distant seeds. *Benth. Lond. Journ.* 1, 497, *A. atomiphylla, Burch. Trav.* 1, 341.

HAB. Kloof Valley, interior of the Cape. *Burchell*, 1685.

A shrub 8–16 feet high, with soft, pale green foliage. Spines very slender, straight, spreading. Leaves looking to the eye as if simply pectinato-pinnate, 8–19-jugate; but really bipinnate, the pinnæ 18–24-jugate. Leaflets very minute, lying very close together, as if cohering. Heads globose, on long peduncles. *Burchell.*

5. A. hebeclada (DC. Prod. 2, 461); twigs, petioles, and peduncles patently tomento-hispid; stipular-spines subulate-conic, short, recurved, tomentose; pinnæ 3–7-jugate, with small, jugal glands; leaflets 12–15-jugate, pubescent, becoming glabrate, linear; peduncles bracteate above the base or in the middle; legume *oblong, obtuse, very thick*, coriaceous, yellowish, tomentose and obliquely striate, at first pulpy within, then hollow. *Benth. Lond. Journ.* 1. *p*, 499 *and* 5, *p.* 95. *Zey.!* 569.

HAB. Ongeluk's Fontein, *Burchell*, 2267. Vaal River and Aapjes River, *Burke & Zeyher!* (Herb. Hk., Sd.)

All the younger parts hairy. Leaflets 3 lines long, ½ line wide, ciliolate or glabrate. Peduncles 1½–2 inches long, the bracts deciduous, minute. Legumes 2–3 inches long, turgid, ochraceous.

6. A. arabica (Willd. sp. 4, 1085); tomentoso-pubescent; stipular-spines long or short, subulate or robust, at length white, straight, or subrecurved; pinnæ 4–8-jugate, with scutelliform glands and often a large petiolar gland; leaflets 10–20-jugate, oblong-linear, obtuse, green, glabrous or ciliate; peduncles axillary, bracteate in the middle; heads globose; legume *flat, linear, moniliform*, tomentose, at length glabrescent, coriaceous, pulpy within. *Benth. L. J.* 1, 500. *E. Mey. Comm. p.* 168.

HAB. Near Port Natal, *Drege, Krauss!* (Herb. Hk., D., Sd.)

A native of North Africa and Arabia, producing the *gum-arabic* of commerce. The Natal specimens belong to Bentham's var. *β. Kraussiana*, and have generally long spines, and a deeply crenate, but scarcely moniliform, tomentose pod. The peduncles, besides the terminal head, have sometimes flowers at the medial bracts.

7. A. hirtella (E. Mey. Comm. p. 167); twigs, petioles, and leaflets minutely hairy; spines straight, subulate; pinnæ 6–8-jugate, with small glands; leaflets 10–15-jugate, oblong-linear; peduncles glabrescent, bracteate in the middle; legume stipitate, narrow-linear, subfalcate, flat, glabrous. *Benth. Lond. Journ.* 1, *p.* 502.

HAB. Between the Omcomas and Omblas, *Drege!* (Herb. Sd.)

I have not seen a legume; the foliage is like that of *A. Arabica, β.*

8. A. horrida (Willd. Sp. 4, 1082); glabrous; stipular spines short or long (on the older twigs longest), straight, ivory-white; twigs, peduncles, and petioles angular; pinnæ 2–5-jugate, with small glands; leaflets 5–12-jugate, oblong-linear, obtuse or subacute; peduncles bracteate in the middle, the upper ones fasciculate-racemose; petals with revolute points; legume long, linear, flat, falcate, glabrous, coriaceous. *Benth. l. c. p.* 502. *A. Capensis, Burch. E. & Z.* 1695.

HAB. Common throughout the Colony. (Herb. Hk., Sd., D.)

This is the common *Doorn-boom*, or Karro-doorn, of the colonists. Spines ½ inch to 2–3 inches long, very sharp. Leaflets 3–4 lines long, 1–1½ wide. Legume 4 inches long, 3 lines wide, with straight or irregularly sinuous margins.

9. A. robusta (Burch. Trav. 2, p. 442); glabrous; stipular spines valid, short, or some long, white; pinnæ 2–4-jugate, with 1–2 jugal glands, and *no petiolar;* leaflets 8–13-jugate, *oblong-linear*, obtuse; peduncles axillary, bracteated below the middle; heads globose, legume straight or subfalcate, lanceolate-oblong, acute, coriaceous, flat, at length convex, obliquely veiny, tapering at base into a short stipe. *Benth. Lond. Journ.* 1, *p.* 501, *and* 5, *p.* 96. *A. clavigera, E. Mey? Comm. p.* 168.

HAB. Interior of the Cape, *Burchell.* Rhinoster Kop and Magaliesberg, *Burke & Zeyher!* (Herb. Hk., D., Sd.)

A tree 20–30 feet high, with a very thick trunk, and thick branches and twigs. Bark of the twigs and branches a reddish brown. Leaflets deep green above, pale beneath and somewhat veiny, 3½–4 lines long, 1½ wide. Unripe legumes 2–2½ inches long, ¾ inch wide, 1–2-seeded. Native name *Mókwi* or *Mokála-mókwi.*

10. A. Natalitia (E. Mey. Comm. p. 167); glabrous; stipular spines small, subulate, white; pinnæ 4–7-jugate, with glands between the pairs, and *a large shield-like petiolar-gland;* leaflets 12–30-jugate, *narrow-linear,* obtuse; peduncles axillary, fascicled, bracteated below the middle; heads globose; legume . . . ? *Benth. l. c.* 502 *and vol.* 5, *p.* 97. *Krauss, No.* 66.

HAB. About Port Natal, *Drege, Gueinzius! Krauss! &c.* (Herb. Hk., D., Sd.)
Bark of the twigs and branches very pale, whitish. Leaflets drying dark, 3 lines long, ½ line wide. *Bentham* suspects that this is the same as *A. robusta,* but the specimens look different. This has much smaller leaflets, well-marked jugal, large petiolar gland, and pale bark. *A clavigera,* E. Mey., which I have not seen, seems to agree better with *A. robusta.*

11. A. detinens (Burch. Trav. 1, 310); glabrous; prickles in pairs just below the nodes, recurved; petiole nearly unarmed; pinnæ about 3-jugate; leaflets unijugate, obliquely obovate, very obtuse; heads loose, subglobose; pedicels equalling the calyx; calyx truncate, one-third as long as the corolla; legume oval, flat, membranous, few-seeded. *Bth. l. c. p.* 507.

HAB. Kloof Valley, *Burchell.*
A shrub, 4–8 feet high; prickles very short. See woodcut in Burchell's Travels. 1, p. 349.

12. A. ferox (Benth.! Lond. Journ. 5, p. 97); young twigs and petioles patently hispidulous; prickles in pairs or threes just below the nodes, recurved; petiole aculeate; glands minute; pinnæ 4–6-jugate; leaflets 3–6-jugate, obliquely obovate oblong, very obtuse; heads subglobose, shortly pedunculate; legume oblong or broadly-linear, acute, flat, membranous, glabrous, few-seeded. *Zey!* 570.

HAB. Magaliesberg, *Burke & Zeyher!* (Herb. Hk., D., Sd.)
Bark dark ashen-grey, that of the older twigs rugose. Leaflets pale, 4 lines long, 1½ wide. Legume 2½–3 inches long, nearly an inch wide, 2–3-seeded.

13. A. Burkei (Benth.! Lond. Journ. 5, p. 98); young twigs and petioles patently hispid; prickles in pairs just below the nodes, recurved; petiole unarmed or armed, and with a petiolar gland below the pairs; pinnæ 3–6-jugate; leaflets 5–8-jugate, *obliquely oblong or obovate,* obtuse or mucronulate, the younger villous at margin; spikes tufted, loose, peduncled, rather longer than the leaf; legume ? *Zey!* 571.

HAB. Magaliesberg, *Burke & Zeyher!* (Herb. Hk., D., Sd.)
Bark dark ashey-brown, rugged. Leaflets drying pale, 4 lines long, 1½–2 lines wide, the uppermost largest and obovate. Spikes 2–3 inches long. Legume not seen; supposed to be membranous by *Bentham.*

14. A. caffra (Willd. Sp. 4, p. 1078); subglabrous, the reddish-brown twigs, petioles, and peduncles minutely puberulous; prickles in pairs below the nodes, recurved, or none; pinnæ 8–14-pair, the unarmed petiole with a gland; leaflets 15–30-jugate, *narrow-linear,* glabrous or nearly so; calyx rather shorter than the corolla; legume linear, flat, bivalve. *Benth. Lond. Journ.* 1, *p.* 509. *DC. p.* 459. *E & Z.!* 1694. *A. fallax, E. Mey.! Zey.!* 595.

HAB. Eastern districts, from Uitenhage to Port Natal, common. (Hb. Hk., D., Sd.)
Spines small and often absent. Leaves 4–6 inches long, the pinnæ 2½ inches long; pinnules 4 lines long, ½ line wide, paler beneath. Spikes often 3–4 inches long.

Legume 3–3½ inches long, 3–4 lines wide, 6–8-seeded, often irregularly sinuous at the margin. *A. multijuga, Meisn. in Hook. Lond. Journ.* 2, *p.* 105, seems from the dedescription to be referable to this.

15. A. eriadenia (Benth.! Lond. Journ. 5, p. 98); prickles scattered, small, straight or recurved, few; twigs, petioles, and peduncles tomentose; stipules membranous, semicordate, acuminate, villous, deciduous; pinnæ 6–10-jugate, the one or two petiolar, and the few jugal glands conical-tubercular, villous; leaflets about 20-jugate, obliquely linear, appressedly silky beneath; spikes elongate, loose, the uppermost in a raceme; flowers subsessile, glabrous; calyx half as long as the corolla. *Zey.!* 568.

HAB. Crocodile River, Magaliesberg, *Burke & Zeyher!* (Herb. Hk., D., Sd.)

Twigs pale ash-coloured. Stipules 6–7 lines long, 2 lines wide. Leaflets 2½–3 lines long, ½ line wide, pale, sometimes silky at the edges only. Petiolar gland very prominent. Legumes unknown.

16. A. Kraussiana (Meisn.! Lond. Journ. 2, p. 103); scandent; prickles scattered, very small, numerous; twigs, petioles, and peduncles minutely downy or glabrate; petioles armed; pinnæ 3–5-jugate, with convex glands; leaflets 6–12-pair, obliquely oblong, obtuse, mucronulate, glabrous, shining above; heads globose, the peduncles racemoso-paniculate; ovary stipitate, pubescent; legume broadly-linear, membranous, flat, glabrous, straight, several-seeded, stipitate. *Bth.! l. c. p.* 505.

HAB. Port Natal, *T. Williamson! Krauss! Gueinzius!* (Herb. Hk., D.)

A slender shrub, with flexuous or twining branches. Petioles mostly aculeate. Leaflets 4–5 lines long, 2 lines wide, paler beneath. Inflorescence disposed in terminal panicles. Legumes 3–4 inches long, the margin slightly sinuous. Prickles much smaller and leaflets larger than in *A. pennata.*

17. A. pennata (Willd. Sp. 4, 1090); scandent; prickles numerous on the twigs and petioles, short, recurved; twigs and petioles thinly tomentose, at length glabrate; pinnæ 8–20-jugate, with petiolar and sometimes jugal glands; leaflets more than 30-jugate, narrow-linear, glabrous or ciliate; heads globose, panicled; calyx nearly equalling the corolla; ovary stipitate, villous; legume glabrous or minutely puberulous, membranous, flat. *Benth.! Lond. Journ.* 1, *p.* 516. *E. Mey.! Comm. p.* 169. *DC. Prod.* 2. 464. *Burm. Zeyl. t.* 1.

HAB. Steep mountain rocks near the Mission Station, Omgaziana, *Drege!* Natal, *Gueinzius!* (Herb. D.)

A slender, half climbing shrub, with large, finely divided, fernlike leaves. Leaves 5–6 inches long; pinnæ 2–3 inches; leaflets 2½–3 lines long, ½ line wide.

Imperfectly known Species.

A. spinosa (E. Mey.! Comm. p. 170); branches and twigs spiny, rigid, divaricate; leaves from the side of the spines, 3–5, tufted, bipinnate, 4–5-jugate; stipitate glands between the first and last or between every pair; leaflets 18–24-jugate, subfalcate-linear, glabrous, shining, acute-angled at the base behind, acute at the anterior margin at the point; stipules membranous, falcato-subulate, soft; *flowers and fruit unknown. E. Mey. l. c.*

HAB. Port Natal, *Drege.*

A. Litakunensis (Burch. Trav. 2, 2452); a tree, 40 feet high, called by the inhabitants *Moshu*, with singularly twisted, bivalve pod. *Burch. Cat. Geogr.* 2205

A. stolonifera (Burch. Trav. 2, 241); stems underground, stoloniferous; twigs, leaves, and spines pubescent; leaves bipinnate; pinnæ 3–7-jugate; leaflets oblongo-lanceolate, 7–15-jugate; spines stipular, in pairs, whitish, spreading, with brown, sub-recurved tips; heads axillary, 2–6 together, globose, pedunculate; legume straight, yellow, obliquely striate, hollow. *Cat. Geogr.* 2138.

A. viridiramis (Burch. Trav. 1, 300); a bush 3–4 feet high, with flexuous, green branches. Stipular-spines recurved, very short; gemmæ woolly-white; leaves small, conjugato-pinnate; pinnæ 6–8-jugate; leaflets oval, close-set. *Cat. Geogr.* 1586.

LXXXI. ALBIZZIA, Duraz.

Flowers mostly bisexual. *Calyx* campanulate or tubular, 5-toothed. *Corolla* monopetalous, funnel-shaped. *Stamens* indefinite, often numerous, united at base into a tube. *Legume* flat, dry, membranous or papery, with thin margins, either indehiscent or dehiscent. *Benth. in Lond. Journ.* 3, *p.* 84.

Unarmed trees or shrubs. Leaves bipinnate. Glands as in *Acacia*. Flowers in heads or spikes, mostly handsome, with long, white or rosy, rarely purple, feathery bundles of stamens. Name of barbarous origin.

ANALYSIS OF THE SOUTH AFRICAN SPECIES.

Pinnæ 2–4-jugate; leaflets falcate-oblong, obtuse, convex (1) **Forbesii.**
Pinnæ 12–14-jugate; leafl. lin.-falc. acute, white beneath (2) **pallida.**

1. **A. Forbesii** (Benth.! Lond. Journ. 3, p. 92); twigs and petioles thinly velvetty-pubescent; stipules small, deciduous; pinnæ 2–4-*jugate;* leaflets 10–15-jugate, *falcate-oblong, obtuse, or mucronulate,* concave beneath, with a submarginal nerve, rufo-sericeous on both sides, at length glabrate above; petioles with a gland above the base and one between the terminal pair; peduncles longish, axillary; heads many-flowered; flowers sessile, rufo-sericeous; calyx ⅓ of corolla; ovary sessile, glabrescent. *Benth. l. c.*

HAB. Delagoa Bay, *Forbes!* (Herb. Hk.)

Bark dark-coloured; pubescence yellowish or foxy. Leaflets 3 lines long, 1½ wide, the margins slightly revolute, drying dark.

2. **A. pallida** (Harv.); twigs and petioles thinly velvetty-pubescent; stipules subulate; pinnæ 12–14-jugate; leaflets 20–24-jugate, *linear-falcate, acute,* with a submarginal nerve, glabrous above, *very pale and thinly silky beneath;* petioles with a gland above the base and one between the terminal pair; peduncles axillary or racemose; heads 15–20-flowered; flowers sessile, fulvo-sericeous; calyx one-third of corolla; ovary sessile, glabrous.

HAB. Banks of the Chobe, Lake Ngami and adjoining forests, *J. McCabe!* (Herb. Hk.)

A shrub about 10 feet high. Petioles 4–5 inches long; leaflets 3 lines long, ½ line wide, ciliolate, glaucescent above, nearly white beneath. Peduncles 2 inches long.

LXXXII. ZYGIA, P. Browne.

Calyx tubular, 5-toothed. *Corolla* funnel-shaped-tubular, shortly 5-

lobed. *Stamens* very numerous, connate in a tube much longer than the corolla, spirally twisted in the bud, free at the summit only. *Legume* as in *Albizzia. Benth. in Lond. Journ.* 3, *p.* 92.

Shrubs or trees with the foliage of *Albizzia*, from which this genus differs in the long staminal tube. Name, ζυγος, a *yoke;* because the stamens are joined together in a long tube.

1. **Z. fastigiata** (E. Mey. Comm. p. 165); twigs and petioles rusty-tomentose; pinnæ 5–6-jugate; leaflets 8–15-jugate, obliquely trapezoid-oblong, puberulous, becoming glabrate above, pale and pubescent beneath, the upper ones smaller. *Benth.! l. c.* 93.

HAB. Between Omsamculo and Omcomas and Port Natal, *Drege! Krauss.* (Herb. Hk., Bth., Sd., D.)

A tree 15–20 feet high. Petioles 5–6 inches long, with a large, oblong gland above the base and a small, round gland between the terminal pair. Leaflets 4–5 lines long; 2½ lines wide. Peduncles axillary, and in a terminal corymbose-raceme, 2–3 inches long. Legumes 5 inches long, nearly an inch wide, obtuse, substipitate, flat, glabrous, many-seeded. A native also of Senegambia.

ORDER XLIX. **ROSACEÆ**, Juss.

(By W. H. HARVEY).

Calyx free or partially adnate with the ovary, its tube short or long, expanded or closed; limb mostly regular, 3–4–5 parted, occasionally with a second external row of segments or adnate-bracts, alternating with the proper segments. *Petals* as many as the calyx-lobes or none, spreading, mostly equal. *Stamens* inserted in the throat of the calyx, indefinite, rarely definite, many or few; filaments filiform, free. *Ovary* apocarpous (except in *Grielum*); *carpels* indefinite or definite, rarely only one, uniovulate, biovulate, or pluriovulate; ovules anatropous. *Styles* one to each carpel, terminal or lateral; stigmata simple or feathery. *Fruit* various; usually of dry *achenia*, naked or enclosed in the calyx-tube; in *Rubus*, of succulent, aggregated drupelets; in *Grielum* a plurilocular capsule (in *Spiræa* and its allies, follicular; in *Pyrus* and its allies a pome; in *Prunus*, &c., a drupe). *Seeds* without albumen; embryo straight, with fleshy cotyledons.

Herbs, shrubs, or trees, most abundant in the temperate and colder parts of the northern hemisphere, few tropical, and few in the south temperate zone. Leaves alternate, pinnately or digitately compound or parted, sometimes simple and entire. Stipules mostly present, adnate to the base of the petiole. Flowers variously disposed, rarely unisexual. To this important Order belong most of the garden fruits of Europe, as apples, pears, peaches, nectarines, apricots, plums, cherries, strawberries, and raspberries, as well as the "Garden's Queen," the rose, which is the type of the group. The Cape Flora possesses very few, and these more of the nature of weeds than flowers. The genus *Grielum*, placed here for want of a better location, looks more like a *Geranium* externally, but has perigynous petals and stamens, &c.

TABLE OF THE SOUTH AFRICAN GENERA.

Sub-Order 1. **DRYADEÆ.** *Calyx* expanded. *Ovary* apocarpous; *carpels* numerous, uniovulate, crowded on a convex, conical or columnar resceptacle.

Shrubs. *Calyx* 5-parted, not bracteate. *Fruit* succulent.

I. **Rubus.**

Herbaceous plants. *Calyx* 5-parted, with 5 external bracts or secondary lobes. Fruit dry.

II. **Potentilla.** Receptacle conical. Carpels without tails.

III. **Geum.** Receptacle columnar. Carpels hairy, with long, bristle-like, twisted tails.

Sub-Order 2. **SANGUISORBEÆ.** *Calyx-tube* turbinate or urceolate, contracted in the throat. *Carpels* 1–4, uniovulate, separate, concealed within the calyx-tube, which is hardened in fruit into a pseudo-pericarp.

Flowers with petals, yellow.

IV. **Leucosidea.** *Calyx-tube* unarmed; limb 10-parted in two rows, the outer lobes small.

V. **Agrimonia.** *Calyx-tube* armed with hooked bristles; limb simple, 5-parted.

Flowers without petals.

VI. **Acæna.** *Calyx-tube* armed with hooked bristles. *Suffruticose.*

VII. **Alchemilla.** *Calyx-tube* unarmed; limb 8-parted, in two rows. Stamens 1–4. *Herbaceous.*

VIII. **Poterium.** *Calyx-tube* unarmed; limb 4-parted. Stamens 20–30. *Herbaceous.*

IX. **Cliffortia.** *Calyx-tube* unarmed; limb 3-parted. Flowers unisexual. Stamens 8–40. Stigma feathery. *Shrubs.*

Sub-Order 3 ? **NEURADEÆ.** Flowers perfect. *Calyx-tube* concrete with the ovary. *Petals* 5, convolute. *Stamens* 10. *Carpels* 5–10, concrete into a plurilocular capsule. Seeds solitary, pendulous.

X. **Grielum.** Herbs with hoary, multi-partite leaves, and large, yellow flowers.

I. RUBUS, L.

Calyx-tube expanded, short; limb 5–parted, without bracts, imbricate. *Petals* 5, crumpled, deciduous. *Stamens* indefinite, inserted on the calyx. *Carpels* indefinite, on a convex receptacle, uniovulate; styles subterminal, filiform; stigmas simple. *Fruit* of many little *drupes*, aggregated on a dry conical receptacle. *Radicle* superior. *Endl. Gen.* 6360. *DC. Prod.* 2, *p.* 556.

Shrubs, rarely herbs, mostly trailing and arching, and armed with sharp prickles, common in temperate latitudes, rare within the tropics and in the southern hemisphere. Leaves either simple, digitate, or impari-pinnate. Stipules adnate to the petiole. Flowers terminal or axillary, commonly panicled, rarely solitary. Name from the Celtic, *rub*, red; the fruit of several (as the *raspberry*) is red. That of the common bramble or *blackberry* has a deep vinous tint. All have edible and some excellent fruits.

ANALYSIS OF THE SOUTH AFRICAN SPECIES.

Leaves pinnate; the uppermost often ternate; fruit red or yellow
 Peduncles axillary and terminal, about one-flowered ... ... (1) **rosæfolius.**
 Peduncles in a terminal raceme or panicle:
  Leaves beneath glabrous or pubescent, but not albo-tomentose ... ... (2) **pinnatus.**
  Leaves beneath albo-tomentose:—
   Leafl. deeply inciso-lobulate; fruit woolly ... ... (3) **Ludwigii.**
   Leafl. serrate; fruits glabrous, golden ... ... (4) **rigidus.**
Leaves digitate, of 5–3 leaflets; fruit black or deep purple ... (5) **fruticosus.**

1. **R. rosæfolius** (Sm. Ic. ined. t. 60); stem and petioles thinly pilose; leaves pinnate, 7–9-foliolate, green, pubescent or pilose, gland-dotted beneath; leaflets ovate or ovato-lanceolate, acute or acuminate, doubly serrate, the serratures acute; peduncles axillary, about one-flowered, shorter than the leaf; petals shorter than the much-acuminated hairy calyx-segments; fruit rather dry, red, of *very many* small carpels on a cylindrical receptacle. *DC. Prod.* 2, 556. *E. & Z.!* 1711. *Hook. Ic. Pl. t.* 349. *Bot. Mag*, 1783 (*with double flowers*).

HAB. Sides of Table Mountain, facing the town, *E. & Z.!* &c. (Herb. Sd. D. Hk.)
A suberect bush, more or less covered with spreading, soft hairs and sessile resinous glands. Leaves 6–7 inches long, the leaf-pairs 1–1½ inch asunder; leaflets 1½–2 inches long, 1 inch wide. Peduncles 2 or 3 inches long, either axillary or ending short branchlets. Prickles small, straight, pale. Fruit shaped like the raspberry, but not very succulent, and raised on a short gynophore; carpels extremely numerous.

2. R. pinnatus (Willd. Sp. 2, p. 1081); branches, panicle and petioles thinly villoso-tomentose, leaves pinnate, 5–9 foliolate, green, and glabrous or nearly so on both surfaces; leaflets very shortly petiolate, ovate, acute, penninerved beneath, doubly and unequally serrate, the serratures acuminate; the terminal leaflet largest; panicle terminal, many flowered; petals shorter than the taper-pointed, tomentose calyx-lobes; fruit glabrous, golden, of few carpels. *Cham. & Schl. in Linn.* 2, *p.* 19. *E. & Z.!* 1705, *and R. Pappei, E. & Z.!* 1706.

HAB. Hanglip, *Mundt.* Table Mountain sides; also Krakakamma and Adow, Uit., and on the Kat River, *E. & Z.!* Zuureberg Forest, *A. Wyley!* (Herb. Sd., D.)
Stems roundish or slightly angular, the younger ones covered with cobwebby hairs, the older often naked. Leaves 3–6 inches long; leaflets 1–3 inches long, ½–2 inches wide, occasionally slightly hairy beneath, especially on the nerves; the uppermost leaves 3-foliolate. Calyx of the fruit erect or spreading, not reflexed. I cannot distinguish *R. Pappei* from ordinary "*pinnatus*," taking Cham. & Schl.'s description as my guide.

3. R. Ludwigii (E. & Z.! 1710); branches and the hooked prickles glabrous, reddish, young twigs and petioles downy; leaves pinnate, 5–7 foliolate, albo-tomentose beneath; leaflets sessile, ovate-oblong, deeply inciso-lobulate, the lobules triangular-acuminate; the terminal leaflet petioled, often trifid or 3-parted and incised; peduncles short, lateral, few flowered, or in a short terminal raceme; petals broad, shorter than the taper-pointed, tomentose calyx-lobes; fruit albo-tomentose. *R. rhodacantha, E. Mey.*

HAB. Among stones on mountain sides, of the Sturmberge, near the Witte, and Zwartkei Rivers, Caffr., *E. & Z.! Drege!* Schneewberg, *Drege!* (Herb. Sond., D.)
Stem terete, creeping, smooth; branches suberect, sometimes glaucous, mostly reddish-brown. Prickles abundant or few, on the twigs and petioles, but not on the nerves. Leaves 4–5 inches long; leaflets 1–1½ inch long, ½–¾ inch wide, glabrous and deep green above, very white and softly tomentose beneath. Flowers small. Fruits very woolly.

4. R. rigidus (Sm.! in Rees. Cycl. 30, No. 5); branches, panicle, petioles, and undersurfaces of the leaves densely albo-tomentose; lower leaves pinnately 5–foliolate, upper ternate or simple; leaflets broadly ovate, acute or obtuse, serrate or doubly serrate, glabrous or pilose above; panicle terminal, contracted, many flowered; calyx-segments ovato-lanceolate; petals obovate: fruit golden or fulvous, glabrous. *R. discolor, E. Mey!*

VAR. α. **chrysocarpus**; minutely glandular; prickles smaller and fewer; leaves glabrous above; calyx segments narrower. *R. chrysocarpus, Ch. & Schl. Linn.* 2, *p.* 17. *E. & Z.!* 1708. *Zey.!* 2450, *ex pte.*

VAR. β. **Mundtii**; without glands; prickles larger and more numerous; leaves sparsely pilose above; calyx-segments rather broader. *R. Mundtii, Ch. & Schl. l. c. p.* 18. *E. & Z.* 1709. *Zey!* 2450, *ex pte.* & 572.

HAB. In bushy places and by river banks, in many places from Tulbagh through Swellendam, and George, eastward to Albany, and in Caffraria. *Mundt, E. & Z.! Drege! Mrs. Barber, &c. &c.* (Herb. Sond., T.C.D., Lin. Soc.)

The upper leaves are commonly 3-foliolate, the pair subsessile, the terminal an inch removed; when 5, the lowest pair is an inch from the upper. Leafl. 1½–2½ inches long, 1–1½ wide. The glands are never plentiful and often wanting. The differences between the two varieties seem to me to be of trifling moment; the fruit in both, judging from dried specimens, seems to be pale. I am indebted to Mr. Kippist for the verification with the original specimens in Linnæus's Herbarium.

5. R. fruticosus, var. Bergii (Ch. & Schl. Linn. 2, p. 16); stem arching, villoso-pubescent or glabrate; prickles slender, straightish or hooked, pale, numerous on the inflorescence and twigs; leaves digitately 5–3 foliolate, glabrous or thinly pubescent above, naked or tomentose beneath; leaflets ovate or rhomboid, sharply serrate, the medial largest; panicle corymboso-fastigiate; calyx-segments ovate, acuminate, tomentose; fruit glabrous, juicy, black, of few carpels. *R. fruticosus, E. & Z.!* 1703 *& R. Bergii, E. & Z.!* 1704.

HAB. About the Lion and Table Mountains, and near Klapmuts, *Stell.! E. & Z.!* Common near Rondebosh, Newlands and Protea, *W.H.H.* (Herb. Sond.)

The common bramble or blackberry of the Cape. It varies, as elsewhere, in pubescence, shape of leaflets, prickles, and other minor characters.

Doubtful Species.

Ecklon and Zeyher's No. 1707, "*R. rigidus*" (not of Smith), of which a specimen exists in Herb. Sonder, is near *R. pinnatas*, but more copiously pubescent, with densely villous twigs and petioles. The inflorescence looks depauperated, and the whole plant has the aspect of a 'drawn' specimen. I therefore pass it by. Among *E. & Z.'s* specimens of *R. rosæfolius* is one which looks almost intermediate between *R. pinnatus* and *R. rigidus*, having the habit and foliage of the former, but the more copious, softer, but not canous pubescence of the latter. Its panicle is many flowered, but only partially developed.

II. POTENTILLA, L.

Calyx-tube short, concave, open; *limb* 4–5 parted, 4–5 bracteate, persistent, the segments valvate in æstivation. *Petals* 4–5, deciduous. *Stamens* indefinite, inserted with the petals, perigynous. *Carpels* indefinite, on a convex receptacle, uniovulate; styles lateral, stigmata simple. *Achenia* dry, on a convex, dry, hairy receptacle, sessile. *Radicle* superior. *Endl. Gen.* 6363. *DC. Prod.* 2, *p.* 571.

A large genus of herbs or suffrutices, rarely shrubs, natives chiefly of the temperate and colder zones of the northern hemisphere; very few passing the tropic of Capricorn. Leaves alternate, digitate or pinnate-partite; leaf segments toothed or cut, mostly pubescent. Stipules adnate to the petiole. Peduncles in the forks of the stem, opposite the leaves, or terminal, one flowered, often corymbose. Flowers yellow or white, rarely red. Name *potens, powerful;* in allusion to the properties, which are strongly astringent.

1. P. Gariepensis (E. Mey!); stems herbaceous, diffuse, villous; leaves pinnati-partite, the upper tripartite; segments obovate-oblong, inciso-dentate, glabrate above, thinly piloso beneath; stipules short, ovate; peduncles opposite, and shorter than the leaves; calyx-segments lanceolate, acute, shorter than the ovate-oblong bracts; petals . . . . ? carpels furrowed, glabrous.

HAB. Near Verleptpram on the Gariep, *Drege!* (Herb. D., Sd.)

Stems 2 feet long, weak, decumbent or ascending, angular and pale. Leaves subdistant, very pale green; leaf-segments an inch long, not ½ inch wide, cuneate at base, deeply 7–11 toothed or lobulate. Peduncles about an inch long. Flowers small.

III. GEUM, L.

Calyx-tube short, concave, open; limb 5-parted, 5-bracted, persistent, the segments valvate in æstivation. *Petals* 5, deciduous. *Stamens* indefinite, inserted with the petals, perigynous. *Carpels* indefinite, on a columnar receptacle, uniovulate; styles terminal, inflexed or sharply bent; stigmas simple. *Achenia* on an elongated receptacle, tailed with the hardened, awnlike, hooked or curled styles. *Radicle* inferior. *Endl. Gen.* 6386. *DC. Prod.* 2, *p.* 550.

Herbaceous plants, common in the north temperate zones, rare in the southern hemisphere. Radical leaves unequally pinnati-partite, the terminal segment mostly much larger than the rest; cauline small or depauperated, trifid. Stipules adnate; flowers terminating the branches, subcorymbose, yellow, or red. Name from γευω, *to taste well;* the roots of some are pleasantly aromatic; all are astringents.

1. G. Capense (Thunb. Prod. p. 91); stem tomentulose, erect; radical leaves villoso-pilose, pinnatisect, the terminal lobe very large, cordate-ovate or subrotund; cauline few and small, tripartite, incised; petals roundish-obovate, longer than the calyx; awns of the fruit twisted in the middle, glabrous upwards. *Thunb. Fl. Cap. p.* 428. *DC. l. c.* 553. *E. & Z.!* 1702. *Harv. Thes. t.* 18.

HAB. Rietvalley, *Thunberg.* Mountains round Grahamstown and various places in Albany and Caffraria, *E. & Z.! Drege! &c.* (Herb. Hk., Sd., D.)

Root fascicled, perennial. Radical leaves numerous, 5–6 inches long, the terminal lobe 2–3 inches long, 1½–2 inches wide; the rest very small, unequal, 2–5 lines long and wide. Stems 1–2 feet high, laxly branched, nearly naked, or with a few small, depauperated leaves. Flowers handsome, bright yellow, 1–1½ inch across, erect. Carpels very hairy.

IV. LEUCOSIDEA, E. & Z.

Flowers complete. *Calyx-tube* obconic, constricted in the throat, with an annular disc; limb 10-parted in two rows, persistent, outer lobes short, ovate, inner lanceolate, acuminate, with valvate æstivation. *Petals* 5, obovate, deciduous. *Stamens* 10–12, inserted on the annular disc. *Carpels* 2–3, enclosed in the calyx-tube; *styles* as many, filiform, exserted, terminal; *stigmata* subclavate, channelled, hook-pointed. *Achenia* membranous, utricular, enclosed in the hardened calyx-tube, subsolitary. *Endl. Gen.* 6375.

A densely leafy shrub, the "*Dwa-dwa*" of the natives, who use it as an astringent medicine. "The woody branches are very inflammable, and eagerly sought after by the Kaffir women, for lighting their fires," *Mrs. F. W. Barber.* The name is compounded of λευκος, *white,* and ιδεα, *a resemblance;* because the pubescence is white.

1. L. sericea (E. & Z.! 1716.)

HAB. Mountain sides, Kat River, and Chumieberg, *E. & Z.!* Zwartkei River, *Mrs. F. W. Barber!* Orange River, *Burke & Zeyher!* (Herb, Sd., D., Hk.)

A shrub, 10–12 feet high, with flexuous branches and loose exfoliating bark. Twigs densely leafy. Stipules membranous, broad, amplexicaul, adnate to the base of the petiole. Leaves petiolate, pinnati-partite, with 2–3 pair of pinnæ; pinnæ obovate-oblong, inciso-dentate, dark green above, white and silky beneath, the lower pairs smaller, and sometimes with a pair or two of minute leaflets interposed.

Racemes dense, cylindrical, terminating short leafy ramuli; bracts membranous, oblong, blunt, under each pedicel, and two bracteoles at the base of the calyx tube. Calyx densely silky, its five inner lobes petaloid at base within. Petals greenish yellow, shorter than the inner calyx-lobes, glabrous, narrow, obovate. Stamens shorter than the petals.

V. AGRIMONIA, L.

Calyx naked at base, the tube turbinate, armed beneath the limb with many hooked bristles, constricted in the throat with an annular ring; limb 5-parted, with subimbricate æstivation, persistent, at length connivent. *Petals* 5, deciduous. Stamens 12–20 inserted on the annular disc. *Carpels* 2, uniovulate, enclosed in the calyx-tube; styles terminal, exserted; stigmas dilated. *Achenia* one or two, enclosed in the hardened and densely hook-bristled calyx-tube. *Endl. Gen.* 6368. *DC. Prod.* 2, *p.* 587.

Perennial herbs, natives chiefly of the north temperate zones. Leaves alternate, imparipinnate. Stipules large, adnate. Flowers in terminal spicate racemes, small, yellow. Pedicels bracteate at base, bibracteolate in the middle. Name, a corruption of *Argemone,* an ancient name for some such plant. English name *Agrimony.*

1. A. Eupatoria (L.); var. **Capensis** (Harv.); stem and petioles softly hirsute; leaves interruptedly pinnate; leaflets 7–9, ovate-oblong, with minute ones between, coarsely toothed, tomentose-pubescent beneath; stipules with a few coarse teeth at base, broadly semi-cordate, acuminate; bracts equalling or exceeding the flower, the lower often leafy; flowers subsessile; petals twice as long as the calyx-limb. *A. Eupatoria, E. & Z.!* 1712 *and A. repens, E. & Z.!* 1713. *A. bracteosa, E. Mey! in Herb. Drege. A. Nepalensis, Don.*

HAB, Near Balfour and Philipstown, Kat River, and on the slopes of Winterberg, *E. & Z.!* Between Keiskamma and Buffel River, and on the Wit-Bergen, *Drege!* Wittedrift, Plettenberg Bay, *Dr. Pappe!* (Herb. Sd., D.)

Stems 2–3 feet high, robust, angular, densely hairy. Leaves 6–8 inches long; leaflets 2–2½ inches long, 1–1½ inch broad, passing from oval to lanceolate. Stipules either quite entire, or variously toothed. *E. & Z.'s* 1712, is a weakly grown, 1713, a strong-growing state (from nearly the same locality) of what appears to me a mere local variety of *A. Europæa,* a species found in Europe, Asia, and North America, in all which countries it varies in size, pubescence, &c.

VI. ACAENA, Vahl.

Flowers bisexual. *Calyx-tube* oblong, echinate or smooth, compressed, 3–4–5-angled, the angles armed with hooked bristles, constricted in the throat; limb 4, rarely 3–5-parted, persistent. *Petals* none. *Stamens* 2–5, inserted on the throat of the calyx. *Carpels* 1–2, enclosed in the calyx-tube, uniovulate; styles terminal, short; stigma pencilled. *Achenia* hidden in the hardened, hook-bristled calyx-tube. *Endl. Gen.* 6372. *DC. Prod.* 2, 592. *Ancistrum, Forst.*

Herbs or suffrutices, frequent in the temperate and cold regions of the southern hemisphere, rare in central America. Leaves imparipinnati-partite, the segments incised. Stipules adnate. Flowers small, green, in terminal or axillary spikes or fascicles. Name ακαινα, a *thorn;* from the prickles on the calyx.

ANALYSIS OF THE SOUTH AFRICAN SPECIES.

Lvs. rosulate; flowers laxly spiked; calyx-bristles very numerous (1) **latebrosa.**
Leaves scattered; flowers in globose heads; cal.-bristles 2–4 ... (2) **sarmentosa.**

1. A. latebrosa (Ait. Hort. Kew. 1, p. 16); stem short, root-like, procumbent; leaves radical, rosulate, softly villous; leaflets in 9–10 pairs, oblong, inciso-serrate; peduncle scape-like, with 2–3 small, distant, leafy bracts; spike laxly several flowered, elongate; calyx-tube densely woolly, and armed with many, dispersed, barbed bristles. *DC. Prod.* 2, *p.* 592. *E. & Z.!* 1717. *Ancistrum decumbens, Th. Fl. Cap. p.* 31. *Agrimonia decumbens, Linn. f. Suppl.* 251.

HAB. Roggeveld, *Thunberg*. Bontjeskraal and Babylon's Toorensberg, Caledon, and near River Zonderende and Breede R. *E. & Z.!* Modderfonteinsberg, 4000–5000 f. *Drege!* (Herb. Sd., D.)

Stem underground or prostrate, simple or branched, the short branches or crowns ending in a tuft of many leaves. Leaves 4 or 5 inches long, clothed with long, soft hairs; leaf-pairs 9–10, the lowest leaflets smallest, the upper gradually larger, opposite or alternate, 3–6 lines long, 3–4 lines wide, deeply incised. Peduncle 1–1½ foot long, the sparsely flowered spike 6–8 inches long. Calyx lobes obovate or spathulate, glabrate above. Wool of the calyx white.

2. A. sarmentosa (Carm? Lin. Trans. 11, p. 20); stem slender, trailing, and creeping; leaves scattered, leaflets in 3–5 pairs, sharply serrate, nearly glabrous above, silky beneath; heads globose; calyx-tube obconical, villous, armed with 2–4 barbed bristles. *DC. l. c. E. & Z.!* 1718.

HAB. Hott. Holl. Berg, *E. & Z.!* (Herb. Sond.)

A slender plant, with the inflorescence and nearly the foliage of *Poterium Sanguisorba*. Leaves 3 inches long, with a pair of lanceolate, adnate stipules at base, a longish petiole, very small or depauperated lower leaflets; the upper 6–7 lines long, 3 lines wide. Peduncles 5–6 inches long, bearing a head 4–5 lines in diameter. Cal.-lobes ovato-lanceolate. Whether this be Carmichael's plant, I cannot say.

VII. ALCHEMILLA, Tournef.

Flowers bisexual. *Calyx-tube* urceolate, constricted in the throat with an annular disc; limb 8-parted, in two rows, the outer lobes shorter, sometimes very small, with imbricate æstivation, deciduous. *Petals* none. *Stamens* 1–4, inserted on the annular disc. *Carpels* 1–4, in the base of the calyx-tube, substipitate, uniovulate; styles basal, filiform; stigmata capitellate. *Achenia* 1–2, in the calyx-tube. *Endl. Gen.* 6370. *DC. Prod.* 2, *p.* 589.

Small herbaceous plants, annual or perennial, natives of the temperate zone, universally dispersed. Leaves alternate, flabelliform or reniform in outline, lobed or parted, or simply crenate, rarely pinnati-partite. Stipules adnate. Flowers minute, green, subcorymbose or clustered. Name from the Arabic *alkemelyeh*.

ANALYSIS OF THE SOUTH AFRICAN SPECIES.

Villous; leaves reniform, short-stalked, crenate, obscurely lobed ... (1) **Capensis.**
Glabrescent; lvs. long-stalked, deeply 5–7 lobed, the lobes toothed above ... ... (2) **elongata.**

1. A. Capensis (Thunb. Cap. 153); stems trailing, filiform, villous; leaves short-petioled, reniform, with 3–5, very shallow, rounded, bluntly crenate lobes, villoso-pilose; racemes axillary, interrupted, with leafy bracts. *DC. Prod.* 2, 589. *Lam. Ill. t.* 86, *f.* 2. *E. & Z.!* 1714.

HAB. Subalpine places, about Table Mountain; Vanstaadensberg, Uit., and on the Winterberg, *E. & Z.!* Blauwberg and Tigerberg; and near Bontjes River, Zuureberg, *Drege!* (Herb. Sd., D., Hk., Th.)

Stems 1–2 feet long, many from the crown, spreading widely. Petioles but half inch long, leaves once and half broader than long, sometimes evidently, sometimes obscurely lobed, thin, pale green, sometimes sparingly, generally copiously villous. Stipules broadly ovate, crenate. Racemes simple, longer than the leaves.

2. A. elongata (E. & Z.! 1715); stems trailing, angular, with appressed pubescence; leaves long-petioled, fanshaped, deeply 5–7 lobed or palmatifid, glabrous or thinly puberulous beneath, the lobes obtuse, coarsely crenato-serrate near the apex; panicles axillary, much branched, corymbose. *A. palmata, E. Mey! in Herb. Drege.*

HAB. Klipplaat River and Katriviersberg, *E. & Z.!* Witbergen, *Drege!* (Herb. Sd., D.)

Much more glabrous and more robust than *A. Capensis*, with palmatifid long-stalked leaves. Stems 2 feet long or more. Petioles 2–2½ inches long; leaves 1½ as broad as long, the lobes cut at least half way to the base, round topped.

VIII. POTERIUM, L.

Flowers polygamous or monœcious, the females in the upper part of the spike. *Calyx-tube* turbinate, constricted in the throat with an annular disc; limb 4-parted, the segments imbricate. *Petals* none. *Stamens* 20–30, on the annular disc. *Carpels* 2–3, enclosed in the calyx-tube, uniovulate; *styles* terminal, exserted; stigma pencilled. *Achenia* concealed in the hardened or fleshy, 4-angled calyx-tube. *Endl. Gen.* 6374. *DC. Prod.* 2, *p.* 594.

Herbs, suffrutices, or shrubs, natives of the warmer parts of the north temperate zone, sometimes spiny. Leaves imparipinnate; leaflets serrate. Stipules adnate. Flowers small, in dense terminal spikes, bracteate at base, and bibracteolate. Name poterium, *a drinking vessel* or drink; formerly an ingredient in cool tankards.

1. P. sanguisorba (Linn. Sp. 1411); herbaceous, the angular stems and leaves glabrous; leaflets ovate or roundish, sharply toothed; lower flowers of the globose heads male, upper female. *E. Bot. t.* 860.

HAB. About Simon's Bay, *C. Wright*. (Herb. T.C.D.)

Introduced from Europe. Leaves chiefly radical, of many leaflets. Stems 12–18 inches high, with few, distant, and smaller leaves, branched; each branch ending in a globular head of flowers.

IX. CLIFFORTIA, L.

Flowers diœcious. *Calyx-tube* urceolate; limb 3-parted (rarely 4-parted). *Petals* none. *Male: stamens* 30–40 or fewer, inserted in the throat of the calyx; filaments very slender. *Female: carpels* 2, enclosed in the calyx tube, uniovulate; style lateral; stigmata long, bearded and feathery. *Achenia* one or two, membranous, enclosed in the hardened and variously sculptured, rarely baccate, calyx-tube. *Endl. Gen.* 6379. *DC. Prod.* 2, *p.* 595.

Small shrubs or suffrutices, natives of South Africa. Leaves properly digitately 3-foliolate, often appearing simple or unifoliolate, either from the confluence of the three leaflets into one, or from the lateral leaflets being very minute or abortive; rarely bifoliolate, the medial leaflets disappearing. Stipules adnate with the petiole, sometimes sheathing the stem. Flowers axillary, small and green, subsessile. Name in honour of George Cliffort, a Dutch gentleman, a great cultivator of plants, and one of Linnæus's earliest patrons.

ANALYSIS OF THE SPECIES.

I. **Multinerviæ.** Leaflets solitary, *many nerved at base;* stipules simple. (Sp. 1-8

Lvs. either sharply few-toothed or incised, or quite entire.

Lvs. elliptic-oblong, 3-9 toothed, or cordate and entire, smooth edged ... ... (1) **ilicifolia.**

Lvs. lanceolate-oblong, 3-toothed, rough edged, quite *glabrous* ... ... (2) **intermedia.**

Lvs. shortly lanceolate, acuminate, concave, entire or toothed, rough edged, downy when young ... ... (3) **ruscifolia.**

Lvs. linear-lanceolate, 1-2 inches long, distantly 2-4 toothed, glabrous ... ... (6) **Meyeriana.**

Lvs. linear-lanceolate, 1-2 inches long, squarrose, quite entire, glabrous ... ... (7) **Dregeana.**

Leaves spinoso-ciliate, but not toothed or incised:

Lvs. broadly cordate, concave, patent or recurved, subsessile ... ... (4) **cordifolia.**

Lvs. broadly oblongo-lanceolate, flat, subsessile ... ... (5) **grandifolia.**

Lvs. narrow-linear, grass-like, flat, petiolate ... ... (8) **graminea.**

II. **Dichopteræ.** Leaflets solitary, *one-nerved;* stipules bifid. (Sp. 9-11.)

Lvs. cordate-ovate, plaited, crenate-serrate; the crenatures mucronate ... ... (9) **odorata.**

Lvs. linear-oblong or lanceolate, narrowed at base, serrate beyond the middle ... ... (10) **ferruginea.**

Lvs. narrow-wedge shaped, truncate, 3-5 toothed at the apex ... ... (11) **cuneata.**

III. **Bifoliolæ.** Leaves bifoliolate; leaflets *many-nerved,* orbicular or reniform, close-pressed together; stipules minute, (Sp. 12-13.)

Leafl. immersedly 5-7 nerved and netted, spinuloso-denticulate ... ... (12) **crenata.**

Leafl. prominently many-nerved, fanlike, scarcely rough-edged ... ... (13) **pulchella.**

IV. **Trifoliolæ.** Leaves either all trifoliolate, or some or all unifoliolate, scattered or tufted; leaflets *one-nerved.* Stipules simple, minute or more or less developed. (Sp. 14-38.)

(*a.*) Leaflets dissimilar; the medial equal-sided, obovate or obcordate, or 3-toothed or trifid; the lateral oblique, entire or toothed.

Glabrous:—(stipules minute or obsolete)

Medial leaflet deeply obcordate, mucronulate .. (14) **obcordata.**

Medial and lateral leafl. broadly obovate, obtuse, mucronulate ... ... (15) **obovata.**

Medial and lateral leaflets spathulate-oblong, white-edged ... ... (16) **marginata.**

Medial and lateral leafl. cuneate, all 3-toothed or 3-lobed ... ... (17) **triloba.**

Roughly villous and pilose; stipules *minute;* medial leafl. 2-5 toothed ... ... (18) **polygonifolia, β.**

Villous or pubescent, *at least the twigs;* stipules deltoid, lanceolate or subulate, conspicuous:

Robust, villous or hirsute; leafl. *villous,* the medial sharply 3-toothed ... ... (19) **octandra.**

Robust, *glabrous* (except the twigs and stipules); medial leafl. bluntly 3-toothed ... ... (22) **dentata.**

Much branched; leafl. pubescent or ciliate, *small,* the medial sharply 3-toothed ... ... (21) **filicaulis.**

Slender, trailing; twigs filiform, elongate:

Lvs. trifoliolate; leafl. glabrous, the medial obcordate, 1-toothed ... ... (20) **gracilis.**

Lvs. *unifoliolate;* leafl. hairy, the medial sharply 3-cuspidate ... ... (23) **tricuspidata.**

(β.) Leaflets similar, *entire;* lanceolate, linear, or obovate, *flat or flattish.*
Leafl. villoso-pubescent or silky:
Roughly villoso-pilose; leafl. ovate or lanceolate (18) **polygonifolia,** *a.*
Rusty tomentose; lvs. scattered; leafl. cuneate-spathulate, concave ... ... (24) **concavifolia.**
Silky and silvery; leafl. linear or spathulate, obtuse, flat above ... ... (25) **sericea.**
Leafl. glabrous:—
Leaflets linear-lanceolate, very acute and rough edged ... ... (26) **strobilifera.**
Leaflets linear-*falcate,* squarrose, acute or mucronate:
Lvs. *petiolate;* leafl. with reflexed margins (27) **falcata.**
Lvs. subsessile; leafl. quite flat, tapering to the base ... ... (28) **drepanoides.**
Leafl. cuneato-spathulate or obovate or linear, obtuse or subacute:
Leafl. obovate to linear-cuneate, flat or flattish ... ... (29) **serpyllifolia.**
Leafl. narrow-linear, with reflexed margins (30) **linearifolia.**
(γ.) Leafl. similar, linear or lanceolate, *with strongly revolute margins.*
Leafl. glabrous, linear, obtuse; lvs. mostly unifoliolate (31) **ericæfolia.**
Leafl. villous, lanceol.-linear or lanceolate, obtuse, 3–4 lines long ... ... (32) **eriocephalina.**
Leafl. villous, linear, mucronate, 6–8 lines long or more ... ... (33) **sarmentosa.**
(δ.) Leaflets similar, subulate or linear, channelled or *concave above,* round-backed or keeled.
Leaves trifoliolate: leaflets pubescent on both sides, acute ... ... (34) **polycephala.**
Leaves trifoliolate; leafl. glabrous, *pungent-mucronate:*
Leafl. subulate, channelled above; twigs puberulous ... ... (35) **juniperina.**
Leafl. linear-filiform, flat above or furrowed ... (36) **filifolia.**
Leaves trifoliolate; leafl. glabrous, linear, obtuse or mucronulate:
Leafl. 1–2 lines long, obtuse; fruits *globose, berry-like, very smooth* ... ... (37) **baccans.**
Leafl. 4–5 lines long, mucronate; fruits oblong, 6-ribbed and rugulose ... ... (38) **teretifolia.**
Leaves unifoliolate; leafl. glabrous and glossy, pungent, *complicate* ... ... (39) **pungens.**

I. MULTINERVIÆ. (Sp. 1–8.)

1. C. ilicifolia (Linn. Sp. 1469); glabrous; leaves cordate-ovate or elliptic-oblong, many-nerved, rigid, amplexicaul, pungently 3–9 toothed or entire; the margin smooth; stipules sheathing, shortly subulate or aristate. *Thunb. Cap.* 436, *DC. Prod.* 2, *p.* 595. *E. & Z.!* 1719. *Dill. Elth. fig.* 35. *Linn. Hort. Cliff. t.* 30. *Drege,* 6830, 6831. *C. rubricaulis, Presl.*

VAR. β. **cordifolia**; leaves quite entire or 3-toothed, cordate. *C. cordifolia, Lam. DC. l. c.*

VAR. γ. **incisa**; leaves narrow-oblong, deeply inciso-lobate, the lobes broadly subulate. *Drege!* 9538.

HAB. Langekloof, *Thunberg!* Banks of Zwartkops R. and Vanstaadensberg, Uit., *E. & Z.! Drege!* Silverberg, *W.H.H.* Camtous R. *Dr. Gill!* (Herb. Th., D. Bth., Hk., Sd.)

A rigid, much branched bush, 1–3 feet high. Branches flexuous, closely imbricate, with flat or concave leaves. Leaves 5–7 lines long, 4–6 lines wide, glaucescent. Sepals 5–7 striate, rigid, acute. The ordinary form varies occasionally with perfectly entire leaves, and often with 3-toothed; and so joins on to var. β, which has no other distinctive character. Var. γ. has much narrower and more deeply incised leaves.

2. C. intermedia (E. & Z.! 1721); glabrous; leaves lanceolate-oblong, 5–7 nerved, rigid, pungent-mucronate, sharply 3-toothed at the apex; the margin scaberulous; stipules sheathing, shortly subulate. *Drege! No.* 6829.

HAB. East side of Table Mountain above Constantia, *E. & Z.!* Dutoitskloof, *Drege!* (Herb. Sond.)

Perhaps only a variety of *C. ruscifolia*, but the leaves are longer and broader, and less acuminate. I should have supposed it to be *C. tridentata*, Willd., but that the leaves are quite glabrous. Leaves 1–1¼ inch long, 3–4 lines wide, scarcely or not at all amplexicaul, erecto-patent, flat.

3. C. ruscifolia (Linn. Sp. 1469); young parts villous, becoming glabrous; leaves lanceolate or ovate-lanceolate, acuminate, 5–7 nerved, rigid, sessile, pungent-mucronate, concave, scaberulous at the margin, entire or sharply tridentate; stipules amplexicaul, shortly subulate or abortive. *Thunb. Cap. p.* 435. *DC. l. c. E. & Z.!* 1720. *Drege,* 6832.

VAR. β. tridentata; almost all the leaves sharply tricuspidate, silky beneath. *Zey.!* 2451. *C. tridentata, Willd.! DC. Prod. l. c.*

HAB. Western districts, very common. (Herb. D., Sd., Bth., Hk., &c.)

A very densely leafy, much branched bush, with more or less persistent pubescence, occasionally quite hoary, and sometimes perfectly glabrous. Leaves very closely set or fascicled, ½ inch long, 2–4 lines wide, generally much acuminate, very sharp.

4. C. cordifolia (E. Mey.! non Lam.); glabrous; leaves broadly cordate, pungent-acuminate, many nerved, rigid, amplexicaul, spinoso-ciliate; stipules sheathing, shortly subulate.

HAB. Dutoitskloof, 1000–2000 f., *Drege!* (Herb. Hk., Bth., D.)

A virgate shrub, with pale brown bark. Leaves horizontally patent, often infolded, nearly an inch long, 9–10 lines wide, closely set, but not imbricating. The margins are set with many rigidly spinous cilia.

5. C. grandifolia (E. & Z.! 1722); glabrous; leaves broadly oblongo-lanceolate, many nerved, rigid, amplexicaul, imbricate, *flat,* pungent-mucronate and *spinoso-ciliate;* stipules setaceo-subulate. *Zey.!* 2455.

HAB. Mountain sides, Puspas Valley, Swell., *E. & Z.!* Voormansbosch, *Zeyher!* (Herb. Sd., D., Hk., Bth.)

Robust, not much branched, very densely imbricated with leaves. Leaves 2–4 inches long, ¾–1 inch broad, shining above, 9–12 nerved, erecto-patent. Flowers ½ inch long; sepals coriaceous, boat-shaped, cuspidate.

6. C. Meyeriana (Presl.!); glabrous; leaves linear-lanceolate, 5–7 nerved, rigid, patent or squarrose, concave, pungent-mucronate and sharply 2–4 toothed; calyx of the fruit ribbed; stipules subulate. *Drege,* 1127.

HAB. Gnadendal, *Drege!* (Herb. Hook., Sond.)

Suffruticose, closely leafy; the leaves often recurved, 1½–2 inches long, 3–4 lines wide, distantly spinous-toothed and much acuminate. Stipules elongate, very slender.

7. C. Dregeana (Presl.!); glabrous; leaves linear-lanceolate, 5–7 nerved, rigid, patent or squarrose, concave or involute, pungent-mucronate, very entire, with a thickened, scaberulous or smooth margin; stipules setaceo-subulate. *Drege!* 1126, 2927.

HAB. Dutoitskloof, and between Bergvalei and Langevalei, *Drege!* (Herb. Sd., Hook., D.)

Very like *C. Meyeriana*, and perhaps merely an entire-leaved variety. Leaves 1–1½ inch long, 2–3 lines wide, mostly convolute, the upper ones, stipules and male flowers, often ferruginous.

8. C. graminea (Linn. f. suppl. 429); glabrous; leaves linear, acuminate, 3–5 nerved, flexible, erect, flat, aristate-mucronate, spinoso-ciliolate, *petiolate;* petiole flat and broad, winged by the adnate stipules, whose points are subulate. *DC. l. c. p.* 595. *Sieb.!* 40, 86. *E. & Z.!* 1723.

HAB. Common in the Western districts, in moist places, among *Restios;* also near Grahamstown, *E. & Z.!* (Herb. Sd., D., Hk., Bth., &c.)

A suffrutex with closely set and singularly grass-like leaves, which are 2–4 inches long or more, and 2–3 lines wide, on petioles varying from ½ inch to 2 inches long. The marginal cilia are very erect or appressed. The subulate points of the stipules are either squarrose or erect.

2. **DICHOPTERÆ.** (Sp. 9–11.)

9. C. odorata (Linn. f. suppl. 431); variably villous, tomentose, or glabrate; leaves subsessile, cordate-ovate, obtuse, plaited, crenato-serrate, the serratures mucronate or aristate; nerves prominent beneath. *DC. l. c. p.* 595. *Thunb.! Cap.* 436.

VAR. α. **vera**; leaves coriaceous, the younger villous or lanose, the old glabrate, *E. & Z.!* 1724. *Drege,* 6827.

VAR. β. **hypoleuca**; leaves coriaceous, *densely albo-tomentose beneath.* *C. hirsuta. E. & Z.!* 1726. *Drege!* 6828.

VAR. γ. **reticulata**; leaves membranaceous, cordate-ovate or subrotund, glabrate, mucronato-crenate. *C. reticulata, E. & Z.!* 1725.

HAB. α. Mountains round Capetown; Hott. Hollandsberg, and near Cape Recief, Uit., *E. & Z.!* &c. Klapmuts, Stell. β. Riv. Zonderende, Swell., *E. & Z.!* Sir Lowry's pass, *Zey.!* 2454. Gnadendal, *Drege!* γ. Winterhoeksberg, Tulbagh, *E. & Z.!* (Herb. Th., Hk., Bth., D., Sd.)

Diffuse or trailing, laxly branched, varying in the size and rigidity of the leaf, and in pubescence, but always known by its plaited, cordate and strongly penninerved leaves. In the common form (*vera*) the pubescence is sometimes very copious and sometimes scanty, but never *persistent;* in β. the undersides of the leaves are white with short tomentum; in other characters this is the same as α. Var. γ. is more slender and weak in all parts, with thinner leaves and larger crenatures; but its characters are scarcely tangible.

10. C. ferruginea (Linn. f. Suppl. 429); mostly glabrous; leaves linear-oblong or lanceolate, tapering to the base, obtuse or acute, spinoso-serrate beyond the middle. *DC. l. c.* 595. *C. berberifolia, Lam. Dict.* 1, 48. *C. serrata, Thunb. Cap. p.* 436.

VAR. β. **flexuosa**; stems flexuous; leaves smaller and narrower, squarrose or recurved, often subentire. *C. flexuosa, E. Mey.!* *Sieb. No.* 87.

VAR. γ. **villosa**; villous; leaves oblong or obovate, scarcely acute at base. *Drege!* 6835.

HAB. Cape flats and western districts, common. (Herb. Hk., Bth., D., Sd.)

Diffuse or trailing, the branches glossy brown (except in γ.), more or less flexuous. Leaves fascicled, very variable in size and shape, but mostly narrowed to the base; the larger 1 inch long, ¼–½ inch wide; the smaller half inch long, and 1–2 lines wide. *Drege's* 6836, of which I have seen but a scrap, may belong to this; but the leaves are either quite entire or tridentate.

11. C. cuneata (Ait. Hort. Kew. 3, p. 413); glabrous; leaves narrow-cuneate, truncate, coarsely 3–5 toothed at the apex, veiny. *DC. l. c.* 595. *E. & Z.!* 1728.

HAB. Klapmuts, Stell., *E. & Z.!* Dutoitskloof and Gnadendal, *Drege!* (Herb. Hk., Bth., D., Sd.)

A robust, much branched, virgate, densely leafy shrub, 3–6 feet high. Leaves sessile, erect, 1–1½ inch long, a line wide at base, gradually widening to the abruptly toothed summit, which is 3–4 lines wide; the veins obvious. Stipules very small, toothlike. Fruit calyces ribbed.

III. **BIFOLIOLÆ.** (Sp. 12–13.)

12. C. crenata (Linn. f. Suppl. 430); leaflets in pairs, close-pressed together, flat, imbricating, orbicular-reniform, immersedly 5–7 nerved and netted-veined, spinuloso-denticulate. *Thunb. Fl. Cap. p.* 437. *DC. l. c. p.* 596. *E. & Z.!* 1729. *Harv. Thes. t.* 95.

HAB. Mountains near Hex River, *Thunberg! Drege!* Kochmanskloof, *E. & Z.!* South side of Genadendal mountain, *Dr. Roser!* (Herb. Th., Hk., Bth., D., Sd.)

A shrub, 3–4 feet high, not much branched; branches 1–2 feet long, undivided, imbricated with distichous, bifoliolate leaves. Leaflets ¾–1 inch broad, glabrous and somewhat glaucous, with a cartilaginous, scabrously denticulate margin. Sepals ovate, reticulate.

13. C. pulchella (Linn. f. Suppl. 430); leafl. in pairs, close-pressed together, flat, imbricating, orbicular, prominently and flabellately many-nerved, minutely scaberulous at the margin. *Thunb. Cap. p.* 437. *DC. l. c.*

HAB. Hartequaskloof, near Safrankraal, *Thunberg!* S. Africa, *Burman! Forsyth!* (Herb. Bth., Sd.)

A small, much branched shrub, 2 feet high, nearly allied to *C. crenata*, but with smaller, more strongly and closely nerved, and entire (though *rough*) edged leaflets. Leaflets 3–4 lines long, 4–5 lines wide, drying fulvous, veined like the frondlets of an *Adiantum*. The most elegant of the genus and one of the rarest.

IV. **TRIFOLIOLÆ.** (Sp. 14–39.)

14. C. obcordata (Linn. f. Suppl. 429); leaves trifoliolate; leaflets glabrous, flat, with immersed veins, scaberulous at the margin, the lateral ones obovate-oblong, roundish or reniform, the medial deeply obcordate, mucronulate. *Thunb.! Cap.* 437. *DC. l. c.* 596. *E. & Z.!* 1734. *C. obliqua, Spreng. DC. l. c.*

HAB. Common in the Cape District. (Herb. Th., Hk., Bth. D., Sd.)

A much branched and ramulous shrub, 2–4 feet high. Leaves scattered or crowded; leaflets ¼–½ inch long and broad. The terminal leaflet is pretty constant in form, but varies much in size, being largest when the lateral are smallest, and much depauperated when these are large. These latter vary in form as well as size, passing from narrow-obovate to broadly reniform, thus becoming almost like those of *C. crenata!* It is worthy of remark that, in such leaves the edge is rougher than in the normal form. I presume they are what Sprengel describes as *C. obliqua.* Stipules minute, toothlike, or obsolete.

15. C. obovata (E. Mey.!); glabrous, slender, ramulous; leaves trifoliolate and fascicled; leaflets one-nerved beneath, veinless, flat, glaucescent, smooth-edged, shortly and broadly obovate, obtuse, the lateral ones oblique, the medial equal-sided, mucronulate; stipules obsolete.

HAB. Gnadendal, *Drege!* (Herb, Hk., Bth., Sd., D.)

A small, divaricately branched or flexuous and twiggy bush, glabrous in all parts. Leaves tufted at short intervals; the outermost at least 3-leaved. Leaflets 2–2½ lines long, 1–1½ wide, quite even above, immersedly 1-nerved beneath. Young fruits glabrous, striate, not ribbed, crowned with 3 lanceolate calyx-lobes. With broader and more obovate leaflets than *C. marginata,* to which it seems allied. It may also be compared to a miniature *C. obcordata,* but wants the strongly obcordate medial leaflet.

16. C. marginata (E. & Z.! 1740); glabrous, slender, diffuse, ramulous; leaves trifoliolate and fascicled; leaflets nerve-keeled beneath, veinless, concave, *white-edged,* spathulate-oblong, subacute, nearly equal, the lateral slightly oblique; stipules obsolete.

HAB. Shrubby hill sides at Somerset, Hott. Holland, Stell., *E. & Z.!* (Herb. Sond., Bth.)

1½–2 feet high, diffuse; twigs flexuous or zig-zag, 6–8 inches long; leaf-tufts about an inch apart, the outer leaf 3-foliolate. Leaflets 2–3 lines long, 1 line wide, channelled or half-complicate, with a variably wide membranous and pellucid margin. Fruits glabrous, striate.

17. C. triloba (Harv.); glabrous or puberulous; leaves *shortly petiolate,* trifoliolate; leaflets shortly and broadly cuneate, truncate, immersedly veiny, all deeply 3-toothed or lobed, with subreflexed margins, pale beneath; stipules toothlike, spreading. *C. dentata, E. & Z.!* 1735, *non Willd.*

HAB. Heathy ground on mountain sides near Brackfontein, Clanw., *E. & Z.!* Aug. (Herb. Bth., Sd., D.)

A slender shrub, with flexuous twigs and reddish bark, perfectly glabrous, except the young parts, which are thinly clothed with deciduous, *microscopic,* appressed, whitish hairs. The leaflets resemble those of *C. trifoliata,* but are neither scabrous nor rugose, and there is always an evident *petiole,* prolonged from 1–3 lines beyond the small, rigid stipules. Leaves scattered or tufted; leaflets 4–5 lines long, 3 lines wide at top.

18. C. polygonifolia (Linn. Sp. 1470); roughly villous and pilose; leaves sessile, trifoliolate; leaflets one-nerved, villous or pilose, scaberulous, piloso-ciliate, ovate or lanceolate, the medial similar to the lateral or cuneate and sharply 2–3 fid; margins subreflexed; stipules minute, toothlike. *Drege,* 6826.

VAR. α. **ternata**; leaflets all subequal and uniform, ovate, oval oblong, or lanceolate. *C. ternata, Linn. f. Suppl.* 430. *DC. l. c.* 596. *Hort. Cliff. t.* 32. *Thunb. Cap.* 438. *E. & Z.!* 1732. *Sieb.* 91. *Drege!* 6840. *Zey.* 2661.

VAR. β. **trifoliata**; leaflets mostly dissimilar, the medial more or less cuneate, 2–3 toothed or lobed, the lateral oblong or lanceolate, entire, but sometimes cuneate and 3-lobed. *C. trifoliata, Linn. Sp.* 1470. *Pluck. Alm. t.* 319, *f.* 4. *DC. l. c. E. & Z.!* 1733. *Drege!* 6826, 6822? 6825.

HAB. Common throughout the western districts, both varieties. (Herb. D., Hk., Bth., Sd., &c.)

An erect, strong growing, much branched, and densely ramulous, roughly pubescent bush, 2–4 feet high. Leaves crowded, very generally fascicled. Leaflets very variable in shape, 3–6 lines long, ½–4 lines wide. After examining large suits of specimens, I find it impossible to keep up the two for msabove indicated, except as *tolerably* constant varieties. Specimens *true* to either descriptive phrase may readily be found; but strictly intermediate forms are common also. Some specimens before me, which would pass for "*C. ternata,*" have a few bidentate or 3-dentate medial leaves; and other specimens of "*C. trifoliata*" vary occasionally with all uniform entire leaves. Some again have very broad, some very narrow leaflets, &c.

19. C. octandra (Ch. & Sch. in Linn. 9, 350); much branched, diffuse, villous or hirsute; stipules withering, broadly subulate, leaf-like, one-nerved, spreading; leaves trifoliolate, leaflets villous, coriaceous, with immersed veins, the lateral ones oblong or lanceolate, entire or toothed, the medial cuneate, obovate, sharply 3-toothed. *Ch. & Sch. in Linn.* 8, 55. *Herb. Un. It.* 190. *E. & Z.!* 1730. *C. serpyllifolia, E. Mey.! non Ch. & Sch.*

HAB. Near the Salt River, Cape; also on the Zwarteberg, near Caledon Baths, *E. & Z.!* Kleinriviersberg, *Zey,!* 2459. Koratra, *Drege!* (Herb. Hk., Bth., Sd., D.)

Not unlike *C. polygonifolia*, var. *trifoliata* in foliage, pubescence, and general aspect, but readily known by the large, persistent, but withering stipules. Stems sometimes shaggy, with long, white, soft hairs; sometimes denuded. The bark is rough and loose on old stems. Leaflets 3-5 lines long, 1-2 lines wide, softly villous when young, afterwards rough.

20. C. gracilis (Harv.); stems slender, trailing, pubescent; stipules membranous, lanceolate; leaves trifoliolate; leaflets glabrous, flat, veiny, the lateral ones obliquely oval or obovate, subrepand, the medial obcordate, unidentate. *C. dentata, E. Mey! non Willd.*

HAB. Dutoitskloof, 2-3000 f,, *Drege!* (Herb. Bth,, Sd.)

This little plant comes nearest to *C. filicaulis*, but has more membranous and veiny leaflets, differing also in shape and venation. Stem 6-12 inches long, flexuous. Leaflets 2-3 lines long and broad, pale green.

21. C. filicaulis (Ch. & Sch. in Linn. 2. p. 33); stems much branched, prostrate; twigs pubescent; stipules membranous, lanceolate or broadly subulate; leaves trifoliolate; leaflets ciliate or pubescent, coriaceous, with immersed veins, the lateral ones oblong, entire, the medial obovate, sharply 3-toothed or 3-lobed. *E. & Z,!* 1731.

HAB. Near Tulbagh, *Mundt.!* (Herb. Sond.)

Stems woody, densely ramulous. Leaves tufted or scattered. Leaflets 2 lines long and wide, often pale beneath, and with slightly reflexed margins, variable in pubescence, of a deep green above. Stipules rufescent, conspicuous. Of this I have only seen a single specimen. The *C. filicaulis* of Hb. Drege, our *C. tricuspidata*, seems distinct.

22. C. dentata (Willd! 4, p. 842); stems much branched, prostrate, ramulous; twigs pubescent; stipules membranous, broadly subulate, one-nerved; leaves trifoliolate; leaflets *glabrous*, coriaceous, veiny, the the lateral ones obovate, entire or 1-2 toothed, the medial obovate, bluntly 3-lobed, the lobes mucronulate. *DC. Prod.* 2, *p.* 596.

HAB. S. Africa, *Heyne!* Devil's Mt. Capetown, *Dr. Alexander Prior!* (Herb. r. Berol., D.)

Stems 1-2 feet long, woody, spreading on the ground, closely branched and ramulous; twigs short, patent. Leaves mostly tufted, close; leaflets 2-3 lines long and broad, the medial one broadly obovate, with three broad, short, terminal lobes and sometimes a pair of lateral teeth; the lateral either obliquely one-toothed or 2-3-toothed, or entire. Stipules and flowers reddish. Dr. Prior's specimens quite agree with the original described by Willdenow. Though growing close to Capetown, this species seems to have escaped the notice of almost all collectors.

23. C. tricuspidata (Harv.); stems slender, trailing, pubescent or villous; stipules triangular, one-nerved; leaves *unifoliolate*, tufted; leaflet loosely hairy, flat, membranous, veiny, elliptical or obovate, sharply tricuspidate; the points recurved, pungent; stamens few; sepals obovate, cuspidate. *C. filicaulis, E. Mey.! in Herb. Drege, non Schl.*

HAB. Dutoitskloof, 2–3000 f. and Drakensteensberg, 4–5000 f. *Drege.* (Herb. Sd., Bth., D.)

Stems 6–12 inches long, filiform, trailing; the branches long, subsimple, flexuous. Pubescence loose, at first copious, but deciduous. Leaves generally 3 or more in small tufts, always unifoliolate; the leaflet 3–4 lines long, 2 lines wide, most commonly with three, but now and then but two strongly recurved, pungent, terminal teeth. Stipules twice as long as broad, adnate, scarcely amplexicaul. With the aspect of *C. filicaulis*, but seemingly well distinguished by difference in foliage and stipulation.

24. C. concavifolia (E. & Z.! 1739); robust, much branched, ramulous, ferrugineo-tomentose; leaves trifoliolate, sessile, scattered; leaflets one-nerved beneath, veinless, concave, thick, silky pubescent on both sides, cuneato-spathulate, acute, nearly equal; stipules obsolete.

HAB. Sides of the Zwarteberg Mountains, Caledon, *E. & Z.!* (Herb. Sond.)

A dwarf, but strong growing bush, a foot or so in height, very densely branched, and minutely twiggy, the old twigs warted with leaf-bases: all the twigs thickly clothed with rusty tomentum. Leaves sessile: leaflets 2 lines long, scarcely 1 line wide, with a thick, prominent midrib, and blunt margins: the upper surface not very concave.

25. C. sericea (E. & Z.! 1746); robust, erect, twiggy, albo-sericeous; leaves on short, membrane-winged petioles, trifoliolate and tufted; leaflets one-nerved beneath, veinless, flat above, thick, silky on both sides, linear or spathulate, obtuse, callous-tipped, the broader with subrevolute margins; stipules adnate, with a setaceo-subulate, excurrent point.

HAB. Rocky places near the Tulbagh Waterfall, *E. & Z.!* (Herb. Sond.)

1½–2 feet high; all parts silky and silvery. Twigs 4–6 inches long, very erect; the nodes ½–¾ inch apart. Stipulated petiole of the outer leaf sheathing at base, 1–2 lines long; leaflets 3–4 lines long, ½–1 line wide, the margin but little reflexed. Stipule points persistent, rather rigid.

26. C. strobilifera (Linn. Syst. 749); robust, glabrous, with virgate twigs; leaves tufted, unifoliolate, the primary often abortive, the stipules remaining; leaflets linear-lanceolate, subaristate, one-nerved beneath, rough-edged, rigid; stipules amplexicaul, with subulate points; fruits striate. *Thunb.! Cap. p.* 435. *E. & Z.!* 1753. *Pluck. Alm. t.* 275, *f.* 2. *DC. l. c.* 596. *Drege,* 6833, 6834?

HAB. Common throughout the colony and through Caffraria to Port Natal. (Herb. D., Sd., &c.)

A large shrub, 6–10 feet high, much branched and twiggy. The rameal leaves are either much reduced in size and quickly deciduous, or abortive altogether, their stipules, adnate to a petiole, remaining, and prolonged beyond the petiole into two awl-shaped points. Leaflets clustered in the axils of these stipules 1–1½–2 inches long, 1–2 lines wide, flattish, The specific name alludes to conelike *galls,* composed of broad, imbricated scales, common on the branches and twigs.

27. C. falcata (Linn. f. Suppl. 431); diffuse, ramulous, with tomentulose twigs; leaves *shortly petiolate,* fascicled, 3-foliolate; leaflets one-nerved beneath, flat above, *with reflexed margins,* glabrous, linear-falcate, acute, squarrose, subequal or the medial shorter; stipules membranous, subulate, basal. *DC. l. c.* 596. *Thunb. Cap. p.* 436. *E. & Z.!* 1742. *Sieb. No.* 89.

HAB. Cape Flats and near Simonstown, *E. & Z.!* &c. Bases of Table and Devil's Mountain, Capetown, *Drege!* (Herb. Sd., Hk., Bth., D.)

A small, flexuous bush, 1–1½ foot high, glabrous, except on the twigs and young branches. Leaves tufted; all on evident, channelled, unwinged petioles 2–4 lines long, with small membranous stipules at their base. Leaflets 2–6 lines long, not a line wide, the longer ones strongly falcate, especially when dry; the margins thickened and reflexed. Colour a dull green, inclining to rusty.

28. C. drepanoides (E. & Z.! 1741); flexuous, with puberulous twigs; leaves subsessile or *very* shortly petioled, fascicled, 3-foliolate; leaflets faintly one-nerved beneath, flat, glabrous, linear, *attenuate at base*, mucronate, subfalcate, nearly equal or the medial smaller; stipules deltoid, bristle-pointed. *C. falcata, Spreng. in Herb. Zey.* 81. *Zey.!* 2458.

HAB. Winterhoeksberg, Uit., *E. & Z.!* Koega River, *Zeyher!* (Herb. Sd., D., Hk.)

Not much branched, 8–12 inches long, leafy. Leaflets 10–12 lines long, 1–1½ broad, perfectly flat on both sides, the longest more or less curved, all squarrose. This has much longer, broader, and flatter leaves than *C. falcata*, shorter or obsolete petioles, and broader and more rigid stipules.

29. C. serpyllifolia (Ch. & Schl. in Linn. 2, p. 34); twigs tomentose or pubescent; leaves fascicled, sessile; leaflets glabrous, subacute or obtuse, cuneato-spathulate or obovate, flat, midribbed beneath, the broader ones also penninerved; stipules setaceous.

VAR. α. **pennineryis**; leaflets broadly obovate, attenuate at base, entire or subtridentate, pale and penninerved beneath; twigs thinly downy. *C. propinqua, E. & Z.!* 1736, *and C. serpyllifolia, β. E. & Z.!* *Drege*, 6817, 6818.

VAR. β. **Chamissonis**; leaflets cuneato-spathulate or linear-cuneate, flat or flattish, faintly penninerved or nearly nerveless beneath, 3–6 lines long; twigs tomentose. *C. serpyll. α. E. & Z.!* *Drege*, 6819, 6820. *C. complanata, E. Mey.*

VAR. γ. **polyphylla**; leaflets small, very blunt, obovate or cuneate-linear, with slightly revolute margins, midribbed beneath, usually veinless, sometimes obscurely veined. *C. polyphylla, E. & Z.!* 1738, *also C. serpyll. var. γ. E. & Z.!*

HAB. Alpine and subalpine situations in the scrub. α. Cederberge, Clanw., and Puspas Valley, Swell., *E. & Z.!* β. Zwarteberg, Caledon, *E. & Z.!* Dutoitskloof, *Drege!* γ. Wagenmacher's Valley, *Mundt!* Moll River, Grahamstown, *E. & Z.!* (Herb. Sd., Bth., D., Hk.)

Vars. β. and γ. are erect, much branched, densely leafy bushes, 2–3 feet high; α. is smaller, more diffuse, or flexuous, and more glabrous. Typical specimens of the three look quite distinct, but many intermediate stages connect the broadest and most nerved leaflets with the narrowest and nerveless. Var. γ. in the same locality varies with narrow, almost linear, and broad, distinctly obovate leaflets; some of the Grahamstown specimens precisely resemble those of Mundt, from Wagenmacher's Valley. Leaflets 2–6–7 lines long, 1–3 wide.

30. C. linearifolia (E. & Z.! 1749); erect, ramulous, with villoso-tomentose, at length naked twigs; leaves trifoliolate and tufted, subsessile; leaflets linear, with reflexed margins, one-nerved beneath, obtuse, callous-tipped, glabrous; stipules membranous, amplexicaul, toothed, villoso-ciliate. *Drege*, 5381, 2353, and 6843.

HAB. Zwartkops River, Uit.; Mountains near Grahamstown and at Klipplaat River, near Silo, Caffr. *E. & Z.!* Between Strandfontein and Matjesfontein; and at Port Natal, *Drege! Dr. Sutherland!* (Herb. Sd., Hk., D., Bth.)

An erect, somewhat virgate and ramulous, straight-stemmed bush, with close-pressed branches, well covered with leaves. Leaftufts close; leaflets 2–3–4 lines long, ¼–½ line wide, varying from exactly linear to linear-oblong, dull green. The narrow leaved forms resemble *C. ericæfolia*, but differ in the tomentose twigs and prominent midrib; the wider leaved come very near *C. serpyllifolia, γ. polyphylla*, but have less expanded, more decidedly reflexed-edged leaflets.

31. C. ericæfolia (Linn. f. Suppl. 430); erect, ramulous, with glabrous twigs; leaves tufted, mostly unifoliolate; leaflets petiolulate, shortly linear, with revolute margins, furrowed beneath, obtuse, glabrous; stipules membranous, sheathing, shortly toothed, ciliate. *DC. l. c.* 596. *E. & Z.!* 1748. *Thunb. Cap. p.* 433.

HAB. Between Capetown and False Bay, *Thunberg!* Cape Flats near Doornhoogde, *E. & Z.!* (Herb. Sd., Bth.)

An erect, much branched, and ramulous shrub, 2–3 feet high, glabrous, except the stipules; the twigs with smooth, reddish brown bark, the old branches cinereous and rough. Leaflets alternate, ¼ inch apart, with brown, fringed stipules; leaflets erect, on very short petioles, 1–2 lines long, ½ line wide, convex above, the edge so rolled back as to cover the whole under surface, leaving a narrow furrow.

32. C. eriocephalina (Cham. & Schl. in Linn. 6. 349); robust, erect, ramulous, villoso-pubescent; leaves on *very short*, membrane-winged petioles, trifoliolate and tufted; leaflets veinless, *convex above, with strongly revolute margins*, lanceolate-linear or lanceolate, obtuse, coriaceous, villoso-pubescent, afterwards glabrate; stipules short and toothlike. *E. & Z.!* 1745. *Also C. phylicoides, E. & Z.!* 1744. *Herb. Un. It.* 192. *C. ericæfolia, E. Mey.! in Herb. Drege, and Drege* 6841.

HAB. Moist places round Table Mt., and on the Winterhoek, Tulbagh, *E. & Z.!* Dutoitskloof, *Drege!* (Herb. Sd., Bth.)

A robust, softly pubescent shrub, 2–3 feet high, much and densely branched. Leaves closely set, more or less fascicled. Stipulated petioles 1 line long, rust coloured; leaflets 3–4 lines long, ½–1 line wide, sometimes with completely revolute sides, concealing the whole under surface; sometimes partly open beneath; the midrib rarely visible. All the younger leaves are softly hairy; the older often smooth and glossy. *E. & Z.'s C. phylicoides* has rather wider and more open leaves, but does not otherwise differ. *Drege's* 6841 may be either a young plant or the growth from an old root, after burning over; its leaves are still broader, flatter, and less coriaceous.

33. C. sarmentosa (Linn. Mant. 299.); diffuse, not much branched, villoso-pubescent; leaves subsessile, trifoliolate and fascicled; leaflets linear-terete, *with revolute margins*, calloso-mucronate, villous (the older often glabrate); stipules membranous, toothlike. *DC. l. c.* 596. *E. & Z.!* 1743. *Thunb. Cap.* 439. *Drege,* 6844. *Zey.!* 534.

HAB. Cape Flats and foot of Muysenberg, &c., *E. & Z.!* (Herb., Sd. D. Hk. Bth.)

A straggling suffrutex, 1–2 feet high, with a few long, subsimple, virgate branches, all parts (except old branches and leaves) softly hairy. Leaflets filiform, by the rolling back of the lamina, 6–8 lines long or more, straight or curved, or squarrose, the young ones almost hoary. Stipules brown, hairy, with small points.

34. C. polycephala (E. Mey.!); robust, ramulous; twigs puberulous; leaves tufted, trifoliolate, subsessile; leaflets linear, acute, subpungent, flattish above, nerve-keeled beneath, *appressedly pubescent on both sides;* stipules subulate, amplexicaul.

HAB. Gnadenthal, 2000–3000 f. *Drege!* (Herb. Hk., Bth., D., Sd.)

Very few and imperfect specimens seen. It seems to differ from *C. juniperina* in the shorter and flatter (occasionally furrowed), less obviously mucronate and regularly pubescent leaflets. Leafl. 3–4 lines long, ½ line wide. Fruits glossy brown, with distant, broadish ribs.

35. C. juniperina (Linn. f. Suppl. 430); robust, diffuse or divaricately much branched or ramulous; twigs puberulous; leaves trifoliolate and tufted, subsessile; leaflets linear-subulate, mucronate, incurved, keeled beneath, somewhat channelled above, glabrous or scaberulous; stipules amplexicaul, subulate; fruits *varying from nearly smooth to striate ribbed, wing-ribbed, tuberculate and muricate!! DC. l. c.* 596. *Thunb. Cap.* 434. *E. & Z.!* 1750. *C. laricina, E. Mey.!* & 6837.

VAR. *α.* **vulgaris**; leaflets 6–9 lines long, smooth or rough-edged; fruits striate or ribbed. *C. laricina, a.*

VAR. *β.* **brevifolia**; leaflets 3–5 lines long; branches divaricate; fruits . . . *C. junip. β. E. & Z.!*

VAR. *γ.* **pterocarpa**; leafl. 4–6 lines long, straight, scaberulous; fruits with raised, rough, sharp ribs.

VAR. *δ.* **tuberculata**; leafl. 4–6 lines long; fruits tuberculated.

VAR. *ε.* **muricata**; leafl. 4–6 lines long; fruits muricated! *C. laricina, b., E. Mey. and Drege!* 6839.

HAB. Common about Capetown and in the Western Districts. *β.* above the source of the Kat River, on the Winterberg, *E. & Z.!* *δ.* Witsenberg, *Zeyher!* *ε.* Dutoitskloof and Drakensteenberg, *Drege!* (Herb. Sd., Hk., Bth., D.)

A slight growing, diffuse or depressed, densely leafy, much branched bush, 1–2 feet high. Branches curved; leaflets close. Leaflets often rusty, sometimes pale green, 3–9 or 10 lines long, ¼–½ line wide, somewhat narrowed at base, and tapering to a pungent mucro. The leaflets do not vary much, except in length and smoothness, but the *fruit* is remarkably variable. I have ventured to indicate some of the principal forms as varieties, but I fear they are very inconstant; on the same branch the degree of furrowing is variable, and sometimes the ribs have smooth, sometimes tuberculated interspaces. Var. *ε.* has such distinct looking fruits, that but for *δ*, I should probably have made it a species!

36. C. filifolia (Linn. f. Suppl. 430); robust, diffusely branched, and ramulous; twigs glabrous or puberulous; leaves trifoliolate and tufted, subsessile; leaflets linear-filiform, subtrigonous, mucronate, incurved, keeled beneath, lightly furrowed or flat above, glabrous or scaberulous; stipules amplexicaul, subulate; fruits nearly smooth, striate or ribbed. *E. & Z.! No.* 1751. *Thunb.! Cap. p.* 434. *C. leptophylla, E. & Z.!* 1752. *Zey.!* 2717 & 573. *Drege,* 6838, 6846.

VAR. *β.* **subsetacea** (E. & Z.!); twigs puberulous; leaflets short (2–3 lines long), pale green, very slender. *Drege!* 6845.

HAB. Cape Flats and about Table Mountain; also about Hauwhoek Pass and Caledon, &c. *E. & Z.!* Breede River, *Burke and Zeyher!* (Herb. Hk., Bth., Sd., D.)

Very similar to *C. juniperina,* but usually with glabrous twigs and much more slender, less tapering, more suddenly mucronate, not so channelled or quite flat above leaflets. Still puzzlingly intermediate states occur, and I fear the two are not sufficiently distinct. I do not see anything to keep *C. leptophylla, E. & Z!* apart from the ordinary state; the roughness or smoothness of the margin is very inconstant. Stems 6–12 inches long or more; leaflets commonly 6–8 lines long, except in *β*, which is smaller in all parts.

37. C. baccans (Harv.); nearly glabrous, ramulous; leaves trifoliolate, tufted; leaflets shortly linear-semiterete, obtuse, flattish above, round-backed, glabrous; stipules obsolete, toothlike; fruits *globose, berry-like, very smooth,* crowned with the ovate calyx-lobes. *Drege,* 552.

HAB. Hexriviersberg, 3000–4000 f., Hexriv. kloof, 1000–2000., *Drege!* (Herb. Hk., Bth., Sd.)

Something like *C. teretifolia* in miniature; but the fruit is remarkably unlike that of any other species, Judging from the dried specimens, it seems to have been almost drupaceous. Leaflets 1½–2 lines long. Fruit reddish, glossy, longer than the subtending leaves.

38. C. teretifolia (Thunb. Prod. 93); glabrous, *virgate;* leaves sessile, trifoliolate and fascicled; leaflets semiterete (flattish above, round backed), linear, incurved, mucronate; stipules obsolete, tooth-like; fruits oblong, 6-ribbed and rugulose. *Thunb. Cap. p.* 433. *E. & Z.!* 1747. *DC. l. c. C. teretifolia, β. tenuior, E. Mey.*

HAB. Piquetberg, *Thunberg!* Brackfontein and Olifant's R., *E. & Z.!* Between Krom River and Berg Valley, *Drege!* (Herb. Sd., Bth., D., Hk.)

A tall, virgate, slightly branched shrub, with very erect, simple branches, not ramulous. Leaftufts ½–1 inch apart, alternate, Leaflets 4–5 lines long, not ½ line in diameter, carnose, nerveless, and veinless, with a short, acute mucro, smooth edged. Fruit finely wrinkled between the six smooth ribs.

39. C. pungens (Presl. Epimel. 202); glabrous, robust, much branched and ramulous; leaves tufted, unifoliolate; leaflets patent or recurved, shortly linear, pungent-mucronate, *complicate*, glabrous and glossy; stipules shortly amplexicaul, toothlike. *C. teretifolia, E. Mey. ex pte.*

HAB. Gnadendal, Mt., 4000–5000 f., *Drege!* (Herb. Hk., Bth., Sd., D.)

A much branched twiggy bush, 1–2 feet high, or perhaps more. Leaftufts close. All the leaves unifoliolate. Leaflets 2–3 lines long, of thick substance, infolded and deeply channelled, squarrose, with a long sharp point. This differs from *C. teretifolia* in habit, in the unifoliolate leaves and infolded leaflets, and longer mucro, &c. Fruit smooth and glossy.

X. GRIELUM, Linn.

Calyx-tube short, at length concrete with the ovary; limb 5-lobed, the segments nearly valvate in æstivation. *Petals* 5, inserted in the throat of the calyx, alternate with the segments, large, obovate, convolute. *Stamens* 10, inserted with the petals. *Carpels* 5–10, in the base of the calyx, concreting with the tube of the calyx and with each other, uniovulate; styles 5–10, filiform, short; stigma capitate. *Capsule* depressed, 5–10 celled, the cells at length opening in the axis, one-seeded. *Endl. Gen.* 6402. *DC. Prod.* 2, *p.* 549.

South African herbs or scarcely suffruticose plants, growing in sandy places, and in salt ground. Leaves alternate, hoary, pinnately decompound. Peduncles axillary, 1-flowered. Flowers large, yellow. Name *γρῆιος, old;* because the leaves are hoary.

ANALYSIS OF THE SPECIES.

Leaves rigid, multi-partite; the narrow lobes mucronate; sepals narrow-lanceolate ... ... (1) **tenuifolium.**
Leaves soft, pinnatifid; the broadish lobes obtuse:
  Stem and upper sides of leaves laxly woolly; sepals ovate, acuminate ... ... (2) **humifusum.**
  Whole plant very woolly; sepals broadly deltoid, obtuse or subacute ... ... (3) **obtusifolium.**

1. G. tenuifolium (Linn. Gen. 578); leaves bipinnately multifid; pinnæ 2–3, alternate, cut into 3 or more narrow-linear, *callous-mucronate* segments, cobwebby above, cano-tomentose with reflexed edges beneath; calyx-segments *lanceolate-acuminate*, at first cobwebby, then glabrous. *DC. l. c.* 549. *Sw. Ger. t.* 171. *Burm. Afr. t.* 53. *Th. Fl. Cap.* 509. *E. & Z.! No.* 455. *E. Mey.! in Hb. Drege.*

VAR. β. **patens**; leaf-segments broader, more rigid, patent or divaricate. *G. humifusum, E. Mey.! in Hb. Drege, non Thunb.*

HAB. Zwartland, *Thunberg!* Ried Valley, Cape, and near Saldanha Bay, *E. & Z.* Salt R., *W.H.H.* Groenekloof and between Bergvalley and Langevalley, *Drege?* Krumriver, *Zeyher!* (var. β.) (Herb. Hk., Bth., Sd., D.)

Root filiform. Stems many from the crown, procumbent, 8–12 inches long, densely leafy and branched. Leaves 1½ inch long, of which half is petiole; the multifid lamina fan-shaped in outline. Peduncles 1–2 inches long. Flowers large and bright yellow, the petals broadly obovate, over an inch in length and breadth.

2. G. humifusum (Thunb. Cap. 509); leaves pinnatifid or sub-bipinnatifid, pinnati-sections 5–6, broadly linear, obtuse, simple or lobulate toward the point, cobwebby above, cano-tomentose with sub-reflexed edges beneath; calyx segments *ovate, acute or acuminate,* persistently tomentose. *DC. l. c. E. & Z.!* 456. *G. flagelliforme, E. Mey.*

VAR. β. **parviflorum**; leaves and flowers smaller. *Drege!* 7516.

HAB. Sandy places near Saldanha Bay, and Karroo near Beaufort, *E. & Z.!* Olifant Riv., Holriver; also between Kaus, Natvoet and Doornport, and near Verleptpram on the Gariep, *Drege!* (Herb. Hk., Bth., Sd., D.)

Stems trailing, 1–2 feet long, angular, cobwebby, becoming glabrate. Leaves scattered or somewhat fascicled, 2 inches long, of which less than half is petiole; the lamina in outline, ovate or ovato-lanceolate; segments 2–3 lines wide. Flowers much smaller than in *G. tenuifolium;* the petals less than an inch long, bright yellow; in β. ½ inch long.

3. G. obtusifolium (E. Mey.!); stems and whole plant densely covered with white wool; leaves pinnatifid, pinnati-sections 5–6, linear-oblong, short and very obtuse, the lowest sometimes 1–2-lobulate; calyx-segments broadly deltoid, very short, obtuse or subacute, densely and persistently woolly. *Zey.!* 165.

HAB. Nieuweveld, between Brackriver and Uitvlugt; and Silverfontein, *Drege!* Springbokkeel, *Zeyher!* Namaqualand, *Wyley!* (Herb. Hk., Bth., Sd., D.)

Very near *G. humifusum,* but much more woolly, with smaller, less divided leaves, and much shorter and broader calyx-lobes. Flowers nearly as in *G. humifusum.*

ORDER L. SAXIFRAGACEÆ, DC.

(By W. H. HARVEY.)

Calyx 5-cleft (rarely 3–4–7–10-cleft), regular, adnate or free; the sepals rarely separate, commonly cohering below into a tube; the limb mostly persistent, sometimes enlarged in fruit. *Petals* as many as the calyx-lobes and alternate with them, rarely wanting. *Stamens* inserted with the petals, either as many and alternate with them, or twice as many: rarely fewer or indefinitely numerous; anthers 2-celled, splitting. *Ovary* either free or more or less adnate with the calyx-tube, of 2 (rarely of 3 or 5) distinct, or more or less cohering, or altogether connate carpels, whose inflexed edges form the complete or incomplete dissepiments: *placentæ* on the inflexed edges of the carpels; *ovules* anatropous, indefinite, rarely few or subsolitary. *Styles* as many as the carpels and terminal, distinct or imperfectly (rarely completely) connate; stigmata simple. *Fruit* capsular, splitting at maturity into its carpellary elements, each carpel opening on its ventral suture. *Seeds* with fleshy albumen, very rarely exalbuminous.

Herbs, shrubs, or trees, very various in aspect; the *Saxifrageæ* abundant throughout the temperate and colder regions of the globe, chiefly of the northern hemisphere; the *Cunonieæ* and *Escalonieæ* chiefly found in South America and Australia, with a few outlying species in tropical and southern Africa. Foliage various in the different suborders. Flowers perfect, regular, cymose or racemose, rarely of large size, but often brightly coloured and abundant. This is a very large and very undefinable Order, allied on the one hand to *Rosaceæ*, from which the albuminous seeds nearly always distinguish it, and on the other to *Crassulaceæ* and *Ribesiaceæ*. None are particularly useful to mankind.

TABLE OF THE SOUTH AFRICAN GENERA.

Sub-Order 1. SAXIFRAGEÆ. Herbaceous or suffruticose plants with alternate or opposite, exstipulate leaves.

I. **Vahlia.**—*Ovary* inferior; *styles* 2, spreading. *Leaves* opposite, simple, linear.

Sub-Order 2. CUNONIEÆ. Shrubs or trees, with *opposite*, compound (or simple) leaves and *interpetiolar stipules*.

II. **Cunonia.**—*Flowers* racemose. *Petals* entire. *Leaves* pinnate.
III. **Platylophus.**—*Flowers* panicled. *Petals* trifid. *Leaves* trifoliolate.

Sub-Order 3. ESCALONIEÆ. Shrubs or trees, with *alternate*, simple, *exstipulate* leaves. *Stamens* as many as the petals, in a single row.

IV. **Montinia.**—Flowers diœcious, 4-cleft, white, terminal, or corymbose. A glabrous *bush* with lanceolate, entire leaves.

V. **Choristylis.**—Flowers 5-cleft, minute and green, in axillary, short panicles. A *shrub* with ovate, serrate, strongly nerved and veined leaves.

Sub-Order 4. ? GREYIEÆ.—Shrubs or trees with alternate, exstipulate leaves and sheathing petioles. *Stamens* in two rows; those of the outer row abortive, of the inner perfect, exserted, *twice* as many as the petals.

VI. **Greyia.** Flowers racemose, bright crimson; stamens much exserted.

I. VAHLIA, Thunb.

Calyx-tube adhering to the ovary; limb 5-parted, persistent, with valvate æstivation. *Petals* 5, spreading, entire, epigynous. *Stamens* 5. *Ovary* inferior, one-celled, with two mult-ovulate placentæ, pendulous from the apex of the cavity. *Styles* 2, spreading; stigmata capitate. *Capsule* membranous, opening between the styles. *Seeds* minute, very numerous. *DC. Prod.* 4, *p.* 53. *Endl. Gen.* 4631.

African or Asiatic herbs or suffruticose, small plants, glabrescent or villous. Leaves opposite, exstipulate, linear or lanceolate, entire. Pedicels axillary, 2-flowered; or in pairs, 1-flowered. Flowers white. Name in honour of Martin Vahl, Professor of Botany at Copenhagen, and author of many botanical works.

1. **V. Capensis** (Thunb. Cap. p. 246); many-stemmed, branched from the base, pubescent or glabrescent; leaves linear or narrow-lanceolate; peduncles 2-flowered, shorter than the leaves; capsules turbinate. *DC. Prod.* 4, *p.* 53. *E. & Z.!* 1764. *Russelia Capensis, Linn. f.*

HAB. Verkeerde Valley, *Thunberg*. Duiker Valley, Cape, and the Kamiesberg, *E. & Z.!* Wolf R., *Burke & Zeyher!* Zwartland, *Wallich*. Many localities to the north of Capetown, near the west coast, *Drege!* Namaqualand, *A. Wyley!* (Herb. D. Sd. Hk.)

Stems 6 inches to 2 feet long, woody and much-branched below, herbaceous above; the twigs 6–12 inches long, simple. Pubescence either scanty or dense and woolly. Leaf-pairs 1–1½ inch apart; leaves 8–10 lines to 1½ inch long, either very narrow-linear with revolute margins, or flat and lanceolate. Peduncles axillary, either very short or 2–6 lines long, forked, bearing 2 flowers. Calyx-lobes lanceolate, longer than the pubescent tube; petals white, shorter than the calyx-lobes.

II. CUNONIA, Linn.

Calyx free, 5-parted, deciduous. *Petals* 5, oblong, entire. *Stamens* 10. *Ovary* free, conical, 2-celled, with mult-ovulate sutural placentæ.

Styles 2, diverging; stigmata simple. *Capsule* conical, 2-horned, 2-celled, separating, from base to apex, from a free, placentiferous column. *Seeds* numerous, compressed, with a narrow, membranous wing. *DC. Prod.* 4, *p.* 12. *Endl. Gen.* 4662.

Only one species known, a South African shrub, the "*Roode Elseboom*" of the colonists. The generic name is in honour of John C. Cuno of Amsterdam, who described his own garden in Dutch, in 1750.

1. C. Capensis (Linn. Sp. 569). *DC. Prod.* 4, *p.* 12. *Lam. Ill. t.* 371. *Lodd. Cab. Bot. t.* 826. *E. & Z.!* 2151.

HAB. Common in moist, woody places, throughout the colony. (Herb. D. Hk. Sd.)

A large shrub or middle-sized tree, 10–50 feet high, glabrous in all parts. Leaves pinnate, 2–3-jugate, on longish, opposite petioles. Leaflets petiolulate, 2-4 inches long, oblongo-lanceolate, sharply serrate, coriaceous, netted-veined. Stipules broadly ovate, deciduous. Racemes opposite, cylindrical. 4–8 inches long, densely many-flowered. Flowers small, white, very numerous, with much exserted stamens. Capsules dark-brown, crowned with the divergent, hornlike styles.

III. PLATYLOPHUS, Don.

Calyx free, 4- (rarely 5-) parted, persistent, with valvate æstivation. *Petals* 4–5, trifid. *Stamens* 8–10, inserted on the outer edge of a hypogynous, fleshy, urceolate disc. *Ovary* free, 2-celled; ovules 2 in each cell, collateral, pendulous. *Styles* 2; stigmata simple. *Capsule* turgid at base, compressed above, membranous, 2-celled, at length splitting; the cells one-seeded. *Don. in Edin. New. Phil. Journ. IX. Endl. Gen.* 4653.

A South African tree, separated by Don from *Weinmannia*, chiefly on account of its trifid petals, and imperfectly dehiscent capsule. The name is derived from πλατυς, *broad;* and λοφος, *crest.* Colonial name "*Witte Elseboom.*"

1. P. trifoliatus (Don.). *Weinmannia trifoliata,* Thunb. *Prod.* 77, *Fl. Cap.* 384. *DC. Prod.* 4, *p.* 9. *E. & Z.!* 2152.

HAB. Waterfall, near Tulbagh, Hex Rivier; Langekloof; and the Vanstaadenberg, *E. & Z.!* Paarl, *Rev. W. Elliott!* Knysna, *Dr. Pappe!* Dutoit's Kloof, &c. *Drege!* (Herb. D., Hk., Sd.)

An umbrageous tree, 40–50 feet high, glabrous in all parts. Bark of the twigs smooth, dark-coloured. Leaf-pairs 2–4 inches apart; leaves on long petioles, trifoliolate; leaflets 3–5 inches long, lanceolate, acute, denticulate, netted-veined, bright green. Stipules small, deciduous. Panicles opposite, on long, naked peduncles, ovate, much-branched, many-flowered. Flowers small, almost always 4-parted; the petals shorter than the calyx, inserted outside the disc, but not adhering to it, as do the stamens. Capsule papery, inflated at base, netted, long remaining closed. Seeds dark-brown, oblong. Embryo straight, not much shorter than the fleshy albumen.

IV. MONTINIA.

Flowers by abortion dioecious. *Calyx-tube* adnate with the ovary; limb short, persistent, 4-cleft. *Petals* 4, ovate, deciduous, with imbricate æstivation. *Disc* (in the male fl.) fleshy, 4-angled. *Stamens* 4, alternating with the petals; filaments short; anthers adnate, opening longitudinally. *Ovary* inferior, imperfectly 2-celled, with parietal, fleshy, multovulate placentæ, filling the greater part of the cavity. *Style* single, short; *stigma* large, capitate, bipartite, the lobes deeply emarginate. *Capsule* ligneous, crowned by the style and calyx-limb, 2-celled, splitting through the centre into two diverging valves. The placentæ confluent, finally free. *Seeds* 4–6 in each cell, compressed, imbricate, with a broad, marginal wing; *testa* membranous; *embryo* large, flat, lying in thin, fleshy albumen, *cotyledons* ovate, *radicle* elongate. *Harv. Gen. S. A. Pl. p.* 101. *DC. Prod.* 3, *p.* 35. *Endl. Gen.* 6123.

A glabrous, South African shrub, usually referred by botanists to *Onagrarieæ*, but, as I think, taking into account the structure of its ovary, the dehiscence of the capsule, the imbricated petals, the slightly albuminous seeds, and the general habit, more nearly allied to *Escalonieæ*. The name is in honour of Laurence Montin, an obscure Swedish botanist.

1. M. acris (Linn. f. Suppl. 427). *Thunb. Fl. Cap.* 142. *DC. Prod.* 3, 35. *E. & Z.!* 1757. *M. caryophyllacea, Thunb. Act. Lund.* 1, *p.* 108. *Sm. Spicil. t.* 15. *Burm. Afr. t.* 90, *f.* 1–2. *M. frutescens, Gaertn.*

HAB. Dry ground, throughout the colony and in Namaqualand, common. (Herb. Hk., D., Sd.)

An erect, rigid, twiggy, glabrous, and somewhat glaucous bush, 1–2 feet high. Ramuli erect, compressed or angular, pale. Leaves varying from oblong to lanceolate and linear, entire, margined, one-nerved, veinless. Flowers small, white, the males in terminal few-flowered, corymbose cymes, the female generally solitary, one at the end of each branch or of short, corymbose branchlets. Capsules oblong, an inch or more long, 4–5 lines wide, at length splitting through the centre into two boatshaped pieces.

V. CHORISTYLIS, Harv.

Flowers polygamous. *Calyx-tube* obconic, adnate with the ovary, limb 5-cleft, persistent. *Petals* 5, inserted on the margin of the calyx-tube, longer than the calyx-lobes, sessile, entire, with valvate æstivation, persistent. *Stamens* 5, alternate with the petals and inserted with them; filaments short; anthers ovate, 2-celled, slitting. *Ovary* 2-celled, multi-ovulate, with axile placentæ. *Styles* 2, short, connate at first, then widely diverging; *stigmata* capitate. *Capsule* half-inferior, its conical, acuminate apex encircled by the persistent calyx-limb and petals, dehiscing scepticidally through the styles. *Seeds* oblong, subincurved, with prominent raphe and leathery testa; embryo . . . . . ? *Harv. in Hook. Lond. Journ. Vol.* I, *p.* 19.

Only one species known. The name is compounded of *χωρις*, *separately* and *στυλος*, a *style*.

1. Choristylis rhamnoides (Harv. l. c.). *Bœobotrys rufescens, E. Mey.! in Herb. Drege. Mœsa palustris, Hochst. in pl. Krauss.*

HAB. Near the Berlin Mission Station, Katberg, *Rev. J. Brownlee! E. & Z.!* Between Omtendo and Omsamculo, *Drege!* Port Natal, *Krauss!* (Herb. D. Sd.)

A leafy shrub, 7–8 feet high or more. Leaves 1½–2 inches long, alternate, petiolate, ovate or oblong, acute, sharply serrate, each serrature mucronate with a minute gland, penni-nerved, the nerves prominent on the paler under-surface. Panicles axillary, shorter than the leaves, much-branched. Flowers small and green; petals broadly subulate, pubescent on both sides. Anthers hairy, with a fleshy connective, often barren and then much reduced in size. Capsule 2–3 lines long.

VI. GREYIA, Hook. & Harv.

Calyx free, 5-parted, short, persistent, with obtuse lobes, imbricated in æstivation. *Petals* 5, oblong, sessile, deciduous, imbricate. *Stamens* inserted in the base of the calyx, subhypogynous, in two rows; those of the outer row 10, without anthers, united at base into a fleshy cup, their very short filaments crowned with a peltate gland; those of the inner row 10, fertile, free, alternating with the barren exterior stamens, much exserted, with subulate filaments, and ovate, erect, short, didymous, splitting anthers. *Ovary* free, deeply 5-furrowed, formed of 5 induplicate-valvate carpels, 1-celled, tapering at apex into a subulate, exserted style; ovules sutural, indefinite. *Capsule* 5-lobed; its carpels follicular, papery, slightly cohering at the sutures; *seeds* minute, with membranous testa, copious, fleshy albumen, and a straight embryo. *Harv. in Proc. Dubl. Univ. Zool. & Bot. Assn. Vol.* 1, *p.* 138, *t.* 13, 14.

A middle-sized tree, with alternate, simple, and exstipulate leaves. Petioles dilated at base and amplexicaul. Racemes terminal, densely many-flowered. Flowers crimson and very handsome. The generic name is in honour of Sir George Grey, K.C.B., Governor-General of the Cape Colony.

1. G. Sutherlandi (Hook. & Harv. MSS.). *Harv. Thes. Cap. t.* 1.

HAB. Rocky, exposed, mountain places near Port Natal, from 2000f. to 6000f. elevation, *Dr. Sutherland.* (Herb Hk., D., Sd.)

A small tree or large shrub, with light porous wood and gray bark. Branches and twigs leafy near the point, bare below; flowering branches naked for a space below the raceme. *Leaves* on long petioles, subrotund, deeply cordate at base, 2–4 inches in diameter, multilobulate and crenate, glabrous, but minutely glandular on the surface. Racemes very dense, 2–4 inches long, many-flowered. Pedicels glabrous, ¾ inch long, with a lanceolate bract at base. Calyx continuous with the pedicel. Petals broadly oblong, sessile, of a thick, glossy substance and bright crimson colour, thrice as long as the calyx, ciliolate. Capsule deeply 5-lobed, almost resolved into 5 follicles.

ORDER LI. BRUNIACEÆ. R. Br.

(By W. SONDER).

Flowers perfect, small, regular. *Calyx*-tube connate with the ovary, or very rarely free; limb 5- rarely 4-cleft, imbricate. *Petals* 5 (or 4), free or cohering into a monopetalous, epigynous, or perigynous corolla, imbricate in æstivation *Stamens* as many as the petals and alternate with them; filaments free or adnate to the base of the petals; anthers erect or incumbent, 2-celled, introrse, opening by slits. *Ovary* more or less inferior, rarely free, 1–3- rarely 5-celled; ovules pendulous, solitary, or two collateral ones in each cell, very rarely (in *Thamnea*) about 10; styles 2–3, distinct or more or less connate. *Fruit* dry, indehiscent or capsular, mostly dicoccous and crowned by the calyx-limb. *Seeds* with copious albumen; embryo minute, straight, next the hilum.

Heathlike shrubs and suffrutices, all natives of South Africa. Leaves small, glabrous or hairy, acerose, rarely ovate, very entire, sessile or subsessile, crowded and mostly imbricated, with a discoloured or withered, callous tip (*ustulate*). Stipules none. Flowers minute and white, rarely red, sessile, spiked or capitate, rarely solitary and axillary. This Order is closely related to *Saxifragaceæ* on the one part, and to *Hamamelideæ* on the other. The habit is peculiar.

TABLE OF THE GENERA.

Fruit 1-seeded, mostly indehiscent.

I. **Berzelia.**—*Fruit* indehiscent, cuneate, 1-seeded. *Ovary* 1-celled, 1-ovuled. *Style* 1. (Flowers in dense heads).

II. **Tittmannia.**—*Fruit* indehiscent, sphærical, 1-seeded. *Ovary* 2-celled, cells, 2-seeded. *Style* 1. (Flowers axillary).

III. **Brunia.**—*Fruit* indehiscent, rarely 2-valved, 1-seeded from abortion. *Ovary* 2-celled, cells 1–2-seeded. *Styles* 2. (Flowers in heads or panicled).

Fruit dicoccous.

Ovary 2-celled, 2–4-ovuled.

IV. **Lonchostoma.**—*Ovary* 4-ovuled. *Corolla* monopetalous, 5-cleft. (Flowers in terminal, leafy spikes).

V. **Linconia.**—*Ovary* 4-ovuled. *Styles* 2. *Petals* free. *Anther*-connective with a conical appendage at top. (Flowers axillary in leafy spikes).

VI. **Berardia.**—*Ovary* 2-ovuled. *Styles* 2. *Petals* free or somewhat cohering into a tube at the base. *Anther* connective without appendage. (Flowers in dense heads.)

VII. **Staavia.**—*Ovary* 2-ovuled. *Style* 1. *Petals* free. (Flowers in heads.)

Ovary 6–10-ovuled. Calyx-segments scarious, imbricate.

VIII. **Audouinia.**—*Ovary* 3-celled, 6-ovuled. *Style* trigonous. (Flowers in spike-like, terminal heads.)

IX. **Thamnea.**—*Ovary* 1- (or 5-?) celled, about 10-ovuled. *Style* cylindrical. (Flowers solitary, terminal.)

I. BERZELIA, Brogn.

Calyx adhering to the ovary, segments 5, rarely 4, unequal, gibbous. *Petals* 5, rarely 4, free. *Stamens* 5, rarely 4, longer than the petals. *Ovary* half-inferior, oblique, 1-celled, 1-ovuled. *Style* simple, terminated by a small, subconical stigma. *Fruit* indehiscent, gibbous. *Brogn. Mem. p.* 14. *Endl. Gen. No.* 4596. *Bruniae spec. Linn. Wendl.*

Small heath-like shrubs, with short, somewhat trigonal, imbricate or spreading leaves. Heads of flowers naked, usually crowded at the top of the branches, with 3 bracteæ at the base of each flower, inferior bractea larger, clavate and callous at top. Named in honour of Berzelius, the famous Swedish chemist.

ANALYSIS OF THE SPECIES.

Heads the size of a pea:
- lvs. subulate, wider at base, heads racemose … … (1) **alopecuroides.**
- lvs. subulate, attenuate at base, heads racemose … … (2) **commutata.**
- lvs. filiform-triquetrous, heads panicled … … … (4) **lanuginosa.**

Heads the size of a nut:
- lvs. linear-trigonal, heads corymbose … … … (3) **intermedia.**
- lvs. linear-lanceolate, heads racemose … … … (6) **squarrosa.**
- lvs. ovate or ovate-lanceolate, heads subcorymbose … (5) **abrotanoides.**
- lvs. subcordate … … … … .. … (7) **cordifolia.**

1. B. alopecuroides (Sond.); smooth, leaves *sessile*, 6-ranked, *imbricate, incurved*, subulate, trigonal, obtuse, ustulate-apiculate, *wider at base;* heads of flowers ovate-globose, the size of a small pea, forming racemes at the tops of the lateral branches. *Brunia alopecuroides, Thunb.? dissert. p.* 6. *R. & Sch. syst. veg. V. p.* 411.

HAB. South Africa, *Thunberg!* October. (Herb. Thunb. Holm.)

A shrub 2–3 feet high, with the aspect of a Dacrydium. Branchlets purple, leafy. Leaves 1½–2 lines long or shorter. Racemes 2–3 inches long. Head of flowers on leafy, 2–3 lines long branchlets, 2–3 lines in diameter. Petals spathulate.

2. B. commutata (Sond.); smooth; leaves *petiolate, spreading* or *recurved*, subulate, obtuse, apiculate, *attenuated* at the base, flat above, keeled beneath; heads of flowers globose, the size of a small pea, terminal, and axillary, *racemose. B. comosa, E. & Z.! No.* 1051. *excl. syn.*

HAB. In the channel of the Zwartkopsrivier. *E. & Z.! Zey.* 2644, *Dec.* (Herb. Sd.)

A much-branched shrub. Leaves 4 lines long, ½ line wide, the uppermost 2 lines long. Racemes 1–2 inches long. Heads of flowers on scaly, 1–3 lines long pedicels. Petals spathulate, spreading. Anthers ovate.

Very similar to the var δ. of *B. abrotanoides*, differs by the longer petioled leaves, much smaller, racemose heads, and smaller anthers.

3. B. intermedia (Schlecht ! Linn. 6, p. 188); smooth or somewhat hairy, leaves *very short-petioled,* spreading or recurved, linear-trigonal, rather *tetragonal* toward the apex, callous at the tip; heads of flowers *size of a nut,* forming a *corymb* on the lateral branches. *E. & Z. ! No.* 1053. *B. Wendlandiana, E. & Z. ! No.* 1052. *B. ericoides, E. & Z. ! No.* 1054. *Brunia paleacea Wendl. ! coll. t.* 21.

VAR. β. Leaves erect, imbricate. *Brunia alopecuroidea, E. & Z. ! No.* 1067. *non Thunb.*

HAB. Mountains, Duyvelsbosch, and in Langekloof near Puspasvalley, Swellendam, Van Staadensrivierberge, *E. & Z. ! Zey.* 2645. Georgetown, *Dr. Pappe;* Howison's Poort, *H. Hutton.* Var. β, near Palmietrivier and Hanglipp, *E. & Z. !* Lowrypass, *Drege,* 6866, Dec., fruct. mat Aug. (Herb. Sond. D. Wendl.)

Near *B. lanuginosa,* but more robust, the young branches a little hairy, not villous, the leaves thicker, (not longer) and spreading, often reflexed, the heads 3–4 times larger and corymbose, the peduncles 1 inch long or longer, leafy, and the flowers somewhat larger. Bracteæ obovate, unguiculate, ustulate at top, at length incurved. Petals white, oblong, spathulate. Stamens 4 or 5, twice longer than the petals. Fruit 1 line long.

4. B. lanuginosa (Brogn. l. c. p. 16, t. 1, f. 1); branches fastigiate, *young ones villous;* leaves sessile, imbricate, erect or spreading, *linear-filiform, triquetrous,* obtuse, apiculate, rather pilose; heads of flowers *the size of a pea,* at the tops of the lateral branches, disposed in a fastigiate panicle. *Brunia lanuginosa. Linn. spec. p.* 288. *Berg. Cap. p.* 60. *Pluk. phyt. t.* 318, *f.* 4. *Wendl. coll. t.* 11. *Lodd. Bot. Cab. t.* 572. *Sieb. fl. cap. exs. No.* 56. *Drege, No.* 6860. *Zeyher,* 2642.

VAR. β. **longifolia;** branches loose, leaves longer, spreading, incurved, ciliate-pilose. *Brunia superba, Don. hort. cant. Willd. ! spec.* 1, *p.* 1143.

VAR. γ. **glabra;** branches and leaves smooth. *Brunia lanuginosa et comosa, Thunb. ! dissert. p.* 4 *and* 5, *fl. cap. p.* 205. *B. tenuifolia, Willd. ! Denksch. Acad. Munchen,* 1808, *p.* 129, *t.* 5, *f.* 2. *Berg. lanuginosa, E. & Z. ! No.* 1050. *Herb. Un. Itin. No.* 140. *Drege, Nos.* 6857, 6858, 6861, 6895.

HAB. Moist places in mountains near Capetown, in Hottentottsholland, Caledon, and Swellendam, Oct. (Herb. Thunb., Holm., Wendl., Willd., Hk., D., Sd.)

A greyish shrub, 3 feet or more in height, branches virgate, very leafy. Leaves straight, sulcate above, 2–3 lines, in var. β. 4–5 lines long. Heads about 3 lines in diameter; peduncles 2–4 lines long, villous, scaly. Bracteæ spathulate, smooth, at top callous. Petals suberect, oblong-lanceolate. Anthers oblong.

5. B. abrotanoides (Brogn. l. c. p. 15); branches glabrous or somewhat hairy; leaves sessile or very shortly petiolate, spreading, *ovate or ovate-lanceolate, subtrigonal,* ustulate at the apex; heads of flowers the *size of a filbert,* terminal, crowded, subcorymbose. *Brunia abrotanoides, Linn. spec. p.* 288. *Berg. cap. p.* 59. *Thunb. ! fl. cap. p.* 207. *Sieb. fl. cap. exs. No.* 57. *Herb. Un. itin. No.* 138, 139. *E. & Z. ! No.* 1059.

VAR. α. **glabra;** branches, as well as the erect or spreading leaves, quite glabrous.

VAR. β. **pilosa;** branches and leaves pilose. *B. brevifolia, E. & Z. ! No.* 1058. *Drege,* 6864.

VAR. γ. **reflexa;** leaves oblong or ovate-lanceolate, reflexed, imbricate. *Brunia squarrosa, Swartz. in Herb. Holm. non Thunb. Drege,* 6863, a.

VAR. δ. **lanceolata;** leaves spreading, the lower lanceolate, 3–4 lines long, the upper ovate. *Burmann. Afr. t.* 100, *f.* 1. *Brun. abrotanoides, Wendl. coll. t.* 45. *Berg. formosa, E. & Z. ! No.* 1060.

VAR. ε. **parvifolia;** leaves spreading, trigonal, small, 1 line long; heads smaller. *Brunia deusta. Thunb. ! diss. p.* 4. *Fl. cap. p.* 205. *Drege,* 6863, b.

HAB. Plains near Capetown. Var. β. mountains near Tulbagh and Capetown; vars. γ. and ε. Franschehoek; var. δ. Table Mountain, Aug.–Dec. (Herb. Thunb. Holm. Sond. D.)

1–3 feet high, branches purplish, the uppermost fastigiate, very leafy. Leaves about 2–3 lines long, the upper smaller. Heads globose, the young ones elliptic. Flowers 4-cleft, 4-androus; or 5-cleft, 5-androus. Petals white, spathulate. Anthers oblong. Fruit glabrous, gibbous.

6. B. squarrosa (Sond.); branches glabrous, younger ones *cobwebbed;* leaves petiolate, spreading or recurved, *linear-lanceolate*, trigonal, ustulate at the apex, glabrous, younger ones pilose; heads of flowers size of a nut, terminal or axillary, racemose. *Brunia squarrosa, Thunb.! dissert. p.* 5, (1804). *B. rubra, Willd.! Denkschrift. Acad. Muench.* (1808), *t.* 4, *f.* 1. *Spreng. syst.* 1, *p.* 782. *B. arachnoidea, Wendl. coll. t.* 62. (1810). *E. & Z.! No.* 1057. *B. ericoides, Wendl. l. c t.* 5, *f.* (*non bona*). *B. plumosa, Lam. Enc.* 1, *p.* 475? *B. superba, Reichenb. hort. bot. t.* 100, *excl. syn., analysis incorrect. Heterodon superbum, Meisn. gen.* 72 (52).

VAR. β. **glabra**; branches and leaves glabrous. *B. superba, E. & Z.! No.* 1056.

VAR. γ. **reflexa**; branchlets villous; leaves crowded, imbricate, incurved, erect or reflexed, rigid, lanceolate, rather tetragonal towards the apex, younger ones pilose or villous. *B. rubra, Schlecht.! Linn.* 6, *p.* 189.

HAB. Mountain rivers in the districts of Worcester, Caledon, and Swellendam. *Thunb. E. & Z.! Dr. Pappe, Zey.* 2641, 2643. *Drege,* 6842. (Herb. Thunb., Willd., Wendl., Sond., D.)

A shrub with greyish or yellowish branches; branchlets often verticillate. Leaves flat or subcanaliculate above, keeled beneath, ½–1 inch long, about ¾ line wide. Heads globose, 4–6 lines in diameter; peduncle 5–6 lines, in fruit often an inch long, in var. γ, much shorter, villous, and scaly. Receptacle hairy. Calyx 5-dentate, never 10-dentate as described by Reichenbach, the teeth blunt, unequal, 3 are longer and gibbous. Petals oblong, spathulate. Anthers ovate, cells diverging at the base. Style simple. Fruit about 1 line long.

7. B. cordifolia (Schlecht. l. c. p. 189); branches erect, younger ones pubescent; leaves very short, petiolate, spreading or reflexed, *subcordate-ovate*, bluntish, callous at the tip, flat above, keeled beneath, smooth; heads of flowers the size of a nut, solitary on the tops of the branches, corymbose. *E. & Z.! No.* 1061.

HAB. Near Mount Potberg, Swellendam, *E. & Z.! Mundt!* Oct. (Herb. Hk. Sd.)

2 feet or more in height, with di-tri-chotomous branches. Petioles large, persistent. Leaves tipped with an obtuse, black mucro, with a pale margin, coriaceous, about 3 lines long, 2–2½ lines wide. Capitules about 8–12 in a corymb; peduncles scaly, ½–1 inch long. Receptacle villous. Bracteæ obtuse, cucullate, callous-mucronate, somewhat hairy. Petals spreading, oblong-spathulate, with a two-crested claw. Anthers linear-oblong.

II. TITTMANNIA, Brogn.

Calyx with a spherical tube, wrinkled and glandular on the outside, adnate to the ovary, 5-cleft, with scarious, erect segments. *Petals* with the claws 2-keeled on the inside, and with ovate-roundish, spreading lamina. *Ovary* inferior, spherical, 2-celled, with a membranous dissepiment, free at the edges; cells 2-seeded. *Ovules* pendulous, fixed to the dissepiment. *Style* simple, conical; stigma bidentate. *Brogn. l. c. p.* 29, *t.* 4, *f.* 2. *Endl. gen.* 4603. *Moesslera Reichenb. consp.* 160. *Meisn. gen.* 72 (52).

A small shrub with fastigiate, subumbellate branches. Leaves linear, subcylindrical, wrinkled, incurved, erect, imbricate, callous at the apex. Flowers axillary, approximate towards the tops of the branches, bent to one side, and calyculated at the base by short, scarious scales. Name in honor of J. A. Tittman, a botanist.

1. T. laxa (Sond.); *Brunia laxa Thunb.! fl. cap. p.* 206. *T. lateriflora, Brogn. l. c. p.* 30. *Moesslera lateriflora, E. & Z.!* 1086.

HAB. Rocky places near the Waterfall, Tulbagh, and near Kochmannskloof, Swellendam, *E. & Z.!* Nieuwekloof, 2–3000! and Drakensteensberge, 4–5000! *Drege,* Oct.–Dec. (Herb. Th., D., Sd., Hk.)

1–2 feet high, with the aspect of a *Juniper.* Branches virgate, filiform, glabrous. Leaves imbricate or erectly-spreading, 1 line long. Flowers about 1½ line long, white.

III. BRUNIA, Linn.

Calyx adhering to the ovary. *Petals* ovate or spathulate. *Ovary* half-inferior, 2-celled, cells 1–2-seeded. *Styles* 2, diverging at the apex. *Fruit* indehiscent, rarely septicidal-dehiscent; 1-seeded from abortion. *Brogn. l. c. p.* 16. *Endl. Gen. No.* 4597.

Shrubs more or less branched, with the branches in whorls, erect or spreading. Leaves small, closely imbricate, and flowers capitate in Sect. I.; but in Sect. II. the leaves are larger, often myrtle-like, spreading, and the flowers in panicles. Flowers furnished with 3 bracteæ each, or sometimes deficient of the two lateral ones. Named in honour of Cornelius Brun, a botanical traveller.

ANALYSIS OF THE SPECIES.

Stamens exserted:
- lvs. subulate, trigonal, glabrous ... ... (1) **nodiflora.**
- lvs. linear, convex beneath, subglabrous ... ... (2) **laevis.**
- lvs. linear-lanceolate, convex beneath, hairy, villous-ciliate ... ... (3) **macrocephala.**

Stamens enclosed:

Calyx and ovary glabrous:
- lvs. subcordate, pubescent on both sides ... ... (5) **cordata.**
- lvs. oblong-sublanceolate, glabrous ... ... (6) **racemosa.**
- lvs. lanceolate-subulate, glabrous, heads globose ... (9) **virgata.**
- lvs. subulate-trigonal, glabrous, heads ovate-globose (11) **alopecuroides.**
- lvs. linear, obtuse, racemes linear ... ... (4) **pinifolia.**

Calyx and ovary villous:
- lvs. petiolate, lanceolate, flat, villous ... ... (7) **villosa.**
- lvs. sessile, lanceolate, concave, glabrous above ... (8) **squalida.**
- lvs. petiolate, linear, trigonal, glabrous ... ... (10) **staavioides.**

Sect. I. **Eu-brunia**; calyx hairy, with spathulate segments. Petals somewhat spathulate. Stamens exserted, unequal. Ovary 2-celled, cells 2-seeded. Fruit crowned by the permanent calyx, stamens, and petals. (sp. 1–3).

1. B. nodiflora (Linn. spec. p. 288); leaves *lanceolate-subulate, trigonal,* acute, smooth, incurved, closely imbricate; heads of flowers globose, size of a cherry, on the top of the branches. *Berg.! cap. p.* 54. *Thunb.! fl. cap. p.* 205. *Brogn. l. c. p.* 17, *E. & Z.! No.* 1062. *Wendl. coll. t.* 35.

HAB. Dry, elevated places near Capetown, and in the districts of Caledon, Stellenbosch, Swellendam, Worcester, and Uitenhage, April–Dec. (Herb. Hm. Th. Sd. &c.)

Shrub 2–3 feet high, much-branched, branches greyish-brown, smooth; branchlets often verticillate, leafy. Leaves on the branches 1 line long. Heads villous-tomentose. Bracteolæ spathulate, villous. Calyx very villous, segments longer than the tube. Petals bicristate at base, limb patent. Stamens unequal, 2 much longer. Fruit coriaceous, indehiscent.

2. B. laevis (Thunb.! l. c. p. 204); leaves *linear,* obtuse, apiculate, *bicarinate above, convex beneath,* puberulous or *subglabrous,* imbricate; heads of flowers globose, size of a large nut. *Br. globosa, E. & Z.! No.* 1063. *B. superba, plant. Krauss.!*

HAB. Mountain sides, Baviaansberg, near Gnadenthal and Zwarteberg, *Pappe, E. & Z.!* Klynriviersberge, *Zey.!* 2640. Ataquaskloof, Nieuweskloof, and Ylandskloof, *Drege,* 6854, a, b, Aug.–April. (Herb. Th. Hm. Vind. D. Sd.)

Very like the preceding, but differing by the greyish colour, more incurved, round-backed, not keeled leaves, and larger heads. Leaves 2–1 line long. Heads of flowers tomentose; bracteæ with a black mucro; bracteoles, as the calyx, white-villous on the outside. Petals oblong-spathulate, narrowed into a claw, little longer than the calyx, glabrous. 1 or 2 of the filaments longer. Ovary very villous. Fruit indehiscent.

3. B. macrocephala (Willd.! l. c. p. 132, t. 6, f. 1); leaves *linear-lanceolate,* crowded, bicarinate above, *convex* beneath, *hairy, villous-ciliate;* heads of flowers globose, very large. *Spreng. syst.* 1, *p.* 782.

HAB. Cape, Herb. Willd. Inferior regions, Niven in Herb. Sond.

Two feet or more high, with the aspect of a *Phylica,* more robust than the preceding. Branches erect, verticillate, as well as the branchlets, pubescent. Leaves 4–5 lines long, ½ line wide, with involute margins and ustulate at the apex. Flowers unknown.

Sect. II. **Beckea.** Calyx generally with scarious, smooth segments. Petals ovate, inclosing the stamens. Ovary 2-celled; cells 1- or 2-seeded. Fruit crowned by the calyx. Petals and stamens mostly caducous. (sp. 4–10).

4. B. pinifolia (Brogn. l. c. p. 19, t. 1, f. 2); branches quite glabrous; leaves erectly-spreading or recurved, sessile, linear, obtuse, 1-nerved, quite smooth, coriaceous, flat; panicle terminal, composed of simple, linear racemes. *Phylica pinifolia, L. fil. suppl. p.* 153. *Thunb. fl. cap. p.* 202. *DC. Prod.* 2, *p.* 73.

VAR. *α.* flowers approximate, as long as the bracteæ; panicle composed of dense racemes. *B. pinifolia, Brogn. Beckea thyrsophora, E. & Z.! No.* 1069.

VAR. *β.* flowers somewhat distant, shorter than the bracteæ; panicle composed of lax racemes. *Beckea Africana, Burm, Prod.* 12, *E. & Z.! No.* 1068. *Brunia pinifolia and Linconia tamariscina, E. Meyer.*

HAB. Mountains, Olifantsrivier, Clanwilliam, *Thunb., E. & Z.! Niven;* var. *β.* near Palmietrivier, *E. & Z.;* Dutoitskloof and Ezelsbank, 3–4000f., *Drege.* (Herb. Th., D., Sd.)

Small, erect shrub, branches fastigiate, filiform. Leaves ustulate at the apex, 6–8 lines long, ¾–1 line wide. Racemes about 6 lines long. Inferior bractea foliaceous. Flowers whitish, not a line long. Calyx obtuse. Petals obovate. Fruit 2-celled, but 1-seeded; a spongeous placenta filling the cavity of the smaller sterile cell.

5. B. cordata (Sond.); branches virgate, *villous;* leaves imbricate, *ovate-subcordate,* acute or bluntish, *pubescent on both sides;* panicle terminal, composed of oblong racemes; calyx obtuse, glabrous; fruit rugulose-papillate. *Beckea cordata, Burm. Prod. p.* 12, *E. & Z.! No.* 1072. *Phylica imbricata, Thunb.! fl. Cap. p.* 202. *B. racemosa, Brogn. l. c. p.* 18. *excl. syn. Linn. et Th.*

HAB. Mountains, Hottentottsholland near Grietjesgat, *E. & Z.!* Dutoitskloof, 1–4000f., *Drege!* Oct.–Jan. (Herb. Th., D., Sd.)

2–3 feet high, branches spreading, brown-purplish, the upper filiform. Leaves 5-nerved, ovate, short, acuminate, villous-ciliate, inferior 6–8 lines long, 4 lines wide, superior twice smaller. Racemes 4–6 lines long or shorter; panicle leafy, dense or lax. Flowers about ½ line long, equalling or shorter than the inferior, foliaceous, concave and ustulate bractea. Fruit reddish, crowned with the whitish calyx.

6. B. racemosa (Sond.); branches virgate, *glabrous*, younger puberulous; leaves erectly-spreading, subimbricate, *ovate-oblong* or *ovate-lanceolate*, obtuse, *narrowed or obtuse at base*, coriaceous, *glabrous*; panicle terminal, composed of ovate or subglobose racemes; calyx *obtuse*, *glabrous*; fruit rugulose, papillate. *Phylica racemosa, Thunb.! fl. Cap. p.* 202. *Beckea laurifolia, E. & Z.! No.* 1070. *Br. laurifolia, Walp.*

VAR. *β*, leaves ovate, obtuse at base, smaller. *B. racemosa, E. & Z.! No.* 1071.

HAB. Mountains, Tulbagh near Waterfall, *Thunberg!* Puspasvalley and Langekloof, *E. & Z.! Drege*, 6856. Grootvadersbosch, *Zey.!* 2226. Var. *β*, near Hexrivier, Worcester, *E. & Z.!* Vanstadensberg, *Zeyher!* 2225, Oct.–Feb. (Herb. Th., Hm., D., Hk., Sd.)

Very like the preceding. It differs by the mostly glabrous branches, and smaller, more oblong, coriaceous leaves, and mostly roundish racemes. Leaves on very short petioles, about 5–6 lines long, 1½–2 lines wide, the upper much smaller and often recurved. Racemes 1–2 lines; panicles pyramidal, about an inch long. Flowers and fruit as in *B. cordata.*

7. B. villosa (E. Meyer); branches *filiform*, *villous*; leaves *petiolate*, erect-spreading, *lanceolate*, flat, *villous and ciliate*, with a black mucro; spike subglobose, solitary, or loose panicles; calyx *acute*, *villous*; fruit striate, villous. *Raspalia villosa, Presl.*

HAB. Mount Blaauwberg, on rocks, 4–5000f., *Drege.* Dec.–Jan. (Herb. Sond.)

Branches long, reddish-brown. Leaves about 3 lines long, ½ line wide, the upper smaller, not imbricate, but longer than the internodes. Spike headlike, the size of a small pea, sometimes ovate. Inferior bracts foliaceous, the bracteolæ half as long as the calyx. Flowers about ½ line long, white. Petals obovate-oblong, longer than the acuminate calyx. Stamens not exserted. Fruit one-seeded, seed oblong, glabrous.

8. B. squalida (Sond.); branches *short*, villous; leaves *sessile*, imbricate, lanceolate, bluntish, *concave*, *glabrous* above, villous and ciliate beneath; spikes subglobose, terminal, solitary; calyx *subacute*, as well as the ovary, *villous. Diosma squalida, E. Meyer.*

HAB. Dutoitskloof, 2–3000f., *Drege.* Oct.–Jan. (Herb. Sd.)

A small shrub, branches 2–3-chomotous, ultimate about 1 inch long, leafy. Leaves 5-farious, about 2 lines long, with a black mucro, villous, at length glabrous on the under surface. Spikes the size of a pea, at the top of leafy branches. Inferior bract foliaceous; bracteolæ ovate, villous-barbate, somewhat shorter than the calyx. Flowers whitish, about ¾ line long. Limb of the calyx shorter than the tube. Petals obovate, narrowed at the base, longer than the calyx. Stamens not exserted. Ovary not papillose, 2-celled, cells 1-seeded.

9. B. virgata (Brogn. l. c. p. 20); branches erect, slender, in whorls, ultimate filiform, fastigiate, *tomentose, at length glabrous*; leaves sessile, closely pressed to the stem, *lanceolate-subulate*, acute, ustulate at the apex, channelled, *glabrous*; heads of flowers terminal, globose; calyx and petals *obtuse*, equal, *glabrous* as well as the ovary. *E. & Z.! No.* 1065.

VAR. *β*. **robustior**; branchlets woolly; leaves carinate; heads of flowers larger. *B. verticillata, Thunb.! fl. Cap. p.* 206. *E. & Z.! No.* 1066.

HAB. Mountains near Puspasvalley, Swellendam, *E. & Z.! Niven.* Var. *β*, Tulbagh, *Thunberg.* Rivier Zonderende, *E. & Z.! Zey.!* 2652. Sept. (Herb. D. Hk. Sd.)

Shrub, 1–2 feet high, much-branched, ultimate branches 1–2 inches long. Leaves adpressed, but not, or only in the ultimate branchlet, imbricate, 1 line, in var. *β*, 2–3 lines long, quite smooth, shining. Heads about the size of a small pea, few-flowered, in var. *β*. twice larger. Flowers whitish, about 1 line long, with 3 bracts, inferior bracteæ foliaceous, lateral ones (bracteoles) opposite, cuneate, shorter and smaller than the flower. Limb of the calyx longer than the tube. Petals obovate-oblong. Ovary cuneate.

10. B. staavioides (Sond.); branches virgate, 2–3-chotomous, *glabrous;* leaves *petiolate, recurved-spreading, linear, trigonal,* obtuse, ustulate, mucronulate, glabrous; heads of flowers terminal and axillary; calyx *villous,* acute, shorter than the glabrous petals; ovary villous. *B. capitellata, E. Mey ! non Thunb.! Raspalia capitella, Presl.*

HAB. Mount Blauwberg, 3–5000f., *Drege!* Nov. (Herb. D. Sd.)

A slender shrub, very like the preceding, but well distinguished by the trigonal, recurved leaves and the axillary heads. Branches reddish. Leaves 3–4 lines long, ⅓ line wide, often bicarinate above. Petiole adpressed, ½ line long. Heads involucrated by some short leaves, the size of a small pea, mostly at the tops of very short, lateral branches, disposed in a leafy raceme. Flowers white, about ½ line long. Bracteolæ setaceous, villous, shorter than the calyx. Petals obovate, narrowed at base. Stamens included. Ripe fruit glabrous, 1 line long, 1-seeded. Seed oblong, shining.

Doubtful Species.

B. alopecuroides (Brogn l. c. p. 19, excl. synon.); branches slender, glabrous; leaves *subulate, trigonal,* acute, smooth, imbricate, *incurved,* ustulate at the apex; heads of flowers terminal, *ovate-globose,* dense, naked, smaller than a pea; bracteæ shorter than the flowers; styles 2.

HAB. Cape. (Herb. Burmann, n. v.)

Perhaps the same as *Berzelia alopecuroides,* but Thunberg's plant has only one style, not two.

IV. LONCHOSTOMA, Wickstroem.

Calyx adhering to the half-inferior ovary, 5-cleft. *Corolla* monopetalous, tubular, 5-cleft, segments unguiculate. *Stamens* 5, very short, inserted in the mouth of the corolla; anthers oblong, bursting inwards, 2 celled, opening lengthwise, cells diverging at the base. *Ovary* pubescent, 2-celled, cells 2-ovulate; ovule pendulous, fixed to the upper part of dissepiment. *Styles* 2 or 1. *Fruit* capsular, dehiscent from the base, 2–4-valved. *Seeds* 4, ovoid, reticulate, attached by a short, thick funicle; embryo small, orthotropus, in the apex of a copious, fleshy albumen; radicle superior. *Wickst. in Kongl. Vetensc. Acad. Handl. St.* 2. 1818. *Gravenhorstia Nees. Esenb. in Lindl. Introd. Ed.* 2, *p.* 439. *Endl. Gen.* 3877, 4606.

Branched shrubs. Leaves alternate, sessile, imbricate, concave, coriaceous. Flowers axillary, disposed in leafy, oblong, or headlike spikes at the top of the branches, bibracteate, reddish or rose-colour. Name compounded of λονχη, a *lance,* and στωμα, a *mouth,* alluding to the lance-shaped sepals and petals of one of the species.

ANALYSIS OF THE SPECIES.

Calyx-segments and lobes of the corolla ovate, obtuse ... (1) **obtusiflorum.**
Calyx-segments and lobes of the corolla lanceolate, acute:
  Styles 2, filiform, exserted ... ... (2) **acutiflorum.**
  Style 1, short ... ... (3) **monostylis.**

1. L. obtusiflorum (Wickstr.! l. c. t. 10, f. 2); leaves oval, hirsute beneath, at length glabrescent; bracteæ *oblong,* obtuse, apiculate; calyx-segments *ovate, obtuse;* limb of the corolla *ovate, obtuse;* styles 2, subclavate, *included. Passerina pentandra, Thunb.! Prod. p.* 76. *Gnidia pentandra, Thunb.! diss. fruct. part. variet. p. post. sect. pr. p.* 19.

HAB. Koude Bockevelde, *Thunberg, Ekeberg.* (Herb. Berg. & Thunb.)

1–2 feet high. Branches terete, mostly trichotomous, glabrous, younger ones hirsute. Leaves with a black, incurved mucro, ciliate, glabrous above, 3–3½ lines long, 1½ line wide. Spike leafy, ovate. Flowers rose-coloured, rather longer than the leaves. Bracteæ ciliate, 1 line long, ½ line wide, membranaceous at the margins and with a pubescent carina. Calyx-segments adpressed to the corolla, imbricate, ciliate, about 1 line long. Corolla glabrous, 3–4 lines, limb about 1 line long. Stamens not longer than the tube; anthers oblong, yellow. Style 1 line long.

2. L. acutiflorum (Wickstr.! l. c. t. 10, f. 1); leaves ovate, hairy beneath, or glabrescent, bracteæ *cuneiform;* calyx-segments *lanceolate, acute;* limb of the corolla *ovate-lanceolate,* acute; *styles* 2, filiform, *exserted. DC. Prod.*

HAB. Cape, *Dr. Hornstedt.* (Herb. Berg. & Swartz.)

Near the preceding, distinguished by larger leaves and flowers. Leaves ovate, upper ones ovate-lanceolate, bluntish, with a short, black mucro, the older glabrate, 5–6 lines long, 2–3 lines wide. Spike foliaceous, terminal, headlike; flowers longer than the leaves, rose-coloured. Bracteæ about 1½ line long, ½ line wide. Calyx-segments 3 lines long. Corolla 6–7 lines, limb about 2 lines long. Stamens a little exceeding the mouth, filaments ½ line long; anthers linear. Styles as long as the corolla, pubescent at base.

3. L. monostylis (Sond.); leaves ovate or oblong-lanceolate, hirsute beneath, or glabrescent; bracteæ *linear-setaceous;* calyx-segments *linear-lanceolate;* limb of the corolla ovate-lanceolate, acute; *style* 1, *short. Gravenhorstia fastigiata, Nees. Esenb.! in Herb. E. & Z. Lonchost. acutiflorum, E. & Z.! No.* 1084, *non Wickstroem. Peliotes detrita, E. Mey. in Herb. Drege.*

HAB. Mountain sides in Hottentottsholland near Palmietrivier, *E. & Z. & Pappe.* Dutoitskloof, 2–3000f., *Drege.* Oct.–Jan. (Herb. D., Sd., &c.)

A small, greyish-villous half-shrub, 1–2 feet high, often much-branched, branches 3-chotomous or whorled. Leaves about 3 lines long, 1 line wide, the upper ones smaller, 5-farious-imbricate, with a black mucro. Spike about ½ inch long. Flowers exceeding the leaves. Bracteæ pubescent, 2 lines, calyx-segments about 3 lines long. Corolla smooth, 4 lines long, limb convolute, as long as the tube. Anthers oblong. Ovary with 4 ovules. Style about 1 line long; stigma sublobed. Fruit about 1 line long, crowned by the calyx. Very like *L. obtusiflorum.*

V. LINCONIA, L.

Calyx adhering to the ovary, with a 5-cleft limb; segments short, membranaceous, smooth. *Petals* oblong, convolute, inclosing the stamens; cells of anthers diverging at the base; connective of anthers with a conical appendage at the top. *Ovary* half-inferior, 2-celled, cells 2-seeded. *Styles* 2. *Fruit* bicoccous. *Swartz. in Berl. Mag.* 1810, *p.* 85. *Brogn. l. c. p.* 26. *Endl. Gen. No.* 4601.

Heathlike shrubs with numerous, erect, fastigiate branches. Leaves spirally inserted on all sides of the branches, spreading or loosely imbricate, on very short stalks, coriaceous, quite smooth, or a little fringed on the margins, marked with a prominent nerve, ustulate at the apex. Flowers axillary, in the axils of the upper leaves, the whole forming a crowded, leafy spike, each flower involucrated by 4 or 5 bracteæ, which are about the length of the calyx. Name in memory of some forgotten person!

ANALYSIS OF THE SPECIES.

Lvs. linear, rough-edged. Flowers rather longer than the lvs. ... (1) **alopecuroides.**
Lvs. oblong, or linear-oblong, smooth-edged. Fl. as long as lvs. (2) **cuspidata.**
Lvs. ovate-oblong, rough-edged. Fl. rather longer than leaves (3) **thymifolia.**

1. L. alopecuroides (L. Mant. p. 216); leaves short, petiolate, imbricate or spreading a little, *linear*, obtuse, ustulate at top, flat or subcanaliculate above, nerve-keeled beneath, *roughish at the margins;* flowers a *little longer* than the leaves; bracteæ ovate, acute; petals *linear-oblong. Thunb.! Prod. p.* 48. *Swartz.! in Berl. Mag.* 1810, *p.* 86, *t.* 4. *DC. Prod.* 2, *p.* 45. *Brogn. l. c. p.* 27, *t.* 3, *f.* 3. *L. cuspidata, E. & Z.! No.* 1083, *non Swartz. Ericeæ. No.* 318, *Herb. E. & Z.*

HAB. Mountains on Krumrivier, Uitenhage; Gouritzrivier in Kannaland, George, Stellenbosch, and Caledon, *Thunb.! E. & Z.!* Aug.–Dec. (Herb. Th. Hm. Vd. D. Sd.)

About 2 feet high. Ultimate branches glabrous or hairy. Leaves 6–8 lines long, 1 line wide; petiole ½ line long. Flowers about 6 lines long. Bracteæ 4–6, carinate, 3 lines long, ciliate, longer than the calyx. Petals white. Stamens as long as the petals; filaments linear, compressed. Ovary glabrous. Fruit about 2 lines long, 2-seeded. Seeds shining, punctulate, 1 line long.

2. L. cuspidata (Swartz.! l. c. p. 284, t. 7, f. 1); leaves short, petiolate, spreading a little, *oblong or linear-oblong*, obtuse, ustulate at the apex, flat or subcanaliculate above, nerve-keeled beneath, *smooth at the margins;* flowers *as long* as the leaves; bracteæ elliptical, equalling the calyx; petals *oblong. Diosma cuspidata, Thunb.! fl. Cap. p.* 227.

HAB. South Africa, *Masson* in Herb. *Thunberg, Niven* in Herb. Sond.

A foot or more in height. Petioles persistent. Leaves often recurved, 3–4 lines long. Flowers 3 lines long. Bracteæ 4-ciliolate. Calyx-teeth very short. Petals white, rather longer than the stamens. Distinguished from the preceding by the smaller leaves and flowers, from the following by the more robust habit, twice longer, thinner leaves.

3. L. thymifolia (Swartz.! l. c. p. 86, t. 4); leaves *very short, petiolate*, imbricate-spreading, *ovate-oblong*, obtuse, ustulate at the top, flattish above, nerve-keeled and carinate beneath, *roughish at the margins;* flowers a little longer than the leaves; bracteæ elliptic, equalling the calyx; petals oblong. *Diosma deusta, Thunb.! l. c. p.* 224.

HAB. Cape, *Masson* in Herb. *Thunb.! & Swartz.;* mountains, Appelskraal near River Zonderende, *E. & Z., Zey.!* 2651. Sept. (Herb. Holm., Sond.)

A dwarf shrub, much-branched, branchlets crowded. Leaves 1½–2 lines long, ¾ line wide, the upper ovate, about 1 line long. Spike about half an inch long. Bracteæ, calyx, and petals as in *L. cuspidata.*

VI. BERARDIA, Brogn. (expte.)

Calyx adhering to the ovary at the base, but free at the apex, 5-cleft. *Petals* 5, free or cohering into a tube at the base. *Stamens* 5; anther connective without an appendage at the top. *Ovary* 2-celled; cells 1-ovuled. *Styles* 2. *Fruit* bicoccous. *Berardia and Raspalia, Brogn. l. c. Bruniæ spec. Thunberg.*

Small shrubs, with erect, fastigiate branches, alternate, opposite or whorled, short branchlets. Leaves small, subulate or rhomboid, keeled, close-pressed, covering the stem on every side, mostly spirally inserted. Heads of flowers solitary, at the tops of the branches. Named in honour of M. Berard, Professor of Chemistry at Montpelier.

ANALYSIS OF THE SPECIES.

A. Petals cohering into a tube at base, at length free.
lower bractea twice as long as the flower, segments of calyx villous; anthers ovate ... ... (1) **paleacea.**

lower bractea longer than the flower, segments of calyx glabrous ; anthers linear-oblong ... ... (2) **affinis.**

B. Petals quite free.

α. **macrocephalæ.** Heads large.

Capitula with a common involucre.

heads of flowers turbinate, branches glabrous ... (3) **lævis.**

heads of flowers globose, branches villous ... (4) **sphærocephala.**

Capitula without common involucre ... ... (5) **globosa.**

β. **microcephalæ.** Heads small.

Stamens enclosed.

lvs. obovate-elliptic, obtuse, convex beneath, puberous or glabrous ... ... (10) **phylicoides.**

lvs. rhomb.-ovate, obt.-keeled beneath, glabrous (6) **microphylla.**

lvs. trig.-ovate, acutely-keeled beneath, downy (7) **angulata.**

lvs. ovate, obtuse, canaliculate beneath, downy or glabrous ... ... (8) **affinis.**

lvs. ovate, mucronulate, canaliculate beneath, rough, with rigid hairs ... ... (9) **aspera.**

Stamens exserted ... ... (11) **Dregeana.**

1. B. paleacea (Brogn. l. c. p. 25, t. 3, f. 2); leaves subulate, acute, short, closely pressed to the stem, glabrous or subciliate, ustulate; heads of flowers corymbose; lower bractea *twice as long* as the flower, subulate, ciliolate at the base; segments of calyx shorter than the petals, *villous;* anthers *ovate. Brunia paleacea, Berg.! Cap. p.* 56. *Thunb.! Prod. p.* 41. *fl. Cap. p.* 206. *Willd. l. c. t.* 3, *f.* 1. *E. & Z.! No.* 1080.

HAB. Mountain sides, Hottentottsholland near Palmietrivier, *E. & Z.!* Zwarteberg near Caledon, *Zey.!* 2649. Dutoitskloof, *Drege.* Nov.–Jan. (Herb. Holm., Willd., Sond., &c.)

Shrub 1–2 feet. Leaves sessile, 5-farious, imbricate, subincurved, 1–2 lines long. Heads the size of a cherry. Exterior leaves of the involucre short, interior 4–6 lines long. Flowers 2 lines, fruit 1 line long.

2. B. affinis (Brogn. l. c.); leaves subulate, acute, closely pressed to the branches, glabrous or a little fringed; lower bractea subulate, glabrous, *longer* than the flowers; segments of calyx shorter than the petals, *glabrous;* anthers *linear-oblong. Linconia capitata, Banks-herb.*

HAB. Cape. (Herb. Banks, n. v.)

3. B. lævis (E. Meyer in Herb. Drege); heads of flowers *turbinate,* corymbose; scales of the involucre foliaceous, lanceolate, obtuse, glabrous, ciliate, the interior *longer than the flowers;* branches *glabrous;* leaves *linear-lanceolate,* carinate, incurved, glabrous or ciliolate; segments of calyx subulate, very villous at the top, shorter than the spathulate, smooth petals.

HAB. Rocky places near Gnadenthal, 3–4000f., *Drege.* Oct. (Herb. D. Sd.)

Leaves 1½–2 lines long. Capitulum about 6 lines long. Leaflets of the involucre in 5–6 rows, twice as large as those of *B. paleacea.* Bracteolæ as long as the petals. Flowers about 3 lines long. Filaments smooth; anthers not seen. Fruit smooth, 1 line long; valves of the cocci at length bifid.

4. B. sphærocephala (Sond.); heads of flowers solitary or aggregated, *globose;* involucre *shorter* than the head; branches *villous;* leaves *lanceolate,* carinate, ustulate, villous-ciliate at the margins and *pubescent on the nerve;* segments of calyx subulate, villous, ustulate, rather shorter than the spathulate, smooth petals. *Brunia microcephala, E. Meyer. non Willd.*

HAB. Dutoitskloof, 3–4000f., *Drege.* Oct.–Jan. (Herb. D. Sd.)

A robust shrub, resembling *Brunia lævis, Thunb.*; distinguished from the preceding by the silky pubescence, larger heads, &c. Leaves multifarious-imbricate, somewhat larger than in *B. lævis*, 1½–2 lines long. Flowering heads about an inch in diameter. Flowers 3 lines long. Bracteolæ not villous. Stamens exserted. Fruit dicoccous, valves at length bifid.

5. B. globosa (Sond.); heads of flowers solitary, scales of the involucre leafy, cuneate, cuspidate, ciliolate, equalling the flowers; leaves subulate, trigonal, incurved, acute, ciliate, or glabrous; bracteolæ and segments of calyx ciliolate or subglabrous, a little longer than the linear-clavate petals. *Brunia globosa, Thunb.! fl. Cap. p.* 205. *B. fragarioides, Willd.! l. c. p.* 128. *Spreng. syst.* 1, *p.* 782. *Berardia fragarioides, Schlecht. Linn.* 6, *p.* 190. *Brun. nodiflora? Drege*, 1908.

HAB. Mountains, Hottentottshollandberge near Palmietrivier, *E. & Z. Zey.* 2650. (Herb. Th., Hm., D., Sd.)

Shrub with the habit of *Br. nodiflora*, much-branched, upper branches filiform. Leaves 4–5-farious, imbricate, 2 lines, in the branches, 1 line long. Heads the size of a large nut or a small walnut. Involucres about 3 lines long. Receptacle hairy. Bracteæ as long as the calyx, sphacelate at the top. Stamens exserted. Styles 2, as long as the petals. Fruit dicoccous. Seeds 2, oblong, black.

6. B. microphylla (Sond.); branchlets *a little spreading*, short, lanate; leaves sessile, 4-farious, imbricate, *rhomboid-ovate*, obtuse, apiculate, *obtuse-keeled, glabrous*, the younger ciliolate; heads of flowers globose, at base involucrated, petals obovate, smooth; ovary *hairy*. *Brunia microphylla, Thunb.! fl. Cap. p.* 207. *Raspalia microphylla, Brogn. l. c. p.* 22, *t.* 3, *f.* 1. *E. & Z.! No.* 1073. *R. teres, E. Meyer.*

HAB. Interior regions, *Thunb.* Hottentottsholland near Palmietrivier, *E. & Z.!* Gnadenthal, 3–4000f., *Drege, Krauss.* Oct. (Herb. Th., Hm., D., Sd., Vd.)

A foot or more in height; branches virgate; ultimate an inch long or shorter. Leaves ½–1 line long and wide, spirally inserted. Heads woolly, the size of a pea, solitary or aggregate. Flowers minute. Calyx adherent to the ovary, villous as well as the bracteolæ. Petals white, with patent limb. Fruit dicoccous.

7. B. angulata (Sond.); branchlets short, *secundate*, lanate, leaves sessile, 4-farious, imbricate, *trigonous-ovate*, subacute, ustulate, *acutely-keeled, downy*, at length glabrate; heads of flowers globose, at base involucrated; petals obovate-oblong, smooth; ovary *very villous. Raspalia angulata, E. Meyer. R. struthioloides, Presl.*

HAB. Rocky places, Gnadenthal and Dutoitskloof, 3–4000f., *Drege*, 6868. Oct. (Herb. D., Hk., Sd.)

Very like the preceding, but distinguished by the more greyish colour, puberous, acutely-keeled leaves, yellowish flowers, and long, hairy ovary. The leafy branches are exceedingly acutely 4-angled; in *B. microphylla*, blunt, 4-angled.

8. B. affinis (Sond.); branchlets somewhat spreading, short, puberous; leaves sessile, 4–5-farious, imbricate, *ovate obtuse, impressed-canaliculate beneath*, downy, at length glabrate; heads of flowers at base involucrated; bracteæ with *long cilia, fringed;* petals glabrous.

HAB. Wupperthal, 1500–2000f., *Drege*, 6867. Dec. (Herb. Sond.)

Like *B. microphylla*, distinguished by obtuse, 1 line long, canaliculate leaves, and imbriate-villous heads. Ovary pubescent. Stamens not seen.

9. B. aspera (Sond.); branchlets somewhat spreading, short, hairy; leaves *short-petiolate, ternate,* imbricate, incurvo-adpressed, *ovate-oblong,* or *ovate, mucronulate,* impressed-canaliculate, *rough* with short, rigid hairs; heads of flowers at base involucrated, bracteæ *rigid-ciliate. Raspalia aspera, E. Meyer.*

HAB. Between Kromrivier and Pietersfontein, sandy places. *Drege, July.* (Hb. Sd.)
Similar to the preceding. Leaves about 1 line long, 3–4 in a whorl. Heads the size of a pea. Flowers not seen.

10. B. phylicoides (Brogn. l. c. p. 25); branches dichotomous, *woolly-tomentose*; branchlets short; leaves sessile, 4–5-farious, imbricate, *obovate-elliptical,* obtuse, callous, *concave,* with *convolute margins,* puberous, at length glabrous, shining; heads of flowers solitary or corymbose, lanate, tomentose, involucrated *to the middle;* calyx and petals *adpressed-hairy* beneath. *Brunia passerinoides, Schlecht. Linn.* 6, *p.* 190. *E. & Z.! No.* 1064.

VAR. *β.* **robusta**; branches more robust, leaves twice larger, obovate-oblong, heads larger. *Brunia phylicoides, Thunb.! fl. Cap. p.* 207. *B. deusta, Willd.! l. c. p.* 127, *t.* 7, *f.* 2, *non Thunb. Phylica squamosa, Willd.! herb.*

HAB. Mountain sides, Zwarteberg, Caledon, *E. & Z.!* Var. *β*, interior regions, *Thunb. & Niven.* Aug. (Herb. Thunb., Willd., Sond.)
A greyish shrub, 2 feet or more in height. Leaves 1 line long or smaller; in var. *β*, 2–2½ lines long, 1 line wide. Heads the size of a pea, greyish-white, in var. *β.* twice as large. Bracteæ black-apiculate, villous, a little shorter than the flower, the laterals smaller. Flowers villous, about 1½ lines long. Petals oblong. Filaments smooth. Ovary superior, very villous. Styles 2, short. Fruit dicoccous.

11. B. Dregeana (Sond.); silky-pubescent, much-branched, branchlets very short; leaves sessile, imbricate, lanceolate, concave, at top ustulate; heads of flowers solitary, at base involucrated; calyx, as well as the oblong petals, silky beneath. *Brunia phylicoides, E. Meyer non Thunb.! Raspalia phylicoides, Presl.*

HAB. Ezelsbank, 3–4000f., *Drege.* Dec. (Herb. D., Sd.)
Branches brown-purplish, pubescent; branchlets very leafy. Leaves about 2 lines long, the upper 1 line long, glabrous above, silky-downy beneath. Heads the size of a pea. Flowers 1 line long. Bracteolæ setaceous, silky, shorter than the calyx. Stamens often twice longer than the yellowish petals. Fruit dicoccous.

VII. STAAVIA, Thunb.

Calyx adhering to the bottom of the ovary, free at the top, ending in 5 subulate, callous lobes. *Petals* free. *Ovary* half-inferior, 2-celled, cells 1-seeded. *Style* simple. *Stigma* 2-lobed. *Fruit* dicoccous. *Brogn. Mem. p.* 22. *Endl. Gen. No.* 4599. *Bruniæ spec. Linn.*

Small shrubs, with linear, spreading leaves, which are callous at the apex. Flowers collected into terminal, disk-like heads, involucrated by numerous, mostly shining, whitish bracteæ, which are either longer or shorter than the leaves. Name in memory of Martin Staaf, a correspondent of Linnæus.

ANALYSIS OF THE SPECIES.

Flowers glutinous ... ... (1) **glutinosa.**
Flowers not glutinous.
  involucre longer than the flowers, coloured:
    leaves linear; heads solitary ... ... (2) **radiata.**
    leaves lanceolate, heads aggregated ... ... (6) **Zeyheri.**

involucre shorter or equalling the flowers, leafy or a little coloured.
capitulum woolly, segments of cal. longer than the petals (4) **globosa.**
capitulum glabrous or a little hairy, segments of calyx shorter than the petals.
branches glabrous, lvs. obl.-linear, trigonal, imbricate (3) **nuda.**
branches puberous, leaves oblong-lanceolate, flat above, spreading or reflexed ... ... (5) **capitella.**

1. S. glutinosa (Thunb.! fl. Cap. p. 207); branches and leaves *quite smooth;* leaves approximate, erect or spreading, linear, *trigonal,* obtuse, callous, mucronulate; heads of flowers usually solitary, bracteæ of involucre erect or stiffly spreading, white, *much longer* than the flowers; flowers *agglutinated with resinous juice. Brogn. l. c. p.* 22, *E. & Z.!* 1074. *Herb. Un. itin. No.* 766. *Brunia glutinosa, Linn. Mant.* 210. *Berg. Cap. p.* 57. *Wendl. coll.* 1, *t.* 22. *St. glutinosa ? et glaucescens, E. Meyer.! in Herb. Drege.*

HAB. On Table Mountain. Oct. (Herb. Th., Sd., &c.)

Shrub 3 feet or more in height. Branches subverticillate, fastigiate, brown-purplish. Leaves small-linear, 6–8 lines long, the uppermost smaller. Petiole adpressed, 1 line long. Heads the size of a cherry. Leaflets of the involucre carinate-trigonous, whitish, at the base greenish, 6 lines long, with a black mucro.

2. S. radiata (Thunb. l. c.); young branches and leaves *pilose;* leaves spreading or deflexed, *linear-acute,* slightly keeled, mucronate; heads of flowers corymbose; bracteas of involucre membranaceous, mucronate, arched, deflexed, white, *a little longer* than the flowers. *Brogn. l. c. tab.* 2, *f.* 2. *Wendl. t.* 82. *E. & Z.! No.* 1075. *Herb. Un. itin. No.* 767. *Zey.!* 726. *Phylica radiata Linn. spec. p.* 283. *Brunia radiata, L. Mant. p.* 209. *Pluk. Mant. t.* 452, *f.* 7. *St. pinifolia, Willd.! l. c. p.* 133, *t.* 3, *f.* 2.

VAR. *β.* **ericetorum** (E. & Z.!); leaves and heads smaller. *Herb. Un. itin. No.* 768.

VAR. *γ.* **glabrata**; branches glabrous, leaves at the base ciliolate or glabrous. *S. Dregeana Presl. St. nuda, E. & Z.!* 1078, *ex pte. Drege,* 6873.

HAB. Cape flats and on mountains near Capetown, Hottentottsholland, &c., very common. Vars. *β.* and *γ.* Cape flats and in Wupperthal and Drakensteenberge. (Herb. Thunb., Willd., Sond., &c.)

Small shrub, 1–2 feet, branches 2–3-chotomous, virgate. Leaves petiolate, 2–3 lines long, a little canaliculate above. Heads the size of a large pea, sometimes smaller or larger. Interior leaflets of the involucre white, a little shorter than the heads.

3. S. nuda (Brogn. l. c. p. 23); branches fastigiate, and, as well as the leaves, *glabrous;* leaves *oblong-linear, short, trigonal,* erect, imbricate; heads of flowers solitary, terminal; involucre *shorter or equal in length* to the leaves, and of *the same colour;* segments of calyx twice *shorter* than the obovate petals. *Brunia verticillata, E. Meyer in Herb. Drege.*

HAB. Dutoitskloof, *Drege.* Oct.–Jan. (Herb. D., Sd.)

Branches verticillate, the ultimate short, filiform. Leaves petiolate, 1–1½ line long, ¼ line wide. Heads the size of a small pea Flowers about 1 line long. Ovary subglabrous. Style glabrous. Fruit dicoccous.

4. S. globosa (Sond.); branches fastigiate, 2–3-chotomous or verticillate, *villous;* leaves *oblong-lanceolate,* nerve-keeled, *ciliate* at the margin, tipped with a short, black mucro; heads of flowers solitary, globose, woolly; involucre *imbricate, shorter* than the leaves and of the *same* *colour*, segments of calyx subulate, *villous*, *longer* than the petals. *Phylica trichotoma et globosa, Thunb.! fl. Cap. p.* 201, 204. *P. elongata, herb. Willd.! St. ciliata, Brogn. l. c. p.* 24, *excl. syn.*

HAB. Sandy places, *Thunberg*. Gnadenthal, Herb. Sond. (Herb. Th., Wld., Sd.)

A small shrub with the aspect of a *Phylica*, 1–2 feet high, erect; the ultimate branches 1–2 inches long. Inferior leaves longer, petiolate, the petiole 1 line long; upper subsessile, subimbricate, 3–2 lines long, slightly concave above. Heads the size of a small cherry. Leaflets of the involucre not different from the upper leaves. Bracteolæ 2, opposite, setaceous, very villous, as long as the flowers. Petals obovate-oblong. Stamens short.

5. S. capitella (Sond.); branches fastigiate, subverticillate, puberous; leaves spreading or reflexed, oblong-lanceolate, obtuse, apiculate, keeled, flat above, at the base ciliate, heads of flowers terminal; interior leaflets of involucre somewhat *spreading, subcolorate, as long as the flowers*; segments of calyx subulate, *hairy, shorter* than the petals. *Brunia capitella, Thunb.! fl. Cap. p.* 206. *S. rupestris, nuda (ex pte.) et ciliata, E. & Z.!* 1077, 1078, 1079.

VAR. β, **composita**; heads of flowers aggregate, forming an ovate spike or corymb. *S. adenandraefolia, E. & Z.!* 1076.

HAB. Mountain sides near Hemel en Aarde and Klynrivier, *E. & Z.!* Zwarteberg, Caledon, *Zey.!* 2647. Var. β. Hauhoeksberge, Stellenbosch, Zwarteberg, *E. & Z.!* Gnadenthal, *Drege*, 6855. Aug.–Oct. (Herb. Th., Hm., Hk., D., Sd.)

Habit of *S. radiata*, 1–3 feet high. Leaves short, petiolate, 1½–2 lines long, ½ line wide. Heads globose, the size of a pea. Leaflets of involucre at base ciliate. Bracteolæ 2, opposite, setaceous, at top hairy. Petals obovate-oblong. Style 1.

6. S. Zeyheri (Sond.); branches pubescent; leaves a little spreading, *lanceolate, channelled above*, keeled beneath, at the top triquetrous, ustulate, glabrous; heads of flowers *aggregated*, forming a dense, ovate, or globose *spike*; exterior scales of involucre foliaceous, interior spreading, coloured, *twice as long* as the flowers; calyx-segments subglabrous; petals at top hairy.

HAB. Rocks on mountains near Appelskraal, Rivier Zonderende. *Zey.!* 2648. Sept. (Herb. Sond.)

Habit of a *Phylica* or of *Stilbe Pinastra*, much more robust than the preceding. Branches subverticillate. Leaves 6–8 lines long, 1 line wide. Petiole adpressed, 1 line long. Spike ½–1 inch long, or twice as long and branched at the base. Heads few-flowered. Exterior scales ciliate at margin, interior yellowish or whitish, channelled, with a black mucro, at the wider base ciliate. Bracteolæ 2, setaceous, hairy, as long as the flowers. Calyx-segments shorter than the oblong, obtuse petals. Stamens equalling the calyx. Style 1. Fruit dicoccous.

VIII. AUDOUINIA, Brogn.

Calyx adhering to the ovary, 5-cleft; segments large, imbricate. *Petals* with a long, 2-keeled claw, and a spreading, roundish limb. *Stamens* included. *Ovary* half-inferior, 3-celled; cells 2-ovuled. *Style* simple, trigonous, terminated by 3 small, papilliform stigmas. *Endl. Gen. No.* 4602.

A heathlike shrub with erect branches and subverticillate, mostly fastigiate branchlets. Leaves spirally inserted, sessile, imbricate, linear, trigonous, bisulcate beneath, scabrous. Flowers crimson, crowded into oblong, spikelike, terminal heads. Named in honour of M. Audouinia, a celebrated entomologist.

1. A. capitata (Brogn. l. c. p. 28, t. 4, f. 1); *E. & Z.! No.* 1085. *Diosma capitata, Thunb.! Prod. p.* 43. *Linn. Mant p.* 210.

HAB. Mountains between Nordhoek and False Bay, *Thunberg;* near Muysenberg, Simonstown, Hanglipp, and Hemel en Aarde, *E. & Z.! Dr. Pappe, Zey.!* 2653. April–Aug. (Herb. Th., Sd., &c.)

About 2–3 feet high. Leaves 2–4 lines long. Flowers about 4 lines long.

IX. **THAMNEA**, Brogn.

Calyx adhering to the ovary at the base, but free at the apex, divided into 5 lanceolate, smooth, scarious, imbricate segments. *Petals* with 2-keeled claws, and an ovate, spreading limb. *Stamens* included. *Ovary* inferior, covered by a fleshy disk, 1- (or 5–2-) celled, many-seeded. *Ovule* hanging from the apex of the column. *Style* simple. *Stigma* entire. *Endl. No.* 4604.

A small shrub with filiform, erect, fastigiate branches. Leaves very small, somewhat rhomboidal, short, blunt-keeled, closely-pressed, spirally inserted; upper ones a little longer than the rest, forming an involucre to the flower. Flowers solitary, terminal, white. Name from θαμνος, a *shrub*.

1. **T. uniflora** (Sol. MSS. Brogn. l. c. p. 30, t. 4, f. 3.)

HAB. Cape of Good Hope, *Masson* in Herb. Banks. (n. v.)

ORDER LII. **HAMAMELIDEÆ**, R. BR.

(By W. SONDER).

Flowers perfect or diclinous, small, regular, in heads or spikes. *Calyx*-tube more or less adnate to the ovary; limb 4–5-cleft or obsolete. *Petals* 4–5, rarely none, inserted in the throat of the calyx or epigynous, more or less valvate in æstivation, deciduous. *Stamens* usually twice as many as the petals, those opposite to them mostly sterile; rarely indefinite: anthers erect, two-celled, each cell either opening by an introrse valve, or slitting at the side. *Ovary* more or less inferior, 2-celled or incompletely 2-celled; ovules solitary, pendulous from the apex of the dissepiment; rarely numerous, and then all but the lowest abortive; styles 2, distinct or 2 sessile stigmata. *Capsule* 2-celled, 2-valved, 2-seeded, loculicidal; or indehiscent *nuts*. *Seed* pendulous, with copious albumen and a straight embryo; radicle next the hilum.

Trees or shrubs, natives of North America, China and Japan, India, Persia, Madagascar, and S. Africa. Leaves mostly alternate (opposite in *Grubbia*), petiolate, simple, penninerved, entire or toothed. Stipules minute, deciduous. Flowers small, white or pink, mostly bracteate. This Order is obviously nearly allied to Saxifragaceæ and Bruniaceæ, from both of which it differs in the structure and dehiscence of the anthers.

TABLE OF THE SOUTH AFRICAN GENERA.

I. **Trichocladus.**—*Flowers* diclinous, spicate. *Petals* 5, linear-clavate, very long. *Stamens* 5, alternate with the petals.

II. **Grubbia.**—*Flowers* perfect, capitate. *Petals* 4, ovate, minute, hairy. *Stamens* 8.

I. **TRICHOCLADUS**, Pers.

Flowers, by abortion, monœcious or diœcious; the female flowers apetalous. *Calyx* 5-cleft, adnate to the base of the ovary, persistent, with valvate æstivation. *Petals* 5, linear-clavate, (much longer than the calyx), their margins revolute, valvate in æstivation. *Stamens* 5, alternate with the petals; filaments short, fusiform; anthers erect, adnate, with valvular dehiscence. *Styles* 2, spreading. *Capsule* didymous, 2-celled; endocarp 2-valved, separating from the two-valved sarcocarp. *Seeds* solitary, pendulous; embryo orthotropous, lying in copious, fleshy albumen; cotyledons broadly ovate, flat; radicle long. *DC. Prod.* 4 *p.* 269.

Small shrubs, natives of S. Africa, with opposite or alternate, ovate or oblong leaves, and densely hairy or pubescent twigs and branches. Flowers white, in dense terminal spikes. Name from *θρίξ, a hair,* and *κλαδος, a branch*:—"Hairy-branch."

ANALYSIS OF THE SPECIES.

Branches and leaves opposite ... ... (1) **crinitus.**
Branches and leaves alternate ... ... (2) **ellipticus.**

1. T. crinitus (Pers. ench. 2 p. 597); branchlets and petioles *opposite*, hirsute; leaves ovate, acute or acuminate, auriculate or obtuse at base, glabrous above, densely clothed with brown stellate hairs below, at length glabrate. *DC. Prod.* 4 *p.* 269. *E. & Z!* 2269. *Dahlia crinita Thunb.! prod. p.* 1, *Act. Soc. Hist. nat. Hafn.*, 2 *vol.* 1, *p.* 133, *s.* 4. *T. vittatus Meisn! in herb. Krauss.*

HAB. Forests in Houtniquas, *Thunberg;* Krakakamma and Vanstadensrivier, Plettenbergsbay, K'nysna, Grootvadersbosch, *E. & Z. Zeyh.* 2657. *Drege* 2311, *Mundt, Alexander Prior, Krauss.* Oct.–Jan. (Herb. D. Sond).

Branches slightly compressed. Petioles 4–6 lines long, rarely longer, as well as the ramuli often blackish villous. Leaves with parallel nerves, prominent on the under surface, 2–6 inches long, 1–3 inches wide. Head of flowers on short peduncles many flowered, sphærical or ovoid, as large as a hazel nut. Male flower: calyx segments obtuse, hirsute, 1 line long, recurved. Petals (white ?) narrow linear, attenuate at base, with involute margins, about 5 lines long. Stamens 2 lines long, the erect, mucronulate anthers opening with 2 oblong lateral valves, equalling the thick filaments. Styles 2, abortive. Female flower: calyx lobes about 1½ lines long. Petals none. Styles 2, divaricate, hirsute at base; stigma punctiform. Capsule hairy, about 4 lines long. Seed white, 2½ lines long. The specimens collected by E. & Z., Drege and Mundt. are diœcious, one in Herb. D. has monœcious flowers.

2. T. ellipticus (E. & Z.! 2270), branchlets and petioles *alternate*, pubescent; leaves elliptic-oblong, acuminate, cuneate at base, glabrous above, reddish or whitish-tomentose beneath.

HAB. In the forests on Bosjesmansrivier, near Philipstown and Balfour, Ceded Territory, *E. & Z.*, District of George, *Mundt., Mrs. F. W. Barber. Drege* 2311. Sept.–Oct. (Herb. D. Sond.)

More slender than the preceding, and distinguished by alternate branches and tomentose, sublepidate leaves. Petioles 2–3 lines, leaves 2–3½ inches long, 3¼–1½ inch wide, shining above. Head of flowers somewhat smaller than in *T. crinitus.* Calyx with shorter lobes; petals 3–4 lines long; stamens very short. Female flower not seen. Capsule 3 lines long. One of the E. Z. specimens has monœcious flowers, the rest is diœcious ?

T. verticillatus (E. & Z.! 2271) is *Bowkeria triphylla, Harv. Thes. Cap. t.* 37 (*Scrophulariaceæ*).

II. GRUBBIA, Berg.

Flowers perfect, capitate, in a diphyllous involucre. *Calyx* adnate with the ovary; limb abortive, truncate. *Petals* 4, epigynous, deciduous, ovate, hairy outside, valvate in æstivation. *Stamens* 8, the 4 alternate with the petals rather longer than the others, all slightly adnate to the base of the petals; anthers 2-celled, minute, roundish, opening lengthwise by introrse valves. *Ovary* inferior, covered by an annular disc, when young (ex *Dne*) 2-celled, with a single pendulous ovule in each cell; afterwards, by a rupture of the septum, falsely one-celled, with an ovule pendulous, as if from the apex of a free central columnar placenta. *Style* very short; stigma bifid. *Nuts* laterally connate, one-seeded, crowned by the disc and style. *Seed* with a straight cylindrical embryo, lying in fleshy albumen; radicle superior, much longer than the narrow, appressed cotyledons. *DC. Prod. XIV. p.* 617.

Small, much branched, South African shrubs, with the habit of *Phylica*. Branches opposite, with swollen nodes, as if jointed. Leaves opposite, exstipulate, with revolute margins. Flowers axillary, minute, three or more united or soldered together, in a bracteate capitulum. Named in honour of Michael Grubb, a Swedish patron of Botany.

ANALYSIS OF THE SPECIES.

Flower-heads 3–2-flowered: lvs. linear or lin-lanceolate:
  Bracts hemispherical, *bifid*, compressed below:
    Branches tomentose or hirsute; lvs. subsessile ... (1) **rosmarinifolia.**
    Branches minutely downy; lvs. petiolate ... ... (2) **pinifolia.**
  Bracts ovate, *undivided*, not compressed ... ... (3) **hirsuta.**
Flower-heads 15–20-flowered; leaves lanceolate ... ... (4) **stricta.**

SEC. I. **Grubbia.** Berg. Klotzsch, Linnæa, 1838, p. 378. DC. l. c. *Ophira* Burm. non Lam. Fruit, consisting of 3, rarely 2 hard, laterally connate nuts, surrounded by 2 scarious bracts. Flowers externally covered with long white hairs.

1. G. rosmarinifolia (Berg.! pl. Cap. p. 90, s. 2), branches *tomentose or hirsute;* leaves *subsessile, linear-lanceolate,* with revolute margins, *hairy and scabrous above,* tomentose beneath; bracts *hemispherical, smooth, bifid, compressed below the fissure;* nuts equalling the bracts, smooth but hairy on the superior margins. *Thunb.! Fl. Cap. p.* 373. *Ophira stricta L. mant. sec. p.* 229.

HAB. Mountains near Cape Town, Hottentottsholland, in the districts of Stellenbosch, Caledon and George, etc. *Zeyh.* 2654. *Drege*! 161. Oct.–Jan. (Herb. Holm. D. Sond.)

Shrub 1–3 feet high with virgate branches. Branchlets terete or somewhat angled, hirsute or glabrescent. Leaves opposite or by abortion of branchlets verticillate, 4–5 lines long, ½–1 line wide; in other specimens 6 lines long, 1½–2 lines wide, subcordate or subauriculate at base, valvate above; petiole not conspicuous, or at most ⅓ line long. Flowers sessile, 1 line long. Bracts chestnut-coloured, rather shorter than the flowers. Fruit 1 line long and thick, about 1½ line wide, enclosed by the bracts, brownish; disk as well as the margins of the perianth hispid. It varies with hairy and subglabrous branches, hirsute or subglabrous, but always scabrous leaves.

2. G. pinifolia (Sond.) branches *minutely downy;* leaves *petiolate, narrow-linear,* obtuse, with revolute margins, *glabrous and smooth above,* shortly pubescent beneath; flowers equalling or shorter than the petiole; bracts hemispherical, smooth, bifid, compressed below the fissure; nuts. . . . . ?

HAB. Mountains near Grietjesgat, Stellenb. 2–4000 feet. *E. & Z.!* Jan. (Herb. Sd.)

Not unlike a small leaved specimen of *G. stricta.* From the preceding it is distinguished by the evidently petiolate, very small and longer leaves. Branches glabrescent. Leaves 1 inch long, ½ line wide, not dilated at the base; petiole 1 line long. Flowers exactly as in *G. rosmarinifolia;* fruit unknown.

3. **G. hirsuta** (E. Meyer) *branches and leaves villous;* leaves subsessile, *linear-lanceolate,* with revolute margins; flowers longer than the petiole; bracts *ovate, smoothish, individed, and not compressed,* twice or thrice as short as the minutely pubescent nuts. *DC. l.* c.

HAB. Mountains in Wupperthal, *Drege.* (Herb. Sond. D.)

Very like *G. rosmarinifolia* but differs by the yellowish appearance and the adpressed hairs. Branchlets often very short. Leaves 3 lines long, ¾ line wide; petiole very short, hispid. Flowers not seen. Fruit compressed, about 1½ line wide, 1 line long and thick, very thinly pubescent on the whole surface.

SEC. II. **Strobilocarpus.** Klotzsh, l. c. p. 380. DC. l. c. *Ophira* Lam. ill. s. 293. non Burm. Many hard nuts, united into a subglobose syncarpium, included at the base by 2 short foliaceous bracts. Flowers externally pubescent.

4. **G. stricta** (DC. l. c.) branches quadrangular and striate, adpressed hairy; leaves petiolate, linear-lanceolate, with revolute margins, glabrous but tuberculate above, silky-pubescent beneath,; the lower ones often much larger, ovate-lanceolate; syncarpium 15–20-flowered, ovoid, when ripe globose; nuts covered with the large, adnate, crustaceous disk. *Taxus tomentosa. Thunb. ! Fl. Cap. p.* 547. *Ophira stricta Herb. Montin. Zeyh. p.* 2650. *Lam. l. c. non. Burm. Strobilocarpus diversifolius Klotzsch. l. c.. G. latifolia Schnizl. Ic. fam. nat fasc.* 13, *p.* 108.

HAB. Mountains in the districts of Cape, Stellenbosch, George and Uitenhage. Oct.–Jan. (Herb. Holm. Thunb. Dubl. Sond.)

An erect, greyish shrub; branches virgate, glabrescent. Leaves 1–1½ inches long, 1½–2 lines wide, with prominent middle-nerve; the lower sometimes 2 inches long, 5–7 lines wide, and evidently 3-nerved; green and sulcate on the upper, yellowish-silky on the under surface; petiole 1–2 lines long. Flowering syncarpium 2–2½ lines long, yellowish. Anthers exactly as in *G. rosmarinifolia.* Fruit about 3 lines long, hard. Lower bracts acuminate, about 1 line long; inner ones (2–4) very minute, fugacious.

ORDER LIII. CRASSULACEÆ, D. C

(By W. H. HARVEY).

Flowers perfect, regular. *Calyx* free, usually 5–4-cleft or parted, (rarely 3–20 parted), the segments imbricate, persistent. *Petals* inserted in the bottom of the calyx, as many as its lobes and alternate with them, free or more or less cohering in a monopetalous corolla, long-persistent, imbricate in æstivation. *Stamens* inserted with the petals, free or adnate to them, as many as the petals and alternate, or twice as many; filaments subulate; anthers 2-celled, splitting. *Squamæ* one at the base of each carpel, sometimes wanting. *Carpels* as many as the petals and opposite them, mostly distinct, each tapering into a style. *Fruit* apocarpous, of several *follicles,* one or many seeded. *Seeds* with a straight, cylindrical embryo, lying in thin fleshy albumen; radicle next the hilum.

Herbaceous or half-shrubby plants, almost always with succulent stems and foliage. Leaves opposite or alternate, fleshy, simple, mostly entire, (rarely ternate or imparipinnate), exstipulate. Flowers very generally in cymes, which are spreading, or dense and subcapitate, sometimes imperfectly umbellate, often corymbose; more rarely in racemes or spikes, or axillary and solitary: often showy, crimson, white or yellow, or of some intermediate colour. A large Order, commonly inhabiting dry places in the warmer temperate zone of the eastern hemisphere; much rarer in America; very abundant in S. Africa. Some of the smaller and less succulent species are found in marshes and on damp ground, and even floating on ponds and in rivulets. In affinity these plants seem nearly allied to *Saxifragaceæ*. Many are cultivated for ornamental purposes.

TABLE OF THE SOUTH AFRICAN GENERA.

Tribe 1. ISOSTEMONES. *Stamens* as many as the petals.

* *Sepals and petals* 4.

I. **Helophytum.**—*Ovules* solitary in each carpel.
II. **Bulliarda.**—*Ovules* several in each carpel.

** *Sepals and petals* 5, *or rarely* 6–9.

III. **Dinacria.**—*Calyx* 5-fid, campanulate. *Petals* clawed, connate at base. *Carpels* with a *hornlike crest* at the back of each style.
IV. **Grammanthes.**—*Calyx* 5-fid, campanulate. *Corolla* salver-shaped, with a short tube.
V. **Crassula.**—*Calyx* 5-parted, stellate or erect. *Petals* free or connate at base, lanceolate or panduriform, sometimes mucronate, or gland-tipped.
VI. **Rochea.**—*Calyx* 5-parted or cleft. *Corolla* salver-shaped, its tube longer than the calyx. *Anthers* subsessile in the throat of the corolla.

Tribe 2. DIPLOSTEMONES. *Stamens* twice as many as the petals.

VII. **Cotyledon.**—*Calyx* 5-parted. *Corolla* 5-lobed.
VIII. **Kalanchoe.**—*Calyx* 4-parted, sepals lanceolate. *Corolla* 4-lobed.
IX. **Bryophyllum.**—*Calyx* inflated, shortly 4-lobed. *Corolla* 4-lobed.

I. HELOPHYTUM, E. & Z.

Calyx 4-cleft or 4-toothed. *Petals* 4, roundish or obovate, spreading. *Stamens* 4, shorter than the petals. *Squamæ* cuneate, truncate. *Carpels* 4; ovules solitary; style short. *Follicles* one-seeded. *E. & Z. Enum. p.* 288.

Water or marsh plants, with weak, filiform, erect or floating, simple or slightly branched stems. Leaves opposite, subdistant, linear or spathulate or subrotund. Flowers axillary, pedicellate, either solitary or in cymules; small, white. Name from ἑλος, a *marsh*, and φυτον, a *plant*.

TABLE OF THE SPECIES.

Stem filiform, slender; flowers solitary, pedicelled, axillary. (1) **natans.**
Stem swollen, hollow; flowers in subsessile, axillary cymules. (2) **inane.**

1. H. natans (E. & Z.! 1843); glabrous; stem filiform, weak, subsimple (mostly floating); lower internodes distant, with linear or spathulate, obtuse or subacute, flat leaves; upper approximate, with spathulate or obovate or subrotund leaves; peduncles axillary, one-flowered, setaceous; calyx 4-toothed, half as long as the subrotund or obovate, spreading petals; stamens shorter than the petals. *Crassula natans, Th. Cap, p.* 281. *Tillæa capensis, Linn. f. sup.* 129.

VAR. α. **fluitans**; stems long and floating; lower leaves linear-elongate, very far apart; upper spathulate. *Hel. fluitans, E. & Z.!* 1844, *ex pte & H. natans*, 1843, *ex pte. H. filiforme, E. & Z.!* 1844, *Drege*, 6876. *Crassula natans, E. Mey.! in Heb. Drege.*

VAR. β. **obovata**; lower leaves spathulate, upper obovate. *H. fluitans, var. obovatum, E. & Z.!* Drege, 6877, 6878.

VAR. γ. **amphibia**; in marshy places, inundated or terrestrial; stems short or longish; leaves equalling or exceeding the internodes, the medial and lower linear, squarrose, the upper spathulate; petals obovate, tapering at base. *Hel. reflexum, E. & Z.!* 1846. *Zey.!* 2510, 2513. Drege, 9540.

VAR. δ. **filiformis**; erect or decumbent; all the leaves linear or linear-oblong. *Bulliarda filiformis, E. & Z.* 1850. *B. capensis, E. M. B. elatinoides, E. & Z.* 1849. *Zey.!* 634. Drege 6883.

HAB. In marshy places, ponds and running streams throughout the colony. (Herb. Sd., D., Hk., Bth.)

Varying with the depth of water and its stillness or fluency. The floating leaves are usually obovate, the submerged linear, and when growing in streams very long and narrow. Stems 3 inches to 3 feet in length, simple or very remotely branched. Leaves from 2 lines to upwards of an inch long, and from ½ line to 2 lines wide. Flowers 1 line wide, white or pale rosy; the anthers red or dark. Petals very obtuse, concave. Carpels obovate.

2. H. inane (E. & Z.! 1847); glabrous; stem terete, swollen (hollow?), subsimple, with distant internodes; leaves shorter than the internodes, connate, ovate, ovato-lanceolate, lanceolate or linear, obtuse or acute, flat; cymules shortly pedunculate; axillary and terminal, several flowered; flowers pedicellate; calyx bluntly 4-lobed, half as long as the ovate petals; stamens shorter than the petals. *Crassula inanis, Thunb. p.* 282, *non E. Mey.! Hel. inane, E. & Z.!* 1847. *Drege!* 6879. *Zey!* 2509.

HAB. In marshes and ditches near the Zwartkops R., Uit., *E. & Z.!* Klein-Drackenstein, near the Berg River; and on the Zuureberg, *Drege!* Grahamstown, *Genl. Bolton!* (Herb. Sd., D., Hk.)

Stems 6–12 inches long or more, according to the wetness or depth of the marsh, 1–2 lines in diameter, pale or strawcolour, weak. Internodes 1–2 inches apart. Leaves very variable in length and breadth, but always broader at base than apex, 3 lines to 1½ inch long. Cymules on peduncles 2–3 lines long, either corymbulose or racemulose. There is often, also, from the same axil as the cymule a 1-flowered, setaceous peduncle. Flowers 1–1½ line in diameter, white. Petals very blunt, one nerved, obovate or spathulate. Carpel oblong, one seeded. Squama cuneate, truncate. Filaments slender; anthers subrotund.

II. BULLIARDA, DC.

Calyx 4-lobed or 4-parted. *Petals* 4, ovate or lanceolate, spreading. *Stamens* 4, shorter than the petals. *Squamæ* linear or cuneate. *Carpels* 4; ovules numerous; style short; *follicles* many-seeded. *DC. Prod.* 3 *p.* 382.

Small, herbaceous plants, mostly annuals, growing in moist places. Stems di-trichotomous. Leaves opposite, linear or obovate. Flowers axillary or in terminal cymes, pedicellate, small, white. Name in honour of *M. Bulliard*, a French botanist,

TABLE OF THE SOUTH AFRICAN SPECIES.

Leaves linear or subulate, acute or subacute:
- Calyx short, with very blunt and shallow lobes ... ... (1) **Vaillantii.**
- Calyx as long as petals, with lanceolate, acute lobes ... (2) **trichotoma.**

Leaves obovate, spathulate or flabelliform, very blunt:
- Calyx-segments with sharp interspaces ... ... (3) **brevifolia.**
- Calyx-lobes with rounded and wide interspaces ... ... (4) **alpina.**

Leaves ovato-lanceolate, acute; sepals and petals lanceolate (5) **Dregei.**

1. B. Vaillantii (DC. Pl. Grass, t. 74); erect or decumbent, dichotomous; leaves linear or subulate, subacute, patent; pedicels equalling or exceeding the leaves; calyx half as long as the ovate petals, with 4 very blunt and shallow lobes. *DC. Prod.* 3 *p.* 382. *E. & Z.!* 1848.

VAR. β. **subulata**; leaves subulate, acute, very slender; stems decumbent. Zey! 634, *a*.

HAB. Margins of small ponds near the Zwartkops River, Uit. *E. & Z.*! Var. β. Buffeljagdrivier, *Zey*. (Herb. Sd.)

Annual, 1–2 inches high, densely much branched. Leaves shorter than the internodes, 2–3 lines long, channelled above. Flowers ½ line long, white. Leaves in β. almost setaceous.

2. B. trichotoma (E.&Z.! 1851); spreading, di-trichotomous; leaves linear or subulate, subacute, patent; pedicels equalling or exceeding the leaves; calyx equalling or exceeding the deltoid petals, deeply 4-parted, with lanceolate, acute segments *Zey! 2511, 75! Drege! 6883.*

HAB. Moist places round Table Mt., *E. & Z.*! Simonstown, *C. Wright*! 550. Draakensteensberg, *Drege*! (Herb. Sd. D.)

A small annual, very like the preceding, but readily known by its deeply parted calyx.

3. B. brevifolia (E.&Z.! 1852); spreading, di-trichotomous; leaves very short, fleshy, linear-obovate, or subrotund, blunt; pedicels filiform, much longer than the leaves; calyx ⅔ as long as the ovate petals, deeply 4-parted, the segments oblong, obtuse, with acute interspaces. *Drege*, 9884, *also Crassula inanis, E. Mey! (ex pte) in Herb. Drege. Zey!* 635.

HAB. Wet spots round Capetown, *E. & Z.*! Greenpoint, *W. H. H.* Simons Bay, *C. Wright*, 549, 555. (Herb. Sd., D.. Hk.)

A small annual, 1–3 inches high, much branched from the base; branches patent. Internodes much longer than the leaves. Leaves 1–2 lines long. Pedicels 5–6 lines long; flowers white, 1 line long.

4. B. alpina (Harv.); spreading, dichotomous; leaves *somewhat petioled*, flat, obovate-spathulate or flabelliform, very obtuse; pedicels filiform, about equal to the leaves; calyx somewhat shorter than the ovate petals, deeply 4-lobed, the lobes oblong, very blunt, roughish, with rounded interspaces. *Petrogeton alpinum, E. & Z.!* 1858. *Crassula umbella, E. Mey! in Hb. Drege.*

HAB. Summit of Table Mt., *E. & Z.*! Hexrivierskloof, *Drege*! (Herb. Sd.)

A minute annual, scarcely an inch high. The leaves are broader and flatter than in *B. brevifolia*, more obviously petioled, and somewhat longer. The calyx lobes are separated by rounded sinuses, and rather longer in proportion to the petals.

5. B. Dregei (Harv.); stem filiform, simple below, dichotomous above, with distant nodes; leaves ovato-lanceolate, acute, rough-edged, flat, thin, translucent when dry, veiny; flowers in a loose, terminal cyme; pedicels filiform; calyx about as long as the concave, lanceolate, acute petals, deeply 4-parted, the segments lanceolate, acute. *Crassula prostrata, E. Mey! in Herb. Drege.*

HAB. Between the Omsamwubo and Omsamcaba, Caffr. *Drege*! (Herb. Sd. Hk.)

This has quite the aspect of a *Crassula*, particularly of *C. centauroides* in miniature. Stems 6–10 inches long, simple for ¾ their length; internodes 1–2 inches long. Leaves 3–5 lines long, 2–3 wide, with purple veins. Flowers 1–1½ line long.

III. DINACRIA, Harv.

Calyx deeply 5-cleft. *Petals* 5, slightly connate at base, with broad, erect claws and spreading or recurved limbs. *Stamens* 5, shorter than the petals. *Carpels* 5, pluri-ovulate, each with a short, dorsal horn at the summit, behind the style; styles short, subulate. *Squamæ* narrow-cuneate, truncate. *Follicles* several seeded.

A small annual, trichotomously branched, with distant nodes. Leaves opposite, obovate or oblong, blunt, fleshy. Cymes corymboso-glomerulate, terminal; flowers small, white. The generic name is compounded of *δις*, *two* and *ακρος*, *a point;* referring to the *apparently* forked apex of the carpel, by which this little plant is known from all others of the Order.

D. filiformis (Harv.); *Grammanthes filiformis, E. & Z.!* 1938. *Zey!* 2517, 637. *Crassula capillacea, b. E. Mey.* (*excl. litt. a.*)

Hab. Sandy places, on hill sides near the Tulbagh Waterfall, *E. & Z.!* Between Capetown and Stellenbosch, *Dr. Pappe!* River Zonderende, and near Driefonteyn, *Zeyher!* (Herb. Sd., Hk.)

Stem 2–3 inches high, trichotomously much branched. Leaves obovate, ovate, or oblong, 3–4 lines long. Corymbs dense, but all the flowers distinctly pedicellate 1–1½ lines long, pale yellow. Limb of the petals elliptic-oblong, blunt. Styles very short. Habit like that of *Crassula glomerata and C. glabra.*

IV. GRAMMANTHES, DC.

Calyx campanulate, semi-quinque-fid. *Corolla* gamopetalous, the tube equalling the calyx, limb 5–6-lobed, spreading. *Stamens* 5–6, adnate to the corolla-tube, shortly exserted. *Carpels* 5–6, pluri-ovulate, with subulate styles, *Squamæ* very minute or obsolete. *Follicles* many-seeded. *DC. Prod.* 3. *p.* 392.

A small, erect, dichotomously branched glabrous and somewhat glaucous annual, with rigid, filiform stems; opposite, distant, fleshy, oblong, ovate or sublinear leaves, and cymoso-paniculate inflorescence. Flowers orange or yellow, or creamy white, each petal (in the full coloured varieties) with a darker mark shaped like the letter V; whence the generic name, from *γραμμα*, a *letter* and *ανθος*, a *flower*.

1. G. gentianoides (DC. Prod. 3. p. 393). *Crassula gentianoides, Lam. Dict.* 2 *p.* 175. *C. retroflexa, Thunb. Cap. p.* 282. *Pluk. Mant. t.* 415. *f.* 6. *C. dichotoma, Linn. Ait. Kew.* 1. *p.* 392.

Var. *α.* **vera**; leaves ovate-oblong; flowers half inch long; calyx-lobes recurved at the point; limb of the petals ovate-oblong, subacute, one-third longer than the stamens. *E. & Z.!* 1934. *G. cæsia, E. Mey. and G. flava, b., E. Mey.*

Var. *β.* **chloræflora**; leaves oblong or linear; fl. ½–¾ inch long; calyx-lobes recurved at point; limb of the petals ovate-lanceolate, subacute, twice as long as the stamens. *E. & Z.!* 1934. *Zey.!* 652.

Var. *γ.* **sebæoides**; leaves oblongo-lanceolate or linear; fl. 4–5 lines long; calyx-lobes *incurved* at the point, very short; limb of the petals broadly lanceolate, subacute, twice as long as the stamens. *G. sebæoides, E. & Z.!* 1936. *G. flava, a. E. Mey.!*

Var. *δ.* **media**; leaves oblong; fl. 2–3 lines long; calyx-lobes erect, very short and blunt; limb of the petals ovate-oblong, blunt, twice as long as the stamens. *Zey.!* 2572.

Var: *ε.* **depressa**; leaves oblong-linear; fl. 1–2 lines long; calyx-lobes erect, blunt; limb of the petals ovate-oblong, blunt, ⅓ longer than the stamens. *G. depressa, E. & Z.!* 1937.

Hab. In sandy ground, throughout the western districs. Var. *α.* about Capetown, Zwartland, and Groenekloof. Var. *β.* Brackfontein, Clanw, and on the seashore near Hott. Holland. Var. *γ.* Mountains of Tulbagh. Var. *δ.* at the River Zonderende, *Zeyher.!* Var. *ε.* Swellendam, *Mundt.* (Herb. Sd., Hk., D. &c.)

Very variable in size, like other annuals of sandy ground, the larger specimens

6 inches high, the smaller 3–2–1 inch. Flowers varying from ¾ inch long to 1½ lines, generally yellow, sometimes pale primrose tint. The leaves are 2–8 lines long, 1–4 wide, a pair at each fork of the stem, the upper ones depauperated. The characters given to the above varieties are by no means constant, and probably intermediate stages between all may easily be found.

V. CRASSULA, L.

Calyx 5-parted or deeply 5-cleft, rarely 6–9 parted, stellate or erect. *Petals* 5, (rarely 6–9), free or connate below, spreading or erect, or erect with recurved points, ovate, obovate, oblong or panduriform, or lanceolate, either simple at the apex or mucronate or gland-tipped. *Stamens* 5 (rarely 6–9) shorter than the petals. *Squamæ* various. *Follicles* several seeded. *DC. Prod.* 3. *p.* 383. *Also, Septas, Lin. DC. l. c. Globulea, Haw. DC. p.* 390. *Curtogyne, Haw. DC. p.* 392. *Rochea, Sect.* 1. *Danielia, DC. l. c. p.* 393. *Sarcolipes, E. & Z.! p.* 290, *Petrogeton, E. & Z.! p.* 291. *Tetraphyle, E. & Z.! p.* 292. *Pyrgosea. E. & Z.! p.* 298. *Sphæritis, E. & Z.! p.* 299. *Thisantha, E. & Z.! p.* 302. *Rochea, E. & Z.! p.* 304.

Shrubby, suffruticose or herbaceous succulents, sometimes annual, very variable in habit and size. Leaves opposite, very generally connate, broad or narrow, flat or semiterete, more or less fleshy, entire, rarely petiolate, frequently cartilagineo-ciliate, either glabrous or pubescent or scaly. Flowers mostly small, white, red or rarely yellow, in cymes or cymules, sometimes solitary, sometimes subumbellate or capitate. The petals vary much in shape and in degree of cohesion, and the genus has (as appears from the copious list of synonyms given above) been subdivided into several "genera" distinguished one from another by differences in the petals. The most obvious of the groups of species so segregated are retained in the 6 sections into which I have divided the genus. I regret being obliged to leave so many species of older authors undetermined; but as most of them have been named in gardens, have never been figured, are not contained in any Herbarium, and have been scarcely more than indicated by the curt descriptive phrases of Haworth, it is quite impossible to make them out satisfactorily. Even Thunberg, who describes 57 species in his Flora, has preserved specimens of but 12 in his Herbarium! The generic name is a diminutive of *crassus*, thick; referring to the succulent or fleshy foliage.

ANALYSIS OF THE SPECIES.

I. **Eu-crassula.** Petals ovate, obovate, oblong or lanceolate, *but not taper-pointed*, often dorsally mucronulate, spreading or reflexed. (Sp. 1–74).

1. Latifoliæ.—Succulent, branching shrubs, with sessile or subconnate, *broad*, flat, fleshy, glabrous, smooth edged leaves, Cymes stalked, corymbose or panicled.

Leaves connate or subconnate. Calyx-lobes shortly lanceolate.
  Lvs. roundish-obovate, obtuse, mucronulate; fl. rosy ... (1) **arborescens.**
  Lvs. narrow-obovate, subacute or acuminate; fl. white (2) **lactea.**
Leaves distinct at base, obliquely obovate; cal-lobes broadly deltoid; flowers red ... ... (3) **portulacea.**

2. Glaucinæ.—Succulent, subsimple shrubs, with connate, *pulverulent-glaucous*, falcate or lanceolate, smooth-edged leaves. Cymes corymbose, densely much branched,

  Lvs. oblong, obliquely falcate, subobtuse ... ... (4) **falcata.**
  Lvs. lanceolate, acuminate, concave above ... ... (5) **perfoliata.**

3. Perfilatæ.—Slender, branching suffrutices, with *connato-perfoliate, roundish or ovate*, smooth-edged or ciliate, glabrous leaves. Cymes corymbose.
  Lvs. roundish or ovate, smooth-edged; cymes terminal, subsessile, dense ... ... (6) **perfossa.**

Lvs. ovate, acute or acuminate, cartilagineo-ciliate; infl. thyrsoid ... ... (7) **perforata.**
Lvs. small, ovato-trigonal, minutely ciliolate; cymes terminal, few flowered ... ... (8) **divaricata.**

4. **Subulares.**—Slender, branching shrubs or suffrutices, with connate or subdistinct, *fleshy, linear-triquetrous* or awlshaped, acute or obtuse, *glabrous* leaves. Cymes corymbose, mostly pedunculate.

Lvs. *linear*-trigonous, 3–4 times as long as thick, *very blunt* ... (9) **brevifolia.**
Lvs. *subulate*-trigonous or cultrate, *acute:*
Lvs. 2–3 inches long, 3–4 lines thick, closely set; petals obovate-oblong ... ... (10) **ramosa.**
Lvs. 1–¾ inches long, ½–2 lines thick, *subdistant:*
Lvs. flattened on both sides ... ... (14) **biplanata.**
Lvs sub-terete or trigonous; bracts *minute*, toothlike.
Lvs. *much* longer than internodes; petals ovate subacute ... ... (11) **tetragona.**
Lvs. not much exceeding internodes; petals oblong-obovate ... ... (13) **acutifolia.**
Lvs. subterete or trigonous; bracts subulate, acute (12) **fruticulosa.**
Lvs. ½ inch long or less, very closely set or imbricate:
Stem dichotomous; lvs. subulate; peduncle elongate, many flowered ... ... (15) **densifolia.**
Stem dichot.; lvs. subulate; cymes *subsessile*, few-flowered ... ... (16) **sarcocaulis.**
Stem branched from base; lvs. cultrate; peduncle short, laxly few-flowered ... ... (17) **alpestris.**

5. **Marginales.**—Virgate suffrutices, simple or branched, laxly leafy. Leaves connate or connato-vaginate, narrow, or oblong, or obovate, *cartilagineo-ciliate*, the surface glabrous or sparsely setose. Cymes corymbose or panicled.

Styles very short or scarcely any: (stem simple, virgate).
Glabrous; calyx-lobes smooth-edged; flowers yellow (18) **vaginata.**
Setose; cal.-lobes ciliate; flowers red ... ... (19) **rubicunda.**
Styles subulate; stem shrubby or suffruticose:
Lvs. narrow-linear or subulate, acute or subacute:
Cymes peduncled; calyx-lobes short, ovate-oblong, obtuse; flowers small ... ... (20) **cymosa.**
Cymes subsessile; calyx-lobes lanceolate-subulate; petals reflexed ... ... (21) **flava.**
Lvs. oblong, tongue-shaped or spathulate; stem decumbent at base, with ascending, simple branches, glabrous.
Lower leaves 1–1½ inch long, 3–4 lines wide ... (22) **undulata.**
Lower leaves 2–2½ inches long, 4–5 lines wide ... (23) **dejecta.**
Lvs. ovate or oblong; stem erect, simple, scabrous above (24) **albiflora.**
Lvs. shortly-obovate; stem diffuse, much branched, glabrous ... ... (25) **rubricaulis.**
Styles subulate; stem herbaceous, erect or decumbent:
Stem more or less albo-setose or hispid:
Erect; lvs. obovate, subrotund or oblong, blunt; cal.-lobes acuminate ... ... (26) **stachyera.**
Erect; lvs. oblongo-lanceolate, subacute; cal.-lobes acute (28) **Meyeri.**
Trailing or decumbent, small; lvs. subrotund or obovate (27) **lasiantha.**
Stem quite glabrous, erect, subsimple; cymes loose, pedunc. (29) **crenulata.**

6. **Squamulosæ.**—Suffrutices or herbs, mostly branching. Stems and foliage more or less clothed with bristles or inflated hairs, or spreading or reflexed scaly-hairs, or scabrous. Cymes corymbose, many or few-flowered.

(1) Stem and foliage densely clothed with *reflexed* flattened hairs or *scales:*
Erect, simple, or corymbose with virgate branches; leaves subulate ... ... (30) **squamulosa.**

Diffuse or decumbent :
Scales copious, broad ; cymes glomerate, with subsessile flowers :
Lvs. linear-lanceolate ; suffruticose ... ... (31) **scabra.**
Lvs. narrow-linear ; suffruticose ... ... (32) **scabrella.**
Lvs. ovate-oblong or ovate ; stem weak, half herbaceous ... ... (34) **Dregeana.**
Scales fewer and very small ; cymes corymbose ; flowers pedicellate ... ... (33) **pruinosa.**
(2) Stem minutely scabrous ; lvs. narrow, scabrous at back and margin :
Cymes much branched, corymbose, many flowered ; fl. pedicellate ... ... (35) **sediflora.**
Cymules simple, 3–5 flowered ; fl. subsessile ... ... (36) **Whiteheadii**
(3) Stem and foliage roughly hispid or albo-pubescent :
Erect, with radical and cauline, oblong or ovato-lanceolate hispid leaves ... ... (37) **setulosa.**
Diffuse, albo-pubescent ; lvs. short, ovato-lanceolate or ovate ... ... (38) **lanuginosa.**
Erect, dwarf, simple ; lvs. fleshy, oblong, obtuse ; cymes 3–5 flowered ... ... (39) **exilis.**

7. Petiolares.—Succulent suffrutices, with glabrous, *distinctly petioled*, ovate, cordate or reniform, crenato-serrate or entire leaves. Cymes laxly trichotomous.

Lvs. cordate-reniform, quite entire and smooth at the margin ... ... (40) **cordata.**
Lvs. broadly cordate, crenate ... ... (41) **spathulata.**
Lvs. ovate, acute, crenato-serrate, on *short* petioles ... (42) **sarmentosa.**

8. Thyrsoideæ.—Erect, succulent herbs, with radical, subrosulate, and depauperated or obsolete cauline leaves. Leaves glabrous or ciliate. Cymules many, in an interrupted thyrsus.

Cymules *capitate* or very dense, *sessile* in the axils of short leafy-bracts :
Rad. leaves oblong or lanceolate, cartilagineo-ciliate or smooth-edged ... ... (43) **Turrita.**
Rad. leaves cuneate, truncate, bearded with long, white hairs ... ... (45) **barbata.**
Cymules *loosely corymbose,* in the axils of leafy bracts ... (44) **corymbulosa.**

9. Rosulares.—Herbaceous, with rosulate, flat, radical leaves and scapelike flowering stems. Leaves cartilagineo-ciliate. Cymes many, in a thyrsus.

Emitting runners ; leaves spathulate-obovate or oblong, obtuse ; fl. subsessile ... ... (46) **orbicularis.**
Without runners ; leaves oblongo-lanceolate, acute ; fl. pedicellate ... ... (47) **rosularis.**

10. Imbricatæ.—Branching suffrutices, with *closely imbricated*, 4-ranked, short leaves. Cymes terminal, corymbulose.

Lvs. deltoid, very fleshy, glaucous ; cymes pedunculate, loosely panicled ... ... (49) **deltoidea.**
Lvs. ovate or ovato-lanceolate, flat, with reflexed edges ; cymes sessile, 3–8 flowered ... ... (48) **ericoides.**

11. Lycopodioides.—Branching suffrutices or herbs, closely covered throughout with small or minute, 4-ranked, imbricate or spreading leaves. Flowers minute, *axillary,* either solitary and *subsessile,* or in axillary fascicles.

Stem woody ; fl. solitary, or 2–3 together, axillary, subsessile.
Axils gemmiferous ; leaves loosely imbricate, ovate or subrotund ... ... (51) **lycopodioides.**
Axils non-gemmiferous ; leaves closely appressed, cordate-ovate, acute ... ... (50) **anguina.**

Stem herbaceous, often annual.; flowers in axillary fascicles:
Lvs., sepals, and petals *much acuminate* or hair pointed (53) **campestris.**
Lvs. ovate or sublanceolate, acute; petals oblong, acute (52) **muscosa.**
Lvs. subulate, taper-pointed; petals oblong, acute ... (54) **subulata**
Lvs. ovato-lanceol. tapering to a blunt point; petals acute (55) **parvula.**
Lvs. rough-edged, ovato-lanceolate, taper-pointed; petals blunt ... ... (56) **bergioides.**

12. Glomeratæ.—Much branched, di-trichotomous annuals, rarely perennial, with small, fleshy leaves. Flowers minute, solitary, or in tufts in the forks of the stem, and glomerato-corymbulose at the ends of the branches.

Dichotomous, *fastigiate*; flowers in the forks solitary:
Lvs. narrow-lanceolate; cal.-lobes equalling or exceeding corolla. ... ... (57) **glomerata.**
Lvs. narrow-linear; cal.-lobes shorter than corolla ... (58) **glabra.**
Di-trichotomous, *prostrate*, with white, angular and compressed branches ... ... (60) **albicaulis.**
Irregularly branched; fl. *in tufts* at the nodes of the stem (59) **decumbens.**

13. Filipedes.—Diffuse or decumbent, weak-stemmed, perennial herbs, with glabrous or pubescent, membranous or fleshy, entire, leaves. Flowers on slender, axillary pedicels, the lower solitary, the upper tufted or subumbellate.

Leaves flat or flattish, *glabrous.*
Lvs. linear-lanceolate, convex beneath; calyx-lobes linear, blunt ... ... (61) **expansa.**
Lvs. cordate or ovate, or oblong-obovate; calyx-lobes lanceolate, very acute ... ... (62) **centauroides.**
Leaves flat, *pubescent.*
Lvs. ovato-lanceolate, acute; calyx-lobes very acute, longer than corolla ... ... (63) **brachypetala**
Lvs. obovate, obtuse; calyx-lobes blunt, half as long as the obovate-oblong petals ... ... (64) **diaphana.**
Lvs. obovate, obtuse; calyx-lobes subacute, equalling the ovate, acute petals ... ... (65) **Sarcolipes.**
Leaves very thick and fleshy, *glabrous.*
Lvs. connate, oblong, obtuse; calyx-lobes and petals obtuse ... ... (66) **peploides.**
Lvs. subglobose or ellipsoidal; calyx-lobes short, obtuse; petals ovate ... ... (67) **dasyphylla.**

14. Crenato-lobatæ.—Slender, weak-stemmed, branching herbs, with *petiolate*, repando-lobate or toothed, glabrous, thin leaves. Flowers loosely cymose or panicled.

Leaves on very long petioles; flowers loosely cymose.
Lvs. reniform, repando-lobulate, calyx-lobes oblong, very blunt ... ... (68) **dentata.**
Lvs. reniform, dentato-lobulate, cal-lobes linear, acute (69) **patens.**
Lvs. subrotund, nearly entire; calyx-lobes blunt; pedicels very long and thread-like ... ... (70) **nivalis.**
Leaves on short or shortish petioles, subentire; flowers in an interrupted thyrsus ... ... (71) **nemorosa.**

15. Tuberosæ.—Root tuberous. Stem herbaceous, simple, with proximate leaf-pairs. Leaves sessile or perfoliate, crenate or subentire. Cymes pedunculate, panicled or umbellate. Flowers sometimes 6–7–9 merous.

Leaves *perfectly confluent* into an orbicular, perfoliate disc (72) **Umbella.**
Leaves subconnate at the base only:
Lvs. roundish flabelliform; cymes laxly panicled, with spreading branches ... ... (73) **flabellifolia.**
Lvs. cordato-reniform; cymes umbellate or corymbulose; petals suberect ... ... (74) **Saxifraga.**

Lvs. cuneate at base, roundish-flabelliform; cymes umbellate; petals lanceolate, spreading, free... ... (75) **Septas.**

II. **Pyramidella.**—Petals much longer than the calyx, tapering above into long, *lanceolate*, channelled points. Leaves imbricated, 4 ranked. (Sp. 76–79).

Stem quite simple:
- Lvs. ovate or deltoid, acute, with reflexed margins; cymes sessile ... ... (76) **pyramidalis.**
- Lvs. orbicular, very obtuse, all closely imbricatad; cymes sessile ... ... (77) **columnaris.**
- Lvs. orbicular, the lower closely imbricated, upper subdistant; cymes subsessile ... ... (78) **semiorbicularis.**

Stem multifid, densely imbricated with broad-based, very acute leaves ... ... (79) **multiceps.**

III. **Sphæritis.**—Petals panduriform, tapering above into narrow-*subulate*, channelled points. (Sp. 80–88).

Inflorescence a single, terminal, corymbose or capitate cyme:
- Virgate, scaberulous; lvs. subul.-acuminate, serrulate (80) **Sphæritis.**
- Virgate, cano-puberulous; lvs. lanceolate-oblong .. (81) **incana.**
- Decumbent, leafy below; lvs. broadly spathul., *ciliate* (84) **ciliata.**
- Dichotomous, *glabrous;* leaves oblong-spathulate, smooth-edged ... ... (83) **clavifolia.**
- Dwarf, simple; leaves oblong, obtuse, *fleshy;* peduncle *hispid* ... ... (88) **hirtipes.**

Inflorescence either a panicle or an interrupted thyrsus:
- Virgate, glabrous; lvs. linear-trigonous, smooth-edged (82) **virgata.**
- Virgate, cano-puberulous; lvs. lanceolate-oblong ... (81) **incana.**
- Erect, hispid; lvs. oblong or obovate, shaggy, with reflexed bristles ... ... (85) **tomentosa.**
- Erect, with subrosulate, roundish, *hispid*, and ciliate lower leaves .. .. (86) **interrupta.**
- Erect, with subrosulate, oblong-obovate, *glabrous*, ciliate leaves ... ... (87) **glabrifolia.**

IV. **Margarella.**—Petals panduriform, *suddenly contracted* at the apex into a gland-like, channelled mucro. (Sp. 89–91).

- Lvs. linear-subulate, subacute, punctulate ... ... (89) **margaritifera**
- Lvs. minute, semiterete, ovate or oblong, scaberulous (90) **subaphylla.**
- Lvs. shortly spathulate or obovate, retrorsely hispid (91) **biconvexa.**

V. **Pachyacris.**—Petals nearly free, lanceolate, with a thickened, *triquetrous*, gland-like point. (Sp. 92).

Whole plant densely hispido-pubescent; lvs. subulate (92) **trachysantha.**

VI. **Globulea.**—Petals panduriform, having an oblong or ovate, fleshy gland immediately behind the blunt apex. Suffrutices with crowded or subrosulate lower leaves, and nearly naked, or barely leafy, flowering stems. (Sp. 93–99).

Lower leaves obovate-oblong, lanceolate-oblong, or cultrate:
- Leaves pubescent-canescent ... ... (97) **canescens.**
- Leaves glabrous, *cartilagineo-ciliate* ... ... (96) **obvallata.**
- Lvs. glabrous, smooth-edged or minutely ciliolate:
  - Stem erect, laxly leafy; leaves obovate-oblong; infl. panicled ... ... (93) **cultrata.**
  - Stem ascending, with lateral, rooting branches, laxly leafy; infl. corymbose ... ... (94) **radicans.**

Stem obsolete; radical leaves rosulate, obovate or subrotund, rigid ... ... (95) **platyphylla.**
Lower leaves semiterete-subulate, channelled above:
Lvs. subpubescent ... ... (98) **nudicaulis.**
Lvs. glabrous ... ... (99) **sulcata.**

Section I. **EU-CRASSULA.** (Sp. 1–75.)

1. C. arborescens (Willd. Sp. 1. p. 1554); stem tall, shrubby, erect, terete; leaves sub-connate, *roundish obovate, obtuse,* mucronulate, fleshy, flat, glaucous, punctate above, glabrous, smooth-edged; cymes panicled, oblong, pedunculate, trichotomous; calyx-lobes very short, lanceolate, keeled, glabrous; petals spreading, slightly connate at base, lanceolate, acute, mucronate, concave, one-nerved; styles subulate; squamæ minute, sessile, obcordate. *DC. Prod.* 3. *p.* 383. *E. & Z.! No.* 1875. *C. Cotyledon, Curt. Bot. Mag. t.* 384. *Jacq. Bot. Misc. t.* 19. *Cotyledon arborescens, Mill. Dict.*

Hab. Among shrubs on the hills near Zwartkops R., Uit. *E. & Z.!* (Herb. Sond).

A large shrub, 8–10 feet high, with robust stems and branches, and fleshy leaves. Leaves 1½–2½ inches long, and 1–2 inches wide. Panicles terminal, 3–6 inches long, with opposite, spreading, trichotomous branches. Bracts small, fleshy, ovate, acute. Calyx ¼ or ⅕ as long as the stellate rose-red petals. Stamens slightly adnate at base, nearly as long as the petals; anthers small and short. Carpels tapering.

2. C. lactea (Ait. Hort. Kew. 1 p. 496); stem shrubby, flexuous, short; leaves connate, *narrow-obovate, subacute or acuminate,* narrowed at base, fleshy, flat, glabrous, punctate within the smooth margin; cyme panicled, oblong, pedunculate, trichotomous; calyx-lobes very short, lanceolate, keeled, glabrous; petals spreading, nearly free, lanceolate, acute, mucronate, concave, one-nerved; styles subulate; squamæ minute, obcordate. *DC. Prod. l. c. Pl. Grass. t.* 37. *Sm. Exot. t.* 33. *Bot. Mag. t.* 1771. *Jacq. Schœnbr. t.* 430. *Thunb. Cap. p.* 289. *E. & Z.!* 1877.

Hab. Among shrubs; Zoutpanshoogde, near Zwartkops Rivier, Uit., *E. & Z.!* (Herb. Sd., Hk., Bth.)

A shrub, 1–2 feet high. Flowers white. Very similar to the preceding in detail, save that the leaves are narrower, more acute, and taper more at base; they are 2½–3 inches long, 1–1½ inch wide, very decidedly connate. Is not this *C. argentea,* L?

3. C. portulacea (Lam. Dict. 2. p. 172); stem tall, shrubby, erect, terete, robust, the branches jointed; leaves opposite, *distinct at base, obliquely obovate,* acute or subacute, fleshy, glabrous, shining, dotted, smooth-edged; cyme corymboso-paniculate, trichotomous, shortly pedunculate; calyx *cup-like, its shallow lobes broadly deltoid,* glabrous; petals nearly free, spreading, oblongo-lanceolate, mucronate, concave; styles subulate; squamæ minute, obcordate. *DC. Prod. l. c. E. & Z.!* 1876. *DC. Pl. Grass. t.* 79. *C. obliqua, Ait. Cotyledon ovata, Mill. Dict. C. articulata, Zuc. Zey.* 2536.

Hab. Among shrubs on the hills and fields near the Zwartkops River, Uit. *E. & Z.!* Aasvogelsberg and Zwaanepoelspoortberg, *Drege!* 6890. (Heb. Sd., D., Hk. Bth.)

A large, succulent shrub, 10–12 feet high, much branched. Leaves very thick, falling off separately, 1–1½ inch long, ¾–1 inch wide. Flowers rosy. Calyx very small, 5 toothed, with wide, shallow spaces between the teeth. The roots are eaten by the Hottentots under the name "T'Karckay," *E. & Z.*

4. C. falcata (Willd. En. 341); stem succulent, suffruticose, simple, pulverulent; leaves connate at base, thick, glaucous, pulverulent, *oblong, obliquely falcate, subobtuse*, the upper-ones degenerating to bracts; cyme corymbiform, trichotomous, much branched; calyx-lobes short, hispido-canescent, ovate or oblong, blunt; petals connate at base, linear-lanceolate, subobtuse, not much longer than the stamens; styles subulate; squamæ minute. *Bot. Mag. t.* 2035. *Rochea falcata, DC. Pl. Grass. t.* 103. *Prod.* 3 *p.* 393. *E & Z.!* 1944. *Larochea falcata, Haw. Syn.* 50. *Tratt. Thes. t.* 20. *C. obliqua, Andr. Rep. t.* 414.

HAB. Hills near Zwartkops R., Uit. and Bothasberg, Grahamstown, *E. & Z.! Drege,* 6918. *Genl. Bolton, &c.* Natal, 1500–3000f. *Dr. Sutherland!* (Heb. Sd., D., Hk. &c.)

A robust, succulent, suffrutex, densely leafy, the leaves diminishing in size upwards. Lower leaves 3–4 inches long, 1–1½ inch wide, strongly reflexed. The surface of leaves and stem is finely granulated or closely and minutely papillate, giving a frosted appearance; the branches of panicle and apex of the stem are finely hispido-pubescent. Flowers bright crimson, rarely white.

5. C. perfoliata (Linn. Sp. 404); stem succulent, suffruticose, simple, pulverulent; leaves connate at base, thick, glaucous, pulverulent, *lanceolate-acuminate, concave above,* the upper ones degenerating; cyme corymbiform, trichotomous, much branched; calyx-lobes short, hispido-canescent, ovato-lanceolate; petals connate at base, linear-lanceolate, subobtuse, not much exceeding the stamens; styles subulate; squamæ minute. *Rochea perfoliata, DC. l. c.* 3. 393. *E. & Z.!* 1945.

β. **albiflora**; flowers white. *Pl. Grass. t.* 13. *Hort. Elth. fig.* 113. *Mill. ic. t.* 108.

HAB. Woods near the Zwartkops R., Uit. *E. & Z.!* (Heb. Sd., D., Hk. &c.)

Very like the preceding except in foliage. Leaves 4–6 inches long, erecto-patent, tapering to a narrow point. Flowers crimson or white.

6. C. perfossa (Lam. Dict. 2. p. 173); stem shrubby, slender, spreading, glabrous; leaves *connato-perfoliate, roundish or ovate,* subacute, thick, glabrous, punctate above, glaucous, *smooth-edged;* cymes terminal, subsessile, dense, oblong or globose, with subulate bracts; calyx-lobes very short, ovate, glabrous; petals connate at base, oblong, obtuse, mucronulate; styles shortly subulate. *DC. l. c.* 3 *p.* 385. *Pl. Grass. t.* 25. *Jacq. Schoenb. t.* 432. *E. & Z.!* 1889. *C. perfilata, Scop. C. punctata, Mill., C. coronata, Don.*

HAB. Among stones, west side of Table Mountain, *E. & Z.!* Nieuweveld, between Brakrivier and Uitvlugt, *Drege!* 6891. (Herb. Sd., Bth.)

Stem 1–2 feet high, woody, dichotomous, the branches short. Leaf pairs close; leaves ¾–1 inch long, ½–¾ inch wide, very fleshy. Infl. dense. Flowers small. Petals reflexed or revolute, dorsally mucronulate, sub-panduriform.

7. C. perforata (Linn. f. Suppl. 190); stem shrubby, slender, erect, subsimple; leaves *connato-perfoliate, ovate, acute or acuminate,* patent or squarrose, subdistant, *cartilagineo-ciliate,* green; cymes in a long, interrupted, contracted panicle (thyrsus), the branches corymbose, dense, with tooth-like bracts; calyx short, ovate-oblong, glabrous, keeled; petals connate at base, oblong, submucronulate; styles shortly subulate. *Thunb. Cap. p.* 287. *DC. l. c.* 3. *p.* 385. *E. & Z.!* 1838. *Zey.* 2534. *C. Anthurus, E. Mey!*

HAB. Woods by the Zwartkops River, *E. & Z.!* Camdebosberg and Fish River, *Drege!* (Herb. Sd., D., Hk.)

The slender branches are often 2 feet long or more, with subdistant or distant leaf pairs, glabrous and angular. Leaves scarcely uncial, ½ inch wide, impunctate. The cartilaginous fringe is generally conspicuous, but sometimes deficient or obsolete. Flowers small. Thyrsus 6–12 inches long, its branches 1 inch long.

8. C. divaricata (E. & Z.! 1891); shrubby, much branched from the base, divaricate, spreading, sub-scaberulous; leaves shorter than the internodes, patent, connato-perfoliate, *small, ovato-trigonal,* fleshy, minutely cartilagineo-ciliate at the margin, acute; cymes terminal, subsessile, forked, 6–7-flowered; flowers minutely pedicellate; calyx-lobes lanceolate, glabrous, keeled; petals ?

HAB. In the Karroo, behind the Langekloof, George, *E. & Z.!* (Herb. Sond.)

A small ligneous fruticulus, 3–6 inches high, with a very thick root. Leaves 2–3 lines long, 1–1½ wide, horizontal, decussate. Perfect flowers not seen. A single imperfect specimen only exists in Hb. Ecklon.

9. C. brevifolia (Harv.); stem shrubby, slender, dichotomous, diffuse; leaves sub-connate, *linear-trigonous, scarcely longer than the internodes,* very thick, *obtuse,* flat above, glabrous; cymes corymbose, terminal, shortly pedunculate, with tooth-like bracts; calyx-lobes short, ovate, blunt, fleshy, glabrous; petals spreading, nearly free, oblong, obtuse, scarcely mucronulate; styles subulate; squamæ membranous, linear.

HAB. Lislap and Springbokkeel, *Zeyher!* 661. (Herb. Hk., Bth., Sd.)

A small bush, 1 foot or more high, with woody stems and short, thick leaves, like those of a *Mesembryanthemum.* Leaf-pairs 6–8 lines apart; leaves 6–8 lines long, 2 lines thick, very blunt. Nearly allied to *C. perfossa,* but differs in the shape of the leaves and their less evident connation. Flowers small (and white ?)

10. C. ramosa (Ait ? fide E. & Z.); stem shrubby, branched at base, diffuse or decumbent; leaves *connato-perfoliate, subulate-trigonous, much longer than the internodes,* very thick, slightly channelled above, glabrous; cymes terminal, pedunculate, trichotomous, corymbose, densely much branched, with toothlike bracts; calyx-lobes very short, ovate, blunt, fleshy, glabrous; petals spreading, nearly free, *obovate-oblong,* bluntish; styles subulate; squamæ short, membranous. *E. & Z.!* 1878. *DC. ? l. c. p.* 384. *Rochea perfoliata, var. glaberrima, E. Mey! in Hb. Drege.*

HAB. Near Louisfontein, Clanwilliam, *E. & Z.!* Between Natvoet and the Gariep, *Drege!* (Herb. Sond.)

Stem short ? woody at base, closely leafy. Leaves 2–3 inches long, 3–4 lines in diameter, very fleshy, tapering to a sharpish point. Peduncle 2–3 inches long, di-trichotomous; the cyme flat-topped, many flowered. Flowers 1–1½ lines long. I cannot tell whether or not this be Aiton's plant; it does not seem to be Thunberg's. My description is taken from Ecklon's and Drege's specimens.

11. C. tetragona (Linn. Sp. 404); stem erect, shrubby, terete, branched; leaves decussately sub-connate, subulate-trigonous, much longer than the internodes, fleshy, flattish above, incurved, acute, glabrous; cyme terminal, pedunculate, corymboso-paniculate, with toothlike bracts; calyx-lobes short, ovate, bluntish, keeled; petals connate at base, spreading, *ovate, subacute;* styles shortly subulate; squamæ minute, emarginate. *DC. l. c. p.* 384. *E. & Z.!* 1879. *Zey!* 2533. *DC. Pl. Grass. t.* 19.

HAB. Woods near Zwartkops R., and Zoutpanshoodge, *E. & Z.!* (Herb. Sd., D., Hk.)

Stem 1–2 feet high, glabrous, corymbosely branched; branches suberect. Leaf pairs 3–6 lines apart; leaves 1½–1¾ inch long. 2–2½ lines thick, widely spreading at base, falcato-incurved. Peduncle 2–3 inches long; panicle much branched, the divisions densely cymose. Flowers small and white.

12. C. fruticulosa (Linn ?); stem suffruticose, spreading, branching; leaves connato-perfoliate, fleshy, subulate, acuminate, glabrous, longer than the internodes; cymes terminal, pedunculate, corymbose, with subulate bracts, few-flowered; calyx-lobes one-third of petals, broadly lanceolate, keeled, subacute; petals connate at base, oblong, subacute, suberect; styles subulate. *Zey! Z. n. N. n. E.* 109. 2.

HAB. Uitenhage? *Zeyher!* (Herb. Sond.)

This has the foliage nearly of *C. acutifolia,* but larger flowers and proportionably longer petals. Flowers white, 2–2½ lines long. Leaves an inch or 1¼ inch long, shrivelling. The specimens are much broken.

13. C. acutifolia (Lam. Dict. 2. p. 175); stem suffruticose, (erect or decumbent), branching; leaves opposite, fleshy, tereti-subulate, acute, patent, glabrous, not much exceeding the internodes; cymes terminal, pedunculate, corymboso-paniculate, with toothlike bracts; calyx-lobes short, oblong, blunt, thick, keeled; petals connate at base, *oblong-obovate,* thin, spreading; styles shortly subulate; squamæ minute. *DC. l. c.* 384. *E. & Z.!* 1882. *Zey?* 2532. *Drege!* 6907. *b.*

VAR. *β.* **radicans;** stem decumbent, rooting, subherbaceous, short; peduncle often forked below the cyme. *Zey!* 2531. *Drege!* 6907. *a. DC. Pl. Grass. t.* 2.

HAB. Hills, &c., near Zwartkops River, *E. & Z.!* both varieties. *β.* Mountain rocks near Welgelegen, *Drege!* (Herb. Sd., Hk., D.)

The erect form resembles *C. tetragona,* but is smaller, with shorter leaves; our var. *β.*, very like DeCandolle's figure, is subherbaceous and diffuse, with larger leaves. Leaves ½–1 inch, in *β.* 1¼ inch long, thick and fleshy. Peduncles 2–6 inches long. Flowers minute, white, densely crowded, in *β.* more lax.

14. C. biplanata (Haw. Phil. Mag. 1824. p. 186); stem suffruticose, erect, with spreading branches; leaves suberect, smooth, subulate, acute, *flat on both sides,* channelled beneath. *DC. l. c.* 384.

HAB. Cape, *Haworth.* (Unknown to us).

I fear to quote *E. & Z.!* 1880 (Herb. Sond.) and which looks very like *C. acutifolia,* as it is impossible to say, from the dried specimen, whether or not the leaves were flat; they seem to have been fleshy.

15. C. densifolia (Harv.); stem suffruticose, dichotomous, fleshy, with *very short internodes;* leaves closely-set, subimbricate, connate, *broadly subulate-trigonous,* fleshy, acute, spreading, glabrous; cymes terminal, *on long peduncles,* corymboso-paniculate, with minute, toothlike bracts; calyx-lobes short, ovate, keeled, blunt; petals slightly connate, oblong-obovate, spreading; styles shortly subulate. *C. bibracteata, E. & Z.!* 1881. (*Vix Haw ?*)

HAB. Sandy and stony ground on the sides of the Devil's Mt., Capetown, *E. & Z.!* (Herb. Sond.)

A small, robust, corymboso-dichotomous undershrub, 8–12 inches high, with very closely placed, shortly subulate leaves. Internodes 2–3 lines long; leaves 5–6 lines. Peduncles 3–4 inches long, of 4–5 joints, with very minute bracts. Flowers white, minute, 1–1½ line long.

16. C. sarcocaulis (E. & Z.! 1884); stem erect, dichotomous, robust, fleshy, with very short internodes, glabrous or scabrous; leaves subconnate, closely set, subulate-trigonous or subterete, glabrous or scabrous, subacute; cymes terminal, *subsessile, few flowered,* corymbulose, with toothlike bracts; calyx-lobes very short, linear, or ovate-oblong, obtuse, keeled; petals nearly free, oblongo-spathulate, erect; stamens thick; ovary contracted at base, styles shortly subulate. *Drege!* 6905. *Zey!* 2535.

VAR. β. **scaberula**; stems, leaves and pedicels scabrous; flowers white.

HAB. Mountain sides near Silo, *E. & Z.!* Witbergen, 5000 ft. *Drege!* Kommandoskaal, *Zeyher!* β. Mountain tops in Graaf-Reinet, *Mrs. F. W. Barber!* (Herb. Sd, Hk., D.)

A stout, corymbose little bush, 6–8 inches high, much branched. Leaves 4–6 lines long, spreading. Flowers, in α, rosy; in β. white, 2–3 lines long.

17. C. alpestris (Thunb?); stem suffruticose, short, branched, with short internodes, nearly glabrous; leaves connate, closely set, *shortly cultrate,* fleshy, keeled, acute or subacute, glabrous; cymes shortly pedunculate, few flowered, corymbose, with toothlike bracts; flowers pedicellate; calyx-lobes not half as long as corolla, oblong-deltoid, subacute, keeled, glabrous; petals oblong, nearly free; styles subulate; squamæ minute, emarginate. *Thunb. Cap.* 285? *fide E. Mey! in Herb. Drege, litt. α.*

HAB. Nieuweveld, between Zakrivierspoort and Leeuwenfontein, 3–4000 ft. *Drege!* Spitskop, 10,250 ft., *Dr. Atherstone!* (Herb. Sd., Hk., Bth.)

Stems 3–4 inches high, chiefly branched from the base, the branches leafy to an inch from the summit. Leaves ½–¾ inch long, 2–3 lines wide, the narrower ones subulate. Peduncles 1 inch long or less; cymes little divided, 7–12-flowered. Flowers white, 2½ lines long. Drege's specimen lettered "*b.*" seems different.

18. C. vaginata (E. & Z.! 1903); stem herbaceous, glabrous, erect, simple, virgate, leafy; leaves *vaginato-perfoliate,* oblongo-lanceolate or lanceolate, the upper ones gradually smaller and more attenuate, all flat, carnoso-coriaceous, sparsely pilose or glabrous, cartilagineo-ciliate; cymes densely corymbose, much branched, subsessile; calyx-lobes half as long as corolla, subulato-lanceolate, acute, *glabrous;* petals suberect, oblong, concave, fleshy; styles very short and squamæ minute. *C. ciliata β, acutifolia, E. Mey! Cyrtogyne, sp. n., Benth! in Pl. Plant.* 82.

HAB. Near Philipstown, Kat R., *E. & Z.!* Between Zandplatt and Komga, *Drege!* Near Grahamstown, *Col. Peddie! Gen. Bolton!* (Herb. Sd., Bth., Hk., D.)

Stem 1–3 feet high, quite simple or merely dichotomous near the summit, glabrous or sparsely setulose. Leaf-pairs 1–2 inches apart; lower leaves 4–6 inches long upper shorter, the uppermost 1–2 inches long. Cymes very large and flat-topped, much branched. Flowers 1–2 lines long, *bright yellow.*

19. C. rubicunda (E. Mey!); stem herbaceous, *setose,* erect, simple, virgate, leafy; leaves connato-perfoliate, oblongo-lanceolate or lanceolate, acute or acuminate, flat, carnoso-coriaceous, glabrous or setose, cartilagineo-ciliate; cymes densely corymbose, much branched, subsessile; calyx-lobes ⅔ of corolla, subulato-lanceolate, acute, *ciliate* and keeled; petals suberect, oblong, concave, acute, fleshy; filaments thick; styles very short and squamæ minute. *Also Globulea stricta, E. Mey!*

HAB. Between the Omtendo and Omsamwubo, and at Natal, *Drege!* Ikubalo, Natal, 2000 ft. *Dr. Sutherland!* (Herb. Sond., Hk.)

Very similar in habit to *C. vaginata*, but with a rough, sometimes a very rough stem, closer leaf-pairs, longer and more tapering leaves, ciliated calyces and *bright red* flowers. Stem 2–3 feet high. Leaves 4–6–8 inches long. Corymb. 4–6 inches across. Flowers 2 lines long.

20. C. cymosa (Linn. Mant. 222); stem suffruticose, erect, branched below, glabrous, leafy, the branches long and simple, laxly leafy upwards; leaves connato-vaginate, *narrow-linear, subacute or obtuse*, flattish, spreading, cartilagineo-ciliate, glabrous; cymes pedunculate, forked or branching, laxly paniculate, with linear bracts; calyx-lobes *short, ovate-oblong, subobtuse*, fleshy, round-backed, glabrous; petals connate at base, oblongo-spathulate, subobtuse, revolute; styles subulate; squamæ truncate. *Bergius, Cap. p.* 84. *Thunb. Cap. p.* 284. *C. subulata, E. & Z.!* 1904, *not of Thunb! Drege!* 6893, 6894.

HAB. Common about Capetown; also near Tulbagh, *E. & Z.! Drege! &c.* Pikeneerskloof, *Zey!* 663. (Herb. Sd., Hk., D., Bth.)

1–1½ feet high, closely leafy below, laxly above, with several erect, simple branches. Leaves 1–1½ inch long, 1 line wide, their common sheath 1–2 lines long. Cymes loosely branched; flowers 2 lines long, white. Bergius's description, above quoted, is full, and very well agrees with our specimens.

21. C. flava (Linn. Mant.); stem suffruticose, *erect*, simple or branched below, glabrous, leafy, virgate; leaves connato-vaginate, *lanceolate or subulate, acute or acuminate*, flat, erect, longer than the internodes, cartilagineo-ciliate, glabrous; cymes subsessile, forked, much branched, corymboso-paniculate or fasciculate, with subulate bracts; calyx-lobes *lanceolato-subulate*, acute, keeled, scabrid-edged; petals nearly free, spathulate, erecto-patent or recurved, mucronulate; styles subulate. *Rochea flava, DC. Curtogyne flava, E. & Z.!* 1942 *& C. Burmanniana, E. & Z.!* 1943. *C. virgata, E. Mey!*

HAB. Hills round Capetown, &c., common. Klynriversberg, *E. & Z.!* (Herb. Sd., Hk., D., Bth.)

1–2 feet high, robust, often quite simple; the branches, when present, are erect and virgate, laxly leafy upwards. Leaves 1½–2 inches long, 2–4 lines wide below, narrowed upwards. Cymes dense. Flowers creamy yellow, the corolla ⅓ longer than the calyx, scarcely hypocrateriform, 3–4 lines long.

22. C. undulata (Haw. Syn. 53); stem suffruticose, flexuous, or at base decumbent, then ascending, with erect, simple branches, laxly leafy upwards; leaves connato-vaginate, *oblong or oblongo-spathulate*, the lower broadest, obtuse, flat, spreading, cartilagineo-ciliate, glabrous; cymes sub-pedunculate, forked, much branched, corymboso-paniculate, with linear bracts; calyx-lobes ½ as long as petals, lanceolate, acute, keeled, glabrous; petals shortly connate, spathulate, erecto-patent or recurved, mucronate; styles subulate. *Curtogyne undulata, DC. l. c. p.* 392. *E. & Z.!* 1940. *Zey!* 2884.

HAB. Base of the mountains round Capetown, on dry ground, common. (Herb. Sd., Hk., D., Bth.)

Woody and commonly decumbent at base, throwing up many erect branches; the barren branches closely leafy with broad leaves, the flowering laxly leafy. Lower leaves 1–1½ inch long, 3–4 lines wide; upper ⅓–⅔ inch long, 2–1 lines wide. Cymes much branched; flowers creamy white, 3–4 lines long. Closely resembles the following, but is smaller in all its parts.

23. C. dejecta (Jacq. Schoenbr. t. 433); stem suffruticose, branched at base or simple, ascending-erect, *tall*, *robust*, densely leafy below, laxly leafy upwards; leaves connate, *broadly* oblong or tongue-shaped, obtuse, flat, spreading, cartilagineo-ciliate, glabrous; cymes sub-pedunculate, forked, much branched, corymboso-paniculate, with linear oblong-bracts; calyx-lobes ⅔ of corolla, lanceolate, acute, keeled, glabrous; petals nearly free, oblongo-lanceolate, erecto-patent or recurved, mucronate; styles subulate. *C. undata, Haw. Suppl. Curtogyne dejecta, DC. l. c.* 392. *E. & Z.!* 1939. *Zey!* 668.

HAB. Near Tulbagh Waterfall, *E. & Z.!* Riebeckskasteel and Rhinosterkloof, *Zeyher!* (Herb. Sd., Hk.)

Stems 2 feet high, robust, surculi densely leafy. Leaves tongue-shaped, the lower ones 2–2½ inches long, 4–5 lines wide; the upper shorter, but scarcely narrower. Cyme much branched, flat-topped. Flowers creamy white, 3–4 lines long. Like a very luxuriant form of the preceding.

24. C. albiflora (Bot. Mag. t. 2391); stem succulent, suffruticose, simple, mostly glabrous except under the *uppermost* nodes; leaves connate at base, *ovate or oblong, acute or obtuse, spreading, glabrous, cartilagineo-ciliate;* cyme corymbiform, densely much branched with scabrous branches; calyx-lobes *lanceolate, acute, glabrous,* smooth-edged; petals slightly connate at base, linear oblong, dorsally mucronate, not much exceeding the stamens; styles subulate; squamæ minute. *Rochea albiflora, DC. Prod.* 3. *p.* 393. *E. & Z!* 1941. *C. dejecta, Drege!*

HAB. Frenchhoek and Drackenstein, *E. & Z.! Verreaux.* Tulbaghskloof, *Zey!* 669. Paarl. *Rev. W. Elliott.* (Herb. T.C.D, Hook, Sd.)

Stem robust, 1–2 feet high, densely leafy; with reflexed, appressed bristles beneath the uppermost nodes, sometimes scabrous throughout. Leaves decidedly connate, 1–1½ inch long, ½–¾ inch wide, squarrose. Cyme very dense, flat-topped, 3–5 inches wide. Flowers white, 4–5 lines long, calyx ½ as long as petals.

25. C. rubricaulis (E. & Z.! 1892); stem suffruticose, *flexuous, diffuse, branching,* closely leafy below, laxly leafy upwards, glabrous; leaves connate, *shortly-obovate,* narrowed at base, obtuse, the uppermost oblongo-spathulate, *minutely* cartilagineo-ciliate, glabrous; cymes subpedunculate, trichotomous, corymboso-paniculate, lax, with linear bracts; calyx-lobes two-thirds of corolla, keeled, *ovato-lanceolate, rough-edged;* petals nearly free, obovato-spathulate, subacute, erecto-patent; styles conniving, shortly subulate.

HAB. Stony places, on the Vanstaadensberg, *E. & Z.!* (Herb. Sond., Hook).

6–12 inches high, much branched and more or less spreading, glabrous or scabrous. Leaves scarcely inch long, 6–7 lines wide above the middle, their fringe very short and fine. Flowers short, 2 lines long.

26. C. stachyera (E. & Z.! 1897); stem herbaceous, erect, simple, (or branched from the base), *albo-hirsute,* leafy throughout; leaves subconnate, the lower *broadly obovate or subrotund,* the upper *elliptic-oblong or oblong,* all subobtuse, flat, thinnish, densely cartilagineo-ciliate, glabrous or sparsely hispid; cymules densely few flowered, disposed either *in a long, leafy spike* or *interruptedly corymbulose*, with oblong bracts; calyx-lobes nearly equalling corolla, lanceolate acuminate, keeled, rough-edged; petals suberect, connate at base, oblongo-spathulate, subapiculate, styles shortly subulate.

VAR. β. **rotundifolia**; leaves subrotund; cymules in a lax raceme. *C. perforata, E. Mey! in Herb. Drege, non Thunb.*

VAR. γ. **pulchella**; dwarf, 2–3 inches high; leaves 3–4 lines long; cymes corymbose; calyx ⅔ of corolla.

HAB. Wet places on the Winterberg, *E. & Z.!* Somerset, *Dr. Atherstone*, 172. Maasstrom, *H. Hutton.* β. at Enon, *Drege!* γ. Rovelo-hills, Natal, *Dr. Sutherland!* in Herb. Hook. (Herb. Sond., Hk., D.)

Stem 6–12 inches high or more. Leaf-pairs 1–1½ inch apart. Leaves about inch long, ¾ inch wide, the uppermost narrower and gradually shorter, when dry membranous and pellucid. The inflorescence varies from a dense spike to a broken corymb; the flowers are subsessile, or very short-stalked, white, 2–3 lines long. *Drege's* 6888, without flowers, may be a dwarf state of this species; or may be an allied and undescribed one.

27. C. lasiantha (E. Mey.!); stem slender, herbaceous, trailing or decumbent, albo-setose, leafy throughout; leaves opposite, *subrotund or broadly obovate*, very obtuse, thin and flat, densely cartilagineo-ciliate, glabrous, or the upper ones setose; cymes shortly pedunculate, corymbose; flowers subsessile; calyx-lobes linear, subcarinate, covered with white bristles; petals connate at base, one-nerved, apiculate; styles shortly subulate; squamæ emarginate.

HAB. Winterhoeksberg, *Drege!* (Herb. Hk., Sd., D.)

Stem 6–8 inches or more long, slender, branched, flexuous; the leaf-pairs scarcely ½ inch apart. Leaves 4–6 lines long, 3–4 lines wide, bordered with white bristles. Flowers small, in dense corymbose fascicles. Near *C. stachyera* but differs in habit, inflorescence, and calyx.

28. C. Meyeri (Harv.); stem herbaceous, erect, retrorsely pilose; leaves opposite, longer than the internodes, spreading, *oblongo-lanceolate, subacute*, flat, thinnish, densely cartilagineo-ciliate, glabrous; cymules densely many-flowered, fasciculate, disposed in a leafy raceme, each fascicle shortly pedunculate, with oblong bracts; calyx-lobes half as long as the recurved corolla, lanceolate, rough at edge and keel; petals nearly free, oblongo-spathulate, mucronulate; styles shortly subulate; squamæ minute, waxy. *C. capitellata, E. Mey.! in Herb. Drege, vix Linn.*

HAB. Between the Omsamculo and Omcomas, *Drege!* (Herb. Sond.)

Stem a foot or more in height, simple? as thick as a goose-quill, tapering upwards, leafy throughout; the internodes about ½ inch apart, more or less hispid with long, swollen bristles. Leaves 1¼–1½ inch long, 3–4 lines wide, membranaceous when dry. Flowers white, 2–3 lines long. This does not agree with the character given of *C. capitellata*, Linn. & Thunb.

29. C. crenulata (Linn.?); stem herbaceous, erect, subsimple, glabrous; leaves opposite, oblongo-lanceolate, obtuse, *narrowed or attenuated at base*, the lower much longer than the internodes, the upper degenerating, flat, thinnish, glabrous, *finely cartilagineo-crenulate*, the crenatures papillate; cymes *pedunculate*, trichotomous, laxly corymbose, much branched, with tooth-like bracts; calyx-lobes very short, acute, glabrous, margined; petals nearly free, oblongo-spathulate, mucronate, scarcely exceeding the stamens; styles subulate; squamæ very small. *DC. ? l. c. p.* 388. *Thunb. Cap. p.* 287 ?

HAB. Vanstaadensberg, Uit. *C. Zeyher!* 2530. (Herb. Sd., D., Bth., Hk.)

A tall, subsimple (?) succulent herb, 2 feet high or more. Lower leaf-pairs an inch apart, upper 2–3 inches. Leaves 3–4 inches long, ½ inch wide, membranous when dry, shrinking; the margin minutely crenato-denticulate. Cyme at length 4 inches in diameter, flat-topped, with long pedicels. The naked part of the stem or peduncle is 6–8 inches long, with a single pair of depauperated leaves in the middle.

30. C. squamulosa (Willd. ? Suppl. 15); stem suffruticose, erect, branching, corymbose, retrorsely scaly; leaves *connate-perfoliate*, subulate-attenuate, squarrose, channelled, retrorsely scaly; cymes subsessile, *densely* corymbose, the lower flowers stalked, upper subsessile; bracts subulate; calyx-lobes ½ of corolla, linear-lanceolate, acute, retrorsely scaly; petals connate at base, lanceolate-spathulate, mucronate, spreading or revolute; styles subulate, equalling the roughish ovary. *DC. l. c.* 385. *E. & Z.!* 1885. *Globulea mesembryanthoides, E. Mey.! in Herb. Drege!*

HAB. Tulbagh Waterfall and Winterhoek, *E. & Z.!* Gamke River, 666, and Riebeckskasteel, *Zey!* 667. Between the Paarl and Pont, near the Berg River, *Drege!* (Herb. Sd., Hk., D.)

1–1½ foot high, erect, with virgate, simple branches. Leaf-pairs nearly an inch apart, the bases of the leaves forming a cup round the stem; leaves 1–1½ inch long, erect at the insertion, curving back and often revolute. Cymes 2–3 inches in diameter. Flowers white (?) 4–5 lines long.

31. C. scabra (Linn. Sp. 405); stem suffruticose, ascending or diffuse, terete, branched, retrorsely scaly; leaves subconnate, patent or reflexed, *linear-lanceolate*, acute, flat, retrorsely scaly on both sides, the uppermost degenerating; cymes subsessile, *corymboso-fasciculate*, with subulate bracts and *subsessile* flowers; calyx-lobes half as long as the corolla, ovato-lanceolate, mucronate, retrorsely scaly or glabrate; petals connate at base, linear-oblong, subspathulate, recurved or revolute, submucronulate; styles subulate, ovaries rough. *DC. l. c. p.* 384. *E. & Z.!* 1886. *Dil. Elth. fig.* 117.

HAB. Hill sides round Capetown, *E. & Z.! Villett. W.H.H. &c.* (Herb. Sd. D. Hk.)

A foot high, flexuous or spreading, branched near the base, every part densely clothed with reflexed, whitish, membranous, swollen hairs or linear scales. Leaves ¾–1 inch long, 2 lines wide, horizontal or squarrose. Cymes almost capitate. Corolla 3 lines long or more, revolute, white (?). Calyx sometimes quite smooth!

31. C. scabrella (Haw. ? rev. suc. 11); stem suffruticose, flexuous, diffuse, slender, terete, much branched, retrorsely scaly; leaves subconnate, patent or squarrose, *narrow-linear*, acute, channelled, retrorsely scaly; cymes subsessile, few-flowered, corymboso-fasciculate, with subulate bracts and subsessile flowers; calyx-lobes ⅔ of corolla, lanceolate, mucronate, retrorsely scaly; petals connate at base, spathulate, recurved or revolute, apiculate; styles subulate. *DC. l. c.* 384. *E. & Z.!* 1887.

HAB. Sandy places at the foot of Table Mt., *E. & Z.! W.H.H., McGillivray! Drege!* 6904, &c. (Herb. Sd., D., Hk.)

Much smaller in all parts than *C. scabra*, less densely squamulose, with narrower and more linear leaves; otherwise very similar, and perhaps a mere dwarfed variety.

33. C. pruinosa (Linn. Mant. 60); stem suffruticose, slender, dichotomous, retrorsely squamulose; leaves subconnate, patent, *linear-subulate, subtrigonous, laxly squamulose;* cymes subsessile, few-flowered, corymbulose, with subulate bracts and *pedicellate* flowers; calyx-lobes half as long as corolla, linear-lanceolate, *sparsely scabrous;* petals linear-spathulate, recurved or revolute; mucronulate; styles subulate. *Zey!* 665. *Drege!* 6906.

HAB. At the 24 Rivers, *Zeyher!* Boschkloof and Blauwberg, *Dregê!* (Herb. Sd., Hk., Bth.)

A much branched, corymbose, *laxly and minutely* scaly suffrutex, 6–12 inches high; the level topped branches ending in small cymes of 3–6–8 *pedicellate* flowers. Leaves ½–¾ inch long, slender, scarcely equalling the internodes. Pedicels 2–3 lines long; corolla 5 lines, hypocrateriform or revolute (cream coloured?). Less scaly than *C. scabrella,* with a different inflorescence and much larger flowers.

34. C. Dregeana (Harv.); stem sub-herbaceous, slender, diffuse, branching, retrorsely hispido-squamose; leaves connate, patent, *short, ovate-oblong or ovate, subacute, flat,* retrorsely squamose; cymes sessile, few-flowered, *capitate,* with subulate bracts and *subsessile* flowers; calyx-lobes ⅔ of corolla, lanceolate, acute, *ciliate at back and keel;* petals oblong, mucronate, recurved; styles subulate; squamæ minute, emarginate. *C. squamulosa, E. Mey.! in Herb. Drege.*

HAB. Between Omsamculo and Omcomas, *Drege!* (Herb. Sd., Hk., D., Bth.)

A slender, weak-growing species with leaves, proportionately broader and shorter than others of this group, and small headlike tufts of whitish flowers. Leaves 3–4 lines long, 2–2½ broad, horizontal. Heads 3–6–8-flowered. Corolla 3–4 lines long.

35. C. sediflora (E. & Z.!); stem suffruticose, diffuse or decumbent, with ascending branches, *minutely scaberulous;* leaves connate, patent, *linear-lanceolate,* flat, subconcave above, *thinnish,* subobtuse, *scabrous at back and margins;* cymes subsessile, much branched, corymbose, with subulate or toothlike bracts and *pedicellate* flowers; calyx lobes ½ of corolla, lanceolate, glabrous, fleshy, keeled; petals connate at base, oblong, blunt or submucronulate, spreading; styles shortly subulate; squamæ small, fleshy. *Pyrgosea sediflora, E. & Z.!* 1909.

HAB. Sides of Kat River, Berg. *E. & Z.!* (Herb. Sd., D.)

Many stemmed, forming wide patches; stems 1–2 feet long, the erect, lateral branches 4–6 inches long, leafy to the summit. Leaves squarrose, 6–8 lines long, 1 line wide, green. Flowers small, 1–1½ line long, white. Scarcely belonging to this section.

36. C. Whiteheadii (Harv.); stem suffruticose, erect, somewhat *fastigiately much branched,* terete, *minutely albo-strigillose;* leaves subconnate, erecto-patent, *shorter than the internodes,* the lower linear, the upper shorter and broader, sub-ovate, all keeled or trigonous, *ciliate at the margin, otherwise glabrous* or sparsely hispidulous; *cymules sessile, 3–5 flowered,* corymbose, with tooth-like bracts and *subsessile* flowers; calyx-lobes ⅔ of corolla, ovate-oblong, subacute, keeled, ciliate and hispidulous; petals connate, oblong-subpanduriform, recurved or revolute; styles shortly subulate.

HAB. Ezel's Fonteyn, Namaqualand, *Rev. H. Whitehead!* (Herb. D., Sd.)

Stems woody, 4–6 inches high, much branched below, all the branches erect and level-topped, reddish-brown. Leaves $1\frac{1}{2}$–$2\frac{1}{2}$ lines long, $\frac{1}{3}$ line wide, mostly shorter than the internodes. Cymules ending all the branches. Flowers 2 lines long, white.

37. C. setulosa (Harv.); stem herbaceous, erect, subsimple, *rigidly hispid*, leaf-pairs subdistant; lower leaves oblong or obovate, upper ovato-lanceolate, all membranous, rigidly hispid on one or both surfaces and ciliate, (or glabrous except the margin); cymes corymboso-paniculate, many flowered; calyx-lobes ovate, acute, $\frac{2}{3}$ of suberect corolla, ciliate with or without dorsal bristles; petals sub-connate at base, oblong, concave, contracted in the middle; styles shortly subulate; scales minute. *Zey.!* 650.

HAB. Doornkop, *Burke & Zeyher!* 401 (650). (Herb. Hk., Sd., D., Bth.)

Radical leaves subrosulate, 2–4 pairs close together. Stems 4–8 inches high, one or more from the same root, simple or branched near the base. Panicle trichotomous or forked, spreading; its divisions densely corymbose. Flowers 1–$1\frac{1}{2}$ line long, white. The whole plant generally clothed with spreading bristles; but the larger leaves sometimes quite bare; sometimes clothed on one side only.

38. C. lanuginosa (Harv.); herbaceous, diffuse, branching, *clothed with short, patent, white pubescence;* leaf-pairs lax; leaves sub-distinct, short, ovato-lanceolate or ovate, acute, sub-oblique, spreading; cymes terminal, subsessile, few flowered, forked; flowers pedicellate; calyx-lobes lanceolate, acute, strigose; petals one-third longer than the calyx, connate at base, oblong, contracted at the sides; stigma subsessile. *C. strigosa, Drege, non L. & C.* 6901. *Drege.*

HAB. Gaatje, near the Stormberg, 5000f. and Nieuwe Hantam, 4500–5000f. *Drege!* (Herb. Sd., Hk., Bth.)

A small, branching, weak-stemmed plant, 3–6 inches long, everywhere equally albo-pubescent. Leaves 3–4 lines long, 1–2 lines wide, when dry somewhat cultriform. Peduncle very short, once or twice forking or unilaterally cymose; the cyme 6–12-flowered. Flowers 1 line long,

39. C. exilis (Harv.); stem short, erect, succulent, simple, closely leafy, scabrous; leaves connate at base, *oblong, very thick and fleshy, blunt*, minutely scabro-pubescent; peduncles terminal, short, slender, pubescent, bearing a 3–5-flowered, corymbulose cyme; flowers subsessile; calyx-lobes ovate-oblong, blunt, pubescent, half as long as the recurved, obovate petals; styles subulate; squamæ minute, fleshy.

HAB. Namaqualand, *Rev. H. Whitehead!* (Herb. T.C.D.)

Stems one or several from the same root, 1–2 inches high, 2–3 lines in diameter. Leaf-pairs 2–3 lines apart. Leaves $\frac{1}{2}$ inch long, 2–3 lines wide, apparently semi-terete. Pubescence minute, but copious, white. Flower stem about an inch high; flowers $1\frac{1}{2}$ line long, white. Remarkable for the disproportion between its clumsy stems and fleshy leaves, and the slender peduncle and small size.

40. C. cordata (Ait. Hort. Kew. 1, p. 396); stem shrubby, slender; leaves opposite, *petiolate, cordate-reniform*, obtuse, cuneate at base, dotted above, glabrous, *quite entire*, smooth-edged; cymes pedunculate, panicled, laxly trichotomous with cordate bracts; calyx-lobes short, glabrous, ovate, acute, one-nerved; petals free, spreading, lanceolate; styles subulate; squamæ minute. *DC.l c.*3, 386. *E. & Z.!* 1898. *DC. Pl. Grass. t.* 121. *Jacq. Schœnb. t.*431. *Willd. Sp.* 1, *p.* 153 (*non Thunb.*) *C. perfossa, E. Mey.! in Herb. Drege.*

HAB. Among shrubs, &c., near the Zwartkops R., Uit., *E. & Z.!* *Zey.!* 2528. Near the Fish R., and Glenfilling, *Drege!* (Herb. Sd., D., Hk., Bth.)

A slender, succulent, shrubby-plant, 1–3 feet high, erect or diffuse, sometimes rooting at the nodes. Leaf-pairs ½–1 inch apart; petioles 2–8 lines long, cuneate; leaves ¾–1 inch in diameter, conspicuously dotted. Panicles very lax, on longish, terminal peduncles. Flowers small, white.

41. C. spathulata (Thunb.! Fl. Cap. 293); stem suffruticose, slender and weak, spreading, branched below; branches long and simple, 4-angled; leaves opposite, *petiolate, broadly cordate,* subacute, cuneate at base, glabrous, *crenate;* cymes pedunculate, corymboso-paniculate, di-trichotomous, with tooth-like bracts; calyx-lobes very short, glabrous, oblongo-lanceolate, keeled; petals free, spreading, lanceolate; styles subulate; squamæ minute. *DC. l. c. p.* 386. *Pl. Grass. t.* 49. *E. & Z.!* 1899. *Zey.!* 2529. *C. lucida Lam. C. cordata, Lodd. Cab. t.* 359.

HAB. Shrubby places near the Zwartkops R., *E. & Z.!* Zoutpanshoodge and near the Bushman's R., *Zeyher!* (Herb. Sd., Hk., D.)

More slender and trailing than the preceding, with a shorter, more corymbose, and denser panicle, and readily known by the crenate leaves. Flowers flesh-coloured.

42. C. sarmentosa (Harv.); stem suffruticose, sarmentose, trailing or climbing, very long, simple; leaves opposite, *shortly petiolate,* the uppermost subsessile, *ovate, acute,* glabrous, shining, *crenato-serrate;* cymes subsessile, laxly panicled, trichotomous, with linear bracts; calyx-lobes very short, lanceolate, cartilage-edged; petals nearly free, spreading, linear-lanceolate, acute; styles subulate, attenuate; squamæ minute. *C. ovata, E. Mey.! MS. in Herb. Drege.*

HAB. Hills of Omblas, near Natal, 500f. *Drege!*—Cult. in England. (*v. v.* cult.; and in Herb. Sond.)

Stem 10–20 feet long or more, simple, scandent. Leaf-pairs 2–4 inches apart; petioles 2–5 lines long; lamina 1–1¾ inch long, ¾–1 inch wide. The serratures are minute, but sharply cut, and the edge is cartilaginous. Branches of the cyme long and widely spreading. Cultivated at Kew and by Mr. Wilson Saunders. Its vine-like stems, if permitted to grow, would trail round a large conservatory. As the *sarmentose* habit is very unusual among Crassulaceous plants, while ovate leaves are common, I have ventured to change Meyer's manuscript name.

43. C. Turrita (Thunb. Fl. Cap. 285); herbaceous; radical leaves subrosulate, spreading, oblong or lanceolate, acute or subacute, glabrous, cartilagineo-ciliate; stem simple, terete, leafy below, floriferous above; cymules *capitato-fasciculate, sessile* in the axils of opposite, leafy bracts, forming an interrupted thyrsus; calyx-lobes ⅔ of corolla, glabrous, keeled; petals oblong, fleshy, concave, dorsally subumbonate; stigma sessile; squamæ minute. *DC. l. c.* 388. *Jacq. Schoenbr. t.* 52. *Pyrgosea Turrita, Haw. E. & Z.!* 1905. *Zey.!* 2543, 2544.

VAR. β. **latifolia**; leaves broadly oblong or obovate, subobtuse, short. *Pyrgosea pyramidalis, Zey!* 2546, 2545. *C. thyrsiflora, litt. b. & d. Drege!*

HAB. Fields near the Zwartkops R., Uit., *E. & Z.!* Olifants R., *Thunb.* β. near the Zwartkops; also on Amsterdamvlakte, Algoa Bay, *Zey.!* (Herb. Sd., D., Hk.)

Leaves crowded near the base of the stem, 2–3 inches long, ¾–1 inch wide, varying from lanceolate to oblong and obovate. Stem 6 inches to nearly 2 feet high, tapering with depauperated leaves. Flowers in a leafy, spicate thyrsus; the tufts sessile. Corolla 1 line long.

44. C. corymbulosa (Link. Enum. 1, p. 301); herbaceous; leaves opposite, decussate, in proximate pairs, all cauline, diminishing upwards, lanceolate, acute or acuminate, glabrous, either smooth-edged or papillato-ciliolate or ciliate; stem erect, simple or with lateral, floriferous branches; cymules *loosely corymbulose*, in the axils of opposite, leafy bracts, forming a long thyrsus; calyx-lobes small, deltoid, glabrous; petals nearly free, oblong, thin, concave, dorsally subumbonate; stigma subsessile; squamæ fleshy, emarginate. *Link. & Otto. Abild. t.* 16. *fide E. & Z.!* 1906. *Zey.!* 2541, 2542. *Also Pyrg. thyrsiflora, E. & Z.!* 1907 *& P. aloides, E. & Z.!* 1908.

HAB. Zoutpanshoodge, Uit. *E. & Z.!* Kommandoskraal, Zondag R., *Zey.!* (Herb. Sd., D., Hk.)

Root biennial? stem 6–12 inches high, leafy throughout, ending in a thyrsus of small, white, loosely corymbulose flowers. The stem and edges of leaves are usually smooth; but sometimes the leaves are either papillate or ciliate on the edges, and the stem, especially in its upper half, more or less densely clothed with deflexed cartilaginous hairs. Flowers scarcely 1 line long. *C. acuminata, E. Mey.!* in Herb. Drege, according to a poor specimen in Herb. Sonder, is very near this species if not the same.

45. C. barbata (Thunb. Cap. 292); herbaceous; radical leaves subrosulate, spreading, membranous when dry, *cuneate-flabelliform, bearded along the truncate apex with long white hairs;* stem simple, scapelike, with opposite subdistant leaf-scales below, angular, floriferous above; cymules capitato-fasciculate, sessile in the axils of opposite, scale-like bracts, forming a long spicate-thyrsus; calyx-lobes short, ovate, nerved, glabrous; petals *connate* at base, oblong, sub-spathulate, suberect with recurved points; stigma sessile; squamæ shortly cuneate. *DC. Prod. l. c.* 388. *Zey.!* 655. *Burke,* 461.

HAB. Rocky places of the Hantum Mts., in the Roggeveld, *Thunberg.* Geelbeck Rivier, *Burke & Zeyher!* (Herb. Hk. Sd.)

The specimens are in an advanced state, with withered leaves. Leaves numerous, 1–2 inches long, with a spathulate petiole, expanding into a shortly cuneate lamina, 1–1½ inch broad, abruptly cut off at top, and fringed with rigid hairs 2–4 lines long. Flowering stem 12–18 inches high, more than half of it occupied by the spiked inflorescence, whose tufts are ½–1 inch apart. A remarkable species.

46. C. orbicularis (Linn. Mant. 361); herbaceous, and *emitting runners;* leaves radical, horizontally spreading, imbricating, rosulate, *spathulate-obovate, or oblong, obtuse,* carnoso-coriaceous, flat, glabrous, cartilagineo-ciliate; peduncle scapelike, leafless; cymes densely fascicled, in an interrupted thyrsus, fascicles opposite, shortly peduncled; flowers *subsessile,* bracteate; calyx-lobes half of corolla, oblong or ovatolanceolate, ciliate; petals obovate-oblong, apiculate; styles very short. *DC. l. c.* 389. *E. & Z.!* 1900. *Dill. Elth, fig.* 118. *DC. Pl. Grass t.* 43. *C. sedoides, Mill. C. hemisphærica, E. Mey.! in Herb. Drege. C. thyrsiflora, litt. e., Drege?*

HAB. Dry hills near the Gauritz R., Swell., *E. & Z.!* Nieuweveldt, near Beaufort, *Drege!* (Herb. Sond.)

Crown throwing out lateral runners and offsets. Leaves spreading in a circle, the undermost 2 inches long, the overlying ones gradually smaller, all more or less obovate, obtuse or scarcely subacute, fringed with cartilaginous cilia. Peduncle 6–8 inches long, with 3–5 pairs of opposite fascicled cymules. Flowers creamy white, 1–1¹ line long.

47. C. rosularis (Haw. Rev. p. 13); herbaceous, *without runners;* leaves radical, sub-horizontally spreading, imbricate, rosulate, *oblongo-lanceolate or spathulate or strap-shaped, acute or subacute,* carnoso-coriaceous, flat, glabrous, cartilagineo-ciliate; peduncles scapelike, leafless or with depauperated, distant, opposite leaves; cymes in a thyrsus or a branching panicle; flowers *pedicellate;* calyx-lobes short, oblong, ciliolate; petals free, obovate, apiculate; styles short. *DC. l. c.* 389. *E. & Z.!* 1901. *Zey.!* 2539. *Drege!* 6897.

HAB. Woods near the Zwartkops, Uit. and Kat-river Settlement, Caffr., *E & Z.!* Adow, *Zey!* Klein and Groot Fish R., *Drege!* Near Grahamstown, *Mr. Hutton!* Natal, *Dr. Sutherland!* (Herb. Sd. Hk., D., Bth.)

Very like the preceding but with much longer and sharper leaves, taller flower-stem and more branching inflorescence; each flower on a distinct, short pedicel. Leaves 3–5 inches long or more, ½, ¾–1½ inch broad, mostly acute, sometimes acuminate, rarely blunt. Flowers small, white.

48. C. ericoides (Haw. Phil. Mag. 1825); stem suffruticose, erect, dichotomous, fastigiate, naked below, *closely imbricated with leaves* above; leaves opposite, *ovate or ovato-lanceolate,* subcordate at base, erect, flat, with sub-recurved margins, glabrous; cymules sessile, terminal, 3–8-flowered; calyx-lobes nearly equalling the corolla, linear, obtuse, glabrous; petals nearly free, oblong, acute, concave, nerve-keeled; styles subulate; squamæ minute. *DC. l. c. p.* 385. *Tetraphyle furcata, E. & Z.!* 1866. *Zey.!* 2521, 2522, 2523. *Drege!* 6903.

HAB. Sandy flats between Krakakamma and Vanstaadensberg; also Quaggavlakte and Adow, and near Grahamstown, *E. & Z.!* Cradockstadt; Zoutpanshoodge, Uit.; and Winterhoeksberg, *Zeyher!* Aasvogelsberg, *Drege!* (Herb. Sd., Hk., D., Bth.)

A much branched, densely leafy suffrutex, with annulated, slightly pubescent or glabrate stems or branches. Leaves quadrifariously imbricated in closely alternating pairs, 3–4 lines long, 2–3 lines wide, subacute. Flowers few at the ends of the leafy branches, white, 2 lines long.

49. C. deltoidea (Linn. f. Suppl. 189); suffruticose, dwarf, branching from the crown; leaves closely imbricated, 4-ranked, deltoid, fleshy, keeled, spreading, pulverulent-glaucous; peduncles terminal, filiform, elongate, pulverulent; cymes loosely panicled, or corymbose; calyx-lobes ovate, scabrido-pulverulent; petals nearly free, spreading, oblong, bluntish, submucronulate; stigma sessile; squamæ minute. *Zey.?* 659. *DC. l. c.* 386? *Thunb. Cap.* 288?

HAB. Lislap, *Zeyher! Wallich!* (Herb. Sd., Hk., Bth. D.)

Branches several from the crown, 2–3 inches long, completely imbricated with fleshy, spreading, decussate, short leaves. Leaves ½–¾ inch long, ovato-deltoid, flattened above, (?) deeply keeled, bluish white. Peduncles 3–4 inches long, forked or trifid, slender. Flowers small, pedicellate.

50. C. anguina (Harv.); stem suffruticose, flexuous, irregularly branched, *closely imbricated* throughout with minute leaves; leaves 4-ranked, *without axillary gemmæ, appressed,* cordate-ovate, acute, *flattish-subconvex,* glabrous; flowers minute, axillary; calyx-lobes nearly equalling corolla, lanceolate, acute, glabrous; petals connate, erecto-patent, oblong, acute, concave; styles shortly subulate. *Zey.!* 641.

Hab. Grootreit, Harteveld, *Zeyher!* Modderfontein, *Rev. H. Whitehead.* (Herb. Sd., Hk., D.)

Very near *C. lycopodioides*, but larger, with the leaves closely appressed, like the scales on the back of a snake, or like tiles in a pavement.

51. C. lycopodioides (Lam. Dict. 2, 173); stem suffruticose, flexuous, irregularly branched, densely covered with minute leaves throughout; leaves 4-ranked, *usually with gemmæ in the axils*, densely crowded or *loosely overlapping*, ovate or deltoid, or subrotund, acute or obtuse, very fleshy and convex, glabrous; flowers minute, axillary, solitary, or 2–3 together; calyx-lobes shortly lanceolate, acute, glabrous; petals connate, erecto-patent, oblong, acute, concave; styles shortly subulate. *DC. l. c.* 385. *C. imbricata, Ait. C. muscosa, Thunb. ex. pte. C. lycioides, E. Mey. ! Tetraphyle lycopodioides, E. & Z. !* 1870. *Zey. !* 2519, 2520, 643.

Var. β. polpodacea; more slender, with smaller leaves. *T. polpodacea, E. & Z. !* 1869. *Zey. !* 2519. *ex pte.*

Var. γ. obtusifolia; leaves subrotund or deltoid, *mostly* obtuse. *T. littoralis & T. propinqua, E. & Z.!* 1867, 1868. *Zey!* 639, 640, 642, 647. *C. muscosa, Drege!*

Hab. Var. α. & β. frequent in Uitenhage and Albany; also at Gauritz River, *E. & Z. !* Heerelogement, *Zey!* var. γ. Kamiesberge, Lislap, Bitterfontein, Saldanha Bay and other localities of the North West, *E. & Z. !* Blaauwberg, *Drege!* (Herb. Sd., D., Hk.)

Stems 1–2 feet long, brittle, spreading or suberect, or decumbent. Leaves 1–2 lines long, with spreading points and mostly with minute axillary leaftufts. Flowers 1 line long. I cannot separate, by readily assignable characters, the four Ecklonian species here united. Those from the western districts have commonly more fleshy, less imbricated and blunter leaves (var. γ.), but at Heerelogement both forms were found by Zeyher. The comparative slenderness and robustness is equally variable.

52. C. muscosa (Linn. Sp. 405); stem herbaceous, thread-like, branching, leafy throughout; leaves opposite and with axillary leaf-tufts, ovate or ovato-lanceolate, acute or acuminate, glabrous; flowers minute, axillary 2-3 or several together, minutely pedicellate; calyx-lobes *equalling* corolla, subulato-lanceolate, acute; petals connate, *oblong, acute*; styles shortly and abruptly subulate. *Thunb. Cap. p.* 281, *ex pte. Tetraphyle muscosa, E. & Z. !* 1872. *Zey.!* 638, 646.

Hab. Among stones, Mts. round Capetown, *E. & Z. ! Gueinzius!* Stellenbosch, *Mund. !* Langspruit, Betchuanaland, *Zey. !* Dornkopf., *Burke!* (Herb. Sd. Hk. D.)

A minute plant, spreading or suberect, irregularly branched, 2–4 inches high. Leaves 1–2 lines long, shorter than the internodes. Flowers 1 line long, sometimes crowded. This has the habit of *Tillæa verticillaris.* It seems to be annual.

53. C. campestris (E. & Z. ! 1873); stem herbaceous, thread-like, erect or decumbent, branching, leafy throughout; leaves connate at base, often with axillary leaf-tufts, ovato-lanceolate or lanceolate, acuminate, almost hairpointed; flowers minute in axillary dense clusters, minutely pedicellate; calyx-lobes *longer than* corolla, subulato-lanceolate, very acute; petals connate, *lanceolate-acuminate or hair-pointed*; styles abruptly subulate. *Also C. lanceolata, E. & Z. !* 1874. *Drege!* 6910. *Zey. !* 2514, 2516.

Hab. Fields near the Zwartkops R., and Krakakamma, *E. & Z. !* Buffeljagdsrivier, *Zeyher!* Compasberg and Nieuweveld, *Drege!* Simonsbay, *C. Wright!* (Herb. Sd., Hk., D.)

A small annual 2–4 inches high, simple or branched. Leaves 2–4 lines long, ½ line wide. Flowers 1 line long; the calyx much longer than the delicate corolla. Known from *C. muscosa* by the *very* taper-pointed leaves, sepals and *petals*. *C. lanceolata, E. & Z.!* has larger leaves than the original specimens of *C. campestris*, but their character varies and both plants grow together.

54. C. subulata (Hook. Ic. t. 590. non Lin. nec E. & Z.); stem herbaceous, *erect*, slender, branched from the base, the branches virgate, leafy throughout; leaves connate at base, erect, or spreading, *subulate, taper-pointed*, concave with inflexed edges, rigid, thin; flowers minute, in axillary clusters, minutely pedicellate; calyx-lobes about equalling the corolla, subulato-lanceolate, acute; petals connate, oblong or ovate-oblong, acute; styles abruptly subulate; squamæ cuneate. *Zey.!* 644, 645. *Drège!* 6909.

HAB. Caledon R. and Doornkopf, *Burke and Zeyher!* Between Kraairiver and the Witberg, *Drege!* Rovelo Hills, Natal, *Dr. Sutherland!* (Herb. Hk., Sd., D.)

Perennial, slightly ligneous at base; the simple or slightly branched stems 6–12 inches high. Leaves 4–6 lines long, ½–1 line wide, pale green. Flowers in dense fascicles much shorter than the leaves.

55. C. parvula (E. & Z.! 1871, ex pte.); stem herbaceous, erect, slender, branched below, leafy throughout; leaves opposite and with axillary leaf-tufts, ovato-lanceolate, tapering to a *narrow bluntish point*, glabrous; flowers minute, axillary, few together, subsessile; calyx-lobes equalling corolla, subulato-lanceolate, acute; petals connate, oblong, acute; styles abruptly subulate.

HAB. Sides of mts. near Grahamstown, *E. & Z.! Gen. Bolton! Mr. H. Hutton!* (Herb. Sd., D.)

Scarcely different from *C. muscosa*, but more robust, with longer and blunter leaves. Leaves 2–3 lines long. Stem ½ line in diameter at base, rigid, and somewhat woody. Ecklon's specimens in Herb. Sonder partly belong to *C. lycopodioides*.

56. C. bergioides (Harv.); stem herbaceous, annual, erect, simple or branched, densely clothed with leaves throughout; leaves connate, ovato-lanceolate or lanceolate, taper-pointed, very acute, *rough-edged*, flat, rigid, *crowded* or imbricated; flowers small, in axillary tufts, shorter than the leaves; calyx-lobes longer than the corolla, lanceolate-acuminate, *rough-edged;* petals connate, *ovate-oblŏng, blunt;* styles shortly filiform; squamæ cuneate. *Zey.!* 2575.

HAB. Breede River, by Kenko, *Zeyher!* (Herb. Sond.)

A small annual; our specimens are 2 inches high and quite simple. Leaves 4–5 lines long, 1–1½ line wide, erecto-patent, with a cartilaginous scabrous or denticulate edge. Flowers 1–1½ line long, hidden among the leaves. Squamæ with a narrow stipe and flat top. Styles very short. Known from others of this group by the rough-edged leaves and blunt petals.

57. C. glomerata (Linn. Mant. 60); stem annual, many times dichotomous, fastigiate, scabrous or glabrous; leaves connate, linear-lanceolate or linear, obtuse, glabrous or scaberulous; flowers minute, *solitary in the forks of the stem, with dense sub-capitate fascicles of flowers ending the branches;* calyx-lobes equalling or exceeding the corolla, oblong or ovate-oblong, *fleshy, mostly blunt*, often scabrous-pointed; petals connate at base, oblong or ovate, subacute; styles shortly filiform; squamæ linear. *DC. l. c. p.* 389. *Pl. Grass. t.* 57. *E. & Z.!* 1929. *also C. strigosa, Lam.? E. & Z.!* 1932. *Drege!* 6881, 6882. *Zey!* 636.

VAR. β. **patens**; calyx-lobes *ovato-lanceolate, acute or mucronate. Thisantha patens, E. & Z.!* 1930.

HAB. Sandy ground round Capetown and in the Western Districts, common. (Herb. Hk., Sd., D.)

A much branched annual, 2–4 inches high, every branch ending in a tuft of small flowers. Leaves 3–4 lines long, 1–2 lines wide, sometimes ovato-lanceolate. Our var. β. merely differs in the calyx-lobes; it is scarcely worth separating.

58. C. glabra (Haw. Syn. 58); stem annual, many times dichotomous, fastigiate, glabrous or nearly so; leaves connate, linear or linear-lanceolate, obtuse, glabrous; flowers minute, *solitary in the forks of the stem, the terminal ones cymoso-paniculate* or subglomerate, all minutely pedicellate; calyx-lobes *shorter* than the corolla, ovate-oblong, obtuse, glabrous, fleshy; petals connate at base, oblong or laterally concave, obtuse; styles shortly subulate; ovary roughish; squamæ cuneate. *DC. l. c. p.* 389. *E. & Z.!* 1931. *Drege! ex pte. C. capillacea, α. E. Mey!*

HAB. Moist sandy places on the Cape flats and elsewhere. (Herb. Sd., D., Hk.)

Like *C. glomerata*, but more laxly panicled, with narrower leaves and shorter calyx-lobes. Stems 2–4 inches high; leaves 2–3 lines long, ½ line wide. "*C. capillacea, b.*" *E. Mey! is Dinacria filiformis.*

59. C. decumbens (E. & Z.! 1933, non Thunb.); stem annual, diffuse or procumbent, irregularly branched, filiform, glabrous; leaves connate, linear-fleshy, obtuse, glabrous; flowers minute, *in tufts at the nodes*, the terminal ones *cymose or subumbellate*, all on filiform pedicels; calyx-lobes *longer* than the corolla, *linear, blunt*, glabrous, fleshy; petals connate at base, oblong, obtuse; styles shortly subulate; squamæ cuneate. *C. glabra, e., E. Mey! in Hb. Drege.*

HAB. Sandy spots at Greenpoint, and near Saldanha Bay, *E. & Z.!* Simons Bay, *C. Wright!* 552, 561. Berg River, *Zey!* 651. (Herb. Sd., D., Hk.)

With the habit of an *Adenogramma*. It is known from *C. glabra* by the irregular branching, the tufts of *usually* long stalked but occasionally *subsessile* flowers at the nodes, and the much longer and more linear calyx-lobes. Stems 3–4 inches long, very weak and slender. Leaves 2–3 lines long, ¼ line wide. Flowers 1 line long. *Thunberg's* "*C. decumbens*," according to his Herbarium is *Bulliarda trichotoma, E. & Z.!*

60. C. albicaulis (Harv.); perennial,? glabrous; stem *prostrate*, di-trichotomous, much branched, *branches divaricating, angular, compressed, very pale* (*or white*); leaves opposite, very patent, linear or linear-oblong or ovate-oblong, sublanceolate, obtuse; flowers minute, *on angular pedicels*, solitary in the forks, the uppermost cymose or subumbellate; calyx-lobes ⅔ of corolla, linear, blunt, with round interspaces, keeled; petals subconnate at base, *ovate*, obtuse; styles shortly subulate; squamæ cuneate.

HAB. Ezel's Fonteyn, Namaqualand, *Rev. H. Whitehead!* (Herb. T.C.D.)

Root thick, somewhat woody. Stems many from the crown, spreading 6–8 inches to all sides, many times decompound, drying of an ivory whiteness and rather rigid. Leaf-pairs about half inch apart toward the ends of the branches, an inch or more below, usually only at the forks of the stem. Leaves 5–8 lines long, 1–2 wide, mostly linear, some inclining to ovate or lanceolate. Pedicels 3–4 lines long, swollen upwards. Flowers 1 line long, white. Petals exactly ovate, about one-third longer than the stamens.

61. C. expansa (Ait. Kew. 1 p. 390); stem herbaceous, scarcely ligneous at base, irregularly dichotomous, much branched, diffuse, glabrous; the branches filiform; leaves subconnate, spreading or recurved, *linear-lanceolate, acute* or subacute, fleshy, convex beneath, glabrous; flowers on long thread-like pedicels, axillary, solitary, or the terminal subcymose; calyx-lobes nearly as long as the spreading corolla, linear, blunt, with obtuse interspaces; petals connate at base, elliptic-oblong, subacute; styles shortly subulate; squamæ shortly cuneate. *DC. l c.* 387. *C. filicaulis, E. & Z.!* 1883. *Zey!* 2524, 653, 2525. *C. expansa*, and *C. parviflora, E. Mey! in Hb. Drege.*

HAB. Near the Zwartkops R., Uit. and Gauritz R., Swell. *E. & Z.!* Breede Riv. and Hassagaiskloof; also at Lislap, *Zeyher!* Between Coega and Zondag Rivers; also Los Tafelberg, Natal, *Drege!* Port Natal, *Dr. Sutherland! Gueinzius!* (Herb. Sd., D., Hk., Bth.)

Biennial or annual? Stems 4–12 inches long or more, widely spreading and much divided, pale, leafy throughout. Leaves ½ inch apart, ½–1 inch long, 1–2 lines wide, rarely wider, mostly acute, shrinking when dry. Pedicels ½–1½ inches long, very slender. Flowers 1–2 lines long. A widely distributed species. *Dr. Gueinzius'* specimens (in Hb. Hooker) have much larger leaves than usual, being 1 inch long, 3 lines wide: otherwise the plant is the same as Natal specimens of the ordinary size.

62. C. centauroides (Linn. Sp. 404); stem herbaceous, distantly forked, diffuse or prostrate, 4-angled, *glabrous;* leaves connato-perfoliate, either cordate-ovate, ovate, elliptical, oblong or obovate, obtuse or acute or mucronate, flat, thinnish, (pellucid when dry), margined and often dotted within the margin, quite entire or crenato-denticulate; flowers on filiform pedicels, the lower axillary, the upper in a terminal sessile or pedunculate umbel or fascicle; calyx-lobes lanceolate, acuminate, keeled, glabrous; petals nearly free, spreading, oblongo-or ovato-lanceolate, acute, concave; styles shortly subulate. *DC. l. c. p.* 386. *Bot. Mag. t.* 1765. *Dill. Elth. t.* 100. *f.* 119. *E. & Z.!* 1893. *and C. pellucida*, 1895. *Zey!* 654. *C. minima, E. & Z.!* 1896, *non Thunb.*

VAR. β. **marginalis**; leaves cordate or ovate, subsessile, mucronate, dotted within the margin. *C. marginalis, DC. l. c., Jacq. Schœnbr. t.* 471. *Zey!* 2525, 2527. *Drege!* 6889. *E. & Z.!* 1894. *C. prostrata, E. Mey! in Hb. Drege.*

HAB. Wet rocks, &c., round Table Mountain, near the summit and on the Winterhoeksberg, Tulbagh, *Thunberg! W.H.H. &c.* Simons Bay, *C. Wright!* 559. β. in Uitenhage; Albany and on to Port Natal, in similar situations. *E. & Z.! Drege! Mrs. Barber! Dr. Sutherland, &c.* (Herb. Thunb., Sd., Hk., D., &c.)

Stems perennial, very weak and rooting at the nodes, 1–3 feet long, the forks 6–8 inches apart. Leaves very variable in shape, even on the same stem, the lower leaves being often obovate and very obtuse, the upper cordate-ovate and acute. Sometimes all the leaves are cordate-ovate; sometimes ovato-lanceolate and even acuminate, When dry they are pellucid and veiny, often with linear purple lines, but as often without them. I cannot regard *C. marginalis* as more than a local variety: and scarcely that, for Thunberg's specimen of *C. centauroides* from Table Mt.., belongs to it. "*C. dichotoma*" of Herb. Thunb. is a form of this species, with *lanceolato-spathulate*, acute leaves.

63. C. brachypetala (E. Mey !); stem herbaceous, distantly forked, diffuse, 4-angled, *pubescent;* leaves sub-connate at base, the lower subpetiolate, *oblongo or ovato-lanceolate, distant, acute, pubescent*, flat, thinnish, margined and dotted within the margin; flowers on filiform pedicels, in terminal umbels or fascicles; calyx-lobes subulato-lanceolate, acute, keeled, glabrous, *setose along the keel, longer than* the concave, acute, oblong or ovate-oblong petals; styles shortly subulate.

HAB. Between the Omsamwubo and Omsamcaba, *Drege!* (Herb. Sd., Bth.)

Stems 2 or more feet long, very weak and probably trailing, patently hispido-pubescent as well as the leaves. Leaf-pairs 2–3 inches apart; leaves 1–1½ inches long, 5–6 lines wide. Flowers 2–3 lines long. Allied to *C. centauroides.*

64. C. diaphana (E. Mey!); stem herbaceous, forked, diffuse, *slender*, pubescent; leaves distant, subconnate, obovate, narrowed to the base, obtuse, pubescent, flat, thin, entire; flowers on filiform pedicels, the lower solitary, the upper loosely fascicled-subumbellate; calyx-lobes *half as long* as the corolla, hispido-pubescent, oblong, obtuse; petals *connate, obovate-oblong;* styles shortly subulate. *C. diaphana, litt. b., Herb. Drege!*

HAB. Between Nieuwekloof and Slangenkeuvel, *Drege!* (Herb. Sd., D.)

A slender herb, 3–6 inches long, pubescent in all parts. Leaf-pairs 1–2 inches apart; the leaves ½ inch long, ¼ inch wide, membranous when dry. Flowers 1 line long. Near *C. Sarcolipes*, but with different flowers.

65. C. Sarcolipes (Harv.); stem herbaceous, annual, forked, diffuse, slender, pubescent; leaves subconnate, elliptical or obovate, narrowed at base, obtuse, thinly pubescent, flat, thin, entire; flowers on filiform pedicels, solitary in the forks of the stem or axils, the uppermost subumbellate; calyx-lobes *equalling* the stellate corolla, lanceolate-oblong, subacute, pubescent; petals *scarcely connate* at base, *ovate, acute;* styles shortly subulate; squamæ linear. *Sarcolipes pubescens, E. & Z.!* 1853. *C. diaphana, litt. a., Drege!*

HAB. In wet places, Brakfontein, Clanw., *E. & Z.!* Simons Bay, *C, Wright!* 560. Piquetberg, *Drege!* (Herb. Sd., D. Hk., Bth.)

A small, weak growing herb, 2–4 inches high. Leaf-pairs ½–1 inch apart; leaves ½–¾, scarcely 1 inch long, 3–5 lines wide. Flowers scarcely 1 line long, on hair-like pedicels 6–8 lines long. This agrees well enough with Thunberg's description of *C. pellucida, Fl. Cap.* p. 282, but as that name has generally been given to a state of *C. centauroides*, I let it drop.

66. C. peploides (Harv.); herbaceous, succulent, decumbent, branched from the base, leafy throughout; leaves connato-perfoliate, oblong, obtuse, fleshy, blunt, glabrous; flowers on slender pedicels, terminal and from the upper axils, few; calyx-lobes oblong, glabrous, fleshy, nearly equalling the oblong petals. *Drege!* 6880.

HAB. Witbergen, 7000–8000f. *Drege!* (Herb. Sond.)

A small perennial species, with a fibrous root and many short, leafy stems from the crown, rooting at the nodes, with the general aspect of *Arenaria peploides.* Leaves 3–5 lines long, 1–2 wide, spreading. Cymules very imperfect, reduced to 1–3 small flowers. Mature corolla not seen.

67. C. dasyphylla (Harv.); small, herbaceous, glabrous, procumbent, with flexuous branches; leaves subglobose or ellipsoidal, fleshy, small, very obtuse, punctate; pedicels terminal or axillary, filiform, one-flowered, short; calyx-lobes oblong, very obtuse, round-backed; petals nearly free, ovate or oblong, bluntish; stigma subsessile. *Drege!* 6885.

HAB Winterveld, between New Year's Fountain and Ezelsfont. 3000-4000f. *Drege!* Cradock, and Gamke R., *Zeyher!* (fragments). (Herb. Sd., Hk,, Bth., D.)

A small plant, with stems 2–3 inches long, resembling *Sedum dasyphyllum* in habit and foliage. Branches jointed, slender. Leaves 1–1½ lines in diameter. Flowers small, on pedicels 2–3 lines long. Fragments of a nearly similar but stronger plant were gathered by Zeyher at Bitterfontein; they differ from Zeyher's plant above quoted in having a large woody root and somewhat larger leaves, but are too imperfect for description.

68. C. dentata (Thunb! Cap. 293); stem herbaceous, simple or forked, with distant nodes; leaves opposite or *subfasciculate, on very long petioles, reniform, at base either cordate or cuneate,* repando-crenate or lobulate, or subentire, thinnish, membranous when dry, glabrous; flowers *patently cymose,* on slender pedicels, the lower ones sometimes crowded in the axils; calyx lobes short, *blunt, oblong,* glabrous; petals free, ovato-lanceolate, acute. *E. & Z.!* 1857; *also Petrogeton typicum, E. & Z.!* 1854. *Drege!* 6886.

VAR. *a.* **minor**; smaller in all parts, with subentire, roundish or flabelliform leaves, and few flowers. *C. minima, Thunb! l. c.* 293.

HAB. Rocky mountain clefts, Bockeveld and Rodesand, and Ribek-Kasteel, *Thunberg!* Table Mountain; also mountains near Hex River, Worces. and Cederbergen, Clanw., *E. & Z.!* Nieuwekloof, *Drege!* Witsenberg at Tigerkloof, Tulbagh, *Dr. Pappe!* (Herb. Thunb., Hk., D., Sd.)

A weak, soft, succulent herb, with something the habit of *Chrysosplenium oppositifolium,* variable in size, 3–6 inches long or more. Petioles 1–3 inches long, slender. Leaves ½–1 inch wide, shorter than their width, occasionally smaller or larger: in the smaller forms subentire; in the larger with a few wide crenatures or very shallow lobules. Flowers 2 lines long, starlike, in a more or less developed cyme.

69. C. patens (E. & Z.! 1855); stem herbaceous, simple or forked, with distant nodes, slender and pellucid; leaves opposite or subfasciculate, on very long petioles, *roundish-reniform,* obtuse at base, *repando-dentate or lobulate,* thin and membranous when dry, glabrous; flowers in a spreading, slightly divided cyme, on long slender pedicels; calyx-lobes very short, *linear-oblong, acute, one-nerved;* petals free, ovato-lanceolate, acute, styles shortly subulate. *Petrogeton patens, E. & Z. l. c. Drege,* 6887 *!*

HAB. Near the Tulbagh Waterfall, *E. & Z.!* Drakensteenberg, *Drege!* (Hb. Sd.)

Scarcely differing from the preceding except by the calyx and the more sharply lobulate or subincised leaves, and laxer inflorescence.

70. C. nivalis (E. & Z.! 1860); stem herbaceous, simple or forked, with distant nodes, slender and pellucid; leaves opposite or subfasciculate, on long petioles, *subrotund, entire, or obscurely repand,* thin and membranous when dry, glabrous; flowers on very long threadlike pedicels, axillary or terminal, or 2–3 together in an imperfect cyme; calyx-lobes *nearly equalling* the petals, thin, *elliptic-oblong, obtuse,* nerved; petals ovate or ovate-elliptical, acute; styles shortly subulate. *Petrogeton nivale, E. & Z.! l. c.*

HAB. Mountain rocks of the Winterberg, Kaffr., *E. & Z.!* (Herb. Sond,)

Resembles the weaker states of *C. dentata*; but differs in inflorescence. Leaves ¼–½ inch wide; petioles ½–1 inch. Stem 3–4 inches long. Pedicels uncial.

71. C. nemorosa (E. & Z.! 1859); stem herbaceous, filiform, with distant nodes; leaves opposite, *on short or shortish petioles*, roundish-subreniform, entire or faintly repand, thin and membranous when dry, glabrous; flowers in an interrupted, racemose panicle, the pedicels threadlike, two or more together, the *terminal umbellate;* calyx-lobes ⅔ of corolla, ovate, subacute; petals lanceolate, acute or acuminate; styles shortly subulate. *Petrogeton nemorosum, E. & Z.! Zey.!* 2518. *C. cordata. E. Mey.!*

HAB. Shady places near the Zwartkops Riv., *E. & Z.!* Between Enon and the Zuureberg, *Drege!* (Herb. Sd., D., Hk.)

A very small plant, with shorter petioles and more racemose flowers than the rest. Drege's specimens have however longish petioles. Stem 2-4 inches high, including the raceme. Leaves ½ inch diameter or less.

72. C. Umbella (Jacq. Ic. Rar. t. 352); root tuberous; stem erect, simple or oppositely branched, each division crowned by a pair of leaves, *perfectly confluent into an orbicular, entire or subentire perfoliate disc;* cymes pedunculate, *panicled,* with spreading branches and pedicellate flowers; calyx-lobes short, deltoid, subacute; petals free, ovate-oblong, acute; styles shortly subulate. *Tratt. Tab. t.* 253. *Septas Umbella, DC. l. c.* 383.

HAB. Modderfontein and Zilverfontein, *Drige! Rev. H. Whitehead!* (Herb. Sd., Hk., D., Bth.)

A very remarkable plant, at once known by its leaves united into a circular disc, which is sometimes 6 inches or more in diameter, though often much smaller. In floral characters it resembles *C. flabellifolia.*

73. C. flabellifolia (Harv.); root tuberous; stem erect, simple, crowned by 2 (rarely 4) horizontally patent leaves; leaves connate and subcuneate at base, *roundish flabelliform,* crenato-lobulate, thinly fleshy, glabrous; cymes pedunculate, *laxly panicled, with spreading branches* and long stalked flowers; calyx-lobes *very short,* bluntly deltoid; petals free, membranous, *ovate-oblong,* acute; styles shortly subulate; squamæ oblong. *Petrogeton Umbella, E. & Z.!* 1856, *excl. syn.*

HAB. Heathy ground on mountains near Brackfontein, Clanw., *E. & Z.!* (Herb. Sd., D., Bth.)

Stem 4-6 inches high, of one internode. Leaves if 4, rosulate (the two upper smaller). 2-3 inches wide, 1½-2 inches long, multicrenate. Peduncle 4-6 inches long, much branched. Flowers small, 1-1½ line long.

74. C. Saxifraga (Harv.); root tuberous; stem erect, simple, crowned by 2 (rarely 4) horizontally patent leaves; leaves subsessile and connate at base, *broadly cordato-reniform,* coriaceo-carnose, crenato-lobulate, glabrous; cymes on *long, naked peduncles*; simple or forked, *umbellate or corymbose;* calyx-lobes ⅓ of corolla, lanceolate, subacute; petals *connate at base, erecto-patent, sub-recurved,* oblong, acute or subacute; styles subulate; squamæ very minute. *Septas globifera, E. & Z.!* 1862, *excl. Syn. Septas,* 918, *Drege!*

HAB. Mountain sides, Steenberge, near Muysenberg, *E. & Z.! W.H.H.!* Between Driekoppen, Bokkeveld and Hex river, and in Dutoits kloof, *Drege!* Albany, *Mrs. F. W. Barber!*, Port Elizabeth, *Mrs. Holland!* 35. (Herb. Sd., Hk., D.)

With the habit and foliage of *C. Septas* and *C. flabellifolia* it differs from both in its flowers. The leaves are 1-3 inches broad, ¾-2 inches long, almost exactly reniform, sometimes doubly crenate. The cymes in weakly grown plants are but 3-6 flowered; in strongly grown they are much branched, in a dense, almost fasciculate corymb, bearing 50–100 flowers or more. In this case the corolla is rather smaller than when there are few flowers. Petals with recurved tips, 2–3 lines long, white, with a rosy tint.

75. C. Septas (Thunb.! Cap. 291); root tuberous; stem erect, with 1–2 internodes, simple, crowned by 2–4 horizontally patent leaves; leaves connate, *cuneate at base*, roundish-flabelliform, coriaceo-carnose, crenate, glabrous; cymes on long naked peduncles, simple or branched, *sub-umbellate*, few or many flowered, with long, slender pedicels; calyx-lobes ¼ of corolla, lanceolate; *petals* 6–7–9 *free*, *stellate*, broadly lanceolate, acute or subacute; *stamens* 6–7–9; styles subulate; squamæ very small and fleshy. *Septas Capensis, Linn. DC. l. c. p.* 383. *E. & Z.!* 1861. *Lam. Ill. t.* 276. *S. globifera, Bot. Mag. t.* 1472.

HAB. Moist places on mountain sides, round Capetown and on Hott. Holland, frequent. (Herb. Thunb., Sd., Hk., D., &c.)

Stem 2–4 inches high. Leaves 1–3 inches long, 1–2 inches wide. Peduncle scapelike, 4–8 inches high. Petals white or rosy, 3–5 lines long. Cyme very variable in composition: *S. globifera* is merely a very luxuriant, garden variety.

Section II. **PYRAMIDELLA.** (Sp. 76–79.)

76. C. pyramidalis (Linn. f. Suppl. 189); stem suffruticose, erect, simple, imbricated with leaves throughout; leaves 4-ranked, most closely imbricating, connate at base, broadly *ovate or deltoid, acute or subacute, with strongly reflexed margins*, somewhat keeled, glabrous; cymes densely capitate, many flowered, sessile, terminal or axillary; calyx-lobes linear, obtuse, round-backed; ciliate; petals connate below, tapering above into long, lanceolate channelled points, much longer than the calyx; stigma subsessile; squamæ cuneate, stipitate, bright orange. *DC. l. c. p.* 388. *Thunb.! Cap.* 287. *E. & Z.!* 1863, *also Tetraphyle quadrangula, E. & Z.!* 1864.

HAB. Olifants River, *Thunberg*. In the Karroo between Uitenhage and Graaf Reynet, and mountain sides near Klipplaat river, *E. & Z.!* Driekoppen, and Zwaanepoelsportberge, *Drege!* Gamke River, *Burke and Zeyher!* 656. (Herb. Sd., Hk., D., Bth.)

Stem 3–8 inches high, forming, with the closely imbricated leaves, a sharply four-angled prism of nearly equal diameter throughout, or gradually widening upwards. Heads of flowers mostly terminal; in luxuriant specimens also lateral, in all cases sessile. Points of the petals very long. E. & Z's. *Tetr. quadrangula* is founded on old specimens with shrivelled leaves and withered flowers.

77. C. columnaris (Linn. f. Suppl. 191); stem short, erect, simple, imbricated with leaves throughout; leaves connate, 4-ranked, closely imbricating, *orbicular, fleshy, very obtuse, with inflexed, ciliate margins*; cymes capitate, terminal, sessile, densely many flowered; calyx-lobes linear, obtuse, ciliate; petals connate below, tapering above into long, lanceolate, channelled points, much longer than the calyx; stigma subsessile; squamæ cuneate, stipitate, orange. *DC. l. c.* 385. *Thunb.! Cap.* 291. *Zey.!* 657.

HAB. Under Roggeveld, *Thunberg*. Hex Rivier'skloof, *Drege!* Zoutkloof, *Burke and Zeyher.!* Namaqualand, *Rev. H. Whitehead!* (Herb. Hk., Sd., D.)

2–4 inches high, closely covered with leaves and crowned with a globose fascicle of flowers, the whole plant resembling one of the *Balanophoreæ*. Leaves nearly an inch broad, not quite so long, more or less inrolled. Flowers white.

78. C. semiorbicularis (E. & Z. ! 1890); stem erect, simple, densely imbricated with leaves below, laxly leafy above, glabrous; leaves connate, the lower ones imbricating, broadly orbicular, fleshy, very obtuse, hispidulous, with inflexed, ciliate margins, the upper depauperated, roundish-ovate, concave; cymes *corymbose*, densely trichotomous, bracteate; calyx-lobes linear, obtuse, ciliate; petals connate below, tapering above into long, lanceolate channelled points; stigma thick, sessile; squamæ flabelliform, stipitate. *C. columnaris, var. β. elongata, E. Mey. ! Zey. !* 658.

HAB. Kamiesberg, Namaqualand, *E. & Z. !* Olifants R., *Zeyher !* 658, Ebenezer, and near Mierenkasteel and Zwartdoorn River, *Drege !* (Herb, Sd., Hk.)

Very similar to *C. columnaris*, and perhaps only a smaller, but taller and more caulescent variety, with looser inflorescence. Stem 3–4 inches high, about ⅓ imbricated with leaves, ⅔ with subdistant nodes.

79. C. multiceps (Harv.); stem suffruticose, cæspitose, multifid, densely imbricated throughout with leaves; leaves connato-vaginate, from a broad base shortly subulate or lanceolate-attenuate, fleshy, with subrecurved margins, glabrous; cymules few flowered, capitate, sessile, terminating the leafy branches; calyx-lobes linear, ciliolate; petals connate below, tapering above into long, lanceolate points; stigma subsessile; squamæ flabelliform, stipitate. *Zey. !* 660.

HAB. Elandsfontein, *Zeyher !* (Herb. Hk. Sd.)

2–3 inches high, dividing from the crown into many, short, forked, corymbose branches, closely leafy throughout like those of *Lycopodium Selago*, the narrow points of the leaves spreading or squarrose. Flowers 3–5, at the ends of the branches, white; the narrow lobes of the corolla twice or thrice as long as the calyx.

Section III. **SPHÆRITIS.** (Sp. 80–88.)

80. C. Sphæritis (Harv.); stem slender, suffruticose, erect, with virgate branches, laxly hispid (chiefly in two opposite lines), with subdistant leaf-pairs; leaves subulate, acuminate or acute, flattish or keeled, *cartilagineo-serrate or ciliate;* cymes sub-capitate, terminal; calyx-lobes ½ of corolla, linear, subacute, entire or denticulate; petals *gradually tapering* into a subulate, channelled apex; stigma subsessile; squamæ linear, truncate. *Sphæritis typica, E. & Z. !* 1910, *S. stenophylla, E. & Z !* 1911, *and S. muricata, E. & Z. !* 1912. *Drege !* 6894, 6908. *Zey. !* 2550. *C. fruticulosa ? Drege !*

HAB. Hills near the Zwartkops R., Uit.; also near Tulbagh; and on the Devil's Mt., Capetown, *E. & Z. !* Adow, and Ebenezer, *Drege !* (Herb. Sd., Hk., D.)

Stems 12–18 inches high, with the habit and foliage of *C. cymosa*, but with different inflorescence and petals. The three Ecklonian species here united are not distinguishable when dry; the leaves vary from broader to narrower, flatter to more convex, and also in the regularity and strength of the marginal toothlets or cilia. Sometimes nearly glabrous and exciliate. *C. fruticulosa*, E. Mey. ! in Herb. Drege, seems to be a dwarf variety.

81. C. incana (E. & Z. ! 1917); everywhere *minutely cano-puberulous;* stem suffruticose, with slender, virgate branches and sub-distant leaf-pairs; leaves *lanceolate-oblong, subacute,* fleshy, convex beneath, the *uppermost* depauperated and *subovate;* cymes capitate, terminal or in a brachiate panicle; calyx-lobes ½ the corolla, ovate or broadly lanceolate, keeled, downy and ciliate; petals panduriform, ribbed and keeled, *gradually* tapering into a channelled point; stigma subsessile; squamæ truncate. *Sphæritis incana, E. & Z.! Crassula pubescens, E. Mey.! in Heb. Drege!*

HAB. In the Karroo, between Beaufort and Graaf Reynet, *Drege! E. & Z.* (Herb. Sd., Hk., Bth.)

More canescent than *C. margaritifera,* with different petals.

82. C. virgata (Harv.) glabrous; stem suffruticose, erect, with virgate branches, angular toward the summit; leaf-pairs distant; leaves sub-distinct, fleshy, linear-trigonous, flat or channelled above, keeled beneath, subobtuse, with smooth margin, the upper ones depauperated; cymules capitate, in a terminal panicle or thyrsus, pedicellate; calyx-lobes ovate or deltoid, bluntly keeled, rough-edged; petals connate, tapering into a channelled point; stigma subsessile; squamæ oblong.

HAB. Pikenier'skloof, *Zeyher!* 664. (Herb. Sd. Hk.)

Ramification unknown. Branches 1–2 feet long, rodlike; the nodes 2–3 inches apart. Lower leaves 1½ inch long, 2–3 lines thick; upper smaller, the uppermost reduced to scales. Flowers sessile, 6–12 in each little head. The foliage is that of several Mesembryanthemums.

83. C. clavifolia (E. Mey.!); glabrous, stem suffruticose, dichotomous, erect, leafy; leaves subconnate at base, oblongo-spathulate, fleshy, tapering below, obtuse, with a smooth margin; peduncles elongate, with two or three distant pairs of ciliolate leaf-scales; cymes terminal, capitate, with ciliolate bracts; calyx-lobes linear, pubescent and ciliate, obtuse; petals connate below, gradually tapering into a channelled point; stigma subsessile; squamæ oblong. *Globulea clavifolia, E.Mey.! in Herb. Drege.*

HAB. Kromrivier and by Welgelegen, 3–4000f. *Drege!* (Herb. Sd. Hk. Bth. D.)

A small species, with the habit and foliage of *C. radicans,* but with more capitate inflorescence and very different petals. Heads of flowers 3–5 lines across; flowers 1–2 lines long.

84. C. ciliata (Linn. Sp. 405); stem short, robust, suffruticose, decumbent, closely leafy below and slightly branched, glabrous, the floriferous branches with distant leaf-pairs and smaller leaves; leaves connate at base, obovate-spathulate or oblong, obtuse, flat, glabrous, cartilagineo-ciliate; cymes corymbose, simple or forked, many flowered, terminal; calyx-lobes ⅔ of reflexed corolla, broadly lanceolate, keeled, acute, with rough edges, glabrous; petals gradually tapering into a broadish, channelled and thickened apex, connate beneath; stigma subsessile; squamæ cuneate, large. *DC. l. c. p.* 387. *E. & Z.!* 1902. *Dill. Elth. t.* 98. *f.* 116. *DC. Pl. Grass. t.* 7. *Zey.!* 2537, 2538.

HAB. Dry hills round Capetown, *E. & Z.! W.H.H., &c.* Paarlberg, *Drege!* Zwartkops river, *Zeyher!* (Herb. Sd., Hk., D., Bth.)

Root woody. Stem forked once or twice, the divisions closely leafy for 3–6 inches; then lengthened into slender, distantly leafy, erect, virgate flowering branches. Cymes many flowered, dense or spreading, 1–3 inches across. Flowers small, cream-coloured. The tapering point of the petals is sometimes very narrow, sometimes broader.

85. C. tomentosa (Linn. f. Suppl. 190); everywhere *densely clothed* *with rigid, reflexed bristly-hairs ;* stem erect, simple, robust, densely leafy at base, with distant leaf-pairs above, the upper leaves small or abortive ; radical leaves broadly oblong or obovate, flat, obtuse, the cauline narrow-oblong, all densely hirsuto-setose and ciliate ; cymules capitate, subsessile at the nodes, forming a long interrupted spiked-thyrsus or spike ; calyx-lobes linear, obtuse, hispid and ciliate, round-backed ; petals gradually tapering into a channelled apex ; stigma subsessile ; squamæ cuneate, short. *DC. l. c.* 387. *Thunb. Cap. p.* 287. *Sphæritis setigera, E. & Z.!* 1921, *and S. tomentosa,* 1920 *!*

HAB. Muysenberg ; and on barren hills near Gauritz R., Swell., *E. & Z. !* Namaqualand, *V. Schlicht.* Modderfonteyn. *Rev. H. Whitehead.* Pikenier's kloof, *Zeyher !* (Herb. Sd.)

Root woody. Stem 1–2 feet high, robust, virgate, tapering upwards. Leaves crowded round the base ; the upper ones diminishing to bracts. Cymules or verticillasters densely many flowered. *Zeyher's* specimen has shorter and less copious bristles, but scarcely differs otherwise. In *Mr. Whitehead's* specimen the bristles on the stem are very fine and close pressed.

86. C. interrupta (E. Mey. !) everywhere hirsuto-pubescent ; stem short, or scarcely any, closely leafy ; leaves subrosulate and almost radical, oblate or subrotund, very obtuse, *hispid and ciliate ;* flowering branches slender and virgate, with distant nodes, scapelike ; cymules capitate, in an interrupted spiked-thyrsus ; calyx-lobes obovate-oblong, blunt, round-backed, hispid ; petals gradually tapering into a broadish, channelled point ; styles short and thick ; squamæ small.

HAB. Zilverfontein, 2000–3000f. Sep.–Oct. *Drege !* (Herb. Sd., Hk., Bth.)

A small, hoary species, with nearly rosulate subradical leaves and a scapelike flowering branch 3–6 inches high, with 3–4 nodes, an inch or two apart. Cauline leaves depauperated. Radical ½–¾ inch long, as broad or broader. Thyrsus of 3–5 verticillasters.

87. C. glabrifolia (Harv.) stem short, scarcely any, closely leafy ; leaves subrosulate and almost radical, *oblong-obovate, obtuse, ciliate, otherwise glabrous ;* flowering branches slender and virgate, with distant nodes, scapelike, canescent ; cymules capitate, in an interrupted spiked-thyrsus ; calyx-lobes oblong, bluntish, round-backed, cano-pubescent ; petals gradually tapering into a broadish, channelled point ; styles short and thick, squamæ small.

HAB. Namaqualand, *Mr. Andrew Wyley !* (Herb. T.C.D.)

Stem ½–1 inch long. Lowest leaves 1½ inches long, an inch wide at top, connate at base, 2–3 pairs crowded together ; upper oblong or sublinear. Scapelike stem 6–8 inches high, striate, canescent with minute pubescence, with 1–2 distant pair of depauperated leaves, ending in an interrupted, compound spike 3–5 inches long. Cymules few flowered, sessile in the axils of short bracts. Allied to *C. interrupta,* but with very different foliage.

88. C. hirtipes (Harv.) ; dwarf ; stem short, erect, succulent, closely leafy ; leaves obovate-oblong, obtuse, fleshy, glabrous ; peduncle slender, *patently hispid,* ending in a capitate, few flowered cyme ; calyx-lobes ovate, densely hispid, half as long as the petals ; petals connate at base, the ovate limb gradually tapering into a broadish, channelled point ; styles short and thick. *Drege,* 6900.

HAB. Ebenezer Mission Station, near the mouth of the Olifant R., Clanw., *Drege !* (Herb. Sond.)

Stem 1–2 inches high, densely clothed with leaves. Leaves ½ inch long, 2–3 lines wide, succulent. Peduncle 1–2 inches long, rough with stiff, spreading, dark hairs or bristles, bearing a flat-topped dense cyme. Flowers 1–1½ lines long. Described from a very imperfect specimen.

Section IV. **MARGARELLA.** (Sp. 89–91).

89. C. margaritifera (E. & Z.! 1913); stem slender, suffruticose, erect with virgate branches, *microscopically puberulent*, with subdistant leaf-pairs; leaves linear-trigonous or nearly subulate, subacute, fleshy, minutely punctulate; cymes sub-capitate, terminal or corymbose; calyx-lobes ⅔ of corolla, *linear, truncate and thickened at point, serrulate*, keeled; petals connate, panduriform, *suddenly contracted* into a fleshy, furrowed apex; stigma subsessile; squamæ cuneate. *Sphæritis margaritifera, E. & Z.! Zey.* 2549, 2551. *Drege!* 6912.

HAB. Fields near the Zwartkops River and Bethelsdorf, *E. & Z.! Drege!* (Herb. Sd., Hk.)

Stem 12–16 inches high, divided near the base into many simple, curved branches, each 12–14 inches long. Lower leaf-pairs ¼–½ inch apart, upper 1½–2 inches. Leaves uncial, mostly acute, minutely dotted with white points. Flowers small.

90. C. subaphylla (E. & Z.! 1916); stem slender, suffruticose, erect or spreading, with virgate branches, puberulous or pubescent, with distant leaf-pairs; leaves connate, *minute*, fleshy, semiterete-ovate or oblong, scaberulous or puberulous, obtuse; cymes *sub-capitate*, terminal or panicled, with subsessile flowers; calyx-lobes ½ of corolla, ovate-oblong, keeled, scabrous or pubescent, blunt; petals subconnate, panduriform, *suddenly* acuminate into a channelled and complicate apex; stigma subsessile; squamæ emarginate. *Sphæritis subaphylla, E. & Z.!*

VAR. β. **puberula**; more evidently pubescent. *Sph. puberula, E. & Z.!* 1919.

HAB. Dry hills near the Gauritz River, Swell., *E. & Z.!* (Herb. Sd.)

Stems 12–14 inches high, branched near the base, with simple flexuous branches. Leaf-pairs ½–1 inch apart; leaves 2–3 lines long, 1–1½ line wide. Heads of flowers small, solitary or in a brachiate panicle. Flowers pale. The pubescence is sometimes very minute, sometimes copious; in other respects the two forms indicated agree.

91. C. biconvexa (E. & Z.! 1918); stem *depressed*, suffruticose, *branching and cæspitose*, closely leafy; leaves *shortly spathulate or obovate*, convex on both sides, *small, retrorsely hispid*, obtuse; peduncle scapelike, naked; cymes capitate, terminal, with subsessile flowers; calyx-lobes ½ of corolla, hispid, *oblongo*-linear, blunt, keeled; petals panduriform, suddenly acuminate into a channelled and complicate apex; stigma subsessile; squamæ cuneate. *Sphæritis biconvexa, E. & Z.!*

HAB. Near the Gauritz River on dry hills. *E. & Z.!* (Herb. Sond.)

Stem 2–3 inches high, closely branched and matted. Leaves 3–5 lines long, 2–3 lines wide. Peduncle 4–6 inches long, threadlike, bearing a small head of flowers, naked or with one or two distant pairs of dwindled leaves. Of this I have only seen a solitary specimen.

Section V. **PACHYACRIS.** (Sp. 92.)

92. C. trachysantha (E. & Z.! 1915); the suffruticose, branching, slender stem, the peduncles, leaves and calyces *densely hispido-pubescent;*

leaves subulate, semiterete, acute, the upper ones in distant pairs; cymes pedunculate, much branched, corymboso-fasciculate; flowers subsessile; calyx-lobes lanceolate-linear, subacute, round-backed; *petals lanceolate, rough-edged, tipped with a triquetrous, fleshy gland;* stigma sessile; squamæ emarginate. *Sphæritis trachysantha, E. & Z.!* and *Sph. paucifolia, E. & Z.!* 1914. *Zey.!* 2547, 2548. *Drege!* 6902.

HAB. Zwartehoogdens, near Grahamstown and dry hills at Zwartkops R., *E. & Z.!* Howison's Poort, *H. Hutton!* Enon, *Drege!* (Herb. Sd., D., Hk., Bth.)

An erect or spreading, half herbaceous succulent, densely hirsute in all parts, 12–18 inches high. Peduncles terminal, forked or trifid, the branches simple or again divided; inflorescence flat-topped. The fleshy glandular apex of the petals is ridged in front, and very prominent. The petals are scarcely connate at base. Possibly this is *Globulea mesembryanthemoides*, Haw.? Leaves sometimes short and subobtuse, ¾–½ inch long.

Section VI. **GLOBULEA.** (Sp. 93–99.)

93. C. cultrata (Linn. Sp. 2.405); stem *erect*, suffruticose, *subsimple*, laxly leafy; leaves connate, obovate elliptical, or obovate-oblong, curved, flattish, obtuse or subacute, glabrous, smooth-edged or ciliolate; peduncle elongate, with distant leaf-scales, pubescent, *panicled* above, the cymules dense with subsessile flowers; calyx-lobes linear, blunt, pubescent, keeled; petals connate at base, panduriform, with an ovate dorsal gland below the apex; stigma nearly sessile; squamæ truncate. *DC. l. c.* 391. *Sims. Bot. Mag. t.* 1940. *Dill. Elth. f.* 114. *Globulea cultrata, Haw. E. & Z.!* 1922. *Zey.!* 2556, 2557, 2558.

HAB. Shrubby places near the Zwartkops R., *E. & Z.!* Olifants R., *Thunberg.* (Herb. Sd., D., Hk.)

Stem 2–3 feet high including the leafy peduncle, mostly simple. Leaves variable in size, sometimes 1–1½ inches long and ½ inch wide, sometimes 2–2½ inches long and ¾–1 inch wide. Panicle either closely thyrsoid or much branched and spreading. Flowers small and green.

94. C. radicans (Haw.); stem suffruticose, ascending-erect, *with spreading lateral branches that take root beneath,* laxly leafy; leaves connate, lanceolate-oblong or lanceolate, narrow, cultrate, flattish, subacute or acute, glabrous, smooth-edged; peduncle elongate, slender, with distant leaf-scales, glabrescent, *cymoso-corymbose* at the summit, the cymules dense; calyx-lobes linear, blunt, pubescent, keeled; petals connate at base, panduriform, with an oblong, subapical dorsal gland; stigma nearly sessile; squamæ emarginate. *DC. l. c.* 391. *Globulea radicans, Haw. E. & Z.!* 1923. *Zey.!* 2552, 2559.

HAB. Woody places near the Zwartkops R., *E. & Z.!* (Herb. Sd., Hk., D.)

Smaller, more branching and more diffuse than *C. cultrata*, with much smaller leaves and a different inflorescence. Leaves seldom more than an inch long, and 3–4 lines wide. Cyme trifid, flat-topped, few flowered and seldom more compound.

95. C. platyphylla (Harv.); stem scarcely any, densely leafy; leaves sub-radical, rosulate, *broadly obovate or subrotund, very obtuse,* flat, rigid, *glabrous, smooth-edged;* peduncle scapelike, puberulous, furrowed, with a few minute leaf-scales, and ending in a long subspicate thyrsus; cymules capitate-subsessile, pubescent; calyx-lobes linear, blunt, pubescent, round-backed; petals connate at base, panduriform, with an ovate, subapical, dorsal gland; stigma subsessile; squamæ truncate.

HAB. South Africa, *Drege!* 6896. (Herb. Sond.)

A single specimen only seen. Leaves 1–1¼ inch long, ¾–1 inch wide, very rigid when dry. Scape 8–10 inches high, the upper half floriferous.

96. C. obvallata (Linn. Mant. 61); stem short, fleshy, simple, densely leafy; leaves (subradical) lanceolate-oblong or obliquely cultrate, subacute or obtuse, *glabrous*, cartilagineo-ciliate; peduncle scapelike, panicled with a few leaf-scales; cymules densely fascicled, pubescent; calyx-lobes puberulent, ciliate, oblong, blunt; petals subconnate, panduriform, with an oblong, dorsal gland below the apex; stigma sessile, capitate; squamæ oblong. *DC. l. c.* 391. *Pl. Grass. t.* 61. *Globulea obvallata, Haw. G. capitata, E. & Z.!* 1924. *Zey.!* 2554.

HAB. Coegakopje &c., by the Zwartkops River, *E. & Z.!* Dutoitskloof, *Drege!* 6916. (Herb. Sd., D., Hk.)

Stem 3–6 inches long, quite covered with leaves. Leaves 2–2½ inches long, ½–¾ inch wide, rigid, carnoso-cartilaginous. Panicle thyrsoid or much branched, the tufts of flowers very dense. Flowers small and green.

97. C. canescens (Schult. Syst. 6. 374); stem short or none, fleshy, densely leafy; leaves radical, oblong-obovate, lanceolate-oblong or lanceolate, cultrate, subacute or obtuse, *pubescenti-canescent*, ciliolate; peduncle scapelike, pubescent, panicled or thyrsoid, with a few leaf-scales; cymules densely capitate, pubescent; calyx-lobes pubescent, ciliate, oblong or linear, very blunt, convex; petals subconnate, panduriform, with an ovate or globose, dorsal gland below the apex; stigma subsessile, capitate; squamæ oblong. *DC. l. c.* 391. *Globulea canescens, Haw. E. & Z.!* 1926, also *G. obvallata, E. & Z.!* 1925.

VAR. α. **latifolia**; leaves *broad*, cultrate-obovate or oblong. *G. obvallata, E. & Z.!*

VAR. β. **angustifolia**; leaves narrow, more or less lanceolate or linear.

HAB. Near the Zwartkops R., *E. & Z.! Zey.!* 2553. *Drege!* 6913. (Herb. Sd., D., Hk.)

Known from *C. obvallata* by its pubescence. Narrow and broad leaved varieties grow together, and I cannot, in the dry state at least, find limits between them.

98. C. nudicaulis (Linn. Sp. 405); stem short or none, densely leafy; leaves subradical, conferto-rosulate, *semiterete, subulate, acute, sub-pubescent;* peduncle scapelike, pubescent, panicled or thyrsoid, with a few distant leaf-scales; cymules densely capitate, pubescent; calyx-lobes ciliate, subglabrous, linear-oblong, blunt; petals subconnate, panduriform, concave, with a globose dorsal gland below the apex; stigma subsessile; squamæ truncate-emarginate. *DC. l. c. p.* 391. *Pl. Grass. t.* 133. *Dill. Elth. f.* 115. *Globulea nudicaulis, Haw. E. & Z.!* 1924. *Zey.!* 662.

HAB. Sandy places, Capeflats near Rietvalley, *E. & Z.!* Klipfontein, *Zey.!* Clanwilliam to Boschkloof, *Drege!* (Herb. Sd., Bth.)

The long-leaved forms are readily known; but those with shorter and broader leaves approach the narrower-leaved states of *C. canescens.* The pubescence however is much more scanty. Leaves 2–5–6 inches long, ¼–½ inch wide, tapering upwards. Stem 1–2 feet high. Flowers small and greenish.

99. C. sulcata (Haw.); nearly stemless; leaves incurved, subulate, semiterete, channelled, shining green, glabrous; peduncle scapelike, panicled, with densely capitate cymules; calyx-lobes pubescent and ciliate, linear-oblong, blunt; petals panduriform, with globose dorsal gland; stigma sessile. *DC. l. c.* 391. *Globulea sulcata, Haw. E. & Z.!* 1928

HAB. Near the Zwartkops, Uit. *E. & Z.!* (Herb. Sond.)

Very like *C. nudicaulis* but glabrous.

Species of Section GLOBULEA *unknown to us.*

Globulea atropurpurea (Haw.); leaves obliquely cuneate-obovate, dark purple; flowering stem scapelike, very long, panicled. *DC. l. c.* 391.

Globulea lingua (Haw.); leaves elongate, loriform, between cultrate and lanceolate, ciliated as well as the calyx; scape panicled. *DC. l. c.* 391.

Globulea lingula (Haw.); similar to *G. lingua*, but half the size.

Globulea capitata (Haw.); leaves ventricosely lanceolate-cultrate, biconvex, imbricately decussate, the younger canous. *DC. l. c.* 391.

Globulea impressa (Haw.); stemless; leaves strap-like, lanceolate, green, impresso-punctate, the dots large, scattered, numerous. *DC. l. c.* 391.

Globulea paniculata (Haw.); stemless; leaves strap-like, acuminate, full-green, minutely impresso-punctate; branches of the panicle spiked. *DC. l. c.* 392.

Globulea hispida (Haw.); leaves crowded, straplike, acuminate, convex beneath, hispid; stem suffruticose, hispid. *DC. l. c.* 392.

Globulea mesembryanthemoides (Haw.); stem suffruticose, bushy, erect; the subulate leaves, branches, ramuli and calyces hispid. *DC. l. c.* 392.

Globulea subincana (Haw.); stem suffruticose, ascending; leaves semiterete, subulate, acute, spreading with incurved points, as well as the branchlets softly canescent. *DC. l. c.* 392.

Globulea mollis (Haw.); leaves semicylindrical, acute, gibbous beneath, suberect, minutely subtomentose; cymes terminal, compound. *DC. l. c.* 392.

Doubtful Species of Section EU-CRASSULA *and species unknown to us.*

C. argentea (Linn. Suppl. 188); *DC. Prod.* 3. *p.* 383, seems by description to be the same as *C. lactea.*

C. telephioides (Haw. rev. suc. p. 9); stems herbaceous ? erect; leaves obovate-oblong, amplexicaul, minutely punctato-crenulate below; flowers cymose. *D.C. l. c. p.* 384. Leaves 3 inches long, 18 lines wide. Petals pale rosy. Squamæ quadrate. In habit very like *Sedum Telephium.*

C. rotundifolia (Haw. Ph. Mag 31. 188); subherbaceous or perennial, erect; leaves petiolate, subrotund, firm, few-toothed; the lowest entire. *DC. l. c.* Stems simple, terete, thick, green, nearly as in *S. Telephium* but shorter.

C. revolvens (Haw. Ph. Mag. 1824. p. 188); stem suffruticose, slender, somewhat branched; branches erect; leaves linear, small, acute, revolute-reflexed or arched, subdistant. *DC. l. c.* 384, near *C. fruticulosa.*

C. bibracteata (Haw. l. c.); effuso-decumbent, rooting at the nodes; leaves subulate, spreading, flat or furrowed above; bracteæ on the common peduncle two. *DC. l. c.* 384. Flowers white; anthers sulphur-coloured, turning brown. Allied to *C. acutifolia.*

C. filicaulis (Haw. l. c); effuse, dichotomous; leaves spreading; subrecurved, lanceolate-subulate, smooth, convex beneath; branches filiform, rooting at the nodes. *DC. l. c.* 384. Flowers white, subcymose; anthers yellow. Allied to the preceding.

C. bullulata (Haw. l. c.); leaves between strap-shaped and lanceolate, as well as the stem minutely roughened with white inflated hairs; flowers cymose. *DC. l. c.* 385. Flowers yellow, allied to *C. scabra.*

C. muricata (Thunb. Fl. Cap. 283); stem frutescent, erect; branches 4-angled; leaves connate, trigonous, ciliate-scabrid, obtuse; flowers subumbellate. *DC. l. c.* 385. Differs from *C. tetragona* by the erect stem and scabrous leaves.

C. vestita (Linn. f. Suppl. 188); leaves connate, deltoid, obtuse, very entire, covered with white powder, the upper ones very close together; flowers terminal, capitate. *DC. l. c.* 385. Stem 6–8 inches high, suberect, branching, naked at base; flowers yellow, sessile, crowded at the ends of the branches.

C. prostrata (Thunb. Cap. 282); stem herbaceous, decumbent, pellucid, glabrous; leaves lanceolate, acute; flowers subumbellate. *DC. l. c.* 386. Probably a form of *C. centauroides.*

C. corallina (Linn. f. Suppl. 188); leaves opposite, deltoid, obtuse, close-placed, dotted; flowers umbellate-corymbose; stems dichotomous, erect. *DC. l. c.* 386. Stems uncial; leaves suborbicular, powdery at the point, 1–2 lines long, longer than the internodes.

C. pubescens (Linn. f. l. c. 191); leaves connate, ovate, acute, fleshy, villous, spreading; stem erect, branched, glabrous; flowers corymbose, small, white; calyx downy. *DC. l. c.* 386.

C. (Turgosea) linguæfolia (Haw. Misc. Nat. 175); lower leaves distinct, opposite, tongue-shaped, ciliate, pubescent; stem simple, leafy; flowers greenish white, whorled, crowded, sessile. *DC. l. c.* 386. Perhaps a variety of *C. tomentosa.*

C. concinnella (Haw. Ph. Mag. 1822. p. 381); leaves obovate, very densely fringed with white cilia. *DC. l. c.* 387.

C. cotyledonis (Linn. f.); radical leaves connate, oblong, obtuse, tomentose, ciliate; stem nearly naked, herbaceous, 4-angled; floral leaves lanceolate; fascicles of flowers corymbose. *Thunb. Cap.* 289. *DC. l. c.* 387. Stem simple, erect, a foot high.

C. spicata (Linn. f. Suppl. 189); radical leaves glabrous, connate, linear-subulate; stem erect, herbaceous, nearly naked; flowers whorled. *Thunb. Cap.* 284. *DC. l. c.* 387.

C. **hirta** (Thunb. Cap. 284); radical leaves lanceolate, hairy; stem herbaceous, erect, nearly naked, pubescent; heads of flowers whorled. *DC. l. c.* 387. Near *C. spicata.*

C. **capitellata** (Linn. f. Suppl. 190); leaves connate, oblong, glabrous, cartilagineo-ciliate, spreading, longer than the internodes; heads of flowers whorled. *Thunb. Cap.* 286. *DC. l. c.* 387.

C. **hemisphærica** (Thunb. Cap. 292); lowest leaves connate, roundish, hemispherically imbricated, cartilagineo-ciliate, stem nearly naked; tufts of flowers spicato-paniculate. *DC. l. c.* 387. Floral leaves very small; flowers small, white.

C. **thyrsiflora** (Linn. f. Suppl. 190); leaves perfoliate, ovate, obtuse, ciliate, glabrous, erecto-patent; thyrsus spiked, branched. *DC. l. c.* 387.

C. **obovata** (Haw. Suppl. 17); leaves opposite, decussate, obovate, ciliate, minutely impresso-punctate; stem hispidulous; flowers axillary, spicato-thyrsoid. *DC. l. c.* 387.

C. **aloides** (Ait. Hort. Kew. 1 p. 304); stem simple, hairy; leaves ovate or spathulato-lanceolate, distinct, ciliate, impresso-punctate; heads axillary, in a thyrsus. *DC. l. c.* 388.

C. **punctata** (Linn. Sp. 406); stem simple, smooth; leaves opposite, ovate, punctate, ciliate, the lower ones oblong; corymbs axillary, very short. *DC. l. c.* 388.

C. **ramuliflora** (Link. enum. l. p. 301); stem shrubby, rough with reflexed hairs; leaves opposite, obovate, acute, subconnate, ciliate; axillary branches few-flowered; petals lanceolate, erect, spreading, white. *DC. l. c.* 388.

C. **montana** (Linn. f.); leaves connate, ovate, acute, the radical close-placed, the cauline distant; flowers whorled in the upper axils, the ultimate capitate. *DC. l. c.* 388. Stem filiform, simple, 3 in. high.

C. **cephalophora** (Linn. f.); radical leaves connate, linear-oblong, obtuse, entire; stem nearly naked, erect; heads opposite, stalked. *DC. l. e.* 388.

C. **debilis** (Thunb. Cap. 280); stem herbaceous, trichotomous, erect; leaves opposite, glabrous, crowded, subterete, concave, papulose; flowers pedicellate; petals linear. *DC. l. c.* 388. Allied to *C. glabra?*

C. **rupestris** (Linn. f.); leaves connate, ovate, very entire, glabrous, crowded, convex-keeled beneath; corymb trichotomous, fastigiate, much branched. *DC. l. c.* 388.

C. **tecta** (Linn. f.); sub-radical leaves connate, ovate, obtuse, imbricate, cartilagineo-ciliate, powdery; scape nearly naked, filiform; flowers sessile, capitate. *DC. l. c.* 388.

C. **minima** (Thunb. Cap. 292); glabrous, nearly stemless; leaves petioled, roundish, entire; peduncles subradical, one-flowered. *DC. l. c.* 388.

C. **neglecta** (Schultz.); stem herbaceous; leaves petioled, cordate, glabrous; flowers solitary. *DC. l. c.* 389.

C. diffusa (Ait.) stem herbaceous; leaves oblong, tapering at base, crenate; peduncles axillary, solitary. *DC. l. c.* 390.

C. subulata (Linn. Mant. 360); stem herbaceous, branched; leaves opposite, subulate, terete, spreading; flowers capitate. *DC. l. c.* 390.

C. sylvatica (Lichst.); stem herbaceous, dichotomous, strigoso-hispid; leaves obovate-oblong, strigose at margin and base; flowers terminal and axillary, solitary. *DC. l. c.* 390.

C. ascendens (Thunb.); stem suffruticose, decumbent; branches erect, above filiform and naked; leaves connate, triquetrous, entire, spreading, glabrous; corymb compound. *DC. l. c.* 390.

VI. ROCHEA, DC.

Calyx 5-parted or deeply 5-cleft. *Corolla* (more or less perfectly) gamopetalous, salver-shaped, its tube longer than the calyx, limb 5-parted, spreading. *Stamens* 5, adnate with the claws of the petals; the anthers subsessile at the throat of the tube. *Carpels* 5, pluriovulate; styles conniving, subulate or clavate. *Squamæ* very minute. *Follicles* many seeded. *DC. Prod.* 3. *p.* 393, *excl. sect.* 1.

Shrubby or half-shrubby succulents, known from *Crassula* by the salver-shaped corolla, with a tube much longer than the calyx. Leaves connate or vaginate at base, bordered with cartilaginous cilia, as are also the sepals. Flowers crimson, rosy, white or pale yellow. Name in honour of *M. de la Roche*, a French botanist.

ANALYSIS OF THE SPECIES.

Stem erect; flowers in subcapitate, many flowered cymes:
- Lvs. ovate-oblong or obovate; flowers crimson, 1½–2 inches long ... ... (1) **coccinea.**
- Lvs. oblongo-lanceolate; flowers uncial, rosy and white, variable ... ... (2) **versicolor.**
- Lvs. linear-lanceolate or subulate, channelled; fl. yellow or cream coloured ... ... (4) **odoratissima.**

Stem decumbent; flowers solitary or few together, white, turning rosy ... ... (3) **jasminea.**

1. R. coccinea (DC. Pl. Grass. t. 1); shrubby, robust; stem erect, subsimple, imbricated with leaves; leaves connate at base, *ovate-oblong or obovate*, acute or subacute; cymes corymboso-capitate, many flowered; sepals uncial, linear-lanceolate; limb of the petals ovate or ovate-oblong, acute. *DC. l. c. p.* 394. *E. & Z.!* 1946. *Crassula coccinea, Linn. Bot. Mag. t.* 495. *Kalosanthes coccinea, Haw.*

HAB. Among stones at the summit of Table Mt., common. (Herb. D., Sd., Hk.)

Stem 1–2 feet high, 2–4 lines in diameter. Leaves tetrastichous, very closely imbricating, 1–1½ inches long, ¾–1 inch wide, the lowest ones often narrower than the upper, all ciliate. Flowers bright scarlet, 1½–2 inches long; the calyx uncial. Often cultivated in England.

2. R. versicolor (DC. Prod. 3. p. 394); suffruticose; stem erect, branched, imbricated with leaves; leaves connato-vaginate, *oblongo-lanceolate*, acute or subacute; cymes corymboso-capitate, several flowered; sepals semiuncial, connate below, keeled, lanceolate, two-thirds as long as the tube of the corolla; limb of the petals lanceolate-oblong, subacute. *Crassula versicolor, Burch. Bot. Reg. t.* 320. *Bot. Mag. t.* 2356. *R. media, DC. ? l. c. E. & Z. !* 1947. *Kalosanthes versicolor, and K. media ? Haw.*

HAB. Table Mt., *Burchell.* At Paradise, east side of Table Mt., *Dr. Wallich !* Dry hills near Zoutendals valley at Hassaquaskloof, Swell., *E. & Z. !* (Herb. Sd., T.C.D., Hk.)

Stem half woody, 1–2 feet high, branched below; branches erect, simple, 10–12 inches long. Leaves 1–1½ inch long, gradually attenuate upwards, cartilagineo-ciliate, flat. Flowers bright red externally, within white except for a rosy margin, at length wholly suffused with red, not quite so long as those of *R. coccinea.* Tube of the corolla about an inch long; the calyx rather more than half an inch. Flowers sweet-scented in the evening. Cultivated in England.

3. **R. jasminea** (DC. Prod. 3, 394); stem suffruticose, branched, decumbent, the branches ascending or erect, simple, closely leafy; leaves connate at base, narrow-oblong or spathulate, blunt, spreading or squarrose; flowers terminal, solitary or few together, sessile; sepals semi-uncial, lanceolate-linear; limb of the petals elliptic-oblong, blunt. *E. & Z. !* 1951. *Crass. jasminea, Bot. Mag. t.* 2178. *Lodd. Cab. t.* 1049. *Kalosanthes jasminea, Haw. Rochea microphylla, E. Mey. !*

HAB. On the Winterhoeksberg, Worcest., *E. & Z. !* Dutoitskloof, *Drege !* (Herb. Sd., D.)

Stems 6–12 inches long, branched near the base, diffuse. Leaves ½–¾ inch long, 1–2 lines wide, cartilagineo-ciliate. Corolla twice as long as the calyx, opening white, becoming rosy, 1½ inch long. Cultivated in England.

4. **R. odoratissima** (D C.); suffruticose; stem erect, scabrous, branched; branches erect, virgate, closely or laxly leafy; leaves connato-vaginate, erecto-patent, *linear-lanceolate or subulate,* channelled, acute or subacute; cymes capitate, many flowered; petals semi-uncial, connate below, lanceolate; limb of the petals lanceolate, subacute. *E. & Z. !* 1850, *also* 1848 *and* 1849. *Crass. odoratissima, Andr. Rep. t.* 26. *Jacq. Schoenbr. t.* 434. *Kalosanthes odor. Haw. Drege,* 6898.

HAB. Rocky and dry ground. Round Capetown, common. Zwarteberg and Kleinriviersberg, *E. & Z. !* &c. (Herb. Sd., D, Hk., &c.)

12–18 inches high, much branched or subsimple, the branches corymbose, each ending in a head of flowers, Leaves 1–1½ inches long, 1–1½ lines wide, taper-pointed, erect or spreading. Flowers about an inch long, pale-yellow or creamy-white, sweet-scented, sometimes rosy. *E. & Z.'s* specimens of "*R. versicolor*" and "*R. bicolor,*" in Herb. Sond., are undistinguishable from this.

Imperfectly known Species.

R. biconvexa (DC.); leaves narrow-linear, distinctly convex at each surface. *Kalosanthes biconvexa, Haw.*

R. flava (Haw.); leaves connato-vaginate, linear; cyme terminal; stem shrubby. *Burm. Afr. t.* 23, *f.* 3. *Crassula flava, Linn.* Perhaps *R. odoratissima ?*

R. fascicularis (Shultz.)) leaves connato-vaginate, linear-lanceolate; flowers fascicled; calyx-lobes lanceolate, acute, ciliate. Corolla nearly that of *C. coccinea,* but shorter. *Crass. fascicularis, Lam.* Probably a mere variety of *C. coccinea.*

R. media (DC.); leaves oblongo-lanceolate; connato-amplexicaul; flowers variable in colour. How does this differ from *R. versicolor ?*

VII. COTYLEDON, L.

Calyx 5-parted, much shorter than the tube of the corolla. *Corolla* gamopetalous, with an ovate or cylindrical 5-angled tube, and a spreading or reflexed and revolute limb, spirally twisted in the bud. *Stamens* 10, attached to the base of the tube of the corolla, exserted or subincluded. *Squamæ* oval. *Carpels* 5, many ovuled; styles subulate; stigmata subcapitate. *Follicles* many seeded. *DC. Prod.* 3, *p.* 396.

Shrubby or half-shrubby, or herbaceous succulents, natives of South Africa. Leaves entire, sessile or subsessile, opposite or scattered. Flowers showy, either in cymoid panicles or spicato-racemose, pedunculate; peduncles mostly terminal. Name from κοτυλη, a *cavity*; referring to the cup-like leaves of some species.

ANALYSIS OF THE SPECIES.

Sect. I. **Paniculatæ.** Inflorescence a branching, corymbose-cyme or panicle; flowers conspicuously pedicellate. (Sp. 1–19.)

(1) Leaves opposite:
  Leaves glabrous, green or powdery:
    Majores: stem robust; peduncle 1–2 feet high:
      Lvs. roundish-obovate or oblong-obovate, flat (1) **orbiculata.**
      Lvs. cuneate-oblong, subcuspidate, concave ... (2) **coruscans.**
      Lvs. broadly linear or strap-shaped, obtuse, concave ... ... (3) **purpurea.**
      Lvs. subterete, elongate, subacute ... ... (4) **decussata.**
    Minores: stem slender; pedunc. 6–12 inches long:
      Lvs. ovoid-oblong, fleshy, acute; stem decumbent ... ... (5) **papillaris.**
      Lvs. obovate-cuneate, obtuse; pedunc. 2–3-fl. (6) **ramosissima.**
      Lvs. obovate-cuneate, acute; pedunc. several flowered ... ... (7) **Meyeri.**
      Lvs. cuneate, *tapering much at base*, acute; peduncs. several flowered, puberulous above (8) **gracilis.**
  Leaves pubescent, hirsute or tomentose:
    Robust; lvs. obovate-cuneate, obtuse ... ... (9) **cuneata.**
    Slender; lvs, ovate-oblong, subacute, (small) ... (10) **tomentosa.**
    Robust; lvs. sub-cylindrical, long, acute ... ... (11) **teretifolia.**
(2) Leaves scattered or tufted, never opposite:—
  Inflorescence loosely panicled; flowers *nodding* or *pendulous*, subsecund:
    Panicle glabrous:
      flowers 9–12 lines long ... ... (12) **fascicularis.**
      flowers 5–6 lines long ... ... (13) **Eckloniana.**
    Panicle and flowers viscoso-pubescent ... ... (14) **Wallichii.**
  Inflorescence corymbose or racemose; flowers *erect.*
    Peduncles simple, elongate, corymb. at the summit:
      Corymb much branched, spreading; fl. uncial (15) **cacalioides.**
      Pedunc. few fl.; corolla curved, 1½–2½ in. long (16) **tuberculosa.**
      Pedunc. zig-zag, several flowered; fl. 8–10 lines long ... ... (18) **ventricosa.**
    Peduncles simple, *short,* few flowered; flowers glabrous, ¾ inch long ... ... (17) **racemosa.**
    Peduncles divaricately much branched, zig-zag, interlaced, hardening and persistent ... ... (19) **reticulata.**

Sect. II. **Spicatæ.** Inflorescence an undivided or rarely forked spike or raceme; flowers subsessile or on very short pedicels, erect. (Sp. 20–23.)

Leaves, peduncles, and flowers *glabrous* :
Leaves obovate or subrotund or oblong, flat ... (20) **hemisphærica.**
Leaves *subpetiolate*, narrow-cuneate, curled at the summit ... . (21) **cristata.**
Leaves sub-cylindrical, narrowed toward each end (23) **mamillaris.**
Lvs. peduncles and flowers *pubescent* ; lvs. flabelliform (22) **Zeyheri.**

Section I. **PANICULATÆ.** (Sp. 1–19.)

1. C. orbiculata (Linn. Sp. 614); leaves opposite, glabrous, powdery and glaucous, *roundish-obovate or oblong-obovate*, cuneate at base, obtuse, mucronate; peduncles terminal, very long, glabrous, loosely panicled, many flowered; flowers glabrous, tube of the corolla 4–5 times as long as the calyx, 2–2½ times as long as the limb. *DC. l. c.* 396. *Pl. Grass. t.* 76. *Bot. Mag. t.* 321. *E. & Z.!* 1957. *C. crassifolia, E. & Z.!* 1956. *C. oblonga, E. & Z.!* 1958. *Zey.!* 2566, 2567, 672. *Drege!* 6925.

HAB. N. & W sides of Table Mt.; near the Zwartkops R., Uit. and at Konabshoogde, Caffr., *E. & Z.!* Paarlberg and Weltevrede, Gamke R., *Drege!* (Herb. Sd., Hk.)

Stem robust, branching, bushy. Leaves 2–4 inches long, 1–2½ inches wide, varying much in shape from oblong to broadly obovate, usually tipped with an abrupt point. Peduncle 2 feet long, ending in a spreading panicle. Tube of the corolla ¾–1 inch long.

2. C. coruscans (Haw. Suppl. 28); leaves opposite, decussate, cuneate-oblong, concave, with thickened margin, sharply mucronate, white-powdery; peduncles elongate, corymboso-paniculate at the apex, glabrous; pedicels elongate, pendulous; tube of the corolla 4–5 times as long as the calyx, about equalling the lanceolate-acute limbs of the petals. *DC. l. c. p.* 396. *Bot. Mag. t.* 2601. *Lodd. Cab. t.* 1030. *C. canalifolia, Haw. E. & Z.!* 1962. *Drege!* 6928. *C. ungulata, E. & Z.!* 1963.

HAB. Karroo between Langekloof and Zwarteberg, Graaf Reynet, and near Gauritz R., George, *E. & Z.!* Nieuweveld, between Brack R. and Uitvlugt, and near Rhinosterkopf, *Drege!* (Herb. Sond.)

Stem . . . . Leaf-pairs ½–1 inch apart. Leaves 1½–2 inches long, 4–5 lines wide, narrowed to the base. Peduncle 12–18 inches long, ending in a 9–12 flowered forked cyme. Flowers nodding, 1–1¼ inch long, red. Calyx-lobes very short, deltoid, acute, with wide interspaces. I cannot distinguish *E. & Z.*'s specimens distributed under the name "*C. ungulata,* Lam." from those of their "*C. coruscans*" here described.

3. C. purpurea (Thunb.! Cap. p. 396); leaves opposite, concave, broadly linear or strap-shaped, obtuse, sub-cuspidate, glabrous; peduncles elongate, laxly corymbose; flowers on long pedicels, nodding, glabrous; tube of the corolla 4–5 times as long as the calyx, and longer than the linear-oblong, mucronate limbs of the petals. *DC. Prod.* 3. *p.* 397.

HAB. Common on hills and mountain sides about Capetown and elsewhere, *Thunberg!* (Herb. Thunb.)

Stem herbaceous, terete, glabrous, erect, a foot high. Leaves sessile, 3–4 inches long, half inch wide, blunt, with a minute projecting apex. Peduncle 6–12 inches long, sub-dichotomous; pedicels uncial, "compressed." Corolla uncial, dull red. Near *C. coruscans,* but with much longer and differently shaped leaves. Though stated by Thunberg to be "common round Capetown and elsewhere," no collector save himself has sent it to Europe.

4. C. decussata (Sims. Bot. Mag. t. 2518); glabrous; leaves opposite, decussate, sub-terete, elongate, subacute or obtuse, glaucous; peduncles elongate, corymboso-paniculate; pedicels elongate, nodding; tube of the corolla 4–5 times as long as the calyx, rather longer than the lanceolate-oblong, acute limbs of the petals. *Lindl. Bot. Reg. t.* 915. *C. papillaris, Haw. Suppl. p.* 21. *E. & Z.!* 1964, *non Thunb.*

HAB. Dry places on mountain sides, Kamiesberge, Namaqualand, *E. & Z.!* (Herb. Sd.)

Stem erect, not much branched, leafy below. Leaf-pairs close together. Leaves 2–2½ inches long, 2–3 lines in diameter, thick and fleshy, sub-cylindrical, flattish above. Peduncles 12–18 inches long, corymbose at the summit, many flowered; pedicels uncial. Corolla about an inch long, red. The leaves are shorter than in *C. teretifolia*, but, except for the pubescence, otherwise similar.

5. C. ramosissima (Haw. Suppl. 25); stem much branched, flexuous, sub-dichotomous; leaves opposite, glabrous, squamuloso-farinose, obovate-cuneate, obtuse; peduncles terminal, short, 2–3-flowered; flowers glabrous; tube of the corolla twice as long as the calyx, equalling the limb. *DC. Prod.* 3. *p.* 396. *E. & Z.!* 1959. *Zey.!* 2565. *Drege,* 6927.

HAB. Zoutpanhoogde, Zwartkops R., *E. & Z.!* District of George, *Herb. Eckl.* (Herb. Sd., Hk., D.)

Stem bushy, 1–2 feet high, 2–3 lines in diameter, with an ash-coloured bark. Leaves 1–1½ inches long, ½–¾ inch wide, whitish, with powdery scales. Peduncle 2–3 inches long, commonly 2-flowered. Tube of the corolla about ½ inch long, 3–4 lines in diameter.

6. C. papillaris (Thunb.! Cap. p. 397); stem slender, decumbent, branched from the base, the branches flexuous, ascending, simple; leaves opposite, decussate, *thick and fleshy, ovoid-oblong*, narrowed at base, acute, glabrous; peduncle elongate, slender, viscoso-puberulous above, cymose at the apex, few flowered; pedicels elongate, pendulous; tube of the corolla thrice as long as the calyx, about equalling the oblong, acute limbs of the petals. *DC. l. c.* 397, *excl. syn. C. angulata, var. foliis minoribus, E. Mey.! in Herb. Drege.*

HAB. In the Karroo, near Camenasie, at the Olifants R., *Thunberg!* Nieuweveld, on hills and mountain sides, near Bokpoort, 3500–4500f., *Drege!* (Herb. Thunb.! Sond.)

Branches several from the crown of the root, herbaceous, 4–6 inches long, a line in diameter, with a chestnut-coloured bark, glabrous or microscopically puberulous. Leaves half an inch apart or rather more, scarcely an inch long, 3–4 lines wide, drying into a nearly fusiform shape, suddenly contracted at their insertion, and acute or apiculate. Peduncle 6 inches long; cyme scarcely branched, 4–5-flowered. Flowers half an inch long. *Thunberg's* specimen in Herb. Upsal is very imperfect, but agrees (so far as it goes) with *Drege's* plant here described.

7. C. Meyeri (Harv.); glabrous; stem slender, branched; leaves opposite, approximate, obovate, cuneate at base, acute, fleshy; peduncles terminal, elongate, corymboso-paniculate, several flowered; pedicels longish, drooping; tube of the (small) corolla twice as long as the calyx, about equalling the lanceolate limbs of the petals. *C. cuneata, E. Mey.! in Herb. Drege, non Thunb.*

HAB. Sternbergspruit, District of Albert, *Drege!* (Herb. Sond.)

Stem 4–6 inches high, 2 lines in diameter, somewhat corymbosely branched; branches 2–3 inches long, leafy, ending in peduncles 6–12 inches long. Leaves about an inch long, 6–7 lines wide. Flowers half inch long, including the limb. Described from a solitary and rather imperfect specimen.

8. C. gracilis (Harv.); stem slender; leaves opposite, cuneate or cuneato-obovate, *tapering much at base, sub-petiolate,* acute or sub-cuspidate, flat, fleshy, glabrous; peduncles terminal, corymboso-paniculate, several flowered, viscidulo-puberulous above; pedicels longish, drooping; tube of the (small) puberulous corolla twice as long as the calyx, and shorter than the lanceolate-acuminate limb of the petals. *Zey.!* 2564.

HAB. Riet-rivier, Tarka, *Zeyher!* (Herb. Sd., Hk.)

Stem 4–6 inches high (?), a line in diameter. Leaves 1–1½ inches long, between cuneate and spathulate, greatly narrowed toward the base, 3–6 lines wide toward the apex, coriaceous when dry and not very fleshy. Peduncles 8–10 inches long; the young ones and the flowers minutely puberulous and viscid. Flowers ½–¾ inch long, including the reflexed limb.

9. C. cuneata (Thunb! Cap. 395); leaves opposite, hispid, oblongo-obovate, cuneate at base, subundulate, obtuse; peduncles terminal, very long, pubescent, loosely panicled, many flowered; panicle, calyx and corolla viscidly hirsute; tube of the corolla twice as long as the calyx, about as long as the limb. *DC. Prod.* 3. *p.* 398. *C. undulata, E. & Z.!* 1960 and *C. cuneata, E. & Z.!* 1961.

HAB. Dry ground, Kamiesberg, Namaqualand; and between Gauritz R., and Cangoberge, George, *E. & Z.!* (Herb. Sd., Thunb.)

Stem robust, half an inch in diameter, closely leafy for 3–6 inches above the base, prolonged into a peduncle 2 feet long, ending in a spreading or condensed panicle. Leaves 3–5 inches long, 2–3 inches wide, very thick. Tube of the corolla 5–6 lines long, and about as much in diameter, the lanceolate lobes as long or longer.

10. C. tomentosa (Harv.); stem slender; leaves opposite, decussate, ovato-oblong, thick and fleshy, sub-petiolate, *densely tomentose,* as are also peduncles and calyx; peduncle elongate, sub-corymbose at the summit, few flowered; pedicels longish, nodding; tube of the pubescent corolla twice as long as the calyx, equalling the lanceolate limbs of the petals. *Zey.! Z. n. N. n. E. Crass.* 3. 108. 5.

HAB. Grootrivier and Trompeterspoort, Uitenhage. *Zeyher!* (Herb. Sond.)

Stem 4–6 inches long, 1½–2 lines in diameter, laxly leafy, prolonged into a naked peduncle 4–8 inches long, bearing a slightly branched, corymbose cyme of 4–6 flowers. Leaves about an inch long, half inch wide, probably convex, certainly fleshy, thickly clothed with woolly hairs. Flowers about ½–¾ inch long, red?

11. C. teretifolia (Thunb! Prod. 83); leaves opposite, *sub-terete, elongate, acute* or cuspidate, densely hirsute (or subglabrous); peduncles elongate, corymboso-paniculate; pedicels elongate; calyx and corolla hirsute; tube of the corolla not quite twice as long as the calyx, shorter than the lanceolate-acuminate limbs of the petals. *Thunb. Cap. p.* 397. *DC. Prod.* 3. *p.* 397. *E. & Z.!* 1965. *Zey.!* 2563.

VAR. *β.* **subglaber**; leaves subglabrous; peduncles and flowers minutely pubescent. *Zey!* 2562.

HAB. Dry hills between Coega and Zondags River, *E. & Z.! Drege!* Near the Zwartkops River, both varieties, *Zeyher!* (Herb. Thunb., Hk., Sd., D.)

Stem suffruticose, 6–8 inches high, simple or branched from the base, 3–4 lines in diameter. Leaves 4–5 inches long, 3–4 lines wide, tapering at base, except in var. β, densely covered with short, patent hairs. Peduncles 12–18 inches high, bearing a many-flowered corymb. Tube of the corolla ½ inch, limb 6–8 lines long. Var. β. differs only in less copiousness of pubescence.

12. C. fascicularis (Ait. Kew. vol. 2, p. 106); leaves crowded toward the end of the branches, scattered, cuneate-obovate, obtuse, flat, fleshy, glabrous; peduncles elongate, panicled, *the branches of the panicle alternate, sub-distant*, patent, scorpioid, glabrous; flowers shortly pedicellate, subsecund, nodding; calyx and corolla minutely puberulous; tube of the corolla more than twice as long as the calyx, rather longer than the lanceolate limbs of the petals. *DC. Prod. p.* 397. *E. & Z.!* 1966. *Zey.!* 673. *Drege!* 6926. *C. paniculata, Thunb.! Cap. p.* 396. *Burm. Afr. Pl. t.* 18. *C. tardiflorum, Bonpl. nav. t.* 37.

HAB. In the Karroo, beyond Hartequa's kloof and in Canna Land, *Thunberg!* Kochman's kloof and Gauritz R., *E. & Z.!* Boschkloof, *Drege!* Blankenberg, Zwartland, *Zeyher!* (Herb. Thunb., Hk., D., Sd.)

Stem very thick and fleshy, 1–2 feet high, little branched; branches short and thick, tubercled with prominent leaf-scars. Leaves 2–3 inches long, 1–1½ wide, tapering into a cuneate base, deciduous. Peduncles 1–2 feet high, more than half occupied by the wide panicle, whose branches are an inch apart at their insertion, and though 6–8 inches long, seldom more than once forked. Flowers dull reddish. with a greenish 5-angled tube, 9–12 lines long.

13. C. Eckloniana (Harv.); leaves crowded toward the end of the branches, scattered, (of unknown form); peduncles elongate, panicled, the branches of the panicle alternate, patent, simple or forked, scorpioid, glabrous; flowers shortly pedicellate, subsecund, nodding; calyx and corolla glabrous; tube of the corolla twice or thrice as long as the calyx, longer than the lanceolate-oblong limbs of the petals. *C. cacalioides, E. & Z.!* 1967, *not of Thunb.*

HAB. Dry places on mountain sides, Kamiesberg, Namaqualand, *E. & Z.!* (Herb. Sond.)

This, though nearly allied to *C. fascicularis*, differs in the much more slender peduncles, looser and more racemose panicle, and smaller, glabrous flowers. The specimens seen are without leaves; the leaf-scars are tubercular, closely spiral on nearly conical ends of a fleshy stem, an inch or more in diameter. Peduncles 2 feet high, 1½–2 lines in diameter below, becoming very slender upwards; branches of the panicle racemose, 4–6 inches long. Flowers ½ inch long.

14. C. Wallichii (Harv.); leaves scattered, (of unknown form); *peduncles, panicle, calyx and corolla viscoso-pubescent*; peduncles elongate, panicled, the branches of the panicle alternate, patent, simple or forked, scorpioid; flowers shortly pedicellate, subsecund, nodding; tube of the corolla 1½ to twice as long as the calyx, rather longer than the oblong, acute limbs of the petals.

HAB. Elandsberg, *Dr. Wallich!* North sides of Snowy Mts., *Burke!* Cape, *Villette* in Hb. Hook. (Herb. Hook., D.)

Allied to *C. fascicularis*, but differs in pubescence, smaller flowers and larger calyx, in proportion to corolla. Stem and leaves unknown; leaf-scars on the peduncle scattered. Flowers perhaps yellowish? half inch long. Peduncles 1–2 feet long.

15. C. cacalioides (Linn. f. Suppl. 242); leaves crowded toward the ends of the branches, tereti-filiform, acute, glabrous; peduncles terminal, elongate, terete, glabrous or hispid, loosely panicled, corymbose, many flowered; flowers erect; panicle, calyx and corolla viscoso-puberulous; tube of the corolla 5-angled, 4–5 times as long as the calyx, rather longer than the narrow-oblong, obtuse-mucronate limb of the petals. *Thunb. ! Fl. Cap.* 397. *DC. Prod.* 3. 397. *Drege !* 9542. *Zey. !* 2569.

HAB. Near Olifants Bath, *Thunberg !* Ataquaskloof, *Drege !* Kuureboomfontyn, Olifants R., and near Kenko R., *Zeyher !* (Herb. Thunb., Sd., Hk.)

Stem short, fleshy, branched from the base; branches 2–6 inches long, closely covered with spirally disposed, prominent leaf-scars. Leaves generally deciduous before flowering, 2–3 inches long, 1–2 lines in diameter, tapering to a point. Flower-stem 1–2 feet high, with depauperated, spirally inserted leaves below, leafless above, ending in a much branched, corymbose panicle. Flowers yellow, turning orange. Corolla-tube uncial, sharply 5-angled. *Drege's* 9542 and *Zeyher's* 2569, precisely agree with the specimen in Herb. *Thunb.*

16. C. tuberculosa (Lam. Dict. 2, p. 139); leaves scattered, subcylindrical, linear or linear-oblong, acute, glabrous; old leaf-scars tubercular; peduncles elongate, angular, laxly beset with filiform, depauperised leaves, cymoso-racemose at the summit, few flowered; pedicels, calyx and corolla viscoso-pubescent; flowers erect; tube of the corolla curved, 4–6 times as long as the calyx, twice as long as the lanceolate-oblong limbs of the petals. *DC. l. c. p.* 397. *E. & Z. !* 1969. *Zey. !* 2568. *Drege !* 6924.

HAB. Hills round Capetown, *E. & Z. ! W.H.H. !* Bufiljagdsriver, *Zeyher !* Between the little and great Fish R., *Drege !* (Herb. Sd., D., Hk.)

Stem short, subsimple, fleshy, closely covered with spirally inserted leaves, which are continued at intervals of about an inch, in a bractlike form, along the peduncle to its summit. Flowers 4–10, in a simple raceme or slightly branched panicle. Corolla 1½–2½ inches long, orange red, with a spreading limb.

17. C. racemosa (E. Mey !); leaves scattered, closely covering the short, fleshy stem, linear-terete, from a sheathing base, subacute, pubescent; peduncles not much longer than the leaves, lateral, numerous, set with a few scattered, *membranous*, subulate bracts, subcorymbose at the summit, few flowered, puberulous; flowers on longish pedicels, erect, *glabrous;* tube of the corolla not much exceeding the *lanceolate* sepals, longer than the ovate-oblong limbs of the petals.

HAB. Between Kaus, Natvoet and Doornport, and near Verleptpram, near the mouth of the Gariep, *Drege !* (Herb. Sd., Hk., D.)

Stem 4–6 inches long, ½ inch in diameter, completely clothed with fleshy leaves. Leaves spreading, 2–2½ inches long, 2–3 lines wide, squarrose when dry. Peduncles 2–3 inches long, with 2–4 membranous, scattered bracts, 4–6 flowered. Flowers yellow ? calyx and corolla more membranous than in other species; the sepals 5 lines long. Corolla, including the erect limb, about ¾ inch long.

18. C. ventricosa (Burm. Pr. Cap. Fl. p. 13); leaves scattered, crowded at the apex of the fleshy stem, linear-terete, elongate, acute, glabrous; leaf-scars tubercular; peduncles elongate, terminal, flexuous, sparsely set with depauperated, pubescent leaves, cymoso-corymbose at the apex, viscoso-pubescent, as are also the calyx and corolla; flowers on long pedicels, erect; tube of the corolla 2–3 times longer than the spreading calyx, longer than the linear-lanceolate, taper-pointed limbs of the petals. *DC. l. c. p.* 397, *E. & Z.!* 1968. *Burm. Afr. Dec. t.* 21. *f.* 1.

VAR. β. **alpina**; peduncles 3–4 inches high, few flowered; corolla uncial.

HAB. Karroo, between Langekloof and Zwarteberge, in Graaf Reynet, *E. & Z.!* β. Elandsberg, *Dr. Wallich!* (Herb. Sond., Hk., D.)

Stem succulent, short. Cauline leaves glabrous, 2–3 inches long, a line in diameter, on prominent tubercles crowded together. Peduncle terminal, 10–12 inches long, 1 line in diameter, somewhat zig-zag, with subulate, uncial bracts at intervals of 8–10 lines throughout its length. Cyme 10–12-flowered, lax, corymbose. Flowers 8–10 lines long, greenish-yellow. Var. β. differs in the shorter peduncles and larger flowers, but in other respects agrees with the normal form.

19. C. reticulata (Thunb. Cap. p. 393); stem short and thick, subsimple; leaves fascicled on wartlike abortive ramuli and at the apex of the stem, small, terete, furrowed above, acute or mucronate; peduncles *divaricately multifid, sub-dichotomous or zig-zag*, persistent; flowers on long, slender, rigid, spreading pedicels, suberect; pedicels, calyx and corolla viscoso-puberulous; tube of the corolla thrice as long as the calyx, limb short, reflexed, pubescent within. *DC. l. c. p.* 398. *Zey.!* 674.

HAB. In the Karroo, beyond Hartequaskloof, *Thunberg!* Springbok-keel, *Zeyher!* Boschjemans-karroo, *Drege!* (Herb. Thunb., Sd., Hk.)

Stem 6–8 inches high, 1–2 inches in diameter, simple or once or twice divided, smooth or covered with wartlike abortive branchlets. Leaves 2–4 lines long, 1 line in diameter, almost fusiform, several in a tuft. Peduncles very numerous and much branched, intricately interlaced (many of them barren ?), hardening after the fall of the flowers and persistent as a *mop* of much branched, spreading spines. The young parts are viscidulous, the older glabrous. Corolla 4–5 lines long, pale. A very remarkable plant.

Section II. **SPICATÆ.** (Sp. 20–23.)

20. C. hemisphærica (Linn. Sp. 614); stem short, leafy; leaves scattered, approximate, *broadly obovate or subrotund*, very obtuse or subacute or mucronulate, flat, thick and fleshy, glabrous; peduncle elongate, simple or rarely forked; flowers spicato-racemose, subsessile or shortly pedicellate, erect or erecto-patent; corolla tubular, much longer than the calyx, with a short, spreading limb. *Dill. Elth. t.* 95. *f.* 111. *DC. Pl. Grass. t.* 87. *E. & Z.!* 1970, *also C. rotundifolia, E. & Z.!* 1971. *C. rhombifolia, E. & Z.!* 1972 *and C. maculata, E. & Z.!* 1973. *C. triflora, Thunb. Fl. Cap. p.* 396. *Zey.!* 2570, 2572. *Drege!* 6821.

HAB. Zekoriver, *Thunberg!* Lion Mt. Capetown, Onderbokkeveld, Clanw.; between Krakakamma and Vanstadensberg, and on dry hills near the Zwartkops R., Uit., *E. & Z.!* Winterveld and Nieuweveld, and Camdeboosberg, *Drege!* Lislap and on the Onrust R., *Zey.!* (Herb. Thunb., Sd., Hk., D.)

Stem a few inches high, thick and succulent, closely covered with leaves. Leaves 1–2 inches long, 1–1½ wide, varying from obovate to oblong and nearly circular, not much tapering at base, sometimes with a minute mucro. Peduncle 8–14 inches high, about half of it occupied by the inflorescence. Flowers solitary or 2–3 together; pedicels 2–4 lines long, rarely 6–8 lines, thickened upwards. Corolla 6–7 lines long.

21. C. cristata (Haw. Phil. Mag. 1827. p. 123); nearly stemless; radical leaves petiolate, much attenuated at base, narrow-cuneate, *abrupt and crispato-undulate* at the summit, flat, fleshy, glabrous;

peduncle elongate, simple; flowers spicato-racemose, subsessile, erect; corolla tubular, much longer than the calyx, with a short spreading limb. *DC. l. c. p.* 399. *E. & Z.!* 1974. *C. clavifolia? Haw. l. c.*

HAB. Dry hills at the *Zwartkops R.*, Uit., *E. & Z.!* (Herb. Sond.)

Crown of the root shaggy, with rigid, red, curled bristles. Leaves 1–1½ inches long, very much attenuated at base into a more or less obvious petiole, the limb cuneate, but varying in breadth from 2 to 6–8 lines; the narrower forms answer to the description of Haworth's *C. clavifolia.* Peduncle slender, 6–8 inches long, about half of it occupied by inflorescence. Flowers as in *C. hemisphærica*, but rather smaller.

22. C. Zeyheri (Harv.); leaves, peduncles and flowers *pubescent*; stem ascending; leaves scattered, approximate, flabelliform, contracted in the middle and tapering into a long, cuneate base, shortly petiolate, rounded and crispato-undulate at the summit, flat, thinly-fleshy; peduncle elongate, simple; flowers spicato-racemose, subsessile, erect; corolla tubular, much longer than the calyx, with a short, spreading limb. *Zey.!* 2571.

HAB. At the Kinko River, *Zeyher!* (Herb. Sond., Hook.)

Stems 3–4 inches long, half recumbent, rooting at the nodes. Leaves approaching in pairs but not opposite, 1½ inch long, the limb subrotund, the base narrow wedge-shaped. Flowers as in *C. hemisphærica.*

23. C. mamillaris (Linn. f. Suppl. 242); stem short or scarcely any; leaves crowded round the apex, or scattered on the short stem, terete, somewhat fusiform, narrowed to both ends, obtuse, glabrous; peduncle elongate, simple; flowers spicato-racemose, subsessile, erect; corolla tubular, much longer than the calyx, with a short, spreading limb. *Thunb.! Fl. Cap. p.* 397. *DC. l. c. p.* 398. *C. filicaulis, E. & Z.!* 1975. *Zey.!* 2897.

HAB. Olifantsbad, *Thunberg!* Kamiesberg, Namaqualand, *E. & Z.!* Springbokkeel, *Zeyher!* (Herb. Sond.)

Stem (according to Thunberg) "creeping and rooting, branched." Leaves 1½–2 inches long, 2–8 lines in diameter, sub-cylindrical, somewhat tapering to each extremity. Flowers as in *C. hemisphærica.*

Imperfectly known and doubtful species.

C. undulata (Haw. Suppl. 50); leaves opposite, rhomb-obovate, with a point, pale green, the oldest very large and thick, with a red point, margined, the younger undulate. *DC. l. c. p.* 396. Probably a form of *C. orbiculata.*

C. ungulata (Lam. Dict. 2. 139); leaves opposite, semi-cylindrical, channelled, glabrous, purple at the callous point; flowers sub-paniculate, glabrous; stem erect. Seems to be the same as *C. coruscans.*

C. curviflora (Sims. Bot. Mag. t. 2044); leaves scattered, semi-cylindrical, glabrous; old leaf-scars prominent; flowers panicled, nodding; calyces spreading; tube of the corolla 5-angled, curved. Seemingly a garden variety of *C. tuberculosa*, with the pedicels "twisted in a fantastical manner."

C. spuria (Linn. Sp. 614?); leaves subradical, terete, oblong, fleshy, obtuse, narrower at base; stem very short and thick; peduncle erect;

naked; flowers panicled. *Burm. Afr. t.* 19. *f.* 1. *Pluk. alm. t.* 323. *f.* 1. *DC. l. c. p.* 397.

C. maculata (Salm.-Dyck); leaves scattered, ovato-spathulate, subauriculate at base, fleshy, shining, marked on both sides with dark red spots; flowers spiked, subalternate, with a spreading limb; stem suffruticose. A var. of *C. hemisphærica?*

C. rhombifolia (Haw.); leaves approximate, obovate-rhomboid, mucronate, powdery; stem branched, robust, decumbent. *DC. l. c.* 398. Allied to *C. hemisphærica*, but more branched and dwarf. Flowers unknown.

C. jasminiflora (Salm.-Dyck); leaves crowded, green, rhomboid-spathulate, fleshy; stem dwarf; peduncle branched; flowers erect, with a green tube, and a revolute purple and white limb; pedicels long, thickened upwards. *DC. l. c.*

C. caryophyllacea (N. L. Burm.); leaves aggregate, ovate, thick, flat, glaucous; flowers panicled on long pedicels, erect; stem branched. *DC. l. c. Burm. Afr. Dec. t.* 17.

C. mucronata (Lam.); leaves subradical, oval, flat, undulate at margin, mucronate; stem short, branched; flowers loosely panicled, erect. *DC. l. c.*

C. dichotoma (Haw.); leaves channelled; cyme dichotomous, pubescent, with spinous bracts; tube of the corolla swollen, limb folded back. *DC. l. c.*

C. parvula (Burch.); leaves oval, compressed, thick; panicle dichotomous; pedicels erect, very long, capillary; stem erect. *DC. l. c.*

C. trigyna (Burch.); stemless; leaves glabrous, flat, fleshy, cuneate-oval or subrotund; flowers erect, alternate on a simple, rarely bifid scape. *DC. l. c.* Carpels 3.

VIII. KALANCHOE, Adans.

Calyx 4-parted, the sepals scarcely cohering at base, small, acute. *Corolla* monopetalous, salver-shaped, with an urceolate tube, and a 4-parted, spreading limb. *Stamens* 8, adnate to the base of the tube of the corolla. *Squamæ* 4, linear or oblong. *Carpels* 4, with subulate styles. *Follicles* 4, many-seeded. *DC. Prod.* 3. *p.* 394.

Succulent suffrutices, with opposite, toothed or entire, or irregularly pinnatifid, fleshy leaves. Flowers in panicled cymes, yellow, red or cream colour. Natives of Asia and Africa. The name is from the Chinese term for one of the species.

ANALYSIS OF THE SOUTH AFRICAN SPECIES.

Inflorescence an oblong, close panicle or thyrsus … … (6) **thyrsiflora.**
Inflorescence corymbose or loosely panicled
  Whole plant, save the corolla, hispid … … … … … (3) **hirta.**
  Whole plant glabrous:
    Leaves narrowed to the base, but not *petiolate:*
      Lvs. fleshy, obovate or spathulate; sepals subulate … … … … … … … … (1) **rotundifolia.**
      Lvs. membranous, obovate-oblong; sepals deltoid … … … … … … … … (4) **oblongifolia.**

Leaves conspicuously petioled :
Lvs. crenate ; panicle not much branched ; sepals subulate ... ... (2) **crenata.**
Lvs. subentire ; panicle supra-decompound ; sepals deltoid ... ... (5) **paniculata.**

1. K. rotundifolia (Haw. Phil. Mag. 1825. p. 31); glabrous ; stem slender, laxly leafy below, naked upwards ; leaves roundish-obovate, obovate or spathulate, subentire or crenulate, tapering at base, subpetiolate ; cymes trichotomous, panicled, the subdivisions flat-topped ; bracts and sepals small, subulate ; lobes of the corolla narrow-lanceolate, acute. *DC. l. c. p.* 395. *E. & Z.!* 1952. *Zey.!* 2561. *Drege!* 6920.

HAB. Hills near the Zwartkops R., Uit., *E. & Z.!* Zoutpanshoogde, *Zey.!* Grassy Hills at Adow, and Klein Winterhoek, in a valley between Zoutpans and Enon, *Drege!* Howisons Poort, *H. Hutton!* (Herb. D., Sd,, Hk.)

Stems 1–3 feet high, leafy below ; the upper internodes 5–8 inches apart, with smaller and narrower leaves. Lower leaves 1–2 inches long, ¾–1 inch wide, often quite entire, fleshy. Cymes flat-topped, simple or trichotomously panicled, the lateral divisons on long common peduncles. Pedicels 3–4 lines long. Calyx 1 line. Corolla 4–5 lines long, orange or deep yellow. I am uncertain whether this be Haworth's plant or not.

2. K. crenata (Haw. Syn. p. 109) ; glabrous ; stem robust, laxly leafy, the upper nodes distant ; leaves *conspicuously petioled*, oblong or ovate, *coarsely crenate ;* cyme trichotomous, panicled, the subdivisions flat-topped ; bracts and *sepals subulate ;* lobes of the corolla lanceolate, acute. *DC. l. c.* 395. *E. & Z.!* 1654. *Bot. Mag. t.* 1436. *Andr. Rep. t.* 21.

HAB. Mountain sides near Philipstown, Caffr., *E. & Z.!* Kreilis Country, *H. Bowker* 72*!* Between the Kei and the Gekau, *Drege!* (Herb. Sd., Hk., D.)

Stems 2–3 feet or more high, 3–5 lines in diameter, the nodes 3–6 inches apart. Lower leaves wanting on our specimens ; cauline leaves 2–3 inches long, 1–2 inches broad, on petioles 1–1½ inch in length, bluntly dentate. Cymes somewhat flat-topped, dense ; the pedicels 2–3 lines long. Corolla 5 lines long, bright yellow. A much larger plant than *K. rotundifolia*, with long petioles and more evidently toothed leaves.

3. K. hirta (Harv.) ; stem, leaves, panicle and calyces *densely hispid*, with short spreading rigid hairs ; leaves conspicuously petioled, ovate or oblong, repando-crenate ; cyme trichotomous, panicled, the subdivisions flat-topped ; sepals ovate, acute ; lobes of the corolla lanceolate.

HAB. Olifantshoek, Uit., *Zeyher!* (Herb. Sond.)

Allied to *K. crenata* from which it differs in pubescence and the form of the sepals. Corolla golden yellow. In pubescence this agrees with *K. lanceolata*, Pers., but differs in foliage.

4. K. oblongifolia (Harv.) ; glabrous ; stem robust, leafy ; leaves obovate-oblong, obtuse, narrowed at base, sessile, very entire, membranous ; cymes trichotomous, panicled, the subdivisions flat-topped ; bracts toothlike ; sepals deltoid, acute ; lobes of the corolla ovato-lanceolate, acute.

HAB. Hopetown District, *Mr. Andrew Wyley!* (Herb. T.C.D.)

With the foliage of *K. thyrsiflora* this has the inflorescence of *K. crenata* and the sepals of *K. hirta*. It is (apparently) a weaker growing plant than any of these, with more membranous leaves. Flowers yellow? Described from an imperfect specimen.

5. **K. paniculata** (Harv.); glabrous; stem robust, rigid, naked upwards; lower leaves oblong, obtuse, *subentire*, tapering at base into a *broad petiole;* upper leaves . . . . ; cyme several times trichotomous and panicled, the partial and general panicles flat-topped; bracts subulate; sepals *ovate or deltoid*, acute; lobes of the corolla ovate, acute. *Zey.!* 671.

HAB. Vetrivier, *Burke and Zeyher!* Hb. (Herb. Hk., Sd.)

Remarkable for its very large, spreading, flat-topped panicle which is 6-8 inches or more in diameter. The stem is more rigid than in *K. crenata*, and may be 3–4 feet high. I have only seen a single leaf which, including the petiole, is 6½ inches long, and 2½ wide in its widest part.

6. **K. thyrsiflora** (Harv.); glabrous; stem robust, leafy; leaves oblongo-spathulate, obtuse, tapering at base, sessile, very entire; cymes *short-stalked, combined into a compact, oblong panicle or thyrsus;* bracts minute; pedicels equalling or exceeding the corolla; sepals ovate or ovato-lanceolate; *lobes of the corolla ovate, obtuse. K. alternans, E. & Z.!* 1953, (*not of Pers. ?*)

HAB. Near the sources of the Kat River, *E. & Z.!* Rhinoster Kopf and Vaal River, *Zey.!* 670. Districts of Cradock and Queenstown, on rocky hill sides, *Mrs. F. W. Barber*, 221! (Herb. Hk., Sd., D.)

Stem 2½–4 feet high, simple, leafy throughout, the upper leaves equalling or exceeding the internodes. Leaves pale green, 4–6 inches long or perhaps longer, the cuneate base half-amplexicaul. Thyrsus 9–12 inches long, cylindrical, densely many flowered. Corolla orange. Quite distinct in inflorescence and the obtuse corolla-lobes from other S. African species. *E. & Z.* refer it to *K. alternans*, Pers., but it hardly agrees with the character given of that species.

IX. BRYOPHYLLUM, Salisb.

Calyx inflated, 4-cleft nearly to the middle, the lobes valvate in æstivation. *Corolla* gamopetalous, with a long, sub-cylindrical tube, bluntly 4-angled at base, and a 4-parted, spreading limb. *Stamens* 8, adnate to the base of the tube of the corolla. *Squamæ* 4, oblong. *Carpels* 4, with subulate styles. *Follicles* many-seeded. *DC. Prod. p.* 395.

Succulent suffrutices. Leaves opposite, fleshy, petioled; either imparipinnate, or by abortion of the lateral segments simple. Pinnæ crenate. Cymes panicled, terminal; flowers yellow, changing to red. Name from *βρυω, to sprout*, and *φυλλον, a leaf;* young plants sprout from the notches in the leaves.

1. **B. tubiflorum** (Harv.); leaves (unknown); corolla thrice or four times as long as the sharply 4-cleft calyx, its segments broadly oblong, very blunt or truncate; stamens as long as the tube of the corolla. *Kalanchoe Delagoensis, E. & Z.!* 1955.

HAB. Delagoa Bay, *Forbes!* (Herb. Sond.)

Of this very remarkable plant a portion of a denuded branch, and part of a dense, probably thyrsoid, inflorescence exist in Herb. Sonder. The internodes are scarcely an inch long, and there are 4 cicatrices, indicating *whorled* leaves, at each node. Calyx 3 lines long. Corolla uncial, bright red, its lobes almost square, 2½ lines long.

ORDER LIV. PORTULACEÆ, Juss.

(BY W. SONDER.)

Flowers perfect, regular. *Calyx* free, or nearly so, deciduous or persistent, 2-leaved, the sepals imbricate, sometimes coloured within. *Petals* 4–6, inserted in the base of the calyx, free or partially connate, very delicate, quickly twisting together and dissolving in decay. *Stamens* mostly definite, as many, or 2–4 times as many as the petals, rarely more, the outermost opposite the petals and attached to their claws; filaments subulate; anthers 2-celled, introrse. *Ovary* sessile, free, one-celled, of 3 or more carpels; ovules amphitropal, rarely solitary, inserted by long cords on a free central placenta; style terminal or none; stigmata as many as the carpels, linear, rarely confluent and capitate. *Fruit* usually a dehiscent capsule; rarely indehiscent and nutlike. *Seeds* lenticular or reniform, with floury albumen. *Embryo* eccentric, curved round the margin of the seed, the radicle next the hilum.

Mostly herbaceous, rarely suffruticose or shrubby plants, very generally with more or less succulent foliage and stems. Leaves alternate or sub-opposite, quite simple or entire, one-nerved or nerveless, without lateral veins, sessile or short-stalked. Stipules none or membranous, often lacerated. Flowers in terminal or axillary cymes, sometimes racemose or tufted; rarely solitary. Natives usually of the temperate and colder zones, in all parts of the world. None are of much use. *Portulaca oleracea* is a potherb, and many others may be similarly used: all are innoxious. Some have esculent fleshy roots, as *Claytonia tuberosa* of Siberia.

TABLE OF THE SOUTH AFRICAN GENERA.

I. **Portulaca.**—*Capsule* circumscissile, one-celled; seeds numerous.

II. **Anacampseros.**—*Capsule* 3-valved, dehiscing longitudinally; valves often bifid. Seeds winged, numerous.

III. **Talinum.**—*Capsule* 3-valved, dehiscing longitudinally. Seeds wingless, numerous.

IV. **Portulacaria.**—*Fruit* 3-winged, indehiscent, 1-seeded.

I. PORTULACA, Tournef.

Calyx bipartite, the tube cohering with the ovary below. *Petals* 4–6, with the 8–20 stamens inserted on the calyx, fugacious. *Ovary* roundish. *Style* 3–8 parted. *Capsule* subglobose, 1-celled, dehiscing transversely about the middle. *Seeds* numerous, affixed to a central placenta. *DC. l. c. Endl. gen. n.* 5174.

Low, herbaceous, fleshy herbs, with scattered quite entire leaves. Name from porto *to carry*, and lac, *milk*; plants milky.

TABLE OF SOUTH AFRICAN SPECIES.

Flowers yellow.	Axils of the leaves naked	(1)	**oleracea.**
Flowers yellow.	Axils of the leaves and joints hairy ...	(2)	**quadrifida.**
Flowers purple.	Axils of the leaves and joints hairy ...	(3)	**pilosa.**

1. **P. oleracea** (Linn. Spec. 638); annual, diffuse, *very smooth;* leaves *obovate* or *cuneiform;* flowers sessile; sepals keeled; petals 5; stamens 7–12; style 5-partite. *DC. Pl. Grass. t.* 123. *Schkuhr, Handb. t.* 130. *P. oleracea et sativa, Haw.*

HAB. Cultivated and waste grounds, "Common Purslane," Aug.–Jan. (Herb. Sd. &c.)

2. P. quadrifida (Linn. Mant. 78); annual, diffuse, creeping; the axils of leaves and joints pilose; leaves *elliptic-oblong*, flat; flowers terminal, nearly sessile, surrounded by four leaves; petals 4; stamens 8–12; style filiform, 4-cleft at the apex. *Thunb. Fl. Cap.* 399, *Jacq. Coll.* 2. 356, *t.* 17. *f.* 2. *DC. l. c. P. meridiana. Linn. Suppl.* 248.

HAB. Uitvlugt near Steelkloof, and between Limoenfontein, Brakvalley and Buffelrivier, 3–5000f. *Drege!* Near Cradock, *Burke & Zey.!* 607. Jan.–Mar. (Herb. Thunb., Vind., D., Sond.)

Root thick. Stem branched, 2–4 inches long, often reddish. Leaves about 1 or 1½ lines long. Flowers small.

3. P. pilosa (Lin. Spec. 639); annual, diffuse, the joints with long hairs; leaves *linear-lanceolate* or *linear*, convex on the back, about equal in length to the axillary hairs; floral leaves in whorls; flowers sessile, crowded at the tops of the branches, surrounded by long hairs; petals ovate, obtuse, retuse, a little longer than the calyx; stamens 15–25; stigmas 5–6. *Commel. hort. Amst.* 1. *t.* 5. *Ker. Bot. Reg. t.* 792. *DC. l. c.* 354.

HAB. Cape, Herb. Hook.; Port Natal, *Gueinzius* in Herb. Sond. (Native of South America.)

Root tuberous. Stem terete, smooth. Leaves subulate, 6–8 lines long, smooth. 2–8 flowers in the head, expanding from 10–12 o'clock in the morning, if the sun is out.

II. ANACAMPSEROS, Sims.

Sepals 2, opposite, oblong, subconcrete at base. *Petals* 5, very fugacious. *Stamens* 15–20, filaments distinct, inserted in the bottom of the calyx with the petals, and adhering to them. *Style* filiform, trifid at the apex. *Capsule* conical, 1-celled, 3-valved, the valves often longitudinally divided, and then apparently 6-valved. *Seeds* numerous, winged, affixed to a central placenta. *DC. l. c.* 355. *Endl. gen. n.* 5176.

Very dwarf undershrubs. Leaves roundish, ovate or lanceolate, fleshy, sometimes very minute. Stipules membranaceous, larger than leaves, or forming hair-like subscarious fascicles in the axils of the leaves. Bracts membranaceous, usually lobed into setaceous segments. Flowers sessile, involucrated or on elongated racemose peduncles, expanding only in the heat of the sun. Name from **ανακαμπτω**, *to cause return*, and **ερος**, *love*.

ANALYSIS OF THE SOUTH AFRICAN SPECIES.

SEC. I. **Avonia,** *E. Mey.!* Flowers terminal, *sessile or subsessile,* involucrated. Sepals sub-persistent. Seeds angular. Stipules membranous, densely imbricated, often bearded at the base, much larger than the hemispherical, fleshy, concave-convex leaves. (Sp. 1–3)

Stipules tongue-shaped, *woolly-bearded* at base; flowers shorter than the involucre ... ... (1) **papyracea.**
Stipules broadly ovate or ovate-triangular, not bearded at base:
Stipules *entire;* flowers 2–4 times longer than the involucre ... ... (2) **quinaria.**
Stipules *lacerate;* flowers equalling the involucre ... (3) **ustulata.**

SEC. II. **Telephiastrum,** *Dill.* Flowers *on scapelike peduncles,* solitary or racemose. Sepals deciduous. Seeds winged. Stipules hair-like, axillary. (Sp. 4–9.)

Axillary stipular-hairs shorter than the leaves:
Leaves obtuse, glabrous ... ... (4) **Telephiastrum.**
Leaves acute or acuminate, cobwebbed ... ... (5) **arachnoides.**
Axillary stipular hairs longer than the leaves:

Leaves ovate-globose or ovate :
Lvs cobwebbed, rather rugged above ... ... (6) **filamentosa.**
Lvs. densely clothed with long woolly hairs ... (7) **lanigera.**
Leaves lanceolate or very narrow :
Lvs. lanceolate ... ... (8) **lanceolata.**
Lvs. narrow-lanceolate ... ... (9) **angustifolia.**

Section I. **AVONIA**, E. Mey. ! (Sp. 1-3.)

1. A. (Avonia) papyracea (E. Mey. ! in Herb. Drege); caudex very short, much divided; stems simple, cylindrical or globular; stipules multifarious, imbricated, scarious, snow-white, *linguiform*, roundish at top, quite entire, transversely wrinkled, woolly-barbate at base, 5–10 times longer than the thick leaves; flowers sessile, solitary, much *shorter* than the involucre. *Fenzl, Wien. Annal.* 1839. *Feb p.* 295.

HAB. Hills in the Great Carroo, near Bloedrivier, 2–2500f. *Drege!* Gamkarivier, in Carroo, *Burke and Zey.!* 649. April. (Herb. Vind., Hook., D., Sd.)

Stems ½–2 inches long, 3–4 lines in diameter. Stipules 3–4 lines long, 1½ lines wide. Flowers included in the uppermost stipules; peduncle smooth, about ½ line long. Petals roundish, yellow, 2 lines long, surpassing the calyx. Capsule 2 lines long.

2. A. (Avonia) quinaria (E. Mey. in Herb. Drege); caudex very short, much divided; stems sub-terminal, numerous, sterile and flower-bearing, undivided, very short, as thick as a pigeon's quill; stipules scarious, white, *broad-ovate, entire*, not bearded at the base, spirally and closely 5-farious, imbricated, much longer than the very minute semi-orbicular flattish leaves; involucral leaves larger, sphacelate at top; flowers *exserted*, 2–4 times *longer* than the involucre. *Fenzl, l. c.*

HAB. Dry flats near Lislap, Boshmannskarroo, 3–3500f. *Drege!* Nov. (Herb. Vind., Sd.)

Caudex ⅓–1 inch long, fleshy. Sterile stems 1–2 lines, fertile 4–6 lines long, terete, silver-coloured. Stipules ½ line, the involucre 1 line long. Flowers purple, about 3 lines long. Sepals ovate, obtuse.

3. A. (Avonia) ustulata (E. Mey. in Herb. Drege); caudex much branched from the base; branches tortuose, densely covered with whitish, globular, ovoid or cylindrical buds of closely imbricated scales (stipules); stipules *broad, ovate-triangular, acute, lacerate* at the margins, not bearded at the base, sphacelate and patent at the point, those of the involucre larger, quite entire; flowers *equalling* the involucre. *Fenzl, l. c. Tetraphyle corallina, E. & Z.!* 1865. excl. synon.

HAB. Hills in the Great Carroo near Bloedrivier, 2–2500f. *Drege!* in Carro near Zwarteruggens, between Uitenhage and Graafreynet, *E. & Z.!* Gamkarivier, *Burke and Zeyher!* April. (Herb. Vind., Hook., D., Sd.)

3–4 inches high. Caudex 3–4 lines in diameter; primary branches as thick as a goose-quill, about 1 inch long, dichotomously divided, branches shorter. Sterile stems budlike, mostly globular, 1–2 lines long; fertile ones 4–6 lines long, as thick as a pigeon's quill, sometimes with one or a few similar branches. Stipules very minute, closely adpressed, those of the fertile stems with a patent rusty-brown but white-margined apex. Involucre 1 line long, twice shorter than the 3-valved capsule.

Section II. **TELEPHIASTRUM**, Dill. (Sp. 4-9.)

4. A. Telephiastrum (DC. l. c.); leaves approximate, ovate or sub-orbicular, *obtuse, glabrous;* axillary hairs filamentous, *shorter* than the leaves; racemes few-flowered, racemose or sub-panicled. *Pl. Grass. t.* 3. *Portulaca Anacampseros, Linn. Spec.* 639. *Talinum Anacampseros, Willd. Spec.* 2. 862. *Rulingia Anacampseros, Ehrh. Beyt.* 3. 133. *A. Telephiastrum et intermedia, Don, Gen. Hist.* 3 *p.* 75. *A. rotundifolia, Bot. Cab. t.* 591.

HAB. Nieuweveldsbergen, near Beaufort, 3–5000f. Witbergen, 7–8000f. *Drege!* Stony hills near Gariprivier, *Zey. Portul.* 2. Nov.–Jan. (Herb. Vind., Sd., D.)

Stem 1–2 inches long. Leaves very fleshy, roundish, ovate or cuneate, about 3–4 lines long and wide, in cultivated specimens larger. Peduncle in the wild plant 1–2 inches long, with 2 or 4 flowers; in the garden specimens often 4–6 inches long, panicled, with many and larger flowers. Sepals 4–6 lines long. Petals large, reddish.

5. A. arachnoides (Sims. Bot. Mag. 1368); leaves ovate, *acute or acuminated,* green, shining, *cobwebbed;* axillary hairs filamentous, *shorter* than the leaves; racemes simple. *E. & Z.!* 1800. *Portulaca trigona, Thunb.! Fl. Cap. p.* 399. *Herbar. fol. II. specimens,* n. 1–3. *P. arachnoides, Haw. Misc.* 142. *Rulingia arachn. Haw. Syn.* 125. *Talinum arachn. Ait. Kew.* 2 *v.* 3. 149.

VAR. β. **rubens**; more robust; leaves often recurved at the apex, as well as the peduncles purplish; flowers a little larger. *Burm. Afric. t.* 30. *A. rubens, DC. l. c. Port. rubens, Haw. Misc.* 142. *Rulingia Haw. Syn.* 125. *P. trigona, Thunb.! Herbar. fol. II. specimen n.* 4.

VAR. γ. **grandiflora**; leaves more crowded, somewhat larger and recurved, purplish as well as the peduncles and flowers. *A. rufescens, DC. l. c. Rulingia, ruf. Haw. Suppl. pl. succ.* 64.

HAB. Carroo near Kayserkuylsrivier, Gondsrivier, Cannaland near Olifantsrivier, *Thunb.!* In the districts of Uitenhage and Albany, *Drege! E. & Z.! Zey.!* 2484. Nov.–Jan. (Herb. Thunb., Vind., Sond.)

Next the preceding, but the leaves are smaller, (2–3 lines long) acute and cobwebbed, the flowers smaller, excepted in var. γ. Stipules woolly, mixed with longer yellowish threads. Scape 2–4 inches high, with several pairs of scarious bracts, bearing 2–3 long peduncled flowers. Petals in var. α. white, oblong. Seeds small, winged.

6. A. filamentosa (Sims. Bot. Mag. 1367); leaves *ovate-globose,* gibbous on both sides and *cobwebbed,* rather *rugged* above; stipulaceous hairs ramentaceous, *straight, longer* than the leaves; raceme simple. *DC. l. c. E. & Z.!* 1799. *Portulaca filamentosa, Haw. Misc.* 142. *Rulingia, Haw. Syn.* 125. *Talinum Ait. l. c.*

HAB. Stony places in Karroo, beyond Hartequaskloof, *Masson;* Gauritzrivier, *E. & Z.!* Schiloh, Klipplaatrivier, 4000f. *Drege!* Gamkarivier, *Burke!* Uitvlugt, *Zey.!* 606. Zwartskopsrivier, *Zey.!* 2483. Dec.–Jan. (Herb. Vind., Hk., D., Sd.)

Very similar to the preceding, but distinguished by the long ramentaceous stipules and larger rose-coloured flowers. Leaves 2–3 lines long, very thick. Scape few-flowered, 2–3 inches long. Petals oblong.

7. A. lanigera (Burch. Cat. Geog. n. 2169); leaves *ovate,* obtuse, small, very *densely clothed with long wool;* axillary hairs ramentaceous, *slightly curled,* much longer than the leaves; raceme simple, few-flowered. *Burch. Trav.* 2. 333. *DC. l. c. E. & Z.!* 1801. *Portulaca trigona, Thunb.! Herbar. fol. I.*

HAB. Bachapin, *Burchell;* Kamisberge, Namaqualand, *E. & Z.!* Uitvlugt, Ramos, *Zey.!* 605. Zilverfontein, 2–3000f. *Drege!* Oct.–Nov. (Herb. Thunb., Vind., Hook., D., Sd.)

Caudex short. Stems ½–1 inch, beset with many rows of minute (1–2 lines long) thick leaves, the whole involved by a dense white wool. Stipules yellow or a little reddish, 5–6 lines long. Scape 1–3 inches long, 2–4 flowered. Flowers rose-coloured. Petals about 3 lines long, obtuse, surpassing the green calyx.

Very like *A. filamentosa*, but more densely leafy, the leaves in many rows and woolly-tomentose, the stipules not so straight and the flowers smaller.

8. A. lanceolata (DC. l. c.); leaves *lanceolate*, fleshy, glabrous, convex beneath; axillary hairs very long; scape leafy, short, generally 1-flowered. *Portulaca lanceolata, Haw. Syn.* 126.

HAB. Cape, *Haworth*, (unknown to me.)

Stems very short. Calyx reddish. Petals reddish. Seeds almost 3-winged. It seems only a variety of *A. telephiastrum.*

9. A. angustifolia (DC. l. c.); leaves fleshy, *narrow-lanceolate*, expanded; stem short, branched. *Rulingia angustifolia, Haw. Rev.* 60.

HAB. Cape, *Haworth.*

Very like the preceding, but smaller.

III. TALINUM, Juss.

Sepals 2, ovate, distinct and free, deciduous. *Petals* 5, ephemeral. *Stamens* 10–30, inserted with the petals in the bottom of the calyx, and often slightly attached to them. *Style* filiform, hispid at the apex. *Capsule* 3-valved, 1 celled, many-seeded. *Seeds* wingless, attached to a central placenta. *DC. l. c.* 356. *Endl. gen. n.* 5178.

Herbaceous or suffrutescent fleshy plants. Leaves alternate, quite entire, exstipulate. Flowers cymose, racemose or panicled, fugacious. Name, probably from *θαλια, a green branch.*

1. T. caffrum (E. Z. ! 1802); smooth, much branched; branches alternate, erect or patent; leaves oblong-linear or linear-lanceolate, mucronulate, with revolute margins; flowers solitary in the axils of the upper leaves, erect, in fruit reflexed. *Portulaca caffra, Thunb. Fl. Cap.* 399. *Anacampseros ramosa, E. Mey. !*

VAR. β. **minus**; branches short, 2–3 inches high; leaves linear-oblong. *T. minus, E. & Z. !* 1803.

HAB. Caffraria, *Thunberg !* stony places in mountains near Silo, Klipplaatrivier, Tambukiland, *E. & Z. !* Los-Tafelberg and Zwartekey, Uitvlugt near Steelkloof; Camdeboo near Hamerkuil, 3–5000f. *Drege !* Crocodillrivier, *Zey. !* 610. Var. β. Katriviersberg, Ceded Territory, *E. & Z. !* Crocodillrivier, *Zey. !* 609. Nov.–Dec. (Herb. Holm. Hook, D., Sd.)

A perennial herb, ½–1 foot high; branches terete, mostly undivided. Leaves on very short petioles, 8–12 lines long, 1–1½ lines wide, in var. β. 5–6 lines long. Flowering peduncles as long as the leaves or shorter, in fruit longer, (1 inch) and thick below the apex, bi-bracteolate in the middle. Sepals ovate, acute, 2 lines long, shorter than the yellow (Thunb.) broad ovate petals, about as long as the stamens. Capsule 3 lines long. Seeds sub-compressed, concentric, striated, black, the hilum and strophiolum white.

IV. PORTULACARIA, Jacq.

Sepals 2, persistent, membranous. *Petals* 5, persistent, equal, obovate, hypogynous. *Stamens* 5–7, inserted with the petals, but disposed without respect to their number (perhaps 10, of which 3–5 are abortive.) *Anthers* short, often barren. *Ovary* ovate, triquetrous. *Style* none.

Stigmas 3, patulous, muricato-glandular above. *Fruit* 3-quetrous, 3-winged, indehiscent, 1-seeded. *DC. l. c.* 360. *Endl. Gen. n.* 3175.

Glabrous, fleshy shrubs or small trees, natives of South Africa. Leaves opposite, roundish, flat, deciduous. Peduncles opposite, denticulate, compressed; pedicels 1-flowered, 3 rising from each notch in the peduncle. Flowers very small, rose-coloured, Name altered from *Portulaca.*

ANALYSIS OF THE SPECIES.

Branches *opposite*; peduncles compressed, *branched*, ... (1) **afra.**
Branches *dichotomous*; peduncles angular, *unbranched*, ... (2) **namaquensis.**

1. P. afra (Jacq. Coll. 1, 160. t. 22); branches *opposite*, smooth; leaves rising from the opposite nodes, obovate-roundish, flat; peduncles compressed, *branched*, branches opposite; pedicels ternate. *DC. Pl. Grass. t.* 132. *E. & Z.!* 1804. *Claytonia Portulacaria, Linn. Mant.* 221. *Lam. ill. t.* 144. *f.* 2. *Crassula Portulacastrum, Linn. spec.* 406. *Portulaca fruticosa, Thunb.! Fl. Cap.* 399. *excl. syn. Dill. Elth.* 1. *t.* 101. *f.* 120.

HAB. In Karroo, in the districts of Uitenhage, Graafreynet and Albany. Nov.–Dec. (Herb. Thunb. Jacq. Sd. D. etc.)

Speckboom of the Colonists. Small tree, 10–12 feet high. Branches articulate. Leaves 4–6 lines long. Flowers on short, delicate (2–3 lines) pedicels, at the base bracteated by very minute scales. Petals about 1 line long.

2. P. namaquensis (Sond.); branches *dichotomous*, all over beset with scattered leaf and flower-bearing nodes; leaves minute, very fleshy, obovate, geminate, sessile on the nodes, deciduous; peduncles inserted between the leaves, angular, *not branched*; pedicels solitary, geminate or ternate.

HAB. Namaqualand, *Dr. Atherstone, A. Wyley, Esq.*, Dec. (Herb. Hook, D.)

4–5 feet high. Branches greyish-green, the ultimate as thick as a goose-quill. Leaves 2 lines long. Peduncles 6–8 lines long. Pedicels involucrated by some minute ovate bracts, about 2 lines long. Calyx 2-phyllous, three times shorter than the 5 rose-coloured, obovate, near 1 line long petals. Stamens 5; filaments linear. Anthers oblong, emarginate at both ends. Ovary ovate. Style short.

ORDER LV. MESEMBRYACEÆ, Lindl.

(By W. SONDER.)

Flowers perfect and regular. *Calyx* gamosepalous, 4–5 cleft, its tube adnate with the ovary, or free. *Petals* indefinite (in *Mesembryanthemum*) or none, marcescent or deliquescent. *Stamens* perigynous, definite or indefinite; filaments slender; anthers 2-celled, introrse. *Ovary* inferior or superior, 2–5–20-celled; ovules numerous or few or solitary, on long cords, attached to the base or inner angle of the cell, amphitropal. *Styles* or *stigmas* as many as the carpels. *Fruit* capsular, variously dehiscent; rarely nucamentaceous and indehiscent. *Seeds* lenticular or reniform, with floury albumen. *Embryo* excentric, curved round the margin of the seed, the radicle next the hilum.

Small shrubs, undershrubs or herbs, with opposite or alternate exstipulate, undivided, usually fleshy or thickened leaves, flat, terete or triangular. Flowers terminal or axillary, in cymes or solitary, often very showy, sometimes minute and inconspicuous. Very numerous in S. Africa, especially in the Karroo districts; thinly scattered over the warmer parts of the temperate zone. None are of much use, except some species of *Tetragonia*, used as pot-herbs.

TABLE OF THE SOUTH AFRICAN GENERA.

* *Petals very numerous, linear.*

I. **Mesembryanthemum.**

** *Petals none.*

† *Stamens indefinite, numerous.*

II. **Tetragonia.**—*Calyx* 4, rarely 3-cleft. *Fruit* inferior, angular, indehiscent. *Seeds* solitary.

III. **Aizoon.**—*Calyx* 5-cleft. *Stigmas* 5, thick. *Capsule* superior, 5-celled, many-seeded.

IV. **Acrosanthes.**—*Calyx* 5-cleft. *Stigmas* 2, filiform. *Capsule* superior, one-celled, 1–2 seeded.

V. **Diplochonium.**—*Calyx* 5-cleft. *Stigmas* 2, filiform. *Capsule* superior, 2-celled, many seeded.

†† *Stamens definite,* 5, 8 *or rarely* 10.

VI. **Galenia.**—*Stamens* 8–10, *in pairs* alternating with the calyx-lobes. *Styles* 2–5. *Capsule* 2–5 (or by abortion 1) celled, loculicidal.

VII. **Plinthus.**—*Stamens* 5, alternate with the calyx-lobes. *Style* 3-partite. *Capsule* 3-celled, 3 valved, loculicidal.

VIII. **Trianthema.**—*Stamens* 5–10, rarely more. *Stigmata* 2 or 1. *Capsule* 2-celled, circumscissile.

* I. MESEMBRYANTHEMUM. L.

Calyx 5, rarely 2–8 parted, its tube adnate with the ovary, lobes unequal, usually leaf-like. *Petals* very numerous, linear, in one, or frequently in many rows, united at the base. *Stamens* innumerable, in many rows united at base. *Ovary* 4–20, but usually 5-celled. *Stigmas* 4–20, usually 5. *Capsule* 5, many-celled, dehiscing in a starlike manner at the summit. *Seeds* innumerable. *DC. l. c. p.* 415. *Endl. Gen.* 5163.

Shrubs or herbs, almost all natives of the Cape of Good Hope, abounding throughout the arid plains and sands of the whole country to the south of the Orange River and west of the Great Fish River. To the east of the Fish River the species are few. Outlying species occur in the Isle of Bourbon, in North Africa, on the Mediterranean coasts of Europe and on the coasts of Australia, Chili and Peru. Leaves usually opposite, thick, fleshy, trigonal, terete or flat. Flowers mostly terminating the branches, white, red or yellow, the greater part opening in the heat of the sun, very few in the evening. The capsules are tightly closed in dry weather and open naturally after rain. If thrown into water until it become thoroughly soaked and then removed, an old capsule will open out its capillary valves, radiating from a centre like a star; and will close them again when dry. This experiment may be repeated several times without destroying their remarkable hygrometric property. Name from *μεσημβρια, mid-day*, and *ανθεμον, a flower.*

☞ *For Synoptical Table of Sections, see next page.*

---

* Owing to the number of imperfectly known species in this most difficult genus, it has been found impracticable to prepare an analytical table of the species. It is hoped, however, that the subjoined Synoptical Table, and key to the arrangement, of the sections, 65 in number, under which the species are distributed, will greatly assist the student. I am indebted to the late Prince Salm Dyck for the use of his manuscript characters of the sections, of which I have largely availed myself: the whole of the species however have been personally worked out, so far as the material at my disposal permitted.—W. S.

- **EPAPULOSA.**
  - Stemless or nearly so . . . **I. SUBACAULIA.** (Sp. 1—76.)
    - § 1, **Sphaeroidea.**
    - § 2, **Subquadrifolia.**
    - § 3, **Rostrata.**
    - § 4, **Aloidea.**
    - § 5, **Ringentia.**
    - § 6, **Dolabriformia.**
    - § 7, **Difformia.**
    - § 8, **Linguaeformia.**
    - § 9, **Gibbosa.**
    - § 10, **Calamiformia.**
    - § 11, **Teretifolia.**
    - § 12, **Bellidiflora.**
    - § 13, **Acuta.**
    - (§ 14, **Macrorhiza**)
    - 1 Spec. Isle of Bourbon.
  - Stem erect or prostrate.
    - Leaves triquetrous or subtriquetrous.
      - Leaves distinct or nearly so. **II. TRIQUETRA.** (Sp. 77—138.)
        - § 15, **Corniculata.**
        - § 16, **Pugioniformia.**
        - § 17, **Sarmentosa.**
        - § 18, **Reptantia.**
        - § 19, **Acinaciformia.**
        - § 20, **Rubricaulia.**
        - § 21, **Heteropetala.**
        - § 22, **Bracteata.**
        - § 23, **Virgata.**
        - § 24, **Virentia.**
        - § 25, **Aurea.**
        - § 26, **Blanda.**
        - § 27, **Amoena.**
        - § 28, **Dilatata.**
        - § 29, **Falcata.**
        - § 30, **Deltoidea.**
        - § 31, **Forficata.**
      - Leaves connate or sheathing. **III. PERFOLIATA.** (Sp. 139—165.)
        - § 32, **Geminata.**
        - § 33, **Uncinata.**
        - § 34, **Microphylla.**
        - § 35, **Rostellata.**
        - § 36, **Vaginata.**
        - § 37, **Tumidula.**
        - § 38, **Crocea.**
    - Leaves terete or semi-terete. **IV. TERETIUSCULA.** (Sp. 166—211.)
      - § 39, **Veruculata.**
      - § 40, **Haworthiana.**
      - § 41, **Spinosa.**
      - § 42, **Cymbaeformia.**
      - § 43, **Defoliata.**
      - § 44, **Splendentia.**
      - § 45, **Juncea.**
      - § 46, **Tenuifolia.**
      - § 47, **Adunca.**
- **PAPULOSA.**
  - Leaves terete or semiterete . . **V. PAPILLOSA.** (Sp. 212—265.)
    - § 48, **Scabrida.**
    - § 49, **Trichotoma.**
    - § 50, **Aspericaulia.**
    - § 51, **Hispida.**
    - § 52, **Barbata.**
    - § 53, **Echinata.**
    - § 54, **Spinulifera.**
    - § 55, **Moniliformia.**
    - § 56, **Crassulina.**
    - § 57, **Geniculiflora.**
    - § 58, **Nodiflora.**
  - Leaves flat . . . . **VI. PLANIFOLIA.** (Sp. 266—293.)
    - § 59, **Scaposa.**
    - § 60, **Platyphylla.**
    - § 61, **Cordifolia.**
    - § 62, **Expansa.**
    - § 63, **Relaxata.**
    - § 64, **Tripolia.**
    - § 65, **Helianthoidea.**

KEY TO THE ARRANGEMENT OF THE SECTIONS.

I. EPAPULOSA.

I. SUBACAULIA.

Fleshy corpuscula ; each plant consisting of two minute leaves united into a globe ... ... (1) **Sphaeroidea.**
Leaves distinct.
Flowers white or reddish.
Leaves (4–6) decussate, entire, obtuse, flat above, convex beneath ... ... (2) **Subquadrifolia.**
Leaves (4–6) divergent, subterete ... ... (11) **Teretifolia.**
Leaves (numerous) tumid, difform, 1 abbreviate gibbous, 1 larger subovate ... ... (9) **Gibbosa.**
Leaves (numerous) cylindrical, blunt, erect. Flowers whitish ... ... (10) **Calamiformia.**
Leaves (numerous) semiterete, subtriquetrous, apex entire. Flowers reddish... ... ... (13) **Acuta.**
Leaves (numerous) triquetrous-compressed, apex dentate. Flowers white ... ... (12) **Bellidiflora.**
Flowers yellow.
Leaves unequal or difform. Stigmas 8 or 9–10.
Leaves obliquely-decussate, semicylindrical, with incrassate or attenuated apex, often toothed (7) **Difformia.**
Leaves distichous, linguæform, one side thicker, obliquely keeled at the apex ... ... (8) **Linguæformia.**
Leaves equal.
Flowers solitary. Stigmas 4, 5 or 6.
Leaves (4–6) semiterete, attenuated, keeled at the apex. Flowers pedunculate ... ... (3) **Rostrata.**
Leaves (4–6) triquetrous, larger in the middle, bluntish, mucronate, entire, tuberculated. Flowers sessile ... ... (4) **Aloidea.**
Leaves (4–6) semiterete, near the apex dilated and triquetrous, angles often dentate or ciliate (5) **Ringentia.**
Flowers ternate
Leaves compressed, the carinal angle much dilated ... ... (6) **Dolabriformia.**

II. TRIQUETRA.

Flowers yellow. Stigmas 10–20.
Leaves cylindrical, more or less triquetrous, not much elongated, punctate ... ... (15) **Corniculata.**
Leaves more or less triquetrous, very long, without dots (16) **Pugioniformia.**
Flowers yellow. Stigmas about 5.
Leaves triquetrous, glaucous ; peduncles without bracts (25) **Aurea.**
Leaves acinaciform, with smooth angles ... ... **M. edule L.** (§ 19)
Leaves subacinaciform, with serrulated angles ... ... **M. serratum L.** (§ 20)
Flowers white or reddish. Stigmas mostly 5 (rarely 5–10).
Flowers solitary.
Stem short, nodulose, prostrate, rooting ... ... (18) **Reptantia.**
Stem erect or decumbent, not rooting :
Branches angular, leaves connate, acinaciform with smooth angles, flowers large ... ... (19) **Acinaciformia.**
Branches angular or terete, leaves connate, subacinaciform, with serrulated angles, fl. large (20) **Rubricaulia.**
Branches angular, lax, leaves very connate, triquetrous-compressed, obtuse, toothed below the apex, flowers large ... ... (31) **Forficata.**
Branches 2-edged, leaves subconnate, compressed-

triquetrous, erect, sub-recurved ; flowers small ; peduncles with 2 bracts in the middle ... ... (23) **Virgata.**

Branches 2-edged ; leaves distinct, compressed-triquetrous, hooked at the apex, scabrous-punctate, peduncles with 2 or 4 bracts near the calyx (22) **Bracteata.**

Flowers ternate or geminate (by abortion rarely solitary)

Branches elongated, sarmentose or rooting ... ... (17) **Sarmentosa.**

Branches not sarmentose or rooting :

Leaves deltoid-triquetrous, dilated and retuse at apex, with toothed angles ... ... (30) **Deltoidea.**

Leaves compressed triquetrous, with dilated keel near the recurved apex, not toothed ... ... (28) **Dilatata.**

Leaves attenuated not dilated :

Leaves falcate-recurved with obtuse angles (29) **Falcata.**

Leaves not falcate :

Leaves triquetrous, carinal angles drawn out, often lacerate ; branches 2-edged, petals biform ... ... (21) **Heteropetala.**

Leaves subconnate, triquetrous subacinaciform, with smooth angles, ; branches subtriquetrous ... ... (24) **Virentia.**

Leaves subtriquetrous, elongated, acute, very smooth ; branches erect, rigid (26) **Blanda.**

Leaves subconnate, triquetrous, with equal sides, gradually attenuated ; branches suberect ... ... (27) **Amoena.**

III. Perfoliata.

Dwarf subshrub with dichotomous branches ; leaves turgid, triquetrous, whitish, smooth, with cartilaginous margins ; flowers unknown ... ... (32) **Geminata.**

Flowers cymose or paniculated :

Flowers subpaniculated ; leaves short, triquetrous, carinal angle rough ... ... (36 **Vaginata.**

Flowers subcymose ; leaves elongated, triquetrous, angles smooth ... ... (37) **Tumidula.**

Flowers terminal, solitary :

Dwarf, procumbent ; leaves beaked-connate, subulate ; flowers pale reddish ... ... (35) **Rostellata.**

Dwarf, erect or procumbent ; leaves minute, triquetrous, aristate, with large dots ... ... (34) **Microphylla.**

Erect ; sheaths of leaves short ; leaves 3-gonous, semicylindrical, weak, sebaceous, without dots ; flowers yellow or croceous ... ... (38) **Crocea.**

Erect, rigid ; sheaths of leaves long ; leaves abbreviate, uncinate or elongate, compressed with toothed carinal angle ; flowers small, reddish ... ... (33) **Uncinata.**

IV. Teretiuscula.

Flowers 3-nate or biternate on terminal spines or on spinous peduncles ; leaves triquetrous, terete, glaucescent, punctate ... ... (41) **Spinosa.**

Flowers dichotomous, 3-nate or 2-ternate, not on spines or spinous peduncles ; leaves cylindrical, glaucous, deciduous, without dots ... ... (43) **Defoliata.**

Flowers solitary or ternate :

Branches continuous, not articulate :

Flowers yellow ; leaves cylindrical, obtuse, mucronate, weak, without dots ... ... (39) **Veruculata.**

Flowers red ; leaves subcylindrical, subulate, glaucous, dotted ; branches decussate ... ... (40) **Haworthiana.**

Flowers yellow or scarlet ; leaves linear-elongate, subterete or compressed, not hooked at the apex ; branches slender ... . (46) **Tenuifolia.**
Flowers reddish, small ; leaves subcylindrical-subulate, incurved and hooked at the apex ; branches flexuous ... ... (47) **Adunca.**
Flowers white ; leaves subcylindrical, spreading-recurved, without dots ; branches erect ; calyx lobes foliaceous ... ... (44) **Splendentia.**
Flowers reddish or yellow ; leaves cymbiform, turgid-triquetrous ... . (42) **Cymbiformia.**
Stem and branches subarticulate ; lvs. linear-subulate (45) **Juncea.**

II. PAPULOSA.

V. Papillosa.

Annual herbs ... ... (58) **Nodiflora.**
Perennial herbs or shrubs :
Branches nodose-moniliform ; joints depressed globose ... (55) **Moniliformia.**
Branches subarticulate, joints or internodes cylindrical (57) **Geniculiflora.**
Branches not articulate :
Leaves barbate at the apex ... ... (52) **Barbata.**
Leaves not barbate :
Branches setiferous ... ... (51) **Hispida.**
Branches rough ... ... (50) **Aspericaulia.**
Branches smooth (bluntish papillate) :
Leaves echinulate or hispid all over ... ... (53) **Echinata.**
Leaves punctate-scabrous ; fl. violaceous ... (48) **Scabrida.**
Leaves minutely papulose :
Leaves semicylindrical or triquetrous-compressed ; flowers trichotomous, white or reddish ; branches sub-woody (49) **Trichotoma.**
Leaves cylindrical, sub-canaliculate ; flowers ternate, greenish ; calyx-lobes elongated ; branches fleshy ... ... (54) **Spinulifera.**
Leaves linear-semiterete, acute, often canaliculate ; flowers white or reddish ; calyx-lobes elongated. Small herbs ... ... (56) **Crassulina.**

VI. Planifolia.

Stigmas 5.
Leaves cordate ... ... (61) **Cordifolia.**
Leaves spathulate, ovate or lanceolate, very papulose ; root biennial or annual ... ... (60) **Platyphylla.**
Leaves spathulate-lanceolate, epapulose ; root biennial ... (64) **Tripolia.**
Leaves ovate-lanceolate, flat, subcarinate, marcescent ; calyx-lobes very unequal. Subshrub ... ... (62) **Expansa.**
Leaves oblong-lanceolate, glaucous, not marcescent ; calyx-lobes subequal. Subshrub ... ... (63) **Relaxata.**
Leaves linear or cuneiform ; flowers on long, mostly radical peduncles ; root annual ... ... (59) **Scaposa.**
Stigmas 10–20.
Leaves flat, lanceolate or spathulate, attenuated at the base ; stem herbaceous ; root annual ... ... (65) **Helianthoidea.**

Series I. EPAPULOSA.—Stem and leaves not papulose. (Groups I.–IV.)

Group I. Subacaulia.—Root perennial. Stems wanting or very short. Leaves variable in form, but not flat. (§§. 1–14.)

§. 1. *SPHÆROIDEA*, Salm Dyck (Minima et Sphæroidea, Haw.); *plants* stemless, forming fleshy corpuscula, or somewhat caulescent, consisting of united corpuscula; *leaves* opposite, very blunt, joined even to the apex into a globe, but separating at length at the apex, and becoming marcescent, but sheathing at the base. *Flowers* sessile, solitary, central. *Calyx* 4–6-cleft. *Petals* generally joined into a tube, reddish or whitish. *Stigmas* 4–6. (Sp. 1–10.)

1. M. minutum (Haw. Obs. 126, Syn. 202); stemless, obconical, *glaucous, without spots;* flowers long, tubular, pale reddish; ovary inclosed. *Sims. Bot. Mag. t.* 1376. *DC. l. c.* 417. *M. nuciforme, Haw. Obs.* 129, *Syn.* 204.

HAB. Gamkarivier, *Burke and Zeyher*, 693. (Herb. D., Sd.)

Plant hardly the size of a common bean. Sheath often with purple dots from the middle to the base. Petals spreading, tube about 3 lines long, shorter than the limb.

2. M. perpusillum (Haw. Rev. 82); stemless, obconical, *green,* with strong *confluent branched dots;* ovary inclosed. *DC. l. c.*

HAB. Cape of Good Hope.

Plant ½ inch, nearly allied to the following, but the colour of the flower is deeper, and the offsets more numerous.

3. M. minimum (Haw. Obs. 126. Syn. 203); stemless, *obconical,* glaucescent, with confluent rather branched dots; flowers *whitish;* petals connate at base; *ovary exserted. Petiv. gaz. t.* 39. *f.* 3? *ex Haw.*

HAB. Cape of Good Hope.

Plant ½ inch; when cultivated, often somewhat caulescent. Offsets fewer from this than from the other allied species, Limb of the corolla very patent.

4. M. truncatellum (Haw. Misc. 22. Syn. 203); stemless, *much depressed* and rather glaucous, with the dots rather distinct; flowers *straw-coloured;* ovary exserted. *DC. l. c.*

HAB. Cape of Good Hope.

Plant ½ inch. Calyx 5-cleft.

5. M. obcordellum (Haw. Misc. 21. Syn. 203); stemless, obconical, glaucescent, with distinct or confluent purple dots; *flowers substipitate,* white; calyx 5-cleft; *petals free;* ovary inclosed. *Sims. Bot. Mag. t.* 1647. *E. & Z.!* 1976. *Salm. Dyck. Monog. fasc.* 6. *t.* 1.

HAB. Karroo, between Beaufort and Graafreynet, near June. (Herb. Sd.)

¾ inch. Sheaths pale, the uppermost equalling or shorter than the corpusculum, irregularly dentate or lacerate. Corpusculum often purple at the cuneate base, plane-convex above. Flower very small, rising from the ciliolate fissure. Peduncle inclosed. Petals about 2 lines long, recurved. Styles 5.

6. M. obconellum (Haw. Misc. 21. Syn. 203); stemless, *obconical,* glaucous green, with prominent confluent purple dots or lines; sheaths membranaceous; flowers *subsessile,* white; *petals connate at the base;* ovary inclosed. *Salm. Dyck. Monog. fasc.* 1. *t.* 1.

HAB. Cape of Good Hope.

Very like the preceding, from which it differs by the prominent dots, nearly sessile flowers and subconnate petals. Calyx 5-cleft.

7. M. uvaeforme (Haw. Rev. 84); stemless or subcaulescent, nearly globose; sheaths *densely imbricate, thick, transversely wrinkled,* obtuse, the uppermost about equalling the pale green, prominently dotted corpuscula; flowers exserted, tubular; ovary inclosed. *Lycoperdastrum, etc. Burm. Afr. t.* 10. *f.* 2.

HAB. Knaus near Lislap, Betchuana territory, on rocks, 3000f. May. *Zeyher!* (Herb. Sd.)

Greyish-brown. A stemless plant, ½ inch, caulescent 1–1½ inch long; 4–6 rising from a branched woody root. Corpusculum the size of a hazelnut. Calyx 6-cleft, inclosed with the turbinate ovarium. Petals purplish when dried, connate at the base, 5–6 lines long. Half ripe fruit as large as a pea, covered with the pointed, purplish dotted calyx-lobes.

8. M. truncatum (Thunb.! Nov. Act. Ephem. nat. curios. Vol. 8, p. 5, App.); stemless or caulescent, *obconical, exactly truncate;* sheaths *thinly membranaceous,* the uppermost *entire,* dotted at the margin, a little longer than the retuse, glabrous, dotted corpusculum; flower pedunculate, exserted; calyx 4-cleft; ovary inclosed. *Flor. Cap.* 412. *excl. syn. Burm. DC. l. c. M. turbiniforme. Haw. Rev.* 84. *Burch. trav.* 1, 310.

HAB. Rocks in Camenasie Karroo, in Bockland and near Hexriver, *Thunberg;* Zandvalley, *Burchell;* Knaus, Betchuana territory, *Zeyher,* 2954. Jan.–May. (Herb. Thunb. Sd.)

Root fibrous, perennial. A stemless plant, ½ inch, caulescent 1–1½ inches, and often with 1 or 2 lateral branches. Sheaths pale yellowish when dry, shining, the uppermost including 1 or 2 corpuscula, 5–6 lines long, truncate. Corpuscula about 3 lines long. Flower exserted, the compressed peduncle inclosed in the fissure. Petals (whitish?) united at the base, about 2 or 3 lines long. Half ripe fruit as large as a small pea, hemisphærical, covered with the 4, pointed, purple-dotted calyx-lobes.

9. M. fimbriatum (Sond.); stemless or somewhat caulescent, obovate; sheaths membranaceous, imbricate, the uppermost cuneate, white but purplish near the base, *deeply lacerate-fimbriate;* corpusculum subglobose, *punctate, glabrous,* but ciliolate at the fissure, half as long as the sheaths.

HAB. Gamkariver, May. *Zeyher!* (Herb. Sd.)

Many plants, ½–¾ inch high, from the perennial root. Upper vagina about 3 lines long, the fimbriae 1 line long. Flowers unknown. Easily known by the lacerate vagina.

10. M. fibulæforme (Haw. Misc. 22. Syn. 203); stemless or somewhat caulescent, *rather canescent,* depressed; sheaths membranaceous, the uppermost whitish, cuneate, *irregularly toothed;* corpusculum *very thinly pubescent,* ciliate at the fissure, *not punctate,* shorter than the vagina; styles 6. *DC. l. c.* 417.

HAB. Cape of Good Hope, *Scholl.;* Namaqualand, *Zeyher!* (Herb. Vind. Sd.)

Root perennial. Stems numerous, ½–1 inch high, greyish, as thick as a pigeon's quill. The upper vagina marcescent, white. Corpuscula the size of a small pea. Petals unknown.

§. 2. *SUBQUADRIFOLIA,* Salm. Dyck. DC. (Semiovata and obtusa, Haw.) *Plants* almost stemless. *Leaves* 4–6, decussate, quite entire, obtuse, flat above but pustulate at the base, convex beneath. *Flowers* solitary, nearly sessile, reddish or whitish. *Calyx* 4–6-cleft. *Stigmas* 4–6. (Sp. 11–13.)

11. M. testiculatum (Jacq. fragm. 20, 73. t. 12. f. 2.); stemless; leaves 4–8, *unequal, whitish,* smooth, rather erect, ovate or oblong-ovate, flat above, convex beneath, quite entire. *M. testiculare, Thunb.! Fl. Cap.* 412. *Haw. Syn.* 265. *M. octophyllum, Haw. Rev.* 85.

HAB. Near Olifantsriver, *Thunberg!* Between Droogekraal and Hollriver, Gamka-river, *Zeyher!* 701. Jan.–Sept. (Herb. Thunb. Sd.)

Root perennial. Leaves 4 or 6, rarely 8, connate at base, very fleshy, about 1 inch long. Peduncle very short, with 2 leafy bracteas near the calyx. Flowers white or reddish, 1–2 inches in diameter. Calyx 6-cleft.

12. M. fissum (Haw. Obs. 134. Syn. 205); plant almost stemless; leaves *equal,* half-terete, very blunt, *glaucescent. DC. l. c.* 418.

HAB. Cape of Good Hope.

Old stem 1–2 inches high, with very short, alternate branches. Flowers unknown.

13. M. obtusum (Haw. Misc. 25. Syn. 206); plant almost stemless, *green;* leaves *unequal,* semiterete, *acinaciform,* obtuse; flower sessile; calyx 6-lobed. *M. fissoides, Haw. Obs.* 135.

HAB. Cape of Good Hope.

Old stem 2 inches high. Peduncle very short, with 2 leafy bracts. Flowers pale red. Calyx thick, lobes subequal, blunt. Petals an inch long. Filaments white. Styles 6, recurved, white.

§. 3. *ROSTRATA,* Haw. DC. *Plants* stemless or nearly so. *Leaves* 4–6, suberect, vaginate-connate, semiterete, attenuate, somewhat carinate at the apex. *Flowers* solitary, pedunculate, yellow. *Calyx* 2–8-cleft. *Stigmas* 8–12, rarely more. (Sp. 14–21.)

14. M. ramulosum (Haw. Misc. 29. Syn. 215); young plant nearly stemless; old stem 3 inches high, branched, prostrate; leaves subulate, obtuse, pustulate inside at the base, when old expanded; scape terete, bracteate at the base; calyx 5-cleft. *DC. l. c.* 421. *M. rostratoides, Haw. Obs.* 154.

HAB. Cape of Good Hope.

Flowers as in *M. caninum.* Styles 9, equalling the stamens. This is the smallest of all the section.

15. M. bifidum (Haw. Misc. 29. Syn. 212); plant almost stemless; leaves subulate, glaucous, obtuse, *with many dots;* scape nearly terminal, 1-flowered; calyx *bifid. DC. l. c.*

HAB. Cape of Good Hope.

Calyx-lobes unequal. Allied to *M. quadrifidum,* but weaker, and the leaves are shorter and blunter.

16. M. quadrifidum (Haw. Misc. 28. Syn. 212); plant almost stemless, at length branched; leaves subulate, obtuse, hoary-glaucous, marked *by a few dots* towards the apex; scape terminal, 1-flowered, longer than the leaves; calyx 4-cleft. *DC. l. c.*

HAB. Cape of Good Hope.

Old stems 2–3 inches long. Styles about the length of the stamens.

17. M. robustum (Haw. Misc. 28. Syn. 211); stem robust, a little branched, short, decumbent; leaves *subulate*, obtuse, *dotted*, *pustulate inside* at the base. *DC. l. c.*

HAB. Cape of Good Hope.
Flowers unknown.

18. M. denticulatum (Haw. Obs. 149. Syn. 215); stemless; leaves very glaucous, *subulately triquetrous*, compressed, dilately keeled at the apex; keel usually *denticulated*; scape *bibracteate*, 1-flowered; styles 12–15. *DC. l. c. M. difforme, Thunb.! Fl. Cap.* 423.

VAR. α. **canum** (Haw. Obs. 149); leaves canescent from minute down.

VAR. β **glaucum** (Haw. Obs. 151); leaves glaucous-white, rather dilated at both ends, a little toothed.

VAR. γ. **candidissimum** (Haw. l. c.); leaves whitish, elongated, a little toothed, compressed on both sides.

HAB. Karroo, between Olifantsriver and Bockland, in Hantum and Roggefeldt. Oct.–Nov. *Thunberg!*
Plant 2–3 inches or more high. Root perennial. Lower leaves marcescent, whitish, upper ones trigonous, acutely green. Peduncle or scape short, angular.

19. M. multipunctatum (Salm Dyck, Monog. fasc. 1. t. 2.); plant almost stemless, much branched; leaves elongated, *semiterete*, obtuse, keeled at the apex, glaucous-green, *with very numerous pellucid dots*, scape elongated *without bracts*; calyx 4-cleft.

HAB. Cape of Good Hope (v. v.)
Leaves 4–6, fleshy, 2–2½ inches long, about 3 lines wide, spreading, when young erect adpressed, with prominent dots, flat above and pustulate at the base, convex beneath, but compressed-keeled at the apex. Peduncle 3 inches long. Calyx-lobes unequal, 2 smaller. Petals in 3 series, about ½ inch long. Styles 8. Capsule 8-locular.

20. M. rostratum (Linn. Spec. 696); plant almost stemless, branched; leaves *subulate*, *elongated*, subtriquetrous, *acute*, subglaucous, dotted; scape elongated, *bibracteate*; calyx 4–5-cleft.

VAR. α. **longebracteatum**; bracteas 2, as long or longer than the scape. M. rostratum, *Haw. DC. Salm.-Dyck. l. c. fasc.* 1, *t.* 3. *E. & Z.!* 1978.

VAR. β. **brevibracteatum** (Salm Dyck, l. c. t. 4); bracts 4, shorter than the scape. *M. bibracteatum* (*Haw. Syn.* 215. *DC. l. c.*)

HAB. In Karroo, between Olifantsriver and Bockland, *Thunberg*; in Zwartland and near Saldanha Bay, *E. & Z.!* var. β. Gamkariver, *Zeyher!* 692. (Herb. Sd.)
Leaves 4–6, in the cultivated plant 4–5 inches long, 5–6 lines wide, in the wild plant 1–2 inches long, flat above, but pustulate at the base, convex beneath at the base, subtriquetrous at the apex. Peduncle or scape 2–4 inches long; bracteas foliaceous. Calyx-lobes unequal. Petals in 3 series. Styles 8–10. Distinguished from the preceding by the long, attenuated leaves.

21. M. purpurascens (Salm Dyck, Obs. Bot. ann. 1822); plant almost stemless, branched; leaves dotted, smooth, gibbous inside at the base, of a bluish-glaucous colour, *obtuse*, triquetrous at the apex; *keel usually extended*; *sheaths purplish*. *DC. l. c.*

HAB. Cape of Good Hope.

§. 4. *ALOIDEA*, DC. (Aloidea et Magnipunctata, Haw.); *stemless. Leaves* 4–6, triquetrous, gradually thickened from the base to the middle, attenuate at the apex, bluntish, mucronulate, keeled beneath, the angles entire, marked on both sides by tubercles. *Flowers* solitary, subsessile, yellow. *Calyx* 4–5-cleft. (Sp. 22–26.)

22. M. nobile (Haw. in Phil. Mag. 1823. p. 381); stemless or nearly so; leaves subelongate, *triquetrously clavate*, obtuse, somewhat recurved, the angles rotundate, *rather concave above*, marked by large elevated *tubercles;* peduncle bibracteate; calyx 6-cleft. *DC. l. c.* 419. *Salm. Dyck. Monog. fasc.* 4. *f.* 1. *M. magnipunctatum γ. affine. Haw. Rev.* 87. *M. compactum, Ait. Kew. vol.* 2. *p.* 191 ?

HAB. Gamkariver, May. *Zeyher!* 688.

Leaves 4–6, connate at the base, 2–2½ inches long, 6–7 lines wide. Flowers 2 inches in diameter. Peduncle 4–6 lines long, with 2 lanceolate foliaceous bracteas. Calyx subglobose. Petals linear, whitish at the base. Styles 16, erect, equalling the stamens.

23. M. magnipunctatum (Haw. Rev. p. 86); stemless; leaves perfect, usually about 4, large, *clavately triquetrous*, very thick, glaucescent, *flat above*, keeled beneath, obtuse at the apex, marked with very *large numerous dots;* flowers *sessile*. *DC. l. c. M. magnipunctatum, Haw. Suppl.* 87.

HAB. Cape of Good Hope.

24. M. canum (Haw. Obs. 158. Syn. 219); stemless; leaves *hoary*, semiterete at the base, attenuated, gibbously keeled at the apex. *DC. l. c.*

HAB. Cape of Good Hope.

Very like *M. magnipunctatum*, but distinguished by its hoary aspect. Flowers unknown.

25. M. albinatum (Haw. Phil. Mag., Aug. 1826, p. 126); stemless; leaves *acinaciformly triquetrous* upwards, with a recurved mucro, full of scattered, rather elevated, whitish dots; flowers sessile. *M. albipunctatum, Haw. l. c.*

HAB. Cape of Good Hope.

Root perennial, tufted. Leaves decussate, green, spreading. Stamens erect, spreading.

26. M. aloides (Haw. Suppl. 88); stemless or nearly so; leaves erecto-patent, entire, *semiterete*, green, *white-dotted*, acute upwards, rather concave above, carinately triquetrous and mucronate at the apex; flowers sessile; calyx 5-cleft. *DC. l. c. Salm. Dyck. l. c. fasc.* 3. *f.* 1.

HAB. Cape of Good Hope.

Old plant tufted. Root fleshy, fusiform. Leaves 6–8, connate at the base, 2 inches long, and at the base 4 lines wide. Flowers 1 inch in diameter. Calyx turbinate, lobes equal in length, 3 of them membranaceous on the margins. Petals in one series. Stamens erect. Styles 10.

§. 5. *RINGENTIA*, DC. (Ringentia et Scapigera, Haw.) *Plants* stemless or nearly so. *Leaves* 4–6, semiterete at the base, gradually thickening to the top and triquetrous; lateral angles ciliate-dentate, carinal angles mostly entire. Flowers solitary, yellow. Calyx 4–5-cleft. Stigmas 4–5, rarely more. (Sp. 27–37.)

* *Sessiliflora.*—Flower sessile or on a short peduncle.

27. M. tigrinum (Haw. Obs. 164. Syn. 216); plant almost stemless; leaves *ovate-cordate*, glaucous-green, *marbled with white*, carinate-convex beneath, carinal angle very entire, lateral ones dentate, teeth ciliated with many long hairs; flowers subsessile. *E. & Z.!* 1982. *Bot. Reg. t.* 260. *Salm. Dyck. Monog. fasc.* 1. *t.* 5. *M. ringens, Thunb.! herb.* α. (*ex parte*).

HAB. In Karroo in Onderste Roggeveld, *Thunberg!* Mount Bothasberg near Vishriver, Albany, Jan. (Herb. Thunb. Sd.)

Leaves rhomboid-dilated, 1–1½ inches long, 8–12 lines wide, very fleshy, with 20–24 ciliated teeth on the margin. Calyx turbinate, 5-cleft. Petals in 2 or 3 series. Styles 5. Distinguished from the following by the larger leaves.

28. M. felinum (Haw. Obs. 161. Syn. 216); nearly stemless; leaves *triquetrous, rhomboid-lanceolate*, glaucescent, *obsoletely white-dotted*, carinate-convex beneath, carinal angle very entire, lateral ones with 8 ciliate teeth; flowers subsessile. *DC. Pl. Grass. t.* 152. *E. & Z.!* 1981. *Salm. Dyck. l. c. fasc.* 1, *t.* 6. *Dill. Elth. t.* 187. *f.* 230. *M. ringens, Thunb.! herb. ex pte.*

HAB. Karroo in Bockland, Roggeveld, *Thunberg!* Zwartkopsrivier, *Zeyher!* 2580. Feb.–Nov. (Herb. Thunb. D., Sd.)

Leaves 1½ inches long, 6–8 lines wide, with 6–10, rigid, recurved teeth. Flowers as large as in the foregoing. Calyx globose, 5-fid. Petals 2–3-seriate. Styles 5.

29. M. lupinum (Haw. Phil. Mag. 64–111); nearly stemless; leaves triquetrous, *lanceolate, green, without dots*, smooth above, carinate-convex and prominently punctate beneath; carinal angle very entire, lateral ones with *numerous long ciliated* teeth; flowers subsessile. *DC. l. c.* 419. *Salm. Dyck. l. c. fasc.* 6. *t.* 2.

HAB. Cape of Good Hope (v.v.)

Leaves 1–1½ inches long, at the base 9 lines wide, with 16–20 lateral, subulate, long ciliated reversed teeth. Calyx turbinate, 5-fid. Petals 2–3-seriate. Styles 5. Very like *M. felinum*, but differs by the greener, from the base attenuated leaves with longer and more numerous teeth, and by the much smaller flowers.

30. M. murinum (Haw. Obs. 165. Syn. 217); nearly stemless, much branched; leaves *elongate-rhomboid*, tuberculate-punctate, half cylindrical at the base, triquetrous at top, carinal angle at *the apex denticulate*, lateral ones with 4–6, *short, acute, subciliated teeth;* flowers on short peduncles.. *DC. l. c. Salm. Dyck. l. c. fasc.* 5. *f.* 1. *M. ringens, Thunb.! herb.* β. (*ex pte.*)

HAB. Karroo in Onderste Bokke'veld, *Thunberg!*

Leaves 12–14 lines long, 4 lines wide, acute, glaucescent, a little thickened towards the apex. Flowers small. Calyx turbinate. Petals in many series. Styles 5, short. Distinguished from the preceding by the short denticulated leaves with denticulate carinal angle.

31. M. mustellinum (Salm. Dyck. Obs. p. 9); plant almost stemless, branched; leaves triquetrous, *gradually thickening towards the apex*, obtuse, greenish, punctate-papulose, pustulate-gibbous on the inside at the base, carinal angle *dilated, very entire*, lateral ones *denticulated above the middle;* flowers on short peduncles. *Monog. fasc.* 1. *t.* 7.

HAB. Bitterfontein, Bechuana territory, April, *Zeyher!* (Herb. Sd.

Leaves 8–10 lines long, 4–5 lines wide, spathulate, flattish above. Flowers small.

Calyx turbinate, 5-fid. Petals in many series. Styles 5, short. In the wild specimens the leaves are shorter, 4–6 lines long, more roundish at the apex, and often only cartilagineous-dentate or entire.

32. M. ermininum (Haw. Phil. Mag. Aug. 1826, p. 126); nearly stemless, branched; leaves triquetrous, a little thickened towards the apex, glaucescent, *wrinkled* from numerous dots, carinal angle very entire, subdilated, *lateral ones with short teeth* at the apex; flowers on short peduncles. *Salm Dyck, l. c. fasc.* 1. *t.* 8.

HAB. Cape of Good Hope.

Leaves 8–9 lines long, 3–4 lines wide, bluntish; lateral angles with 6–8 very short teeth. Flowers small. Calyx turbinate. Petals in many series, setaceous. Styles 5, short. Different from *M. mustellinum* in the leaves being smaller, glaucous, punctate but not gibbous, tuberculate on the inside at the base.

33. M. musculinum (Haw. Phil. Mag. 1826. p. 228); stem short, branches elongate, prostrate; leaves *triquetrous*, glaucous-green, *pellucid-punctate*, gibbous, pustulate on the inside at the base, *margins and keel usually bearing but one tooth each;* flowers on short peduncles. *Salm Dyck, l. c. fasc.* 1. *t.* 9.

HAB. Cape of Good Hope (v.v)

Branches angular, sometimes 6 inches long. Leaves about 8 lines long, 2–3 lines wide, bluntish, lateral angles with 2–4 very short teeth. Flowers small. Calyx 5-cleft. Petals lanceolate, about 1 line wide. Styles 5.

34. M. agninum (Haw. Phil. Mag. 1826. p. 127); stem very short, branched; leaves *oblong, canescent, wrinkled from dots*, flat and pustulate at the base above, carinate-convex beneath, lateral angles somewhat *toothed or entire* towards the apex; flowers on short peduncles; calyx 6-cleft. *Salm Dyck, l. c. fasc.* 3. *t.* 2.

VAR. *α.* **denticulatum**; leaves evidently toothed at top.

VAR. *β.* **integrifolium**; leaves smaller, entire.

HAB. Cape of Good Hope (v.v.)

Leaves with a large white pustule on the inside at the base, 1½–2 inches long, at the base 5–6 lines wide, gradually attenuated towards the apex, bluntish; the margin in var. *α.* with 3–5 teeth on each side. Flowers 1 inch in diameter; peduncle compressed, 1 inch long. Calyx sub-compressed, with 6 triquetrous, elongated lobes. Petals linear, in many series. Styles 6.

** *Scapigera.*—Flower pedunculate; peduncle bibracteate.

35. M. caninum (Haw. Obs. 159. Syn. 217); stem very short, branched; leaves glaucous, carinately-triquetrous, *subclavate*, incurved towards the apex, carinal angle entire, *lateral ones somewhat toothed;* peduncles longer than the leaves; calyx 5-fid. *Salm Dyck, l. c. fasc.* 3. *t.* 3.

VAR. *α.* **pluridentatum**; leaves smaller, lateral angles denticulate. *M. caninum, DC. Pl. Grass. t.* 95. *M. ringens and caninum, Linn. Spec.* 298. *Dill. Elth.* 241 *&* 188. *f.* 231.

VAR. *β.* **paucidentatum**; leaves larger; lateral angles obscurely toothed. *M. vulpinum, Haw. Rev.* 88.

HAB. Cape of Good Hope (v.v.)

Leaves very thick, not punctate, 2 inches (in var. *β.* 3) long, 7–8 lines wide, acute, flat above, triquetrous-convex beneath. Peduncle 4 inches long, thickened at top, with 2 foliaceous, vaginate-connate bracteas. Calyx turbinate. Petals in many series. Styles 5, subulate.

36. M. albidum (Linn. Spec. 699); stemless, smooth, whitish; leaves thick, *subulate, triquetrous,* flat above, obtuse, *with an acumen,* semiterete at the base, *quite entire;* flower large, peduncle longer than the leaves; calyx 5-fid. *Thunb.! Fl. Cap.* 423. *Dill. Elth. t.* 189. *f.* 232. *Sims. Bot. Mag. t.* 1821. *Salm Dyck, l. c. fasc.* 5. *t.* 283.

HAB. Namaqualand, *Drege! A. Wyley!* (Herb. Thunb., D., Sd.)

Leaves (4–6) vaginate-connate, about 2 inches long, 6–8 lines wide, attenuated, mucronate, very thick, not punctate. Peduncle as long or longer than the leaves, thickened below the punctate calyx, with 2 foliaceous bracts at the base. Calyx-lobes subequal, 3 with a membranaceous margin. Petals an inch long, linear. Styles 6–15, erect. Garden specimens are larger, 6 inches or more high, the leaves 3 inches long, sometimes on the lateral angles with one or two short teeth; the peduncle often with two or three flowers, but calyx and petals are not different. *M. hybridum, Haw.* is quite the same.

37. M. namaquense (Sond.); stem very short, branched; leaves thick, smooth, whitish, punctate or wrinkled, *acute but not mucronate,* a *little channelled above, keeled and triquetrous beneath,* quite entire; peduncle 3–4 times longer than the leaves; calyx 5-fid. *M. difforme, Thunb.! Fl. Cap. p.* 423. *non Linn.*

HAB. Olifantsriver, in Bockland and Hantum, *Thunberg!* Namaqualand, *Drege! A. Wyley!* (Herb. Thunb., D., Sd.)

A small, well distinguished species. Branches ½–1 inch long. Leaves about 4, not including some decayed at the base, long, connate, 5–6 lines long, about 3 lines wide, the inferior ones white-punctate, scabrous, the superior wrinkled, when dry channelled above. Peduncle with 2 foliaceous, punctate, small bracts at the base. Three of the calyx-lobes with a large, white-membranaceous margin. Petals 6–8 lines long, linear, gold-coloured. Style 5, short.

§. 6. *DOLABRIFORMIA,* DC. (Dolabriformia and Carinata, Haw.) *Plants* stemlss or on short stems. *Leaves* entire, decussate, with a keeled, gibbous angle. *Flowers* ternate, yellow; peduncle bibracteate. *Calyx* 5-fid. Stigmas 5. (Sp. 38–42.)

38. M. scapiger (Haw. Phil. Mag. Dec. 1824. p. 423); plant almost stemless; leaves elongate, carinately triquetrous, green, a little roughish and dotted on the margins, carinal angle *much compressed and produced; scape 2-edged. Salm Dyck, l. c. fasc.* 1. *t.* 10.

HAB. Cape of Good Hope.

Radical leaves 10–12, erecto-patent, 3–4 inches long, 5–7 lines wide at the base, cartilagineous at the apex, the two opposite unequal, one gradually attenuated, scarcely dilated below the apex, the other with very prominent carinal angle. Scape 3 or 5-flowered, 2–3 inches long; bracts short, subleafy. Flowers middle-sized. Styles 5, filiform.

39. M. multiceps (Salm Dyck, Monog. fasc. 6. t. 3); stemless, tufted; leaves elongate, attenuated at both ends, *very smooth,* shining, green, flat above, keeled beneath, triquetrous-compressed and *acuminate at the apex; scape subterete. M. bibracteatum, E. & Z.!* 1980.

HAB. On fields near the Zwartkopsriver and Koegariver, Howisonsport, &c., *Zeyher!* 2577. July–Dec. (Herb. D., Sd.)

Leaves 1–1½ inches long, ¾–4 lines wide, cultivated larger, attenuated near the base, gradually narrowed at the apex and not much dilated on the carinal angle. Peduncle sometimes 1-flowered, as long as the leaves, but generally there is a 3-flowered, rarely 5-flowered subcompressed scape, 2 or 3 lines longer than the leaves.

Bracts foliaceous, pedicels ½–1 inch long, thickened upwards. Calyx-lobes 5, very acuminate, 4 lines long. Petals ½ inch long. Much smaller than the preceding, and with smaller flowers, but perhaps only a variety.

40. M. rhomboideum (Salm. Dyck. l. c. fasc. 5, t. 4); stemless; leaves smooth, *rhomboid-dilated*, green, *white-dotted*, flat-concave above, convex and carinate-keeled beneath, *rotunded-obtuse at the apex*, angles white-margined; scape terete. *M. scapiger, E. & Z.!* 1984.

HAB. On the fields by the *Zwartkopsriver*, Coega and *Zondagsriver*, *Zey.!* 2579. Janr. (Herb. D., Sd.)

8–10 radical leaves, 1–2 inches long, about 6–8 lines wide, subglaucous-green, half cylindrical from the base to the middle, slightly concave above, compressed-keeled beneath from the middle to the apex. Scape 1–2 inches long. Flowers biternate. bracts 1–2 lines long. Pedicels ½–1 inch, with or without bracts. Calyx 5-cleft, lobes ovate acute. Petals not much longer than the calyx. Styles 5.

41. M. dolabriforme (Linn. Spec. 699); caulescent; leaves glaucous, dotted, exactly *dolabriform*, i. e. depressed at the base and compressed at the apex, obtuse and somewhat emarginate; scape abbreviate, compressed; pedicels short. *Dill. Elth. t.* 191. *f.* 237. *DC. Pl. Grass. t.* 6. *Curt. Bot. Mag. t.* 32. *Salm. Dyck. l. c. fasc.* 3. *t.* 4.

HAB. Cape of Good Hope (v.v.)

Stem in the young plant very short, but when old ½–1 foot. Leaves connate, 1–1½ inches long, at the base 2 lines, at the top 6–7 lines wide. Flowers ternate or quinate. Bracts subleafy. Calyx 5-cleft. Petals lanceolate, ½ inch long. Styles 5.

42. M. carinans (Haw. Rev. 90); nearly stemless; leaves elongated, *subincurved* and spreading, *semiterete at the base*, compressed at the apex, and dilated into a keel, whitish and dotted. *DC. l. c.* 423. *E. & Z.!* 1983? *M. canum, Salm. Dyck. obs. bot.* 20. *non Haw.*

HAB. Cape of Good Hope.

Flowers ternate, exactly as in *M. dolabriforme*, and expanding in the evening. The specimen of *E. & Z.!* collected between Beaufort and Graafreynet is incomplete, and seems not to be different from *M. dolabriforme*.

§. 7. *DIFFORMIA*, Salm. Dyck. (Cruciata and Difformia, Haw. Linguæf. β. DC.) *Stem* very short or erect, branched. *Leaves* unequal, obliquely decussate, half-cylindrical, dilated or attenuate at the apex, variously difformed and often furnished with 1–2 fleshy teeth. *Flowers* solitary, large, yellow, pedunculate or sessile. *Calyx* 4-cleft. *Stigmas* 8. (Sp. 43–52.)

43. M. bidentatum (Haw. Suppl. 89. Rev. 103); nearly *stemless*; leaves semi-cylindrical, thick, soft, oblique, and compressed at the apex, and difformed, bearing 2 *large, almost opposite, fleshy teeth* in the middle. *Salm. Dyck. l. c. fasc.* 2. *f.* 1. *Dill. Elth. p.* 252. *f.* 241. *(excl. f.* 242.*)*

HAB. Cape of Good Hope (v.v.)

Branches prostrate, Leaves unequal, about 3–4 lines wide, one 2 inches long, the opposite 1–1½ inches long, very smooth, green, punctate. Peduncle 1½ inches long, thickened upwards. Calyx turbinate. Petals uniseriate, obtuse, denticulate. Styles short, ramentaceous.

44. M. semicylindricum (Haw. Obs. 238. Syn. 228.); *caulescent;* leaves semi-cylindrical, *narrow-tongue-shaped*, bullate-punctate, oblique and subcompressed at the blunt apex, furnished in the middle *with* 1 *or* 2 *obsolete teeth or tubercles. Flowers on short pedicels. Salm Dyck, l. c. fasc.* 1. *t.* 11. *M. difforme Lin.* ? *Willd.* ? *Dill. Elth. p.* 252. *f.* 242.

HAB. Cape of Good Hope.

Stem erect, rigid, ½ foot or higher, bifarious, branched. Leaves difformed, one of them 2 inches, the opposite 12–15 lines long, 3–4 lines wide. Peduncle 1 inch. Calyx turbinate. Petals entire.

45. M. difforme (Haw. Rev. 103. Syn. 228); rather caulescent; leaves obliquely cruciate, semi-cylindrical, narrow-tongue-shaped, punctate, *not dentate or tubercled in the middle,* difformed, one shorter, bluntish and mucronulate, the other longer, attenuated, compressed, and obliquely keeled at the blunt apex. *Salm Dyck, l. c. fasc.* 1. *t.* 12. *DC. l. c. E. & Z.!* 1985, *but not of Linn. or Thunb.*

HAB. In Karroo, near Olifantsrivierbad, Clanwilliam, Oct. (Herb. Sd.).

Stem short-branched. Leaves ½ inch wide, the longer 3, the other 2–2½ inches long. Peduncle 1 inch. Petals entire. Styles 8–9.

46. M. bigibberatum (Haw. Phil. Mag. Nov. 1825. p. 329); rather caulescent; leaves obliquely somewhat cruciate, small, semi-cylindrical, scarcely tongue shaped, very smooth, usually with 2 gibbosities, *subequal,* one *attenuated and acute,* the other *keeled-dilated, obtuse* at the apex. *Salm Dyck, l. c. fasc.* 6. *f.* 4.

HAB. Cape of Good Hope (v. v.)

Twice smaller than the preceding, leaves smaller, sub-equal. It varies with or without gibbosities on the middle. Peduncle 8–10 lines long. Petals denticulate.

47. M. praepingue (Haw. Rev. 95. Syn. 222); nearly stemless; leaves obliquely tongue-shaped, semiterete, thick, very smooth, pale green, *when young ciliated with pubescence,* much difformed, one abbreviate, subacute, the other dilated, compressed, *keeled with incurved point;* flowers nearly sessile; *calyx ciliated* on the angles. *Salm Dyck, l. c. fasc.* 5. *t.* 5.

HAB. Cape of Good Hope (v. v.)

Very like *M. difforme,* but easily known by the triquetrous-compressed, ciliated calyx. Opposite leaves about 1½–2½ inches long, 6 lines wide. Peduncle ½ inch long. Petals biseriate, denticulate.

48. M. angustum (Haw. Obs. 176. Syn. 222); nearly stemless; *leaves obliquely distichous,* linear-tongue-shaped, semi-cylindrical, very long, *obliquely keeled at the apex,* subincurved, punctate, subequal, one acute, the other *uncinate;* flowers *subsessile, calyx glabrous. Salm.-Dyck. l. c, fasc.* 5. *t.* 6. (*var. pallidum, Haw.*)

VAR. *β.* **heterophyllum** (Salm Dyck, l. c. t. 7); leaves longer and thicker, evidently incurved at the apex. *M. heterophyllum,* Haw. Rev. 102.

HAB. Cape of Good Hope (v. v.)

Distinguished from the preceding by the less decussate more tongue-shaped and not ciliated leaves. Calyx subtrigonous-turbinate. Petals biseriate, acute.

49. M. cruciatum (Haw. Obs. 173. Syn. 224); nearly stemless; leaves cruciate, incrassate, linear-tongue-shaped, semi-cylindrical, very soft, obliquely keeled at the apex, subequal, *one attenuated,* the other *compressed dilated;* flowers *on long peduncles. Salm Dyck, l. c. fasc.* 2. *t.* 2.

HAB. Cape of Good Hope.

Stem short, somewhat branched. Leaves very thick, curved upwards, subpunctate, green, 3–4 inches long, 8–9 lines wide. Peduncle thickened upwards, 2 inches long. Calyx turbinate. Petals biseriate, bluntish. It differs from all the preceding by more semi-cylindrical, attenuated leaves, less differmed at the apex.

50. M. Salmii (Haw. Suppl. 89. Rev. 100); nearly stemless; leaves decussate, obliquely semi-cylindrical, attenuate, one *acute*, the other *oblique and blunt; flowers sessile. Link. and Otto. Abbild. Gen. fasc.* 8. 9. 95. *t.* 44. *Salm Dyck, l. c. fasc.* 4. *t.* 2.

VAR. β. **elongatum** (Salm Dyck); leaves somewhat smaller and longer.

VAR. γ. **semicruciatum** (Salm Dyck); leaves flattish, rigid and shorter, obliquely cuneate.

HAB. Cape of Good Hope (v. v.)

Leaves 6–8, patent, 3–4 inches long, 8–9 lines wide, flattish above, with a white spot at the inside of the base. Calyx compressed, obconical. Petals uniseriate, obtuse, denticulate.

51. M. taurinum (Haw. Syn. 224. Rev. 100); planta lmost stemless; leaves disposed in two rows, obliquely cruciate, *semiterete, obtuse,* very thick, of a yellowish green colour, *incurved;* flowers sessile. *DC. l. c.* 422.

HAB. Cape of Good Hope (v. v.)

Habit of *M. cruciatum,* but much larger. Old stem branched at the base, half a foot high. Leaves about one finger long, flattish above, often oblique at the apex; young ones always incurved. Calyx bifid (Haw.), segments unequal. Styles 8, ramentaceous.

52. M. surrectum (Haw. Rev. 101); leaves decussate, erectish or spreading, more or less semiterete, subulate, acute, soft, usually pustulate at the base; ovarium exserted, somewhat pedunculate.

HAB. Cape of Good Hope.

§. 8. *LINGUAEFORMIA,* Haw. (Linguaef. α. disticha. DC.) *Stemless* or nearly so. *Leaves* differmed, distichous, tongue-shaped, one of the margins thicker than the other, obliquely carinulate, often uncinate-incurved at the apex; flowers solitary, large, sessile or pedunculate. *Calyx* 3–5-fid. *Stigmas* 8–10. (Sp. 53–62.)

(Perhaps all the species of this section are varieties of one plant.)

53. M. scalpratum (Haw. Obs. 187. Syn. 220); nearly stemless; leaves sloped down much, *scalprate, very broad,* one of the margins thicker, pustulate inside at the base; flowers sessile; calyx 4-fid; petals emar ginate. *Salm Dyck. l. c. fasc.* 4. *t.* 3. *M. linguiforme, Linn. Dill. Elth. t.* 183. *fig.* 224.

VAR. β. **angustius** (*Salm.-Dyck. l. c.*); leaves smaller, more attenuated at the apex.

HAB. Cape of Good Hope.

Leaves 3–4 inches long, 1½–2 inches wide, subequal in length, obtuse, patent, green, not punctate. Calyx obconical, compressed. Petals uniseriate, broadly-linear. Styles 10–11. ramentaceous.

54. M. fragrans (Salm Dyck, l. c. fasc. 4. t. 4); nearly stemless; leaves sloped down, *obliquely tongue-shaped,* thick, one side rather convex and obtuse at the apex, the other side *thrown out into a keel;* flowers sessile;

petals obtuse, crenulate. Haw. Revis. 95. *Link. et Otto. Abbild. Gen. fasc.* 8. *p.* 93. *t.* 43. *DC. l. c.* 421.

HAB. Cape of Good Hope.

Nearly allied to *M. scalpratum*, but the leaves are shorter and thicker, and the flowers larger and fragrant. Leaves 2–3 inches long and 1 inch wide, patent. Flowers 3 inches in diameter. Calyx 4-fid, obconical, compressed. Petals uniseriate, broadly-linear. Styles 10.

55. M. grandiflorum (Haw. Phil. Mag. Nov. 1826. p. 328); nearly stemless; leaves sloped down, *broadly tongue-shaped*, long, thick, with a large pustule on the inside of the base, one side oblique cultrate, the other *subuncinate;* flowers sessile, very large; petals obtuse, crenulate. *Salm Dyck, l. c. fasc.* 4. *t.* 5.

HAB. Cape of Good Hope (v. v.)

Leaves 4–6 inches long, 1–1½ inch wide, a little attenuated at the apex, pale green. Flowers almost scentless. Calyx globose, subcompressed, 4-fid. Petals 2–3 lines broad. It differs from the preceding by the larger, flatter, exactly linguae-formed, more soft leaves, and from the following by the robust habit and particularly by the very large flower.

56. M. adscendens (Haw. Syn. 220. Rev. 96); stemless or nearly so, *very proliferous;* leaves distichous, sloped down, adscendent, broadly-tongue-shaped, *flattish on both sides, very blunt*, green; flowers pedunculate; peduncle and calyx *pustulate. Salm Dyck, l. c. fasc.* 5. *t.* 8. *Dill. Elth. p.* 237. *t.* 226.

HAB. Cape of Good Hope.

About 8–10 leaves, a little fleshy, subequal, 2½ inches long, 9–10 lines wide. Peduncle 1 inch long or longer, subtriquetrous at the apex. Petals subbiseriate, narrow, acute. Styles 10. It differs from the preceding by more flattish (not excavate above) leaves, not thicker at one side and blunt, scarcely uncinate; from the following by the proliferous stem.

57. M. cultratum (Salm Dyck, Obs. 1820. p. 7); nearly stemless; leaves distichous, sloped down, exactly tongue-shaped, thick, *cultrate at the margin and blunt apex;* flowers pedunculate; peduncle subtriquetrous. *Monog. fasc.* 5. *t.* 9. *E. & Z.!* 1986.

VAR. β. **perviride** (Salm Dyck, l. c.); leaves saturate green. *M. medium*, Haw. Suppl. 88.

HAB. On the fields near the Zwartkopsriver. Oct. (Herb. D. Sd.)

Leaves 3 inches long, 10–12 lines wide, subdeflexed, soft, two-edged at the margins, subequal. Peduncle 1 inch long or longer. Petals patent, subbiseriate, acute. Styles 11–13.

58. M. uncatum (Salm Dyck, l. c. fasc. 5. t. 10); nearly stemless; leaves distichous, patent, subincurvate, *narrow tongue-shaped*, thick, flattish above, obliquely convex beneath, *incurvate-uncinate at the apex;* flowers *short-pedunculate; petals revolute. M. longum, ε. uncatum. Haw. Rev.* 97.

HAB. Cape of Good Hope.

Leaves 2½ inches long, 8 lines wide, incrassate at the apex. Peduncle ½ inch. Calyx compressed globose. Petals subbiseriate, acute. Style 10.

59. M. depressum (Haw. Misc. p. 33. Syn. 221); nearly stemless, prostrate; leaves narrow-tongue-shaped, *recurved-depressed*, obtuse or variously incurved, acute at the apex; flowers *long pedunculate; petals sub-recurved. Salm Dyck, l. c. t.* 11.

VAR. β. **lividum** (Haw. Rev. p. 99); leaves of a livid rufous colour.

HAB. Cape of Good Hope (v, v.)

Leaves 2½ inches long, 9–10 lines wide, flattish above, obliquely convex beneath. Peduncle about 1½ inch, subcompressed. Petals subbiseriate. Styles 11.

60. M. linguaeforme (Haw. Obs. 188. Syn. 221); stemless or nearly so; leaves *unequally tongue-shaped,* distichous, *subfalcate-deflexed,* when young sloped down, when old depressed, flattish above, *obliquely attenuated,* obtuse or often keeled; flower *short-pedunculate. Salm.-Dyck. l. c. fasc.* 6. *t.* 5. *M. obliquum, Willd. Spec.* 2. 1027. *excl. fig. Dill. M. lucidum. Haw. Rev.* 95.

VAR. β. **latum** (Salm Dyck, l. c. t. 6); leaves shorter, very blunt and thicker at the apex, flower subsessile. *M. latum, Haw. Rev.* 98.

HAB. Cape of Good Hope (v. v.)

Leaves broadly tongue-shaped, 3 inches long, 1 inch wide at the base, 5–6 lines below the apex. Peduncle 8–10 lines long. Petals subbiseriate, lanceolate. Styles 8.

61. M. longum (Haw. Obs. 177) nearly stemless, subprostrate; leaves sloped down, *elongated* tongue-shaped, shining, deep green, *obliquely acute at the apex;* flowers long-pedunculate; petals acute. *M. linguaeforme, DC. pl. grass. t.* 71.

VAR. β. **declive** (Haw. Rev. 96); leaves very long, arcuate-deflexed; peduncle shorter. *Salm Dyck, l. c. fasc.* 5. *t.* 12. *M. longum, β. flaccidum.*

HAB. Cape of Good Hope.

Leaves 3–4 inches long, 10–12 lines wide at the base, somewhat attenuated, distichous. Peduncle 2 inches long, Petals subbiseriate. Styles 9.

62. M. pustulatum (Haw. Suppl. 88. Rev. 96); stemless or nearly so; leaves distichous, narrow tongue-shaped, adscending, elongated, subattenuate, blunt, furnished with large pustules on the inside at the base; flower long-pedunculate. *Salm.-Dyck, l. c. fasc.* 5. *t.* 13.

VAR. β. **lividum** (Salm Dyck, l. c. t. 14); leaves shorter, of a livid colour. *Haw. Rev.* 96.

HAB. Cape of Good Hope.

Leaves 3–4 inches long, 8–10 lines wide. Peduncle 2 inches or more long, slightly compressed. Petals subbiseriate, acutish. Styles 9.

§. 9. *GIBBOSA,* Haw. DC. subcaulescent; *stem* short, decumbent, branched. *Leaves* connate a great way above the base, large, unequal, one short, gibbous, the other somewhat larger, subovate. *Flowers* sessile, or on short pedicels, small, reddish. *Calyx* 6-cleft. *Stigmas* 6. (Sp. 63–66.)

63. M. gibbosum (Haw. Obs. 137. Syn. 226); nearly stemless; leaves of a yellowish green colour, spreading, *ovate,* semi-cylindrical, very rarely keeled at the apex; *peduncles short,* 2-edged.

HAB. Cape of Good Hope.

Calyx lobes unequal. Petals reddish, with paler eyes. Stigmas very short.

64. M. luteo-viride (Haw. Syn. 226); stem short, prostrate, weak;

leaves *oblong*, semi-cylindrical, triquetrous at the apex, *greenish yellow; flowers sessile. M. perviride, β. Haw. Misc.* 37.

HAB. Cape of Good Hope.

Stem 1–2 inches. Calyx 2-edged at the base, lobes unequal. Petals numerous, broad linear, reddish, marked by a deeper line each. Styles very short, at length spreading. Perhaps only a variety of *M. perviride.*

65. M. perviride (Haw. Obs. 186. Syn. 227); stem weak, prostrate; leaves semi-cylindrically-triquetrous or somewhat ovate, *very green;* pedicels very short, 2-edged.

HAB. Cape of Good Hope.

Stem 2–3 inches long. Leaves shorter and broader than in *M. luteo-viride.* Calyx small. Petals paler than those of *M. gibbosum.*

66. M. pubescens (Haw. Obs. 138. Syn. 227); plant almost stemless; old stem weak, prostrate; leaves *silky-pubescent*, semi-cylindrical, oblique at the apex.

HAB. Cape of Good Hope.

Flowers unknown. It differs from the two preceding species in being silky and downy.

§. 10. *CALAMIFORMIA*, Haw. DC. *Stemless* or nearly so. *Leaves* numerous, erect, terete, bluntish. *Flowers* in short peduncles, of a dirty white colour. *Calyx* 5-cleft. *Stigmas* 8–10. (Sp. 67–69.)

67. M. calamiforme (Linn. Spec. 690); leaves subulate, nearly terete, glaucescent dotted, flat above, obtuse, mucronulate. *Haw. Syn.* 208. *DC. pl. grass. t.* 5. *Salm Dyck, monog. fasc.* 4. *t.* 6. *Dill. Elth. t.* 186. *f.* 228.

HAB. Cape of Good Hope (v. v.)

Stem 1–2 inches long. Leaves 2–2½ inches long, 4 lines in diameter. Peduncle 1 inch. Calyx subglobose, 3 of the lobes leafy and longer. Petals in many rows, linear-lanceolate. Styles 7–8.

68. M. obsubulatum (Haw. Misc. 26. Syn. 208); stemless, or nearly so; leaves inversely subulate or *gradually thickening towards the apex*, thick, obtuse, greenish, punctulate. DC. l. c. 424.

HAB. Cape of Good Hope.

Very like *M. calamiforme.* Flowers unknown.

69. M. digitiforme (Thunb. Nov. Ephem. Nat. Cur. v. 8. p. 6. App.); stemless or nearly so; leaves 3–4, approximate, *terete, finger-shaped, obtuse*, smooth,soft; flower *subsessile. Fl. cap.* 412. *M. digitatum, Ait. Kew.* 2. 181. *Haw. Syn.* 211.

HAB. in Karroo, between Olifantsrivier and Bocklandsberg. Oct. Novb. *Thunberg* (but now wanting in herb. Thunb.)

Root fasciculate. Old stem decumbent, brown. Leaves about one finger long, broad at the base. Calyx 5-cleft, lobes obtuse. Petals linear, white.

§. 11. *TERETIFOLIA*, Haw. DC. *Stemless* or subcaulescent. *Leaves* (4–6) nearly terete, spreading. *Flowers* pedunculate, pale red. *Calyx* 4-cleft. *Stigmas* 8–12. (Sp. 70–72.)

70. M. cylindricum (Haw. Obs. 411. Syn. 209); almost stemless;

leaves bluntly triquetrous, *rather glaucous*, dotted, when young more glaucous and *more triquetrous;* peduncles compressed at the base, bibracteate. *DC. l. c.* 424.

HAB. Cape of Good Hope.

Old stems 2 inches long, crowdedly branched. Leaves 3 inches long. Peduncles 1-2 inches long. Bracteas large, leafy. Calyx lobes unequal, very blunt. Petals saturate-reddish.

71. M. teretifolium (Haw. Syn. 210); subcaulescent; leaves *nearly terete* or cylindrical, *greenish*, rather dotted, but when young polished, very green *and semiterete;* peduncle *nearly terete, bibracteate. DC. l. c. M. cylindricum, β. Haw. Misc.* 27.

HAB. Cape of Good Hope (v. v.)

Branches procumbent. Leaves 4 inches long. Peduncle 2 inches long. Bracts leafy. Petals often emarginate, reddish, white at the base. Styles about 10. Perhaps a variety of the preceding.

72. M. teretiusculum (Haw. Obs. 410. Misc. 27. excl. Syn.); stemless; leaves *bluntly triquetrous*, firm, thick, green, dotted. *DC. l. c.*

HAB. Cape of Good Hope.

Leaves 2 inches long. Flowers unknown.

§. 12. *BELLIDIFLORA*, Haw. DC. *Subcaulescent. Leaves* triquetrous, compressed, spreading-incurvate, acute at the angles, toothed at the apex. *Flowers* solitary, pedunculate. *Petals* white, with a purple middle-rib. *Calyx* 5-cleft. *Stigmas* numerous, hair-formed. (Sp. 73.)

73. M. bellidiflorum (Linn. Spec. 590); caudex short, suffruticose; leaves triquetrous, compressed, denticulate at the apex; peduncle bibracteate at the base.

VAR. *α.* **glaucum** (Salm Dyck, l. c. fasc. 3. t. 5); leaves spreading, glaucous, compressed-triquetrous, 3-fariously denticulate at the apex. *M. bellidiflorum, Linn. Lam. Haw. Dill. Elth.* 244. *f.* 233.

VAR. *β.* **viride** (Haw. Salm Dyck, l. c. t. 6); leaves suberect, pale green, toothed beneath on the keel, lateral angles entire. *M. bellidiflorum, Spreng. Thunb. fl. cap.* 418. *et herbar. !*

VAR. *γ.* **subulatum** (Haw. Salm Dyck, l. c. t. 7); leaves spreading, glaucesent, attenuate-triquetrous, the carinal angle dentate at the apex. *M. subulatum, Haw. Syn.* 208. *M. bellid. β. simplex. DC. pl. grass. t.* 41.

HAB. Cape of Good Hope.

Leaves decussate, connate, 2 inches, in var *γ* 1 inch long. Flowers in var. *α* 1½ inch, in var. *β* et *γ* 1 inch in diameter. Peduncle in var. *α* 2 inches, in var. *β* 1 inch, in var. *γ* ½ inch long. Calyx lobes broad, subequal. Petals biseriate, narrow lanceolate, bifid or acute. Styles 5, ramentaceous.

M. Burmanni, Haw. et DC. founded on Burmann's t. 25, is not different from *M. serrulatum*, Haw.

§. 13. *ACUTA*, Haw. DC. *Stemless* or nearly so. *Leaves* semiterete, subulate-acute, subtriquetrous at the apex, full of pellucid dots. *Flowers* pedunculate, purple. *Calyx* 5-cleft. *Stigmas* 8–10. (Sp. 74–76.)

74. M. acutum (Haw. Misc. 26. Syn. 207); stemless; leaves semicylindrical, *triquetrous, acute* at the apex, glaucous-green, full of pellucid dots, finely wrinkled; flowers long-pedunculate, peduncles bibracteate at at the base. *Salm Dyck, l. c. fasc.* 1. *t.* 13. *M. subulatoides, Haw. Obs.* 141. *M. subrostratum, Willd. Enum.* 529. *excl. Syn. M. rostroides, Haw.*

HAB. in Karroo, near Caledon, *Zeyh.*! 2578; Gamkariver, *Zeyh.*! 691. (Herb. Sd.)

Leaves aggregate, connate at the base, 2–2½ inches long, 3–5 lines wide, attenuated from the base, flat above, angles entire, sometimes a little cartilagineous, acutish. Peduncles 2–3 inches long, terete. Calyx hemisphærical, lobes acute. Petals uniseriate, serrulate at the apex.

75. M. diminutum (Haw. Misc. 26. Syn. 230); nearly stemless, smooth, shining; leaves semiterete, *obsoletely triquetrous*, terminating in a *white point at the apex*, full of pellucid dots, flat above. *M. corniculatum, Haw. Obs.* 226. *excl. Syn.* M. *loreum, Linn. Spec.*

VAR. β. cauliculatum (Haw. Suppl. 90); stem half erect; leaves longer and with larger dots. Perhaps only an old plant.

HAB. Cape of Good Hope.

3 or 4 times smaller than *M. acutum.* Flowers red or purplish.

76. M. punctatum (Haw. Obs. 411. Rev. 107); stemless, smooth; leaves semiterete, *triquetrous* at the top, flat above, full of pellucid dots, pale green, furnished with a minute, white point at the apex. *DC. l. c.* 425.

HAB. Cape of Good Hope.

Perhaps only a variety of *M. diminutum.* Flowers unknown.

Group II. TRIQUETRA.—Stem fruticose or suffruticose, erect, decumbent or rooting. Leaves more or less triquetrous, distinct or connate at the base. (§§. 15–31.)

§. 14. *CORNICULATA*, Haw. DC. *Caudex* branched, prostrate, substrumose at the knots. *Leaves* more or less crowded at the node, elongated, exactly triquetrous or cylindrical-triquetrous, in- or re-curvate. *Flowers* pedunculate, yellow, in one species whitish, with purple lines. *Calyx* 5-cleft. *Stigmas* 10–20. (Sp. 77–84.)

77. M. reptans (Ait. Kew. 2. 241); stems *filiform*, very slender, creeping; leaves crowded, incurvate-erect, *triquetrous*, acute, glaucous, *scabrous* from pellucid dots; petals yellow. *Haw. Syn.* 242. *Salm.-Dyck. l. c. fasc.* 6. *t.* 7. *E. & Z.!* 1988. M. *crassifolium, Thunb.! fl. cap.* 421. *non. Linn.* M. *debile (Haw. Phil. Mag.* 1826. 331). *Zeyher*, 2591. 2610.

HAB. sandy places near Capetown, Zoutriver, and in Swartland. June–July. (Herb. Thunb. D. Sd.)

Stem and branches prostrate, ½–1 foot or longer, angulate. Leaves 8–12 lines long, 2–3 lines wide in the middle, attenuated at the base, acute and mucronulate. Peduncle 1–2 inches long, bibracteate. Calyx turbinate, ½ inch long, 2 or 3 lobes acuminate, the others membranaceous at the margins. Petals somewhat longer than the calyx. Specimens from rocks of Lions Mountain are more filiform, the leaves half an inch long, and the flowers smaller and externally reddish; but there are intermediate forms from various localities.

78. M. diversifolium (Haw. Misc. 38. Syn. 230); stem very short; branches *sarmentose;* leaves crowded, exactly decussate, *semi-cylindrical*, very unequal in length, glaucous-green, *rugulose*, triquetrous-compressed at top; petals yellow. *Salm Dyck, l. c. fasc.* ii. *t.* 3. *M. loreum, Dill. Elth. t.* 200. *f.* 255. *Haw. Rev.* 108. *M. corniculatum, β. diversiphyllum*, Haw. Willd.

VAR. β. **congestum** (Salm Dyck. l. c. t. 4); leaves very numerous, crowded, the lower ones often the longest. *Dill. Elth. t.* 198. *f.* 252. *M. diversifolium, E. & Z.!* 1987.

HAB. Sandy places near Greenpoint, Bethelsdorfe, and Cradockstadt. July (Herb. D. Sd.)

Branches ½ foot or longer. Leaves capitate, flattish above, convex beneath, lower ones 2–3 inches, upper ones 1–½ inch long, at the base 3–4 lines wide, proliferous from the axils. Peduncle 2–3 inches long, with 2 foliaceous bracts. Calyx depressed-globose. Petals ½ inch long, in var. β. purplish on the under side near the apex. Styles 14–17.

79. M. purpureo-album (Haw. Phil. Mag. Nov. 1826. p. 329); stem elongate, branches short, prostrate, leafy; leaves vaginate, terete-semi-cylindrical, subtriquetrous at the apex, very green, *full of little dots*, upper ones crowded; petals *whitish, with purple lines. Salm Dyck, l. c. fasc.* 2. *t.* 5.

HAB. Cape of Good Hope.

Stem 1 foot, branches 2–4 inches long, yellowish. Leaves bluntish, purplish-mucronulate, 1½ inches long, 2 lines wide. Peduncle 2 inches long. Calyx depressed-globose. Petals uniseriate, twice or thrice longer than the calyx. Styles 16–18.

80. M. laeve (Thunb.! Nov. Oct. Nat. Cur. v. 8. p. 16. Ap.); stem prostrate; branches short, erect, terete; leaves connate, elongate, *bluntish-triquetrous, subincurved*, rather acute, punctate, green; petals *yellow*, reddish-lineate on the outside; styles 10–14. *M. dubium, Haw. Syn.* 231. *Salm Dyck, l. c. t.* 6. *M. decipiens, Haw. Rev.* 110. *M. Thunbergii, Haw. M. crassifolium, corniculatum et spectabile. E. & Z.!* 1995. 1989. 2017.

HAB. Sandy places near Saldanhabay, Tigermountain, Sondagsriver, and in Swartland. July–Sept. (Herb. Thunb. D. Sd.)

Stem 1 foot or more, subglaucous. Leaves 1½–2½ inches long, 3–4 lines broad at the base, flat above, green or subglaucous, purplish at the sheathing base. Peduncle 1–1½ inches long, with 2 short sheathing bracts. Calyx 5-cleft, depressed-globose. Petals twice longer than the calyx. Capsule subglobose, the size of a small hazle nut, dehiscing in 14 spreading valves. It comes very near *M. veruculoides, Sond.* (168), from which it differs by subtriquetrous leaves and solitary flowers. *Zeyh.* 2586 seems to be the same, but the fruit-bearing specimens are too imperfect.

81. M. corniculatum (Linn. Spec. 676); stems spreading, angular, with distant nodes; leaves rather crowded, triquetrously semi-cylindrical, *very long*, glaucous, incurved, *blunt, mucronulate;* petals yellow, *emarginate;* stigmas purple. *Dill. Elth. f.* 253 *et* 254. *DC. pl. grass. t.* 108. *Salm Dyck, l. c. t.* 7. *M. loreum, Linn. Hort. Cliff.*

HAB. Cape of Good Hope (v. v.)

Stem 1–1½ foot, with decumbent, flexuous branches. Leaves 3–4 inches long, 3–4 lines wide, minutely punctate, sheathing at the base. Peduncle 2–4 inches long; bracts elongate, foliaceous. Calyx 5–6-cleft. Petals pluriseriate, about twice longer than the calyx. Styles 13–14.

82. M. procumbens (Haw. Rev. 111); stems flexuous, *procumbent;* leaves by pairs, recurved, corniculate, *semi-cylindrically-triquetrous*, glaucescent. *DC. l. c.* 426.

HAB. Cape of Good Hope.

Allied to the following, but the leaves are shorter and more expanded.

83. M. tricolorum (Haw. Obs. 233. Syn. 332); stem prostrate; branches distant; leaves *exactly cylindrical*, acute, green; petals acute, yellow, *blood-coloured at the base*; anthers brown. *Salm Dyck, l. c. fasc.* 1. *t.* 14. *M. stramineum, Willd. Enum.* 233.

HAB. Cape of Good Hope (v. v.)

Stem flexuous, 1 foot long. Leaves subvaginate, 2–3 inches long, 2–3 lines in diameter, minutely punctate. Peduncle 1½–2 inches long, bibracteate. Calyx depressed-globose, 5–4-cleft, twice shorter than the spreading acute petals. Styles 18–20, adpressed to the ovary, ramentaceous, yellow.

84. M. validum (Haw. Phil. Mag. Nov. 1826. 329); stem procumbent; branches flexuous; leaves elongate, *triquetrous-compressed*, semiterete at the base, bluntish and mucronulate at the apex, erect, green; *petals and anthers yellow. Salm Dyck, l. c. t.* 15.

HAB. Cape of Good Hope.

Stem 2 feet or longer. Leaves crowded, connate, 2–3 inches long, at the base 3–4 lines wide, minutely punctate. Peduncle 2–3 inches, bibracteate. Petals twice longer than the calyx. Styles 17–20, ramentaceous. Distinguished by the robust leaves.

§. 16. *PUGIONIFORMIA*, Salm Dyck, (Capitata, Haw. DC.) *Stem* erect or procumbent. *Leaves* crowded at the tops of the branches, alternate, very long, exactly or subtriquetrous, without dots. *Flowers* solitary, long, pedunculate, yellow, large. *Peduncle* without bracts. *Calyx* 5-cleft, lobes elongated. *Petals* ciliated at the base. *Stigmas* 10–20. (Sp. 85–89.)

* *Stem annual.*

85. M. elongatum (Haw. Obs. 236. Syn. 223); root tuberous; stem prostrate, herbaceous, annual; leaves very long, semi-cylindrical, channelled or semiterete. *M. pugioniforme, DC. pl grass. t.* 72. *Salm Dyck, l. c. fasc.* 2. *t.* 8.

VAR. β. minus; flowers a little smaller, petals hardly ciliated. *Bot. Reg. t.* 493.

HAB. Cape of Good Hope.

Root thick, tuberous. Stem 1 foot or longer. Leaves 4–6 inches long, 3 lines wide, dilated at the base, in var. α. channelled, in var β. subconvex near the apex. Peduncle 5–6 inches long. Calyx urceolate. Petals linear, somewhat longer than the calyx-lobes.

** *Caudex perennial.*

86. M. brevicaule (Haw. Rev. 113); caudex suffruticose, undivided, erect; leaves crowded, very long, triquetrous, *green*; calyx lobes *as long as the petals. Salm Dyck, l. c. fasc.* 3. *t.* 8.

HAB. Cape of Good Hope.

Caudex 4–5 inches long. Leaves 4–5 inches long, 2 lines wide, linear elongate, subcanaliculate. Peduncle 4–5 inches long. Petals about 1 inch long, scarcely ciliate at the base, equalling the 2 longer calyx lobes. Styles 12, filiform, erect.

87. M. capitatum (Haw. Syn. 228); caudex suffruticose, simple or branched; leaves crowded, very long, triquetrous, scarcely *canaliculate above, glaucescent*; calyx lobes *slightly longer than the petals*; styles 15, recurvate. *Salm Dyck. l. c. fasc.* 4. *t.* 7. *E. & Z.!* 1991.

HAB. Sandy places in the Cape flats, also near Tulbagh. Nov. (Herb. D. Sd.)

Stem ½–1 foot. Leaves equilaterally triquetrous, 5–7 inches long, 3 lines wide, purplish at the larger base. Calyx urceolate. Petals linear, attenuate. Perhaps a variety of the following.

88. M. pugioniforme (Linn. Spec. 699); caudex suffruticose, simple or branched; leaves crowded, very long, *excavate-triquetrous*, dilated near the middle, *glaucous;* calyx lobes *twice longer* than the petals; styles 13, erect-recurved. *Dill. Elth. t.* 210. *f.* 269. *Thunb.! fl. cap.* 424. *E. & Z.!* 1990. *Salm Dyck, l. c. fasc.* 5. *t.* 15.

HAB. Sandy places in the Cape flats, in Verlooren Valley, etc. Oct.-Nov. (Herb. Thunb. Sd.)

Caudex ½-1 foot. Leaves 7-8 inches long, in the middle 6 lines wide, trinerved, purplish at the base. Peduncle 4-5 inches long. Calyx lobes elongate, subequal. Petals in many rows.

98. M. corruscans (Haw. Suppl. 90. Rev. 113); stem suffruticose; leaves crowded at the top, alternate, dagger-shaped, long, *glittering. DC. l. c.* 426.

HAB. Cape of Good Hope.

Flowers large, yellow. Seemingly a variety of *M. pugioniforme.*

§. 17. *SARMENTOSA,* Salm Dyck, Haw. DC. *Stem* shrubby, angular, branches straight or sarmentaceous. *Leaves* opposite, connate, elongate, acute, triquetrous, with serrulated margins. *Flowers* usually ternate, small, pedunculate. *Peduncles* bracteate at the base and in the middle. *Petals* white or rose-coloured, with a red dorsal line. *Calyx* 5-cleft. *Stigmas* 5. (Sp. 90-94.)

90. M. Schollii (Salm Dyck, Obs. 1820. p. 10); stem *diffuse,* with the branches sarmentaceous-decumbent; leaves *spreading-recurvate,* dotted, elongate, attenuate-triquetrous, *serrulated* on the angles; pedicels bigeminate or ternate. *Monog. fasc.* 3. *t.* 9. *M. recurvum, Haw. Suppl.* 90. *M. aduncum, Jacq. Fragm. t.* 51. *f.* 2. *M. multiflorum, E. & Z.!* 1026.

HAB. On rocks in mountains. Table-mountain. Feb. (Herb. Sd.)

Stem greyish, when young compressed. Leaves glaucous, rigid, 2-3 inches long, 5-6 lines wide at the base, smaller in the branches, cartilagineous-serrulate. Pedicels ½-1 inch long. Calyx lobes subequal. Petals about ½ inch, in the cultivated plant not or scarcely larger than in the wild specimen.

91. M. rigidicaule (Haw. Rev. 116); stem and branches *erect,* floriferous, subdecumbent; leaves *erect-spreading,* dotted, *elongate,* triquetrous, equal-sided, with *roughish* margins; pedicels bigeminate or ternate. *Salm Dyck, l. c. t.* 10.

HAB. Langevalley, June. *Zeyher* (Herb. Sd.)

Near the preceding, but more erect, the leaves not recurvate, not evidently serrulate, and the flowers a little smaller. Leaves 2-3 inches long, about 3 lines wide, glaucous. Petals 4-5 lines long.

92. M. sarmentosum (Haw. Syn. 238); stem *diffuse;* branches *prostrate, sarmentaceous, rooting at the nodes;* leaves erectish, triquetrous, equalsided, *mucronulate,* roughish at the margins; pedicels bigeminate or ternate. *Salm Dyck. l. c t.* 11. *M. filamentosum, E. & Z.!* 2001. *M. scabrum, Thunb.! herb. I.*

VAR. β. **rigidius** (Salm Dyck, l. c.); more robust, internodes shorter; leaves slender. *M. simile, Haw. Rev.* 115.

HAB. Sandy places, Cape Flats, in Zwartland. Oct.-Nov. (Herb. Thunb. Sd.)
Stem and branches often 2 feet or more long; flowering stem erect, 3-4 inches long. Leaves fasciculate, mostly 1 inch, sometimes 1½-2 inches long, 3 lines wide, obtuse, with often recurved point, more roughish near the apex. Upper leaves and bracteas ½ inch or shorter. Pedicels ½-1 inch, incrassate at the apex. Calyx 2-2½ lines. Petals not much longer. Ripe capsule turbinate as in the preceding.

93. M. geminiflorum (Haw. Rev. 114); stem diffuse; branches slender, creeping; leaves erect-spreading, triquetrous, equalsided, *acute, recurved* at the apex, roughish at the margins; pedicels bigeminate or ternate. *Salm Dyck, l. c. fasc.* 1. *t.* 16. *M. geminatum, Jacq. Fragm. t.* 50.

HAB. Cape of Good Hope.
Very like the preceding, and only distinguished by the slender, not sarmentaceous branches, more attenuated and recurved leaves. Stems 2 feet. Leaves 2 inches long, 2 lines wide. Petals scarcely 3 lines long.

94. M. laxum (Willd. Enum. 536); stem loose, diffuse, shrubby; branches creeping, very slender; leaves connate, *compressed,* triquetrous, *more green than the others,* tubercularly dotted, usually shorter than the internodes; margins and keels finely denticulated. *Haw. Rev.* 115.

HAB. Cape of Good Hope.
Flowers reddish.

§. 18. *REPTANTIA,* Salm Dyck. (Humillima, Haw. DC.) *Stem* suffruticose, short, prostrate, nodulose, rooting at the nodes. *Leaves* subconnate, opposite, triquetrous, with smooth margins. *Flowers* solitary, pedunculate, reddish or whitish. *Calyx* 5-cleft. *Stigmas* 5. [M. clavellatum, Haw. and M. Australe, Ait. in New Holland.] (Sp. 95-96.)

95. M. crassifolium (Linn. Spec. 693); stem semiterete, creeping; leaves obtuse, triquetrous, mucronulate, dotless, very green, smooth, semi-cylindrical at the base; peduncles a little compressed, without bracts, short. *Haw. Syn.* 241. *Salm Dyck, l. c. fasc.* 1. *t.* 18.

HAB. Sandy places Uylenkraal. *Zeyher* (Herb. Sd.)
Stem prostrate, very long, 1 foot or longer, terete. Leaves fasciculate, erecto-patent, 1 inch long, 1½-2 lines wide, half cylindrical at the base, triquetrous at the apex. Peduncles 1 inch long, thickened upwards. Calyx subturbinate, twice shorter than the purplish petals. Very like *M. læve, Thunberg,* but the leaves are nearly twice smaller.

96. M. dunense (Sond.); stems prostrate, adscending; leaves triquetrous, equal-sided, acute, *dotted,* glaucous-green, smooth; peduncles with *two large flattish bracts;* calyx lobes acuminate, triquetrous, longer than the petals.

HAB. Sea shore, near Cape Town, *Ecklon, Dr. Pappe.* (Herb. D. Sd.)
Stems 3-4 inches long, purplish. Leaves erectish, about 1 inch long, 3 lines wide, wider at the base, with dispersed dots. Bracts somewhat larger than the leaves, dotted, flattish, but attenuated in a triquetrous apex, longer than the compressed peduncle. Calyx very cuneate, lobes ½ inch long, dotted, 2 or 3 at the base, with a brownish, large, membranaceous margin. Petals white, linear, twice shorter than

the calyx. Capsule glaucous, with 5 valves. The figure of *Dillanius, t.* 201. *f.* 257. more resembles this species than the foregoing.

§. 19. *ACINACIFORMIA* (Salm Dyck.) *Stem* suffruticose, robust, angular, decumbent. *Leaves* connate, triquetrous, acinaciform, thick, with smooth margins. *Flowers* solitary, large, reddish (or, as in *M. edule.* yellow). *Calyx* 5-cleft, lobes unequal. *Stigmas* 6–10. *Fruit* fleshy, [M. aequilaterale, Haw. M. Rossi, Haw. M. virescens, Haw. M. glaucescens, Haw. and M. abbreviatum, Haw. in new Holland.] (Sp. 97–98.)

97. M. acinaciforme (Linn. Spec. 695); stem angular, procumbent; leaves subglaucescent, acinaciform, compressed, *carinal angle much dilated*, cartilagineous, entire or subundulate-scabrous; peduncle with 2 large bracts; petals purple; stigmas 14. *Dill. Elth. f.* 270–271. *Andr. Rep. t.* 508. *M. laevigatum, Haw. Syn.* 233. *M. rubrocinctum, E. & Z.!* 1999. *Lindl. bot. Reg. t.* 1732. *M. subalatum, Haw. Syn.* 235.

HAB. Sandy flats, near Cape Town, and Zwartkopsriver. July–Nov. (Herb. Sd.)

Stem 2–4 feet, articulate, young branches much compressed. Leaves 2–3 inches long, 6–8 lines wide, subincurved, much compressed, the carinal angle mostly entire, or near the apex subscabrous. Flowers the largest in the genus. Peduncle compressed, 2-edged, about 2 inches long. Calyx lobes subequal. Petals in many rows, lanceolate. Fruit eatable, Hottentot-figs or T'gaukum. A small red line at the keel is the only but variable difference between M. rubrocinctum and the true acinaciforme.

98. M. edule (Linn. Spec. 698); stem angular, expanded; leaves *subequally triquetrous*, subincurved, carinal angle serrulated; peduncle without bracts; petals yellow or purple; stigmas 8. *Dill. Elth. t.* 272. *Seb. thes.* 1. *t.* 19. *f.* 6. *M. acinaciforme, E. & Z.!* 1997. *Zeyher n.* 2575. *M. edule, Thunb.! et E. & Z.!* 1998. *Pappe. Flor. cap. med. p.* 16.

HAB. Very common in the sandy tracts of the colony. July–Dec. (Herb. Thunb. D. Sd.)

Often confounded with the preceding, but very different. Leaves 3–4 inches long, 3–6 lines wide, the carinal angle scarcely dilated. Peduncle 1 inch long, shorter than the uppermost leaves. Calyx turbinate. Flower large. The eatable fruit also called Hottentot-figs, *Zuure, or Paarde Vigen.*

§. 20. *RUBRICAULIA*, Salm Dyck, Haw. DC. *Stem* suffruticose, branches erect, hardly decumbent, usually reddish. *Leaves* connate, triquetrous, rather acinaciform, serrulated at the angles. *Flowers* solitary, pedunculate, deep red (or, in *M. serratum*, yellow.) *Calyx* 5-cleft. *Stigmas* 5–8. (Sp. 99–102.)

99. M. serratum (Linn. Spec. 696); stem *erect*, branched; leaves opposite, subconnate, triquetrous, *subulate, elongate*, dotted, serrated at the angles or only at the keel; flowers *yellow*, with purple lines, calyx lobes subulate, *longer than the petals. Dill. Elth. f.* 238.

HAB. Cape of Good Hope, *Dr. Pappe.* (Herb. D.)

One foot or higher; stem reddish, terete. Leaves 3–4 inches long, flattish at the base, and about 3 lines wide, with prominent dots. Internodes 1–2 inches long. Peduncle shorter than the uppermost leaves, without bracts. Calyx turbinate, lobes subequal, subulate-triquetrous, 1 and 1½ inch long. Petals yellow, in the upper part with small purplish lines, about twice shorter than the calyx.

100. M. filamentosum (L. Spec. 694); stem subterete, branches short, decumbent; leaves triquetrous, *subacinaciform*, crowded, thick, dotted, longer than the internodes, roughly serrulated at the angles; flowers *purplish;* calyx lobes *twice shorter than the acute petals. Dillen. Elth. t.* 212. *f.* 273. *Salm Dyck, l. c. fasc.* 5. *t.* 14.

HAB. Cape of Good Hope.

Stem angular. Leaves 1½ inches long, 3 lines wide, equally triquetrous, the carinal angle in the middle subdilated. Peduncle 1 inch, with two large bracts. Calyx lobes subtriquetrous. Petals linear-lanceolate.

101. M. serrulatum (Haw. Misc. 77. Syn. 239.); stem shrubby, when young erect; branches *ascending;* leaves triquetrous, subacinaciform, rather glaucous, thick, with cartilaginous, minutely serrulated angles; flowers purplish; petals *bidentate, a little longer* than the calyx-lobes. *Salm Dyck, l. c. t.* 15.

HAB. Cape of Good Hope (v.v.)

Nearly allied to the foregoing, but the flowers are much smaller on longer peduncles, with smaller bracts, and the petals cuneate and emarginate. Leaves 1½ inches long, usually longer than the internodes. Peduncle 2½–3 inches long.

102. M. rubricaule (Haw. Syn. 239); stem and branches *erect*, mostly reddish; leaves subcompressed *triquetrous, subincurved*, glaucous-green, with cartilaginous, serrulated angles; flowers purplish; petals lanceolate, *acute*, slightly longer than the calyx-lobes. *Salm Dyck, l. c. t.* 16.

HAB. Cape of Good Hope.

By the straight reddish stem and not emarginate petals it is distinguished from *M. serrulatum.* Leaves 1–1½ inches long, not acinaciform, usually shorter than the internodes.

§. 21. *HETEROPETALA*, Salm Dyck. (Megacephala, Haw. Forficata, DC.) *Stem* frutescent, branches adscendent, 2-angled. *Leaves* opposite, triquetrous, compressed, with the carinal angle drawn out, sometimes lacerately-toothed. *Flowers* ternately disposed, lateral ones often abortive, pedunculate, reddish or whitish; *petals* biformed, subulate and linear-lanceolate. *Calyx* 5-cleft. *Stigmas* 5. (Sp. 103–107.)

103. M. lacerum (Haw. Rev. 119); stem erect, branches erectly spreading; leaves rather acinaciform, acutely triquetrous, glaucous; carinal angle lacerately-toothed; calyx compressed, with lacerate margins; petals numerous, linear-lanceolate, *longer than* the calyx. *Salm Dyck, l. c. fasc.* 4. *t.* 9. *M. carinatum, Vent. Malm. t.* 109. *M. gladiatum, Jacq.! hort. Vind. t.* 111. *M. acinaciforme, DC. Pl. Grass. t.* 89. *M. falcatum, Thunb.! Fl. Cap.* 422.

HAB. Cape of Good Hope. (Herb. Thunb. Vind.)

Stem 2–3 feet high. Leaves 1½–2 inches long, 4–5 lines wide, mucronate. Flowers often solitary, large, 2-edged. Calyx subturbinate, lobes triquetrous. Petals covering the stamens; inner ones short, the exterior longer than the calyx, rose-coloured.

104. M. heteropetalum (Haw. Syn. 294. Misc. 67); stem and branches erect-spreading; leaves glaucous, *subfalcate*, compressed-triquetrous, *carinal angle lacerate; petals unequal, shorter than the calyx. Salm Dyck. l. c. fasc.* 3. *t.* 17.

HAB. Cape of Good Hope (v.v.)

Branches subflexuous. Leaves in sterile branches crowded, in fertile ones distant, 12–14 lines long, 4 lines wide, acinaciform, mucronulate, punctate, lateral angles entire. Flowers small, solitary, rarely geminate. Calyx globose, lobes large, thick, triquetrous, erect. Petals pale red or whitish, interior ones very short.

105. M. mutabile (Haw. Obs. 377. Syn. 294); stem and branches subtortuous, erect; leaves nearly distinct, crowded, glaucous-green, compressed-triquetrous, acute, incurved, *carinal angle entire*, cartilaginous; petals *linear-subulate*, *a little longer than* the calyx. *Salm Dyck, l. c. fasc.* 4. *t.* 10. *M. tricolor*, *fascq. hort. Schoenb. t.* 440. *M. glaucinum, Haw. Rev.* 132. *M. forficatum, Jacq. hort.-Vind.* 1. *t.* 26. *M. filamentosum, DC. Pl. Grass. t.* 60.

HAB. In district of Uitenhage, *Ecklon!* Langevalley, Rhinosterkop, dist. Beaufort, et Gamkariver, April–June, *Zey.* 682, 689, (Herb. Sd.)

Branches straw-coloured or reddish, old ones terete. Leaves 6–8 lines long, 2–3 lines wide, mucronate, and the angles often cartilaginous undulate, but always entire. Flowers mostly solitary, peduncle thickened upwards. Calyx subglobose, about 4 lines long. Petals rose-coloured, interior ones much shorter, pale yellowish.

106. M. inclaudens (Haw. Rev. 133); stem and branches tortuous, spreading; leaves subconnate, crowded, green, compressed-triquetrous, acute, *acinaciform, carinal angle much dilated*, entire or subscabrous; petals *subspathulate, longer than the calyx. Andr. Repos. t.* 384. *M. mutabile and inclaudens, E & Z.!* 2004 *et* 2005. *Salm Dyck, l. c. fasc.* 3. *t.* 18.

HAB. Rocks in Hottentotshollandskloof, Oct. Cape flats, *Zeyher!* 2918. (Hb. Sd.)

Leaves 7–10 lines long, at the dilated apex 3–4 lines wide, with large dots. Peduncles 1–2, about 1½–2 inches long, rarely without bracts. Calyx turbinate, lobes triquetrous, about twice longer than the calyx-lobes; the interior very short and and small, purplish. Distinguished from the preceding by the subdeltoid acinaciform leaves and the broad petals.

107. M. Dregeanum (Sond.); stem and branches erect; leaves subconnate, crowded, glaucous-green, compressed-triquetrous, mucronate, acinaciform, carinal angle much dilated, entire or scarcely denticulate, dotted; petals (white) *linear-subulate, three times longer than the calyx.* M. *strictum et cymbifolium, E. & Z.!* 2011, 2012.

HAB. Sandy and stony places at Tulbagh and Vogelvalley, Worcester, *E. & Z.! Drege!* Dec. (Herb. Sd.)

Stem 2 feet or higher, purplish or red, subpruinose, terete-angulate; branches compressed, ultimate 2-edged, reddish. Leaves smaller than in any other species of this section, 3–4 lines long, 1½–2 lines broad, with large dots, much compressed. Flowers terete or solitary; the upper lateral, short branchlets often terminated by a flower. Terminal peduncles ¾–1 inch long, inferior ones (branchlets) 1½–2½ inches. Calyx subturbinate, dotted, 2 of the lobes triquetrous, green, the rest larger, obtuse, subcoloured. Petals when dry pale yellowish, 8–9 lines long, acute. Capsule 4 lines long, 5-valved.

§. 22. *BRACTEATA*, Salm Dyck. (Haw.) DC. *Stem* suffrutescent, branches erect, much compressed. *Leaves* distinct, compressed-triquetrous, hooked at the apex, subrecurved, more or less scabrous from dots. *Flowers* pedunculate, girded by 2–4 broadly ovate, keeled bracts, which generally clasp the calyx, solitary, reddish, always expanded; inner *petals* thread-like. *Calyx* 5-cleft. *Stigmas* 5. (Sp. 108–114.)

108. M. gracile (Haw. Rev. 144); stem and branches very slender, straight; leaves green, *triquetrous, with equal sides* mucronate, recurved at the apex; peduncle with 2 bracts in the middle and 2 at the top; petals narrow-linear, very spreading. *Salm Dyck, l. c. fasc.* 4. *t.* 11. *M. stellatum, Haw. Misc.* 91. *M. ternifolium, I. herb. Thunberg!*

HAB. Sandy places near Capetown. (Herb. Thunb. D. Sd.)

Stem 2 feet, branches smooth. Leaves 8–11 lines long, 1 line wide, pellucid-dotted. Peduncle 1½–2 inches long; bracts ovate, acute. Calyx obconical. Inner petals very short, yellow; outer biseriate, reddish. The slender habit and straight branches distinguish this species from the others of the section.

109. M. anceps (Haw. Syn. 289. Rev. 143); branches erecto-patent; leaves somewhat spreading, green, *acinaciformly triquetrous*, sides rather membranous below, with large, elevated dots, mucronate, recurved at the apex; peduncle *at the top* with 4 bracts, petals narrow, straight. *Salm Dyck, l. c. t.* 12. *M. lacerum, E. & Z. !* 2000. *M. bracteatum, E. & Z. !* 2006, *ex parte. Herb. Un. atino,* 513. *ex pte.*

HAB. Stony places near Tablemountain, and Swellendam. (Herb. Sd.)

Stem 2–3 feet. Leaves 10–12 lines long, 1 line wide, pellucid-dotted. Bracts as in the preceding, but near or clasping the calyx. Petals a little longer as in *M. gracile.*

110. M. asperum (Haw. Rev. 145); stem and branches erect; leaves spreading, longish, glaucous-green, triquetrous, *subequalsided*, full of pellucid dots, *very scabrous*, hooked at the apex; peduncle with 2 bracts *in the middle;* petals lanceolate, patent-recurved. *Salm Dyck, l. c. t.* 13.

HAB. Cape of Good Hope.

Distinguished from the foregoing by longer, very scabrous leaves and bractless calyx. Stem 2 feet or more. Leaves 12–18 lines long, 1 line wide. Petals longer and wider at the top. It is also closely allied to *M. scabrum*, but differs by the large bracts and not punctate-scabrous calyx.

111. M. compressum (Haw. Obs. 326. Syn. 289); stem and branches erect, spreading; leaves glaucescent, triquetrous, *with equal sides, somewhat scabrous* from dots, recurved and mucronate at the apex; peduncle with 2 bracts *at the top*; petals lanceolate, erect-recurved. *Salm Dyck, l. c. t.* 14.

HAB. Cape of Good Hope.

Distinct from *M. asperum* by the more slender branches, shorter, less scabrous leaves, and bracts at the top of the peduncle. Except the glaucous colour, I cannot find any difference from *M. bracteatum*. Leaves 10–12 lines long, 1 line wide. Peduncle 1 inch long. Calyx obconical. Petals as in the preceding, reddish.

112. M. bracteatum (Ait. Hort. Kew. 2 p. 185); stem and branches erect; branches of a reddish brown colour; leaves compressed, triquetrous, *with equal sides*, green, dotted, *nearly smooth*, recurved and mucronate at the apex; peduncle with 2 or 4, broadly ovate, keeled bracts *at the top;* petals lanceolate, spreading. *Haw. Syn.* 289. *Lodd. Bot. Cab.* 251. *Herb. Un. itin.* 513. *ex pte. M. bracteatum, E. & Z. !* 2006. *ex pte. M. gracile, E. & Z. !* 2010. *ex pte.*

HAB. Stony places on the north side of Tablemountain, and in Hottentotholland, March–May. (Herb. D., Sd.)

Stem 1–2 feet, much branched. Leaves 8–12 lines long, 1 line wide. Bracts hooked at the apex, large, equal, much dotted, membranous at the margins, clasping the punctate calyx. Petals the size of those of *M. compressum*, reddish, with many purplish lines.

113. M. patulum (Haw. Syn. 334); branches numerous, *diffuse, patent-reflexed;* leaves suberect, glaucous-green, linear, triquetrous, equal-sided, roughish with dots, recurved and mucronate at the apex; peduncle with 2 bracts *on the middle* and 2 *at the top*; petals lanceolate, attenuate at the base, erect, recurved. *Salm Dyck, l. c. t.* 15. *E. & Z.!* 2009. *M. gracile, E. & Z.!* 2010. *ex pte. M. incurvum, E. Mey. in Heb. Drege.*

HAB. Stony places in distr. Stellenbosch and near Tablebay, Mar. (Herb. Sd.)

By the spreading or subrecurved, slender branches, smaller leaves, and by the bracts, from which the pair in the middle of the peduncle is smaller than those at the top, it is easily known from *M. gracile* and *bracteatum*, Leaves 6–12 lines long, ¾–1 line wide. Flower somewhat smaller than in M. bracteatum. Petals ½ inch long, reddish, with darker lines. *M. patulum, E. & Z.!* 2008, with greyish, dotted-scabrous branches, many dotted subscabrous, short leaves, and very short peduncles with many approximate bracts, seems to be a distinct species, but the specimens are insufficient.

114. M. radiatum (Haw. Obs. 232. Syn. 289); stem erect, branches *erectish*, canescent, smooth; leaves *very glaucous*, triquetrous, equal-sided, attenuated and hooked at the apex, prominently dotted or wrinkled; peduncle with 2 bracts above the middle, and 2 at the top; petals spreading, lanceolate, attenuated at the base. *E. & Z.!* 2007. *Herb. Un. itin.* 520. *M. scabrum, fol.* 2. *herb. Thunberg.*

HAB. Stony places near Lionsmountain and Greenpoint, Jan.–Feb. (Herb. Thunb. Sd.)

By the grey-bluish colour of the whole plant, it differs from the other species of this section; from *M. patulum* especially it is distinguished by the equal not heterogeneous bracts. Stem ½–1½ feet high, branches often crowded, 2-edged. Leaves 6–8 lines long, 1 line wide. Bracts ovate with recurved apex, somewhat membranous at the base. Petals as in *M. bracteatum.* The figure 249 of Dillenius, cited by Haworth, is a very bad one.

§ 23. *VIRGATA,* (Haw.) *Stem* suffrutescent, branches erect, virgate, 2-edged. *Leaves* subconnate, distant, compressed-triquetrous, erect-recurved, punctate. *Flowers* pedunculate, peduncle with 2 bracts, solitary, reddish, small. *Calyx* 5-cleft. *Stigmas* 5. (Sp. 115–117.)

115. M. virgatum (Haw. Syn. 290); stem weak; branches twiggy; leaves glaucescent, *triquetrous*, compressed, recurved at the mucronate apex; flowers solitary; peduncle with 2 bracts. *Salm Dyck, l. c. fasc.* 2. *t.* 9. *M. compressum, Haw. Obs. App.* 416.

HAB. Cape of Good Hope.

Stem 2–3 feet; branches rigid, straight. Leaves 6–9 lines long, 1 line wide; carinal angle scarcely dilated. Peduncle 1 inch long, with 2 connate, cymbiform, 3–4 lines long, bracts in the middle. Flowers small. Calyx obconical. Petals subuniseriate, about 4 lines long.

116. M. congestum (Salm Dyck, l. c. fasc. 6. t. 8); stem and branches straight; leaves glaucous, *subacinaciformly-triquetrous*, much compressed, uncinate at the apex; flower solitary; peduncle with 2 leafy bracts. *M. heteropetalum, E. & Z.!* 2003. *non Haw.*

VAR. β. Stem and branches diffuse, flexuous. *M. glaucinum, E. & Z.!* 2002. *non Haw.*

HAB. Near Mt. Bothasberg, Vishriver, Albany; var. β. near Zwartkopsriver; stony places near Heerelogemont, Kammapur, *Zey.!* 2926. Bethelsdorp, *Zey.!* 2595; sea shores near Cape Receif, *Zey.!* 2588. (Herb. Sd.)

Very near the preceding, but the leaves are larger, (7–12 lines long, 2–3 lines wide) the carinal angle more dilated, the flowers a little larger and the bracts nearly as large as the leaves. Stem 2–3 feet. Leaves dotted or wrinkled. Peduncle 2 inches long, thickened above. Capsule when ripe glaucous, 5-valved. One of *Zeyher's* specimens has both solitary and ternate flowers.

117. **M. cymbifolium** (Haw. in Till. Phil. Mag. 1824, vol. 64, p. 424); stem shrubby, erectish; branches few, 2-edged, hoary; leaves trigonal, *boat-shaped, obtuse,* pale green, with large dots. *DC. l. c.* 437.

HAB. Cape of Good Hope.
Flowers unknown.

§. 24. *VIRENTIA*, Salm Dyck. *Stem* suffrutescent, branches erect, rigid. *Leaves* subconnate, subacinaciform, thick, green, with smooth angles. *Flowers* ternate, by abortion geminate or solitary, large, pedunculate; *peduncle* with 2 thick, keeled bracts; *petals* reddish. *Calyx* 5–6-cleft. *Stigmas* 5–6. (Sp. 118.)

118. **M. virens** (Haw. Rev. 121); leaves distant, compressed-triquetrous, bluntish, mucronulate; flowers ternate, hexagynous; calyx 6-cleft. *Salm Dyck, l. c. fasc.* 3. *t.* 19.

HAB. Sandy places between Bethelsdorp and Cradockstadt, *Zeyher!* 2587. Jan. (Herb. D., Sd.)

Stem weak, 1–2 feet, with spreading branches and compressed branchlets. Leaves about 1 inch long, 3 lines wide, subacinaciform, dotted. Peduncles ½–1 inch long, the middle mostly ebracteate. Calyx subturbinate. Petals twice longer than the the calyx. Capsule 6-locular.

§ 25. *AUREA*, Haw. DC. *Stem* suffruticose, branches erect. *Leaves* distinct, spreading, elongated, bluntish, triquetrous, glaucous. *Flowers* solitary, large, yellow or copper-coloured, long-pedunculate; peduncle without bracts. *Calyx* 5-cleft. *Stigmas* 5. (Sp. 119–121.)

119. **M. glaucum** (Linn. Spec. 696); stem and branches erect, subcompressed; leaves subconnate, *triquetrous,* much compressed, glaucous, *roughish* from dots, carinal angle cartilaginous-serrulate; flowers *sulphur-yellow. Dill. Elth. t.* 196. *f.* 248. *DC. Pl. Grass. t.* 146. *Salm Dyck, fasc.* 3. *t.* 20.

VAR. β. **tortuosum** (Salm Dyck, l. c.); leaves smaller, branches more slender, tortuous.

HAB. Cape of Good Hope (v.v.)

Stem 2 feet or more. Leaves 9–14 lines long, about 3 lines wide, blunt, mucronate. Petals subbiseriate, 1 inch long. Stigmas ramentaceous.

120. **M. aurantiacum** (Haw. Misc. 84. Syn. 264); branches fastigiate, subcompressed; leaves subconnate, *bluntly triquetrous, smooth,* glaucous; flowers *orange-coloured. E. & Z.!* 2014. *Salm Dyck, fasc.* 1. *t.* 21. *M. verruculatum, Thunb. herb. ex parte. M. glaucum, Thunb. herb. ex pte. M. aurantium, Willd. M. glaucoides, Haw.*

HAB. Sandy places in Cape Flats, Doornhoogte, Rietvalley, Vygekraal, Aug.–Oct. (Herb. Thunb., D., Sd.)

Stem ½–1½ feet, erect or decumbent at base, much branched. Leaves 6–12 lines long, ½–2 lines wide, acutish, prominently dotted. Peduncle thickened above. Calyx turbinate. Petals subtriseriate, about 8 lines long.

121. M. aureum (Linn. Syst. nat. ed. 10. p. 1050); branches erect, subcompressed; leaves subconnate, *cylindrically triquetrous*, smooth, glaucous, bluntish, mucronate; flowers large, *golden-coloured*. *Bot. Mag. t.* 262. *DC. Pl. Grass. t.* 11. *E. & Z.!* 2015. *Salm Dyck. l. c. t.* 22.

HAB. Sandy places near Saldanha bay, Aug.–Sep. (Herb. D. Sd.)

More robust than the preceding, leaves æquilaterally triquetrous, with convex sides, 1½–2 inches long, 3 lines wide. Flowers 2 inches in diameter. Petals in many series. Capsule obconical, 5-valved. It varies, but very rarely, with ternate or geminate peduncles.

§. 26. *BLANDA*, (Haw. Conferta, DC.) *Stem* fruticose, branches erect, rigid. *Leaves* connate, subtriquetrous, elongate, acute, very smooth. *Flowers* ternate, by abortion geminate or solitary, large, whitish or pale rose-coloured, pedunculate; lateral *peduncles* bracteate, the intermediate bractless. *Calyx* 5-cleft. *Stigmas* 5. (Sp. 122–124.)

122. M. blandum (Haw. Suppl. 95. Rev. 147); branches ascending; leaves compressed, triquetrous, *with equal sides*, elongated, narrow, acutish, smooth; peduncles *subequal;* petals *spreading*, *straight*, *pale rose-coloured*, bidentate. *Bot. Reg. t.* 582. *Bot. Cab. t.* 599. *Salm Dyck, fasc.* 4. *t.* 16.

HAB. Cape of Good Hope.

Stem 2 feet. Branches numerous. Leaves distant, 1½–2 inches long, 1½ lines wide, minutely dotted. Peduncle 2 inches long, subcompressed, scarcely thickened above. Petals twice longer than the calyx-lobes.

123. M. curviflorum (Haw. Rev. 147); branches erect, straight; leaves triquetrous-compressed, the *carinal angle a little dilated* below the apex, elongated, acute, smooth; peduncles *clavate;* petals *incurvate*, *white*, bluntish. *Salm Dyck*, 2. *t.* 10.

HAB. Cape of Good Hope.

More robust, leaves thicker (2–3 inches long, 2 lines wide), less crowded. Peduncles much thickened at the apex, and incurved. White petals distinguish this species from the foregoing. Stem 2–3 feet high.

124. M. turbinatum (Jacq. Hort. Vind. t. 476); stem branched, diffuse; leaves glaucous, elongated, acute, triquetrous, crowded; flowers on long peduncles, reddish; ovarium *contracted into a neck beneath the calyx*. *DC. l. c.* 436.

HAB. Cape of Good Hope.

Petals numerous, linear, much spreading.

§. 27. *AMOENA*, Salm Dyck. (Eximia. Haw. Conferta, DC.) *Stem* suffruticose, branches erectish or adscendent. *Leaves* crowded, subconnate, triquetrous, gradually attenuated, elongated, acute. *Flowers* ternate, by abortion geminate or solitary, large, showy, reddish, pedunculate; *peduncles* bracteate. *Calyx* 5-cleft. *Stigmas* 5–6. (Sp. 125–128.)

125. M. conspicuum (Haw. Syn. 240); branches tortuous, adscending; leaves crowded, green, incurved, erect, *triquetrous, attenuate, acute;* flowers and filaments purplish. *Salm Dyck, fasc.* 2 *t.* 11.

HAB. Cape of Good Hope (v.v.)

Stem 1–1½ feet, rigid. Leaves 2–2½ inches long, 2 lines wide. Floriferous branches erect, with distant leaves. Peduncles about 3 inches long. Calyx turbinate. Petals subtriseriate, 9 lines long, beautifully red. Styles 5, thick, acutish, suberect.

126. M. amoenum (Salm Dyck, in DC. Prod. 3. 436); branches suberect; leaves crowded, *green,* incurved, erect, *cylindrically-triquetrous, bluntish, mucronulate;* flowers purplish, filaments white. *Monog.* 2. *t.* 12.

HAB. Near Grahamstown, Nov. (Herb. D., Sd.)

Nearly allied to the preceding, and only distinguished by the shorter stem and horter (1–1½ inches long, 2 lines wide) cylindrically-trigonous, nearly subclavate, bluntish leaves. Peduncles 8–12 inches, petals 6–7 lines long. Styles 5, patent, acute.

127. M. spectabile (Haw. Obs. 385. Syn. 240); branches ascending; leaves crowded, glaucous, incurved, patent, *triquetrous, attenuate,* mucronate; flowers purplish, filaments white. *Bot. Mag. t.* 396. *DC. Pl. Grass. t.* 153. *Salm Dyck, l. c. t.* 13.

HAB. Cape of Good Hope (v.v.)

Stem prostrate, floriferous, elongate. Leaves keeled, 2–3 inches long. 3 lines wide. Peduncles bracteate in the middle and above, 3–6 inches long. Calyx turbinate. Petals spreading, 1 inch long, the inner shorter. Styles 5, erect, deltoid or obovate at the apex.

128. M. formosum (Haw. Rev. 145); sterile branches very short, floriferous ones elongated; leaves crowded, subdistinct, incurved-patent, *triquetrous,* elongated, *bluntish,* mucronulate, *green;* petals purplish; filaments white. *Salm Dyck, fasc.* 3. *t.* 21.

HAB. Cape of Good Hope.

Differs from *M. amoenum* by the prostrate flowering branches, carinate-triquetrous, thicker leaves, and short peduncles; from *M. spectabile* by the green, not glaucous, leaves, ternate, short peduncled, much smaller flowers. Leaves 2 inches long, 3 lines wide. Peduncles rigid, 1–1½ inch long. Petals bidentate. Styles 5, erect-spreading, thick, acute, ramentaceous.

§. 28. *DILATATA,* Haw. *Stem* fruticose, branches erect-spreading. *Leaves* crowded, glaucous, much dotted, compressed-triquetrous, attenuate at the base, much dilated and recurved above the middle. *Flowers* small, pale, rose-coloured, mostly solitary. *Calyx* 5-cleft. *Stigmas* 5. (Sp. 129.)

129. M. dilatatum (Haw. Syn. 303); branches subflexuous, rigid; leaves spreading, acute-recurvate, attenuated at the base, dilated above the middle, triquetrous-compressed, angles obtuse, pellucid-punctate; flowers solitary. *Salm Dyck, fasc.* 6. *t.* 9.

HAB. Cape of Good Hope.

Stem woody, 1½ inch; branches yellowish. Leaves 1–1½ inches long, 3 lines wide, obtuse. Peduncle tender, 1 inch long, with 2 small, foliaceous bracts. Calyx obconical. Petals uniseriate, spreading, subrecurved, lanceolate, twice longer than the calyx-lobes. Styles 5, thick, acute.

§. 29. *FALCATA*, DC. (Lunata and Pallidiflora, Haw.) *Stem* suffruticose as well as the branches suberect, flexuous or divaricate. *Leaves* crowded, glaucous, triquetrous or subtriquetrous-compressed, falcate, with obtuse, smooth angles. *Flowers* ternate, or in a 5-flowered cyme, rose-coloured, long pedunculate, peduncle furnished with 2–4 bracts. *Calyx* 5-fid. *Stigmas* 5. (Sp. 130–134.)

130. M. falciforme (Haw. Syn. 299); branches spreading; leaves much crowded, triquetrous, falcate, *angles acute*, the carinal *acinaciformly dilated*, with numerous, large, prominent, dots; flowers ternate or solitary, *showy*. *DC. l. c.* 433. *Salm Dyck, fasc.* 1. *t.* 23.

HAB. Cape of Good Hope (v.v.)

Stem 1–2 feet, branches angular. Leaves 6–9 lines long, 2 lines wide, mucronate. Peduncle thickened upwards, 2–3 inches long; bracts small, leafy. Flowers an inch and a-half in diameter. Calyx turbinate. Petals in many series, lanceolate. Filaments white Styles 5, short, acute, ramentaceous.

131. M. falcatum (Linn. Spec. 694); much branched, branches filiform; leaves *minute*, crowded, subtriquetrous-compressed, *subfalcate, with obtuse angles*, mucronulate, attenuated on both ends, *pellucid-punctate*; flowers ternate, *small*. *Dill. Elth. t.* 213. *f.* 275, 276. *Salm Dyck, fasc.* 3. *t* 22.

HAB. Cape of Good Hope.

Leaves 2–3 lines long, 1 line wide; dots scattered, large. Peduncles very slender, 1½–2 inches long. Flowers 6–8 lines diameter. Petals biseriate, bidentate. Styles 5, subulate.

132. M. lunatum (Willd. Enum. 538); branches suberect, flexuous; leaves crowded, subtriquetrous-compressed, incurvedly *half moon shaped*; *angles obtuse*, the carinal dilated, very glaucous, dotless; flowers ternate, or bigeminate, small. *DC. l. c.* 433. *Salm Dyck, fasc.* 1. *t.* 24. *M. falcatum, lunatum et falciforme, E. & Z.!* 2019–2021.

HAB. Stony places on mountain sides near Brackfontein and Vierentwintig Rivieren, *Clanwilliam*. June–Sept. (Herb. Sd.)

Leaves thick, 5–6 lines long, 2 lines wide, obtuse, mucronulate. Flowers ternate, bigeminate or cymose, 5-nate. Peduncle 1 inch, pedicels ½–¾ inch long. Petals about ½ inch long, pale rose-coloured, acute, when dry whitish. Nearly intermediate between *M. falciforme et falcatum.*

133. M. maximum (Haw. Obs. 402. Syn. 292); stem woody, erect, bushy; leaves crowded, large, very much compressed, triquetrous, incurvedly half-moon shaped, very *glaucous, obtuse*, half-stem-clasping, full of pellucid dots; flowers *small*. *DC. l. c.* 433.

HAB. Cape of Good Hope.

Peduncles with 2 bracts. Calyx 5-cleft. Petals reddish.

134. M. roseum (Willd. Enum. 535); branches spreading, leafy; leaves attenuate on both ends; *incurved*, glaucous, compressed-triquetrous, *the carinal angle dilated above the middle*, mucronulate, punctate; flowers ternate or geminate, *showy*. *Salm Dyck, fasc.* 5, *t.* 18. *M. multiradiatum, Jacq. Fragm. t.* 53. *f.* 1. *M. incurvum, var. roseum, DC.*

VAR. β. **confertum** (Salm Dyck, l. c.); branches subtortuous and leaves more crowded. *M. incurvum et decumbens, Haw. Syn.* 300.

HAB. Cape of Good Hope (v.v.)

Stem 1½–2 feet high. Leaves 12–14 lines long, 2 lines wide, subdistinct. Peduncle about 2 inches long, thickened above; bracts small, leafy. Calyx turbinate. Petals pale rose-coloured, biseriate, emarginate, about 8–9 lines long. Styles 5, short, acute. In var. β. the branches and leaves are more crowded and the flowers of a deeper red.

§. 30. *DELTOIDEA*, Salm Dyck, DC. (Muricata, Haw.) *Stem* suffruticose, branches erect, spreading. *Leaves* subconnate, crowded, glaucous, deltoid-triquetrous, attenuate at the base, dilated, retuse at the apex, with the angles muricately toothed, flowers ternate, rose-coloured, small, sweet-scented, pedunculate, peduncles furnished with 2–4 leafy bracts. *Calyx* 5-cleft. *Stigmas* 5. (Sp. 135–137.)

135. M. caulescens (Mill. dict ed. 8, p. 12); leaves incurvate-erect, glaucous, rather long, triquetrously deltoid, acutish, with the sides obtuse, toothed and *the keel entire;* petals obtuse, emarginate. *Dill. Elth. t.* 195. *f.* 243, 244. *Salm Dyck, fasc.* 3. *t.* 23. *M. deltoides* β, *simplex. DC. Pl. Grass. n.* 53.

HAB. Cape of Good Hope (v.v.)

Shrub 1½ feet. Leaves 6–9 lines long, at the base 2 lines, below the apex 4 lines wide, without dots, on the lateral angles with 2 or 3 short teeth, often red-marginate. Flowers 6–8 lines in diameter, ternate or often solitary. Pedicels about 4 lines long. Petals subspathulate, obtuse or erose. Easily distinguished from the two following by the larger leaves with entire keel.

136. M. deltoides (Mill. l. c. p. 13); leaves incurvate-erect, glaucous, deltoid, *trifariously toothed; peduncles elongate;* petals acute. *Dill. Elth. t.* 195. *f.* 245. *DC. Pl. Grass. t.* 53. *Salm Dyck, l. c. t.* 24.

HAB. Witsenberg, Decemb. *Zey. !* 694. (Herb. D., Sd.)

Stem suberect, branches reddish brown. Leaves 5–6 lines long, below the apex 3–4 lines wide, not dotted; teeth on the 3 angles acute. Flowers ternate or cymose-tri-ternate, Peduncle 1–1½ inches long, pedicels shorter. Bracts often entire. Flowers ½ inch in diameter.

137. M. muricatum (Haw. Obs. 364. Syn. 297); leaves incurvate-erect, very glaucous, deltoid, *trifariously muricate-dentate; peduncles short;* petals acute. *Dill. Elth. t.* 195. *f.* 246. *Salm Dyck, l. c. t.* 25. *M. deltoides, Linn. Ait. E. & Z. !* 2022.

HAB. Mountain sides near Tulbagh, Worcester. Sept.–Nov. (Herb. Sd.)

Scarcely distinguished from the preceding. Whole plant bluish or greyish-blue. Leaves 3–5 lines long, below the apex 3 lines wide, the angles with several short, mucronulate teeth. Flowers ternate or tri-ternate. Peduncle 3–4 lines long. Flowers as in the preceding; in the wild specimens a little smaller.

§. 31. *FORFICATA*, Haw. DC. *Stem* suffruticose, with the branches angulate, lax. *Leaves* triquetrous-compressed, long connate, decurrent, obtuse, carinal angle toothed at the apex. *Flowers* solitary, terminal, long pedunculate, reddish. *Calyx* 5-cleft. *Stigmas* 5. (Sp. 138.)

138. M. forficatum (Linn. Spec. 695); branches decumbent, 2-edged; leaves erect, much triquetrous-compressed, green, without dots, at the rounded apex denticulate. *Haw. Syn.* 280. *Salm Dyck, fasc.* 1. *t.* 25. *M. filamentosum* β. *anceps. DC. Pl. Grass. p.* 60.

HAB. Cape of Good Hope.

About 1 foot high; branches flexuous. Leaves rather distant, 1 inch long, 4–5 lines wide; carinal angle much compressed. Flowers showy, about 1½ inches in diameter. Peduncle 1–1½ inches long. Calyx subglobose. Petals subuniseriate, red, with a darker line from the base to the middle.

Group III. PERFOLIATA.—Stem fruticose or suffruticose, mostly erect. Leaves vaginate-connate, more or less triquetrous. (§§. 32–38.)

§. 32. *GEMINATA*, Haw. *Stem* suffruticose, dwarf, as well as the branches procumbent, dichotomous. *Leaves* connate a long way, turgid-triquetrous, whitish, smooth, with cartilaginous, entire margins. *Flowers* unknown. (Sp. 139–140.)

139. M. geminatum (Haw. Misc. 92. Syn. 280); branchlets dichotomous, ascending; leaves triquetrous, erect, glaucous, smooth, cartilaginous at the margins.

HAB. Cape of Good Hope.
Flowers unknown, probably white.

140. M. marginatum (Haw. Obs. 412. Syn. 294); branches erect; leaves triquetrous, *rather acinaciform*, glaucous, with whitened margins.

HAB. Cape of Good Hope.
Stem 4 inches high. Leaves small.

§. 33. *UNCINATA*, Salm Dyck, DC. (Uncinata et Lineolata, Haw.) *Stem* fruticose, erect, as well as the branches rigid. *Leaves* sheathing, sheaths fleshy, covering the internodes, limb of the leaves abbreviate, solid, uncinate; or elongate, compressed, the carinal angle toothed or serrulate. *Flowers* at the tops of the branches, solitary, short peduncled, rather small, reddish. *Calyx* 5-cleft. *Stigmas* 5–8. (Sp. 141–147.)

* Leaves retuse, not compressed.

141. M. perfoliatum (Haw. Misc. 92. Syn. 281); erect, with few, straight branches; leaves sheathing at the base, rather decurrent, dotted, *whitish*, abbreviate, *triquetrous*, mucronate, hard; keel 1–2-*toothed* beneath near the apex. *Bradl. Succ. Dec.* 3. *f.* 26. *dextra. Salm Dyck, fasc.* 2. *t.* 14. *M. perfoliatum, β. monacanthum, DC. Prod.* 3. 430.

HAB. Cape of Good Hope.
Stem 2 feet or higher; branches simple. Leaves distant, erect-spreading, obtuse-triquetrous, mucronate, with nearly convex sides, 8–9 lines long, 3 lines wide at the base; carinal angle with 1, rarely 2 short teeth. Peduncle 3–4 lines long. Flower about ¾ inch in diameter. Styles 5–6, subulate, erect.

142. M. viride (Haw. Syn. 283); erect, with straight branches; leaves sheathing at the base, rather decurrent, *green, subtriquetrous-cylindraceous*, elongate, incurvate-erect, uncinately recurved at the apex, *quite entire*. *Salm Dyck. l. c. t.* 15.

HAB. Cape of Good Hope.
Leaves 8–9 lines long, 2 lines wide at the base, shorter than the sheaths. Peduncle 3–4 lines long. Flowers middle-sized, pale red. Styles 7–8, very short, erect, subulate. It differs from the preceding by the green colour, slender branches, and smaller, entire leaves.

143. M. uncinatum (Will. Dict. ed. 8. n. 18); stem tortuose, diffuse; leaves sheathing at the base, rather decurrent, glaucous-green, punctate, *short and equal-sided, triquetrous,* mucronulate, often furnished with 1 or 2 spines underneath at the apex. *Burm. Afr. t.* 26. *f.* 3. *DC. Pl. Grass. t.* 54. *Salm Dyck, fasc.* 6, *t.* 10. *M. edentulum, Haw. Rev.* 125.

HAB. Karro, between Olifantsriver and Bocklandsberg, *Thunberg!* Springbokkeël, Mar. *Zey.!* 703 and 2956. (Herb. Thunb. D., Sd.)

Stem erect, diffuse, much branched, branches sub-compressed, greenish, floriferous short. Leaves spreading, 3 lines long, 1½–2 lines wide, in the wild specimens often as broad as long, fleshy, rugose when dry, three times shorter than the internodes, at the keel near the apex with or without a short tooth. Flowers on very short, thick peduncles, half an inch in diameter, rose-coloured. Styles 5.

144. M. uncinellum (Haw. Rev. 125); stem erect, diffuse; leaves sheathing at the base, rather decurrent, glaucous, punctate, short, *triquetrous subrecurved,* mostly *trifariously denticulate* at the apex. *Salm Dyck, fasc.* 5, *t.* 19. *M. uncinatum var. minor. Salm Dyck, Catal. Dill. Elth. t.* 193. *fasc.* 239.

HAB. Karro in Zoutpanshoogde near Zwartkopsriver, Oct. *Zey.!* 2598. Gamkariver, *Zey.!* 685 et 688. Port Natal, *Miss Owen.* (Herb. D., Sd.)

Distinguished from *M. uncinatum* by a little longer, tridenticulate, recurved leaves, internodes not much shorter than the leaves. Leaves 4 lines long, 1 line wide, crowded. Flowers as in *M. uncinatum.* It varies with subentire or unidentate leaves.

** Leaves elongate, compressed.

145. M. semidentatum (Salm Dyck, Obs. p. 9); branches *few, erect,* simple, straight; leaves *distant,* sheathing, erecto-spreading, equalsided-triquetrous, compressed, whitish, dotted, bluntish, mucronulate, keel furnished *with* 2–4 *teeth* near the top. *Monog. fasc.* 1. *t.* 26.

HAB. Gamkariver, *Zeyher!* (Herb. Sd.)

Stem 2–3 feet. Sheaths compressed, long. Leaves 12–15 lines long, 2–3 lines wide; the carinal angle with 2 or 4 recurved teeth. Peduncle ½ inch long, compressed, thickened upwards. Flowers middle sized. Petals very narrow. Styles 5, subulate.

146. M. unidens (Haw. Phil. Mag. 1826, p. 331); branches *numerous,* rigid, *spreading;* leaves *crowded,* sheathing, erecto-recurved, compressed-triquetrous, elongate, whitish, dotted, attenuated-mucronate, keel generally with *one tooth* near the top. *Salm Dyck, l. c. fasc.* 6. *t.* 11. *M. rigidicaule, E. & Z.!* 1992.

HAB. Stony places on Mt. Bothasberg near Vischrivier, Albany, June. (Hb. Sd.)

Flowering branches short. Leaves 9–12 lines long, 2 lines wide, twice longer than the internodes. Peduncle 6–9 lines long. Flowers rose-coloured, 6 lines in diameter. Styles 5, linear, spreading.

147. M. lineolatum (Haw. Rev. 130); stem *depressed;* branches numerous, spreading; leaves much crowded, sheathing, triquetrous, rigid, mucronate; one of a pair *incurved,* with dilated, roughish carinal angle, the other *recurved,* subuncinate; sheaths with a short impressed line. *Salm Dyck, fasc.* 2. *t.* 16.

VAR. β. **minus** (Haw.); leaves shorter and glaucescent.

VAR. γ. **nitens** (Haw.); leaves shining green.

HAB. Hills near Zwarteberg & Babylons Toorensberg, Caledon. July. (Herb. Sd.)
Leaves 6-10 lines long, 2 lines wide, triquetrous, angles not dentate. Flowers as in the preceding. Peduncle 6 lines long. Petals purplish with a dark dorsal line.

§. 34. *MICROPHYLLA*, Salm Dyck, Haw. DC. *Stem* suffrutescent, short, much branched, as well as the branchlets divaricate. *Leaves* minute, connate, triquetrous, aristate, with large and pellucid dots. *Flowers* on the tops of the branches solitary, small, reddish. *Calyx* 5-cleft. *Stigmas* 5. (Sp. 148–152.)

148. M. pulchellum (Haw. Misc. 72. Syn. 298); stems and branches decumbent, tortuous, terete; leaves minute, triquetrous, with equal convex sides, *somewhat boat-shaped, aristate-mucronate, glaucous*, ciliated with pubescence on the angles, especially on the keel. *Salm Dyck, fasc.* 2. 17. *M. canescens, Haw. Rev.* 135.

HAB. Cape of Good Hope.
Stem 1 foot. Branches retroflex. Leaves crowded on the tops of the branches, erecto-spreading, 4 lines long, 1½ lines wide. Peduncle ½–¾ inch, thickened upwards. Calyx turbinate, lobes subequal. Petals about ½ inch long, acute, pale rose-coloured. Styles 5, erect, filiform, longer than the stamens.

149. M. microphyllum (Haw. Obs. 417. Syn. 297); stem short, as well as the branches slender, crowded; leaves minute, subconnate, triquetrous, *bluntish, mucronulate, green, shining*, pustulate at the base on the inside, carinal angle *subconvex* and very entire beneath. *Salm Dyck, fasc.* 6. *t.* 12.

HAB. Cape of Good Hope.
Branches 3–4 inches high. Leaves very spreading, 2 lines long, 1 line wide. Peduncles 4–6 lines long. Calyx obconical, lobes subequal. Petals twice longer than the calyx, rose-coloured, whitish at the base. Styles 5, filiform, shorter than the stamens.

150. M. aristulatum (Sond.); stem very short or none; branches *long, prostrate, compressed*, flowering branchlets very short, leafy; leaves much crowded, connate, erectish, triquetrous, equal-sided, (when dry somewhat canaliculate above), *glaucous*, acute, *mucronate-aristate*, angles *acute*, very thinly *ciliate*. *M. forficatum, E. & Z.!* 2023.

HAB. Stony places on the sides of Lionsmountain. August. (Herb. Sd.)
Branches 1 foot or more long, creeping, somewhat rooting at the nodes, glabrous. Flowering branchlets half an inch long. Leaves much crowded in the axils, 3–4 lines long, line wide, acute, dotted, mucro recurved. Peduncle shorter than the leafy bracts, about 2–3 lines long. Calyx dotted, lobes subequal, aristate, 2½ lines long, with a small membranous margin. Petals pale red (when dry) scarcely longer than the calyx. Styles 5, subulate, as long as the stamens. Distinguished by the creeping branches, acute angles, and very small flowers. In habit is like *M. Rostellum*.

151. M. mucronatum (Haw. Misc. 73. Syn. 297); stem very short, erect, much branched; leaves oblong-ovate, triquetrous at the apex, connate at the base, *glaucescent*, terminated in a *white mucro*, coarsely dotted.

HAB. Cape of Good Hope.
Stem 1–3 inches. Leaves 3 lines long. Flowers unknown.

152. M. pigmaeum (Haw. Suppl. 99); stem very short, branched; leaves connate at the base, oblong-ovate, *semiterete, awnless*, in winter united nearly to the top.

HAB. Cape of Good Hope.
Flowers unknown.

§. 35. *ROSTELLATA*, Haw. DC. *Stem* suffrutescent, dwarf, much branched, as well as the branches prostrate. *Leaves* connate, terete, subulate, recurved. *Flowers* at the top of the branches, solitary, white, tipped with red. *Calyx* 5-cleft. *Stigmas* 5. (Sp. 153.)

153. M. Rostellum (Salm Dyck, fasc. 2. t. 18); stem and branches prostrate, rigid; leaves beaked, connate, semiterete, subulate, recurved, dotted, glaucous-green; flowers white. *M. rostellatum, DC. l. c.* 430.

HAB. Cape of Good Hope.
½–1 foot. Branches terete. Leaves 4–6, vaginate-connate at the base, 6–9 lines long, 2 lines wide, triquetrous at the apex. Peduncle clavate, 4–6 lines long. Calyx-lobes subequal, acute. Petals somewhat longer than the calyx-lobes. Styles 5, thickish.

§ 36. *VAGINATA*, Salm Dyck (Paniculata, Haw. DC.) *Stem* fruticose, erect, branches rigid. *Leaves* crowded, sheathing-connate, with the longitudinal lines of the sheaths more or less distinct, triquetrous, short, the carinal angle scabrous. *Flowers* at the top of the flowering branches, panicled, small, white; *peduncles* short, bracteate. *Calyx* 5-cleft. *Stigmas* 5. (Sp. 154–159.)

154. M. tenellum (Haw. Obs. 315. Syn. 283); branches erect, *filiform*; leaves distant, much shorter than the internodes, green, rather spreading, minute, triquetrous, *acute, recurvate at the apex*, carinal angles scabrous. *Salm Dyck, fasc.* 5. *t.* 20. *M. uncinatum, E. & Z.!* 2024.

HAB. Karro on hills near the Gauritzriver, Swellendam, Dec. (Herb. Sd.)
Stem 1 foot or higher, with slender branches. Leaves 3–4 lines long, ¾ line wide. Flowers half an inch in diameter. Peduncle short, compressed; bracts leafy. Calyx turbinate, lobes subequal. Petals uniseriate. Styles 5, short, acute.

155. M. rigidum (Haw. Misc. 95. Syn. 283); branches erect, spreading, very *stiff*; leaves shorter than the internodes, green, horizontal, minute, triquetrous, *bluntish, mucronulate*, as well as the keel scabrous *at the apex. Salm Dyck, fasc.* 6. *t.* 13.

HAB. Karro on hills near the *Zwartkopsriver, Zeyher!* 2597. Gamkariver, *Zey.!* 690. (Herb. Sd.)
Very near the preceding, but the branches are more robust and rigid; the leaves thicker, more obtuse, short, mucronulate. Internodes 6 lines, leaves 4–5 lines long. Petals about 3 lines long.

156. M. parviflorum (Haw. Misc. 95. Syn. 284); stem and branches erect; leaves erectish, *glabrous*; keel finely serrulated. *DC. l. c.* 432.

HAB. Cape of Good Hope.
Leaves half an inch long. Peduncles bracteate even to the calyx. Flowers white, small, 3 lines.

157. M. vaginatum (Haw. Misc. 95. Syn. 284); stem and branches erect, rigid; leaves about as long as the internodes, green, *spreading, linear-triquetrous, hamate-recurved* at the apex, *smooth*, but the angles roughish near the top. *Salm Dyck, l. c. t.* 14. *M. curtum, γ. minus. M. hamatam, Willd. Haw.*

HAB. Cape of Good Hope (v.v.)

Shrub 2 feet. Leaves sheathing-connate, 6 lines long, equal-sided-triquetrous. sometimes quite smooth. Peduncle very short, compressed. Petals 3–4 lines long. Styles 5, erect, subulate.

158. M. acutangula (Haw. Phil. Mag. 64. p. 424); branches erect, spreading, rigid; leaves *about as long* as the internodes, glaucous-green, *incurvate-erect*, triquetrous, compressed near the apex, carinal angle a little dilated, *scabrous. Salm Dyck, fasc.* 5. *t.* 21.

HAB. Cape of Good Hope.

Distinguished from *M. vaginatum* by the incurvate-erect, not recurved, more scabrous leaves. Internodes and leaves 6 lines long. Petals 4 lines long.

159. M. curtum (Haw. Rev. 126. Syn. 334); branches somewhat spreading, rigid; leaves *longer than the internodes*, glaucous-green, spreading, triquetrous, *attenuate, acute*, carinal angle scabrous. *Salm Dyck, l. c. t.* 22. *M. imbricatum, E. & Z.!* 2925.

HAB. Stony places near Saldanhabay, Aug.–Sept. (Herb. Sd.)

Stem 2 feet. Leaves 6–8 lines long, acute, a little recurved at the apex; the internodes 3–4 lines long, a little dilated upwards. Peduncle compressed. Flowers 9–10 lines in diameter.

§. 37. *TUMIDULA* Haw., Salm Dyck, (Paniculata DC.) *Stem* fruticose, erect, with the branches rigid. *Leaves* subdistant, sheathing-connate, the sheaths tumid, abbreviate; limb elongate, triquetrous, with smooth angles. *Flowers* at the top of the flowering branches, subcymose, small, white or pale rose-coloured; *peduncles* bracteate. *Calyx* 5-clsft. *Stigmas* 5. (Sp. 160–162.)

160. M. multiflorum (Haw. Misc. 96. Syn. 285); leaves distant, *longer than the internodes*, smooth, glaucous-green, subtriquetrous, linear-elongate; sheaths *scarcely tumid;* flowers *white;* cyme many-flowered. *M. imbricatum, Haw. Salm Dyck, fasc.* 5, *t.* 23. *et fasc.* 6. *t.* 15. *M. patens, Willd. M. foliosum, E. & Z.!* 2027.

HAB. Karro near Gauritzriver, Swellendam, Dec. (Herb. Sd.)

Stem 2–3 feet, branches straight, terete, greenish. Leaves 2–3 inches long, 2 lines wide near the sheath, bluntish, mucronulate. Cyme bearing often more than 20 flowers; pedicels short, thick. Flowers half an inch in diameter.

161. M. tumidulum (Haw. Syn. 286); leaves distant, *shorter or equalling the internodes*, smooth, glaucous-green, subtriquetrous, *linear-elongate*, erect-recurved; sheaths very tumid, abbreviate; flowers *rose-coloured*, cymose. *Salm Dyck, fasc.* 5. 24.

VAR. β. **foliosum**; leaves longer, more crowded. *M. foliosum, Haw. Syn.* 130.

HAB. Driefonteyn, *Zeyher!* 698. (Herb. D., Sd.)

Branches when young often purplish. Leaves 1–2 inches long, 2 lines wide near the sheaths, minutely punctate. Lower pedicels of the cyme elongate, slender, bracteate in the middle, bracts very large, white-margined. Calyx scarcely turbinate. Petals longer than in the preceding and following.

162. M. umbellatum (Linn. Spec. 481); leaves distant, *longer* than the internodes, smooth, glaucous-green, *subcylindraceous*, blunt, mucronulate; sheaths tumid, abbreviate; flowers *white*, *umbelled*. *Dill. Elth. t.* 208. *Salm Dyck, fasc.* 6. *t.* 16. *Thunb.! Fl. Cap.* 414. *ex pte. E. & Z.!* 2028. *M. anomalum, Willd. Enum.* 531.

HAB. Sandy places, Capeflats, Kaeberg, Gnadenthal, Vygekraal, Heerelogement, etc. *Zey.!* 697, 699. (Herb. Thunb. D., Sd.)

Stem robust, 2–3 feet. Leaves 2–3 inches, 2–3 lines wide, subtriquetrous at the apex, with a red mucro. Flowers umbellate or corymbose, often numerous, white or purple.

§. 38. *CROCEA*, Salm Dyck. (Sebacea, Haw. Veruculata, DC.) *Stem* fruticose, erect, branched. *Leaves* somewhat crowded, connate, sheaths short, trigonous-semiterete, soft, sebaceous, mealy-glaucous, without dots. *Flowers* terminal, solitary, yellow or croceous, long peduncled. *Calyx* 4-cleft. *Stigmas* 8. (Sp. 163–165.)

163. M. luteum (Haw. Phil. Mag. Aug. 1826, p. 128); stem erect; leaves semicylindrical, attenuate, subtriquetrous at the apex, *acutish*, subsebaceous, glaucous; petals *acute*, *yellow*. *Salm Dyck, fasc,* 3. *t.* 26. *M. purpureo-croceum, β. flavo-croceum. Haw.*

HAB. Cape of Good Hope.

Stem 1–2 feet. Leaves erecto-patent, 1–1½ inches long and 2 lines wide, yellowish green. Peduncle circ. 1 inch long. Calyx subturbinate, 2 of the lobes triquetrous, 2 shorter. Petals uniseriate, acute, yellow, when older croceous. Stigmas 8, erect, ramentaceous.

164. M. croceum (Jacq. fragm. t. 59. f. 2); stem erect; leaves semicylindrical, turgid, scarcely triquetrous-compressed at the apex, *obtuse*, *subsebaceous*, mealy-pruinose; petals *erose*, *croceous* above, more or less purplish beneath. *Salm Dyck, l. c. t.* 27. *M. insititium, Willd. Enum.* 536. *M. purpureo-croceum, Haw. Misc.* 81. *E. & Z.* 2029. *M. glaucum, E. & Z.!* 2013.

HAB. Stony places on mountain sides near Olifantsriver, Clanwilliam, Oct. (Herb. Sd.)

Very near the preceding, differs by the thicker stem, more whitish, bluntish leaves and erose, croceous petals. Leaves 1–1½ inches long, about 3 lines in diameter. Petals biseriate. Stigmas 8–9, subglobose.

165. M. luteolum (Haw. Phil. Mag. 1826, p. 129); leaves crowded, acute at the apex, and a little recurved; branches slender and dense; flowers small, yellow.

HAB. Cape of Good Hope.

It differs from *M. luteum* in the leaves and in the more dwarf stature. Flowers more numerous than in *M. luteum.*

Group IV. TERETIUSCULA.—Stem fruticose or suffruticose, erect or nearly so. Leaves distinct, rarely connate at the base, terete or semicylindrical or turgid-trigonous. (§§. 39–47.)

§. 39. *VERUCULATA*, Salm Dyck, DC. (Sebacea, Haw.) *Stem* fruticose, with spreading branches. *Leaves* crowded, cylindraceous, soft, dotless, very glaucous, obtuse, mucronulate. *Flowers* ternate, rarely solitary, small, yellow, or croceous, sweet-scented, short pedunculate;

peduncles bracteate. Calyx 5, rarely 4-cleft. *Stigmas* 5, 4 or 8. (Sp. 166–168.)

166. M. veruculatum (Linn. Spec. 696); leaves fasciculate, incurvate, subsebaceous, mealy-pruinose, subtrigonous, cylindrical, mucronate; flowers *yellow*, mostly ternate, *subsessile*; *calyx* 5-*cleft*; *stigmas* 5. *Dill. Elth. t.* 203. *f.* 259. *DC. Pl. Grass. t.* 36. *Salm Dyck, fasc.* 3. *t.* 28. *E. & Z.!* 2030.

HAB. Mountain sides near Brackfonteyn, Clanwilliam, Oct.–Nov. (Herb Sd.)

Stem 1 foot or more, branches tortuose. Leaves connate, 1–1½ inches long, 3–4 lines wide, arcuate-incurvate, when young erect, obsoletely trigonous, with an evident, purplish mucro. Pedicels 2–4 lines long, at the base and in the middle with 2 leafy bracts. Calyx hemispherical, lobes subequal, with broadly membranacrous margins. Petals yellow, scarcely longer than the calyx.

167. M. monticolum (Sond.); leaves fasciculate incurvate, subsebaceous, mealy-pruinose, *obtusely trigonous*, mucronate; flowers *croceous*, ternate, *pedunculate*, lateral peduncles enclosed by 2 connate bracts, the intermediate bractless; calyx 4-cleft; *stigmas* 8.

HAB. Stoofkraal, *Zeyh.!* Mar. ((Herb. Sd.)

Shrub with woody, tortuous, short branches. Leaves connate, a little sheathing, 1 inch long, 3 lines wide, when dry evidently trigonous. Flowers the size of those of *M. veruculatum*, or a little larger. Peduncles 5–6 lines long, compressed; bracts of the two lateral peduncles sheathing from the base to the middle, the upper or free part leafy, trigonal, acute, equalling the flowers. Calyx-lobes subequal, keeled. Styles subulate, longer than the stamens.

168. M. veruculoides (Sond); leaves fasciculate, erectish, scarcely incurvate, soft, mealy-pruinose, *cylindrical, flattish above*, obtuse, with a very short mucro; flowers *croceous*, ternate, rarely solitary, pedunculate; lateral peduncles bracteate *at the middle*, the intermediate bractless; *calyx* 4–5-*cleft*; *stigmas* 5. *M. veruculatum*, β. *Herb. Thunb.!*

VAR. **minus**; leaves smaller, flowers mostly solitary. *M. pruinosum, E. & Z.!* 2110. *non Thunb.!*

HAB. In Hantum, *Thunberg*; 'Kamus, Feb.; Droogekraal, Hartveld, June, *Zeyher!* var. β. fields near Zwartkopsriver, *Zey.!* 2585; Rhinosterkop, dist. Beaufort, *Zeyher!* 684; Roggeveld, *A. Wyley*. (Herb. Thunb., D., Sond.)

Stem procumbent, terete, soft, glabrous. Leaves connate, 1–1½ inches long, 3 lines in diameter, obsoletely trigonous. Peduncles subequal, 1 inch long, compressed, thickened above. Bracteas vaginate, limb leafy, as long as the peduncle. Calyx-lobes unequal, 2 longer and blunt. Petals a little longer than the calyx, linear, obtuse. Capsule turbinate, subangulate, 8–10 valved. Var. β. is smaller, more depressed, leaves ¾–1 inch long, 2 lines wide, the flowers are not different. Very similar to *M. laeve, Thunb.!*

§. 40. *HAWORTHIANA*, DC. (Corallina, Haw.) *Stem* fruticose, erect, branches decussate. *Leaves* subcylindrical, elongate, more or less subulate, glaucous, punctate. *Flowers* solitary (or in **M. productum** ternate), showy, reddish, long peduncled; peduncles bracteate. *Calyx* 5-cleft. *Stigmas* 5. (Sp. 169–173.)

169. M. Haworthii (Don. Hort. Cantab. 66); stem and branches erect; leaves somewhat crowded, subdistinct, incrassate, semicylindrical, *subcompressed at the top, attenuated on both ends*, subincurved, spreading, glaucous, smooth; flowers large, purplish. *Salm Dyck, fasc.* 1. *t.* 27.

HAB. Cape of Good Hope.

Stem 2 feet high; branches decussate. Leaves 1–1½ inches long, 2–3 lines wide, bluntish, mucronulate. Flowers about 3 inches in diameter. Peduncles 1–1½ inches long, thickened upwards; bracts leafy. Calyx turbinate. Petals subtriseriate, broad-lanceolate. Styles 5, very short, roundish.

170. M. coralliflorum (Salm Dyck, l. c. t. 28); stem and branches erect; leaves *distant, subconnate, clavate-elongate*, subcylindrical, spreading, in- or re-curvate, glaucous, smooth; flowers *on very long peduncles. M. corallinum, Haw. Rev.* 154. *excl. syn. Thunb. M. laeve, Haw. misc.* 64?

HAB. Cape of Good Hope.

Distinct from the preceding by the more slender and branched stems; subclavate, at the apex not attenuated leaves; longer peduncled, somewhat smaller flowers and uniseriate petals. Leaves about 2 inches long, 1½ lines wide. Peduncle 4–6 inches long. Styles 5 or 4, acutish.

171. M. stipulaceum (Linn. Spec. 693); stem and branches erect; leaves *crowded, subdistinct, linear-elongate, semiterete*, spreading, recurved, very glaucous, smooth; axils very proliferous; flowers purplish. *Dill. Elth. t.* 209 *f.* 267, 268. *Salm Dyck, l. c. t.* 29. *E. & Z.!* 2031. *M. laeve, E. & Z.!* 2033. *ea pte.*

HAB. Near Gauritzriver, Swellendam, and in dist. Uitenhage, Dec. (Herb. D. Sd.)

Distinct from *M. coralliflorum* by a shorter stem, more crowded, at the apex not incrassate leaves, smaller flowers and effuse stamens. Stem 1–1½ inches high. Leaves 1½–2 inches long, 1–2 lines wide. Peduncles solitary, rarely subternate, 1½–2 inches long. Calyx scarcely turbinate. Petals subtriseriate, narrow lanceolate, about twice shorter than in *M. Haworthii.* Styles erect, acute.

172. M. productum (Haw. Phil. Mag. Dec. 1834. p. 425); stem and branches erect; leaves crowded, subdistinct, elongate, *semiterete*, erect-incurved, *glaucous*, smooth; flowers *bigeminate or ternate*, pale rose-coloured; calyx-lobes *elongate*, 2 longer *equalling the petals. Salm Dyck, fasc.* 2. *t.* 19. *M. tetragonum, E. & Z.!* 2035. *M. tenuifolium, Thunb.! herb.* 3.

VAR. β. **lepidum** (Salm Dyck, l. c.); stem higher; flowers whitish.

HAB. Karro hills near Zwartkopsriver; Winterhoeksberg and near Bethelsdorp, *Zeyher!* 2594; Albany, *Williamson.* Var. β. in the Capeflats, Nov. (Herb. Thunb. D., Sd.)

Stem 1–2 inches high. Leaves 1–1½ inches long, 1½ lines wide, terete, half cylindrical, bluntish mucronulate. Peduncles 1–1½ inches long, thickened above, rarely solitary. Calyx lobes unequal, 2 of them longer, subcylindrical, acute, nearly as long or a little longer than the lanceolate petals. Styles 5, clavate, acute. By the inflorescence, the elongated calyx-lobes and smaller flowers it is easily distinguished from the preceding.

173. M. Zeyheri (Salm Dyck, fasc. 5. t. 25); stem and branches erect, subflexuous, leaves much crowded, subdistinct, elongate, terete, incurved, erect, attenuated on both ends, acute, very smooth and *green*; flowers *solitary, purplish*; calyx-lobes *broad*, much *shorter than the petals. M. Haworthii, E. & Z.!* 2032.

HAB. In fields near Zwartkopsriver, Oct. (Herb. Sd.)
Stem straight, 1½ foot. Leaves 1–1½ inch long, 1 line wide. Peduncle 2 inches long, thickened upward. Calyx-lobes subequal, 2 larger with membranaceous margins. Petals purplish-violet, spathulate-lanceolate, bluntish, about 10 lines long. Styles 5, acute.

§. 41. *SPINOSA*, Salm Dyck, Haw. DC. *Stem* fruticose, erect; branches rigid. *Leaves* triquetrous-terete, dotted, glaucescent. *Flowers* on terminal spines, ternate or biternate, numerous, or ternate, and the peduncles after flowering spinescent, reddish, small. *Calyx* 5-cleft. *Stigmas* 5. (Sp. 174–175.)

174. M. spinosum (Linn. Spec. 693); stem, erect; branches hard, dichotomous, *spinose after flowering*; leaves nearly distinct, teretely-triquetrous, dotted. *Dill. Elth. t*, 208. *f*. 265. *Salm. Dyck, fasc*. 5. *t*. 26. *Thunb. Fl. Cap*. 420. *E. & Z.* ! 2034.

HAB. In Karro, Olifantsriver, Beaufort and Graafreynet, Rhinosterkop, *Zeyher* ! 675, Oct.–April. (Herb. Thunb., D., Sd.)
Stem 2 feet or higher, branches spreading. Leaves 6–12 lines long, 1½ lines wide, in the wild plant generally smaller, (3–6 lines long), blunt, mucronulate. Peduncles tripartite, lateral branches triacanthous, flowering. Pedicels short. Flowers ½ inch in diameter or smaller.

175. M. mucroniferum (Haw. Phil. Mag. 1823. 381); stem erect; branches straight; leaves glaucescent, spreading, bluntly triquetrous, mucronulate; flowers disposed by threes; *peduncles permanent* after flowering and spinose. *M. pulverulentum, Willd. Enum. p*. 583.

HAB. Cape of Good Hope.
Stem 1 foot and higher.

§. 42. *CYMBIFORMIA*, Salm Dyck, Haw. DC. *Stem* suffrutescent, dwarf, branches often decussate. *Leaves* distinct, turgidly-trigonous, obtuse, cymbiform. *Flowers* solitary, reddish or yellow. (Sp. 176–180.)

176. M. Lehmanni (E. & Z. ! 1996); branches *compressed*, ascending; leaves spreading, subconnate, turgidly-triquetrous, subcymbiform, *without dots*, very smooth, glaucous; flowers terminal, solitary, on short peduncles; *calyx* 6-cleft, lobes turgid, keeled.

HAB. Karro-like hills near Zwartkopsriver, April. *Zey.* ! 2576.
Stems ½–1 foot. Leaves when dry rugose, 8–12 lines long, 3–4 lines wide, about as long as the internodes, blunt, with a minute mucro. Flowers 1–1½ inches in diameter. Peduncle about 1 inch long, compressed, with 2 leaf-like bracts near the calyx. Lobes of the calyx broad, subequal. Petals in many rows, interior ones shorter, pale yellow. Styles 6. Capsule when ripe, subangulate, glaucous, about 4 lines long, 6-locular.

177. M. molle (Ait. Kew. 2. 192); branches crowded, 2-*edged*, decumbent; leaves spreading, turgidly-triquetrous, firm, canescent, with the margins blunt, and *lined with dots*; flowers terminal, solitary, peduncled; *calyx* 5-*cleft*. *Haw. Syn*. 262.

HAB. Cape of Good Hope (v.v.)
Subshrub 1 foot. Peduncles subterete, 1½ inches long, with 2 leafy bracts near the calyx. Lobes of the calyx small, 2 membranaceous. Corolla ½ inch in diameter, pale-reddish. Filaments spreading, purple but white at the base. Styles 5, acute.

178. M. strictum (Haw. Misc. 82. Syn. 262); stem woody, branched, very stiff, straight; leaves *triquetrous*, obtuse, expanded, glaucescent, beset *with large dots. DC. l. c.* 437.

HAB. Cape of Good Hope.
Shrub 2–3 feet high, has never yet flowered in the gardens; but from a specimen of it received from the Cape by Haworth, the flowers are said to be showy, and yellow.

179. M. trichotomum (Thunb.! Nov. Act. Nat. Cur. v. 8. p. 14. Ap.); stem *erect;* branches trichotomous, spreading, fastigiate, *subterete*, glabrous; leaves spreading, connate, cylindrical, trigonous, obtuse, subcymbiform, quite smooth and glabrous; flowers terminal, solitary, sessile; *calyx* 4-*cleft*, lobes unequal, 2 longer, leaf-like; styles 4, very short. *Fl. Cap.* 419.

HAB. Karro between Olifantsriver and Bocklandsberg, Oct.–Nov. (Herb. Th.)
Shrub 1 foot or more in height, rigid. Ultimate branches 2–3 inches long. Leaves remote, about ½ inch long, 1 line broad, young ones 2 in each axil, 3–4 lines long, 1½ lines broad when dry, yellowish green, very minutely punctate. Shorter lobes of calyx carinate, obtuse, with membranous margins. Petals linear, spreading, purple, interior ones shorter, white, (Thunb.!) Only one specimen with imperfect flowers in herb. Thunberg.

180. M. sessile (Thunb.! Nov. Oct. Nat. Curios. v. 8. p. 14. App.); stem erect; branches spreading; leaves *minute, trigonous-globose*, blunt, *subconcave above*, smooth, dotted; flower solitary, on a very short peduncle; calyx 5-cleft. *Fl. Cap.* 419. *M. cymbiforme, Haw. Obs.* 264. *Syn.* 263.

HAB. Karro between Olifantsriver and Bocklandsberg, Oct. (Herb. Thunb.)
Stem 1 foot or higher. Branches subflexuous, ultimate very short, flower bearing. Leaves connate, 4–6, crowded, 1 line long and wide, subglobose, flattish above, gibbous beneath. Peduncle 2–4 lines long with 2 leafy bracts. Flowers are wanting in herb. Thunberg; from description in Fl. Cap. they are reddish, and the calyx-lobes rotundate-obtuse. Ripe capsule conical, 5-locular.

§. 43. *DEFOLIATA*, Salm Dyck. (Noctiflora, Haw. DC.) *Stem* suffruticose, slender, rigid, sparingly branched. *Leaves* remote, cylindraceous, glaucous, without dots, soon falling off. *Flowers* by threes or biternately cymose, often expanding in the evening, white, yellowish or violet. *Calyx* turbinate, 4-cleft. *Stigmas* 4. (Sp. 181–183.)

181. M. defoliatum (Haw. Misc. 83); stem erect; branches terete; leaves spreading, *subcylindraceous*, blunt, soon deciduous; flowers dichotomously cymose or biternate, *scentless;* peduncles *very short*, clavate; ovarium terete. *Salm Dyck, fasc.* 3. *t.* 29. *M. clavatum, Jacq.! Hort. Schoenb. t.* 108. *M. horizontale, Haw. Syn.* 261.

HAB. Cape of Good Hope.
Shrub 1 foot or more. Leaves distinct, 1–1½ inches long, 2–3 lines wide. Central flower subsessile, without bracts; lateral ones very short pedunculate, with 2 short bracts near the calyx. Lobes of the calyx unequal, 2 are longer and half cylindrical. Petals a little longer than the calyx, white or straw-coloured.

182. M. noctiflorum (Linn. Spec. 689); stem erect; branches *terete*, leaves spreading, *cylindraceous*, blunt, soon deciduous; flowers biternate-cymose, *fragrant ;* peduncles *long,* thickened above ; ovarium *terete.*

VAR. *α.* **phœniceum** (Haw. Rev. 179) ; flowers white inside, and scarlet outside. *Dill. Elth. t.* 206. *f.* 262. *Salm Dyck, fasc.* 4, *t.* 17. *DC. Pl. Grass. t.* 10.

VAR. *β.* **stramineum** (Haw. Rev. 179) ; flowers white inside, and straw-coloured outside. *Dill. Elth. t.* 206. *f.* 263.

VAR. *γ.* **fulvum** (Salm Dyck, fasc. 6. t. 17) ; flowers mostly ternate, white inside and fulvous outside. *M. fulvum, DC. l. c.* 445.

HAB. Springbokkeel, *Zeyher!* 700. (Herb. Sd.)

Distinguished from *M. defoliatum* by somewhat thicker leaves, long peduncled flowers, obtuse calyx-lobes and a sweet-scented, twice larger flower. Stem 2–3 feet high. Leaves 1–1½ inch long, 3 lines wide. Peduncles and pedicels 1 inch long, terete, intermediate without bracts, lateral ones with two leafy bracts in the middle. Flower 1½ inch in diameter. Calyx subturbinate, 2 lobes longer, obtuse. Styles 4 subulate.

183. M. tetragonum (Thunb. ! Fl. Cap. 426) ; stem erect as well as the branches *subtetragonal ;* leaves spreading subcylindrical, obtuse deciduous ; flowers ternate or biternate-cymose ; peduncles longish *compressed,* thickened above ; ovarium *tetragonal. M. fasciculatum, Th. ! l. c.*

HAB. in Hantam, *Thunberg ;* Springbokkeel, Komseep, *Zeyher* 2955, 702. Mart. (Herb. Thunb. Sd.)

Stem 1 foot high, more branched than the foregoing, branches patent, pale green straw-coloured. Leaves subincurved, 6–12 lines long, 2–3 lines wide, very obtuse, flattish above. Peduncles about 1 inch long or shorter, intermediate bractless, lateral ones near the middle, with 2 leafy bracts. Flowers as large as those of *M. defoliatum,* yellowish (Thunberg). Calyx lobes obtuse, 2 longer. Styles 4 subulate. Ripe capsule angulate, turbinate.

§. 44. *SPLENDENTIA,* Salm Dyck, DC. (Digitiflora, Haw.) *Stem* suffruticose, erect, branched. *Leaves* crowded, distinct, subcylindrical, spreading-recurved, without dots, when young sulcate above. *Flowers* solitary, or rarely ternate, white, middle-sized. *Calyx* 4–5 cleft, lobes leafy. *Stigmas* 4–5. (Sp. 184–191.)

184. M. sulcatum (Haw. Rev. 173) ; stem and branches erect, straight ; leaves crowded, linear-subulate, bluntish, green, when young erecto-incurvate, *canaliculate,* when old expanded, subterete ; calyx lobes *unequal ;* flowers very pale rose-coloured. *Salm Dyck, fasc.* 3. *f.* 30.

HAB. Cape of Good Hope.

Shrub 2 feet. Leaves ½–1 inch long, 1 line wide, mucronulate. Peduncles about 1 inch long, very minutely papillate. Calyx lobes 5, subulate, three of them shorter. Petals 8–10 lines long. Styles 5, subulate. Perhaps a variety of the following.

185. M. splendens (Linn. Spec. 689) ; stem and branches *flexuous-erect ;* leaves crowded, *semiterete,* glaucous-green, when young erect, recurved at the apex, when old very patent, bluntish ; calyx lobes *equal,* abbreviate, subulate ; flowers white. *Dill. Elth. t.* 204. *f.* 260. *DC. Pl. Grass. t.* 35. *Salm Dyck, fasc.* 6. *t.* 4.

HAB. Cape of Good Hope (v. v.)

Leaves 6–12 lines long, 2 lines wide, the axils very proliferous. Peduncle short, thickened above, a little papillate as well as the subclavate calyx. Petals bidentate, 6 lines long. Styles 5.

186. M. fastigiatum (Haw. Rev. 173. Syn. 256); stem and branches *straight, fastigiate;* leaves crowded, semiterete, obtuse, glaucous-green, when young spreading, when old very patent; calyx lobes subequal, abbreviate, *bluntish;* flowers whitish. *Salm Dyck, l. c. t.* 19.

HAB. Cape of Good Hope.
Stem 2 feet, branches minutely papillose. Leaves 5–8 lines long, scarcely 1 line wide, attenuate at the base. Peduncle short. Petals acute, linear. Styles 4–5.

187. M. acuminatum (Haw. Phil. Mag. 1824. p. 426); stem erect; branches *flexuous-patent;* leaves crowded, semiterete, *attenuate, mucronulate,* green, when young erect-incurved, when old spreading-recurved; calyx lobes *unequal,* abbreviate, *acute;* flowers pale straw-coloured. *Salm Dyck, l. c. t.* 20.

HAB. Cape of Good Hope.
Leaves 6–10 lines long, 1 line wide. Peduncles 1 inch long, minutely papillate, without bracts. Calyx 5-cleft, two of the lobes longer, leafy. Petals about ½ inch long, linear-lanceolate. Styles 5, subulate. It comes very near *M. sulcatum.*

188. M. albicaule (Haw. Phil. Mag. 1826. p. 331); stem erect; branches flexuous, divergent; leaves remote, semiterete, *acute,* glaucous-green, when young erect-recurved, when old very patent and recurved; calyx lobes subequal, *much elongated,* acute; flowers pale yellow.

HAB. Cape of Good Hope.
Leaves 6–10 lines long, 1 line wide. Peduncle short, thick. Petals as in *M. splendens,* of which it only seems a variety.

189. M. umbelliflorum (Jacq. Willd. Enum. 534); stem and branches erect, flexile; leaves crowded, curvate-spreading, depressed-terete, *obtuse, subglaucous,* in the floriferous branches linear-elongate, in sterile branches shorter, subclavate; calyx lobes subequal, acutish; flowers *subpaniculate, violet-white. Salm Dyck, fasc.* 3. *t.* 31.

HAB. Cape of Good Hope.
Very near *M. fastigiatum,* and only distinguished by the laxer stem and branches, less crowded leaves, and the inflorescence. Leaves ½–1 inch long, often proliferous from the axils. Flowers rarely solitary. Calyx turbinate-clavate, papillate. Petals acute, about 6–8 lines long.

190. M. flexuosum (Haw. Rev. 172. Syn. 257); stem and branches erect-spreading, *flexuous;* leaves crowded, curvate-patent, depressed-terete, *mucronulate, green, shining,* in the flowering branches elongated, in the sterile branches short, sublanceolate; calyx lobes subulate, elongate, subequal; flowers white, a little straw-coloured. *Salm Dyck, l. c. t.* 32.

HAB. Cape of Good Hope.
By the slender, flexuous stem and branches, very green, shining leaves, it differs from the preceding. Leaves 6–12 lines long, 1 line wide. Peduncle ½–1 inch long, thickened above, 1-flowered. Calyx clavate, smooth, spreading. Petals 8–10 lines long, acute, nearly white. Styles 5 short.

191. M. longistylum (DC. Pl. Grass. t. 156); branches elongated; leaves distinct, when young *linear-filiform,* but at length becoming *a little keeled, acute,* and *minutely papulose;* peduncles 1-flowered; calyx

5-cleft, 2 or 3 of the acutish lobes having hyaline margins; styles 5, *exceeding the stamens. M. pallens, Jacq.! Hort. Schœnb.* 3. *t.* 279. *not of Ait. M. pallescens*, β. *Haw. Rev.* 174.

VAR. β. **purpurascens** (DC. Prod. 3. 446); flowers purplish. *M. pallescens a, Haw. Rev.* 174. *M. reflexum*, β. *Haw. Misc.* 64.

HAB. Cape of Good Hope. (Herb. Jacquin.)

Leaves 6–10 lines long. Flowers about 10 lines in diameter, white, rose-coloured or violet at the apex.

§. 45. *JUNCEA*, Haw. DC. *Stem* suffruticose, erect. *Branches* herbaceous. *Leaves* small, linear-subulate, dotless, deciduous when dry; whence the stems appear to be articulated by the cicatrices of the fallen leaves. *Flowers* small, pedunculate, often ternate, white or reddish. (Sp. 192–200.)

a. tetramerous. (Sp. 192–193.)

192. M. junceum (Haw. Misc. 175. Syn. 255); smooth; branches suberect, terete, articulate, not contracted at the joints; leaves very spreading, *semicylindraceous, linear*, acute, subcanaliculate, glaucous-green; flowers terminal, dichotomous, subcymose; lobes of calyx 4, 2 longer, subulate; *petals longer than the calyx;* styles 4; capsule 4-valved. *Salm Dyck, fasc.* 2. *t.* 20. *M. articulatum, Thunb.! Fl. Cap.* 416. *ex pte. M junceum, E. & Z.!* 2043.

VAR. β. **pauciflorum**; branches elongated, few-flowered at the top. *M. coralloides, E. & Z.!* 2041. *excl. Syn.*

HAB. Karro, in Zwartsland, and near Olifantriver, and Gauritzriver, Bosjesveld, Springbokkeel, *Zeyh.* 2957. (Herb. Thunb. D. Sd.)

Stem 2 feet, woody. Branches herbaceous, rugose when dry. Internodes terete, very unequal. Leaves ½–1 inch long, in the branches much shorter, 1–1½ inch wide. Flowers subunilateral, short pedicellate. Pedicels thickened above. Calyx turbinate, 2 lobes subulate, 2 larger, with membranaceous margins. Petals a little longer than the calyx, pale rose-coloured or nearly white. In var. β the branches are mostly secundate, ½–1 foot high, with 2–4 flowers, and nearly leafless.

193. M. micranthum (Haw. Syn. 257); smooth, branches herbaceous, very numerous and slender, terete, subarticulate; leaves crowded, erect, distinct, subterete, *attenuated at both ends*, pale green; flowers on the top of the short branchlets, solitary or ternate; lobes of calyx 4, 2 elongated, *longer than the petals;* styles 4; capsule 4-valved. *Salm Dyck, l. c. t.* 22. *M. parviflorum, Jacq. Hort. Schœnb.* 3. *t.* 278. *M. tenue, Haw. Rev.* 175. *M. aduncum et flexifolium, E. & Z.!* 2050-2051. *M. junceum, Herb. Drege.*

HAB. Sandy places near Rietvalley, near Gnadenthal; Hassaquaskloof. *Zeyher* 2615. Oct.–Nov. (Herb. D. Sd.)

Branches filiforme, the ultimate flowerbearing ½ inch or shorter. Leaves about 3 lines long, ½ line wide, acute, incurved. Flowers very minute, about 3 lines long, subsessile. Calyx turbinate. Petals linear, obtuse, very white.

β. pentamerous. (Sp. 194–200.)

194. M. bicorne (Sond.); quite smooth; branches herbaceous, very numerous and slender, terete, *subarticulate;* leaves crowded, erect, spreading, *distinct*, subterete, attenuated at both ends, pale green; flowers solitary or ternate; lobes of calyx 5, 2 much elongated, subulate, longer than the petals; styles 5; capsule 5-valved. *M. micranthum et tenue, E. & Z.!* 2044-2045.

HAB. Fields near Zwartkopsriver. Dec.-Jan. *Zeyh.* 2616. (Herb. D. Sd.)

Very near the preceding, but the leaves mostly longer (½–1 inch), the flowers pedunculate, the calyx less turbinate, the capsules globose. Flowers about 2 lines long; the two calyx lobes as long or longer than the calyx tube. Petals white.

195. M. granulicaule (Haw. Phil. Mag. 1824. V. 64. p. 427); stem and branches *puberous-scabrous;* branches herbaceous, erectish, terete, articulate, not contracted at the joints; leaves *subconnate*, spreading, linear-cylindraceous, acute, pale green; flowers solitary on the top of the very short, subunilateral branchlets, sessile or short-pedicellate; lobes of calyx 5, 2 *obtuse*, a little longer and about equalling the petals; styles 5; capsule 5-valved. *E. & Z.!* 2042. *Salm Dyck, l. c. t.* 21. *M. articulatum, Thunb.! Fl. Cap. ex pte.*

HAB. Karro-like-hills, near Zwartkopsriver, *Zeyh.* 2617; Dickkopulakte, Vishriver, *Zeyh.* 705; Olifantsriver, Gauritzriver, and Nieuwefeld, Beaufort. Nov.-Jan. (Herb. Thunb. Sd.)

Shrub 1–1½ foot, smaller than *M. junceum*, but larger than *M. micranthum*, from all others of the section distinguished by the scabrous pubescence, consisting of very minute, acute, rigid hairs or papulæ, not of dots, as described by the authors. Leaves 3–6 lines long, larger at the base. Flowers 3 lines long. Calyx subturbinate. Petals obtuse, white.

196. M. simile (Sond.); *quite smooth;* stem adscending or erect; branches subunilateral, herbaceous, terete, articulate, not contracted at the joints; leaves subconnate, spreading or incurvate, semicylindrical acute, glaucous-green; flowers cymose, rarely subsolitary, on short peduncles; lobes of calyx 5, 3 subulate and much longer, nearly equalling the petals; styles 5; capsule 5-valved. *M. fulvum, E. & Z.!* 2037, *not of Haw.*

VAR. β. **Namaquense**; more glaucous, joints of the branches longer, flowers somewhat smaller, cymose, paniculate.

HAB. Fields near Zwartkopsriver, *Zeyh.* 2618. var. β, Namaqualand, *V. Schlicht*, *A. Wyley.* (Herb. D. Sd.)

Stem often very long. Branches mostly short. Joints very unequal, 2 lines–1 inch. Leaves as in *M. junceum*; the flowers also very similar, but white. Calyx lobes ovate-lanceolate, at length subequal. Pedicels thick, in var. β longer and thinner.

197. M. Schlichtianum (Sond.); quite smooth; branches erect, soft, *subspongious*, terete, articulate, somewhat contracted at the joints; internodes 4–6 lines longer than broad; flowers distinct, incurvate-erect, *subcylindrical, attenuated at both ends;* flowers terminal, cymose, subsessile; lobes of calyx 5, equal, acutish, equalling the petals; styles 5.

HAB. Namaqualand, *v. Schlicht.* (Herb. Sd.)

A very distinct species, of which only a few branches were collected. The whole plant greyish-white. Internodes ½ inch long; the leaves of the same length. Ovarium about 1 line long, subglobose, not turbinate, very short, pedicellate. Petals white linear, equalling or scarcely longer than the (¾ line long) calyx lobes.

198. M. corallinum (Thunb.! Nov. Oct. Nat. Curios. V. 8. p. 12);

quite smooth; stem woody, much branched; branches herbaceous, terete, articulate, *much contracted at the joints, internodes oval, twice as long as broad;* leaves distinct, erect-incurved, semiterete, *bluntish;* flowers terminal, sessile, solitary; lobes of calyx 5; petals white. *Fl. Cap.* 416.

HAB. Karro, in Bockland, and near Olifantsriver. Oct. (Herb. Thunb.)

Easily known by the thick, woody, much branched stem and the oval articles. Stem ½ foot. Branches spreading, 1 line broad. Leaves a little longer than the articles. There are no flowers in herb. Thunb.; from the description the calyx is 5-cleft, lobes erect, subterete, the petals spreading, linear, white.

199. M. ciliatum (Thunb.! R. c. p. 11. App.); stem and branches smooth; branches secundate, erect-spreading, terete, articulate, *not contracted at the joints;* leaves connate, a little sheathing, *sheaths at the base ciliated with long deflexed hairs;* leaves subcylindrical, obtuse, *papillose;* flowers terminal, short, peduncled, subcymose; calyx lobes 5, equal, obtuse, subpapulose, nearly equalling the white petals. *Flor. Cap.* 416.

HAB. Karro, between Olifantsriver and Bocklandsberg. Oct.–Nov. (Herb. Thunb.)

Stem 1 foot. Branches fastigiate. Articles cylindrical. 4–6 lines long. Sheaths of the leaves nearly 1 line long, ciliated, with 1–1½ line long hairs at the base, not at the top. Leaves as long as the articles or longer. Flowers on the tops of the branchlets; peduncles 1–1½ lines long. Calyx 3 lines long, a little shorter than the very narrow-linear petals. Capsule as large as a small pea, 5-valved.

200. M. rapaceum (Jacq. Fragm. 43. t. 52. f. 1); root tuberous; stem elongated, herbaceous; branches terete and somewhat articulated; leaves distant, terete, obtuse, dotted, spreading; peduncles 1-flowered; lobes of calyx filiform; styles 5, spreading, subulate.

HAB. Cape of Good Hope.

Leaves 8–10 lines long, 1 line wide. Flowers snow-white, 9–10 lines in diameter. Peduncles 1 inch long, with 2 bracts in the middle. Lobes of calyx nearly equal or very unequal.

§. 46. *TENUIFOLIA,* Salm Dyck, Haw. DC. *Stem* fruticose, branches slender, effuse. *Leaves* elongated, linear, nearly terete or subcompressed, punctate, in one species not punctate. *Flowers* solitary or ternate, showy, yellow or scarlet, long pedunculate. *Peduncles* bracteate. *Calyx* 5-cleft. *Stigmas* 5. (Sp. 201–206.)

201. M. coccineum (Haw. Obs. 247. Syn. 265); stem and branches *erect, straight;* leaves semicylindrical-triquetrous, blunt, mucronulate, glaucescent; peduncles smooth at the base; petals *scarlet. Lodd. Bot. Cab. t.* 1033. *DC. Pl. Grass. t.* 83. *Salm Dyck, fasc.* 3. *t.* 33. *M. bicolorum, Curt. Bot. Mag. t.* 59. *M. tenuifolium, E. & Z.!* 2048. *Zeyh.* 696.

HAB. Sandy places, Cape Flats, Saldanhabay, etc. May–Sept. (Herb. D. Sd.)

Stem 1–3 feet. Leaves semicylindrical, subtriquetrous-compressed near the apex, 6–16 lines long, ¾–1 line wide, with prominent dots. Flowers solitary or ternate. Peduncles 1 inch or longer, thickened upwards. Calyx dotted-scabrous, lobes subequal. Petals scarlet on both surfaces.

202. M. variabile (Haw. Syn. 266); stem and branches *effuse,* slender; leaves semicylindrical-triquetrous, acutish, glaucescent, punctate; peduncles smooth at the base ; petals *variable*, yellow, at length becoming reddish. *Salm Dyck, l. c. t.* 34. *E. & Z.!* 2049. *Zeyh.* 2612.

HAB. Sandy places near the sea shore, Saldanhabay. Oct. (Herb. Thunb. Sd.)
From the foregoing it is only distinct by more spreading or effuse branches and a different colour of the petals.

203. M. bicolorum (Linn. Spec. 695); stem and branches erect; leaves subtriquetrous, erect, *acute, green; peduncles and calcyes papulose-scabrous;* petals yellow inside and scarlet outside *Dill. Elth. t.* 202. *f.* 288. *Salm Dyck, fasc.* 4. *t.* 18. *E. & Z.!* 2046. *M. coccineum and emarginatum, E. & Z.!* 2047 *and* 2057. *M. tenuifolium, Thunb.! herb. ex pte.*

HAB. Sandy places in Cape Flats and on the sides of Tablemountain, near Brackfontein. May–Sept. (Herb. Thunb., D., Sd.)
Distinguished from *M. coccineum* by the scabrous-papulose peduncle and calyx, and yellow flowers ; the dots on the leaves are also more evident. Calyx-lobes unequal.

204. M. inæquale (Haw. Syn. 266); stem and branches slender, *effusely decumbent;* leaves semicylindrical-triquetrous, green; fructiferous peduncles compressedly clavate ; petals *croceous* with a red line on the outside. *Salm Dyck, l. c. t.* 19.

HAB. Cape of Good Hope.
Very near *M. bicolorum*, and perhaps a variety. Branches decumbent, peduncle longer and more slender, flowers paler on the outside. Calyx-lobes longer than the membrane ; in *M. bicolorum* not longer.

205. M. tenuifolium (Linn. Spec. 693); stem and branches effuse-procumbent; leaves *linear-semiterete, subulate, punctate,* as well as the peduncles *smooth;* flowers *scarlet. Dill. Elth. t.* 201. *f.* 256. *DC. Pl. Grass. t.* 82. *Salm Dyck, l. c. t.* 21.

VAR. *β.* **minus**; stem and branches elongate, prostrate, often creeping or rooting ; branches erect, short ; leaves ½–1 inch long.

HAB. Near Capetown, var. *β.*, in the Cape Flats. (Herb. Sd.)
Branches numerous, decumbent. Leaves longer than the internodes, 1½–2 inches long, 1 line wide, pellucid-punctate. Flowers solitary; peduncle 1–2 inches long. Calyx a little scabrous, lobes acute. It varies with suberect stem, and very rarely with 4-cleft calyx and 4-valved capsule.

206. M. stenum (Haw. Phil. Mag. 1831. 420) ; stem and branches slender, effuse, flexuous ; leaves incurvate-erect, subterete, mucronate, attenuate at the base, glaucescent, *without dots;* peduncles compressed ; *petals rose-coloured. Salm Dyck, l. c. t.* 20.

HAB. Cape of Good Hope.
Much branched. Leaves numerous, 6–15 lines long, 1 line wide. Flowers ternate, by abortion solitary, about an inch in diameter. Peduncles 1 inch long, the lateral bracteate. Different from the other species of the section by the rosy flowers and not punctate leaves.

§. 47. *ADUNCA*, Salm Dyck, Haw. DC. *Stem* suffruticose, dwarf, branches flexuous, suberect or prostrate. *Leaves* crowded at the top of the branches, subcylindraceous, subulate, patent, usually incurved, hooked at the apex. *Flowers* solitary, small, reddish, pedunculate ; *peduncles* bracteate. *Calyx* 5-cleft, in one species 4-cleft. *Stigmas* 5, very rarely 4. (Sp. 207–211.)

207. M. spinforme (Haw. Misc. 87. Syn. 291); stem and branches *erect;* leaves *distant,* subconnate, cylindrically-subulate, *incurvate-erect,* recurved at the apex; peduncles and keels of the bracts rather scabrous; styles purple. *Salm Dyck, fasc.* 1. *t.* 30.

HAB. Lions Mountain, May, 1838, *Dr. Harvey.* (Herb. D.)

Stem about 1 foot, woody, rigid; branches greyish-brown. Leaves unequal, 1–2 inches long, 1–2 lines wide, subpunctate. Peduncle 1 inch long, with 2 small, acute bracts in the middle. Calyx-lobes subequal. Petals 4 lines long, bluish.

208. M. curvifolium (Haw. Misc. 88. Syn. 290); branches *divaricate,* crowded, flexuose; leaves *crowded,* connate, cylindrically-subulate, *spreading-incurvate,* recurved at the apex; flowers short pedunculate; *calyx clavate, bracteate at the base;* styles *purple.*

VAR. *α.* **majus** (Salm Dyck, l. c. t. 31); *M. ceratophyllum, Willd. Enum. Suppl.* 36.

VAR. *β.* **minus** (Salm Dyck, l. c.); *M. flexifolium, Haw. Rev.* 153. *M. aduncum, Willd. Enum.* 534.

HAB. Cape of Good Hope.

Stem ½ foot or higher, diffuse. Leaves in var. *α.* 12–15 lines long, 2 lines wide, in var. *β.* a little shorter, acuminate, subflexuous, green, obsoletely punctate. Peduncle 4–5 lines; calyx 6–7 lines long. Petals bluntish, nearly white at the base. Distinguished from the preceding by shorter, more acuminate leaves and twice longer, shorter peduncled flowers.

209. M. aduncum (Haw. Misc. 87. Syn. 291); branches *suberect,* slender, crowded, flexuous, *smooth;* leaves crowded, connate, cylindrically subulate, spreading incurved, much recurved at the apex; flowers pedunculate; peduncle *with 2 bracts in the middle;* calyx *obconical;* styles *yellowish. Salm Dyck, l. c. t.* 32. *M. splendens, E. & Z.!* 2038.

HAB. Near Vankampsbay and on Lionsmountain, July. (Herb. Sd.)

Shrub ½ foot, much branched. Leaves 6–10 lines long, 1 line wide, subflexuous, scarcely punctate. Flowers small; peduncle 9–12 lines long. Petals about 4 lines long, reddish, paler or whitish at the base. Styles 5, subulate, erect. Much smaller than *M. curvifolium,* leaves more recurvate and smaller, calyx not clavate and petals much shorter.

210. M. inconspicuum (Haw. Phil. Mag. 1826. p. 128); much branched, branches *divaricate,* slender, punctate, *rough;* leaves subdistant, connate, *compressed-semiterete,* shining, spreading, *uncinate-mucronulate* at the apex; peduncle short, papillose, *ebracteate. Salm Dyck, fasc.* 6. *t.* 22.

HAB. Cape of Good Hope.

Shrub 1 foot, diffusely branched. Leaves 5–7 lines long, ½ line wide, green, minutely punctate. Peduncle 6–8 lines long. Calyx obconical, very small, 4-cleft. Petals reddish, 3 lines long. Styles 4, erect.

211. M. filicaule (Haw Misc. 88. Syn. 291); stem and branches filiform, *prostrate, creeping;* leaves crowded, connate, cylindrically-subulate, spreading-incurved, recurvate at the apex; flowers *very long peduncled;* peduncles *bracteate at the base;* styles yellowish-green. *Salm Dyck, fasc.* 1. *t.* 33.

HAB. Cape of Good Hope.

The smallest of the section and very distinct by the filiform, creeping or prostrate, weak stems and elongated (2 inches long) peduncles. Leaves about 1 inch long.

Calyx obconical, small. Petals reddish with a deeper colour on the outside, 4 lines long.

Series II. PAPULOSA.—Stem and leaves more or less beset with glittering papillæ. (Groups V.-VI.)

Group. V. PAPILLOSA.—Stem fruticose or suffruticose, as well as the branches woody or fleshy, often rough or hispid. Leaves triquetrous or terete, beset with scabrous or glittering dots. (§§. 48–58.)

§. 48. *SCABRIDA*, Haw. DC. *Stem* fruticose or suffruticose, with slender branches. *Leaves* more or less triquetrous-compressed, elongated, punctate-scabrous. *Flowers* ternate, reddish-violet, pedunculate; *peduncles* bracteate. *Calyx* 5-cleft. *Stigmas* 5. (Sp. 212–218.)

212. M. glomeratum (Linn. Spec. 694); suffruticose, branches slender, compressed, *erect*, crowded; leaves spreading, linear, compressed, semiterete, *a little incurved, green*, with large, prominent dots; flowers mostly biternate. *Salm Dyck, fasc.* 6. *t.* 23. *Dill. Elth. t.* 213. *f.* 274. *M. inflexum, Haw. Rev.* 138. *M. glomeratum et polyanthum, E. & Z.!* 2053, 2054. *Herb. Un. itin,* 518. *M. tenuifolium, Thunb. herb. ex pte. Drege* 6998.

VAR. *β.* **majus**; leaves and flowers larger. *M. laeve, E. & Z.!* 2033.

HAB. Near Capetown in the flats, and on Table and Lionsmountain, and in district of Tulbagh, var. *β*, at Seapoint near Adow, Uitenhage, Oct.–Nov. (Herb. Thunb., D., Sd.)

Stem 1–1½ foot high; young branches compressed, purplish. Leaves 6–9 lines long, 1 line wide, in var. *β.* 1½ lines wide, subattenuated at the base, acute. Flowers often subpaniculate, middle sized, pedicels about 1 inch long. Calyx obconical, 3 of the lobes with a large membranaceous margin. Petals narrow-lanceolate, acute, twice, in var. *β.* three times longer than the calyx-lobes, rose-coloured.

213. M. polyanthum (Haw. Syn. 270); stem and branches slender, *spreading, diffuse*, flexuose; leaves much crowded, erect-spreading, *very narrow*, compressed-semiterete, *bluish-glaucous*, prominently dotted; flowers very numerous, ternately panicled. *Salm Dyck, l. c. t.* 24. *M. imbricans, Haw. Rev.* 139. *M. flexile, Haw. Rev.* 140. *M. violaceum, E. & Z.!* 2056.

HAB. Mountain sides at Tulbagh near the cataract, Nov. (Herb. D., Sd.)

Subshrub 1 foot, much branched. Leaves 4–8 lines long, ½ line wide, subtriquetrous-compressed at the top. Flowers ternate, biternate or paniculate, peduncles 1½–2 inches long, with two bracts in the middle. Calyx turbinate, lobes reflexed. Petals biseriate, about ½ inch long, obtuse or bidentate, rosy-violet.

214. M. violaceum (DC. Pl. Grass. t. 84); stem and branches *erect*; leaves crowded, spreading-incurved, compressed-semiterete, *a little incrassate* at the apex, bluntish, mucronulate, glaucous, prominently dotted; flowers *numerous, very long peduncled*, ternately panicled; petals *entire*. *Salm Dyck, l. c. t.* 25. *M. polyphyllum, Haw. Rev.* 141. *M. puniceum, Jacq. Hort. Schoenb. t.* 442.

HAB. Cape of Good Hope.

By the more robust, erect stem, less diffuse branches, longer and a little incrassate leaves distinct from the preceding. Leaves 5–12 lines long, scarcely 1 line wide at the apex. Peduncles 3 inches long. Petals twice longer than the calyx, violet.

215. M. emarginatum (Linn. Spec. p. 692); stem erect, branches expanded, flexuose; leaves subremote, linear-elongated, spreading, in- or re-curved, semiterete, subcompressed at the apex, acutish, glaucous, *scabrous-punctate;* flowers very long pedunculate, *ternate;* petals *emarginate. Dill. Elth. t.* 197. *f.* 250. *Salm Dyck, l. c. t.* 26.

HAB. Cape of Good Hope.

Stem 2 feet high. Flowers less numerous, very long pedunculate, and petals bidentate distinguish it from *M. violaceum.*

216. M. elegans (Jacq. Hort. Schoenb. 4. t. 436); stem suffruticose; branches compressed, decumbent or deflexed, bark whitish or red; leaves *rather triquetrous,* narrow, very glaucous, scabrous; flowers numerous, mostly panicled; lobes of calyx reflexed; petals pale reddish, *entire;* stamens collected. *M. retroflexum, emarginatoides, leptaleum et deflexum, Haw. M. incurvum, E. & Z.!* 2018. *M. Thunbergii, E. & Z.!* 2036. *ex pte. M. flexuosum, longistylum and versicolor, E. & Z.!* 2039, 2040, 2055. *M. tenuifolium, Thunb.! herb. ex pte. Herb. Un. itin.* 512.

HAB. Sandy places and rocks, Table and Lionsmountain, Hottentottsholland, Swellendam, Worcester and Uitenhage. (Herb. Thunb., Sd.)

Shrub ½–1 foot or higher, much branched. Leaves crowded, 6 lines long, ½–¾ line wide. Peduncle 1 inch or longer. Flowers reddish or whitish, sometimes pale red with darker lines. Petals ½ inch long. Smaller and more slender than the preceding.

217. M. versicolor (Haw. Misc. 17. Syn. 268); stem shrubby, branched; leaves almost triquetrous, glaucescent, scabrous from the warts; lobes of calyx ovate, acuminated; petals somewhat bidentate at the apex; stamens collected. *DC. l. c. p.* 434.

HAB. Cape of Good Hope.

Petals variable, of a shining white or silvery colour, but when closed in the morning and evening they are pale-reddish. Probably a variety of the polymorphous *M. elegans.*

218. M. scabrum (Linn. Spec. 992); stem and branches erect, *straight;* leaves *linear-elongated,* spreading-recurved, triquetrous-compressed, bluntish, green, *as well as the calyx very rough from shining warts;* flowers ternate; petals often crenated at the apex; stamens collected. *Dill. Elth. t.* 197. *f.* 251. *Salm Dyck, l. c. t.* 27. *M. emarginatum, E. & Z.!* 2057. *Herb. Un. itin.* 516. *ex pte.*

HAB. Stony places on the sides of Lionsmountain and Tablemountain, Feb.–June. *Zey.!* 2589. (Herb. D., Sd.)

Stem. 1–1½ foot. Leaves 9–15 lines long, 1 line wide. Flowers ternate, or by abortion solitary. Peduncle 1½–2 inches, thickened and warted near the calyx. Petals 2 or three times longer than the calyx, narrower than in the cultivated plant and not crenated, obtuse or acute, reddish.

§ 49. *TRICHOTOMA,* Haw. DC. *Stem* fruticose or suffruticose, fleshy, at length woody, erect, often with tuberous roots. *Leaves* semicylindrical or triquetrous-compressed, minutely papulose. *Flowers* disposed by threes, corymbose, small, white or reddish. *Calyx* 4–5-cleft. *Stigmas* 4–5. (Sp. 219–223.)

219. M. tuberosum (Linn. Spec. 693); root tuberous, hard; stem erect; branches *diffuse, tortuose;* leaves rather triquetrous-compressed, incurvate-spreading, recurved at the apex, green, beset with very minute papulæ; flowers trichotomous, subcorymbose, lateral peduncles biternate; *petals red. Dill. Elth.* 275. *f.* 264. *DC. Pl. Grass. t.* 78. *Salm Dyck, fasc.* 6. *t.* 28. *M. umbellatum, fol.* δ. *and* ε. *herb. Thunb. E. & Z.!* 2058.

HAB. Mountain sides, Bothasberg near Vischriver, in Karro between Beaufort and Graafreynet; Zwartskopriver, Rhinosterkop, *Zey.!* 679. 2608, Dec.–July. (Herb. Thunb., D., Sd.)

Root very large, globose. Stem woody, much branched, about 1 foot high. Leaves crowded, 6–10 lines long, 1 line wide. Peduncles ternate, intermediate without bract, lateral ones with small bracts. Pedicels 2–3 lines long, persistent and subspinescent. Calyx subglobose, as large as a pea. Petals 2 lines long, acute.

220. M. megarhizum (Don. Gen. Hist. v. 3. 145); *root tuberous, hard;* stem erect, with the branches *straight;* leaves rather triquetrous, compressed, incurvate-spreading, recurved at the apex, green, beset *with very minute papulæ;* flowers trichotomous, subcorymbose, lateral peduncles biternate; *petals white. M. macrorbizum, Haw. Phil. Mag.* 1826. 332, *not of DC. Salm Dyck, fasc.* 4. *t.* 22.

HAB. Cape of Good Hope.

Very like *M. tuberosum*, but differs in the principal stem being more equal in thickness, straight and erect branches, and in the flowers being white.

221. M. subincanum (Haw. Phil. Mag. Dec. 1824, p. 427); *root woody, branched, fibrous;* stem firm, erect, branched; leaves connate, spreading, trigonal-compressed, rather canescent, soft, *without dots or papulæ*, recurved and mucronulate at the apex; flowers trichotomous, subcorymbose, fragrant, white. *Salm Dyck, fasc.* 2. *t.* 23. *M. testaceum, et brachiatum, E. & Z.!* 2059, 2060.

HAB. Karro-like hills near the Zwartkopsriver, Rhinosterkop, dist. Beaufort and Albany, *Zey.!* 678. 681, 686, 2604. Nov.–April. (Herb. Thunb., D., Sd.)

Most nearly allied to the preceding, but generally a little larger, the stem 1½ feet high, leaves 1 inch long, 1½–2 lines wide, not papulose, but clothed with a very minute greyish down, the flowers somewhat larger, petals about 5 lines long.

222. M. testaceum (Haw. Suppl. 97. Rev. 178); stem shrubby, erect; branches often declinate, glabrous; leaves connate, spreading, *triquetrous*, compressed, glaucous-green, *minutely punctate-papulose*, recurved and mucronulate at the apex; flowers terminal, ternate, corymbose or in trichotomous umbels; flowers of a coppery-colour. *M. geminiflorum, E. & Z.!* 1991.

HAB. Karro-like hills near Zwartkopsriver, *Zey.!* 2584. Jan.–April. (Hb. D. Sd.)

Leaves 6–12 lines long, 1–1½ lines wide, more triquetrous than in the preceding; flowers the size of *M. subincanum*, often only ternate; lateral pedicels bracteate in the middle; lobes of calyx subequal, acute; petals a little longer than the calyx.

223. M. Ecklonis (Salm Dyck, l. c. fasc. 6. t. 29); root woody, branched; stem and branches *pubescent*, at length glabrate; leaves depressed-triquetrous, attenuated on both ends, spreading-recurved, mucronulate, minutely papillate and *hairy;* flowers white, subtrichotomous; calyx 5-cleft, 2 lobes elongated. *M. lanceum et villosum, E. & Z.!* 2105, 2106.

HAB. Woods near Adow, Zwartkopsriver, *Zey !* 2603. Sept.–Oct. (Herb. Thunb. D. Sd.)

Branched from the base, 1½ foot high, when young fleshy. Leaves subconnate, 6–10 lines long, 3–4 lines wide in the middle, flattish or a little canaliculate above, obtusely keeled beneath. Flowers ternate or biternate, rarely by abortion subsolitary; lateral peduncles with 2 leafy bracts. Flowers ½ inch in diameter. Petals biseriate, as long as the two longer lobes of the calyx.

§. 50. *ASPERIUSCULA*, Haw. DC. *Stem* fruticose or suffruticose; branches erect, slender, rough. *Leaves* distant; cylindraceous or triquetrous-compressed, glittering from papulæ. *Flowers* solitary, reddish, yellow or copper-coloured, small or large. *Calyx* 5–6-cleft. *Stigmas* 5–6. (Sp. 224–233.)

* *Parviflora.*—Flowers reddish or yellow. (Sp. 224–229.)

224. M. pulverulentum (Haw. Syn. 272); stem erect, branches crowded; leaves cylindrically-triquetrous, obtuse, dotted with white, powdery, scabrous; *calyx 6-cleft;* petals reddish.

HAB. Cape of Good Hope.

Flowers almost like those of *M. barbatum*, whitish at the smaller base. Filaments erect. Styles 6, recurved at the apex.

225. M. asperulum (Salm Dyck, l. c. fasc. 5. t. 28); stem erect; branches numerous, slender, straight, rough; leaves linear-elongate, crowded, semicylindrically-triquetrous, incurvate-spreading, hooked and mucronulate at the apex, papulose; *flowers lateral; calyx* 4-*cleft.*

HAB. Cape of Good Hope.

Shrub 1½ foot. Leaves 6–10 lines long, about 1 line wide, green, minutely papillate. Peduncle 5 lines long, thickened above, papillose. Petals uniseriate, 5 lines long, pale rose-coloured, with a darker dorsal line. Stamens erect.

226. M. parvifolium (Haw. Rev. 184); stem suberect; branches diffuse, filiform, rough; leaves minute, crowded, expanded, *triquetrous, subcymbæform,* papulose; *flowers terminal;* calyx 5-cleft; petals minute, deep purple. *Salm Dyck, fasc.* 3. *t.* 35. *M. pulverulentum, E. & Z.!* 2061. *ex pte.*

HAB. Rhinosterkop, *Zeyher,* 638; Namaqualand, *A. Wyley!* Karro-like hills between Gauritzriver and Langekloof, *George.* Dec. (Herb. Sd.)

The smallest of the section, ½–¾ foot high. Leaves 2 lines long and wide, a little dilated at the keel, glittering. Peduncles ½ inch long. Calyx with small, acute, subleafy, equal lobes. Petals about 3 lines long. Exterior filaments spreading. Styles 5, recurvate.

227. M. brevifolium (Ait. Kew. v. 2. 188); stem erect; branches numerous, erecto-diffuse, slender, rough; leaves small, crowded, spreading, *triquetrous-compressed,* very blunt, papulose; flowers *lateral;* calyx 5-cleft; filaments reddish, barbate at base. *Salm Dyck, fasc.* 4. *t.* 23. *M. erigeriflorum, Jacq. Hort. Schoenb. t.* 477. *M. lateriflorum, Red. pl. grass. M. subglobosum et parvifolium, E. & Z. !* 2062, 2063. *M. capillare, Thunb.! herb. M. hispidum,* δ. *Thunb.! herb. M. subglobosum, Haw. Syn.* 273.

HAB. Karro-like hills near Uitenhage, Olifantsriver, and on Gauritzriver, Swellendam, Sept.–Dec. (Herb. Thunb., Jacq., D., Sd.)

1 foot or higher, much branched, branches adscendent. Leaves 2–5 lines long, 1–1½ lines wide, glittering from papulæ. Peduncle 3–5 lines long, glittering as well as the obtuse calyx. Petals 4 lines long. Stamens erect. Styles 5, short.

228. M. obliquum (Haw. Rev. 183); stem erect; branches spreading, filiform, rough; leaves much distant, *cylindrical*, obtuse, glittering from papulæ, with one of the pair deflexed, the other opposite it ascending. *Salm Dyck, fasc.* 2. *t.* 24. *Bot. Reg. t.* 863. *M. brevifolium, E. & Z.!* 2065.

HAB. Karro-like hills near Zwartkopsriver, *Zeyher*, 2605, 2606. Nov.–Jan. (Herb. D., Sd.)

Very distinct by the longer (6–8 lines) reflexed and adscending leaves. Branches when old, smooth, erect or more or less spreading. Peduncles ½–1 inch, in the cultivated plant twice as long. Calyx as large as a large pea, obconical, 5-cleft, lobes subequal. Petals purple, 4 lines long. Styles 5, filiform.

229. M. flavum (Haw. Rev. 183); stem dwarf, erectish; branches very slender, scabrous; leaves much crowded, *nearly terete*, rather attenuated on both ends, glittering from papulæ, *subincurved or variously bent;* calyx obconical, 5-cleft, lobes equal, obtuse; petals *yellow. Salm Dyck, l. c. t.* 25. *M. obliquum, E. & Z. !* 2064.

HAB. Karro-like hills near Zwartkopsriver, *Zey. !* 2590 *ex pte.* June–Oct. (Herb. D., Sd.)

Small shrub. Leaves 3–4 lines long, about 1 line wide. Peduncles 1½–2 inches long, filiform, papillate, glittering, without bract. Calyx turbinate, 2 lines long, glittering; petals nearly twice as long, subbiseriate. Capsule obconical, depressed, umbilicate, 5-valved.

* *Grandiflora.*—Flowers croceous. (Sp. 230–233.)

230. M. collinum (Sond.); stem and branches erect, *straight*, slender, rough; leaves distant, *erect*, subtrigonous-cylindraceous, obtuse, attenuate at the base, papillate; flowers middle-sized croceous; petals *linear-subulate. M. micans, Thunb. ! Fl. Cap.* 426. *M. flavum, E. & Z.!* 2066.

HAB. Karro-like hills in Bockefeldt, *Thunberg ;* near Gauritzriver, *E. & Z. !* Dec. (Herb. Thunb., Sd.)

Shrub 1 foot, with filiform, purplish branchlets. Leaves 5–8 lines long, ½–¾ line wide, minutely papulate, a little larger near the apex. Peduncles 1½–2 inches long, ebracteate. Flowers about 6 lines in diameter. Calyx glittering, 5-cleft, 5-horned. Petals in many rows, very narrow. Filaments yellow, much smaller than the following.

231. M. micans (Linn. Spec. 696); stem erect; branches elongate, *erect-spreading*, slender, rough; leaves distant, *spreading*, semicylindrical, obtuse, *subrecurved* at the apex, papillate; flowers large, purple-croceous; petals *lanceolate. Dill. Elth. t.* 215. *f.* 282. *DC. Pl. Grass. t.* 167. *Salm Dyck, fasc.* 4. *t.* 24. *M. hispidum, var. γ. herb. Thunb. ! M. micans, E. & Z. !* 2068.

HAB. Stony places near Puspasvalley, Swellendam, *E. & Z. !* Genadenthal, *Dr. Roser;* Rietkuil, *Zey. !* 2613. Sept.–Oct. (Herb. Thunb., D., Sd.)

Shrub 2–3 feet high; branches papillate-scabrous. Leaves 8–12 lines long, 1–2 lines wide, obtuse, with a short obtuse recurved mucro. Flowers 1 inch and more in diameter. Calyx glittering with 2 longer, subulate lobes. Petals when dry nearly blackish.

232. M. speciosum (Haw. Syn. 270); stem erect; branches elongate, erect-spreading, slender, rough; leaves much distant, much spreading, *turgid-cylindraceous*, abbreviate, acutish, papillate; flowers very large, deep scarlet; petals spathulate-lanceolate. *Salm Dyck, fasc.* 6. *t.* 30. *E. & Z.!* 2067.

HAB. Karro-like hills near Gauritzriver, Swellendam. Dec. (Herb. Sd.)

Larger than *M. micans*, branches more divergent, leaves 6–8 lines long, 3 lines wide. Calyx with 5 subequal lobes. Petals greenish at the base, nearly 1 inch long.

233. M. maculatum (Haw. Syn. 272); stems erect, covered with rough spots; leaves expanded, remote, obtuse, semicylindrical and rather compressed, papulose. *M. micans, var. β. Haw. Misc.* 98.

HAB. Cape of Good Hope.

Nearly allied to *M. micans*, but the stem is higher, more erect, the branches less rough, more slender and filiform, the leaves shorter, scarcely incurved, but more horizontal. Flowers unknown.

§. 51. *HISPIDA*, DC. (Hispicaulia, Haw.) *Stem* suffruticose, woody at the base, diffuse; branches slender, divergent, hispid from bristles. *Leaves* cylindraceous, papulose. *Flowers* solitary, reddish or white; peduncles hispid. *Calyx* 5-cleft. *Stigmas* 5. (Sp. 234–241.)

234. M. pruinosum (Thunb.! Nov. Act. Nat. Cur. v. 8. p. 17. App.); branches erect, spreading, terete, when young densely beset with setiferous papulæ, when old only papulose; leaves trigonous-cylindrical, obtuse, papulose; flowers axillary, solitary or by threes terminal, lateral pedicels bracteate; calyx setiferous, 5-cleft, nearly equalling the petals. *Fl. Cap. p.* 425.

HAB. Karro near Lurisriver and in Cannaland, *Thunb.!* near Zwartkopsriver, *Zeyher!* 2596. Dec.–Jan. (Herb. Thunb., D., Sd.)

Subshrub ½–1 foot, decumbent; lower branches nearly smooth, upper ones whitish-punctate from papulæ, in the young state terminated by a rigid bristle. Leaves fleshy, thick, approximate or remote, 5–6 lines long, 2 lines wide, rarely longer; young ones setiferous. Peduncles 2–4 lines long, thickened above. Calyx about 2–3 lines long, lobes erect, thick, subequal. Petals reddish as it seems in the dry specimens, not yellow, as described by Thunberg. Capsule obconical, 5-valved on a compressed 4–6 lines long peduncle.

235. M. striatum (Haw. Syn. 275); stem and branches suberect, setose; leaves *turgid-cylindraceous*, obtuse, glittering from *setiferous papulæ;* calyx woolly, with subequal, leafy lobes; petals rose-coloured, purple, striate; stamens collected. *Salm Dyck, fasc.* 2. *t.* 26. *M. striatum and hirtellum, E. & Z.!* 2072, 2073.

VAR. *α.* **roseum** (DC. l. c. 441); petals pale rose-coloured, with a deeper coloured line in the middle. *M. striatum, Haw. l. c. M. hispidum, γ. Linn. Dill. Elth. f.* 281. *M. hispidum, fol. ε. herb. Thunb.! Zey.!* 2593. *ex pte.*

VAR. *β.* **pallidum** (DC. l. c.); petals white, with a red line at the base. *M. striatum, DC. Pl. Grass. t.* 132.

VAR. *γ.* **hispifolium** (Salm Dyck, l. c.); papulæ of the leaves beset with reversed hairs. *M hispifolium, Haw Rev.* 198. *M. tuberculatum, DC. l, c.*

HAB. In fields near Zwartkopsriver; between Zwarteberg and Klynriviersberge, Caledon; near Tulbagh and on Olifantsriver. Sept.–Nov. (Herb. Thunb., D., Sd.)

Stem 1–2 feet, as well as the branches beset with very spreading bristles. Leaves 6–12 lines long, 1½ line in diameter. Peduncles setose, 1–3 inches long, filiform Calyx subcampanulate, 2 of the lobes a little shorter, with membranaceous margins.

Petals subbiseriate, lanceolate, 5–6 lines or a little longer than the calyx-lobes. Styles 5, short, erect.

236. M. attenuatum (Haw. Rev. 188); stem and branches short, adscending, slender, setose; leaves crowded, *linear-cylindraceous, obtuse*, glittering from *crystalline papulæ;* calyx-lobes *subequal;* petals white, reddish-striate; stamens collected. *Salm Dyck, l. c. t.* 27. *E. & Z. !* 2074.

HAB. Sandy places on the sea shore near Cape Agulhas. Nov. (Herb. Sd.)

Near the preceding but much smaller, stem and branches shorter, more slender, decumbent, and the flowers somewhat smaller. Leaves 5–8 lines long, 1 line wide. Lobes of calyx leafy, 2 membranaceous on the margins, about twice shorter than the petals. Styles, 5, short, recurvate.

237. M. calycinum (Haw. Rev. 187); stem and branches erect-effuse, very slender, setose; leaves distant, linear-cylindrical, *subattenuated at the apex*, bluntish, glittering from very minute papulæ; calyx-lobes unequal, 2 *of them very elongated*, cylindrical, leafy; petals white. *Salm Dyck, l. c. t.* 28. *M. hispidum, fol. β. herb. Thunb. ! and Fl. Cap.* 418.

HAB. Sandy fields near Olifantsriver. Oct.–Nov. (Herb. Thunb., D,)

Branches subfiliform. Leaves often crenate-recurved, 6–9 lines long, 1 line wide. Peduncles 6–10 lines long. Calyx subcampanulate, the longer lobes equalling the acute, white petals.

238. M. candens (Haw. Rev, 186); stem and branches effuse, ascending, very slender, *minutely setose;* leaves crowded, *linear-cylindrical,* blunt, subattenuate at the base, glittering and *canescent* from minute papulæ; lobes of calyx *subequal;* petals *white. Salm Dyck, fasc.* 4. *t.* 25. *M. hispidum, fol. α. herb. Thunb.! and Flor. Cap.* 418. *M. candens, E. & Z.!* 2076.

HAB. Sea shore near Cape Recief, Uitenhage, Jan.–Feb. (Herb. Thunb., D., Sd.)

Stem diffuse or prostrate, often rooting, branches 1–2 feet or longer, subflexuous, filiform, scabrous from very short setae. Leaves 4–6 lines long, scarcely 1 line wide. Peduncles terminating the short branches, 4–8 lines long. Calyx obconical, 3 lines long, a little shorter than the uniseriate, acute, white petals. Styles subulate, ramentaceous, equalling the stamens. It varies with greener and subglaucous leaves when cultivated.

239. M. subcompressum (Haw. Phil. Mag. Dec. 1826. p. 131); stem erect; branches slender, diffuse, when young rather pilose; leaves *compressed-semiterete,* very blunt, *greenish-canescent,* glittering from very minute papulæ; lobes of calyx deflexed, 2 subleafy, 3–4 shorter, flat, with membranaceous margins; petals *purplish. Salm Dyck, fasc.* 5. *t.* 29.

HAB. Cape of Good Hope.

About 2 feet high. Lower part of the branches scabrous-punctate, but not beset with very short hairs as the upper part. Leaves remote, 6–9 lines long, 1–2 lines wide. Peduncle 1–2 inches long, filiform, papillate. Calyx obconical, 5–6 cleft, lobes a little shorter than the spreading petals. Styles 6, filiform. Capsule 6-locular.

240. M. hispidum (Linn. Spec. 691); stem, branches, as well as the peduncles *erect,* hispid; leaves *cylindraceous, very obtuse,* glittering from crystalline papillæ; petals large, purplish; *stamens effuse*; styles exserted. *Dill. Elth. t.* 214. *f.* 277 *et* 278. *Salm Dyck, fasc.* 4. *t.* 26. *Thunb. Fl. Cap.* 418. *M. hispidum et Boerhavii, E. & Z.!* 2069, 2070.

HAB. Mountains near Olifantsriver and between Hauwhoek and Caledon. Oct.–Nov. (Herb. Thunb., Sd.)

In habit very near *M. micans.* Stem 1–2 feet; branches with reflexed bristles. Leaves 5–10 lines long, 1½ line wide, exactly cylindrical, green. Peduncle elongated. Calyx obconical, lobes equal, short, bluntish. Petals uniseriate, subspathulate, much longer than the calyx. *M. hirtellum,* Haw. Obs. 284, is the same, with pale red flowers.

241. M. floribundum (Haw. Syn. 274); stem and branches tortuous, *subdecumbent,* setose; leaves cylindrical, obtuse, glittering from minute papulæ; flowers *axillary, small;* petals rose-coloured; stamens effuse; styles exserted. *Salm Dyck, fasc.* 3. *t.* 36. *E. & Z.!* 2071. *M. tuberculatum, E. & Z.!* 2075. *M. torquatum & furfureum, Haw. M. hispidum, β. pallidum, Willd. Dill. Elth. t.* 214. *f.* 279–280.

VAR. *β.* **erectius**; Salm Dyck, l. c. Branches more erect, and flowers smaller.

HAB. Karro-like hills, near Gauritzriver, Hassaquaskloof, and Breederiver, *E. & Z.*; Zwartkopriver, *Zeyh.* 2593; Rhinosterkop, *Zeyh.* 677; Namaqualand, *A. Wyley.* Sept.–Dec. (Herb. D. Sd.)

Branches ½ foot or shorter. Leaves mostly arcuate-curvate, 5–10 lines long, 1 line wide, a little thicker near the apex. Peduncles 1–1½ inch long or shorter. Calyx subturbinate, lobes equal, obtuse. Petals uniseriate, twice longer than the calyx. Styles 5, subulate.

§. 52. *BARBATA,* Salm Dyck, Haw. DC. *Stem* suffruticose; branches erectly-decumbent, diffuse or crowded. *Leaves* near terete, subpapulose, thick, bearded at the apex by radiating hairs. *Flowers* solitary, reddish. *Calyx* turbinate, papulose, 5–8-cleft, with as many blackish-green tubercles on the torus, and bearded lobes. *Stigmas* 5–8. (Sp. 242–247.)

242. M. barbatum (Linn. Spec. 691); stem and branches diffuse, decumbent, *smooth;* leaves remote, erect-spreading, turgid-semicylindraceous, *ending in* 5–6 *radiating hairs;* base of calyx glabrous, lobes unequal; petals entire. *Dill. Elth. t.* 190. *f.* 234. *Salm Dyck, fasc.* 4. *t.* 27. *E. & Z.!* 2078. *DC. Pl. Grass. t.* 28. *M. stelligerum, Haw. Phil. Mag. Jul.* 1824.

HAB. Hills in Zwartland. Nov. *E. & Z.* (Herb. Sd.)

About 1–1½ foot high. Leaves 5–6 lines long, 2 lines wide, pellucid-papulose, green. Flowers axillary and terminal. Calyx turbinate, 5–6-cleft, two of the lobes longer. Petals uniseriate, twice or three times longer than the calyx, acute.

243. M. intonsum (Haw. Phil. Mag. 1824. p. 62); stem and branches slender, erectly-deflexed, *hispid;* leaves remote, spreading-recurved, turgid-semicylindraceous, *attenuated,* echinate-papulose, *ending in* 8–10 nutbrown *radiating* hairs at the apex; lobes of calyx *unequal. Salm Dyck, fasc.* 2 *t.* 29. *M. bulbosum, E. & Z.!* 2081.

VAR. *a.* **rubicundum**; flowers reddish.

VAR. *β.* **album**; flowers white, but at length becoming reddish as they fade *Haw. l. c.*

HAB. Fields near Zwartkopsriver and Zondagsriver. *Zeyh.* 2581. July–Oct. (Herb. Sd.)

Very distinct, by the slender stem, attenuated, recurvate, hispid-papulose leaves, and the brownish hairs at the apex. Leaves 6–7 lines long, 2 lines wide. Flowers terminal on short peduncle. Calyx setiferous, two of the lobes longer. Petals sub-biseriate, interior row shorter.

244. M. bulbosum (Haw. Phil. Mag. Dec. 1824. p. 428); root tuberous; stem and branches diffuse, *pubescent*, at length subglabrous; leaves crowded, *horizontal*, subterete, minutely echinate-papulose, terminating *in* 10–15 *radiating white* hairs; lobes of calyx *equal*. *Salm Dyck, l. c. t.* 30. *M. intonsum, E. & Z.!* 2080.

HAB. Karro-like hills, near Port Elizabeth. Oct.–Nov. (Herb. D. Sd.)

Much branched, ½ foot or higher. Leaves connate, 4–5 lines long, 1½ lines wide. Peduncles short. Calyx papulose, lobes short. Petals reddish, ⅓ inch long, sub-uniseriate, entire.

245. M. stelligerum (Haw. Rev. 190); stem and branches diffuse, tortuose-suberect, *smooth;* leaves crowded, very spreading, nearly cylindrical, papulose, terminating in 5–10 radiating white or brownish hairs; lobes of calyx equal; petals *bidentate. M. barbatum, Bot. Mag. t.* 70. *DC. l. c.* 440. *M. stelligerum. Salm Dyck, fasc.* 5. *t.* 30.

HAB. Cape of Good Hope.

In habit near *M. barbatum*, but differing by more crowded, very spreading leaves, equal calyx lobes, and bidentate petals. Leaves 4–5 lines long, 2 lines wide, obtuse. Flowers terminal and axillary. Petals uniserial, reddish. *M. stelligerum, E. & Z.!* 2083, is a very different plant, nearly intermediate between *M. barbatum* and *intonsum.*

246. M. stellatum (Mill. Dict. ed. 8. n. 14); stem and branches short, fleshy, tufted; leaves much crowded, *glaucous*, nearly *semiterete, papillose-scabrous*, terminating in *many radiating* hairs at the apex; calyx 6–8-cleft as well as the peduncles hairy. *Dill. Elth. t.* 199. *f.* 235. 1, 2, 3. *DC. Pl. Grass. t.* 29. *E. & Z.!* 2082. *Salm Dyck, fasc.* 6. *t.* 31. *M. barbatum, β. Linn.. M. hirsutum, Spreng.*

HAB. Karro-like-hills, near Zwartkopsriver, Coega, and Zondagsriver. *Zeyh.* 2583. Oct.–Janr. (Herb. D. Sd.)

About 3–4 inches high. Leaves 3–4 lines long, 1 line wide, scarcely attenuated at the apex, radiating hairs (12–16), white. Flowers solitary, peduncle 4–6 lines long. Calyx campanulate, 3 lines long, lobes obtuse, equal. Petals uniseriate, narrow, reddish-violet.

247. M. densum (Haw. Syn. 279); stem and branches short, fleshy, tufted; leaves much crowded, *flattish above, convex beneath*, beset with glittering papulæ, terminating in many radiating hairs at the apex, rather ciliated at the base; calyx 6-cleft, as well as the peduncles *very hairy. Sims. Bot. Mag. t.* 1220. *Dill. Elth. t.* 190. *f.* 236. *Salm Dyck, fasc.* 6. *t.* 32. *E. & Z.!* 2079. *M. barbatum, β. densum, Linn.*

HAB. Karro, near Olifantsriver, Clanwilliam, and near Zwartkopsriver. *Zeyh.* 2582, ex pte. (Herb. Sd.)

It differs from *M. stellatum* by longer, greenish and larger, turbinate, hirsute calyx. Leaves 6 lines long, 2 lines wide, turgid, a little recurved; radiating hairs (20–25), white. Peduncle 8–12 lines long, hirsute. Calyx 5 lines long, lobes equal, obtuse. Petals uniseriate, narrow, reddish-violet.

§. 53. *ECHINATA*, Salm Dyck, Haw. *Stem* suffruticose, short or nearly wanting; branches erect-tortuose or prostrate, strumose-nodulose. *Leaves* oblong-ovate, distinct, echinate or hispid. *Flowers* solitary, white or yellow. *Calyx* 4–5-cleft. *Stigmas* 4–5. (Sp. 248–249.)

248. M. strumosum (Haw. Rev. 190); stem and branches short, decumbent, fleshy, strumose; leaves much crowded, *depressedly-cylindrical, attenuated at both ends*, hispid all over; calyx 5-cleft, lobes nearly equal, obtuse; petals emarginate. *Salm Dyck, fasc.* 5. *t.* 29.

HAB. Cape of Good Hope.

Stem wanting or very short. Leaves spreading, 6–9 lines long, 1–1½ lines wide. Peduncle 4–6 lines long, papillate. Calyx subglobose, pilose. Petals pale-yellow or white, with a red dorsal line, 3–4 lines long. Styles 5, short, acute.

249. M. echinatum (Ait. Kew. 2. 194); stem erect, branches diffuse, erectish-tortuose papillate; leaves turgid, *ovate or oblong ovate*, gibbous, echinaceous-hispid from glittering papulæ; calyx 4–5-cleft, lobes *unequal;* petals entire. *DC. Pl. Grass. t.* 24. *Salm Dyck, fasc.* 4. *t.* 28. *E. & Z.!* 2084. *M. setosum, Moench.*

HAB. Sandy places near Zoutpan, Zwartkopsriver, *Zeyh.* 2907; Karro-like-hills, between Gauritzriver and Langekloof, George. Nov.–Dec. (Herb. D. Sd.)

Stem 1 foot, branches woody. Leaves 3–6 lines long, about 3–4 lines wide, gibbous beneath, subtriquetrous, crowded. Peduncles very short, papulose-hispid. Calyx 4 or 5-cleft, a little shorter than the white or yellowish petals. Capsule 4–6-valved.

§. 54. *SPINULIFERA*, Haw. DC. *Stem* suffruticose, often strumose at the base. *Branches* fleshy, papulose: when old they appear as if they were spiny from the permanent remains of the dried leaves. *Leaves* cylindraceous, somewhat channelled, glittering from papulæ. *Flowers* mostly ternate, greenish-yellow or greenish red. *Calyx* 5-cleft, lobes often elongated. *Stigmas* 5. (Sp. 250–256.)

250. M. nitidum (Haw. Syn. 253); stem and branches erect; leaves semiterete, *subattenuate*, obtuse, channelled above, green, glittering from papulæ; flowers *ternate*, long-pedunculate, *yellow;* lobes of calyx elongated, subulate; styles *short*. *Salm Dyck, fasc.* 4. *t.* 29. *M. brachiatum, DC. Pl. Grass. t.* 129. *M. salmoneum, E. & Z.!* 2088. *non. Haw. Zeyh.* 2599.

HAB. Fields near Zwartkopsriver. Dec. (Herb. D. Sd.)

Stem 1–2 feet, young branches papillate, old ones terete, glabrous. Leaves 8–12 lines long, 1–2 lines wide. Flowers ternate, rarely solitary, ½–1 inch long, papillate, intermediate without bracts. Calyx turbinate, lobes equal, or 2–3 shorter, equalling or a little longer than the yellow petals.

251. M. decussatum (Thunb.! Prod. 68); stem and branches erect; leaves semiterete, *scarcely attenuate*, bluntish, slightly sulcate above, green, papulose; flowers terminal, *subsolitary, white;* lobes of calyx unequal, two of them longer, subfoliaceous, obtuse; styles *long*. *M. brachiatum, Ait. Kew. ed.* 1. *v.* 2. 19?

HAB. Karro-like-hills, near Olifantsriver, and in Bocklands. Oct.–Dec. (Herb. Thunb.)

Stem 2 feet. Branches opposite, erect-spreading, subfastigiate, terete, greyish, papulose. Leaves 1 inch long, 1 line wide, upper ones gradually shorter. Flowers smaller than in the preceding. Peduncle 1–2 lines long. Calyx subturbinate, papulose, glittering, 5-cleft, 3 lines long; lobes erect, two longer and more obtuse. Petals equalling the calyx, or a little longer. Styles filiform, as long as the stamens.

252. M. auratum (Sond.); stem substrumose, with the branches erect, crystalline-papillose; leaves connate, terete, subcanaliculate above, *very obtuse*, papulose; flowers terminal, ternate or solitary, *yellow*, lobes of calyx unequal, three of them longer, cylindraceous, obtuse; styles *long*. M. *aureum*, *Thunb.! Fl. Cap.* 425. *non Linn. nec Haw.*

HAB. Karro, between Olifantsriver and Bocklandsberg, *Thunberg;* Springbokkeel and Kamos, *Zeyher*. Oct.–Feb. (Herb. Thunb. Sd.)

Near the preceding. Stem more robust, 2–3 feet high; colour of the branches not greyish but yellowish. Leaves shorter and thicker (6–8 lines long, 1½ line wide), the rudiments of old remaining leaves very spinous, in *M. decussatum* wanting or not spinous. Peduncles longer, intermediate ½–1 inch long without bracts, lateral ones longer, bibracteate. Flowers golden-yellow. Calyx 5-cleft, 6 lines long or longer; lobes thicker, as long as the petals. Capsule turbinate, 5-valved, ½ inch long.

253. M. spinuliferum (Haw. Rev. 176. Syn. 252); stem very thick, strumose; branches incrassate, tortuose-erect; leaves connate, semicylindrical, canaliculate, *attenuate*, *acute*, green, crystalline-papulose; flowers ternate, pedunculate, *pale-straw-coloured; calyx* 4-*cleft;* styles *very short*. *Salm Dyck, fasc.* 4. *t.* 30.

HAB. Cape of Good Hope (v. v.)

Stem 1 foot high; old branches smooth. Leaves 1–1½ inch long, 2–3 lines wide at the base, gradually attenuate, not persistent and spinous. Peduncles 6–9 lines long, thick, ebracteate. Calyx about ½ inch long; lobes unequal or subequal. Petals somewhat longer. Capsule 4-valved.

253. M. grossum (Haw. Syn. 252); stem strumose, as well as the branches ascending; leaves connate, semicylindrical, canaliculate at the base, flattish near the attenuate *bluntish apex*, green, crystalline-papulose; flowers ternate, *pale-reddish; calyx* 5-*cleft*, lobes unequal, two longer, cylindraceous; styles very short. *Salm Dyck, l. c. t.* 31. M. *rapaceum, E. & Z.!* 2086.

HAB. Fields near Zwartkopsriver. Oct. (Hb. Sd.)

It differs from the very similar *M. spinuliferum* by much shorter stem and branches, subspinous persistent rudiments of leaves, bracteate lateral peduncles, and rosy flowers. *M. grossum, E. & Z.!* 2085, comes very near, but differs in several points. The specimen is insufficient.

255. M. longispinulum (Haw. Phil. Mag. Dec. 1824. p. 426); stem and branches elongate, *filiform*, prostrate or creeping; leaves connate, in the flowering, upper branches *mostly alternate*, semicylindrical, subcanaliculate, *attenuate, acute*, greenish, as well as the branches and peduncles, crystalline-papulose; peduncles 1-flowered, but *often irregularly branched, and then* 3–7 *flowered;* flowers pale-yellow; calyx 5-cleft, lobes elongated, acute, nearly as long as the petals; styles shorter than the stamens. *Salm Dyck, l. c. t.* 32. M. *calycinum, E. & Z.* 2077.

HAB. Sandy places in Cape Flats near Rietvalley, Sept.–Nov. (Herb. Sd.)

Stem and branches often several feet long, filiform; young branches short, with very crowded leaves. Remnants of leaves only on the higher branches, flexible, not spinose. Leaves about 1 inch long, 1–1½ lines wide at the base. Short branches mostly 1 flowered, larger ones with 2–7, often panicled flowers. Calyx turbinate, lobes subequal or unequal, often as long as the tube.

256. M. viridiflorum (Ait. Kew. ed. 1. v. 2. 196); stem incrassate; branches *nodulose,* erect-decumbent; leaves *opposite,* connate, semi-terete, *bluntish,* glittering, papulose hairy; flowers solitary or subternate; calyx 5-cleft, *hairy,* lobes subequal, leafy; petals narrow-linear, subfiliform, green or greenish-red; styles very short. *Bot. Cab. t.* 326. *DC. Pl. Grass. t.* 159. *Jacq. Fragm. t.* 52. *f.* 2. *Salm Dyck, fasc.* 5, *t.* 32. *M. tenuiflorum, Jacq. Fragm. t.* 32. *f.* 3.

HAB. Cape of Good Hope.

Branches 1 foot or longer. Leaves distant, erect-recurved, 1–1½ inch long, 2–3 lines wide, subcanaliculate or flattish above: subspinous remnant on the old branches not frequent. Peduncles thick, papillate, 4–6 lines long. Petals as long as the longer cylindrical calyx-lobes.

§. 55. *MONILIFORMIA,* Haw. DC. *Stem* very short, branches nodose-moniliform, leafless in the summer. *Leaves* produced in autumn, 2, connate at the base in a green globule, elongate, semiterete, glittering-papulose, marcescent and deciduous. *Flowers* solitary, whitish. *Calyx* 4–6-cleft. *Stigmas* 7–8. (Sp. 257–258.)

257. M. moniliforme (Thunb. ! Nov. Oct. Nat. Cur. v. 8 p. 7. App.); stem very short, as well as the branches articulate-moniliform, articles depressed-globose; terminal pair of leaves joined into a spherical form; the following ones half-terete, subulate, very long, green, and somewhat recurved. *Fl. Cap.* 413.

HAB. Hills near Olifantsriver, Sept. (Herb. Thunb.)

Branches in Herb. Thunberg woody, 3–4 inches long, branchlets opposite or alternate, subdistichous, 2–1 inch long, upper ones shorter, consisting of terete articles about 4–6 lines in diameter, and 1 line long. Leaves and flowers are wanting. Peduncle, as described, angulate, one flowered, erect, terminal, 2 inches long. Calyx 4-cleft. Petals snow-white. Styles 7, ex Haw.

258. M. pusiforme (Haw. Misc. 23. Syn. 207); leaves full of crystalline papulæ; the first two united into the form of a pea; the following 2 semiterete; stem much branched and very dwarf.

HAB. Cape of Good Hope.

Flowers unknown. The first leaves are produced in Autumn, and the second in winter.

§. 56. *CRASSULINA,* Salm Dyck, Haw. (Spinulifera, DC.) *Stem* suberect or prostrate, branches effuse, filiform. *Leaves* linear-semiterete, acute, often canaliculate, papulose. *Flowers* solitary or ternate, whitish or pale rose-coloured. *Calyx* 4–5-cleft, lobes elongated. *Stigmas* 4–5. (Sp. 259–262.)

259. M. canaliculatum (Haw. Misc. 77. Syn. 253); stem and branches prostrate, very slender; leaves *linear-elongate,* spreading or reflexed, convex beneath, canaliculate above, green, crystalline-papulose; *flowers solitary,* flesh-coloured; calyx 4-cleft, *turbinate. Salm Dyck, fasc.* 5, *t.* 33. *M. reflexum, Willd.*

HAB. Karro-like hills near Zwartkopsriver, Feb. *Zey.!* 2614. (Herb. Sd.)

Root fleshy, somewhat tuberous. Branches 2–4 inches high. Leaves subdistant, 10–12 lines long, 1 line wide. Peduncle short, thickened upwards. Calyx 4; rarely 5-cleft, 7–8 lines long, lobes unequal. Petals uniseriate, as long as the longest calyx lobe. Styles 5, erect, subulate.

260. M. salmoneum (Haw. Rev. 176); stem and branches very slender, somewhat creeping; leaves crowded, spreading, *narrow-lanceolate*, obtuse, convex beneath, flattish, sulcate above, green, *glittering from minute papulæ; peduncle subtrichotomous;* flowers yellowish, rose-coloured at the apex; calyx *subglobose, 5-cleft, Salm Dyck, fasc.* 2. *t.* 23.

HAB. Cape of Good Hope.

Very similar to *M. canaliculatum*, but differing by its shorter (6–8 lines long) on both ends attenuated minute papillose leaves, globose calyx, and pluriseriate, differently coloured petals. Branches some inches long. Peduncle trichotomous, dichotomous or rarely 1-flowered. Calyx-lobes short, subequal, three times shorter than the petals.

261. M. crassulinum (DC. l. c. 445); stem and branches filiform, prostrate; leaves distant, spreading, narrow-lanceolate, subacute, convex beneath, subsulcate above, *scarcely papillate*; peduncles *subsolitary;* calyx subglobose, 5-cleft; petals *white. Salm Dyck, l. c. t.* 34. *M. crassuloides, Haw. Rev.* 170.

HAB. Karro-like hills in Hassaquaskloof, Sept. *Zeyher!* 2611. (Herb. Sd.)

It differs from *M. salmoneum* by more distant, scarcely papillate leaves and smaller white flowers, not 1 inch, but only ½ inch in diameter. Peduncles filiform, without bracts. Petals subuniseriate.

262. M. incomptum (Haw. Suppl. 96); stem and branches very slender, suberect, with elongated internodes; leaves erect, spreading, subulate, subacute, convex on both surfaces, subglaucescent, minutely papillate; peduncles ternate, bracteate; petals white. *Salm Dyck, l. c. t.* 35.

VAR. β. **Ecklonis** (Salm Dyck, fasc. 6. t. 33); stem and branches more diffuse, flexuose; petals at the base and stamens rose-coloured. *M. incomptum, E. & Z.!* 2087.

HAB. Hills near Adow, Uitenhage, Aug.–Sept. (Herb. Sd.)

½–1½ foot, weak, very distantly nodulose on the nodes with fasciculate leaves; young branches a little papulose. Leaves 6–10 lines long, 1–1½ line wide, semiterete, subulate. Peduncles by abortion 2 or 1, subpapillate, 1–2 inches long. Flowers ½ inch in diameter. Calyx campanulate with acute lobes, a little shorter than the petals.

§. 57. *GENICULIFLORA*, DC. (Cylindracea, Haw.) *Stem* erect, as well as the branches herbaceous, but woody at the base. *Leaves* distinct, semiterete, papulose, marcescent-deciduous. *Flowers* ternate, by abortion often solitary in the forks of the branches, yellowish. *Calyx* 4-cleft. *Stigmas* 4. (Sp. 263.)

263. M. geniculiflorum (Linn. Spec. 688); stem frutescent; branches subarticulate, divaricate, papillose; leaves remote, spreading, linear-semicylindrical; flowers mostly solitary and axillary in the forks of the branches. *Dill. Elth. t.* 205. *f.* 261. *DC. Pl. Grass. t.* 17. *Salm Dyck, fasc.* 5. *t.* 34. *E. & Z.!* 2089.

HAB. Sandy places near Olifantsriver, Clanwilliam; near Papendorf, Oct.–Nov. (Herb. D., Sd.)

Stem 2 feet or more high. Leaves 8–14 lines long, 1–1½ line wide. Pedicels short, thickened above, lateral ones foliate. Calyx turbinate, papillose, 2 lobes elongated and about as long as the petals.

§. 58. *NODIFLORA*, DC. (Cylindracea, Haw.) *Root* annual. *Stem* herbaceous, branched. *Leaves* subterete or linear, papulose. *Flowers* axillary, subsessile. *Calyx* 4–5-cleft, lobes longer than the very minute, white petals. *Stigmas* 4–5. (Sp. 264–265.)

264. M. nodiflorum (Linn. Spec. 687); stem nearly erect or diffuse; leaves opposite or alternate, semiterete or subterete, obtuse, sometimes a little cilated at the base; flowers sessile or short-pedunculate; calyx 4–5-cleft, lobes unequal, *much longer than the minute petals. DC. Pl. Grass. t.* 88. *Thunb.! Fl. Cap.* 413. *M. copticum, Linn. Spec.* 688. *M. apetalum, Linn. f. Suppl.* 258. *M. copticum, Jacq. hort. Vind.* 3. *t.* 6. *M. nodiflorum et apetalum, E. & Z.!* 2090, 2091.

HAB. Sandy places in Cape Flats; near Olifantsriver, near Grootvadersbosh, in Zwartland, *Zey.!* 2620, Oct.–Nov. (Herb. Thunb. Jacq. D. Sd.)

Much branched, 2 inches 1 foot high, greyish-green. Branches terete, more or less papulose, branchlets mostly secundate. Leaves ½–1 inch long, ¼–1 line wide. Flowers axillary or subterminal. Calyx turbinate, lobes subulate, as long as the tube. A very polymorphous plant. I cannot distinguish the specimens from Sicily, Barbary, Egypt, from the Cape plant.

265. M. caducum (Ait. Kew. 2. p. 179); stem and branches erectish or diffuse, as well as the leaves and calyxes papulose; leaves alternate or opposite, semiterete, obtuse; flowers sessile, in the axil of a leaf, or terminal, girded by a pair of leaves; calyx 5-cleft, lobes unequal, *a little longer* than the whitish petals. *DC. l. c.* 447. *E. & Z.!* 2092.

HAB. Sandy places near Buffaloriver, Clanwilliam, Nov. (Herb. Sd.)

Only distinguished from *M. nodiflorum* by the flower with larger petals. Branches mostly alternate.

Group. VI. PLANIFOLIA.—Stem slightly woody, fleshy, or herbaceous. Leaves flat, more or less papulose. (§§. 59–65.)

§. 59. *SCAPOSA*, DC. (Limpida, Haw.) *Root* annual, nearly stemless. *Leaves* almost radical, linear or cuneiform, papulose. *Flowers* solitary, rising from a radical peduncle, reddish, yellow or white. *Calyx* 5-cleft. *Stigmas* 5. (Sp. 266–268.

266. M. pyropaeum (Haw. Suppl. 19); plant almost stemless, branched from the base; branches subpapulose; leaves connate, *linear-elongate, semicylindrical*, obtuse, canaliculate near the base; flowers solitary, pedunculate; peduncle and calyx papillose. *Salm Dyck, fasc.* 6 *t.* 34. *M. tricolor, Willd. Enum,* 530. *Sims. Bot. Mag. t.* 2144. *M. gramineum, Haw. M. claviforme, DC. Prod.* 3. 448. *M. clavatum, Haw. Petiv. Gaz. t.* 88. *f.* 7.

VAR. β. **roseum** (Haw); flowers rose-coloured.

VAR. γ. **album** (Haw.); flowers white. *M. lineare, Thunb.! Fl. Cap.* 411. *M. apetalum, Thunb.! l. c.* 417.

HAB. Sandy places in Cape Flats ; in Zwartland and Groenekloof, Sept. (Herb. Thunb., D., Sd.)

2–3 inches high. Leaves radical or rising from the base of the very short stem, crowded on the top of the branches, dilated at the base, opposite, linear or somewhat broader near the apex, 1–2 inches long, ½–1 line wide. Peduncle radical or terminating the branches, ½–1½ inch long. Calyx-lobes unequal, 2 or 3, often 4–6 lines long, and equalling the petals. Capsule, when ripe, depressed, globose.

267. M. criniflorum (Houtt. Pfl. Syst. 2 D. t. 53); plant almost stemless or branched from the base, herbaceous; leaves *obovate or cuneiform, flat,* obtuse, rather scabrous from papulæ; peduncles solitary, longer than the leaves, radical or rising from the top of the branches, pilose-papulose. *Linn. Suppl.* 259. *Thunb.! Fl. Cap.* 411. *M. cuneifolium, Jacq. Icon. rar.* 3. *t.* 288. *DC. Pl. Grass. t.* 134. *M. spathulatum, Willd. M. limpidum, Ait. M. criniflorum and cuneifolium, E. & Z.!* 2094, 2095. *Herb. un itin.* 511. *Zey.!* 709.

HAB. Sandy places in Cape Flats, Tablemountain, Saldanhabay, &c. Aug.–Sept. (Herb. Thunb., Jacq., D., Sd.)

Near the preceding, but differing by much broader, flat and scabrous papulose leaves and larger flowers. Leaves 1–3 inches long, 3–4 lines wide, attenuated in the petiole with larger base; crowded if rising from the top of the branches. Peduncles 1–3 inches long. Calyx with 2 or 3 elongated, obtuse lobes, shorter than the pale rose-coloured, red, or whitish narrow petals. It varies with narrow leaves and resembles *M. pyropaeum.*

268. M. papulosum (Linn. f. Suppl. 259); nearly stemless; branched, decumbent, terete, papulose; leaves opposite, spathulately-oblong, *acute,* papulose; peduncles solitary, terminal, *shorter* than the leaves; calyx lobes unequal; flowers yellow. *M. sabulosum, E. & Z.!* 2096. *non. Thunb.!*

HAB. Sandy places near Brackfontein, Clanwilliam, Aug. (Herb. Sd.)

Root annual, stem ½–1 inch, with spreading, decumbent 1–3 inches long, terete branches. Leaves opposite in the lower part, crowded at the top of the branches; lower ones about 2 inches long, 4 lines wide, attenuate in a linear petiole; leaves of the branches smaller, 1–2 inches long, 2–3 lines wide, subscabrous from pellucid papulæ. Peduncles 4–6 lines long, minutely papulose. Calyx subglobose, the size of a large pea; 2 or 3 of the lobes longer, equalling or a little surpassing the yellow petals. *M. oligandrum Kze,* Del sem. hort. Lips. 1845 is a depauperated state of *M. papulosum.*

§. 60. *PLATYPHYLLA,* Haw. DC. *Root* annual or biennial. *Stem* herbaceous. *Leaves* flat, variable in form, papulose as well as the branches. *Flowers* of various colours. *Calyx* 5-cleft. *Stigmas* 5. (Sp. 269–278.)

269. M. crystallinum (Linn. Spec. 688); diffusely procumbent, herbaceous, covered with *large,* white, glittering papulæ; leaves opposite, or in the branches alternate, stem clasping, ovate or spathulate, undulated; flowers axillary, almost sessile; calyx 5-cleft, tube campanulate, terete, lobes ovate, retuse and acute. *Dill. Elth. t.* 130. *f.* 221. *Thunb.! Fl. Cap.* 413. *DC. Pl. Grass. t.* 128. *Pappe. Prod. Flor. Cap. med. p.* 16.

VAR, β. **grandiflorum** (E. & Z. ! 2097); flowers larger on longer peduncles.

HAB. Sandy places frequent in the neighbourhood of Capetown, especially in the flats near Rietvalley, Sondagsriver, Olifantsriver; var. β. Karro-like places in Gauritzriver and near Uitenhage, *Zey.!* 711 *et* 2626. (Herb. Thunb., D., Sd.)

The ice-plant is also a native of Greece and the Canary Islands. The annual plant is called by Haworth *M. glaciale*, the biennial *M. crystallimum*. Flowers white or rose-coloured.

270. M. angulatum (Thunb.! Fl. Cap. 426); stem and branches angulose, herbaceous, procumbent as well as the leaves, peduncles and calyxes covered with *minute*, white, glittering papulæ; leaves opposite, or in the branches alternate, obovate or spathulate-rhomboid; flowers pedunculate, terminal and axillary; calyx 5-cleft, tube *pentagonal*, lobes unequal, 2 or 3 longer, spathulate, obtuse. *M. crystallophanes, E. & Z.!* 2099. *Salm Dyck, fasc.* 5. *t.* 35. *Zey.!* 2625.

VAR. β. **ovatum**; leaves ovate, or obovate obtuse or acute. *M. ovatum, Thunb.! Fl. Cap.* 417. *M. elongatum, E. & Z.!* 2098. *excl. Syn. Zey.!* 2623.

VAR. γ. **gracile**; stem very slender, leaves and flowers smaller. *M. angulatum, var.* γ. *Thunb.! Herb.*

HAB. Fields near Zwartkopsriver and Sondagsriver. Var. β. in the same localities and near Howisons Poort. Var. γ. on hills near Bethelsdorp, Sep.–Dec. (Herb. Thunb., D., Sd.)

Stem 1–2 feet, not so robust as in *M. crystallinum*. Leaves 1–2 inches long, 6–12 lines wide, attenuated in a broad-linear channelled petiole. Peduncle angulate, 2–6 lines long. Calyx papillose, longer lobes often equalling the whitish petals. In var. γ. the stem and branches are much thinner from a greyish-white colour, the leaves ½ inch long, upper ones oblong-spathulate, and the smaller flowers on very slender peduncles.

271. M. puberulum (Haw. Phil. Mag. Sept. 1831. p. 419); stem branched, procumbent, papillose; floriferous branches and margins of leaves pubescent; leaves opposite or alternate, obovate-spathulate, channelled, keeled; peduncles subcylindrical. *M. papulosum, E. & Z.!* 2100.

HAB. Sandy places on the Sondagsriver near Graafreynet. Sept. (Herb. Sd,)

The specimen of E. & Z.! is a span high with spreading branches. Leaves ½–¾ inch long, 3–1 line wide. Peduncles solitary, 6 lines long. Calyx-lobes unequal, 3 lobes rather longer and larger. Petals nearly as long as the calyx, whitish, of pale rose-coloured in the dry plant.

272. M. sessiliflorum (Ait. Kew. 3. 193); branches divaricate; leaves flat, spathulate as well as the stems beset with papulæ; flowers sessile, yellow. *Haw. Syn.* 247.

HAB. Cape of Good Hope.

273. M. Aitonis (Jacq. Hort. Vind. t. 7.) branches decumbent, angular; leaves opposite or alternate, ovate-spathulate, papulose; pedicels short; calyxes angular. *Haw. Misc.* 48.

HAB. Cape of Good Hope.

Flowers expanding in the evening, pale-reddish, about the size of those of *M. cordifolium.* Calyx-lobes very unequal. Styles 5, erect, recurved at the apex.

274. M. papuliferum (DC. l. c. 448); nearly stemless; branches 2–4, opposite, divided, fastigiate, gradually thickened, as well as the leaves and calyxes papulose; leaves connate, ovate, obtuse; flowers terminal, sessile, 1–3 together, white; calyx 5-cleft. *M. fastigiatum, Thunb.! Fl. Cap.* 413. *non Haw.*

HAB. On hills in very dry places near Olifantsriver, Oct. *Thunberg!*

Root annual. Branches some inches high; branchlets very short. Leaves ½–1 inch long. Calyx-lobes ovate, obtuse, purplish. The plant is now wanting in herb. Thunberg.

275. M. lanceolatum (Haw. Misc. 45. Rev. 159); stems decumbent; leaves alternate, lanceolate, bluntish, papulose; calyx and peduncles beset with crystalline dots; flowers white. *M. Volkameri, Haw. Obs.* 426.

VAR. β. **roseum** (Haw. Rev. 159); flowers reddish; leaves lanceolate-spath late.

HAB. Cape of Good Hope.

276. M. clandestinum (Haw. Phil. Mag. 1826. 129); stem very short, prostrate, subwoody; branches ascendent, terete, herbaceous; leaves ovate-lanceolate, acute, papillose-shining; flowers terminal, paniculate or subsolitary, small; petals very minute, white. *Salm Dyck, fasc.* 6. *t.* 36. *M. sessiliflorum,* β. *album, Haw. Rev.* 158.

HAB. Cape of Good Hope.

Branches ½–1 foot, green, minutely papillose. Leaves distant, thickish, 4–10 lines long, 2–5 lines wide, attenuated in a short petiole. Peduncles short, mostly cymose or dichotomously panicled. Calyx 2 lines long, 2 lobes longer, obtuse. Petals uniseriate, scarcely conspicuous. Styles 5, short, acute.

277. M. lanceum (Thunb. ! Fl. Cap. 417); stem erect, subtetragonal, subpapulose; leaves *sessile, lanceolate,* acutish, flat, papulose, erect-spreading; flowers terminal, subternate, middle-sized, white.

HAB. Cape of Good Hope. (Herb. Thunberg.)

There are two stems? or branches in herb. Thunb. ! about 1 foot high, subflexuose erect, a little branched at the top. Leaves opposite, subconnate and larger at the base, gradually attenuated, about 1½ inch long, 4 lines wide, somewhat longer than the internodes, upper ones shorter, acute. Intermediate flower short-peduncled, without bracts, lateral ones 8–12 lines long, bibracteate. Calyx turbinate, papulose, ½ inch long, lobec acute, subequal, about as long or shorter than the narrow petals. It comes very near *M. expansum,* L. which has the leaves attenuated at the base.

278. M. pinnatifidum (Linn. f. Suppl. 260); stems diffuse; leaves obovate, *lyrate pinnatifid,* lobes obtuse; flowers axillary, solitary, pedunculate; petals as long as the calyx, yellow. *Bot. Mag. t.* 67. *DC. Pl. Grass. t.* 142. *Salm Dyck, fasc.* 6. *t.* 35. *Thunb.! Fl. Cap.* 427. *E. & Z.!* 2101.

HAB. Sandy places in mountains near the cataract of Tulbagh, Sept.–Oct. (Herb. Th. Sd.)

Root annual. Stem dichotomous, branched from the base, terete, papillose. Leaves 1–2 inches long, terminal lobe the largest, minutely papillose. Peduncle in the forks of the branches, ½–1 inch long. Calyx 5-cleft, small. Petals subtriseriate. Styles 5.

§ 61. *CORDIFOLIA,* DC. (Platyphylla, Haw.) *Stem* suffruticose, branches herbaceous, prostrate. *Leaves* petiolate, flat, cordate-ovate, papulose. *Flowers* pedunculate, solitary, purple. *Calyx* 4-cleft. *Stigmas* 4. (Sp. 279.)

279. M. cordifolium (Linn. f. Suppl. 26c); stems diffuse; leaves opposite, flat, petiolate, ovate-cordate, rather papulose; peduncles terminal or rather lateral on the elongated branches; calyx obconical, 2 lobes flat, large, 2 subulate. *Jacq. Icon. rar.* 3. 3. *t.* 487. *DC. Pl. Grass. t.* 102. *Salm Dyck, fasc.* 4. *t.* 31. *E. & Z. !* 2102.

HAB. Woods near Zeekoriver, Thunb.; Zwartkopsriver, near Adow, and on Fishriver, *E. & Z. ! Zeyh.* 706, 2621. Sept.–Dec. (Herb. Thunb. Sd.)

1–2 feet. Stem and branches minutely papulose. Leaves 6–12 lines long and nearly wide. Peduncles 4–8 lines; calyx tube about 4 lines long, longer lobes equalling the tube or longer. Petals short, linear.

§. 62. *EXPANSA*, DC. (Planifolia, Haw.) *Stem* and branches suffruticose, diffuse, or procumbent, leafy at top. *Leaves* flat, ovate-lanceolate, subcarinate, papulose, marcescent, nerves and veins persistent. *Flowers* ternate or biternate in a terminal, elongated, thick peduncle, whitish or pale yellow. *Calyx* 4–5-cleft, lobes very unequal. *Stigmas* 4–5. (Sp. 280–285.)

280. M. anatomicum (Haw. Syn. 249); stem much branched, decumbent, as well as the branches slender, tortuose; leaves connate, crowded, *erectish, oblong-lanceolate*, acute, flattish, crystalline-glittering, when dead persistent; flowers ternate; calyx 4-cleft. *Salm Dyck, fasc.* 4. 32. *M. emarcidum, Thunb. ! Nov. Act. Nat. Cur. v.* 8. *p.* 9. *App. Fl. Cap.* 415. *M. tortuosum, E. & Z. !* 2103.

HAB. Karro, in Bockland, and near Gauritzriver. Nov.–Dec. (Herb. Thunb. Sd.)

Stem ½–1 foot. Leaves ¾–1 inch long, 4–5 lines wide, when old the nerves and a pellucid epiderme alone remain persistent. Flowers large. Peduncles 1–1½ inch, without bracts. Calyx pyriform, 4, rarely 5-cleft (Thunb.) Petals white, very narrow, as long as the longer calyx lobes. Styles 4, filiform.

281. M. expansum (Linn. Spec. 697); stem diffuse; branches lax, reflexed; leaves connate, *recurved, much-spreading, broad lanceolate*, acute, attenuated at the base, flat, keeled by the prominent middle nerve, rather glittering; flowers bigeminate; *calyx 5-cleft*, three of the lobes very large, two subulate. *Salm Dyck, fasc.* 1. *t.* 34. *M. tortuosum, C. Pl. Grass. t.* 94. *Dill. Elth. t.* 182. *f.* 223.

HAB. Cape of Good Hope (v, v.)

Branches ½–1 foot. Leaves 1–1½ inch long, 4–6 lines wide; when young thickish, green, minutely papillate, when old marcescent and membranaceous, dried nerves persistent. Flowers large. Pedicels 4–6 lines long, bracteated. Pale yellowish petals, as long as the longer calyx-lobes. Styles 5, short.

282. M. tortuosum (Linn. Spec. 697); stem divaricate; branches lax, procumbent; leaves connate, erect-recurved, *ovate-lanceolate*, rather *concave above, carinate-convex beneath*; flowers subternate; calyx 5-cleft, two of the lobes very large, two subulate, acute. *Dill Elth. t.* 181. *f.* 222. *Thunb. ! Fl. Cap.* 427. *Salm Dyck, fasc.* 2. *t.* 36. *M. varians, Haw. Syn.* 249, *according to the very bad figure. Petiv. Gaz. t.* 78. *f.* 10.

HAB. Cape of Good Hope, in Karro. (Herb. Thunb.)
Branches elongated, foliaceous at top. Leaves in the flowering branches distant, 1 inch long, 4 lines wide, attenuated on both ends; when old marcescent and membranaceous. Flowers large, white. Calyx turbinate, papulose, twice shorter than the narrow-linear petals. Styles 5, short.

283. M. concavum (Haw. Rev. 168); stem suberect, much branched; branches slender, tortuous; leaves connate, erect, crowded, *lanceolate-acute*, rather concave above, convex beneath, crystalline-glittering; flowers ternate; calyx 4-cleft. *Salm Dyck, fasc.* 4. *t.* 33.

HAB. Cape of Good Hope.
Very similar to *M. tortuosum*, but much slenderer and the calyx 4-fid. Leaves 1 inch long, 3–4 lines wide. Petals white, twice longer than the calyx.

284. M. crassicaule (Haw. Phil. Mag. 1824. 425); stem very short, thick; branches procumbent, papillose; leaves connate, erectish-recurvate, *linear-lanceolate*, acute, shining, papulose, *subcanaliculate above*, convex beneath; flowers bigeminate; calyx 5-cleft, three of the lobes larger, two subulate. *Salm Dyck, l. c. t.* 34.

HAB. Cape of Good Hope (v. v.)
Branches subherbaceous, ½–1 foot long. Leaves fleshy, 1–2 inches long, 3–5 lines wide, upper ones much shorter. Peduncles 1 inch or longer, bracteate. Calyx pyriform, papulose; lobes twice shorter than the straw-coloured linear petals. Styles short.

285. M. humifusum (Ait. Kew. 2. 179); stem suffruticose, trailing; leaves stem-clasping, spathulate, keeled, scabrous from conical papulæ; petals very minute.

HAB. Cape of Good Hope.
Corolla white.

§. 63. *RELAXATA*, Salm Dyck (Expansa, DC. Planifolia, Haw.) *Stem* incrassate, diffuse; branches flexuose, adscending. *Leaves* flat, oblong-lanceolate, glaucous, with minute papulæ. *Flowers* ternate or about 5, by abortion often solitary, reddish or white. *Calyx* 5-cleft, lobes subequal, without papulæ. *Stigmas* 5. (Sp. 286–287.)

286. M. relaxatum (Willd. Enum. Suppl. 36); stem diffuse; branches ascendent, flexuose; leaves stem-clasping, distinct, *linear-lanceolate*, bluntish, erecto-patent, glaucescent, subcanaliculate above, *obtuse-carinate* beneath; flowers purplish. *Salm Dyck, fasc.* 1. *t.* 35. *E. & Z.!* 2108.

HAB. Rocky places near Fort Beaufort or Katriver. July. (Herb. Sd.)
Branches 1 foot or more, terete. Leaves 1–2 inches long, 3–4 lines wide, scarcely attenuated at the base, minutely papulose. Flowers 3 or 5, large. Calyx turbinate; lobes subequal, twice shorter than the linear-lanceolate petals. Styles as long as the stamens.

287. M. pallens (Ait. Kew. v. 2. 182); stem diffuse; branches decumbent, flexuose; leaves stem-clasping, distinct, *oblong-lanceolate*, acute, recurvate-spreading, glaucous, minutely papulose, subcanaliculate above, *costate-carinate* beneath; flowers *white*. *E. & Z.!* 2104. *Salm Dyck, l. c. t.* 36. *M. expansum*, *DC. Pl. Grass. t.* 47. *M. loratum*,

Haw. Rev. 168. *E. & Z.!* 2169. *M. angulatum, E. & Z.!* 2107. *Zeyh.* 2600. 2601. 2602.

HAB. Fields near Zwartkopsriver, mountain sides of Bothasberg, near Vishriver, and at Katriver. Oct.–Jan. (Herb. D. Sd.)

The whole plant paler or more glaucous than the preceding; the white petals are not longer than the calyx lobes. The wild specimens are smaller and more erect than in the cultivated; the panicle is also larger but the flowers smaller.

§. 64. *TRIPOLIA*, DC. (Planifolia, Haw.) *Root* biennial. *Stem* herbaceous, lax. *Leaves* flat, spathulate-lanceolate, opposite, without papulæ, shining. *Flowers* terminal, peduncled, white. *Calyx* pentagonal, 8-cleft. *Stigmas* 5. (Sp. 288.)

288. M. Tripolium (Linn. Spec. 690); stem herbaceous, loose, simple; leaves spathulate-lanceolate, dotless, shining, almost destitute of papulæ; flowers pedunculate. *Dill. Elth. t.* 179. *f.* 220. *Bradl. Suec.* 5. *p.* 14. *t.* 47. *Salm Dyck, fasc.* 4. *t.* 36. *E. & Z.!* 2111. *M. expansum, Herb. Thunb.!*

HAB. Sandy places in the flats near Rietvalley. Oct. (Herb. Thunb. D. Sd.)

About 1 foot high. Leaves in sterile branches crowded, in floriferous distant, 2–4 inches long, 6–8 lines wide, bluntish, much attenuated at the base, minutely papulose. Peduncles 2–3 inches long, thickened above, without bracts. Calyx turbinate, denticulate at the angles; lobes shorter than the linear petals.

§. 65. *HELIANTHOIDEA*, DC. (Pomeridiana and Hymenogyne, Haw.) *Root* annual. *Stem* herbaceous, branched. *Leaves* flat, lanceolate or spathulate, attenuate at the base, subpapulose. *Flowers* long-peduncled, yellow. *Calyx* hemispherical, often rather angular at the base, with five elongated lobes. *Ovarium* depressed. *Stigmas* 10–20. (Sp. 289–293.)

289. M. pomeridianum (Linn. Spec. 698); annual; stem erectish or diffuse, as well as the branches, peduncles and calyxes hairy; leaves spathulate or spathulate-lanceolate, attenuated in a canaliculate petiole, flat, smooth, *ciliated;* peduncles elongated; lobes of the *hemispherical* calyx unequal, two leafy and longer than the petals. *Bot. Mag. t.* 540. *Jacq. Icon. rar.* 3. *t.* 489. *E. & Z.!* 2112. *Salm Dyck, fasc.* 4. *t.* 36. *M. calendulaceum, Haw. Rev.* 161. *M. Candollii, E. & Z.!* 2114. *non. Haw.*

VAR. β. **glabrum** (Haw. Rev. 160); plant glabrous or nearly so, upper leaves a little ciliated, petals as long or a little longer than the calyx. *Andr. Rep. t.* 57.

HAB. Sandy places in the Cape Flats, common. Oct. (Herb. D. Sd.)

Stem simple or dichotomous, branches adscending, 3 inches–1 foot long. Leaves 2–4 inches long, 6–12 lines wide, in small specimens, or in the upper part of the branches smaller. Peduncles 2–5 inches long. Longer lobes of the calyx ½–1 inch long. Petals pluriseriate, linear-lanceolate. Capsule large, flat above, about 16-locular.

290. M. helianthoides (Ait. Kew. 198); annual; stem erectish, as well as the branches and peduncles hairy; leaves oblong-spathulate or spathulate-lanceolate, attenuated at the base, flat, smooth, *glabrous;* peduncles elongated; lobes of the *subpentagonal* calyx unequal, as long or a little shorter than the petals. *DC. Pl. Grass. t.* 135. *M. pilosum,*

Haw. Rev. 161. *M. calendulaceum, Haw. Misc.* 47. *Breyn. Cent. t.* 79. *Moris. Hist. Sect.* 12. *t.* 6. *f.* 13.

VAR. *β.* **glabrum**; plant glabrous or nearly so.

HAB. Cape of Good Hope; var. *β.* sandy flats between Klippfontein and Predikstoel. Nov. *Zeyh.* 707. (Herb. Sd.)

Leaves somewhat smaller than in *M. pomeridianum*, and not ciliated, and the calyxes glabrous with subangular not exactly hemispherical tubes. Stem about 1 foot high. Stigmas 12–18.

291. M. flaccidum (Jacq. Hort. Vind. t. 475); biennial; leaves lanceolate, acute, flat, glabrous, quite entire; peduncles 1-flowered, erectish, glabrous, very long. *DC. l. c.* 450.

HAB. Cape of Good Hope.

Petals linear, acute on both ends. Styles 5! Perhaps not belonging to this section.

292. M. sabulosum (Thunb.! Nov. Act. Nat. Cur. v. 8. p. 17. App.); *nearly stemless, quite glabrous;* branches ascending, terete, at top with crowded leaves; leaves opposite, radical ones larger, *oblong* or *oblong-spathulate*, petiolate, *acutish*, minutely papillose; peduncles longish; lobes of the hemispherical subangulate calyx subequal, as long as the linear, yellow petals. *M. calendulaceum, E. & Z.!* 2113.

HAB. Sandy places near Saldanhabay and in Zwartland. Aug.–Oct. (Herb. Thunb. Sd.)

Habit of *M. criniflorum*, but easily distinguished by the yellow petals and radiating, numerous stigmas. Root filiform, annual. Stem about ½ inch, branches 1–2 inches long. Limb of the radical leaves about 1 inch long, 3–4 lines wide, attenuated in a broad-linear, 6 lines long petiole. Upper leaves ¾–1 inch. Peduncles 2–4 inches long. Calyx lobes foliaceous, three of them with large membranaceous margins. Petals about 6 lines long. Stigmas 10–12, not 5, as described by Thunberg. Capsule depressed, globose, twice smaller than in *M. pomeridianum.*

293. M. glabrum (Ait. Kew. 2. 198); leaves petiolate, spathulately-lanceolate, dilated at the base, and as well as the branches glabrous; base of calyx hemispherical, with linear, unequal lobes; stigmas usually 12, united into a tube at the base. *Hymenogyne glabra, Haw. Rev.* 192.

HAB. Cape of Good Hope.

Habit almost of *M. helianthoides.* Corolla straw-coloured; petals rufescent at the base; sterile filaments copper-coloured. Probably the same as *M. sabulosum.*

SPECIES NOT SUFFICIENTLY KNOWN.

M. campestre (Burch. Trav. 1. 259). Erect, 1½ foot high. Flowers rose-coloured. Allied to *M. pulchellum, Haw.*

HAB. in Roggeveld.

M. arboriforme (Burch. Trav. 1. 343). Shrub 1–2 feet high, branched, with the trunk mostly simple. Cymes 8-times dichotomous. Flowers minute, of a testaceous colour. Species allied to *M. parvifolium, Haw.*

HAB. Gattikamma.

M. coriarium (Burch. Trav. 1, 243). Used in tanning leather by the Hottentotts. Allied to *M. uncinatum*.

M. magnipunctum (Burch. Trav. 1. 272). Only the name.

M. humile (Haw. Misc. 80). Founded on the very bad figure of Petiv. Gaz. t. 88. f. 8.

M. graniforme (Haw. Misc. 82). Founded on the figure. Bradl. succ. t. 20, can be united with many others of Haworth's species.

M. guiganense (Klotzsch in Schœnb. Reise.) The description is too imperfect to determine the section.

Excluded from the genus *Mesembryanthemum*.

M. filiforme, Thunb. ! herb. is not a *Mesembryanthemum,* but *Aizoon* or *Galenia*.

M. crispum, Haw. (*M. crispatum,* Haw.) Founded on Pet. Gaz. t. 88, f. 5, is doubtless a *Composita*.

M. ? villosum (Linn. Spec. 692), is probably a species of *Aizoon*.

II. TETRAGONIA, L.

(By Prof. E. Fenzl.)

Calyx 4-cleft, rarely 3–5-cleft, its tube adhering to the 4–5-horned ovary, lobes coloured within. Petals wanting. Stamens variable in number, solitary, or in fascicles, anthers oblong or linear. Ovary 3–9-celled, by abortion 1–2-celled; cells 1-ovulate. Styles as many as ovary-cells, very short. Drupe or long nut, winged or horned, indehiscent, 1–9-locular. Seeds solitary in each cell. DC. Prod. 3, 451. Endl. gen.-n, 5164. Fenzl. Wien. Annal. 1839.

Herbs or subshrubs with alternate, flat, fleshy, undivided, usually quite entire leaves and axillary, sessile or stalked flowers. Name, from τετρα, four; and γωνια, an angle.

ANALYSIS OF THE SOUTH AFRICAN SPECIES.

I. **Tetragonoides,** DC. Stamens about as many as the calyx lobes.
Fruit topshaped, with 3–4 unequal wings ... ... (1) **microptera.**
Fruit ovoid, 3–4 angled, echinate ... ... (2) **echinata.**
II. **Tetragonocarpus,** Comm. Stamens more than the calyx lobes.
1. Pterigonia.—Fruit with 3–4 wings.
Pedunculares. Decumbent herbs. Flowers on long peduncles.
Prostrate:
Papulose-hirsute; lvs. rhomb-ovate or elliptical; calyx lobes 3–4 lines long (3) **herbacea.**
Papulose-pruinose; lvs. spathulate-oblong; calyx-lobes 1–2 lines long; styles as long as calyx ... ... (4) **portulacoides.**
Decumbent or ascending, papulose-hirsute,
Lvs. obovate or oblong, *rounded*; calyx-lobes 1–2 lines long; styles shorter ... (5) **nigrescens.**
Lvs. oblong or lanceolate ... ... (6) **halimoides.**
** *Chenopodinæ.* Annuals, with axillary, sessile, clustered flowers.
Calyx-lobes triangular-ovate; stam. 15-20; styles 3–5 ... ... (7) **chenopodioides.**

Calyx-lobes linear ; stam. 5–8 ; style 1 ... (8) **galenioides.**
*** *Macranthæ.* Calyx-lobes in bloom 4–6 lines long ; a *hairy* undershrub, with petioled, oblong or lanceolate leaves ... ... (9) **hirsuta.**
**** *Fruticulosæ.* Calyx-lobes in bloom under 4 lines long ; shrubby or half shrubby, glabrous or papulose, but *not hairy* :
Leaves petiolate,
Stem erect ; calyx-lobes ovate, *obtuse*,
Flowers solitary, remote, sessile, (12) **verrucosa.**
Flowers 2–3 together, pedicellate, (in a leafy raceme) ... ... (18) **arbuscula.**
Stem decumbent,
Flowers in a leafy raceme ; calyx-lobes ovate, subacute ... ... (13) **spicata.**
Fl. in a lax, leafless raceme ; calyx-lobes linear ... ... (16) **calycina.**
Leaves sessile or subsessile,
Decumbent, rough with papillæ,
Lvs. broadly ovate, subsessile, flat ... ... (10) **decumbens.**
Lvs. oblong or lanceolate, obtuse, with revolute margins ... (11) **Zeyheri.**
Erect or suberect,
Papulose ; pruinose or canous,
Lvs. sessile, oval-oblong or linear. Anthers *oval* ... (19) **robusta.**
Lvs. tapering at base, subpetiolate. Anthers *linear* (20) **sarcophylla.**
Glaucous. Lvs. lanceolate ; cal.-lobes ovate, acute. Fruit 2–3 celled ... ... (21) **glauca.**
Subglabrous ; leaves lanceolate or linear,
Calyx-lobes linear, blunt. Fruit orbicular. Style 1 (15) **psiloptera.**
Cal.-lobes ovate, obtuse. Fr. obovate, 3–4-celled ... (14) **fruticosa.**
Cal.-lobes ovate, bluntish. Fr. orbicular. Style 1... (17) **distorta.**

2. HAPLOGONIA.—Fruit wingless, evidently ribbed.
Fruit 3-4-locular. Styles 3–4 ... ... (22) **Haworthii.**
Fruit one-celled. Styles 1–2 ... ... (23) **saligna.**

Section I. **TETRAGONOIDES**, DC. (Sp. 1–2.)

1. T. microptera (Fenzl.), annual, herbaceous, glabrous, diffuse ; leaves deltoid-ovate or ovate-oblong, obtuse, attenuated in a short petiole ; flowers axillary, *glomerate-sessile*, very minute, 1–5-androus ; fruit 1–2 lin. long, *turbinate*, crystalline-papulose, 3–4-quetrous at the base, substipitate, *unequally* 3–4-*winged*, wings alternating with 3–4 tubercles or very small wings ; larger wings obverse-triangular, 1–2-dentate or entire.

VAR. α. **trisperma** ; flowers 4–5-androus, 3-gynous, a few 1–2-androus, 1-gynous ; fruit 3–1-locular, wings unequal, one or two of them often bilobed, bidentate or truncate.

VAR. γ. **monosperma** ; flowers 1–3-androus, 1–2-gynous, a few 4–5 androus ; fruit 1–2-locular, wings mostly very entire.

HAB. Stony places near Zilverfontein, Namaqualand. *Drege!* 2932, 7023, 7029. Var. β. near the Garip, *Drege!* 7026, and Ebenezar, *Drege!* 7035; Bitterfontein, Betchuana territory, *Zeyher* 716. (Herb. Vind. Sd.)

Stem and primary branches ½–1 foot, or only 2–5 inches high. Leaves very variable in size, 1–2 inches or 3–8 lines long, as well as the flowers green. It is distinguished from the following only by the winged, not echinate fruit.

2. T. echinata (Ait. Kew. 2. 177); herbaceous, glabrous, diffuse; leaves rhomboid-ovate or oblong, petiolate, petiole very short; flowers axillary *pedicellate,* solitary or 2–4 subglomerate, very minute; fruit *ovoid,* truncate at the base, 3–4-*angular,* 3–4-locular, crystalline-papulose, *echinate by* very patent *or reflexed cornicles. Haw. Misc.* 123. *DC. Prod.* 3. 432. *Pl. grass, t.* 113, *E. & Z.!* 2116.

HAB. Karrolike hills near Zwartkops, and Sondagsriver, *E. & Z., Zeyh.!* 2627. Albany, *Drege,* 7028. (Herb. Vind. D. Sd.)

Section II. **TETRAGONOCARPUS,** Commel. (Sp. 3–23.)

3. T. herbacea (Linn. spec. 687); root (rhizoma?) tuberous or fusiform-strumose; stems herbaceous, prostrate, as well as the pedicels and flowers *papulose-hirsute;* leaves ½–1½ inch long, 2–6 lines wide, rhomboid-ovate, elliptic or oblong, attenuated in the petiole, crystalline-papulose; flowers axillary on very long pedicels, the lower solitary, terminal ones 3–9 corymbose or subumbellate; lobes of calyx broad ovate or oblong, 1½–3½ lines long; stamens very numerous; fruit pendulous *oblong,* drupaceous, fleshy, 4-winged, 3–4 often by abortion 1–2 seeded. *Commel. hort. Amstel. II. t.* 102, *Haw. Misc.* 122, *DC. l. c.* 452. *E. Z.!* 2115.

HAB. Sandy places in the Cape Flats and on mountains near Capetown, on Hexriver, Bergvallei, and Krom river. *E. & Z., Pappe. Zeyh.* 715 *b. Krauss. Drege!* 394, 2932, 7027. (Herb. Vind. D. Sd.)

Stems very lax, simple or more or less branched, ½–1 foot. Pedicels filiform, in flower ½–2 inches long, ⅓ line thick. It differs from the two following especially by twice or thrice larger flowers, with longer and thicker pedicels, of which the lower are solitary, the uppermost umbellate, and by papulose-hirsute, not minutely crystalline-pruinous or pubescent branches and pedicels.

4. T. portulacoides (Fenzl.); root (rhizoma?) fusiform; stems herbaceous, thin, filiform, prostrate, virgately branched, as well as the leaves, pedicels and calyxes *papulose-pruinose;* leaves 8–18 lines long, 2–5 lines wide; ovoid or spathulate-oblong, obtuse, attenuated in the petiole; flowers axillary, lower pedicels 2–3-nate, superior 4–7 umbellate; pedicels filiform, lax, 4–12 lines long; lobes of calyx ovate or oblong, 1¼–2 lines long; stamens very numerous; *styles as long as the calyx;* fruit pendulous *obovate-oblong,* at *length suborbicular,* 3–5 locular, 8–10 lin. long, 4-winged, *wings membranaceous. T. chenopodioides et nigrescens, E. & Z.!* plant ex sicc. ex parte.

HAB. Sandy places in Cape Flats, Rietvalley and Duykervalley, and on mountains near Capetown, *Thunberg, Ecklon, Zeyher, Drege!* Kuilsriver and near Muizenberg, *Pappe.* (Herb. Vind. D., Sd.)

Distinguished from T. herbacea by the tender and smooth stem and leaves and smaller flowers; from T. nigrescens by the same character and the style. (The two specimens in herb. Thunberg belong to this, not to the preceding species. Sond.)

5. T. nigrescens (E & Z.! 2117); root (rhizoma?) fusiform-oblong or strumose-subbranched; stem herbaceous, virgate-branched, procumbent and ascendent, fleshy, pruinose or papulose-hirsute, *from the middle flowerbearing;* lower leaves 4–1 inch long, 1¾ inch–4 lines wide, *obovate or oblong, rotundate,* superior ones smaller, rhomboid-ovate or broad ovoid, elliptic or oblong, attenuated in the petiole; flowers axillary, terminal pedicels 3–7, cymose-umbellate, filiform, 4–12 lines long; lobes of calyx ovate or oblong, 1–3 lines long; stamens 15–25; styles 3–4 *evidently shorter than the calyx;* fruit 4-winged.

VAR. α. **hirsuta**; pedicels as well as the calyces dense papulose-hirsute.

VAR. β. **hirta**; pedicels and calyx papulose hairy. *T. heterophylla E & Z.!* 2119.

VAR. γ. **pruinosa**; pedicels and calyx beset with minute papulae. *T. herbacea,* Fenzl. *olim in herb. Drege n.* 2932. a.

VAR. δ. **maritima,** *Sond.;* branches very short; leaves aggregated at top, obovate, longish petiolate; pedicels ½–¾ inch long; as well as the calyx beset with minute papulæ; fruit 4-winged, 2 wings larger.

HAB. Sandy places near Grootepost, *Zeyher;* var. β. near Brackfontein, Clanwilliam, *E. & Z.!;* near Grootepost and in Piquetbergen, 1–4000 ft. *Zeyh.!* 715. a.; var. γ. rocky mountain places near Kasparskloof, *Drege;* var. δ. sea shore near Cape Recief, *Zeyher.* 715. (Herb. Vind. Sd.)

VAR. α. looks very like *T. herbacea,* but differs by the fascicles of flowers, beginning from the lower part or the middle of stem and branches. Var. γ. has smaller leaves (about 4 lines wide, lamina as long as the petiole), subsolitary pedicels, 2 or 3 styles, 4–5-locular ovarium and a suborbicular fruit about 4 lines long.

6. T. halimoides (Fenzl.); stems herbaceous, decumbent, branched, papulose-hirsute, from the middle floriferous; leaves 2–1 inch long, 9–3 lines wide, *oblong or lanceolate, subacute,* attenuated in the petiole, papulose, when dry subleprose; flowers axillary, flowering pedicels 1½–5 lines, in fruit 6 12 lines long, terminal ones sometimes aggregate, subhirsute; lobes of calyx ovate, 1–1½ lin. long; stamens very numerous; styles 3–5; fruit *suborbicular,* 10–16 *lines long,* deeply emarginate at the apex, 4-winged, wings pergameous, smooth, often alternating with tubercles or very minute wings. *T. heterophylla, E & Z.!* ex parte.

HAB. On the Bergriver, Zwartland, and on Mount Paardeberg, *E. & Z.;* Hexriver, Tulbagh, *Drege* 7025. b. g. (Herb. Vind. Sd.)

It is distinguished from var. γ. of *T. nigrescens* by the indument, lower solitary flowers, and the large fruit.

7. T. chenopodioides (E & Z.! 2118); papulose, stem decumbent, slender, branched from the base; leaves subspathulate-elliptic or lanceolate bluntish, more attenuated at the base than at the apex; flowers subsessile, at length very short pedicellate, disposed in an interrupted leafy spike; pedicels dense papulose; calyx lobes yellowish, *broad triangular-ovate,* acutish, 1 line long, tube 4-costate; *stamens* 15–20; *styles* 3–5, very long; fruit (probably) winged.

HAB. near Saldanhabay, *E. & Z.!* (Herd. Sd.)

It resembles small specimens of *Chenopodium polyspermum.* From *T. portulacoides and galenioides* it differs by the slender stem, very minute, aggregate flowers, short, only 2–4 lines long, fruit-bearing pedicels and very long styles.

8. T. galenioides (Fenzl.); papulose-hirsute, diffuse, branched; leaves elliptical acutish or obtuse, lower ones long-petiolate; flowers glomerate, rarely subsolitary, very minute, yellow, hirsute, at the top of the branchlets fasciculate or spicate; calyx mostly 4-parted, *lobes linear; stamens* 5–8; *styles* 1; fruit sessile, monospermous, orbicular, 3–5 lines long, 4-winged, wings alternating with 4 acute triquetrous tubercles.

HAB. Rocky places, Uienvallei, Onderbokkeveld and between Boschkloof and Honigvallei, near Mount Blauwberg, 2500–3000 ft., *Drege* 7034. (Herb. Vind. Sd.)

In habit it comes near *T. crystallina*, stems ¾–1 foot, much branched. Larger leaves often 3 inches, the uppermost only 3–4 lines long, lanceolate. Flowers about 1–1½ lines long. Fruit when dry, cribose-punctate with papulae.

9. T. hirsuta (Linn. fil. Suppl. 258); suffruticose, prostrate; floriferous branches erect, hirsute; leaves elliptical, oblong or lanceolate, and attenuated at the base, petiolate, on the margins or on both sides hirsute or denudate and papulose; flowers axillary, 3–5 laxly glomerate on short pedicels, forming a leafy or leafless interrupted raceme; calyx 4–6 lines long, lobes oval or oblong, mostly hirsute; stamens very numerous; style deeply 3–5-parted; fruit 4-winged, 4-locular. *Thunb.! fl. cap.* 408.

VAR. α. **hirsutissima**; the whole plant densely hirsute by long, simple, horizontal, when dry, pellucid-paleaceous hairs. *T. hirsuta, Haw. Misc.* 119. *DC. l. c.*

VAR. β. **denudata**; leaves denudate, papulose, lobes of calyx subglabrous, but at the top and on the tube hirsute.

HAB. Sandy hills in Zwartland near Olifantsriver, etc., *Thunb. Zeyh.* 713; Brackfontein and Heerenlogement, *E. & Z.! Pappe.*; Cederbergen, Langevallei and Bergvallei, *Drege*, 7024. 7025. 7032. 7052. (Herb. Thunb. Vind. D. Sd.)

Branches at length woody, branchlets 3–12 inches, sometimes fasciculate. Leaves acute or obtuse or rotundate, 1–3½ inches long; hairs often 1 line long. Styles always connate at the base. Wings of the fruit according to Thunberg's description, crispate, villous. (Fruit in Thunberg's herbarium orbicular, 1 inch long and broad, beset with scattered hairs; the 4 wings large, membranaceous, shining, veined, in the dry state somewhat undulated, 4-locular. Sond.)

10. T. decumbens (Mill. Dict. n. 2); suffrutescent, decumbent, branches woody at the base, annual ones elongate, simple or with short branchlets at the base, rough from papulæ; *leaves broad-ovate, cuneate, subsessile,* or ovate, rhomboid-ovate or oblong, attenuated in a short, broad petiole, superior ones spathulate-oblong, *flat;* flowers axillary, 3–5, aggregated, on unequal pedicels, disposed in a very long, interrupted-leafy raceme; leafy bracts *ovate or spathulate;* fruit 6–9 lines long, suborbicular, 4–5-winged, papulose-leprose, wings often alternate with tubercles.

VAR. α. **obovata**; leaves mostly obovate, subsessile or sessile, larger ones more than 2 inches long and often 1 inch wide. *T. obovata, Haw. Rev.* 73. *DC. l. c. E. & Z.!* 2126.

VAR. β. **ovalifolia**; leaves mostly oval or rhomboid-obovate, ½–1½ inch long, 4–6 lines wide. *T. decumbens, Mill. Ic. t.* 263. *f.* 1. *Haw. Misc.* 121. *DC. l. c. var.* γ. *Pl. Grass. t.* 23.

VAR. γ. **oblongifolia**; lower leaves mostly oblong, attenuated in a petiole. *T. decumbens, Krauss. exs.*

HAB. Sandy places, Table Bay and Zwartkopsriver, *E. & Z.! Drege.* 7033; var. β. Table Bay and Simons Bay, *Eckl. Pappe, C. Wright;* var. γ. Zwartevalley, George, *Krauss.;* Zwartkopsriver, *Zeyh.* 2631. (Herb. Vind. D. Sd.)

Bi-triennial, 1–1½ foot high, primary branches as thick as a goose's quill, with or without axillary fasciculate branchlets, leaves broadly-petiolate, rotundate.

11. T. Zeyheri (Fenzl.); suffruticose, decumbent, much branched, rough from large papulæ; primary branches elongate, thick, flexuose, secondary numerous, approximate, leafy, often fasciculate-branched; *leaves oblong* or *lanceolate, obtuse, attenuated at the base*, sessile, with revolute or reflexed margins, when young linear; flowers axillary, 1–3 aggregate, disposed in a mostly leafy raceme, interrupted below, but dense above; leafy bracts *sublinear*, generally as long as the flowers; fruit 6–9 lines long, suborbicular, papulose-leprose, 4–5-winged, *wings coriaceous-woody*, often alternate with small or obsolete tubercles. *T. tetrapteris, E. & Z.!* 2125, *non Haw. Aizoon perfoliatum, Thunb.! Fl. Cap.* 411.

HAB. Sandy hills near Rietvalley, Thunb. E. & Z.! (Herb. Thunb. Jacq. Sd.)
Nearly allied to *T. decumbens*, and only distinguished by a higher, more woody stem, with numerous, very leafy branches, oblong, cuneate leaves, with revolute margins. Primary branches thicker than a goose's quill, angulate. Larger leaves 12–15 lines long, 5–7 lines wide, upper ones 2–3 times smaller. Pedicels 6 lines, upper ones 3–1 line long. (In herb. *Thunb.* is only a single branch, but it agrees pretty well with the specimens collected by *Zeyher*. Sond.)

12. T. verrucosa (Fenzl.); suffruticose, erect, papulose-crystalline; branches thick, rigid, at length woody; lower leaves mostly ovate-oblong, the rest oblong, lanceolate or linear, obtuse; *petioles long-persistent, winged; flowers* axillary, *sessile, solitary, remote;* lobes of calyx *broad-ovate,* very blunt; stamens very numerous; fruit sessile, orbicular, 1 inch long, 3-*locular*, 3-*winged, wings dotted*, alternate with 3 triangular tubercles.

VAR. *α.* **latifolia**; leafy bracts oblong or lanceolate.

VAR. *β.* **angustifolia**; superior leaves as well as the leafy bracts broadly linear, with revolute margins.

HAB. Karrolike-hills, near Ebenezer, Clanwilliam, *Drege*, 7031. (Herb. Vind. Sd.)
Subshrub ½–1 foot, woody at the base, branches subtortuous, with a spongy pith, as thick as a pigeon's quill. Leaves 1–1½ inch long, very fleshy. Calyx with 3–4 lobes.

13. T. spicata (Linn. fil. Suppl. 258); suffruticose, branched at the decumbent base; branches ascendent or erect, virgate, papulose, glabrous or puberous; leaves pruinose, petiolate, ovate, oblong or lanceolate, obtuse, acute or acuminate, rotundate or rhomboid at the base, flat, upper ones subsessile, linear with revolute margins; flowers axillary, 3–7 aggregated, at length longish-pedicellate, disposed in a very long, interrupted, leafy raceme; lobes of calyx ovate, acutish; fruit 4-winged, *turbinate*, broader than long, retuse at the apex, acutely keeled between the smooth wings.

VAR. *α.* **latifolia**; branches and branchlets mostly elongate, upper leaves ovate or ovate-oblong; racemes mostly elongate, interrupted. *T. decumbens, E. & Z.!* 2121. *non Mill. Zeyh.* 2629.

VAR. *β.* **angustifolia**; branchlets numerous, more abbreviate and fasciculate; upper leaves subsessile or sessile, ovate-oblong or lanceolate; racemes often dense-flowered. *T. spicata, L. Thunb.! Fl. Cap.* 409, *E. & Z.!* 2122. *T. tetrapteris, Haw. Misc.* 121? *T. saligna, Zeyh.* 2628, *b. non Fenzl.*

HAB. Fields near Zwartkopsriver, at Rondebosh and Saldanha-bay, Roodebloen, *E. & Z.!* Howison's Poort, *H. Hutton;* Groenriver, *Drege*, 2931. Var. *β.* Zwartkopsriver and Kenkoriver, *E. & Z.!* Paarl and Pardeberg, *Drege,* 7043, 7046, 7047; Tulbagh, *Dr. Pappe.* (Herb. Thunb. Vind. D. Sd.)

A very polymorphous species—the var. *α.* nearly allied to *T. decumbens,* var. *β.* to *T. fruticosa.* It differs from the first by a more shrubby habit, ovate or ovate-oblong, small petiolate leaves; from *T. fruticosa* by the abrupt petiolate leaves, with rotundate base, and by the twice smaller fruit. Primary flowering branches 3–12 inches long. Larger leaves 1–2 inches long, 4–9 lines wide. The fascicles of flowers often somewhat compound. Flowers very often hermaphrodite-monœcious, pedicels of the sterile flowers 2–3 lines, of the fertile flowers 4–7 lines long, as well as the calyx often densely papulose-puberulous. Stamens very numerous. Styles 2–4, in the sterile flower mostly 1. Fruit 3–5 lines long, 4–6 lines wide.

14. T. fruticosa (Linn. Spec. 687); suffruticose, erect or ascending, much branched, branches virgate, branchlets fasciculate; *leaves sessile or subsessile,* lanceolate or linear, obtuse, attenuated at the base, with more or less revolute margins, papulose; flowers axillary, yellow, 2–5 aggregated, pedicellate, disposed in an elongated bracteolate or leafless raceme; pedicels 2–8 lines long as well as the calyx papulose or pruinose subhirsute; lobes of calyx *broad, ovate, very blunt;* fruit 1½ inch or larger, *obovate, 4-winged,* 3–4-locular; wings very large, membranaceous, papulose, alternate with four long tubercles. *Thunb.! Fl. Cap.* 408.

VAR. *α.* **lanceolata**; leaves much attenuated at the base, subsessile, a few revolute at the margins, larger ones lanceolate or broad-linear. *T. fruticosa, DC. l. c. Comm. Hort. Amst. II. t.* 103. *Zeyher,* 2628 *a.*

VAR. *β.* **linearis**; leaves sessile, with revolute or conduplicate-revolute margins. *T. linearis, Haw. Rev.* 73? *DC. l. c. T. fruticosa, E. & Z.!* 2124. *Mill. gc. t.* 263. *f.* 2. *Zeyh.* 2630. *Herb. Un. itin. n.* 789.

HAB. Hills near Zwartkopsriver and Zondagsriver, *E. & Z.!* Olifantsriver, *Drege,* 7038, 7042, 7050. Var. *β.* near Cape-Town, and in the districts of Zwellendam, Uitenhage, Albany, and in Namaqualand, *E. & Z.! Pappe, Drege,* 7039, 7045, 7051. (Herb. Thunb. Vind. D. Sd.)

A large under-shrub woody at the base. Flowering branches ½–1½ foot high. Leaves ½–1½ inch long and ½–3 lines wide. It differs from *T. spicata* and its varieties by the very large fruit and the leaves not rounded at the base, but gradually attenuated in a short petiole, or perfectly sessile.

15. T. psiloptera (Fenzl.); herbaceous but suffruticose at the base, erect; leaves carnose, sessile, lanceolate, acutish, attenuate at the base; leafy bracts linear, with subrevolute margins; flowers axillary, ternate or 5-nate, subracemose; pedicels 1½–2½ lines long, densely papulose; lobes of calyx *linear, bluntish; stamens* 10–15; style 1; fruit *orbicular,* 4–6 *lines long,* 4-winged, *very smooth;* tubercles none.

HAB. Mount Giftberg, Clanwilliam, 1500–2000 ft., *Drege,* 7052. (Herb. Vind. Sd.)

½–1 foot high, with the habit of *T. verrucosa,* branches as thick as a raven's quill. The axillary fascicles of flowers form often minute racemes. Lobes of calyx 1½–2½ lines long.

16. T. calycina (Fenzl.); suffruticose, diffuse, much branched, pruinose-papulose; flowering branches undivided; leaves carnose, lanceolate or linear-lanceolate, acutish, *attenuated in a distinct petiole,* margins revolute; flowers axillary, solitary or geminate, disposed in a lax, nearly leafless raceme; pedicels 1 inch or shorter; lobes of calyx *linear,* *bluntish*, 2–2½ lines long; stamens very numerous; fruit 8–10 lines long, suborbicular, with four large wings, *not costate*, 3–4 *locular*.

HAB. Cape, *Drege*, 7059. (Herb. Vind. Sd.)

Decumbent subshrub; annual branches as large as a raven's quill. Leaves 1 inch or shorter, 2½–1 line wide, scattered, with a few fascicles in the axils. From *T. fruticosa* it is distinguished by the distinctly petiolate leaves and linear not broadly ovate calyx segments.

17. T. distorta (Fenzl.); frutescent, much branched, squarrose-distort, nearly smooth; *leaves subsessile*, linear-lanceolate or linear, bluntish, conduplicate-revolute, papulose-pruinose; flowers axillary, 2–3-nate, or solitary; pedicels 2½–4 lines long; lobes of the pruinose calyx *ovate*, bluntish, 1–1½ line long; *stamens* 12-15; *style* 1, *filiform*, elongate, arcuate; fruit papulose-pruinose, at length very smooth, 8–10 lines long, orbicular, *unilocular*, with four wings and another four alternate small ones on triangular tubercles.

HAB. Between Ebenezer and Giftberg, near Olifantsriver, Clanwilliam, and in Little Namaqualand, *Drege*, 7044. (Herb. Vind. Sd.)

A small, stiff shrub, very squarrose, in habit nearer to the following than to the preceding. Leaves generally 6 lines long, 1 line wide or narrower.

18. T. arbuscula (Fenzl.); suffruticose, much branched, *glabrous*; branches *diffuse*, woody, virgate; branchlets abbreviate, leafy; leaves *ovate-oblong*, lanceolate or linear, obtuse, *attenuated in a short petiole*, *flat*, not revolute at the margins; flowers 2–3, pedicellate in the axils of the leaves, disposed in a long leafy raceme; lobes of calyx broad-ovate, very blunt; *stamens* 20–30; *styles* 2–3, a little longer than the calyx; fruit ½ inch, shining, *obovate-orbicular*, 2–3-*locular*, with 3 or 4 wings and 4 alternate, obsolete costæ.

VAR. α. **linearis**; leaves linear or lanceolate, 4–10 lines long, ½–2 lines wide; pedicels of flowers 2½–4 lines long.

VAR. β. **latifolia**; leaves obovate-oblong, 4–10 lines long, 2–4 lines wide, uppermost lanceolate; pedicels of flowers 1–3 lines long.

HAB. Sneeuwbergen, and near Steelkloof, 4-5000 ft. *Drege*, 659, 7036, 7040; Gamkariver, *Burke and Zeyher*, 712. Var. β. Gamkariver, *Drege*, 7048. (Herb. Vind. D. Sd.)

Leaves 4–10 lines long. Calyx-lobes mostly 1½ line long. Allied to *T. fruticosa*; differs by the woody branches and branchlets, smaller flowers and fruit, and more exserted styles.

19. T. robusta (Fenzl.); frutescent, erect, much branched; branches *very straight*, virgate, branchlets many-leaved, papulose-pruinose; leaves crowded, sessile, ovate, oblong or sublinear, obtuse, with *revolute or conduplicate margins*; flowers very numerous, solitary or 3–5-nate, pedicellate, in the axils of the leaves disposed in a longish raceme, yellow, minute, hermaphrodite-monœcious; lobes of the 3–4-parted calyx ovate, fertile 4–6-*androus*, 2–4-*gynous*, sterile 8–12-androus, 1-gynous, or by abortion of the ovary male; *anthers oval, not linear*; fruit *orbicular*, 6–8 lines long, 2–4-locular, 3–4-winged, very smooth, and without costæ.

HAB. Namaqualand, between Koussie-river and Zilverfontein, 2000 ft. *Drege*, 7041. *A. Wyley*. (Herb. Vind. D. Sd.)

Distinct from *T. fruticosa* by the minute, very short pedicelled, 4–12-androus flowers and ovate anthers. Shrub about 1½ foot high. Leaves 1 inch or shorter, 1–2 lines wide. Pedicels ½–1 line long. Lobes of calyx acute or obtuse, yellow, 1–1½ line long.

20. T. sarcophylla (Fenzl.); frutescent, much branched, erect, *papulose-canous;* branches rigid, thickish, many-leaved; leaves ½ inch or smaller, carnose, obovate, obovate-oblong or oblong-linear, gradually attenuated at the base, subpetiolate, with revolute margins; flowers axillary, 1–5-nate, very short-pedicelled, forming a spike at the top of the branches, hermaphrodite-monœcious, lower ones 5–8-androus, 2–3-gynous, superior ones sterile, larger, 12–20-androus, 1-gynous or only male; lobes of the 4-fid calyx *oval, rotundate; anthers linear;* fruit (probably) winged.

VAR. β. **glabrata,** Sond.; branches minutely papulose; leaves glabrate, obovate, cuneate, petiole 1–3 lines long, dilated at the base; flowers short-pedicellate, axillary; stamens nearly 15; ovary papulose, 3–4-locular; style 3 or 1; fruit 3–4-winged, cuneate at the base, emarginate at the apex; wings membranaceous, alternate with 3 or 4 often obsolete tubercles.

HAB. Stony places of Mount Kendo, in Groote Zwartebergen, 3–4000 ft. *Drege,* 8018; var. β. Springbokkeel, *Zeyher,* 714. (Herb. Vind. D. Sd.)

Subshrub erect, ½–1 foot high, very rigid, primary branches as thick as a goose-quill. Pedicels and calyx-lobes nearly 1 line long. Stem leaves more than 4 lines long. In var. β. the lower leaves 8–10 lines long, 3–4 lines broad, the calyx rarely 5-parted, lobes ovate or oval, the fruit about 4–6 lines long and broad.

21. T. glauca (Fenzl.); suffruticose, much branched, erect? glaucous, branches virgate, elongate, rather straight, subherbaceous; leaves *subsessile, lanceolate,* bluntish or acute, flat, with subrevolute margins, the uppermost linear; flowers axillary, 3–6-nate, *long pedicellate,* disposed in a lax, interrupted, sometimes subcompound spike, near the apex leafless; lobes of calyx *ovate, acutish;* stamens very numerous; fruit 2–3-locular, *elliptic,* 4-winged, wings alternate with four obsolete costæ. *T. linearis,* E. & Z.! 2133. *excl. specim. circa Zwellendam lectis ad T. salignam spectantibus.*

HAB. Karrolike-hills, between Gauritzriver and Langekloof, dist. George, *E. & Z.! Drege,* 7030. (Herb. Vind. Sd.)

Branches ½–1 foot or longer, very smooth, glaucous or whitish. Leaves 1–2½ inches long, 3–6 lines wide. Pedicels 2½–10 lines long, capillary. Common axillary peduncle 1–6 lines long. Calyx-lobes 1–1½ line long. Fruit (unripe) 4 lines long.

22. T. Haworthii (Fenzl); suffruticose, branches ascendent or erect, angular, obsoletely papulose; leaves very short, petiolate, lanceolate, elliptic or ovate-lanceolate, carnose, subpapulose, with subdeflexed margins; flowers axillary, small, 2–3-nate, on short, unequal pedicels; lobes of the 4-parted calyx rotundate-ovate; stamens very numerous; *styles* 3–4; fruit subglobose, 3–4-*locular,* not winged, with 7–8 very blunt angles. *T. fruticosa, Haw. Misc.* 120 (*excl. omnibus Syn. ad T. fruticosam, L. spect.*)

HAB. Cape of Good Hope. Haworth.

Perhaps only a 3–4-gynous form of the following.

23. T. saligna (Fenzl.); suffruticose, decumbent, divaricately branched; branches ascendent, virgate, greyish, smooth; leaves subfleshy, distant, short, petiolate, lanceolate or linear-lanceolate, 1–2 inches long, with subrevolute margins, minutely papulose, upper ones sessile, linear, obtuse; flowers 3–8 axillary, glomerate, sessile or subsessile, glomeruli (clusters?) disposed along the branches; calyx-lobes ovate-roundish, stamens very numerous; *style* 1, *rarely* 2, *filiform;* fruit subglobose, 2½–3 lines long, 1-*locular*, 8-*angulate*, angles obtuse or acutish. *T. linearis, E. & Z.!* 2123, *partim.*

HAB. Karrolike-hills, near the River Zonderende and Breederiver, Zwellendam, *E. & Z.!* Zwartebeestkraal, between Olifantsriver and Bergriver, Clanwilliam, *Drege*, 7037. (Herb. Vind. Sd.)

Subshrub 1–2 feet or higher, with the aspect of *Atriplex angustifolia;* primary branches nearly as thick as a goose-quill. Larger leaves 2–4 lines, small ones ¾–1½ line wide. Pedicels ½–1 line long. Lobes of calyx ¾–1 line long.

III. AIZOON, L.

Calyx 5-parted, coloured and petaloid within. *Petals* none. *Stamens* about 20, inserted at the bottom of the calyx, and disposed in 3–5 tufts. *Ovary* free, 5-angled, 5-celled; cells with 2 or many ovules. *Styles* or *stigmas* 5. *Capsule* 5-celled, dehiscing at the apex in a stellate manner; cells 2–10-seeded. *Seeds* pyriform or subreniform, brown, shining, striate. *DC. l. c.* 3, *p.* 453. *Endl. Gen.* 5165. *Fenzl. l. c.* 288.

Humble herbs or subshrubs. Leaves fleshy, quite entire. Flowers sessile in the axils of the leaves or forks of the stems, rarely pedicellate. Name from ἀεὶ, always, and ζωος, alive.

ANALYSIS OF THE SPECIES.

Leaves alternate.

Stem villous; leaves obtuse, pubescent ... ... (1) **Canariense.**
Stem hirsute; leaves acute or mucronate, hirsute ... ... (3) **glinoides.**
Stem canous-tomentose; leaves glaucous-tomentose .. .. (4) **rigidum.**
Stem and leaves papulose ... ... (2) **galenioides.**

Leaves opposite.

Stem appressed hairy as well as the ovate leaves ... ... (5) **Zeyheri.**
Stem not sarmentose, as well as the lanceolate or spathulate leaves tomentose... ... ... (6) **paniculatum.**
Stem sarmentose, as well as the linear-subulate leaves glabrous or hairy ... ... (7) **sarmentosum.**

1. A. Canariense (Linn. Spec. 700); stems herbaceous, procumbent, much branched, *villous;* leaves alternate *obovate-cuneiform*, petiolate, *pubescent;* flowers sessile, subadnate at the origin of the branches; stigmas 5, thick. *DC. Pl. Grass. t.* 136. *Glinus crystallinus, Forsk. Descript.* 95. *t.* 14. *Veslingia cauliflora, Mœnch.*

VAR *β.* **denudata,** Sond.; stem and leaves subglabrous; leaves long-petiolate, punctate, ciliate. *A. spathulatum, E. & Z.!* 2128.

HAB. Fields near Bitterfontyn, *Zeyh.* 718; var. *β.* sandy places near Brackfontyn, Clanwilliam, *E. & Z.! Drege*, 7063. (Herb. Vind. Sd.)

Stem or primary branches ½ foot long. Leaves ½–1 inch, spathulate, obtuse, attenuated in a petiole nearly as long or longer than the lamina. Calyx 5-angled, lobes 1 line long or longer. Capsule much depressed.

2. A. galenioides (Fenzl. in herb. Drege); stems herbaceous, ascendent-erect, as well as the leaves and calyces *densely beset with papulæ;* leaves alternate or subopposite, spathulate-oblong, cuneate or shortly petiolate; flowers axillary, sessile; styles 5, as long as the calyx.

HAB. South Africa, *Drege,* 7060. (Herb. Vind. Sd.)

Nearly ½ foot high, glittering from papulæ. Branches terete, primary opposite, upper ones alternate. Leaves ¾–1 inch long. 2–3 lines wide, much smaller in the branchlets, 3–4 lines long, 1 line wide. Flowers 1 line long, lobes ovate, acute. Capsule 5-angled, twice smaller than in *A. Canariense.*

3. A. glinoides (Linn. fil. Suppl. 261); stems herbaceous, elongate, procumbent, as well as the leaves and calyces *hirsute with long white hairs;* leaves alternate, *obovate, mucronate or acute,* petiolate; flowers axillary, sessile; lobes of the calyx ovate-acuminate; styles 5 short. *Thunb.! Fl. Cap.* 410. *A. hirsutum, E. & Z.!* 2130. *Fenzl.! in herb Drege, Zeyher,* 2632.

HAB. Karrolike places near Zwartkopsriver, Onaggasvlakt, Grahamstown, Koegariver, *E. & Z.! Col. Bolton, Drege, Krauss. T. Williamson,* and near Port Natal, *Drege.* April–Sept. (Herb. Thunb. Vind. D. Sd.)

Stem 1–2 feet long, subflexuose, terete. Ultimate branches short. Leaves about 1 inch long, 4–8 lines wide, on the branches smaller, attenuated in the short petiole. Lobes of the spreading calyx 2 lines long, white or yellowish-white, and glabrous above. Stamens very numerous, shorter than the calyx-lobes. Capsule when ripe much depressed in the middle, 5-angled, papulose, 3 lines in diameter.

4. A. rigidum (Linn. fil. Suppl. 261); stem suffrutescent, as well as the alternate branches *canous-tomentose;* leaves *alternate, flat, obovate-acute,* attenuated in a short petiole, covered with *appressed, glaucous tomentum;* flowers axillary, sessile, canescent; lobes of the calyx ovate-lanceolate; styles 5, filiform. *Thunb.! Fl. Cap.* 409. *E. & Z.!* 2127. *Zeyh.* 2634 *a.*

VAR. *β.* **angustifolia,** Sd.; leaves oblong-spathulate, acute or acuminate, covered with white, subsilky tomentum. *A. argenteum, E. & Z.!* 2129. *A. sericeum, Fenzl. in herb. Drege, Zeyh.* 2634 *c.*

HAB. Karrolike-hills near Zoutriver, Caledon, Gauritzriver, Swellendam, and between Uitenhage and Graafreynet; var. *β.* on the sea shore, near Seaview, Cape Recief. Oct.–February. (Herb. Thunb. Vind. D. Sd.)

Stem ½–1 foot or longer, with appressed, rarely a little spreading tomentum; branches terete, spreading. Leaves 6–8 lines long, 2–2½ lines wide, in some specimens about 1 inch long, 5 lines broad, upper ones often crowded and always smaller; in var *β.* 1–2 lines broad, and usually more acuminate at the apex. Flowers secundate, a little smaller than in *A. glinoides,* tube often with spreading tomentum, lobes 2½ lines long, appressed-tomentose. Stamens numerous. Capsule 5-angled, papulose; cells many-seeded.

5. A. Zeyheri (Sond.); suffruticose; branches alternate, short, as well as the leaves and calyces covered with appressed, white hairs; leaves *opposite,* very short-petiolate, *ovate, acutish, concave above, obtusely-keeled beneath;* flowers axillary, sessile; calyx-lobes ovate, acuminate; styles 5, filiform; cells of the capsule 2-seeded.

A small procumbent shrub, with woody, erect, short, divided branches. Branchlets about 1–2 inches long; internodes as long or shorter than the obtusely carinate white leaves. Petiole ½–1 line, leaves or lamina 2–2½ lines long, 2 lines wide. Flowers 1 line long, smaller than in all the other species. Capsule papulose, scarcely 5-angled, depressed in the middle. Seeds 10, blackish, agreeing in size with those of *A. rigidum.*

6. A. paniculatum (Linn. Spec. 700); stem herbaceous, decumbent or erectish, *tomentose*, branched; leaves opposite, *linear-lanceolate, lanceolate or subspathulate*, acute, attenuated at the base, tomentose; flowers sessile, peduncles trichotomous, panicled. *Thunb.! Fl. Cap.* 410. *E. & Z.!* 2141. *A. tomentosum, Lam. Enc. Meth.* 3. 418.

HAB. Sandy places in the Cape Flats, near Saldanhabay, Simonsbay. Aug-Oct. (Herb. Thunb., Vind., D., Sd.)

From several inches to 1 foot or more high, greyish, tomentose. Branches alternate or opposite. Lower leaves 2 inches long, 2–3 lines wide, upper ones about 1 inch long, 1–2 lines broad or smaller. Panicle terminal, few or many flowered. Flowers with 2 leafy bracts, sessile, or the lateral short-pedunculate. Calyx-lobes 5, rarely 4, appressed-hairy on the outside, glabrous and pale-yellowish on the inside, 2–3 lines long. Stamens 15 or more. Styles 5, filiform. Capsule 5-angled depressed. It varies with rose-coloured and with larger flowers.

7. A. sarmentosum (Linn. fil. Suppl. 260); stems suffruticose, diffuse, *sarmentose, glabrous;* branches subfiliform, appressed-villous towards the apex; leaves opposite, *linear-subulate, rather connate;* flowers 3 or ternately-panicled on the top of the branches, with 2 longer leafy bracts. *Thunb. Fl. Cap.* 416. *Burm. Afr. t.* 26. *f.* 2. *Herb. Un. Itin. n.* 11. *E. & Z.!* 2140. *Zeyher* 721. *A. stellatum, Lam. l. c. Mesembryanth. hexaphyllum, Haw. Rev.* 168. *ex syn. Burm.*

VAR. β. **strigosum** (E. & Z.! l. c.); stems, branches and calyces strigose-scabrous.

VAR. γ. **hirsutum** (E. Z.! l. c.); branches and calyces hirsute with long spreading hairs.

HAB. Mountains near Capetown and in the Cape Flats. Var. β. Zwarteberg, Caledon, *E. & Z.!* Paarlberg, *Drege!* Var. γ. hills near the cataract of Tulbagh, *E. & Z.! Pappe;* Nieuwekloof, *Drege!* July–Nov. (Herb. Thunb., Vind., D., Sd.)

Stems many from the root, ½–1 foot, often rooting. Branches alternate or opposite. Leaves subfiliform and flat above, acute, ¾–1 inch long. Flowers rarely solitary or geminate, usually ternate or ternately-compound, the intermediate sessile. Calyx stellate, appressed-villous on the outside, glabrous and white inside, ovate-lanceolate, or lanceolate, 3–4 lines long. Stamens more than 20, twice shorter than the calyx, equalling the 5 filiform styles. Capsule shorter than the calyx, obtusely 5-angled, depressed-globose, many seeded. Var. γ is usually more robust and erect, stem more woody and rarely sarmentose, with glabrous or pilose leaves, but is united by var. β. with the typical form.

IV. ACROSANTHES, E. & Z.

Calyx 5-parted, tube short, infundibuliform, lobes subfleshy, keeled, coloured on the inside, acuminate, erect. *Petals* wanting. *Stamens* 10–40, 2-seriate and in many parcels on the top of the calyx-tube, the exterior longer, alternate with the calyx-lobes; filaments capillary; anthers linear. *Ovary* 2-celled, 2-ovulate. *Stigmas* 2, filiform. *Capsule* subglobose, included in the persistent calyx, 1-celled, 2-valved. *Seeds* 1–2, affixed at the base, globose-reniform, lacunose-tuberculate, estrophiolate. *E. & Z.! Enum.* 328. *Fenzl. l. c. Endl. gen.* 5191. *Trianthema spec. Thunb.!*

Decumbent, dichotomous. quite glabrous subshrubs. Leaves subconnate, opposite or by abortion of branchlets in fours. Flowers axillary, or in the forks, solitary, pedunculate. Name from *ακρος, the summit,* and *ανθος a flower.*

ANALYSIS OF THE SPECIES.

Flowers subsessile :
Leaves lanceolate or nearly so ; stamens 9–10 ... (2) **humifusa.**
Flowers on peduncles, longer than the calyx :
Leaves oblong, acute or sublanceolate ; stamens 17–23 ... ... (1) **anceps.**
Leaves linear-lanceolate ; stamens 11–21 ... ... (3) **angustifolia.**
Leaves terete, filiform ; stamens 28–40 ... ... (4) **teretifolia.**

1. **A. anceps** (Sond.) ; stem procumbent ; branches terete, fistulose ; leaves flat, opposite or 4–6 verticillate, *oblong, acute at both ends,* or sublanceolate ; flowers axillary, solitary ; *pedicels rather longer than the calyx ;* lobes of the ovoid calyx subequal, ovate, mucronulate ; *stamens* 17–23. *Trianthema anceps. Thunb. ! Fl. Cap.* 390. *Acros. fistulosa, E. & Z. !* 2146.

HAB. Sandy hills near Heerelogement, Clanwilliam, *E. & Z. ! Drege !* Tulbaghskloof, *Zeyher!* Oct. Herb. Thunb., Sd.)

Branches often 1–2 feet, woody, branchlets spreading. Leaves 4–10 lines long, 1½–3 lines wide, very patent, when dry rigid, papulose at the margins. Pedicels 2-edged, at length recurved. Calyx tube short, longer lobes 2–2½ lines long.

2. **A. humifusa** (Sond.) ; prostrate, tufted ; branches short ; leaves flat, *ovate-lanceolate, lanceolate,* lanceolate-linear, attenuated at the base, mucronulate, those of the short branches imbricate-aggregated, of the longer branches distant ; flowers axillary, *subsessile ;* exterior calyx-lobes larger, lanceolate ; *stamens* 9–10. *Trianthema humifusa, Thunb. ! Fl. Cap.* 389. *Acros. decandra, Fenzl. l. c.* 270. *Didaste pentandra, E. Meyer !*

HAB. High plains of the Cederbergen, Clanwilliam, 3500–4000f. *Drege !* Koude Bokkeveldt and Hexriver, *Thunberg,* Jan. (Herb. Thunb., D., Sd.)

Branchlets 1–2 inches long. Leaves spreading, 2–4 lines long, ½–1 line wide. Internodes in the sterile branches 3–6 lines long. Flowers sometimes very shortly pedicellate, yellowish-green, but purplish above the middle. Calyx-lobes keeled, mucronate. Filaments purple. Very like the following, and only distinguished by the subsessile flowers and fewer stamens.

3. **A. angustifolia** (E. & Z. ! 2147) ; decumbent, branches diffuse, elongated ; leaves opposite, verticillate, *lanceolate-linear or linear,* cuspidate, with denticulate-scabrous margins ; flowers axillary ; *pedicels twice longer than the calyx ;* exterior lobes of the calyx longer ; *stamens* 11–21. *Didaste decandra, E. Meyer !*

VAR. *α.* **do-decandra** (Fenzl.) ; stamens 11–14 ; leaves of the branchlets approximate, longer than the internodes.

VAR. *β.* **icosandra** (Fenzl.) ; stamens 16–21 ; leaves of the branchlets distant, as long as the internodes.

HAB. Var. *α.* rocks on Piquetberg, Clanwilliam, 15–2000f., *Drege ;* var. *β.* near the cataract of Tulbagh, *E. & Z. !* Witsenberg, *Zeyher !* 721, Nov.–Dec. (Herb. D., Sd.)

Much branched, ½–2 feet ; branches virgate, subterete, greyish, young ones 2-edged or semiterete, testaceous. Leaves 4–6 lines long, ½–1½ line wide, erect or spreading, a little keeled, mucronate. Flowers yellowish, pedicels 2-edged. Calyx in flower campanulate, 2 lines long, in fruit 3 lines long, evidently keeled, purplish-sphacelate at the apex, white above. Capsule 1–2-seeded, ⅓ shorter than the calyx.

4. A. teretifolia (E. & Z. ! 2148) ; procumbent ; branches diffuse or ascending, lax or rigid, virgate, nearly terete ; leaves opposite, or 4–5 verticillate, *terete-filiform*, acutish, sphacelate-mucronate ; flowers axillary, or in the forks solitary, pedicellate ; lobes of the calyx *ovate ; stamens* 28–40. *Didaste icosandra, E. Meyer !*

HAB. Stony places, Mount Zwarteberg, Caledon ; near Tulbagh, Worcester, *E. & Z. !* Mount Paarlberg, 1500–2000f., *Drege !* Nov.–Dec. (Herb. D., Sd.)

Habit of the preceding. Root thick, 3–4 inches long. Stems many, 6–10 inches long ; branches short. Leaves ½–1 inch long, ½ line broad. Flowers as long as the pedicels or twice shorter, 1½–2½ lines long. Calyx-lobes nearly twice longer than the turbinate tube. Capsule a little shorter than the calyx.

V. DIPLOCHONIUM, Fenzl.

Calyx 5-parted, tube infundibuliform, lobes petaloid-membranaceous. *Petals* none. *Stamens* 40–70, inserted (not in sets) on the top of the calyx tube. *Anthers* didymous, ellipsoid, versatile. *Ovary* free, 2-celled. *Cells* many-ovulate. *Styles* 2, filiform. *Capsule* ellipsoid-oblong, 2-celled, transversely dehiscent in the middle, opercle obtusely conical ; the epicarpium of the lower segment separating from the endocarpium, *Seeds* subreniform, blackish, smooth, shining, with a large hilum. *Fenzl. in Endl. nov. stirp. decad. n.* 65. *Endl. Gen.* 5169.

A fleshy, papulose, dichotomous herb. Leaves opposite, fleshy, quite entire, flat, obovate and elliptical, attenuated at the base, with revolute margins. Flowers sessile in the forks of the branches, the uppermost axillary. Name from διπλοος, *double*, and χονος, *a beaker ;* in reference to the separation of the endocarpium from the persistent-epicarpium.

1. D. sesuvioides (Fenzl. l. c.)

HAB. Rocky places on the Garip, *Drege !* Namaqualand, *A. Wyley !* (Herb. Vind., D., Sd.)

Herb probably prostrate, with the habit of *Sesuvium Portulacastrum*, glabrous, beset with dispersed, often whitish granules ; branches terete, as thick as a raven's quill. Leaves acute or obtuse, about 1 inch long, 3–4 lines wide, the upper ones smaller. Flowers 6 lines long, tube 2 lines long, lobes much spreading, 4 lines long, ovate, mucronate, petaloid, herbaceous in the middle. Stamens somewhat shorter than the calyx.

VI. GALENIA, L.

Calyx 4 or 5-parted, coloured within. *Petals* none. *Stamens* 8 or 10, in 4 or 5 sets, alternate with the sepals, of unequal length (a long and short stamen in the axil of each sepal) ; anthers versatile, didymous, cells turgid, longitudinally dehiscing. *Ovary* free, 2–5-celled, by abortion often 1-celled ; cells 1-ovulate. *Styles* or stigmas 2–5. *Capsule* 2–5 celled, by abortion sometimes 1-celled, 3–5-sulcate or 2-edged, dehiscent, or, if unilocular, indehiscent. *Seeds* solitary, pyriform or subreniform, brown, shining, striate, affixed to long funiculi, rising from the base of the cell ; radicle superior, next the hilum ; embryo on the outside of a farinaceous albumen, slightly curved. *Harv. Gen. p.* 123. *Fenzl. l. c. Kolleria, Aizoonis species, and Sialoides, E. & Z. ! Endl. Gen. n.* 5166.

Herbaceous or suffruticose, much branched plants, with alternate or opposite, simple, subfleshy, quite entire leaves and minute, axillary, sessile, rarely subpedicellate flowers. Named after Claudius Galenus, a Roman naturalist.

ANALYSIS OF THE SPECIES.

Subgenus I. **Kolleria.** Calyx 4–5-parted. Ovary mostly 3–5-celled. Styles 3–5. Flowers cymoso-spicate. (Sp. 1–9.)

Decumbent or prostrate perennial herbs or suffrutices:
- Villous and hirsute with white hairs, in all parts ... (1) **secunda.**
- Clothed with diaphanous, hairlike *scales*:
  - Upper and lower leaves of similar form, *obovate* ... (4) **sarcophylla.**
  - Lower leaves oblong-spathulate; upper *linear-spathulate* ... (5) **portulacea.**
- Clothed with appressed, rounded papillæ ... (6) **papulosa.**
- Appressedly hairy; lvs. obovate or spathulate; cal. 5-parted; styles 3–5 ... (2) **spathulata.**
- Thinly pubescent; calyx 4-parted; styles 4.
  - Lvs. obovate or spathulate ... (3) **pallens.**
  - Lvs. lanceolate or oblong-lanceolate ... (9) **herniariæfolia.**
- Glabrous or nearly so; lvs. obovate or spathulate; cal. 5-parted ... (8) **humifusa.**

Erect, shrubby; thinly pubescent; lvs. obovate-lanceolate; cal. 5-parted ... (7) **affinis.**

Subgenus II. **Eugalenia.** Calyx 4-parted. Ovary 2 or 1-celled. Styles 2. Flowers panicled. (Sp. 10–18.)

Covered with diaphanous scales:
- Scales lanceolate, spreading... ... (10) **squamulosa.**
- Scales short, appressed; lvs. obovate-cuneate ... (13) **crystallina.**
- Scales short, appressed; lvs. suborbicular ... (11) **Dregeana.**

Covered with minute papillæ, except on the old branches (12) **pruinosa.**

Hairy with stiff hairs; lvs. lanceolate-linear ... (18) **hispidissima.**

Silky-pubescent; leaves oblong-cuneate or lanceolate erect and shrubby ... (14) **fruticosa.**

Glabrous and glaucous; leaves linear-filiform, 4–9 lines long; stem decumbent ... (16) **glauca.**

Glabrous (not glaucous):
- Prostrate; leaves linear-clavate, squarrose, 1–2 lines long ... (15) **procumbens.**
- Erect; lvs. linear-lanceolate or spathulate, uncial (17) **Africana.**

Subgen. I. **Kolleria** (Presl.) Ovary 4–5, very rarely 3-celled. Styles 3–5. Capsule dehiscent. Prostrate perennial herbs or subshurbs; flowers cymose-spiked, subsecundate, in the axils of leaves, 3–5-gynous, sometimes intermixed with a few 2-gynous. Calyx 5-parted. (Sp. 1–9.)

1. G. secunda (Sond); stem herbaceous, decumbent, elongate as well as the branches and leaves *villous or hirsute with white hairs;* branches alternate, diffuse; branchlets short, leafy; leaves alternate, obovate, or obovate-spathulate, attenuated at the base, acute; flowers sessile, axillary, solitary, subsecundate. *Aizoon secundum, Thunb.! Fl. Cap.* 410. *Aizoon glinoides, elongatum, propinquum, et contaminatum, E. & Z.!* 2131, 2232, 2134, 2135. *Zeyher!* 2633.

VAR. β. **strigulosa** (Sond); stem, branches and leaves strigose, leaves oblong-spathulate, acute or subacuminate. *G. aizoides, Fenzl. in herb. Drege.*

HAB. Near Amsterdammbatterie and Gorrichoogde, *Dr. Pappe;* Zwartland, *Thunberg;* in fields and Karro-like hills in the districts of Uitenhage and Swellendam, *E. & Z.!* var. β. near Zwartkopsriver and in Sneeuwberge. *Drege!* Aug.–Dec. (Herb. Thunb., Vind., D., Sd.)

Stems 1–2 feet; branches gradually shorter. Leaves, when young, often opposite, on short petioles, 6–10 lines long, 2–4 lines wide, the upper ones much smaller and more approximate, often recurved at the apex, 1–3 lines long. Flowers in the ultimate branches appressed-hairy or hirsute, 1 line long. Calyx mostly 5-parted. Styles 3–5. In *A. contaminatum, E. & Z.!* the indument is more appressed and the leaves subtomentose, but *A. propinquum* is the intermediate form.

2. **G. spathulata** (Fenzl. in Herb. Drege) stem and branches herbaceous, procumbent, as well as the leaves *covered with short appressed hairs;* leaves alternate, short-petiolate, obovate-spathulate, subcomplicate, obtuse, mucronate or acute, often recurved at the apex; flowers sessile, axillary, solitary, 5-parted; *styles* 3–5. *Aizoon pubescens, E. & Z.!* 2133. *G. heterophylla, Fenzl. Zeyher,* 2635.

HAB. Fields near the Zwartkopsriver, *E. & Z.;* near Grahamstown, *Col. Bolton;* Kendo, Roodesand, and Buffelriver, *Drege.* Octob.–Jan. (Herb. Vind. D. Sd.)

Very nearly allied to *G. secunda,* and only distinguished by the more greenish or subglaucous colour and more distant leaves in the branches. The whole plant is very thinly pubescent, the young leaves opposite, old ones about ½ inch long, 2–3 lines wide, the upper smaller. Flowers usually more distant than in the preceding, 1 line long, pubescent on the outside.

3. **G. pallens** (Fenzl.); *very thinly puberous, glaucous or pale-yellowish;* stem suffruticose, procumbent, branches diffuse or ascendent; leaves alternate, petiolate, *obovate-spathulate,* subcomplicate, with recurved apex, young ones opposite; flowers axillary, sessile, solitary, *terminal ones mostly ternate,* central sessile, lateral ones short-pedicellate; *calyx* 4-*parted; styles* 4. *Kolleria pallens and glauca, E. & Z.!* 2144. 2145. *Zeyher* 2636.

HAB. Fields near the Zwartkopsriver, *E. & Z.;* Dec. (Herb. Vind. D. Sd.)

Habit and foliage exactly as in *G. spathulata,* from which it differs by the scarcely conspicuous, sometimes wanting pubescence and tetramerous flowers, very rarely mixed with a 5-parted calyx; 5 styles.

4. **G. sarcophylla** (Fenzl.); perennial, *squamulose-pubescent;* prostrate or suberect, primary branches opposite; leaves *obovate* or spathulate, obtuse, fleshy, papulose, *upper ones smaller but similar,* complicate; flowers axillary, sessile, solitary, in the ultimate branchlets approximate, *subspicate;* calyx 5-cleft; styles 3–4.

HAB. Springbokkeel, Feb., *Zeyher,* 719. (Herb. Vind. Sd.)

Whole plant greyish or pale, covered with appressed, hairlike papulæ, ½–1 foot high. Branches alternate, spreading, terete. Leaves 4–6 lines long, 2–3 lines wide, attenuated in a short petiole, upper ones 2–1 line long. Ultimate or flowering branchlets very short. Flowers alternate, scarcely 1 line long, pubescent on the outside. It comes very near *G. crystallina,* and is perhaps a 3–4 styled form. From *G. portulacoides Fenzl.* it differs by the thick, uniform leaves.

5. **G. portulacacea** (Fenzl. in herb. Drege); perennial, *squamulose-puberous,* stems herbaceous, prostrate or suberect; branches alternate, primary ones often opposite; leaves alternate, *oblong-spathulate,* flat, obtuse, papulose, upper *ones much smaller, linear-spathulate,* complicate; flowers axillary, sessile, solitary, *remote;* calyx 5-cleft; *styles* 3.

HAB. At Driekoppen and Jakkalsfontyn, *Drege,* Sept. (Herb. Vind. Sd.)

Nearly allied to the preceding, but the leaves are smaller and flat, ¾–1 inch long, 2 lines wide, attenuated at the base; those of the flowering branchlets 2–3 lines long, ½–1 line broad; flowers not approximate, otherwise not differing.

6. **G. papulosa** (Sond.); *densely clothed with appressed round papulæ;*

stem decumbent suffruticose ; branches alternate, rarely opposite ; leaves alternate, *obovate-cuneate*, obtuse; flowers solitary, axillary, sessile ; calyx 5-cleft, papulose ; *styles mostly* 5. *Aizoon papulosum E. & Z. !* 2137.

HAB. Karrolike hills on Gauritzriver near Grootriver, Swellendam. Dec.

Habit of *G. crystallina*, and *sarcophylla*, from the first it differs by a 5-parted calyx and 5, rarely 3 styles ; from *G. sarcophylla*, by a very different indument, not difformed leaves, and not subspiked flowers. Stem about 1 foot long. Leaves 4–6 lines long, 2–3 lines broad, floral ones nearly twice smaller, but obovate. Flowers 1 line long.

7. G. affinis (Sond.); *fruticose, erect*, branches alternate, virgate, spreading, glabrous, ultimate short, as well as the leaves *very thinly and appressedly pubescent ;* leaves alternate, small, obovate-lanceolate, complicate, with acute, recurved points ; flowers axillary, solitary, sessile, terminal ones mostly ternate, central sessile, lateral ones short-pedicellate ; calyx 5-cleft ; styles 3–4, rarely 5. *Aizoon fruticosum, E. & Z. !* 2139. *non. Thunb. !*

HAB. Hills between Hassaquaskloof and Breederiver, Swellendam, *E. & Z.* ; Bitterfontyn, *Zey. !* 724. (Herb. Vind. D. Sd.)

A woody greyish shrub, several feet high ; primary branches as thick as a goose's quill, ultimate very short and leafy. Leaves crowded, 3–5 lines long, 1 line wide, attenuated in a short petiole. Flowers secundate, 1 line long. Calyx appressed-hairy on the outside, longer than the stamens. *G. fruticosa* is very similar, but is distinguished by more silky pubescent leaves, 4-parted calyx, and 2 styles.

8. G. humifusa (Fenzl. in herb. Drege) ; *suffruticose, diffuse, quite glabrous*, or with a few appressed hairs on the young branches ; leaves alternate, *spathulate or obovate*, petiolate, upper one oblong-spathulate or oblong, attenuated at the base, smaller, subcomplicate, subacute ; flowers axillary, solitary, sessile, secundate ; calyx 5-cleft ; styles 3, rarely 4. *Kolleria collina, E. & Z.!* 2143.

HAB. Near Karakuis (III. B.) 1500–200 ft., and near the Garip, *Drege ;* Karrolike hills near Breederiver, and on the Zwarteberg, Caledon, *Zeyh. !* 2637 ; Bitterfontein, *Zeyher*, 722. (Herb. Vind. D. Sd.)

Many stems from the woody root, very pale-greyish or whiteish, ½–1½ ft. long, with alternate, slightly spreading, similar, terete branches. Leaves 6–8 lines long, lower ones 2–3 lines broad, upper ones about 1 line wide, those in the axils or of the ultimate very short branches, 1–1½ line long. Calyx 1 line long, glabrous. From the preceding, different by the diffuse, weaker stems, glabrous leaves and calyx ; from the following, by the 5-parted, 3-gynous flowers.

9. G. herniariæfolia (Fenzl.) ; suffruticose, diffuse, much branched ; branches alternate, the ultimate as well as the leaves and calyces on the outside, covered with a very thin, scarcely conspicuous indument ; leaves oblong-lanceolate or lanceolate-bluntish, attenuated at the base, subcomplicate ; flowers axillary, sessile, solitary, secundate ; calyx 4-parted ; styles 4. *Kolleria herniariefolia, Presl. Symb. bot.* 1, *p.* 24, *t.* 14. *E. & Z. !* 2142., *Aizoon microphyllum Bartl. ! Linnæa, Vol.* 7, 541. *A. Herniaria, Reichbch. in Sieb. herb. fl. cap.* 164.

HAB. Sandy places and hills in the Cape Flats near Rietvalley and Constantia ; Salt River, Decemb. (Herb. Vind. D. Sd.)

Stems 1–2 feet, prostrate, woody at the base, branches spreading ; ultimate very short and leafy. Leaves 3–4 lines long, ½–1 line wide, the uppermost smaller, often only one line long. Calyx nearly glabrous, scarcely 1 line long, white inside. Stamens 8, a little shorter than the calyx. Capsule depressed, 4-valved.

Subgenus II. **Engalenia,** Fenzl. Ovary 2-celled, or by abortion 1-celled. Styles 2. Capsule dehiscent or indehiscent. Prostrate or erectish, perennial herbs or shrubs ; flowers cymose-panicled, in the axil of a minute leaf, 2-gynous, very rarely mixed with a few 3-4-gynous. Calyx 4-parted. (Sp. 10-18.)

10. G. squamulosa (Fenzl.) ; densely covered with *spreading, lanceolate* squamulæ ; stem suffruticose decumbent, branches erectish alternate or subopposite ; leaves *alternate,* in the young branches opposite, *obovate or obovate-cuneate,* obtuse, flat ; flowers sessile, disposed in terminal, lax, *cymose-panicles. Aizoon squamulosum, E. & Z.!* 2136.

HAB. Sandy places between Mount Kamiesberg and Orange River, Namaqualand, *E. & Z.!* Dec. (Herb. Sd.)

Whole plant greyish-yellowish, about 1 foot high ; branches short. Leaves 4-6 lines long, 2 lines wide, on the terminal branches sometimes suborbicular, 3 lines long and broad. Panicles subsecundate about 1 inch. Flowers in the axils of a minute, roundish leaf, as large as the 1 line long calyx.

11. G. Dregeana (Fenzl.) ; covered with *pellucid squamulæ ;* stem suffruticose, erect ; branches opposite or alternate ; leaves opposite, rarely subalternate, *suborbicular,* short petioled ; flowers sessile, united into *glomerate cymes,* disposed in *a large terminal panicle.*

HAB. Plains near the Garip, *Drege,* Sept. (Herb. Vind. D. Sd.)

More robust than the preceding, stem as thick as a goose's quill ; branches erect-spreading, pale yellowish, dichotomous at top. Leaves 6-8 lines long, 6-7 lines broad, squamulose and papulate, upper ones 3-4 lines long and wide. Cymes of the panicle ¼-½ inch. Flowers 1 line long, bracteated by a minute leaf. Calyx squamulose.

12. G. pruinosa (Sond.) ; branches, leaves, and calyx *pruinose,* beset *with minute papulæ ;* stem suffruticose erect ; branches opposite ; leaves opposite, *obovate-cuneate,* obtuse ; flowers sessile, disposed in a terminal cymose panicle. *G. papulosa, Fenzl. excl. Syn. Aiz. papulosum E. & Z. ! Tephras papulosa, E. Mey. in herb. Drege.*

HAB. Plains near Koussie, Buffelriver and on the Garip, *Drege.* Sept. (Herb. Vind. D. Sd.)

Branches 1 foot long, white, as well as the spreading branchlets. Leaves on the branches 4-6 lines long, 2 lines broad. Panicle dichotomous, spreading. Flowers rarely 1 line long, in the axil of an equal or a little larger leaf. Easily known by the white colour of stem and branches.

13. G. crystallina (Fenzl.) ; covered with *pellucid appressed squamulæ ;* stem suffruticose, decumbent ; branches alternate, erectish ; leaves alternate, oblong-cuneate, upper ones obovate-cuneate, obtuse ; flowers sessile, disposed in *secundate subcymose* panicles. *Aizoon crystallinum, E. & Z.!* 2138.

HAB. Sandy places near Valleyfontein, Clanwilliam, Nov., *E. & Z.!* (Herb. Vind. Sd.)

Primary branches more than a foot long, terete, greyish, or grey-white as the whole plant. Leaves 6-8 lines long, 2-4 lines wide, much attenuated at the base, upper ones smaller, those of the young branches often opposite. Flowers ¾ line long, bracteated by a *minute* leaf. It comes very near *G. squamulosa* and *G. sarcophylla,* from the first it differs by the appressed, not lanceolate-spreading squamulæ and more approximate, smaller flowers ; from the latter by the 4-parted calyx and 2 styles.

14. G. fruticosa (Sond.) ; *young branches, leaves and calyx silky-pubes-*

cent ; stem fruticose erect, much branched, branches and leaves opposite ; leaves *oblong-cuneate* or oblong-lanceolate, upper ones smaller, complicate; flowers sessile, disposed in a terminal panicle. *Aizoon fruticosum Thunb. ! fl. cap.* 410. *G. sericea et salsoloides, Fenzl. in herb. Drege.*

HAB. In Zwartland, Thunberg ; *Scholl. in herb. Jack. ;* hills near Ebenezer, and between Hexriver and Buffelriver, *Drege ;* Bitterfontyn, Betchuanavald, *Zeyh. !* 723. (Herb. Thunb. Vind. Sd.)

Greyish shrub, 1 to several feet high, woody. Branches rigid, virgate. Leaves silky from adpressed hairs, ½–1 inch long, 1–2 lines wide, upper ones crowded and smaller, and recurved. Panicle spreading, dichotomously divided, 1 to several inches long. Flowers ½ line long, in the axil of an equal or rarely somewhat larger leaf.

15. G. procumbens (Linn. fil. Suppl. p. 227) ; glabrous, stem fruticose, decumbent ; branches and leaves opposite, *short, linear-clavate, canaliculate, squarrose-recurved ;* flowers sessile and short-pedunculate, cymose-paniculate at the top of the branches. *Thunb. ! fl. cap.* 384.

HAB. In Hantam, *Thunberg ;* Tarka, *Zeyher ;* Nieuweveld, dist. Beaufort, *Drege.* Nov.-Dec. (Herb. Thunb. Vind. D. Sd.)

A low, much branched shrub ; upper part of the branches and leaves with very thin, scarcely conspicuous indument, often quite glabrous. Leaves 1–2 lines long, ½ line wide, on the young branches mostly 3–4 lines long, spreading-recurved, with impressed line above. Flowers ½ line long ; the central sessile, lateral ones subpedicellate. Styles 2.

16. G. glauca (Sond.); glabrous, suffruticose, subdecumbent, stems filiform, branches and leaves opposite ; leaves connate at the base, linear-filiform, acute, subfleshy ; flowers ternately-cymose-paniculated, the central sessile, the lateral shortly pedicellate ; calyx shorter than the leafy bract. *Sialodes glauca, E. & Z. !* 2149. *Gal. Ecklonis Walp. Rep.* 2, 232.

HAB. Sandy places at the base of the Winterhoeks mountains, near Tulbagh, Worcester, Sept., *E. & Z. ;* Predikstoel, *Zeyher.* (Herb. Vind. D. Sd.)

Stems several from the root, 3–8 inches long, often reddish. Internodes 1 inch or longer. Leaves mostly crowded, 4–9 lines long, ½ line wide. Panicle terminal. Calyx 1 line long, twice shorter than the leafy bract ; lobes acuminate, white.

17. G. Africana (Linn. Spec. 515); glabrous, fruticose, erect ; branches and leaves opposite ; leaves linear or linear-lanceolate, attenuated at the base, carinate above, subfleshy ; flowers sessile, disposed in a terminal, dichotomous, large panicle. *G. linearis Thunb. ! fl. cap. p.* 384. *Sieb. herb. fl. cap. n.* 351. *Herb. Un. itin. n.* 309. *Zeyh.* 2638.

VAR. β. **halimifolia** ; leaves spathulate-lanceolate, 1–2 lines broad, 1 inch long. *G. halimifolia, Fenzl. in herb. Drege.*

HAB. Fields and among shrubs in the Cape Flats, Hottentottsholland, Breederivierspoort in Wupperthal, and in the distr. of Uitenhage ; var. β. in Langevalei and on Tarkariver. (Herb. Thunb. Vind. D. Sd.)

Stem 3 feet and more high. Branches woody, terete, greyish, erect, ultimate subfiliform, virgate. Leaves mostly crowded, 1 inch long, 1 line wide, sometimes longer, in the branches shorter and narrower. Panicle often very large. Flowers ½ line long, equalling the bracts, rarely shorter.

18. G. hispidissima (Fenzl. in herb. Drege.) ; stem fruticose, prostrate ; branches densely clothed with long stiff, spreading hairs ; leaves opposite, lanceolate-linear, as well as the calyx hispid; flowers sessile, disposed in dense cymose panicles.

HAB. Cape (station not given), *Drege.!* 7055. (Herb. Vind. Sd.)

A very distinct species. Stem 1 foot or more long; branches ascending short. Leaves of the branches 4–6 lines long, ¾–1 line wide, hispid, at length subglabrous, thickish. Calyx nearly 1 line long, white, with spreading white setulæ on the outside.

VII. **PLINTHUS**, Fenzl.

Calyx tubulose, 5-parted, lobes erect, subequal, coloured within. *Petals* none. *Stamens* 5, inserted in the lower part of the calyx, alternate with the lobes; filaments exserted. *Ovary* 3-celled, cells with 1 pendulous ovule. *Style* 3-partite. *Capsule* ovoid, densely papillose, rotundate, not depressed, 3-celled, loculicidal, 3-valved, cells one-seeded. *Seeds* pyriform, shining, striate. *Embryo* uncinate. *Fenzl. in nov. stirp. decad. n.* 60. *Endl. Gen.* 5167.

Small shrub, humifuse, branched, imbricated-leafy. Leaves very minute, ovate-triquetrous, opposite and alternate, without stipulæ. Flowers sessile, alternate, hidden in the axil of the leaves, with 1 or 2 equal, leafy, bracteoles. Name from πλινθος, a tile; so called in reference to the imbricated leaves.

1. **P. cryptocarpus** (Fenzl. l. c.)

HAB. Near Rietpoort, Nieuweveld, 3000–3500 ft., Nov., *Drege.* (Herb. Vind. Sd.)

Stems 1–3 inches long. Branches subunilateral, 3–4 lines long, as well as the leaves covered with appressed, silky hairs. Leaves densely 3–5-farious, imbricated, acutish, ¾–1½ line long. Flowers ½–¾ line long. Calyx puberous on the outside, yellowish inside. Stamens nearly hypogynous.

ORDER LVI. **CACTEÆ**, DC.

(BY W. H. HARVEY.)

Flowers perfect, regular. *Calyx* and *corolla* confounded together, in a many-leaved perianth. *Calyx*-tube attached to the ovary, sometimes much produced beyond it; sepals numerous, in few or many rows, the innermost petaloid. *Petals* also usually in several rows, of delicate texture, the outer ones confounded with the inner sepals, marcescent or deciduous. *Stamens* indefinite; filaments filiform; anthers 2-celled, versatile. *Ovary* inferior, unilocular, with numerous ovules on 3 or more parietal placentæ; style terminal, filiform; stigmas as many as the placentæ. *Fruit* succulent, one celled, many-seeded; seeds lying in pulp, usually without albumen; embryo straight, curved, or spiral, with the radicle next the hilum.

Succulent shrubs, very varied in form, almost exclusively natives of the American continent, though several species (chiefly of the genus *Opuntia*) are now naturalized in the warmer parts of the old world. Leaves very generally wanting, or reduced to minute scales or spines; the functions of a leaf being discharged by the green bark of the succulent stems. The genus *Pereskia*, however, possesses large, petioled, deciduous leaves of ordinary structure. The flowers are solitary, terminal, or axillary, and of large or small size, often very showy.

I. **RHIPSALIS**, Gaertn.

Tube of the perianth not produced beyond the ovary, the limb rotate, of 12–18 short, scale-like parts, the outer sepaloid, the inner petaloid.

Stamens numerous, about equalling the perianth. *Style* filiform; stigmas 3–6, radiating. *Berry* globose, smooth, crowned with the dried up limb of the perianth. *Seeds* numerous, exalbuminous; cotyledons short, acute. *Endl. Gen.* 5160.

Slender, epiphytical succulents, growing on forest trees, with whip-like or expanded and leaf-like, often articulate stems, and small lateral flowers. Name from ῥιψ, a *willow-branch;* from the long, flexuous branches.

1. **R. cassytha?** (Gaern.; Pfeiff. Enum. Cact. p. 133); stem erect or creeping, at length woody; branches slender, green, terete, pendulous, more or less verticillate, remotely scaly, obtuse. *DC. Prodr.* 3. *p.* 476. *Bot. Mag. t.* 3080. *Cactus flagelliformis, E. Mey. in Herb. Drege.*

HAB. Between the Omtata and Omsamwubo, *Drege.* Caffirland, *J. Backhouse.*

Whether or not *Drege's* plant belong to the common *R. cassytha,* which is found in the Mauritius as well as in the West Indies, or to a new species I cannot say. *Mr. Backhouse,* travelling in Caffirland in 1838, observed a *Rhipsalis* on trees, which no doubt was the same as what *Drege* collected. We have not seen specimens from either collector.

ORDER LVII. BEGONIACEÆ, R. Br.

(BY W. SONDER.)

Flowers unisexual, monœcious. *Male: Perianth* petaloid, 4-leaved, the two outer sepals larger, roundish. *Stamens* indefinitely numerous, occupying the centre of the flower; filaments short; anthers extrorse, 2-celled, the cells adnate to the edges of a thickened connective. *Female: Perianth* corolloid, with a 3-winged tube adhering to the ovary, the limb 4–9 parted, with imbricate æstivation. *Ovary* inferior, 3-celled; ovules very numerous, attached to axile placentæ. *Stigmas* 3, subsessile, bifid, incrassated, tortuous or capitate. *Capsule* membranous, crowned with the withered perianth limb, 3-winged, 3-celled, opening by slits at the base of each wing; seeds very numerous, minute, oblong, exalbuminous; cotyledons very short; radicle long, next the hilum.

Herbaceous or half-shrubby plants, with succulent stems and foliage. Branches swollen at the nodes. Leaves alternate, petiolate, simple, palmate-nerved, entire or lobed, very generally unequal-sided or semi-cordate at base, variously toothed, often covered with membranous scales and brightly coloured on the under surface. *Stipules* lateral, membranous, free. Inflorescence cymose. Natives chiefly of the tropics of Asia and America; very few African. The roots are astringent and slightly bitter, and occasionally employed in medicine.

I. BEGONIA, L.

Capsule opening by arched or longitudinal slits along the face of the loculi. *Placentæ* from the inner angle of the loculi. *A. DC. in An. Sc. Nat. Ser.* 4. *Vol. XI. p.* 119.

A large genus, abundant in Asia and America, but rare in Africa, where until lately they were unknown. Many are deservedly favourites in cultivation as ornamental stove-plants. The name is in honor of Michael Begon, a French patron of botany in the 17th century.

ANALYSIS OF THE SOUTH AFRICAN SPECIES.

Stems herbaceous:
  Leaves reniform-cordate; stipules ovate, obtuse, mucronulate; wings of the fruit subequal ... ... (1) **Dregei.**

Leaves reniform-cordate; stipules lanceolate-acuminate; wings of the fruit subequal ... ... (2) **Caffra.**
Leaves semicordate-acuminate; 2 wings of the fruit larger (3) **Natalensis.**
Stem woody at the base:
Leaves palmately 3-4-lobed, lobes lanceolate; wings of the fruit equal ... ... (4) **suffruticosa.**

1. B. Dregei (Otto & Dietr. Gartenzeitg. IV. 357); stem fleshy-nodose; leaves petiolate, unequal-sided, reniform-cordate, coarsely *angulate-serrate,* very smooth, shining; stipules *ovate, obtuse, mucronulate;* cymes axillary, pedunculate, few-flowered; flowers snow-white; wings of the capsule subequal, acutangular, truncate at the apex. *B. parvifolia, Graham, Bot. Mag. t.* 3720. *Augustia Dregei, Klotzsch, Begoniac. p.* 80, *t.*

HAB. Near Port Natal, *Drege.* (Herb. Sond.)

Stem 1–3 feet. Leaves ¾–1½ inch long. 1–2½ inches wide. Petioles 1–2 inches long, in the upper leaves shorter. Stipules 3–4 lines long, nearly 2 lines broad. Cymes longer than the petioles. Bracts greenish, roundish-obovate. Wings of the capsule 6–9 lines long, 2–3 lines broad.

2. B. Caffra (Meisn. Linn. 14.501); stem fleshy-nodose, branched; leaves petiolate, unequalsided-reniform-cordate, *angulate-lobed,* bluntish-serrate, acute or acuminate; stipules *lanceolate-acuminate;* cymes dichotomous, pedunculate, axillary, 4-flowered; *flowers whitish;* bracts orbicular-ovate; wings of the capsule *subequal,* acutangular, truncate at the apex. *B. sinuata, E. Meyer. Graham. Bot. Mag. t.* 3731, *not of Wallich. Augustia Caffra, Klotzsch. l. c.*

HAB. Near Port Natal, *Drege* (v.v.) (Herb. Sond. D.)

Stem 2–3 feet. Leaves 2–2½ inches long, 3–4 inches broad, with red nerves beneath. Petioles 2–3 inches long. Stipules from a sublobate, large base, acuminated, 7 lines long, at the base 3–4 lines wide. Pedicels 2 inches. Bracts broad-ovate, blunt. Wings of the fruit near the apex 3 lines broad.

3. B. Natalensis (Hook.! Bot. Mag. t. 4841); tuberous, glabrous; stem fleshy, at the base thick, nodose-articulate, branched; leaves unequal, *semicordate, acuminate,* lobate, and coarsely auriculate-serrate, acute, whitish-spotted above; cymes pedunculated, axillary, 4–6-flowered; *flowers pale-rosy;* petals of the male flower rhomboid-orbicular, of the female rhomboid-ovate; fruit 3-winged; 2 *wings larger,* acutish-angular, one shorter, obtuse-angled. *Augustia Natalensis, Kl. l. c. VIII. B.*

HAB. Port Natal, Capt. Garden; *Gueinzius,* 210. (Herb. Hook. Sond.)

Stem 1–1½ foot high, green, reddish-spotted. Leaves 1–3 inches long, 3–4 inches wide, with a red middle nerve. Petioles ½–1½ inch long, reddish. Pedicel 1–1½ inch. Fruit 8–12 lines long, larger wings 2–4 lines wide.

4. B. suffruticosa (Meisn. Linn. 14.502); tuberous, slender, very smooth; stem flexuous-erect, *woody at the base;* stipules ovate-oblong, acute, entire; leaves oblique, *palmately 3–4-lobed,* lobes *unequal, lanceolate, pinnate-incised,* dentate or entire; cymes axillary, pedunculate, few-flowered; capsules ovate-triangular, reticulate-veined, truncate at the apex, wings equal, obtuse-angled at the apex.

HAB. Near Port Natal, *Drege, Gueinzius.* (Herb. Sond.)

Very near and perhaps only a variety of the preceding. Petioles 4 lines–1½ inch long. Leaves 1–1½ inch long, 8–12 lines broad. Stipules 2 lines long. Pedicels ¾–1½ inch long. Wings of the capsule much veined, 8 lines long, 3 lines broad.

ORDER LVIII. **CUCURBITACEÆ**, Juss.

(BY W. SONDER.)

Flowers monœcious or diœcious. *Calyx*-tube adnate to the ovary, and sometimes produced beyond its summit, in the male-flowers short and mostly campanulate; limb 5-lobed, with imbricate æstivation. *Corolla* (very rarely of separate petals) usually monopetalous, rising from the summit of the calyx-tube, with which it seems continuous, rotate or campanulate, 5-lobed. *Stamens* inserted in the bottom of the calyx, 5, rarely 3 or 2, free or united wholly or in part; anthers extrorse, adnate, 1–2-locular, linear, usually very long and flexuous. *Ovary* inferior, of 3–5 united carpels, at first unilocular with prominent, but revolute, parietal placentæ; afterwards (by the union of the placentæ into a central column and the adherence of their revolute edges to the walls of the ovary) 6–10-celled; ovules numerous, anatropous, pseudo-parietal. *Style* terminal, short, 3–5-cleft or parted; stigmas thickened, lobed or fimbriate. *Fruit* a gourd or berry, dry, or fleshy or juicy; usually by the dissolving of the septa into pulp, unilocular, many-seeded. *Seeds* flat or convex, with a succulent or membranous envelope, exalbuminous; embryo with leafy and veined cotyledons, orthotropous.

Herbaceous or half-shrubby, rarely shrubby plants, natives chiefly of the tropics and of the warmer parts of the temperate zones. Stems usually prostrate or climbing. Leaves alternate, petioled, simple, palmate-nerved, entire or palmatifid, or variously lobed, mostly cordate at base. Tendrils formed out of a lateral stipule. Flowers often of large size, either solitary, tufted, racemose, or panicled, usually white or yellow, sometimes red. To this Order belong melons, gourds, cucumbers, and vegetable-marrow, &c. Many however have highly acrid and poisonous or powerfully cathartic fruits. Of the former class *Momordica Elaterium* (the spurting cucumber) is one of the most virulent examples, and of the latter the *Citrullus Colocynthis*, the source of the drug colocynth. The fruit of *C. vulgaris* (bitter-apple or wild water-melon) is a useful colonial substitute for the drug.

TABLE OF THE SOUTH AFRICAN GENERA.

* Stamens 5, free. Anther-cells linear, straight.

I. **Coniandra.**

** Stamens 3. Anther-cells linear, straight.

II. **Zehneria.** Connective without conical appendage.

III. **Mukia.** Connective terminated by a conical appendage.

*** Stamens 3; two of the filaments bearing a 2-celled, the third a 1-celled anther. Anther-cells flexuous or gyrous.

a. Flowers coetaneous. (appearing with the leaves).

IV. **Lagenaria.** Petals white. Gourd fleshy, indehiscent. Seeds with a tumid border. Male peduncle without bract.

V. **Luffa.** Petals yellow. Gourd at length dry and fibrous. Male peduncle without bract.

VI. **Momordica.** Petals white or yellow. Gourd 3-valved, fleshy, prickly. Male peduncle with a large bract.

VII. **Cephalandra.** Corolla 5-parted, yellow. Many-seeded, oblong berry. Anthers combined to a globose head.

VIII. **Citrullus.** Corolla 5-parted, yellow. Many-seeded gourd. Anther-connective without terminal appendage. Seeds with obtuse margin.

IX. **Cucumis.** Corolla 5-parted, yellow. Many-seeded gourd. Anther-connective terminated by an appendicula. Seeds not margined.

β. Flowers precocious (appearing before the leaves).

X. **Pisosperma.**

I. CONIANDRA, Schrad.

Flowers monœcious or diœcious. *Male: Calyx* campanulate, 5-fid. *Corolla* with 5-parted, spreading limb. *Stamens* 5, inserted on the throat of the perianth, free or triadelphous; *filaments* short and thick; *anthers* anterior, linear-oblong, straight, one-celled. *Female: Calyx* and *corolla* as in the male flower. *Stigma* usually capitate, tri-lobed. *Fruit* berried, beaked, pseudo-trilocular, few-seeded. *Seeds* obovate. *Coniandra and Cyrtonema, Scrad. in E. & Z.! Enum. pp.* 275, 276. *Endl. Gen.* 5124, 5125.

Herbaceous, climbing plants, with a tuberous root. Tendrils simple. Leaves palmate- or digitate-partite, rarely reniform, often scabrid. Flowers very small, green, or a little yellowish; the male racemose, female much shorter, pedunculate, solitary. Name from *κωνος*, a *cone*, and *ανηρ,ανδρος* a *man;* the anther bearing part of the stamens conniving or conical.

ANALYSIS OF THE SPECIES.

Flowers monœcious:
- Leaves reniform, crenate-dentate, pubescent ... ... (6) **mollis.**
- Leaves reniform-cordate and trilobed, glabrous ... ... (5) **Thunbergii.**
- Leaves digitate-partite:
  - segments pinnatifid, lobes flat, smooth ... ... (2) **Africana.**
  - segments pinnatifid, lobes scabrous, with recurved margins (4) **punctulata.**
  - segments bifid or trifid, lobes flat, linear-cuneate ... ... (1) **digitata.**
  - segments bifid or trifid, lobes obovate-cuneate ... ... (3) **glauca.**

Flowers diœcious ... ... (7) **Zeyheri.**

1. C. digitata (Sond.); leaves petiolate, orbiculate, digitate-partite; segments *linear or linear-cuneate, lateral ones deeply bifid or trifid; the middle trifid or undivided,* obtuse, mucronate; uppermost leaves often *3-foliate,* with linear-cuneate leaflets; peduncle of the male flowers as long as the leaf, racemose or panicled; calyx-lobes with very short bristles; fruit oval, beaked, 4–6-seeded. *Bryonia digitata, Thunb.! Fl. Cap.* 35. *Cyrtonema digitata, Schrad.! E. & Z.*, 1780, *partim. Drege,* 8187.

HAB. In fields among bushes near the Zwartkopsriver. Feb.–March. *E. & Z., Drege.* (Herb. D., Hk., Sd.)

Quite glabrous. Stem sulcate-angulate. Petiole 3–4 lines long. Segments of the larger leaves 1½–2 inches long, 1½–2 lines broad; of the upper leaves about 1 inch long, or shorter. Male peduncle capillary, with 4–8 short pedicels. Fruit yellow, nearly 6 lines long.

2. C. Africana (Sond.); leaves petiolate, *triangulate,* digitate, 5-partite; segments *pinnatifid,* lobes lanceolate or linear-lanceolate, mucronate, entire or dentate, the terminal longer; peduncle of the male flowers as long as the leaf or longer, racemose at top; calyx-lobes with *long, subulate bristles;* fruit roundish, shortly beaked. *Bryonia Africana, Linn.! Spec.* 1438. *B. dissecta, Thunb.! l. c.* 36. *B. multifida et pinnatifida, E. Meyer in herb. Drege. Coniand. pinnatisect, Schrad.! E. & Z.* 1776. *B. pinnatifida, Burch. trav.* 1. 547.

HAB. Near Gauritzriver, *E. & Z.!* Orange and Vaalriver, *Zey.!* 601. Great Vetriver and near Wanderfontyn, Betchld. *Zey.!* 600, 599; Nieuwehantum, 4–5000ft.; between Basche and Morley, and near Omsamwubo, 1–2000ft., *Drege.* Dec.–March. (Herb. Holm., D., Hk., Sd.)

Habit of the preceding. Petiole 3–4 lines long. Leaves mostly 3-partite, middle segment 1½–2 inches long, usually with 8–12 horizontal, 1–2 lines broad lobes, the lower of which 4–8 lines long, the uppermost toothlike; lateral segments bifid at the base, 1–1½ inch long, outer lobe bifid, inner with 6–8, spreading or horizontal, gradually shorter lobes. Upper leaves much smaller. Peduncles as in *C. digitata.* Anthers free. Ripe fruit about 4 lines long, scarlet.

3. C. glauca (Schrader l. c.); leaves *petiolate, orbiculate,* digitate, 3–5-partite; segments *flat, obovate-cuneate,* dentate, mucronate or acute; segments of the superior leaves smaller, with lanceolate-cuneate lobes; peduncle of the male flowers elongate, pedicels subumbellate; calyx-lobes *with short bristles;* fruit ovate, beaked, 3–4-seeded. *Bryonia Africana et grossulariæfolia, E. Meyer, Zey.!* 604, 2476.

VAR. β. **dissecta**; segments evidently dentate or inciso-serrate. *C. dissecta, Schrad.! E. & Z.!* 1777, *excl. syn. Zeyher,* 2475.

HAB. Among bushes in distr. Uitenhage near Adow and on Coega and Zwartkopsriver, *E. & Z., Drege;* Howisonspoort, *Hutton;* Port Natal, *Drege.* Feb. (Herb. D., Sd.)

Habit of *C. digitata,* but generally smaller. Leaves about 1 inch in diameter; lower ones often 3-lobed or half 5-lobed, with shortly toothed lobes; segments of the middle and upper leaves subflabellate, cuneate, dentate or lobed; segments of the uppermost leaves smaller. Peduncles and flowers as in the preceding. Fruit about 4 lines long. Many of the specimens distributed by *E. & Z.* under the name of *C. Zeyheri,* belong to *C. glauca. Drege,* 8188 seems not to be different from *C. glauca.*

4. C. punctulata (Sond.); leaves *sessile or shortly petiolate,* digitate, 3-partite; segments pinnatifid or bipinnatifid, lobes oblong or obovate-oblong, obtuse, the terminal mucronulate, *recurved at the margins, scabrous, white-punctate above;* peduncles of the male flowers elongate, racemose at top; calyx shortly aristate; fruit roundish, shortly beaked, 2–3-seeded.

VAR. β. **tenuiloba**; leaves short-petiolate; lobes short, lanceolate or subulate, mucronulate.

HAB. Springbokkeel, *Burke, Zey.,* 603. Var. β, near Orangeriver and Rhinosterkop, distr. Beaufort, *Zey.!* 602 & 602?; Uitvlugt, *Drege,* 8189. Jan.–April. (Herb. Hk., Sd.)

Near *C. digitata,* but the leaves are subsessile, and the segments remotely pinnatifid and narrower. Leaves of var. β, about ½–¾ inch in diameter. Racemes 3–5-flowered. Fruit 3–4 lines long.

5. C. Thunbergii (Sond.); glabrous leaves petiolate, scabrous, *reniform-cordate, subangulate or trilobed;* lobes ovate or cuneate-obovate, obtuse, mucronulate, lateral ones often sub-bilobed; peduncles of the male flowers elongate, *glabrous,* racemose or subpanicled; fruit 3–5-seeded, ovate, beaked.

VAR. α, all the leaves reniform-cordate, more or less angulate, or the upper ones often 3-lobed. *Bryonia triloba, Thunb.! l. c.* 34. *Cyrtonema triloba, Schrad.! E. & Z.,* 1778. *Sicyos angulata, Berg.! cap. p.* 352.

VAR. β, all the leaves trilobate or palmate 5-lobed, or the lower ones reniform-cordate, subangulate. *B. Africana, Thunb.! l. c.* 352, *non Linn. B. triloba, Drege, exs. Cyrtonema latiloba, Schrad.! E. & Z.,* 1781.

HAB. Cape Flats and in mountains near Capetown; var. β, Bosjesmansriver, Houtbay, Cape Recief, *Thunb., Drege, E. & Z., Harv.* Dec.-Jan. (Herb. Thunb., Holm., D., Sd.)

Stems and petioles glabrous or with scattered hairs. Petioles 3–4 lines long. Leaves coriaceous, ¾–1 inch long and broad, palmately nerved. Raceme 6–10-flowered, pedicels 2–1 line long. Male flower 2 lines, fruit 5–6 lines long, yellow. What *E. & Z.* described as a green fruit, is a production of insects.

6. C. mollis (Sond.); leaves petiolate, *reniform, crenate-dentate, mucronulate, pubescent on both sides;* peduncles of the male flowers elongated, *pubescent,* racemose at top; fruit ovate, beaked, 4–6-seeded. *Cyrtonema molle, Kunze in Linn. Vol.* 20, *p.* 49.

HAB. Mountain places among shrubs near Uitenhage. March. *Zey.!* 2480. (Herb. Sd.)

Stem very long, as thick as a pigeon's quill, angulate-sulcate, pubescent, when old subglabrous. Petiole ½–1 inch long, densely pubescent. Leaves 1½–2 inches long, 2–2½ inches broad, rarely sublobate, dentate. Male peduncle 2–3 inches long, with 6–10, capillary, 2–3 lines long pedicels. Flowers very small. Fruit on a very short peduncle, ½–¾ inch long, yellow or a little reddish when dry.

7. C. Zeyheri (Schrad.!) glabrous; leaves petiolate, palmate, 3–5-partite; segments of the lower, obovate-cuneate, shortly 3-lobed; of the upper, oblong-cuneate or lanceolate-cuneate, evidently 3-lobed, lobes obtuse, mucronulate; peduncles of the male flowers elongate, racemose; stamens 3-adelphous; fruit short-peduncled, ovate, beaked, 4–6-seeded. *E. & Z.,* 1775. *Cyrton. sphenoloba, Schrad.! E. & Z.,* 1779. *C. digitata, E. & Z.! (pars inferior) Zeyher,* 2471, 2472, 2473.

VAR. β. **angustiloba**; inferior leaves reniform-3-lobed, upper ones palmately 3-partite, with lanceolate or sublinear, bifid or trifid segments; flowers and fruit as in var. α. *Zey.!* 2474, & 592.

HAB. Among bushes in fields near the Zwartkopsriver, and hills near Boshmansriver, *E. & Z.;* var. β, Komandokraal and near Zondagriver, *Zeyher;* Albany, *T. Williamson.* Rhinosterkop near Beaufort, *Burke & Zeyher.* April–Sept. (Herb. D., Hk., Sd.)

Quite glabrous or a little pubescent at the base in var. β. Petiole 3–4 lines long. Lower leaves 3-partite, middle segments about 6–10 lines long, 6–8 lines broad, sub-3-lobed, lateral segments shortly bilobed, lobes subacute, 3-dentate or entire; upper leaves smaller, often more compound, segments narrower, evidently 3 or 2-lobed. Male peduncle 1 inch long, with 4–6, bracteated, 1–2 lines long pedicels. Flowers very small. Fruit about ½ inch long. Var. β, is perhaps a distinct species.

II. ZEHNERIA, Endl.

Flowers monœcious or diœcious. *Male: Calyx* campanulate or rugulose, 5-dentate. *Corolla* with 5-parted, very spreading limb. *Stamens* 3, inserted at the base of the corolla, free, filiform-cylindraceous, terminated by a subcordate connective, along each margin of which at the back are attached the linear, straight, 1-celled, free or cohering anthers. *Female: Calyx* and *corolla* as in the male flowers. *Stigma* cap-shaped, somewhat trilobed. *Fruit* berried, oblong-fusiform, ovate or globose, with a subcoriaceous pericarp, pseudo-trilocular or bilocular, few-seeded. Seeds compressed. *Endl. Prod. Norfolk,* (1833), *p.* 69. *Gen. n.* 5127. *Pilogyne Schrad. in E. & Z. Enum. p.* 277.

Herbaceous plants with a tuberous rhizome, simple tendrils and lobed leaves. Flowers small, white; the male racemose; the female umbellate or on simple peduncles. Fruit globose, as large as a pea, rarely larger. Seeds small, much-compressed, obovate.

ANALYSIS OF THE SOUTH AFRICAN SPECIES.

§ I. PILOGYNE.—Peduncles without bracts. Perianth campanulate. Fruit 2–4-seeded.

Quite glabrous and smooth. Leaves cordate, quite entire ... (5) **hederacea.**
Stem glabrous. Leaves scabrous.
cordate, acute, dentate ... ... (1) **scabra.**
cordate, obtuse, dentate ... ... (2) **cordata.**

Stem pubescent. Leaves
sub-5-lobed, serrate-dentate. Fruit glabrous ... ... (3) **velutina.**
3-angulate-cordate, denticulate. Fruit pubescent ... (4) **obtusiloba.**

§ II. BRACTEARIA.—Peducles with a large bract. Perianth tubulose. Fruit 2-seeded.

Bracts ciliate. Leaves
palmately 3–5-lobed, lobes ovate. Male peduncle shorter than the petiole ... ... (6) **Garcini.**
palmately 3-partite, lobes oblong-lanceolate. Male peduncle longer than the petiole ... ... (7) **pectinata.**
Bracts dentate. Leaves palmately 5-partite, lobes lin.-lanceol. (8) **Wyleyana.**

§ III. PLEIOSPERMION.—Peduncles without bract. Perianth tubulose. Fruit 6–9-seeded.

Lobes of the leaves lanceolate. Fruit oval, acute ... ... (9) **macrosperma.**
Lobes of the leaves lin.-lanceol. Fruit subglobose, apiculate (10) **debilis.**

1. Z. scabra (Sond.); stem glabrous, young branches pilose; leaves *cordate, angulate, acute,* dentate, callous-punctate and scabrous above, pilose or hispid by short hairs on the veins beneath, lower ones sub-5-lobed, upper ones 3-lobed, middle lobes subacuminate. *Bryonia scabra, Thunb.! fl. cap.* 34. *B. punctata, Thunb.! l. c. B. angulata, Thunb.! l. c.* 35. *B. Maderaspatana, Berg.! cap.* 351. *Pilogyne sauvis, affinis, Ecklonii, cuspidata et membranacea, Schrad.! E. & Z.*, 1782, 1783, 1785, 1786, 1787. *Herb. Un. itin. n.* 141. *Zeyher,* 597, 598, 2479.

VAR. β, **glabrata**; quite glabrous, the leaves scabrously punctate above. *Bryonia dentata, E. Mey. in herb. Drege. Zeyher,* 2478.

VAR. γ. **peduncularis**; branches and leaves rough with short, stiff hairs; peduncles 2–3 inches, pedicels 3–4 lines long, pubescent; male flowers a little larger.

HAB. Near Capetown, in the distr. of Uitenhage, Albany, George, and in Betchuanaland; var. β, near Uitenhage and Port Natal, *Drege;* var. γ, valley of Tarka-river, *Mrs. Barber.* Natal, *Gerr. & McKen.!* 551. May–Dec. (Herb. Th., Hm., D., Hk., Sd.)

Stem angulate-sulcate. Petiole 4–12 lines long. Leaves mucronate-dentate, paler beneath, as long as broad, or somewhat longer; lower ones 1–2½ inches long, 3 or 5-lobed; upper ones gradually smaller, usually trilobed, with an acuminate, aristate middle lobe. Peduncles solitary or aggregated, uncial or longer, at top with many (10–20) densely racemose, puberous, 1–1½ line long pedicels. Female flowers often solitary on shorter peduncles. Flowers puberous or glabrous. Fruit glabrous, minutely punctate. Seeds 1 line long.

2. Z. cordata (Sond.); glabrous, young branches pilose; leaves scabrous above, hairy on the veins beneath, lower ones *broad-triangular-cordate, obtuse or emarginate,* mucronulate-dentate; upper ones subtrilobed, obtuse or subacute, denticulate; fruit glabrous. *Bryonia cordata, Thunb.! fl. cap.* 34. *Pilogyne dilatata, Schrad.! E. & Z.*, 1784.

HAB. Woods at Olifantshoek, Uitenhage, *E. & Z.* Sept. (Herb. Thunb., Sd.)

Near the preceding, flowers and fruit the same, but the leaves are more broad-cordate, obtuse, lower ones 2 inches broad, 1½ inch long, upper ones uncial or smaller and acute, cordate or truncate at the base. Peduncle about ½ inch long, with 6–12, very short pedicels. Female peduncles 4 lines long.

3. Z. velutina (Schrad.!) stem and petioles *pubescent;* leaves cordate, sub-5-lobed, *unequally serrate-dentate,* mucronate, *very scabrous above, densely pubescent beneath;* peduncles of the female flowers elongated, umbellate, or very short and 1-flowered; *fruit glabrous. E. & Z.!* 1788. *Bryonia scabra, var. Drege, herb.*

HAB. Woods on the Katriver-mountains, Caffraria, *E. & Z.*; Zuureberge, April, *Drege.* (Herb. D., Sd.)

Leaves about as long as broad, 1–2-uncial, distinctly 5-lobed, lobes short. Flowers pubescent. Fruit as in *Z. scabra*, from which it is only distinguished by the serrated densely pubescent leaves. *Bryonia lævis, Thunb.! fl. cap.* 35, now wanting in his herbarium is perhaps a glabrous variety.

4. **Z. obtusiloba** (Sond.); *whole plant pubescent*; leaves *triangulate-cordate, subangulate,* with 2 obtuse lobes at the base, minutely-mucronulate-dentate, *not scabrous above;* peduncles of the male flower elongated, racemose, of the female short, and 1-flowered; fruit *pubescent. Bryonia obtusiloba, E. Meyer in herb. Drege.*

HAB. Woods near Port Natal, *Drege. Gerrard & McKen.*, 559. (Herb. D. Sd.)

Leaves 1½–2 inches long and wide, with soft hairs on both sides, entire or 3–5-lobate, lobes short, often obsolete on the margins, with very short teeth, obtuse or subacute. Male peduncles longer than the half-uncial petiole. Perianth campanulate. Fruit about 4 lines in diameter. Seeds 3, ovate, with a small membranaceous margin.

5. **Z. hederacea** (Sond.); *quite glabrous and smooth;* leaves cordate, acute, mucronate, *quite entire,* trilobed or obsoletely 5-lobed, lateral lobes short, obtuse, with or without a mucro; in the uppermost leaves acute; peduncles of the male flower elongate, racemose, few-flowered, as well as the pedicels capillary.

HAB. Kromriver in woods, March, *Drege.* (Herb. Sd.)

Easily distinguished by the smoothness of all parts, and the entire, not dentate or serrate leaves. Stems as in the others, lower leaves about 2 inches long and broad or a little broader, palmately nerved, lateral lobes roundish or obtuse, 3-angular. Peduncle 1 inch or longer; pedicels 3–6 lines long, not densely aggregated as in the preceding. Female flowers unknown.

6. **Z. Garcini** (Stocks in Hook. Kew. Journ. 4, 149); leaves petiolate, *palmate-3–5-lobed,* lobes ovate, mucronate-toothed, sprinkled on both sides with minute, rigid bristles; male peduncles shorter than the petiole, with a cordate, ciliate bract; fruit inverse, reniform, 2-seeded; seeds oblong, thickest at the margin. *Harv. thes. cap. t.* 96. *Bryonia Garcini, Willd. spec.* 4, 623. *W. & Arn. Prod.* 344. *Sicyos Garcini, Linn. Burm. Ind. t.* 57. *f.* 3.

HAB. Near Port Natal, *J. Sanderson.* Kreili's country, *H. Bowker.* (Herb. D., Sd.)

Stem several feet long, at first hispidulous, then quite glabrous. Leaves 2–3 inches apart, spreading; petiole scarcely uncial; lamina broader than long, 2–2½ inches long, about 3 inches broad, scabrous, 3–5-lobed, lobes ovate, coarsely toothed. Male peduncle about 2-flowered. Male perianth nearly ½ inch long, with a cylindrical tube, and horizontally patent, acuminate limb. Fruit (of an Indian specimen) 3 lines broad, 2 lines long, glabrous, on a very short peduncle. Seeds 2, compressed.

7. **Z. pectinata** (Sond.); monœcious; leaves petiolate or deeply palmate, 5-partite, scabrous on both sides, lateral segments oblongo-lanceolate or oblong, deeply bifid, 2–3-dentate at the apex; the middle segment cuneate-acute, 3–5-dentate; bracts cordate, ciliate; male peduncle longer than the petiole; the female much shorter, pedunculated; (unripe) fruit ovate. *Bryonia pectinata, E. Meyer in herb. Drege.*

HAB. Buffelriver, Jan., *Drege.* Port Natal, *Miss Owen.* (Herb. D., Sd.)

Stem as in *Z. Garcini.* Petioles 3–6 lines long. Leaves very scabrous by whitish bristles, in some specimens digitate- in other palmate-parted, the intermediate segment about 1 inch long, 3-5 lines broad, mucronulate or acute, with several acute teeth, rarely trilobed; the lateral segments 8-10 lines long, bifid, obovate or oblong-cuneate; lobes dentate, rarely entire Bracts as long as the petiole. Male peduncle uncial. Perianth a little smaller than in *Z. Garcini*. Stamens 3, with cohering, straight anthers. Unripe fruit ovate or oblong-ovate, about 4 lines long.

8. Z. Wyleyana (Sond.); monœcious? quite glabrous; leaves petiolate, deeply palmate, 5-partite, smooth; segments linear-lanceolate, quite entire, the lower ones often bifid, the middle longer; bracts broad-cordate, dentate; male peduncles 2–3-aggregated, nearly as long as the bract.

HAB. Namaqualand, *A. Wyley*. (Herb. D., Sd.)

Stem very long, angulate-sulcate. Petiole 3-6 lines long. Tendrils simple, elongate. Leaves glaucous green; middle lobe 1¼–1½ inch long, 2 lines wide, lateral ones a little shorter. Bracts 6–7 lines long, 8 lines broad, with short, acute teeth, not ciliated. Tube of the male perianth 4 lines, lobes 2 lines long. Stamens exactly as in *Z. Garcini*. Female plant unknown.

9. Z. macrocarpa (Sond.); perennial; stem and branches angulate, glabrous; leaves palmately-digitate, 5 lobed, scabrous; *lobes lanceolate*, entire or dentate; the lower often 2-lobed; peduncles solitary; perianth tubulose with linear-lanceolate, spreading lobes; ovary glabrous; fruit *ovate, acute;* seeds scarcely compressed, ovate, in two rows.

HAB. Grassy places among shrubs above Mooyeriver, *Burke* 290, *Zey.!* 579. Dec. (Herb. Hk., Sd.)

Prostrate, about 2 feet long. Tendrils simple, elongate. Petioles 4–6 lines long. Lobes of the leaves about 1 inch long, 1–3, rarely 4 lines, broad, acute; the lower ones shorter. Pedicels ¾–1 inch. Tube ½ inch, lobes 4–6 lines long; calyx-teeth very short. Stamens 3, inserted near the base of the tube. Anthers connate in a tube. Fruit 2 inches long, 1 inch broad. Seeds 5 lines long. *Drege* 8190 seems to be the same, but I have not seen the fruit.

10. Z. debilis (Sond.); perennial; stem and branches sulcate-angulate, glabrous; leaves short, pedicellate, palmate-digitate, 5-lobed, scabrous; *lobes linear-lanceolate*, entire or with a few acute teeth; peduncles solitary; perianth tubulose, with lanceolate, spreading lobes; ovary glabrous, acute; fruit *subglobose, apiculate*, spuriously 3-locular; seeds ovate, compressed.

HAB. Karro-like places at Uitvlugt near *Andr. Burger's*, Dec., and at Rhinosterkop, near Beaufort, April, *Burke*, 141. *Zeyher*, 577. (Herb. Hk., Sd.)

Stem and branches nearly as in the preceding. Tendrils elongate. Leaves about 1 inch long or smaller; the 3 middle lobes often with 2–4 acute teeth. Peduncles 3–4 lines long. Tube of the flower nearly 6 lines, lobes 2 lines long. Stamens as in *Z. macrocarpa*. Fruit 6–7 lines, seeds 3½ lines long.

III. MUKIA, Arnott.

Flowers monœcious. *Male: Calyx* campanulate, 5-dentate. *Corolla* with 5-parted, obtuse limb. *Stamens* 3, inserted at the base of the corolla, free; anthers extrorse linear, straight, cohering, terminated by a short, conical appendage of the connective. *Female: Calyx* and *corolla* as in the male flowers. *Ovary* ovoid, setulose, with 2–3 placentæ. *Style* short, with a fleshy, annular disc at the base; stigmas 3, erect. *Fruit* berried, globose, smooth or echinulate, few-seeded. *Seeds* oblong-oval, subcompressed, surrounded by a broad or narrow zone. *Arn. in Hook. Journ. Bot. III. p.* 276. *Endl. n.* 5130. *Naudin in Ann. Scienc. Nat.* 1859. *Bot. p.* 141. *Bryonia spec. Linn.*

Annual, very scabrous, climbing herbs with simple tendrils, and angulate or lobed leaves. Flowers small, yellow, the male fascicled, the female solitary or aggregated. Fruit greenish, at length red. Name from Murra (-Peri) Rheed. Hort. Malabar ?

1. **M. scabrella** (Arn. l. c.); *Wight. Illustr. Ind. Bot. II. t.* 105. *Bryonia scabrella, Linn. fil. Suppl.* 424. *W. & Arn. Prod.* 345. *B. micropoda, E. Meyer in herb. Drege.*

HAB. Near Port Natal, *Drege.* (Herb. Hk., Sd.)

Stem angulate. Petiole uncial or longer. Leaves cordate, 5-angled or 3-5-lobed, middle segment triangular, very shortly toothed, hispid on both sides, as well as the stem and petioles, 1½-2 inches long and broad. Flowers short-peduncled. Fruit as large as a large pea. Seeds rugose, or elevated-punctate, surrounded by a narrow zone, exactly as in the Indian specimens. *M. leiosperma,* Arnott! from an authentic specimen in herb. *Hooker,* is only distinguished from *M. scabrella* by perfectly smooth seeds with a broad tumid zone.

IV. LAGENARIA, Seringe.

Flowers monœcious. *Calyx* campanulate; *segments* subulate or broadish. *Corolla* (white) petals 5, obovate, springing from within the margin of the calyx. *Male: Stamens* 3; *anthers* subsessile, triadelphous, cells very flexuose. *Style* scarcely any. *Stigmas* 3, subsessile, thick, 2-lobed. *Pepo* fleshy, indehiscent. *Seeds* numerous, obovate, compressed, with a tumid border. *Mem. Soc. d'hist. Geneve III.* 2, *p.* 1. *DC. Prod.* 3, 299. *Endl. Gen. n.* 5136.

Climbing, annual herbs, softly pubescent. Tendrils 2-cleft; leaves cordate, nearly entire; flowers solitary or fascicled, axillary; fruit often very large, pyriform or subclavate. Name from *lagena,* a bottle; form of fruit of some of the species.

ANALYSIS OF THE SOUTH AFRICAN SPECIES.

Leaves suborbicular, cordate at base, glaucous ... ... (1) **vulgaris.**
Leaves sagittate, acuminate, scabrous ... ... (3) **sagittata.**

1. **L. vulgaris** (Ser. l. c.); leaves large, suborbiculate, cordate at the base, entire or obsoletely lobed, denticulate, somewhat glaucous, with 2 glands at the base; flowers fascicled; fruit pubescent, at length nearly glabrous and very smooth. *W. & Arn. Prod.* 341. *L. idolatrica, Ser. l. c. Naudin. l. c. p.* 91. *Cucurbita Lagenaria, Linn.*

HAB. Spontaneous in the gardens in Caffraria and Tambokiland, *E. & Z.* Dec.–Jan.

2? **L. sagittata** (Harv. mst.); diœcious; stem glabrous, sulcate; tendrils simple; leaves scabrous, with short, appressed hairs, sagittate, lower ones larger, dentate or acutely lobed, upper ones entire; peduncles fascicled, 1-flowered; flowers very small.

HAB. Port Natal, *J. Sanderson.* (Herb. Hk., D.)

Many filiform stems from a tuberous, woody root-stock, 1–1½ foot long. Petiole 4–6 lines long. Leaves in the female specimen 2 inches long, about 1 inch broad, ovate, acuminate, with several lanceolate teeth; the lanceolate, basal lobes entire or bidentate. Leaves of the male specimens exactly sagittate, lanceolate or linear-lanceolate, very entire, 1 inch long, 4–3 lines broad. Peduncles ½–1 inch long. Male flower about 3 lines long; calyx campanulate, with short, subulate teeth; corolla 5-parted, lobes ovate-lanceolate. Anthers cohering, gyrose, subsessile, shorter than the corolla. Female flower a little smaller; calyx and corolla the same; ovary oblong, glabrous, attenuated at the apex. Stigma (only in one flower seen) bilobed. Fruit unknown.

V. LUFFA, Tournef. Cav.

Flowers monœcious or rarely diœcious. *Calyx* 5-toothed, tube in the male campanulate or turbinate, in the female oblong-clavate. *Petals* 5, somewhat deciduous. *Male: Stamens* 3; 2-bilocular, deeply bipartite, the third unilocular; anthers very flexuous. *Female: Stamens* more or less abortive. *Style* 3-cleft; *stigmas* reniform or bipartite. *Pepo* becoming at length dry and fibrous within, usually opening by the fall or decay of a lid or stopple at the apex, sometimes indehiscent. *Seeds* broad-oval, flattened. *W. & Arn. l. c.* 343. *Endl. Gen.* 5134. *Naudin Ann. Scienc. nat. 4th Series, Bot. XII. p.* 118.

HAB. Herbaceous, climbing herbs with angulate stems. 2–7-partite tendrils, palmate-lobed leaves, and large yellow flowers. Male peduncles racemose, female 1-flowered. Name from Louff, the Arabic name of *L. Ægyptiaca.*

ANALYSIS OF THE SOUTH AFRICAN SPECIES.

Stem glabrous; leaves palmately 3–5-lobed, lobes ovate, acute (1) **sphærica.**
Stem pubescent; leaves palmately 5-lobed, lobes obovate, obtuse (2) **Caledonica.**

1. L. sphærica (Sond.); stem sulcate, *glabrous;* leaves triangulate, palmately 3–5-lobate, at the base subcuneate, biglandulose, *lobes ovate, acute,* repand-mucronate-dentate, or the lower ones bifid, upper side slightly scabrous, under with very short hairs and paler; male racemes long peduncled, 2–6-flowered; calyx campanulate, with short, subulate teeth; petals large, obovate. *Lagenaria sphærica, E. Meyer in herb. Drege.*

HAB. Between Omsamculo and Omcomas, *Drege;* Port Natal, *Krauss,* 89. *Plant* 103. *Gerrard & McKen.,* 556. April. (Herb. Hk, D., Sd.)

Tendrils bifid. Petiole 2-uncial, at the base of the cuneate leaf with 2 acutish glands. Leaves about 4 inches long and broad, deeply lobed; lobes toothed or lobate-dentate, especially in the smaller upper leaves. Peduncles as long as the leaves, glabrous. Calyx about 4 lines long, glabrous. Petals 1–1¼ inch, much veined. Stamens 3–4, distinct; anthers 2–3 lines long. Female plant not seen.

2. L. Caledonica (Sond.); monœcious; stem sulcate, as well as the petioles with short, *yellowish pubescence;* leaves palmately-5-lobate, without glands at the base; *lobes obovate, obtuse,* shortly mucronate-dentate, upper side glabrous, under with short, appressed hairs on the nerves and veins; male racemes long peduncled, 2–5-flowered; calyx turbinate, with subulate teeth, as well as the peduncle beset with long, articulated, yellowish hairs; petals large; female flowers short-peduncled; fruit (half-ripe) oblong, tapering to each end, reddish, smooth, and glabrous.

HAB. Rocks on Wolvekop near Caledon river, Dec. *Burke,* 305; *Zeyher,* 589. (Herb. Hk., Sd.)

Stem several feet long. Tendril bifid. Petiole uncial. Leaves 3 inches long and wide, middle lobe about 1½ inch broad, scarcely paler on the under side. Male peduncles solitary or fascicled, ½–1 foot, pedicels 6–3 lines long. Calyx-tube ⅓ inch, lobes 3 lines long. Petals twice or thrice longer than the calyx. Filaments 3, distinct, very short; anthers cohering, flexuose, about 2 lines long. Peduncle of the female flower 3–4 lines long. Young fruit somewhat fleshy, when half-ripe, about 3 inches long, 1 inch broad.

VI. MOMORDICA, L.

Flowers monœcious or diœcious. Male peduncle with a large sessile bractea. *Calyx* 5-cleft, with a very short tube. *Petals* 5, much longer than the calyx-segments. *Stamens* 3, one-dimidiate; *anthers* flexuous, free or connate. *Ovary* with 3 placentæ; *ovules* horizontal. *Style* with 3 stigmata. *Pepo* fleshy, not fibrous, prickly, bursting when ripe, with or without elastic force. *Seeds* compressed, enveloped in a fleshy arillus, reticulated. *DC. Prod.* 3, 311. *W. & Arn. Prod.* 348. *Endl. Gen.* 5133. *Naudin. l. c. p.* 129.

Annual or perennial, climbing herbs, with petiolate, lobed or compound leaves, simple, rarely 2-fid tendrils, and yellow or white flowers. Name from *mordeo*, to bite; the seeds have the appearance of being bitten.

ANALYSIS OF THE SPECIES.

Leaves reniform-cordate	(4) **cordifolia.**
Leaves palmately-5-lobed; bracteole cordate, toothed	(1) **Balsamina.**
Leaves palmately-5-lobed; bracteole reniform, quite entire ...	(2) **involucrata.**
Leaves bipinnate	(3) **clematidea.**

1. M. Balsamina (Linn. Spec. 1453); annual, stem glabrous; leaves palmately-5-lobed, *deeply acute-toothed*, glabrous, shining; male peduncles 1-flowered; with a *toothed-cordate* bracteole above the middle; *calyx lobes acute;* fruit roundish-ovate, attenuated at both ends, tubercled, bursting irregularly and laterally; seeds with a red arillus. *W. & Arn. Prod.* 348. *Lam. Ill. t.* 794, *f.* 1. *Blackw. herb.* 6, *t.* 539, *a. b. M. gariepensis, E. Meyer.*

HAB. On the Gariep near Verleptpram, *Drege; Namaqualand, A. Wyley; Magalisberg* and Vaalriver, *Burke, Zeyher,* 594, 595. Sept.–May. (Herb. Hk. D. Sd.)

Leaves 1–1½ inches long and wide; upper ones smaller. Tendrils simple. Peduncles 1-2-uncial. Bracteole about 3 lines broad, nearly as long as the pedicel; in the female peduncle below the middle. Petals ½ inch long, twice longer than the calyx. Fruit orange-coloured, as large as a walnut.

2. M. involucrata (E. Meyer in herb. Drege); stem glabrous; leaves *palmately-5-lobed, sinuate-dentate, teeth obtuse,* mucronulate, glabrous; male peduncles 1-flowered, with a very large, *reniform, quite entire* bracteole, close to the flower; *calyx lobes roundish-obtuse,* 2–3 times shorter than the petals; fruit roundish-ovate, attenuated at both ends, tubercled, bursting irregularly and laterally; seeds with a red arillus.

HAB. Port Natal, *Drege, Krauss,* (90), *Dr. Grant, Hewitson, Gueinzius.* April. (Herb. Hk. D. Sd.)

Closely allied to the preceding, the leaves are very similar, but the teeth are blunt, with a short mucro, and about 1½–2 inches long and wide; the flowers are a little larger, the bracteole much larger, nearly ½ inch broad, 3 lines broad and much longer than the pedicels. Stamens 3, free; anthers cohering. The fruit, of the same size, seems to be smoother.

3. M. clematidea (Sond.); quite smooth and glabrous; stem angulate; tendrils simple or bifid; leaves *bipinnate,* partial petioles sulcate above the middle, at the apex 3 or 5-parted, lateral ones alternately branched, leaflets ovate or suborbiculate dentate, mucronulate or inciso-serrate; male peduncle 1-flowered, with a quite entire, cordate bract near the calyx; petals 3 times longer than the ovate calyx-lobes.

Hab. Among shrubs on the Crocodile river, Magalisberg, *Burke*, 357, *Zeyher*, 578. Dec. (Herb. Hk. Sd.)

Stem angulate-sulcate. Common petiole ½–¾ inch long. Lower leaves 3–4 inches long and broad. Leaflets 4–10 lines long, obtuse at the base. Male peduncle 2 inches long. Bracts and flower as in *M. involucrata*. Filaments 3, free, very short, with a gland on each side near the apex. Anther cells flexuose, not cohering. Female flower not seen.

4. M. cordifolia (Sond.); stem glabrous; leaves long, petiolate, *reniform-cordate*, acute, dentate, mucronulate, a little scabrous above, with short hairs on the nerves and veins beneath; *male peduncle umbellate*, with a puberous bract at the base of the pedicels; female peduncle 1-flowered; calyx-lobes roundish-obtuse, fimbriate at the margin, 3 times shorter than the petals; ovary globose, densely muricate. *Cucumis ? cordifolius, E. Meyer.*

Hab. Between Omtata and Omsamwubo, Feb., *Drege*. Port Natal, *Dr. Krauss*, 47. *Gerr. & McKen.*, 1560. (Herb. Hk. D. Sd.)

Tendrils simple or bifid. Petiole 2–3 inches long. Larger leaves 3 inches long and broad, a little paler beneath. Male peduncle as long or longer than the leaves. Bracteole in our specimens truncate, obtuse, subcuneate, 3–4 lines long. Pedicels 3–6, one inch or longer, as well as the calyx-lobes powdery. Calyx-tube glandular, muricate. White petals 1 inch long, about 7–8 lines broad. Anthers connate. Female peduncles often not shorter than the male. Style filiform with a very thick and lobed stigma. Fruit (nearly half-ripe) as large as a hazlenut, muricate-echinate.

Momordica latana, Thunb. ! fl. cap. 36, is *Citrullus vulgaris* (*amarus Schrad.*)

VII. CEPHALANDRA, Schrader.

Flowers diœcious. *Male: Calyx* short, campanulate, 5-toothed. *Corolla* 5-parted, flattish, the laciniæ recurved at the apex. *Filaments* inserted at the base of the corolla, 3, free, but the connective connate, and united into a subglobose, antheriferous capitulum. *Anthers* distinct, flexuous, two of which are 2-celled, the third 1-celled. *Female: Calyx* and *corolla* as in the male flower. *Ovary* oblong, with 3 placentas. *Style* with thick, lobed stigmata. *Fruit* berried, many-seeded, smooth. *Seeds* compressed, obliquely subattenuated at the base. *E. & Z. ! Enum. p.* 280. *Endl. Gen.* 5142.

Climbing, herbaceous, perennial plants, with 5-lobed leaves, simple tendrils, golden or yellowish flowers, and purple fruit. Male peduncle subumbellate or racemose; the female shorter and 1-flowered. Name from κεφαλη, a *head*, and ανηρ, *man*; the anthers are united to a head.

ANALYSIS OF THE SPECIES.

Glabrous; leaves long-petioled, 5-lobed; lobes ovate, acuminate ... (4) **palmata.**
leaves short-petioled, 5-lobed; lobes ovate or oblong obtuse ... ... (1) **quinqueloba.**
Pubescent; leaves petioled, 5-lobed; lobes oblong or oblong-lanceolate, callous-denticulate ... ... (2) **pubescens.**
Glabrous; leaves sessile, 5-lobed; lobes ovate, acute ... ... (3) **sessilifolia.**

1. C. quinqueloba (Schrad. ! l. c.); *diœcious, glabrous, glaucescent;* leaves shortly petiolate, palmately 5-lobed, lobes divaricate, *ovate or ovate-oblong*, obtuse mucronulate, entire or with a few remote teeth; the sinus rounded; male peduncles umbellate, rarely 1-flowered. *Bryonia quinqueloba, Thunb. ! fl. cap.* 35. *Momordica quinqueloba, E. Meyer.*

HAB. Woods in Uitenhage, Albany, and Kaffirland. Feb.–April. (Herb. Th. Holm. Hk. D. Sd.)

Stem angulate. Petiole 2–3 lines long. Leaves papillose-scabrous above, lower 3–4 inches long and broad, middle lobe about 2 inches long, 1 inch wide, lateral ones shorter; upper leaves gradually smaller, often subsessile. Male peduncles solitary, or 2–3 in the axils of the upper leaves, about 1 inch long, terminated by one flower or by 3–6 subumbellate, 3–6 line long pedicels. Flowers ½ inch or a little larger. Calyx-teeth subulate. Petals veined, acute. Stamens short. Female peduncles solitary or geminate, 1 inch long. Ovary oblong, attenuated at both ends. Fruit as large as a pigeon's egg, acute, containing 10–16 obovate seeds.

2. C. pubescens (Sond.); *pubescent;* leaves petiolate, scabrous, palmately 5-lobed, lobes spreading, *oblong or oblong-lanceolate,* mucronulate, *callous-denticulate;* male peduncles elongated, racemose, rarely 1-flowered.

HAB. On the Magalisriver, *Burke,* 408, *Zeyher,* 588. Dec. (Herb. Hk., Sd.)

Habit of *C. quinqueloba.* Petioles 4–6 lines long, as well as the stem and peduncles shortly pubescent. Leaves deeply palmate-partite, in the preceding species only to the middle, the lower nearly 3 inches long, 2½ inches wide; the middle lobe 2 inches long, 6–8 lines broad, lateral lobes evidently shorter; all the lobes punctate-scabrous. Male peduncle 2–3 inches long, from the middle with racemose, 4–6 lines long, pubescent pedicels, without bracts. Flower ½ inch, and stamens as in *C. quinqueloba.* Female peduncle 3–4 lines long. Ovary oblong, attenuated at both ends.

3. C. sessilifolia (Sond.); glabrous; *leaves sessile,* punctate, scabrous at the margins, palmately 5-lobed, lobes spreading, ovate-lanceolate, acute, mucronate, coarsely dentate or the middle trifid; upper leaves with lanceolate, acuminate, paucidentate or trifid lobes; male peduncles solitary, 1-flowered or subumbellate. *Bryonia ? lagenaria, E. Meyer.*

HAB. Near Mooyeriver and Vaalriver on Rhinosterkop, *Burke,* 289, *Zeyher,* 580. Nieuwe Hantum, 4–5000ft., *Drege.* Dec.–Feb. (Herb. Hk. Sd.)

Glaucous-green, much-branched. Lower leaves 1½–2 inches long and broad, at the middle 5-lobed; middle lobes 8–12 lines long, 8 lines wide, 5-dentate or the lower teeth attenuated into 4 lines long, acute lobes; lateral lobes dentate and shorter. Upper leaves 3-fid, or, when 5-parted, the lobes often very small, 2–3 lines broad, 1–1½ inch long. Peduncle about uncial. Flowers and stamens as in *C. quinqueloba.* Fruit 2–2½ inches long, nearly 1 inch broad, acute, with 16–20 seeds.

4. C. palmata (Sond.); *quite smooth and glabrous;* leaves *long-pedunculate,* palmately 5-lobed, lobes *ovate, acuminate,* the margins denticulate by distant calli; the 2 lower lobes very short; male peduncles elongated, racemose; the female shorter, 1-flowered. *Momordica palmata, E. Meyer in herb. Drege.*

HAB. Near Port Natal, *Drege.* April. (Herb. Hk. Sd.)

Tendrils simple, rarely bifid. Petiole 1½–2 inches long. Leaves 3–4 inches long and broad. The middle lobe about 2 inches long, 1 inch wide, the 2 lateral lobes 1½ inch, the lower much shorter. Male peduncle often as long as the leaf, above the middle racemose, pedicels ½–1 inch. Calyx and stamens as in *C. quinqueloba,* Petals uncial. Fruit ovate, acute, in size and colour agreeing with that of *C. quinqueloba;* seeds also the same.

VIII. CITRULLUS, Schrad.

Flowers monœcious. *Male: Calyx* campanulate, deeply 5-fid. *Corolla* 5-partite, flattish. *Stamens* 3, inserted at the base of the corolla, two bilocular, deeply partite, the third unilocular, connectivum without terminal appendage. *Anther*-cells linear, flexuous. *Female: Calyx* and *corolla* as in the male flower. *Ovary* with 3 placentas, ovoid, villous or smooth. *Style* trifid. *Stigmas* 3, thick. *Fruit* a globose, rarely oblong, 3 or pseudo 6-celled, many-seeded pepo. *Seeds* oval, compressed, with obtuse margins. *E. & Z.! Enum. p.* 279.

Annual or perennial, prostrate herbs. Tendrils bifid, rarely trifid or undivided. Leaves deeply 3–5-lobed, lobes lobulate or dissect, with rounded sinus. Peduncles axillary, 1-flowered. Flowers yellow. Pepo with fleshy or spongious, white, yellowish, reddish or purple, sweet or bitter, pulp. Name unexplained.

1. C. vulgaris (Schrad. l. c.); root annual; leaves stalked; the upper ones 3-parted; middle segment sinuated, pinnatifid; lateral ones 2-fid; lobes obovate-rotundate-obtuse, scabrid; radical leaves 5-parted; fruit elliptico-globose, glabrous, when young often woolly. *Naudin. l. c. p.* 100. *Cucurbita Citrullus, Linn. Citrullus caffer, Schrad. l. c. (fruit sweet). Citrullus amarus, Schrad.! l. c. Cucumis Colocynthis, Thunb.! herbar. var.* γ *et Drege herb. Momordica lanata, Thunb.! fl. Cap.* 36. *(fruit bitter.)*

HAB. In the sands of the Cape downs, near Tigerberg and Rietvalley, and in similar localities, *Somerset;* Gamkariver, *Zeyher*, 587. Jan.–May. (Herb. Th., Hm., Hk., D., Sd.)

Stem woolly or pubescent. Radical leaves often 6 inches long, 5 inches wide, on a petiole of the same length; upper ones gradually smaller, scabrous above, hairy beneath or nearly glabrous. Pepo the size of an apple or of a child's head; when edible or sweet it is called water-melon or Kaffir water-melon; when bitter, it is the *bitter apple* or *wild water-melon* of the colonists. The pulp of the latter may be used like that of Colocynth; conf. *Pappe Flor. Cap. med. p.* 14.

IX. CUCUMIS, Linn.

Flowers monœcious or diœcious. *Male: Calyx* campanulate, 5-fid. *Corolla* patent, limb 5-partite. *Stamens* 3, inserted at the base of the corolla, one-dimidiate; *anthers* posterior, linear, gyrose, terminated by a papillose, bilobed connectivum. *Female: Calyx* and *corolla* as in the male. *Stigmas* 3, thick. *Fruit* a pepo or gourd, 3 or spuriously 6-celled, many-seeded. *Seeds* oval, compressed, not margined. *Schrad. Naudin.! Endl. Gen.* 5137.

Herbaceous, annual or perennial, scabrous p ınts, with succulent stems, simple, rarely wanting tendrils, and angular or lobed leaves. Flowers axillary, solitary or fascicled, yellow. Name from κικυος a cucumber?

ANALYSIS OF THE SOUTH AFRICAN SPECIES.

Tendrils none or soon deciduous. Stem erect ... ... (9) **rigidus.**
Tendrils simple. Stems prostrate.
  Fruit spinous or muricate.
    Fruit with large, conical spines.
      Leaves cordate, angulate, or subtrilobed ... ... (1) **metuliferus.**
      Leaves deeply palmately 5-lobed ... ... (6) **Naudinianus.**
    Fruit with short, thin, often weak, spinelike bristles.
      Annual,
        Leaves 3–5-lobed, lobes rotundate-obtuse; ovary oblong; pepo ovoid ... ... (2) **Africanus.**
        Leaves 3–7-lobed, lobes rotundate-obtuse; ovary roundish; pepo globose ... ... (3) **myriocarpus.**
        Leaves 3–5-lobed, middle lobe lanceolate-acute; pepo pyriform-globose ... ... (4) **Zeyheri.**
      Perennial,

Leaves deeply 5-7-lobed, lobes lobulate and dentate; flowers monoecious; pepo ovoid ... ... (5) **dissectifolius.**
Leaves deeply 5-7-lobed, lobes linear, entire; flowers dioecious; pepo ovoid ... ... (7) **heptadactylus.**
Fruit pubescent or glabrous, not spinous or muricate ... (8) **hirsutus.**

1. C. metuliferus (E. Meyer in herb. Drege); annual, branched, deeply green; stem and branches angulate, hispid; leaves long, petiolate, *palmately sub-3-lobed, dentate, with cordate base;* lobes angulate, denticulate, the terminal acute, mucronate; pepo oblong, bluntish-trigonous, at both ends obtuse, armed with thick, conical, sharp spines. *Naud. l. c.* 10.

HAB. Omsamwubo, near the river, *Drege.* Feb. (Herb. Sd.)

Branches long. Leaves 2-3 inches long and broad, or in the upper often longer than broad and more acute, with shorter, lateral lobes. Male flowers nearly as in *C. Melo.* Fruit of the wild specimens 4 inches long, 1½-1¾ inch in diameter, with red spines, about 12-20 in a pepo, ⅛ inch long, terminated by a conical, very hard apex. Pulp pale green; taste of that of *C. sativus, (Naudin).* Seeds 2½ lines long.

2. C. Africanus (Linn. fil. Suppl. 423); annual, green, scabrous everywhere; branches angulate; *leaves deeply* 3- *or* 5-*lobed,* lobes entire or sublobed, denticulate, as well as the sinus rotundate, middle lobe obovate, longer than the lateral ones; *ovary oblong, muricate-echinate,* on a slender peduncle; *pepo ovoid,* densely beset with short, but sharp spines.

VAR. *α*, leaves mostly trilobed or the lateral ones 2-lobed. *C. Africanus, Thunb.! fl. cap.* 36. *Drege herb. a, b, c. C. prophetarum d. herb. Drege, and b. partim.*

VAR. *β*, leaves mostly 5-lobed, lateral lobes often deeply 2-lobed, the terminal sub-3-lobed. *C. Africanus, E. & Z.!* 1794. *Naudin. l. c.*

HAB. Var. *α.* on the Gariep, in the distr. of George and near Omtata, *Drege;* Caledon river, *Zeyh.* 584. Var. *β.* Mount Winterberg, Caffirland, and in the gardens near Capetown, *E. & Z.* and *Dr. Pappe;* Port Natal, *Miss Owen.* Sept.-Jan. (Herb. Th. Hk. D. Sd.)

Stem much branched. Leaves on longish petioles, in var. *α.* 1¼-1¾ inch long, 1-1½ inch wide, the upper smaller; in var. *β.* 1½-2½ inches long, and very similar to those of *C. Anguria,* L. Male flowers fascicled, very small, much shorter than the hispid petiole; female flowers on longer peduncles. Fruit 1½ inch long, ¾ or nearly 1 inch broad. Spines 2 lines long; the ripe fruit sometimes denudate or only tubercled by the remaining base of the spines. Seeds nearly 2 lines long. *C. arenarius,* Schrad.! E. & Z.! 1795, founded on a single specimen, is a depauperated state of *C. Africanus. C. arenarius, Arn. and Planch. in herb. Hook.* is *C. myriocarpus,* Naud. *C. arenarius, Schum. and Thom.!* is a quite different plant; the branches are hispid, tendrils very long, leaves about 2 inches long, 15 lines broad, bluntish-5-lobed, the ovate middle lobe 1 inch long, the four lateral ones short but equal; the whole leaf on both sides subsilky by appressed hairs, a little hispid on the nerves beneath; flowers very small, fascicled and short peduncles, the tube hirsute; the fruit unknown. N. 4919 of *Drege's* collection has some resemblance to this, but the flowers are much larger, the tendrils very short, and the branches pubescent.

3. C. myriocarpus (Naudin.! l. c. 22); annual, green, scabrous; branches angulate-striate; leaves long petiolate, palmately 3, 5, 7-lobed, lobes and sinus rotundate, middle lobe larger, hispidulous-scabrous beneath; *ovary roundish, densely muricate;* pepo *subglobose,* beset with weakly prickles, caducous; peduncle slender. *C. prophetarum, Thunb.! fl. cap.* 36. *Jacq. hort. Vind. t.* 9. *Blackw. herb. t.* 589. *E. & Z.!* 1793, *and herb. Drege b. ex parte. C. Colocynthis α, herb. Thunb.*

HAB. On Tablemountain, near Genadenthal; at Buffelfontein, and in the districts of Uitenhage and Albany. (Herb. Th. Hk. D. Sd.)

Leaves 1½–2 inches long and wide, upper ones much smaller, usually on very long petioles, on the margins denticulate, with cordate base, subglabrous above, the middle lobe roundish, cuneate, more or less 3-lobed. Peduncles fascicled, much shorter than the hispid petiole. Flowers very small. Fruit usually very numerous, round or nearly so, ¾–1 inch in diameter, densely or sparingly beset with bristle-like, 1 line long spines. According to Naudin the true *C. prophetarum, Linn.* is a different species.

4. C. Zeyheri (Sond.); annual, pale green; stem and branches angulate, scabrous; leaves palmately 3–5-lobed, lobes denticulate, lateral ones shortly obtuse, the middle elongated, lanceolate, more or less 3-lobed, acute, very scabrous on both sides; *pepo pyriform-globose*, beset with very short, weak prickles.

HAB. Gamkariver, May, *Zey.!* 582; Magalisberg, Nov., *Zey.!* 583. (Herb. Sd.)

Stem 1–2 feet, branched, on the 5 angles scabrous by minute, whitish hooks. Tendrils very short. Petiole 4–6 lines long, scabrous. Leaves 12–15 lines long, 8–10 lines broad, middle lobe cuneate at the base, 10–12 lines long, 3–4 lines wide. Fruit 12–14 lines long, 10–12 lines in diameter, with 1 line long prickles. Seeds not margined. A similar plant is collected by *Gueinzius* (398) at Port Natal, the leaves are twice larger, very green, scarcely scabrous above, the tendrils long, the ovary oval, very prickly. If the fruit is not different, it may be a variety of *C. Zeyheri.*

5. C. dissectifolius (Naudin. l. c. 23); perennial, monœcious, pale green, stem and branches very long, angulate, scabrous; tendrils elongate; leaves *deeply palmately 5–7-lobed;* lobes lanceolate, lobulate and dentate, acute, the middle longer, subglabrous above, very scabrous beneath; pepo ovoid, *with very short, weak prickles.*

HAB. Sandy places near Moojeriver, *Burke*, 276, *Zey.*, 585. Jan. (Herb. Hk. Sd.)

Root perennial (*Zeyher msc.*). Stem often 4–6 feet long, creeping, hooked scabrous on the 5 angles, especially on the branches. Tendrils 2–3 inches, petioles about 1 inch long, hispid-scabrous. Leaves 1½–2 inches long and broad, all the lobes lobulate and inciso-dentate, the middle acutely 5-lobed. Male flowers fascicled, nearly ½ inch; female ones solitary. Fruit 10 lines long, 8 lines in diameter; prickles 1–1½ line long. Seeds oblong, not margined. It comes very near the preceding, but the leaves are more compound and the lateral lobes longer and acute.

6. C. Naudinianus (Sond.); perennial, diœcious; stem and branches very long, angulate, smooth or a little scabrous-hairy on the young branches; tendrils short, spinous; leaves deeply palmately 5-lobed, scabrous on both sides, lobes narrow-lanceolate, pinnatifid, and dentate, lower ones very short, the middle the longest; male flowers campanulate; pepo ovoid, *armed with large, conical spines. C. dissectifolius, Naudin. l. c. partim.*

HAB. Sandy, grassy places near the Moojeriver and Magalisberg, *Burke*, 488; *Zeyher*, 586. Dec. (Herb. Hk. Sd.)

Stem much branched, as thick as a pigeon's quill, quite smooth and glabrous or with short hairs near the nodes. Tendrils ½ inch long, or shorter on the young branches, straight. Petioles 6–8 lines long, scabrous, and purplish as the prominent nerves of the leaves. Middle lobe about 1½ inch long, 1½–2½ lines broad, with 4–6 horizontal lobes, the lower of which are 3–4 lines long or shorter in the upper leaves, the intermediate lobes divaricate, somewhat shorter, dentate, or lobes like the middle lobes; the lower ones generally very short, angulate-toothed. Peduncle ½ inch. Calyx of the male flower broad-campanulate, with lanceolate, recurved lobes. Petals pubescent outwards. Stamens and anthers as in the genus. Fruit

1¾ inch long, 1 inch in diameter, not fasciated or striated; spines numerous, about 3 lines long, at the base 2–3 lines wide. Seeds nearly 5 lines long, not margined.

7. **C. heptadactylus** (Naud. l. c. 24); perennial, diœcious, greyish; stem and branches angulate, hispid, and scabrous; leaves short, petiolate, *palmate-digitate; lobes* 5–7, *linear, elongate, entire,* acute, with revolute margins, very scabrous; male flowers subtubulose; pepo ovoid, beset with *short, weakly bristles.*

HAB. Sandy places, Vanderwaltsfontyn, distr. Colesberg, *Burke,* 139, *Zeyher,* 591. on the Caledonriver, *Burke,* 7, *Zeyher* 590; Winterfeld, distr. Beaufort, *Drege,* 8183 (*not collected at Port Natal, as indicated by Naudin*). Jan.–Feb. (Herb. Hk. Sd.)

Many stems from the whitish root, 1–4 feet long, flexuous, sulcate-angulate, with a few branches, covered as well as the other parts with short and stiff, spreading hairs. Tendrils short. Petioles in the lower leaves 4–8 lines, in the upper 2–3 lines long. Limb of the leaves 2–3 lines; the lobes unequal, the middle from 1½–3 inches long, 1–2 lines broad, the lower ones much shorter. Upper surface glabrous except the middle nerve. Flowers fascicled or subracemose, very small. Calyx lobes short, subulate, erect, spreading. Corolla hispidulous. Pepo 14–15 lines long, 9–10 lines in diameter; the bristle-like spines, 1–1½ line long.

8. **C. hirsutus** (Sond.); perennial, diœcious, hirsute; branches elongate, sulcate-striate; leaves short-petiolate, scabrous, ovate-cordate, dentate, inciso-serrate or 3–5-lobed, lobes acute, the middle much longer; flowers subsolitary on slender peduncles; pepo subglobose, hirsute, at length glabrous or nearly so.

HAB. Sandy places near Wonderfontyn and Moojeriver, Betchuanaland, *Burke,* 297, *Zeyher,* 581. Dec. Near Port Natal, *Krauss.* 91. (Herb. Hk. D. Sd.)

Stem as thick as a pigeon's quill, prostrate as the similar branches, several feet long. Tendrils short, in some branches often wanting. Petioles 2–3 lines long, hirsute. Leaves 1½–2½ inches long, 1 inch broad, hirsute, and very scabrous, with shortly dentate margins, entire or lobed; lateral lobes short, rarely ½ inch long, the middle lobe uncial or longer, ovate-lanceolate, rarely obtuse. Upper leaves of the branches twice or thrice shorter, usually entire and not scabrous. Peduncles about 1 inch long. Flowers subcampanulate, ½ inch, lobes of the corolla oblong. Ovary densely hirsute. Fruit 1 inch long and broad, on a longer peduncle. Seeds about 4 lines long, subcompressed, not margined. The edible fruit is acidulous, *Zeyher.*

9. **C. rigidus** (E. Meyer in herb. Drege); perennial, stem erect, sulcate-angulate, alternately branched, very scabrous and appressedly hairy; leaves petiolate, 3-lobed, scabrous and whitish by adpressed hairs, lobes ovate or obovate, obtuse, short-dentate, the middle lobe somewhat longer; flowers axillary, pedunculate, scabrous-hairy outwards; fruit ovoid, beset with longish, subulate spines.

HAB. On the Gariep, *Drege.* Namaqualand, *A. Wyley.* Sept. (Herb. D. Sd.)

Whole plant greyish-white. Stem several feet high, as thick as a goose's quill; branches spreading. Petioles uncial, thick, and rigid. Leaves 1 inch long and broad; the lateral lobes 4–5 lines long, the middle about ½ inch long, 4–6 lines wide, on the margins and at the apex repand-dentate. Male flowers unknown, the females pale yellow. Petioles short, in fruit ½ inch long, thick. Ovary ovate, spinous-scabrous. Calyx campanulate, lobes subulate, 1 line long, green. Style very short. Stigma thick, 3-lobed. Fruit 1½ inch long, 1 inch in diameter, red or scarlet, the spines 2 lines long. Seeds compressed, not margined.

Cucumis spec. Drege, 8182, *without flower and fruit, may be the true C. prophetarum.*

Cucumis spec. Gueinzius, 397, *without female flowers and fruit, seems to be a new species.*

Bryonia spec. Drege, 8185, *with very small, reniform, and trilobate puberulous leaves, and racemose flowers, is perhaps a Coniandra.*

Bryonia acutangula, Thunb. ! fl. cap. 35, *and herbar. is a species of Senecio.*

X. PISOSPERMA, Sond.

Flowers monœcious, much aggregated, on radical branches; the male on longish, racemose, 1-flowered pedicels; the female solitary, on shorter pedicels. *Male: Calyx* 5-fid, tube subcampanulate, lobes lanceolate. *Petals* 5, oblong. *Stamens* 3, short. *Anther*—cells flexuose, cohering, without appendage. *Female: Calyx* and *corolla* as in the male. *Style* 1; *stigma* thick, lobed. *Fruit* subbaccate, pseudo-trilocular, subglobose, apiculate, 6–12-seeded. *Seeds* round, subcompressed, with a tumid margin.

A herbaceous, perennial, subscabrous plant, with tuberous root and precocious flowers. Leafy branches prostrate, rising from the short radical flower-bearing branches, when the fruit begins to ripen. Tendrils simple. Leaves petiolate, palmate-digitate, 5-lobed; lobes linear, the middle elongated. Flowers small, pale yellow, striped with green, and very thinly pubescent. Name from *πισος*, a *pea*, and *σπερμα*, a *seed*.

1. P. Capense (Sond.)

HAB. Nieuwejaarspoint and Caledonriver, *Zeyher*, 593, and *Cucurbit.* 1; Camdeboosberg, *Drege*, 8188; Zwartekey River, *Mrs. F. W. Barber.* Oct.–Jan. (Herb. D., Hk., Sd.)

Leafy branches 1 foot or longer, prostrate, sulcate-angulate, hairy-scabrous, at length subglabrous. Petioles 3–4 lines long, shorter than the tendrils. Middle lobe of the leaves 1–2-uncial, 1–1½ line wide, acute, very entire; the intermediate lobes twice shorter; the lower ones very short, often bilobed. Radical flowering stem 1–2-uncial; branches with 10–16, filiform, uncial pedicels, each of which has a small, subulate bract. Flowers 4–5 lines long. Ripe fruit the size of a large hazlenut, pubescent, at length nearly glabrous. Seeds as large as a pea.

ORDER LIX. PASSIFLOREÆ, Juss.

(BY W. H. HARVEY.)

Flowers perfect or unisexual. *Perianth* (consisting either wholly of *calyx*, or of *calyx* and *corolla* soldered together) monophyllous, free, the tube long or short, sometimes scarcely any, the limb, if in a single row, 3–4–5-cleft, if double, 8–10-parted, the outer segments herbaceous, the inner more or less petaloid. *Corona-staminea* occupying the bottom of the perianth, annular, fimbriated or entire, sometimes consisting of fleshy glands, always exterior in insertion to the stamens. *Stamens* as many or twice as many as the lobes of the perianth, rarely subindefinite, monadelphous or free; anthers introrse, either versatile or adnate, bilocular. *Ovary* stipitate or rarely subsessile, free, unilocular; ovules many or few, on 3–5 parietal placentæ, pendulous, orthotropous. *Styles* or stigmata as many as the carpels. *Fruit* either a succulent berry or a 3–5-valved capsule, usually many-seeded. *Seeds* on long seed-cords, mostly arillate, with a furrowed and ridged seed-coat, albuminous; embryo orthotropous, with flat, leafy cotyledons.

Herbaceous or suffruticose, rarely shrubby plants, mostly climbers, natives chiefly of the warmer parts of America, with outlying genera and species in Africa, Asia, and Australia. Leaves mostly simple, entire or variously lobed, rarely imparipinnate, alternate, petioled or sessile. Stipules in pairs at the base of the petioles, sometimes wanting. Tendrils, when present, axillary, formed out of abortive peduncles. The type of this Order is the well-known genus *Passiflora* or Passion-Flower, no African species of which has yet been discovered, though *P. cærulea* (a native of South America) is now almost naturalised and apparently wild in some parts of our colony. The few South African species known have small or minute, greenish flowers.

TABLE OF THE SOUTH AFRICAN GENERA.

* *Lobes of the perianth in two rows, those of the inner row like petals. No involucel.*

I. **Tryphostemma.**—Each row of the perianth of 3 segments. Corona-staminea annular, double, the outer fringed.

II. **Modecca.** Each row of the perianth of 5 segments. Corona-staminea obsolete or wanting.

** *Lobes of the perianth (or calyx) in a single row. Involucral bracts present or absent, subtending the perianth.*

III. **Ceratiosicyos.**—A vine-like climber. Perianth 5- (or 4-) fid. Male flowers racemose, involucrate; females solitary, without involucre.

IV. **Acharia.**—A suberect, small herb. Perianth 3- (or 4-) fid. Male and female flowers involucrate, neither racemose.

I. TRYPHOSTEMMA, Harv.

Flowers hermaphrodite. Tube of the *perianth* short, conical; limb 6-parted, in two rows, the three inner segments unequal, two of them larger, herbaceous and albomarginate, the third linear and petaloid. *Corona-staminea* perigynous, annular, double; the outer fimbriated, the inner entire or crenulate, bearing the stamens. *Stamens* 5, attached to the interior *corona;* filaments subulate; anthers erect, sagittate, 2-celled. *Ovary* subsessile, unilocular; ovules few, on three or four parietal placentæ. *Styles* 3–4, filiform; stigmas capitate. *Capsule* shortly stipitate, membranous, 3–4-valved, few-seeded; seeds pendulous, enclosed in a membranous arillus, areolate-corrugate; embryo not seen.

But one species known. The name is compounded of τρυφος, *a delicate fragment,* and στέμμα, *a crown;* in allusion to the depauperated condition of the crown of rays in this miniature Passion-flower.

T. Sandersoni (Harv. Thes. t. 51).

HAB. Port Natal, *J. Sanderson,* No. 59 and 440. (Herb. D. Hk. Sd.)

Root perennial, woody. Stems numerous, 4–12 inches high or more, erect, quite simple, angular, ribbed and furrowed. Leaves alternate, quite sessile or shortly petiolate, ovate or ovato-lanceolate, acute, 1–2½ inches long, ½–1 inch wide, distantly ciliate-dentate, glabrous, reticulated with veins. Stipules small, subulate, free. Peduncles axillary, as long as the leaves or shorter, 2–3-flowered, with one or more bracteoles at the base of each pedicel. Flowers 2–5 lines in diameter, greenish, with purple dots outside. The outer lobes of the perianth and two of the inner, are broadly ovate or ovate-oblong; the third inner lobe is much smaller, narrower, and more petaloid than the rest. Specimens recently received from *Mr. Sanderson* differ from that figured in the Thesaurus in being taller and stouter, with evidently petioled, longer, and comparatively narrower leaves, and somewhat larger flowers.

II. MODECCA, Lam.

Flowers unisexual. *Perianth* without involucre, double; the *outer* (or *calyx*) tubular-conical, campanulate or subrotate, more or less deeply 4–5-cleft; the *inner* (or *corolla*) of 4–5, ovate, oblong, or linear petals, smaller than the calycine lobes and inserted either at the summit, or far beneath the summit of the calyx-tube. MALE: *Stamens* 4–5, inserted in the bottom of the calyx and opposite its lobes; filaments subulate, connate in a ring at base; anthers introrse, 2-celled, erect. A rudiment of an ovary. FEMALE: Abortive filaments 5, subulate, surrounding the ovary, sometimes wanting. *Ovary* stipitate or subsessile, unilocular; ovules numerous, on 3 parietal placentæ. *Stigmata* subsessile, dilated. *Capsule* fleshy (leathery when dry), subglobose, 3-valved, many-seeded; seeds arillate, areolate-corrugate; embryo in fleshy albumen. *Endl. Gen. No.* 5130.

Herbaceous or shrubby plants, mostly climbing, natives of Asia and Africa. Leaves alternate, undivided or lobed, the petioles biglandular at the apex. Stipules obsolete or none. Peduncles axillary, branched, the medial branch tendriliferous. Flowers small and greenish. *Modecca* is the native Indian name for one of the species.

ANALYSIS OF THE SOUTH AFRICAN SPECIES.

Calyx-tube conico-campanulate; limb 5-cleft. *Petioles short.*
Leaves lanceolate-linear, obtuse, undivided ... ... (1) **Paschanthus.**
Leaves deeply 3–5-lobed, the lobes incised ... ... (2) **digitata.**
Calyx nearly rotate, 5-parted. Leaves on long petioles, bluntly 3-lobed ... ... (3) **gummifera.**

1. M. Paschanthus (Harv.); stem scarcely climbing; leaves subsessile, *lanceolate-linear, obtuse, distantly repand,* glabrous, glaucous, semi-complicate, reticulated, having two large glands at the apex of the petiole, and one beneath each of the marginal inæqualities. *Paschanthus repandus, Burch. Trav.* 1, *p.* 533. *DC. Prod.* 3, *p.* 336.

HAB. In the interior of South Africa, lat. 29°.20′, long 23°.43′, *Burchell, Cat. Geogr.* No. 2036 *and* 2486, 2. At Motito, Feb. 1842. (Herb. Hk.)

Stem about 2 feet high. Leaves of tender substance but thickish (much of the substance of a cabbage-leaf) subglaucous, elongate-lanceolate, the margins repand and reddish. Peduncles axillary, cirrhose. Flowers polygamous, ochraceous. Calyx tubular, 5-cleft. Petals 5, small, lanceolate, inserted between the divisions of the calyx. Filaments 5, inserted near the bottom of the calyx; anthers linear. Ovary stipitate; style very short; stigma lacero-capitate. Capsule 1-celled, ovate, inflated, 3–6-seeded, 3-valved, purple-rosy. Seeds ovate, inclosed in a scarlet arillus. Such is *Burchell's* account. The specimen in Herb. Hook. above referred to, and which alone I have seen, is in fruit only, but in foliage and other characters it so nearly agrees with *Burchell's* description that I feel little doubt of its identity with his plant. By the description of the flower, as given by *Burchell,* I cannot see how this species differs generically from *Modecca:* it probably belongs, as does the following, to the subgenus *Blepharanthes.*

2. M. digitata (Harv. Thes. t. 12); stem herbaceous, climbing, angularly-striate; leaves on short petioles *digitately 3–5-parted, the lobes pinnatifid,* glabrous, margined, with two glands at the apex of the petiole, and glands beneath the sinuses of the lamina; racemes few-flowered, equalling the petiole, sometimes cirrhiferous; calyx funnel-shaped, tapering at base; petals lanceolate, inserted toward the base of the calyx-tube, included.

HAB. In the Zulu country, *Miss Owen!* Macallisberg, *Burke!* (Herb. Hk. D.)

The female flowers are unknown. For a full description see *Thes. Cap.* above quoted.

3. M.? gummifera (Harv.); stem shrubby, extensively climbing, striated; leaves *on long petioles*, abrupt at base, *bluntly 3-lobed*, the lobes short, round-topped or emarginate, very entire, glabrous, membranous, paler, nigro-punctulate and veiny beneath, with two glands at the apex of the petiole; peduncles (of the female flowers) axillary, much shorter than the petiole, 1-flowered, those of the fruit elongate and cirrhiferous; calyx rotate, its segments ovate, nigro-punctate; petals narrow-linear, minute, inserted in the sinus between the calyx-lobes; staminodia none; ovary subsessile; stigmas 3, expanded, fimbriate. *Passiflorearum species, Drege! No.* 5211.

HAB. Omsamculo and Omcomas, *Drege.!* Natal, *Sanderson*, 555. Common round D'Urban, *Gerrard & McKen!* (Herb. D., Sd., Hk.)

A woody climber, rising to the tops of trees. "Stems green, striated, vine-like, 2–3 inches in diameter, resembling green snakes," (*Gerr.*). Petioles 2–6 inches long, slender; leaves 1–4 inches long, 1½–5 inches wide, the blunt lobes separated by wide, rounded sinuses; sometimes the lateral lobes are obsolete. "The Kafirs use a claret-coloured, gummy substance obtained from this plant to paint their faces," *Gerrard*. "Also used by them as an emetic, *Sanderson*." Until the male flowers are known the genus of this plant cannot be perfectly determined. If it be a *Modecca* it will perhaps be referred to the subgenus *Microblepharis*, Arn.

III. CERATIOSICYOS, Nees.

Flowers monœcious. MALE: *Perianth* campanulate, 4–5-lobed, subtended by 4–5, slender, involucral bracts. *Stamens* inserted in the base of the perianth, free, as many as its lobes and alternate with them; filaments dilated upwards; anthers adnate to a clavate connective, the cells slightly separated, introrse. *Glands* as many as the stamens and alternating with them, oblong, fleshy. FEMALE: *Perianth* as in the male, but destitute of involucre, marcescent. *Glands* as in the male, but smaller, opposite the lobes of the perianth. *Ovary* stipitate, unilocular; ovules numerous, on 4–5 parietal placentæ. *Stigmata* 4–5, subsessile, channelled, bilobed. *Capsule* siliquæform, 4–5-valved, several-seeded; seeds with a fleshy integument; embryo in the axis of fleshy albumen. *Endl. Gen. No.* 5106.

A slender, herbaceous, nearly glabrous climher, with palmately 5–7-lobed, membranaceous leaves, and axillary, greenish flowers; the male flowers in racemes, the female solitary. The name is compounded of κερατιον, a *pod* or siliqua, and σικυος, a *cucumber*; in reference to the aspect of the fruit.

C. Ecklonii (Nees. in E. & Z. Enum. No. 1797); *Harv. in An. Nat. Hist. 1st Ser. vol. 3, p.* 421, *t.* 10. *Modecca septemloba, E. Meyer.! in Herb. Drege.*

HAB. In woods of Uitenhage, Albany, and Caffraria, *E. & Z.! Drege!* Grahamstown, *General Bolton!* Port Elizabeth, *Mrs. Holland!* Port Natal, *Gueinzius! Gerrard & McKen!* (Herb. D., Hk., Sd.)

Root perennial. Stems several feet long, slender, pale, twining round other plants. Leaves on long petioles, exstipulate, cordate at base, deeply 3, 5, 7-lobed, the lobes acuminate and sharply serrate. Flowers of one or both sexes axillary; the males in 3–6-flowered, pedunculate racemes, on slender, filiform pedicels bracteolate at base; the female generally solitary, on a simple peduncle. Perianth 4–5 lines long, greenish, veiny. *Glands* waxy. Stamens spathulate. Capsule 2–3 inches long, 4–5-angled, 2–4 lines wide, tapering to both extremities, on a long stipes, the perianth remaining till the seeds are nearly ripe.

IV. ACHARIA, Thunb.

Flowers monœcious. MALE: *Perianth* campanulate, 3–4-lobed, subtended by 3–4 involucral bracts. *Stamens* adnate to the perianth for more than half their length, as many as its lobes and alternate with them; filaments dilated upwards, subexserted; anthers adnate to a broad, bilobed connective, didymous, the cells separated, introrse. *Glands* 3–4, fleshy, in the base of the perianth, alternating with the stamens. FEMALE: *Perianth* as in the male, but enlarged in fruit, persistent. *Ovary* subsessile, with 3 glands at base, unilocular; ovules few, on 3–4 parietal placentæ. *Style* 3–4-fid; stigmas 3–4-channelled, 2-lobed. *Capsule* shortly stipitate, membranous, 3–4-valved, few-seeded; seeds pendulous, with a small arillus; embryo cylindrical, in the axis of fleshy albumen. *Endl. Gen. No.* 5107.

A small, herbaceous, thinly pubescent plant, with branching stems, alternate, petioled, 3-lobed and cut leaves, and small, green, axillary flowers. The name is in honour of Erick Acharius, a celebrated Swedish botanist, author of a system of lichenology and of several descriptive works on lichens.

1. A. tragioides (Thunb. Prodr. p. 14, cum ic.); *Thunb.! Cap. p.* 37. *Arn. & Harv. in An. Nat. Hist. 1st ser. vol.* 3, *p.* 420, *t.* 9.

HAB. Shady places in the forests of Uitenhage and Albany, frequent. (Herb. D., Hk., Sd., &c.)

Root perennial, woody. Stems numerous, erect or ascending, simple or branched, angular, pubescent. Leaves on longish petioles, exstipulate, scattered, deeply 3-lobed, the lobes coarsely toothed or cut, pubescent. Flowers axillary, 2 or more (of one or both sexes) together, shortly pedicellate, cernuous. Perianth usually 3-fid, occasionally 4-fid. The connectives of the anthers are broadly spathulate and emarginate, and the anther-cells so far separate as to appear like 2 anthers. The pollen-case is inflated, and granulated or gland-toothed externally. A full analysis of the flower will be found in the An. Nat. Hist., as above quoted.

ORDER LX. LOASACEÆ.

(BY W. H. HARVEY.)

Flowers perfect, regular. *Calyx*-tube adnate to the ovary, frequently ribbed; the limb 4–5-parted, persistent or rarely deciduous. *Petals* inserted in the throat of the calyx, deciduous, rarely as many as its lobes and alternate with them, usually twice as many, in a double row, those of the outer row larger, concave, shortly clawed, induplicate-valvate in the bud, or rarely flat, sessile, and twisted in æstivation; those of the inner row much smaller, often resembling abortive stamens. *Stamens* inserted with the petals, mostly polyadelphous, in parcels opposite the petals; filaments filiform; anthers introrse, 2-celled. *Ovary* inferior, unilocular; ovules numerous, on 3–5 parietal placentæ, pendulous, anatropous. *Style* simple; stigma undivided or 3–5-fid. *Capsule* crowned with the persistent limb of the calyx, very rarely fleshy and indehiscent, usually opening in 3–5 valves. *Seeds* albuminous; embryo orthotropous.

Erect or twining, herbaceous or suffruticose plants, almost all natives of America. Leaves opposite or alternate, often lobed, and rough with stinging or rigid hairs, exstipulate. Flowers yellow or orange, often showy, solitary or many together. Many are ornamental plants, but none of much use in medicine or the arts. Only one is South African.

I. KISSENIA, R. Br.

Calyx-tube 10-ribbed; limb 5-parted, the lobes equal, enlarged in fruit, persistent. *Petals* 10, deciduous, inserted at the summit of the calyx-tube; 5 outer alternating with the calyx-lobes, roundish, concave; 5 inner, opposite the calyx-lobes, smaller, ligulate, angularly bent. *Stamens* indefinite, those of the outer row barren, with cordate bases. *Ovary* turbinate, 3-celled; cells uni-ovulate. *Fissenia, Endl. Gen. Suppl. II. p.* 76. *Cnidone, E. Mey. MSS.*

A very remarkable plant, the only *Loasacea* yet known on the African continent. It was originally discovered in Arabia by a traveller named *Kissen*, to whose memory Dr. R. Brown inscribed the genus in MSS. in the British Museum. Endlicher, who first published a generic character, miscalled it *Fissenia*, under which name it is figured in Thesaurus Capensis. For the correct spelling now given I am indebted to my friend Dr. T. Anderson, who has carefully compared the Arabian with the South African specimens, and finds no difference between them.

1. **K. spathulata** (R. Br. in Herb. Br. Mus.); *Andr. Fl. Aden. p.* 43. *Fissenia capensis, Endl. l. c. Harv. Thes. t.* 98. *Cnidone Mentzelioides, E. Mey. ! in Herb. Drege.*

HAB. Between Verleptpram and the mouth of the Gariep, *Drege!* Aapjes R., *Dr. Atherstone!* Namaqualand, *A. Wyley!* (Herb. Sd., D., Hk.)

Stem robust, rigid, striate, very scabrous and pale, as are all parts of the plant. Leaves alternate, petioled, the lower ones 5–7-lobed, 2–3 inches long, coarsely toothed, ribbed, and veiny, thickish and very rough; the upper smaller and less cut, passing toward the summit into linear or lanceolate bracts. Flowers in a terminal, scorpioid cyme, subsessile. Calyx-lobes much enlarged after flowering, spathulate-oblong, obtuse, 3–5-nerved, much longer than the corolla; tube obconical, shaggy with fulvous, straight hairs. Petals pale yellow or buff, style trifid.

ORDER LXI. **ONAGRARIEÆ.**

(BY W. H. HARVEY.)

Flowers perfect, mostly regular (rarely irregular). *Calyx*-tube adnate with the ovary, frequently produced beyond its apex; limb 4-parted, rarely 2–3-parted, the lobes valvate in æstivation, persistent or deciduous. *Petals* inserted in the throat of the calyx, rarely absent, as many as its lobes and alternate with them, more or less clawed, twisted in æstivation. *Stamens* inserted with the petals, as many or twice as many, some occasionally sterile; filaments filiform; anthers 2-celled, introrse; pollen bluntly triangular. *Ovary* inferior, mostly 4-celled; ovules mostly numerous (rarely solitary) on axile placentæ, anatropous. *Style* filiform; stigma 2–4-lobed. *Fruit* either a 4-valved capsule or a berry, rarely nutlike, 4–2-celled. *Seeds* exalbuminous; embryo orthotropous.

Herbaceous plants or shrubs, dispersed over the globe, but most abundant in the temperate zones, east and west, of the northern hemisphere, particularly of the new world. Leaves opposite or alternate, exstipulate, simple, entire or variously lobed and cut. Flowers either axillary or in racemes or spikes, often showy. None are remarkably useful. The numerous species and garden varieties of *Fuchsia*, and *Oenothera*, both genera chiefly American, are much cultivated for ornament. *Montinia*, usually referred to this Order, will be found under Saxifragaceæ.

TABLE OF THE SOUTH AFRICAN GENERA.

* *Limb of the calyx persistent. Capsule septicidal.*

I. **Jussiæa.**—Stamens *twice* as many as the calyx-lobes.
II. **Ludwigia.**—Stamens as many as the calyx-lobes.

** *Limb of the calyx deciduous. Capsule loculicidal.*

III. **Œnothera.**—Calyx-tube much produced beyond the ovary. Seeds naked.
IV. **Epilobium.**—Calyx-tube not produced beyond the ovary. Seeds with a tuft of hairs at one end.

I. JUSSIÆA, L.

Calyx-tube not produced beyond the ovary; the limb 4–5-parted persistent. *Petals* 4–5. *Stamens* 8–10. *Stigma* capitate. *Capsule* 4–5-celled, crowned by the calyx segments and opening longitudinally between the ribs. *Seeds* numerous, small, without any appendage. *Endl. Gen.* 6109.

Herbaceous or shrubby, rarely arborescent plants, chiefly natives of marshes in tropical and subtropical America, with a few species in Asia and Africa. Leaves alternate, mostly quite entire. Flowers axillary, solitary, sessile or shortly pedicellate, yellow or white. Named in honour of the illustrious Jussieu, the restorer of the natural system of Botany.

ANALYSIS OF THE SOUTH AFRICAN SPECIES.

Flower 4-parted. Stem erect. Lvs. linear or lanceolate-linear (1) **angustifolia.**
Flower 5-parted. Stem decumbent or floating. Leaves much attenuated at the base ... ... (2) **fluitans.**

1. J. angustifolia (Lam. Dict. 3, p. 331. Ill. t. 280. f. 3); stem herbaceous, *erect*, laxly pilose; leaves *subsessile*, linear-lanceolate or linear, acute at both ends, minutely hispidulous on both surfaces; flowers on very short pedicels, 4-*cleft;* calyx-lobes acuminate, ovato-lanceolate, 3–5-nerved, tube 4-angled, elongate. *DC. Prodr.* 3. *p.* 55.

VAR. *β*, **linearis**; leaves very narrow, nearly linear. *J. linearis, Hochst. in Pl. Krauss, No.* 73, *not of Willd.*

HAB. Port Natal, *Mr. Hewitson!* *β*. Natal, *Krauss!* Nototi, *Gerrard and McKen.* (Herb. D.)

Stem 2–3 feet high, much branched. Leaves 3–4 inches long, 2–5 lines wide, with a thickish midrib, and slender pennate nerves. Pedicels 2–3 lines long; ripe capsule an inch or rather more in length. Calyx-lobes very acute. *Mr. Hewitson's* specimens agree well with East Indian ones distributed by *Drs. Hooker and Thomson.* It seems to be a common East Indian species.

2. J. fluitans (Hochst.); stem *procumbent* (or floating), subsimple, in the upper part more or less villous; leaves lanceolate or oblongo-lanceolate, *tapering much at base into a petiole*, acute or obtuse at the apex, sparsely pilose or glabrous; flowers on shorter or longer pedicels, 5-*cleft;* calyx-lobes lanceolate, acute, pilose or glabrous, tube elongate. *J. alternifolia, E. Mey.*

HAB. *γ*. Between Omtata, Omsamculo and Omcomas, *Drege!* Port Natal, *T. Williamson! Gueinzius,* 459*! Krauss! No.* 36. (Herb. D., Sd.)

Very nearly related to *J. repens*, but usually with longer and narrower, more acute leaves, and shorter pedicels. But *Dr. Gueinzius'* specimens are intermediate in both these respects. Leaves, including petiole, 3–4 inches long, ¼–½ inch wide. Flowers yellow.

II. LUDWIGIA, L.

Characters as in JUSSIÆA, but stamens 4–5. *Petals* wanting in *L. palustris. Endl. Gen.* 6110, 6111.

Herbaceous, aquatic or marsh plants, with the habit of Jussiæa, natives of the four quarters of the globe. The name is in honour of C. G. Ludwig, once professor of medicine at Leipsic, and author of numerous botanical works.

ANALYSIS OF THE SOUTH AFRICAN SPECIES.

Stem procumbent; leaves opposite; fl. apetalous ... ... (1) **palustris.**
Stem erect; leaves alternate; fl. 4-petalled ... ... (2) **jussæoides.**

1. L. palustris (Ell. Car. Vol. 1. p. 211); stem *procumbent*, creeping, glabrous; leaves *opposite*, ovate, acute, tapering at base into a petiole, glabrous; flowers axillary, solitary, sessile, *without petals;* calyx-lobes 4, ovate. *Isnardia palustris, Linn. Sp.* 175. *DC. Prod.* 3, *p.* 61. *E. Bot. t.* 2593. *E. & Z.! No.* 1763.

HAB. In ditches and marshy places. Near the baths at Kochmanskloof, Swell., *E. & Z.!* King William's Town, Caffr., *Rev. J. Brownlee!* Macallisberg, *Burke and Zeyher!* (Herb. D., Sd., &c.)

Stems 1–2 feet long, subsimple, or with a few erect branches. Leaves, including petiole, 1–1½ inch long. Flowers 2–3 lines long. A native also of Europe, Asia, and America.

2. L. jussæoides (Lam. Dict. 3, p. 588); stem herbaceous, *erect*, nearly glabrous; leaves *alternate*, tapering at base into a petiole, lanceolate or ovato-lanceolate, acute or acuminate, scaberulous, especially along the margin and nerves; flowers shortly pedicellate, 4-cleft; 4-petalled; calyx lobes lanceolate, acuminate, 3-nerved, tube bluntly 4-angled, elongate, slender. *Jussiæa cylindrocarpa, Boiv.! No.* 3412.

HAB. On the Nototi, Natal, *W. T. Gerrard!* Mayotte, *Boivin!* (Herb. T.C.D.)

Stem tall, branching, dark-coloured, bluntly 4 angled. Leaves, including the petiole, 4–6 inches long, ½–1 inch wide, penninerved, thin. Flowers on *Mr. Gerrard's* specimen on short lateral branchlets, racemulose, one or more from the axil of a small floral leaf. Calyx-tube, in fruit, about an inch long, scarcely a line in diameter.

III. ŒNOTHERA, L.

Calyx-tube much produced beyond the ovary, deciduous; limb 4-parted. *Petals* 4, obcordate. *Stamens* 8. *Stigma* 4-lobed or capitate. *Capsule* various in form and texture, 4-celled, 4-valved, many-seeded. *Seeds* naked. *Endl. Gen.* 6115.

A very large genus of biennial herbs or suffrutices, common throughout North and South America, from which continents some species have become naturalized in Europe and Asia, and two have taken effectual footing in South Africa. Radical leaves mostly rosulate; cauline alternate, entire or denticulate, sometimes sinuate or pinnatifid. Flowers axillary, solitary, or forming a terminal, leafy spike, very generally opening in the evening; whence the popular name of "*Evening Primrose*" given to these plants. The generic name *Oenothera* is derived from *οινος, wine,* and *θηραω, to hunt;* the roots of *Œ. biennis* were formerly eaten as incentives to wine-drinking.

ANALYSIS OF THE SOUTH AFRICAN SPECIES.

Cauline leaves sessile, the uppermost half-amplexicaul ... ... (1) **biennis.**
Cauline leaves petioled, tapering to base and apex ... ... (2) **nocturna.**

1. Œ. biennis (Linn. Sp. 492); stem erect, simple, hirsute; radical leaves rosulate, oblongo-lanceolate, acute, tapering at base, cauline ovato-lanceolate, sessile, the uppermost short and subamplexical, all repando-denticulate; flowers in a terminal, leafy spike; tube of the calyx twice or thrice as long as the ovary or as the segments; stamens somewhat declined; capsules oblong-linear, bluntly 4-sided, 4-ribbed. *Tor. and Gray. Fl. Bor. Amer.* 1, *p.* 492. *Bot. Mag. t.* 2048. *E. Bot. t.* 1534. *Œ. villosa, Thunb. Cap. p.* 373. *E. & Z!* 1761.

Hab. Naturalised from North America. In fields near Klapmuts, *E. & Z.!* Fields near Steandal, Tulbagh, *Dr. Pappe!* Cape Flats, *J. Sturk!* (Herb. D. Sd. &c.)

Stem 1–4 feet high, in the Cape specimens very hairy, with long, soft hairs. Flowers pale yellow: the calyx-tube 3–4 inches long. This plant, the "*Evening Primrose*," has long been naturalized in Europe, and probably brought by the first settlers to South Africa. In Thunberg's time it had already become so wild as to be even then mistaken for an indigenous species.

2. Œ. nocturna (Jacq.? Ic. Rar. t. 455); stem erect, simple or branched, pubescent; cauline leaves lanceolate, acute or acuminate, tapering at base into a short petiole, the uppermost subsessile, all sinuato-denticulate, the lower ones erose, or near the base almost runcinato-pinnatifid; flowers in a terminal, leafy spike; tube of the calyx twice as long as the ovary, or as the segments; stamens erect; capsules oblong-linear, very bluntly 4-sided or subterete, not obviously ribbed. *DC. Prodr.* 3. *p.* 47? *Œ. erosa, Lehm.! ind. Sem. Hort. Hamb.* 1820. *E. & Z.!* 1762.

Hab. Probably of South American (Chilian ?) origin. Naturalized in fields and waste places near Rondebosch, *E. & Z.!* (Herb. Sd., D.)

Stem 2–3 feet high, simple or with several lateral, virgate branches from the axils of the upper leaves. Whole plant softly pubescent. Leaves 3–4 inches long, all but the uppermost much attenuated at base, variably dentate, either repand, sinuate or erose, the lowest ones frequently deeply and sharply sinuate. Flowers smaller than in *Œ. biennis*, of a deeper yellow, changing to reddish in decay.

IV. EPILOBIUM, L.

Calyx-tube not produced beyond the ovary; limb deeply 4-lobed or 4-parted, deciduous. *Petals* 4, obovate or obcordate. *Stamens* 8. *Stigma* clavate or 4-lobed. *Capsule* linear, 4-sided, 4-celled, 4-valved, loculicidal. *Seeds* with a tuft of hairs at the chalaza-end. *Endl. Gen.* 6121.

Herbaceous plants or suffrutices, natives of the temperate zones, chiefly in the northern hemisphere. Leaves alternate or opposite, entire or serrulate; flowers axillary, solitary or in terminal spikes, purple or rosy: very rarely yellow. Name from επι, *upon*, and λοβος, *a pod;* a flower growing on a pod.

ANALYSIS OF THE SOUTH AFRICAN SPECIES.

Stigma 4-lobed:
- leaves sessile, softly hairy, lanceolate ... ... (1) **hirsutum.**
- leaves subpetiolate, ovate-oblong, puberulous ... ... (2) **flavescens.**

Stigma undivided; leaves sessile, lanceolate, subglabrous ... (3) **tetragonum.**

1. E. hirsutum (Linn.); stem tall, erect, much branched, terete, softly hairy and villous; leaves opposite and alternate, *villous*, lanceolate or oblongo-lanceolate, *sessile or half-clasping*, unequally and rather sharply serrulate; stigma *deeply 4-lobed, its lobes strongly revolute. DC. Prodr.* 3, *p.* 42. *E. Bot. t.* 838. *E. villosum, Thunb. Cap. p.* 374. *DC. l. c. Drege!* 6851. *Zey.!* 545.

Hab. Moist places and by river-banks. Districts of Cape, Uitenhage, and Albany, *E. & Z.!* Worcester, *Pappe!* Stellenbosch, *W. H. H.* Dutoitskloof, Kamdebo, Sneeuwberg; and several localities in Caffraria, *Drege!* Natal, *Sanderson!* (Herb. Sd., D.)

Stem 3–5 feet high, robust, pyramidal, with many lateral branches. Pubescence copious, soft, and somewhat hoary, but variable in amount and in the length of the hairs. Leaves mostly lanceolate; the lower ones opposite, broader, and more oblong. Flowers bright purple. I cannot separate this from the European *E. hirsutum;* Cape specimens differ as much among themselves, in hairiness, shape, and size of leaves and size of flower, as any of them do from the European plant. *Seringe* (DC. l. c.) chiefly relies in distinguishing *E. villosum* from *E. hirsutum*, on the stigma, which he states to be "somewhat thicker and more convolute" in *E. villosum*.

2. E. flavescens (E. Mey.! in Herb. Drege); stem erect, simple, virgate, terete, *puberulous*; leaves (except the 2–3 lowest pairs which are opposite and subsessile) alternate, *minutely petiolate*, rounded at base, ovate-oblong or ovato-lanceolate, distantly repando-dentate, puberulous; stigma 4-*lobed, its lobes oblong, erecto-patent*; pedicels of the fruit much longer than the leaves. *E. montanum, E. & Z.!* 1759 (*not of Linn.*)

HAB. On the Winterberg, Kaffr., *E. & Z.!* Between Zandplaat and Komga, and between the Omsamwubo and Omsamcaba, *Drege*. Water courses in Kreili's country, *Mrs. F. W. Barber*, 285. Natal, *Krauss!* 154. (Herb. Sd., D.)

Stems 1–2 feet high, in all our specimens quite simple. Leaves 1–1½ inch long, ½–¾ inch wide, longer than the internodes; petioles 1–2 lines long. Fruit-pedicels 1½ to twice as long as the floral leaf. Flowers a creamy white? Nearly allied to *E. montanum*, but the leaves are more closely placed, and none but the lowest opposite, and the fruit-stalks are proportionally much longer.

3. E. tetragonum (Linn.); stem erect, branched, 4-angled, nearly glabrous or minutely puberulous; leaves opposite and alternate, sessile, lanceolate or linear-lanceolate, acute, repando-dentate, glabrous or nearly so; stigma *club-shaped, undivided*; pedicels of the fruit equalling the floral-leaf or longer. *DC. Prodr.* 3, *p.* 43. *E. Bot. t.* 1948. *E. obscurum, E. & Z.!* 1760. *E. Dregeanum, E. Mey.! in Hb. Drege.*

HAB. Moist places at Rietvalley and Doornhoogde, Cape, *E. & Z.!* Cape, *Capt. Carmichael!* Zwartkops R., *Zeyher!* Winterveld, 3000f.; Sternbergspruit; and on the Witberg, 6000–7000f.; also in Dutoitskloof, *Drege!* (Herb. Sd., D.)

Stem 2–3 feet high, with many lateral, erect, virgate branches. Pubescence, if any, very scanty and minute. Leaves closely placed, sometimes mostly opposite, in other specimens mostly scattered, rather closely placed, 1½–3 inches long, ¼–½ inch wide. Stem mostly 4-angled, at least toward the summit of the internodes. Flowers smaller than in *E. flavescens*, purplish-pink. A very variable and widely dispersed plant, common to most parts of the temperate zones, north and south.

ORDER LXII. COMBRETACEÆ.

(BY W. SONDER.)

Flowers regular, perfect or unisexual. *Calyx*-tube adnate with the ovary; limb 4–5-parted, valvate in æstivation, rarely persistent. *Petals* (sometimes wanting altogether and often very minute) inserted on the summit of the calyx-tube. *Stamens* inserted within the petals, as many, or twice, rarely thrice as many; the filaments subulate; anthers introrse, 2-celled. *Ovary* inferior, generally crowned by a fleshy or woolly disc, unilocular; ovules definite, 2–4, rarely 5, pendulous from the apex of the cavity. *Style* single; stigma undivided. *Fruit* drupaceous, mostly longitudinally 4–5-winged. *Seed* mostly solitary, filling the cavity of the fruit, exalbuminous; embryo orthotropous, with leafy, spirally twisted, folded, or flat cotyledons.

Trees or shrubs, sometimes climbers. Leaves alternate or opposite, simple, entire penninerved, petiolate, exstipulate. Flowers often of small size, in spikes, racemes or heads, naked or bracteated. The species are numerous in the tropics of both hemispheres; a few straggling into the warmer parts of the temperate zone. None of the South African species are found to the west of Uitenhage.

TABLE OF THE SOUTH AFRICAN GENERA.

Tribe 1. TERMINALIEÆ. Flowers without petals. Cotyledons spirally twisted.

I. **Terminalia.**

Tribe 2. COMBRETEÆ. Petals 4–5. Cotyledons usually thick, plano-convex or irregularly and longitudinally plaited, rarely thin, leafy and intricately folded.

II. **Combretum.**—Calyx short, 4-toothed. Petals 4. Stamens 8.
III. **Poivrea.**—Calyx short, 5-lobed. Petals 5. Stamens 10.
IV. **Quisqualis.**—Calyx tubular, very long and slender, 5-toothed. Petals 5. Stamens 10.

I. TERMINALIA, L.

Flowers often polygamous from abortion. Limb of the calyx deciduous, campanulate, 5-cleft; lobes acute. *Petals* wanting. *Stamens* 10, in a double row, longer than the calyx. *Ovary* 2–3-ovuled. *Style* filiform, acutish. *Drupe* not crowned by the calyx, usually dry, indehiscent, 1-seeded. *Seeds* almond-like. *DC. l. c.* 3. 10. *Endl. Gen. n.* 6076.

Trees or shrubs. Leaves alternate or rarely opposite, sometimes crowded toward the extremities of the branches. Flowers spiked, spikes often racemose or panicled, bisexual in the lower part of the spike, male in the upper. Name from terminus, *end;* leaves and spikes at the ends of the branches.

1. **T. sericea** (Burch. Cat. Geog. Afr. Austr. n. 2399); leaves alternate, crowded at the tops of the branches, oblong, tapering at base and shortly petiolate, mucronulate, quite entire, clothed with silky, appressed hairs on both surfaces; spikes shorter than the leaves, pedunculate, ovate or oblong, silky; drupe broad-winged, reddish. *DC. l. c.* 13.

HAB. On the Aapjesriver, Dec., *Zey.!* 548. (Herb. Sd., D., Hk.)

Branches glabrous, dichotomous or trichotomous. Leaves oblong or obovate-oblong, narrowing into a short petiole, 2½–3 inches long, nearly an inch broad. Spikes pedunculate, shorter than the leaves. Flowers small. Drupe glabrous, 1–1½ inch long, 9–10 lines broad. Seed ovate, 4 lines long.

II. COMBRETUM, L.

Calyx funnel-shaped; tube as short as or longer than the ovary; limb campanulate, 4-lobed, deciduous. *Petals* 4, inserted between the lobes of the calyx. *Stamens* 8, in two rows, exserted. *Ovary* 2–5-ovuled. *Style* exserted, acute. *Fruit* 4-winged, 1-celled, 1-seeded, indehiscent. *Seed* pedulous. *DC. l. c.* 18. *Endl. Gen. n.* 1087.

Shrubs or trees, more or less scandent. Leaves often opposite, quite entire. Spikes terminal and axillary, sometimes panicled. Derivation of the name unknown.

TABLE OF THE SOUTH AFRICAN SPECIES.

Leaves elliptical, ovate, obovate or broadly cordate:
  Adult leaves densely velvetty-tomentose ... ... (5) **holosericeum.**
  Adult leaves glabrate (at least on the upper surface):
    Inflorescence globose (pseudo-capitate).
      Calyx-teeth blunt, ciliolate... ... ... (1) **glomeruliflorum.**
      Calyx-teeth acute, glabrous ... ... (2) **erythrophyllum.**
    Inflorescence oblong or cylindrical (racemose).
      Branches glabrous, the twigs minutely pubescent:

Lvs. obovate, cuneate at base ; racemes shorter than leaves ... ... (4) **Kraussii.**
Lvs. elliptic, recurved at point ; racemes equalling leaves ... ... (6) **apiculatum.**
Branches velvetty or densely tomentose :
Petals bearded ; fruit 9–10 lines long, with moderate wings ... ... (3) **Gueinzii.**
Petals glabrous ; fruit 2 inches long, with very wide wings ... ... (7) **Zeyheri.**
Leaves lanceolate or oblongo-lanceolate, acute :
Branches velvetty ; leaves pubescent beneath, about the nerves :
Lvs. lanceolate, minutely petiolate ... ... (8) **riparium.**
Lvs. *broadly* oblongo-lanceolate, *conspicuously* petiolate ... ... (9) **Sonderi.**
Branches subglabrous ; leaves glabrous beneath ... (10) **salicifolium.**

1. C. glomeruliflorum (Sond. in Linnæa, vol. xxiii., p. 42) ; branches unarmed, spreading, glabrous ; the twigs very short, pubescent ; leaves opposite, petiolate, elliptic, acute at both ends, quite entire, glabrous above, *pubescent beneath ;* spikes axillary, solitary, subcapitate, shorter than the leaves ; calyx campanulate, *with blunt, ciliolate teeth ;* petals spathulate, glabrous ; stamens 8 ; filaments exserted.

HAB. Port Natal, *Gueinzius*, 62, 565, *Dr. Sutherland !* (Herb. Sd., D.)
An erect shrub, with opposite, terete, greyish, smooth branches. Leaves reddish, about 2 inches long, 10 lines broad. Petiole 3 lines long. Spikes nearly as long as the semi-uncial, puberulous pedicels. Calyx 1 line long, at length glabrous. Petals yellow, glabrous as well as the filaments. Anthers oblong. Style short. Fruit not seen.

2. C. erythrophyllum (Sond. l. c. 43) ; branches unarmed, glabrous, young ones pubescent ; leaves alternate or opposite, petiolate, acutely ovate, quite entire, *glabrous ;* spikes axillary, solitary, capitate, shorter than the leaves ; *calyx campanulate, with acute, glabrous teeth ;* petals obovate, unguiculate, glabrous ; stamens 8 ; filaments exserted. *Terminalia ? erythrophylla, Burch. trav.* 1, 400. *DC. l. c.* 13.

HAB. On the banks of the Ky-gariep, *Burchell ;* woods on Crocodileriver, *Zeyher*, 550. August. (Herb. Sd., D.)
" A large tree of picturesque growth and thin foliage, called by the Hottentotts of Klarwater " Roodeblat," on account of the beautiful crimson colour which the leaves assume at the autumnal season, or rather season of fading : in which circumstance it remarkably agrees with the Indian almond (*Terminalia Catappa*). It grows to the height of 40 feet, with several crooked, spreading trunks, from 1–2 feet in diameter, covered with a smooth, white or pale-green bark," *Burchell.* It comes very near *C. glomeruliflorum*, but differs by somewhat larger, glabrous, red leaves and the calyx.

3. C. Gueinzii (Sond. l. c. 43) ; unarmed ; branches terete, *velvetty ;* leaves opposite, short-petioled, elliptic or obovate, acutish or obtuse-mucronulate, entire, *often cordate at the base*, young ones appressed-pubescent, when old glabrous above, ferrugineous, lepidote and reticulated beneath ; spikes oblong, axillary, as long as the leaves, *rhachis and calyx hairy and lepidote ;* flowers 8-androus, bracteated ; petals ciliate-bearded ; fruit sub-pedicellate, *elliptic*, 4-winged ; wings scarious, shining, *slightly broader* than the lanceolate body of the fruit.

HAB. Port Natal, *Gueinzius*, 567. Attercliffe, *Sanderson!* 249. *Gerrard and McKen!* (Herb. Sd. D.)

Branches greyish. Leaves 3–4 inches long, 2–2½ inches wide, shining above. Petiole 3 lines long. Raceme-like spikes 2 inches long. Bracts subulate, minute. Calyx acute, nearly 1 line long, hairy. Petals minute, obovate, unguiculate, yellow. Filaments exserted. Style as long or longer than the stamens. Fruit obtuse at both ends, 9 lines long, 6 lines broad; wings yellowish, transversely striated, with crenulate margins, lepidote.

4. C. Kraussii (Hochst.! pl. Krauss, 58); branches *glabrous;* leaves opposite, short-petiolate, obovate or obovate-oblong, obtuse, mucronulate or acutish, *cuneate at the base,* quite entire and *glabrous,* pale olivaceous and reticulated beneath; spikes oblong, axillary, usually shorter than the leaves, *glabrous;* flowers 8-androus; petals minute; fruit *suborbicular,* emarginate at the base and apex, 4-winged; wings scarious, *twice broader* than the lanceolate body of the fruit. *Sond. l. c. C. lucidum, E. Meyer. non Blume.*

HAB. Woods near Port Natal, *Krauss.* 253, *Drege, Gueinzius,* 566, *Plant,* 27. Oct.–Feb. (Herb. S., D.)

Shrub or tree 15–20 feet high, with greyish branches. Leaves 2–3 inches long, 1½–2 inches broad. Spike 2 inches, flowers sessile, not pedunculate as in the preceding. Fruit 8 lines long and broad, wings transversely striated.

5. C. holosericeum (Sond. l. c. 44); arborescent, unarmed, branchlets terete, as well as the leaves and spikes *fulvous-silky;* leaves opposite, very shortly-petiolate, broad-ovate, subcordate, acute, very entire, on both sides densely and softly velvetty with yellowish-brown hairs; spikes oblong-cylindrical, axillary, solitary, shorter than the leaves; flowers 8-androus; calyx cup-shaped; petals obtuse, ciliate; fruit subsessile, elliptic, 4-winged, wings semiorbicular, thinly pubescent, not or scarcely broader than the lanceolate body of the fruit. *Harv. Thes. Cap. p.* 47, *t.* 74.

HAB. Magalisberg, *Burke, Zeyher,* 575. June. (Herb. D., Sd.)

A small tree, with glabrate branches, and opposite, densely velvetty twigs. Leaves 2½–3½ inches long, 2–2½ inches broad, with minutely recurved margins, the lower surface densely netted with prominent veinlets between the parallel primary veins. Spikes shortly pedunculate, 1½–2 inches long. Flowers minute. Calyx with 4 shallow, broad teeth, separated by rounded interspaces. Petals very minute, obovate. Stamens exserted. Fruit 9–10 lines long, obtuse at each end (when young, somewhat taper-pointed); the wings subentire at margin, yellowish, cross-striate, 2 lines wide in the middle.

6. C. apiculatum (Sond. l. c. 45); erect, unarmed; branches glabrous; leaves opposite, shortly-petiolate, *elliptic or oblong, recurvate-apiculate,* glabrous on both sides, lepidote, reddish; raceme-like spikes axillary, solitary, as long as the leaves, subglabrous; flowers 8-androus; calyx campanulate; petals bearded-ciliate; fruit 4-winged, subemarginate at both ends; wings lunate, shining, glabrous, broader than the oblong-lanceolate, lepidote body of the fruit.

HAB. Magalisberg, *Zeyher,* 553. Oct. (flower), Jan. (fruit). (Herb. Sd., D.)

A small, much-branched tree. Branches opposite, greyish-yellowish, young ones at top viscous. Leaves reticulate, 2½–3 inches long, 14–18 lines broad, when young subviscous. Racemes 1½–2 inches long; peduncle and rachis glabrous. Calyx glabrous, with 5 short, ciliolate teeth. Petals obovate, minute, yellow. Stamens exserted, as long as the style. Fruit cordate-ovate, 10 lines long, 9 lines wide, golden-yellow; the wings a little larger at the base, with subundulate margins; the pedicel 3½ lines long.

7. **C. Zeyheri** (Sond. l. c. 46); arborescent, unarmed; branches, petioles, young leaves, and inflorescence softly pubescent; adult leaves oblong-elliptic, obtuse, subemarginate, glabrate, reticulate-veined, shining above; spikes oblong, axillary, solitary; calyx campanulate; petals glabrous; stamens 8, rarely 12 or 16; fruit very large, petiolate, roundish-elliptical, emarginate at both ends, glabrous, 4-winged; wings semiorbicular, shining, twice as wide as the oblong-lanceolate body of the fruit. *Harv. Thes. Cap. p.* 48, *t.* 75.

HAB. Magalisberg, *Burke & Zeyher*, 552. (Herb. D., Sd.)

A tree, 20–30 feet high. Leaves on petioles 2–4 lines long, 2–3 inches long, 14–20 lines broad. Spike shorter than the leaf or equalling it, densely velvetty, cylindrical, many-flowered. Flowers with a minute bract. Calyx shortly 4-toothed, 1 line long. Petals minute, on short claws, ovato-trapeziform. Style equalling the stamens. Fruit 2–2¼ inches long, and nearly as wide; the wings 8 lines wide, papery in substance, cross-striate, and easily splitting in the direction of the striæ.

8. **C. riparium** (Sond. l. c. 47); erect, unarmed, branches pubescent; leaves opposite, very short-petiolate, oblong-lanceolate, acuminate, narrowed at the base, glabrous above, reticulated and pubescent between the nerves beneath; racemes axillary, solitary; fruit pedicellate, elliptic, 4-winged; wings lunate, glabrous, shining, broader than the lanceolate, powdery body of the fruit.

HAB. On the Magalisriver, *Zey.!* 549. July. (Herb. Sd., D.)

Branches terete, young ones subangulate, very thinly pubescent. Leaves 2½–4 inches long, 10–12 lines broad, quite glabrous above, tomentose on the nerves beneath; petiole 1 line long. Flowers unknown. Fruit on a pedicel 2½ lines long, 6 lines long, 5 lines wide, subacute, obtuse at the base; the wings golden, cross-striated.

9. **C. Sonderi** (Gerr.! Mss.); arborescent, unarmed, the twigs and petioles and peduncles *densely velvetty; leaves conspicuously* petiolate, *broadly oblongo-lanceolate*, (4–7 inches long), acute or acuminate, membranaceous, glabrous and minutely punctate above, pubescent, especially on the nerves beneath, penninerved and finely reticulated; flowers unknown; fruit in axillary, short racemes, pedicellate, elliptical, 4-winged; wings lunate, glabrous, shining, rather wider than the smooth and even body of the fruit.

HAB. On the Nototi R., near Natal, *W. T. Gerrard!* (Herb. T.C.D.)

A large tree, "the handsomest of the South African species (*Gerr.*) Leaves often 6 or 7 inches long on the young shoots, 2–3 inches wide, of a thin substance and bright green colour, mostly tapering to an acute point, opposite. Petioles semi-uncial. Fruit on a pedicel 4–5 lines long; 6–7 lines long, 5 lines wide, emarginate; the wings yellowish, cross-striated. Allied to *C. riparium*, but with much larger and broader leaves, longer petioles and pedicels, and larger fruit.

10. **C. salicifolium** (E. Meyer. in herb. Drege); erect, unarmed, branches glabrous; leaves opposite, petiolate, lanceolate, quite glabrous on both sides, glaucescent; racemose spike capitate, shorter than the leaves, peduncle pubescent; fruit pedicellate, elliptic or sub-orbicular, subemarginate, 4-winged; wings shining, as wide as the oblong-lanceolate body of the fruit. *Dodonæa caffra, conglomerata et dubia, E. & Z.!* 421–423. *Zey.* 551. *Drege*, 6849.

Hab. On rivers in the districts of Uitenhage, Albany, and in Caffraria, *E. & Z.! Drege*. Oct.–Feb. (Herb. Sd.)

Tree, 20–50 feet high. Branches terete, young ones puberulous. Leaves 3 inches long, 8 lines broad, glaucous above, pale-green, at length reddish beneath; petioles 2 lines long. Flowers polygamous. Male-flowers capitate-racemose. Female ones racemose; racemes 3–6-flowered, on a 4–6 lines long peduncle; pedicles 2 lines long. Fruit 7–9 lines long, 6–7 lines wide, or in some specimens smaller and suborbicular.

III. POIVREA, Comm.

Limb of the calyx infundibuliform, 5-lobed, deciduous. *Petals* 5. *Stamens* 10, protruded. *Ovary* 2–3-ovuled. *Style* filiform, protruded, acute. *Fruit* oval or oblong, or 5-winged. *Seed* solitary, pendulous, 5-angled. *Cotyledons* convolute. *Endl. Gen. n.* 6086.

Usually climbing shrubs. Leaves opposite or alternate, quite entire. Spikes axillary and terminal. Bracteoles solitary under the flowers. Name in honour of N. Poivre, Intendant of the Mauritius.

1. P. bracteosa (Hochst.! in pl. Krauss.); unarmed; branches glabrous; leaves opposite or ternate, shortly petiolate, ovate or ovate-oblong, acute at both ends or obtuse at the base, glabrous; spikes on axillary branches, nodding; bracts large, pedicellate, oval, as long or longer than the calyx, green; calyx 5-toothed; petals oblong, puberulous outwards; stamens exserted; fruit wingless. *Codonocroton triphyllum, E. Meyer. in herb. Drege.*

Hab. Between Omtata and Omsamwubo, 1–2000ft., *Drege!* Common, near Port Natal, *Krauss.! Gueinzius*, 103. Oct. (Herb. Sd., D.)

Fruit 8–10 feet high. Petioles 2 lines long, puberulous above. Leaves 2½–3 inches long, 1–1½ inch broad, veined, paler beneath, acutish or with a short, obtuse acumen. Racemes terminating the lateral (and terminal?) branches, about 1 inch long. Bracts foliaceous, ovate, acute. Pedicels 1 line long, as well as the oblong ovary and calyx minutely puberulous. Calyx campanulate, with 5 acute, 1 line long teeth. Petals unguiculate, oblong, reddish, pubescent outwards, 4 lines long. Stamens exserted, glabrous; anthers elliptic. Ovary 1-celled. Fruit oval or slightly obovate, indistinctly 5-angled, glabrous, 1-seeded. Called "Hiccup-nut" in the colony.

IV. QUISQUALIS, Rumph.

Tube of the *calyx* slender, produced much beyond the ovary, deciduous; limb 5-lobed. *Petals* 5, oblong or roundish, obtuse, longer than the calyx-teeth, imbricate in æstivation. *Stamens* 10, inserted within the throat of the calyx, those opposite the petals longest. *Ovary* 4-ovuled. *Style* filiform, exserted, its base adhering to the calyx-tube. *Drupe* dry, 5-furrowed and 5-ribbed, one seeded. *Seed* pendulous, 5-angled. *Cotyledons* plano-convex. *Endl. Gen. No.* 6089.

Shrubs, natives chiefly of the tropics of Asia and Africa. Branches often twining. Leaves opposite or alternate, entire. Spikes axillary or terminal, bracteate. The name is compounded of *quis*, who, and *qualis*, what kind; the older botanists did not know what to make of it.

1. Q. parviflora (Gerr. Mss.); twigs, petioles and young leaves hispid with patent, rusty pubescence; leaves on short petioles, membranous, reticulated, oval-oblong, acuminate, glabrescent above, pubescent on the nerves and margins beneath; bracts oblongo-lanceolate, persistent; *calyx-tube uncial*, its lobes deltoid, acute; petals *subrotund*, hispidulous on the inner surface; anthers globose, *subsessile*.

HAB. On dry rocks. Umhtoti, Natal, *W. T. Gerrard!* (Herb. T.C.D.)

A shrub, with the aspect of *Q. indica*, but with very much smaller flowers. Calyx tube about an inch long, clavate; lobes ½-line long, tipped with rufous bristles. Petals 1–1½ lines long and nearly as wide, silky on the outside, minutely hispidulous within. Filaments not ½ line long, the anthers of the longer stamens in the throat of the calyx; those of the shorter quite included.

ORDER LXIII. **RHIZOPHOREÆ.**

(BY W. SONDER.)

Flowers perfect, regular. *Calyx*-tube adnate to the ovary, wholly or in part; limb 4–12-parted, persistent, with valvate æstivation. *Petals* as many as the lobes of the calyx, inserted on a fleshy ring within the calyx-tube, sessile, either entire and flat, or bifid and inflexed at base, the lobes entire or laciniate. *Stamens* inserted with the petals, twice or thrice their number, rarely many times, usually in pairs opposite the petals; filaments subulate; anthers 2-celled, introrse, erect, slitting. *Ovary* inferior or half-inferior, 2–4-celled, with ovules in pairs; or very rarely unilocular, with 6 ovules; style filiform or conical; stigma entire or 2–4-toothed; ovules anatropous. *Fruit* coriaceous, crowned with the persistent calyx-limb, abortively unilocular and 1-seeded; seed exalbuminous, germinating before it falls; radicle very long, issuing through the summit of the fruit.

Trees or shrubs, natives of muddy sea-shores and estuaries in the tropics and warmer parts of the temperate zone. Branches opposite; twigs 4-angled. Leaves opposite, petioled, simple, very entire, coriaceous, penninerved. Stipules interpetiolar, on each side one, convolute, deciduous. Inflorescence terminal or axillary, cymose or capitate. Under the name of *Mangroves*, the shrubs of this family form tangled thickets on the muddy seashore, the seeds germinating before they fall, the long radicle issuing as a thread from the fruit, reaching the mud beneath before it loosens its hold above. Mangroves are among the few shrubs that vegetate in sea-water. The bark is astringent, and may be used as a febrifuge.

TABLE OF THE SOUTH AFRICAN GENERA.

I. **Rhizophora.**—Flowers with 4 petals.
II. **Bruguiera.**—Flowers with 10–13 petals.

I. **RHIZOPHORA,** Linn.

Calyx-tube obovate, adhering to the ovary; limb divided into 4, oblong, persistent segments. *Petals* 4, oblong, coriaceous, emarginate, conduplicate, and when young embracing the alternate stamens, the margins each with a double row of long, woolly hairs. *Stamens* 8; anthers nearly sessile, large, linear-oblong. *Ovary* 2-celled, with 2 ovules in each cell. *Style* conical, short, 2-furrowed. *Stigma* 2-dentate. *Fruit* ovate or oblong, crowned near the base with the persistent segments of the calyx, longer than the tube, at length perforated at the apex by the radicle of the germinating embryo. *Lam. Ill. t.* 396. *W. & Arn. Prod.* 310. *Endl. Gen. n.* 6098.

Trees, with quite entire leaves and axillary inflorescence. Name from ῥιζα, a *root*, and φορεω, *to bear.*

1. **R. mucronata** (Lam. Dict. 6, p. 169); leaves petiolate, oval, abruptly acuminated; racemes nodding, dichotomous.

HAB. On the seashore near Port Natal, *Drege, Krauss.* April–Aug. (Hb. Sd. D.)

II. BRUGUIERA, Lam.

Calyx-tube turbinate, adhering to the ovary; limb divided into 5–13, persistent segments. *Petals* as many as the calycine segments, oblong, bifid, coriaceous, conduplicate, each embracing two stamens, woolly along the margin. *Stamens* twice as many as petals, and inserted by pairs opposite to them; filaments unequal, half the length of the petals; anthers linear or sagittate. *Ovary* 2–4-celled; ovules 2 in each cell. *Style* nearly the length of the stamens. *Stigma* 2–4-toothed. *Fruit* contained within the tube of the calyx, crowned at the apex by its segments, at length perforated by the germinating embryo. *Lam. Ill. t.* 397. *W. & Arn. Prod.* 311. *Endl. Gen. n.* 6101.

Trees or shrubs, with quite entire leaves and axillary inflorescence. Named after the French botanist, Bruguiere.

1. B. gymnorhiza (Lam. l. c.); leaves ovate-oblong, acuminate at both ends; peduncles solitary, 1-flowered, drooping; calyx about 12-cleft; segments linear-acuminated, triquetrous toward the point; petals 2-lobed, acute, 2–4-setose, and a longish setula in the sinus. *B. Capensis, Wightii, and Rheedii. Blume Mus. Lugd. Bat.* 137. *Rhizophora gymnorhiza, L.*

HAB. Near Port Natal, *Drege, Krauss, Gueinzius.* July. (Herb. Sd., D.)

Tree 12–15 feet. Leaves in our specimens 4 inches long, 1½–2 inches broad, on longish petioles. Flowers reddish-yellow. Calyx glabrous, lobes 7–8 lines long, 1 line wide. Petals nearly as long as the calyx, hirsute at the base, subpilose at the margins; the setula in the sinus a little shorter than the petals. Germinating radicle 3–4 inches long, cylindraceous, obtuse, when dry scarcely striated.

ORDER LXIV. LYTHRARIEÆ.

(By W. SONDER.)

Flowers perfect, rarely irregular. *Calyx* free, persistent, tubular or campanulate, the tube nerved or ribbed, the limb few or many-toothed, the teeth in one or two rows, with valvate æstivation. *Petals* (rarely wanting) inserted at the summit of the calyx-tube, alternating with its teeth, or with those of the inner row, when they are doubled, imbricated in æstivation, tender in substance, deciduous. *Stamens* inserted about the middle, or toward the bottom of the calyx-tube, as many as the petals and alternate with them, rarely fewer, or twice or thrice as many, in one or more rows; filaments filiform; anthers introrse, bilocular, erect or incumbent, opening lengthwise. *Ovary* free, sessile or substipitate, 2–3–4–5–6-celled, sometimes imperfectly unilocular; ovules numerous (rarely few) on axile placentæ. *Style* simple, terminal; stigma simple or emarginate. *Capsule* membranous or woody, inclosed in the persistent base of the calyx, either opening by valves, circumscissile or irregularly bursting. *Seeds* exalbuminous; embryo orthotropous.

Herbs, shrubs or trees, few in number, but widely diffused throughout the temperate zones; much more numerous and arborescent within the tropics, especially of America. Leaves opposite or whorled, or often on the same stem alternate, simple, penninerved, entire, petioled or sessile, sometimes gland-dotted, exstipulate.

Flowers either solitary or axillary, or in tufts or cymes, or spicato-racemose, rarely panicled, purple or white, sometimes showy. None are remarkably useful, unless we include among useful products the Henna (prepared from *Lawsonia alba*), universally used by the ladies of Egypt and in the East for dyeing their nails and hair.

TABLE OF THE SOUTH AFRICAN GENERA.

I. **Ammannia.**—Calyx bracteolated at the base, campanulate, 8–14-toothed. Petals 4–7 or wanting. Capsule 4-celled, or when ripe only 1-celled.

II. **Lythrum.**—Calyx bracteolated at the base, cylindrical, 8–12-toothed. Petals 4–6. Capsule 2-celled.

III. **Nesaea.**—Calyx not bracteolated at the base, hemispherical-campanulate, 8–12-lobed. Petals 4–6. Ovary 4-celled.

I. AMMANNIA, Linn.

Calyx bracteolated at the base, more or less campanulate, with 4–7, erect, flat teeth, and 4–7, horn-formed, spreading, smaller ones rising from the sinuses. *Petals* 4–7, alternating with the erect teeth of the calyx. *Stamens* as many or twice as many as the calycine lobes. *Ovary* 2-3-4-celled. *Style* shortish or elongated. *Stigma* capitate. *Capsule* ovate-globose, membranous, either bursting transversally, the upper part falling away with the style, or opening by valves. *Seeds* numerous, attached to thick, central placentas. *Lam. Ill. t.* 77. *DC. Prod.* 3. 77. *Endl. Gen. n.* 6146.

Herbaceous plants, growing in wet soil or in water, all nearly quite glabrous. Stem usually 4-angled. Leaves opposite, quite entire. Flowers small, axillary, sessile or short-peduncled, usually pink or red. Named after John Ammann, once professor of botany at Petersburg.

ANALYSIS OF THE SOUTH AFRICAN SPECIES.

Stem erect, simple; lvs. lanceolate; fl. in axillary corymbs ... (1) **pusilla.**
Stem diffuse much-branched; lvs. oblong; fl. subsolitary ... (2) **anagalloides.**

1. A. pusilla (Sond. in Linnæa, Vol. 23, p. 40); annual; stem *erect, simple*, quadrangular, glabrous; leaves *lanceolate*, sessile with auriculate base, paler beneath; axillary *corymbs* 3–7-flowered, pedunculate, bracteolate; flowers 4-petalous, 4-androus.

HAB. Wet places near Sandriver, May, *Zeyher*. (Herb. Sd.)

2-3-uncial. Leaves usually reflexed, a little scabrous above, 3–4 lines long, 1 line wide. Corymbs of the lower axils on longer peduncles, about as long as the leaves. Calyx nearly 1 line long, shortly 8-toothed. Petals 4, minute, when dry, blueish. Capsule roundish, red-brown, terminated by a style of the same length, 1-celled, many-seeded. It comes near *A. auriculata*, DC.

2. A. anagalloides (Sond. l. c.); stem *prostrate*, at the base *much-branched;* branches quadrangular, a little scabrous; leaves *oblong*, acute, upper ones oblong-lanceolate, mucronulate, sessile with cordate base, hispidulous on both surfaces, at length glabrous; flowers *subsolitary*, shortly-pedunculate, apetalous, 4-androus.

Wet places on Rhinosterkop near Vaalriver, May, *Zeyher*, 541. (Herb. Sd., D.)

Annual, the opposite branches 3–4 inches long. Leaves green on both surfaces, the lower 6 lines long, 2–3 lines wide, the upper ones smaller. Flowers solitary or geminate. Calyx glabrous, $\frac{3}{4}$ line long, with 4 erect, bluntish, mucronulate, and 4 (from the sinuses) subulate, acuminate teeth. Capsule terminated by the filiform style, 1-celled. Nearly allied to *A. aspera*, Guill. and Perrot.

II. LYTHRUM, Linn.

Calyx bracteated at the base, cylindrical, striated, with 8–12 teeth, of which from 4–6 are broader than the rest, and erect; the others smaller and spreading. *Petals* 4–6, inserted in the orifice of the calyx, alternating with its erect teeth. *Stamens* inserted in the middle, or at the base of the tube of the calyx, twice as numerous as the petals, or occasionally fewer. *Style* filiform; stigma capitate. *Capsule* oblong, included in the calyx, 2-celled, many-seeded. *Placentas* thick, adnate to the dissepiment. *Endl. Gen. n.* 1649.

Herbs or suffruticose plants, with entire leaves and axillary, purple or purplish flowers. Name from λυθρον, *black-blood*; from the purple colour of the flowers.

ANALYSIS OF THE SPECIES.

Leaves bluntish; peduncles axillary, subsessile, 1-flowered; stamens 6 ... ... (1) **hyssopifolium**.
Leaves bluntish; peduncles axillary, nearly as long as the leaves, 1-flowered; stamens 12 ... ... (3) **rigidulum**.
Leaves acute; peduncles axillary-corymbose; stamens 4 ... (2) **sagittæfolium**.

1. L. hyssopifolium (Linn. spec. 642); leaves linear-lanceolate, bluntish; lower ones opposite and often oblong, upper ones alternate; *flowers axillary, solitary, nearly sessile*, each with a pair of very small bracts at the base; stamens 6. *Jacq. flor Austr. t.* 133. *Smith Engl. Bot. t.* 292. *L. thymifolium, hyssopif. et tenellum, E. & Z.!* 1769–1771. *Herb. Un. itin.* 495.

VAR. β. **acutifolium,** DC. Prod. 3, p. 82, leaves all or the upper ones acutish. *L. thymifolium, Hoffm. fl. germ.* 213, *but not of Linn.*

VAR. γ. **latifolium,** all the leaves oblong, at the base obtuse or attenuated. *L. tenellum, Thunb.! fl. cap.* 400. *L. hyssopif. var.* δ *? tenellum, DC. l. c.*

HAB. Wet places or in rivulets near Capetown, in Uitenhage, Stellenbosch, Caledon, &c. Nov.–Jan. (Herb. Th., Sd., D., &c.)

Stem erect or at the base prostrate, 1–1½ inch high, much-branched, glabrous. Lower or larger leaves 8–10 lines long, 2 lines wide, upper ones 4–6 lines long, 1¾ line broad. Calyx about 2 lines long. Petals 6, small, light purple. Capsule oblong. *L. tenellum, Thunb.*, differs only from the typical form by the larger (6–8 lines long, 2 lines broad) leaves, which are attenuated at the base in young specimens as gathered by *Thunberg*.

2. L. sagittæfolium (Sond. in Linnæa, Vol. 23, p. 41); stem suffruticose, pubescent-scabrid; leaves sessile, erect-adpressed, oblong-lanceolate, acute, sagittate at the base, with recurved margins, scabrous above, smooth beneath; *corymbs axillary, pedunculate*, longer than the leaves, 3–5-flowered; flowers 4-petalous, 4-androus; pedicels and calyx minutely pubescent.

HAB. Boggy places near Magalisberg, Nov., *Zeyher*, 543. (Herb. Sd., D.)

A small shrub. Stem purplish, branches virgate. Leaves 6 lines long, near the base 2 lines wide, those of the branches subimbricate. Peduncle of the corymbs 2–3 lines long; pedicels 1 line long, bracteolated. Calyx striated, 1½ line long, 4 teeth acutish, 4 minute, corniform. Petals obovate, light-purple, 2 lines long. Filaments exserted, glabrous as the style. Capsule covered by the calyx, glabrous, 1-celled.

3. L. rigidulum (Sond. l. c. p. 42); stem suffruticulose, at the base branched; branches and opposite leaves scabrid; leaves sessile, oblong-lanceolate, bluntish, with auriculate-cordate base; *peduncles axillary, solitary*, 1–2-flowered, as long as the leaves; calyx 12-striated; petals 6; stamens 12.

HAB. Aapjesriver, Oct., *Burke, Zeyher*, 542, near Ladysmith, *W. T. Gerrard!* (Herb. D., Sd.)

Stems herbaceous, 3-4 inches long, erectish, at length nearly smooth. Leaves erect, 3-4 lines long, 1½ line wide. Pedicels at the base 2-bracteolate. Calyx 2 lines long, 12-toothed. Petals purplish-red. Stamens unequal, glabrous as the style. Capsule oblong, twice shorter than the style. [The upper leaves are often alternate, *W.H.H.*]

III. NESÆA, Comm.

Calyx hemispherically-campanulate, bractless at the base, 8–12-lobed; the inner 4 or 6 lobes erect, the outer or those from the sinus spreading and horn-formed. *Petals* 4–6, alternating with the erect lobes. *Stamens* 8–12, nearly equal. *Ovary* sessile, almost globose, 4-celled. *Capsule* covered by the calyx. *Seeds* minute, wingless. *DC. Prod.* 3, 90. *Endl. Gen. n.* 6147.

Herbaceous plants. Leaves lanceolate or oblong, nearly sessile, obtuse or acute. Peduncles longish, 3-flowered or capitate-manyflowered at the apex, with 2 larger bracts and 4 minute ones at the origin of the pedicels. Name from Nesæa, in mythology, a sea-nymph.

1. N. floribunda (Sond.); stem herbaceous, erect; branches pubescent with spreading hairs; leaves opposite, sessile, oblong, or lanceolate appressed-pubescent, at length subglabrous; flowers capitate; peduncles pubescent, as long or longer than the many-flowered capitula. *Tolypeuma floridum, E. Meyer.*

HAB. On the Omblasriver, near Port Natal, *Drege.* April. (Herb. Sd., D.)

Branches erect-spreading, terete. Leaves quite entire, 1½ inch long, 4 lines broad, others 1 inch long, 5–6 lines broad. Peduncles in the axils of the upper leaves, flowering shorter than in fruit. Capitulum at the base, with 2 ovate-acuminate, pubescent, leafy bracts, a little longer than the heads. The bracts at the base of the pedicels (not at the calyx) linear-ciliate at the apex, shorter or nearly as long as the calyx. Tube of the campanulate calyx glabrous, 1 line long, with 5 connivent, short, subtriangular, and 5 erect, subulate, ciliate teeth. Style exserted, flexuous. Petals not seen.

ORDER LXV. MELASTOMACEÆ.

(By W. SONDER.)

Flowers perfect, regular. *Calyx*-tube enclosing the ovary, either quite adnate to it, or attached by its ribs to the ovary, leaving interspaces, or rarely quite free; limb 4–6-parted or subentire, with valvate æstivation. *Petals* inserted at the summit of the calyx-tube, alternate with its lobes, expanded, shortly clawed, twisted in æstivation. *Stamens* inserted with the petals, twice their number, either all perfect or those opposite the petals abortive; filaments bent inwards in æstivation; anthers terminal, 2-celled, (hidden during æstivation in the interstices between the calyx and ovary), almost always opening by terminal pores; the connective most frequently prolonged downwards below the cells and articulated with the filament. *Ovary* either free or adnate, plurilocular; ovules numerous, on axile placentæ, anatropous.

Style simple; stigma undivided. *Fruit* capsular or fleshy; seeds exalbuminous.

Trees, shrubs, suffrutices or rarely herbaceous, annuals or perennials. Leaves opposite, one sometimes smaller than the opposing, simple, entire, very generally 3-ribbed (3–5–7–9-ribbed), with transverse, connecting nerves; rarely penninerved, always without pellucid dots, exstipulate. Flowers in cymes or panicles, rarely solitary, brightly coloured. These plants are most abundant in the tropical or subtropical regions of America, a few extending in North America to the parallel of 40°; they are much less frequent in tropical Asia and Africa, and in South Africa are only known in the vicinity of Natal. Many are cultivated in European gardens as ornamental plants. None are particularly useful.

I. OSBECKIA, Linn.

Calyx-tube ovate, usually covered with stellate bristles or pubescence; limb 4–5-cleft, with appendages between the lobes springing from the outside. *Petals* 4–5. *Stamens* 8–10; filaments glabrous; anthers nearly equal and similar to each other, shortly rostrate or very rarely truncated, opening by a single, terminal pore; the connectivum with two short auricles at the base. *Ovary* covered with bristles at the apex. *Capsule* 4–5-celled. *Seeds* cochleate; hilum orbicular, at the base. *Lam. Ill. t.* 283. *Endl. Gen. n.* 6221.

Herbs, or usually subshrubs. Leaves quite entire or minutely serrulated, 3–5-nerved. Flowers terminal. Named in honour of Peter Osbeck, a Swedish clergyman and naturalist.

ANALYSIS OF THE SOUTH AFRICAN SPECIES.

Flowers panicled. Stem and leaves pubescent or villous:
Lvs. quite entire, oblong-lanceolate, stellate-pubescent above ... ... (1) **Umlaasiana.**
Lvs. serrulate, ovate-acuminate, appressed-villous above (2) **eximia.**
Flowers capitate; stem and lvs. hispid, with rigid, patent, yellow hairs ... ... (3) **phæotricha.**

1. O. Umlaasiana (Hochst.! pl. Krauss.); stem erect, quadrangular, covered with short, brown, stellate pubescence; leaves opposite, very short-petiolate, *oblong or oblong-lanceolate, quite entire*, 5-nerved, subcordate at the base, *dotted by stellate hairs above*, greyish pubescent beneath; flowers racemose-panicled; calyx greyish-puberulous, tube ovate-globose, lobes ovate-lanceolate; appendages very minute, subulate; stamens 10, unequal, the anthers of the longer ones equalling the petals. *O. canescens, E. Mey. in herb. Drege, not of Graham.*

HAB. Near Port Natal, *Drege, Plant, Gueinzius*, 137, 393. Jan.–Feb. (Herb. Sd., D., Hk.)

Several feet high, from the habit of *Lythrum Salicaria*, L. Branches erect, spreading. Petioles 1–2 lines long. Leaves a little scabrous, the lower 2 inches long, 6–8 lines broad, the upper smaller. Panicle terminal, oblong, more or less compound; the racemes equalling or somewhat longer than the leaves. Calyx-tube about 3 lines long; segments glabrous on the inner side, nearly as long as the tube. Petals rotundate-obtuse, purple, more than twice as long as the calyx-lobes. Anthers 3 lines long. Capsule ovate, glabrous, 5-celled.

2. O. eximia (Sond. in Linnæa, Vol. 23, p. 48); stem erect, quadrangular, *pubescent;* leaves opposite, petiolate, *ovate, acuminate, minutely serrulate,* subcordate at the base, *appressed-villous above,* subtomentose and paler beneath; *panicle terminal, scabrous;* calyx covered with white, fascicled, setose hairs; tube oblong, in fruit urceolate, lobes lanceolate, pectinate-ciliate, deciduous; appendages linear, palmate-ciliate; stamens 10, unequal, the anthers of the longer ones equalling the petals.

HAB. Near Port Natal, *Gueinzius*, 145, 492. *Gerrard & McKen!* (Herb. Sd., D.) Stem densely covered with fulvous, fascicled hairs. Leaves 4 inches long, 1½ inch broad, 5-nerved, subsilky on the upper, greyish on the under-surface; petiole 6–8 lines long. Panicle about 4 inches long, the primary branches opposite or ternate. Bracts acuminate, puberulous, as long as the peduncles. Pedicels 1–2 lines long. Petals rotundate, purple, 1 inch long. Filaments glabrous, connectivum biauriculate at the base; anthers 1-porose. Fruit-bearing calyx 5 lines long. Capsule roundish, silky at the apex, 5-celled.

3. O. phæotricha (Hochst.! l. c.); suffruticulose; stems erect or ascendent, *hispid with rigid, yellowish hairs;* leaves opposite, very short-petiolate or the upper ones subsessile, *ovate-oblong or sublanceolate,* 3–5-nerved, remotely serrulated, *hispid; flowers terminal, capitate-aggregate,* bracteated; calyx setose, with 4, pectinate-ciliate lobes; appendages linear, short, palmate-ciliate; stamens 8. *O. Simsii, E. Meyer. in herb. Drege.*

VAR. β. **debilis**, Sond. Stems glabrous or hispid at the apex; leaves oblong or oblong-lanceolate, glabrous or the upper ones hispidulous; flowers a little smaller. *O. debilis, Sond. l. c.* 48.

HAB. On the plains near Port Natal, *Drege, Krauss., Gueinzius*, 395 et 494. Var. β, muddy places on Magalisberg, *Zeyher*, 538. Dec.–April. (Herb. Sd. D. Hk.) Stem from 3 inches to 1–1½ foot, a few-branched or simple, purplish. Leaves 8–12 lines long, 5–6 lines, in var. β. 3–4 lines broad, acute. Flowers 3–8, united into a head, very short-petiolated. Calyx ovate, 2 lines long, lobes ovate or oblong. Petals 4, rotundate-obtuse, nearly ½ inch long, purple. Ripe capsule as large as a small pea, setose at the apex, 4-celled.

(GENUS ALLIED TO MELASTOMACEÆ.)

OLINIEÆ, W. Arnott.

(By W. SONDER.)

OLINIA, Thunb.

Calyx tubulose, adhering to the ovary, with 5, rarely 4, minute teeth. *Petals* 5, very rarely 4, inserted in the throat of the calyx. *Scales* 5, minute, obovate, alternate with the petals. *Stamens* 5; filaments very short, adnate to the calyx below the scales; anthers subglobose, 2-celled, introrse; connectivum thick. *Ovary* inferior, 4–5-celled; cells 3-ovulate; ovules pendulous, uniseriate, affixed to a central placenta. *Style* subulate; stigma obtuse. *Berry* drupaceous, elliptic or subglobose, truncate, 3–4-celled; putamens elongate, incurved, mostly (by abortion) 1-seeded. *Seed* oval. *Embryo* without albumen, spirally rolled; cotyledons scarcely distinct. *Klotzsch, in Linn. Klotzsch & Otto, Icon. pl. rar. hort. Berol.* 1, *p.* 6. *Endl. Gen. n.* 6272.

A glabrous shrub or tree, with 4-angled, patent branches; opposite, petiolate, coriaceous, green and shining above, penninerved, quite entire, not punctate leaves; terminal and axillary, densely cymose, subtrichotomous panicles of small, white flowers. Two opposite, obovate, mucronate, ciliate, white, deciduous bracts at the base of a flower. Fruit scarlet-red.

1. O. cymosa (Thunb.! fl. cap. 194); *Sideroxylon cymosum, Linn. fil. Suppl.* 152.

VAR. α. **latifolia**; leaves obovate or broad-ovate, obtuse, subemarginate, and apiculate, sometimes acute, cuneate at the base. *O. cymosa, Thunb. l. c. Klotzsch. l. c.* 60, *t.* 24.

VAR. β, **intermedia**; leaves elliptic, acute or subacuminate at both ends, rarely obtuse and apiculate at the apex. *O. Capensis, Klotzsch. l. c.* 6, *t.* 3.

VAR. δ. **acuminata**; leaves oblong, acuminate at both ends, obtuse and mucronulate at the apex; flowers generally a little smaller. *O. acuminata, K. l. c.* 53, *t.* 21.

HAB. About Tablemountain and in the districts of Stellenbosch, Caledon, George, Uitenhage, and Albany, *Zeyher,* 2464, 2465, *Ecklon, Drege, Thunb.* Magalisberg. *Zey.*, 308. June–Dec. (Herb. Th., Hm., D. Sd.)

Leaves undulated or flat, a little reflexed at the margins, 1–2½ inches long, paler beneath, with short petioles. Cymes terminal, and in the upper axils, shorter than the leaves. Bracts white, oblong-linear, ciliolate at the margins, 2½ lines long. Calyx about 2–3 lines long, greenish-white, with nearly obsolete teeth. Petals white, spathulate, acute, twice shorter than the calyx-tube. Scales incurved, pilose. Fruit the size of a small hazle-nut.

ORDER LXVI. MYRTACEÆ.

(By W. SONDER.)

Flowers perfect, regular, *Calyx*-tube adnate to the ovary, either wholly or in part; limb 4–5 or many-cleft or parted, imbricated (or valvate) in æstivation. *Petals* rarely wanting, inserted on the fleshy margin of the calyx-tube, alternate with its lobes, imbricate or twisted in æstivation. *Stamens* indefinite or rarely definite, inserted with the petals; filaments free or polyadelphous; anthers introrse, slitting. *Ovary* inferior or half-inferior, sometimes unilocular, with one or few ovules; most usually 2 or many-celled, with numerous ovules. *Style* simple; stigma undivided. *Fruit* either a succulent berry or a dry capsule; sometimes dry and indehiscent. *Seeds* without albumen.

Trees or shrubs, very rarely herbs. Leaves usually opposite, rarely alternate or whorled, entire, penninerved, with an intra-marginal vein, almost always pellucid-dotted, exstipulate. Flowers either axillary and solitary, or in axillary or terminal cymes, corymbs or panicles, or sometimes capitate or spiked. A very large Order, extremely abundant in South America and Australia; less common in Asia, very thinly scattered over Africa, chiefly tropical, with a few outlying species in the temperate zones. Many valuable spices, as cloves, allspice, &c.; and many fruits, as the guava, pomegranate, ugni, &c., are products of these plants. The bark in all is astringent, and the foliage of most yields an aromatic essential oil.

TABLE OF THE SOUTH AFRICAN GENERA.

Tribe 1. LEPTOSPERMEÆ, DC. *Fruit* dry, many-celled, dehiscent. *Seeds* exarillate.

I. **Metrosideros.**—Stamens free. Capsule 2–3-celled.

Tribe 2. MYRTEÆ, DC. *Fruit* fleshy, baccate, many-seeded. *Leaves* full of pellucid dots.

II. **Syzygium.**—Limb of calyx almost entire or repandly-lobed. Petals 4–5, concrete, falling off in the shape of a calyptra.

III. **Eugenia.**—Limb of calyx deeply 4- rarely 5-parted. Petals 4, very rarely 5, not concrete.

Tribe 3. BARRINGTONIEÆ, DC. *Fruit* baccate or dry, valveless, many-celled. *Leaves* without pellucid dots.

IV. **Barringtonia.**

I. METROSIDEROS, R. Brown.

Calyx-tube adhering to the ovary, not angular ; limb 5-cleft. *Stamens* 20–30, free, very long, exserted. *Style* filiform ; stigma simple. *Capsule* 2–3-celled, cells many-seeded. *Seeds* wingless. *DC. Prod.* 3. 224. *Endl. Gen. n.* 6303.

Trees or shrubs. Leaves opposite or alternate. Flowers not adnate to the branches, on axillary, umbellate peduncles. Name from μητρα, *the heart of a tree*, and σιδηρος, *iron* ; the wood of these trees is very hard.

1. M. angustifolia (Smith. Linn. transact. 3, 268); branches tetragonal ; leaves opposite, linear-lanceolate, naked ; peduncles axillary, umbellate ; bracts lanceolate, glabrous. *Myrtus angustifolia, L. Thunb.! fl. cap.* 408. *E. & Z.!* 1773. *Houtt. Pflanz. Syst.* 3. *t.* 25, *f.* 2.

Hab. Sides of rivers in various parts of the colony, in the districts of Stellenbosch, Worcester, Caledon, &c. Jan.–March. (Herb. Sd., D., &c.)

Tree 20 feet. Leaves 2–3 inches long, 3–4 lines broad. Umbels densely flowered, subcapitate, much shorter than the leaves. Flowers yellowish. Fruit 1 line long.

II. SYZYGIUM, Gaertn.

Tube of the calyx obovate ; limb nearly entire or repandly lobed. *Petals* 4–5, roundish, joined into a calyptra, and falling off either in that state from the calyx, or immediately after expansion. *Stamens* numerous, distinct. *Ovary* 2-celled, with few ovules in each cell. *Style* 1. *Stigma* simple. *Berry* 1-celled, 1- or few-seeded. *Seed* globose. *Cotyledons* large, fleshy, nearly hemispherical ; radicle small, inserted between the cotyledons below their middle, and concealed by them. *DC. l. c.* 259. *Endl. Gen. n.* 6320.

Trees or shrubs. Leaves opposite, quite entire, glabrous. Peduncles axillary and terminal, cymose or corymbose. Name from συζυγος, *coupled* ; in allusion to the manner in which the branches and leaves are united by pairs.

1. S. cordatum (Hochst. ! pl. Krauss.); arborescent, glabrous; leaves subsessile, suborbiculate or elliptic-cordate, quite entire, coriaceous ; much veined, paler beneath ; cymes terminal, many-flowered, the branches quadrangular ; calyx very short 4-tooth ; style longer than the stamens. *Jambosa cyminifera, E. Meyer. in herb. Drege.*

Hab. Woods on the rivers near the seashore from Omtendo to Port Natal, *Drege, Gueinzius, Krauss, Plant.* (71) Oct.–Feb. (Herb. Sd., D.)

A tree 30–40 feet. Branches spreading, tetragonous, as well as the leaves opposite Petioles 1 line long. Leaves about 2½ inches long, 2–2¼ inches wide, roundish or with a short, obtuse apex ; with prominent veins on the under-surface. Cyme 3-chotomous, leafy at the base, 3–4 inches high and wide. Calyx 2 lines long, turbinate. Stamens exserted. Fruit the size of a small cherry, acidulous.

III. EUGENIA, Linn.

Calyx-tube nearly globose, limb divided down to the ovary into 4, rarely 5, segments. *Petals* 4, rarely 5. *Stamens* numerous, distinct. *Ovary* 2-celled, cells many ovulate. *Berry* nearly globose, crowned by the calyx, when ripe 1-celled, but rarely 2-celled. *Seeds* 1–2, large, roundish. *Cotyledons* very thick, and conferruminated. *Radicle* very short, hardly distinguishable. *DC. l. c.* 262. *Myrtus spec. Swartz. Endl. gen. n.* 6323.

Trees or shrubs. Leaves opposite, quite entire, pellucid-dotted. Peduncles axillary or terminal, solitary or several together, simple and 1-flowered, or racemose-cymose. Named in honour of Prince Eugene of Savoy, who was a protector of botany.

ANALYSIS OF THE SOUTH AFRICAN SPECIES.

Peduncles cymose or racemose, shorter than the leaves:
- Leaves elliptic, acute, membranaceous ... ... (1) **Natalitia.**
- Leaves elliptic-lanceolate, coriaceous ... ... (2) **Zeyheri.**

Peduncles at the apex 3-flowered, as long as the leaves ... (3) **Albanensis.**
Peduncles 1-flowered, about as long as the (small) leaves ... (4) **Capensis.**
Peduncles 1-flowered, 4 times shorter than the (large) leaves ... (5) **Gueinzii.**

1. E. Natalitia (Sond.); glabrous, much-branched; branches greyish-white, young ones subangulate; leaves shortly-petiolate, opposite, *membranaceous, much veined and dotted, elliptic,* acute at both ends, with recurved margins; peduncles axillary, cymose or racemose, much shorter than the leaves; calyx-tube glabrous.

HAB. Port Natal, *Gueinzius*, 60, 568. *Gerrard & McKen*, 707. (Hb. Hk. Sd. D.)

Erect shrub, with opposite, erect, spreading branches. Petiole 1 line long. Leaves 1½ inch long, nearly 1 inch wide, others 1½ inch long, 6–9 lines broad, with short, obtuse apex, shining above, pale beneath and reticulated, with an intra-marginal vein. Cymes or racemes about ½ inch long, often intermixed with several 1-flowered, 3–4 lines long, bibracteate peduncles. Calyx-lobes rotundate-obtuse, ¾ line long. Petals 4, twice longer than the calyx, dotted, concave, rounded, a little acuminate at the apex. Fruit unknown.

2. E. Zeyheri (Harv. Gen. Sth. Afr. 416); *glabrous*, much-branched; branches greyish-white, young ones subangulate; leaves shortly petiolate, opposite, *coriaceous, not veined, elliptic-lanceolate,* with recurved margins; peduncles axillary, cymose or racemose, much shorter than the leaves; calyx-tube glabrous. *Zeyher*, 2467.

HAB. Woods on Vanstadesrivier, Krakakamma, and near Howisonsport, *E. & Z.!* *Drege*, 5366, a. Dec.–May. (Herb. D., Sd.)

Shrub or tree, 10–15 feet high. Petiole 1 line long. Leaves thick and not conspicuously dotted, attenuated at each end, but obtuse at the apex, 1–1½ inch long, 4–7 lines broad, shining above, paler beneath. Peduncles cymosely or racemosely 3–7-flowered, intermixed sometimes with a few 1-flowered peduncles; pedicels with 2 minute bracts near the calyx. Petals obovate. Berry 1–2- very rarely 3-seeded, as large as a small cherry.

3. E. Albanensis (Sond.); branches at top *appressed-hairy;* leaves on very short petioles, opposite or ternate, coriaceous, not veined, *ovate or elliptic-oblong,* with recurved margins; peduncles solitary, axillary, at the apex *bibracteate and 3-flowered;* pedicels *adpressed-hairy* as well as the 4- or 5-parted calyx.

HAB. Hills on the Great Vishriver, *Zeyher, Memecyl.* 1; between Kovi and Kapriver, *Drege*, 5366 b; near Somerset, *Bowker*. Nov.–Dec. (Herb. Hk., Sd., D.)

Dwarf shrub, ½–1 foot, a few-branched; branches purplish. Petiole ½ line long. Leaves opposite, rarely alternate and ternate, 8–9 lines long, 4–5 lines broad, acutish, upper ones ovate-oblong, when dry olive-green or blueish-green and shining above, pale fulvous-beneath, only the young ones conspicuously dotted. Peduncle compressed, glabrous, a little thicker upwards; pedicels 1–2 lines long, the lateral ones divaricate, equalling the leafy, ovate bracts. Calyx at the base with 2 minute bracts; lobes 4 or 5, obtuse. Petals 4 or 5, twice longer, obovate. Stamens inserted in a large disc. Fruit not seen.

4. E. Capensis (Harv. l. c.); *quite glabrous*, much-branched; leaves (small) on very short petioles, opposite, coriaceous, a little veined, *elliptical or suborbiculate*, with recurved margins; peduncles axillary, *1-flowered, solitary, geminate or ternate, about as long as the leaves*, glabrous as well as the calyx; berry globose, usually 1-seeded. *Memecylon Capense, E. & Z.!* 1772.

VAR. *β*, **major**; leaves and flowers larger.

HAB. Sandy downs near the Bosjesmansriver, and between Zwartkops and Koegariver, *E. & Z.! Zey.!* 2466. Var. *β*. sandy hills between Omtendo and Omsamculo, *Drege*. Feb.-April. (Herb. D., Hk., Sd.)

Shrub 4–6 feet high, with opposite or aggregated, greyish branches. Leaves pale green, 6–8 lines long, 5–6 lines broad, in var. *β*. 10–12 lines long, 8 lines broad. Petiole ½ line long, sulcate. Peduncles about ½ inch. Petals 2 lines, in var. *β*. 3 lines long, obovate.

5. E. Gueinzii (Sond.); quite glabrous; branches terete, when young subcompressed; leaves (large) on very short petioles, opposite, coriaceous, slightly veined and dotted beneath, elliptic or broad-ovate, subcordate at the base, with recurved margins; peduncles axillary, geminate, 1-flowered, 4 *times shorter than the leaves;* calyx-lobes obtuse, twice shorter than the obovate petals.

HAB. Port Natal, *Gueinzius*. (Herb. Hk., Sd.)

Erect shrub. Branches purplish. Petiole 1 line long. Leaves 1½–2 inches long, 1¼–1½ inch broad, shining above, much paler beneath. Peduncles 4–5 lines long. Calyx-lobes much dotted. Petals unguiculate, 3 lines long. Fruit unknown.

IV. BARRINGTONIA, Forst.

Calyx-tube ovate; limb 2–3–4-parted; lobes ovate, obtuse, concave, persistent. *Petals* 4, coriaceous, attached to the ring at the base of the stamens. *Stamens* numerous, in several rows; filaments filiform, long, distinct, combined at the base into a short ring; all bearing anthers. *Ovary* 2–4-celled, surmounted by an urceolus sheathing the base of the style; ovules 2–6 in each cell. *Style* filiform; stigma simple. *Fruit* fleshy, more or less 4-angled, crowned by the limb of the calyx, 1-celled. *Seed* solitary. *Embryo* large, fleshy, not separable into cotyledons and radicle, formed of two concentric, homogeneous, combined layers. *W. & Arnott. prodr.* 333. *Stravadium, Juss.*

Trees. Leaves crowded about the ends of the branches, opposite or verticillate, obovate, quite entire or crenated or serrated, without pellucid dots. Flowers racemose. Name in honour of Dr. Barrington.

1. B. racemosa (Roxb. fl. Ind. 2. 634); leaves cuneate-oblong, shortly acuminate, serrulated or crenulated; flowers forming a long, pendulous raceme; pedicels short; calyx 3–4-cleft; fruit ovate; endocarp fibrous. *W. & Arn. l. c. Blume in DC. Prod.* 3. 288. *B Caffra, E. Meyer. in herb. Drege.*

HAB. Near Port Natal, *Drege, Gueinzius*, 459, 542, 575. (Herb. Sd.)

Leaves ½–1 foot long, 3–5 inches broad. Raceme 1 foot or longer. Pedicels 3 lines, calyx 6 lines long, with ovate lobes. Petals ovate, 10–12 lines long. Stamens longer than the petals, but shorter than the style. Fruit 2 inches long, ½ inch wide. Seed the size of a walnut. The South African specimens are not different from those collected by *Zollinger in Java*, except that in the latter the pedicels are as long as the calyx or a little longer.

ORDER LXVII. UMBELLIFERÆ, Juss.

(By W. SONDER).

Flowers usually perfect, in umbels, small. *Calyx* adhering to the ovary; its margin 5-toothed or obsolete. *Petals* 5, inserted on the outside of a fleshy, epigynous disc *(stylopodium)*, mostly with inflexed points, the inflexed portion connate with the middle vein of the lamina; æstivation slightly imbricate or valvate. *Stamens* 5, alternate with the petals; anthers 2-celled. *Ovary* of 2-carpels, 2-celled; ovules solitary, pendulous; styles 2, distinct. *Fruit* dry, consisting of two easily separable carpels *(mericarps)*, which cohere by their inner face *(commissure)* to a common, filiform axis *(carpophore)*, but at maturity separate from it and are for a time pendulous from its summit: each mericarp is indehiscent, marked with 5 longitudinal *(primary)* ribs, one opposite each petal and each stamen, and often also with 4 *(secondary)* intermediate ribs, the ribs being separated by furrows. In the substance of the pericarp are linear, longitudinal oil-vessels *(vittæ)*, most commonly opposite the furrows, *(valleculæ)* sometimes opposite the ribs, and sometimes wanting altogether. *Albumen* copious, horny, with a minute embryo in its base.

A very large and most natural Order of herbaceous, or rarely shrubby plants, common throughout the temperate zones, rare within the tropics. Leaves alternate, very rarely opposite, usually with sheathing petioles, pinnately or ternately divided, often cut into capillary segments, rarely entire. Flowers in umbels or rarely capitate, with or without involucre. Many garden vegetables, as the carrot, parsnip, parsly, celery, &c., and several poisonous plants, of which the Hemlock *(conium)* is the most famous, belong to this Order. The drugs asafœtida, ammoniacum, galbanum, &c.; and the carminative seeds caraway, anise, dill, cummin, coriander, &c., are also products of umbelliferous plants. The generic characters of many can only be well examined when the fruit is ripe or nearly so; this, together with the uniformity of floral structure throughout the order, and the minute differences that require to be noted, render the study of these plants very difficult to the student. The peculiar terms used in the following descriptions are given in italics in the above character, immediately after the explanation of each term.

ANALYTICAL TABLE OF THE SOUTH AFRICAN TRIBES AND GENERA.

Sub-Order I. ORTHOSPERMEÆ. *Albumen* (as seen in a cross section of the fruit) flat or nearly so on its inner face (next the commissure).

Umbels simple or imperfect:

1. **HYDROCOTYLEÆ**: Fruit laterally compressed ... ... **1. Hydrocotyle.**

2. **SANICULEÆ**: Fruit ovato-globose, cross section circular:
   Fruit covered with hooked bristles ... ... **2. Sanicula.**
   Fruit tuberculated ... ... **3. Alepidea.**

Umbels compound or perfect:

3. **AMMINEÆ**: Fruit laterally compressed or didymous: (IV.–XIII.)
   Mericarps equal; *leaves* much cut or divided:
      Carpophore distinct, entire (not bipartite):
         Fruit roundish, didymous. Petals roundish **4. Apium.**
         Fruit ovate or oblong. Petals ovate ... ... **6. Helosciadium.**
      Carpophore bipartite:
         Furrows of the fruit uni-vittate:
            Margin of the calyx obsolete:
               Petals roundish, apiculate, entire ... **5. Petroselinum.**
               Petals obcordate, apiculate ... ... **8. Carum.**

Margin of the calyx 5-toothed :
Petals deeply emarginate, white. Ft. ovate or oblong ... ... 7. **Ptychotis.**
Petals obovate, entire, involute, yellowish. Fruit roundish ... ... 11. **Rhyticarpus.**
Furrows of the fruit with many vittæ :
Fruit ovate. Neither involucre, nor involucels ... ... 9. **Pimpinella.**
Fruit subdidymous. Involucre and involucels present ... ... 10. **Sium.**
Mericarps equal ; leaves undivided, quite entire ... ... 12. **Bupleurum.**
Mericarps unequal ; leaves entire, 3-lobed or 3-parted 13. **Heteromorpha.**

4. **SESELINEÆ** : cross section of the fruit circular or nearly so, or the mericarps slightly compressed at back (commissure broad :—(XIV.-XXIII.)
Vittæ under the ribs of the fruit ; none inthe furrows 14. **Lichtensteinia.**
Vittæ in the furrows of the fruit :
Mericarps unequal ... ... 15. **Anesorhiza.**
Mericarps equal :
Mericarps hispid or scaly ... ... 19. **Deverra.**
Mericarps glabrous :
Calyx margin enlarged after flowering. Carpophore indistinct ... ... 16. **Œnanthe.**
Calyx margin unchanged. Carpophore distinct and free :
Ribs (of fruit) obtuse, filiform, the lateral wider. Fl. white ... ... 21. **Polemannia.**
Ribs prominent, bluntly-keeled, the lateral wider. Fl. yellow ... ... 18. **Foeniculum.**
Ribs prominent, filiform, equal ... 20. **Sesili.**
Ribs prominent, sharp, winglike, equal ... ... 17. **Glia.**
Ribs thick, rounded, *corky*, winglike 22. **Stenosemis.**
Ribs membranaceous, winglike ... 23. **Cnidium.**

5. **ANGELICEÆ** : Fruit much compressed dorsally, having a *double wing* on each side. Raphe central ... ... 24. **Levisticum.**

6. **PEUCEDANEÆ** : Fruit much compressed dorsally, with a *single* acute or thickened wing on each side :—(XXV.—XXX.)
Fruit with 5 conspicuous dorsal ribs, and vittæ in the furrows.
Dorsal ribs slender, filiform :
5 ribs equidistant and equally filiform :
Margin of fruit broad. Petals emarginate 25. **Peucedanum.**
Margin of fruit narrow. Petals entire 26. **Bubon.**
5 ribs equidistant ; the 3 intermediate acutely keeled ... ... 27. **Anethum.**
3 ribs equidistant ; 2 lateral distant, marginal ... ... 28. **Pastinaca.**
Dorsal ribs thick, keeled, tubercled or flexuous ... 29. **Capnophyllum.**
Fruit without dorsal ribs, hairy ; margins thickened. No vittæ ... ... 30. **Pappea.**

7. **DAUCINEÆ** : Fruit somewhat compressed dorsally. Mericarps with 5 primary, bristly, and 4 secondary, prickly ridges ... ... 31. **Daucus.**

Sub-Order II. CAMPYLOSPERMEÆ. Albumen with a longitudinal furrow along its inner face (a cross section of the fruit showing it concave on the side next the commissure).

8. **CAUCALINEÆ** : Fruit laterally compressed or subterete, lateral primary ridges on the commissure, the dorsal (primary and secondary) bristly or setose ... ... 32. **Torilis.**

9. **SMYRNEÆ**: Fruit turgid, often laterally compressed. Ribs sometimes obliterated :—(XXXIII.—XXXV.)
 Diœcious. Fruit adnate to a large, spinous involucre ... **33. Arctopus.**
 Fruit not involucrate.
  Fruit sub-compressed dorsally; dorsal ribs wingless, lateral very small in the commisure. Furrows multi-vittate ... ... **34. Hermas.**
  Fruit ovate, laterally compressed; ribs 5, equal, undulato-crenate. Furrows without vittæ ... ... **35. Conium.**

Sub-Order I. ORTHOSPERMEÆ. DC. Prodr. 4, 58. Albumen flat on the inner side; neither involute nor convolute. (Gen. I.—XXXI.)

I. HYDROCOTYLE, Tourn.

Calyx-tube subcompressed, limb with an obsolete margin. *Petals* ovate, entire, acute, with a straight apex. *Fruit* flatly compressed from the sides, bi-scutate. *Mericarps* without vittae; the 5 ribs nearly filiform, the carinal and lateral ones usually obsolete, and the 2 intermediate ones joined. *Seed* carinately compressed. *Endl. gen. n.* 4355.

Usually slender, bog herbs, rarely subshrubs. Flowers sessile or pedicellate, white. Umbel usually 3-flowered but monocarpous, 2 of the flowers being sterile. Involucre in anthesis 4-leaved, but when fruit-bearing 2-leaved. Name from ὑδωρ, *water*, and κοτυλη, *a cavity*, in reference to the plants growing in moist situations, and to the leaves often being hollowed like a bowl.

ANALYSIS OF THE SOUTH AFRICAN SPECIES.

Leaves peltate:
 peduncles bearing interrupted whorls of flowers ... (1) **verticillata.**
 peduncles bearing a terminal umbel ... ... (2) **Bonariensis.**
Leaves cordate or orbicularly reniform (not peltate):
 umbels with 3–4 fertile flowers ... ... (3) **Asiatica.**
 umbels with 1 fertile flower (monocarpous)
  Leaves orbicularly reniform:
   stem glabrous; leaves crenately toothed, woolly beneath .. .. (4) **eriantha.**
   stem glabrous; leaves sharp-toothed, glabrous ... . (5) **calliodus.**
   stem and leaves villous ... ... (6) **flexuosa.**
  Leaves 3–7 toothed or angled, cordate, subreniform, as well as the stem villous ... ... (7) **hederaefolia.**
Leaves lanceolate, ovate, or linear (not cordate).
 Leaves not toothed or lobed:
  lvs. ovate or elliptic, acute, as well as the stem villous ... ... (8) **villosa.**
  lvs. oval, oblong or lanceolate, acute or acuminate at both ends, 3-nerved, with the stem glabrous or subtomentose ... ... (15) **Centella.**
  lvs. linear-lanceolate, attenuate at the base, 5 nerved, with uncinate apex; quite glabrous ... (16) **debilis.**
  lvs. linear-filiform ... ... (17) **virgata.**
 Leaves toothed or lobed:
  Involucre of fruit 4–6 leaved
   stem short, with the leaves tomentose; leaves obovate-cuneiform, bluntish, 7-toothed ... (9) **Solandra.**
   stem elongate, with the leaves villous-tomentose; leaves oblong-cuneate, 9–11-toothed (10) **hermanniaefolia.**
   stem short with the leaves villous-tomentose; leaves narrow-cuneiform, 3-dentate ... ... (14) **tridentata.**

Involucre of fruit 2-leaved:
petals villous; stem puberous; leaves ovate-acute, 5-toothed; peduncles solitary, very short ... ... (11) **Dregeana.**
petals hairy on back; stem subtomentose; leaves ovate, 3-5-toothed, or subentire; umbel sessile, many-flowered ... ... (13) **montana.**
petals glabrous; stem glabrous; leaves cuneate-ovate, 3-lobed ... ... (12) **triloba.**

1. H. verticillata (Thunb.! diss. de Hydroc. 1798, t. 3,) leaves peltate, orbicular, doubly crenated, 11-nerved, glabrous as well as the petioles; flowers disposed in many, subdistant whorls, ultimate ones umbellate; fruit rather attenuated at the base, coloured. *Flor. cap.* 252. *E. & Z.!* 2154. *H. interrupta, Muehl. Catal. p.* 10, (1813). *H. vulgaris, α, communis, Cham. & Schlecht Linnæa, vol.* 1, 356. *H. vulgaris, Thunb. fl. cap. and E. & Z.* 2153. *Zeyh.* 2659–2660.

HAB. Marshy, boggy places, and on the margins of rivulets near Capetown, Zwartkopsriver, Howison's Poort, Olifantsriver, etc. Oct.–April. (Herb. Thunb. Hk. D. Sd.)

Stems rooting at the nodes. Petioles 2–6 inches long. Leaves orbicular, 1–2 uncial. Peduncles shorter than the petioles. Umbels in the young plant capitate, about 5-flowered, as in *H. vulgaris.* Fruit about 1 line broad, ¾ line long, usually reddish-brown; mericarps with 1 or 2 ribs on each side.

2. H. Bonariensis (Lam. Dict. 3. 147); leaves peltate, orbicular, doubly-crenated, 15–20-nerved, glabrous as well as the petioles; *scapes umbellate at the apex and umbellately-branched;* flowers disposed in interrupted whorls along the branchlets. *Cham. & Schlecht. l. c. H. Caffra, Meisn.! in pl. Krauss. H. multiflora, Ruiz. & Pav. fl. Peruv.* 3. *p.* 24. *t.* 246. *f. a.*

HAB. Near Port Natal, *Drege, Krauss.* 127. *Gueinz.* 534. Delagoa Bay, *Forbes.* Dec.–April. (Herb. Hk. D. Sd.)

Stem rooting. Petioles 2 inches, or 1 foot or longer. Leaves perfectly round or with an incision at the base; young ones dentate-crenated; when old often 2–2½ inches in diameter. Peduncles as long or shorter than the petioles. Umbel compound of 4–12 rays ½–2 inches long; the rays simple or dichotomous; flowers interruptedly-verticillate; pedicels very short. Fruit 1-line broad, coloured, scarcely emarginate at the base and apex, mericarps with 1–2 ribs on each side.

3. H. Asiatica (Linn. Spec. 234,) subvillous or glabrescent; leaves orbicularly reniform, or reniform-cordate, crenated, 7–9 nerved; petioles and peduncles in fascicles, pubescent; umbels capitate, on short peduncles, 3–4 flowered, *all the flowers fertile;* fruit orbicular, furnished with 4 ribs on both sides, *much longer than the involucre. Rich. Hydr. n.* 15, *f.* 11. *Cham. & Schlect, l. c.* 365. *H. pallida, DC. l. c.* 63. *H. brevipes, E. Mey. in herb. Drege. H. Asiatica et ficariodes, E. & Z.!* 2155, 2156. *Herb. Un. Itin.* 405. *Zeyh.* 2661. *H. ficarioides Meisn. in herb. Krauss.*

VAR. β, **repanda**, leaves reniform-cordate, repandly-toothed. *H. repanda,* Pers. *Ench. t.* 102. *Rich. Hydr. n.* 13, *f.* 14. *H. ficarioides, Rich. l. c. n.* 12, *f.* 12.

HAB. In humid places on the sides of Table Mountain, in the Cape flats, on the rivers from Zwartkopriver to Port Natal. Var. β, in the same localities and near Bethelsdorf. Dec.–April. (Herb. Hk. D. Sd.)

A very variable plant. Stem much-creeping and stoloniferous. Petioles 1 inch or shorter, in other specimens from 2–6 inches long. Leaves thin or coriaceous, larger and rather woolly at the base, with crenated, dentate or nearly quite entire margins; in small specimens 3–4 lines, in the largest 2–2¼ inches broad at the base, with rotundate or subtruncate sinus; usually a little broader than long, but sometimes and particularly in var. β. a little longer than broad. Peduncles 2 lines, ½ or 1 inch long, terminated by 3 subsessile or short-pedicelled, fertile flowers. Fruit 1 line long, when ripe reticulated, involucrated by 2 shorter ovate leaves.

4. H. eriantha (Rich. Hydr. n. 18. f. b.); *suffruticose*, stem sarmentaceous, elongated, much-branched, branches ascendent; leaves nearly orbicularly reniform, *crenately-toothed*, woolly beneath; petioles dilated and woolly at the base; peduncles aggregated, as long or shorter than the leaves, villous, with a 3-flowered, *monocarpous* umbel; fruit obovate-cordate, *a little shorter or as long as the ovate, acuminate involucre. Cham. et Schlecht. l. c. Sieb. herb. fl. cap. n.* 247. *Herb. Un. Itin.* 404. *E. & Z.!* 2157. *H. cuspidata, Willd. H. reniformis, Spreng. H. asiatica, Thunb.! fl. cap.* 252. *Zey.* 2661. *b.*

VAR. γ. **glabrata**; nearly glabrous, peduncles villous. *H. pallida, E. & Z.!* 2158. *H. eriantha, E. Mey. in Herb. Drege.*

HAB. Rocky places on Tablemountain; Klynriviersberge, Caledon; Puspasvalley; and on mountains near Voormansbosch, Port Natal; var. γ. in Hottentottsholland. Sep.–Feb. (Herb. Thunb. Hk. D. Sd.)

Branches woody, terete, purplish. Leaves fascicled or villous, on short or long petioles, somewhat broader than long, 1–2 uncial, paler beneath, much reticulated, usually coriaceous, crenated and mucronulate. Peduncles ½–1 inch long. Involucre hairy, 4 or 5-leaved, 2 or 3 falling off with the sterile flowers. Fruit 2 lines long and broad; the mericarps with 3 or 4 ribs on each side. It differs from *H. Asiatica* by the suffruticose stem, mucronately-toothed leaves, and twice larger fruit.

5. H. calliodus (Cham. & Schlecht. l. c. 371); suffruticose, *glabrous;* stems elongated, dichotomous; leaves orbicularly-reniform, *coarsely, sharply, and unequally toothed;* peduncles aggregated, glabrous or hardly pubescent, shorter than the leaves, with a 3-flowered monocarpous umbel; fruit obovate, cordate, *twice longer than the ovate-involucre. DC. l. c.* 64. *E. & Z.!* 2159. *Zey.* 2663.

HAB. Among shrubs, mountains Tradouw, *Mundt & Maire;* Mount Baviensberg near Gnadenthal and Voormansbosch, *E. & Z., Pappe;* Dutoitskloof and Paarlberg, *Drege.* Oct.–Feb. (Herb. Beral. Hk. D. Sd.)

Nearly allied to *H. eriantha*, but easily distinguished by its smoothness, by usually smaller, thinner, scarcely-reticulated leaves, and the shorter involucre. Stems and primary branches woody, 2–3 feet, not rooting. Leaves 5–7-nerved, rarely 1 inch long and broad, with cordate or subtruncate base; the margin with sharp, often incurved teeth. Petioles dilated at the base, 2–3 inches long. Peduncles ½–1 inch, filiform, puberulous or glabrous. Fruit nearly the same as in the preceding, but a little smaller.

6. H. flexuosa (E. & Z.! 2160); suffruticose, *quite villous;* stems elongated, branched; leaves cordate or orbicularly-reniform, *coarsely many-toothed;* peduncles aggregated, shorter than the leaves, with a 3-flowered, monocarpous umbel; fruit glabrous, orbicularly-cordate, *twice shorter than the villous, ovate, acuminate involucre. Zey.* 2662.

HAB. Sandy-stony places in Hottentottsholland, near Palmietriver, *E. & Z.!* Feb. (Herb. Hk. Sd.)

Easily known by the yellowish or greyish soft indument. Leaves ¾–1 inch broad, with 10–18 sharp, erect, 1–2 lines long, ovate teeth. Peduncles ½–1 inch. Involucre, when young, 4-leaved, subulate; fruitbearing, consisting of 2 striated, acuminated leaves, 3 lines long. Mericarp on each side with 3 or 4 ribs, reticulated or undulated.

7. **H. hederæfolia** (Burch. Trav. 1. 46); perennial, decumbent, villous; leaves *cordately-reniform, coarsely 3–7-toothed;* teeth or lobules subequal, acute, entire; peduncles geminate or ternate, shorter than the leaves, with a 3-flowered, monocarpous umbel; fruit glabrous, suborbicularly-cordate, *as long or a little shorter than the villous, ovate-acuminate* involucre. *DC. l. c.* 70. *Herb. Un. Itin.* 407. *E. & Z.* 2161. *H. macrodus, Spreng. Cham. & Schlecht. l. c. H. moschata, Spreng. non Thunb.*

HAB. On the Table Mountain, *Burchell, E. & Z., Mundt.* Nov.–Jan. (Herb. Reg. Berol. Sd.)

Much smaller than *H. flexuosa*, herbaceous, not woody at the base. Petioles filiform, 1–2 inches long. Leaves 6–8 lines broad, 4–6 lines long, usually 5-toothed or lobed, the lobes 3-angulate, the middle sometimes larger; greenish, with impressed nerves above; on both sides appressed villous by short hairs, at length subglabrous. Peduncles 4–6 lines long. Fruit as in *H. flexuosa*, from which it is chiefly distinguished by the scarcely-villous, 5-angled-leaves.

8. **H. villosa** (Linn. fil. Suppl. 175); suffruticose, whole plant more or less villous; stem decumbent, branched; branches herbaceous; leaves *ovate or elliptic, acute,* with short straight or uncinate point, entire, 3–5-nerved; peduncles solitary or aggregated, shorter than the leaves, with a 3-flowered, monocarpous umbel; petals villous; fruit obovate-cordate, glabrous, longer than the villous, ovate-acuminate, 2-leaved involucre. *Centella villósa, Linn. Spec.* 1393. *Mercurialis Afra. Linn. Mant.* 298.

VAR. *α*, **minor**; villous or glabrescent; leaves about 4–6 lines long, 3–4 lines wide, obtuse at the base. *H. villosa, Thunb.! Herb. var. α. E. & Z.!* 2162, *et var. lanceolata. H. mollissima, E. Mey. in Herb. Drege. H. uncinata, Turcz! Bullet. de Moscou, vol.* 20. 1847. *Drege,* 7610, 7615, 7611.

VAR. *β*, **major**; villous or glabrescent; leaves 8–12 lines broad, 7–10 lines long; lower ones sometimes subcordate. *H. villosa, Thunb.! Herb. var. β. H. villosa, E. Mey. E. & Z.! var β. latifolia.*

HAB. Rocks on Table Mountain, at Tulbagh near Waterfall, Piquetberg, Blauwberg, Winterhoeksberg, and at Olifantsriver; var *β*, in the same locality and in Dutoitskloof. Oct.–Jan. (Herb. Thunb. Hk. D. Sd.)

Stem and branches filiform, ½–1 foot. Leaves flat or undulated at the margins. Petioles ½–2 inches long. Peduncles 6–12 lines long, villous. Involucre in anthesis 4-leaved. Petals very small, villous outward. Fruit 2 lines long and broad, compressed; mericarps wrinkled, on each side with 3 ribs.

9. **H. Solandra** (Linn. fil. Suppl. 176); perennial, whole plant tomentose; stem and branches short; leaves petiolate, obovately-cuneiform, bluntly 7-toothed at the apex; peduncles subsolitary, *nearly as long or longer than the leaves,* with a 5–6-flowered, monocarpous umbel; *petals glabrous;* fruit elliptic-cordate, tomentose, a little *longer* than the 4–6-leaved, lanceolate involucre. *Lam. Ill. t.* 188. *E. & Z.!* 2163. *Herb. Un. Itin.* 408. *H. tomentosa, Thunb.! fl. cap.* 250. *Sieb. fl. cap. exs. n.* 140. *Solandra Capensis, Linn. Spec.* 1407. *excl. syn.*

VAR. *β*. **longifolia** (DC. l. c. 69); peduncles shorter than the cuneate-elongated leaves.

HAB. Mountains near Capetown and in Cape Flats, Paarlberg, Van Kampsbay, near Tulbagh, etc. Var. β. near Saldanhabay, Driekop, Brackfontein. Mar.–Oct. (Herb. Thunb. Hk. Sd. D.)

Root or subterraneous stem long, branched. Stems very short, rarely 1–2 inches long, densely leafy. Leaves much aggregated, about ½ inch long, 4 lines broad, sometimes smaller, but in var. β. 8–10 lines long and more cuneated; with 7, rarely 5, acute teeth. Petioles much dilated at the base, ½–1 inch. Peduncles filiform. Flowers white, the sterile pedicellate, the fertile sessile. Petals ovate, ¾ line long. Fruit compressed, 2½ lines long, truncate, wrinkled; mericarps with 3 prominent ribs on each side, in var. β. sometimes subglabrous.

10. H. hermanniæfolia (E. & Z.! 2164); suffruticose, whole plant *tomentose-villous;* stem elongate, branched; leaves petiolate, oblong-cuneate, 3-nerved, with 9–11 short teeth at the blunt apex; peduncles *very short,* solitary, with a 5–6-flowered, monocarpous umbel; *petals villous;* fruit elliptic-cordate, hairy, a little *shorter* than the 6-leaved ovate, acuminate involucre. *Eckl. in S. Afr. Quart. Journ.* 1830, *p.* 375.

VAR. β brevifolia (E. & Z.! l. c.); leaves cuneate-spathulate.

VAR. γ. littoralis; leaves linear-oblong, cuneate, 3-dentate or acute. *H. littoralis, E. & Z.!* 2168.

HAB. Sandy flats between Krakakamma and Van Stadensriviersberge, *E. & Z., Drege,* 7613; var. β. near the Zwartkopsriver, *Zey.* 2665; var. γ. on the sea shore near Cape Agulhas, *Mundt.* July–Jan. (Herb. Hk. Sd.)

Readily distinguished from the preceding by the longer leaves with shorter and usually more numerous teeth, and by the subsessile umbels with hairy petals. Root very thick. Stem and primary branches woody. Petioles 1–2 inches long. Leaves 1–1½ inch long, 3–6 lines wide; in var. β. 8–9 lines long, 5–6 lines broad; in var. γ. 1 inch long, 3 lines broad. Peduncles solitary, rarely geminate, 1–2 lines long. Flowers equalling the involucre, the fertile sessile. Fruit 2–2½ lines long and broad; mericarps wrinkled, with 3 ribs on each side.

11. H. Dregeana (Sond.); suffruticose, puberulous; stems very short, with many fascicles of petiolate leaves at top; petioles dilated at the base, longer than the *ovate, shortly-acuminate, at the base subcuneate,* 5-*toothed leaves;* peduncles solitary, *very short,* with a 3-flowered, monocarpous umbel; petals villous; fruit orbicularly-cordate, puberulous or subglabrous, a little shorter than the lanceolate, acuminate 2-leaved involucre.

HAB. Near Ezelsfontyn and on the Roodeberg, 3–4000 ft., Nov., *Drege.* (Hb. Sd.)

Stems about uncial. Petioles 6–8 lines long. Leaves rather coriaceous, obsoletely 3-nerved, minutely puberulous, at length subglabrous, 4–6 lines long, 2–3 lines broad at the margin, with about 5 acute teeth. Peduncles at the base of the leaves, scarcely 1 line long. Fruit 2 lines long and broad, wrinkled; mericarps with 2 ribs on each side. It has the habit of *H. villosa,* but is quite different.

12. H. triloba (Thunb.! diss. de Hydr. p. 6. t. 3); suffruticose, *glabrous;* stem erect, branched; leaves cuneate-ovate, trilobed at the apex, 3–5-nerved, lobes or teeth acute; peduncles much aggregated, puberulous, at length glabrous, shorter than the petioles, *with a* 3–4-*flowered,* monocarpous umbel; petals glabrous; fruit obovate, glabrous, longer than the ovate, acuminate, 2-leaved involucre. *Flor. Cap.* 250. *Rich. Monog,* 65. *f.* 36. *E. & Z.!* 2166. *Zey.* 2664.

HAB. Rocks on Klynriviersberge, Caledon, Hauhoeksberge, Hottentottshollandberge, and on the Table Mountain, *E. & Z.! Harvey, Masson.* July–Sept. (Herb. Thunb. Hk. D. Sd.)

Stem and branches woody, ½–1½ feet, erect. Petioles 1–2 inches long. Leaves about 1 inch long, 6–8 lines broad, sometimes narrower, attenuated at the base; teeth or lobes at the apex 1–3 lines long. Peduncles 4–6 lines long. Flowers very small. Fruit 1½ lines long, compressed; mericarps with 3 ribs on each side; furrows flat, not wrinkled.

13. H. montana (Cham. & Schlecht. l. c. 374); suffruticose, *subtomentose*, at length glabrescent; stems suberect; branches spreading, elongate; leaves *ovate or elliptic-oblong, acute*, entire or acutely-3–5-toothed at the apex, coriaceous, 3–5-nerved, obtuse or subcuneate at the base; those of the branches smaller, sublanceolate, much shorter, petiolated; umbels *sessile, many-flowered*, monocarpous; petals of the fertile flowers hairy on back; fruit obovate-orbicular, glabrous, shorter or longer than the villous, ovate, cuspidate, 2-leaved involucre. *H. difformis, E. & Z.!* 2165. *Zey.* 2666.

HAB. Mountains near Langekloof and Tradow, *Mundt & Maire;* Hottentottshollandberge near Grietjesgat on Palmietriver, near Klynriver, and between Gnadenthal and Mount Zwarteberge, *E. & Z.! May.* Aug. (Herb. Berol. Hk. Sd.)

Root thick and woody. Stem and branches striate, purplish, several feet long. Leaves much aggregated at the base and on the nodes, on carinate, 2–5-uncial petioles, usually 1–2 inches long, 8–14 lines broad; old ones sometimes nearly 4 inches long, 2 inches broad, very thick and with revolute margins; those of the upper branches nearly sessile, 1–½ inch long, 4–2 lines wide. Umbels capitate, woolly, 10–20-flowered; sterile flowers on filiform, villous pedicels, the fertile quite sessile. Petals white and violaceous, obtuse. Fruit 2 lines long and broad, quite glabrous; mericarps with 2 ribs on each side, very prominent as well as the dorsal rib; furrows flattish or rugose. The lower leaves are similar to those of *H. triloba*, but the base is ovate, obtuse, rarely subcuneate. The specimens collected by *Mundt. & Maire* are not very perfect.

14. H. tridentata (Linn. fil. Suppl. 176); suffruticose, villous-tomentose, at length glabrescent; stems and branches short, ascending; leaves petiolate, *oblong-cuneiform or obversely-lanceolate, narrow, with 3 acute teeth or lobes at the apex*, 1–3-nerved, coriaceous; peduncles solitary or geminate, with a very short 4–5-flowered, monocarpous umbel; petals villous; fruit cordate-orbicular, glabrous, shorter than the villous, ovate, acuminate, 4-leaved involucre. *Thunb.! Dissert. p.* 6. *t.* 1. *Flor. Cap.* 250. *Rich. Monog, n.* 60. *f.* 37. *E. & Z.!* 2167.

HAB. Sandy places near Duikervalley and Doornhoogde, and near Swellendam, *Thunb. E. & Z. Drege, Pappe.* Sep.–Dec. (Herb. Thunb. Hk. D. Sd.)

Root very long and woody. Stem and branches 2–4 inches, tomentose. Leaves fasciculate, 2 lines broad, ½–1 inch long, attenuated into a short or longish petiole, dilated at the base, 3-nerved, the lateral ones often obsolete; when young tomentose-villous, at length nearly glabrous, with 3, rarely 2 teeth at the apex, very rarely acute, not toothed. Peduncles 1 line long or shorter. Involucre 3 lines long. Pedicels of the sterile flowers glabrous. Fruit sessile, 2 lines long and broad; mericarps with 2 lateral ribs on each side; furrows flattish, a little rugose.

15. H. Centella (Cham. & Schlecht. l. c. 375); suffruticose, subtomentose or glabrous; stem and branches terete or subangulate; leaves petiolate, *oval, oblong, or lanceolate, acute or acuminate at both ends*, 3-nerved; peduncles fascicled, filiform, with a 3–5-flowered, monocarpous umbel; petals glabrous, rarely subpilose; fruit obcordate, glabrous or pubescent, longer than the ovate, acute or acuminate, 2-leaved involucre.

VAR. α. **latifolia** (Cham. & Schlecht.! l. c.); leaves on longish petioles, elliptic or cuneiform-oval, acute or acuminate at each end, 3-, rarely 5-nerved (1–2 inches long, 6–8 lines broad), the upper smaller.

αα. *Woolly or subtomentose. H. villosa, Thunb.! herb. var. β. H. glabrata, β. subtomentosa, E. & Z.! Drege,* 7614.

ββ. *Glabrous. H. falcata, E. & Z.!* 2175. *H. glabrata, E. & Z.!* 2172 *(incl. var. minore). H. Centella, β. latifolia and γ. plantaginea, E. Mey. in herb. Drege. H. bupleurifolia, Rich. Monog.* 67. *f.* 39.

γγ. *tridentata;* leaves sometimes at top 2 or 3-dentate. *H. bupleurifolia, E. & Z.! var. β.*

VAR. β. **plantaginea**; leaves on longish or short petioles, oblong or oblong-lanceolate, acute or acuminate at both ends, 3-nerved (8–18 lines long, 2–4 lines broad), the upper smaller.

αα. *Subtomentose or subhirsute. H. plantaginea, Spreng. Grundz. der Pflanzenk. t.* 8. *f.* 5–7. *H. Cent. var. lasiocarpa. Cham. & Schl.! H. montana, E. Meyer in herb. D. non Cham. & Schl. (villosissima). Drege,* 7616.

ββ. *Glabrous. H. glabrata, Thunb.! fl. cap.* 251. *Centella glabrata, Linn. Amoen. Acad.* 6. *p.* 112. *H. bupleurifolia and plantaginea, E. & Z.!* 2173, 2174. *H. Cent, var. α. glabra, E. Mey.*

VAR. γ. **coriacea**; quite glabrous, or in the axils woolly; stem and branches short; leaves much aggregated, linear-lanceolate, cuneate, thick, coriaceous, with obsolete nerves, 1–2 inches long, 1½–2 lines broad. *H. montana and rupestris, E. & Z.!* 2169, 2170.

VAR. δ. **linifolia**; subtomentose or glabrous; branches elongate; leaves linear-lanceolate, cuneate, 3-nerved, the lower 3–4 inches, upper ones 1–2 inches long, 1–2 lines broad.

αα. *rigida;* stem erect as well as the leaves rigid. *H. linifolia, Thunb. fl. cap.* 250. *H. rigescens, E. & Z.!* 2179.

ββ. *flaccida H. linifolia, E. & Z.!* 2180. *H. falcata, var. β. & γ. E. & Z.!* 2175. *H. affinis, E. & Z.!* 2177. *H. fusca, E. & Z.!* 2178, *ex pte. H. cent. var. linifolia, E. Meyer. H. virgata, c. & d. E. Mey. Drege,* 7618. *Zeyh.* 2658.

γγ. *verticillata*; leaves much aggregated on the nodes, linear, attenuated at the base, 6–9 lines long, ¾ line broad. *H. linearis, E. Mey.*

HAB. Sandy and stony places in mountains throughout the Colony. (Herb. Thunb. Berol. Hk. D. Sd.)

Very variable in form and habit. Polygamo-monoecious or dioecious. Stem terete, striate-angulate, usually elongate; internodes without leaves. Peduncles 4–10 lines long, villous or glabrous. Umbel 1–5-flowered, but only 1 flower is fertile. Involucre in anthesis, 4, rarely 5-leaved. Petals 4 or 5, usually glabrous. Fruit about 2 lines broad; the mericarps with 5 filiform ribs; furrows rugulose.

16. H. debilis (E. & Z! 2176); suffruticose, quite glabrous; stem erect, *quadrangular;* branches elongate, filiform; leaves longish-petiolate, elongated-linear, *attenuate at the base,* quite entire, *5-nerved and with uncinate apex;* peduncle aggregated, capillary, 1–3-flowered; fruit elliptic, small, glabrous, twice longer than the ovate, acute, 2-leaved involucre.

HAB. Among shrubs and high grass on the Van Stadensriviersberge, Uitenhage. Aug. *E. & Z.!* (Herb. Sd.)

Distinguished from the smallest forms of the preceding by the 4-edged stem, longer, filiform branches, and uncinate leaves. Lower leaves 4–5 inches long, 2 lines broad, incrassate on the margins, and with 5 parallel, very prominent nerves on the under surface; upper leaves 2–1 inch long, 1–½ line broad, at the apex incurved-hamate. Peduncles 4–6 lines long. Fruit 1 line long and broad, compressed; mericarps with 5 filiform ribs.

17. H. virgata (Linn. fil. Suppl. 176); suffruticose, erect, branched, lanuginose-villous, or quite glabrous; leaves *linear-filiform,* quite entire; peduncles aggregated, rarely solitary, filiform, short, 1–5-flowered; fruit obovate or suborbicular, longer than the ovate, acute or acuminate, 2-leaved involucre. *Cham. & Schlecht.! l. c.* 379.

VAR. α. **glaberrima** (DC. l. c. 69); erect, branched, quite glabrous; flowers polygamous, monoecious. *H. virgata, Thunb.! fl. cap.* 50. *Lam.! ill. t.* 188. *f.* 3. *E. & Z.!* 2183. *H. filicaulis, E. & Z.!* 2184. *H. fusca, E. & Z.! ex pte. Zeyh.* 727, 2668, *a. Drege,* 7620, 7652, 7625, 7626, 9543.

VAR. β. **lanuginosa** (Cham. & Schlect. l. c.); erect, virgate, lanuginose-villous, polygamo-monoecious; leaves broader, sulcate, but not flat. *H. lanuginosa, E. & Z.!* 2182.

VAR. γ. **macrocarpa** (Cham. & Schl. l. c.); virgate, glabrous, polygamo-dioecious; fruit larger. *H. macrocarpa, Rich. Monog.* 67, *f.* 40. *E. & Z.!* 2181. *Zey.* 2668, *b. Drege,* 1839, 7623, 7624. Herb. Un. Itin. 7.

VAR. δ. **nana** (Cham. & Schl. l. c.); plant short, glabrous, polygamo-dioecious or monoecious; fruit small. *H. alpina et trichophylla, E. & Z.!* 2171, 2185.

HAB. In plains and on hills throughout the Colony; var β. near the Gauritzriver, Swellendam; var δ, on Tablemountain and Devilsmountain, and above the Waterfall near Tulbagh. (Herb. Thunb. Hk. D. Sd.)

Stem from ½–2 feet high, terete, dichotomously-branched. Leaves terete, acute, dilated at the base, 2 or more at the nodes; lower and intermediate ones 2–5 inches long, ½–1 line broad, the upper gradually smaller. Peduncles 2–6 lines long, in the upper, often woolly axils, capillary. Umbels very small, the fertile monocarpous, the male ones sometimes 4–8-flowered. Petals glabrous. Fruit more or less rugulose, 1½–2 lines long, 1½ line broad, in var. γ, 3 lines long, 2–3 lines broad; in var. δ, suborbicular, about 1 line long and wide. Mericarps with 5 filiform ribs. The species is easily distinguished from all the others by the filiform or terete leaves.

II. SANICULA, Linn.

Calyx-tube bristly; its margin 5-cleft, leafy. *Petals* erect, conniving, obovate, their apices inflexed. *Fruit* subglobose, not spontaneously bipartite. *Mericarps* with obsolete ridges, many-vittate, densely covered with hooked bristles. *Carpophore* indistinct. *Seeds* semiglobose. *DC. l. c. p.* 84. *Endl. Gen. n.* 4382.

Perennial herbs. Leaves radical, petiolate, palmate-lobed; lobes cuneated, deeply toothed at the apex. Stem naked or sparingly leafy. Flowers in dense heads, sessile. General umbel with few rays. Leaves of involucre few, lobed. Umbellules of many rays. Leaves of involucel many, entire. Flowers, male, female and hermaphrodite in the same umbel. Name from sano, *to heal or cure.*

1. S. Europæa (Linn. Spec. 339); leaves radical, palmate-parted; lobes trifid, serrate-toothed; flowers polygamous, all nearly sessile, disposed in umbellules; lobes of calyx denticulated. *Schkuhr. Handb. t.* 60. *Engl. Bot.* 98. *S. Europ. var. Capensis, Cham. & Schl. Linn.* 1, 352. *S. Canadensis, Thunb.! Fl. Cap.* 254, *excl. syn. S. Capensis, E. & Z.!* 2186. *Schimper Abyss. sect.* 2 *n.* 1127.

HAB. Woods and groves, districts of Worcester, Swellendam, George, etc. Oct.–Jan. (Herb. Hk. Sd.)

Stems 1–3 feet high. Flowers white or tinged with red, disposed in little heads. I have been unable to detect a character by which to separate this plant from the European specimens.

II. ALEPIDEA, La Roche.

Calyx-tube glabrous or muricately-tuberculate; lobes erect, leafy. *Petals* inflexed. *Fruit* ovate, somewhat compressed and contracted from the sides, its transverse section nearly circular. *Mericarps* without vittæ, but having 5 filiform or elevated-obtuse, inflated ribs. *Carpophore* united to the mericarps, but at length free, undivided. *Seed* semiterete. *Endl. Gen. n.* 4385.

Herbaceous, glabrous plants, with the habit of some *Eryngium.* Stems nearly naked or leafy, branched, umbellate at the apex. Leaves not lobed, oblong, ciliately-toothed with spinescent bristles. Universal umbels irregular, surrounded by variable involucra; partial umbels regular like those of *Astrantia,* but the flowers are sessile and surrounded by an involucel of about 10 unequal leaves, which are connate at the base,·coriaceous and coloured inwards. Name from α, *privitive,* and λεπις, a *scale:* these plants are glabrous.

ANALYSIS OF THE S. AFRICAN SPECIES.

Fruit densely muricate; leaves 2–3 inches long... ... ... (1) **ciliaris.**
Fruit quite or nearly smooth; leaves 7–15 inches long ... (2) **amatymbica.**

1. A. ciliaris (L. Roche, Eryng. p. 19. t. 1); radical leaves petiolate, *oblong,* obtuse; cauline leaves smaller, stem—clasping; fruit densely *muricate-tuberculate. Astrantia ciliaris, Linn. fil. Suppl.* 177. *Thunb. Fl. Cap.* 196. *Jasione Capensis, Berg. Act. Ups.* 3. *p.* 187. *t.* 10.

VAR. α. leaves oblong, truncate, or unequal-sided, or subcordate at the base. *A. serrata. E. & Z.!* 2188. *A. ciliaris, Drege, coll. Zeyh.* 2669.

VAR. β. leaves oblong or elliptic-oblong, more or less cordate at the base. *A. ciliaris et var. latifolia, E. & Z.!* 2187. *A. cordata, E. Mey. in Herb. Drege. Zey.* 2669.

HAB. Hills and mountains (1–8000 ft.) in the districts of Uitenhage, Albany and George, on Katberg, Witbergen and Zuurebergen. Oct.–Feb. (Herb. Th. Hk. D. Sd.)

Root fibrous. Stem 1–1½ foot, erect, striate, somewhat branched. Petioles of the radical leaves 1–2 inches long, carinate, its leaves 1½–3 inches long, ½–1 inch broad, coriaceous, toothed and ciliate; stem-leaves gradually smaller. Universal umbel of few rays. Involucel 4–6 lines in diameter, usually consisting of 5 ovate, mucronate, and 5 smaller, ovate-lanceolate leaflets. Flowers aggregate on a flat receptacle, white. Fruit subcompressed, 1 line long, crowned with the persistent calyx-lobes and the erect styles. Stylopodium depressed. Mericarps with 5, often obsolete, obtuse ribs. Albumen flat in front.

2. A. Amatymbica (*E. & Z.!* 2189); radical leaves oblong, or linear-oblong, *cuneate,* petiolate; cauline leaves smaller, stem-clasping; fruit broadly-ovate, *quite smooth, or at the apex tuberculated. A. longifolia, E. Mey.*

VAR. β, **cordata**; radical leaves long—petiolate, cordate at the base.

HAB. Top of Mount Winterberg, Ceded Territory, *E. & Z.!* between Buffelriver and Key, *Drege;* Port Natal, *Gueinzius, Plant.*

VAR β. Dornkop, near Sandriver, *Zey.* 728. (Herb Hk. Sd. D.)

Stem 2–4 feet high, fistular, sulcate. Radical leaves 7–15 inches long, 1½–2½ inches broad, toothed and ciliate; stem-leaves numerous, gradually smaller, the lower ones sometimes nearly as large as the radical, but not cuneate. Umbels numerous, forming a large panicle. Involucel as in the preceding, but a little larger, the leaflets unequal, lanceolate, pale or reddish inwards. Fruit when ripe 2 lines long and broad, contracted on the sides. Mericarps broad-ovate, with 5 obtuse, subinflated ribs. *A. peduncalaris, Steud.! in Rich. Tent. fl. Abyss.;* seems not to be a distinct species.

IV. APIUM, Linn.

Margin of the *calyx* obsolete. *Petals* roundish, entire. *Stylopodium* depressed. *Fruit* roundish, laterally contracted, didymous. *Mericarps* with equal, filiform ridges, the lateral on the margins. *Furrows* 1-

vittate, the outer frequently 2–3-vittate. *Carpophore* undivided. *Seeds* gibbous, convex, flattish in front. *DC. l. c. p.* 100, *Nees ab Esenb. Gen. pl. Germ. fasc.* 26, *t.* 7. *Endl. Gen. n.* 4393.

Herbaceous. Stem furrowed, branched. Leaves pinnate, leaflets cuneiform, cut. Umbels nearly sessile, on axillary branchlets, or at the apex of the stems. Involucre and involucel wanting. Flowers greenish-white. Name from *Apone*, Celtic for water; habitation of plants.

1. A. graveolens (Linn. Spec. 379); glabrous; leaves pinnate, upper ones ternate, leaflets cuneate, incised and toothed at the apex; petals with the point closely involute. *Engl. Bot. t.* 1210. *Hayn. Arzn. Gew.* 7, *t.* 24. *A. graveolens et decumbens, E. & Z.!* 2190, 2191. *Smyrnium laterale, Thunb.! Fl. Cap.* 259. *Helosciadium Ruta, DC. l. c.* 106 ? *Zeyh.* 267c. *Drege*, 9544.

HAB. On the sea-shore near Rietvalley, Algoabay, Zwartkopsriver, Albany, Port Natal. Dec.–Feb. (Herb. Thunb. Hk. D. Sd.)

The wild state of the *celery* is usually smaller and more decumbent than the plant in cultivation. The specimens from South Africa agree perfectly with those from the sea-shore of Germany.

V. PETROSELINUM, Hoffm.

Margin of the *calyx* obsolete. *Petals* roundish, incurved, entire, scarcely emarginate, with a narrow incurved point. *Stylopodium* short, conical, subcrenulate. *Styles* diverging. *Fruit* ovate, laterally compressed, subdidymous. *Mericarps* with 5-filiform equal ridges, the lateral marginal. *Furrows* with single vittæ. *Commissure* bivittate. *Carpophore* bipartite. *Seeds* gibbous, convex, flattish in front. *DC. l. c.* 102. *Nees ab Esenb. l. c. t.* 8. *Endl. Gen. n.* 4394.

Branched, glabrous herbs. Leaves decompound, with cuneated segments. Involucra few-leaved; involucels many-leaved. Flowers white or greenish, uniform: those in the disc of the umbel frequently sterile. Name from *πετρος, a rock*, and *σελινον, parsley*.

1. P. sativum (Hoffm. Umb. 78, t. 1, f. 7); stem erect, angular; leaves decompound, shining; lower leaflets ovate, cuneate, trifidly toothed; upper ones lanceolate, nearly entire; leaves of involucel filiform. *Apium Petroselinum, Linn. Hayn. Arg. Gen.* 7, *t.* 23.

HAB. In cultivated grounds, naturalized from Europe. "Parsley."

VI. HELOSCIADIUM, Koch.

Margin of *calyx* 5-toothed, or obsolete. *Petals* ovate, entire, with a straight or incurved apiculus. *Fruit* ovate or oblong, compressed from the sides. *Mericarps* with 5 filiform, prominent, equal ridges, the lateral marginal. *Furrows* with single vittæ. *Carpophore* distinct, entire. *Seeds* gibbously or teretely convex, flattish in front. *Nees ab Esenb. l. c. t.* 10. *Endl. Gen. n.* 4397.

Herbs with prostrate creeping stems and white flowers. Name from *ἑλος*, a *marsh*, and *σκιαδιον*, an *umbel*.

ANALYSIS OF THE S. AFRICAN SPECIES.

Glabrous, perennial, prostrate; fruit glabrous ... ... (1) **repens.**
Hairy, annual, erect; fruit hairy ... ... (2) **Capense.**

1. H. repens (Koch. DC. l. c. 105); *glabrous, perennial;* stem *prostrate,* rooting; leaves pinnate; leaflets roundish-ovate, unequally and acutely inciso serrate; umbels on long peduncles, opposite to the leaves; involucre of 2–4 ovate-lanceolate, permanent leaves; fruit *glabrous. Sium repens Linn. fil. Fl. Dan.* 1514. *Eng. Bot. t.* 1431. *E. & Z.!* 2195.

HAB. Marshy places in Zwartland, *Brehm.* (Herb. Sd.)

Stem ½–1 foot long, quite prostrate. Leaves petiolate, 1–2 inches; leaflets 2–3 lines long and wide, sessile. Involucre shorter than the (3–6) rays; involucels as long or longer than the pedicels. Flowers very small.

2. H. Capense (E. & Z.! 2196); *pubescent, annual;* stem *erect,* branched; leaves long, petiolate, ternately-partite; lobes 2–3-fid, linear-oblong, acute; umbels axillary, with 3–5 rays; umbellules 1–3-flowered, involucel few-leaved or wanting; fruit ovate-oblong, *hairy.*

HAB. Rocky places on the hills by the Coegariver, Uitenhage. Oct. *E. & Z.!* (Herb. Hk. Sd.)

Stem ½–1 foot, terete. Petiole of the radical leaves 1–2 inches long, of the upper leaves shorter. Lobes of the leaves 2–3 lines long, ¾–1 line broad, entire or bifid. Rays of the umbel unequal, the longer 6–12, the shortest 1–2 lines long, capillary. Umbellules of 2–4 flowers, pedicels scarcely ½ line long. Flowers very small. Ripe fruit 1 line long, a little contracted at the margins. Stylopodium depressed; styles very short. Ridges obtuse. Commisure with 2 vittæ.

VII. PTYCHOTIS, Koch.

Margin of *calyx* 5-toothed. *Petals* obovate, bifid, or deeply emarginate, with a long inflexed point proceeding from the sinus. *Fruit* compressed, ovate or oblong. *Mericarps* with 5 equal, filiform, primary ridges, the lateral ones marginal. *Interstices* with single vittæ. *Carpophore* bi-partite. *Seeds* terete or gibbously convex, flattish in front. *DC. l. c. p.* 107. *Endl. Gen. n.* 4400.

Annual or bi-ennial herbs. Cauline leaves multifidly capillaceous. Umbels axillary, compound; involucre wanting or few-leaved; involucel of several leaves. Flowers white. Name from πτυχη, a *plait,* and ους, ωτος, an *ear*; the petals have (usually) a plait in the middle, emitting a little ear or segment. (The European species of *Ptychotis* are characterised by a large gland they bear on the back of the petals; I never saw it in the exotic species.)

ANALYSIS OF THE S. AFRICAN SPECIES.

1. **Euptychotis.** Involucre wanting. Fruit smooth.
   - Segments of leaves ovate-oblong or oblong; umbels pedunculate; fruit oval ... ... (1) **Meisneri.**
   - Segments of leaves linear, acute; umbels sessile; fruit roundish-ovate ... ... (2) **tenuis.**
   - Segments of leaves filiform; umbels pedunculate; fruit ovate ... ... (3) **caruifolia.**
2. **Trachyspermum.** Involucre present. Fruit muricated.
   - Stem hispidulous; fruit ovate ... ... (4) **hispida.**
   - Stem glabrous; fruit broader than long ... ... (5) **didyma.**

Sect. 1. EUPTYCHOTIS, DC. l. c. 108. Universal involucrum wanting. Fruit smooth, not muricated. (Sp. 1–3)

1. P. Meisneri (Sond.); smooth; stem erect, branched; leaves bipinnate, lobes pinnatifid, segments *ovate-oblong* or *oblong, acute,* asperulous on the margin; umbels opposite to the leaves, *pedunculate;*

leaflets of involucel few, subulate, as long or shorter than the pedicels, persistent; fruit *oval. Petroselinum humile, Meisn.! in Hook. Lond. Journ. Bot. vol.* ii. *p.* 531.

HAB. In the plains near Port Natal, *Krauss,* 418; *Drege,* 9545. Oct. (Herb. Meisn. D. Sd.)

Root thin, perpendicular. Stem ½–1 foot high, striated. Leaves petiolate; petiole at the base with a white-margined sheath; limb subtriangular, about 2–3 inches long and broad; leaflets ovate like those of *Anthriscus Cerefolium.* Umbel with 5–6 rays, shorter than the peduncle. Pedicels 2 lines long. Petals deeply emarginate. Fruit line long, broadest in the middle. Stylopodium short. Styles reflexed. Commissure with 2 vittæ.

2. **P. tenuis** (Sond.) annual, glabrous; stem erect, dichotomously branched; leaves bipinnate, lobes deeply pinnatifid, *segments linear, acute;* umbels opposite to the leaves, *sessile;* involucel 1–2 leafed, much shorter than the smooth pedicels; fruit *roundish-ovate,* as well as the ovary quite smooth.

HAB. Hills between Buffeljagdriver and Rietkuil, Oct.–Nov. *Zeyh.* 2672. (Herb. Sd.)

A small herb about a span high. Radical leaves on longer petioles. Segments linear-filiform, 1–1½ line long. Umbel with 4–5 rays, ½-inch long. Pedicels 1½ line long. Fruit nearly 1 line long and broad, contracted from the sides. Stylopodium minute, conical. Styles very short. Valliculæ with one elevated vitta each. Commissure, 2-vittate.

3. **P. caruifolia** (Sond.) annual, glabrous; stem erect, dichotomously much branched; leaves cut into numerous *filiform* segments: umbels opposite to the leaves, *pedunculate;* involucel 1–2-leaved, much shorter than the *scabrid* pedicels; *ovary muricated* on the ridges; fruit *ovate,* smooth.

HAB. Near Riebekkasteel, Nov. *Zeyher,* 729. (Herb. Hk. Sd.)

Root whitish. Stem 1–2½ feet, terete, purplish and glaucous, with many spreading branches. Leaves nearly as in *P. Coptica,* bipinnate; lobes cut in setaceous, 2–3 lines long, segments. Umbel with 5–8 scabrid rays, on shorter or subequal peduncle. Pedicels 2 lines long. Fruit a little compressed on the sides; 1 line long. Stylopodium subdepressed. Styles short, spreading. Mericarps with 5 filiform ridges. Valleculæ 1 vittate; the commissure with 2 vittæ. Carpophore 2 partite.

Sect. 2. TRACHYSPERMUM, DC. l. c. Universal involucre composed of a few linear entire or trifid leaves. Fruit muricated. (Sp. 4–5.)

4. **P. hispida** (Sond.) annual; stem decumbent or erect, branched, as well as the leaves and peduncles scabrid or hispidulous; leaves subtripinnate, segments filiform; umbels opposite to the leaves, pedunculate, with 5–8 rays; *fruit ovate, acute. Sium hispidum, Thunb.! fl. c.* 261. *Trachyspermum rigens, E. & Z.!* 2197. *(excl. synon.) Phymatis cyminoides, E. Meyer. Herb Un. Itin. n.* 212. *Zeyh.* 2697, 2673. *Drege,* 4645.

HAB. Sandy places near Capetown; in Hottenthold. near Tulbagh, Hassaquaskloof, and on the Zwartkopsriver. Sept.–Jan. (Herb. Th. Hk. D. Sd.)

Root long, subsimple. Stem several inches, or 1–1½ foot, terete; branches striated. Leaves 2–, or nearly 3-pinnate, the radical ones longer, petiolate, 1–3 inches long. Segments 1–2 lines long, setaceous. Leaves of involucre usually pinnated or compound. Rays of umbel ½–1 inch. Involucel few leaved, twice shorter than the pedicels. Petals white. Fruit 1 line long, when ripe nearly as broad as long, contracted from the sides. Mericarps with 5 filiform ribs. Valleculæ with 1 elevated vitta; the commissure 2 vittate. Styles diverging. It varies with glabrous stem and leaves, hispidulous or minutely muricated fruit.

5. P. didyma (Sond.) annual; stem decumbent, branched, glabrous; leaves bi- or subtripinnate; segments filiform; umbels opposite to the leaves, pedunculate, with 10–16 rays; fruit *didymous*, broader than long. *Carpophyllum Jacquini, α, herb. Drege!*

HAB. Woods near Tulbaghskloof, Nov. *Zeyh.* 730. (Herb. Hk. Sd.)

Stem and leaves as in the preceding, but well distinguished by the many-rayed umbels and didymous fruit. Involucre of umbel pinnatifid. Rays 6–7 lines long, sometimes a little scabrid. Leaves of involucel 2–4, subulate, shorter than the pedicels. Flowers very small. Fruit ½ line long, ¾ line broad, densely muricated, much contracted from the sides. Mericarps roundish-ovate, with 5 bluntish ribs. Stylopodium and styles very short.

VIII. CARUM. Linn.

Margin of *calyx* obsolete or nearly so. *Petals* obovate or elliptic, emarginate, with an inflexed point. *Stylopodium* depressed or shortly conical; styles divaricate. *Fruit* ovate or oblong, contracted from the sides. *Mericarps* with 5 filiform, equal ribs, the lateral marginating. *Furrows* with 1 vitta; commisure bi-vittate. *Carpophore* free, forked or bipartite. *Seed* terete-convex, flattish in front. *Koch. Umb.* 121. *Nees. ab. Es. gen. fl. germ. fasc.* 26, *n.* 15. *Endl. gen. n.* 4406.

Biennial or perennial, glabrous herbs, much branched. Leaves pinnate, segments or leaflets, multifid. Both the involucra and involucels variable, often wanting. Flowers white. Name from *Caria*, in Asia Minor; the native country of the *Caraway*.

1. C. Capense (Sond.) root fleshy, stem branched, radical leaves tripinnate, leaflets filiform, setaceously acuminated, fastigiate; cauline leaves sheathy, without limb. *Apium, radice crassa, etc. Burm. Afr. t.* 72, *f.* 1. *Anethum Capense, Thunb! fl. cap.* 262. *Fœniculum Capense, DC. l. c.* 142. *Chamarea Capensis et Caffra, E. & Z.!* 2220, 2221. *Drege.* 7635. *Zeyher,* 736, 739, 2682.

HAB. Flats between Constantia and Hotthld.; karroolike places and mountains in the distr. of Stellenbosch, Albany, Uitenhage, Caffraria, Namaqualand, &c. Apr.–July. (Herb. Thunb. Hk. D. Sd.)

Root solitary or geminate, oblong, aromatic, called by the colonists "Fenkelwortel." Radical leaves aggregated; petiole 2–3 inches, leaf 3–5 inches long; the two primary lateral branches about twice shorter than the middle pinnæ; the ultimate pinnulæ opposite with tridichotomously divided, numerous, setaceous lobes, 1–2 lines long. Umbel 4–7 rayed; rays unequal, ½–1½ inch long. Umbellules 8–12 flowered, pedicels often violaceous, 2–4 lines longer than the involucel. Fruit ovate, 1½ line long, ovate, much contracted at the sides.

IX. PIMPINELLA (Linn.)

Margin of the *calyx* obsolete. *Petals* obovate, emarginate, with the point long and inflexed. *Fruit* contracted laterally, ovate, crowned with the reflexed style, whose bases are much swollen, smooth or hairy. *Mericarps* with 5 filiform, equal ridges, the lateral marginal. *Furrows* with many vittæ; *Commissure* 2 vittate; *carpophore* free, bifid. *Seed* gibbous, flattish in front. *DC. l. c. Nees. ab Esenb. l. c. n.* 16. *Endl. gen. n.* 4410.

Herbs with radical, pinnatifid leaves, the segments roundish, dentate or cut, rarely entire; cauline leaves more finely divided. Involucres wanting. Flowers white or yellow. Name from *bipinnata*, twice pinnate; the leaves are often so.

ANALYSIS OF THE S. AFRICAN SPECIES.

Radical leaves bi-pinnatisect ... ... (1) **Stadensis.**
Radical leaves cordate-suborbicular, inciso-serrate ... ... (2) **Caffra.**

1. P. Stadensis (Harv. gen. 135.) erect, branched, *glabrous, but minutely downy above the middle;* leaves *bipinnatisect*, lobes pinnatifid and trifid; segments sublinear, cuspidate, spreading; upper leaves pinnatifid or trifid or nearly entire; the radical ones long petiolated; involucrum wanting; involucels 1–2 leaved or wanting; rays and pedicels as well as the ovary pubescent; fruit, when ripe, *broadly ovate*, hairy or subglabrous. *Anisum Stadense, E. & Z!* 2199.

HAB. Among shrubs in the Vanstadens mountains, Uitenhage, *E. & Z!* Feb. (Herb. Sd.)

Habit of a small *P. dissecta*. Root seemingly biennial. Stem 1–1½ foot, terete, striate. Radical leaves about 1 inch long, ¾ inch wide, on a twice longer, subfiliform petiole,; segments 1–2 lines long, the terminal often longer, nearly 1 line broad. Umbel with 6–8 uncial rays. Involucels capillary, shorter than the pedicels. Flowers small, white. Fruit 1 line long, with short, conical stylopodium, and spreading, short styles. Mericarps with 5 prominent ribs.

2. P. Caffra (Harv. l. c.) erect, branched, minutely *downy from the base;* radical leaves longish-petiolate, *cordate-suborbicular*, inciso-serrate; lower cauline leaves pinnately lobed, lobes linear-lanceolate or cuneate; uppermost leaves trifid or undivided linear; involucrum and involucel wanting; rays and pedicels pubescent, *ovary glabrous* or nearly so; fruit *ovate-oblong.*

HAB. Grassy places on Mount Katriviersberg, Ceded Territory; Mohlamba range, Natal, *Dr Sutherland, Drege,* 7628. March. (Herb. Hk. Sd.)

With the habit of the foregoing, 1–2 feet high; easily distinguished by the short pubescence of the whole plant, and the lower leaves. The umbels and flowers nearly the same, but the ovary and fruit quite glabrous or sometimes beset with a few hairs. The stylopodium is depressed. Ripe mericarps are unknown.

X. SIUM, Linn.

Margin of the *calyx* 5-toothed or obsolete. *Petals* obcordate, with an inflexed point. *Stylopodium* depressed or shortly conical. *Styles* divaricated or recurved. *Fruit* laterally compressed or contracted, and subdidymous; *mericarps* with 5 filiform, equal, obtuse ridges; furrows and commissure with many vittæ. *Carpophore* bipartite. *Seed* subterete. *Koch. Umbell.* 117. *DC. l. c.* 124.

Mostly aquatic herbs. Leaves pinnate or pinnatifid. Umbels terminal, manyparted, surrounded by many-leaved involucra. Flowers white. Name from *siu*, water, in Celtic; habitation of most of the species.

1. S. (Berula) Thunbergii (DC. l. c. 125); root fibrous, stoloniferous; stem erect, striated; branches angular; leaves pinnate; leaflets ovate, acute, regularly and callously serrated; umbels pedunculate, lateral or terminal; involucre and involucels many-leaved. *E. & Z.!* 2200. *S. angustifolium, Thunb.! Fl. Cap.* 260. *Herb. Un. Itin. n.* 69. *Zeyh.* 2674.

HAB. Marshy spots near Zeekoevalley, in the bed of the Zwartkopsriver, and in similar localities in the district of Albany, and near Port Natal. Feb.–Mar. (Herb. Thunb. Hk. D. Sd.)

Stem 1–3 feet high. The root called *Tandpynwortel* (Tooth-ache-root) by the colonists, is renowned for its allaying tooth-ache when held in the mouth or chewed. *Pappe, Fl. Cap. med.* 18.

XI. RHYTICARPUS, Sond.

Margin of *calyx* 5-toothed. *Petals* obovate, entire, involute, the apex acute or acuminate. *Fruit* roundish, laterally compressed, crowned with the conical stylopodium and the short styles. *Mericarps* rugose, with 5 filiform ridges, the lateral marginal; furrows with single, commissure with 2 vittæ. *Carpophore* bipartite. *Seeds* convex, flattish in front.

Perennial, glabrous herbs. Stem erect, glaucous. Petioles ternately branched. Leaves pinnatisect, lobes cuneate or subulate; ramifications of petiole articulate. Umbels compound. Involucres and involucels of many small leaves. Flowers yellowish, monœcious or diœcious. Name from ῥυτις, a *wrinkle*, and καρπος, *fruit.*

ANALYSIS OF THE SPECIES.

Leaf-lobes *cuneate*, incised or toothed ... ... (1) **rugosus.**
Leaf-lobes long, *terete*, furrowed ... ... (2) **Ecklonis.**

1. R. rugosus (Sond.); radical leaves 2–3 pinnatisect, *lobes cuneate, incised or toothed, or oblong-lanceolate, trifid, or quite entire;* cauline leaves ternately compound, lobes linear-terete, acute, short; uppermost leaves tripartite or undivided; leaves of involucre and involucel short, linear-subulate; fruit roundish-ovate, laterally contracted, rugose. *Conium rugosum, Thunb.! Fl. Cap.* 253. *Sium paniculatum, Thunb.! l. c.* 261. *Trinia Uitenhagensis, E. & Z.!* 2193. *Lepisma paniculatum, E. Mey. (ex parte) Zeyh.* 2671.

HAB. On the fields near the Zwartkopsriver, *E. & Z.!* Zondagsriver, Enon, and on the Giftberg, *Drege;* in the Hantum mountains, *Thunberg.* Nov.–June. (Herb. Thunb. Hk. Sd.)

Root blackish. Stem 2–3 feet high, terete, striate; pruinose, spotted with purple, much branched above. Leaves glaucous, much aggregated at the base of the stem. Petioles 2–4 inches long, terete, ternately compound. Larger leaflets generally trifid, lobes 2-3 lines broad, incised or acutely 3-4-toothed; other leaflets with lanceolate lobes, 1 line broad; those of the upper leaves more distant, thicker, subulate, carinate, 3–1 line long. Umbels on longish peduncles, with 6–8 glaucous rays 1–1½ inch long. Leaves of involucre (6–9) about 2 lines long. Calyx teeth acute, as long as the stylopodium. Fruit 1½ lines long and broad, when ripe, didymous. *Sium paniculatum, Thunb.!* is the flowering, *Conium rugosum* the fruit-bearing plant

2. R. Ecklonis (Sond.); lower leaves *biternate; lobes elongate, terete, acute, sulcate, rigid;* the terminal longer; upper leaves trifid or undivided; leaves of involucre and involucel short, subulate; fruit broad-ovate, compressed, rugose. *Trinia Swellendammensis, E. & Z.!* 2194. *Bupleurum acerosum, E. Meyer.*

HAB. In Karro, between Kochmanskloof and Gauritzriver, E. & Z.; near Gnadenthal, Drege. Oct.–Nov. (Herb. Hk. Sd.)

Nearly allied to the preceding, but easily distinguished by the terete, striate-sulcate ramifications of the terete petiole. Umbels with 5–7 rays; involucre and involucel 4–5-leaved; flowers and fruit as in the foregoing. This comes near *Bupleurum difforme*, but I never found a leaf or phyllodium without lobes.

XII. BUPLEURUM, Linn.

Margin of the *calyx* obsolete. *Petals* roundish, entire, involute at the apex, which is broad and retuse. *Fruit* laterally compressed or somewhat didymous, crowned with the depressed stylopodium. *Mericarps* with 5 acute, winged, filiform, or obsolete ridges, the lateral marginal. *Furrows* with or without vittæ, smooth or granulate. *Seed* teretely convex, flattish in front. *DC. l. c.* 127. *Nees ab Esenb. l. c. n.* 18. *Endl. Gen. n.* 4414.

Herbaceous or shrubby glabrous plants. Leaves rarely divided, usually from the abortion of the limb and dilatation of the petiole changed into phyllodia, with quite entire margins. Umbels compound. Involucres various. Flowers yellow. Name from βους, *an ox*, and πλευρον, *a side;* so called from a supposed bad quality in swelling kine that feed on some species of the genus.

ANALYSIS OF THE S. AFRICAN SPECIES.

Herbaceous; leaves linear-subulate, nerved; fruit oblong ... (1) **Mundtii.**
Shrubby; leaves filiform, rushlike; fruit obovoid ... ... (2) **difforme.**

1. B. Mundtii (Cham. & Schlecht.! in Linnæa I. 384); perennial, glabrous, much branched; leaves *linear-subulate, nerved;* radical ones tapering into the petioles; umbels 5–8-rayed; involucels of 5 narrow lanceolate leaves, which are equal in length to the umbellules; fruit *oblong*, smooth. *E. & Z.!* 2201. *Zeyh.* 2676. *B. Baldense, E. Mey. in Herb. Drege, non Host. Sium filifolium, β. Thunb.! herb. B. falcatum, β. Africanus, Berg. Cap.* 76.

HAB. Mountains near Fort Beaufort, Katriviersberge, Zuureberge, Uitvlugt, and on downs near Zwartkops, and Koegariver, and Port Natal. Jan.–Feb. (Herb. Thunb. reg. Berol. Hk. D. Sd.)

Stem 1–2 feet high, flexuous, striate. Leaves much aggregated at the base, 6–8 inches long, 1–2 lines broad, or in other specimens subsetaceous; cauline ones not attenuated at the base, gradually smaller. Flowers yellow. Involucre of 3, rarely 5 unequal, lanceolate leaves. Rays 1–1½ inch long. Umbellules 8–10-flowered. Fruit nearly 2 lines long, 1 line broad. Mericarps with very prominent, paler ridges. Furrows with several vittæ. Stylopodium depressed; styles recurved. *B. Mundtii*, as described by *Cham. & Schlechtendal*, is a depauperate state, with 2 or 3 rays in the umbels.

2. B. difforme (Linn. Spec. 343); *shrubby*, erect, branched; leaves *filiform, variable on the same branch, simple or ternate;* leaflets undivided or cut; umbels 8–20-rayed; leaves of involucre linear, acute, much shorter than the peduncles; fruit *obovoid*, tubercled, rugulose. *Thunb. Fl. Cap.* 248. *Burm. Afr. t.* 71, *f.* 1. *E. & Z.!* 2202. *Tenoria difformis, Sprengl. Oenanthe exaltata, Thunb.! l. c.* 254. *Zeyh.* 2475.

HAB. Mountains in the districts of Cape, Swellendam, Worcester, Uitenhage, and Caffraria. Feb. (Herb. Thunb. Hk. D. Sd.)

A rigid, glabrous shrub, 2 feet and more high. Branches forming a large, leafless panicle. Leaves aggregated, often ½–1 foot long; young ones on the lower part of the stem composed of many small, flat leaflets, finely cut, of a sea-green colour; these leaves soon fall off, and the upper part of the branches are closely covered with long, rush-like, angled leaves, coming out in clusters from each joint. Rays of umbel uncial or longer. Involucre and involucel about 5-leaved. Calyx with acute teeth. Fruit when ripe 2–2½ lines long, 1½ line broad, subcompressed. Mericarps with blunt ridges; furrows and commissure bivittate.

XIII. HETEROMORPHA, Cham. & Schlecht.

Margin of the *calyx* 5-toothed. *Petals* roundish, entire, involute, the apex broad, retuse. *Fruit* obovato-pyriform, 3-winged; mericarps of 2 forms, the outer one 2-winged, the inner one 3-winged, wings decurrent from the teeth of the calyx. *Furrows* with solitary vittæ. *Commissure* bivittate. *Linnœa,* 1, 385. *DC. l. c.* 127. *Endl. Gen. n.* 4415.

A glabrous shrub with the habit of *Bupleurum fruticosum.* Leaves petiolate, ovate or oblong, rarely somewhat triple-nerved, sometimes quite entire, sometimes 3-lobed, tripartite or ternate. Umbels of many rays. Involucra and involucels of many short leaves. Flowers yellow, as in *Bupleurum.* Name from ἑτερος, *diverse,* and μορφη, *form.*

1. H. arborescens (Cham. & Schl. l. c.)

VAR. α. **integrifolia**; leaves ovate or oblong, obtuse, mucronate, or acute. *Bupleurum arborescens, Linn. Thunb. Fl. Cap.* 247; *E. & Z.!* 2203. *Tenoria arborescens, Zeyh.* 2677. *H. Abyssinica, Hochst.! Schimp. n.* 1844.

VAR. β. **trifoliata**; leaves ternate. *Bupl. trifoliatum, Wendl. & Bartl. Beyt.* 2, *p.* 13. *E. & Z.!* 2204.

VAR. γ. **collina**; leaves entire, oval-oblong, obtuse, mucronate, coriaceous, pinnately-veined, with thicker margins. *H. collina, E. & Z.* 2205.

HAB. Among shrubs on the Krum, Loeri, and Zwartkopsriver, Uitenhage; in the district of Albany, George, and in Caffraria. Jan.–Aug. (Herb. Hk. D. Sd.)

Branches panicled. Leaves quite entire or remotely crenate, 1–3 inches long; in var. γ, smaller and more coriaceous. In *Drege's* specimens, *n.* 7630, the leaves are entire, bifid, trifid, and ternate. Fruit 3 lines long, with small wings, crowned with a conical stylopodium. Carpophore bipartite.

XIV. LICHTENSTEINIA, Cham. & Schlecht.

Margin of *calyx* 5-toothed. *Petals* elliptic, drawn out into a long inflexed point, reaching nearly to their base. *Fruit* nearly terete, variable in length, crowned with the erect, calycine teeth, surmounted by the short spreading styles with conical bases. *Mericarps* smooth, 5-ridged, ridges filiform, equal, the lateral marginal; vittæ large, solitary under each ridge, but none in the commissure nor furrows. *Carpophore* bipartite. *Linn.* 1, 394. *DC. l. c.* 135. *Endl. Gen. n.* 4416.

Perennial herbs, yielding an aromatic juice. Radical leaves cut. Stem erect, naked, branched, furnished with sheating scales or abortive leaves. Terminal umbels compressed, fertile, lateral ones often sterile. Involucra and involucels of many short marcescent leaves. Flowers white. Name in honour of the late Prof. von Lichtenstein, a celebrated botanist.

ANALYSIS OF THE SPECIES.

Leaves ovate or obovate, undivided or lacerate, setacously-serrate	(1) **lacera.**
Leaves ovate or obovate, undivided or 3-lobed, lobes mucronately-toothed	(2) **latifolia.**
Lvs. tripartite; segments lanceolate or oblong-lanceolate	(3) **trifida.**
Lvs. tripartite; segments linear, 2 or 3-pinnatisect	(4) **crassijuga.**
Lvs. ternate; segments interruptedly-pinnate; umbels 8–12-rayed; mericarps ovate	(5) **interrupta.**
Lvs. interruptedly 2-pinnate; umbels 3–5-rayed; mericarps oblong	(6) **Beiliana.**

1. L. lacera (Cham. & Schlecht. l. c.); leaves ovate or obovate, coriaceous, scabrous from short pubescence, *undivided or lacerate,* irregularly serrated; serratures *setaceously cuspidate;* fruit oblong, a little compressed. *Hermas rudissima, Richb. in Spreng. syst.* 4. 118. *Sieb. Flor. Cap. exs. n.* 213. *E. & Z.!* 2206.

VAR. β. **pinnatifida**; leaves smooth, pinnatifid-lobed.

HAB. On mountains near Capetown, and near Clasenbosch and Constantia. Jan.–Feb. (Herb. reg. Berol. Hk. D. Sd.)

Stem fistulose, 3 feet and more in height, sulcate, glabrous, branched. Petioles very large, 3–5-inches long. Leaves ½–1 foot long, 4–9 inches broad or smaller, much-veined, and with large branched middle nerve; the margin with spreading, often ½–1 inch long serratures, terminated by a longish mucro. Umbel many-rayed. Leaves of involucre (circ. 10) 3–4 lines long. Umbellules about 10–12-flowered; pedicels a little longer than the involucel. Flowers white. Fruit when ripe 5 lines long, 2 lines broad. Stylopodium conical. Mericarps with 5 filiform ridges.

2. L. latifolia (E. & Z.! 2207); leaves shortly petiolate, *orbicular or obovate,* coriaceous, smooth and glabrous, *undivided or 3-lobed,* lobes obovate, *margined with short callous-mucronate teeth;* fruit ovate, nearly terete.

HAB. Among shrubs on the Van Stadensriviersberge, Uitenhage. Feb. *E. & Z.!* (Herb. Sd.)

Habit of the preceding, distinguished by the minutely-toothed leaves and fruit. Petioles 1–2 inches long. Leaves 8–12 inches long, or smaller, with prominent nerves and veins. Umbels as in *L. lacera,* or with somewhat longer rays. Umbellules 12–16-flowered. Calyx-teeth acute, persistent. Fruit 3 lines long.

3. L. trifida (Cham. & Schl.! l. c. 396); leaves long-petiolate, *trifid to the base,* rarely 2 or 4-fid; segments *lanceolate or oblong-lanceolate,* toothed, glabrous, rarely hairy, coriaceous; fruit ovate-oblong. *E. & Z.!* 2208. *L. runcinata, E. Mey. in Herb. Drege.*

VAR. β, **palmata**; leaves 3–5-fid, the middle segment trifid. *L. palmata, DC. l. c.* 135, (with hairy leaves).

VAR. γ, **pinnatifida**; leaves 3-fid, the segments pinnatifid.

HAB. Var. α, in Schurfdeberg, *Mundt. & Maire;* Paarlberg, *Drege;* Zwarteberg, *E. & Z.;* var. β, Piquetberg, *Drege,* 7633; var. γ, Cape Flats, *E. & Z.* Jan.–Feb. (Herb. reg. Berol. Hk. Sd.)

Stem 2–5 feet, sulcate-striate. Petioles 4–6 inches long, 6–9 lines broad. Radical leaves 5–10 inches long, the margins decurrent on the petiole, the middle segment longer, 2–3 inches broad, shortly and callously toothed, rarely incised or inciso-dentate, acute or acuminate, rarely obtuse, entire or 3-lobed. One of the primary leaves often undivided, obovate. Umbels many-rayed, nearly the same as in *L. lacera,* and also the fruit, but the latter is sometimes a little smaller and more ovate. The var. γ has narrower segments of leaves (4–8 lines broad) with uncial or shorter, acutely-toothed, horizontal, lateral lobes.

4. L. crassijuga (E. Meyer); leaves longish-petiolate, trifid to the base; segments *linear, bi or tri-pinnatisect;* lobes lanceolate-acuminate.

HAB. Langevallei. Feb. *Drege.* (Herb. Hk. Sd.)

Petioles 4–5 inches long, 4–6 lines broad. Leaves 6–10 inches long. Segments 2 lines broad, with horizontally-spreading marginal lobes, and short, often minute, subulate teeth. Lobes pinnately divided or simple, 2–4 lines long. Umbels with 6–9 rays, 2–2½ inches long. Involucre and involucels as in the preceding. Fruit, when young, ovate.

5. L. interrupta (E. Mey.); leaves petiolate, *ternate; segments decursively and interruptedly pinnate;* leaflets ovate, unequally inciso-serrate or subpinnatifid, glabrous; *umbels* 8–12-*rayed; fruit roundish-ovate;*

mericarps ovate. Oenanthe interrupta, Thunb.! Prod. 50. *Fl. Cap.* 253. *O. obscura, Spreng.! Lichtenst. pyrethrifolia, Cham. & Schlecht.! l. c.* 397. *L. Sprengeliana, E. & Z.!* 2212. *Physospermum terebinthaceum, E. Mey.*

HAB. Zwartland, *Thunb.;* Ruytersbosh, *Mundt. & Maire;* Zwartkopsriver, *E. & Z., Zeyh.* 2680, *Drege,* 7631, 7632; Buffelriver and near Port Natal, *Drege, Krauss.* Nov.–Feb. (Herb. Thunb. reg. Berol. Hk. D. Sd.)

Stem 1–2½ feet, subangulate or terete, striate, somewhat branched. Radical leaves on short or longish, 1–2 lines broad petioles. Leaves 3–4 inches long, 2–3 inches broad, ternate, the lateral segments at the base with an ovate, sessile, pinnatifid leaflet; the middle segment usually twice longer than the quite similar lateral ones; rhachis winged, dentate; leaflets with mucronulate serratures. Sheaths at the ramifications of the stem without limb or with a very short one. Involucre and involucels about 6-leaved; leaflets of involucre 4–6 lines long. Rays of umbel 1½–2-uncial. Fruit 2 lines long and broad; mericarps a little contracted at the sides with roundish back; ribs filiform, obtuse; commissure with a longitudinal furrow.

6. L. Beiliana (E. & Z.! 2211); leaves petiolate, decursively and interruptedly pinnate or bipinnatisect; leaflets ovate, cuneate, serrato-ciliate or inciso-serrate, glabrous; *umbels 3–5-rayed; fruit ovate-oblong; mericarps oblong. L. pyrethrifolia, E. Mey. non Cham. & Schl. Oenanthe interrupta, Thunb.! Herb. ex prte. not of Fl. Cap. Zeyh.* 732, 740.

VAR. β, **simplicior**; leaves decursively pinnate; lobes 3–9, mostly larger, elliptic or elliptic-ovate, minutely and bluntish-toothed or ciliate. *L. pyrethrifolia and inebrians, E. & Z.!* 2209, 2210. *L. triradiata, E. Mey.*

HAB. In the districts of Cape, Worcester, Stellenbosh and George. Nov.–Feb. (Herb. Thunb. Hk. D. Sd.)

Distinguished at first sight from *L. interrupta* by the less compound leaves and few-rayed umbels. Stem 1–3 feet, with spreading branches. Leaves on short or very long petioles. Rachis winged, toothed. Lobes 6–12 lines long, 3–6 lines broad, sharply and ciliolate-serrated; in var. β, nearly quite entire, ciliate or crenate, or shortly and bluntish-toothed, mostly coriaceous and often 1½ inch long, 8 lines wide. Involucral leaves 3–4, lanceolate. Rays of umbel 1½–3-uncial. Umbellules 9–12-rayed, with 5 or 6 subulate involucel-leaflets. Fruit 2½ lines long, 1½ line broad. Mericarps a little contracted at the sides, with roundish back and obtuse ridges.

XV. ANNESORHIZA, Cham. & Schlecht.

Margin of the *calyx* 5-toothed, persistent. *Petals* elliptical, acuminate, more or less emarginate, with an inflexed point. *Fruit* 5-angled, prismatic, crowned by the calyx and inflexed styles. *Mericarps* convex on the back, unequal, one of them 3-winged, the central dorsal rib and the two lateral being wing-like, the two intermediate filiform; the other 4-winged, the central dorsal being filiform, the lateral and intermediate ridges winged; (in one species the mericarps are equal or subequal, with prominent, scarcely-winged ridges). *Furrows* with single vittæ. *Commissure* bivittate. *Carpophore* bipartite. *Linnæa, vol.* 1, 398. *DC. l. c.* 139. *Endl. Gen. n.* 4420.

Biennial or perennial herbs, known by the colonial name *Anyswartel.* Stem erect. Radical leaves petiolate, pinnati-partite, cauline small and scale-like. Umbels with many rays. Involucres many-leaved. Flowers white. Name from *αννησον, anise, ῥιζα, a root.*

ANALYSIS OF THE SPECIES.

1. Biennial, with 1–2 roots:
   Fruit oblong, 2½–3 lines long; mericarps unequal … … … (1) **Capensis.**

Fruit oblong, much cuneated, 4–5 lines long; mericarps equal ... (2) **filicaulis.**
Fruit oblong, scarcely cuneated, 2–2½ lines long; mericarps unequal ... ... (3) **montana.**
2. Perennial, with 5 or more roots:
Leaves glabrous:
Fruit oblong-cuneate, 6 lines long; mericarps unequal... (4) **macrocarpa.**
Fruit ovate-oblong, 2–3 lines long; mericarps equal ... (5) **elata.**
Leaves hairy:
Lobes of leaves pinnatifid-incised; umbels 5–6-rayed ... (6) **villosa.**
Lobes of leaves ovate, obtuse, toothed; umbels 3–5-rayed (7) **hirsuta.**

1. Capensis (Cham. & Schlecht.! l. c.); glabrous; leaves tripinnatifid, segments trifid or pinnatifid, *lobes spreading, lanceolate-subulate;* umbels 3–8-rayed; involucre 3–5-leaved, deciduous; fruit *oblong,* a little narrower at the base; *mericarps unequal. E. & Z.!* 2213. *Chaerophyllum Capense, Thunb.! Fl. Cap.* 253. *Myrrhis Capensis. Spreng. Spec. Umb.* 132.

HAB. Near Lurisriver, *Thunb.;* in mountains near Capetown and in Hottentottsholland, *Mundt. & Maire, E. & Z.* Dec.–June. (Herb. Thunb. reg. Berol. Hook. D.)

Root (rarely 2) fusiform, a finger long, much corrugated when dry. Stem 1–1½ feet high, branched, terete, leafless. Leaves radical, nearly triangular, about 4 inches long and wide, on longish petioles; ultimate lobes 1–2 lines long, ½ line broad, quite entire, rugulose, acute or mucronulate. Rays of umbel 1–2 inches long. Leaves of involucre ovate-lanceolate, 2–3 lines long; leaflets of involucel as long or shorter than the (10–16) pedicels. Fruit 2½–3 lines long, straw-coloured. Stylopodium short, conical, often purplish. The root is called by the colonists *Vlackte Anyswortel.* The leaves are incorrectly described by Ecklon and Zeyher.

2. A. filicaulis (E. & Z.! 2216); glabrous; leaves . . . . ; umbels 3–5-rayed; rays unequal; leaves of involucre 2–3, very short; umbellules 5–12-radiate; fruit oblong, *much cuneated; mericarps nearly equal.*

HAB. Sandy places near Olifantriver, Clanwilliam. Jan. *E. & Z.!* (Herb. Sd.)

Stem 2–2½ inches long, filiform, branched above. Radical leaves unknown, as well as the root. Scales appressed, cuspidate. Rays of umbel filiform, 1 inch long, some of them twice or thrice shorter. Leaflets of involucel subulate. Fruit 4–5 lines long. Stylopodium longer than in the preceding, and styles erect-spreading. The plant is only known by a few imperfect specimens.

3. A. montana (E. & Z.! 2214); quite glabrous; leaves bi- or tripinnate, segments pinnatifid; *lobes ovate-oblong,* shortly dentate, mucronulate, the terminal rotundate 3-lobed; umbels 3–5-rayed, rays subequal; leaves of involucre 5, lanceolate; umbellules 18–24-radiate; fruit oblong, scarcely cuneated; mericarps *unequal. Acroglyphe flexuosa, E. Meyer, ex parte.*

HAB. In mountains near Capetown and in Zwartland, *E. & Z.;* Wetkamp, *Zeyh.;* Bergriver, between Paarl and Pont, *Drege.* Feb.–April. (Herb. Hk. Sd.)

Root solitary or geminate, fusiform, 2 inches to ½ foot long, a finger thick, called by the colonists *Berg-Anyswortel.* Stem terete, 2–3 feet at the base of the branches, with whitish margined, cuspidate scales. Leaves petiolate, the 3 primary branches 3–5 inches long; the ultimate lobes about 6 lines long, 4 lines broad, paler and reticulated above. Rays of umbel 1 inch long. Fruit 2–2½ lines long, with depressed stylopodium and spreading styles.

4. A. macrocarpa (E. & Z.! 2219); quite glabrous; stem striated, branched above, leaves ternately 3–4-pinnatisect; segments multipartite, divaricate; *lobes minute, linear-subulate;* umbels 5–8-rayed, rays unequal; leaves of involucre 4, ovate-acuminate; umbellules 25–30-flowered; involucels 7-leaved, as long as the pedicels; *fruit oblong, cuneate; mericarps unequal. Zeyh.* 2681.

HAB. Sandy hills on the Zwartkopsriver, *E. & Z.* Jan.–May. (Herb. Hk. D. Sd.)

Roots numerous (10–20), elongated, fusiform, bluntish, 4 or 5-sided, 6–8 inches long, 2–3 lines in diameter. Stem 3–4 feet high. Radical leaves long petiolate, 1 foot long and broad; divisions of petiole naked, the ultimate only leafy; segments crowded, lobes 1 line long. Rays of umbel 2–3 inches, pedicels 1½ lines long. Fruit 6 lines long; mericarps largely winged; stylopodium conical.

A. spuria, (E. & Z.! 2217) of which the roots and leaves are unknown, is only distinguished from *A. macrocarpa* by the many (12–28) rayed umbels; stem, flowers, and fruits are quite the same. It grows in the Cape flats, near Doornhoogde, *E & Z.*; Muysenberg, *W. H. Harvey.*

5. A. elata (E. & Z.! 2218); quite glabrous; stem striated, much branched; leaves. . . ; umbels 3–5-rayed; rays subequal; leaves of involucre 5; umbellules 20–30-flowered; leaflets of involucel 8–10, linear-subulate, shorter than the pedicels; fruit *ovate-oblong; mericarps equal,* with prominent, but not winged, ridges. *Pimpinella Capensis, Thunb. Fl. cap.* 260.

HAB. On Tablemountain near Tokay, and on the Zwarteberg, Caledon, *E. & Z.* Dec.–Feb. (Herb. Sd.)

Five and more roots, 3 inches or more long, terete, corrugated when dry, 2–3 lines in diameter. Stem 4–5 feet, branched from the middle, branches fastigiate. Rays of umbel 2–4 inches; pedicels 1½–2 lines long. Flowers small, crowded. Fruit 2 lines long, 1 line broad. Calyx-teeth short, acute. Stylopodium short, conical. Styles recurved. *Pimpinella Capensis* is now wanting in herb. Thunberg.

6. A. villosa (Sond.); stem erect, striate, glabrous; leaves *villous,* ternately-tripinnatifid, lobes *pinnatifid-incised, or inciso-serrate,* rugulose; umbels 5–6-rayed; rays unequal; leaves of involucre 5–6; umbellules 20–30-flowered; leaflets of involucel 5–7, shorter than the pedicels; fruit (unripe) cuneate. *Sium villosum, Thunb. Fl. Cap.* 51. *Acroglyphe hispida, E. Meyer.*

HAB. Near Ezelbank, in sandy localities, 3000 ft. Dec.—*Drege.* (Hb. Tb. Sd.)

Many terete, 4–6 inches long roots, as thick as a goose's quill. Stem 2 feet or more high; branches spreading, subfastigiate. Petioles of the radical leaves 1–3 inches long, sulcate, very hairy. Leaves 3–4 inches long and wide; segments ovate, 3–4 lines long, 2–3 lines broad, deeply serrate or incised. Rays of umbel 2–3 inches long; leaves of involucre ovate, acute, with scarious margins. Half-ripe fruit 2 lines long; mericarps with wing-like ridges. Stylopodium short, conical. Styles diverging.

7. A. hirsuta (E. & Z.! 2215); stem erect, striated, glabrous; leaves hairy, ternately tripinnatifid; the secondary ramifications of the petioles with winged rhachis; *lobes ovate, obtuse, mucronulate, serrato-dentate,* the terminal trilobed; umbels 3–5-rayed; rays subequal; leaves of involucre 4–5, ovate-lanceolate; umbellules 20–30-flowered; leaflets of involucel 7–8, shorter than the pedicels; fruit oblong, cuneate; mericarps unequal. *Acroglyphe flexuosa, E. Mey., ex parte.*

HAB. Mountains of Hottentottsholland, *E. & Z.*; Cape flats, *Wallich.*; between Paarl and Pont, *Drege.* Nov.–Jan. (Herb. Hk. D. Sd.)

Root as in the preceding. Stem 2–3 feet high. Leaves on long petioles, very similar to those of *A. montana,* but the teeth are more numerous and sharper, and the petiole and leaf are hairy. Ultimate lobes 6-8 lines long, 4-6 lines broad, reticulated. Rays of umbel 4-6 inches long. Fruit 4 lines long. Stylopodium short, conical; styles spreading. It is distinguished from *A. villosa* by the less divided, larger lobes, with shortly dentate margins.

XVI. OENANTHE, Linn.

Margin of the *calyx* 5-toothed, persistent, after flowering enlarged. *Petals* obovate, emarginate, with inflexed points. *Stylopodium* conical. *Fruit* cylindrical-ovate, crowned with the long, erect styles. *Mericarps* with 5, obtuse, rather convex ridges, the lateral marginal, a little broader. *Furrows* with single vittæ. *Carpophore* indistinct. *Lam. Ill. t.* 203. *DC. l. c.* 136. *Endl. Gen. n.* 4418.

Smooth, usually aquatic herbs. Umbels compound. Flowers white. Name of *οινος, wine,* and *ανθος, a flower.*

1. O. filiformis (Lam. Dict. 4, 520); annual, glabrous; stem erect, flexuous, striated, branched; leaves simple, filiform, linear, or linear-lanceolate, the radical attenuated into a long petiole; umbels 3–7-rayed; involucre and involucels of 3–5 subulate leaves; fruit cylindrical-prismatic. *Sium filifolium, Thunb.! Fl. Cap.* 260 *(var. α, in herbar). O. filiformis, Sieb. Fl. Cap. exs. n.* 209. *Herb. Un. Itin. n.* 54, *et* 761. *Zeyher,* 734, 735.

VAR. *α*, **erecta**; stem erect; 1–2 feet, evidently striated; umbels with 5–10 rays.

VAR. *β*, **humilis**; stem filiform, finely striated, mostly diffuse, or with divaricating branches, 3–8 inches high; umbels with 3–5 rays.

VAR. *γ*, **latifolia**; stem erect; radical leaves oblong or linear-oblong, cauline ones linear-lanceolate. *O. Dregeana, E. Meyer.*

HAB. Sandy and rocky places in the Cape flats and in mountains near Capetown, Paarlberg, Tulbagh, 24-Riviers, and Olifantsriver; var. *β*, Dutoitskloof, *Drege.* Nov.–Jan. (Herb. Thunb. Hk. D. Sd.)

Root fusiform or tuberous-incrassate, nearly 1 inch long. Stem dichotomously branched. Radical leaves much aggregated, several inches or 1 foot long; the lamina sometimes scarcely broader than the petioles, but usually ½ to 1½ line broad, 3-nerved, bluntish or acute. Cauline leaves shorter, linear, broader at the base. Involucral leaves 4–6 lines long, lanceolate, subulate. Rays of umbel 4–6 lines long. Umbellules 8–12 flowered; leaflets of an involucel equalling the pedicels. Flowers white. Fruit 1½–2 lines long, crowned by the capillary, spreading calyx teeth, and the longer, diverging styles. Mericarps with prominent ridges. The var. *γ* seems at first sight to be a different species, but there are intermediate forms; the leaves are ½–1 inch long, 2–3 lines broad, the cauline gradually smaller; flowers and fruit as in var. *α*.

XVII. GLIA, Sond.

Margin of the *calyx* 5-toothed, teeth triangular, acute, persistent; after flowering not enlarged. *Petals* obovate, subemarginate, with inflexed, lanceolate point. *Fruit* ovate-oblong, subterete, crowned with the conical stylopodium and spreading styles. *Mericarps* with 5 equal, sharp, nearly wing-like ridges, the lateral marginal. *Furrows* with single vittæ. *Commissure* 2-vittate. *Carpophorum* bipartite. *Seed* semiterete, flattened in front.

A perennial glabrous herb, with erect stem, pinnate leaves, compound umbels, and whitish or greenish flowers. Involucre and involucel many-leaved. Name from *gli;* the plant is so called by the Hottentots, who prepare from the roots an inebriating liquor.

1. **G. gummifera** (Sond.) leaves aggregated, dimorphous, the radical with obovate, trifid, or trilobed, serrate segments, the cauline with linear or subulate segments, the uppermost linear, entire. *Bubon gummiferum L. Comm. hort. Amst.* 2, *t.* 58. *E. & Z.!* 2250. *Herb. Un. Itin. n.* 563. *Peucedanum Caledonicum, E. & Z.!* 2233. *Krauss,* 1180. *Œnanthe inebrians et tenuifolia, Thunb.! fl. cap.* 253. *Lichtensteinia inebrians, E. Meyer. L. pyrethrifolia, DC. l. c.* 135. *Sieb. Fl. Cap. exs. n.* 211. *Lepisma paniculatum, E. Meyer, ex parte.*

HAB. On Platteklip, Table Mountain and Cape Flats, Paarlberg, Dutoitskloof, Ezelsbank, Klynriviersberge, and near Tulbagh. Oct.–Jan. (Hb. Thun. Hk. D. Sd.)

Root perpendicular, a finger thick, when dry, blackish, subwoody. Stem 2–5 feet high, leafy, terete, striate, branched or panicled above. Lower or radical leaves on longish, subangulate or striate petioles; petiole and its ramifications articulate and sulcate. Leaves pinnate or bipinnate; segments obovate-cuneate, incised and toothed-serrate, rugulose, 6–12 lines long, 4–8 lines broad; terminal lobe 3-fid, the lateral ones often bi- or trifid. The following stem-leaves often longer and more cut, with narrower ovate or oblong-cuneate, pinnatifid, or serrate segments; the upper leaves with linear-lanceolate or linear-subulate lobes, 6–2 lines long. Umbels with 10–16 rays, 1–2 inches long. Involucre of many lanceolate-subulate leaves, 3–4 lines long. Involucels shorter than the pedicels. Fruit about 3 lines long. *Œnanthe tenuifolia,* Herb. Thunb. is the same as *O. inebrians,* but without radical leaves.

XVIII. FOENICULUM, Adans.

Margin of the *calyx* tumid, obsolete, toothless. *Petals* roundish, entire, involute, with a subquadrate, retuse point. *Fruit* in a transverse section nearly circular. *Mericarps* with 5 prominent, obtusely-keeled ribs, the lateral marginal and a little broader. *Furrows* with single vittæ. *Commissure* bi-vittate. *Seed* semicylindrical. *DC. l. c.* 142. *Endl. Gen. n.* 4425.

Biennial or perennial herbs. Root fusiform. Stem terete, branched. Leaves triply-pinnate, decompound, with linear, setaceous leaflets. Involucra and involucels almost wanting. Flowers yellow. Name from *fœnum,* hay; the smell of the plant resembling that of hay.

1. **F. officinale** (All. Ped. 2. 25); radical leaves rather distinct; leaves all tripinnate, with capillary, elongated leaflets; umbels 6–10-rayed. *E. & Z.!* 2223. *Zeyh.* 2683. *Anethum Fœniculum, L. Hayn. Arz. Gew. vol.* 7. *t.* 18.

HAB. Naturalised in various parts of the Colony, near Capetown, on hills near Zwartkopsriver and Van Stadensriver. "*Fennel.*" Dec.–May. (Herb. D. Sd.)

XIX. DEVERRA, DC.

Margin of the *calyx* obsolete, toothless. *Petals* ovate, acuminate, with inflexed points. *Styles* short, at length spreading. *Fruit* ovate or roundish, laterally subcompressed, covered with patent scales or hairs. *Mericarps* semiterete, with often obsolete ridges; solitary vittæ in the furrows, and two in the commissure. *Carpophore* bipartite. *DC. l. c.* 143. *Endl. Gen. n.* 4427.

Aromatic, nearly leafless, glaucous, rigid, broom-like subshrubs. Petioles sheathing, permanent; limbs of leaves wanting in the adult plants or nearly so; but in young plants they are small and 3-parted, with linear lobes. Umbels of few rays. Involucra and involucel, 4–6-leaved, deciduous. Flowers white.

ANALYSIS OF THE S. AFRICAN SPECIES.

Fruit hairy ... ... (1) **aphylla.**
Fruit tuberculated ... ... (2) **Burchellii.**

1. D. aphylla (DC. l. c. 143); stem erect; sheaths leafless, or the lower ones furnished with simple, entire or petiolate, trifoliolate leaves, the segments of which are linear-filiform; *fruit very hispid. Bubon aphyllum, Cham. & Schlecht. Linnæa,* 1389. *D. aphylla, E. & Z.!* 2224. *Zeyher,* 738, 2685.

VAR. *β*, **denudata**; fruit sparingly hispid. *D. Burchellii, E. & Z.!* 2225. *excl. syn.*

HAB. In mountains in the districts of Clanwilliam, Beaufort, Uitenhage, Graafreynet, in Caffraria and Little Namaqualand, 1–4000 feet. Oct.–July. (Herb. Sd. Hk. D.)

Root woody. Stem 1–3 feet high, terete, striate, often flexuous, panicled above. Leaves of the lower sheaths wanting or filiform, terminating the large sheaths, ½–1 inch long, or 3-foliolate, with a short filiform petiole, the lobes ½–1½ inch long. Umbels with 3–9 rays, ¾–1 inch long. Involucre ovate-lanceolate, whitish-margined. Umbellules 6–12-flowered; pedicels, when young, shorter than the 5-leaved involucel, but fruitbearing longer. Fruit ovate, 1½ lines long, very villous, in var *β*. sparingly villous or muricate. Mericarps with filiform, obtuse ridges, more evident in var. *β*, the lateral marginal, equal. Stylopodium short, conical, glabrous. Styles spreading.

2. D. Burchellii (Sond.); stem erect; sheaths leafless or with simple, entire or petiolate, trifoliolate leaves, the segments of which are linear-filiform; *fruit tuberculate. D. aphylla, β, Burchellii. DC. l. c.*

HAB. Near Kapockfontyn, district of Graafreynet, *Zeyh.* 738. *b.*; Wonderfontyn near Vaalriver, and Magalisberg, *Zeyh.* 737. Dec.–Feb. (Herb. Hk. D. Sd.)

Stem, leaves and umbels as in the preceding, from which it is distinguished by the ripe fruit, not covered with hairs but with elevated granules or tubercles.

XX. SESELI, Linn.

Margin of *calyx* 5-toothed; teeth short, thickish, sometimes obliterated. *Petals* ovate, with inflexed points, emarginate or nearly entire. *Fruit* oval or oblong, its transverse section nearly terete, crowned by the reflexed styles. *Mericarps* with 5 prominent, filiform or elongated, thick ridges, the lateral marginal and often a little broader. *Furrows* with single vittæ, the outer rarely 2-vittate. *Commissure* 2-vittate, rarely 4-vittate. *Seeds* semiterete. *DC. l. c.* 144. *Endl. Gen. n.* 4430.

Biennial or perennial herbs, with pinnatifid or decompound leaves. Involucre scarcely any; involucel of many leaves. Flowers white, rarely yellow. Name from *Seycelyous*, the Arabic name of an umbelliferous plant.

ANALYSIS OF THE SOUTH AFRICAN SPECIES.

Leaves 2-pinnatipartite; segments incised; involucel cupuliform ... (1) **Caffrum.**
Leaves 3-pinnatisect, segments multifid, lobes subulate; involucel 4-6-leaved... ... ... (3) **asperum.**
Leaves 3-foliolate, leaflets linear-lanceolate ... ... (2) **Natalense.**

1. S. (Hippomarathroides) caffrum (Meisn.! in Hook. Lond. Journ. Bot. vol. 2. 533); quite glabrous; stem terete, stiff, striate, with a few branches above the middle; radical leaves aggregated, *bipinnatipartite;* segments cuneate, pinnatifid and inciso-dentate; teeth mucronulate; stem-leaves reduced to a large vagina; involucre 2-leaved or wanting;

involucel *cupuliform*, semi-5-6-parted, with acute, deciduous lobes; rays of umbel elongated; pedicels 2-3 times longer than the involucel; fruit smooth.

HAB. Near Port Natal, *Krauss*, 403; on hills near Adow, Uitenhage, and near Philipstown, Ceded Territory, *Ecklon*; on the Kowieriver, Albany, *Zeyher, n. N. n. E. Umbell.* 4. Aug.-Oct. (Herb. D. Meisn. Sd.)

Root woody. Stem 1-2 feet high. Radical leaves petiolate. Petiole as long or a little shorter than the leaf, canaliculate above, striate-sulcate beneath. Leaves about 3 inches long and broad, the segments or leaflets 6-12 lines long, obovate or ovate. Rays of umbel 2-3 inches long. Umbellules 12-16-flowered. Flowers white. Stylopodium depressed. Styles spreading. Ripe fruit unknown.

2. S. (Euseseli) Natalense (Sond.); quite glabrous; stem terete, fistular, multistriate, branched above the middle; radical and the lower stem-leaves petiolate, *trifoliolate;* leaflets linear-lanceolate, 3-nerved, quite entire; upper leaves quite entire, linear; involucre with 4-5 unequal leaves; leaves of involucel *distinct;* fruit smooth.

HAB. On the Tableland near Port Natal, *Krauss*, 433; *Gueinzius*, 535. (Herb. D. Sd.)

Perennial herb, 2-3 feet high, easily known by the trifoliolate leaves. Stem finely striated. Segments of the leaves unequal, the middle somewhat longer, about 2 inches long, 2 lines broad, on a petiole of the same length. Cauline leaves few. Flowers white. Rays of umbel (6-10) unequal, the longer 1-1½ inches. Leaves of involucre setaceous, acuminated, 4-1½ lines long. Involucel of 4-5 subulate leaflets. Fruit, when half ripe, oblong, with 5 equal, filiform ridges. Stylopodium depressed. Styles short.

3. S. (Euseseli) asperum (Sond.); glabrous; stem terete, striate, branched; *leaves sub-3-pinnatisect; segments multifid; lobes abbreviate, subulate, sulcate above;* involucre and involucel 4-6-leaved; peduncles, pedicels, as well as the fruit, subscabrous. *Sium asperum, Thunb.! Fl. Cap.* 21?.

HAB. Sea shore near Capetown, *Thunb., Zeyh.* March. (Herb. Thunb. Sd.)

Stem 1-1½ ft., erect. Leaves petiolate, the lower larger, 3-4 inches long. Ultimate lobes 1-2 lines long, acute. Cauline leaves much smaller. Rays of umbel 8-12, ¾-1 inch long. Leaves of involucre and involucel 1-2 lines long, with scarious margins. Flowers white. Fruit 1 line long, nearly terete, a little compressed from the sides. Mericarps with prominent, obtuse ridges, with the valleculæ muriculate, scabrous, at length nearly smooth. Valleculæ 1-vittate. Commissure with 2 vittæ.

XXI. POLEMANNIA, E. & Z.

Margin of *calyx* shortly 5-toothed, permanent. *Petals* elliptical, entire, with acuminate, inflexed point. *Fruit* oblong, its transverse section nearly cylindrical, crowned with the depressed-conical stylopodium and short styles. *Mericarps* with 5 prominent, obtuse ridges, the lateral marginal a little larger. *Furrows* with 1, commissure with 2, vittæ. *Carpophore* bipartite. *Seed* semicylindrical, flattened in front. *E. & Z.! Enum.* 347. *Endl. Gen. n.* 4431.

Glabrous shrubs, with petiolate, ternate or ternately-pinnate leaves, the leaflets wedge-shaped, trifid or undivided, multiradiate umbels and umbellules and white flowers. Named after P. H. Polemann of Capetown.

ANALYSIS OF THE SPECIES.

Leaves tripartite or 3-lobed; umbels many-rayed ... ... (1) **grossulariæfolia.**
Leaves triternate bipinnatifid; umbels few-rayed ... ... (2) **verticillata.**

1. P. grossulariæfolia (E. & Z.! l. c. n. 2227); leaves cuneiform, trifoliolate, tripartite or subtrilobed; lobes cuneate-rotundate, mucronate or dentate, with diaphanous margins; umbels terminal, with 12–16 unequal rays; involucre none or few-leaved, deciduous. *Lepisma verticillatum, d. E. Mey. in Herb. Drege.*

HAB. Rocky places among shrubs on Mount Chumiberg, Caffraria, *E. & Z.*; on the Katberg, 4–5000 ft., *Drege.* Nov.–April. (Herb. Sd.)

Shrub 8–10 feet high, with purplish branches. Leaves alternate, solitary or aggregated, 8–12 lines long and broad. Lobes 3-fid or 3-dentate, reticulate-veined. Petiole sulcate, as long or a little longer than the leaf. Involucre of some petiolate, mostly undivided, obovate or cuneate leaves. Rays of umbel 6–12 lines long. Involucel of some subulate, deciduous leaflets. Umbellules 10–15-flowered. Fruit nearly 2 lines long.

2. P. verticillata (Sond.); leaves triternate or ternately-bipinnate; segments cuneate, 3-fid; lobes tridentate, mucronulate; umbels subverticillate, with 4–8 unequal rays; involucre of some filiform leaves.

HAB. On rocks near the Gariep, and near Mierenkasteel, 500–1000 ft., Sept., *Drege*; near Kammapus, *Zeyher.* (Herb. Hk. Sd.)

Easily known by the more slender stem, more compound leaves, and elongated, often not umbellately-aggregated rays. Leaves solitary or aggregated; petiole 1–1½ inch; the leaf 1½–2 inches long and broad, more compound in the sterile than in the flowering branch. Lobes much-cuneate, 3–5 lines long, the terminal lobes a little longer. Rays of umbel 2–3 inches long, sometimes dispersed on the branch and not verticillate; the terminal umbel perfect. Involucel of 6–8 subulate leaflets, twice shorter than the pedicels. Ripe fruit unknown.

XXII. STENOSEMIS, E. Meyer.

Margin of *calyx* minutely 5-toothed. *Petals* obcordate, deeply emarginate, with subulate, inflexed point. *Fruit* roundish-ovate, its transverse section terete. *Mericarps* with 5 large, rather winged ribs; the 3 dorsal roundish-obtuse, corky; lateral ones marginating, a little more dilated and sharp-winged. *Furrows* with single vittæ. *Commissure* 2-vittate. *Carpophore* bipartite. *Seed* convex, flat in front.

Perennial, glabrous herbs, with erect, sulcate stem and branches, and petiolate, ternately-pinnate leaves; leaflets filiform or linear. Umbels terminal, compound. Involucre and involucel of many lanceolate leaves. Flowers white. Name probably from στενος, *narrow*, and σημα, in the sense of a *petal.*

ANALYSIS OF THE SPECIES.

Leaves triternate; lobules elongate ... ... (1) **Caffra.**
Leaves tripinnate; lobules short ... ... (2) **angustifolia.**

1. S. Caffra (Sond.); leaves triternate; ultimate lobes elongated, terete, acute, sulcate, undivided or trifid; upper ones similar, but smaller. *S. teretifolia, E. Mey. Krubera Caffra, E. & Z.!* 2253.

HAB. Rocks on Mount Bothasberg near Grahamstown, *E. & Z.*; between Kaprivier and Vishrivier, *Drege.* Jun.–July. (Herb. Hk. Sd.)

Root woody. Stem 2–3 feet high, a goose's quill thick. Branches erect, spreading. Lower leaves on longish (4–6 inches), terete, sulcate petioles. Segments 1–3 inches long, the ultimate, if again divided, shorter. Umbel with 10–14 unequal rays, ½–¾ inch long, sulcate, twice longer than the ovate-lanceolate leaves of involucre. Fruit, when ripe, 3 lines long, 2½ lines broad. Mericarps with thick wings; the lateral ones compressed. Stylopodium depressed, much shorter than the deflexed styles.

2. S. angustifolia (E. Mey.); leaves ternately 3-pinnate; *lobes short, linear*, sulcate beneath; upper ones smaller.

HAB. Stony hills between Morley and Omtata, Feb., *Drege*. (Herb. Sd.)
Stem and branches, flowers and fruit of the preceding, but the petioles are shorter (2–3 inches long), and the leaves more compound. Lobes of the leaves 3–4 lines long, ½ line broad, flat above, but sulcate and elevated-nerved beneath.

XXIII. CNIDIUM, Cusson.

Margin of the *calyx* obsolete or with very short teeth. *Petals* obovate or ovate, emarginate, with inflexed points. Transverse section of the *fruit* subcircular. *Mericarps* with 5 equal, winged ridges; the lateral marginal. *Furrows* univittate. *Commissure* 2-vittate. *Carpophore* bipartite. *Seed* semicylindrical, flattened in front. *DC. l. c. p.* 152. *Endl. Gen. n.* 4436. *Heteroptilis, E. Mey. Meisn. in Hk. Lond. Jour. Bot. vol.* 2, *p.* 534.

Decumbent or erect, perennial or suffruticose herbs. Leaves variable, mostly pinnatifid or multifid. Umbels terminal or lateral. Flowers white or rosy. *Cnidium* was the ancient name of Orach, a potherb.

ANALYSIS OF THE SOUTH AFRICAN SPECIES.

Leaves tripinnatisect or ternately decompound ... ... (1) **suffruticosum.**
Leaves reniform-cordate, the upper trifid ... ... (2) **Kraussianum.**

1. C. suffruticosum (Cham. & Schl.! Linn. 1. 387); quite glabrous; stem prostrate or erect, striate, branched; leaves fleshy, *bipinnatisect or ternately-decompound;* leaflets short, rather trifid, obtuse or acute; umbels, terminal and lateral, many-rayed; involucre and involucel many-leaved; fruit with thickish-membranaceous, subequal wings. *E. & Z.!* 2228. *Conium suffruticosum, Berg.! Cap.* 77. *Conium rigens, Thunb.! Prod.* 50. *Fl. Cap.* 253. *Sium patulum, Thunb.! Fl. Cap.* 261. *Heteroptilis arenaria, E. Mey. Ligusticum Capense, DC. l. c.* 159. *Athamantha Capensis, Burm. Fl. Cap. p.* 7. *Zeyh.* 2686.

HAB. On the sea shore and on sandy hills from Tablebay to Port Natal. Dec.–April. (Herb. Th. reg. Berol. Hk. D. Sd.)
Suffruticose, stem and branches terete, or the latter subangulate, subflexuose, ½–1 foot or more long. Leaves petiolate, duplicate-pinnate, pinnæ oblong, bluntish, incised. Petiole about as long as the leaf. Umbels with 9–16 rays, ¾–1 inch long. Leaves of involucre and involucel lanceolate, short. Flowers white. Ripe fruit 2–2½ lines long and broad. Stylopodium minute, conical, shorter than the reflexed styles. Wings of the mericarps subequal, or 1 or 2 of the dorsal wings somewhat smaller. The whole plant is usually glaucous, rarely green; the lobes of the leaves vary—obtuse, acute or acuminate, approximate or remote.

2. C. Kraussianum (Sond.); stem erect, as well as the leaves and peduncles minutely downy; *lower leaves reniform-cordate*, nearly as long or a little longer than the broad petiole, duplicate-serrate or toothed; upper leaves gradually smaller, 3-lobed, with entire or toothed lobes; umbels terminal, 10–12-rayed; involucre 1–2-leaved; involucel few leaved; fruit glabrous, roundish, winged. *Fœniculum? Kraussianum, Meisn.! in Hook. Lond. Journ. Bot. vol.* 2, 532.

VAR. β, **elatior**; downy; lower leaves cordate-acute, mucronately-toothed, twice shorter than the petiole; upper ones 5- or 3- partite, with lanceolate or linear segments, or undivided.

VAR. γ, **glabrata**; lower leaves cordate or reniform-cordate; upper ones subcordate or truncate at the base, or ovate-cuneate, sharply-serrated, rarely divided; stem and leaves glabrous; peduncle and rays of umbel downy. *Pimpinella cordata, E. Mey.*

HAB. In the plains near Port Natal, *Krauss*, 140; var. β, near Port Natal, *Gueinzius;* var. γ, between Omtendo and Omsamculo, *Drege.* Nov.–Feb. (Herb. Meisn. Hook. D. Sd.)

Root fusiform, simple. Stem terete, faintly striated, 1–1½ foot high, branched above the middle. Petiole of the lower leaves 3–2 lines broad, 2–1 inch long. Leaves 1½–2 inches long and broad, with a deep and broad sinus; the upper ones 3-fid, with cuneate lobes. Rays of umbel nearly 1 inch long, 3 times longer than the linear involucral leaves. Pedicels 3 lines long. Flowers white. Half-ripe fruit with winged ribs, as in *C. suffruticosum.* Furrows 1-vittate; commissure with 2 vittæ. Styles longer than the conical, short stylopodium. Var. β is 3 feet or more high; the lower leaves 2–2½ inches long, 18–20 lines broad, with equal, acute teeth; the upper leaves with elongated, entire, or toothed segments; the uppermost reduced to sheaths, without lamina. Umbels 16–20-rayed. Young fruit glabrous, ripe unknown. Var. γ is 1–1½ foot high, glabrous, but puberulous at the upper part of the stem and on the rays of umbel. Lower leaves as in var. α. and β. on very long or short, broad or narrow, petioles, duplicate-toothed or subserrate; the upper smaller and gradually narrower, serrate, rarely 3-fid, with linear lobes. Umbel with 12–16 rays. Flowers white. Ripe fruit unknown.

XXIV. LEVISTICUM, Koch.

Margin of *calyx* obsolete or with short teeth. *Petals* incurved, entire, with an acute point. *Fruit* compressed from the back, having 2 wings on each side. *Mericarps* with 5 winged ribs; the wings of the lateral ribs usually twice the breadth of the others; vittæ 1 in each furrow, and 2–4 in the commissure. *Carpophore* bipartite. *Seed* convex on the back, and flattish in front. *Koch. Umb.* 101. *f.* 41. *Endl. Gen. n.* 4453. *Species of Ligusticum, Linnæus.*

Strong, perennial, smooth and glabrous herbs. Stems erect, terete. Leaves pinnately divided; leaflets obovate-cuneated, toothed. Involucre and involucels of many leaves. Flowers yellow or yellowish. Name from *levo,* to *assuage;* said to relieve flatulency.

1. **L. grandiflorum** (Sond.); radical leaves bipinnate; lower stem-leaves pinnate; segments pinnatifid or inciso-lobed; lobes obovate or suborbiculate-cuneate, rugose, much veined and serrate-toothed; umbels with 5–8 elongated rays; involucre and involucel 4–6-leaved; fruit with short wings, the lateral ones scarcely broader. *Sium grandiflorum, Thunb.! Fl. Cap.* 260. *Bubon pimpinellifolium, E. & Z.!* 2251.

HAB. Stony places in Zwartland, *Thunberg, Wallich;* Riebeckkasteel, *Zeyh.* 733; Vierentwintig Rivieren, Worcester, *E. & Z.* Nov. (Herb. Thunb. Hk. D. Sd.)

Radical and lower leaves 1 foot and more long. Pinnulæ or segments 2–1 inch long, often rhomboid, more or less lobed or pinnatifid-incised; on the whole margin with sharp or mucronate teeth. Stem branched, striate, pruinose, 2 feet or more in height. Umbels terminal. Rays unequal, 1–3 inches, sometimes ½ foot in length. Leaves of involucre and involucel ovate, acuminate. Umbellules 10–12-flowered. Petals 1 line long, reddish-veined. Calyx-teeth short, acute. Stylopodium conical, with short, spreading styles. Fruit 4–5 lines long, 2 lines broad. Valleculæ with 1 large vitta. Commissure 2-vittate. The fruit agrees well with that of *Ligusticum Scoticum,* but is a little broader, and is distinguished by the solitary vittæ and the central, not marginal, raphe.

XXV. PEUCEDANUM, Koch.

Margin of the *calyx* 5-toothed or nearly obsolete. *Petals* obovate, emarginate or subentire, with inflexed points. *Fruit* dorsally compressed, flat or lenticular, with a dilated, flattened margin. *Mericarps* with subequidistant ribs, the 3 intermediate or dorsal ones filiform, the 2 lateral confounded with the dilated margin. *Furrows* with single vittæ, the lateral sometimes 2-vittate. *Commissure* 2-vittate. *Carpophore* 2-partite. *Seed* flat in front. *Koch. Umb.* 92. *f.* 28. *et* 29. *DC. l. c.* 176. *Peucedanum et Dregea, E. & Z.! Peucedanum et Sciothamnus, Endl. Gen.* 4463, 4464.

Glabrous, perennial herbs or shrubs. Leaves simply pinnate, ternately-sect or multifid. Umbels terminal. Involucres and involucels many-leaved. Flowers white or yellow. Name from *πευκη, a pine*, and *δαυος, parched;* so called on account of the strong smell, which resembles resin.

ANALYSIS OF THE S. AFRICAN SPECIES.

1. Shrubs:
   - Leaflets entire.
     - Umbels with many rays. Fruit obtuse at both ends ... (1) **Capense.**
     - Umbels with 4–6 rays. Fruit emarginate at both ends ... (2) **abbreviatum.**
   - Leaflets cuneate, 3–5-toothed or incised ... (3) **Ecklonianum.**
   - Leaflets or lobes linear-subulate ... (4) **striatum.**
2. Perennial herbs, with depressed stylopodium:
   - Lobes of leaves terete, angustate, pungent ... (10) **pungens.**
   - Lobes of leaves linear-subulate, capillary or filiform.
     - Leaves triternate ... (8) **triternatum.**
     - Leaves trichotomously decompound:
       - Lobes capillary or sulcate, refracted ... (5) **capillaceum.**
       - Lobes subulate, erect or spreading, not refracted:
         - Lobes subulate, sulcate, triquetrous; fruit 3 lines long; stem sulcate-striate ... (6) **Ferulaceum.**
         - Lobes long, subulate, sulcate, triquetrous; fruit 5–6 lines long; stem sulcate ... (7) **Sieberianum.**
         - Lobes filiform, sulcate; stem finely striate ... (9) **lateriflorum.**
   - Lobes of leaves ovate or oblong, or linear-oblong:
     - Radical leaves 3-nately sub-2-pinnate; lobes acute, serrate-toothed; fruit 5–6 lines long ... (11) **connatum.**
     - Radical leaves pinnate, lobes cuneate, mucronately toothed; fruit 4 lines long ... (12) **platycarpum.**
     - Radical leaves 2-pinnatisect; lobes linear-oblong, acute ... (13) **magalismontanum.**
3. Perennial herbs with conical stylopodium:
   - Stem deeply sulcate, lobes of leaves lanceolate ... (16) **sulcatum.**
   - Stem striate:
     - Umbel 20–30-rayed; fruit obtuse at each end; lobes of leaves linear, spreading ... (14) **Cynorhiza.**
     - Umbel 12–20-rayed; fruit elliptical, emarginate at both ends; lobes of leaves lanceolate ... (15) **Zeyheri.**
     - Umbel 10–16-rayed; fruit obovate, emarginate at the apex; lobes of leaves subcapillary, secundate ... (17) **millefolium.**

Sect. I. Shrubs. *Dregea*, E. & Z.! SCIOTHAMNUS, Endl. (Sp. 1–4.)

1. P. Capense (Sond.); fruticose; leaves pinnate, bipinnate or subtripinnate; leaflets ovate, oblong, or lanceolate, with revolute or thickened margins; umbels many-rayed; *fruit elliptic-oblong or obovate, obtuse at both ends.*

VAR. *α*, **latifolium**; leaflets ovate or oblong, mucronate; umbels with 20–40 rays. *Laserpitium Capense, Thunb.! Fl. Cap.* 256. *Dregea Capensis, E. & Z.* 2240. *P. rigidum, E. Meyer.*

VAR. β, **lanceolatum**; leaflets lanceolate, mucronate; umbels 10–24-rayed. *P. virgatum, Cham. & Schlecht.! Linnæa I.* 392. *P. frutescens et lanceolatum, E. Mey. Dregea virgata, E. & Z.!* 2241, *Zeyh.* 2691.

HAB. Hills and mountains in the districts of Uitenhage, Albany, George, Graafreynet and Caffraria, and near Port Natal. March–June. (Herb. reg. Berol. Hk. D. Sd.)

Shrub several feet high, branches terete, sulcate-striate, yellowish-brown, few-leaved on the upper part. Lower leaves larger; leaflets in var. α 1½–2 inches long, ½–1 inch broad; in var. β ¾–1 inch long, 1–2 or 3 lines wide, sessile or shortly petiolate, penninerved, entire, rarely 2 or 3-lobed. Petiole adnate to the back of the 4-6 lines broad sheaths, angulate, striate, its ramifications articulated with the rachis. Upper leaves much smaller and less divided, often reduced to sheaths without limb. Rays of umbel in var. α 1½–2 inches, in var. β 1–1½ inches, long. Leaves of involucre (5–10) lanceolate, 4–8 lines long. Involucel similar, twice shorter than the pedicels. Flowers yellow. Fruit 3 lines long, 2 lines broad, quite flat on the dilated margin; the 3 dorsal ribs filiform, elevated. Stylopodium and styles very short.

2. P. abbreviatum (E. Meyer); fruticose; leaves pinnate; leaflets lanceolate, mucronate, attenuated at the base, *subfalcate*, with revolute margins, sessile, the lower ones petiolate, binate or ternate; umbels with 4–6 rays; *fruit elliptical, emarginate at both ends.*

HAB. Camdeboosberg, stony places, 4–5000 ft. Jan.—*Drege.* (Herb. Sd.)

Allied to *P. Capense*, but a much smaller shrub with smaller leaves; the leaflets 4–6 lines long, 1 line broad; the rachis articulate. Involucre and involucel of few lanceolate leaves. Rays of umbel 8–12 lines, of umbellules 1½–2 lines long. Fruit 2½–3 lines long.

3. P. Ecklonianum (Sond.); fruticose; leaves *pinnate; leaflets cuneate, 3–5-dentate or incised,* with revolute margins, teeth acute, spreading; umbels 8–10-rayed; fruit obovate. *Dregea montana, E. & Z.!* 2242.

HAB. On the Winterhoekberge, near the Elandriver, Uitenhage, March. *E. & Z. Zeyher*, 2689. (Herb. Hk., Sd.)

About 1–1½ foot high. Stem often a little rough. Leaves short petiolate, 1 inch long, rhachis articulate; leaflets 4–6 lines, 1–1½ line broad; the lateral teeth ½–1 line long, often recurved. Involucre and involucel of 4–6 small leaves. Rays of umbel 4–6 lines; of umbellules 1 line long. Flowers yellow. Fruit 2 lines long. Dorsal ribs filiform, elevated.

4. P. striatum (Sond.); fruticose; leaves rigid, *tripinnatisect; lobes linear-subulate,* mucronulate, sulcate; umbels 6–8-rayed; fruit elliptic. *Seseli striatum, Thunb.! Fl. Cap.* 259. *Dregea collina, E. & Z.!* 2243.

HAB. Near Buffeljagdrivier, *Mundt. in herb., E. & Z.;* Rietkuil and Kafferkuilsrivier, and between Zwarteberg and Rivierszondereinde. *Zeyher*, 2684 *et* 2682. Sept.–Dec. (Herb. Hk. Sd.)

A dwarf shrub, leafy, with green or purplish striate branches. Leaves with a very short petiole, adnate at the back of the broad, whitish-margined vagina, about 1 inch long, 2–3-pinnatisect; segments 2–3-partite; lobes 1 line long, often recurved, sulcate as well as the articulate rhachis. Uppermost leaves without limb. Involucre and involucels of 4-5 subulate leaves. Rays of umbel 4–6 lines long. Flowers yellowish-green. Ripe fruit unknown.

Sect. 2. Perennial herbs. Stylopodium depressed. PEUCEDANUM, Koch. (Sp. 4–13.)

5. P. capillaceum (Thunb.! Fl. Cap. 257); stem erect, terete, sulcate-striate, naked, a little branched; radical leaves long-petiolate, trichotomously supra-decompound; segments capillaceous, refracted, sulcate; involucre and involucel 5–8-leaved; *fruit large, elliptic-oblong. E. & Z.!* 2236.

VAR. β, **rigidum**; segments rigid, longer and thicker. *P. rigidum, E. & Z.* 2237. *Zeyh.* 2687. *Drege,* 7640.

HAB. Mountains near Zoete Melksvalley, *Thunb.;* River Zondereinde, and near Genadenthal, *E. & Z.;* var. β on the same localities, and on the Loari and Van Stadensriver, *E. & Z., Drege.* Dec.–April. (Herb. Thunb. Hk. Sd.)

Stem 1–2 feet. Petiole of the radical leaves ½–1 foot, sulcate above, all the divisions refracted at the base; the primary about 1 inch long, the following gradually smaller and articulated, the whole leaf nearly as long and large (4–6 inches); ultimate segments in var. α capillary, 1–4 lines, in var. β 5–8 lines; in the largest form 1 inch long or longer, and nearly ½ line broad, unisulcate above, bisulcate beneath, thence triquetrous. Umbel 10–16-rayed; rays 1–1½ inch long. Flowers yellow. Fruit when ripe 5 lines long, 3½ lines broad, the flat margin nearly 1 line broad. Valleculæ 1-vittate; commissure 2-vittate.

6. P. ferulaceum (Thunb.! herbar.); stem erect, terete, sulcate-striate, branched, leafy; *leaves very short-petiolate,* trichotomously decompound; segments subulate, sulcate-triquetrous; involucre and involucel 5–6-leaved; fruit small, obovate or obovate-oblong. *Oenanthe ferulacea, Thunb.! Fl. Cap.* 253. *P. tenuifolium, E. & Z.!* 2231. *Herb. Un. Itin. n.* 564 *et* 565. *Lichtensteinia pyrethrifolia, DC. Prod.* 4, 135. *Sieb. Fl. Cap. exs. n.* 211, 222. *P. elongatum, E. Mey. Oenanthe seseloides, Presl. bot. Bem.* 74.

VAR. β, **Stadense**; segments divaricate. *P. Stadense, E. & Z.!* 2232, *Zeyh.* 2688, *Drege,* 7643.

HAB. On Table- and Devilsmountain, Dutoitskloof, and in Hottentholld.; var β. Vanstadensriviersberge, and near Grahamstown. Oct.–Jan. (Herb. Thunb. Hk. D. Sd.)

Near the foregoing; the stem is usually higher (3–4 feet) and leafy, only the upper part is naked or nearly so. Petiole adnate to the broad, uncial vagina; lower leaves about 4 inches long, 2–3 inches broad; the segments 2–6 lines long, erect; in var. β spreading, but not refract. Umbels and flowers as in *P. capillaceum,* but the fruit is 3 lines long, 2 lines broad, the flat margin ⅔ line broad.

7. P. Sieberianum (Sond); stem erect, terete, sulcate, branched, leafy at the base; *leaves long-petiolate,* trichotomously decompound; segments *erect,* elongated, subulate, *triquetrous;* involucre and involucel 5–10-leaved; *fruit large, elliptical. P. capillaceum DC. l. c. Sieb.! Fl. Cap. exs. n.* 212. *Ferula stricta, Spr. E. Mey. in herb. Drege. P. ferulaceum, E. & Z.!* 2234, *excl. syn.*

HAB. Rocks on the Kasteel and Tablemountain, and near Tulbagh, *Sieber, Wallich E. & Z.;* Dutoitskloof, *Drege.* Dec.–Feb. (Herb. Hk., Sd.)

Distinguished from *P. capillaceum* by the higher, stiff stems, erect, not refracted segments and ramifications of petiole and larger umbel. Stem a finger thick. Leaves radical and on the lower part of the stem. Petiole terete, striate, shorter or as long as the 4–12 inches long leaf. Segments of leaves 1 inch long, sometimes shorter or longer. Rays of umbel 16–24, unequal, 1–2 inches long. Umbellules many-flowered. Flowers yellow. Fruit 5–6 lines long, 3–3½ lines broad; the flat margin 1 line broad.

8. P. triternatum (E. & Z.! 2235); stem erect, terete, finely striate, simple or with a few branches at top, naked; radical leaves petiolate, *triternate; segments linear-filiform,* acute; umbels 5–6-rayed; involucre and involucel 5–6-rayed; fruit small, ovate-oblong.

HAB. Rocks in Hotthldbergen., near Palmietrivier, *E. & Z.* Jan. (Herb. Sd.)

Stem 2–3 feet high, nearly filiform. Petiole of the radical leaves filiform, terete, sulcate, equalling the 2–3 uncial, simply biternate leaf, the segments of which are 1–1½ inch long. Rays of umbel very unequal, ½–1½ inch long. The small flowers seem to be yellow. Fruit 2½–3 lines long, 1½–2 lines broad; the flat margin nearly ½ line broad.

9. P.? lateriflorum (Sd.); stem erect, terete, striate, branched above the middle, naked; leaves radical, petiolate, 4–5 *times triternate;* segments spreading, filiform, sulcate; umbels often proliferous, 5–8-rayed; involucre and involucel many-leaved; fruit . . . *Athamantha lateriflora, E. & Z.!* 2229.

HAB. On the Kamiesberge, Namaqualand; Nov.; *E. & Z., Rev. H. Whitehead.* (Herb. Sd. D.)

Stem straight, 3–4 feet high. Petiole of the radical leaves 2–6 inches long, terete, sulcate; its ramifications similar, gradually narrower, ultimate segments 1–1½ inch, in other leaves only 2–6 lines long and then capillary. All the branches terminated by umbels. Rays very unequal, some of them scarcely 6 lines, others 4 inches long. Umbellules 8–12-flowered, pedicels unequal, 1–6 lines long. Leaves of involucre and involucel lanceolate, 2–1½ lines long. Fruit when very young obovate. Stylopodium depressed; styles short; ripe fruit unknown. It has exactly the habit of the preceding, but much divided leaves.

10. P. pungens (E. Meyer); stem erect, much branched, sulcate-striate, leafy, but naked above the middle; leaves sessile, ternately-pinnate; rhachis terete, striate, as well as the attenuated spinous segments; umbels many-rayed; involucre and involucel 6–8-leaved; fruit small, obovate. *Seseli striatum, E. & Z.!* 2226, *non Thunb.*

HAB. Hills between Potrivier and Langehoogde, Caledon, *E. & Z.;* near Gnadenthal, *Drege.* Aug.–Nov. (Herb. Hk. Sd.)

Stem several feet high. Leaves aggregated at the base or to the middle. Petiole adnate to the back of the large, 1–2 uncial vagina. Pinnæ or segments articulated with the rachis. Lower pinnæ 3-partite, the upper 3-fid, 2-fid, or undivided, at the base as thick as a pigeon's, the rhachis as a goose's, quill; attenuated like a needle, and pungent, the longer 1–2 inches, the shorter 4–6 lines, long. Umbel 15–20-rayed, rays 1 inch long. Umbellules many-flowered. Fruit 3 lines long, 2 lines broad; the flattened margin narrow, nearly ⅓ line broad. Valleculæ 1-vittate; commissure 2-vittate.

11. P. connatum (E. Meyer); stem erect, striate, simple or slightly branched at top; leaves radical, *ternately sub-bipinnate;* segments ovate, cuneate, trifid or 5-fid, *lobes acute, mucronately serrate-toothed;* umbels many-rayed; involucre 1-leaved or wanting; involucel of 4–6 large, membranaceous leaflets; fruit *obovate, large;* stylopodium *short, conical;* calyx-teeth nearly obsolete.

HAB. Grassy places, and on the sea-shore from Omtendo to Port Natal, *Drege.* Feb.–April. (Herb. Hk. Sd.)

Stem stiff, 2–3 feet high. Radical leaves on 1–3 uncial petioles. Leaves 3–6 inches long; segments pinnatifid-incised, ½–1 inch long, 6–8 lines wide, the upper ones smaller, and, as well as the lobes of the larger segments, commonly 3-dentate or sub-3-lobed. Cauline leaves none, or one at the base, similar to the radical. Rays of umbel unequal, 1–2 inches long. Flowers yellow. Fruit 5–6 lines long, 3 lines broad, the dorsal ribs scarcely prominent; the flat margin ½ line broad. Styles diverging, as long as the stylopodium. In habit it comes near the broader leaved form of *Peuc. Oreoselinum.*

12. P. platycarpum (E. Mey.); stem erect, striate, branched; leaves radical, *pinnate;* pinnæ deeply pinnatifid, *lobes cuneate, mucronate-toothed;* umbels 8–10-rayed; involucre 5–7-leaved, deciduous; involucel of 4–7 small, linear leaflets; fruit *elliptical, middle-sized;* stylopodium *depressed;* calyx-teeth very minute.

HAB. On the Katriver and Kachu- or Geelhoutriver, 2–3000 feet. Nov.–Jan. *Drege.* (Herb. Sd.)

Radical leaves 2–3 inches long, 1½–2 inches wide, 2–4 times shorter than the terete petiole. Segments about ½ inch long, cuneate as the usually 3-dentate lobes. Stem 2–3 feet high, with leafy sheaths at the base of the ramifications. Leaves of involucre and involucel 1–2 lines long. Rays of umbel unequal, 1–2-uncial. Flowers wanting. Fruit 4 lines long, 3 lines broad, much compressed; the flat margin ¾ lines broad. Dorsal ribs obtuse. Valleculæ 1-vittate; and commissure 2-vittate, as in the preceding; the vittæ superficial. It belongs to the section *Selinoides,* of DC.

13. P. magalismontanum (Sonder); glaucous-green, stem erect, branched; radical leaves petiolate, *bipinnatisect, segments deeply pinnatifid, lobes linear-oblong, acute,* entire or bifid; umbel 6–12-rayed; involucre and involucel wanting; fruit obovate, with a dilated, rather convex, margin; stylopodium depressed; calyx-teeth very minute.

HAB. Grassy fields on the Magalisberg, and near Vaalriver; *Burke,* 277; *Zeyher,* 118, 744. Nov.–Jan. (Herb. Hk. D. Sd.)

Petiole of radical leaves ½–4 inches long. Leaves 4–6 inches long, the middle pinna twice longer than the 2 lowest; the segments sessile, or the lower pair short petiolate; lobes 1–3 lines long, ½ line wide, the terminal mostly 3-fid, the lateral entire or bifid. Stem 1½–2 feet high, with 1 pinnate leaf or quite naked, with sheaths at the ramifications. Rays of umbel unequal, 1–3 inches long. Flowers yellow. Fruit 4–5 lines long, 2–3 lines broad, much compressed. Ribs filiform, obtuse. Valleculæ 1-vittate; commissure 2-vittate; margin ¾ line broad, more thickened than in the other species. In the form of the leaves it resembles *P. alsaticum.*

Sect. 3. Perennial herbs with a large, fleshy root. Leaves all radical. Involucre and involucel many-leaved. Calyx-teeth minute, acute. Stylopodium conical. Margin of mericarps very broad, diaphanous. Dorsal ribs 5, filiform, obtuse, at equal distances. Vittæ 1 in each furrow, as well as the 2 commissural vittæ, superficial. *Cynorhiza,* E. & Z.! Enum. p. 351. This section agrees nearly with *Imperatoria* and *Selinoides,* DC.; from the latter it is distinguished by broader margins and conical stylopodium; from *Imperatoria* by the involucre and short calyx-teeth. The raphe is marginal, not central, as indicated by *E. & Z.;* the furrows are always 1-vittate, not 2-vittate, as described by *Endlicher.* (Sp. 14–17.)

14. P. Cynorhiza (Sond.); stem erect, terete, striate; leaves petiolate, *with 3-angular circumference,* tripinnatisect, segments pinnatifid; lobes linear or oblong-linear, acute; umbels 20–30-rayed; rays unequal; leaves of involucre and involucel ovate, cuspidate, or lanceolate, deciduous; fruit elliptical or ovate, *obtuse at each end,* crowned by the conical stylopodium. *Cynorhiza typica, E. & Z.!* 2244.

HAB. Sandy hills near the Zwartkopsriver, *E. & Z.; Zeyh.* 2693. Jan.–April. (Herb. Hk. Sd.)

Stem 1–1½ feet high. Petiole 2–3 inches long, 3-partite. Leaf 4–5 inches long and broad. Lobes 2–3 lines long, ¾–1 line wide, those of the lateral segments often bifid, and the terminal sometimes 3-fid. Rays of umbel 1½–2 inches long. Flowers white, E. & Z. Fruit 4–5 lines long, 3–4 lines broad; the margins 1 line broad.

15. P. Zeyheri (Sond.); stem erect, terete, *striate;* leaves petiolate, *with oblong-triangular circumference,* quadri-pinnatisect; segments pinnatifid or ternately 3-fid; lobes lanceolate; umbels 12–20-rayed; rays unequal; leaves of involucre and involucel ovate, cuspidate, or lanceolate, deciduous; fruit elliptical or elliptical-obovate, *emarginate at both ends,* the conical stylopodium in the notch. *C. montana, E. & Z. !* 2245.

HAB. Stony places near Coegakopje at the Zwartkopsriver, *Zeyh.* 2692; *Drege,* 7641, *d.* Oct.–Jan. (Herb. Hk. D. Sd.)

Stem 2–3 feet high and more, branched as in the foregoing; the leaves are often 2 feet long, the ramifications of the petiole more spreading and naked at the base; the segments larger and often tripartite, and the lobes in the perfectly developed leaves 4–6 lines long, 1–2 lines broad, with acuminate point. The umbels with 1½–2½ uncial rays. Fruit 4–5 lines long, 4 lines broad. *Cynorhiza? alta,* E. & Z.! 2246. *Drege,* 7641, *a, c, e,* seems to consist of gigantic specimens of *P. Zeyheri;* the leaves are not different, the fruit is unknown.

16. P. sulcatum (Sond.); stem erect, terete, *deeply sulcate;* leaves petiolate, with oblong-triangular circumference, quadri-pinnatisect; segments pinnatifid or twice or ternately 3-fid; *lobes spreading, lanceolate;* umbels 30–40-rayed; rays subequal; leaves of involucre and involucel ovate, cuspidate; fruit (when young) obovate-oblong, broadly-margined; stylopodium conical. *C. ? sulcata, E. & Z.!* 2247.

HAB. Mount Kamiesberg, Namaqualand, *Zeyher.* (Herb. Sd.)

Well distinguished by the deeply sulcate stem. Leaves as large as in *P. Zeyheri,* the lobes 1 line broad. Rays of umbel 2½–3 inches long. Ripe fruit unknown.

17. P. millefolium (Sd.); stem erect, terete, *striate,* slightly branched; leaves petiolate, with oblong circumference, 4–5 times pinnatisect, supra-decompound; lobes erect, *secundate, linear-subulate,* short and crowded; rhachis and primary ramifications muricate, lobes glabrous; cauline leaves wanting, or one much smaller; umbels terminal, with 10–16 unequal rays; leaves of involucre lanceolate-acuminate, of involucel linear; fruit obovate, deeply emarginate at the apex, with short conical style in the notch. *Ferula meifolia, E. & Z. !* 2230. *Analyrium millefolium, E. Meyer.*

HAB. Sandy places near Vierentwintig Rivieren, *E. & Z., Drege;* Cape Flats, *Zeyher.* Nov. (Herb. Hk. Sd.)

Stem 1–3 feet high. Leaves ½–1 foot long, on a short or 3-uncial broad petiole. Rhachis striate at the base, 2 lines broad; the primary branches 1–2 inches long; the multifid segment 5–3 lines long, the lobes 1 line long, subcapillary. Rays of the solitary umbels 1–2 inches; pedicels 2–4 lines long. Flowers white. Fruit 5–8 lines long, 4–6 lines broad, truncate or obtuse at the cuneate base; margin 1 line broad. In the form of the leaves it has a great resemblance to *Peuced. meifolium,* Boiss.!

XXVI. BUBON, Linn.

Margin of the *calyx* obsolete. *Petals* obovate, entire, with an acute, involute point. *Fruit* dorsally compressed, lenticular, girded by a narrow, flattened margin. *Mericarps* with 5 ribs at equal distances; the 3 intermediate ones filiform; the 2 lateral ones confounded in the complanate margin. Vittæ broad, solitary in each furrow, and rarely (in one species) under the jugum. *Commissure* 2-vittate. *Carpophore* bipartite. *Seed* rather convex, flat in front. *Koch, Umb.* 95, *DC. l. c.* 134; *Agasyllis Spec. Spreng. Endl. gen.* 4466.

Glabrous shrubs, with a resinous smell. Stems terete, branched. Leaves ternately compound, leaflets veined. Umbels of many rays. Involucre and involucels of many, linear leaves. Flowers greenish-yellow. Name from βουβων, the *groin*, or a tumour, which this herb was supposed to cure. This genus is very nearly allied to *Peucedanum*, from which it is only distinguished by the narrower margin of the fruit, and the petals.

ANALYSIS OF THE SPECIES.

Lobes of leaves rhomboid, cuneated ... ... (1) **Galbanum.**
Lobes of leaves elongated, lanceolate, acute, green ... ... (2) **tenuifolium.**
Lobes of leaves elongated, linear-lanceolate, glaucous beneath; umbels many-rayed; leaves of involucre lanceolate ... ... (3) **hypoleucum.**
Lobes of leaves abbreviate, linear-subulate, glaucous beneath; umbels 8–12-rayed; leaves of involucre linear-setaceous ... ... (4) **montanum.**
Lobes of leaves elongate, filiform, glaucous; umbels many-rayed (5) **Capense.**

1. B. Galbanum (Linn. Spec. 364); *leaves petiolate, pinnate*, triternate, glaucous; *segments cuneated, rhomboidal, toothed or incised*, terminal ones 3-lobed. *Thunb. Fl. Cap.* 258. *Jacq. Hort. Vind.* 3, *t.* 36. *Sims. Bot. Mag. t.* 2489. *Pappe. Fl. Cap. med.* 18. *E. & Z.!* 2248.

HAB. Moist places, and in the ravines of mountains all over the colony. Sept.–Jan. (Herb. Sd. etc.)

Stem leafy, 6–8 feet high. Leaves variable; segments sometimes 2–3 inches long, 1½–2 inches broad (var. Tulbaghica, E. & Z.); commonly about 1 inch long, and cuneate; in other specimens elongate-cuneate, inciso-serrate, or subpinnatifid. Rays of umbel often very numerous, 100 or more. Fruit 2–2½ lines long, 1½ line broad.

2. B. tenuifolium (Sond.); *leaves petiolate*, 2–3-*pinnatisect; segments pinnatifid* and trifid; *lobes lanceolate*, acute, with revolute margins, *green on both surfaces; fruit elliptic or ovate*, with filiform, bluntish ridges, and a very narrow margin. *Peucedanum tenuifolium, Thunb. Fl. Cap.* 257. *Oreoselinum uliginosum, E. & Z.!* 2238. *Bubon? multiradiatum, E. Meyer, c, d, e.*

HAB. Mountains near Capetown, *Thunberg & Drege;* Krumriver, *E. & Z.;* Langekloof and Klipriver, *Drege.* Nov.–Jan. (Herb. Thunb. Hk. Sd.)

Several feet high, flexuose, very leafy. Leaves ½–⅔ foot long, ternately pinnate, decompound; lobes 2–4 lines long, ½–¾ line broad. Umbel 30–40-rayed; rays 1½–2 inches long. Leaves of involucre 3–4 lines long, lanceolate, acuminate. Fruit 2 lines long, 1 line broad. Mericarps with univittate furrows and bivittate commissure; the juga without vittæ.

3. B. hypoleucum (Meisn.! in Hook. Lond. Journ. Bot. v. ii. 536); leaves petiolate, 2–3-pinnatisect; segments pinnatifid and trifid; *lobes linear or linear-lanceolate, acute*, with revolute margins, green above, *glaucous beneath; umbels many-rayed;* leaves of involucre lanceolate-acuminate; *fruit oblong*, with very obtuse, filiform ridges, and a narrow convex-flattened margin. *B. gummiferum, Drege, Herb. ex pte. Oreoselinum uliginosum, β. glaucum, E. & Z.! l. c.*

HAB. Near Duyvels and Voormannsbosch, Swellendam, *E. & Z.;* Bavianskloof, near Gnadenthal, *Drege, Krauss.* Oct.–Dec. (Herb. Hk. Sd.)

Very like the preceding; the whole plant glaucous, the leaves with twice or 3 times longer, 1 line broad lobes; stem and umbels are the same. The fruit is 3½–4 lines long, 1½ line broad, the ridges more prominent, and the margin a little broader. As already observed by *Dr. Meisner*, there are not only vittæ in the valleculæ, but also under each jugum.

4. B. montanum (Sond.); leaves petiolate, 3–4-pinnatisect; segments pinnatifid; *lobes abbreviate, linear-subulate,* with revolute margins, glaucous beneath; *umbels* 8–12-*rayed;* leaves of involucre linear-setaceous; fruit oblong, or ovate-oblong, with filiform, obtuse ridges, and a narrow convex-flattened margin. *B. tenuifolium, E. Meyer.*

HAB. Dutoitskloof, *Drege.* Oct.–Jan. (Herb. Hk. Sd.)

Perhaps this species is a variety of the foregoing, but the habit is very different. The stem is much slenderer, 3–4 feet, the leaves about 1 foot in length; the sections with very delicate, nearly capillary, 1–2 lines long lacinulæ. Rays of umbel 2 inches long. Involucre and involucel capillary, attenuated. Fruit 3–3½ lines long, 1½ line broad. Mericarps with univittate furrows, but the juga always without vittæ.

5. B. Capense (Sond.); leaves petiolate, *ternately decompound; lobes elongate,* linear-filiform, undivided or pinnatifid, with similar lacinulæ; umbels 40–50-rayed; fruit ovate-oblong, with filiform, obtuse ridges, and a very narrow margin. *Oreoselinum Capense, E. & Z.!* 2239. *Bubon multiradiatum, E. Mey. a. b.*

HAB. Mountains near Klapmuts, Stellenbosh, and on Bergriver near Paarl, *E. & Z., Drege.* Oct.–Dec. (Herb. Hk. Sd.)

Habit and leaves of a true *Peucedanum.* Stem and leaves glaucous-pruinose. Lower leaves often more than a foot long and broad; stem-leaves similar, but smaller. Terminal lobes 1–2 inches long, ⅓ line broad, sulcate; when divided, with remote, alternate lacinulæ. Rays of umbel 2–2½ inches long. Leaves of involucre lanceolate. Fruit 2½–3 lines long, 1–1¼ line broad. Mericarps with 1-vittate valleculæ and 2-vittate commissures; the juga without vittæ.

(Species not sufficiently known.)

B. lævigatum (Ait. Hort. Kew. ed. 1, vol. 1, 352. ed. 2, vl. 2, 146); stem frutescent; leaves bipinnate; leaflets lanceolate, bluntly and obsoletely crenated; fruit glabrous. *Ferula lævigata, Spreng. Umb. Spec.* 88.

HAB. South Africa.

Umbel depauperate. Flowers yellow. Fruit thick, solid, with 3 dorsal, obtuse ribs (ex Spreng.) Seemingly a species of *Peucedanum* (Dregea).

XXVII. ANETHUM, Linn.

Margin of the *calyx* obsolete. *Petals* roundish, entire, involute, with a square, retuse point. *Fruit* dorsally compressed, lenticular, with a flattened margin. *Mericarps* with filiform, equidistant ridges, the 3 dorsal acutely carinate, the 2 lateral subobsolete, confounded with the margin. *Vittæ* broad, solitary in each furrow and completely filling it, 2 in the commissure. *DC. l. c.* 185. *Endl. Gen. n.* 4467.

Annual, glabrous herbs, with decomposed leaves and yellow flowers, without involucres. This is the *ανηθον* of Theophrastes; derived from *ανω, upwards,* and *θεω, to run,* alluding to the quick growth of this plant.

1. A. graveolens (L. Spec. 377); lobes of the leaves elongate, linear-filiform; fruit elliptic, with a flat margin. *Hayne Arz. Gew.* 7. *t.* 17.

HAB. In corn-fields and in cultivated grounds, escaped from gardens.

XXVIII. PASTINACA, Linn.

Margin of the *calyx* obsolete or minutely toothed. *Petals* roundish, entire, involute, the involute part broad and retuse. *Fruit* flat-compressed dorsally, surrounded by a dilated, flattened margin. *Mericarps* with very slender ridges; the dorsal and 2 intermediate ones equidistant, the lateral near the outer edge of the dilated margin. *Vittæ* linear, scarcely shorter than the ridges, solitary in each furrow, 2 or more on the commissure. *Carpophore* bipartite. *Seed* flattened. *DC. l. c.* 188. *Endl. Gen. n.* 4473.

Herbaceous plants with a fusiform and often fleshy root. Leaves pinnated, the segments toothed, cut or lobed. Umbels compound. 1 involucre and involucel wanting or few-leaved. Flowers usually yellow. Name from *pastinum*, a dibble; in reference to the form of the root.

1. **P. Capensis** (Sond.); stem deep-furrowed; leaves pinnate, nearly glabrous above, clothed with short pubescence beneath; lateral pinnæ subsessile, entire or bifid, the terminal 3-lobed; segments ovate, acute, dentate; involucre 1-, involucel 2-leaved; calyx-teeth obsolete; fruit orbicular, glabrous; lateral vittæ close to the intermediate ridges; commissure with 2–4 vittæ.

HAB. Cape (special locality not indicated), *Drege*, 7627. (Herb. Sd.)

Root of *P. sativa*. Stem 2–3 feet high, angulate-sulcate, glabrous, much-branched. Leaves similar to those of *P. ligusticifolia*, W. & A., but the segments not serrate, but with short, mucronulate teeth, 1½–2 inches long, 1–1½ inch broad. Petiole downy, sulcate above, multistriate beneath. Rays of umbel (10–16) unequal, 1–2 inches long; pedicels (16–20) a little hairy. Flowers yellow. Fruit 2 lines long and broad or a little larger. Commissure with 2 vittæ, or with 4, two of which are larger. It is nearly allied to *P. divaricata* and *lucida*, Gouan.; but it is distinguished by the fruit and stem. *P. lucida*, Linn., that I have not seen, must be a quite different plant.

XXIX. CAPNOPHYLLUM, Gaertn.

Margin of the *calyx* obsolete. *Petals* oblong, subemarginate, with an acute, inflexed point. *Fruit* lenticularly compressed, girded by a flattened, dilated margin. *Mericarps* having the 3 dorsal ridges thick, carinate, flexuose or tuberculated; the 2 lateral passing into the dilated margin. *Furrows* with single vittæ. *Commissure* 2-vittate. *Seed* rather convex, but flat in front. *Koch. Umb. p.* 95. *n.* 44. *in add. DC. l. c.* 187. *Endl. Gen. n.* 4470.

Annual herbs, with glaucous, multifid leaves almost like those of *Fumaria*. Umbels opposite the leaves or nearly terminal, many-rayed. Involucre and involucels composed of 3–6 leaves, with membranous edges. Flowers white. Name from *καπνος*, the Greek name for *Fumaria*, and *φυλλον, a leaf*.

1. **C. Africanum** (Koch. l. c.); umbels of 3–10 rays; involucre and involucels of 3–6 leaves; mericarps with tuberculated ridges. *Conium Africanum, Linn. Mant.* 352. *Thunb.! Fl. Cap.* 257. *Jacq. Hort. Vind. t.* 194. *E. & Z.!* 2252. *C. Africanum et Jacquini, DC. l. c. Actinocladus cinerascens, E. Mey.! Ind. Sem. Hort. Bot. Regiomont.* 1847. *Zeyh.* 742.

VAR. β, **leiocarpon**; mericarps with flexuose, not tuberculated ridges.

HAB. Sandy places in the Cape Flats and near Van Kampsbay; var. β, near Capetown, *Drege*, 6243. Sep.–Nov. (Herb. Thunb. Hk. D. Sd.)

An erect or prostrate herb, 1–1½ foot. Leaves 2–3-pinnate; segments multifid; lobes short, subcapillary, often thickish; cauline leaves smaller, shorter petiolate. Rays of the shortly pedunculate umbel ¾–1½ inch; pedicels of the many-flowered umbellules 2–4 lines long. Some of the flowers of an umbellule usually sterile. Fruit 3–4 lines long, 2–2½ lines broad, glaucous.

XXX. PAPPEA, Sond. & Harv. (non Eck. & Zey.)

Flowers all hermaphrodite, fertile, regular. *Margin* of the calyx obsolete. *Petals* glabrous, ovate, shortly acuminate, furnished externally with a longitudinal medial fold, dorsally biconvex, keeled on the inside, with an acute, incurved point. *Styles* 2, broad-based, short; stigmata terminal, capitellate. *Stylopodium* depressed. *Fruit* dorsally compressed, lenticular, crowned with the styles, pilose externally, consisting of 2 mericarps. *Mericarps* somewhat convex at back, and covered with hairs of two kinds, one shorter, the other longer, clavate, minutely tuberculate, and patent. *Ribs* none, except the marginal, which form a broad, densely villoso-ciliate margin, furnished within with a series of oil-cells. *Commissure* flat, even, glabrous, completely joined at the margin. *Raphe* marginal, at one side. *Vittæ* none (save the above-mentioned oil-cells). *Carpophore* reaching to the middle of the mericarps, bipartite from the base, persistent after the fall of the mericarps, swelling when moistened (as if gelatinous). *Albumen* rather convex at back, flat in front.

An annual, prostrate, many-stemmed, glaucous herb, glabrous in all parts except on the fruit. Root filiform, simple. Stems 1–2 inches long, somewhat branched. Leaves (nearly resembling those of a *Fumaria*), irregularly, subternately cut, the radical with longer petioles, about uncial, including the petiole. Lobes capillary, obtuse, about 1 line long. Umbels at the base of the stem, at the origins of the branches, and also at their apices, solitary, sessile. Umbellules few, 4–6-rayed, unequally pedicellate. Pedicels about 1 line long. Involucre and involucels similar to the leaves, equalling the flowers. Flowers white, 1 line long. Fruit 1–1½ line long and wide. Having (vol. 1, p. 241) been compelled to unite the *Pappea* of Eck. & Zey. with *Sapindus*, we gladly seize the opportunity, now afforded us, of dedicating to our valued friend Dr. Ludwig Pappe, Colonial Botanist, the very remarkable plant here described, and which constitutes a genus of whose distinctness from all others there can be no question.

1. **Pappea Capensis** (Sond. & Harv., non E. & Z.!)

Hab. Nieuwejaarsspruit, between the Gariep and Caledon river, near the foot of the Witberg, 4–5000 feet. Oct. *Zeyher!* (Herb. Sond. D.)

XXXI. DAUCUS, Linn.

Margin of *calyx* 5-toothed. *Petals* obcordate, with an inflexed lobe, exterior usually larger and bifid. *Fruit* dorsally compressed. *Mericarps* with bristly, primary ridges; secondary ridges equal-winged, with 1 row of spines. *Furrows* with single vittæ under the secondary ridges. *Endl. Gen. n.* 4497.

Herbaceous, often biennial plants. Leaves 2–3-pinnated. Involucra of several trifid or pinnatifid leaves; involucels of many entire or trifid leaflets. Flowers white or yellow; the central one often fleshy and sterile. Δαυκος, of Dioscorides, is said to be from δαιω, *to make hot;* from its supposed effect in medicine.

1. **D. Carota** (Linn. Spec. 348); stem hispid; leaves 2- or 3- pinnated; leaflets pinnatifid; lobes lanceolate, cuspidate; leaflets of the involucre pinnated, about the length of the umbel; umbel with a solitary, coloured, abortive, central flower, concave when in seed. *Smith. Engl. Bot. t.* 1174.

Hab. Borders of fields and in gardens, probably escaped from culture, *Carrot.*

Sub-Order II. **Campylospermeæ** (DC. l. c. 215); albumen involute, or marked by a longitudinal furrow or channel on the inner side. (*Gen.* xxxii., xxxv.)

XXXII. TORILIS, Spreng.

Calyx-teeth 5, triangular-lanceolate, acute, persistent. *Petals* obovate, emarginate, with an inflexed point, the outer larger, bifid. *Fruit* contracted from the sides. *Mericarps* with 5 primary setulose ridges, 3 dorsal and 2 lateral (placed on the commissure); the secondary ridges represented by rows of bristles (hook-pointed) filling the whole furrow. *Vittæ* solitary under the secondary bristles. *Carpophore* setaceous, bifid. *Seed* with its margins inflexed. *DC. l. c.* 219. *Koch, Umb.* 80, *t.* 15. *Endl. Gen. n.* 4503.

Annual herbs, with multifid leaves, beset with short appressed hairs, which are retrorse on the stems, and erect on the rays of the umbels. Umbels opposite the leaves. Involucra wanting, or of 1–5 leaves. Involucels of 5–8 lanceolate, ciliated leaves. Fl. white, those in the disk of the umbellules male. Name unexplained

1. T. Africana (Spreng. in Schult. Syst. 6, 486); umbels on long peduncles, 2–4-fid; involucra wanting; umbellules 3–6-flowered, with 4–5-leaved involucel. *E. & Z.!* 2254. *Caucalis Africana, Thunb.! Fl. Cap.* 256. *C. Capensis, Lam. Herb. Un. Itin. n.* 805. *Zeyh.* 2696.

HAB. In cultivated ground, common. Oct.–Dec. (Herb. Sd. etc.)

Plant 1–1½ foot, erect, leafy. Leaves pinnately decompound; lobes short, lanceolate, dentate. Fruit ovate, 2–2½ lines long; its prickles shorter than the breadth of the fruit.

XXXIII. ARCTOPUS, Linn.

Flowers polygamo-dioecious. Margin of *calyx* 5-toothed. *Petals* lanceolate, with an incurved, acute, entire acumen. *Male flower:* stamens twice as long as the corolla, perfect. *Stylopodium* flat. *Styles* 2, very short, deciduous. *Ovary* abortive. *Female;* stamens none. *Styles* divaricating, with thick bases. *Fruit* ovate, acute or rostrate, crowned with the calyx, its lower half adnate with the involucre, marked with a furrow, not separable into two parts, but bilocular, with one of the cells abortive. *Seed* 1, roundish, convex at one side, furrowed on the other. *Lam. Ill. t.* 865. *DC. l. c.* 236. *Endl. Gen. n.* 4524.

Perennial, stemless plants, with stellate, ciliate leaves, close pressed to the ground. Male umbels compound, pedunculate, sterile, but mixed with a few female flowers. Umbellules nearly globose. Involucrum usually of 5–7 leaves, which are joined together after flowering. Female umbels sessile, fertile, surrounded by the 4, rarely 5, concrete leaves of the involucrum, which are coriaceous, reticulated, and spiny toothed, girding the fruit. Petals white. Name from *αρκτος, a bear,* and *πους, a foot.*

ANALYSIS OF THE SPECIES.

Female involucre ovate, cuspidate, carinate, laterally 3-spinous (1) **echinatus.**
Female involucre orbicular, flat, quite entire, with 1 spine at the apex ... ... (2) **monacanthus.**
Female involucre orbicular, with involute, quite entire margins, and 3–5 bristle-like spines at the apex ... ... (3) **Dregei.**

1. A. echinatus (Linn. Spec. ed. 2, v. 2, 1512); petiole flat; leaves suborbicular, trifid; lobes toothed, ciliate-spinous; female involucre *ovate, carinate on back, spinous-cuspidate on each side, with 3 large spines; fruit ovate, shortly rostrate, densely beset with erect, spiny bristles. Thunb.! Fl. Cap.* 255. *Pappe, Fl. Cap. med.* 19. *E. & Z.!* 2255. *Burm. Afr.* 1, *t.* 1. *Sieb. Herb. Fl. Cap. n.* 141. *Herb. Un. Itin. n.* 42.

Hab. Common in sandy places in the districts of Cape, Stellenbosch, rarely in Albany and Uitenhage. May–Aug. (Herb. Sd. D. etc.)

Root long and thick. Leaves expanded, the inner smaller, suborbicular, or subcuneate, or obovate, about 2–3 inches long and broad, slightly incised-3-fid; segments often again 3-fid or 3-dentate. Lobes dentate, with long yellowish cilia; at the incisions with 1 or several longish spines; glabrous on both sides, thickish-veined beneath. Petioles 1–2 inches long, 3–4 lines broad Male pedunculate umbel, sometimes with a lateral branch. Rays 10–20, 1–1½ inch long. Involucre of many broad, linear, spinous leaves. Umbellule many-flowered. Pedicels as long as the many-leaved, spinous involucel. Male flower 1 line long. Female umbels shortly pedunculate. Universal involucre as in the male, mostly broader. Leaves of the partial involucre 6 lines long, 3–4 lines broad, convex-carinate. Terminal and lateral spines 3–4 lines long. Petals minute, erect, acute. Styles elongated. Fruit 4–5 lines long, 1-seeded, the abortive cell very narrow, to separate when fully ripe. Dorsal ribs not conspicuous. Vittæ none.

2. A. monacanthus (Carmichael!); petiole flat; leaves suborbicular, cuneate, trifid; lobes toothed, ciliate-spinous; female involucre orbicular, *with 1 spine at the round apex*, quite entire at the flat margins; *fruit ovate, long-rostrate*, sparingly beset with short bristles.

Hab. Cape, *Captain Carmichael;* near Somerset in Hottentottsholland, *E. & Z.;* between Paarl and Pont, on stony hills, *Drege*, 7648. Sept. (Herb. Sd. Hk. D.)

Habit and leaves as in the preceding, leaves often larger. Female umbel subsessile; rays 1–1½-uncial. Partial involucre much larger than in *A. echinatus*, 1–1¾ inch long and broad, much reticulated, with only 1 short, but hard spine. The fruit is 6–8 lines long, the beak of which is nearly as long as the fruit. Dorsal ribs obliterated. Male plant unknown.

3. A. Dregei (Sond.); petiole flat; leaves suborbicular, cuneate, trifid; lobes toothed, subspinous-ciliated; female involucre orbicular, a little emarginate, *with 3 or 5 short bristle-like spines at the apex*, quite entire at the involute margins; fruit *broad-ovate, acute, not rostrate, quite smooth* or scarcely tuberculated at back.

Hab. Sandy hills, Agter de Paarl, *Drege*, 7649. Oct. (Herb. Sd.)

Distinguished from the preceding by the leaves, usually ciliate, with shorter bristles; smaller partial involucre (6 lines long and broad) and the unarmed (3 lines long, 2½ lines broad) fruit without any beak. The female umbel is shortly pedunculate as in *A. monacanthus;* the peduncle ½–¾ inch long, with ovate, acuminate involucre and several short rays, nearly as long as the pedicels. Partial involucre less veined than in the foregoing; the spines at the apex line long, and removed 1 line from each other. Fruit much striated at the back. Vittæ none. Male plant unknown.

XXXIV. HERMAS, Linn.

Margin of the *calyx* 5-parted, leafy, persistent. *Petals* oval-oblong, acute, carinate, entire, equal. *Fruit* ovate. *Mericarps* somewhat inflated, dorsally compressed, 5-ridged, 1 dorsal exserted, 2 intermedial larger, and the 2 lateral very small, in the commissure. *Furrows* broad, with many vittæ. *Carpophore* undivided. *Seed* not adnate to its segment, elliptic, subconcave on its inner face. *Lam. Ill. t.* 851. *DC. l. c.* 241. *Endl. Gen. n.* 4530.

Large herbaceous or usually suffruticose plants, with simple, subdentate leaves, compound, many-rayed, and nearly globose umbels. Involucre many-leaved; involucel of about 3 leaves. Lateral umbels sterile, male. Flowers white or purplish. Name of unknown meaning.

ANALYSIS OF THE SPECIES.

Radical or lower leaves petiolate:
Leaves ovate or obovate, ciliated ... ... (5) **ciliata.**
Leaves oval or oblong, serrated ... ... (1) **gigantea.**
Leaves ovate-cordate, crenately-toothed ... ... (3) **capitata.**
Leaves ovate, cuneate at the base, 5–7-toothed ... ... (4) **quinquedentata.**
Lower leaves stem clasping, obovate-oblong, subcordate ... (2) **villosa.**

1. H. gigantea (Linn. fil. Suppl. 435); *radical leaves on long petioles, oval or oblong,* somewhat serrated, clothed with dense soft tomentum on both surfaces; stem branched. *Thunb.! Nov. Act. Petr. v.* 14, 529, *t.* 11. *Fl. Cap.* 249. *E. & Z.!* 2256. *Bupleurum giganteum, Thunb. Prod.* 50.

HAB. Mountains near Capetown, Tulbagh in Winterhoeksberg, Duyvels and Voormannsbosch, Swellendam, etc. Jan.–March. (Herb. Holm. Sd. D., etc.)

Radical leaves ½–1 foot long, about 3–4 inches broad, obtuse, with obtuse or narrow base, serrulated, or from the dense tomentum, often entire, woolly as well as the petioles. Stem terete, glabrous, purplish, 1–2 feet high, naked or with some very small leaves at the base of the branches. Umbel with many uncial rays, the lateral globose. Leaves of involucre lanceolate. Flowers purplish according to Thunberg. The plant is called "Tundelbloom" by the colonists, who use the wool scraped from the leaves for tinder.

2. H. villosa (Thunb.! Nov. Act. Petrop. v. xiv. 531); leaves *stem-clasping, ovate-oblong,* acute, subcordate at the base, toothed, glabrous above, whitish-tomentose beneath; stem tomentose between the leaves, glabrous above. *Flor. Cap.* 249. *E. & Z.!* 2257. *Herb. Un. Itin. n.* 397. *Sieb. n.* 214. *Burm. Afr. t.* 71, *f.* 2. *Bupleurum villosum, Thunb. Prod.* 50.

HAB. Mountains near Capetown, *Thunb. E. & Z. Drege, Sieber;* Platteklipp, *Zeyh.* 3020. Jan.–May. (Herb. Thunb. Holm. Sd. etc.)

Stem 2–3 feet or more in height; leafy on the lower parts. Leaves sessile or very short petiolate (*H. depauperata,* Linn.), 3–4 inches long, 1½–2 inches broad on the margins, with many short teeth. Umbel with many ½–1 uncial rays. Leaves of involucre oblong, acute. Outer rays of the umbellules sterile or abortive. Fruit 2 lines long and broad.

3. H. capitata (Linn. fil. Suppl. 435); leaves *radical, petiolate, ovate-cordate,* crenately-toothed, glabrous above, whitish-tomentose beneath; stem tomentose or glabrous at the apex, leafless. *Thunb. l. c. E. & Z.!* 2260. *Bupleurum capitatum, Thunb. Prod.* 50.

VAR. β, **minima**; stem filiform (2–4 inches), leaves and umbels smaller. *H. minima,* E. & Z.! 2261.

HAB. Rocks on the top of Tablemountain; var. β, Van Staadensriviersberge, Uitenhage, *E. & Z.* Jan.–Feb. (Herb. Thunb. Holm. Sd. Hk. D.)

Stem about 1 foot high, with 1 or 2 short, umbelliferous branches. Leaves only radical, 1 inch long, 9–10 lines wide, often a little smaller, rarely larger, on a filiform, equal or longer petiole; quite glabrous above or rather villous in the younger leaves of the var. β. Stem leaves none, but a small scale at the base of the branches. Umbels subcapitate; rays in fruit nearly ½ inch long. Leaves of involucre about 7. Flowers white. Fruit 1 line long.

4. H. quinquedendata (Linn. fil. Suppl. 436); leaves *from the lower part of the stem,* petiolate, *ovate, cuneated at the base,* 5–7-toothed, glabrous above, whitish-tomentose beneath; stem tomentose between the leaves, glabrous above. *Thunb. l. c. E. & Z.!* 2262. *H. quercifolia, E. & Z.!* 2263. *Bupleurum quinquedentatum, Thunb. Prod.* 50.

HAB. Top of Tablemountain, *Thunb.;* near Tulbagh, and on Zwarteberg, near Gnadenthal, *E. & Z.;* between Nieuwekloof and Ylandskloof, *Drege.* Dec.–Feb. (Herb. Hk. Sd.)

Near the foregoing, but the leaves are not radical, but rise from the stem, and are aggregated on shorter petioles, more oblong and attenuated at the base, 8–16 lines long, 3–8 lines broad; the stem is filiform, naked, simple or with a few short branches, and the subglobose umbels smaller. The involucre of the umbel is 4–5-leaved. Flowers white. In *H. quinquedentata*, E. & Z., the margins of the leaves are more revolute, and the teeth acute or sharp; in *H. quercifolia*, E. & Z., the margins are not revolute, and the teeth bluntish.

5. H. ciliata (Linn. fil. Suppl. 436); leaves *radical*, petiolate, *ovate or obovate, obtuse, ciliated*, glabrous above, whitish-tomentose beneath; stem glabrous, a little branched, leafless. *Thunb.! l. c. Burm. Afr. t.* 72, *f.* 3. *E. & Z.!* 2258. *H. Uitenhagensis, E. & Z.!* 2259.

HAB. Mountains, Hottholld., near Somerset and Vanstadensriviersberge, *Thunb. E. & Z.;* Dutoitskloof, *Drege.* Jan.–Feb. (Herb. Sd. Hk.)

Stem 1–2 feet. Leaves 2–3 inches long, 1–1½ inch broad, shortly ciliate. Petiole shorter or as long as the leaf. Rays of umbel ¾–1 inch, twice longer than the many-leaved involucre. Flowers white. Fruit nearly 2 lines in length.

XXXV. CONIUM, Linn.

Margin of the *calyx* obsolete. *Petals* obcordate, submarginate, with a short inflexed point. *Fruit* ovate, compressed from the sides. *Mericarps* with 5 prominent, equal, subundulated or undulately-crenulated ribs; lateral ones marginating. *Furrows* with many striæ, but without vittæ. *Carpophore* bifid at the apex. *Seed* with a deep, narrow furrow, and as if it were complicate. *Koch, Umb.* 135, *f.* 40. *Endl. Gen. n.* 4532.

Biennial, poisonous herbs. Stem terete, branched. Leaves decompound. Involucre of 3–5 leaves; involucels dimidiate. Flowers white, all fertile. Name said to be from, *κονια, dust.*

1. C. chærophylloides (E. & Z.! 2264); stem at the base, as well as the petiole and its primary ramification, scabrid by short hairs; leaves 3–4-pinnate, glabrous; segments deeply pinnatifid, lacinulæ oblong or linear-acute; involucre of 4–6 short leaves; fruit with very prominent, subundulated, not crenulated ridges. *Seseli chærophylloides, Thunb. Fl. Cap.* 259.

HAB. Woods in the districts of Albany and Uitenhage, *E. & Z.;* Klipplaatriver and Nieuweveld, between Rhinosterkop and Gangefontein, 3–4000 ft., *Drege;* Caledonriver and near the Gariep, *Burke & Zeyher, n.* 745. Oct.–Nov. (Herb. Thunb. Hk. D. Sd.)

Stem much branched, round, striated, hollow. Leaves like those of *C. maculatum;* radical ones very large. Segments ovate, incised, or deeply pinnatifid; lacinulæ entire, acute, or with one or a few serratures. Umbels with 8–12 rays, 1–2 inches long, a little scabrous at the apex. Involucels of few short, submembranaccous leaflets. Flowers white. Fruit when ripe 2 lines long, with pale ridges. Stylopodium depressed. Styles short, divergent. There is a variety with more dissect leaves collected by *E. & Z.* and *Dr. Pappe* on the Zwarteberg near Caledon.

Drege n. 7636, 7639, 7644 are imperfect specimens of new or undescribed umbelliferous plants.

ORDER LXVIII. ARALIACEÆ, Juss.

(BY W. SONDER.)

Flowers perfect or unisexual, mostly umbellate. *Calyx*-tube adnate to the ovary; limb entire or toothed. *Petals* 5–16, alternate with the teeth of the calyx, epigynous, valvate in æstivation. *Stamens* as many, rarely twice as many as the petals, inserted on the outside of a fleshy, epigynous disc; anthers 2-celled. *Ovary* inferior, of two or several cells; ovules solitary, pendulous; styles or stigmas as many as the cells of the ovary. *Fruit* fleshy or nearly dry, 2–16-celled, crowned by the limb of the calyx; endocarp crustaceous. *Albumen* copious, horny, with a minute, basal embryo.

Trees and shrubs, more rarely herbaceous plants, natives chiefly of the warmer zones. Leaves alternate, simple or compound, digitate, pedate, or pinnate. Best known from *Umbelliferæ* by the fleshy fruit. There is but one S. African genus of the Order.

I. CUSSONIA, Thunb.

Margin of the *calyx* short, with 5–7 acute teeth, or entire. *Petals* 5–7. *Stamens* as many as the petals, and alternate with them. *Ovary* turbinate, surmounted with a broad disc (stylopodium). *Styles* 2–3, short, erect, distinct, approximate. *Fruit* 2–3 celled, roundish, somewhat fleshy. *Fl. Cap.* 247. *DC. Prod.* 4, 255. *Endl. Gen. n.* 4552.

Shrubs or trees. Stem thick and somewhat succulent. Leaves long-petiolate, approximate, glabrous, glossy, palmate or digitate, with 5–9 one-nerved, entire or lobed leaflets. Flowers small. Name in honour of Peter Cusson, once professor of botany in the university of Montpellier.

ANALYSIS OF THE SPECIES.

Leaves palmato-partite ... ... (1) **Natalensis.**
Leaves peltato-digitate:
  Flowers in simple spikes. Leafl. petiolulate.
    Leaflets lanceolate; flowers sessile ... ... (2) **spicata.**
    Leaflets obverse lanceolate; flowers pedicellate ... ... (3) **Kraussii.**
  Flowers in panicled spikes; leaflets shortly petiolulate ... ... (4) **paniculatum.**
  Flowers in simple racemes; leaflets sessile ... ... (5) **thyrsiflora.**
  Flowers umbellate ... ... (6) **umbellifera.**

1. C. Natalensis (Sond.); leaves *palmato-partite*, lobes lanceolate-acuminate, serrate; spikes elongate; rhachis from the base laxly beset with sessile or subsessile flowers; calyx obsoletely 5–6-toothed; ovary roundish; styles two, spreading.

HAB. Port Natal, *Gueinzius*. (Herb. Sd.)

Petiole terete, striate, about 1 foot long. Leaves 5-lobed or sometimes 3-lobed, with deeply bifid, lateral lobes, united by an half-uncial lamina; lobes 3–4 inches long, 1–1½ inch broad, with revolute, shortly-serrated margins, attenuated in a long apex, somewhat narrowed at the entire base, much reticulated; the lower lobes equal or a little smaller. Spikes not pedunculate, aggregate, ½ foot long; rhachis beset with minute, ovate, cuspidate scales, in the axil of which are placed the flowers. Ovary roundish. The half-ripe (?) fruit 2 lines long, a little fleshy, 2-celled.

2. C. spicata (Thunb! Nov. Act. Ups. 3, 212, t. 13); leaves *peltato-digitate;* leaflets (about 7) *lanceolate*, petiolulate, coarsely serrate or toothed from the middle, entire or trifid; spikes oblong, cylindrical, pedunculate; rhachis densely beset *with exactly sessile* flowers; calyx nearly entire; ovary roundish; styles 2–3, short. *E. & Z.!* 2266.

HAB. In the districts of Uitenhage, Albany, Caffraria, and near Port Natal Nov.–Dec. (Herb. Sd. D., etc.)

Tree, 15 feet high, with the aspect of a Palm, called by the colonists "Samareelboom, Nojesboom." Leaflets 7–9, rarely 5, on longish petioles, the lower ones often simple, the intermediate mostly ternate, the upper ones 3-nate or pinnate, with decurrent pinnæ, 3–5 inches long, 1–1½ inches broad, with revolute margins, serrate or toothed at the lanceolate, terminal lobes, rarely quite entire. Spikes 8–12, umbellate, on a very long, common peduncle; partial peduncle as long or shorter than the 2–3 uncial spikes. Flowers spirally disposed along the rhachis in 6–10 series. A specimen with very short-petiolate or subsessile leaves, collected by Zeyher at Kœgaskoppe (n. 1003) agrees well with *C. triptera Colla. Hort. Ripul. p.* 43, *t.* 26.

3. **C. Kraussii** (Hochst. in pl. Krauss.); leaves peltato-digitate; leaflets (7–9) petiolulate, *obverse-lanceolate*, shortly acuminate or cuspidate, ternate or pinnately-incised, *quite entire* or a little toothed at the apex; spikes ovate-oblong; rhachis rather densely beset *with very short-pedicellate* flowers; calyx acutely 5–6 toothed; ovary obovate; styles 2–3, short.

HAB. Port Natal, *Gueinzius*. (Herb. Sd.)

Leaves as in the preceding, but the leaflets are not attenuated at the apex, but obovate, shortly, or mucronately acuminate; the intermediate and upper ones pinnate, at the apex ternate, with much spreading, lateral leaflets. Spikes 1 inch long, twice or 3 times shorter than the peduncle. Pedicels ½ line long, 3 times shorter than the linear lanceolate, scarious bract. Calyx nearly as long as the petals.

4. **C. paniculata** (E. & Z.! 2267); leaves peltato-digitate; leaflets (7–9) *lanceolate-acuminate*, attenuated in a short petiole, quite entire or serrate or pinnatifid-incised, coriaceous, with revolute margins; *spikes paniculated;* calyx nearly entire; fruit globose; styles 2.

HAB. Nieuweveldsberge and Rhinosterkop, distr. Beaufort, *Drege, Zeyher*, 746. Stormberg, *A. Wyley*. Jan.–Feb. (Herb. D. Sd.)

Shrub 10–12 feet high. Leaflets 6–8 inches long, 1–1½ inch broad, more coriaceous than in the preceding and following; very often quite entire, or with a few sharp teeth, rarely deeply incised or pinnatifid, and the upper leaflet ternate. Panicles (or racemes) aggregated, 1 foot or more in length; the branches or peduncles spreading, 1–2 inches long at the base, with short bracteolæ. Spikes 1–2 inches long, dense. Fruit the size of a large pea, crowned with the spreading styles.

5. **C. thyrsiflora** (Thunb. l. c. t. 12); leaves peltato-digitate, leaflets (circ. 5) *sessile, obovate, cuneate*, obtuse, truncate or mucronate or shortly acuminate, quite entire or a little toothed, upper ones sometimes 3-sect; *flowers pedicellate, disposed in an oblong raceme;* calyx 5-toothed; fruit roundish; styles 2–3. *Fl. Cap.* 247. *E & Z.!* 2265. *Jacq. fil. Eclog.* 1, 89, *t.* 61. *C. thyrsoidea Pers. Ench.* 1, 98. *Drege*, 7607. *Zeyher*, 2698.

HAB. In Tablemountain, Houtbay; distr. of Uitenhage, George, etc. June–Sept. (Herb. Sd. D., etc.)

Leaflets 2–3 inches long, 1–1½ inch broad, without distinct petiole, coriaceous; margins revolute, usually simple, rarely some of them jointed, the lowest joints dilated at end into smaller lobes, or ternate at top like the leaflets of *C. spicata*. Racemes 2–4 inches long, dense-flowered, about the length of the peduncle; 8–12 aggregated to an umbel. Pedicels 1–2 lines long at the base, with a lanceolate, scarious bracteole. Calyx mucronate-toothed. Styles in the wild specimens usually 2.

6. C. umbellifera (Sond. in Linnæa, v. 23, p. 49); leaves peltato-digitate; leaflets (circ. 5) *petiolulate, oblong or elliptic-oblong, cuneate,* undulate-serrate or quite entire, emarginate at the apex; *umbels terminal, proliferous, paniculated;* umbellules with many rays; ovary hemispherical or subturbinate; styles 2, very short; fruit roundish. *Dietr. Flor. Univ. fasc.* 9 (1856), *t.* 90. *C. paniculata, E. Meyer, non E. & Z.!*

HAB. In woods near Port Natal and in grassy hills between Omsamculo and Omcomas, *Drege, Gueinzius.* (Herb. Sd. D.,)

Tree 30–40 feet in height; diameter of the stem 1–1½ foot; bark resinous; wood very soft. Petiole ½–1 foot; the intermediate of the petiolules 1½ inch, the lateral ones shorter. Leaflets 5 or rarely 3, coriaceous, 3–6 inches long, 1–2 inches broad. shining above. Umbels pedunculate. Rays of the terminal umbel somewhat longer than those of the lateral ones. Involucre wanting. Pedicels 3–4 lines long. Calyx minutely-toothed; ovary 1 line long.

ORDER LXIX. CORNEÆ, DC.

(BY W. H. HARVEY).

Flowers bisexual or unisexual, small, regular. *Calyx*-tube adnate to the ovary; limb 4-toothed. *Petals* 4, epigynous, with valvate æstivation. *Stamens* 4, alternate with the petals, inserted round the margin of a fleshy, epigynous disc. *Ovary* inferior, 2–4-celled; ovules solitary, pendulous, anatropal; style single. *Fruit* fleshy and juicy, with a 2–4-celled nucleus. *Embryo* in the axis of fleshy albumen.

Trees or shrubs, natives of the temperate zone, and chiefly of the northern hemisphere. Leaves almost always opposite, entire or toothed, penninerved, exstipulate. Flowers in heads, umbels or cymes, mostly white.

I. CURTISIA, Ait.

Calyx-tube turbinate, 4-angled; limb 4-parted. *Petals* 4, oblong, valvate. *Stamens* 4, inserted with the petals, and alternate with them; filaments subulate; anthers versatile, short, didymous. *Ovary* 4-celled (occasionally 3-celled), crowned with a hairy disc; style single; stigmata 3–4. *Fruit* thinly fleshy, with a bony, 4-celled (or 2–3-celled) nut; seeds one in each cell, pendulous. *DC. Prod.* 2, *p.* 12. *Endl. Gen.* 4577.

But one species, the well-known *Hassagaywood.* The name, *Curtisia,* is in honor of W. Curtis, the founder and for a long time the proprietor of "Curtis's Botanical Magazine."

1. C. faginea (Ait. Hort. Kew. 1, p. 162); *DC. Prod. l. c. Lam. Ill. t.* 71. *Thunb. Fl. Cap. p.* 141. *Burm. Dec. Afr. p.* 235, *t.* 82.

HAB. In forests throughout the colony; also in Kaffirland. (Herb. D. Sd. &c.)

A tree, 20–40 feet high, with dark-coloured, smooth bark. Leaves ovate or oblong, acute, coarsely-toothed, penninerved, glabrous and glossy on the upper surface, tomentose beneath; petioles semi-uncial. Twigs, petioles and inflorescence tomentose and rusty. Panicles terminal, trichotomous, much-branched; flowers minute, pubescent. "The wood is solid, extremely tough, heavy, close-grained, very durable, and resembles plain mahogany." *Pappe, Silv. Cap. p.* 17.

ORDER LXX. HALORAGEÆ, R. Br.

(BY W. H. HARVEY).

Flowers minute, bisexual or unisexual. *Calyx*-tube adnate to the ovary; limb 2–3–4–toothed or entire. *Petals* 2-3-4 (or more), epigynous, with valvate or imbricate æstivation, deciduous. *Stamens* as many or twice as many as the petals and inserted with them, rarely fewer. *Ovary* inferior, 1–3–4-celled; ovules solitary or in fours, pendulous. *Styles* (if any) separate, very short; stigmata villous or feathery. *Fruit* nut-like or fleshy, 1–2–3–4-celled. *Seeds* pendulous; embryo in the axis of fleshy albumen.

Herbaceous plants or small shrubs, growing in water or in very wet soil. Leaves entire or toothed, or the submerged ones pectinate, opposite, whorled or alternate. Flowers inconspicuous, axillary or panicled. Dispersed throughout the temperate zones; the shrubby species are chiefly Australian. In *Gunnera* the stamens are opposite the petals; in the other genera they are alternate to them, when petals are present.

TABLE OE THE SOUTH AFRICAN GENERA.

I. **Gunnera.**—*Leaves* radical, *on long petioles,* reniform. *Flowers* in compound spikes, on a common, scapelike peduncle.

II. **Serpicula.**—*Leaves* sessile, alternate or opposite, entire or toothed. *Flowers* axillary.

III. **Myriophyllum.**—*Leaves* whorled, pectinato-partite. *Flowers* in terminal, interrupted spikes.

I. GUNNERA, L.

Flowers unisexual or bisexual. *Calyx*-tube terete or angular; limb 2–3-lobed. *Petals* 2 or none, coriaceous, deciduous. *Stamens* 1–2, opposite the petals; anthers innate, 2-celled. *Ovary* uni-locular, uni-ovulate; stigmata 2, elongate, simple, papillose. *Fruit* succulent, with a bony endocarp. *Albumen* copious; embryo very minute. *Hook. f. Fl. Tasm.* 1, *p.* 124.

Nearly stemless, herbaceous plants, with alternate, petioled, cordate or reniform, many-nerved leaves. Flowers in compound spikes or capitate, on simple or branching scapes, minute, green. A curious genus of few species, natives of Java, Tasmania, New Zealand, the Society and Sandwich Islands, Peruvian Andes, Chile and Fuegia, besides the solitary South African species on which the genus was founded. The name is in honor of Bishop Gunner, a Norwegian botanist of the last century, author of a "*Flora Norvegica.*"

1. G. perpensa (Linn.); stemless, monœcious; leaves on long petioles, broadly-reniform, minutely-toothed, scabrous, pedately many-nerved, reticulate; scapes simple, ending in a long, compound spike (or thyrsus), the spikelets of female flowers occupying the lower half of the spike. *Thunb.! Fl. Cap. p.* 32. *Bot. Mag. t.* 2376.

HAB. In wet ditches throughout the colony. (Herb. T.C.D. &c.)

Leaves radical, on petioles 12–18 inches high, the lamina 6–12 inches broad. All the younger parts pubescent. Scapes at length 2–3 feet high, two-thirds of the length occupied by the inflorescence. Flowers of both sexes in numerous slender *spikes,* distributed along a common peduncle; the males in the upper, the females in the lower spikes, neither bracteolate. Male flower: calyx 2-toothed. Petals 2, spathulate, entire, deciduous. Stamens 2, opposite the petals; anthers subsessile, 2-celled, opening laterally. Female flowers: calyx-tube compressed; limb minutely 4-toothed. Petals none. Stigmata 2, spreading, papillate. Fruit succulent, minute.

II. SERPICULA, L.

Flowers unisexual. *Male: calyx* minute, 4-toothed. *Petals* 4, oblong, concave. *Stamens* 4, alternate with the petals, or 8. *Female: calyx*-tube 4-angled, 8-ribbed; limb 4-toothed. *Petals* and *Stamens* none. *Stigmata* 4, long and feathery. *Ovary* 1-celled; ovules 4, pendulous. *Fruit* 8-ribbed, fleshy, with a bony endocarp, 1-seeded. *DC. Prod.* 3, *p.* 65. *Endl. Gen.* 6136.

Small, herbaceous, creeping plants, natives of warm countries. Leaves alternate or opposite, entire or toothed, sessile. Flowers axillary; the males on slender pedicels, the female subsessile. Name from *serpo*, to creep.

1. **S. repens** (Linn. Mant. 124); leaves alternate or the lower ones opposite, linear-lanceolate or oval, entire or 2–3-toothed, scabrous at the margin; male flowers tetrandrous. *DC. Prod.* 3, *p.* 64. *S. rubicunda, Burch. DC. l. c. E. & Z.!* 1765, 1766.

HAB. In moist places throughout the colony, common. (Herb. D. Sd. &c.)

Very variable in size, in pubescence, and in the shape and toothing of the leaves. Stems many from the crown, subsimple, decumbent, rooting on the under side, glabrous or hairy. Leaves 4–7 lines long, 1–3 lines wide, entire or toothed, opposite and alternate on the same branch.

III. MYRIOPHYLLUM, L.

Flowers unisexual, rarely bisexual. *Male: calyx* 4-parted. *Petals* 4, ovate, caducous. *Stamens* 4–6–8. *Female: calyx*-limb 4-parted. *Petals* none, or very small and caducous. *Stamens* abortive or none. *Ovary* 4-celled; stigmata 4, long, compressed, papulose on the inner surface. *Fruit* 4-lobed, of 4 nut-like, indehiscent carpels. *DC. Prod.* 3, *p.* 69. *Endl. Gen.* 6135.

Herbaceous water plants, natives of all climates. Leaves opposite or whorled, the submerged ones pectinato-pinnate. Flowers axillary, solitary, sessile; sometimes in the axils of depauperated floral leaves, and then forming leafy spikes. Name from *μυριος, a thousand,* and *φυλλον, a leaf.*

1. **M. spicatum** (Linn. Spec. 1409); leaves whorled, pectinate, the lobes opposite, capillary; flowers in an interrupted, terminal spike, the subtending bracts shorter than the flowers. *DC. l. c. E. & Z.!* 1767. *Drege,* 8805. *E. Bot. t.* 83. *Fl. Dan. t.* 681.

HAB. In the Zwartkopsriver, *E. & Z.! Drege!* Kraairiver and the Gariep, *Drege!* (Herb. Sd. D. &c.)

Stems submerged, erect. Leaves 4-stichous, 1 inch long, multipartite; lobes bristle-like. Spikes 2 inches long; the whorls of flowers 7–8 lines apart; bracts very small. A common plant in Europe and North America.

ORDER LXXI. BALANOPHOREÆ, Rich.

(BY W. H. HARVEY).

Flowers unisexual, in dense heads, spikes or panicles. *Male flower: perianth* 3-parted, valvate in æstivation, or none. *Stamens* 1–3 opposite the segments of the perianth. *Female flower: perianth* epigynous, 3-lobed, or obsolete. *Ovary* inferior, 1-celled; ovules solitary, pendulous. *Styles* filiform, 1 or 2, or none. *Fruit* dry or succulent, indehiscent; the seed filling the cavity. *Embryo* minute, in fleshy or friable albumen.

Fleshy, fungous-like root-parasites, inhabiting the tropics of Asia, Africa and America. Leaves reduced to fleshy, coloured scales. Flowers surrounded by bracts, or involucrate or naked. An Order of very doubtful affinity supposed to be related to *Halorageæ* and *Loranthaceæ*.

TABLE OF THE SOUTH AFRICAN GENERA.

I. **Mystropetalon.**—*Flowers* monœcious, in dense spikes.

II. **Sarcophyte.**—*Flowers* diœcious; the males panicled; the females concrete in globose heads.

I. MYSTROPETALON, Harv.

Flowers monœcious, in dense spikes. *Male: perianth* tripartite, bilabiate, the segments with long claws, spathulate, valvate in æstivation, the two posterior connate. *Stamens* 2, opposite to and inserted on the posterior segments of the perianth, conniving; anthers 2-celled, extrorse, opening longitudinally; pollen cubical. *Female: perianth* epigynous, tubular, 3-toothed. *Ovary* seated in a cup-like, fleshy disc or receptacle, uni-ovulate; style filiform, exserted, deciduous; stigma capitate, 3-lobed. *Fruit* subtended by the unchanged receptacle, with a thin, juicy epicarp and a crustaceous endocarp, 1-seeded. *Embryo* very minute, in the base of easily friable albumen. *Harv. in Ann. Nat. Hist. 1st ser. vol. 2, p. 386. Griff. in Linn. Trans. 19, p. 336. Hook. f. in Linn. Trans. vol. 22, p. 29.*

Root parasites. Stem simple, fleshy, densely imbricated with linear-spathulate scales, ending in a dense spike of flowers. Spikes monœcious, the female flowers occupying the lower, the males the upper part of the spike. Bracts, 3 under each flower, 1 anteal, and 2 lateral. The generic name is compounded of *μυστρον, a spoon,* and *πεταλον, a petal;* the segments of the perianth are spoon-shaped.

1. M. Polemanni (Harv. l. c. t. 20); anterior bract *spathulate, with a narrow claw;* female perianth *tubular. Hook. f. l. c.*

Hab. At the Hoouw Hoek Pass, *Mrs. Denys* (v. v.)

Stem about 6 inches high, ¾ inch in diameter. Bracts orange, the anterior ones densely bearded. Flowers bright carmine.

2. M. Thomii (Harv. l. c. t. 19); anterior bract *broadly-oblong;* female perianth *subglobose. Hook. f. l. c. Griff. l. c. p. 336. Balanophora Capensis, E. & Z.! in Herb.*

Hab. About Caledon Baths, *Thom, Ecklon, Polemann!* (v. v.)

Stem 6–8 inches high, ½–1 inch in diameter. Anterior bract of nearly equal breadth throughout, bearded at the apex and along the prominent keel, orange. Flowers a dark, brownish, or dull red.

II. SARCOPHYTE, Sparm.

Flowers diœcious. *Male* flowers panicled; perianth 3-lobed, valvate in æstivation. *Stamens* 3, opposite the lobes; filaments free; anthers multilocular. *Female* flowers in globose heads, densely crowded. *Perianth* none. *Ovaries* seated on a subglobose, common receptacle, becoming concrete, unilocular. *Stigma* sessile, peltate. *Syncarpium* globose, berry-like. *Endl. Gen. 714. Griff. in Linn. Trans. l. c. Hook. f. in Linn. Trans. l. c.*

A root parasite. Stem branching, thick and fleshy, imbricated with scales below, and ending in a panicled inflorescence. Smell offensive, resembling that of rotten fish. The generic name is compounded of *σαρξ, flesh,* and *φυτον, a plant;* a fleshy-plant.

1. **S. sanguinea** (Sparm. Ait. Holm. 37, p. 300, t. 7); *Griff. l. c. p.* 339. *Hook. f. l. c. p.* 37. *Wedd. Ann. Sc. Nat. ser.* 3, *vol.* 14, *t.* 10, *f.* 34–38. *Ichthyosma Wehdemanni, Schl. in Linn. vol.* 2, *t.* 8.

HAB. On the roots of *Ekebergia Capensis* and of *Acacia Caffra* in the districts of Albany and Uitenhage.

Stems 9–10 inches high, an inch or more in diameter, dull flesh-coloured or reddish. Male flowers purplish. The concrete fruit resembles a mulberry.

ORDER LXXII. **LORANTHACEÆ**, DC.

(BY W. H. HARVEY).

Flowers bisexual or unisexual, regular. *Calyx* bracteate at base, adnate; limb short, entire or 4–8-toothed. *Petals* 4–8, separate or more or less cohering, with valvate æstivation. *Stamens* as many as the petals and opposite to them; filaments adhering to the base or claw of the petals; anthers 2-celled. *Ovary* inferior, 1-celled, with 1–3 pendulous ovules; style filiform or none; stigma capitate. *Fruit* a succulent, 1-seeded berry. *Embryo* straight, in the axis of fleshy albumen; radicle superior.

Shrubby plants, almost always parasitical, natives chiefly of the tropics and warmer parts of the temperate zones. Leaves entire, opposite or alternate, coriaceous or fleshy, exstipulate. Flowers axillary or terminal, either solitary or clustered, or in cymes, spikes, racemes or umbels.

TABLE OF THE SOUTH AFRICAN GENERA.

I. **Loranthus.**—*Flowers* bisexual. *Petals* clawed, more or less united in a tubular corolla. *Style* filiform.

II. **Viscum.**—*Flowers* unisexual, *minute*. *Petals* sessile, free or connate at base. *Style* none or very short.

I. **LORANTHUS**, L.

Flowers bisexual. *Calyx*-limb short, truncate or toothed. *Petals* 4–8, with slender claws, more or less united in a tubular corolla. *Stamens* inserted on the claws of the petals; filaments subulate; anthers 2-celled. *Ovule* solitary. *Style* filiform; stigma capitate. *Berry* usually crowned by the limb of the calyx. *DC. Prod.* 4, *p.* 286. *Endl. Gen. No.* 4586.

A very large genus almost wholly tropical or subtropical; particularly numerous in South America. One species is found in Europe. The flowers are generally bright-coloured, yellow, orange or scarlet. *L. oleæfolius* was formerly cultivated in Ludwigsburg Gardens, Capetown. Name from *lorum*, a lash of leather, and *ανθος, a flower.*

ANALYSIS OF THE SOUTH AFRICAN SPECIES.

Unopened corolla cylindrical or clavate (not swollen at base or constricted):
  Corolla densely hirsute or squamulose and scurfy:
    Corolla densely silky with *smooth* hairs ... ... (1) **Dregei.**
    Corolla clothed with brittle, *scabrous* hairs; leaves mealy ... ... (2) **ovalis.**
    Corolla clothed with *flat scales;* leaves mealy ... (3) **glaucus.**

Corolla quite glabrous (or very minutely downy):
  Leaves petiolate:
    Twigs, petioles and peduncles *hispid;* leaves opposite ... ... (4) **Zeyheri.**
    Twigs and all parts *quite glabrous:*
      Lobes of the corolla erect; peduncles axillary, 1-flowered ... ... (5) **Natalitius.**
      Lobes of the corolla spirally revolute; peduncles umbellate ... ... (6) **oleæfolius.**
  Leaves sessile, *cordate at base;* corolla 4–5-petaled (7) **undulatus.**
Unopened corolla strongly constricted above the urceolate or swollen base:
  Flowers *sessile,* in subsessile, 3–5-flowered heads; lobes of corolla reflexed ... ... (8) **Namaquensis.**
  Flowers pedicellate, in pedunculate or subsessile umbels; lobes of corolla erect:
    Leaves *opposite,* ovato-lanceolate, long-petioled; flowers quite *glabrous* ... ... (9) **Kraussianus.**
    Leaves alternate, ovate, long-petioled; flowers *minutely puberulous* ... ... (10) **prunifolius.**
    Leaves alternate, short-petioled; flowers glabrous, *very slender* ... ... (11) **quinquenervius.**

1. L. Dregei (E. & Z.! 2284); the twigs, young leaves and peduncles covered with ferruginous, stellate pubescence; leaves opposite or nearly so, petiolate, elliptic-oblong, obtuse at both ends, penninerved, minutely pulverulent-rugulose; peduncles short, 2–3–4-flowered; flowers sessile; calyx hirsute; *corolla densely clothed with long, silky, erect or appressed, smooth hairs,* subcylindrical, semi-5-cleft, the narrow linear segments at length reflexed. *L. oblongifolius, E. Mey.! in Herb. Drege.*

HAB. Bodasberg, *E. & Z.!* Morley, and between Omtata and Omsamwubo, *Drege!* about D'Urban, Natal, frequent, *Sanderson! Gerrard & McKen! Gueinzius!* &c. (Hb. Sd. D. Hk.)

Robust; the older branches with a rough, greyish-ashen bark, spreading. Leaves pale-green, 2–3 inches long, ¾–1–1¼ inch wide; petioles about ¼ inch long. Peduncles ¼–½ inch long, or very short. Bracts cup-shaped, obliquely-ovate. The pubescence of the young parts looks powdery, on older parts it is stellate, and though it soon rubs off, is sometimes long-persistent on the under surfaces of the leaves. Corolla pale-yellowish green and red, slender, nearly 2 inches long.

2. L. ovalis (E. Mey.!); *branches,* twigs, leaves, and pedicels densely clothed with powdery and *somewhat furry,* glaucous indument; leaves scattered, shortly-petiolate, *oval or elliptical-obovate,* obtuse, acute at base, thick, densely-pulverulent, without conspicuous nerves; peduncles axillary, 1-flowered, solitary or in pairs, very short; bract *oval;* calyx shaggy, truncate; corolla densely clothed with *brittle, deciduous, scabrous (spinuliferous) hairs,* subcylindrical-clavate, 4-lobed, the tube splitting, the short, spoon-shaped lobes reflexed.

HAB. Kaus and Natvoet, Gariep, *Drege!* Namaqualand, *A. Wyley!* (Hb. D. Sd.)

Nearly related to *L. glaucus,* but with broader and more oval leaves, and a different indument, especially that of the flowers. The hairs on the corolla are fully a line long, yellowish or creamy, and curiously *whorled* at short intervals with minute spines. Leaves ¾ inch long, ½ inch wide. Flowers 1½ inch long; peduncles 1–2 lines long.

3. L. glaucus (Thunb. Cap. p. 295); the twigs, leaves, pedicels, and flowers *densely clothed with minute, powdery, glaucous scales;* leaves scattered, very shortly-petiolate, *obovate-oblong or lanceolate-oblong*, obtuse, tapering at base, thick, pulverulent, without conspicuous nerves; peduncles axillary, 1-flowered, 2 or 4 together, very short; bract *linear;* calyx pulverulent, minutely-crenate; corolla clothed with *pulverulent scales*, subcylindrical clavate, slightly swollen at base, 4- (or sometimes 5-?) lobed, the tube splitting, the short spoon-shaped lobes reflexed. *E. & Z.!* 2280, *also L. Burchellii, E. & Z.!* 2281. *Zey.!* 753, 754.

HAB. Saldanhabay, *Thunb.! E. & Z.! Drege!* and near Hexriver, Worcester, and Gaaup, and Winterveld, Beaufort, *E. & Z.!* Heerelogement, Kraus in Bushmansland and in Graafreynet, *Zey.!* Snowy Mts. *Burke!* Ebenezer and other stations in N. W., *Drege!* (Herb. Sd. D. Hk.)

Parasitical on *Rhus*, and on *Lycium*, &c. All parts covered with very persistent, pale scurf, which gradually wears off. Leaves scarcely an inch long, often less, ¼–½ inch wide; petioles 1–2 lines long. Peduncles 2–3 lines long. Corolla 1½ inch long, usually splitting down one side. Petals 4, so far as I have seen; *Thunberg* says 5. Limb of the petals 2 lines long. *L. canescens* (Burch. Trav. 2, p. 90) seems, by description, to be a synonym of this species.

4. **L. Zeyheri** (Harv.); the twigs, petioles and peduncles *hispid, with short, spreading hairs;* leaves opposite, minutely-petiolate, elliptic-oblong or ovate, obtuse or subacute, acute at base, 3-nerved at base and penninerved, glaucous, the younger ones *scabrous and hispidulous*, especially on the lower side, at length nearly glabrate; peduncles axillary, 2–4 together, 1-flowered, *hispid*, short; bract linear, hispid; calyx ciliate, repand; corolla glabrous, subcylindrical, 5-lobed, the tube splitting, the linear-lanceolate clawed lobes erect. *Zey.! No.* 751.

VAR. β, **minor**; leaves and flowers smaller.

HAB. Magallisberg, *Zeyher!* Gamkeriver, *Burke!* (Herb. Hk. Sd. D.)

Robust, with a rough, brown bark; all the young parts clothed with short, whitish, rough pubescence. Leaves 1½–2 inches long, an inch or more wide; petioles 1–2 lines long. Peduncles 2–3 lines long. Flowers 2–2½ inches long, resembling those of *L. Natalitius*. Bract fully as long as the calyx, or a little longer.

5. **L. Natalitius** (Meisn.! in Lond. Journ. Bot. 2, p. 539); *glabrous;* leaves scattered or sub-opposite, shortly-petiolate, ovato- or oblongo-lanceolate, obtuse or subacute, acute or tapering at base, penninerved; peduncles axillary, 4–5 together, 1-flowered, semi-uncial; bract obliquely-cupulate; calyx truncate, crenate; corolla subcylindrical, 5-lobed (occasionally 6-lobed), the tube splitting, the linear-lanceolate, clawed lobes erect. *Gueinzius!* 47 & 545. *Harv. Thes. Cap. t.* 30.

HAB. Common round D'Urban, Natal, *Krauss! Gueinzius! Sanderson!* (Herb. Hk. D. Sd.)

Robust, with rugose, dark-coloured bark. The lower leaves and branches frequently opposite, those of the upper shoots mostly alternate. Leaves 1½–2½ inches long, ¾–1½ wide, yellow-green. Flowers "waxy-white, tipped with yellow, standing erect from the pendulous branches, and called '*Lighted-candles*' by the children at Natal" (*Sand.* in Litt.) The figure above quoted was made from unopened flowers and does not represent the form of the mature corolla perfectly.

6. **L. oleæfolius** (Ch. & Schl. Linn. 3, p. 209); *glabrous;* leaves opposite or scattered, *minutely glauco-pulverulent*, shortly petioled, linear-lanceolate, obtuse, acute at base, faintly penninerved beneath; peduncles axillary, *umbellate*, 3–5-flowered; bract obliquely cupulate,

acute; calyx crenulate; unopened corolla subcylindrical, at length urceolate at base, deeply 5-parted, the narrow-linear lobes reflexed and *spirally revolute. E. & Z.! No.* 2282, (*not of E. Mey.*). *L. Lichtensteinii, Herb. Willd. L. speciosus, Dietr. Lichtensteinia oleæfolia, Wendl. Coll.* 2, *p.* 4, *t.* 39.

VAR. β, **elegans**; leaves sometimes ternate; flowers rather smaller. *L. elegans, Cham. & Schl. l. c. E. & Z.! No.* 2283. *L. croceus, E. Mey.!*

HAB. Parasitic on *Acacia* trees. Olifantsriver, Clanw., *E. & Z.!* Betw. Dweka and Zwart-Bulletye, Langvallei, and at the Boschjemans R., *Drege!* β, Zwartkops and Zondags Rs., Uit., *E. & Z.!* Gamke R., *Zey.!* 752. Grahamstown, *Gen. Bolton!* Berlin, Br. Kaffr., *Mr. D'Urban!* (Herb. Sond., D., Hk.)

Robust, with ashen bark. The younger parts, under a lens, are minutely granulated. Leaves 2–2½ inches long, seldom quite ½ inch wide, tapering more or less to a blunt point. Veins more evident in the dried specimens. Peduncles 2–6 lines long; pedicels 2–3 lines. Bract rusty, ovate-acute. Unopened corolla 1½ inch long, brilliant orange-scarlet, the tips of the petals blackish, at length 5-parted nearly to the base, its lobes strongly revolute. I can scarcely distinguish *L. elegans*, even as a variety.

7. **L. undulatus** (E. Mey.!); glabrous; leaves mostly opposite, *sessile, cordate at base*, linear-oblong or ovate-oblong, very obtuse, undulate, thick, faintly 3-nerved; peduncles terminating short ramuli, 2-flowered; bract oblong, truncate, short; calyx obconic, truncate; unopened corolla suburceolate at base, clavate, at length separating quite to the base into 5 (sometimes 4) spreading or revolute petals.

HAB. Between Natvoet and the Gariep, and near Verleptpram, *Drege!* Namaqualand, *A. Wyley!* (Herb. D. Sd.)

Robust, with ashen back. Leaves 1–2 inches long, varying much in shape, but always very blunt and conspicuously cordate at base. Branches virgate, with many short, floriferous ramuli, 1–2 inches long, with 1–2 pair of leaves. Common peduncle and pedicels both 4–5 lines long. Bract much shorter than the calyx, somewhat ciliate. Petals at length quite separate, broad at base, narrowing in the middle, spathulate at the apex.

8. **L. Namaquensis** (Harv.); the twigs, very young leaves, calyx, and corollas *minutely pulverulent*, otherwise glabrous; leaves opposite or alternate, shortly petiolate, ovate-oblong or oblongo-lanceolate, obtuse, sub-acute at base, thick, penninerved beneath, glaucescent; peduncles axillary, shorter than the calyx, 3–5-flowered; flowers *sessile;* bract obliquely cupshaped; calyx repand or truncate; unopened corolla much constricted above the urceolate, swollen base, obtuse, 5-lobed, the tube splitting, the short, spoonshaped lobes reflexed. *L. oleæfolius, E. Mey.! (not of Ch. and Schl.)*

HAB. Groenrivier, and near Verleptpram, on the Gariep, *Drege!* Namaqualand, *A. Wyley!* Modderfontyn, *Rev. H. Whitehead!* (Herb. Sd. D.)

Robust, with a pale, ashen, or rufous bark. Leaves 1½–2 inches long, ¾–1 inch wide, all but the very young ones quite glabrous and glaucous. Common peduncle 1–2 lines long. Corolla 1–1½ inch long, its blunt apex somewhat 5-umboned; tube splitting down one side nearly to the base.

9. **L. Kraussianus** (Meisn.! in Lond. Journ. Bot. vol. 2, p. 539); quite glabrous *in all parts;* leaves *opposite*, on long petioles, *ovato-lanceolate* or *ovate*, obtuse, 3-nerved at base, and faintly penninerved; peduncles axillary, about 5-flowered; flowers pedicellate; bract *ovate, acute;* calyx repand; unopened corolla much constricted above the

urceolate, swollen base, obtuse, *glabrous*, 5-lobed, the tube splitting, the narrow, spoonshaped lobes erect, pale within.

HAB. Near D'Urban, Natal, *Krauss.! Gueinzius! Gerrard & McKen!* No. 640. (Herb. Sd. D.)

Very nearly related to *L. prunifolius;* but with opposite, narrower, and more lanceolate leaves, which turn black in drying, and perfectly glabrous flowers, ovate bracts, and nearly truncate calyces. The peduncles and pedicels are longer, and the lobes of the corolla shorter. Leaves 2–3 inches long, ¾–1 inch wide, occasionally subacute; peduncles and pedicels 4–5 lines long. Flowers 1½ inches long.

10. L. prunifolius (E. Mey.!); glabrous, *except the minutely viscoso-puberulous pedicels, bracts and flowers;* leaves *mostly alternate,* on long petioles, *broadly-ovate or ovate-oblong,* obtuse, faintly penninerved; peduncles axillary, umbellately 5–10-flowered; flowers pedicellate; bract *oblong,* pubescent; calyx minutely 5-toothed; unopened corolla much constricted above the swollen base, obtuse, *puberulous,* 5-lobed, the narrow, spoonshaped lobes erect, pale within.

HAB. Glenfilling and between the Keiskamma and Buffalo rivers, *Drege!* Albany, *Eck. & Zey.! H. Hutton!* (Herb. D. Sd.)

Bark a dark ashen-grey, rough. Leaves 2–2½ inches long, 1–1¾ wide, those of the flowering branches mostly scattered, *glaucous* (?); petioles ½ inch long. Peduncle 2–3 lines long, several-flowered; pedicels 2–3 lines long. Flowers 1½–2 inches long, the lobes (when dry) cream-coloured within.

11. L. quinquenervius (Hochst. Bot. Zeit. 27, 11, 432); glabrous; leaves alternate, on short petioles, *ovate* or oblong, obtuse, *faintly 5-nerved or nerveless, thick and coriaceous;* umbels axillary, *sessile*, 3–5 or several-flowered; flowers shortly pedicellate; bract obliquely cup-shaped, obtuse; calyx truncate; unopened corolla constricted above the swollen base, *very slender,* splitting to the middle into 5 very narrow, linear, erect lobes. *Walp. Ann.* 5, *p.* 938. *L. tenuiflorus, Harv. (non Hk. f.)*

HAB. Natal, *Hochstetter, T. Williamson! Gerrard & McKen!* 639. (Herb. D.)

Bark ashen-grey, smooth. Leaves 1½–2 inches long, 1–1¼ wide, of a thick substance, with immersed nerves, rather more conspicuous when dry. Flowers 1½ inches long, ½ a line in diameter, very much more slender than in any other Cape species.

II. VISCUM, L.

Flowers unisexual. *Calyx*-limb obsolete. *Petals* 3–4, short, triangular or ovate. *Male fl.:—Anthers* sessile on the face of the petals, opening inwards by several pores. *Fem. fl.:—Stamens* none. *Style* very short or none; stigma capitate. *Ovules* 3. *Berry* 1-seeded. *DC. Prod.* 4, 278. *Endl. Gen. n.* 4584.

Parasitical shrubs, natives of the Old World, and chiefly of the warmer zones. Stems dichotomous or trichotomous, often jointed. Leaves opposite, or none. Flowers minute, greenish, axillary, tufted, umbellate or solitary. The mistletoe (*V. album*) of Europe is the most famous species. Name from *viscus,* birdlime, which is made from the bark of *V. album.*

ANALYSIS OF THE SOUTH AFRICAN SPECIES.

1. Leafy species: (1—6)
   Leaves obovate or oblong, very obtuse, *tapering* at base, or
      subpetiolate:

Twigs hexagonal; female flowers and *smooth* fruit pedunculated ... ... (1) obscurum.
Twigs terete; flowers and *tuberculated* fruit sessile ... (2) obovatum.
Leaves orbicular, ovate or oblong, not tapering at base:
Leaves coriaceous, obviously 3–5-nerved; berries *pedicellate.*
Lvs. elliptical or ovate, somewhat cuneate at base (nearly *uncial*) ... ... (3) pauciflorum.
Lvs. broadly-ovate or suborbicular (¼–½ inch long and wide) ... ... (4) rotundifolium.
Lvs. oblong or lanceolate-oblong, *acute at each end,* patent ... ... (5) tricostatum.
Leaves fleshy, without obvious veins; berries *sessile* ... (6) Crassulæ.
2. Leafless species: (7—11)
Stem much branched and bushy: branches and twigs *terete.*
Berries smooth, mostly in pairs. Branches robust (3–4 lines in diam.) ... ... (7) robustum.
Berries smooth, mostly solitary. Branches slender (1–2 lines in diam.) ... ... (8) Capense.
Berries tuberculated ... ... (9) verrucosum.
Stem much branched and bushy: branches and twigs *two-edged, flattish* ... ... (11) dichotomum.
Stem minute, of one internode; the single vagina bearing 3 sessile flowers ... ... (10) minimum.

1. V. obscurum (Thunb. Prod. 31); old branches terete, the younger ones and twigs hexagonal, articulated; leaves elliptical-*obovate,* or oblong, *tapering at base,* very obtuse, coriaceous, obscurely 3-nerved, glabrous; flowers in threes, the males sessile, the females pedicellate; peduncle of the fruit mostly longer than the white, or yellowish-white, smooth berry. *Fl. Cap. p.* 154. *DC. Prod.* 4, *p.* 285.

VAR. *α,* **longifolium**; leaves narrow-obovate, 1½–2 inches long, much attenuated at base. *V. obscurum, E. & Z.!* 2273.

VAR. *β,* **brevifolium**; leaves broadly-obovate, ¾–1½ inches long, acute at base. *V. rotundifolium, E. & Z.!* 2272. *Zey.!* 2700. *V. pauciflorum, E. Mey. in Herb. Drege. V. obscurum, E. Mey.!*

HAB. Var. *α,* Grootvaders and Duyvelsbosch, Swell., and at the Chumiberg, Kaffr., *E. & Z.!* VAR. *β,* chiefly on willows, in Uitenhage, Albany, and Kaffirland, *E. & Z.!* Kaymansgat, *Drege!* (Herb. Sd. D.)

Much branched, bushy, dichotomous or trichotomous; the young twigs sharply hexagonal. Internodes 1–3 inches long, the nodes somewhat swollen. Leaves variable in length and breadth. Peduncles of the fruit 3–6 lines long. Style ½ line long. This comes very near *V. Orientale* of the East Indies, but has a more evident style, and somewhat different berries.

2. V. obovatum (Harv.); branches and *twigs* terete, articulated; leaves *broadly-obovate,* acute at base, very obtuse, coriaceous, obscurely 3-nerved, glabrous; flowers in pairs or solitary, both sexes sessile; fruits *sessile, tuberculated.*

HAB. Near D'Urban, Port Natal, *Gerrard & McKen!* 659. (Herb. T.C.D.)

Robust, bushy, much branched. Very similar in foliage to *V. obscurum,* var. *β,* but readily known by the terete twigs, and especially by the fruit. Leaves ¾ inch long, more than ½ inch wide.

3. V. pauciflorum (Thunb. Prodr. 31); old branches terete, the younger ones and twigs sharply 6–12-angled; leaves *elliptical or ovate,* sessile, somewhat *cuneate at base,* obtuse or subacute, *carnoso-coriaceous,*

3-nerved, glabrous (or "pubescent"); flowers on minute, 2–3-flowered peduncles, the females minutely pedicellate; pedicel of the fruit shorter than the oblong, smooth, yellowish-white berry. *Thunb. Cap.* 154. *DC. Prod.* 4, *p.* 285. *E. & Z.!* 2274. *Drege, No.* 7650.

VAR. β, **Eucleæ**; leaves subacute or obtuse; berries reddish. *E. & Z.!* 2275.

HAB. Near Tulbagh and on mountains in the Onderbokkeveld, Clanw., *E & Z.!* Paarlberg and in Dutoitskloof, *Drege!* β. Parasitical on *Euclea* near Driefontein in Groenekloofveld, *E. & Z.!* On *Rhus*, at Heerelogement, Clanw., *Zeyh.!* 750. (Herb. Sd.)

Robust, di-trichotomous, brittle, with dark-coloured bark. Internodes 1–1½ inches long, the nodes somewhat swollen. Leaves ¾–1 inch long, 4–7 lines wide, most commonly elliptical, erecto-patent. Pedicel of the fruit 2–3 lines long. Style scarcely any. This has much larger, and thicker, and more oblong leaves than *V. rotundifolium*. *Thunberg* describes them as being "alternate and canescent"; in our plant they are opposite, glabrous, but probably glaucous when fresh, blackish when dry. *E. & Z.'s V. Eucleæ* scarcely differs, but is said to have reddish berries.

4. V. rotundifolium (Thunb. Prodr. 31); old branches terete, the younger ones and twigs sharply hexagonal; leaves *broadly-ovate or subrotund, sessile, broad-based,* obtuse or subacute, coriaceous, glaucous, obscurely 3-nerved, glabrous; flowers on minute, 2–3-flowered peduncles, the females pedicellate; pedicel of the fruit shorter than the ovate, red berry. *Thunb. Cap.* 154. *DC. Prod.* 4, *p.* 279. *V. glaucum, E. & Z.!* 2276. *Drege,* 7651. *Zeyh.!* 2701.

HAB. In woods near the Zwartkops river, Uit., also in Albany and Kaffirland, *E. & Z.!* Klein Winterhoek and near Beaufort, *Drege!* Gamke river, *Burke!* Magalisberg, *J. Sanderson!* Albany, *Hutton!* &c. (Herb. D., Sd., Hk.)

Robust, frequently trichotomous, brittle, with pale bark. Internodes 1–1½ inch long, the nodes swollen. Leaves ¼–½ inch long, sometimes exactly orbicular, but more commonly roundish-ovate, obtuse and acute on the same branch. Pedicels of the fruit 1–2 lines long. Style scarcely any.

5. V. tricostatum (E. Mey.!); old branches terete, the younger ones and twigs sharply hexagonal; leaves *oblong-ovate or lanceolate-oblong,* subsessile, *acute at both ends, horizontally-spreading,* coriaceous, 3-nerved, glabrous, glaucous; flowers on short, 2–3-flowered peduncles, the females pedicellate; pedicel of the fruit about as long as the smooth, reddish berry. *Zeyh.!* 747. *Drege,* 7652, 7651.

HAB. Between Verleptpram and the mouth of the Gariep, *Drege!* Parasitical on willows by the Gariep, *Zey.!* Namaqualand and Hopetown district, *A. Wyley!* (Herb. Sd. Hk. D.)

This has the habit of *V. rotundifolium*, but differently-shaped leaves. It is more slender than *V. pauciflorum*, with pale bark, spreading branches, remarkably patent, more uniformly acute, and evidently glaucous leaves, not turning black in drying. Leaves 5–7 lines long, 2–4 lines wide. Pedicels of the fruit 2 lines long. Style scarcely any.

6. V. Crassulæ (E. & Z.! 2277); branches and twigs *terete,* succulent (rugose when dry); leaves *suborbicular,* sessile, *very thick and fleshy, without obvious nerves;* flowers *sessile,* 3–4 together, axillary; berries *sessile,* oblong, truncate, red, smooth, tipped with a style. *V. Euphorbiæ, E. Mey.!*

HAB. Parasitical on shrubby *Crassulæ* on the Bothasberg, Albany, *E. & Z.!* Zondagsrivier on *Euphorbia* (?), *Drege!* (Herb. Sd. D.)

Very robust, much branched, articulated, and brittle, with swollen nodes; internodes ¾–1 inch long. Leaves 4 lines long, 3 lines wide, very obtuse or rounded, cuneate at base. Berries 2–2½ lines long; style ½ line long.

7. **V. robustum** (E. & Z.! 2279); leafless, much branched, articulated, di-trichotomous; stem and branches terete (very robust), flexuous; leaf-scales connate, shortly ovate, patent, acute, scabrous at the edge; flowers 2–3-together in the axils of the scales, sessile; berries globose, smooth.

HAB. At the T'Kaussi river, Namaqualand, *E. & Z.!* (Herb. Sond.)
Branches 3–4 lines in diameter, very fragile, pale, sulphur-yellow, wrinkled when dry. Leaf-scales (*vaginæ*) patellæform, decussate, their ovate points 1–1½ lines long. Flowers mostly in pairs. Style ½ line long. Perhaps merely a robust variety of *V. Capense*.

8. **V. Capense** (Thunb. Prod. 31); leafless, much branched, articulated, trichotomous; stem and branches terete, the twigs opposite, bluntly 4-angled; leaf-scales connate, short, patellæform, scabrous at edge; flowers solitary or in threes, opposite, sessile, in the axils of the scales; berries globose, minutely pedicellate, *smooth. Thunb. Cap. p.* 154. *DC. Prod.* 4, *p.* 283. *E. & Z.!* 2278. *Zey.!* 749. *V. continuum, E. Mey.! Drege!* 7653.

HAB. Parasitical on trees and shrubs in the Cape, Worcester, Swellendam, and Uitenhage districts, *E. & Z.! Drege! Wallich!* Gamkeriver, *Burke!* Namaqualand, *Wyley!* (Herb. Sd. D.)
Bushy, the lesser branches 1–2 lines in diameter, the twigs more slender, generally opposite and widely-spreading; all of a pale, yellowish-green colour, sometimes turning blackish in drying, wrinkled when dry. Leafless vaginæ short, spreading, decussate. Flowers and fruits mostly solitary and opposite. Style ½ line long.

9? **V. verrucosum** (Harv.); leafless, much-branched, articulated, di-trichotomous; stem and branches terete; leaf-scales connate, short, patellæform; berries in pairs or solitary, globose, *covered with wart-like prominences.*

HAB. Weenen country, Natal, 3–5000f., *Dr. Sutherland!* Magalisberg, *J. Sanderson!* (Herb. Hook. D. Sd.)
More numerous and perfect specimens are required fully to establish this species as distinct from *V. Capense*. All those yet seen have the fruits uniformly tuberculated, and look normal. They may, however, be in a diseased condition.

10. **V. minimum** (Harv.); leafless, *nearly stemless, very minute, univaginate;* the vagina bearing a pair of ovate, connate leaf-scales, and 3 terminal flowers; berries globose, smooth.

HAB. On stems of the succulent *Euphorbiæ*, in Albany, *Mrs. Barber!* (Herb. D.)
Plants consisting of a single terete internode but 2–3 lines in height, crowned with 3 flowers. The large berries 3–4 lines in diameter are, as Mrs. Barber says, "about six times as large as the plant that bears them: when quite ripe the long, cylindrical radicle shoots out, turns round, and plants itself on the *Euphorbia* stem." A very curious and distinct little plant.

11. **V. dichotomum** (Don. Pr. Nep. p. 147); leafless, much-branched, articulated, di-trichotomous; old stems terete, *branches and twigs strongly-compressed and two-edged,* striate or ridged when dry; internodes linear, slightly narrowed at each end, many times longer than

broad, truncate; leaf-scales patellæform; berries globose, minutely pedicellate or sessile, mostly warted. *DC. Prod.* 4, *p.* 284. *V. Nepalense, Spreng. V. anceps, E. Mey.! Zey.! No.* 748.

HAB. Morley, and between the Omtata and Omsamwubo, *Drege!* Magalisberg, *Burke & Zeyher!* Natal, *J. Sanderson!* (Herb. Sd. Hk. D.)

Robust, the older stems nearly terete, with smooth, rather glossy, oblivaceous bark; all the branches and twigs striate, and plano-compressed. Mature internodes 3–4 inches long, 4–5 lines wide; younger ones about uncial, 1½–3 lines wide, yellowish-green. The South African specimens are very similar to those from several parts of India, where this species seems to be common.

END OF VOL. II.

ADDENDA AND CORRIGENDA

TO

THE FIRST VOLUME.

(By W. H. Harvey.)

Page 9, after **Guatteria**, introduce:

III. ANONA, L.

Sepals 3, minute, united at base. *Petals* 6, in two rows, valvate in æstivation; the outer fleshy, triquetrous, hollow at base, or altogether concave. *Stamens* indefinite; connective produced as an oval process beyond the linear, extrorse, contiguous cells of the anthers. *Torus* hemispherical. *Ovaries* numerous, concrescent; styles terminal, oblong. *Ovules* solitary, erect. *Carpels* numerous, united into a many-celled, fleshy, ovoid or roundish fruit. *Seeds* one in each cell, erect, with a shining skin. *Hook. f. & Thoms. Fl. Ind.* 1, *p.* 114. *Endl. Gen. No.* 4723.

African and American trees and shrubs, with edible fruits, commonly called "*Custard-apples.*" Leaves alternate, entire, penninerved, sometimes pellucid-dotted. Peduncles axillary or opposite the leaves, solitary or tufted, one or few-flowered, bracteolate. Flowers greenish or yellowish. *Anona* is the native name of these plants among the aborigines of St. Domingo.

1. **A. Senegalensis** (Pers. Syn. 2, p. 95); leaves broadly-elliptical or ovate, acute or obtuse, rounded or cordate at base, strongly netted-veined and *thinly pubescent* beneath; twigs, petioles, and peduncles pubescent or tomentulose; peduncles 1–3, lateral, 1-flowered. *Rich. Fl. Seneg.* 1, *p.* 5. *DC. Prodr.* 1, *p.* 86. *Deless. Ic. vol.* 1, *t.* 86.

Hab. On the Nototi river, Natal, *W. T. Gerrard!* (Herb. D.)
"A shrub, 6–8 feet high. Fruit edible, 1½–2 inches in diameter, with the flavour of *A. reticulata;* foliage resembling that of *A. Cherimolia.*" *W. T. G.* *Dr. Barter*, who sends it from the Niger river, says, "10 feet high; flowers fleshy, cream-coloured, fragrant; fruit size of an apple, deep orange, when ripe of an apricot flavour, the best of indigenous fruit." As yet I have only seen a single leaf and seeds of the Natal plant; these quite agree with *Dr. Barter's*, and also with specimens from *Dr. Kirk* collected in S. E. tropical Africa, near Moramballa. It appears to be generally distributed throughout tropical Africa, north and south.

Page 67, after **Oncoba Kraussiana,** Pl., introduce:

2. **O. spinosa** (Forsk. Ægypt. p. 103); armed with axillary spines, glabrous; leaves on short petioles, ovate-acuminate, membranaceous, reticulate, denticulate; peduncles terminal, 1-flowered; petals varying

from 5 to 12, denticulate; anthers mucronate. *Lam. Encycl. t.* 471. *Guillm. Fl. Senegamb. t.* 10. *O. monacantha, Steud. Lundia monacantha, Schum. & Thonn.*

HAB. Near Port Natal, *W. T. Gerrard!* Shiré river, *Dr. Kirk.* (Herb. Hk. D.)

A shrub, 6 feet high and more. Stems thick, with whitish, tuberculated bark. Leaves 3–4 inches long, 1½–2½ inches wide, gradually or suddenly acuminated, finely and bluntly toothed, acute or obtuse, in our specimens rounded at base. Spines ½–1½ inch long. Flowers white, 2 inches across, resembling wild roses. Anthers tipped by a fleshy point, sometimes obsolete. Fruit "gourd-like, one-celled, with a solid shell, internally pulpy and many-seeded; well known to the Natal and Zulu Kaffirs, who wear it hung round the neck, and use it as a snuff-box, calling it "Thunga!" *Dr. Kirk's* specimens from the Shiré (in Hb. Hook.) are precisely similar to *Mr. Gerrard's.* Those of *O. monacantha,* from the Niger and Sierra Leone (Hb. Hook.), have rather more coriaceous leaves, less evidently denticulate, and more acute, or even tapering at base, with more evident points to the anthers; the petals are 10.

3. O. Tettensis (Hook. f.); twigs, petioles, foliage, peduncles, and calyx *densely pubescent,* with short, stiff, spreading hairs; leaves short-petioled, rounded at base, oblong, obovate, obtuse, penni-nerved and netted-veined, with slightly reflexed, very entire margins; peduncles lateral and terminal, 1-flowered; petals 10–12; ovary and young fruit *densely tomentose and deeply furrowed;* old fruit glabrous and angular, cuspidate. *Chlanis Tettensis, and Ch. macrophylla, Klotsch. in Peters' Reise nach, Mosamb.* 1, *p.* 145.

HAB. Delagoa bay, *Forbes!* Tette, *Dr. Kirk!* (Herb. Hook. D.)

Readily known by its dense and rigid, though short, somewhat rusty pubescence. *Forbes'* specimens have rather smaller leaves and shorter petioles than those from *Dr. Kirk,* which were collected in the same locality as those described by Klotsch.

Page 68,—The genus "*Phoberos,*" Lour., is the same as **Scolopia,** Schreb, which name is to be substituted. The following new species is to be added:—

4. S. Gerrardi (Harv.); armed with spreading spines; leaves broadly-rhomboid, not tapering at base, obtuse, entire or subrepand; racemes and calyces glabrous; sepals obtuse.

HAB. Nototi river, Natal, *W. T. Gerrard!* (Herb. T.C.D.)

A shrub, 10–12 feet high, with whitish-ashen bark, armed with axillary spines, 2 or more inches in length. Leaves 3–4 inches long, 2–2½ inches wide. It differs from *S. Ecklonii* in the much broader and more rhomboid leaves and the axillary spines; and from *S. Zeyheri* in the perfectly glabrous inflorescence, &c.

Page 70, alter the generic char. of **Aberia** as follows:—

"*Ovary* sessile, on a lobed, fleshy disc, imperfectly 2–6-celled or 1-celled; placentæ prominent, 2–6, each with 2–6 ovules; *styles* 2–6, divergent; stigmas expanded or bifid."

Add the following new species:—

3. A. Caffra (Hk. f. & Harv.); arborescent, *thorny, glabrous;* leaves membranaceous, obovate, obtuse, cuneate at base, 3-nerved and veiny, quite entire, with slightly revolute margins, concolourous; female flowers solitary, on axillary peduncles about twice as long as the leaf-stalks; calyx pubescent, sepals 5–6, oblong, acute, spreading; ovary *ovate, glabrous, with 5–6 divergent styles.*

HAB. Eastern districts and Kaffirland, *R. Hallack, Mrs. Holland,* &c. (Herb. D. Hk.)

A shrub, or small tree, with the habit of *Dovyalis rhamnoides.* Leaves 1½ inch long, ¾–1 inch wide, pale green; petioles 2–3 lines long. Fruit edible, "like a small, yellowish apple." Colonial name, "the Kei apple."

4. **A.? longispina** (Harv.); arborescent, armed with long, divergent spines, glabrous; *branches and spines warted;* leaves coriaceous, rhomboid-ovate or elliptical, obtuse, 5-nerved at base and veiny, quite entire, concolourous; female flowers unknown; male flowers fascicled, shortly pedicellate; calyx 5-parted, tomentose; sepals ovate, acute, nearly valvate.

HAB. Near D'Urban, Natal, *Gerrard & McKen,* No. 541, 542. (Herb. D.)

A shrub or small tree, resembling *Celastrus buxifolius.* Bark dark-coloured, minutely warted. Leaves 2–2½ inches long, 1–1½ inch broad, petioles 2 lines long. Male flowers 6–12 in a tuft, on pedicels 1–2 lines long. Stamens numerous; anthers subglobose.

Page 72, after **Blackwellia rufescens,** introduce:

2. **B. dentata** (Harv.); leaves *on long petioles,* broadly elliptical, or ovate, coarsely and bluntly toothed; panicles axillary, *shorter* than the leaves; pedicels shorter than the calyx-tube; perianth 16–18-parted; stamens 8–9.

HAB. Near Port Natal, *Gerrard & McKen.* (Herb. D.)

Perhaps only a variety, though a strongly marked one, of *B. rufescens.* Petioles uncial. Lamina of the leaves 2–3 inches long, 1½–2½ inches broad, membranaceous. Panicle about twice as long as the petioles, many-flowered.

Page 74, after **Ionidium Caffrum,** introduce:

2*. **I. Natalense** (Harv.); suffruticose; stems erect, *virgate,* pubescent; leaves on very short petioles, ovate-oblong or rhomboid, obtuse or subacute, flat, repando-dentate or sub-entire, when *young puberulous, afterwards glabrous;* stipules subulate; peduncles axillary, 1-flowered; sepals lanceolate, ciliate, and hispid; labellum *somewhat obovate, cuneate at base,* with a very short spur.

HAB. Near Port Natal, *J. Sanderson!* 415, *Gerrard & McKen!* (Herb. Hk. D. Sd.)

By much the largest of the S. African species. Stems 12–18 inches high. Leaves 1½–2½ inches long, ¾–1 inch wide, pale green, the full grown quite glabrous. Peduncles much shorter than the leaves, pubescent. Labellum smaller than in *I. Caffrum,* differently shaped, and much narrower in proportion to its breadth.

Page 113, at the end of **Polygaleæ,** add:

IV. SECURIDACA, Linn.

Sepals 5, unequal, the two lateral *(alæ)* much larger than the rest, winglike, coloured. *Petals* 5, the two lateral adnate to the base of the staminal tube, distinct from the carina, erecto-connivent; carina of equal length, concave-helmetshaped, or 3-lobed; rudiments of the upper petals minute or none. *Stamens* 8, united into a slit tube and hidden within the carina; anthers 2-celled, opening by terminal pores. *Ovary* 1-celled, 1-ovuled; style bent upwards, terete or dilated. *Fruit* samaroid, indehiscent, produced at the apex into a wing. *Benth. & Hook. f. Gen. Pl.* 1, *p.* 138. *Endl. Gen. No.* 5652.

Shrubs or woody climbers. Leaves alternate, mostly entire and bi-glandular. Racemes terminal or axillary, often panicled. Chiefly natives of America, a few Asiatic and African. The name is from securis, *a hatchet,* alluding to the shape of the wing of the fruit.

1. S. oblongifolia (Bth. & H. f. l. c.); shrubby, divaricately branched, the old twigs often spinescent; twigs, petioles, peduncles, and pedicels pubescent or hispid; leaves short-petioled, oblong, obtuse, glabrous and somewhat glaucous, with slightly recurved margins; peduncles terminating short ramuli, racemose, many-flowered; pedicels longer than the flowers; wing nearly thrice as long as the dorsally umbonate fruit. *Lophostylis oblongifolia, Hochst. Fl. Ratisb.* 1842, *n.* 15. *Schimp. Pl. Abyss. No.* 771.

HAB. Delagoa Bay, *Forbes!* (Herb. Hk. D.)

A middle-sized shrub, with pale bark and foliage, the defoliated twigs often hardening into spines. Leaves 1–2 inches long, 4–6 lines wide, obtuse at both ends. Racemes 12 or more flowered; pedicels ½ inch long. Fruit umbonate or bluntly cristate at back, by an abortive second carpel, the wing 1¼–1½ inches long, and ½–¾ inch wide. A native of Abyssinia and Senegambia; found also by *Dr. Kirk* at Moramballa, S. E. Africa, where it is called *"Buaze"*: the "young branches yield an excellent, durable fibre; the seeds a valuable oil," *Livingstone.*

Page 170, after **Pavonia præmorsa**, Willd., introduce:

5. P. urens (Cav. Diss. 3, t. 49, f. 1 & 5, p. 283); herbaceous, erect, the stem, petioles, and leaves setose, with spreading, rigid, subfasciculate, yellow hairs; leaves on very long petioles, 5–7-angled or shortly 5–7-lobed, the lobes acuminate, coarsely toothed; stipules filiform; flowers axillary, tufted, subsessile; involucel of many linear leaflets; calyx densely setose. *Jacq. Ic. Rar. t.* 522. *DC. Prodr.* 1, *p.* 443. *Gerr. & McKen.! No.* 443.

HAB. A kloof near the Tugela R., Natal, *W. T. Gerrard.* (Herb. D.)

Stem 7–8 feet high. Pubescence rigid, bright yellow, copious on the younger parts. Petioles 6–10 inches long. Leaves 5–8 inches long, 4–6 inches wide. Flowers pale rosy, with a deeper centre. A native also of Mauritius and Bourbon.

6. P. odorata (Willd. Sp. 3, p. 837); herbaceous, diffuse or prostrate, the stem, petioles, and peduncles viscidulous, hispid with long, patent hairs; leaves on long petioles, cordate-hastate, bluntly 3-angled or lobed, the middle lobe largest, crenate or subentire, stellate-hispid, especially beneath; stipules filiform; peduncles axillary, slender, one-flowered, equalling the petiole; involucel of 10-12 narrow-linear, rigidly ciliate leaflets, twice as long as the calyx; carpels unarmed. *DC. Prodr.* 1. *p.* 444. *W. & A. Prodr* 1, *p.* 40.

HAB. In and about D'Urban, Natal, *Gerrard and McKen.* (Herb. T.C.D.)

"A prostrate herb, with several stems 1-2 feet long from the same crown. Lvs. various in size and shape, from ¼ to 1½ inches long and broad. Flowers white, more than half inch across: petals semi-transparent, with prominent nerves." *W. T. G. Pav. triloba,* Hochst. in Kotsch. Pl. Nos. 220, 395, seems to be a variety of this: and I fear that Sonder's *Hibiscus leptocalyx,* if afresh examined, may prove to be a synonym also.

Page 176, next **Hibiscus cannabinus,** L., introduce,

20* H. Natalitius (Harv.); annual, thinly sprinkled with simple or stellate hairs; stem and petioles prickly; leaves on long petioles, deeply 5–7-lobed, the lobes acute or acuminate, crenate-toothed, glandless; flowers axillary (small), subsessile; involucel of 9–10 subulate, hispid leaflets, shorter than the ovate, acute, glandless, sparsely pilose calyx-lobes; capsule veiny, ciliate; seeds glabrous, granulated.

HAB. Palmiet R., near the Umgena, Natal, *Gerrard and McKen!* (Herb. D.)

A tall, leafy species, 4–5-feet high, allied in several respects to *H. cannabinus*, but with different foliage and calyx, and much smaller flowers. Petioles 3–4-inches long. Leaves as long as broad; lobes 2 inches long, inch wide, lower surface rather paler. Corolla yellow, with a dark centre, about 1½ inch in diameter. More perfect specimens are needed to establish the species.

Page 177, after **Hibiscus Surattensis,** introduce:

23. H. Gibsoni (Stocks); annual, *glaucous, sparsely setulose;* leaves petiolate, digitate, 3–5-phyllous; leaflets ovato-lanceolate, acute, or acuminate, coarsely serrate, glabrous or sprinkled with a few trifid bristles; stipules subulate; peduncles longer than the leaves, spreading, jointed just beneath the flower; *involucel of* 8–10 *rigid, stellately patent, setulose and pungent leaflets;* calyx deeply parted, its segments ovato-lanceolate, acuminate, 3-ribbed, ciliate; seeds glabrous.

HAB. Damara land, *Miss Elliott!* (Herb. D., Hk.)

A tall, sparingly branched plant, 2–4 feet high? Stems purplish. Foliage remarkably pale. Flowers yellow with a dark purple eye. Stamens few, in interrupted whorls. Stigmas clavate. Remarkable for its star-like involucel. It is a native of the Deccan and of Afghanistan, and was also found by *Dr. Kirk* in S. E. Africa.

***** *Involucel wanting.*

24. H. Elliottiæ (Harv.); suffruticose, finely *stellato-pulverulent and canescent;* branches virgate; leaves petiolate, ovate, crenulate; stipules setaceo-subulate; peduncles axillary, equalling the leaves; *involucel none;* calyx-segments lanceolate, acuminate; petals stellate along the dorsal ridge; ovules 3; seeds (young) *quite glabrous.*

HAB. Damara land, *Miss Elliott!* (Herb. T.C.D.)

A slender undershrub, clothed with very minute, yellowish stellulate pubescence. Leaves 1–1½ inches apart. Petioles 3–4 lines long. Lamina ¾–1 inch long, ½–¾ inch wide. Flowers an inch across, bright scarlet or crimson. Staminal column antheriferous in the middle. Very unlike any Cape *Hibiscus;* but closely related to *H. denudatus* (Benth! Sulp. p. 7. t. 3.), a native of Lower California and New Mexico, from which it scarcely differs save in the more minute pubescence, the smaller flowers, and the glabrous seeds. It is also allied to *H. micranthus*, L., a native of North Africa and tropical Asia; but differs in pubescence, want of involucel, and in the glabrous seeds.

Page 177, after **Paritium,** introduce:

XI. FUGOSIA, Juss.

Involucel 3 or several leaved, often small or deciduous. *Corolla* and *Stamens* as in *Hibiscus*. *Ovary* 3–4 celled; cells 3 or several ovuled; *style* club-shaped, either 3–4 furrowed at the point, or shortly cleft into

3–4 erect lobes. *Capsule* 3–4 celled, loculicidal; seeds subglobose; often pubescent or woolly. *Endl. Gen. No.* 5279. *Benth. & Hook. Gen. Pl.* 1. *p.* 208.

Shrubs or subshrubs, with the habit of *Hibiscus*, chiefly natives of America, one Australian. Leaves entire or lobed. Flowers mostly yellow. The name is in honour of *Bernard Cienfuegos*, a Spanish botanist of the 16th century.

1. F. Gerrardi (Harv.); suffruticose, diffuse or decumbent, sparsely stellulato-pubescent; leaves petiolate, *cordate at base, broadly-ovate or reniform, bluntly 3-lobed or entire*, the lobes mucronulate, nigro-punctate beneath; stipules leafy, ovato-lanceolate or spathulate; calyx campanulate, its lobes oblong, acute, with rounded interspaces; involucel few-leaved, its leaves shorter than the calyx-tube; one-half of each petal glabrous and nigro-punctate, the other half tomentose; seeds thinly woolly.

HAB. On dry plains near Ladysmith, Natal, *W. T. Gerrard*, No. 632. (Hb. D.)
"A trailling, shrubby plant, the branches 3 feet long or more; flowers deep-yellow."—*Gerr.* in litt. Petioles about uncial. Leaves 1–1½ inch long, 1½–2 inches broad, 3–5-nerved, most frequently 3-lobed or 3-angled, the upper ones roundish or reniform. Peduncles axillary, longer than the leaves. Stipules green and leaf-like, tapering at base, midribbed and veiny, ½–¾ inch long, 2–4 lines wide. A very distinct species; most like *F. heterophylla*, Juss., but abundantly different.

2. F. triphylla (Harv.); suffruticose, densely stellato-pulverulent and canescent; leaves petiolate, *triphyllous; leaflets broadly-lanceolate, acute, very entire*, nigro-punctate; stipules subulate, deciduous; calyx campanulate, its small lobes deltoideo-cuspidate, with rounded interspaces; involucel 3-leaved, its lanceolate leaves equalling the calyx-tube; one-half of each petal glabrous and nigro-punctate, the other half thinly stellulate; ovary trilocular; style much longer than the stamens, tricrenate.

HAB. Damaraland, *Miss Elliott*. (Herb. T.C.D.)
A branch only seen. Petioles ¾ inch long. Leaves digitately triphyllous; leaflets 2 inches long, ½ inch wide, both surfaces canescent with extremely minute, stellulate pubescence. Peduncles equalling the petioles. Stipules 2–3 lines long, ½ line wide. Flowers pale-yellow with a dark-purple centre. Very unlike *F. digitata* in pubescence, the shape of leaflets, and especially the involucre. The calyx, involucre, petals, and foliage are nigro-punctate.

Page 185, after **Hermannia decumbens**, W., introduce:

4.* H. Gerrardi (Harv.); procumbent, suffruticose; the stem and petioles hispid with *stipitate-stellate*, rigid hairs; leaves petiolate (very large), oblong, obtuse, cordate at base, green, repando-crenate, laxly clothed on both sides with stipitate and sessile stellate hairs; stipules very broad, *amplexicaul, palmatifid;* flowers in lateral or terminal branched racemes; pedicels slender, equalling the calyx; bracts lanceolate or bifid, with lanceolate lobes; calyx semi-quinquefid, densely stellato-canescent; lobes lanceolate, acute; petals not much exceeding the calyx, the broad, stellate claw equalling the ovate limb; filaments *linear or subulate*, with very narrow wings, stellate-pubescent; ovary turbinate, pubescent.

HAB. Dry rocks near the Mooi river, *W. T. Gerrard!* (Herb. D.)

Stems 2–3 feet long, trailing, subsimple or branched. Petioles about uncial. Leaves 2½–5 inches long, 1½–3 inches wide, of thinnish substance, with prominent nerves and reticulate veins. Pubescence copious and rough, chiefly of *stalked*, stellate hairs, especially on the older parts. Flowers orange-yellow. This has quite the habit and even the cloven stipules of *Mahernia chrysantha*, but the filaments are those of a *Hermannia*, though extremely narrow.

Page 209, after **Mahernia heterophylla,** Cav., introduce:

3.* M. Elliottiana (Harv.); erect, suffruticose; *the stems, leaves, peduncles and calyces densely glandular, with stipitate glands;* leaves spuriously whorled, 2 or 3 or more in each whorl, pinnatifid or sub-bipinnatifid, the rest linear, entire; peduncles much longer than the leaves, 2-flowered; bracts 3–6-parted, the laciniæ linear; calyx deeply parted; the segments narrow-lanceolate, acute, much longer than the calyx, obovate, retuse, tapering at base; anthers bicuspidate.

HAB. Damaraland, *Miss Elliott.* (Herb. T.C.D., Hk.)

Stems 6–12 inches high? copiously glandular and viscidulous in all parts. Leaves ½–¾ inch long; leaf-stipules 3–4 lines long. Peduncles 1½–2 inches long. The flowers are nearly as large as those of *M. grandiflora*, and of similar form and colour; the foliage quite different.

Page 211, after **Mahernia pulchella,** Cav., introduce:

9.* M. vernicata (Burch. Cat. 1461. Trav. 1, p. 278); erect, suffruticose, glabrous, resiniferous; leaves petiolate, inciso-pinnatifid or *sub-bipinnatifid, the bluntly-lobed laciniæ* and the sinuses very obtuse, the margin somewhat inflexed; stipules small, oblong, subacute; peduncles shorter than the leaves, 2-flowered; bracts connate, hood-shaped, incised; calyx semi-quinquefid, resinous-dotted, *the segments ovate, subacute,* half as long as the obovate petals; ovary obovoid, stellato-pubescent. *DC. Prod.* 1, *p.* 496.

HAB. South Africa, *Burchell;* Modderfontyn, Namaqualand, *Rev. H. Whitehead!* (Herb. D.)

Very similar to *M. pulchella*, but more copiously resinous, with a different calyx. In *M. palchella* the calyx-lobes are sensibly acuminate; here they are broader, and barely acute. I describe from *Mr. Whitehead's* specimen.

Page 221, after **Dombeya rotundifolia,** H., introduce:

4. D. densiflora (Planch.!); young branches stellato-tomentose, older glabrous; leaves *subsessile,* suborbicular, *unequally denticulate* or multilobulate, densely stellato-pubescent on both sides, prominently 3–5-nerved and netted beneath; peduncles equalling the leaves, simple or forked, tomentose, each arm densely 6–12-flowered; involucel leaflets narrow-linear, shorter than the bud, deciduous; sepals lanceolate, reflexed, tomentose; ovary tomentose.

HAB. Macallisberg, *Burke & Zeyher!* (Herb. Hk. Sd.)

I formerly confounded this with *D. rotundifolia*, to which it is very nearly allied, but from which it differs in the subsessile, much more densely stellate, thicker, and more acutely toothed leaves, and crowded flowers. Petioles 1–2 lines long. Leaf about an inch long and broad, subacute.

5. D. cymosa (Harv.); young branches thinly stellulate, older glabrous; leaves *on longish, stellato-tomentose petioles, cordate, acute or* *acuminate*, denticulate, minutely and sparsely stellato-pubescent, 5–7-ribbed; peduncles filiform, axillary, equalling the leaves, *cymoso-corymbose, many-flowered*, canescent; invol. leaflets narrow-linear, shorter than the bud, deciduous; ovary tomentose.

HAB. Kreili's country, Kaffraria, *H. Bowker!* 216. (Herb. Hk. D.)

This has flowers of nearly the same size and structure as those of *D. rotundifolia*, with the cordate, acuminate leaves of *D. Natalensis.* Flowers scarcely ½ inch across, white. Pedicels ¼–½ inch long, slender.

6. D. Burgessiæ (Gerr.! MSS.); young branches, petioles, and peduncles *densely villous*; leaves on very long petioles, cordate at base, *shortly and bluntly 3–5-lobed*, softly villoso-tomentose on both sides, especially the under; peduncles equalling the petiole, cymoso-corymbose, many-flowered; invol. leaflets . . . ?; sepals lanceolate, acuminate, villous; ovary tomentose.

HAB. Zululand, *J. M. McKen.* Klip river, Natal, *W. T. Gerrard!* (Herb. T.C.D.)

A much-branched shrub, 8–10 feet high, with beautiful, pale-green foliage, resembling that of *Sparmannia Africana.* Petioles 4–6 inches long; stipules ½ inch, ovato-lanceolate. Leaves 5–7 inches long and broad, 5–7-nerved, with as many shallow and blunt lobes, clothed with long, soft, simple hairs. Flowers 1¼–1¾ inches across, white, pencilled with pink or rosy lines, showy and fragrant. Discovered by *Mr. McKen* in Zululand, and raised by him in the botanic garden, D'Urban. It is named in compliment to Miss Burgess, of Birkenhead.

Page 223, after **Melhania Burchellii**, DC., introduce:

5. M. Damarana (Harv.); diffuse, *densely stellato-tomentose and canescent;* leaves conspicuously petioled, *broadly ovate or oblong*, subcordate at base, obtuse, unequally toothed, with prominent veins beneath; peduncles 1-flowered; invol. leaflets *lanceolate*, shorter than the lanceolate-acuminate, stellato-tomentose calyx lobes.

HAB. Damara land, *Miss Elliott.* (Herb. T.C.D.)

Nearly related to *M. ovata*, Cav. (*M. oblongata*, Hochst.), but with much broader involucral leaflets. Petioles uncial. Leaves 1–1½ inches long, ¾–1 inch wide. Petals expanded, twice as long as the sepals. Filaments of the fertile stamens very short; anthers 3–4 times longer. Staminodia narrow-oblong, subspathulate. Style as long as the staminodia. All parts of the plant densely tomentose, with whitish or yellowish hairs.

Page 225, line 24, for *G. officinalis*, read *G. ocecidentalis.*

Page 225, after **Grewia cana**, introduce:

5*. G. bicolor (Juss. An. Mus. p. 90, t. 50, f. 2); twigs velvetty-canescent; leaves on short petioles, oblong, *acute*, minutely serrulate, minutely puberulous, *becoming glabrous* above, *velvetty and canescent beneath;* peduncles solitary, 2–3-flowered; flower-buds *oblong*, velvetty; sepals 3-nerved, longer than the petals; ovary hairy. *Fl. Senegamb. p.* 96. *DC. Prodr.* 1, *p.* 509. Also *G. Rothii*, DC.; and *G. salvifolia*, Roth.

HAB. Damara land, *Miss Elliott.* (Herb. T.C.D.)

A native of North and East Africa, and of India. Pubescence very short, thin, soft, and whitish. Petiole 1–2 lines long. Leaves 1½–2½ inches long, green above, white beneath, ¾–1 inch wide. Petals cream-coloured?

Page 227, at the end of **Grewia**, add:

Doubtful Species.

Grewia robusta (Burch. Cat. No. 2845), "branches robust, rigid;

leaves small, ovate, obtuse, crenulate, whitish-tomentose beneath; fruit somewhat hairy, 4-lobed; peduncles solitary, opposite the leaves and a little shorter than them, 1–2 flowered; flowers purple." *Bch. Trav.* 2, *p.* 133.

On the descent of the Schneeuweberg, *Burchell.*

Page 233, after **Triaspis**, introduce:

III. TRISTELLATEIA, Thouars.

Calyx 5-parted, with minute glands or none. *Petals* clawed, keeled externally, glabrous, sagittate-ovate. *Stamens* 10, fertile, connate at base, those opposite the petals longer. *Ovary* 3-lobed, the lobes many-crested dorsally; style 1 (the other 2 reduced to papillæ), slender, elongate. *Samaræ* 3, many-winged; the wings narrow, elongate, stellately-patent. *Endl. Gen. No.* 5571. *Benth. & Hook. f. Gen. Pl.* 1, *p.* 258.

Climbing shrubs, with opposite or quaternate leaves; the petioles often 2-glanded, minutely stipulate at base. Racemes terminal and lateral. Flowers yellow. Name from tres, *three*, and stella, *a star*; the wings of the three carpels spread like a star.

1. **T. Madagascariensis** (Poir. Suppl.); leaves elliptical-ovate, obtuse or mucronulate, quite glabrous, the lower ones in fours, the upper opposite; filaments exserted; pedicels bibracteolate below the middle; carpels commonly 7-winged. *Juss. l. c. p.* 241, *t.* 16.

HAB. Delagoa Bay, *Commr. Owen.* (Herb. D. ex Herb. R. Br.)

A nearly glabrous climber. Leaves 1½–2½ inches long, commonly obtuse and exactly oval, sometimes ovate and subacute. Racemes laxly many-flowered; pedicels uncial. Glands of the calyx 2 at the base of each segment, minute. Calyx and pedicels minutely-strigillose.

Page 234, after **Erythroxylon pictum,** E. M., insert:

3. **E. (Sethia) monogynum** (Roxb. Cor. 1, t. 88); leaves oblong-obovate, obtuse, cuneate at base, membranaceous, reticulated, paler beneath, glabrous; stipules broadly-subulate, deciduous; peduncles axillary, 1–3, twice as long as the petiole; calyx-lobes triangular; *styles combined beyond the middle,* longer than the stamens. *Sethia Indica, DC. Prod.* 1, 576.

HAB. Near Natal, *Gerrard & McKen!* (Herb. D.)

A shrub, 8–10 feet high. Leaves 1½–1¾ inch long, ½–¾ inch wide, very acute at base, of thin substance. Petioles 1–2 lines long. Flowers small, white. *W. T. Gerrard's* specimens are very similar in foliage to those from India, where this species is common: they want flowers; the character of the style therefore requires verification.

Page 255, after **Monsonia umbellata,** Harv., insert:

5.* **M. Senegalensis** (Guill. & Perr. Fl. Senegamb. p. 131):—

VAR. **hirsutissima** (Harv.); annual, much-branched, diffuse, *densely hairy with patent, white hairs;* leaves on longish petioles, ovate or cordate, toothed and plaited; stipules membranaceous, lanceolate; peduncles 1- or rarely 2-flowered, short; sepals with a reflexed mucro; petals not much longer than the calyx, emarginate.

HAB. Sandy flats near the Orange R., Namaqualand, *Dr. Atherstone, A. Wyley.* (Herb. Hk. D.)

Much more hairy than the North African *M. Senegalensis*, with which in most other respects it agrees, or than any specimens we have seen of *M. umbellata;* from which species this is chiefly known by its inflorescence.

Page 278, after **Pelargonium Caffrum**, E. & Z., insert :.

66.* P. Bowkeri (Harv.); stem short and succulent; radical leaves on long petioles, *tri-quadri-pinnati-partite*, with a linear-lanceolate outline, the segments *short, filiform, multifid, canescent* with appressed pubescence; ultimate lobes setaceous; stipules lanceolate-acuminate, adnate; scapes longer than the leaves, pubescent; umbel many-flowered, the pedicels about equalling the oblong, villous bracts, and, as well as the calyx, villoso-canescent; petals bipartite, their segments fimbriato-multifid.

HAB. In the Trans-Kei country, *H. Bowker*. (Herb. T.C.D.)

The petioles are 5–6 inches long; the decompound lamina about as long, and not more than 2 inches wide or less, multijugate, each segment as finely divided as a fennel-leaf. The pubescence generally is very short and white; that of the leaves and petioles appressed, of the scapes and calyx spreading. Scapes 12–14 inches high. Calyx-tube 1½ inch long; the lobes 4–5 lines long, obtuse, at length reflexed. Petals twice as long as the calyx-lobes, dark-coloured at base, with yellow, capillary, fringe-like lobules. A very distinct and handsome species.

Page 456, after **Celastrus tenuispinus**, introduce :

9.* C. (Gymnosporia) ruber (Harv.); armed with slender spines; twigs *roughly-puberulous;* leaves on very short petioles, ovate, membranaceous, veiny, sharply-serrate, obtuse, glabrous; cymes on long peduncles, forked, diffuse; bracts and stipules subulate; *petals pink or rosy!*

HAB. On the Nototi river, Natal, *W. T. Gerrard!* (Herb. T.C.D.)

Quite unlike any South African species. Twigs slender and possibly scandent, covered with minute, but rigid, white hairs. Petiole 1–2 lines long. Leaves ¾–1 inch long, ½–⅝ inch wide, broad-based, exactly ovate, of a thin substance. Peduncles longer than the subtending leaves, 1–1½ inch long, slender, smooth, or slightly scabrid. Petals oblong, twice as long as the sepals, rosy or purplish-pink! Anthers short, didymous. Ovary sunk in the disc, 3-celled, with 2 erect ovules in each cell.

Page 526, after **Balsamodendron**, insert :

VIII.* PROTIUM, W. & A.

Character nearly as in *Balsamodendron;* but, *inflorescence* panicled; panicle pedunculate, diffuse. *Benth. & Hook. Gen. Pl.* 1, *p.* 336.

1. P. Africanum (Harv.); glabrous; leaves impari-pinnate or trifoliolate, leaflets oblongo- or ovato- lanceolate, acute at base, taper-pointed, veiny, *serrulate;* panicles *alternately*-branched; calyx cleft beyond the middle.

HAB. Near D Urban, Port Natal, *Gerrard & McKen! No.* 689. (Herb. D. Hk.)

A tree, with ash-like foliage. Leaflets three or five, 1½–2½ inches long, ¾–1 inch broad, tapering *gradually* to a long, acute point, rather paler beneath, serrulated from a short distance above the base. Allied to *P. caudatum*, from which it is known by the narrower and less cuspidate, *serrulated* leaves, the alternately-branched, not dichotomous inflorescence, and the more deeply-parted calyx.

ADDENDA AND CORRIGENDA

TO

THE SECOND VOLUME.

Page 24, the following genus, accidentally omitted, ought to be introduced, between **Amphithalea** and **Cœlidium**:

VI. LATHRIOGYNE, Eck. & Z.

Calyx, ovary and legume as in the one-ovuled *Amphithaleæ*. *Corolla* scarcely longer than the calyx; the carina incurved, rostrate. *Benth. in Lond. Journ.* 2, *p.* 453.

Only one species. The name is compounded of λαθριος, *hidden*, and γυνη, here meaning an *ovary;* because the one-seeded legume is concealed in the hairy calyx.

1. **L. parvifolia** (E. & Z.! No. 1244). *Heudeusa decipiens, E. Mey.! Comm. p.* 153. *Liparia tomentosa, Thunb.! Fl. Cap. p.* 568.

HAB. Hott. Holl. Mts., *E. & Z.!* Zwarteberge, *Drege!* (Herb. Th. Hk. Bth. Sd. D.)

A small, virgate shrub, 12–18 inches high. Branches erect. Leaves 4 lines long, lanceolate, flat, silky canescent. Flowers 2–4, capitate, terminal. Corolla yellow, almost hidden in the very hairy, fulvous calyx.

Page 47, after **Crotalaria Natalitia,** Msn., introduce:

25. C. podocarpa (DC. Prod. 2, p. 133); erect, herbaceous, divaricately-branched, *thinly clothed with long, patent, soft hairs;* stipules leaf-like, one-sided, falcate, ribbed, ciliate; leaflets oblong or *oblongo-lanceolate,* acute, ciliate on margin and midrib; racemes opposite the leaves, 2–6–8-flowered; flowers subdistant; calyx-lobes lanceolate-acuminate, much longer than the tube; legume *stipitate,* many-seeded, glabrous. *Benth. Lond. Journ.* 2, *p.* 589.

HAB. Damaraland, *Miss Elliott!* (Herb. D.)

1–2 feet high, much-branched; the branches pale. Pubescence long, loose, yellowish. Petioles 1–1½ inch long. Leaflets (on our specimens) 1–1½ inch long, the medial longest, *lanceolate,* 2½–3 lines wide. Stipules 6–7 lines long, dimidiate, 1½ line wide. A native also of Tropical and North Africa. It varies with *obovate* leaflets and in amount of pubescence. The calyx-lobes in our specimen are glabrous. Carina long-rostrate.

Miss Elliott has also sent from Damaraland a single specimen of a *Crotalaria,* allied to *C. podocarpa,* differing chiefly in having appressed and rigid pubescence; and in the stipules, which are ***pedately or*** *secundly trifoliolate*, a character so remarkable that I fear to assume it to be normal without further evidence. If this specimen prove to belong to a new species, it may be called **C. diversistipula.**

Page 51, under **Lotononis carnosa,** Bth., introduce:

VAR. β, **condensata** (Harv.); leaflets and leafy stipules linear-lanceolate, acute, the stipules longer than the petioles; racemes terminal, *sub-umbellate, densely several-flowered.*

HAB. Trans-Kei country, *H. Bowker*, No. 107. (Herb. D.)

This may prove to be a distinct species, very closely allied to *L. carnosa*, from which, without further evidence, I do not venture to separate it.

Page 52, after **Lotononis dichilioides,** Sond., introduce:

11.* L. Wrightii (Harv.); herbaceous, prostrate, many-stemmed; stems filiform, subsimple, glabrous, or sparsely appressed-pubescent; leaves (or phyllodia) scattered, *falcato-subulate*, acute, fleshy, compressed, glabrate; stipules in pairs, small, subulate; flowers pedunculate, terminal, or in terminal 2–3-flowered racemes, bibracteate below the calyx; calyx appressedly-pubescent, nearly equalling the glabrous corolla; vexillum small; carina obtuse; legume oblongo-lanceolate, acute, many-seeded, compressed, appressedly-pubescent.

HAB. Mountain sides near Simonstown, *C. Wright!* (Herb. Wright, D.)

Root simple, vertical. Stems many from the crown, 1–1½ foot long, trailing. Leaves an inch or more apart, reduced to phyllodia, which are 1–1½ inch long, laterally-compressed, curved or arched backwards, and scarcely a line wide; the young ones are appressedly-puberulous, the older glabrate. Stipules a line or two in length. Flowers either opposite the leaves or terminal, imperfectly racemose, of a "deep, dark-purple" (*C. W.*) Calyx acute at base, 3 lines long, the lateral lobes lanceolate, the anteal subulate. Petals with longish claws. Ovary multiovulate. Legume an inch or more in length, 2–2½ lines wide, compressed. Staminal tube slit. Very distinct from any other species. The habit is that of a *Hallia.*

Vol. 2, page 68, under **Argyrolobium speciosum,** E. & Z., introduce:

VAR. β, **glaberrimum**; more slender than the normal form, with shorter petioles and more lanceolate leaflets; *the rachis and calyx perfectly glabrous.*

HAB. Kreili's Country, *H. Bowker*. (Herb. D. Hk.)

Said to have "the properties of Spanish liquorice," *Mrs. F. W. Barber.*

Page 69, after **Argyrolobium speciosum,** E. & Z., introduce:

3*. A. Sandersoni (Harv.); glaucous, and nearly glabrous (except the inflorescence); stem erect, bluntly angular, branching; *stipules narrow-subulate*, longer than the petioles; leaflets obovate or oblong, mucronate; racemes elongate, terminal; lower lip of the silky calyx *minutely 3-toothed;* petals glabrous. *Sanderson, No.* 99.

HAB. Flats between Field's and Botha's Hills, Natal, *J. Sanderson!* (Herb. Hk. D.)

Very near *A. speciosum*, but with different stipules and calyx. Legumes 2½ inches long, 1½ line wide, slightly curved, wavy between the seeds.

3. A. Sutherlandi** (Harv.); thinly hairy and ciliate with long, soft yellow hairs; stem erect, angular; stipules *broadly lanceolate, all much longer than the very short petioles;* leaflets obovate-oblong, mucronulate, ciliate; raceme *densely many flowered*, elongate, terminal; lower lip of the calyx *sharply trifid;* petals *quite glabrous.*

HAB. Near Pieter Maritzberg, 2–3,000 ft., *Dr. Sutherland!* (Herb. Hk. D.)
Nearly intermediate between *A. speciosum* and *baptisioides;* having the stipules, calyx and petals of the former, and the pubescence and general aspect of the latter. The inflorescence is more dense than in any of this group.

Page 74, after **Argyrolobium molle,** E. & Z., introduce:

22.* A. lotoides (Harv.); slender, erect, branching, *glabrescent;* stems substrigillose; *stipules broadly ovate or subrotund, shorter than the long petiole;* leaflets broadly obovate, mucronulate, veinless, ciliate along the margin and midrib; peduncles longer than the leaves, umbellately 5–7 flowered; lower lip of the nearly glabrous calyx deeply trifid; corolla glabrous.

HAB. Tyomo River, Kaffraria, *H. Bowker*, No. 366 (Herb. D.)
A small herbaceous plant, 5–6 inches high, with the aspect of *Lotus corniculatus*, nearly glabrous, except for a few scattered hairs and appressed bristles. Petioles ¾–1 inch long. Stipules 3 lines long, 2 lines wide. Leaflets 4 lines long, 2½ wide. Peduncles 1–2 inches long. Vexillum equalling the carina. Legume unknown.

Page 184, after **Indigofera corniculata,** E. M., introduce:

54.* I. Gerrardiana (Harv.); suffruticose, ascending or suberect, branched, *tomentose and canescent;* branches curved, angular; leaves short-petioled, 5–6-jugate, the common petiole recurved, glandless; leaflets *lanceolate*, acute, mucronulate, tomentose, the terminal sessile; stipules small, subulate; racemes laxly many-flowered, elongate, on peduncles (at first) equalling the leaves; calyces albo-tomentose, their segments lanceolate; petals silky with fulvous hairs; legumes *albo-tomentose*, cylindrical, many-seeded, spreading.

HAB. Bushman's River, Natal, *Gerrard and McKen!* 431. (Herb. D.)
Two feet or more high; the stem, foliage and calyces covered with *short*, whitish curled pubescence. Common petiole 1–1½ inches long, the leaf-pairs 2–3 lines apart, without gland-stipells. Leaflets 5–6 lines long, 1 line wide, pubescent on both sides. Stipules almost setaceous. Peduncle 1½–2 inches long, having a raceme which lengthens to 2–2½ inches additional; pedicels 2–3 lines long. Flowers purple, the vexillum and keel tawny without. Legume an inch long, acute.

Page 239, after **Canavalia Bonariensis,** Lindl., introduce:

3. C. gladiata (DC. Prodr. 2, 404); stem voluble, glabrous or downy; leaflets *broadly ovate, mostly acuminate*, rigidly membranous, glabrous or *pubescent underneath;* upper lip of the calyx bilobed, shorter than the lobe, lower trifid; carina not beaked. *Benth! in Mart. Fl. Braz. p.* 178. *Wight, Ic. t.* 753. *Can. Braziliensis, Mart.—Dolichos gladiatus, Linn. Jacq. Ic. Rar. t.* 560. *D. acinaciformis, Jacq. Ic. t.* 559.

HAB. Port Natal, *Gerrard & McKen!* (Herb. D.)
A native also of tropical Africa, India and S. America. Stem extensively climbing, the younger parts pubescent. Petioles 4–6 inches long. Leaflets 4–5 inches long, 2–3–4 inches wide, acute or acuminate, green, prominently nerved and veined beneath; our specimens pubescent underneath, the nerves on both sides hairy. Peduncles 8–12 inches long, several flowered beyond the middle; nodes tubercular. Ovary densely and appressedly hirsute. Legume 4–12 inches long, 1½ inch wide.

Page 241, under **Vigna triloba,** Walp., insert:

VAR. γ. **acutifolia;** leaflets ovate-hastate, *acute*, the medial equal-sided, with obsolete lateral lobes, the lateral leaflets dimidiate-hastate, the outer lobe blunt.

HAB. Near D'Urban, Natal. *Sanderson!* 451. *Gerrard & McKen,* 387. (Herb. D.)
A slenderer plant than the normal form, with thinner leaves, constantly acute, and less obviously 3-lobed. Were it not for such plants as var. β, I should be tempted to regard this as a species.

Page 261, under **Eriosema squarrosum,** Walp., insert:

VAR. ε, **longatum** (Harv.); stems weak, distantly leafy; lower leaves frequently unifoliolate; peduncles very long (8–12 inches) flowering at the summit; pubescence silvery. *Gerrard & McKen, No.* 421.

HAB. Near D'Urban, Natal, *Gerrard & McKen.!* (Herb. D.)
Apparently a specimen drawn up among long grass.

Page 275, after **Bauhinia tomentosa,** L., introduce,

1.* **B.** [*Pauletia*] **Bowkeri** (Harv.); shrubby, unarmed; leaves rounded at base, netted-veined beneath, *glabrous;* leaflets obliquely oblong, 3–4-nerved, obtuse, concrete for a short distance above the base; peduncles terminal, very short, 2–3-flowered; flowers very shortly pedicelled; calyx tube cylindrical, puberulous, its limb spathaceous, reflexed; petals *lanceolate, tapering at base, strongly midribbed and penninerved;* stamens 10, fertile, unequal.

HAB. Along the Basche River, Fort Bowker, Caffraria, *H. Bowker, Esq.* (Herb. D.)
"A tree, covered with white blossoms; would make a good ornamental tree for a garden, *H.B.*—Allied to *B. tomentosa,* from which it may be at once known by its very different petals; those of *B. tomentosa* are broadly obovate, retuse, and neither strongly midribbed (ribbed only at base) nor penninerved. The young twigs, petioles, and nerves of leaves, and the calyx are minutely puberulous. Stipules inconspicuous. Leaflets 1 inch long, 6–7 lines wide; petiole ½ inch long. Petals 1½ inch long, scarcely ½ inch wide.

Page 285, in the Table of Genera, under **Rosaceæ,** introduce (altering the numbers of the 3 Sub-Orders there enumerated to Nos. 2, 3, 4.)

Sub-Order I.—CHRYSOBALANEÆ. *Calyx* tubular or campanulate. *Ovary* composed of one carpel, unilocular or bilocular; mostly adnate to one side of the calyx-tube; ovules 1–2, erect. *Style* lateral or basal. Fruit drupaceous.

I.* PARINARIUM.

Page 286, before **Rubus,** introduce:

I.* PARINARIUM, Juss.

Flowers bisexual. *Calyx*-tube short or long, subequal or unequal-sided; limb 5-parted, subequal, imbricate. *Petals* 5, rarely 4, sessile or clawed, inserted in the throat of the calyx, deciduous. *Stamens* 10 or indefinite, inserted with the petals, shortly connate at base (or united in a unilateral parcel), all perfect, or some barren; filaments filiform; anthers short. *Ovary* adnate at one side to the calyx-tube, exserted, 2-celled (or incompletely so); ovules one in each cell, erect; style basal, filiform, hairy; stigma truncate. *Drupe* ovoid or sphærical, with fibrous or pulpy flesh, and a bony, one-seeded putamen. *Cotyledons* fleshy; radicle very short. *Endl. Gen. No.* 6411.

Trees or shrubs, natives of the tropics of both hemispheres; several in tropical Africa. Leaves alternate, persistent, penninerved, mostly coriaceous, entire, often bi-glandular at base. Stipules subulate or lanceolate. Flowers racemose, corymbose or panicled, 2-bracteolate, white or rosy. Fruits sometimes edible. *Parinari* is the name of *P. montanum*, in Guiana.

1. **P. Capense** (Harv.); a dwarf shrub; twigs and petioles rufo-villous; leaves lanceolate-oblong, obtuse, acute or tapering at base, glabrous above, albo-tomentose and netted with veins beneath; inflorescence corymbulose, shorter than the leaves; bracts ovate; calyx-lobes ovate, acute; petals oblong, sessile; stamens 10, shorter than the calyx-lobes. *Zey.!* 537. *Burke!* 518.

HAB. Aapjes River, *Burke & Zeyher!* (Herb. Hk., Sd., D.)

Apparently a dwarf, but ligneous, branching shrub, 6–12 inches high, the younger parts clothed with foxy hairs. Leaves 2½–4½ inches long, ¾–1 inch broad, coriaceous, white beneath, penninerved and strongly veiny. Panicles 1–1½ inch long, not much branched.

Page 288. Erase the name **Potentilla Gariepensis**, and substitute: (retaining the specific chararcter and remarks)

1 **P. supina** (Linn. Sp. 711); *DC. Prodr.* 2. *p.* 580. *Jacq. Fl. Austr.* 5. *t.* 406. *Lehm. Pot.* 43. *P. Gariepensis, E. Mey.! in Herb. Drege.*

Our S. African *Potentilla* proves, on re-examination, to be merely a small-flowered and weaker-stemmed variety of *P. supina*, Linn.

Page 304, at the end of **Cliffortia**, introduce the following:—

Doubtful Species.

40? C. flabellifolia (Sond. MSS.); glabrous, robust, much-branched and ramulous, procumbent (?); twigs angular, the bark splitting lengthwise; *leaves opposite*, decussate, sessile, cuneate-flabelliform, plaited, 6–7-crenate on the subtruncate apex; stipules one at each side, minute, subulate; flowers unknown.

HAB. Magalisberg, *Zeyher!* (Herb. Sond.)

A small shrub, whose genus cannot at present be satisfactorily ascertained. It has the general aspect of a *Cliffortia*, but the leaves are *opposite!* Leaves very rigid, 6–7 lines long, 3–4 lines wide.

Page 309, at the end of **Saxifragaceæ**, introduce:

BREXIA, Thouars.

Calyx free, 5-cleft, persistent, with short, acute, coriaceous segments, imbricate in æstivation. *Petals* 5, inserted outside the margin of a perigynous ring, coriaceous, oblong, obtuse, imbricate in æstivation. *Stamens* 5, alternate with the petals and inserted with them; filaments subulate; anthers oblong, erect, basifixed, slitting. Annular *disc* thick, adnate to the base of the ovary, with 5 fimbriated lobes. *Ovary* superior, 5-angled, 5-celled, ovules very numerous, on axile placentæ. *Style* very short; stigma 5-lobed. *Fruit* oblong, with a ligneous pericarp, 5-celled, many-seeded. *Embryo* exalbuminous, straight, almond-like; cotyledons fleshy, ovate, obtuse; radicle very short. *Endl. Gen. No.* 4681.

Shrubs, natives of Madagascar and South Africa, with alternate, exstipulate, coriaceous, entire or serrate leaves, and axillary or terminal subumbellate, green flowers. The name is said to be from *βρεξις, rain;* because the foliage may afford shelter in rain.

1. B. Madagascariensis (Lindl. Bot. Reg. t. 730); leaves oblong or *obovate*, quite entire, with revolute margins, netted-veined beneath. *Venana Madagascariensis, Lam. Encycl. t.* 131.

HAB. Delagoa Bay, *Forbes!* (Herb. Hk. D.)

A glabrous and somewhat glaucous shrub. Leaves in our specimens shorter and more obovate than in those from Madagascar, but otherwise similar; 3–4 inches long, 2½–3 inches wide. Ripe fruit 2½ inches long, obscurely 5-angled, tapering to a conical, acute point; the rind thick and woody. Seeds horizontal.

Page 310, line 38, for **Brunia alopecuroides**, *Thunb. ?* read *Thunb. !*
Page 320, line 1, for *B. microcephala*, E.M., read *B. macrocephala*, E.M.
Page 470, under **Aizoon Zeyheri**, Sond., add:

HAB. Bitterfontein, *Zeyher. !* 717. (Herb. Sond.)

Page 479, the following genus, accidentally omitted, ought to be introduced after **Plinthus**:—

VIII. TRIANTHEMA, Lam.

Sepals 5, persistent, united at the base, coloured on the inner surface and mucronate below the apex. *Petals* none. *Stamens* 5 or 10, rarely more, free, inserted on the tube of the calyx. *Anthers* cordate-ovate. *Ovary* ovate. *Styles* (or stigmas) filiform, 1–2, rarely 3. *Capsule* circumscissile below the middle, bilocular; seeds subsolitary in each cell. *DC. l. c.* 3. 353. *Endl. gen. n.* 5168.

Subfleshy herbs, sometimes suffrutescent at the base; leaves opposite, entire, petiolate. Petiole dilated at the base at each side into a stipuloid membrane. Flowers bibracteolate, axillary, sessile, solitary, glomerate or cymose. Name from *τρις, three,* and *ανθος, a flower.*

1. T. crystallina (Vahl. Symb. 1. 32.); perennial, cæspitose, woody at the base; stems prostrate, terete, papulose; leaves *ovate or somewhat spathulate*, opposite, one of them smaller than the other; flowers crowded, axillary; stamens 5; style simple. *Wight & Arn. Prod.* 1. 355. *Papularia crystallina. Forsk. descr.* 69.

VAR. β. *rubens*, stems nearly glabrous, at the apex papulose, reddish as well as the spathulate leaves; flowers cymose-glomerate; tube of the calyx finely striated. *T. rubens, E. Meyer in herb. Drege.*

VAR. γ. *corymbosa*, stems elongate, epapulose or nearly so, pale; leaves oblong-spathulate, evidently petiolate; flowers in loose cymes. *T. corymbosa, E. Meyer in herb. Drege.*

HAB. (var. α. Arabia, East Indies); var. β. on the Garip River, *Drege;* near Springbokkeel, *Zeyher*, 633 partim; var. γ. near Verleptpram on the Garip, *Drege;* Namaqualand, *A. Wyley*, Sept. (Herb. D. Sd.)

Stem from several inches to 1 foot long, filiform, with opposite branches. Leaves 4–6 lines long, 1–1½ line wide; petiole 1–2 lines long, membranaceous. Glomerule of flowers many flowered, usually shorter, rarely equalling the leaf; calyx reddish, 1 line long; lobes acute, a little longer than the tube. Stamens equalling the calyx. Seeds glabrous, reticulate-punctate, exactly as in *T. crystallina.*

2. T. parvifolia (E. Meyer in herb. Drege); perrennial? cæspitose, herbaceous; stems terete, prostrate or diffuse, papulose or subglabrous; leaves *suborbicular*, opposite, one of them generally smaller than the other; flowers ternately aggregated, axillary; stamens 5; style simple.

HAB. Zwartbulletje, stony hills, and on the Gamka River, 2–3000 ft., Drege. Springbokkeel, Zeyh. 633, Feb.-Apr. (Herb. D. Sd.)

It comes very near the preceding, and is only distinguished by the roundish leaves and the 3-flowered, not many flowered glomerules. Stipules very large, inclosing the young flowers. Leaves 1–3 lines long and wide. Flowers 1 line long, tube finely striated.

Page 502, after **Passifloreæ**, insert :—

ORDER LIX.* TURNERACEÆ, DC.

(BY W. H. HARVEY.)

Flowers regular, bisexual. *Calyx* tubular, 5-cleft, with imbricate æstivation. *Petals* 5, alternate with the lobes of the calyx, and inserted on the tube or in the throat, twisted in æstivation, deciduous. *Stamens* 5, inserted below the petals, with which they alternate; filaments subulate, flat; anthers erect, dorsally affixed. *Ovary* free, unilocular; placentæ 3, parietal, multi-ovulate; styles 3, terminal, distinct, opposite the placentæ; stigmata fimbriate. *Fruit* capsular, 3-valved; valves placentiferous. *Seeds* numerous, with fleshy albumen, and a crustaceous, hollow-dotted testa; embryo straight, axile.

Herbs, half-shrubs, or small shrubs, natives chiefly of tropical America; a few African. Leaves alternate, simple, entire or toothed, rarely pinnatifid, often with 2 glands at base. Stipules none. A small Order, of 3 or 4 genera, closely related to *Passifloreæ*.

I. TURNERA, Plum.

Calyx coloured, tubular-funnell-shaped, more or less deeply 5-parted. *Petals* inserted in the throat of the calyx, alternate with its lobes, short-clawed. *Stamens* 5, alternate with the petals. *Styles* undivided; stigmata flabellate-multifid. *Capsule* ovate or oblong, 3-valved. *Endl. No.* 5056.

Suffrutices or small shrubs, chiefly American. *T. ulmifolia* is naturalized throughout the tropics of both hemispheres. The generic name is in honour of William Turner, M.D., Prebendary of York, Canon of Windsor, and Dean of Wells, who died 1568. He was the author of a "New Herbal."

1. T. Capensis (Harv.); dwarf, suffruticose, many-stemmed, densely hirsute; leaves lanceolate-oblong, obtuse, coarsely toothed, tapering at base into a short petiole; peduncles axillary, 1-flowered, free, much shorter than the leaves; calyx deeply 5-parted, laciniæ linear-lanceolate, acute; petals obovate.

HAB. Aapje's river, *Burke & Zeyher!* (Herb. Hook. D.)

Root thick and woody. Stems 4–6 inches high, erect, simple or slightly branched. Every part of the plant thickly clothed with loose, coarse pubescence. Leaves 1–1½ inches long, 4–5 lines wide, very hairy. Peduncles 4–7 lines long, curved or nodding. Flowers small, white? Calyx-tube not one-fourth as long as the lobes. Filaments flat, subulate, scarcely half as long as the petals. Ovary hirsute. Styles longer than the stamens; stigmas expanded, channelled, fimbrato-multifid.

INDEX.

[SYNONYMS IN *italics*.]

PRINTED AND BOUND IN GREAT BRITAIN BY
WILLIAM CLOWES AND SONS, LIMITED
LONDON AND BECCLES

Printed in the United States
By Bookmasters